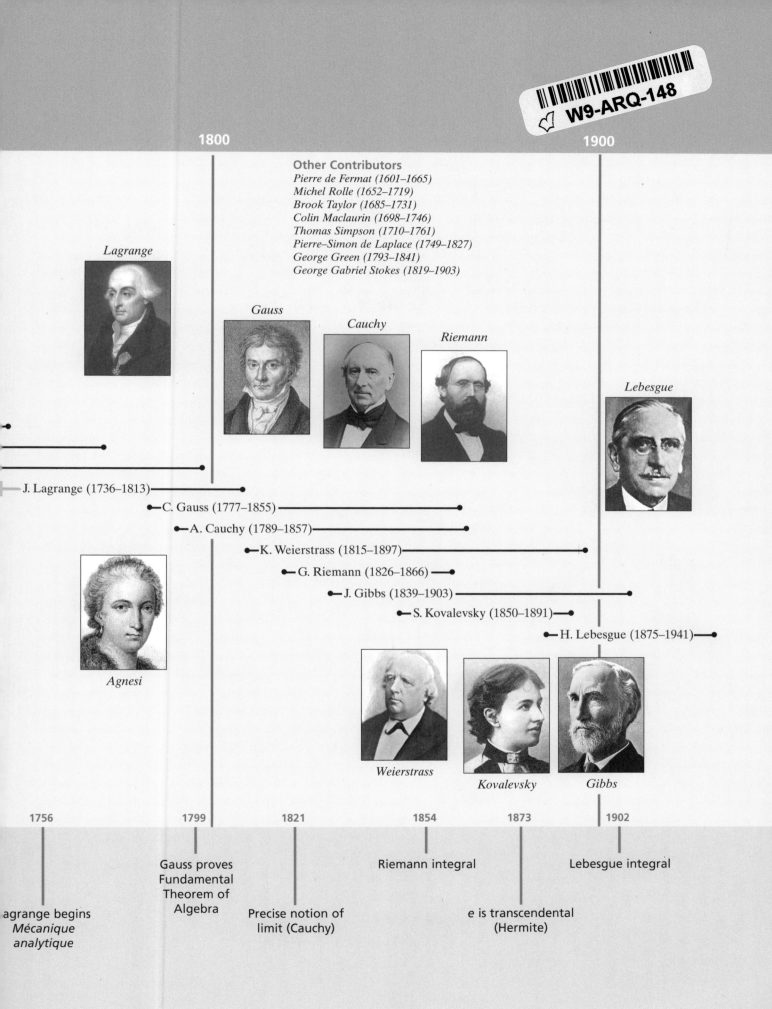

1800

1900

Other Contributors
Pierre de Fermat (1601–1665)
Michel Rolle (1652–1719)
Brook Taylor (1685–1731)
Colin Maclaurin (1698–1746)
Thomas Simpson (1710–1761)
Pierre–Simon de Laplace (1749–1827)
George Green (1793–1841)
George Gabriel Stokes (1819–1903)

Lagrange

Gauss

Cauchy

Riemann

Lebesgue

J. Lagrange (1736–1813)

C. Gauss (1777–1855)

A. Cauchy (1789–1857)

K. Weierstrass (1815–1897)

G. Riemann (1826–1866)

J. Gibbs (1839–1903)

S. Kovalevsky (1850–1891)

H. Lebesgue (1875–1941)

Agnesi

Weierstrass

Kovalevsky

Gibbs

1756

1799

1821

1854

1873

1902

Gauss proves
Fundamental
Theorem of
Algebra

Riemann integral

Lebesgue integral

Lagrange begins
*Mécanique
analytique*

Precise notion of
limit (Cauchy)

e is transcendental
(Hermite)

Calculus

for Biology and Medicine

Third Edition

Calculus

for Biology and Medicine

Claudia Neuhauser

University of Minnesota

Prentice Hall

Boston   Columbus   Indianapolis   New York   San Francisco   Upper Saddle River
Amsterdam   Cape Town   Dubai   London   Madrid   Milan   Munich   Paris   Montréal   Toronto
Delhi   Mexico City   São Paulo   Sydney   Hong Kong   Seoul   Singapore   Taipei   Tokyo

Editor-in-Chief: *Deirdre Lynch*
Executive Editor: *Jennifer Crum*
Senior Project Editor: *Joanne Dill*
Editorial Assistant: *Joanne Wendelken*
Senior Managing Editor: *Karen Wernholm*
Senior Production Supervisor: *Tracy Patruno*
Project Managers: *Rob Merenoff and Lifland et al., Bookmakers*
Manufacturing Buyer: *Debbie Rossi*
Executive Marketing Manager: *Jeff Weidenaar*
Marketing Assistant: *Kendra Bassi*
Art Director: *Heather Scott*
AV Project Manager: *Tom Benfatti*
Manager, Cover Visual Research & Permissions: *Elaine Soares*
Cover/Interior Designer: *Tamara Newnam*
Compositor: *Dennis Kletzing*
Art Studio: *Laserwords Private Limited*
Cover Image: *HI Virus infecting cell / Shutterstock*

Photo credits appear on page C1 and constitute a continuation of the copyright page.

Many of the designations used by manufacturers and sellers to distinguish their products are claimed as trademarks. Where those designations appear in this book, and Pearson was aware of a trademark claim, the designations have been printed in initial caps or all caps.

Library of Congress Cataloging-in-Publication Data

Neuhauser, Claudia, 1962–
  Calculus for biology and medicine / Claudia Neuhauser.--3rd ed.
    p. cm.
  Includes bibliographical references and index.
  ISBN 978-0-321-64468-8 (alk. paper)
  1. Biomathematics--Textbooks. 2. Medicine--Mathematics Textbooks.
I. Title.
  QH323.5.N46 2011
  515--dc22
                                                    2011027223

1 2 3 4 5 6 7 8 9 10—EB—13 12 11 10 09

Prentice Hall
is an imprint of

www.pearsonhighered.com

ISBN 10:   0-321-64468-9
ISBN 13: 978-0-321-64468-8

About the Author

Claudia Neuhauser is Vice Chancellor for Academic Affairs and Director of the Center for Learning Innovation at the University of Minnesota Rochester (UMR). She is a Distinguished McKnight University Professor, Howard Hughes Medical Institute Professor, and Morse-Alumni Distinguished Teaching Professor. She received her Diploma in Mathematics from the Universität Heidelberg (Germany), and a Ph.D. in Mathematics from Cornell University. Before joining UMR in July 2008, she was Professor and Head in the Department of Ecology, Evolution and Behavior at the University of Minnesota Twin Cities, and a faculty member in mathematics departments at the University of Southern California, UW-Madison, University of Minnesota, and UC Davis.

Dr. Neuhauser's research is at the interface of ecology and evolution. She investigates effects of spatial structure on community dynamics, in particular, the effect of competition on the spatial structure of competitors and the effect of symbionts on the spatial distribution of their hosts. In addition, her research in population genetics has resulted in the development of statistical tools for random samples of genes.

In her role as Director of the Center for Learning Innovation at the University of Minnesota Rochester, Dr. Neuhauser is responsible for the development of the Bachelor of Science in Health Sciences. The Center promotes a learner-centered, concept-based learning environment in which ongoing assessment guides and monitors student learning and is the basis for data-driven research on learning. Dr. Neuhauser's interest in furthering the quantitative training of biology undergraduate students has resulted in a textbook titled *Calculus for Biology and Medicine* and a web page called Numb3r5 Count! (http://bioquest.org/numberscount/). In her spare time, she enjoys riding her bike, working out in the gym, and reading history and philosophy.

Contents

Preface

Though it has been several years since the first edition was published, my goal of the first edition of *Calculus for Biology and Medicine* remains true in this edition:

> ***To show students right from the beginning how calculus is used to analyze phenomena in nature without compromising the rigor of the presentation of calculus.***

The result of this goal is a calculus text that has plentiful life and health sciences applications and that provides students with the knowledge and skills necessary to analyze and interpret mathematical models of a diverse array of phenomena in the living world. Since this text is written for college freshmen, the examples were chosen so that no formal training in biology is needed.

The rigor of the text prepares students well for more advanced courses in mathematics and statistics. My hope is that students will find calculus concepts easier to understand and more interesting if they are related to their major and career aspirations.

While the table of contents resembles that of a traditional calculus text, the content does not: abstract calculus concepts are introduced in a biological context and students learn how to transfer and apply these concepts to biological situations. The book does not teach modeling, but students are exposed to and asked to apply numerous models while they begin to see how simple models can capture the essence of natural phenomena.

■ New to this Edition

- ■ Approximately 20% of the problems have been updated and new problems have been added.
- ■ The application problems are now labeled to make it easier to identify the area of application.
- ■ Learning objectives have been added to each chapter to help students structure their learning and help teachers organize their syllabus and class notes.
- ■ Some sections have been rewritten or reorganized in response to users.
- ■ Additional explanations and examples have been added to aid student understanding.
- ■ Where necessary, figures have been added to aid students in visualizing the mathematics.
- ■ The basic organization of the text has remained the same; however, some changes to the organization have been made: Practicing integration and partial fraction decomposition are now in separate sections. The final chapter on probability and statistics has been expanded to include more statistics and more on stochastic processes, and it can now be used for a semester-based course.

■ Features of the Text

Examples and Explanations   Each topic is inspired by biological examples. This motivating introduction is followed by a thorough discussion *outside* of the life science context to enable students to become familiar with both the meaning and the mechanics of the mathematical topic. Finally, biological examples are presented to teach students how to apply the material in a life science context. Examples in the text are completely worked out, and the steps in the calculations are frequently explained in words.

Exercises   Calculus cannot be learned by watching someone do it. Because of this, *Calculus for Biology and Medicine*, Third Edition, provides students with skill-based exercises as well as word problems. Word problems are an integral part of teaching calculus in a life science context. The word problems contained in the text are up-to-date and are adapted from either standard biology texts or original research. The exercises and word problems are at the end of each section and are organized by subsection to help students reference specific subsections of content while completing homework. This also aids instructors in assigning homework problems.

Technology   *Calculus for Biology and Medicine* takes advantage of graphing calculators. This allows students to develop a much better visual understanding of the concepts in calculus. Beyond this, no special software is required.

■ Reflections and Outlook

The process of revising this book gave me good reason to look back and reflect upon how this book came about and to make decisions as to where it should go in the future. The motivation for writing this book more than ten years ago came from my teaching the second quarter of calculus to a large group of students at the University of Minnesota Twin Cities. The course covered standard calculus material, primarily focused on integration techniques, and the students came from diverse majors outside of the physical and engineering sciences. In fact, many of the students were life science majors.

Through interactions with colleagues in the life sciences, I increasingly became aware that much of what was taught in a standard first-year calculus course was of little use to the students in their pursuit of careers in the life and health sciences. Since life science majors rarely take more than a year of calculus, they are left with some rudimentary knowledge of an enormously useful area, but with few skills in applying this knowledge. I was fortunate to be in an environment where I was able to experiment with a different kind of calculus course with the support of my department (School of Mathematics) and departments within the College of Biological Sciences. I developed the course in 1997–98 and taught it for the first time in 1998–99.

Much has happened since then. The life and health sciences have undergone an information revolution. The human genome project was completed in 2003. Many more complete genomes have become available since then. High-throughput technologies, sensor systems, imaging devices, and other new technologies generate data at a breath-taking pace. The data will ultimately enable us to find solutions to challenges in areas as diverse as health, energy, environment, and national security. These revolutionary changes necessitate changes in how we educate future scientists.

National reports on quantitative education of students in the life and health sciences have addressed the needed changes. Foremost, Bio2010, a 2003 report[1] by the National Research Council, examined the undergraduate education of life and health science majors and identified fundamental science and mathematics skills needed to prepare them for a career in biomedical research. More recently, I was a member of a committee that was convened by the Association of the American Medical Colleges (AAMC) and the Howard Hughes Medical Institute (HHMI) and that prepared a report[2] on scientific competencies for medical school graduates and undergraduate students interested in going to medical school. Both reports emphasize the need for undergraduate students to develop *quantitative reasoning skills* to analyze, model, and predict phenomena in the natural world. Increasingly, students must be able to extract information from large data sets. Many of the mathematics concepts listed in Bio2010 are covered in this book, including dynamical systems and probability and statistics.

(1) Bio2010: Transforming Undergraduate Education for Future Research Biologists. Committee on Undergraduate Biology Education to Prepare Research Scientists for the 21st Century, Board of Life Sciences, Division on Earth and Life Studies, the National Research Council of the National Academies. 2003. **(2)** Scientific Foundations for Future Physicians. Report of the AAMC-HHMI Committee. 2009.

I continue to question the way we teach mathematics, statistics, and computation to students in the life and health sciences. We need to prepare our students for this data-rich environment. The traditional way of completing paper-and-pencil exercises and working with moderately-sized data sets (with a focus on developing the skills to master differentiation and integration techniques) is no longer sufficient.

I have been fortunate to receive funding from HHMI as a HHMI Professor to develop quantitative curricula for the life and health sciences. This growing set of resources is available on my web site NUMB3R5 COUNT! (http://bioquest.org/numberscount/). The goal is to take a data-driven approach to algebra, calculus, probability, and statistics. Worksheets and spreadsheets with authentic data sets have been written to supplement this book and enrich this course. It is my goal that students will begin to explore calculus concepts with real data.

■ Chapter Summary

Chapter 1   Basic tools from algebra and trigonometry are summarized in Section 1.1. Section 1.2 contains the basic functions used in this text, including exponential and logarithmic functions. Their graphical properties and their biological relevance are emphasized. Section 1.3 covers log-log and semi-log plots; these are graphical tools that are frequently used in the life sciences. In addition, a section on translating verbal descriptions of biological phenomena into graphs will provide students with skills that they will need when they read biological literature.

Chapter 2   This chapter covers difference equations (or discrete time models) and sequences, which provides a more natural way to explain the need for limits. The chapter ends with classical models of population growth, giving students a first glimpse at how models can aid in understanding biological phenomena.

Chapter 3   Limits and continuity are key concepts for understanding the conceptual parts of calculus. Visual intuition is emphasized before the theory is discussed. The formal definition of limits is at the end of the chapter and can be omitted.

Chapter 4   The geometric definition of a derivative as the slope of a tangent line is given before the formal treatment. After the formal definition of the derivative, differential equations are introduced as models for biological phenomena. Differentiation rules are discussed. These sections give students time to acquaint themselves with the basic rules of differentiation before applications are discussed. Related rates and error propagation, in addition to differential equations, are the main applications.

Chapter 5   This chapter presents biological and more traditional applications of differentiation. Many of the applications are consequences of the mean value theorem. The word problems originate from either biology textbooks or research articles. This use of sources puts the traditional applications (such as extrema, monotonicity, and concavity) in a biological context. Analysis of difference equations is available in an optional section.

Chapter 6   Integration is motivated geometrically. The fundamental theorem of calculus and its consequences are discussed in depth. Both biological and traditional applications of integration are provided before integration techniques are covered.

Chapter 7   This chapter contains integration techniques. However, only the most important techniques are covered. A section on Taylor polynomials is also included in this chapter. The section on using tables of integrals was moved to the end of the chapter and is optional. While computer software is not required in this book, it is likely that students will have access via the web to easy-to-use, free software to calculate integrals. For instance, WolframAlpha (http://www.wolframalpha.com/) allows integration of functions on the web and shows the steps.

Chapter 8 This chapter provides an introduction to differential equations. The treatment is not complete, but it will equip students with both analytical and graphical skills to analyze differential equations. Eigenvalues are introduced early to facilitate the analytical treatment of systems of differential equations in Chapter 11. Many of the differential equations discussed in the text are important models in biology. Though this text is not a modeling text, students will see how differential equations can model biological phenomena and will be able to interpret differential equations. Chapter 8 contains a large number of up-to-date applications of differential equations in biology.

Chapter 9 Matrix algebra is an indispensible tool for every life scientist. The material in this chapter covers the most basic concepts and is tailored to Chapters 10 and 11, where matrix algebra is frequently used. Special emphasis is given to the treatment of eigenvalues and eigenvectors because of their importance in analyzing systems of differential equations.

Chapter 10 This is an introduction to multidimensional calculus. The treatment is brief and tailored to Chapter 11, where systems of differential equations are discussed. The main topics are partial derivatives and linearization of vector-valued functions. The discussions of gradient and diffusion and the section on extrema and Lagrange multipliers are not needed for Chapter 11. If difference equations were covered early in the course, the final section in this chapter provides an introduction to systems of difference equations with many biological examples.

Chapter 11 This material is most relevant for students in the life sciences. Both graphical and analytical tools are developed to enable students to analyze systems of differential equations. The material is divided into linear and nonlinear systems. Understanding the stability of linear systems in terms of vector fields, eigenvectors, and eigenvalues helps students to master the more difficult analysis of nonlinear systems. Theory is explained before applications are given. This sequencing allows students to become familiar with the mechanics before delving into applications. An extensive problem set allows students to experience the power of this modeling tool in a biological context.

Chapter 12 This chapter contains some basic probabilistic and statistical tools. The statistics section has been expanded and more on stochastic processes has been added. The chapter can now be used for a full-semester course in probability and statistics, in particular if supplemented by real data sets, which are available on my web site: NUMB3R5 COUNT! (http://bioquest.org/numberscount/).

■ How to Use This Book

This book contains more material than can be covered in one year. The intent is to allow for more flexibility in the choice of material covered. Sections whose heading says "Optional" can easily be omitted; the material in those sections is not needed in subsequent sections.

The book's content can be arranged so that the course can be taught as a one-semester, two-quarter, two-semester, four-quarter, or three-semester course. Chapters 1–4 must be covered in that order before any of the other sections are covered. In addition to Chapters 1–4, the following sections can be chosen:

One semester—integration emphasis 5.1–5.6, 5.8, 6.1–6.3 (without 6.3.4 and 6.3.5)

One semester—differential equation emphasis 5.1–5.6, 5.8, 6.1, 6.2, 8.2 (without solving any of the differential equations)

Two quarters 5.1–5.6, 5.8, 6.1–6.3 (without 6.3.4 and 6.3.5), Chapter 7, Chapter 8

Two semesters 5.1–5.6, 5.8, 6.1–6.3, Chapters 7, 8, and 9, 10.1–10.4, 10.7, 11.1–11.4 (select two of the subsections in Section 11.4)

Four quarters or three semesters All sections that are not labeled optional; optional sections should be chosen as time permits

One semester—probability emphasis Chapter 1, only Section 2.2.1 and 2.2.2 in Chapter 2, Chapter 3 (except 3.6), Chapter 4, 5.1–5.4, 5.8, 6.1, 6.2, 7.1, 7.2.1, 12.1–12.5 (without 12.5.5), 12.6 (if time permits)

■ Supplements

Online Instructor's Solutions Manual Provides fully worked-out solutions to every textbook exercise, including the Chapter Review problems. Available to download online at the Instructor Resource Center at www.pearsonhighered.com/irc

Student's Solutions Manual Provides fully worked-out solutions to the odd-numbered exercises in the section and Chapter Review problems. ISBN-13: 978-0-321-64492-3, ISBN-10: 0-321-64492-1

■ Acknowledgments

This book would not have been possible without the help of numerous people. The book greatly benefited from critical reviews of an earlier draft.

Youn-Sha Chan, *University of Houston—Downtown*

Jeong-Mi Yoon, *University of Houston—Downtown*

Guillermo Goldsztein, *Georgia Institute of Technology*

Peter Howard, *Texas A&M University*

Petr Lisonek, *Simon Fraser University*

Lawrence Marx, *University of California—Davis*

Thanks to all my students who took this course over the years at the University of Minnesota and the University of California–Davis for their constructive criticism and enthusiasm. I owe a special thanks to George Lobell, my former mathematics editor at Prentice Hall, who made this book possible in the first place; to Eric Frank, my math editor for the second edition; and to the staff at Pearson Education who helped develop this third edition: Deirdre Lynch, Editor-in-Chief; Jennifer Crum, Executive Editor; Joanne Dill, Senior Project Editor; Jeff Weidenaar, Executive Marketing Manager; Heather Scott, Art Director; Bayani DeLeon, Assistant Managing Editor; Rob Merenoff, Project Manager; Tracy Patruno, Senior Production Supervisor; Joanne Wendelken, Editorial Assistant; and Kendra Bassi, Marketing Coordinator.

Finally, I wish to thank my husband, Maury Bramson, who again patiently put up with my long hours of working on the third edition of the book.

Claudia Neuhauser
neuha001@umn.edu
University of Minnesota Rochester

Calculus

for Biology and Medicine

Preview and Review

LEARNING OBJECTIVES
The first two sections of this chapter serve as a review of algebra, trigonometry, and precalculus, material needed to master the topics covered in this book. Section 1.3 reviews graphing functions and introduces the important concept of transforming functions into linear functions. The section includes a subsection on visualizing verbal descriptions of biological phenomena.

A Brief Overview of Calculus

Isaac Newton (1642–1727) and Gottfried Wilhelm Leibniz (1646–1716) are typically credited with the invention of calculus and were the first to develop the subject systematically.

Calculus has two parts: differential and integral calculus. Historically, differential calculus was concerned with finding lines tangent to curves and with calculating extrema (i.e., maxima and minima) of curves. Integral calculus has its roots in attempting to determine the areas of regions bounded by curves or in finding the volumes of solids. The two parts of calculus are closely related: The basic operation of one can be considered the inverse of the other. This result is known as the *fundamental theorem of calculus* and goes back to Newton and Leibniz, who were the first to understand its meaning and to put it to use in solving difficult problems.

Finding tangents, locating extrema, and calculating areas are basic geometric problems, and it may be somewhat surprising that their solution led to the development of methods that are useful in a wide range of scientific fields. The main reason for this historical development is that the slope of a tangent line at a given point is related to how quickly the function changes at that point. Knowing how quickly a function changes at a point opens up the possibility of a dynamic description of biology, such as a description of population growth, the speed at which a chemical reaction proceeds, the firing rate of neurons, and the speed at which an invasive species invades a new habitat. For this reason, calculus has been one of the most powerful tools in the mathematical formulation of scientific concepts. Applications of calculus are not restricted to biology, however; in fact, physics was the driving force in the original development of calculus. In this text we will be concerned primarily with how calculus is used in biology.

In addition to developing the theory of differential and integral calculus, we will consider many examples in which calculus is used to describe or model situations in the biological sciences. The use of quantitative reasoning is becoming increasingly more important in biology—for instance, in modeling interactions among species in a community, describing the activities of neurons, explaining genetic diversity in populations, and predicting the impact of global warming on vegetation. Today, calculus (Chapters 2–11) and probability and statistics (Chapter 12) are among the most important quantitative tools of a biologist.

■ 1.1 Preliminaries

This section reviews some of the concepts and techniques from algebra and trigonometry that are frequently used in calculus. The problems at the end of the section will help you reacquaint yourself with this material.

■ 1.1.1 The Real Numbers

Figure 1.1 The real-number line.

The **real numbers** can most easily be visualized on the **real-number line** (see Figure 1.1), on which numbers are ordered so that if $a < b$, then a is to the left of b. Sets (collections) of real numbers are typically denoted by the capital letters A, B, C, etc. To describe the set A, we write

$$A = \{x : \text{condition}\}$$

where "condition" tells us which numbers are in the set A. The most important sets in calculus are **intervals**. We use the following notations: If $a < b$, then

the **open** interval $(a, b) = \{x : a < x < b\}$

and

the **closed** interval $[a, b] = \{x : a \leq x \leq b\}$

We also use **half-open** intervals:

$$[a, b) = \{x : a \leq x < b\} \qquad \text{and} \qquad (a, b] = \{x : a < x \leq b\}$$

Unbounded intervals are sets of the form $\{x : x > a\}$. Here are the possible cases:

$$[a, \infty) = \{x : x \geq a\}$$
$$(-\infty, a] = \{x : x \leq a\}$$
$$(a, \infty) = \{x : x > a\}$$
$$(-\infty, a) = \{x : x < a\}$$

The symbols "∞" and "$-\infty$" mean "plus infinity" and "minus infinity," respectively. These symbols are *not* real numbers, but are used merely for notational convenience. The real-number line, denoted by **R**, does not have endpoints, and we can write **R** in the following equivalent forms:

$$\mathbf{R} = \{x : -\infty < x < \infty\} = (-\infty, \infty)$$

The location of the number 0 on the real-number line is called the **origin**, and we can measure the distance of the number x to the origin. For instance, -5 is 5 units to the left of the origin. A convenient notation for measuring distances from the origin on the real-number line is the absolute value of a real number.

> **Definition**   The **absolute value** of a real number a, denoted by $|a|$, is
>
> $$|a| = \begin{cases} a & \text{if } a \geq 0 \\ -a & \text{if } a < 0 \end{cases}$$

For example, $|-7| = -(-7) = 7$. We can use absolute values to find the distance between any two numbers x_1 and x_2 as follows:

$$\text{distance between } x_1 \text{ and } x_2 = |x_1 - x_2|$$

Note that $|x_1 - x_2| = |x_2 - x_1|$. To find the distance between -2 and 4, we compute $|-2 - 4| = |-6| = 6$, or $|4 - (-2)| = |4 + 2| = 6$.

We will frequently need to solve equations containing absolute values, for which the following property is useful:

> Let $b \geq 0$. Then
>
> **1.** For $a \geq 0$, $|a| = b$ is equivalent to $a = b$.
> **2.** For $a < 0$, $|a| = b$ is equivalent to $-a = b$.

EXAMPLE 1

Solve $|x - 4| = 2$.

Solution

If $x - 4 \geq 0$, then $x - 4 = 2$ and thus $x = 6$. If $x - 4 < 0$, then $-(x - 4) = 2$ and thus $x = 2$. The solutions, illustrated graphically in Figure 1.2, are therefore $x = 6$ and $x = 2$. The points of intersection of $y = |x - 4|$ and $y = 2$ are at $x = 6$ and $x = 2$. Solving $|x - 4| = 2$ can also be interpreted as finding the two numbers that have distance 2 from 4. ∎

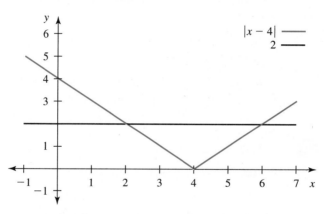

Figure 1.2 The graph of $y = |x - 4|$ and $y = 2$. The points of intersection are at $x = 6$ and $x = 2$.

We write the solution of an equation of the form $|a| = |b|$ as either $a = b$ or $a = -b$, illustrated in the next example.

EXAMPLE 2

Solve $|\frac{3}{2}x - 1| = |\frac{1}{2}x + 1|$.

Solution

Either

$$\frac{3}{2}x - 1 = \frac{1}{2}x + 1 \qquad \text{or} \qquad \frac{3}{2}x - 1 = -\left(\frac{1}{2}x + 1\right)$$

$$x = 2 \qquad\qquad \frac{3}{2}x - 1 = -\frac{1}{2}x - 1$$

$$2x = 0$$

$$x = 0$$

A graphical solution of this example is shown in Figure 1.3. ∎

Returning to Example 1, where we found the two points whose distance from 4 was equal to 2, we can also try to find those points whose distance from 4 is less than (or greater than) 2. This amounts to solving inequalities with absolute values. Looking back at Figure 1.2, we see that the set of x-values whose distance from 4 is less than 2 (i.e., $|x - 4| < 2$) is the interval $(2, 6)$. Similarly, the set of x-values whose distance from 4 is greater than 2 (i.e., $|x - 4| > 2$) is the union of the two intervals $(-\infty, 2)$ and $(6, \infty)$, or $(-\infty, 2) \cup (6, \infty)$.

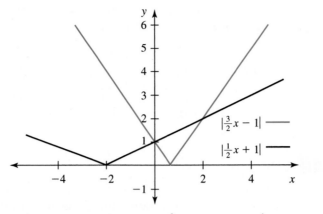

Figure 1.3 The graphs of $y = |\frac{3}{2}x - 1|$ and $y = |\frac{1}{2}x + 1|$.
The points of intersection are at $x = 0$ and $x = 2$.

In general, to solve absolute-value inequalities, the following two properties are useful:

Let $b > 0$. Then

1. $|a| < b$ is equivalent to $-b < a < b$.
2. $|a| > b$ is equivalent to $a > b$ or $a < -b$.

EXAMPLE 3 **(a)** Solve $|2x - 5| < 3$.   **(b)** Solve $|4 - 3x| \geq 2$.

Solution **(a)** We rewrite $|2x - 5| < 3$ as

$$-3 < 2x - 5 < 3$$

Adding 5 to all three parts, we obtain

$$2 < 2x < 8$$

Dividing the result by 2, we find that

$$1 < x < 4$$

The solution is therefore the set $\{x : 1 < x < 4\}$. In interval notation, the solution can be written as the open interval $(1, 4)$.

(b) To solve $|4 - 3x| \geq 2$, we go through the following steps:

$$
\begin{array}{ccc}
4 - 3x \geq 2 & & 4 - 3x \leq -2 \\
-3x \geq -2 & \text{or} & -3x \leq -6 \\
x \leq \dfrac{2}{3} & & x \geq 2
\end{array}
$$

The solution is the set $\{x : x \geq 2 \text{ or } x \leq \frac{2}{3}\}$, or, in interval notation, $(-\infty, \frac{2}{3}] \cup [2, \infty)$.

■

■ 1.1.2 Lines in the Plane

We will frequently encounter situations in which the relationship between quantities can be described by a **linear equation**. For example, when a weight is attached to a helical spring made of some elastic material (and the weight is not too heavy), the relationship between the length y of the spring and the weight x is

$$y = y_0 + kx \tag{1.1}$$

where y_0 denotes the length of the spring when no weight is attached to it and k is a positive constant. Equation (1.1) is an example of a linear equation, and we say that x and y satisfy a linear equation.

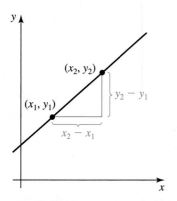

Figure 1.4 The slope of a straight line.

The **standard** form of a linear equation is given by

$$Ax + By + C = 0$$

where A, B, and C are constants, A and B are not both equal to 0, and x and y are the two variables. In algebra, you learned that the graph of a linear equation is a straight line.

If the two points (x_1, y_1) and (x_2, y_2) lie on a straight line, then the **slope** of the line is

$$m = \frac{y_2 - y_1}{x_2 - x_1}$$

(See Figure 1.4.) Two points (or one point and the slope) are sufficient to determine the equation of a straight line.

If you are given one point and the slope, you can use the **point–slope** form of a straight line to write its equation, given by

$$y - y_0 = m(x - x_0)$$

where m is the slope and (x_0, y_0) is a point on the line. If you are given two points, first compute the slope and then use one of the points and the slope to find the equation of the straight line in point–slope form.

Lastly, the most frequently used form of a linear equation is the **slope–intercept** form

$$y = mx + b$$

where m is the slope and b is the y-intercept, which is the point of intersection of the line with the y-axis; the y-intercept has coordinates $(0, b)$.

We summarize these three forms of linear equations in the following box:

$$
\begin{array}{ll}
Ax + By + C = 0 & \text{(Standard Form)} \\
y - y_0 = m(x - x_0) & \text{(Point–Slope Form)} \\
y = mx + b & \text{(Slope–Intercept Form)}
\end{array}
$$

EXAMPLE 4 Determine, in slope–intercept form, the equation of the line passing through $(-2, 1)$ and $(3, -\frac{1}{2})$.

Solution The slope of the line is

$$m = \frac{y_2 - y_1}{x_2 - x_1} = \frac{-\frac{1}{2} - 1}{3 - (-2)} = \frac{-\frac{3}{2}}{5} = -\frac{3}{10}$$

Using the point–slope form with $(-2, 1)$, we find that

$$y - 1 = -\frac{3}{10}(x - (-2))$$

or, in slope–intercept form,

$$y = -\frac{3}{10}x + \frac{2}{5}$$

We could have used the other point, $(3, -\frac{1}{2})$, and obtained the same result. ■

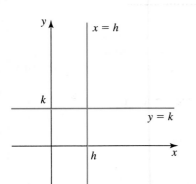

Figure 1.5 The horizontal line $y = k$ and the vertical line $x = h$.

We now recall two special cases that we illustrate in Figure 1.5:

$$
\begin{array}{ll}
y = k & \text{horizontal line (slope 0)} \\
x = h & \text{vertical line (slope undefined)}
\end{array}
$$

In the next example, we show how to determine the slope and the y-intercept of a given straight line.

EXAMPLE 5

Determine the slope and the y-intercept of the line $3y - 2x + 9 = 0$.

Solution

We solve for y in $3y = 2x - 9$. We obtain $y = \frac{2}{3}x - 3$. We can now read off the slope $m = \frac{2}{3}$ and the y-intercept $b = -3$.   ■

When two quantities x and y are linearly related so that

$$y = mx$$

we say that y is **proportional** to x, with m denoting the **constant of proportionality**, and we write

$$y \propto x$$

The symbol $\propto$ is read "is proportional to." If we write Equation (1.1) in the form

$$y - y_0 = kx$$

then the change in length $y - y_0$ is proportional to the attached weight with constant of proportionality k, and we can write

$$y - y_0 \propto x$$

There are two more properties of straight lines we wish to mention. When two lines l_1 and l_2 in the plane have no points in common or are identical, they are called **parallel**, denoted by $l_1 \parallel l_2$. The following criterion is useful in deciding whether two lines are parallel: Two noncoincident lines l_1 and l_2 are parallel ($l_1 \parallel l_2$) if and only if their slopes are identical. For two noncoincident, nonvertical lines l_1 and l_2 with slopes m_1 and m_2, respectively, the criterion becomes

$$l_1 \parallel l_2 \qquad \text{if and only if} \qquad m_1 = m_2$$

Two lines l_1 and l_2 are called **perpendicular** ($l_1 \perp l_2$) if their intersection forms an angle of 90°. The following criterion is useful for deciding whether two lines are perpendicular: Two nonvertical lines are perpendicular if and only if their slopes are negative reciprocals. That is, if l_1 and l_2 are nonvertical lines with slopes m_1 and m_2, then

$$l_1 \perp l_2 \qquad \text{if and only if} \qquad m_1 m_2 = -1$$

We will prove this result in Problem 54 at the end of this section.

■ 1.1.3 Equation of the Circle

A **circle** is the set of all points at a given distance, called the **radius**, from a given point, called the **center**. If r is the distance from (x_0, y_0) to (x, y) (see Figure 1.6), then, using the Pythagorean theorem, we find that

$$r^2 = (x - x_0)^2 + (y - y_0)^2$$

If $r = 1$ and $(x_0, y_0) = (0, 0)$, the circle is called the **unit circle**.

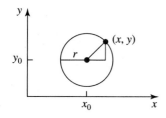

Figure 1.6 Circle with radius r centered at (x_0, y_0).

EXAMPLE 6

Find the equation of the circle with center $(2, 3)$ and passing through $(5, 7)$.

Solution

Using the Pythagorean theorem, we can compute the distance in the plane between $(2, 3)$ and $(5, 7)$:

$$\sqrt{(5 - 2)^2 + (7 - 3)^2} = \sqrt{9 + 16} = 5$$

Thus, this circle has radius 5 and center $(2, 3)$, and its equation is

$$25 = (x - 2)^2 + (y - 3)^2$$ ■

■ 1.1.4 Trigonometry

We will need a few results from trigonometry. Recall that angles are measured in either degrees or radians and that a complete revolution on a unit circle (Figure 1.7) corresponds to $360°$, or 2π. For reasons that will become clear, the radian measure is preferred in calculus. To convert between degree and radian measure, we use the formula

$$\frac{\theta \text{ measured in degrees}}{360°} = \frac{\theta \text{ measured in radians}}{2\pi}$$

For instance, to convert $23°$ into radian measure, we compute

$$\theta = 23° \frac{2\pi}{360°} = 0.401$$

To convert $\frac{\pi}{6}$ into degrees, we compute

$$\theta = \frac{\pi}{6} \frac{360°}{2\pi} = 30°$$

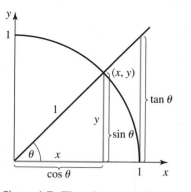

Figure 1.7 The trigonometric functions on a unit circle.

There are four trigonometric functions that you should be familiar with: sine, cosine, tangent, and secant; the other two, cotangent and cosecant, are rarely used. The six are defined on a unit circle (see Figure 1.7) and are abbreviated as sin, cos, tan, sec, cot, and csc, respectively. Recall that a positive angle is measured counterclockwise from the positive x-axis, whereas a negative angle is measured clockwise. The six trigonometric functions are defined as follows:

$$\sin\theta = \frac{y}{1} = y \qquad \csc\theta = \frac{1}{\sin\theta} = \frac{1}{y}$$

$$\cos\theta = \frac{x}{1} = x \qquad \sec\theta = \frac{1}{\cos\theta} = \frac{1}{x}$$

$$\tan\theta = \frac{y}{x} \qquad \cot\theta = \frac{1}{\tan\theta} = \frac{x}{y}$$

There are a number of frequently used trigonometric identities. First, since $\tan\theta = y/x$ with $y = \sin\theta$ and $x = \cos\theta$, it follows that

$$\tan\theta = \frac{\sin\theta}{\cos\theta}$$

Now, applying the Pythagorean theorem to the triangle in Figure 1.7 and using the notation $\sin^2\theta = (\sin\theta)^2$, we find that

$$\sin^2\theta + \cos^2\theta = 1$$

Next, if we divide the preceding identity by $\cos^2\theta$, we obtain

$$\frac{\sin^2\theta}{\cos^2\theta} + 1 = \frac{1}{\cos^2\theta}$$

Using $\tan\theta = \sin\theta/\cos\theta$ and $\sec\theta = 1/\cos\theta$, we can write this as

$$\tan^2\theta + 1 = \sec^2\theta$$

In the next example, we solve a trigonometric equation.

EXAMPLE 7   Solve

$$2\sin\theta\cos\theta = \cos\theta \quad \text{on } [0, 2\pi)$$

Solution   We should not be tempted to cancel $\cos\theta$ on each side; this would cause us to lose solutions. Instead, we bring $\cos\theta$ to the left side and factor $\cos\theta$ to obtain

$$\cos\theta(2\sin\theta - 1) = 0$$

That is,

$$\cos\theta = 0 \quad \text{or} \quad 2\sin\theta - 1 = 0$$

Solving $\cos\theta = 0$, we find that

$$\theta = \frac{\pi}{2} \quad \text{or} \quad \theta = \frac{3\pi}{2}$$

Solving $2\sin\theta - 1 = 0$, we get

$$\sin\theta = \frac{1}{2}$$

which yields

$$\theta = \frac{\pi}{6} \quad \text{or} \quad \theta = \frac{5\pi}{6}$$

The solution set is therefore $\{\frac{\pi}{6}, \frac{\pi}{2}, \frac{5\pi}{6}, \frac{3\pi}{2}\}$.   ■

Figure 1.8 yields the following two identities when we compare the two angles θ and $-\theta$ (a positive angle is measured counterclockwise from the positive x-axis, whereas a negative angle is measured clockwise):

$$\sin(-\theta) = -\sin\theta \quad \text{and} \quad \cos(-\theta) = \cos\theta$$

Some exact trigonometric values are collected in Table 1-1. Of course, $\frac{1}{2}\sqrt{0} = 0$, $\frac{1}{2}\sqrt{1} = \frac{1}{2}$, and $\frac{1}{2}\sqrt{4} = 1$, and you should memorize these simplified values. Rewriting Table 1-1 will make it easier to re-create the table in case you forget the exact values. Using $\tan\theta = \sin\theta/\cos\theta$, you immediately get the values for $\tan\theta$.

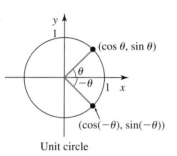

Figure 1.8 Using the unit circle to define trigonometric identities.

TABLE 1-1 Some Exact Trigonometric Values

Angle θ	0 (0°)	$\dfrac{\pi}{6}$ (30°)	$\dfrac{\pi}{4}$ (45°)	$\dfrac{\pi}{3}$ (60°)	$\dfrac{\pi}{2}$ (90°)
$\sin\theta$	$\frac{1}{2}\sqrt{0}$	$\frac{1}{2}\sqrt{1}$	$\frac{1}{2}\sqrt{2}$	$\frac{1}{2}\sqrt{3}$	$\frac{1}{2}\sqrt{4}$
$\cos\theta$	$\frac{1}{2}\sqrt{4}$	$\frac{1}{2}\sqrt{3}$	$\frac{1}{2}\sqrt{2}$	$\frac{1}{2}\sqrt{1}$	$\frac{1}{2}\sqrt{0}$

■ 1.1.5 Exponentials and Logarithms

Exponentials and logarithms are particularly important in biological contexts.
An exponential is an expression of the form

$$a^r$$

where a is called the **base** and r the **exponent**. Unless r is an integer or unless r is a rational number of the form p/q where p is an integer and q is an odd integer, we will assume that a is positive. We summarize some of the properties of an exponential as follows:

$$a^r a^s = a^{r+s} \qquad (ab)^r = a^r b^r$$

$$\frac{a^r}{a^s} = a^{r-s} \qquad \left(\frac{a}{b}\right)^r = \frac{a^r}{b^r}$$

$$a^{-r} = \frac{1}{a^r} \qquad (a^r)^s = a^{rs}$$

EXAMPLE 8 Evaluate the following exponential expressions:

(a) $3^2 3^{5/2} = 3^{2+5/2} = 3^{9/2}$

(b) $\dfrac{2^{-4} 2^3}{2^2} = \dfrac{2^{-1}}{2^2} = 2^{-1-2} = 2^{-3} = \dfrac{1}{2^3} = \dfrac{1}{8}$

(c) $\dfrac{a^k a^{3k}}{a^{5k}} = a^{k+3k-5k} = a^{-k} = \dfrac{1}{a^k}$ ■

Logarithms allow us to solve equations of the form

$$2^x = 8$$

The solution of this equation is $x = 3$, which we can write as

$$x = \log_2 8 = 3$$

In other words, a logarithm is an exponent. The expression

$$\log_a y$$

is the exponent on the base a that yields the number y. Logarithms are defined only for $y > 0$ (where the base is assumed to be positive and different from 1). We have the following correspondence between logarithms and exponentials:

$$x = \log_a y \qquad \text{is equivalent to} \qquad y = a^x$$

EXAMPLE 9 Which real number x satisfies

(a) $\log_3 x = -2$? **(b)** $\log_{1/2} 8 = x$?

Solution **(a)** We write this in the equivalent form

$$x = 3^{-2}$$

Hence,

$$x = \frac{1}{3^2} = \frac{1}{9}$$

(b) We write this in the equivalent form

$$\left(\frac{1}{2}\right)^x = 8$$
$$2^{-x} = 2^3$$
$$2^x = 2^{-3}$$

Setting the exponents equal to each other, we find that $x = -3$. Note that, in order to compare exponents, the bases must be the same. ■

Some important properties of logarithms are as follows:

$$\log_a(xy) = \log_a x + \log_a y$$
$$\log_a\left(\frac{x}{y}\right) = \log_a x - \log_a y$$
$$\log_a x^r = r \log_a x$$

The most important logarithm is the **natural logarithm**, which has the number e as its base. The number e is an irrational number whose value is approximately 2.7182818. The natural logarithm is written $\ln x$; that is, $\log_e x = \ln x$.

EXAMPLE 10   Assume that x and y are positive, and simplify the following expressions:

(a) $\log_3(9x^2) = \log_3 9 + \log_3 x^2 = 2 + 2\log_3 x$

(b) $\log_5 \frac{x^2+3}{5x} = \log_5(x^2 + 3) - \log_5 5 - \log_5 x = \log_5(x^2 + 3) - 1 - \log_5 x$

[Note that $\log_5(x^2 + 3)$ cannot be simplified any further.]

(c) $-\ln\frac{1}{2} = \ln(\frac{1}{2})^{-1} = \ln 2$

(d) $\ln\frac{3x^2}{\sqrt{y}} = \ln 3 + \ln x^2 - \ln\sqrt{y} = \ln 3 + 2\ln x - \frac{1}{2}\ln y$

(In the last step, we used the fact that $\sqrt{y} = y^{1/2}$.)   ■

In algebra, you learned how to solve equations of the form $e^{2x} = 3$ or $\ln(x+1) = 5$. We will need to do this frequently. The key to solving such equations are the two identities

$$\log_a a^x = x \qquad \text{and} \qquad a^{\log_a x} = x$$

The next example illustrates how to use these identities.

EXAMPLE 11   Solve for x.

(a) $e^{2x} = 3$ \qquad (b) $\ln(x + 1) = 5$ \qquad (c) $5^{2x-1} = 2^x$

Solution   (a) To solve $e^{2x} = 3$ for x, we take logarithms to base e on both sides:

$$\ln e^{2x} = \ln 3$$

But $\ln e^{2x} = 2x$; hence,

$$2x = \ln 3, \qquad \text{or} \qquad x = \frac{1}{2}\ln 3$$

(b) To solve $\ln(x + 1) = 5$, we write the equation in exponential form:

$$e^{\ln(x+1)} = e^5$$

This simplifies to

$$x + 1 = e^5, \qquad \text{or} \qquad x = e^5 - 1$$

(c) To solve $5^{2x-1} = 2^x$ for x, we observe that the two bases are different. We therefore cannot compare the exponents directly. Instead, we take logarithms on both sides. Any positive base (different from 1) for the logarithm would work, and we choose base e, since it is the most commonly used base in calculus. Doing so yields

$$\ln 5^{2x-1} = \ln 2^x$$

or, after simplifying,

$$(2x - 1)\ln 5 = x\ln 2$$

Solving for x, we find that

$$2x\ln 5 - x\ln 2 = \ln 5$$
$$x(2\ln 5 - \ln 2) = \ln 5$$

Hence,

$$x = \frac{\ln 5}{2\ln 5 - \ln 2}$$

■

■ 1.1.6 Complex Numbers and Quadratic Equations

The square of a real number is always nonnegative. However, there are situations in which we need to take a square root of a negative number. Since the resulting square root cannot be a real number, we introduce a new symbol, which we denote by i, that will allow us to deal with this case. We set

$$i^2 = -1$$

The symbol i is called the **imaginary unit**. Thus, instead of writing $\sqrt{-17}$, for instance, we can now write $i\sqrt{17}$.

The symbol i allows us to introduce a new number system, the set of **complex numbers**:

A **complex number** is a number of the form

$$z = a + bi$$

where a and b are real numbers. The real number a is the **real part** of $a + bi$, and the real number b is the **imaginary part**.

For instance, $-3 + 7i$ has real part -3 and imaginary part 7, and $2 - 5i$ has real part 2 and imaginary part -5. Since $a + 0i = a$, it follows that the set of real numbers is a subset of the set of complex numbers. Complex numbers of the form bi are called **purely imaginary numbers**.

Two complex numbers are equal if their respective real and imaginary parts are equal; that is,

$$a + bi = c + di \qquad \text{if and only if} \qquad a = c \quad \text{and} \quad b = d$$

To add two complex numbers, we use the following rule:

$$(a + bi) + (c + di) = (a + c) + (b + d)i$$

This rule says that real and imaginary parts are added separately. To calculate the product of two complex numbers, we proceed as follows:

$$\begin{aligned}
(a + bi)(c + di) &= ac + adi + bci + bdi^2 \\
&= ac + (ad + bc)i - bd \\
&= (ac - bd) + (ad + bc)i
\end{aligned}$$

Note that we used $i^2 = -1$ in the penultimate step. There is no need to memorize the product of two complex numbers, since we can always compute it by the distributive law.

EXAMPLE 12

Find

(a) $(2 + 3i) - (5 - 6i)$, (b) $(5 - 3i)(1 + 2i)$.

Solution

(a) $(2 + 3i) - (5 - 6i) = 2 + 3i - 5 + 6i = -3 + 9i$,

(b) $(5 - 3i)(1 + 2i) = 5 + 10i - 3i - 6i^2 = 5 + 7i - (6)(-1) = 11 + 7i$. ■

If $z = a + bi$ is a complex number, its **conjugate**, denoted by $\bar{z}$, is defined as

$$\bar{z} = a - bi$$

For complex numbers z and w, it can be shown (see Problems 113–115) that

$$\overline{(\overline{z})} = z$$
$$\overline{z + w} = \overline{z} + \overline{w}$$
$$\overline{zw} = \overline{z}\,\overline{w}$$

Furthermore, if we multiply a complex number by its conjugate, we find that

$$z\overline{z} = (a + bi)(a - bi) = a^2 - abi + abi - b^2i^2$$
$$= a^2 + b^2$$

That is,

If $z = a + bi$, then
$$z\overline{z} = a^2 + b^2$$

EXAMPLE 13　　Let $z = 3 + 2i$.

(a) Find $\overline{z}$.　　　(b) Compute $z\overline{z}$.

Solution　　(a) $\overline{z} = 3 - 2i$.

(b) $z\overline{z} = (3 + 2i)(3 - 2i) = 9 - 4i^2 = 9 + 4 = 13$.　　■

We encounter complex numbers primarily when we solve quadratic equations. Recall that, to solve

$$ax^2 + bx + c = 0$$

for $a \neq 0$, we use the quadratic formula

$$x_{1,2} = \frac{-b \pm \sqrt{b^2 - 4ac}}{2a}$$

where $x_{1,2}$ refers to the two solutions x_1 (with the "+" sign) and x_2 (with the "−" sign).

EXAMPLE 14　　Solve
$$x^2 + 4x + 5 = 0$$

Solution　　Using the quadratic formula, we obtain

$$x_{1,2} = \frac{-4 \pm \sqrt{4^2 - (4)(1)(5)}}{(2)(1)}$$

$$= \frac{-4 \pm \sqrt{16 - 20}}{2} = \frac{-4 \pm \sqrt{-4}}{2}$$

If we allowed solutions only in the real-number system, we would conclude that $x^2 + 4x + 5 = 0$ has no solutions. But if we allow solutions in the complex number system, we find that

$$x_{1,2} = \frac{-4 \pm \sqrt{4i^2}}{2} = \frac{-4 \pm 2i}{2} = \frac{2(-2 \pm i)}{2} = -2 \pm i$$

That is, $x_1 = -2 + i$ and $x_2 = -2 - i$.　　■

The term $b^2 - 4ac$ under the square root sign in the quadratic formula is called the **discriminant**. If the discriminant is nonnegative, the two solutions of the corresponding quadratic equation are real. (When the discriminant is equal to 0, the two solutions are identical.) If the discriminant is negative, the two solutions are complex conjugates of each other.

EXAMPLE 15   Without solving

$$2x^2 - 3x + 7 = 0$$

what can you say about the solution?

Solution   We compute the discriminant

$$b^2 - 4ac = (-3)^2 - (4)(2)(7) = 9 - 56 = -47 < 0$$

Since the discriminant is negative, the equation $2x^2 - 3x + 7 = 0$ has two complex solutions, which are conjugates of each other.   ■

Section 1.1 Problems

■ **1.1.1**

1. Find the two numbers that have distance 3 from -1 by **(a)** measuring the distances on the real-number line and **(b)** solving an appropriate equation involving an absolute value.

2. Find all pairwise distances between the numbers -5, 2, and 7 by **(a)** measuring the distances on the real-number line and **(b)** computing the distances by using absolute values.

3. Solve the following equations:
(a) $|2x - 4| = 6$ **(b)** $|x - 3| = 2$
(c) $|2x + 3| = 5$ **(d)** $|7 - 3x| = -2$

4. Solve the following equations:
(a) $|2x + 4| = |5x - 2|$ **(b)** $|5 - 3u| = |3 + 2u|$
(c) $|4 + \frac{t}{2}| = |\frac{3}{2}t - 2|$ **(d)** $|2s - 3| = |7 - s|$

5. Solve the following inequalities:
(a) $|5x - 2| \leq 4$ **(b)** $|1 - 3x| > 8$
(c) $|7x + 4| \geq 3$ **(d)** $|6 - 5x| < 7$

6. Solve the following inequalities:
(a) $|2x + 3| < 6$ **(b)** $|3 - 4x| \geq 2$
(c) $|x + 5| \leq 1$ **(d)** $|7 - 2x| < 0$

■ **1.1.2**

In Problems 7–42, determine the equation of the line that satisfies the stated requirements. Put the equation in standard form.

7. The line passing through $(2, 4)$ with slope $-\frac{1}{3}$

8. The line passing through $(1, -2)$ with slope 2

9. The line passing through $(0, -2)$ with slope -3

10. The line passing through $(-3, 5)$ with slope $1/2$

11. The line passing through $(-2, -3)$ and $(1, 4)$

12. The line passing through $(-1, 4)$ and $(2, -\frac{1}{2})$

13. The line passing through $(0, 4)$ and $(3, 0)$

14. The line passing through $(1, -1)$ and $(4, 5)$

15. The horizontal line through $(3, \frac{3}{2})$

16. The horizontal line through $(0, -1)$

17. The vertical line through $(-1, \frac{7}{2})$

18. The vertical line through $(2, -3)$

19. The line with slope 3 and y-intercept $(0, 2)$

20. The line with slope -1 and y-intercept $(0, -3)$

21. The line with slope $1/2$ and y-intercept $(0, 2)$

22. The line with slope $-1/3$ and y-intercept $(0, -1)$

23. The line with slope -2 and x-intercept $(1, 0)$

24. The line with slope 1 and x-intercept $(-2, 0)$

25. The line with slope $-1/4$ and x-intercept $(3, 0)$

26. The line with slope $1/5$ and x-intercept $(-1/2, 0)$

27. The line passing through $(2, -3)$ and parallel to

$$x + 2y - 4 = 0$$

28. The line passing through $(1, 2)$ and parallel to

$$x - 3y - 6 = 0$$

29. The line passing through $(-1, -1)$ and parallel to the line passing through $(0, 1)$ and $(3, 0)$

30. The line passing through $(2, -3)$ and parallel to the line passing through $(0, -1)$ and $(2, 1)$

31. The line passing through $(1, 4)$ and perpendicular to

$$2y - 5x + 7 = 0$$

32. The line passing through $(-1, -1)$ and perpendicular to

$$x - y + 3 = 0$$

33. The line passing through $(5, -1)$ and perpendicular to the line passing through $(-2, 1)$ and $(1, -2)$

34. The line passing through $(4, -1)$ and perpendicular to the line passing through $(-2, 0)$ and $(1, 1)$

35. The line passing through $(4, 2)$ and parallel to the horizontal line passing through $(1, -2)$

36. The line passing through $(-1, 5)$ and parallel to the horizontal line passing through $(2, -1)$

37. The line passing through $(-1, 1)$ and parallel to the vertical line passing through $(2, -1)$

38. The line passing through $(3, 1)$ and parallel to the vertical line passing through $(-1, -2)$

39. The line passing through $(1, -3)$ and perpendicular to the horizontal line passing through $(-1, -1)$

40. The line passing through $(4, 2)$ and perpendicular to the horizontal line passing through $(3, 1)$

41. The line passing through $(7, 3)$ and perpendicular to the vertical line passing through $(-2, 4)$

42. The line passing through $(-2, 5)$ and perpendicular to the vertical line passing through $(1, 4)$

43. To convert a length measured in feet to a length measured in centimeters, we use the facts that a length measured in feet is proportional to a length measured in centimeters and that 1 ft corresponds to 30.5 cm. If x denotes the length measured in ft and y denotes the length measured in cm, then

$$y = 30.5x$$

(a) Explain how to use this relationship.

(b) Use the relationship to convert the following measurements into centimeters:
(i) 6 ft   **(ii)** 3 ft, 2 in   **(iii)** 1 ft, 7 in

(c) Use the relationship to convert the following measurements into ft:
(i) 173 cm   **(ii)** 75 cm   **(iii)** 48 cm

44. (a) To convert the weight of an object from kilograms (kg) to pounds (lb), you use the facts that a weight measured in kilograms is proportional to a weight measured in pounds and that 1 kg corresponds to 2.20 lb. Find an equation that relates weight measured in kilograms to weight measured in pounds.

(b) Use your answer in (a) to convert the following measurements:
(i) 63 lb   **(ii)** 150 lb   **(iii)** 2.5 kg   **(iv)** 140 kg

45. Assume that the distance a car travels is proportional to the time it takes to cover the distance. Find an equation that relates distance and time if it takes the car 15 min to travel 10 mi. What is the constant of proportionality if distance is measured in miles and time is measured in hours?

46. Assume that the number of seeds a plant produces is proportional to its aboveground biomass. Find an equation that relates number of seeds and aboveground biomass if a plant that weighs 217 g has 17 seeds.

47. Experimental study plots are often squares of length 1 m. If 1 ft corresponds to 0.305 m, compute the area of a square plot of length 1 m in ft^2.

48. Large areas are often measured in hectares (ha) or in acres. If 1 ha = 10,000 m^2 and 1 acre = 4046.86 m^2, how many acres is 1 hectare?

49. To convert the volume of a liquid measured in ounces to a volume measured in liters, we use the fact that 1 liter equals 33.81 ounces. Denote by x the volume measured in ounces and by y the volume measured in liters. Assume a linear relationship between these two units of measurements.

(a) Find the equation relating x and y.

(b) A typical soda can contains 12 ounces of liquid. How many liters is this?

50. To convert a distance measured in miles to a distance measured in kilometers, we use the fact that 1 mile equals 1.609 kilometers. Denote by x the distance measured in miles and by y the distance measured in kilometers. Assume a linear relationship between these two units of measurements.

(a) Find an equation relating x and y.

(b) The distance between Minneapolis and Madison is 261 miles. How many kilometers is this?

51. Car speed in many countries is measured in kilometers per hour. In the United States, car speed is measured in miles per hour. To convert between these units, use the fact that 1 mile equals 1.609 kilometers.

(a) The speed limit on many U.S. highways is 55 miles per hour. Convert this number into kilometers per hour.

(b) The recommended speed limit on German highways is 130 kilometers per hour. Convert this number into miles per hour.

To measure temperature, three scales are commonly used: Fahrenheit, Celsius, and Kelvin. These scales are linearly related. We discuss these scales in Problems 52 and 53.

52. (a) The Celsius scale is devised so that 0°C is the freezing point of water (at 1 atmosphere of pressure) and 100°C is the boiling point of water (at 1 atmosphere of pressure). If you are more familiar with the Fahrenheit scale, then you know that water freezes at 32°F and boils at 212°F. Find a linear equation that relates temperature measured in degrees Celsius and temperature measured in degrees Fahrenheit.

(b) The normal body temperature in humans ranges from 97.6°F to 99.6°F. Convert this temperature range into degrees Celsius.

53. (a) The Kelvin (K) scale is an **absolute** scale of temperature. The zero point of the scale (0 K) denotes **absolute zero**, the coldest possible temperature; that is, no body can have a temperature below 0 K. It has been determined experimentally that 0 K corresponds to -273.15°C. If 1 K denotes the same temperature difference as 1°C, find an equation that relates the Kelvin and Celsius scales.

(b) Pure nitrogen and pure oxygen can be produced cheaply by first liquefying purified air and then allowing the temperature of the liquid air to rise slowly. Since nitrogen and oxygen have different boiling points, they are distilled at different temperatures. The boiling point of nitrogen is 77.4 K and of oxygen is 90.2 K. Convert each of these boiling-point temperatures into Celsius. If you solved Problem 52(a), convert the boiling-point temperatures into Fahrenheit as well. Consider the two techniques described for distilling nitrogen and oxygen. Which element gets distilled first?

54. Use the following steps to show that if two nonvertical lines l_1 and l_2 with slopes m_1 and m_2, respectively, are perpendicular, then $m_1 m_2 = -1$: Assume that $m_1 < 0$ and $m_2 > 0$.

(a) Use a graph to show that if θ_1 and θ_2 are the respective angles of inclination of the lines l_1 and l_2, then $\theta_1 = \theta_2 + \frac{\pi}{2}$. (The angle of inclination of a line is the angle $\theta \in [0, \pi)$ between the line and the positively directed x-axis.)

(b) Use the fact that $\tan(\pi - x) = -\tan x$ to show that $m_1 = \tan \theta_1$ and $m_2 = \tan \theta_2$.

(c) Use the fact that $\tan(\frac{\pi}{2} - x) = \cot x$ and $\cot(-x) = -\cot x$ to show that $m_1 = -\cot \theta_2$.

(d) From the latter equation, deduce the truth of the claim set forth at the beginning of this problem.

■ **1.1.3**

55. Find the equation of a circle with center $(-1, 4)$ and radius 3.

56. Find the equation of a circle with center $(2, 3)$ and radius 4.

57. (a) Find the equation of a circle with center $(2, 5)$ and radius 3.

(b) Where does the circle intersect the y-axis?

(c) Does the circle intersect the x-axis? Explain.

58. (a) Find all possible radii of a circle centered at $(3, 6)$ so that the circle intersects only one axis.

(b) Find all possible radii of a circle centered at $(3, 6)$ so that the circle intersects both axes.

59. Find the center and the radius of the circle given by the equation
$$(x - 2)^2 + y^2 = 16$$

60. Find the center and the radius of the circle given by the equation
$$(x + 1)^2 + (y - 3)^2 = 9$$

61. Find the center and the radius of the circle given by the equation
$$0 = x^2 + y^2 - 4x + 2y - 11$$
(To do this, you must complete the squares.)

62. Find the center and the radius of the circle given by the equation
$$x^2 + y^2 + 2x - 4y + 1 = 0$$
(To do this, you must complete the squares.)

■ **1.1.4**

63. (a) Convert $75°$ to radian measure.
(b) Convert $\frac{17}{12}\pi$ to degree measure.

64. (a) Convert $-15°$ to radian measure.
(b) Convert $\frac{3}{4}\pi$ to degree measure.

65. Evaluate the following expressions without using a calculator:
(a) $\sin(-\frac{5\pi}{4})$ **(b)** $\cos(\frac{5\pi}{6})$ **(c)** $\tan(\frac{\pi}{3})$

66. Evaluate the following expressions without using a calculator:
(a) $\sin(\frac{3\pi}{4})$ **(b)** $\cos(-\frac{13\pi}{6})$ **(c)** $\tan(\frac{4\pi}{3})$

67. (a) Find the values of $\alpha \in [0, 2\pi)$ that satisfy
$$\sin \alpha = -\frac{1}{2}\sqrt{3}$$
(b) Find the values of $\alpha \in [0, 2\pi)$ that satisfy
$$\tan \alpha = \sqrt{3}$$

68. (a) Find the values of $\alpha \in [0, 2\pi)$ that satisfy
$$\cos \alpha = -\frac{1}{2}\sqrt{2}$$
(b) Find the values of $\alpha \in [0, 2\pi)$ that satisfy
$$\sec \alpha = 2$$

69. Show that the identity
$$1 + \tan^2 \theta = \sec^2 \theta$$
follows from
$$\sin^2 \theta + \cos^2 \theta = 1$$

70. Show that the identity
$$1 + \cot^2 \theta = \csc^2 \theta$$
follows from
$$\sin^2 \theta + \cos^2 \theta = 1$$

71. Solve $2\cos\theta \sin\theta = \sin\theta$ on $[0, 2\pi)$.
72. Solve $\sec^2 x = \sqrt{3}\tan x + 1$ on $[0, \pi)$.

■ **1.1.5**

73. Evaluate the following exponential expressions:

(a) $4^3 4^{-2/3}$ **(b)** $\frac{3^2 3^{1/2}}{3^{-1/2}}$ **(c)** $\frac{5^k 5^{2k-1}}{5^{1-k}}$

74. Evaluate the following exponential expressions:
(a) $(2^4 2^{-3/2})^2$ **(b)** $\left(\frac{6^{5/2}6^{2/3}}{6^{1/3}}\right)^3$ **(c)** $\left(\frac{3^{-2k+3}}{3^{4+k}}\right)^3$

75. Which real number x satisfies
(a) $\log_4 x = -2$? **(b)** $\log_{1/3} x = -3$? **(c)** $\log_{10} x = -2$?

76. Which real number x satisfies
(a) $\log_{1/2} x = -4$? **(b)** $\log_{1/4} x = 2$? **(c)** $\log_5 x = 3$?

77. Which real number x satisfies
(a) $\log_{1/2} 32 = x$? **(b)** $\log_{1/3} 81 = x$? **(c)** $\log_{10} 0.001 = x$?

78. Which real number x satisfies
(a) $\log_4 64 = x$? **(b)** $\log_{1/5} 625 = x$? **(c)** $\log_{10} 10,000 = x$?

79. Simplify the following expressions:
(a) $-\ln\frac{1}{3}$ **(b)** $\log_4(x^2 - 4)$ **(c)** $\log_2 4^{3x-1}$

80. Simplify the following expressions:
(a) $-\ln\frac{1}{5}$ **(b)** $\ln\frac{x^2-y^2}{\sqrt{x}}$ **(c)** $\log_3 3^{2x+1}$

81. Solve for x.
(a) $e^{3x-1} = 2$ **(b)** $e^{-2x} = 10$ **(c)** $e^{x^2-1} = 10$

82. Solve for x.
(a) $3^x = 81$ **(b)** $9^{2x+1} = 27$ **(c)** $10^{5x} = 1000$

83. Solve for x.
(a) $\ln(x - 3) = 5$ **(b)** $\ln(x + 2) + \ln(x - 2) = 1$
(c) $\log_3 x^2 - \log_3 2x = 2$

84. Solve for x.
(a) $\ln(2x - 3) = 0$ **(b)** $\log_2(1 - x) = 3$
(c) $\ln x^3 - 2\ln x = 1$

■ **1.1.6**

In Problems 85–92, simplify each expression and write it in the standard form $a + bi$.

85. $(3 - 2i) - (-2 + 5i)$ **86.** $(7 + i) - 4$
87. $(4 - 2i) + (9 + 4i)$ **88.** $(6 - 4i) + (2 + 5i)$
89. $3(5 + 3i)$ **90.** $(2 - 3i)(5 + 2i)$
91. $(6 - i)(6 + i)$ **92.** $(-4 - 3i)(4 + 2i)$

In Problems 93–98, let $z = 3 - 2i$, $u = -4 + 3i$, $v = 3 + 5i$, and $w = 1 - i$. Compute the following expressions:

93. $\bar{z}$ **94.** $z + u$ **95.** $\overline{z + v}$
96. $\overline{v - w}$ **97.** $\overline{vw}$ **98.** $\overline{uz}$
99. If $z = a + bi$, find $z + \bar{z}$ and $z - \bar{z}$.
100. If $z = a + bi$, find $\bar{z}$. Use your answer to compute $(\overline{\bar{z}})$, and compare your answer with z.

In Problems 101–106, solve each quadratic equation in the complex number system.

101. $2x^2 - 3x + 2 = 0$ **102.** $3x^2 - 2x + 1 = 0$
103. $-x^2 + x + 2 = 0$ **104.** $-2x^2 + x + 3 = 0$
105. $4x^2 - 3x + 1 = 0$ **106.** $-2x^2 + 4x - 3 = 0$

In Problems 107–112, first determine whether the solutions of each quadratic equation are real or complex without solving the equation. Then solve the equation.

107. $3x^2 - 4x - 7 = 0$ **108.** $3x^2 - 4x + 7 = 0$
109. $-x^2 + 2x - 1 = 0$ **110.** $4x^2 - x + 1 = 0$
111. $3x^2 - 5x + 6 = 0$ **112.** $-x^2 + 7x - 2 = 0$
113. Show $(\overline{\bar{z}}) = z$.
114. Show $\overline{z + w} = \bar{z} + \bar{w}$.
115. Show $\overline{zw} = \bar{z}\,\bar{w}$.

■ 1.2 Elementary Functions

■ 1.2.1 What Is a Function?

Scientific investigations often study relationships between quantities, such as how enzyme activity depends on temperature or how the length of a fish is related to its age. To describe such relationships mathematically, the concept of a function is useful.

The word *function* (or, more precisely, its Latin equivalent *functio*, which means "execution") was introduced by Leibniz in 1694 in order to describe curves. Later, Euler (1707–1783) used it to describe any equation involving variables and constants. The modern definition is much broader and emphasizes the basic idea of expressing relationships between any two sets.

> **Definition** A **function** f is a rule that assigns each element x in the set A exactly one element y in the set B. The element y is called the **image** (or **value**) of x under f and is denoted by $f(x)$ (read "f of x"). The set A is called the **domain** of f, the set B is called the **codomain** of f, and the set $f(A) = \{y : y = f(x) \text{ for some } x \in A\}$ is called the **range** of f.

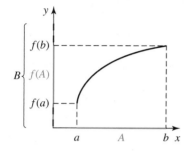

Figure 1.9 A function $f(x)$ with domain A, codomain B, and range $f(A)$.

To define a function, we use the notation

$$f : A \to B$$
$$x \to f(x)$$

where A and B are subsets of the set of real numbers. Frequently, we simply write $y = f(x)$ and call x the **independent** variable and y the **dependent** variable. We can illustrate functions graphically in the x–y plane. In Figure 1.9, we see the graph of $y = f(x)$, with domain A, codomain B, and range $f(A)$.

The function $f(x)$ must be specified; for example, $f(x)$ could be given by a graph as in Figure 1.9, or it could be expressed algebraically, such as $f(x) = x^2$. Note that $f(A) \subset B$, but not every element in the codomain B must be in $f(A)$. For instance, let

$$f : \mathbf{R} \to \mathbf{R}$$
$$x \to x^2$$

The domain of f is $\mathbf{R}$, but the range of f is only $[0, \infty)$ because the square of a real number is nonnegative; that is, $f(\mathbf{R}) = [0, \infty) \neq \mathbf{R}$. Also, the domain of a function need not be the largest possible set on which we can define the function, as $\mathbf{R}$ is in the preceding example. For instance, we could have defined f on a smaller set, such as $[0, 1]$, calling the new function g, given by

$$g : [0, 1] \to \mathbf{R}$$
$$x \to x^2$$

Although the same rule is used for f and g, the two functions are not the same, because their respective domains are different.

> Two functions f and g are **equal** if and only if
>
> **1.** f and g are defined on the same domain, and
> **2.** $f(x) = g(x)$ for all x in the domain.

EXAMPLE 1   Let

$$f_1 : [0, 1] \to \mathbf{R}$$
$$x \to x^2$$
$$f_2 : [0, 1] \to \mathbf{R}$$
$$x \to \sqrt{x^4}$$

and

$$f_3 : \mathbf{R} \to \mathbf{R}$$
$$x \to x^2$$

Determine which of these functions are equal.

Solution Because f_1 and f_2 are defined on the same domain and $f_1(x) = f_2(x) = x^2$ for all $x \in [0, 1]$, it follows that f_1 and f_2 are equal.

Neither f_1 nor f_2 is equal to f_3, because the domain of f_3 is different from the domains of f_1 and f_2. ■

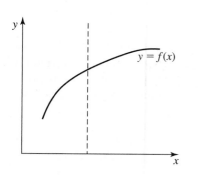

Figure 1.10 The vertical line test shows that the graph of $y = f(x)$ is a function.

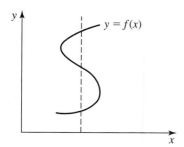

Figure 1.11 The vertical line test shows that the graph of $y = f(x)$ is not a function.

The choices of domains for the functions that we have thus far considered may look somewhat arbitrary (and they are arbitrary in the examples we have seen so far). In applications, however, there is often a natural choice of domain. For instance, if we look at a certain plant response (such as total biomass or the ratio of above to below biomass) as a function of nitrogen concentration in the soil, then, given that nitrogen concentration cannot be negative, the domain for this function could be the set of nonnegative real numbers. As another example, suppose we define a function that depends on the fraction of a population infected with a certain virus; then a natural choice for the domain of this function would be the interval [0, 1] because a fraction of a population must be a number between 0 and 1.

In our definition of a function, we stated that a function is a rule that assigns, to each element $x \in A$, *exactly* one element $y \in B$. When we graph $y = f(x)$ in the x–y plane, there is a simple test to decide whether or not $f(x)$ is a function: If each vertical line intersects the graph of $y = f(x)$ at most once, then $f(x)$ is a function. Figure 1.10 shows the graph of a function: Each vertical line intersects the graph of $y = f(x)$ at most once. The graph of $y = f(x)$ in Figure 1.11 is not a function, since there are x-values that are assigned to more than one y-value, as illustrated by the vertical line that intersects the graph more than once.

Sometimes functions show certain symmetries. For example, in Figure 1.12, $f(x) = x$ is symmetric about the origin; that is, $f(x) = -f(-x)$. In Figure 1.13, $g(x) = x^2$ is symmetric about the y-axis; that is, $g(x) = g(-x)$. In the first case, we say that f is odd; in the second case, that g is even. To check whether a function is even or odd, we use the following definition:

A function $f : A \to B$ is called

1. **even** if $f(x) = f(-x)$ for all $x \in A$, and
2. **odd** if $f(x) = -f(-x)$ for all $x \in A$.

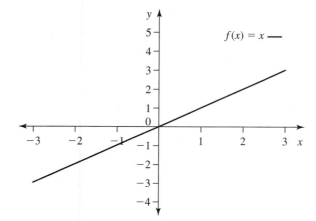

Figure 1.12 The graph of $y = x$ is symmetric about the origin.

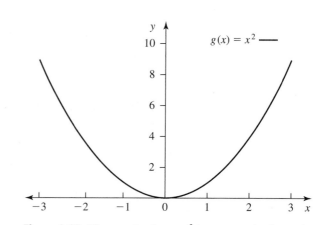

Figure 1.13 The graph of $y = x^2$ is symmetric about the y-axis.

Using this criterion, we can show that $f(x) = x, x \in \mathbf{R}$, is an odd function:

$$-f(-x) = -(-x) = x = f(x) \qquad \text{for all } x \in \mathbf{R}$$

Likewise, to show that $g(x) = x^2, x \in \mathbf{R}$, is an even function, we compute

$$g(-x) = (-x)^2 = x^2 = g(x) \qquad \text{for all } x \in \mathbf{R}$$

We will now look at the case where one quantity is given as a function of another quantity that, in turn, can be written as a function of yet another quantity. To illustrate this situation, suppose we are interested in the abundance of a predator, which depends on the abundance of a herbivore, which, in turn, depends on the abundance of plant biomass. If we denote the plant biomass by x and the herbivore biomass by u, then x and u are related via a function g, namely, $u = g(x)$. Likewise, if we denote the predator biomass by y, then u and y are related via a function f, namely, $y = f(u)$. We can express the predator biomass as a function of the plant biomass by substituting $g(x)$ for u. That is, we find $y = f[g(x)]$. Functions that are defined in such a way are called composite functions.

Definition The **composite function** $f \circ g$ (also called the **composition** of f and g) is defined as

$$(f \circ g)(x) = f[g(x)]$$

for each x in the domain of g for which $g(x)$ is in the domain of f.

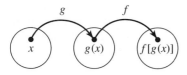

Figure 1.14 The composition of functions.

The composition of functions is illustrated in Figure 1.14. We call g the inner function and f the outer function. The phrase "for each x in the domain of g for which $g(x)$ is in the domain of f" is best explained with the use of Figure 1.14. In order to compute $f(u)$, u needs to be in the domain of f. But since $u = g(x)$, we really require that $g(x)$ be in the domain of f for the values of x we use to compute $g(x)$.

EXAMPLE 2   If $f(x) = \sqrt{x}, x \geq 0$, and $g(x) = x^2 + 1, x \in \mathbf{R}$, find

(a) $(f \circ g)(x)$ and   **(b)** $(g \circ f)(x)$.

Solution   **(a)** To find $(f \circ g)(x)$, we set $f(u) = \sqrt{u}$ and $g(x) = x^2 + 1$. Then

$$y = f(u) = f[g(x)] = f(x^2 + 1) = \sqrt{x^2 + 1}$$

To determine the domain of $f \circ g$, we observe that the domain of the inner function g is $\mathbf{R}$ and its range is $[1, \infty)$. Since the range of g is contained in the domain of the outer function f ($[1, \infty) \subset [0, \infty)$), the domain of $f \circ g$ is $\mathbf{R}$.

(b) To find $(g \circ f)(x)$, we set $g(u) = u^2 + 1$ and $f(x) = \sqrt{x}$. Then

$$y = g(u) = g[f(x)] = g(\sqrt{x}) = (\sqrt{x})^2 + 1 = x + 1$$

To determine the domain of $g \circ f$, we observe that the domain of the inner function f is $[0, \infty)$ and its range is $[0, \infty)$. The range of f is contained in the domain of the outer function g ($[0, \infty) \subset \mathbf{R}$), so the domain of $g \circ f$ is $[0, \infty)$.   ■

In the last example, you should observe that $f \circ g$ is different from $g \circ f$, which implies that the order in which you compose functions is important. The notation $f \circ g$ means that you apply g first and then f. In addition, you should pay attention to the domains of composite functions. In the next example, the domain is harder to find.

EXAMPLE 3   If $f(x) = 2x^2, x \geq 2$, and $g(x) = \sqrt{x}, x \geq 0$, find $(f \circ g)(x)$ together with its domain.

Solution   We compute

$$(f \circ g)(x) = f[g(x)] = f(\sqrt{x}) = 2(\sqrt{x})^2 = 2x$$

This part was not difficult. However, finding the domain of $f \circ g$ is more complicated. The domain of the inner function g is the interval $[0, \infty)$; hence, the range of g is the interval $[0, \infty)$. The domain of f is only $[2, \infty)$, which means that the range of g is *not* contained in the domain of f. We therefore need to restrict the domain of g to ensure that its range is contained in the domain of f. We can choose only values of x such that $g(x) \in [2, \infty)$. Since $g(x) = \sqrt{x}$, we need to restrict x to $[4, \infty)$. Thus, for every $x \in [4, \infty)$, $g(x) \in [2, \infty)$, which is the domain of f. Therefore,

$$(f \circ g)(x) = 2x, \quad x \geq 4$$

See Figure 1.15.   ■

In the subsections that follow, we introduce the basic functions that are used throughout the remainder of this book.

■ 1.2.2 Polynomial Functions

Polynomial functions are the simplest elementary functions.

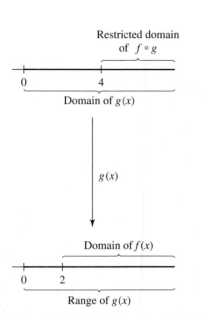

Figure 1.15 Finding the domain of a compositie function: The domain of $g(x)$ must be restricted in Example 3.

> **Definition** A **polynomial** function is a function of the form
>
> $$f(x) = a_0 + a_1 x + a_2 x^2 + \cdots + a_n x^n$$
>
> where n is a nonnegative integer and $a_0, a_1, \ldots, a_n$ are (real-valued) constants with $a_n \neq 0$. The coefficient a_n is called the **leading coefficient**, and n is called the **degree** of the polynomial function. The largest possible domain of f is **R**.

We have already encountered polynomials, namely, the constant function $f(x) = c$, the linear function $f(x) = mx + b$, and the quadratic function $f(x) = ax^2$. The constant, nonzero function has degree 0, the linear function has degree 1, and the quadratic function has degree 2. Other examples are $f(x) = 4x^3 - 3x + 1$, $x \in \mathbf{R}$, which is a polynomial of degree 3, and $f(x) = 2 - x^7$, $x \in \mathbf{R}$, which is a polynomial of degree 7. In Figure 1.16, we display $y = x^n$ for $n = 2$ and 3. Looking at the figure, we see that $y = x^n$ is an even function (i.e., symmetric about the y-axis) when $n = 2$ and an odd function (i.e., symmetric about the origin) when $n = 3$. This property holds in general: $y = x^n$ is an even function when n is even and an odd function when n is odd. We can show this algebraically by using the criterion in Section 1.2.1. (See Problem 28 at the end of this section.)

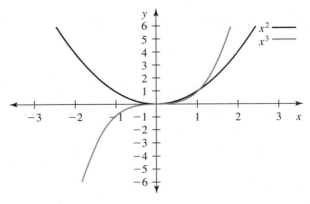

Figure 1.16 The graphs of $y = x^n$ for $n = 2$ and $n = 3$.

Polynomials arise naturally in many situations. We present two examples.

EXAMPLE 4

Suppose that at time 0 an apple begins to drop from a tree that is 64 ft tall. Ignoring air resistance, we can show that at time t (measured in seconds) the apple is at height $h(t)$ (measured in feet) given by

$$h(t) = 64 - 16t^2$$

We assume that the height of the ground level is equal to 0. Show that $h(t)$ is a polynomial and determine its degree. How long will it take the apple to hit the ground? Find an appropriate domain for $h(t)$.

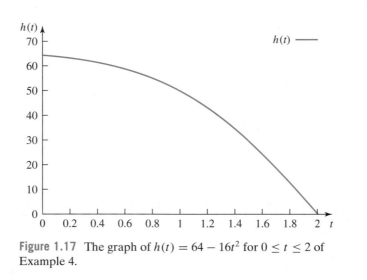

Figure 1.17 The graph of $h(t) = 64 - 16t^2$ for $0 \leq t \leq 2$ of Example 4.

Solution

The function $h(t)$ is a polynomial of degree 2, with $a_0 = 64$, $a_1 = 0$, and $a_2 = -16$. The graph of $h(t)$ is shown in Figure 1.17. The apple will hit the ground when $h(t) = 0$. That is, we must solve the quadratic equation $0 = 64 - 16t^2$ as follows:

$$0 = 64 - 16t^2$$
$$t^2 = \frac{64}{16} = 4$$
$$t = 2 \quad (\text{or } t = -2)$$

Since the apple begins to drop at time $t = 0$, we can ignore the solution $t = -2 < 0$. We find that it takes the apple 2 seconds to hit the ground (ignoring air resistance). Note that because $h(t) \geq 0$ [where $h(t)$ is the height above the ground and the height of the ground level is equal to 0], the range is $[0, 64]$. Because $t \geq 0$, the domain of $h(t)$ is the interval $[0, 2]$.   ■

EXAMPLE 5

A Chemical Reaction Consider the reaction rate of the chemical reaction

$$A + B \longrightarrow AB$$

in which the molecular reactants A and B form the molecular product AB. The rate at which this reaction proceeds depends on how often A and B molecules collide. The **law of mass action** states that the rate at which this reaction proceeds is proportional to the product of the respective concentrations of the reactants. Here, *concentration* means the number of molecules per fixed volume. If we denote the reaction rate by R and the concentration of A and B by [A] and [B], respectively, then the law of mass action says that

$$R \propto [A] \cdot [B]$$

Introducing the proportionality factor k, we obtain

$$R = k[A] \cdot [B]$$

Note that $k > 0$, because [A], [B], and R are positive. We assume now that the reaction occurs in a closed vessel; that is, we add specific amounts of A and B to the vessel at the beginning of the reaction and then let the reaction proceed without further additions.

We can express the concentrations of the reactants A and B during the reaction in terms of their initial concentrations a and b and the concentration of the molecular product [AB]. If $x = [AB]$, then

$$[A] = a - x \quad \text{for } 0 \le x \le a \qquad \text{and} \qquad [B] = b - x \quad \text{for } 0 \le x \le b$$

The concentration of AB cannot exceed either of the concentrations of A and B. (For example, suppose five A molecules and seven B molecules are allowed to react; then a maximum of five AB molecules can result, at which point all of the A molecules are used up and the reaction ceases. The two B molecules left over have no A molecules to react with.) Therefore, we get

$$R(x) = k(a - x)(b - x) \quad \text{for } 0 \le x \le a \text{ and } 0 \le x \le b$$

The condition $0 \le x \le a$ and $0 \le x \le b$ can be written as $0 \le x \le \min(a, b)$, where $\min(a, b)$ denotes the minimum of a and b. To see that $R(x)$ is indeed a polynomial function, we expand the expression for $R(x)$ as

$$R(x) = k(ab - ax - bx + x^2)$$
$$= kx^2 - k(a + b)x + kab$$

for $0 \le x \le \min(a, b)$. We now see that $R(x)$ is a polynomial of degree 2.

A graph of $R(x), 0 \le x \le a$, is shown in Figure 1.18 for the case $a \le b$. (We chose $k = 2, a = 2$, and $b = 5$ in the figure.) Notice that when $x = 0$ (i.e., when no AB molecules have yet formed), the rate at which the reaction proceeds is at a maximum. As more and more AB molecules form and, consequently, the concentrations of the reactants decline, the reaction rate decreases. This should also be intuitively clear: As fewer and fewer A and B molecules are in the vessel, it becomes less and less likely that they will collide to form the molecular product AB. When $x = a = \min(a, b)$, the reaction rate $R(a) = 0$. This is the point at which all A molecules are exhausted and the reaction necessarily ceases. ■

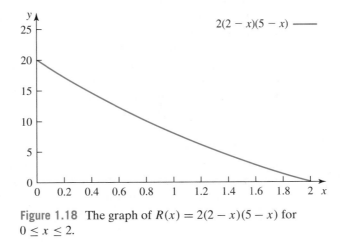

$2(2 - x)(5 - x)$ ——

Figure 1.18 The graph of $R(x) = 2(2 - x)(5 - x)$ for $0 \le x \le 2$.

■ 1.2.3 Rational Functions

Rational functions are built from polynomial functions.

> **Definition** A **rational** function is the quotient of two polynomial functions $p(x)$ and $q(x)$:
>
> $$f(x) = \frac{p(x)}{q(x)} \quad \text{for } q(x) \ne 0$$

Since division by 0 is not allowed, we must exclude those values of x for which $q(x) = 0$. Here are a couple of examples of rational functions, together with their largest possible domains:

$$y = \frac{1}{x}, \quad x \neq 0$$

$$y = \frac{x^2 + 2x - 1}{x - 3}, \quad x \neq 3$$

An important example of a rational function is the hyperbola, together with its largest possible domain:

$$y = \frac{1}{x}, \quad x \neq 0$$

The graph of y is shown in Figure 1.19.

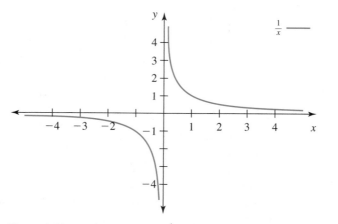

Figure 1.19 The graph of $y = \frac{1}{x}$ for $x \neq 0$.

Throughout this text, we will encounter populations whose sizes change with time. The change in population size is described by the **growth rate**. Roughly speaking, the growth rate tells you how much a population changes during a small time interval. (The growth rate is analogous to the velocity of a car: Velocity is also a rate; it tells you how much the position changes in a small time interval. We will give a precise definition of rates in Section 4.1.) The **per capita** growth rate is the growth rate divided by the population size. The per capita growth rate is also called the **specific** growth rate. The next example introduces a function that is frequently used to describe growth rates.

EXAMPLE 6

Monod Growth Function There is a function that is frequently used to describe the per capita growth rate of organisms when the rate depends on the concentration of some nutrient and becomes saturated for large enough nutrient concentrations. If we denote the concentration of the nutrient by N, then the per capita growth rate $r(N)$ is given by the *Monod growth function*

$$r(N) = \frac{aN}{k + N}, \quad N \geq 0$$

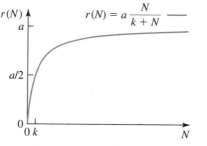

Figure 1.20 The graph of the Monod function $r(N) = a\frac{N}{k+N}$ for $N \geq 0$.

where a and k are positive constants. The graph of $r(N)$ is shown in Figure 1.20; it is a piece of a hyperbola. The graph shows a decelerating rise approaching the saturation level a, which is the maximal specific growth rate. When $N = k$, $r(N) = a/2$; for this reason, k is called the *half-saturation constant*. The growth rate increases with nutrient concentration N; however, doubling the nutrient concentration has a much bigger effect on the growth rate for small values of N than when N is already large. When this type of function is used in biochemistry to describe enzymatic reactions, it is called the Michaelis–Menten function. ■

■ 1.2.4 Power Functions

> **Definition** A **power** function is of the form
>
> $$f(x) = x^r$$
>
> where r is a real number.

Examples of power functions, with their largest possible domains, are

$$y = x^{1/3}, \quad x \in \mathbf{R}$$
$$y = x^{5/2}, \quad x \geq 0$$
$$y = x^{1/2}, \quad x \geq 0$$
$$y = x^{-1/2}, \quad x > 0$$

Polynomials of the form $y = x^n$, $n = 1, 2, \ldots$, are a special case of power functions. Since power functions may involve even roots, as in $y = x^{3/2} = (\sqrt{x})^3$, we frequently need to restrict their domain.

Figure 1.21 compares the power functions $y = x^{5/2}$, $y = x^{1/2}$, and $y = x^{-1/2}$ for $x > 0$. Pay close attention to how the exponent determines the ranking according to size for x between 0 and 1 and for $x > 1$. We find that $x^{5/2} < x^{1/2} < x^{-1/2}$ for $0 < x < 1$, but $x^{5/2} > x^{1/2} > x^{-1/2}$ for $x > 1$.

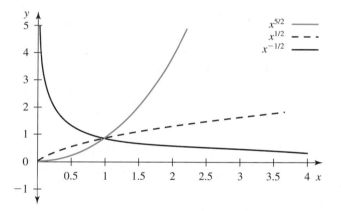

Figure 1.21 Some power functions with rational exponents.

EXAMPLE 7

Power functions are frequently found in "scaling relations" between biological variables (e.g., organ sizes). These are relations of the form

$$y \propto x^r$$

where r is a nonzero real number. That is, y is proportional to some power of x. Recall that we can write this relationship as an equation if we introduce the proportionality factor k:

$$y = kx^r$$

Finding such relationships is the objective of **allometry**. For example, in a study of 45 species of unicellular algae, a relationship between cell volume and cell biomass was sought. It was found [see, for instance, Niklas (1994)] that

$$\text{cell biomass} \propto (\text{cell volume})^{0.794}$$

Most scaling relations are to be interpreted in a statistical sense; they are obtained by fitting a curve to data points. The data points are typically scattered about the fitted curve given by the scaling relation. (See Figure 1.22.)   ■

Figure 1.22 Some data points and the fitted curve of Example 7. (*Note*: The "data points" aren't real.)

The next example relates the volume and the surface area of a cube. This relationship is not to be understood in a statistical sense, because it is an exact relationship resulting from geometric considerations.

EXAMPLE 8

Suppose that we wish to know the scaling relation between the surface area S and the volume V of a cube. The scaling relations of each of these quantities with the length L of the cube are as follows:

$$S \propto L^2 \quad \text{or} \quad S = k_1 L^2$$
$$V \propto L^3 \quad \text{or} \quad V = k_2 L^3$$

Here, k_1 and k_2 denote the constants of proportionality. (We label them with different subscripts to indicate that they might be different.) To express S in terms of V, we must first solve L in terms of V and then substitute L in the equation for S. Because $L = (V/k_2)^{1/3}$, it follows that

$$S = k_1 \left[\left(\frac{V}{k_2} \right)^{1/3} \right]^2 = \frac{k_1}{k_2^{2/3}} V^{2/3}$$

Introducing the constant of proportionality $k = k_1 / k_2^{2/3}$, we find that

$$S = k V^{2/3}, \quad \text{or simply} \quad S \propto V^{2/3}$$

In words, the surface area of a cube scales with the volume in proportion to $V^{2/3}$. We can now ask, for instance, by what factor the surface area increases when we double the volume. When we double the volume, we find that the resulting surface area, denoted by S', is

$$S' = k(2V)^{2/3} = 2^{2/3} \underbrace{k V^{2/3}}_{S}$$

That is, the surface area increases by a factor of $2^{2/3} \approx 1.587$ if we double the volume of the cube. This scaling has implications on heat retention in animals: A larger body has a relatively smaller surface area and will retain more heat. ■

■ 1.2.5 Exponential Functions

In our study of exponential functions, let's first look at an example that illustrates where they occur.

EXAMPLE 9

Exponential Growth Bacteria reproduce asexually by cellular fission, in which the parent cell splits into two daughter cells after duplication of the genetic material. This division may happen as often as every 20 minutes; under ideal conditions, a bacterial colony can double in size in that time.

Let us measure time such that one unit of time corresponds to the doubling time of the colony. If we denote the size of the population at time t by $N(t)$, then the function

$$N(t) = 2^t, \quad t \geq 0$$

has the property of doubling its value every unit of time

$$N(t + 1) = 2^{t+1} = 2 \cdot 2^t = 2N(t) \tag{1.2}$$

The function $N(t) = 2^t$, $t \geq 0$, is an exponential function because the variable t is in the exponent. We call the number 2 the base of the exponential function $N(t) = 2^t$.

We find that when $t = 0$, $N(0) = 1$; that is, there is just one individual in the population at time $t = 0$. If, at time $t = 0$, 40 individuals were present in the population, we would write $N(0) = 40$ and

$$N(t) = 40 \cdot 2^t, \quad t \geq 0 \tag{1.3}$$

You can verify that $N(t)$ in (1.3) also satisfies $N(t + 1) = 2N(t)$.

It is often desirable not to specify the initial number of individuals in the equation describing $N(t)$. This approach has the advantage that the equation for $N(t)$ then describes a more general situation, in the sense that we can use the same equation for different initial population sizes. We often denote the population size at time 0 by N_0 (read "N sub 0") instead of $N(0)$. The equation for $N(t)$ is then

$$N(t) = N_0 2^t, \quad t \geq 0$$

We can verify that $N(0) = N_0 2^0 = N_0$ and that $N(t + 1) = N_0 2^{t+1} = 2(N_0 2^t) = 2N(t)$. ■

The function $f(t) = 2^t$ can be defined for all $t \in \mathbf{R}$; its graph is shown in Figure 1.23.

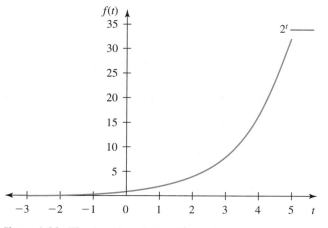

Figure 1.23 The function $f(t) = 2^t$, $t \in \mathbf{R}$.

Here is the definition of an exponential function:

Definition The function f is an **exponential** function with base a if

$$f(x) = a^x$$

where a is a positive constant other than 1. The largest possible domain of f is **R**.

When $a = 1$, $f(x) = 1$ for all values of x. This is a case that will occur in biological examples, but is excluded from the definition since it is simply the constant function.

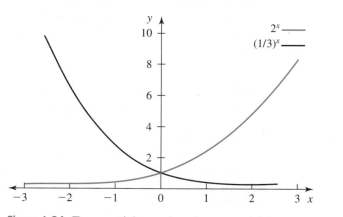

Figure 1.24 Exponential growth and exponential decay.

The basic shape of the exponential function $f(x) = a^x$ depends on the base a; two examples are shown in Figure 1.24. As x increases, the graph of $f(x) = 2^x$ shows a rapid increase, whereas the graph of $f(x) = (1/3)^x$ shows a rapid decrease toward 0. We find the rapid increase whenever $a > 1$ and the rapid decrease whenever $0 < a < 1$. Therefore, we say that we have exponential *growth* when $a > 1$ and exponential *decay* when $0 < a < 1$.

Recall that $a^0 = 1$ and $a^{1/k} = \sqrt[k]{a}$, where k is a positive integer. In Subsection 1.1.5, we summarized the properties of exponentials. Since they are very important, we list them again here:

$$a^r a^s = a^{r+s}$$

$$\frac{a^r}{a^s} = a^{r-s}$$

$$a^{-r} = \frac{1}{a^r}$$

$$\left(a^r\right)^s = a^{rs}$$

In many applications, the exponential function is expressed in terms of the base $e = 2.718\ldots$, which we encountered in Subsection 1.1.5. The number e is called the **natural exponential base**. The exponential function with base e is alternatively written as $\exp(x)$. That is,

$$\exp(x) = e^x$$

The advantage of this alternative form can be seen when we try to write something like $e^{x^2/\sqrt{x^3+1}}$: $\exp(x^2/\sqrt{x^3 + 1})$ is easier to read. More generally, if $g(x)$ is a function in x, then we can write, equivalently,

$$\exp[g(x)] \quad \text{or} \quad e^{g(x)}$$

Bases 2 and 10 are also frequently used; in calculus, however, e will turn out to be the most common base.

The next two examples provide an important application of exponential functions.

EXAMPLE 10

Radioactive Decay Radioactive isotopes such as carbon 14 are used to determine the absolute age of fossils or minerals, establishing an absolute chronology of the geological time scale. This technique was discovered in the early years of the 20th century and is based on the property of certain atoms to transform spontaneously by giving off protons, neutrons, or electrons. The phenomenon, called *radioactive* *decay*, occurs at a constant rate that is independent of environmental conditions. The method was used, for instance, to trace the successive emergence of the Hawaiian islands, from the oldest, Kauai, to the youngest, Hawaii (which is about 100,000 years old).

Carbon 14 is formed high in the atmosphere. It is radioactive and decays into nitrogen (N^{14}). There is an equilibrium between atmospheric carbon 12 (C^{12}) and carbon 14 (C^{14})—a ratio that has been relatively constant over a fairly long period. When plants capture carbon dioxide (CO_2) molecules from the atmosphere and build them into a product (such as cellulose), the initial ratio of C^{14} to C^{12} is the same as that in the atmosphere. Once the plants die, however, their uptake of CO_2 ceases, and the radioactive decay of C^{14} causes the ratio of C^{14} to C^{12} to decline. Because the law of radioactive decay is known, the change in ratio provides an accurate measure of the time since the plants' death.

According to the radioactive decay law, if the amount of C^{14} at time t is denoted by $W(t)$, with $W(0) = W_0$, then

$$W(t) = W_0 e^{-\lambda t}, \quad t \geq 0$$

where $\lambda > 0$ (λ is the lowercase Greek letter lambda) denotes the **decay rate**. The function $W(t) = W_0 e^{-\lambda t}$ is another example of an exponential function. Its graph is shown in Figure 1.25.

Frequently, the decay rate is expressed in terms of the **half-life** of the material, which is the length of time that it takes for half of the material to decay. If we denote this time by T_h, then (see Figure 1.25)

$$W(T_h) = \frac{1}{2} W_0 = W_0 e^{-\lambda T_h}$$

from which we obtain

$$\frac{1}{2} = e^{-\lambda T_h}$$
$$2 = e^{\lambda T_h}$$

Recall from algebra (or Subsection 1.1.5) that, to solve for the exponent λT_h, we must take logarithms on both sides. Since the exponent has base e, we use natural logarithms and find that

$$\ln 2 = \lambda T_h$$

Solving for T_h or λ yields

$$T_h = \frac{\ln 2}{\lambda} \qquad \text{or} \qquad \lambda = \frac{\ln 2}{T_h}$$

It is known that the half-life of C^{14} is 5730 years. Hence,

$$\lambda = \frac{\ln 2}{5730 \text{ years}}$$

Note that the unit "years" appears in the denominator. It is important to carry the units along. When we compute λt in the exponent of $e^{-\lambda t}$, we need to measure t in units of years in order for the units to cancel properly. For example, suppose $t = 2000$ years; then

$$\lambda t = \frac{\ln 2}{5730 \text{ years}} 2000 \text{ years} = \frac{(\ln 2)(2000)}{5730} \approx 0.2419$$

and we see that "years" appears in both the numerator and the denominator and thus can be canceled. ■

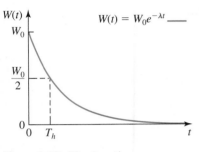

Figure 1.25 The function $W(t) = W_0 e^{-\lambda t}$.

An application of the C^{14} dating method is given in the next example.

EXAMPLE 11

Suppose that, on the basis of their C^{12} content, samples of wood found in an archeological excavation site contain about 23% as much C^{14} as does living plant material. Determine when the wood was cut.

Solution

The ratio of the current amount of C^{14} to the amount of living plant material is expressed as

$$0.23 = \frac{W(t)}{W(0)} = e^{-\lambda t}$$

Taking logarithms (base e) on both sides, we obtain

$$\ln(0.23) = -\lambda t$$

or

$$\lambda t = -\ln(0.23) = \ln \frac{1}{0.23}$$

With $\lambda = \ln 2/(5730 \text{ years})$ from Example 10,

$$t = \frac{5730 \text{ years}}{\ln 2} \ln \frac{1}{0.23}$$

Using a calculator to compute this result, we find that the wood was cut about 12,150 years ago.   ■

■ 1.2.6 Inverse Functions

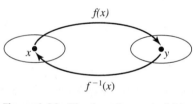

$f(x)$

$x \bullet \qquad \bullet y$

$f^{-1}(x)$

Figure 1.26 The function $y = f(x)$ and its inverse.

Before we can introduce logarithmic functions, we must understand the concept of inverse functions. Roughly speaking, the inverse of a function f reverses the effect of f. That is, if f maps x into $y = f(x)$, then the inverse function, denoted by f^{-1} (read "f inverse"), takes y and maps it back into x. (See Figure 1.26.) Not every function has an inverse: Because an inverse function is a function itself, we require that every value y in the range of f be mapped into exactly one value x. In other words, for a function to have an inverse, it must be that whenever $x_1 \neq x_2$, it follows that $f(x_1) \neq f(x_2)$ or, equivalently, that $f(x_1) = f(x_2)$ implies $x_1 = x_2$. (Recall the definition of a function, in which we required that each element in the domain be assigned to *exactly one* element in the range.)

Functions that have the property "$x_1 \neq x_2$ implies $f(x_1) \neq f(x_2)$" [or, equivalently, "$f(x_1) = f(x_2)$ implies $x_1 = x_2$"] are called **one to one**. If you know what the graph of a particular function looks like over its domain, then it is easy to determine whether or not the function is one to one: If no horizontal line intersects the graph of the function f more than once, then f is one to one. This criterion is called the horizontal line test. We illustrate it in Figures 1.27 and 1.28.

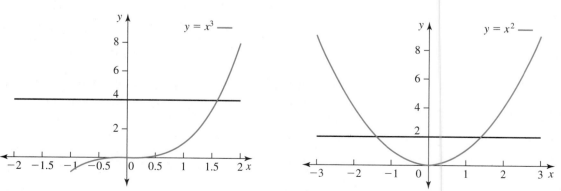

Figure 1.27 Horizontal line test successful.

Figure 1.28 Horizontal line test unsuccessful.

Now consider $y = x^3$ and $y = x^2$, for $x \in \mathbf{R}$. The function $y = x^3, x \in \mathbf{R}$, has an inverse function, because $x_1^3 \neq x_2^3$ whenever $x_1 \neq x_2$. (See Figure 1.27.) The function $y = x^2, x \in \mathbf{R}$, does *not* have an inverse function, because $x_1 \neq x_2$ does not imply $x_1^2 \neq x_2^2$ (or, equivalently, $x_1^2 = x_2^2$ does not imply $x_1 = x_2$; see Figure 1.28). The equation $x_1^2 = x_2^2$ implies only that $|x_1| = |x_2|$. Since x_1 and x_2 can be positive or negative, we cannot simply drop the absolute-value signs. For instance, both -2 and 2 are mapped into 4, and we find that $f(-2) = f(2)$ but $-2 \neq 2$. (Note that $|-2| = |2|$.) To invert this function, we would have to map 4 into -2 and 2, but then it would no longer be a function by our definition. By restricting the domain of $y = x^2$ to, say, $x \geq 0$, we can define an inverse function of $y = x^2, x \geq 0$.

Here is the formal definition of an inverse function:

> **Definition** Let $f : A \to B$ be a one-to-one function with range $f(A)$. The **inverse** function f^{-1} has domain $f(A)$ and range A and is defined by
>
> $$f^{-1}(y) = x \quad \text{if and only if} \quad y = f(x)$$
>
> for all $y \in f(A)$.

EXAMPLE 12

Find the inverse function of $f(x) = x^3 + 1, x \geq 0$.

Solution

First, note that $f(x)$ is one to one. To see this quickly, graph the function and apply the horizontal line test. (See Figure 1.29.) Be aware, though, that unless you know what the graph looks like over its entire domain, the graphical approach can be misleading. To demonstrate it algebraically, start with $f(x_1) = f(x_2)$ and show that this implies $x_1 = x_2$:

$$f(x_1) = f(x_2)$$
$$x_1^3 + 1 = x_2^3 + 1$$
$$x_1^3 = x_2^3$$

Taking the third root on both sides gives $x_1 = x_2$, which tells us that $f(x)$ has an inverse. Now we will find f^{-1}.

To find an inverse function, we follow three steps:

1. Write $y = f(x)$:
$$y = x^3 + 1$$

2. Solve for x:
$$x^3 = y - 1$$
$$x = \sqrt[3]{y - 1}$$

The range of f is $[1, \infty)$, and this range becomes the domain of f^{-1}, so we obtain
$$f^{-1}(y) = \sqrt[3]{y - 1}, \quad y \geq 1$$

Typically, we write functions in terms of x. To do this, we need to interchange x and y in $x = f^{-1}(y)$. This is the third step:

3. Interchange x and y:
$$y = f^{-1}(x) = \sqrt[3]{x - 1}, \ x \geq 1$$

Note that switching x and y in step 3 corresponds to reflecting the graph of $y = f(x)$ about the line $y = x$. The graphs of f and f^{-1} are shown in Figure 1.30. Look at the graphs carefully, and observe how they are related to each other: Each can be obtained from the other by reflection about the line $y = x$. ■

Figure 1.29 The graph of $f(x) = x^3 + 1$ in Example 12. The horizontal line test is successful.

Figure 1.30 Inverse functions.

As mentioned in the beginning of this subsection, the inverse of a function f reverses the effect of f. If we first apply f to x and then f^{-1} to $f(x)$, we obtain the original value x. Likewise, if we first apply f^{-1} to x and then f to $f^{-1}(x)$, we obtain the original value x. That is, if $f : A \to B$ has an inverse f^{-1}, then

$$f^{-1}[f(x)] = x \qquad \text{for all } x \in A$$
$$f[f^{-1}(x)] = x \qquad \text{for all } x \in f(A)$$

[A note of warning: The superscript in f^{-1} does *not* indicate the reciprocal of f (i.e., $1/f$). This difference is further explained in Problem 74 at the end of this section.]

■ 1.2.7 Logarithmic Functions

Recall from algebra (or Subsection 1.1.5) that, to solve the equation

$$e^x = 3$$

for x, you must take logarithms on both sides:

$$x = \ln 3$$

In other words, applying the natural logarithm undoes the operation of raising e to the x power. Thus, the natural logarithm is the inverse of the exponential function, and conversely, the exponential function is the inverse of the logarithmic function.

We will now define the inverse of the exponential function $f(x) = a^x$, $x \in \mathbf{R}$. The base a can be any positive number, except 1.

> **Definition** The inverse of $f(x) = a^x$ is called the **logarithm to base a** and is written $f^{-1}(x) = \log_a x$.

The maximum domain of $f(x) = a^x$ is the set of all real numbers, and its range is the set of all positive numbers. Since the range of f is the domain of f^{-1}, we find that the maximum domain of $f^{-1}(x) = \log_a x$ is the set of positive numbers.

Because $y = \log_a x$ is the inverse function of $y = a^x$, we can find the graph of $y = \log_a x$ by reflecting the graph of $y = a^x$ about the line $y = x$. Recall that the graph of $y = a^x$ had two basic shapes, depending on whether $0 < a < 1$ or $a > 1$. (See Figure 1.24.) Figure 1.31 illustrates the graphs of $y = a^x$ and $y = \log_a x$ when $a > 1$.

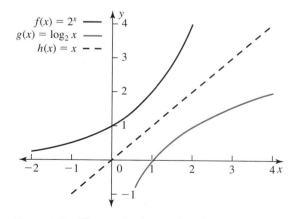

$f(x) = 2^x$ ——
$g(x) = \log_2 x$ ——
$h(x) = x$ - - -

$f(x) = \left(\frac{1}{2}\right)^x$ ——
$g(x) = \log_{1/2} x$ ——
$h(x) = x$ - - -

Figure 1.31 The graph of $y = a^x$ and the graph of $y = \log_a x$ for $a = 2$.

Figure 1.32 The graph of $y = a^x$ and the graph of $y = \log_a x$ for $a = \frac{1}{2}$.

Figure 1.32 shows the graphs of $y = a^x$ and $y = \log_a x$ when $0 < a < 1$.

We can now summarize the relationship between the exponential and the logarithmic functions:

> **1.** $a^{\log_a x} = x$ for $x > 0$
> **2.** $\log_a a^x = x$ for $x \in \mathbf{R}$

It is important to remember that the logarithm is defined only for positive numbers; that is, $y = \log_a x$ is defined only for $x > 0$. The logarithm satisfies the following properties:

$$\log_a(st) = \log_a s + \log_a t$$

$$\log_a\left(\frac{s}{t}\right) = \log_a s - \log_a t$$

$$\log_a s^r = r \log_a s$$

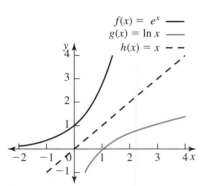

$f(x) = e^x$ ——
$g(x) = \ln x$ ——
$h(x) = x$ – – –

Figure 1.33 The graphs of $y = e^x$ and $y = \ln x$.

The inverse of the exponential function with the natural base e is denoted by $\ln x$ and is called the natural logarithm of x. The graphs of $y = e^x$ and $y = \ln x$ are shown in Figure 1.33. Note that both e^x and $\ln x$ are increasing functions. However, whereas e^x climbs very quickly for large values of x, $\ln x$ increases very slowly for large values of x. Looking at both graphs, we can see that each can be obtained as the reflection of the other about the line $y = x$.

The logarithm to base 10 is frequently written as $\log x$ (i.e., the base of 10 in $\log_{10} x$ is omitted).

EXAMPLE 13 Simplify the following expressions:

(a) $\log_2[8(x - 2)]$ **(b)** $\log_3 9^x$ **(c)** $\ln e^{3x^2+1}$

Solution **(a)** We simplify as follows:

$$\log_2[8(x - 2)] = \log_2 8 + \log_2(x - 2) = 3 + \log_2(x - 2)$$

No further simplification is possible.

(b) Simplifying yields

$$\log_3 9^x = x \log_3 9 = x \log_3 3^2 = 2x$$

The fact that $\log_3 9 = 2$ can be seen in two ways: We can write $9 = 3^2$ and say that applying $\log_3$ undoes raising 3 to the second power (as we did previously), or we can say that $\log_3 9$ denotes the exponent to which we must raise 3 in order to get 9.

(c) We use the fact that $\ln x$ and e^x are inverse functions and find that

$$\ln e^{3x^2+1} = 3x^2 + 1 \qquad\qquad ■$$

Any exponential function with base a can be written as an exponential function with base e. Likewise, any logarithm to base a can be written in terms of the natural logarithm. The following two identities show how:

> $$a^x = \exp[x \ln a]$$
> $$\log_a x = \frac{\ln x}{\ln a}$$

The first identity follows from the fact that exp and ln are inversely related (which implies that $a^x = \exp[\ln a^x]$) and the fact that $\ln a^x = x \ln a$. To understand the second identity, note that

$$y = \log_a x \qquad \text{means} \qquad a^y = x$$

Taking logarithms to base e on both sides of $a^y = x$, we get

$$\ln a^y = \ln x$$

or

$$y \ln a = \ln x$$

Hence,

$$y = \frac{\ln x}{\ln a}$$

EXAMPLE 14   Write the following expressions in terms of base e:

(a) 2^x (b) 10^{x^2+1} (c) $\log_3 x$ (d) $\log_2(3x - 1)$

Solution   (a) $2^x = \exp(\ln 2^x) = \exp(x \ln 2) = e^{x \ln 2}$

(b) $10^{x^2+1} = \exp(\ln 10^{x^2+1}) = \exp\left[(x^2 + 1) \ln 10\right] = e^{(x^2+1) \ln 10}$

(c) $\log_3 x = \dfrac{\ln x}{\ln 3}$

(d) $\log_2(3x - 1) = \dfrac{\ln(3x - 1)}{\ln 2}$ ■

EXAMPLE 15   DNA sequences evolve over time by various processes. One such process is the substitution of one nucleotide for another. The simplest substitution scheme is that of Jukes and Cantor (1969), which assumes that substitutions are equally likely among the four types of nucleotide. In comparing two DNA sequences that have a common origin, it is possible to estimate the number of substitutions per site. Since more than one substitution can occur per site, the number of observed substitutions may be smaller than the number of actual substitutions, particularly when the time of divergence is large. Mathematical models are used to correct for this difference. The proportion p of observed nucleotide differences between two sequences that share a common ancestor can be used to find an estimate of the actual number K of substitutions per site since the time of divergence. According to the substitution scheme of Jukes and Cantor, K and p are related by

$$K = -\frac{3}{4} \ln \left(1 - \frac{4}{3} p \right)$$

provided that p is not too large. Assume that two sequences of length 150 nucleotides differ from each other by 23 nucleotides. Find K.

Solution   The variable p denotes the proportion of observed nucleotide differences, which is $23/150 \approx 0.1533$ in this example. We thus obtain

$$K = -\frac{3}{4} \ln \left(1 - \frac{4}{3} \frac{23}{150} \right) \approx 0.1715$$ ■

■ 1.2.8 Trigonometric Functions

The trigonometric functions are examples of periodic functions.

> **Definition** A function $f(x)$ is **periodic** if there is a positive constant a such that
> $$f(x + a) = f(x)$$
> for all x in the domain of f. If a is the smallest number with this property, we call it the **period** of $f(x)$.

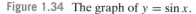

Figure 1.34 The graph of $y = \sin x$.

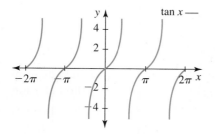

Figure 1.35 The graph of $y = \cos x$.

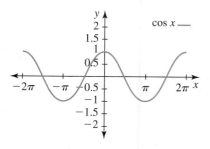

Figure 1.36 The graph of $y = \tan x$.

We begin with the sine and cosine functions. In Subsection 1.1.4, we recalled the definition of sine and cosine on a unit circle. There, $\sin \theta$ and $\cos \theta$ represented trigonometric functions of angles, and θ was measured in degrees or radians. Now we define the trigonometric functions as functions of *real numbers* . For instance, we define $f(x) = \sin x$ for $x \in \mathbf{R}$. The value of $\sin x$ is then, *by definition*, the sine of an angle of x radians (and similarly for all the other trigonometric functions).

The graphs of the sine and cosine functions are shown in Figures 1.34 and 1.35, respectively.

The sine function, $y = \sin x$, is defined for all $x \in \mathbf{R}$. Its range is $-1 \le y \le 1$. Likewise, the cosine function, $y = \cos x$, is defined for all $x \in \mathbf{R}$ with range $-1 \le y \le 1$. Both functions are periodic with period 2π. That is, $\sin(x + 2\pi) = \sin x$ and $\cos(x + 2\pi) = \cos x$. [We also have $\sin(x + 4\pi) = \sin x$, $\sin(x + 6\pi) = \sin x, \ldots$, and $\cos(x + 4\pi) = \cos x$, $\cos(x + 6\pi) = \cos x, \ldots$, but, by convention, we use the smallest possible value to specify the period.] We see from Figures 1.34 and 1.35 that the graph of the cosine function can be obtained by shifting the graph of the sine function a distance of $\pi/2$ units to the left. (We will discuss horizontal shifts of graphs in more detail in the next section.)

To define the tangent function, $y = \tan x$, recall that

$$\tan x = \frac{\sin x}{\cos x}$$

Because $\cos x = 0$ for values of x that are odd integer multiples of $\pi/2$, the domain of $\tan x$ consists of all real numbers with the exception of odd integer multiples of $\pi/2$. The range of $y = \tan x$ is $-\infty < y < \infty$. The graph of $y = \tan x$ is shown in Figure 1.36, from which we see that $\tan x$ is periodic with period π.

The graphs of the remaining three trigonometric functions are shown in Figures 1.37–1.39. Recall that $\sec x = \frac{1}{\cos x}$, $\csc x = \frac{1}{\sin x}$, and $\cot x = \frac{1}{\tan x}$. It follows that the domain of the secant function $y = \sec x$ consists of all real numbers with the exception of odd integer multiples of $\pi/2$; the range is $|y| \ge 1$. The domain of the cosecant function $y = \csc x$ consists of all real numbers with the exception of integer multiples of π; the range is $|y| \ge 1$. The domain of the cotangent function $y = \cot x$ consists of all real numbers with the exception of integer multiples of π; the range is $-\infty < y < \infty$.

Figure 1.37 The graph of $y = \sec x$.

Figure 1.38 The graph of $y = \csc x$.

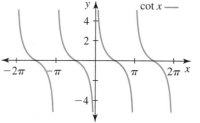

Figure 1.39 The graph of $y = \cot x$.

Since the sine and cosine functions are of particular importance, we now describe them in more detail. Consider the function

$$f(x) = a \sin(kx) \qquad \text{for } x \in \mathbf{R}$$

where a is a real number and $k \ne 0$. Now, $f(x)$ takes on values between $-a$ and a. We call $|a|$ the **amplitude**. The function $f(x)$ is periodic. To find the period p of $f(x)$, we set

$$|k|p = 2\pi \qquad \text{or} \qquad p = \frac{2\pi}{|k|}$$

Because the cosine function can be obtained from the sine function by a horizontal shift, we can define the amplitude and period analogously for the cosine function. That is, $f(x) = a \cos(kx)$ has amplitude $|a|$ and period $p = 2\pi/|k|$.

EXAMPLE 16   Compare

$$f(x) = 3 \sin\left(\frac{\pi}{4} x\right) \qquad \text{and} \qquad g(x) = \sin x$$

Solution   The amplitude of $f(x)$ is 3, whereas the amplitude of $g(x)$ is 1. The period p of $f(x)$ satisfies $\frac{\pi}{4} p = 2\pi$ or $p = 8$, whereas the period of $g(x)$ is 2π. Graphs of $f(x)$ and $g(x)$ are shown in Figure 1.40. ■

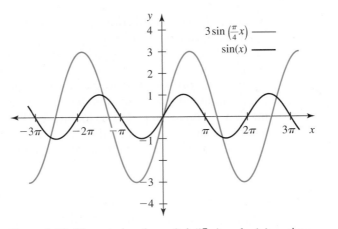

Figure 1.40 The graphs of $y = 3 \sin(\frac{\pi}{4} x)$ and $g(x) = \sin x$ in Example 16.

Remark. A number is called **algebraic** if it is the solution of a polynomial equation with rational coefficients. For instance, $\sqrt{2}$ is algebraic, as it satisfies the equation $x^2 - 2 = 0$. Numbers that are not algebraic are called **transcendental**. For instance, π and e are transcendental.

A similar distinction is made for functions. We call a function $y = f(x)$ algebraic if it is the solution of an equation of the form

$$P_n(x) y^n + \cdots + P_1(x) y + P_0(x) = 0$$

in which the coefficients are polynomial functions in x with rational coefficients. For instance, the function $y = 1/(1 + x)$ is algebraic, as it satisfies the equation $(x + 1) y - 1 = 0$. Here, $P_1(x) = x + 1$ and $P_0(x) = -1$. Other examples of algebraic functions are polynomial functions with rational coefficients and rational functions with rational coefficients.

Functions that are not algebraic are called transcendental. All the trigonometric, exponential, and logarithmic functions that we introduced in this section are transcendental functions.

Section 1.2 Problems

■ **1.2.1**

In Problems 1–4, state the range for the given functions. Graph each function.

1. $f(x) = x^2$, $x \in \mathbf{R}$ **2.** $f(x) = x^2$, $x \in [0, 1]$

3. $f(x) = x^2$, $-1 < x \le 0$ **4.** $f(x) = x^2$, $-\frac{1}{2} < x < \frac{1}{2}$

5. (a) Show that, for $x \ne 1$,

$$\frac{x^2 - 1}{x - 1} = x + 1$$

(b) Are the functions

$$f(x) = \frac{x^2 - 1}{x - 1}, \quad x \ne 1$$

and

$$g(x) = x + 1, \quad x \in \mathbf{R}$$

equal?

6. (a) Show that

$$2|x - 1| = \begin{cases} 2(x - 1) & \text{for } x \ge 1 \\ 2(1 - x) & \text{for } x \le 1 \end{cases}$$

(b) Are the functions

$$f(x) = \begin{cases} 2 - 2x & \text{for } 0 \le x \le 1 \\ 2x - 2 & \text{for } 1 \le x \le 2 \end{cases}$$

and

$$g(x) = 2|x - 1|, \; x \in [0, 2]$$

equal?

In Problems 7–12, sketch the graph of each function and decide in each case whether the function is (i) even, (ii) odd, or (iii) does not show any obvious symmetry. Then use the criteria in Subsection 1.2.1 to check your answers.

7. $f(x) = 2x$ **8.** $f(x) = 3x^2$

9. $f(x) = |3x|$ **10.** $f(x) = 2x + 1$

11. $f(x) = -|x|$ **12.** $f(x) = 3x^3$

13. Suppose that

$$f(x) = x^2, \quad x \in \mathbf{R}$$

and

$$g(x) = 3 + x, \quad x \in \mathbf{R}$$

(a) Show that

$$(f \circ g)(x) = (3 + x)^2, \quad x \in \mathbf{R}$$

(b) Show that

$$(g \circ f)(x) = 3 + x^2, \quad x \in \mathbf{R}$$

14. Suppose that

$$f(x) = x^3, \quad x \in \mathbf{R}$$

and

$$g(x) = 1 - x, \quad x \in \mathbf{R}$$

(a) Show that

$$(f \circ g)(x) = (1 - x)^3, \quad x \in \mathbf{R}$$

(b) Show that

$$(g \circ f)(x) = 1 - x^3, \quad x \in \mathbf{R}$$

15. Suppose that

$$f(x) = 1 - x^2, \quad x \in \mathbf{R}$$

and

$$g(x) = 2x, \quad x \ge 0$$

(a) Find

$$(f \circ g)(x)$$

together with its domain.

(b) Find

$$(g \circ f)(x)$$

together with its domain.

16. Suppose that

$$f(x) = \frac{1}{x + 1}, \quad x \ne -1$$

and

$$g(x) = 2x^2, \quad x \in \mathbf{R}$$

(a) Find $(f \circ g)(x)$.   **(b)** Find $(g \circ f)(x)$.
In both (a) and (b), find the domain.

17. Suppose that

$$f(x) = 3x^2, \quad x \ge 3$$

and

$$g(x) = \sqrt{x}, \quad x \ge 0$$

Find $(f \circ g)(x)$ together with its domain.

18. Suppose that

$$f(x) = x^4, \quad x \ge 3$$

and

$$g(x) = \sqrt{x + 1}, \quad x \ge 3$$

Find $(f \circ g)(x)$ together with its domain.

19. Suppose that $f(x) = x^2, x \ge 0$, and $g(x) = \sqrt{x}, x \ge 0$. Typically, $f \circ g \ne g \circ f$, but this is an example in which the order of composition does not matter. Show that $f \circ g = g \circ f$.

20. Suppose that $f(x) = x^4, x \ge 0$. Find $g(x)$ so that $f \circ g = g \circ f$.

■ **1.2.2**

21. Use a graphing calculator to graph $f(x) = x^2, x \ge 0$, and $g(x) = x^4, x \ge 0$, together. For which values of x is $f(x) > g(x)$, and for which is $f(x) < g(x)$?

22. Use a graphing calculator to graph $f(x) = x^3, x \ge 0$, and $g(x) = x^5, x \ge 0$, together. When is $f(x) > g(x)$, and when is $f(x) < g(x)$?

23. Graph $y = x^n, x \ge 0$, for $n = 1, 2, 3$, and 4 in one coordinate system. Where do the curves intersect?

24. (a) Graph $f(x) = x, x \ge 0$, and $g(x) = x^2, x \ge 0$, together, in one coordinate system.

(b) For which values of x is $f(x) \ge g(x)$, and for which values of x is $f(x) \le g(x)$?

25. (a) Graph $f(x) = x^2$ and $g(x) = x^3$ for $x \ge 0$, together, in one coordinate system.

(b) Show algebraically that

$$x^2 \ge x^3$$

for $0 \le x \le 1$.

(c) Show algebraically that

$$x^2 \le x^3$$

for $x \ge 1$.

26. Show algebraically that if $n \ge m$,

$$x^n \le x^m \quad \text{for } 0 \le x \le 1$$

and

$$x^n \ge x^m \quad \text{for } x \ge 1$$

27. (a) Show that $y = x^2, x \in \mathbf{R}$, is an even function.

(b) Show that $y = x^3, x \in \mathbf{R}$, is an odd function.

28. Show that

(a) $y = x^n, x \in \mathbf{R}$, is an even function when n is an even integer.

(b) $y = x^n, x \in \mathbf{R}$, is an odd function when n is an odd integer.

29. In Example 5 of this section, we considered the chemical reaction

$$A + B \rightarrow AB$$

Assume that initially only A and B are in the reaction vessel and that the initial concentrations are $a = [A] = 3$ and $b = [B] = 4$.

(a) We found that the reaction rate $R(x)$, where x is the concentration of AB, is given by

$$R(x) = k(a - x)(b - x)$$

where a is the initial concentration of A, b is the initial concentration of B, and k is the constant of proportionality. Suppose that the reaction rate $R(x)$ is equal to 9 when the concentration of AB is $x = 1$. Use this relationship to find the reaction rate $R(x)$.

(b) Determine the appropriate domain of $R(x)$, and use a graphing calculator to sketch the graph of $R(x)$.

30. An autocatalytic reaction uses its resulting product for the formation of a new product, as in the reaction

$$A + X \rightarrow X$$

If we assume that this reaction occurs in a closed vessel, then the reaction rate is given by

$$R(x) = kx(a - x)$$

for $0 \leq x \leq a$, where a is the initial concentration of A and x is the concentration of X.

(a) Show that $R(x)$ is a polynomial and determine its degree.

(b) Graph $R(x)$ for $k = 2$ and $a = 6$. Find the value of x at which the reaction rate is maximal.

31. Suppose that a beetle walks up a tree along a straight line at a constant speed of 1 meter per hour. What distance will the beetle have covered after 1 hour, 2 hours, and 3 hours? Write an equation that expresses the distance (in meters) as a function of the time (in hours), and show that this function is a polynomial of degree 1.

32. Suppose that a fungal disease originates in the middle of an orchard, initially affecting only one tree. The disease spreads out radially at a constant speed of 10 feet per day. What area will be affected after 2 days, 4 days, and 8 days? Write an equation that expresses the affected area as a function of time, measured in days, and show that this function is a polynomial of degree 2.

■ **1.2.3**

In Problems 33–36, for each function, find the largest possible domain and determine the range.

33. $f(x) = \dfrac{1}{1 - x}$
34. $f(x) = \dfrac{2x}{(x - 2)(x + 3)}$

35. $f(x) = \dfrac{x - 2}{x^2 - 9}$
36. $f(x) = \dfrac{1}{x^2 + 1}$

37. Compare $y = \frac{1}{x}$ and $y = \frac{1}{x^2}$ for $x > 0$ by graphing the two functions. Where do the curves intersect? Which function is greater for small values of x? for large values of x?

38. Let n and m be two positive integers with $m \leq n$. Answer the following questions about $y = x^{-n}$ and $y = x^{-m}$ for $x > 0$: Where do the curves intersect? Which function is greater for small values of x? for large values of x?

39. Let

$$f(x) = \dfrac{1}{x + 1}, \quad x > -1$$

(a) Use a graphing calculator to graph $f(x)$.

(b) On the basis of the graph in (a), determine the range of $f(x)$.

(c) For which values of x is $f(x) = 2$?

(d) On the basis of the graph in (a), determine how many solutions $f(x) = a$ has, where a is in the range of $f(x)$.

40. Let

$$f(x) = \dfrac{2x}{3 + x}, \quad x \geq 0$$

(a) Use a graphing calculator to graph $f(x)$.

(b) Find the range of $f(x)$.

(c) For which values of x is $f(x) = 1$?

(d) Based on the graph in (a), explain in words why, for any value a in the range of $f(x)$, you can find exactly one value $x \geq 0$ such that $f(x) = a$. Determine x by solving $f(x) = a$.

41. Let

$$f(x) = \dfrac{3x}{1 + x}, \quad x \geq 0$$

(a) Use a graphing calculator to graph $f(x)$.

(b) Find the range of $f(x)$.

(c) For which values of x is $f(x) = 2$?

(d) On the basis of the graph in (a), explain in words why, for any value a in the range of $f(x)$, you can find exactly one value $x \geq 0$ such that $f(x) = a$. Determine x by solving $f(x) = a$.

In Problems 42–44, we discuss the Monod growth function, which was introduced in Example 6 of this section.

42. Use a graphing calculator to investigate the Monod growth function

$$r(N) = \dfrac{aN}{k + N}, \quad N \geq 0$$

where a and k are positive constants.

(a) Graph $r(N)$ for (i) $a = 5$ and $k = 1$, (ii) $a = 5$ and $k = 3$, and (iii) $a = 8$ and $k = 1$. Place all three graphs in one coordinate system.

(b) On the basis of the graphs in (a), describe in words what happens when you change a.

(c) On the basis of the graphs in (a), describe in words what happens when you change k.

43. The Monod growth function $r(N)$ describes growth as a function of nutrient concentration N. Assume that

$$r(N) = 5\dfrac{N}{1 + N}, \quad N \geq 0$$

Find the percentage increase when the nutrient concentration is doubled from $N = 0.1$ to $N = 0.2$. Compare this result with what you find when you double the nutrient concentration from $N = 10$ to $N = 20$. This is an example of *diminishing return*.

44. The Monod growth function $r(N)$ describes growth as a function of nutrient concentration N. Assume that

$$r(N) = a\dfrac{N}{k + N}, \quad N \geq 0$$

where a and k are positive constants.

(a) What happens to $r(N)$ as N increases? Use this relationship to explain why a is called the saturation level.

(b) Show that k is the half-saturation constant; that is, show that if $N = k$, then $r(N) = a/2$.

45. Let

$$f(x) = \dfrac{x^2}{4 + x^2}, \quad x \geq 0$$

(a) Use a graphing calculator to graph $f(x)$.

(b) On the basis of your graph in (a), find the range of $f(x)$.

(c) What happens to $f(x)$ as x gets larger?

46. The function

$$f(x) = \frac{x^n}{b^n + x^n}, \quad x \geq 0$$

where n is a positive integer and b is a positive real number, is used in biochemistry to model reaction rates as a function of the concentration of some reactants.
(a) Use a graphing calculator to graph $f(x)$ for $n = 1, 2,$ and 3 in one coordinate system when $b = 2$.
(b) Where do the three graphs in (a) intersect?
(c) What happens to $f(x)$ as x gets larger?
(d) For an arbitrary positive value of b, show that $f(b) = 1/2$. On the basis of this demonstration and your answer in (c), explain why b is called the half-saturation constant.

■ 1.2.4

In Problems 47–50, use a graphing calculator to sketch the graphs of the functions.

47. $y = x^{3/2}, x \geq 0$ **48.** $y = x^{1/3}, x \geq 0$
49. $y = x^{-1/4}, x > 0$ **50.** $y = 2x^{-7/8}, x > 0$

51. (a) Graph $y = x^{-1/2}, x > 0$, and $y = x^{1/2}, x \geq 0$, together, in one coordinate system.
(b) Show algebraically that

$$x^{-1/2} \geq x^{1/2}$$

for $0 < x \leq 1$.
(c) Show algebraically that

$$x^{-1/2} \leq x^{1/2}$$

for $x \geq 1$.
52. (a) Graph $y = x^{5/2}, x \geq 0$, and $y = x^{1/2}, x \geq 0$, together, in one coordinate system.
(b) Show algebraically that

$$x^{5/2} \leq x^{1/2}$$

for $0 \leq x \leq 1$. (*Hint*: Show that $x^{1/2}/x^{-1/2} = x \leq 1$ for $0 < x \leq 1$.)
(c) Show algebraically that

$$x^{5/2} \geq x^{1/2}$$

for $x \geq 1$.

In Problems 53–56, sketch each scaling relation (Niklas, 1994).

53. In a sample based on 46 species, leaf area was found to be proportional to (stem diameter)$^{1.84}$. On the basis of your graph, as stem diameter increases, does leaf area increase or decrease?
54. In a sample based on 28 species, the volume fraction of spongy mesophyll was found to be proportional to (leaf thickness)$^{-0.49}$. (The spongy mesophyll is part of the internal tissue of a leaf blade.) On the basis of your graph, as leaf thickness increases, does the volume fraction of spongy mesophyll increase or decrease?
55. In a sample of 60 species of trees, wood density was found to be proportional to (breaking strength)$^{0.82}$. On the basis of your graph, does breaking strength increase as wood density increases? or as wood density decreases?
56. Suppose that a cube of length L and volume V has mass M and that $M = 0.35V$. How does the length of the cube depend on its mass?

■ 1.2.5

57. Assume that a population size at time t is $N(t)$ and that

$$N(t) = 2^t, \quad t \geq 0$$

(a) Find the population size for $t = 0, 1, 2, 3,$ and 4.
(b) Graph $N(t)$ for $t \geq 0$.
58. Assume that a population size at time t is $N(t)$ and that

$$N(t) = 40 \cdot 2^t, \quad t \geq 0$$

(a) Find the population size at time $t = 0$.
(b) Show that

$$N(t) = 40e^{t \ln 2}, \quad t \geq 0$$

(c) How long will it take until the population size reaches 1000? [*Hint*: Find t so that $N(t) = 1000$.]
59. The half-life of C^{14} is 5730 years. If a sample of C^{14} has a mass of 20 micrograms at time $t = 0$, how much is left after 2000 years?
60. The half-life of C^{14} is 5730 years. If a sample of C^{14} has a mass of 20 micrograms at time 0, how long will it take until **(a)** 10 grams and **(b)** 5 grams are left?
61. After 7 days, a particular radioactive substance decays to half of its original amount. Find the decay rate of this substance.
62. After 5 days, a particular radioactive substance decays to 37% of its original amount. Find the half-life of this substance.
63. Polonium 210 (Po^{210}) has a half-life of 140 days.
(a) If a sample of Po^{210} has a mass of 300 micrograms, find a formula for the mass after t days.
(b) How long would it take this sample to decay to 20% of its original amount?
(c) Sketch the graph of the amount of mass left after t days.
64. The half-life of C^{14} is 5730 years. Suppose that wood found at an archeological excavation site contains about 35% as much C^{14} (in relation to C^{12}) as does living plant material. Determine when the wood was cut.
65. The half-life of C^{14} is 5730 years. Suppose that wood found at an archeological excavation site is 15,000 years old. How much C^{14} (based on C^{12} content) does the wood contain relative to living plant material?
66. The age of rocks of volcanic origin can be estimated with isotopes of argon 40 (Ar^{40}) and potassium 40 (K^{40}). K^{40} decays into Ar^{40} over time. If a mineral that contains potassium is buried under the right circumstances, argon forms and is trapped. Since argon is driven off when the mineral is heated to very high temperatures, rocks of volcanic origin do not contain argon when they are formed. The amount of argon found in such rocks can therefore be used to determine the age of the rock. Assume that a sample of volcanic rock contains 0.00047% K^{40}. The sample also contains 0.000079% Ar^{40}. How old is the rock? (The decay rate of K^{40} to Ar^{40} is 5.335×10^{-10}/yr.)
67. (*Adapted from Moss, 1980*) Hall (1964) investigated the change in population size of the zooplankton species *Daphnia galeata mendota* in Base Line Lake, Michigan. The population size $N(t)$ at time t was modeled by the equation

$$N(t) = N_0 e^{rt}$$

where N_0 denotes the population size at time 0. The constant r is called the **intrinsic rate of growth**.
(a) Plot $N(t)$ as a function of t if $N_0 = 100$ and $r = 2$. Compare your graph against the graph of $N(t)$ when $N_0 = 100$ and $r = 3$. Which population grows faster?

(b) The constant r is an important quantity because it describes how quickly the population changes. Suppose that you determine the size of the population at the beginning and at the end of a period of length 1, and you find that at the beginning there were 200 individuals and after one unit of time there were 250 individuals. Determine r. [*Hint:* Consider the ratio $N(t+1)/N(t)$.]

68. Fish are indeterminate growers; that is, they grow throughout their lifetime. The growth of fish can be described by the von Bertalanffy function

$$L(x) = L_\infty(1 - e^{-kx})$$

for $x \geq 0$, where $L(x)$ is the length of the fish at age x and k and L_∞ are positive constants.
(a) Use a graphing calculator to graph $L(x)$ for $L_\infty = 20$, for **(i)** $k = 1$ and **(ii)** $k = 0.1$.

(b) For $k = 1$, find x so that the length is 90% of L_∞. Repeat for 99% of L_∞. Can the fish ever attain length L_∞? Interpret the meaning of L_∞.

(c) Compare the graphs obtained in (a). Which growth curve reaches 90% of L_∞ faster? Can you explain what happens to the curve of $L(x)$ when you vary k (for fixed L_∞)?

■ **1.2.6**

69. Which of the following functions is one to one (use the horizontal line test)?
(a) $f(x) = x^2, x \geq 0$ **(b)** $f(x) = x^2, x \in \mathbf{R}$
(c) $f(x) = \frac{1}{x}, x > 0$ **(d)** $f(x) = e^x, x \in \mathbf{R}$
(e) $f(x) = \frac{1}{x^2}, x \neq 0$ **(f)** $f(x) = \frac{1}{x^2}, x > 0$

70. (a) Show that $f(x) = x^3 - 1, x \in \mathbf{R}$, is one to one, and find its inverse together with its domain.
(b) Graph $f(x)$ and $f^{-1}(x)$ in one coordinate system, together with the line $y = x$, and convince yourself that the graph of $f^{-1}(x)$ can be obtained by reflecting the graph of $f(x)$ about the line $y = x$.

71. (a) Show that $f(x) = x^2 + 1, x \geq 0$, is one to one, and find its inverse together with its domain.
(b) Graph $f(x)$ and $f^{-1}(x)$ in one coordinate system, together with the line $y = x$, and convince yourself that the graph of $f^{-1}(x)$ can be obtained by reflecting the graph of $f(x)$ about the line $y = x$.

72. (a) Show that $f(x) = \sqrt{x}, x \geq 0$, is one to one, and find its inverse together with its domain.
(b) Graph $f(x)$ and $f^{-1}(x)$ in one coordinate system, together with the line $y = x$, and convince yourself that the graph of $f^{-1}(x)$ can be obtained by reflecting the graph of $f(x)$ about the line $y = x$.

73. (a) Show that $f(x) = 1/x^3, x > 0$, is one to one, and find its inverse together with its domain.
(b) Graph $f(x)$ and $f^{-1}(x)$ in one coordinate system, together with the line $y = x$, and convince yourself that the graph of $f^{-1}(x)$ can be obtained by reflecting the graph of $f(x)$ about the line $y = x$.

74. The reciprocal of a function $f(x)$ can be written as either $1/f(x)$ or $[f(x)]^{-1}$. The point of this problem is to make clear that a reciprocal of a function has nothing to do with the inverse of a function. As an example, let $f(x) = 2x + 1, x \in \mathbf{R}$. Find both $[f(x)]^{-1}$ and $f^{-1}(x)$, and compare the two functions. Graph all three functions together.

■ **1.2.7**

75. Find the inverse of $f(x) = 3^x, x \in \mathbf{R}$, together with its domain, and graph both functions in the same coordinate system.
76. Find the inverse of $f(x) = 5^x, x \in \mathbf{R}$, together with its domain, and graph both functions in the same coordinate system.
77. Find the inverse of $f(x) = (\frac{1}{4})^x, x \in \mathbf{R}$, together with its domain, and graph both functions in the same coordinate system.
78. Find the inverse of $f(x) = (\frac{1}{3})^x, x \in \mathbf{R}$, together with its domain, and graph both functions in the same coordinate system.
79. Find the inverse of $f(x) = 2^x, x \geq 0$, together with its domain, and graph both functions in the same coordinate system.
80. Find the inverse of $f(x) = (\frac{1}{2})^x, x \geq 0$, together with its domain, and graph both functions in the same coordinate system.
81. Simplify the following expressions:
(a) $2^{5\log_2 x}$ **(b)** $3^{4\log_3 x}$
(c) $5^{5\log_{1/5} x}$ **(d)** $4^{-2\log_2 x}$
(e) $2^{3\log_{1/2} x}$ **(f)** $4^{-\log_{1/2} x}$
82. Simplify the following expressions:
(a) $\log_4 16^x$ **(b)** $\log_2 16^x$
(c) $\log_3 27^x$ **(d)** $\log_{1/2} 4^x$
(e) $\log_{1/2} 8^{-x}$ **(f)** $\log_3 9^{-x}$
83. Simplify the following expressions:
(a) $\ln x^2 + \ln x^3$ **(b)** $\ln x^4 - \ln x^{-2}$
(c) $\ln(x^2 - 1) - \ln(x + 1)$ **(d)** $\ln x^{-1} + \ln x^{-3}$
84. Simplify the following expressions:
(a) $e^{3\ln x}$ **(b)** $e^{-\ln(x^2+1)}$
(c) $e^{-2\ln(1/x)}$ **(d)** $e^{-2\ln x}$
85. Write the following expressions in terms of base e, and simplify:
(a) 3^x **(b)** 4^{x^2-1} **(c)** 2^{-x-1} **(d)** 3^{-4x+1}
86. Write the following expressions in terms of base e:
(a) $\log_2(x^2 - 1)$ **(b)** $\log_3(5x + 1)$
(c) $\log(x + 2)$ **(d)** $\log_2(2x^2 - 1)$
87. Show that the function $y = (1/2)^x$ can be written in the form $y = e^{-\mu x}$, where μ is a positive constant. Determine μ.
88. Show that if $0 < a < 1$, then the function $y = a^x$ can be written in the form $y = e^{-\mu x}$, where μ is a positive constant. Write μ in terms of a.
89. Assume that two DNA sequences of common origin, each of length 300 nucleotides, differ from each other by 47 nucleotides. Use the Jukes and Cantor correction of Example 15 to find an estimate for the number K of substitutions per site.
90. A community measure that takes both species abundance and species richness into account is the Shannon diversity index H. To calculate H, the proportion p_i of species i in the community is used. Assume that the community consists of S species. Then

$$H = -(p_1 \ln p_1 + p_2 \ln p_2 + \cdots + p_S \ln p_S)$$

(a) Assume that $S = 5$ and that all species are equally abundant; that is, $p_1 = p_2 = \cdots = p_5$. Compute H.
(b) Assume that $S = 10$ and that all species are equally abundant; that is, $p_1 = p_2 = \cdots = p_{10}$. Compute H.
(c) A measure of equitability (or evenness) of the species distribution can be measured by dividing the diversity index H by $\ln S$. Compute $H/\ln S$ for $S = 5$ and $S = 10$.
(d) Show that, in general, if there are N species and all species are equally abundant, then

$$\frac{H}{\ln S} = 1$$

■ **1.2.8**

In Problems 91–96, for each given pair of functions, use a graphing calculator to compare the functions. Describe what you see.

91. $y = \sin x$ and $y = 2\sin x$

92. $y = \sin x$ and $y = \sin(2x)$

93. $y = \cos x$ and $y = 2\cos x$

94. $y = \cos x$ and $y = \cos(2x)$

95. $y = \tan x$ and $y = 2\tan x$

96. $y = \tan x$ and $y = \tan(2x)$

97. Let
$$f(x) = 3\sin(4x), \quad x \in \mathbf{R}$$
Find the amplitude and the period of $f(x)$.

98. Let
$$f(x) = -2\sin\left(\frac{x}{2}\right), \quad x \in \mathbf{R}$$
Find the amplitude and the period of $f(x)$.

99. Let
$$f(x) = 4\sin(2\pi x), \quad x \in \mathbf{R}$$
Find the amplitude and the period of $f(x)$.

100. Let
$$f(x) = -\frac{3}{2}\sin\left(\frac{\pi}{3}x\right), \quad x \in \mathbf{R}$$

Find the amplitude and the period of $f(x)$.

101. Let
$$f(x) = 4\cos\left(\frac{x}{4}\right), \quad x \in \mathbf{R}$$
Find the amplitude and the period of $f(x)$.

102. Let
$$f(x) = 7\cos(2x), \quad x \in \mathbf{R}$$
Find the amplitude and the period of $f(x)$.

103. Let
$$f(x) = -3\cos\left(\frac{\pi x}{5}\right), \quad x \in \mathbf{R}$$
Find the amplitude and the period of $f(x)$.

104. Let
$$f(x) = -\frac{2}{3}\cos\left(\frac{3x}{\pi}\right), \quad x \in \mathbf{R}$$
Find the amplitude and the period of $f(x)$.

105. Use the fact that $\sec x = \frac{1}{\cos x}$ to explain why the maximum domain of $y = \sec x$ consists of all real numbers except odd integer multiples of $\pi/2$.

106. Use the fact that $\csc x = \frac{1}{\sin x}$ to explain why the maximum domain of $y = \csc x$ consists of all real numbers except integer multiples of π.

■ 1.3 Graphing

In the preceding section, we introduced the functions most important to our study. You must be able to graph the following functions without a calculator: $y = c, x, x^2$, $x^3, 1/x, e^x, \ln x, \sin x, \cos x, \sec x$, and $\tan x$. This will help you to sketch functions quickly and to come up with an analytical description of a function based on a graph. In this section, you will learn how to obtain new functions from these basic functions and how to graph them. In addition, we will introduce important transformations that are often used to display data graphically.

■ 1.3.1 Graphing and Basic Transformations of Functions

In this subsection, we will recall some basic transformations: vertical and horizontal translations, reflections about $x = 0$ and $y = 0$, and stretching and compressing.

Definition The graph of
$$y = f(x) + a$$
is a **vertical translation** of the graph of $y = f(x)$. If $a > 0$, the graph of $y = f(x)$ is shifted up a units; if $a < 0$, the graph of $y = f(x)$ is shifted down $|a|$ units.

This definition is illustrated in Figure 1.41, where we display $y = x^2$, $y = x^2 + 2$, and $y = x^2 - 2$.

Definition The graph of
$$y = f(x - c)$$
is a **horizontal translation** of the graph of $y = f(x)$. If $c > 0$, the graph of $y = f(x)$ is shifted c units to the right; if $c < 0$, the graph of $y = f(x)$ is shifted $|c|$ units to the left.

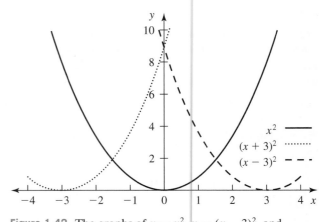

Figure 1.41 The graphs of $y = x^2$, $y = x^2 + 2$ and $y = x^2 - 2$.

Figure 1.42 The graphs of $y = x^2$, $y = (x - 3)^2$, and $y = (x + 3)^2$.

This definition is illustrated in Figure 1.42, where we display $y = x^2$, $y = (x - 3)^2$, and $y = (x + 3)^2$. Note that $y = (x + 3)^2$ is shifted to the left, since $y = (x - (-3))^2$ and therefore $c = -3 < 0$.

Reflections about the x-axis ($y = 0$) and the y-axis ($x = 0$) are illustrated in Figure 1.43. We graph $y = \sqrt{x}$; its reflection about the x-axis, $y = -\sqrt{x}$; and its reflection about the y-axis, $y = \sqrt{-x}$.

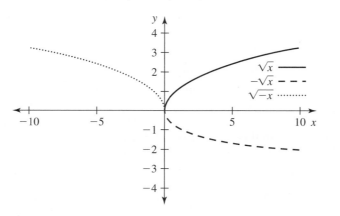

Figure 1.43 Reflections about the x-axis and the y-axis.

Multiplying a function by a factor between 0 and 1 compresses the graph of the function; multiplying a function by a factor greater than 1 stretches the graph of the function. These operations are illustrated in Figure 1.44, where we graph $y = x^2$ and $y = \frac{1}{2}x^2$, and in Figure 1.45, where we graph $y = x^2$ and $y = 2x^2$.

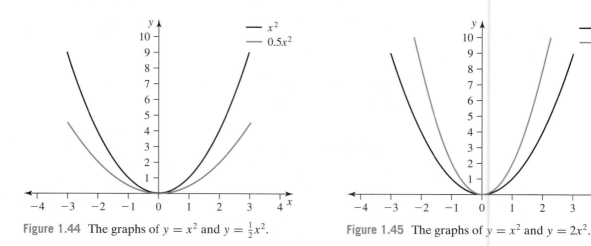

Figure 1.44 The graphs of $y = x^2$ and $y = \frac{1}{2}x^2$.

Figure 1.45 The graphs of $y = x^2$ and $y = 2x^2$.

We illustrate the preceding transformations in the next two examples.

EXAMPLE 1   Explain how the graph of

$$y = 2 \sin \left(x - \frac{\pi}{4} \right) \quad \text{for } x \in \mathbf{R}$$

can be obtained from the graph of $y = \sin x$, $x \in \mathbf{R}$.

Solution   We transform $y = \sin x$ in two steps, illustrated in Figure 1.46. First we shift $y = \sin x$ to the right $\frac{\pi}{4}$ units. This yields $y = \sin(x - \frac{\pi}{4})$. Then we multiply $y = \sin(x - \frac{\pi}{4})$ by 2. This corresponds to stretching $y = \sin(x - \frac{\pi}{4})$ by the factor 2.   ■

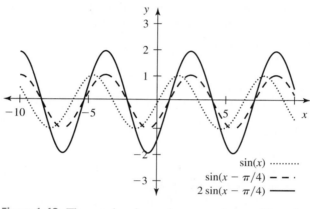

Figure 1.46 The graphs of $y = \sin x$, $y = \sin(x - \frac{\pi}{4})$, and $y = 2 \sin(x - \frac{\pi}{4})$.

EXAMPLE 2   Explain how the graph of

$$y = -\sqrt{x - 3} - 1, \quad x \geq 3$$

can be obtained from the graph of $y = \sqrt{x}$, $x \geq 0$.

Solution   We transform $y = \sqrt{x}$ in three steps, illustrated in Figure 1.47. First we shift $y = \sqrt{x}$ three units to the right and obtain $y = \sqrt{x - 3}$. Then we reflect $y = \sqrt{x - 3}$ about the x-axis, which yields $y = -\sqrt{x - 3}$. Finally, we shift $y = -\sqrt{x - 3}$ down one unit. This is the graph of $y = -\sqrt{x - 3} - 1$.   ■

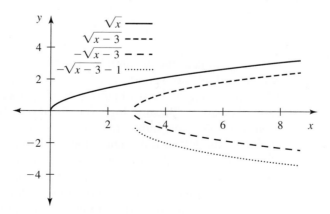

Figure 1.47 The graphs for Example 2.

■ 1.3.2 The Logarithmic Scale

We often encounter sizes that vary over a wide range. For instance, lengths in the metric system are measured in meters (m). (A meter is a bit longer than a yard: 1 meter is equal to 1.0936 yards.) A longer metric unit that is commonly used is a kilometer (km), which is 1000 m. Shorter commonly used metric units are a millimeter (mm), which is 1/1000 of a meter; a micrometer (μm), which is 1/1,000,000 (one-millionth) of a meter; and a nanometer (nm), which is 1/1,000,000,000 (one-billionth) of a meter. Here are some examples of lengths of organisms: A ribosome is about 20 nm ($= 2 \times 10^{-8}$ m), a poxvirus is about 400 nm (4×10^{-7} m), a bacterium is about 1 μm ($= 10^{-6}$ m), a tardigrade (or "water bear") is about 1.2 mm ($= 1.2 \times 10^{-3}$ m), an adult human is about 1.8 m, a blue whale is between 25 and 35 m, the diameter of the earth is 12,755 km ($\approx 1.3 \times 10^{7}$ m), and the average distance from the sun to the earth is about 150 million km ($= 1.5 \times 10^{11}$ m). These sizes are conveniently illustrated with the use of a **logarithmic scale**—a scale according to which multiples of 10 are equally distant (Figure 1.48).

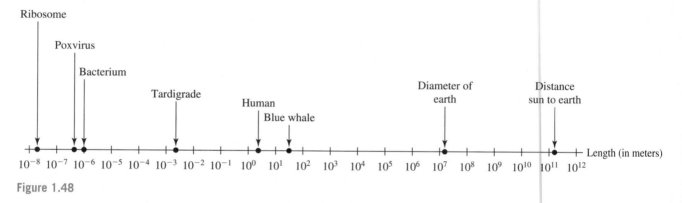

Figure 1.48

When we take logarithms to base 10 of the quantities displayed in Figure 1.48, we find that the transformed scale looks like the arithmetic scale we are familiar with (Figure 1.49). The numbers on the logarithmic scale in Figure 1.49 correspond to exponents.

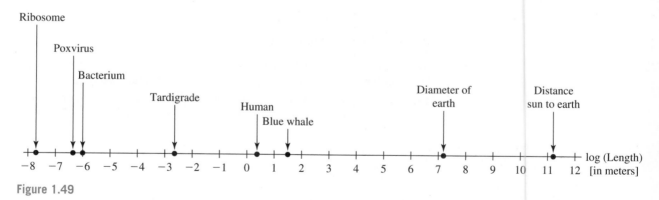

Figure 1.49

Let's look at the two number lines in more detail. The origin of the number line in Figure 1.49 corresponds to the number 1 in Figure 1.48, since $\log 1 = 0$. If we go to the left of 1 on the line in Figure 1.48, we get smaller and smaller numbers, but they are *all* positive. (Because $\log x$ is defined only for $x > 0$, we cannot logarithmically transform negative numbers.) Going to the left of 1 in Figure 1.48 corresponds to going to the left of 0 in Figure 1.49. The negative numbers on the number line in Figure 1.49 correspond to negative exponents; for instance, the -8 in Figure 1.49 means $\log x = -8$, or $x = 10^{-8}$. A similar interpretation holds when we go to the right of 1 in Figure 1.48 (or to the right of 0 in Figure 1.49). A logarithmic scale is typically based on logarithms to base 10, since this base makes conversion between the two representations in Figures 1.48 and 1.49 easier.

The lengths in the preceding examples differed by many factors of 10. Instead of saying that quantities differ by many factors of 10, we will say that they differ by many **orders of magnitude**: If two quantities differ by a factor of 10, they differ by one order of magnitude. (If they differ by a factor of 100, they differ by two orders of magnitude, and so on.) Orders of magnitude are approximate comparisons: A 1.8-m-tall human and a 25-m-long blue whale differ by about one order of magnitude.

EXAMPLE 3 Display the numbers 0.003, 0.1, 0.5, 6, 200, and 4000 on a logarithmic scale.

Solution To display the numbers, we need to take logarithms first:

x	0.003	0.1	0.5	6	200	4000
$\log x$	-2.5229	-1	-0.3010	0.7782	2.3010	3.6021

Since $\log 0.003 = -2.5229$, we find this number 2.5229 units to the left of 0 on the logarithmic scale. Similarly, since $\log 0.1 = -1$, this number is one unit to the left of 0, and $\log 200 = 2.3010$ is 2.3010 units to the right of 0 (Figure 1.50).

Figure 1.50 Example 3.

Figure 1.51 Example 3.

In the biological literature, x rather than $\log x$ is used to label logarithmic number lines. The locations of the numbers are the same; only the labeling changes. That is, 0.003 would be -2.5229 units to the left of the origin of the line (which is now at 1). The line in Figure 1.50 would then look like the line in Figure 1.51. ■

■ 1.3.3 Transformations into Linear Functions

When you look through a biology textbook, you very likely find graphs like the ones in Figures 1.52 and 1.53. In either graph, you see a straight line (with data points scattered about it). In Figure 1.52, the vertical axis is logarithmically transformed and the horizontal axis is on a linear scale; in Figure 1.53, both axes are logarithmically transformed. Why do we display data like this, and what do these graphs mean?

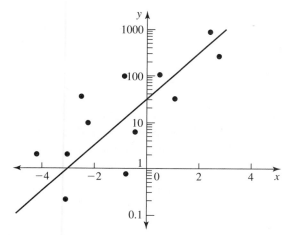

Figure 1.52 A straight line when the vertical axis is logarithmically transformed.

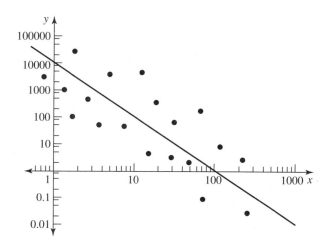

Figure 1.53 A straight line when both axes are logarithmically transformed.

The first question is quick to answer: Straight lines (or linear relationships) are easy to recognize visually. If transforming data results in data points lying along a straight line, we should do the transformation, because, as we will see, this will allow us to obtain a functional relationship between quantities. Now on to the second question: What do these graphs mean?

Exponential Functions Let's look at Figure 1.52 and redraw just the straight line, using $\log y$ (instead of y) on the vertical axis and x on the horizontal axis (Figure 1.54). Set $Y = \log y$ and forget for a moment where the graph came from. We see a linear relationship between Y and x—a relationship of the form

$$Y = c + mx$$

where c is the Y-intercept and m is the slope. We can read these two quantities off of the graph in Figure 1.54:

$$c = 1.5 \qquad m = 0.5$$

That is, we have

$$Y = 1.5 + 0.5x$$

Now, $Y = \log y$, and thus

$$\log y = 1.5 + 0.5x$$

Exponentiating both sides, we find that

$$y = 10^{1.5+0.5x} = 10^{1.5}(10^{0.5})^x$$

Since $10^{1.5} \approx 31.62$ and $10^{0.5} \approx 3.162$, we can write the preceding equation as

$$y = (31.62)(3.162^x) \tag{1.4}$$

Looking at (1.4), we see that it is an exponential function.

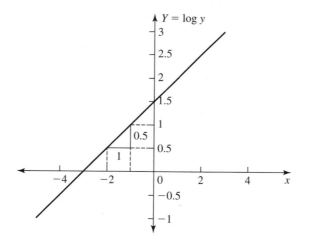

Figure 1.54 Figure 1.52 redrawn. Now the vertical axis is labeled $\log y$.

A graph in which the vertical axis is on a logarithmic scale and the horizontal axis is on a linear scale is called a **log-linear plot** or a **semilog plot**. If we display an exponential function of the form $y = ba^x$ on a semilog plot, a straight line results. To see this, we take logarithms to base 10 on both sides of $y = ba^x$:

$$\log y = \log(ba^x) \tag{1.5}$$

Using the properties of logarithms, we simplify the right-hand side to

$$\log(ba^x) = \log b + \log a^x = \log b + x \log a$$

If we set $Y = \log y$, then (1.5) becomes

$$Y = \log b + (\log a)x \tag{1.6}$$

Comparing this equation with the general form of a linear function $Y = c + mx$, we see that the Y-intercept is $\log b$ and the slope is $\log a$. You do not need to memorize this statement, since you can always do the calculation that resulted in (1.6), but you should memorize the fact that an exponential function results in a straight line on a semilog plot. If $a > 1$, the slope of the line is positive; if $0 < a < 1$, the slope of the line is negative.

EXAMPLE 4   Graph

$$y = 2.5 \cdot 3^x, \quad x \in \mathbf{R}$$

on a semilog plot.

Solution   We take logarithms first:

$$\log y = \log(2.5 \cdot 3^x)$$
$$= \underbrace{\log 2.5}_{\approx 0.3979} + x \underbrace{\log 3}_{\approx 0.4771}$$

The graph is shown in Figure 1.55. Note that the origin of the coordinate system is where $x = 0$ and $y = 1$ (or $\log y = 0$). The labeling on the vertical axis is for y, and we see that the labels are multiples of 10. To find 2.5 on the vertical axis, we use the fact that $\log 2.5 = 0.3979$ and that 2.5 is therefore 0.3979 units above the x-axis, as illustrated in Figure 1.55. (One unit on the vertical axis corresponds to a factor of 10.)   ■

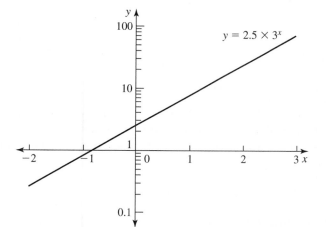

Figure 1.55   The graph of $y = 2.5 \times 3^x$ on a semilog plot.

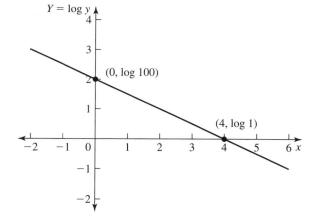

Figure 1.56   The graph for Example 5. The line goes through the points $(0, 2)$ and $(4, 0)$.

EXAMPLE 5   Find the functional relationship between x and y based on the graph in Figure 1.56.

Solution   Figure 1.56 shows a semilog plot. We set $Y = \log y$. Then, in an x–Y graph, the Y-intercept is $\log 100 = 2$, and, using the two points $(0, \log 100)$ and $(4, \log 1)$, we find that the slope of the line is $(\log 1 - \log 100)/(4 - 0) = (0 - 2)/(4 - 0) = -0.5$. Hence,

$$Y = c + mx = 2 - 0.5x$$

Since $Y = \log y$, after exponentiating the linear equation, we obtain

$$y = 10^{2 - 0.5x} = 10^2 (10^{-0.5})^x = (100)(0.3162)^x$$ ■

Power Functions   Let's look back at Figure 1.53. There, both axes are logarithmically transformed. We redraw just the straight line, using $Y = \log y$ on the vertical axis and $X = \log x$ on the horizontal axis (Figure 1.57).

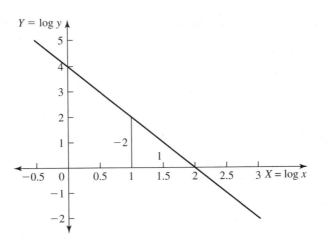

Figure 1.57 Figure 1.53 redrawn. Now the vertical axis is labeled $\log y$ and the horizontal axis is labeled $\log x$.

Using the linear equation $Y = c + mX$, where c is the Y-intercept and m is the slope, we find, from Figure 1.57, that

$$c = 4 \quad \text{and} \quad m = -2$$

With $Y = \log y$ and $X = \log x$, the linear equation becomes

$$\log y = 4 - 2\log x$$

Exponentiating both sides, we get

$$y = 10^{4 - 2\log x} = 10^4 (10^{\log x^{-2}}) = 10^4 x^{-2}$$

The function $y = 10^4 x^{-2}$ is a power function.

A graph in which both the vertical and the horizontal axis are logarithmically scaled is called a **log-log plot** or **double-log plot**. If we display a power function $y = bx^r$ in a double-log plot, a straight line results. To see this, we take logarithms to base 10 on both sides of $y = bx^r$:

$$\log y = \log(bx^r) \tag{1.7}$$

Using the properties of logarithms on the right-hand side of (1.7), we get

$$\log(bx^r) = \log b + \log x^r = \log b + r\log x$$

If we set $Y = \log y$ and $X = \log x$, then (1.7) becomes

$$Y = \log b + rX$$

Comparing this equation with the general form of a linear function, $Y = c + mX$, we see that the Y-intercept is $\log b$ and the slope is r. If $r > 0$, the slope is positive. If $r < 0$, the slope is negative.

EXAMPLE 6

Graph

$$y = 100x^{-2/3}, \quad x > 0$$

on a double-log plot.

Solution

We take logarithms first:

$$\log y = \log(100x^{-2/3}) = \log 100 - \frac{2}{3}\log x$$

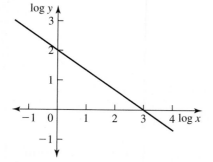

Figure 1.58 The graph $y = 100x^{-2/3}$ on a double-log plot where the axes are labeled $X = \log x$ and $Y = \log y$.

We set $Y = \log y$ and $X = \log x$. Then, with $\log 100 = 2$, we find that

$$Y = 2 - \frac{2}{3}X$$

This is the equation of a straight line with X-intercept 3 and Y-intercept 2 (and thus slope $-2/3$). We graph this function in Figure 1.58, where we have X and Y on the two axes. If we use x and y on the two axes (Figure 1.59), the labels change: The y-intercept is now 100 (corresponding to $\log 100 = 2$) and the x-intercept is 1000 (corresponding to $\log 1000 = 3$). Note that the origin in Figure 1.58 is $X = 0$ and $Y = 0$; the origin in Figure 1.59 is $x = 1$ and $y = 1$. ■

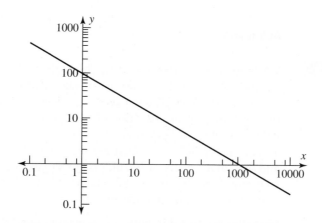

Figure 1.59 The graph $y = 100x^{-2/3}$ on a double-log plot where the axes are labeled x and y.

EXAMPLE 7

Find the functional relationship between x and y on the basis of the graph in Figure 1.60.

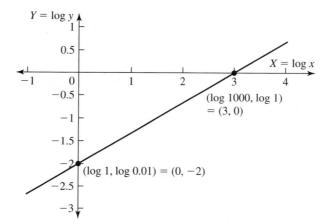

Figure 1.60 The graph of the function for Example 7: The two points on the double-log plot used for finding the relationship are $(\log 1, \log 0.01)$ and $(\log 1000, \log 1)$.

Solution

If $Y = \log y$ and $X = \log x$, then, in an X–Y graph, the Y-intercept is $\log 0.01 = -2$ and, using the two points $(\log 1, \log 0.01)$ and $(\log 1000, \log 1)$, we calculate the slope as

$$\frac{\log 1 - \log 0.01}{\log 1000 - \log 1} = \frac{0 - (-2)}{3 - 0} = \frac{2}{3}$$

Hence, the equation is

$$Y = -2 + \frac{2}{3}X$$

With $Y = \log y$ and $X = \log x$, we find that

$$\log y = -2 + \frac{2}{3}\log x$$

and, after exponentiating both sides of this equation, we get

$$y = 10^{-2 + \frac{2}{3}\log x} = 10^{-2}10^{\log x^{2/3}} = (0.01)x^{2/3}$$

Thus, the functional relationship between x and y is a power function of the form

$$y = (0.01)x^{2/3}$$

■

Applications

EXAMPLE 8

When growing plants at sufficiently high initial densities, we often observe that the number of plants decreases as the size of the plants grows. This property is called *self-thinning*. When the per-plant dry weight of the aboveground biomass is plotted on a log-log plot as a function of the density of survivors, we frequently find that the data lie along a straight line with slope $-3/2$. Assume that, for a particular plant, such a relationship holds for plant densities between 10^2 and 10^4 plants per square meter and that, at a density of 100 plants per square meter, the dry weight per plant is about 10 grams. Find the functional relationship between dry weight and plant density, and graph this function on a log-log plot.

Solution

Since the relationship between density (x) and dry weight (y) follows a straight line with slope $-3/2$ on a log-log plot, we set

$$\log y = C - \frac{3}{2}\log x \quad \text{for } 10^2 \leq x \leq 10^4$$

where C is a constant. To find C, we use the fact that when $x = 100$, $y = 10$. Therefore,

$$\log 10 = C - \frac{3}{2}\log 100$$

or

$$1 = C - \frac{3}{2} \cdot 2, \quad \text{which implies that} \quad C = 4$$

Hence,

$$\log y = 4 - \frac{3}{2}\log x$$

Exponentiating both sides (and remembering that "log" denotes the logarithm to base 10), we find that

$$y = 10^4 x^{-3/2} \quad \text{for } 10^2 \leq x \leq 10^4$$

The graph of this function on a log-log scale is shown in Figure 1.61. ■

EXAMPLE 9

Polonium 210 (Po^{210}) is a radioactive material. To determine the half-life of Po^{210} experimentally, we measure the amount of radioactive material left after time t for various values of t. When we plot the data on a semilog plot, we find that we can fit a straight line to the curve. The slope of the straight line is -0.0022/day. Find the half-life of Po^{210}.

Solution

Radioactive decay follows the equation

$$W(t) = W(0)e^{-\lambda t} \quad \text{for } t \geq 0$$

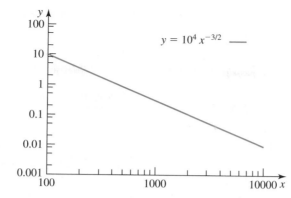

Figure 1.61 The graph of the function for Example 8: The line has slope $-3/2$ and goes through the point $(\log 100, \log 10)$ on a double-log plot.

where $W(t)$ is the amount of radioactive material left after time t. If we log transform this equation, we obtain

$$\log W(t) = \log W(0) - \lambda t \log e$$

Note that we use logarithms to base 10. If we plot $W(t)$ as a function of t on a semilog plot, we obtain a straight line with slope $-\lambda \log e$. Matching this slope with the number given in the example, we obtain

$$\lambda \log e = 0.0022/\text{day}$$

Solving for λ yields

$$\lambda = \frac{1}{\log e} 0.0022/\text{day}$$

To find the half-life T_h, we use the formula (see Subsection 1.2.5)

$$T_h = \frac{\ln 2}{\lambda} = \frac{\ln 2}{0.0022}(\log e) \text{ days}$$

$$\approx 136.8 \text{ days}$$

■

Note that in the preceding example we used logarithms to base 10 to do the log transformation. The radioactive law was given in terms of the natural exponent e. The slope therefore contained the factor $\log e \approx 0.4343$.

EXAMPLE 10   Light intensity in lakes decreases with depth. Denote by $I(z)$ the light intensity at depth z, with $z = 0$ representing the surface. Then the percentage surface radiation at depth z, denoted by $\text{PSR}(z)$, is computed as

$$\text{PSR}(z) = 100\frac{I(z)}{I(0)}$$

When we graph the percentage surface radiation as a function of depth on a semilog plot, a straight line results. An example of such a curve is given in Figure 1.62, where the coordinate system is rotated clockwise by $90°$ so that the depth axis points downward. Derive an equation for $I(z)$ on the basis of the graph.

Solution   We see that the dependent variable, $100I(z)/I(0)$, is logarithmically transformed, whereas the independent variable, z, is on a linear scale. The graph is a straight line. We thus find that

$$\log 100\frac{I(z)}{I(0)} = c + mz$$

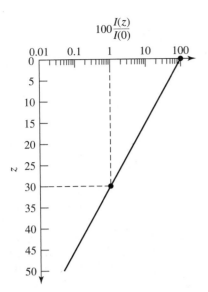

Figure 1.62 The graph of percentage surface radiation as a function of depth.

where c is the intercept on the percentage surface radiation axis and m is the slope. We see that

$$c = \log 100$$

and, using the points (0, 100) and (30, 1), we get

$$m = \frac{\log 100 - \log 1}{0 - 30} = -\frac{2}{30} = -\frac{1}{15}$$

Hence,

$$\log 100 \frac{I(z)}{I(0)} = \log 100 - \frac{1}{15} z$$

The left-hand side simplifies to $\log 100 + \log \frac{I(z)}{I(0)}$. After canceling $\log 100$ on both sides and exponentiating both sides, we find that

$$\frac{I(z)}{I(0)} = 10^{-(1/15)z} = \exp[\ln 10^{-(1/15)z}]$$

Thus

$$I(z) = I(0)e^{-(\frac{1}{15}\ln 10)z}$$

The number $\frac{1}{15} \ln 10$ is called the *vertical attenuation coefficient*. The magnitude of this number tells us how quickly light is absorbed in a lake.   ■

■ 1.3.4 From a Verbal Description to a Graph (Optional)

Being able to sketch a graph on the basis of a verbal explanation of some phenomenon is an extremely useful skill since a graph can summarize a complex situation that can be more easily communicated and remembered. Let's look at an example.

EXAMPLE 11   The following quote in Rosenzweig and Abramsky (1993) relates primary productivity (i.e., the rate at which autotrophs convert light or inorganic chemical energy into chemical energy of organic compounds) to species diversity (i.e., number of species):

> The relationship of primary productivity and species diversity on a regional scale ($10^6 \, \text{km}^2$) is not simple. But within such regions, and perhaps even larger ones, a pattern is emerging: as productivity rises, first diversity increases, then it declines.

If we wanted to translate this verbal description into a graph, we would first determine the independent and the dependent variable. Here, we consider species diversity as a function of primary productivity; hence, primary productivity is the independent variable and species diversity is the dependent variable. We will therefore use a coordinate system whose horizontal axis denotes primary productivity and whose vertical axis denotes species diversity. Since both primary productivity and species diversity are nonnegative, we need to draw only the first quadrant (Figure 1.63).

Going back to the quote, we see that as productivity increases, diversity first increases, then decreases. The graph in Figure 1.64 illustrates this behavior.

The exact shape of the curve cannot be inferred from the quote and will depend on the system studied. For instance, the graph in Figure 1.65 resembles the curve from a study in the Costa Rican forests; as productivity increases, the curve shows an initial increase followed by a decrease in species diversity. The shape of the curve in Figure 1.65 is quantitatively different from the graph in Figure 1.64, but both have the same qualitative features of an initial increase followed by a decrease.

As another example, we will look at the *functional response* of a predator to its prey density.   ■

Figure 1.63 The coordinate system for species diversity as a function of primary productivity can be restricted to the first quadrant.

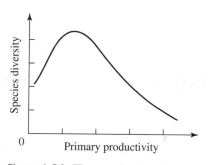

Figure 1.64 The graph of species diversity as a function of primary productivity is hump shaped.

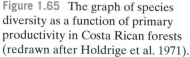

Figure 1.65 The graph of species diversity as a function of primary productivity in Costa Rican forests (redrawn after Holdrige et al. 1971).

EXAMPLE 12

The functional response of a predator to its prey density relates the number of prey consumed per predator (the dependent variable) to the prey density (the independent variable). Holling (1959) introduced three basic types of response. Type 1 describes a response in which the number of prey eaten per predator as a function of prey density rises linearly to a plateau. The type 2 functional response increases at a decelerating rate and eventually levels off. The type 3 functional response is S shaped, or sigmoidal, and also eventually levels off. Now let's translate these three ways into graphs. All graphs will be plotted in coordinate systems, with prey density on the horizontal axis and the number of prey eaten per predator on the vertical axis. Since both prey density and number of prey eaten per predator are nonnegative variables, we need to draw only the first quadrant.

Even though this was not mentioned, we will assume that when the prey density is equal to zero, the number of prey eaten per predator will also be zero, and once the prey density is positive, the number of prey eaten per predator will be positive. This means that the three functional response curves all go through the origin.

The type 1 functional response first increases linearly (i.e., results in a straight line) and then reaches a plateau (stays constant) (See Figure 1.66.)

The type 2 functional response is described as a function that increases at a decelerating rate. This means that the function will increase less quickly as prey density increases (Figure 1.67). In contrast to the type 1 functional response, the type 2 response will continue to increase and approach, but not actually reach, the plateau at a finite value of prey density.

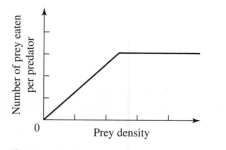

Figure 1.66 The type 1 functional response.

Figure 1.67 The type 2 functional response.

Figure 1.68 The type 3 functional response.

The type 3 functional response is described as sigmoidal. Sigmoidal curves are characterized by an initial accelerating increase followed by an increase at a decelerating rate (Figure 1.68). Similar to the type 2 functional response, the type 3 functional response approaches a plateau as prey density increases, but will not reach the plateau at a finite value of prey density. ■

For each example discussed so far, the functional relationship depended on just one variable, such as the number of prey eaten per predator as a function of prey density. Often, however, a response depends on more than one independent variable.

The next example presents a response that depends on two independent variables, and shows how to draw a graph of this more complex relationship.

EXAMPLE 13 The successful germination of seeds depends on both temperature and humidity. When the humidity is too low, seeds tend not to germinate at all, regardless of the temperature. Germination success is highest for intermediate values of temperature. Finally, seeds tend to germinate better when humidity levels are higher.

One way to translate this information into a graph is to graph germination success as a function of temperature for different levels of humidity. If we measure temperature in Fahrenheit or Celsius, we can restrict the graphs to the first quadrant (Figure 1.69), since the temperature needs to be well above freezing for germination to occur (the temperature at which freezing starts is $0°C$, or $32°F$). Germination success will be between 0 and 100%. To sketch the graphs, it is better not to label the axes beyond what we provided in Figure 1.69, because we do not know the exact numerical response.

There is enough information to provide three graphs: one for low humidity, one for intermediate humidity, and one for high humidity. We will graph them all in one coordinate system, so that it is easier to compare the different responses. The graph for low humidity is a horizontal line where germination success is 0%. For intermediate and high humidity, the graphs are hump shaped, since germination success is highest for intermediate values of temperature. The graph for high humidity is above the graph for intermediate humidity, because seeds tend to germinate better when humidity levels are higher (Figure 1.70). ■

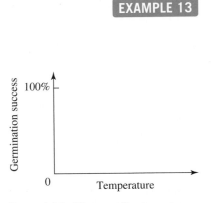

Figure 1.69 The coordinate system for germination success as a function of temperature can be restricted to the first quadrant. Germination success will be between 0 and 100%.

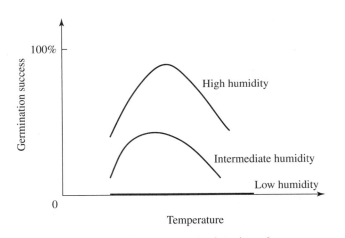

Figure 1.70 Germination success as a function of temperature for three humidity levels (low, intermediate, high).

Section 1.3 Problems

■ 1.3.1

In Problems 1–22, sketch the graph of each function. Do not use a graphing calculator. (Assume the largest possible domain.)

1. $y = x^2 + 1$

2. $y = -(x - 2)^2 + 1$

3. $y = x^3 - 2$

4. $y = -x^4 + 1$

5. $y = -2x^2 - 3$

6. $y = -(3 - x)^2$

7. $y = 3 + 1/x$

8. $y = 1 - 1/x$

9. $y = 1/(x - 1)$

10. $y = 1 + 1/(x + 2)^2$

11. $y = \exp(x - 2)$

12. $y = \exp(-x)$

13. $y = e^{-(x+3)}$

14. $y = 3e^{2x+1}$

15. $y = \ln(x + 1)$

16. $y = \ln(x - 3)$

17. $y = -\ln(x - 1) + 1$

18. $y = -\ln(1 - x)$

19. $y = 2\sin(x + \pi/4)$

20. $y = 0.2\cos(-x)$

21. $y = -\sin(\pi x/2)$

22. $y = -2\cos(\pi x/4)$

23. Explain how the following functions can be obtained from $y = x^2$ by basic transformations:

(a) $y = x^2 - 2$ **(b)** $y = (x - 1)^2 + 1$ **(c)** $y = -2(x + 2)^2$

24. Explain how the following functions can be obtained from $y = x^3$ by basic transformations:

(a) $y = x^3 + 1$ **(b)** $y = (x + 1)^3 - 1$ **(c)** $y = -3(x - 2)^3$

25. Explain how the following functions can be obtained from $y = 1/x$ by basic transformations:

(a) $y = 1 - \dfrac{1}{x}$ **(b)** $y = -\dfrac{1}{x - 1}$ **(c)** $y = \dfrac{x}{x + 1}$

26. Explain how the following functions can be obtained from $y = 1/x^2$ by basic transformations:
(a) $y = \dfrac{1}{x^2} + 1$ **(b)** $y = -\dfrac{1}{(x+1)^2}$ **(c)** $y = -\dfrac{1}{x^2} - 2$

27. Explain how the following functions can be obtained from $y = e^x$ by basic transformations:
(a) $y = 2e^x - 1$ **(b)** $y = -e^{-x}$ **(c)** $y = e^{x-2} + 1$

28. Explain how the following functions can be obtained from $y = e^x$ by basic transformations:
(a) $y = e^{-x} - 1$ **(b)** $y = -e^x + 1$ **(c)** $y = -e^{x-3} - 2$

29. Explain how the following functions can be obtained from $y = \ln x$ by basic transformations:
(a) $y = \ln(x - 1)$ **(b)** $y = -\ln x + 1$ **(c)** $y = \ln(x+3) - 1$

30. Explain how the following functions can be obtained from $y = \ln x$ by basic transformations:
(a) $y = \ln(1 - x)$ **(b)** $y = \ln(2 + x) - 1$
(c) $y = -\ln(2 - x) + 1$

31. Explain how the following functions can be obtained from $y = \sin x$ by basic transformations:
(a) $y = 1 - \sin x$ **(b)** $y = \sin\left(x - \dfrac{\pi}{4}\right)$
(c) $y = -\sin\left(x + \dfrac{\pi}{3}\right)$

32. Explain how the following functions can be obtained from $y = \cos x$ by basic transformations:
(a) $y = 1 + 2\cos x$ **(b)** $y = -\cos\left(x + \dfrac{\pi}{4}\right)$
(c) $y = -\cos\left(\dfrac{\pi}{2} - x\right)$

■ **1.3.2**

33. Find the following numbers on a number line that is on a logarithmic scale (base 10): 0.0002, 0.02, 1, 5, 50, 100, 1000, 8000, and 20000.

34. Find the following numbers on a number line that is on a logarithmic scale (base 10): 0.03, 0.7, 1, 2, 5, 10, 17, 100, 150, and 2000.

35. (a) Find the following numbers on a number line that is on a logarithmic scale (base 10): 10^2, 10^{-3}, 10^{-4}, 10^{-7}, and 10^{-10}.
(b) Can you find 0 on a number line that is on a logarithmic scale?
(c) Can you find negative numbers on a number line that is on a logarithmic scale?

36. (a) Find the following numbers on a number line that is on a logarithmic scale (base 10):
(i) $10^{-3}, 2 \times 10^{-3}, 5 \times 10^{-3}$ **(ii)** $10^{-1}, 2 \times 10^{-1}, 5 \times 10^{-1}$
(iii) $10^2, 2 \times 10^2, 5 \times 10^2$
(b) From your answers to (a), how many units (on a logarithmic scale) is **(i)** 2×10^{-3} from 10^{-3} **(ii)** 2×10^{-1} from 10^{-1} and **(iii)** 2×10^2 from 10^2?
(c) From your answers to (a), how many units (on a logarithmic scale) is **(i)** 5×10^{-3} from 10^{-3} **(ii)** 5×10^{-1} from 10^{-1} and **(iii)** 5×10^2 from 10^2?

In Problems 37–42, insert an appropriate number in the blank space.

37. The longest known species of worms is the earthworm *Microchaetus rappi* of South Africa; in 1937, a 6.7-m-long specimen was collected from the Transvaal. The shortest worm is *Chaetogaster annandalei*, which measures less than 0.51 mm in length. *M. rappi* is ____ order(s) of magnitude longer than *C. annandalei*.

38. Both the La Plata river dolphin (*Pontoporia blainvillei*) and the sperm whale (*Physeter macrocephalus*) belong to the suborder Odontoceti (individuals that have teeth). A La Plata river dolphin weighs between 30 and 50 kg, whereas a sperm whale weighs between 35,000 and 40,000 kg. A sperm whale is ____ order(s) of magnitude heavier than a La Plata river dolphin.

39. Compare a ball of radius 1 cm against a ball of radius 10 cm. The radius of the larger ball is ____ order(s) of magnitude bigger than the radius of the smaller ball. The volume of the larger ball is ____ order(s) of magnitude bigger than the volume of the smaller ball.

40. Compare a square with side length 1 m against a square with side length 100 m. The area of the larger square is ____ order(s) of magnitude larger than the area of the smaller square.

41. The diameter of a typical bacterium is about 0.5 to 1 μm. An exception is the bacterium *Epulopiscium fishelsoni*, which is about 600 μm long and 80 μm wide. The volume of *E. fishelsoni* is about ____ order(s) of magnitude larger than that of a typical bacterium. (*Hint*: Approximate the shape of a typical bacterium by a sphere and the shape of *E. fishelsoni* by a cylinder.)

42. The length of a typical bacterial cell is about one-tenth that of a small eukaryotic cell. Consequently, the cell volume of a bacterium is about ____ order(s) of magnitude smaller than that of a small eukaryotic cell. (*Hint*: Approximate the shapes of both types of cells by spheres.)

■ **1.3.3**

In Problems 43–46, when $\log y$ is graphed as a function of x, a straight line results. Graph straight lines, each given by two points, on a log-linear plot, and determine the functional relationship. (The original x–y coordinates are given.)

43. $(x_1, y_1) = (0, 5), (x_2, y_2) = (3, 1)$
44. $(x_1, y_1) = (-1, 4), (x_2, y_2) = (2, 8)$
45. $(x_1, y_1) = (-2, 3), (x_2, y_2) = (1, 1)$
46. $(x_1, y_1) = (1, 4), (x_2, y_2) = (6, 1)$

In Problems 47–54, use a logarithmic transformation to find a linear relationship between the given quantities and graph the resulting linear relationship on a log-linear plot.

47. $y = 3 \times 10^{-2x}$ **48.** $y = 4 \times 10^{5x}$
49. $y = 2e^{-1.2x}$ **50.** $y = 7e^{3x}$
51. $y = 5 \times 2^{4x}$ **52.** $y = 6 \times 2^{-0.9x}$
53. $y = 4 \times 3^{2x}$ **54.** $y = 5^{-6x}$

In Problems 55–58, when $\log y$ is graphed as a function of $\log x$, a straight line results. Graph straight lines, each given by two points, on a log-log plot, and determine the functional relationship. (The original x–y coordinates are given.)

55. $(x_1, y_1) = (1, 2), (x_2, y_2) = (5, 1)$
56. $(x_1, y_1) = (3, 5), (x_2, y_2) = (1, 5)$
57. $(x_1, y_1) = (4, 2), (x_2, y_2) = (8, 8)$
58. $(x_1, y_1) = (2, 5), (x_2, y_2) = (5, 2)$

In Problems 59–66, use a logarithmic transformation to find a linear relationship between the given quantities and graph the resulting linear relationship on a log-log plot.

59. $y = 2x^5$ **60.** $y = 3x^2$
61. $y = x^6$ **62.** $y = 5x^3$
63. $y = x^{-2}$ **64.** $y = 6x^{-1}$
65. $y = 4x^{-3}$ **66.** $y = 7x^{-5}$

In Problems 67–72, use a logarithmic transformation to find a linear relationship between the given quantities and determine whether a log-log or log-linear plot should be used to graph the resulting linear relationship.

67. $f(x) = 3x^{1.7}$

68. $g(s) = 1.8e^{-0.2s}$

69. $N(t) = 130 \times 2^{1.2t}$

70. $I(u) = 4.8u^{-0.89}$

71. $R(t) = 3.6t^{1.2}$

72. $L(c) = 1.7 \times 10^{2.3c}$

73. The following table is based on a functional relationship between x and y that is either an exponential or a power function:

x	y
1	1.8
2	2.07
4	2.38
10	2.85
20	3.28

Use an appropriate logarithmic transformation and a graph to decide whether the table comes from a power function or an exponential function, and find the functional relationship between x and y.

74. The following table is based on a functional relationship between x and y that is either an exponential or a power function:

x	y
0.5	7.81
1	3.4
1.5	2.09
2	1.48
2.5	1.13

Use an appropriate logarithmic transformation and a graph to decide whether the table comes from a power function or an exponential function, and find the functional relationship between x and y.

75. The following table is based on a functional relationship between x and y that is either an exponential or a power function:

x	y
−1	0.398
−0.5	1.26
0	4
0.5	12.68
1	40.18

Use an appropriate logarithmic transformation and a graph to decide whether the table comes from a power function or an exponential function, and find the functional relationship between x and y.

76. The following table is based on a functional relationship between x and y that is either an exponential or a power function:

x	y
0	3
0.5	2.20
1	1.61
1.5	1.18
2	0.862

Use an appropriate logarithmic transformation and a graph to decide whether the table comes from a power function or an exponential function, and find the functional relationship between x and y.

77. The following table is based on a functional relationship between x and y that is either an exponential or a power function:

x	y
0.1	0.045
0.5	1.33
1	5.7
1.5	13.36
2	24.44

Use an appropriate logarithmic transformation and a graph to decide whether the table comes from a power function or an exponential function, and find the functional relationship between x and y.

78. The following table is based on a functional relationship between x and y that is either an exponential or a power function:

x	y
0.1	1.72
0.5	1.41
1	1.11
1.5	0.872
2	0.685

Use an appropriate logarithmic transformation and a graph to decide whether the table comes from a power function or an exponential function, and find the functional relationship between x and y.

So far, we have always used base 10 for a logarithmic transformation. The reason for this is that our number system is based on base 10 and it is therefore easy to logarithmically transform numbers of the form ..., 0.01, 0.1, 1, 10, 100, 1000, ... when we use base 10. In Problems 79–82, use the indicated base to logarithmically transform each exponential relationship so that a linear relationship results. Then use the indicated base to graph each relationship in a coordinate system whose axes are accordingly transformed so that a straight line results.

79. $y = 2^x$; base 2

80. $y = 3^x$; base 3

81. $y = 2^{-x}$; base 2

82. $y = 3^{-x}$; base 3

83. Suppose that $N(t)$ denotes a population size at time t and satisfies the equation

$$N(t) = 2e^{3t} \quad \text{for } t \geq 0$$

(a) If you graph $N(t)$ as a function of t on a semilog plot, a straight line results. Explain why.

(b) Graph $N(t)$ as a function of t on a semilog plot, and determine the slope of the resulting straight line.

84. Suppose that you follow a population over time. When you plot your data on a semilog plot, a straight line with slope 0.03 results. Furthermore, assume that the population size at time 0 was 20. If $N(t)$ denotes the population size at time t, what function best describes the population size at time t?

85. Species–Area Curves Many studies have shown that the number of species on an island increases with the area of the island. Frequently, the functional relationship between the number of species (S) and the area (A) is approximated by $S = CA^z$, where z is a constant that depends on the particular species and habitat in the study. (Actual values of z range from about 0.2 to 0.35.) Suppose that the best fit to your data points on a log-log scale is a straight line. Is your model $S = CA^z$ an appropriate description of your data? If yes, how would you find z?

86. Michaelis–Menten Equation Enzymes serve as catalysts in many chemical reactions in living systems. The simplest such reactions transform a single substrate into a product with the help of an enzyme. The Michaelis–Menten equation describes the initial velocity of such enzymatically controlled reactions. The equation, which gives the relationship between the initial velocity of the reaction (v_0) and the concentration of the substrate (s_0), is

$$v_0 = \frac{v_{max}s_0}{s_0 + K_m}$$

where v_{max} is the maximum velocity at which the product may be formed and K_m is the Michaelis–Menten constant. Note that this equation has the same form as the Monod growth function.
(a) Show that the Michaelis–Menten equation can be written in the form

$$\frac{1}{v_0} = \frac{K_m}{v_{max}}\frac{1}{s_0} + \frac{1}{v_{max}}$$

This formula is known as the Lineweaver–Burk equation and shows that there is a linear relationship between $1/v_0$ and $1/s_0$.
(b) Sketch the graph of the Lineweaver–Burk equation. Use a coordinate system in which $1/s_0$ is on the horizontal axis and $1/v_0$ is on the vertical axis. Show that the resulting graph is a line that intersects the horizontal axis at $-1/K_m$ and the vertical axis at $1/v_{max}$.
(c) To determine K_m and v_{max}, we measure the initial velocity of the reaction, denoted by v_0, as a function of the concentration of the substrate, denoted by s_0, and fit a straight line through the points in a coordinate system in which the horizontal axis is $1/s_0$ and the vertical axis is $1/v_0$. Explain how to determine K_m and v_{max} from the graph.

(Note that this is an example in which a nonlogarithmic transformation is used to obtain a linear relationship. Since the reciprocals of both quantities of interest are used, the resulting plot is called a double-reciprocal plot.)
87. (*Continuation of Problem 86*) Estimating v_{max} and K_m from the Lineweaver–Burk graph as described in Problem 86 is not always satisfactory. A different transformation typically yields better estimates (Dowd and Riggs, 1965). Show that the Michaelis–Menten equation can be written as

$$\frac{v_0}{s_0} = \frac{v_{max}}{K_m} - \frac{1}{K_m}v_0$$

and explain why this transformation results in a straight line when you graph v_0 on the horizontal axis and $\frac{v_0}{s_0}$ on the vertical axis. Explain how you can estimate v_{max} and K_m from the graph.
88. (*Adapted from Reiss, 1989*) In a case study in which the maximal rates of oxygen consumption (in ml/s) of nine species of wild African mammals (Taylor et al., 1980) were plotted against body mass (in kg) on a log-log plot, it was found that the data points fell on a straight line with slope approximately equal to 0.8 and vertical-axis intercept approximately equal to 0.105. Find an equation that relates maximal oxygen consumption and body mass.
89. (*Adapted from Benton and Harper, 1997*) In vertebrates, embryos and juveniles have large heads relative to their overall body size. As the animal grows older, proportions change; for instance, the ratio of skull length to body length diminishes. That

this is the case not only for living vertebrates, but also for fossil vertebrates, is shown by the following example:

Ichthyosaurs are a group of marine reptiles that appeared in the early Triassic and died out well before the end of the Cretaceous.[1] They were fish shaped and comparable in size to dolphins. In a study of 20 fossil skeletons, the following allometric relationship between skull length S (measured in cm) and backbone length B (measured in cm) was found:

$$S = 1.162B^{0.93}$$

(a) Choose suitable transformations of S and B so that the resulting relationship is linear. Plot the transformed relationship, and find the slope and the y-intercept.

(b) Explain why the allometric equation confirms that juveniles had relatively large heads. (*Hint*: Compute the ratio of S to B for a number of different values of B—say, 10 cm, 100 cm, 500 cm—and compare.)

90. Light intensity in lakes decreases exponentially with depth. If $I(z)$ denotes the light intensity at depth z, with $z = 0$ representing the surface, then

$$I(z) = I(0)e^{-\alpha z}, \quad z \geq 0$$

where α is a positive constant called the *vertical attenuation coefficient*. Figure 1.71 shows the percentage surface radiation, defined as $100I(z)/I(0)$, as a function of depth in different lakes.

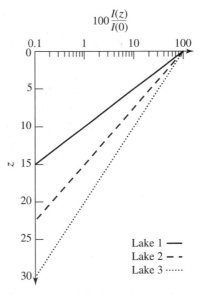

Figure 1.71 Light intensity as a function of depth for Problem 90.

(a) On the basis of the graph, estimate α for each lake.

(b) Reproduce a graph like the one in Figure 1.71 for Lake Constance (Germany) in May ($\alpha = 0.768\,\text{m}^{-1}$) and December ($\alpha = 0.219\,\text{m}^{-1}$) (data from Tilzer et al., 1982).

(c) Explain why the graphs are straight lines.

---

(1) The Triassic is a geological period that began about 248 million years ago and ended about 213 million years ago; the Cretaceous began about 144 million years ago and ended 65 million years ago.

91. The absorption of light in a uniform water column follows an exponential law; that is, the intensity $I(z)$ at depth z is

$$I(z) = I(0)e^{-\alpha z}$$

where $I(0)$ is the intensity at the surface (i.e., when $z = 0$) and α is the *vertical attenuation coefficient*. (We assume here that α is constant. In reality, α depends on the wavelength of the light penetrating the surface.)

(a) Suppose that 10% of the light is absorbed in the uppermost meter. Find α. What are the units of α?

(b) What percentage of the remaining intensity at 1 m is absorbed in the second meter? What percentage of the remaining intensity at 2 m is absorbed in the third meter?

(c) What percentage of the initial intensity remains at 1 m, at 2 m, and at 3 m?

(d) Plot the light intensity as a percentage of the surface intensity on both a linear plot and a log-linear plot.

(e) Relate the slope of the curve on the log-linear plot to the attenuation coefficient α.

(f) The level at which 1% of the surface intensity remains is of biological significance. Approximately, it is the level where algal growth ceases. The zone above this level is called the *euphotic zone*. Express the depth of the euphotic zone as a function of α.

(g) Compare a very clear lake with a milky glacier stream. Is the attenuation coefficient α for the clear lake greater or smaller than the attenuation coefficient α for the milky stream?

92. When plants are grown at high densities, we often observe that the number of plants decreases as plant weights increase (due to plant growth). If we plot the logarithm of the total aboveground dry-weight biomass per plant, $\log w$, against the logarithm of the density of survivors, $\log d$ (base 10), a straight line with slope $-3/2$ results. Find the equation that relates w and d, assuming that $w = 1$ g when $d = 10^3 \, \text{m}^{-2}$.

In Problems 93–98, find each functional relationship on the basis of the given graph.

93. Figure 1.72   **94.** Figure 1.73

95. Figure 1.74   **96.** Figure 1.75

97. Figure 1.76 (*Hint*: This relationship is different from the ones considered so far. The x-axis is logarithmically transformed, but the y-axis is linear.)

98. Figure 1.77 (*Hint*: This relationship is different from the ones considered so far. The x-axis is logarithmically transformed, but the y-axis is linear.)

99. The free energy ΔG expended in transporting an uncharged solute across a membrane from concentration c_1 to one of concentration c_2 follows the equation

$$\Delta G = 2.303 RT \, \log \frac{c_2}{c_1}$$

where $R = 1.99 \, \text{kcal K}^{-1} \text{kmol}^{-1}$ is the universal gas constant and T is temperature measured in kelvins (K). Plot ΔG as a function of the concentration ratio c_2/c_1 when $T = 298$ K (25°C). Use a

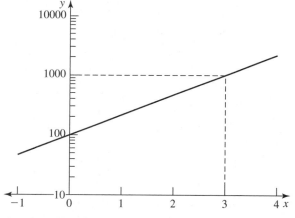

Figure 1.72 Graph for Problem 93.

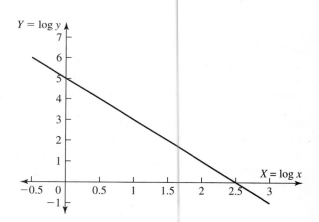

Figure 1.73 Graph for Problem 94.

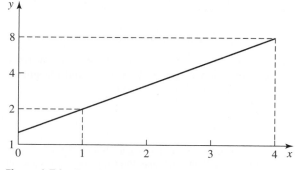

Figure 1.74 Graph for Problem 95.

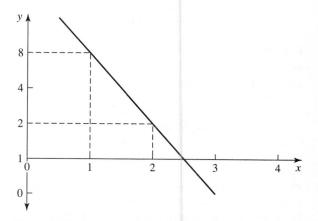

Figure 1.75 Graph for Problem 96.

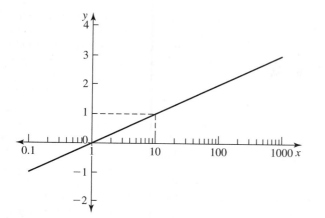

Figure 1.76 Graph for Problem 97.

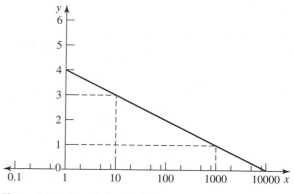

Figure 1.77 Graph for Problem 98.

coordinate system in which the vertical axis is on a linear scale and the horizontal axis is on a logarithmic scale.

100. Logistic Transformation Suppose that

$$f(x) = \frac{1}{1 + e^{-(b+mx)}} \quad (1.8)$$

A function of the form (1.8) is called a **logistic** function. The logistic function was introduced by the Dutch mathematical biologist Verhulst around 1840 to describe the growth of populations with limited food resources. Show that

$$\ln \frac{f(x)}{1 - f(x)} = b + mx \quad (1.9)$$

This transformation is called the *logistic transformation*. It is a standard transformation for linearizing functions of the form (1.8).

■ **1.3.4**

101. Not every study of species richness as a function of productivity produces a hump-shaped curve. Owen (1988) studied rodent assemblages in Texas and found that the number of species was a decreasing function of productivity. Sketch a graph that would describe this situation.

102. Species diversity in a community may be controlled by disturbance frequency. The intermediate disturbance hypothesis states that species diversity is greatest at intermediate disturbance levels. Sketch a graph of species diversity as a function of disturbance level that illustrates this hypothesis.

103. Preston (1962) investigated the dependence of number of bird species on island area in the West Indian islands. He found that the number of bird species increased at a decelerating rate as island area increased. Sketch this relationship.

104. Phytoplankton converts carbon dioxide to organic compounds during photosynthesis. This process requires sunlight. It has been observed that the rate of photosynthesis is a function of light intensity: The rate of photosynthesis increases approximately linearly with light intensity at low intensities, saturates at intermediate levels, and decreases slightly at high intensities. Sketch a graph of the rate of photosynthesis as a function of light intensity.

105. Brown lemming densities in the tundra areas of North America and Eurasia show cyclic behavior: Every three to four years, lemming densities build up very rapidly, and they typically crash the next year. Sketch a graph that describes this situation.

106. Nitrogen productivity can be defined as the amount of dry matter produced per unit of nitrogen per unit of time. Experimental studies suggest that nitrogen productivity increases as a function of light intensity at a decelerating rate. Sketch a graph of nitrogen productivity as a function of light intensity.

107. A study of Borchert's (1994) investigated the relationship between stem water storage and wood density in a number of tree species in Costa Rica. The study showed that water storage is inversely related to wood density; that is, higher wood density corresponds to lower water content. Sketch a graph of water content as a function of wood density that illustrates this situation.

108. Species richness can be a hump-shaped function of productivity. In the same coordinate system, sketch two hump-shaped graphs of species richness as a function of productivity, one in which the maximum occurs at low productivity and one in which the maximum occurs at high productivity.

109. The size distribution of zooplankton in a lake is typically a hump-shaped curve; that is, if the frequency (in percent) of zooplankton is plotted against the body length of zooplankton, a curve that first increases and then decreases results. Brooks and Dodson (1965) studied the effects of introducing a planktivorous fish in a lake. They found that the composition of zooplankton after the fish was introduced shifted to smaller individuals. In the same coordinate system, sketch the size distribution of zooplankton before and after the introduction of the planktivorous fish.

110. *Daphnia* is a genus of zooplankton that comprises a number of species. The body growth rate of *Daphnia* depends on food concentration. A minimum food concentration is required for growth: Below this level, the growth rate is negative; above, it is positive. In a study by Gliwicz (1990), it was found that growth rate is an increasing function of food concentration and that the minimum food concentration required for growth decreases with increasing size of the animal. Sketch two graphs in the same coordinate system, one for a large and one for a small *Daphnia* species, that illustrates this situation.

111. Grant (1982) investigated egg weight as a function of adult body weight among 10 species of Darwin's finches. He found that the relationship between the logarithm of the average egg size and the logarithm of the average body size is linear and that smaller species lay smaller eggs and larger species lay larger eggs. Graph this relationship.

112. Grant et al. (1985) investigated the relationship between mean wing length and mean weight among males of populations of six ground finch species. They found a positive and nearly linear relationship between these two quantities. Graph this relationship.

113. Bohlen et al. (2001) investigated stream nitrate concentration along an elevation gradient at the Hubbard Brook Experimental Forest in New Hampshire. They found that the nitrate concentration in stream water declined with decreasing elevation. Sketch stream nitrate concentration as a function of elevation.

114. In Example 13, we discussed germination success as a function of temperature for varying levels of humidity. We can also consider germination success as a function of humidity for various levels of temperature. Sketch the following graphs of germination success as a function of humidity: one for low temperature, one for intermediate temperature, and one for high temperature.

115. Boulinier et al. (2001) studied the dynamics of forest bird communities. They found that the mean local extinction rate of area-sensitive species declined with mean forest patch size, whereas the mean extinction rate of non-area-sensitive species did not depend on mean forest size. In the same coordinate system, graph the mean extinction rate as a function of mean forest patch size for **(a)** an area-sensitive species and **(b)** a non-area-sensitive species.

116. Dalling et al. (2001) compared net photosynthetic rates of two pioneer trees—*Alseis blackiana* and *Miconia argenta*—as a function of gap size in Barro Colorado Island. They found that net photosynthetic rates (measured on a per-unit basis) increased with gap size for both trees and that the photosynthetic rate for *Miconia argenta* was higher than that for *Alseis blackiana*. In the same coordinate system, graph the net photosynthetic rates as functions of gap size for both tree species.

Chapter 1 Key Terms

Discuss the following definitions and concepts:

1. Real numbers
2. Intervals: open, closed, half-open
3. Absolute value
4. Proportional
5. Lines: standard form, point–slope form, slope–intercept form
6. Parallel and perpendicular lines
7. Circle: radius, center, equation of circle, unit circle
8. Angle: radians, degrees
9. Trigonometric identities
10. Complex numbers: real part, imaginary part
11. Function: domain, codomain, range, image
12. Symmetry of functions: even, odd
13. Composition of functions
14. Polynomial
15. Degree of a polynomial
16. Chemical reaction: law of mass action
17. Rational function
18. Growth rate
19. Specific growth rate and per capita growth rate
20. Monod growth function
21. Power function
22. Allometry and scaling relations
23. Exponential function
24. Exponential growth
25. Natural exponential base
26. Radioactive decay
27. Half-life
28. Inverse function, one to one
29. Logarithmic function
30. Relationship between exponential and logarithmic functions
31. Periodic function
32. Trigonometric function
33. Amplitude, period
34. Translation: horizontal, vertical
35. Logarithmic scale
36. Order of magnitude
37. Logarithmic transformation
38. Log-log plot
39. Semilog plot

Chapter 1 Review Problems

1. Population Growth Suppose that the number of bacteria in a petri dish is given by

$$B(t) = 10,000e^{0.1t}$$

where t is measured in hours.

(a) How many bacteria are present at $t = 0, 1, 2, 3$, and 4?

(b) Find the time t when the number of bacteria reaches 100,000.

2. Population Decline Suppose that a pathogen is introduced into a population of bacteria at time 0. The number of bacteria then declines as

$$B(t) = 25,000e^{-2t}$$

where t is measured in hours.

(a) How many bacteria are left after 3 hours?

(b) How long will it be until only 1% of the initial number of bacteria are left?

3. Chemical Reaction Consider the chemical reaction

$$A + 2B \rightarrow AB_2$$

Assume that the reaction occurs in a closed vessel and that the initial concentrations of A and B are $a = [A]$ and $b = [B]$, respectively.

(a) Explain why the reaction rate $R(x)$ is given by

$$R(x) = k(a - x)(b - 2x)^2$$

where $x = [AB_2]$.

(b) Show that $R(x)$ is a polynomial and determine its degree.

(c) Graph $R(x)$ for the relevant values of x when $a = 5, b = 6$, and $k = 0.3$.

4. History of Mathematics Euclid, a Greek mathematician who lived around 300 B.C., wrote the *Elements*, by far the most important mathematical text of that period. The book, arranged in 13 volumes, is a systematic exposition of most of the mathematical knowledge of that time. In Book III, Euclid discusses the construction of a tangent to a circle at a point P on the circle. To phrase the construction in modern terminology, we draw a straight line through the point P that is perpendicular to the line through the center of the circle and the point P on the circle.

(a) Use this geometric construction to find the equation of the line that is tangent to the unit circle at the point $(\frac{1}{2}\sqrt{3}, \frac{1}{2})$.

(b) Determine the angle θ between the positive x-axis and the tangent line found in (a). What is the relationship between the angle θ and the slope of the tangent line found in (a)?

5. Hypothetical Plants To compare logarithmic and exponential growth, we consider two hypothetical plants that are of the same genus, but that exhibit rather different growth rates. Both plants produce a single leaf whose length continues to increase as long as the plant is alive. One plant is called *Growthus logarithmiensis*; the other one is called *Growthus exponentialis*. The length L (measured in feet) of the leaf of *G. logarithmiensis* at age t (measured in years) is given by

$$L(t) = \ln(t+1), \quad t \geq 0$$

The length E (measured in feet) of the leaf of *G. exponentialis* at age t (measured in years), is given by

$$E(t) = e^t - 1, \quad t \geq 0$$

(a) Find the length of each leaf after 1, 10, 100, and 1000 years.

(b) How long would it take for the leaf of *G. exponentialis* to reach a length of 233,810 mi, the average distance from the earth to the moon? (Note that 1 mi = 5280 ft.) How long would the leaf of *G. logarithmiensis* then be?

(c) How many years would it take the leaf of *G. logarithmiensis* to reach a length of 233,810 mi? Compare this with the length of time since life appeared on earth, about 3500 million years. If *G. logarithmiensis* had appeared 3,500 million years ago, and if there was a plant of this species that had actually survived throughout the entire period, how long would its leaf be today?

(d) Plants started to conquer land only in the late Ordovician period, around 450 million years ago.[2] If both *G. exponentialis* and *G. logarithmiensis* had appeared then, and there was a plant of each species that had actually survived throughout the entire period, how long would their respective leaves be today?

6. Population Growth In Chapter 3 of *The Origin of Species* (Darwin, 1859), Charles Darwin asserts that a "struggle for existence inevitably follows from the high rate at which all organic beings tend to increase. ... Although some species may be now increasing, more or less rapidly, in numbers, all cannot do so, for the world would not hold them." To illustrate this point, he continues as follows:

> There is no exception to the rule that every organic being naturally increases at so high a rate, that, if not destroyed, the earth would soon be covered by the progeny of a single pair. Even slow-breeding man has doubled in twenty-five years, and at this rate, in a few thousand years, there would literally not be standing room for his progeny.

Starting with a single pair, compute the world's population after 1000 years and after 2000 years under Darwin's assumption that the world's population doubles every 25 years, and find the resulting population densities (number of people per square foot). To answer the last part, you need to know that the earth's diameter is about 7900 mi, the surface of a sphere is $4\pi r^2$, where r is the radius of a sphere, and the continents make up about 29% of the earth's surface. (Note that 1 mi = 5280 ft.)

7. Population Growth Assume that a population grows $q\%$ each year. How many years will it take the population to double in size? Give the functional relationship between the doubling time T and the annual percentage increase q. Produce a table that shows the doubling time T as a function of q for $q = 1, 2, \ldots, 10$, and graph T as a function of q. What happens to T as q gets closer to 0?

(2) The Ordovician lasted from about 505 million years ago to about 438 million years ago.

8. Beverton–Holt Recruitment Curve Many organisms show density-dependent mortality. The following is a simple mathematical model that incorporates this effect: Denote the density of parents by N_b and the density of *surviving* offspring by N_a.

(a) Suppose that without density-dependent mortality, the number of surviving offspring per parent is equal to R. Show that if we plot N_b/N_a versus N_b, the result is a horizontal line with y-intercept $1/R$. That is,

$$\frac{N_b}{N_a} = \frac{1}{R}$$

or

$$N_a = R \cdot N_b$$

The constant R is called the **net reproductive rate**.

(b) To include density-dependent mortality, we assume that N_b/N_a is an increasing function of N_b. The simplest way to do this is to assume that the graph of N_b/N_a versus N_b is a straight line with y-intercept $1/R$ and that goes through the point $(K, 1)$. Show that this implies that

$$N_a = \frac{RN_b}{1 + \frac{(R-1)N_b}{K}}$$

This relationship is called the Beverton–Holt recruitment curve.

(c) Explain in words why, for small initial densities N_b, the model described by the Beverton–Holt recruitment curve behaves like the model for density-independent mortality described in (a).

(d) Show that if $N_b = K$, then $N_a = K$. Furthermore, show that $N_b < K$ implies $N_b < N_a < K$ and that $N_b > K$ implies $K < N_a < N_b$. Explain in words what this means. (Note that K is called the *carrying capacity*.)

(e) Plot N_a as a function of N_b for $R = 2$ and $K = 20$. What happens for large values of N_b? Explain in words what this means.

9. Fish Yield (*Adapted from Moss, 1980*) Oglesby (1977) investigated the relationship between annual fish yield (Y) and summer phytoplankton chlorophyll concentration (C). Fish yield was measured in grams dry weight per square meter per year, and the chlorophyll concentration was measured in micrograms per liter. Data from 19 lakes, mostly in the Northern Hemisphere, yielded the following relationship:

$$\log_{10} Y = 1.17 \log_{10} C - 1.92 \tag{1.10}$$

(a) Plot $\log_{10} Y$ as a function of $\log_{10} C$.

(b) Find the relationship between Y and C; that is, write Y as a function of C. Explain the advantage of the log–log transformation resulting in (1.10) versus writing Y as a function of C. [*Hint*: Try to plot Y as a function of C, and compare with your answer in (a).]

(c) Find the predicted yield (Y_p) as a function of the current yield (Y_c) if the current summer phytoplankton chlorophyll concentration were to double.

(d) By what percentage would the summer phytoplankton chlorophyll concentration need to increase to obtain a 10% increase in fish yield?

10. Radioactive Decay (*Adapted from Moss, 1980*) To trace the history of a lake, a sample of mud from a core is taken and dated. One dating method uses radioactive isotopes. The C^{14} method is effective for sediments that are younger than 60,000 years. The $C^{14} : C^{12}$ ratio has been essentially constant in the atmosphere over a long time, and living organisms take up carbon in that ratio.

Upon death, the uptake of carbon ceases and C^{14} decays, which changes the $C^{14}:C^{12}$ ratio according to

$$\left(\frac{C^{14}}{C^{12}}\right)_t = \left(\frac{C^{14}}{C^{12}}\right)_{\text{initial}} e^{-\lambda t}$$

where t is the time since death.

(a) If the $C^{14}:C^{12}$ ratio in the atmosphere is 10^{-12} and the half-life of C^{14} is 5730 years, find an expression for t, the age of the material being dated, as a function of the $C^{14}:C^{12}$ ratio in the material being dated.

(b) Use your answer in (a) to find the age of a mud sample from a core for which the $C^{14}:C^{12}$ ratio is 1.61×10^{-13}.

11. Fossil Coral Growth (*Adapted from Futuyama, 1995, and Dott and Batten, 1976*) Corals deposit a single layer of lime each day. In addition, seasonal fluctuation in the thickness of the layers allows for grouping them into years. In modern corals, we can count 365 layers per year. J. Wells, a paleontologist, counted such growth layers on fossil corals. To his astonishment, he found that Devonian[3] corals that lived about 380 million years ago had about 400 daily layers per year.

(a) Today, the earth rotates about its axis every 24 hours and revolves around the sun every $365\frac{1}{4}$ days. Astronomers have determined that the earth's rotation has slowed down in recent centuries at the rate of about 2 seconds every 100,000 years. That is, 100,000 years ago, a day was 2 seconds shorter than today. Extrapolate the slowdown back to the Devonian, and determine the length of a day and the length of a year back when Wells's corals lived. (*Hint*: The number of hours per year remains constant. Why?)

(b) Find a linear equation that relates geologic time (in million of years) to the number of hours per day at a given time.

(c) Algal stromatolites also show daily layers. A sample of some fossil stromatolites showed 400 to 420 daily layers per year. Use your answer in (b) to date the stromatolites.

12. Tree Growth The height y in feet of a certain tree as a function of age x in years can be approximated by

$$y = 132e^{-20/x}$$

(a) Use a graphing calculator to plot the graph of this function. Describe in words how the tree grows, paying particular attention to questions such as the following: Does the tree grow equally fast over time? What happens when the tree is young? What happens when the tree is old?

(b) How many years will it take for the tree to reach 100 ft in height?

(c) Can the tree ever reach a height of 200 ft? Is there a final height—that is, a maximum height that the tree will eventually reach?

13. Model for Aging The probability that an individual lives beyond age t is called the *survivorship function* and is denoted by $S(t)$. The Weibull model is a popular model in reliability theory and in studies of biological aging. Its survivorship function is described by two parameters, λ and β, and is given by

$$S(t) = \exp[-(\lambda t)^\beta]$$

Mortality data from a *Drosophila melanogaster* population in Dr. Jim Curtsinger's lab at the University of Minnesota were collected and fitted to this model separately for males and females (Pletcher, 1998). The following parameter values were obtained (t was measured in days):

Sex	λ	β
Males	0.019	3.41
Females	0.022	3.24

(a) Use a graphing calculator to sketch the survivorship function for both the female and male populations.

(b) For each population, find the value of t for which the probability of living beyond that age is $1/2$.

(c) If you had a male and a female of this species, which would you expect to live longer?

14. Carbon Isotope Carbon has two stable isotopes: C^{12} and C^{13}. Organic material contains both stable isotopes but the ratio $[C^{13}]:[C^{12}]$ in organic material is smaller than that in inorganic material, reflecting the fact that light carbon (C^{12}) is preferentially taken up by plants during photosynthesis. This process is called *isotopic fractionation* and is measured as

$$\delta^{13}C = \left[\frac{([C^{13}]:[C^{12}])_{\text{sample}}}{([C^{13}]:[C^{12}])_{\text{standard}}} - 1\right]$$

The standard is taken from the isotope ratio in the carbon of belemnite shells found in the Cretaceous Pedee formation of South Carolina. Explain, on the basis of the preceding information, why the following quotation from Krauskopf and Bird (1995) makes sense:

> The low [negative] values of $\delta^{13}C$ in the hydrocarbons of petroleum are one of the important bits of evidence for ascribing the origin of petroleum to the alteration of organic material rather than to condensation of primeval gases from the Earth's interior.

15. Chemical Reaction The speed of an enzymatic reaction is frequently described by the Michaelis–Menten equation

$$v = \frac{ax}{k + x}$$

where v is the velocity of the reaction, x is the concentration of the substrate, a is the maximum reaction velocity, and k is the substrate concentration at which the velocity is half of the maximum velocity. This curve describes how the reaction velocity depends on the substrate concentration.

(a) Show that when $x = k$, the velocity of the reaction is half the maximum velocity.

(b) Show that an 81-fold change in substrate concentration is needed to change the velocity from 10% to 90% of the maximum velocity, regardless of the value of k.

16. Lake Acidification Atmospheric pollutants can cause acidification of lakes (by acid rain). This can be a serious problem for lake organisms; for instance, in fish the ability of hemoglobin to transport oxygen decreases with decreasing pH levels of the water. Experiments with the zooplankton *Daphnia magna* showed a negligible decline in survivorship at pH = 6, but a marked decline in survivorship at pH = 3.5, resulting in no survivors after just eight hours. Illustrate graphically the percentage survivorship as a function of time for pH = 6 and pH = 3.5.

---

(3) The Devonian period lasted from about 408 million years ago to about 360 million years ago.

17. Lake Chemistry The pH level of a lake controls the concentrations of harmless ammonium ions (NH_4^+) and toxic ammonia (NH_3) in the lake. For pH levels below 8, concentrations of NH_4^+ ions are little affected by changes in the pH value, but they decline over many orders of magnitude as pH levels increase beyond pH = 8. By contrast, NH_3 concentrations are negligible at low pH, increase over many orders of magnitude as the pH level increases, and reach a high plateau at about pH = 10 (after which levels of NH_3 are little affected by changes in pH levels). Illustrate the behavior of $[NH_4^+]$ and $[NH_3]$ graphically.

18. Development and Growth Egg development times of the zooplankton *Daphnia longispina* depend on temperature. It takes only about 3 days at 20°C, but almost 20 days at 5°C, for an egg to develop and hatch. When graphed on a log–log plot, egg development time (in days) as a function of temperature (in °C) is a straight line.

(a) Sketch a graph of egg development time as a function of temperature on a log-log plot.

(b) Use the data to find the function that relates egg development time and temperature for *D. longispina*.

(c) Use your answer in (b) to predict egg development time of *D. longispina* at 10°C.

(d) Suppose you measured egg development time in hours and temperature in Fahrenheit. Would you still find a straight line on a log-log plot?

19. Resource Model Organisms consume resources. The rate of resource consumption, denoted by v, depends on resource concentration, denoted by S. The *Blackman* model of resource consumption assumes a linear relationship between resource consumption rate and resource concentration: Below a threshold concentration (S_k), the consumption rate increases linearly with $S = 0$ when $v = 0$; when $S = S_k$, the consumption rate v reaches its maximum value v_{max}; for $S > S_k$, the resource consumption rate stays at the maximum value v_{max}. A function like this, with a sharp transition, cannot be described analytically by just one expression; it needs to be defined piecewise:

$$v = \begin{cases} g(S) & \text{for } 0 \leq S < S_k \\ v_{max} & \text{for } S \geq S_k \end{cases}$$

Find $g(S)$, and graph the resource consumption rate v as a function of resource concentration S.

20. Light Intensity Light intensity in lakes decreases exponentially with depth. If $I(z)$ denotes the light intensity at depth z, with $z = 0$ representing the surface, then

$$I(z) = I(0)e^{-\alpha z}, \quad z \geq 0$$

where α is a positive constant called the *vertical attenuation coefficient*. This coefficient depends on the wavelength of the light and on the amount of dissolved matter and particles in the water. In the following, we assume that the water is pure:

(a) About 65% of red light (720 nm) is absorbed in the first meter. Find α.

(b) About 5% of blue light (475 nm) is absorbed in the first meter. Find α.

(c) Explain in words why a diver would not see red hues a few meters below the surface of a lake.

21. Light Intensity Light intensity in lakes decreases with depth according to the relationship

$$I(z) = I(0)e^{-\alpha z}, \quad z \geq 0$$

where $I(z)$ denotes the light intensity at depth z, $z = 0$ represents the surface, and α is a positive constant denoting the vertical attenuation coefficient. The depth where light intensity is about 1% of the surface light intensity is important for photosynthesis in phytoplankton: Below this level, photosynthesis is insufficient to compensate for respiratory losses. The 1% level is called the *compensation level*. An often used and relatively reliable method for determining the compensation level is the *Secchi disk* method. A Secchi disk is a white disk with radius 10 cm. The disk depth is the depth at which the disk disappears from the viewer. Twice this depth approximately coincides with the compensation level.

(a) Find α for a lake with Secchi disk depth of 9 m.

(b) Find the Secchi disk depth for a lake with $\alpha = 0.473$ m^{-1}.

22. Population Growth Assume that the population size $N(t)$ at time $t \geq 0$ is given by

$$N(t) = N_0 e^{rt}$$

with $N_0 = N(0)$. The parameter r is called the *average annual growth rate*.

(a) Show that

$$r = \ln \frac{N(t+1)}{N(t)} \tag{1.11}$$

Formula (1.11) is used, for instance, by the U.S. Census Bureau to track world population growth.

(b) Suppose a population doubles in size within a single year.

(i) What is the percent increase of the population during that year?

(ii) What is the average annual growth rate in percent during that year, according to (1.11)?

(c) Suppose the average annual growth rate of a population is 1.3%. How many years will it take the population to double in size?

(d) To calculate the doubling time of a growing population with a constant average annual growth rate, we divide the percent average annual growth rate into 70. Apply this "Rule of 70" to (c) and compare your answers. Derive the "Rule of 70."

2

Discrete-Time Models, Sequences, and Difference Equations

■ 2.1 Exponential Growth and Decay

■ 2.1.1 Modeling Population Growth in Discrete Time

Imagine that we observe bacteria that divide every 20 minutes and that, at the start of the experiment, there was one bacterium. How will the number of bacteria change over time? We call the time when we started the observation time 0. At time 0, there is one bacterium. After 20 minutes, the bacterium splits in two, so there are two bacteria at time 20. Twenty minutes later, each of the bacteria splits again, resulting in four bacteria at time 40, and so on (Figure 2.1).

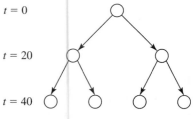

Figure 2.1 Bacteria split every 20 units of time.

We can produce a table that describes the growth of this population:

Time (min)	0	20	40	60	80	100	120
Population size	1	2	4	8	16	32	64

We can simplify the description of the growth of the bacterial population if we measure time in more convenient units. We say that one unit of time equals 20 minutes. Two units of time then corresponds to 40 minutes, three units of time to 60 minutes, and so on. We reproduce the table of population growth with these new units:

Time (20 min)	0	1	2	3	4	5	6
Population size	1	2	4	8	16	32	64

The new time units make it easier to write a general formula for the population size at time t. Denoting by $N(t)$ the population size at time t, where t is now measured in the new units (one unit is equal to 20 minutes), we guess from the second table that

$$N(t) = 2^t, \quad t = 0, 1, 2, \ldots \tag{2.1}$$

We encountered this function in Section 1.2 when we discussed exponential functions. There, the function was defined for all $t \geq 0$, whereas now, the function is defined only for nonnegative integer values. Equation (2.1) allows us to determine the population size at any discrete time t directly, without first calculating the population sizes at all previous time steps. For instance, at time $t = 5$, we find that $N(5) = 2^5 = 32$, as shown in the second table, or, at time $t = 10$, $N(10) = 2^{10} = 1024$. The graph of $N(t) = 2^t$ is shown in Figure 2.2.

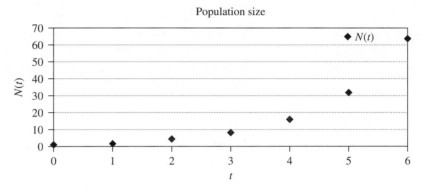

Population size

Figure 2.2 The graph of $N(t) = 2^t$ for $t = 0, 1, 2, \ldots, 6$.

The function $N(t) = 2^t$, $t = 0, 1, 2, \ldots$, is an exponential function, and we call the type of population growth that it represents **exponential growth**. The base 2 reflects the fact that the population size doubles every unit of time.

Instead of $N(t)$, we will often write N_t. The subscript notation is used only for functions $N(t)$ where t is a nonnegative integer. So, instead of writing $N(t) = 2^t$, $t = 0, 1, 2, \ldots$, we can write $N_t = 2^t$, $t = 0, 1, 2, \ldots$.

So far, we assumed that $N(0) = N_0 = 1$. Let's see what N_t looks like if $N_0 = 100$. Regardless of N_0, the population size doubles every unit of time. We obtain the following table, where time is again measured in units of 20 minutes:

Time (20 min)	0	1	2	3	4	5	6
Population size	100	200	400	800	1600	3200	6400

We can guess the general form of N_t with $N_0 = 100$ from the table:

$$N_t = 100 \cdot 2^t, \quad t = 0, 1, 2, \ldots$$

We see that the initial population size $N_0 = 100$ appears as a multiplicative factor in front of the term 2^t. If we do not want to specify a numerical value for the population size N_0 at time 0, we can write

$$N_t = N_0 2^t, \quad t = 0, 1, 2, \ldots$$

We already mentioned that the base 2 indicates that the population size doubles every unit of time. Replacing 2 by another number, we can describe other populations. For instance,

$$N_t = 3^t, \quad t = 0, 1, 2, \ldots$$

describes a population with $N_0 = 1$ and that triples in size every unit of time. The corresponding table is

Time	0	1	2	3	4
Population size	1	3	9	27	81

Now that we have some experience with exponential growth in discrete time, we give the general formula:

$$N_t = N_0 R^t, \quad t = 0, 1, 2, \ldots \tag{2.2}$$

The parameter R is a positive constant called the **growth constant**. The constant N_0 is nonnegative and denotes the population size at time 0. The assumptions $R > 0$ and $N_0 \geq 0$ are made for biological reasons: Negative values for R or N_0 would result in negative population sizes, and $R = 0$ would be uninteresting.

EXAMPLE 1

Suppose a population of cells reproduces every 15 minutes and we measure its size every 30 minutes:

Time (min)	0	30	60	90	120	150	180
Population size	1	4	16	64	256	1024	4096

Write a formula for time $n = 0, 1, 2, \ldots$ when (a) one unit of time is 30 minutes, (b) one unit of time is 60 minutes, and (c) one unit of time is 15 minutes.

Solution

(a) We see from the values listed in the table that when one unit of time is 30 minutes, the population quadruples every unit of time, with $N_0 = 1$. Thus,

$$N_t = 4^t, \quad t = 0, 1, 2, \ldots$$

(b) This time, we see from the values in the table that when one unit of time is equal to 60 minutes, the population grows by a factor of 16 each unit of time. Again, $N_0 = 1$. Hence,

$$N_s = 16^s, \quad s = 0, 1, 2, \ldots$$

We could also have arrived at this answer by noting that the time step in (b) is twice that of the time step in (a). In other words, when one unit of time elapses in (b), two units of time elapse in (a):

t	0	1	2	3	4	5	6
s	0		1		2		3

We find that $t = 2s$. If we substitute $2s$ for t in (a), we find that

$$N_t = 4^t \quad \text{yields} \quad N_s = 4^{2s} = 16^s$$

for $s = 0, 1, 2, \ldots$.

(c) When one unit of time is 15 minutes, and we use the variable $u = 0, 1, 2, \ldots$ to denote time, it follows that $t = u/2$ and

$$N_t = 4^t \quad \text{yields} \quad N_u = 4^{u/2} = 2^u$$

for $u = 0, 1, 2, \ldots$. ■

The function $N_t = N_0 R^t$, $t = 0, 1, 2, \ldots$, is an exponential function. We discussed exponential functions in the previous chapter. There, we looked at $f(x) = a^x$, $x \in \mathbf{R}$. To make the comparison easier, we choose $N_0 = 1$ in (2.2) and restrict the function $f(x) = a^x$ to $x \geq 0$. If we choose the same values for R and a, then the two functions N_t and $f(x)$ use the same rule to compute their values. The difference is in the domain: N_t is defined only for nonnegative integers, whereas $f(x)$ is defined for all nonnegative real numbers. The two functions agree where they are both defined. This can be seen when we graph N_t and $f(x)$ in the same coordinate system for $R = a$ (Figure 2.3).

In Chapter 1, we learned how $f(x) = a^x$, $x \in \mathbf{R}$, behaves for different values of a. We can use this behavior now to describe that of $N_t = N_0 R^t$, $t = 0, 1, 2, \ldots$. In Figure 2.4, we show the function $f(x) = a^x$, $x \geq 0$, for different values of a. Superimposed are the graphs of $N_t = N_0 R^t$, $t = 0, 1, 2, \ldots$, for $R = a$ and $N_0 = 1$.

We see that when $R > 1$, the population size N_t increases indefinitely; when $R = 1$, the population size N_t stays the same for all $t = 0, 1, 2, \ldots$; and when $0 < R < 1$, the population size N_t declines and approaches 0 as t increases. The behavior is the same for other positive initial population sizes ($N_0 > 0$).

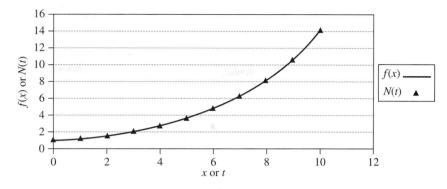

Figure 2.3 The graphs of $f(x) = a^x$, $0 \leq x \leq 10$, and $N(t) = R^t$, $t = 0, 1, 2, \ldots, 10$, when $a = R = 1.3$.

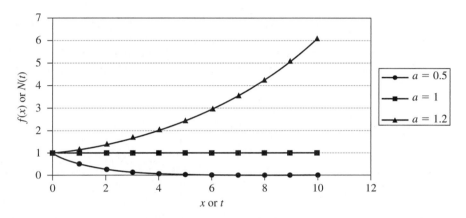

Figure 2.4 The graphs of $f(x) = a^x$, $0 \leq x \leq 10$, and $N(t) = R^t$, $t = 0, 1, 2, \ldots, 10$, for three different values of $a = R$: $a = R = 0.5$, $a = R = 1$, and $a = R = 1.2$.

■ 2.1.2 Recursions

When we constructed the tables for the bacterial population size with $R = 2$ at consecutive time steps, we doubled the population size from time step to time step. In other words, we computed the population size at time $t + 1$ on the basis of the population size at time t, using the equation

$$N_{t+1} = 2N_t \tag{2.3}$$

Equation (2.3) is a rule that is applied repeatedly to go from one time step to the next and is called a **recursion**. We say that Equation (2.3) defines the population size **recursively**.

If we want to use Equation (2.3) to find the population size, say, at time $t = 4$, we need to know the population size at some earlier time, say, time $t = 0$. Let's assume that $N_0 = 1$. Then, applying the recursion (2.3) repeatedly, we find that

$$N_1 = 2N_0 = 2$$
$$N_2 = 2N_1 = 4$$
$$N_3 = 2N_2 = 8$$
$$N_4 = 2N_3 = 16$$

We thus have two equivalent ways to describe this population: For $t = 0, 1, 2, \ldots,$

$$N_t = 2^t \quad \text{is equivalent to} \quad N_{t+1} = 2N_t \quad \text{with } N_0 = 1$$

The recursion for a general value of R is

$$N_{t+1} = RN_t \quad \text{with } N_0 = \text{population size at time 0} \tag{2.4}$$

Applying (2.4) repeatedly, we obtain

$$N_1 = RN_0$$
$$N_2 = RN_1 = R^2 N_0$$
$$N_3 = RN_2 = R^3 N_0$$
$$N_4 = RN_3 = R^4 N_0$$
$$\vdots$$
$$N_t = RN_{t-1} = R^t N_0$$

The two descriptions for $t = 0, 1, 2, \ldots$, namely,

$$N_t = N_0 R^t \quad \text{and} \quad N_{t+1} = RN_t \quad \text{with } N_0 = \text{population size at time 0}$$

are equivalent. We say that $N_t = N_0 R^t$ is a **solution** of the recursion $N_{t+1} = RN_t$ with initial condition N_0 at time 0, since the function $N_t = N_0 R^t$ satisfies the recursion with initial condition $N(0) = N_0$.

We can visualize recursions by plotting N_t on the horizontal axis and N_{t+1} on the vertical axis. The exponential growth recursion

$$N_{t+1} = RN_t \tag{2.5}$$

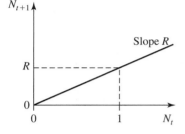

Figure 2.5 The exponential growth recursion $N_{t+1} = RN_t$ when $R > 0$.

is then a straight line through the origin with slope R (Figure 2.5). Since $N_t \geq 0$ for biological reasons, we restrict the graph to the first quadrant.

What does this graph tell us? For any current population size N_t, it allows us to find the population size in the next time step, namely, N_{t+1}. For instance, if $R = 2$ and $N_0 = 1$, then successive population sizes are 1, 2, 4, 8, 16, 32, ... For this choice of N_0, we will never see a population size of, say, 5 or 10. Thus, for a specific choice of N_0, only a selected number of points on the graph $N_{t+1} = RN_t$ will be realized (Figure 2.6). A different choice of initial condition would yield a different set of points.

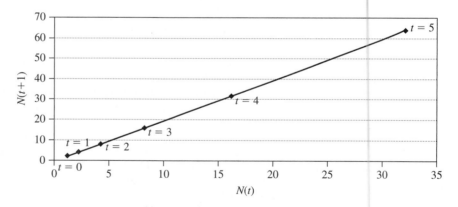

Figure 2.6 Successive population sizes on the graph of the exponential growth recursion when $R = 2$ for $t = 0, 1, 2, \ldots, 5$.

We also see from Figure 2.6 that unless we label the points according to the corresponding t-value, we would not be able to tell at what time a point (N_t, N_{t+1}) was realized. We say that time is *implicit* in this graph. Compare Figure 2.6 with Figure 2.2, in which we graphed N_t as a function of t for the same values of R and N_0; in Figure 2.2, time is *explicit*.

The hallmark of exponential growth is that the ratio of successive population sizes, N_t/N_{t+1}, is constant. When $N_t > 0$ (and hence $N_{t+1} > 0$), it follows from $N_{t+1} = RN_t$ that

$$\frac{N_t}{N_{t+1}} = \frac{1}{R}$$

If the population consists of annual plants, we can interpret the ratio N_t/N_{t+1} as the parent–offspring ratio. If this ratio is constant, parents produce the same number of

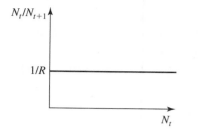

Figure 2.7 The graph of the parent–offspring ratio $\frac{N_t}{N_{t+1}}$ as a function of N_t when $N_t > 0$.

offspring, regardless of the current population density. Such growth is called **density independent**.

When $R > 1$, it follows that $1/R$, the parent–offspring ratio, is less than 1, implying that the number of offspring exceeds the number of parents. Density-independent growth with $R > 1$ results in an ever-increasing population size. This model eventually becomes biologically unrealistic, since any population will sooner or later experience food or habitat limitations that will limit its growth. (We will discuss models that include such limitations in Section 2.3.)

The density independence in exponential growth is reflected in a graph of N_t/N_{t+1} as a function of N_t, which is a horizontal line at level $1/R$ (Figure 2.7).

As before, only a selected number of points are realized on the graph of N_t/N_{t+1} as a function of N_t, and time is implicit in the graph. (See Figure 2.8, with $R = 2$ and $N_0 = 1$.)

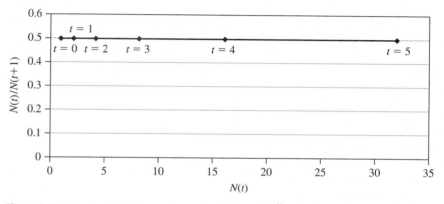

Figure 2.8 The graph of the parent–offspring ratio $\frac{N_t}{N_{t+1}}$ as a function of N_t when $N_t = 1$ and $R = 2$.

Section 2.1 Problems

In Problems 1–4, produce a table for $t = 0, 1, 2, \ldots, 5$ and graph the function N_t.

1. $N_t = 3^t$

2. $N_t = 10 \cdot 2^t$

3. $N_t = \dfrac{25}{4^t}$

4. $N_t = (0.3)(0.9)^t$

In Problems 5–10, give a formula for $N(t)$, $t = 0, 1, 2, \ldots$, on the basis of the information provided.

5. $N_0 = 2$; population doubles every 20 minutes; one unit of time is 20 minutes

6. $N_0 = 4$; population doubles every 40 minutes; one unit of time is 40 minutes

7. $N_0 = 1$; population doubles every 40 minutes; one unit of time is 80 minutes

8. $N_0 = 6$; population doubles every 40 minutes; one unit of time is 60 minutes

9. $N_0 = 2$; population quadruples every 30 minutes; one unit of time is 15 minutes

10. $N_0 = 10$; population quadruples every 20 minutes; one unit of time is 10 minutes

11. Suppose $N_t = 20 \cdot 4^t$, $t = 0, 1, 2, \ldots$, and one unit of time corresponds to 3 hours. Determine the amount of time it takes the population to double in size.

12. Suppose $N_t = 100 \cdot 2^t$, $t = 0, 1, 2, \ldots$, and one unit of time corresponds to 2 hours. Determine the amount of time it takes the population to triple in size.

13. A strain of bacteria reproduces asexually every hour. That is, every hour, each bacterial cell splits into two cells. If, initially, there is one bacterium, find the number of bacterial cells after 1 hour, 2 hours, 3 hours, 4 hours, and 5 hours.

14. A strain of bacteria reproduces asexually every 30 minutes. That is, every 30 minutes, each bacterial cell splits into two cells. If, initially, there is one bacterium, find the number of bacterial cells after 1 hour, 2 hours, 3 hours, 4 hours, and 5 hours.

15. A strain of bacteria reproduces asexually every 23 minutes. That is, every 23 minutes, each bacterial cell splits into two cells. If, initially, there is 1 bacterium, how long will it take until there are 128 bacteria?

16. A strain of bacteria reproduces asexually every 42 minutes. That is, every 42 minutes, each bacterial cell splits into two cells. If, initially, there is 1 bacterium, how long will it take until there are 512 bacteria?

17. A strain of bacteria reproduces asexually every 10 minutes. That is, every 10 minutes, each bacterial cell splits into two cells. If, initially, there are 3 bacteria, how long will it take until there are 96 bacteria?

18. A strain of bacteria reproduces asexually every 50 minutes. That is, every 50 minutes, each bacterial cell splits into two cells. If, initially, there are 10 bacteria, how long will it take until there are 640 bacteria?

19. Find the exponential growth equation for a population that doubles in size every unit of time and that has 40 individuals at time 0.

20. Find the exponential growth equation for a population that doubles in size every unit of time and that has 53 individuals at time 0.

21. Find the exponential growth equation for a population that triples in size every unit of time and that has 20 individuals at time 0.

22. Find the exponential growth equation for a population that triples in size every unit of time and that has 72 individuals at time 0.

23. Find the exponential growth equation for a population that quadruples in size every unit of time and that has five individuals at time 0.

24. Find the exponential growth equation for a population that quadruples in size every unit of time and that has 17 individuals at time 0.

25. Find the recursion for a population that doubles in size every unit of time and that has 20 individuals at time 0.

26. Find the recursion for a population that doubles in size every unit of time and that has 37 individuals at time 0.

27. Find the recursion for a population that triples in size every unit of time and that has 10 individuals at time 0.

28. Find the recursion for a population that triples in size every unit of time and that has 84 individuals at time 0.

29. Find the recursion for a population that quadruples in size every unit of time and that has 30 individuals at time 0.

30. Find the recursion for a population that quadruples in size every unit of time and that has 62 individuals at time 0.

In Problems 31–34, graph the functions $f(x) = a^x$, $x \in [0, \infty)$, and $N_t = R^t$, $t \in \mathbf{N}$, together in one coordinate system for the indicated values of a and R.

31. $a = R = 2$ **32.** $a = R = 3$
33. $a = R = 1/2$ **34.** $a = R = 1/3$

In Problems 35–46, find the population sizes for $t = 0, 1, 2, \ldots, 5$ for each recursion.

35. $N_{t+1} = 2N_t$ with $N_0 = 3$ **36.** $N_{t+1} = 2N_t$ with $N_0 = 5$
37. $N_{t+1} = 3N_t$ with $N_0 = 2$ **38.** $N_{t+1} = 3N_t$ with $N_0 = 7$
39. $N_{t+1} = 5N_t$ with $N_0 = 1$ **40.** $N_{t+1} = 7N_t$ with $N_0 = 4$
41. $N_{t+1} = \frac{1}{2}N_t$ with $N_0 = 1024$
42. $N_{t+1} = \frac{1}{2}N_t$ with $N_0 = 4096$
43. $N_{t+1} = \frac{1}{3}N_t$ with $N_0 = 729$
44. $N_{t+1} = \frac{1}{3}N_t$ with $N_0 = 3645$
45. $N_{t+1} = \frac{1}{5}N_t$ with $N_0 = 31250$
46. $N_{t+1} = \frac{1}{4}N_t$ with $N_0 = 8192$

In Problems 47–58, write N_t as a function of t for each recursion.

47. $N_{t+1} = 2N_t$ with $N_0 = 15$ **48.** $N_{t+1} = 2N_t$ with $N_0 = 7$
49. $N_{t+1} = 3N_t$ with $N_0 = 12$ **50.** $N_{t+1} = 3N_t$ with $N_0 = 3$
51. $N_{t+1} = 4N_t$ with $N_0 = 24$ **52.** $N_{t+1} = 5N_t$ with $N_0 = 17$
53. $N_{t+1} = \frac{1}{2}N_t$ with $N_0 = 5000$
54. $N_{t+1} = \frac{1}{2}N_t$ with $N_0 = 2300$
55. $N_{t+1} = \frac{1}{3}N_t$ with $N_0 = 8000$
56. $N_{t+1} = \frac{1}{3}N_t$ with $N_0 = 3500$
57. $N_{t+1} = \frac{1}{5}N_t$ with $N_0 = 1200$
58. $N_{t+1} = \frac{1}{7}N_t$ with $N_0 = 6400$

In Problems 59–66, graph the line $N_{t+1} = RN_t$ in the N_t–N_{t+1} plane for the indicated value of R and locate the points (N_t, N_{t+1}), $t = 0$, 1, and 2, for the given value of N_0.

59. $R = 2, N_0 = 2$ **60.** $R = 2, N_0 = 3$
61. $R = 3, N_0 = 1$ **62.** $R = 4, N_0 = 2$
63. $R = \frac{1}{2}, N_0 = 16$ **64.** $R = \frac{1}{2}, N_0 = 64$
65. $R = \frac{1}{3}, N_0 = 81$ **66.** $R = \frac{1}{4}, N_0 = 16$

In Problems 67–74, graph the line $\frac{N_t}{N_{t+1}} = \frac{1}{R}$ in the N_t–$\frac{N_t}{N_{t+1}}$ plane for the indicated value of R and locate the points $(N_t, \frac{N_t}{N_{t+1}})$, $t = 0$, 1, 2, for the given value of N_0. Find the parent–offspring ratio.

67. $R = 2, N_0 = 2$ **68.** $R = 2, N_0 = 4$
69. $R = 3, N_0 = 2$ **70.** $R = 4, N_0 = 1$
71. $R = \frac{1}{2}, N_0 = 16$ **72.** $R = \frac{1}{2}, N_0 = 128$
73. $R = \frac{1}{3}, N_0 = 27$ **74.** $R = \frac{1}{4}, N_0 = 64$

75. A bird population lives in a habitat where the number of nesting sites is a limiting factor in population growth. In which of the following cases would you expect that the growth of this bird population over the next few generations could be reasonably well approximated by exponential growth?
(a) All nesting sites are occupied.

(b) The bird population just invaded the habitat, and the population size is still much smaller than the available nesting sites.

(c) In the previous year, a hurricane killed more than 90% of the birds in this habitat.

76. Pollen records show that the number of Scotch pine (*Pinus sylvestris*) grew exponentially for about 500 years after colonization of the Norfolk region of Great Britain about 9500 years ago. Can you find a possible explanation for this growth?

77. Exponential growth generally occurs when population growth is density independent. List conditions under which a population might stop growing exponentially.

■ 2.2 Sequences

■ 2.2.1 What Are Sequences?

Before we explore other discrete-time population models, we need to develop further the theory of functions with domain **N**. The functions are of the form

$$f : \mathbf{N} \to \mathbf{R}$$
$$n \to f(n)$$

When the independent variable denotes time, we will frequently use t instead of n. Tables and graphs are useful tools to illustrate these functions.

EXAMPLE 1

Let

$$f : \mathbf{N} \to \mathbf{R}$$
$$n \to f(n) = \frac{1}{n+1}$$

Produce a table for $n = 0, 1, 2 \ldots, 5$ and graph the function.

Solution

The table is

n	0	1	2	3	4	5
$\frac{1}{n+1}$	1	$\frac{1}{2}$	$\frac{1}{3}$	$\frac{1}{4}$	$\frac{1}{5}$	$\frac{1}{6}$

The graph of this function consists of discrete points (Figure 2.9). On the horizontal axis, we display the variable n; on the vertical axis, the function $f(n)$. Note that we did not connect the points with lines or curves. ■

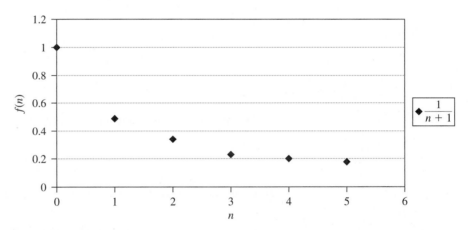

Figure 2.9 The graph of the function $f(n) = \frac{1}{n+1}$ in Example 1.

We can write the function

$$f : \mathbf{N} \to \mathbf{R}$$
$$n \to f(n)$$

as a list of numbers $a_0, a_1, a_2, \ldots$, where $a_n = f(n)$. We refer to this list as a **sequence**. We will write $\{a_n : n \in \mathbf{N}\}$ (or $\{a_n\}$ for short) if we mean the entire sequence. Note that we list the values of the sequence $\{a_n\}$ in order of increasing n:

$$a_0, a_1, a_2, \ldots$$

EXAMPLE 2

The sequence

$$a_n = (-1)^n, \quad n = 0, 1, 2, \ldots$$

takes on values

$$1, -1, 1, -1, 1, \ldots$$ ■

When we see a sequence and recognize a pattern, we can often write an expression for a_n.

EXAMPLE 3

Find a_n for the sequence

$$0, 1, 4, 9, 16, 25, \ldots$$

Solution

Looking at the sequence, we can guess the next terms, namely, 36, 49, 64, and so on. We thus find that

$$a_n = n^2, \quad n = 0, 1, 2, \ldots$$

We do not need to start a sequence at $n = 0$. If we started the sequence at $n = 1$, we would write

$$a_n = (n-1)^2, \quad n = 1, 2, 3, \ldots$$

In either case, it is important to include the domain of the sequence. ■

EXAMPLE 4

Find a_n for the sequence

$$1, -\frac{1}{4}, \frac{1}{9}, -\frac{1}{16}, \frac{1}{25}, \ldots$$

Solution

This sequence has alternating signs: The first term is positive, the second negative, the third positive, and so on. This indicates that we need a factor $(-1)^n$, $n = 0, 1, 2, \ldots$. The numerators are all equal to 1, and the denominators are successive squares of integers, starting with the integer 1. We can thus write

$$a_0 = (-1)^0 \frac{1}{(1)^2} = 1$$

$$a_1 = (-1)^1 \frac{1}{(2)^2} = -\frac{1}{4}$$

$$a_2 = (-1)^2 \frac{1}{(3)^2} = \frac{1}{9}$$

and so on. This set of equations suggests that

$$a_n = (-1)^n \frac{1}{(n+1)^2}, \quad n = 0, 1, 2, \ldots$$

If we wanted to start the sequence at $n = 1$, we could write

$$a_n = (-1)^{n-1} \frac{1}{n^2}, \quad n = 1, 2, 3, \ldots$$

or

$$a_n = (-1)^{n+1} \frac{1}{n^2}, \quad n = 1, 2, 3, \ldots$$

Look carefully at the exponent of (-1). Any of the terms $(-1)^n$, $(-1)^{n-1}$, and $(-1)^{n+1}$ produces alternating signs. Since the first term in the sequence $\{a_n\}$ is positive, we need to use $(-1)^n$ if we start with $n = 0$, and either $(-1)^{n-1}$ or $(-1)^{n+1}$ if we start with $n = 1$. ■

The exponential growth model we considered in the previous section is an example of a sequence. We gave two descriptions, one explicit and the other recursive. These two descriptions can be used for sequences in general. An explicit description is of the form

$$a_n = f(n), \quad n = 0, 1, 2, \ldots$$

where $f(n)$ is a function of n.

A recursive description is of the form

$$a_{n+1} = g(a_n), \quad n = 0, 1, 2, \ldots$$

where $g(a_n)$ is a function of a_n. If, as is shown here, the value of a_{n+1} depends only on the value one time step back, namely, a_n, then the recursion is called a **first-order recursion**. Later in the chapter, we will see an example of a second-order recursion, in which the value of a_{n+1} depends on the values a_n and a_{n-1}—that is, on the values one and two time steps back. To determine the values of successive members of a sequence given in recursive form, we need to specify an initial value a_0 if we start the sequence at $n = 0$ (or a_1 if we start the sequence at $n = 1$).

In the notation of this section, the exponential growth of the previous section is given explicitly by

$$a_n = a_0 R^n, \quad n = 0, 1, 2, \dots$$

and recursively by

$$a_{n+1} = Ra_n, \quad n = 0, 1, 2, \dots$$

Note that, in the recursive definition, the initial value a_0 needs to be specified.

■ 2.2.2 Limits

When studying populations over time, we are often interested in their **long-term behavior**. Specifically, if N_t is the population size at time t, $t = 0, 1, 2, \dots$, we want to know how N_t behaves as t increases, or, more precisely, as t tends to infinity. Using the notation of this section, we want to know the behavior of a_n as n tends to infinity. When we let n tend to infinity, we say that "we take the limit of the sequence a_n as n goes to infinity" and use the shorthand notation

$$\lim_{n \to \infty} a_n \quad \text{or} \quad \lim_{n \to \infty} a_n$$

read as "the limit of a_n as n tends to infinity," in equations. Let's first discuss limits informally to get an idea of what can happen.

EXAMPLE 5

Let

$$a_n = \frac{1}{n + 1}, \quad n = 0, 1, 2, \dots$$

Find $\lim_{n \to \infty} a_n$.

Solution

Plugging successive values of n into a_n, we find that a_n is the sequence

$$1, \frac{1}{2}, \frac{1}{3}, \frac{1}{4}, \frac{1}{5}, \dots$$

and we guess that the terms will approach 0 as n tends to infinity. This is indeed the case, and we will learn shortly how to show that

$$\lim_{n \to \infty} \frac{1}{n + 1} = 0$$

Since the limiting value is a unique number, we say that the limit exists. ■

Note that plugging in successive values of n into a_n is only a heuristic way of determining how a_n behaves as $n \to \infty$.

EXAMPLE 6

Let

$$a_n = (-1)^n, \quad n = 0, 1, 2, \dots$$

Find $\lim_{n \to \infty} a_n$.

Solution

The sequence is of the form

$$1, -1, 1, -1, 1, \dots$$

and we see that its terms alternate between 1 and -1. There is thus no single number we could assign as the limit of a_n as $n \to \infty$. We then say that the limit does not exist. ■

EXAMPLE 7

Let

$$a_n = 2^n, \quad n = 0, 1, 2, \dots$$

Find $\lim_{n \to \infty} a_n$.

Solution   Successive terms of a_n, namely,

$$1, 2, 4, 8, 16, 32, \ldots$$

indicate that the terms continue to grow. Hence, a_n goes to infinity as $n \to \infty$, and we can write $\lim_{n \to \infty} a_n = \infty$. Since infinity ($\infty$) is not a real number, we say that the limit does not exist. ■

Let's look at one more example of a limit that exists before we give a formal definition.

EXAMPLE 8   Find

$$\lim_{n \to \infty} \frac{n+1}{n}$$

Solution   Starting with $n = 1$ and computing successive terms, we find that

$$2, \frac{3}{2}, \frac{4}{3}, \frac{5}{4}, \frac{6}{5}, \ldots$$

We see that the terms get closer and closer to 1, and, indeed,

$$\lim_{n \to \infty} \frac{n+1}{n} = 1$$ ■

The way we solved the first four examples is unsatisfying: We guessed the limiting values. How do we know that our guesses are correct? There is a formal definition of limits that can be used to compute them. However, except in the simplest cases, the formal definition is quite cumbersome to use. Fortunately, there are mathematical laws that build on simple limits (which can be computed from the formal definition). We will first discuss the formal definition (as an optional topic) and then introduce the limit laws.

Formal Definition of Limits (Optional)   Example 8 will motivate the formal definition of limits. When we guessed the limit in Example 8, we realized that successive terms approached 1. This means that no matter how small an interval about 1 we choose, all points must lie in this interval for all sufficiently large values of n. Graphically, the points of the graph of a_n must lie between the two dashed lines in Figure 2.10 for all large enough values of n, no matter how close those lines are to the horizontal line at height 1.

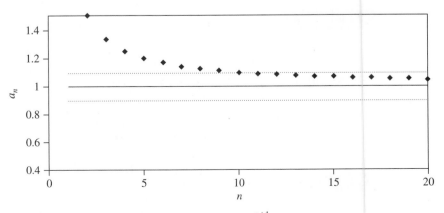

Figure 2.10   Convergence of the sequence $a_n = \frac{n+1}{n}$ to $a = 1$.

Translating this condition into a formal statement for the general case, we arrive at the following definition:

Formal Definition of Limits The sequence $\{a_n\}$ has **limit** a, written as $\lim_{n\to\infty} a_n = a$, if, for every $\epsilon > 0$, there exists an integer N such that

$$|a_n - a| < \epsilon \qquad \text{whenever } n > N$$

If the limit exists, the sequence is called **convergent** and we say that a_n **converges** to a as n tends to infinity. If the sequence has no limit, it is called **divergent**.

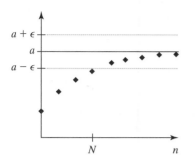

Figure 2.11 An illustration of the formal definition of limits to show convergence of the sequence a_n to a as $n \to \infty$: For all $n > N$, a_n lies in the strip of width 2ϵ and centered at a.

The value of N will typically depend on ϵ: The smaller ϵ is, the larger N is. We illustrate the concept of a converging sequence in Figure 2.11. The horizontal dashed lines are at heights $a + \epsilon$ and $a - \epsilon$, respectively. They form a strip of width 2ϵ centered at the horizontal line at height a. Points a_n within this strip satisfy the inequality $|a_n - a| < \epsilon$. For a sequence to be convergent, we require that *all* points a_n lie in this strip for *all* n sufficiently large (namely, larger than some N).

EXAMPLE 9

Show that

$$\lim_{n\to\infty} \frac{1}{n} = 0$$

Solution

Before we show this for any arbitrary ϵ, let's try to find N for a particular choice of ϵ, say, $\epsilon = 0.03$. We need to find an integer N such that

$$\left|\frac{1}{n} - 0\right| < 0.03 \quad \text{whenever } n > N$$

Solving the inequality $|\frac{1}{n} - 0| < 0.03$ for n positive, we find that

$$\left|\frac{1}{n}\right| < 0.03, \qquad \text{or} \qquad n > \frac{1}{0.03} \approx 33.33$$

The smallest value for N that we can choose is $N = 33$, which is the largest integer less than or equal to $1/0.03$. Successive values for $n > 33$ give us confidence that we are on the right track but don't prove that our choice is correct:

$$a_{34} = \frac{1}{34} \approx 0.0294, \quad a_{35} = \frac{1}{35} \approx 0.0286, \quad \text{and so on}$$

To see that our choice for N works, we need to show that $n > 33$ implies $|1/n| < 0.03$. Now, since n takes on only integer values, $n > 33$ is equivalent to $n \geq 34$, which implies that $1/n \leq 1/34 \approx 0.0294$. Since $n > 0$, we have

$$\left|\frac{1}{n} - 0\right| < 0.03 \quad \text{whenever } n > 33$$

To show that $a_n = \frac{1}{n}$ converges to 0, we need to do the same calculation for any arbitrary ϵ. That is, we need to show that, for every $\epsilon > 0$, we can find an N such that

$$\left|\frac{1}{n} - 0\right| < \epsilon \quad \text{whenever } n > N$$

To find a candidate for N, we solve the inequality $|\frac{1}{n}| < \epsilon$. Since $\frac{1}{n} > 0$, we can drop the absolute-value signs and find

$$\frac{1}{n} < \epsilon, \qquad \text{or} \quad n > \frac{1}{\epsilon}$$

Let's choose N so that $1/N \geq \epsilon$ and $1/(N+1) < \epsilon$, or, equivalently, $N \leq 1/\epsilon$ and $N+1 > 1/\epsilon$. This means that we choose N to be the largest integer less than or equal to $1/\epsilon$. If $n > N$, then $n \geq N+1$, which is equivalent to $1/n \leq 1/(N+1)$. Since N is the largest integer less than or equal to $1/\epsilon$, it follows that $1/n \leq 1/(N+1) < \epsilon \leq 1/N$ for $n > N$. This condition, together with $n > 0$, shows that if N is the largest integer less than or equal to $1/\epsilon$, then

$$\left| \frac{1}{n} - 0 \right| < \epsilon \quad \text{whenever } n > N \qquad ■$$

Limit Laws The formal definition of limits is cumbersome when we want to compute limits in specific examples. Fortunately, there are mathematical laws that facilitate the computation of limits:

> **Limit Laws** If $\lim_{n\to\infty} a_n$ and $\lim_{n\to\infty} b_n$ exist and c is a constant, then
>
> $$\lim_{n\to\infty} (a_n + b_n) = \lim_{n\to\infty} a_n + \lim_{n\to\infty} b_n$$
>
> $$\lim_{n\to\infty} (ca_n) = c \lim_{n\to\infty} a_n$$
>
> $$\lim_{n\to\infty} (a_n b_n) = (\lim_{n\to\infty} a_n)(\lim_{n\to\infty} b_n)$$
>
> $$\lim_{n\to\infty} \frac{a_n}{b_n} = \frac{\lim_{n\to\infty} a_n}{\lim_{n\to\infty} b_n}, \text{ provided } \lim_{n\to\infty} b_n \neq 0$$

Although we do not need to know the formal definition of limits in order to use the limit laws, in the next two examples we will need to know that

$$\lim_{n\to\infty} \frac{1}{n} = 0 \qquad (2.6)$$

which was proved (using the formal definition of limits) in Example 9.

EXAMPLE 10   Find

$$\lim_{n\to\infty} \frac{n+1}{n}$$

Solution   We break $\frac{n+1}{n}$ into a sum of two terms, namely, $1 + \frac{1}{n}$. Since $\lim_{n\to\infty} 1$ and $\lim_{n\to\infty} \frac{1}{n}$ exist [the former is equal to 1 and the latter to 0, according to (2.6)], it follows that

$$\lim_{n\to\infty} \frac{n+1}{n} = \lim_{n\to\infty} \left(1 + \frac{1}{n} \right) = \lim_{n\to\infty} (1) + \lim_{n\to\infty} \frac{1}{n} = 1 + 0 = 1$$

as claimed in Example 8.   ■

EXAMPLE 11   Find

$$\lim_{n\to\infty} \frac{4n^2 - 1}{n^2}$$

Solution   We rewrite a_n:

$$a_n = \frac{4n^2 - 1}{n^2} = 4 - \frac{1}{n^2} = 4 - \frac{1}{n} \cdot \frac{1}{n}$$

Since $\lim_{n\to\infty} 4$ and $\lim_{n\to\infty} \frac{1}{n}$ exist, we have

$$\lim_{n\to\infty} \frac{4n^2 - 1}{n^2} = \lim_{n\to\infty} \left(4 - \frac{1}{n} \cdot \frac{1}{n} \right) = \lim_{n\to\infty} 4 - \left(\lim_{n\to\infty} \frac{1}{n} \right) \left(\lim_{n\to\infty} \frac{1}{n} \right)$$

$$= 4 - 0 \cdot 0 = 4 \qquad ■$$

EXAMPLE 12

Without proof, we will state the long-term behavior of exponential growth. For $R > 0$, exponential growth is given by

$$a_n = a_0 R^n, n = 0, 1, 2, \ldots$$

Figure 2.12 indicates that

$$\lim_{n \to \infty} a_n = \begin{cases} 0 & \text{if } 0 < R < 1 \\ a_0 & \text{if } R = 1 \\ \infty & \text{if } R > 1 \end{cases}$$

Figure 2.12 Exponential growth in Example 12 for three different values of R.

This conclusion can also be shown rigorously by using the formal definition of limits. ■

■ 2.2.3 Recursions

In the previous subsection, we learned how to find $\lim_{n \to \infty} a_n$ when a_n is given explicitly as a function of n. We will now discuss how to find such a limit when a_n is defined recursively.

When we define a first-order sequence $\{a_n\}$ recursively, we express a_{n+1} in terms of a_n and specify a value for a_0. We can then compute successive values of a_n, which might allow us to guess the limit if it exists. In some cases (as in the next example), we can find a solution of the recursion and then determine the limit (if it exists), as in Subsection 2.2.2.

EXAMPLE 13

Compute a_n for $n = 1, 2, \ldots, 5$ when

$$a_{n+1} = \frac{1}{4}a_n + \frac{3}{4} \quad \text{with } a_0 = 2 \tag{2.7}$$

Find a solution of the recursion, and then take a guess at the limiting behavior of the sequence.

Solution

By repeatedly applying the recursion, we find that

$$a_1 = \frac{1}{4}a_0 + \frac{3}{4} = \frac{1}{4} \cdot 2 + \frac{3}{4} = \frac{5}{4} = 1.25$$

$$a_2 = \frac{1}{4}a_1 + \frac{3}{4} = \frac{1}{4} \cdot \frac{5}{4} + \frac{3}{4} = \frac{17}{16} = 1.0625$$

$$a_3 = \frac{1}{4}a_2 + \frac{3}{4} = \frac{1}{4} \cdot \frac{17}{16} + \frac{3}{4} = \frac{65}{64} \approx 1.0156$$

$$a_4 = \frac{1}{4}a_3 + \frac{3}{4} = \frac{1}{4} \cdot \frac{65}{64} + \frac{3}{4} = \frac{257}{256} \approx 1.0039$$

$$a_5 = \frac{1}{4}a_4 + \frac{3}{4} = \frac{1}{4} \cdot \frac{257}{256} + \frac{3}{4} = \frac{1025}{1024} \approx 1.0010$$

There seems to be a pattern, namely, that the denominators are powers of 4 and the numerators are just 1 larger than the denominators. We therefore set

$$a_n = \frac{4^n + 1}{4^n} \tag{2.8}$$

and check whether this is indeed a solution of the recursion. First, we need to check the initial condition: $a_0 = \frac{4^0+1}{4^0} = \frac{2}{1} = 2$. This agrees with the given initial condition. Next, we need to check whether a_n satisfies the recursion. Accordingly, we write

$$a_{n+1} = \frac{4^{n+1} + 1}{4^{n+1}} = 1 + \frac{1}{4 \cdot 4^n} = 1 + \frac{1}{4}\frac{1}{4^n}$$

Now, $a_n = \frac{4^n+1}{4^n}$ implies that $a_n = 1 + \frac{1}{4^n}$, or $\frac{1}{4^n} = a_n - 1$. Using the latter equation and simplifying then yields

$$a_{n+1} = 1 + \frac{1}{4}\frac{1}{4^n} = 1 + \frac{1}{4}(a_n - 1) = \frac{1}{4}a_n + \frac{3}{4}$$

which is the given recursion and thus proves that (2.8) is a solution of (2.7). We can now use (2.8) to find the limit. We have

$$\lim_{n \to \infty} a_n = \lim_{n \to \infty} \frac{4^n + 1}{4^n} = \lim_{n \to \infty} \left(1 + \frac{1}{4^n}\right) = 1$$

since $\lim_{n \to \infty} \frac{1}{4^n} = \lim_{n \to \infty} \left(\frac{1}{4}\right)^n = 0$, according to Example 12. ■

Finding an explicit expression for a_n as in Example 13 is often not a feasible strategy, because solving recursions can be very difficult or even impossible. How, then, can we say anything about the limiting behavior of a recursively defined sequence?

The following procedure will allow us to identify *candidates* for limits: A **fixed point** is a point such that if a_0 is equal to the fixed point, then all successive values of a_n are also equal to the fixed point. In mathematical terms, if we call the fixed point a, then if $a_0 = a$, we have $a_1 = a$, $a_2 = a$, and so on.

Now, if $a_{n+1} = g(a_n)$, then if $a_0 = a$ and a is a fixed point, it follows that $a_1 = g(a_0) = g(a) = a$, $a_2 = g(a_1) = g(a) = a$, and so on. That is, a fixed point satisfies the equation

$$a = g(a) \tag{2.9}$$

We will use (2.9) to find fixed points.

In Example 13, we had the recursion $a_{n+1} = \frac{1}{4}a_n + \frac{3}{4}$. Fixed points for the recursion thus satisfy

$$a = \frac{1}{4}a + \frac{3}{4}$$

Solving this equation for a, we find that $a = 1$. It turns out that in Example 13 the fixed point is also the limiting point. This will not always be the case: A fixed point is only a *candidate* for a limit; a sequence does not have to converge to a given fixed point (unless a_0 is already equal to the fixed point). The next two examples illustrate convergence and nonconvergence, respectively.

EXAMPLE 14

Assume that $\lim_{n \to \infty} a_n$ exists for

$$a_{n+1} = \sqrt{3a_n} \quad \text{with } a_0 = 2$$

Find $\lim_{n \to \infty} a_n$.

Solution

Since the problem tells us that the limit exists, we don't have to worry about existence. The problem that remains is to identify the limit. To do this, we compute the fixed points. We solve

$$a = \sqrt{3a}$$

which has two solutions, namely, $a = 0$ and $a = 3$. When $a_0 = 2$, we have $a_n > 2$ for all $n = 1, 2, 3, \ldots$, so we can exclude $a = 0$ as the limiting value. This leaves only one possibility, and we conclude that

$$\lim_{n \to \infty} a_n = 3$$

Using a calculator, we can find successive values of a_n, which we collect in the following table (accurate to two decimals):

n	0	1	2	3	4	5	6	7
a_n	2	2.45	2.71	2.85	2.92	2.96	2.98	2.99

The tabulated values suggest that the limit is indeed 3. ■

EXAMPLE 15　Let

$$a_{n+1} = \frac{3}{a_n}$$

Find the fixed points of this recursion, and investigate the limiting behavior of a_n when a_0 is not equal to a fixed point.

Solution　To find the fixed points, we need to solve

$$a = \frac{3}{a}$$

This equation is equivalent to $a^2 = 3$; hence, $a = \sqrt{3}$ or $a = -\sqrt{3}$. These are the two fixed points. If $a_0 = \sqrt{3}$, then $a_1 = \sqrt{3}$, $a_2 = \sqrt{3}$, and so on, and likewise, if $a_0 = -\sqrt{3}$, then $a_1 = -\sqrt{3}$, $a_2 = -\sqrt{3}$, and so on.

Let's start with a value that is not equal to one of the fixed points—say, $a_0 = 2$. Using the recursion, we find that

$$a_1 = \frac{3}{a_0} = \frac{3}{2}$$

$$a_2 = \frac{3}{a_1} = \frac{3}{\frac{3}{2}} = 3 \cdot \frac{2}{3} = 2$$

$$a_3 = \frac{3}{a_2} = \frac{3}{2}$$

$$a_4 = \frac{3}{a_3} = \frac{3}{\frac{3}{2}} = 3 \cdot \frac{2}{3} = 2$$

and so on. That is, successive terms alternate between 2 and 3/2. Let's try another initial value, say, $a_0 = -3$. Then

$$a_1 = \frac{3}{a_0} = \frac{3}{-3} = -1$$

$$a_2 = \frac{3}{a_1} = \frac{3}{-1} = -3$$

$$a_3 = \frac{3}{a_2} = \frac{3}{-3} = -1$$

$$a_4 = \frac{3}{a_3} = \frac{3}{-1} = -3$$

and so on. Successive terms now alternate between -3 and -1. Alternating between two values, one of which is the initial value, happens with any initial value that is not one of the fixed points. Specifically, we have

$$a_1 = \frac{3}{a_0} \quad \text{and} \quad a_2 = \frac{3}{a_1} = \frac{3}{\frac{3}{a_0}} = a_0$$

Thus, a_3 is the same as a_1, a_4 is the same as a_2 and hence a_0, and so on. ■

The last two examples illustrate that fixed points are only *candidates* for limits and that, depending on the initial condition, the sequence $\{a_n\}$ may or may not converge to a given fixed point. If we know, however, that a sequence $\{a_n\}$ does converge, then the limit of the sequence must be one of the fixed points.

There is a graphical method for finding fixed points, which we will mention briefly here: If the recursion is of the form $a_{n+1} = g(a_n)$, then a fixed point satisfies $a = g(a)$. This suggests that if we graph $y = g(x)$ and $y = x$ in the same coordinate system, then fixed points are located where the two graphs intersect, as shown in Figure 2.13.

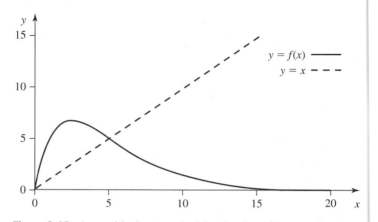

Figure 2.13 A graphical way to find fixed points. (See text for explanation.)

We will return to the relationship between fixed points and limits in Section 5.6, where we will learn methods that allow us to determine whether a sequence converges to a particular fixed point.

Section 2.2 Problems

■ **2.2.1**

In Problems 1–16, determine the values of the sequence $\{a_n\}$ for $n = 0, 1, 2, \ldots, 5$.

1. $a_n = n$

2. $a_n = 3n^2$

3. $a_n = \dfrac{1}{n+2}$

4. $f(n) = \dfrac{1}{1+n^2}$

5. $f(n) = \dfrac{1}{(1+n)^2}$

6. $a_n = \dfrac{1}{\sqrt{n+1}}$

7. $f(n) = (n+1)^2$

8. $f(n) = \sqrt{n+4}$

9. $a_n = (-1)^n n$

10. $a_n = \dfrac{(-1)^n}{(n+1)^2}$

11. $a_n = \dfrac{n^2}{n+1}$

12. $a_n = n^3\sqrt{n+1}$

13. $f(n) = e^{\sqrt{n}}$

14. $f(n) = 3e^{-0.1n}$

15. $f(n) = \left(\dfrac{1}{3}\right)^n$

16. $f(n) = 2^{0.2n}$

In Problems 17–24, find the next four values of the sequence $\{a_n\}$ on the basis of the values of $a_0, a_1, a_2, \ldots, a_5$.

17. $1, 2, 3, 4, 5$

18. $0, 1, \sqrt{2}, \sqrt{3}, \sqrt{4}$

19. $1, \dfrac{1}{4}, \dfrac{1}{9}, \dfrac{1}{16}, \dfrac{1}{25}$

20. $-1, \dfrac{1}{4}, -\dfrac{1}{9}, \dfrac{1}{16}, -\dfrac{1}{25}$

21. $\dfrac{1}{2}, \dfrac{2}{3}, \dfrac{3}{4}, \dfrac{4}{5}, \dfrac{5}{6}$

22. $\dfrac{1}{5}, \dfrac{4}{10}, \dfrac{9}{17}, \dfrac{16}{26}, \dfrac{25}{37},$

23. $\sqrt{1+e}, \sqrt{2+e^2}, \sqrt{3+e^3}, \sqrt{4+e^4}, \sqrt{5+e^5}$

24. $\sin\dfrac{\pi}{2}, -\sin\dfrac{\pi}{4}, \sin\dfrac{\pi}{6}, -\sin\dfrac{\pi}{8}, \sin\dfrac{\pi}{10}$

In Problems 25–36, find an expression for a_n on the basis of the values of $a_0, a_1, a_2, \ldots$.

25. $0, 1, 2, 3, 4, \ldots$

26. $0, 2, 4, 6, 8, \ldots$

27. $1, 2, 4, 8, 16, \ldots$

28. $1, 3, 5, 7, 9, \ldots$

29. $1, \dfrac{1}{3}, \dfrac{1}{9}, \dfrac{1}{27}, \dfrac{1}{81}, \ldots$

30. $\dfrac{1}{3}, \dfrac{2}{5}, \dfrac{3}{7}, \dfrac{4}{9}, \dfrac{5}{11}, \ldots$

31. $-1, 2, -3, 4, -5, \ldots$

32. $2, -4, 6, -8, 10, \ldots$

33. $-\dfrac{1}{2}, \dfrac{1}{3}, -\dfrac{1}{4}, \dfrac{1}{5}, -\dfrac{1}{6}, \ldots$

34. $\dfrac{1}{2}, -\dfrac{1}{8}, \dfrac{1}{18}, -\dfrac{1}{32}, \dfrac{1}{50}, \ldots$

35. $\sin(\pi), \sin(2\pi), \sin(3\pi), \sin(4\pi), \sin(5\pi), \ldots$

36. $-\cos\dfrac{\pi}{2}, \cos\dfrac{\pi}{4}, -\cos\dfrac{\pi}{6}, \cos\dfrac{\pi}{8}, -\cos\dfrac{\pi}{10}, \ldots$

■ **2.2.2**

In Problems 37–44, write the first five terms of the sequence $\{a_n\}$, $n = 0, 1, 2, 3, \ldots$, and find $\lim_{n\to\infty} a_n$.

37. $a_n = \dfrac{1}{n+2}$

38. $a_n = \dfrac{2}{n+1}$

39. $a_n = \dfrac{n}{n+1}$

40. $a_n = \dfrac{2n}{n+2}$

41. $a_n = \dfrac{1}{n^2+1}$

42. $a_n = \dfrac{1}{\sqrt{n+1}}$

43. $a_n = \dfrac{(-1)^n}{n+1}$ **44.** $a_n = \dfrac{(-1)^n}{n^3+3}$

In Problems 45–52, write the first five terms of the sequence $\{a_n\}$, $n = 0, 1, 2, 3, \ldots$, and determine whether $\lim_{n\to\infty} a_n$ exists. If the limit exists, find it.

45. $a_n = \dfrac{n^2}{n+1}$ **46.** $a_n = \dfrac{n^3}{n+1}$

47. $a_n = \sqrt{n}$ **48.** $a_n = n^2$

49. $a_n = 2^n$ **50.** $a_n = \left(\dfrac{1}{2}\right)^n$

51. $a_n = 3^n$ **52.** $a_n = \left(\dfrac{1}{3}\right)^n$

Formal Definition of Limits: In Problems 53–64, $\lim_{n\to\infty} a_n = a$. Find the limit a, and determine N so that $|a_n - a| < \epsilon$ for all $n > N$ for the given value of ϵ.

53. $a_n = \dfrac{1}{n}, \epsilon = 0.01$ **54.** $a_n = \dfrac{1}{n}, \epsilon = 0.02$

55. $a_n = \dfrac{1}{n^2}, \epsilon = 0.01$ **56.** $a_n = \dfrac{1}{n^2}, \epsilon = 0.001$

57. $a_n = \dfrac{1}{\sqrt{n}}, \epsilon = 0.1$ **58.** $a_n = \dfrac{1}{\sqrt{n}}, \epsilon = 0.05$

59. $a_n = \dfrac{(-1)^n}{n}, \epsilon = 0.01$ **60.** $a_n = \dfrac{(-1)^n}{n}, \epsilon = .001$

61. $a_n = \dfrac{n}{n+1}, \epsilon = 0.01$ **62.** $a_n = \dfrac{n+1}{n}, \epsilon = .05$

63. $a_n = \dfrac{n^2}{n^2+1}, \epsilon = 0.01$ **64.** $a_n = \dfrac{n^2}{n^2+1}, \epsilon = .001$

Formal Definition of Limits: In Problems 65–70, use the formal definition of limits to show that $\lim_{n\to\infty} a_n = a$; that is, find N such that for every $\epsilon > 0$, there exists an N such that $|a_n - a| < \epsilon$ whenever $n > N$.

65. $\lim_{n\to\infty} \dfrac{1}{n} = 0$ **66.** $\lim_{n\to\infty} \dfrac{1}{n+1} = 0$

67. $\lim_{n\to\infty} \dfrac{1}{n^2} = 0$ **68.** $\lim_{n\to\infty} \dfrac{1}{n^2+1} = 0$

69. $\lim_{n\to\infty} \dfrac{n+1}{n} = 1$ **70.** $\lim_{n\to\infty} \dfrac{n}{n+1} = 1$

In Problems 71–82, use the limit laws to determine $\lim_{n\to\infty} a_n = a$.

71. $\lim_{n\to\infty} \left(\dfrac{1}{n} + \dfrac{1}{n^2}\right)$ **72.** $\lim_{n\to\infty} \left(\dfrac{2}{n} - \dfrac{1}{n^2+1}\right)$

73. $\lim_{n\to\infty} \left(\dfrac{n+1}{n}\right)$ **74.** $\lim_{n\to\infty} \left(\dfrac{2n-3}{n}\right)$

75. $\lim_{n\to\infty} \left(\dfrac{n^2+1}{n^2}\right)$ **76.** $\lim_{n\to\infty} \left(\dfrac{3n^2-5}{n^2}\right)$

77. $\lim_{n\to\infty} \left(\dfrac{n+1}{n^2-1}\right)$ **78.** $\lim_{n\to\infty} \left(\dfrac{n+2}{n^2-4}\right)$

79. $\lim_{n\to\infty} \left[\left(\dfrac{1}{3}\right)^n + \left(\dfrac{1}{2}\right)^n\right]$ **80.** $\lim_{n\to\infty} \left(3^{-n} - 4^{-n}\right)$

81. $\lim_{n\to\infty} \dfrac{n+2^{-n}}{n}$ **82.** $\lim_{n\to\infty} \dfrac{n+3^{-n}}{n}$

■ 2.2.3

In Problems 83–92, the sequence $\{a_n\}$ is recursively defined. Compute a_n for $n = 1, 2, \ldots, 5$.

83. $a_{n+1} = 2a_n, a_0 = 1$ **84.** $a_{n+1} = 2a_n, a_0 = 3$

85. $a_{n+1} = 3a_n - 2, a_0 = 1$ **86.** $a_{n+1} = 3a_n - 2, a_0 = 2$

87. $a_{n+1} = 4 - 2a_n, a_0 = 5$ **88.** $a_{n+1} = 4 - 2a_n, a_0 = \dfrac{4}{3}$

89. $a_{n+1} = \dfrac{a_n}{1+a_n}, a_0 = 1$ **90.** $a_{n+1} = \dfrac{a_n}{a_n+3}, a_0 = 2$

91. $a_{n+1} = a_n + \dfrac{1}{a_n}, a_0 = 1$ **92.** $a_{n+1} = 5a_n - \dfrac{5}{a_n}, a_0 = 2$

In Problems 93–102, the sequence $\{a_n\}$ is recursively defined. Find all fixed points of $\{a_n\}$.

93. $a_{n+1} = \dfrac{1}{2}a_n + 2$ **94.** $a_{n+1} = \dfrac{1}{3}a_n + \dfrac{4}{3}$

95. $a_{n+1} = \dfrac{2}{5}a_n - \dfrac{9}{5}$ **96.** $a_{n+1} = -\dfrac{1}{3}a_n + \dfrac{1}{4}$

97. $a_{n+1} = \dfrac{4}{a_n}$ **98.** $a_{n+1} = \dfrac{7}{a_n}$

99. $a_{n+1} = \dfrac{2}{a_n+2}$ **100.** $a_{n+1} = \dfrac{3}{a_n-2}$

101. $a_{n+1} = \sqrt{5a_n}$ **102.** $a_{n+1} = \sqrt{7a_n}$

In Problems 103–110, assume that $\lim_{n\to\infty} a_n$ exists. Find all fixed points of $\{a_n\}$, and use a table or other reasoning to guess which fixed point is the limiting value for the given initial condition.

103. $a_{n+1} = \dfrac{1}{2}(a_n + 5), a_0 = 1$

104. $a_{n+1} = \dfrac{1}{3}\left(a_n + \dfrac{1}{9}\right), a_0 = 1$

105. $a_{n+1} = \sqrt{2a_n}, a_0 = 1$

106. $a_{n+1} = \sqrt{2a_n}, a_0 = 0$

107. $a_{n+1} = 2a_n(1 - a_n), a_0 = 0.1$

108. $a_{n+1} = 2a_n(1 - a_n), a_0 = 0$

109. $a_{n+1} = \dfrac{1}{2}\left(a_n + \dfrac{4}{a_n}\right), a_0 = 1$

110. $a_{n+1} = \dfrac{1}{2}\left(a_n + \dfrac{9}{a_n}\right), a_0 = -1$

■ 2.3 More Population Models

The material presented in this section will be revisited in Section 5.6. Section 2.3 can be postponed until then.

An important biological application of sequences consists of models of seasonally breeding populations with nonoverlapping generations where the population size at one generation depends only on the population size of the previous generation. The exponential growth model of Section 2.1 fits into this category. We denote the population size at time t by $N(t)$ or N_t, $t = 0, 1, 2, \ldots$. To model how the population size at generation $t + 1$ is related to the population size at generation

t, we write

$$N_{t+1} = f(N_t) \tag{2.10}$$

where the function f describes the density dependence of the population dynamics.

As explained in Section 2.2, a recursion of the form (2.10) is called a first-order recursion because, to obtain the population size at time $t+1$, only the population size at the previous time step t needs to be known. A recursion is also called a **difference equation** or an **iterated map**. [The name *difference equation* comes from writing the dynamics in the form $N_{t+1} - N_t = g(N_t)$, which allows us to track population size changes from one time step to the next. The name *iterated map* refers to the recursive definition.]

When we study population models, we are frequently interested in asking questions about the long-term behavior of the population, such as, Will the population size reach a constant value? Will it oscillate predictably? or Will it fluctuate widely without any recognizable patterns? We will explore these questions in the examples that follow, in which we will see that discrete-time population models show very rich and complex behavior.

■ 2.3.1 Restricted Population Growth: The Beverton–Holt Recruitment Curve

In Section 2.1, we discussed exponential growth defined by the recursion

$$N_{t+1} = RN_t \quad \text{with } N_0 = \text{population size at time } 0$$

When $R > 1$, the population size will grow indefinitely, provided that $N_0 > 0$. We can understand why this happens if we look at the parent–offspring ratio for $N_t > 0$, N_t/N_{t+1}, which is equal to the constant $1/R$. This means that, regardless of the current population density, the number of offspring per parent is a constant. Such growth, called density-independent growth, is biologically unrealistic. As the size of the population increases, individuals will start to compete with each other for resources, such as food or nesting sites, thereby reducing population growth. We call population growth that depends on population density *density-dependent growth*.

To find a model that incorporates a reduction in growth when the population size gets large, we start with the ratio of successive population sizes in the exponential growth model and assume that N_0 is positive, so that all successive population sizes are positive:

$$\frac{N_t}{N_{t+1}} = \frac{1}{R} \tag{2.11}$$

The ratio N_t/N_{t+1} is a constant. If we graphed this ratio as a function of the current population size N_t, we would obtain a horizontal line in a coordinate system in which N_t is on the horizontal axis and the ratio N_t/N_{t+1} is on the vertical axis (Figure 2.7). Note that as long as the parent–offspring ratio N_t/N_{t+1} is less than 1, the population size increases, since there are fewer parents than offspring. Once the ratio is equal to 1, the population size stays the same from one time step to the next. When the ratio is greater than 1, the population size decreases.

To model the reduction in growth when the population size gets larger, we drop the assumption that the parent–offspring ratio N_t/N_{t+1} is constant and assume instead that the ratio is an increasing function of the population size N_t. That is, we replace the constant $1/R$ in (2.11) by a function that increases with N_t. The simplest such function is linear. Graphically, this is a straight line with positive slope (Figure 2.14). To compare the model with density dependence with the exponential growth model (2.11), we assume that the two models agree when the population sizes are very small. We can achieve this agreement by assuming that the line corresponding to density-dependent growth goes through the point $(0, 1/R)$. To make sure that the population grows at low densities, we also assume that $R > 1$. The population density where the parent–offspring ratio is equal to 1 is of particular importance, since it corresponds to the population size, which does not change from one generation to the next. We call this population size the **carrying capacity** and denote it by K, where

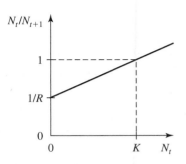

Figure 2.14 Density-dependent growth: The parent–offspring ratio increases as a function of current population size. (Note that this ratio is defined only for $N_t > 0$.)

K is a positive constant. We thus require that the line corresponding to density-dependent growth connect the points $(0, 1/R)$ and $(K, 1)$ in a graph in which N_t is on the horizontal axis and the ratio N_t/N_{t+1} is on the vertical axis (Figure 2.14).

The straight line in Figure 2.14 has slope $(1 - 1/R)/K$ and vertical-axis intercept $1/R$, which yields the equation

$$\frac{N_t}{N_{t+1}} = \frac{1}{R} + \frac{1 - \frac{1}{R}}{K} N_t$$

We solve this equation for N_{t+1} to obtain a recursion. Multiplying both sides by N_{t+1} yields

$$N_t = N_{t+1}\left(\frac{1}{R} + \frac{1 - \frac{1}{R}}{K} N_t\right)$$

which allows us to isolate N_{t+1}:

$$N_{t+1} = \frac{N_t}{\frac{1}{R} + \frac{1 - \frac{1}{R}}{K} N_t}$$

We next simplify the right-hand expression by multiplying numerator and denominator by R:

$$N_{t+1} = \frac{R N_t}{1 + \frac{R-1}{K} N_t} \tag{2.12}$$

This recursion is known as the **Beverton–Holt recruitment curve**.

Using results from Section 2.2, we can compute the fixed points of (2.12). Solving

$$N = \frac{R N}{1 + \frac{R-1}{K} N}$$

for N, we immediately find that $N = 0$. If $N \neq 0$, we divide both sides by N, producing

$$1 = \frac{R}{1 + \frac{R-1}{K} N}$$

Algebraic manipulation then yields

$$1 + \frac{R-1}{K} N = R \quad \text{or} \quad \frac{R-1}{K} N = R - 1$$

from which we solve for N to obtain

$$N = \frac{R-1}{\frac{R-1}{K}} = (R-1)\frac{K}{R-1} = K$$

We thus have two fixed points when $R > 1$: the fixed point $N = 0$, which we call **trivial**, since it corresponds to the absence of the population, and the fixed point $N = K$, which we call **nontrivial**, since it corresponds to a positive population size.

In Figure 2.15, we set $K = 20$ and $R = 1.4$ and plot N_t as a function of t for three different initial population sizes. For clarity, we include the lines that connect successive population sizes. We see from the figure that if $N_0 > 0$, then N_t will eventually approach $K = 20$. (If $N_0 = K$, then $N_t = K$ for all $t = 1, 2, 3, \ldots$, since K is a fixed point.) This is the reason for calling K the carrying capacity. On the basis of on Figure 2.15, we conclude that, when $K > 0$, $R > 1$, and $N_0 > 0$, we have

$$\lim_{t \to \infty} N_t = K$$

At this point, we need to rely on graphs and tables to investigate the long-term behavior of the population. This is a serious limitation, since it restricts our investigations to specific parameter values and we cannot then explore all possible parameter

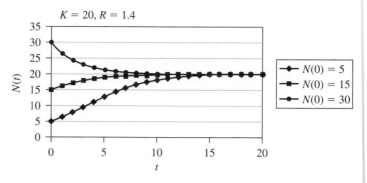

Figure 2.15 The population sizes N_t when $K = 20$ and $R = 1.4$ in the Beverton–Holt recruitment model for different initial population densities.

values. It turns out that this example has the same qualitative behavior for all R and K, provided that $R > 1$ and $K > 0$. In Section 5.6, we will learn analytical methods that will allow us to make general statements (like the one in the previous sentence) about the behavior of discrete-time population models such that that behavior will not depend on tables and graphs. In the next subsection, we will see an example where the behavior depends strongly on the choice of parameters.

■ 2.3.2 The Discrete Logistic Equation

The most popular discrete-time single-species model is the discrete logistic equation, whose recursion is given by

$$N_{t+1} = N_t \left[1 + R \left(1 - \frac{N_t}{K} \right) \right] \tag{2.13}$$

where R and K are positive constants. R is called the **growth parameter** and K is called the **carrying capacity**. The analysis that folllows will explain the terminology. This model of population growth exhibits very complicated dynamics, described in an influential review paper by Robert May (1976).

Before we illustrate its behavior, we will rewrite the model in what is called the *canonical form*. The advantage of this form is that the resulting recursion will be simpler. The algebraic steps presented next are not obvious, but will lead to the canonical form of the discrete logistic equation. We write

$$\begin{aligned} N_{t+1} &= N_t \left[1 + R \left(1 - \frac{N_t}{K} \right) \right] \\ &= N_t \left[1 + R - \frac{R}{K} N_t \right] \\ &= N_t (1 + R) \left[1 - \frac{R}{K(1+R)} N_t \right] \end{aligned}$$

Now, dividing by $1 + R$ yields

$$\frac{1}{1+R} N_{t+1} = N_t \left[1 - \frac{R}{K(1+R)} N_t \right]$$

Let's multiply both sides by R/K (you'll see why in a moment):

$$\frac{R}{K(1+R)} N_{t+1} = \frac{R}{K} N_t \left[1 - \frac{R}{K(1+R)} N_t \right] \tag{2.14}$$

If we define the new variable as

$$x_t = \frac{R}{K(1+R)} N_t \tag{2.15}$$

then

$$\frac{R}{K(1+R)}N_{t+1} = x_{t+1} \qquad \text{and} \qquad \frac{R}{K}N_t = (1+R)x_t$$

and (2.14) becomes

$$x_{t+1} = (1+R)x_t(1-x_t)$$

At this point, the new parameter $r = 1 + R$ is customarily introduced. Note that $r > 1$, since $R > 0$. We thus arrive at the canonical form of the logistic recursion:

$$x_{t+1} = rx_t(1-x_t) \tag{2.16}$$

The advantage of this form is threefold: (1) The recursion (2.16) looks simpler than the original recursion (2.13); (2) instead of two parameters (R and K), there is just one (r); and (3) the quantity x_t is *dimensionless*. The last point needs some explanation. The original variable N_t has units (or dimension) of number of individuals; the parameter K has the same units. Dividing N_t by K in (2.15), we see that the units cancel and we say that the quantity x_t is dimensionless. [The parameter R does not have a dimension, so multiplying N_t/K by $R/(1+R)$ does not introduce any additional units.] A dimensionless variable has the advantage that it has the same numerical value regardless of what the units of measurement are in the original variable. (See Problems 31–34.) The process of making a quantity dimensionless is called **nondimensionalization**.

Let's go back to the discrete logistic equation in its canonical form (2.16) and see what its behavior is. The function $f(x) = rx(1-x)$ is an upside-down parabola, since $r > 1$ (Figure 2.16). We see from the figure that if x is outside of the interval $(0, 1)$, $f(x)$ is nonpositive. Since $x_t = \frac{R}{K(1+R)}N_t$ [see (2.15)], and we want N_t to be positive (it is a population size, after all), we require x_t to be positive. This means that we need to ensure that $x_{t+1} = f(x_t)$ stays within the interval $(0, 1)$. The maximum value of $f(x)$ occurs at $x = 1/2$, and $f(1/2) = r/4$, so, in order to make sure that $f(x_t) \in (0, 1)$, we require that $r/4 < 1$, or $r < 4$. We already require that $r > 1$, since $R > 0$. To summarize, if $1 < r < 4$, then x_t stays within the interval $(0, 1)$ for all $t = 1, 2, 3, \ldots$, provided that $x_0 \in (0, 1)$. In what follows, we will therefore assume that $1 < r < 4$ and $x_0 \in (0, 1)$.

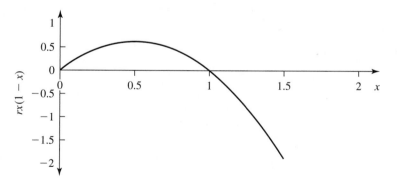

Figure 2.16 A graph of the discrete logistic equation in its canonical form. (Here, $r = 2.5$.)

We first compute fixed points of (2.16). We need to solve

$$x = rx(1-x)$$

Solving immediately yields the solution $x = 0$. If $x \neq 0$, we divide both sides by x and find that

$$1 = r(1-x), \qquad \text{or} \qquad x = 1 - \frac{1}{r}$$

(See Figure 2.17.) Provided that $r > 1$, both fixed points are in $[0, 1)$.

Figure 2.17 A graphical illustration of the fixed points of the discrete logistic equation in its canonical form. The fixed points are where the parabola and the line $y = x$ intersect.

We return to the original variable N_t for a moment to see what $x = 0$ and $x = 1 - 1/r$ mean in terms of N. Since $x = \frac{R}{K(1+R)}N$, the fixed point $x = 0$ corresponds to the fixed point $N = 0$, which is why we call $x = 0$ a trivial equilibrium. When $x = 1 - 1/r$, then, using $r = 1 + R$, we obtain

$$N = \frac{K(1+R)}{R}x = \frac{K(1+R)}{R}\left(1 - \frac{1}{1+R}\right)$$
$$= \frac{K(1+R)}{R}\frac{1+R-1}{1+R} = K$$

so $N = K$ is the other fixed point.

The long-term behavior of the discrete logistic equation is very complicated. We will go through the different cases by simply listing them. Later, in Section 5.6, we will be able, at least to some extent, to understand why this equation has such complicated behavior.

When $1 < r < 3$ and $x_0 \in (0, 1)$, x_t converges to the fixed point $1 - 1/r$ (Figure 2.18). Increasing r to a value between 3 and $3.449\ldots$, we learn that x_t settles into a cycle of period 2 (Figure 2.19). This means that, for large enough times, x_t will oscillate back and forth between a larger and a smaller value. For r between $3.449\ldots$ and $3.544\ldots$, the period doubles: A cycle of period 4 appears for large enough times. The population size now oscillates between the same four values (Figure 2.20). Increasing r continues to double the period: A cycle of period 8 is born when $r = 3.544\ldots$, a cycle of period 16 when $r = 3.564\ldots$, and a cycle of period 32 when $r = 3.567\ldots$. This doubling of the period continues until r reaches a value of about 3.57, when the population pattern becomes **chaotic** (Figure 2.21). The population dynamics seem to be random, although the rules are entirely deterministic! There is no regular pattern we can discern: x_t no longer oscillates between the same values; the dynamics are **aperiodic**. Furthermore, starting from ever so slightly different initial conditions quickly produces very different trajectories (Figure 2.22). This sensitivity to initial conditions is characteristic of chaotic behavior.

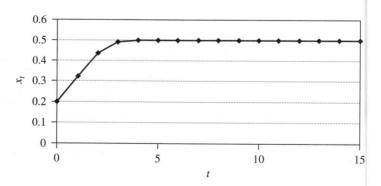

Figure 2.18 A graph of x_t as a function of t when $r = 2$.

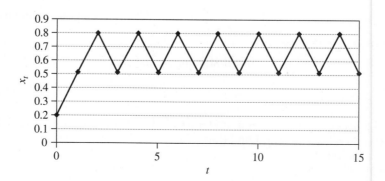

Figure 2.19 A graph of x_t as a function of t when $r = 3.2$.

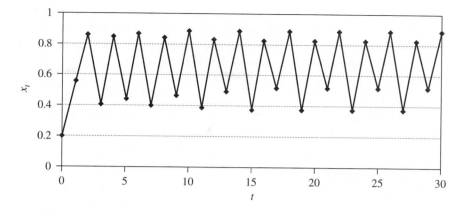

Figure 2.20 A graph of x_t as a function of t when $r = 3.52$.

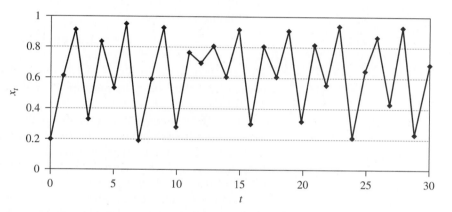

Figure 2.21 A graph of x_t as a function of t when $r = 3.8$.

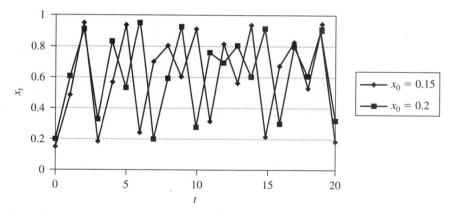

Figure 2.22 Graphs of x_t as a function of t when $r = 3.8$ for two different initial values of x_0.

To obtain biologically sensible results, we needed to restrict both r and x_0. The reason was that if $x_t > 1$, then x_{t+1} is negative. This situation can be easily remedied by changing the dynamics slightly. We discuss such a model in the next subsection.

■ 2.3.3 Ricker's Curve

The discrete logistic map has the biologically unrealistic feature that, unless one restricts the initial population size and the growth parameter, negative population sizes can occur. The reason is that the function $f(x) = rx(1 - x)$ takes on negative values for $x > 1$, so if $x_t > 1$, then $x_{t+1} < 0$. It is not difficult to avoid this problem.

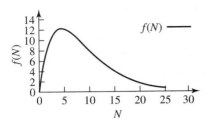

Figure 2.23 Ricker's curve when $R = 2.8$ and $K = 9$.

One example of an iterated map that has the same (desirable) properties as the logistic map but does not admit negative population sizes (provided that the population size at time 0 is positive) is **Ricker's curve**. The recursion, called the **Ricker logistic equation**, is given by

$$N_{t+1} = N_t \exp\left[R\left(1 - \frac{N_t}{K}\right)\right] \tag{2.17}$$

where R and K are positive parameters. As in the discrete logistic model, R is the growth parameter and K is the carrying capacity. The graph of Ricker's curve (Figure 2.23), $f(N_t) = N_t \exp\left[R\left(1 - \frac{N_t}{K}\right)\right]$, is positive for all $N_t > 0$, thus avoiding the problem of negative population sizes we encountered in the discrete logistic equation.

Fixed points of (2.17) satisfy

$$N = N \exp\left[R\left(1 - \frac{N}{K}\right)\right] \tag{2.18}$$

We find the trivial fixed point $N = 0$. If $N \neq 0$, we can divide both sides of (2.18) by N, obtaining

$$1 = \exp\left[R\left(1 - \frac{N}{K}\right)\right]$$

This equation holds if $R(1 - N/K) = 0$, from which it follows that $N = K$. The parameter K has the same meaning as in the discrete logistic equation, namely, that it is the carrying capacity.

The Ricker logistic equation shows the same complex dynamics as the discrete logistic map [convergence to the fixed point for small positive values of R (Figure 2.24), periodic behavior with the period doubling as R increases, and chaotic behavior for larger values of R (Figure 2.25)]. The values of R where the behavior changes are different than in the discrete logistic equation. For instance, the onset of chaos in the discrete logistic equation occurs for $R = 2.570\ldots$, whereas the onset of chaos in the Ricker logistic equation occurs for $R = 2.692\ldots$.

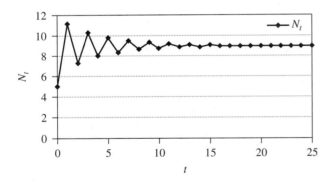

Figure 2.24 The population size N_t as a function of t when $R = 1.8$ and $K = 9$.

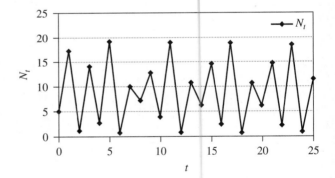

Figure 2.25 The population size N_t as a function of t when $R = 2.8$ and $K = 9$.

■ 2.3.4 Fibonacci Sequences

As a last example in this section, we will look at a **second-order difference equation**—an equation in which N_{t+1} depends on both N_t and N_{t-1}.

A famous example of a second-order difference equation is the **Fibonacci sequence**. The equation comes from the following problem posed in 1202 by Leonardo of Pisa (1175–1250), an Italian mathematician known by the name Fibonacci: How many pairs of rabbits are produced if each pair reproduces one pair of rabbits at age one month and another pair of rabbits at age two months and initially there is one pair of newborn rabbits?

If N_t denotes the number of newborn rabbit pairs at time t (measured in months), then at time 0, there is one pair of rabbits ($N_0 = 1$). At time 1, the pair of rabbits we started with is one month old and produces a pair of newborn rabbits, so $N_1 = 1$.

At time 2, there is one pair of two-month-old rabbits and one pair of one-month-old rabbits. Each pair produces a pair of newborn rabbits, so $N_2 = 2$. At time 3, our original pair of rabbits is now three months old and will stop reproducing; there is then one pair of two-month-old rabbits and two pairs of one-month-old rabbits. Since each pair of one-month-old and two-month-old rabbits produces a pair of newborn rabbits, at time $t = 3$ there will be $2 + 1 = 3$ newborn rabbits. More generally, to find the number of pairs of newborn rabbits, we need to add up the number of pairs of one-month-old rabbits and two-month-old rabbits. The one-month-old rabbits at time $t + 1$ were newborn rabbits at time t; the two-month-old rabbits were newborns at time $t - 1$. So the number of pairs of newborn rabbits at time $t + 1$ is

$$N_{t+1} = N_t + N_{t-1}, \ t = 1, 2, 3, \ldots \quad \text{with } N_0 = 1 \text{ and } N_1 = 1$$

Note that we need to specify N_t for $t = 0$ and $t = 1$ in order to be able to use the recursion. Using the recursion, we find the sequence

$$1, 1, 2, 3, 5, 8, 13, \ldots$$

We see that the number of newborn pairs of rabbits will go to infinity as t tends to infinity, so N_t will not converge to a finite limit. It turns out, however, that the ratio N_{t+1}/N_t converges (although we cannot show this here). We can find a candidate for the limiting value as follows: Start with the recursion

$$N_{t+1} = N_t + N_{t-1}$$

and divide both sides by N_t, yielding

$$\frac{N_{t+1}}{N_t} = 1 + \frac{N_{t-1}}{N_t} \tag{2.19}$$

If we now assume that

$$\lim_{t \to \infty} \frac{N_{t+1}}{N_t} = \lambda$$

(λ is the lowercase Greek letter lambda), which also implies that

$$\lim_{t \to \infty} \frac{N_t}{N_{t-1}} = \lambda$$

then

$$\lim_{t \to \infty} \frac{N_{t-1}}{N_t} = \lim_{t \to \infty} \frac{1}{\frac{N_t}{N_{t-1}}} = \frac{1}{\lim_{t \to \infty} \frac{N_t}{N_{t-1}}} = \frac{1}{\lambda}$$

Taking the limit as $t \to \infty$ in (2.19), we find that

$$\lambda = 1 + \frac{1}{\lambda}$$

which is $\lambda^2 = \lambda + 1$ after multiplying both sides by λ. We thus need to solve

$$\lambda^2 - \lambda - 1 = 0$$

The formula for solving quadratic equations yields

$$\lambda_{1,2} = \frac{1 \pm \sqrt{1+4}}{2} = \frac{1 \pm \sqrt{5}}{2}$$

One solution is positive, the other negative. Only the positive solution is relevant when $N_0 = N_1 = 1$, since then $N_{t+1}/N_t > 0$ for all $t = 0, 1, 2, \ldots$. The ratio

$$\frac{1 + \sqrt{5}}{2} \approx 1.61803$$

is the limit of N_{t+1}/N_t as $t \to \infty$ and is called the **golden mean**.

A rectangle whose sides bear the golden ratio is called a **golden rectangle**; it is thought to be the visually most pleasing proportion a rectangle can have. Golden rectangles were known to the ancient Greeks, who used them to scale the dimensions of their buildings (e.g., the Parthenon). Ratios of successive Fibonacci numbers can be found in nature as well. For instance, the florets on plants such as the sunflower run in spirals, and the ratios of the number of spirals running in opposite directions are often successive Fibonacci numbers.

Section 2.3 Problems

■ 2.3.1

In Problems 1–6, assume that the population growth is described by the Beverton–Holt recruitment curve with growth parameter R and carrying capacity K. For the given values of R and K, graph N_t/N_{t+1} as a function of N_t and find the recursion for the Beverton–Holt recruitment curve.

1. $R = 2, K = 15$

2. $R = 2, K = 50$

3. $R = 1.5, K = 40$

4. $R = 3, K = 120$

5. $R = 2.5, K = 90$

6. $R = 2, K = 150$

In Problems 7–12, assume that the population growth is described by the Beverton–Holt recruitment curve with growth parameter R and carrying capacity K. Find R and K.

7. $N_{t+1} = \dfrac{2N_t}{1 + N_t/20}$

8. $N_{t+1} = \dfrac{3N_t}{1 + 2N_t/40}$

9. $N_{t+1} = \dfrac{1.5N_t}{1 + 0.5N_t/30}$

10. $N_{t+1} = \dfrac{2N_t}{1 + N_t/200}$

11. $N_{t+1} = \dfrac{4N_t}{1 + N_t/150}$

12. $N_{t+1} = \dfrac{5N_t}{1 + N_t/20}$

In Problems 13–18, assume that the population growth is described by the Beverton–Holt recruitment curve with growth parameter R and carrying capacity K. Find all fixed points.

13. $N_{t+1} = \dfrac{4N_t}{1 + N_t/30}$

14. $N_{t+1} = \dfrac{3N_t}{1 + N_t/60}$

15. $N_{t+1} = \dfrac{2N_t}{1 + N_t/30}$

16. $N_{t+1} = \dfrac{2N_t}{1 + N_t/100}$

17. $N_{t+1} = \dfrac{3N_t}{1 + N_t/30}$

18. $N_{t+1} = \dfrac{5N_t}{1 + N_t/120}$

In Problems 19–24, assume that the population growth is described by the Beverton–Holt recruitment curve with growth parameter R and carrying capacity K. Find the population sizes for $t = 1, 2, \ldots,$ 5 and find $\lim_{t \to \infty} N_t$ for the given initial value N_0.

19. $R = 2, K = 10, N_0 = 2$

20. $R = 2, K = 20, N_0 = 5$

21. $R = 3, K = 15, N_0 = 1$

22. $R = 3, K = 30, N_0 = 0$

23. $R = 4, K = 40, N_0 = 3$

24. $R = 4, K = 20, N_0 = 10$

■ 2.3.2

In Problems 25–30, assume that the discrete logistic equation is used with parameters R and K. Write the equation in the canonical form $x_{t+1} = rx_t(1 - x_t)$, and determine r and x_t in terms of R, K, and N_t.

25. $R = 1, K = 10$

26. $R = 1, K = 20$

27. $R = 2, K = 15$

28. $R = 2, K = 20$

29. $R = 2.5, K = 30$

30. $R = 2.5, K = 50$

In Problems 31–34, we will investigate the advantage of dimensionless variables.

31. (a) Let N_t denote the population size at time t and let K denote the carrying capacity. Both quantities are measured in units of number of individuals. Show that $x_t = N_t/K$ is dimensionless.

(b) Let M_t denote the population size at time t and let L denote the carrying capacity. Assume that M_t and L are measured in units of 1000 individuals. Show that $y_t = M_t/L$ is dimensionless.

(c) How are N_t and M_t related? How are K and L related?

(d) Use (c) to find M_t and L if there are 20,000 individuals at time t and the carrying capacity is 5000.

(e) Show that, for the population size and the carrying capacity in (d), $x_t = y_t$.

32. To quantify the spatial structure of a plant population, it might be convenient to introduce a characteristic length scale. This length scale might be characterized by the average dispersal distance of the plant under study. Assume that the characteristic length scale is denoted by L. Denote by x the distance of seeds from their source. Define $z = x/L$. Find z if $x = 100$ cm and $L = 50$ cm, and show that z has the same value if x and L are measured in units of meters instead.

33. Suppose a bacterium divides every 20 minutes, which we call the characteristic time scale and denote by T. Let t be the time elapsed since the beginning of an experiment that involves this bacterium. Define $z = t/T$. Find z if $t = 120$ minutes, and show that z has the same value if t and T are measured in units of hours instead.

34. The time to the most recent common ancestor of a pair of individuals from a randomly mating population depends on the population size. Let t denote the time, measured in units of generations, to the most recent common ancestor, and let T be equal to N generations, where N is the population size of the randomly mating population. Define $z = t/T$. Show that z is dimensionless and that the value of z does not change, regardless of whether t and T are measured in units of generations or in units of, say, years. (Assume that one generation is equal to n years.)

In Problems 35–46, we investigate the behavior of the discrete logistic equation

$$x_{t+1} = rx_t(1 - x_t)$$

Compute x_t for $t = 0, 1, 2, \ldots, 20$ for the given values of r and x_0, and graph x_t as a function of t.

35. $r = 2, x_0 = 0.2$

36. $r = 2, x_0 = 0.1$

37. $r = 2, x_0 = 0.9$

38. $r = 2, x_0 = 0$

39. $r = 3.1, x_0 = 0.5$

40. $r = 3.1, x_0 = 0.1$

41. $r = 3.1, x_0 = 0.9$

42. $r = 3.1, x_0 = 0$

43. $r = 3.8, x_0 = 0.5$

44. $r = 3.8, x_0 = 0.1$

45. $r = 3.8, x_0 = 0.9$

46. $r = 3.8, x_0 = 0$

■ 2.3.3

In Problems 47–50, graph the Ricker's curve

$$N_{t+1} = N_t \exp\left[R\left(1 - \frac{N_t}{K}\right) \right]$$

in the N_t–N_{t+1} plane for the given values of R and K. Find the points of intersection of this graph with the line $N_{t+1} = N_t$.

47. $R = 2$, $K = 10$ **48.** $R = 3$, $K = 15$
49. $R = 2.5$, $K = 12$ **50.** $R = 4$, $K = 20$

In Problems 51–54, we investigate the behavior of the Ricker's curve

$$N_{t+1} = N_t \exp\left[R\left(1 - \frac{N_t}{K}\right) \right]$$

Compute N_t for $t = 1, 2, \ldots, 20$ for the given values of R, K, and N_0, and graph N_t as a function of t.

51. (a) $R = 1$, $K = 20$, $N_0 = 5$
(b) $R = 1$, $K = 20$, $N_0 = 10$
(c) $R = 1$, $K = 20$, $N_0 = 20$ **(d)** $R = 1$, $K = 20$, $N_0 = 0$
52. (a) $R = 1.8$, $K = 20$, $N_0 = 5$
(b) $R = 1.8$, $K = 20$, $N_0 = 10$
(c) $R = 1.8$, $K = 20$, $N_0 = 20$ **(d)** $R = 1.8$, $K = 20$, $N_0 = 0$
53. (a) $R = 2.1$, $K = 20$, $N_0 = 5$
(b) $R = 2.1$, $K = 20$, $N_0 = 10$
(c) $R = 2.1$, $K = 20$, $N_0 = 20$ **(d)** $R = 2.1$, $K = 20$, $N_0 = 0$

54. (a) $R = 2.8$, $K = 20$, $N_0 = 5$
(b) $R = 2.8$, $K = 20$, $N_0 = 10$
(c) $R = 2.8$, $K = 20$, $N_0 = 20$ **(d)** $R = 2.8$, $K = 20$, $N_0 = 0$

■ 2.3.4

55. Compute N_t and N_t/N_{t-1} for $t = 2, 3, 4, \ldots, 20$ when

$$N_{t+1} = N_t + N_{t-1}$$

with $N_0 = 1$ and $N_1 = 1$.

56. Compute N_t and N_t/N_{t-1} for $t = 2, 3, 4, \ldots, 20$ when

$$N_{t+1} = N_t + 2N_{t-1}$$

with $N_0 = 1$ and $N_1 = 1$.

57. In the text, an interpretation of the Fibonacci recursion

$$N_{t+1} = N_t + N_{t-1}$$

is given. Use a similar example to give an interpretation of the recursion

$$N_{t+1} = N_t + 2N_{t-1}$$

58. In the text, an interpretation of the Fibonacci recursion

$$N_{t+1} = N_t + N_{t-1}$$

is given. Use a similar example to give an interpretation of the recursion

$$N_{t+1} = 2N_t + N_{t-1}$$

Chapter 2 Key Terms

Discuss the following definitions and concepts:

1. Exponential growth
2. Growth constant
3. Fixed point
4. Equilibrium
5. Recursion
6. Solution
7. Density independence
8. Sequence
9. First-order recursion
10. Limit
11. Long-term behavior
12. Convergence, divergence
13. Limit laws
14. Difference equation
15. Beverton–Holt recruitment curve
16. Density dependence
17. Carrying capacity
18. Growth parameter
19. Discrete logistic equation
20. Nondimensionalization
21. Periodic behavior
22. Chaos
23. Ricker's curve
24. Fibonacci sequence
25. Golden mean

Chapter 2 Review Problems

In Problems 1–10, find the limits.

1. $\lim_{n\to\infty} 2^{-n}$
2. $\lim_{n\to\infty} 3^n$
3. $\lim_{n\to\infty} 40(1 - 4^{-n})$
4. $\lim_{n\to\infty} \dfrac{2}{1 + 2^{-n}}$
5. $\lim_{n\to\infty} a^n$ when $a > 1$
6. $\lim_{n\to\infty} a^n$ when $0 < a < 1$
7. $\lim_{n\to\infty} \dfrac{n(n+1)}{n^2 - 1}$
8. $\lim_{n\to\infty} \dfrac{n^2 + n - 6}{n - 2}$
9. $\lim_{n\to\infty} \dfrac{\sqrt{n}}{n+1}$
10. $\lim_{n\to\infty} \dfrac{n+1}{\sqrt{n}}$

In Problems 11–14, write a_n explicitly as a function of n on the basis of the first five terms of the sequence a_n, $n = 0, 1, 2, \ldots$.

11. $\dfrac{1}{2}, \dfrac{3}{4}, \dfrac{5}{6}, \dfrac{7}{8}, \dfrac{9}{10}$
12. $\dfrac{2}{2}, \dfrac{6}{4}, \dfrac{12}{8}, \dfrac{20}{16}, \dfrac{30}{32}$
13. $\dfrac{1}{2}, \dfrac{2}{5}, \dfrac{3}{10}, \dfrac{4}{17}, \dfrac{5}{26}$
14. $0, \dfrac{1}{3}, \dfrac{2}{4}, \dfrac{3}{5}, \dfrac{4}{6}$

15. Density-Dependent Growth The Beverton–Holt recruitment curve is given by the recursion

$$N_{t+1} = \frac{RN_t}{1 + \frac{R-1}{K}N_t}$$

where $R > 1$ and $K > 0$. When $N_0 > 0$, $\lim_{t\to\infty} N_t = K$ for all values of $R > 0$. To investigate how R affects the limiting behavior of N_t, find N_t for $t = 1, 2, 3, \ldots, 10$ for $K = 100$ and $N_0 = 20$ when **(a)** $R = 2$, **(b)** $R = 5$, and **(c)** $R = 10$, and plot N_t as a function of t for the three choices of R in one coordinate system.

In Problems 16–18, we discuss population models when the population size at time $t + 1$ depends not only on the population size at time t, but also on the growth conditions at time t, which may vary over time.

16. Temporally Varying Environment The recursion

$$N_{t+1} = R_t N_t$$

describes growth in a temporally varying environment if we interpret R_t as the growth parameter in generation t. A population was followed over 10 years and the population sizes were recorded each year. Use the data provided to find R_t for $t = 0, 1, 2, \ldots, 9$:

t	N_t
0	10
1	15.5
2	15.6
3	10.8
4	15.6
5	32.2
6	95.1
7	103.2
8	165.0
9	418.7
10	15.7

17. Temporally Varying Environment The recursion

$$N_{t+1} = R_t N_t$$

describes growth in a temporally varying environment if we interpret R_t as the growth parameter in generation t. A population was followed over 20 years and the population sizes were recorded every year. The following table provides the population size data and the inferred values of R_t for each of the 20 years:

t	N_t	R_t
0	10.0	2.78
1	27.8	0.29
2	8.10	0.43
3	3.49	0.25
4	0.87	2.90
5	2.52	1.67
6	4.21	1.17
7	4.94	0.69
8	3.39	1.45
9	4.92	1.13
10	5.56	0.08
11	0.45	0.88
12	0.40	2.69
13	1.06	0.36
14	0.38	0.08
15	0.03	2.34
16	0.07	2.13
17	0.15	2.20
18	0.34	2.80
19	0.94	0.29
20	0.28	1.22

The values of N_t indicate that the population heads toward extinction. The long-term behavior of the geometric mean of the growth parameter, denoted by $\hat{R}_t$ (read "R sub t hat"), is defined as

$$\hat{R}_t = \left(R_0 R_1 \cdots R_{t-1}\right)^{1/t}$$

and determines whether the population will go extinct. Specifically, if

$$\lim_{t \to \infty} \hat{R}_t < 1$$

then the population will go extinct. Compute $\hat{R}_t$ for $t = 1, 2, \ldots, 20$.

18. Temporally Varying Environment The recursion

$$N_{t+1} = R_t N_t$$

describes growth in a temporally varying environment if we interpret R_t as the growth parameter in generation t.
(a) Show that

$$N_t = (R_{t-1}R_{t-2}\cdots R_1 R_0)N_0$$

(b) The quantity $\hat{R}_t$ (read "R sub t hat"), defined as

$$\hat{R}_t = (R_{t-1}R_{t-2}\cdots R_1 R_0)^{1/t}$$

is called the **geometric mean**. Show that

$$N_t = (\hat{R}_t)^t N_0$$

(c) The **arithmetic mean** of a sequence of numbers $x_0, x_1, \ldots, x_{n-1}$ is defined as

$$\bar{x}_n = \frac{x_0 + x_1 + \cdots + x_{n-1}}{n}$$

Set $r_t = \ln R_t$ and show that

$$\bar{r}_t = \frac{\ln R_{t-1} + \ln R_{t-2} + \cdots + \ln R_0}{t}$$

(d) Use (c) to show that

$$N_t = N_0 e^{\bar{r}_t t}$$

19. Harvesting Model Let N_t denote the population size at time t, and assume that

$$N_{t+1} = (1-c)N_t \exp\left[R\left(1 - \frac{(1-c)N_t}{K}\right)\right]$$

where R and K are positive constants and c is the fraction harvested. Find N_t for $t = 1, 2, \ldots, 20$ when $R = 1$, $K = 100$, and $N_0 = 50$ for **(a)** $c = 0.1$, **(b)** $c = 0.5$, and **(c)** $c = 0.9$.

20. Harvesting Model Let N_t denote the population size at time t, and assume that

$$N_{t+1} = (1-c)N_t \exp\left[R\left(1 - \frac{(1-c)N_t}{K}\right)\right]$$

where R and K are positive constants and c is the fraction harvested. Find N_t for $t = 1, 2, \ldots, 20$ when $R = 3$, $K = 100$, and $N_0 = 50$ for **(a)** $c = 0.1$, **(b)** $c = 0.5$, and **(c)** $c = 0.9$.

Limits and Continuity

3

LEARNING OBJECTIVES

The two concepts of limits and continuity are fundamental to differential calculus. Specifically, in this chapter we will learn how to

- determine the value of a function $f(x)$ at $x = c$ as x approaches c, both from graphs and mathematical expressions defining the function;
- determine whether a function is continuous or discontinous at a point;
- identify where a function is continuous and where it is discontinuous;
- extract information from continuous functions on the basis of generic properties of such functions.

3.1 Limits

In Chapter 2, we discussed limits of the form $\lim_{n \to \infty} a_n$, where n took on integer values. In this chapter, we will consider limits of the form

$$\lim_{x \to c} f(x) \tag{3.1}$$

where x is now a continuously varying real variable that tends to a fixed value c (which may be finite or infinite). Let's look at an example that will motivate the need for limits of the form (3.1).

Population growth in populations with discrete breeding seasons (as in Chapter 2) can be described by the change in population size from generation to generation. By contrast, in populations that breed continuously, there is no natural time scale such as generations. Instead, we will look at how the population size changes over small time intervals. We denote the population size at time t by $N(t)$, where t is now varying continuously over the interval $[0, \infty)$. We will investigate how the population size changes during the interval $[t, t + h]$, where $h > 0$. The absolute change during this interval, denoted by ΔN, is

$$\Delta N = N(t + h) - N(t)$$

(The symbol Δ indicates that we are taking a difference.) To obtain the change relative to the length of the interval $[t, t + h]$, we divide ΔN by the length of the interval, denoted by Δt, which is $(t + h) - t = h$. We find that

$$\frac{\Delta N}{\Delta t} = \frac{N(t + h) - N(t)}{h}$$

This ratio is called the **average growth rate**.

We see from Figure 3.1 that $\Delta N / \Delta t$ is the slope of the secant line connecting the points $(t, N(t))$ and $(t + h, N(t + h))$. The average growth rate $\Delta N / \Delta t$ depends on the length of the interval Δt. This dependency is illustrated in Figure 3.2, where we see that the slopes of the two secant lines (lines 1 and 2) are different. But we also see that, as we choose smaller and smaller intervals, the secant lines converge to the tangent line at the point $(t, N(t))$ of the graph of $N(t)$ (line 3).

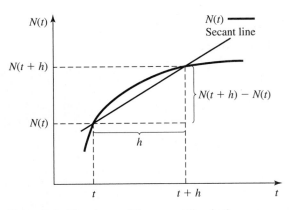

Figure 3.1 The slope of the secant line is the average growth rate.

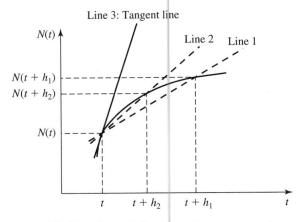

Figure 3.2 The slope of the secant line converges to the slope of the tangent line as the length of the interval $[t, t + h]$ shrinks to 0.

The slope of the tangent line is called the **instantaneous growth rate** and is a convenient way to describe the growth of a continuously breeding population. To obtain this quantity, we need to take a limit; that is, we need to shrink the length of the interval $[t, t + h]$ to 0 by letting h tend to 0. We express this operation as

$$\lim_{h \to 0} \frac{N(t + h) - N(t)}{h} \tag{3.2}$$

In (3.2), we take a limit of a quantity in which a continuously varying variable, namely, h, approaches some fixed value, namely, 0. This is a limit of the form (3.1).

■ 3.1.1 An Informal Discussion of Limits

> **Definition** The "**limit of $f(x)$, as x approaches c, is equal to L**" means that $f(x)$ becomes arbitrarily close to L whenever x is sufficiently close (but not equal) to c. We denote this statement by
>
> $$\lim_{x \to c} f(x) = L$$
>
> or $f(x) \to L$ as $x \to c$.

If $\lim_{x \to c} f(x) = L$ and L is a finite number, we say that the limit **exists** and that $f(x)$ **converges** to L. If the limit does not exist, we say that $f(x)$ **diverges** as x tends to c.

Note that we say that we choose x close, but not equal, to c. That is, when finding the limit of $f(x)$ as x approaches c, we do not simply plug c into $f(x)$. In fact, we will see examples in which $f(x)$ is not even defined at $x = c$. The value of $f(c)$ is irrelevant when we compute the value of $\lim_{x \to c} f(x)$.

Furthermore, when we say "x approaches c," we mean that x approaches c in any fashion. When x approaches c from only one side, we use the notation

$$\lim_{x \to c^+} f(x) \quad \text{when } x \text{ approaches } c \text{ from the right}$$

$$\lim_{x \to c^-} f(x) \quad \text{when } x \text{ approaches } c \text{ from the left}$$

and talk about right-handed and left-handed limits, respectively. The notation "$x \to c^+$" indicates that when x approaches c from the right, values of x are greater than c, and when x approaches c from the left ("$x \to c^-$"), values of x are less than than c. Let's look at some examples.

Limits That Exist

EXAMPLE 1

Define $f(x) = x^2$, $x \in \mathbf{R}$. Find

$$\lim_{x \to 2} f(x)$$

Solution

The graph of $f(x) = x^2$ (see Figure 3.3) immediately shows that the limit of x^2 is 4 as x approaches 2 (from either side). We also suspect this from the following table, where we compute values of x^2 for x close, but not equal, to 2:

x	x^2	x	x^2
1.9	3.61	2.1	4.41
1.99	3.9601	2.01	4.0401
1.999	3.996001	2.001	4.004001
1.9999	3.99960001	2.0001	4.00040001

Note that in the left half of the table we approach $x = 2$ from the left ($x \to 2^-$), whereas in the right half of the table we approach x from the right ($x \to 2^+$).
We find that

$$\lim_{x \to 2} x^2 = 4$$

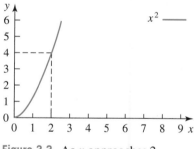

Figure 3.3 As x approaches 2, $f(x) = x^2$ approaches 4.

Since this limit is a finite number, we say that the limit exists and that x^2 converges to 4 as x tends to 2. The fact that $f(x) = x^2$ at $x = 2$ is 4 as well is a nice property that will be introduced and named later. Not all functions are like that. ■

EXAMPLE 2

(a) Define

$$g(x) = \begin{cases} x^2 & \text{if } x \neq 2 \\ 5 & \text{if } x = 2 \end{cases}$$

Find

$$\lim_{x \to 2} g(x)$$

(b) Define $h(x) = x^2$, $x \neq 2$. Find

$$\lim_{x \to 2} h(x)$$

Solution

(a) In computing the value of $\lim_{x \to 2} g(x)$, the value of $g(2)$ is irrelevant. We find, as in Example 1, that

$$\lim_{x \to 2} g(x) = \lim_{x \to 2} x^2 = 4$$

We note that $\lim_{x \to 2} g(x) \neq g(2)$.

(b) To obtain the limit of $h(x)$ as $x \to 2$, the function $h(x)$ need not be defined at $x = 2$. We obtain

$$\lim_{x \to 2} h(x) = \lim_{x \to 2} x^2 = 4$$ ■

EXAMPLE 3 Find

$$\lim_{x \to 3} \frac{x^2 - 9}{x - 3}$$

Solution In the previous two examples, determining the limits did not require any calculations. This is not the case here, since both numerator and denominator tend to 0 as $x \to 3$. We define $f(x) = \dfrac{x^2 - 9}{x - 3}$, $x \neq 3$. Since the denominator of $f(x)$ is equal to 0 when $x = 3$, we exclude $x = 3$ from the domain. When $x \neq 3$, we can simplify the expression, namely,

$$f(x) = \frac{x^2 - 9}{x - 3} = \frac{(x - 3)(x + 3)}{x - 3} = x + 3 \qquad \text{for } x \neq 3$$

We were able to cancel the term $x - 3$ because $x - 3 \neq 0$ for $x \neq 3$ and we assumed that $x \neq 3$. (If we allowed $x = 3$, then canceling $x - 3$ would mean dividing by 0.) The graph of $f(x)$ is a straight line with one point deleted at $x = 3$. (See Figure 3.4.) Taking the limit, we find that

$$\lim_{x \to 3} \frac{x^2 - 9}{x - 3} = \lim_{x \to 3} (x + 3)$$

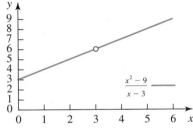

Figure 3.4 The graph of $f(x) = \frac{x^2-9}{x-3}$ is a straight line with the point $(3, 6)$ removed.

Now, using either the graph of $y = x + 3$ for $x \neq 3$ or a table, we suspect that

$$\lim_{x \to 3} (x + 3) = 6$$

We conclude that $\lim_{x \to 3} f(x)$ exists and that $f(x)$ converges to 6 as x tends to 3. Note that $f(x)$ is not defined at $x = 3$. ■

One-Sided Limits To compute one-sided limits, we use the notation

$$\lim_{x \to c^+} f(x) \quad \text{when } x \text{ approaches } c \text{ from the right}$$

$$\lim_{x \to c^-} f(x) \quad \text{when } x \text{ approaches } c \text{ from the left}$$

that was introduced previously.

EXAMPLE 4 Find

$$\lim_{x \to 0} e^{-|x|}$$

Solution We set

$$f(x) = e^{-|x|} = \begin{cases} e^{-x} & \text{for } x > 0 \\ e^{x} & \text{for } x < 0 \end{cases}$$

Figure 3.5 indicates that $\lim_{x \to 0^+} f(x) = \lim_{x \to 0^+} e^{-x} = 1$ and $\lim_{x \to 0^-} f(x) = \lim_{x \to 0^-} e^{x} = 1$. We therefore conclude that

$$\lim_{x \to 0} e^{-|x|} = 1$$ ■

EXAMPLE 5 Find

$$\lim_{x \to 0^+} \frac{|x|}{x} \qquad \text{and} \qquad \lim_{x \to 0^-} \frac{|x|}{x}$$

Solution We set $f(x) = \dfrac{|x|}{x}$, $x \neq 0$. Since $|x| = x$ for $x \geq 0$ and $|x| = -x$ for $x \leq 0$, we find that

$$f(x) = \frac{|x|}{x} = \begin{cases} +1 & \text{for } x > 0 \\ -1 & \text{for } x < 0 \end{cases}$$

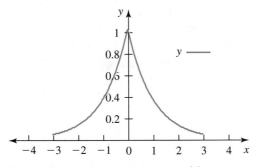

Figure 3.5 The graph of $f(x) = e^{-|x|}$ in Example 4.

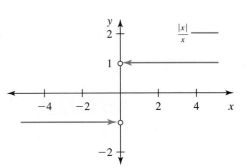

Figure 3.6 The graph of $\frac{|x|}{x}$ in Example 5: The function is not defined at $x = 0$.

The graph of $f(x)$ is shown in Figure 3.6. This function can be used to model a switch, where the value of $f(x)$ switches from -1 to $+1$ as x goes through 0. We see that $f(x)$ converges to 1 as x tends to 0 from the right and that $f(x)$ converges to -1 as x tends to 0 from the left. We can write this property as

$$\lim_{x \to 0^+} \frac{|x|}{x} = 1 \quad \text{and} \quad \lim_{x \to 0^-} \frac{|x|}{x} = -1$$

and observe that the one-sided limits exist.   ■

In Example 5, we computed one-sided limits. Since the right-hand limit differs from the left-hand limit, we conclude that

$$\lim_{x \to 0} \frac{|x|}{x} \quad \text{does not exist}$$

because the phrase "x approaches 0" (or, in symbols, $\lim_{x \to 0}$) means that x approaches 0 in any fashion.

More Limits That Do Not Exist

EXAMPLE 6   Find

$$\lim_{x \to 0} \frac{1}{x^2}$$

Solution

A graph of $f(x) = 1/x^2, x \neq 0$, reveals that $f(x)$ increases without bound as $x \to 0$. (See Figure 3.7.) We also suspect such an increase when we plug in values close to 0. By choosing values sufficiently close to 0, we can get arbitrarily large values of $f(x)$:

x	-0.1	-0.01	-0.001	0.001	0.01	0.1
$f(x)$	100	10,000	10^6	10^6	10,000	100

This table of values indicates that the limit does not exist.   ■

When $\lim_{x \to c} f(x)$ does not exist, we say that $f(x)$ diverges as x tends to c. The divergence in Example 6 was such that the function grew without bound. This is an important case, and we define it in the following box:

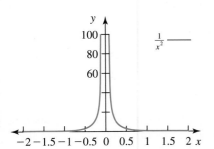

Figure 3.7 The graph of $f(x) = \frac{1}{x^2}$ in Example 6: The function grows without bound as x tends to 0.

$$\lim_{x \to c} f(x) = +\infty \quad \text{if } f(x) \text{ increases without bound as } x \to c$$

$$\lim_{x \to c} f(x) = -\infty \quad \text{if } f(x) \text{ decreases without bound as } x \to c$$

Similar definitions can be given for one-sided limits, which we will need in the next example. Note that when we write $\lim_{x \to c} f(x) = +\infty$ (or $-\infty$), we say that $f(x)$

diverges as $x \to c$. In particular, this means that $\lim_{x \to c} f(x)$ does *not* exist. (The symbols $+\infty$ and $-\infty$ do not refer to real real numbers.) Nevertheless, we write $\lim_{x \to c} f(x) = +\infty$ (or $-\infty$) if $f(x)$ increases (or decreases) without bound as $x \to c$, since it is useful to know when a function does that.

EXAMPLE 7   Find

$$\lim_{x \to 3} \frac{1}{x - 3}$$

Solution   The graph of $f(x) = 1/(x - 3)$, $x \neq 3$, in Figure 3.8 reveals that

$$\lim_{x \to 3^+} \frac{1}{x - 3} = +\infty \quad \text{and} \quad \lim_{x \to 3^-} \frac{1}{x - 3} = -\infty$$

We arrive at the same conclusion when we compute values of $f(x)$ for x close to 3. We see that if x is slightly larger than 3, then $f(x)$ is positive and increases without bound as x approaches 3 from the right. Likewise, if x is slightly smaller than 3, $f(x)$ is negative and decreases without bound as x approaches 3 from the left. We conclude that $f(x)$ diverges as x approaches 3.   ■

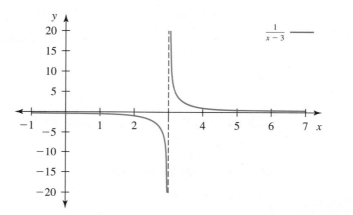

Figure 3.8   The graph of $f(x) = \frac{1}{x-3}$ in Example 7 grows without bound as x approaches 3 from the right and decreases without bound as x approaches 3 from the left.

The next example shows that a function can diverge without having one-sided limits or without going to $+\infty$ or $-\infty$.

EXAMPLE 8   Find

$$\lim_{x \to 0} \sin \frac{\pi}{x}$$

Solution   Simply using a calculator and plugging in values to find limits can yield wrong answers if we do not exercise proper caution. If we produced a table of values of $f(x) = \sin \frac{\pi}{x}$ for $x = 0.1, 0.01, 0.001, \ldots$, we would find that $\sin \frac{\pi}{0.1} = 0$, $\sin \frac{\pi}{0.01} = 0$, $\sin \frac{\pi}{0.001} = 0$, and so on. (Note that we measure angles in radians.) These calculations might prompt us to conclude that the limit of the function is 0. But let's look at its graph, which is shown in Figure 3.9. The graph does not support our calculator-based conclusion.

What we find instead is that the values of $f(x)$ oscillate infinitely often between -1 and $+1$ as $x \to 0$. We can see why as follows: As $x \to 0^+$, the argument in the sine function goes to infinity. (Likewise, as $x \to 0^-$, the argument goes to negative infinity.) That is,

$$\lim_{x \to 0^+} \frac{\pi}{x} = \infty \quad \text{and} \quad \lim_{x \to 0^-} \frac{\pi}{x} = -\infty$$

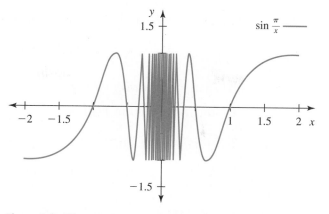

Figure 3.9 The graph of $f(x) = \sin \frac{\pi}{x}$ in Example 8.

As the argument of the sine function goes to $+\infty$ or $-\infty$, the function values oscillate between -1 and $+1$. Therefore, $\sin \frac{\pi}{x}$ continues to oscillate between -1 and $+1$ as $x \to 0$. ∎

The behavior exhibited in Example 8 is called **divergence by oscillation**.

Pitfalls The next example is an interesting one that shows other limitations of using a calculator to compute limits.

EXAMPLE 9 Find

$$\lim_{x \to 0} \frac{\sqrt{x^2 + 16} - 4}{x^2}$$

Solution The graph of $f(x) = \frac{\sqrt{x^2+16}-4}{x^2}$, $x \neq 0$, in Figure 3.10 indicates that the limit exists. So, on the basis of the graph, we conjecture that the limit is equal to 0.125. If, instead, we use a calculator to produce a table of values of $f(x)$ close to 0, something strange seems to happen:

x	0.01	0.001	0.0001	0.00001	0.000001	0.0000001
$f(x)$	0.1249998	0.125	0.125	0.125	0.1	0

As we get closer to 0, we first find that $f(x)$ gets closer to 0.125, but when we get very close to 0, $f(x)$ seems to drop to 0. What is going on? First, before you worry too much, note that $\lim_{x \to 0} f(x) = 0.125$. In the next section, we will learn how to compute this limit without resorting to the (somewhat dubious) help of the calculator. The strange behavior of the calculated values happens because, when x is very small, the difference in the numerator is so close to 0 that the calculator can no longer accurately determine its value. The calculator can compute only a certain number of digits accurately, which is good enough for most cases. Here, however, we need greater accuracy. The same strange thing happens when you try to graph this function on a graphing calculator. When the x range of the viewing window is too small, the graph is no longer accurate. (Try, for instance, $-0.00001 \leq x \leq 0.00001$ and $-0.03 \leq y \leq 0.15$ as the range for the viewing window.) ∎

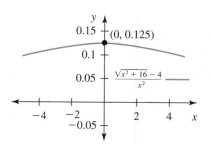

Figure 3.10 The graph of $f(x)$ in Example 9: As x tends to 0, the function approaches 0.125.

At the end of this chapter, we will discuss how limits are formally defined. The formal definition is conceptually similar to the one we used to define limits of the form $\lim_{n \to \infty} a_n$, but we will not use it to compute limits. As in Chapter 2, there are mathematical laws that will allow us to compute limits much more easily.

■ 3.1.2 Limit Laws

We encountered limit laws in Chapter 2. Analogous laws hold for limits of the type $\lim_{x \to c} f(x)$.

> **Limit Laws** Suppose that a is a constant and that
>
> $$\lim_{x \to c} f(x) \quad \text{and} \quad \lim_{x \to c} g(x)$$
>
> exist. Then the following rules hold:
>
> **1.** $\lim\limits_{x \to c} af(x) = a \lim\limits_{x \to c} f(x)$
>
> **2.** $\lim\limits_{x \to c}[f(x) + g(x)] = \lim\limits_{x \to c} f(x) + \lim\limits_{x \to c} g(x)$
>
> **3.** $\lim\limits_{x \to c}[f(x) \cdot g(x)] = \lim\limits_{x \to c} f(x) \cdot \lim\limits_{x \to c} g(x)$
>
> **4.** $\lim\limits_{x \to c} \dfrac{f(x)}{g(x)} = \dfrac{\lim\limits_{x \to c} f(x)}{\lim\limits_{x \to c} g(x)}$ provided that $\lim\limits_{x \to c} g(x) \neq 0$

You are probably easily convinced that

$$\lim_{x \to c} x = c \tag{3.3}$$

In Section 3.6, we will use the formal definition of limits to show that this equation is true. For now, we accept (3.3) as a fact. Starting from that equation, we can use the limit laws to compute limits of polynomials and rational functions.

EXAMPLE 10 Find

$$\lim_{x \to 2}[x^3 + 4x - 1]$$

Solution   Using Rules 1 and 2, we see that this equation becomes

$$\lim_{x \to 2} x^3 + 4 \lim_{x \to 2} x - \lim_{x \to 2} 1$$

provided that the individual limits exist. For the first term, we use Rule 3,

$$\lim_{x \to 2} x^3 = \left(\lim_{x \to 2} x \right) \left(\lim_{x \to 2} x \right) \left(\lim_{x \to 2} x \right)$$

provided that $\lim_{x \to 2} x$ exists. From (3.3), it then follows that $\lim_{x \to 2} x = 2$ and we find that

$$\left(\lim_{x \to 2} x \right) \left(\lim_{x \to 2} x \right) \left(\lim_{x \to 2} x \right) = (2)(2)(2) = 8$$

To compute the second term, we use (3.3) again to obtain $\lim_{x \to 2} x = 2$. For the last term, we find that $\lim_{x \to 2} 1 = 1$. Now that we have shown that the individual limits exist, we can use Rules 1 and 2 to evaluate

$$\lim_{x \to 2}[x^3 + 4x - 1] = \lim_{x \to 2} x^3 + 4 \lim_{x \to 2} x - \lim_{x \to 2} 1 = 8 + (4)(2) - 1 = 15 \qquad ■$$

EXAMPLE 11 Find

$$\lim_{x \to 4} \frac{x^2 + 1}{x - 3}$$

Solution   Using Rule 4, we find that

$$\lim_{x \to 4} \frac{x^2 + 1}{x - 3} = \frac{\lim_{x \to 4}(x^2 + 1)}{\lim_{x \to 4}(x - 3)}$$

provided that the limits in the numerator and denominator exist and the limit in the denominator is not equal to 0. Using Rules 2 and 3 in the numerator, we obtain

$$\lim_{x \to 4}(x^2 + 1) = \left(\lim_{x \to 4} x^2\right) + \left(\lim_{x \to 4} 1\right) = (4)(4) + 1 = 17$$

Breaking up the limit of the sum in the numerator into a sum of limits is justified only after we have shown that the individual limits exist. Using Rules 1 and 2 in the denominator, we get

$$\lim_{x \to 4}(x - 3) = \lim_{x \to 4} x - \lim_{x \to 4} 3 = 4 - 3 = 1$$

Again, using the limit laws is justified only after we have demonstrated that the individual limits exist. Since the limits in both the denominator and the numerator exist and the limit in the denominator is not equal to 0, we obtain

$$\lim_{x \to 4} \frac{x^2 + 1}{x - 3} = \frac{17}{1} = 17$$

■

The computations in Examples 10 and 11 look somewhat awkward, and it appears that what we have done is plug 2 into the expression $x^3 + 4x - 1$ in Example 10 and 4 into the expression $\frac{x^2 + 1}{x - 3}$ in Example 11, even though we emphasized in the informal definition of limits that we are not allowed to simply plug c into $f(x)$ when computing $\lim_{x \to c} f(x)$. But, in essence, we did the calculation

$$\lim_{x \to 2}[x^3 + 4x - 1] = 2^3 + (4)(2) - 1 = 15$$

in Example 10 and the calculation

$$\lim_{x \to 4} \frac{x^2 + 1}{x - 3} = \frac{17}{1} = 17$$

in Example 11.

Even though we made a point that we cannot simply substitute the value c into $f(x)$ when we take the limit $x \to c$ of $f(x)$, the limit laws and (3.3) (which we will prove in Section 3.6) show that we can do just that when we take a limit of a polynomial or a rational function. Let's summarize this property and then look at two more examples that show how to compute limits of polynomials or rational functions by using these results.

If $f(x)$ is a polynomial, then

$$\lim_{x \to c} f(x) = f(c)$$

If $f(x)$ is a rational function

$$f(x) = \frac{p(x)}{q(x)}$$

where $p(x)$ and $q(x)$ are polynomials, and if $q(c) \neq 0$, then

$$\lim_{x \to c} f(x) = \lim_{x \to c} \frac{p(x)}{q(x)} = \frac{p(c)}{q(c)} = f(c)$$

EXAMPLE 12   Find

$$\lim_{x \to 3}[x^2 - 2x + 1]$$

Solution   Since $f(x) = x^2 - 2x + 1$ is a polynomial, it follows that

$$\lim_{x \to 3}[x^2 - 2x + 1] = 9 - 6 + 1 = 4$$

■

EXAMPLE 13   Find

$$\lim_{x \to -1}\frac{2x^3 - x + 5}{x^2 + 3x + 1}$$

Solution   Note that

$$f(x) = \frac{2x^3 - x + 5}{x^2 + 3x + 1}$$

is a rational function that is defined for $x = -1$. (The denominator is not equal to 0 when we substitute $x = -1$.) We find that

$$\lim_{x \to -1}\frac{2x^3 - x + 5}{x^2 + 3x + 1} = \frac{2(-1)^3 - (-1) + 5}{(-1)^2 + 3(-1) + 1} = \frac{4}{-1} = -4$$

■

When you use the limit laws for finding limits of the form

$$\lim_{x \to c}[f(x) + g(x)] \quad \text{or} \quad \lim_{x \to c}[f(x) \cdot g(x)] \quad \text{or} \quad \lim_{x \to c}\frac{f(x)}{g(x)}$$

you need to check first that both $\lim_{x \to c} f(x)$ and $\lim_{x \to c} g(x)$ exist and, in the case of $\lim_{x \to c}\frac{f(x)}{g(x)}$, that $\lim_{x \to c} g(x) \neq 0$. The next two examples illustrate the importance of checking the assumptions in the limit laws before applying them.

EXAMPLE 14   Find

$$\lim_{x \to 0}\frac{\frac{1}{x}}{\frac{1}{x} + 1}$$

Solution   We observe that neither

$$\lim_{x \to 0}\frac{1}{x} \quad \text{nor} \quad \lim_{x \to 0}\left(\frac{1}{x} + 1\right)$$

exist. So we cannot use Rule 4 right away. Multiplying both numerator and denominator by x, however, will help:

$$\lim_{x \to 0}\frac{\frac{1}{x}}{\frac{1}{x} + 1} = \lim_{x \to 0}\frac{1}{1 + x}$$

Now we have a rational function on the right-hand side, and we can plug in 0 because the denominator, $1 + x$, will be different from 0. We get

$$\lim_{x \to 0}\frac{1}{1 + x} = \frac{1}{1 + 0} = 1$$

■

EXAMPLE 15   Find

$$\lim_{x \to 4}\frac{x^2 - 16}{x - 4}$$

Solution   The function $f(x) = \frac{x^2-16}{x-4}$ is a rational function, but since $\lim_{x\to 4}(x-4)=0$, we cannot use Rule 4. Instead, we need to simplify $f(x)$ first:

$$\lim_{x\to 4}\frac{x^2-16}{x-4} = \lim_{x\to 4}\frac{(x-4)(x+4)}{x-4}$$

Because $x \neq 4$, we can cancel $x-4$ in the numerator and denominator, which yields

$$\lim_{x\to 4}(x+4)=8$$

where we used the fact that $x+4$ is a polynomial in computing the limit.   ■

Section 3.1 Problems

■ 3.1.1

In Problems 1–32, use a table or a graph to investigate each limit.

1. $\lim\limits_{x\to 2}(x^2-4x+1)$ **2.** $\lim\limits_{x\to 2}\dfrac{x^2+3}{x+2}$

3. $\lim\limits_{x\to -1}\dfrac{2x}{1+x^2}$ **4.** $\lim\limits_{s\to 2}s(s^2-4)$

5. $\lim\limits_{x\to \pi}3\cos\dfrac{x}{4}$ **6.** $\lim\limits_{t\to \pi/9}\sin(3t)$

7. $\lim\limits_{x\to \pi/2}2\sec\dfrac{x}{3}$ **8.** $\lim\limits_{x\to \pi/2}\tan\dfrac{x-\pi/2}{2}$

9. $\lim\limits_{x\to -2}e^{-x^2/2}$ **10.** $\lim\limits_{x\to 0}\dfrac{e^x+1}{2x+3}$

11. $\lim\limits_{x\to 0}\ln(x+1)$ **12.** $\lim\limits_{t\to e}\ln t^3$

13. $\lim\limits_{x\to 3}\dfrac{x^2-16}{x-4}$ **14.** $\lim\limits_{x\to 2}\dfrac{x^2-4}{x+2}$

15. $\lim\limits_{x\to \pi/2}\sin(2x)$ **16.** $\lim\limits_{x\to \pi/2}\cos(x-\pi)$

17. $\lim\limits_{x\to 0}\dfrac{1}{1+x^2}$ **18.** $\lim\limits_{x\to 0}\dfrac{1}{x^2-1}$

19. $\lim\limits_{x\to 0^+}(1-e^{-x})$ **20.** $\lim\limits_{x\to 0^-}(1+e^x)$

21. $\lim\limits_{x\to 4^-}\dfrac{2}{x-4}$ **22.** $\lim\limits_{x\to 3^+}\dfrac{1}{x-3}$

23. $\lim\limits_{x\to 1^-}\dfrac{2}{1-x}$ **24.** $\lim\limits_{x\to 2^+}\dfrac{4}{2-x}$

25. $\lim\limits_{x\to 1^-}\dfrac{1}{1-x^2}$ **26.** $\lim\limits_{x\to 2^+}\dfrac{2}{x^2-4}$

27. $\lim\limits_{x\to 3}\dfrac{1}{(x-3)^2}$ **28.** $\lim\limits_{x\to 0}\dfrac{1-x^2}{x^2}$

29. $\lim\limits_{x\to 0}\dfrac{\sqrt{x^2+9}-3}{x^2}$ **30.** $\lim\limits_{x\to 0}\dfrac{\sqrt{x^2+4}-2}{x}$

31. $\lim\limits_{x\to 0}\dfrac{1-\sqrt{1-x^2}}{x^2}$ **32.** $\lim\limits_{x\to 0}\dfrac{\sqrt{2-x}-\sqrt{2}}{2x}$

33. Use a table and a graph to find out what happens to

$$f(x)=\frac{2}{x^2}$$

as $x \to \infty$. What happens as $x \to -\infty$? What happens as $x \to 0$?

34. Use a table and a graph to find out what happens to

$$f(x)=\frac{2x}{x-1}$$

as $x \to \infty$. What happens as $x \to -\infty$? What happens as $x \to 1$?

35. Use a graphing calculator to investigate

$$\lim_{x\to 1}\sin\frac{1}{x-1}$$

36. Use a graphing calculator to investigate

$$\lim_{x\to 0}\cos\frac{1}{x}$$

■ 3.1.2

In Problems 37–54, use the limit laws to evaluate each limit.

37. $\lim\limits_{x\to -1}(x^3+7x-1)$ **38.** $\lim\limits_{x\to 2}(3x^4-2x+1)$

39. $\lim\limits_{x\to -5}(4+2x^2)$ **40.** $\lim\limits_{x\to 2}(8x^3-2x+4)$

41. $\lim\limits_{x\to 3}\left(2x^2-\dfrac{1}{x}\right)$ **42.** $\lim\limits_{x\to -2}\left(\dfrac{x^2}{2}-\dfrac{2}{x^2}\right)$

43. $\lim\limits_{x\to -3}\dfrac{x^3-20}{x+1}$ **44.** $\lim\limits_{x\to 1}\dfrac{x^3-1}{x+2}$

45. $\lim\limits_{x\to 3}\dfrac{3x^2+1}{2x-3}$ **46.** $\lim\limits_{x\to -2}\dfrac{1+x}{1-x}$

47. $\lim\limits_{x\to 1}\dfrac{1-x^2}{1-x}$ **48.** $\lim\limits_{u\to 3}\dfrac{9-u^2}{3-u}$

49. $\lim\limits_{x\to 3}\dfrac{x^2-2x-3}{x-3}$ **50.** $\lim\limits_{x\to 1}\dfrac{(x-1)^2}{x^2-1}$

51. $\lim\limits_{x\to 2}\dfrac{2-x}{x^2-4}$ **52.** $\lim\limits_{x\to -4}\dfrac{x+4}{16-x^2}$

53. $\lim\limits_{x\to -2}\dfrac{2x^2+3x-2}{x+2}$ **54.** $\lim\limits_{x\to 1/2}\dfrac{1-x-2x^2}{1-2x}$

■ 3.2 Continuity

■ 3.2.1 What Is Continuity?

Consider the two functions

$$f(x) = \begin{cases} \dfrac{x^2 - 9}{x - 3} & \text{if } x \ne 3 \\ 6 & \text{if } x = 3 \end{cases}$$

and

$$g(x) = \begin{cases} \dfrac{x^2 - 9}{x - 3} & \text{if } x \ne 3 \\ 7 & \text{if } x = 3 \end{cases}$$

We are interested in how these functions behave for x close to 3. Both functions are defined for all $x \in \mathbf{R}$ and are the same for $x \ne 3$. Furthermore, as we saw in Example 3 of Section 3.1,

$$\lim_{x \to 3} f(x) = \lim_{x \to 3} g(x) = \lim_{x \to 3} \frac{x^2 - 9}{x - 3} = 6 \tag{3.4}$$

But the two functions differ at $x = 3$: $f(3) = 6$ and $g(3) = 7$. Comparing these results with (3.4), we see that

$$\lim_{x \to 3} f(x) = f(3) \quad \text{but} \quad \lim_{x \to 3} g(x) \ne g(3)$$

This difference can also be seen graphically [Figures 3.11(a) and 3.11(b)]: Although the graph of $f(x)$ can be drawn without lifting the pencil, in graphing $g(x)$ we need to lift the pencil at $x = 3$, since $\lim_{x \to 3} g(x) \ne g(3)$. We say that the function $f(x)$ is **continuous** at $x = 3$, whereas $g(x)$ is **discontinuous** at $x = 3$. Here is the definition of continuity at a point:

> **Definition** A function f is said to be **continuous** at $x = c$ if
>
> $$\lim_{x \to c} f(x) = f(c)$$

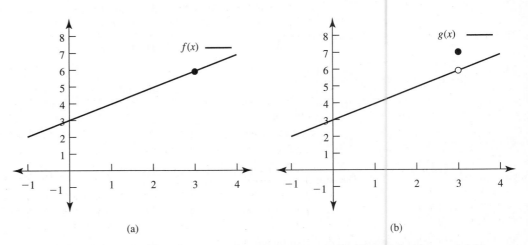

(a)   (b)

Figure 3.11 (a) The graph of $y = f(x)$ is continuous at $x = 3$. (b) The graph of $y = g(x)$ is discontinuous at $x = 3$.

To check whether a function is continuous at $x = c$, we need to check the following three conditions:

1. $f(x)$ is defined at $x = c$.
2. $\lim\limits_{x \to c} f(x)$ exists.
3. $\lim\limits_{x \to c} f(x)$ is equal to $f(c)$.

If any of these three conditions fails, the function is **discontinuous** at $x = c$.

EXAMPLE 1   Show that $f(x) = 2x - 3$, $x \in \mathbf{R}$, is continuous at $x = 1$.

Solution   We must check all three conditions:

1. $f(x)$ is defined at $x = 1$, since $f(1) = 2 \cdot 1 - 3 = -1$.
2. We use the fact that $\lim_{x \to c} x = c$ to conclude that $\lim_{x \to 1} f(x)$ exists.
3. Using the limit laws, we find that $\lim_{x \to 1} f(x) = -1$. This is the same as $f(1)$.

Since all three conditions are satisfied, $f(x) = 2x - 3$ is continuous at $x = 1$. ■

EXAMPLE 2   Let

$$f(x) = \begin{cases} \dfrac{x^2 - x - 6}{x - 3} & \text{if } x \neq 3 \\ a & \text{if } x = 3 \end{cases}$$

and find a so that $f(x)$ is continuous at $x = 3$.

Solution   To compute

$$\lim_{x \to 3} \frac{x^2 - x - 6}{x - 3}$$

we factor the numerator: $x^2 - x - 6 = (x - 3)(x + 2)$. Hence, since $x \neq 3$,

$$\lim_{x \to 3} \frac{x^2 - x - 6}{x - 3} = \lim_{x \to 3} \frac{(x - 3)(x + 2)}{x - 3} = \lim_{x \to 3} (x + 2) = 5$$

To ensure that $f(x)$ is continuous at $x = 3$, we require that

$$\lim_{x \to 3} f(x) = f(3)$$

We therefore need to choose 5 for a. This is the only choice for a that will make $f(x)$ continuous. Any other value of a would result in $f(x)$ being discontinous. ■

The function $y = \frac{x^2 - x - 6}{x - 3}$, $x \neq 3$, is not defined at $x = 3$ and is therefore automatically discontinuous there. (Condition 1 does not hold.) But we saw in Example 2 that we can *remove the discontinuity* by appropriately defining the function at $x = 3$. Still, it is not always possible to remove discontinuities, as the next three examples will show. In the first two, the discontinuity is a jump; that is, both the left-hand and the right-hand limits exist at the point where the jump occurs, but the limits differ. In the third example, the function grows without bound where it is discontinuous.

EXAMPLE 3   The floor function

$$f(x) = \lfloor x \rfloor = \text{the largest integer less than or equal to } x$$

is graphed in Figure 3.12. The closed circles in the figure correspond to endpoints that are contained in the graph of the function, whereas the open circles correspond to endpoints that are not contained in the graph of the function. To explain this function, we compute a few values: $f(2.1) = 2$, $f(2) = 2$, and $f(1.9999) = 1$. The function jumps whenever x is an integer. Let k be an integer; then $f(k) = k$ and

$$\lim_{x \to k^+} f(x) = k, \qquad \lim_{x \to k^-} f(x) = k - 1$$

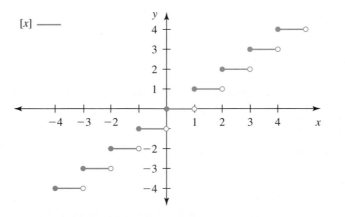

Figure 3.12 The floor function $f(x) = \lfloor x \rfloor$.

That is, only when x approaches an integer from the right is the limit equal to the value of the function. The function is therefore discontinuous at integer values, and the discontinuity cannot be removed. If c is *not* an integer, then $f(x)$ is continuous at $x = c$.   ■

Example 3 motivates the definition of one-sided continuity:

> **Definition** A function f is said to be continuous from the right at $x = c$ if
>
> $$\lim_{x \to c^+} f(x) = f(c)$$
>
> and continuous from the left at $x = c$ if
>
> $$\lim_{x \to c^-} f(x) = f(c)$$

The function $f(x) = \lfloor x \rfloor$, $x \in \mathbf{R}$, of Example 3, is therefore continuous from the right but not from the left. In the next example, the discontinuity is again a jump; however, this time we do not even have one-sided continuity.

EXAMPLE 4   Show that

$$f(x) = \begin{cases} \dfrac{|x|}{x} & \text{if } x \neq 0 \\ 0 & \text{if } x = 0 \end{cases}$$

is discontinuous at $x = 0$ and that the discontinuity cannot be removed.

Solution   The graph of $f(x)$ is shown in Figure 3.13. We can write

$$f(x) = \begin{cases} 1 & \text{for } x > 0 \\ 0 & \text{for } x = 0 \\ -1 & \text{for } x < 0 \end{cases}$$

since $|x| = x$ for $x > 0$ and $|x| = -x$ for $x < 0$. We therefore get

$$\lim_{x \to 0^+} f(x) = 1 \qquad \text{and} \qquad \lim_{x \to 0^-} f(x) = -1$$

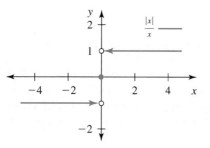

Figure 3.13 The function $f(x) = \frac{|x|}{x}$ is discontinuous at $x = 0$.

The one-sided limits exist, but they are not equal [which implies that $\lim_{x \to 0} f(x)$ does not exist]. When we graph the function, a jump occurs at $x = 0$. (See Figure 3.13.) This function does not exhibit even one-sided continuity because $f(x)$ is neither 1 nor -1 at $x = 0$. There is no way that we could assign a value to $f(0)$ such that the function would be continuous at $x = 0$.   ■

EXAMPLE 5   At which point is the function

$$f(x) = \frac{1}{(x - 4)^2}$$

discontinuous? Can the discontinuity be removed?

Solution   The graph of $f(x)$ is shown in Figure 3.14. The function $f(x)$ cannot be defined for $x = 4$, since $f(x)$ is of the form $\frac{1}{0}$ when $x = 4$. The function is defined for all other values of x. Therefore, we look at $x = 4$. We find that

$$\lim_{x \to 4} \frac{1}{(x - 4)^2} = \infty \quad \text{(limit does not exist)}$$

Because ∞ is not a real number, we cannot assign a value to $f(4)$ such that $f(x)$ would be continuous at $x = 4$. We therefore conclude that $f(x)$ is discontinuous at $x = 4$ and the discontinuity cannot be removed.   ■

Figure 3.14   The function $f(x) = \frac{1}{(x-4)^2}$ is discontinuous at $x = 4$.

■ 3.2.2 Combinations of Continuous Functions

Using the limit laws, we find that the following statements hold for combinations of continuous functions:

Suppose that a is a constant and the functions f and g are continuous at $x = c$. Then the following functions are continuous at $x = c$:

1. $a \cdot f$
2. $f + g$
3. $f \cdot g$
4. $\dfrac{f}{g}$ provided that $g(c) \neq 0$

Proof   We will prove only the second statement. We must show that conditions 1–3 of the previous subsection hold:

1. Note that $[f + g](x) = f(x) + g(x)$. Therefore, $f + g$ is defined at $x = c$ and $[f + g](c) = f(c) + g(c)$.
2. We assumed that f and g are continuous at $x = c$. This means, in particular, that

$$\lim_{x \to c} f(x) \quad \text{and} \quad \lim_{x \to c} g(x)$$

both exist. That is, the hypothesis in the limit laws holds, and we can apply Rule 2 for limits and find that

$$\lim_{x \to c}[f + g](x) = \lim_{x \to c}[f(x) + g(x)] = \lim_{x \to c} f(x) + \lim_{x \to c} g(x) \quad (3.5)$$

In other words, $\lim_{x \to c}[f + g](x)$ exists and condition 2 holds.

3. Since f and g are continuous at $x = c$, it follows that

$$\lim_{x \to c} f(x) = f(c) \qquad \text{and} \qquad \lim_{x \to c} g(x) = g(c) \tag{3.6}$$

Therefore, combining (3.5) and (3.6), we obtain

$$\lim_{x \to c}[f + g](x) = \lim_{x \to c} f(x) + \lim_{x \to c} g(x) = f(c) + g(c)$$

which is equal to $[f + g](c)$ and hence condition 3 holds.

Since we showed that all three conditions hold, it follows that $f + g$ is continuous at $x = c$. The other statements are shown in a similar way, using the limit laws. ■

We say that a function f is continuous on an interval I if f is continuous for all $x \in I$. Note that if I is a closed interval, then continuity at the left (and, respectively, right) endpoint of the interval means continuous from the right (and, respectively, left). Many of the elementary functions are indeed continuous wherever they are defined. For polynomials and rational functions, this statement follows immediately from the fact that certain combinations of continuous functions are continuous. We give a list of the most important cases:

The following functions are continuous wherever they are defined:

1. polynomial functions
2. rational functions
3. power functions
4. trigonometric functions
5. exponential functions of the form a^x, $a > 0$ and $a \neq 1$
6. logarithmic functions of the form $\log_a x$, $a > 0$ and $a \neq 1$

The phrase "wherever they are defined" is crucial. It helps us to identify points where a function might be discontinuous. For instance, the power function $1/x^2$ is defined only for $x \neq 0$, and the logarithmic function $\log_a x$ is defined only for $x > 0$. We will illustrate the six cases cited in the preceding box in the next example, paying particular attention to the phrase "wherever they are defined."

EXAMPLE 6   For which values of $x \in \mathbf{R}$ are the following functions continuous?

(a) $f(x) = 2x^3 - 3x + 1$ (b) $f(x) = \dfrac{x^2 + x + 1}{x - 2}$ (c) $f(x) = x^{1/4}$

(d) $f(x) = 3 \sin x$ (e) $f(x) = \tan x$ (f) $f(x) = 3^x$

(g) $2 \ln(x + 1)$

Solution   (a) $f(x)$ is a polynomial and is defined for all $x \in \mathbf{R}$; it is therefore continuous for all $x \in \mathbf{R}$.

(b) $f(x)$ is a rational function defined for all $x \neq 2$; it is therefore continuous for all $x \neq 2$.

(c) $f(x) = x^{1/4} = \sqrt[4]{x}$ is a power function defined for $x \geq 0$; it is therefore continuous for $x \geq 0$.

(d) $f(x)$ is a trigonometric function. Because $\sin x$ is defined for all $x \in \mathbf{R}$, $3 \sin x$ is continuous for all $x \in \mathbf{R}$.

(e) $f(x)$ is a trigonometric function. The tangent function is defined for all $x \neq \frac{\pi}{2} + k\pi$, where k is an integer; it is therefore continuous for all $x \neq \frac{\pi}{2} + k\pi$, where k is an integer.

(f) $f(x)$ is an exponential function. $f(x) = 3^x$ is defined for all $x \in \mathbf{R}$ and is therefore continuous for all $x \in \mathbf{R}$.

(g) $f(x)$ is a logarithmic function. $f(x) = 2\ln(x+1)$ is defined as long as $x+1 > 0$ or $x > -1$; it is therefore continuous for all $x > -1$. ■

The following result is useful in determining whether a composition of functions is continuous:

Theorem If $g(x)$ is continuous at $x = c$ with $g(c) = L$ and $f(x)$ is continuous at $x = L$, then $(f \circ g)(x)$ is continuous at $x = c$. In particular,

$$\lim_{x \to c}(f \circ g)(x) = \lim_{x \to c} f[g(x)] = f[\lim_{x \to c} g(x)] = f[g(c)] = f(L)$$

To explain this theorem, recall what it means to compute $(f \circ g)(c) = f[g(c)]$. When we compute $f[g(c)]$, we take the value c, compute $g(c)$, and then take the result $g(c)$ and plug it into the function f to obtain $f[g(c)]$. If, at each step, the functions are continuous, the resulting function will be continuous.

EXAMPLE 7

Determine where the following functions are continuous:

(a) $h(x) = e^{-x^2}$ **(b)** $h(x) = \sin\dfrac{\pi}{x}$ **(c)** $h(x) = \dfrac{1}{1 + 2x^{1/3}}$

Solution

(a) Set $g(x) = -x^2$ and $f(x) = e^x$. Then $h(x) = (f \circ g)(x)$. Since $g(x)$ is a polynomial, it is continuous for all $x \in \mathbf{R}$, and the range of $g(x)$ is $(-\infty, 0]$. $f(x)$ is continuous for all values in the range of $g(x)$. [In fact, $f(x)$ is continuous for all $x \in \mathbf{R}$.] It therefore follows that $h(x)$ is continuous for all $x \in \mathbf{R}$.

(b) Set $g(x) = \frac{\pi}{x}$ and $f(x) = \sin x$. $g(x)$ is continuous for all $x \neq 0$. The range of $g(x)$ is the set of all real numbers, excluding 0. $f(x)$ is continuous for all x in the range of $g(x)$. Hence, $h(x)$ is continuous for all $x \neq 0$. Recall that we showed in Example 8 of Section 3.1 that

$$\lim_{x \to 0} \sin\frac{\pi}{x}$$

does not exist. That is, $h(x)$ is discontinuous at $x = 0$.

(c) Set $g(x) = x^{1/3}$ and $f(x) = \frac{1}{1+2x}$. Then $h(x) = (f \circ g)(x)$. $g(x)$ is continuous for all $x \in \mathbf{R}$, since $g(x) = x^{1/3} = \sqrt[3]{x}$ and 3 is an odd integer. The range of $g(x)$ is $(-\infty, \infty)$. $f(x)$ is continuous for all real x different from $-1/2$. Since $g(-\frac{1}{8}) = -\frac{1}{2}$, $h(x)$ is continuous for all real x different from $-1/8$. Another way to see that we need to exclude $-\frac{1}{8}$ from the domain of $h(x)$ is by looking directly at the denominator of $h(x)$. We have $1 + 2x^{1/3} = 0$ when $x = -\frac{1}{8}$. ■

When we compute $\lim_{x \to c} f(x)$ and we know that $f(x)$ is continuous at $x = c$, it follows that $\lim_{x \to c} f(x) = f(c)$. The next three examples illustrate this property.

EXAMPLE 8

Find

$$\lim_{x \to 3} \sin\left(\pi\frac{x^2 - 1}{4}\right)$$

Solution

The function $f(x) = \sin\left(\pi\frac{x^2-1}{4}\right)$ is continuous at $x = 3$. Hence,

$$\lim_{x \to 3} \sin\left(\pi\frac{x^2 - 1}{4}\right) = \sin\left(\pi\frac{9 - 1}{4}\right) = \sin(2\pi) = 0$$ ■

EXAMPLE 9   Find

$$\lim_{x \to 1} \sqrt{2x^3 - 1}$$

Solution   The function $f(x) = \sqrt{2x^3 - 1}$ is continuous at $x = 1$. Thus,

$$\lim_{x \to 1} \sqrt{2x^3 - 1} = \sqrt{(2)(1)^3 - 1} = \sqrt{1} = 1$$ ■

EXAMPLE 10   Find

$$\lim_{x \to 0} e^{x-1}$$

Solution   The function $f(x) = e^{x-1}$ is continuous at $x = 0$. Therefore,

$$\lim_{x \to 0} e^{x-1} = e^{0-1} = e^{-1}$$ ■

We conclude this section by calculating the limit of the expression in Example 9 of Section 3.1.

EXAMPLE 11   Find

$$\lim_{x \to 0} \frac{\sqrt{x^2 + 16} - 4}{x^2}$$

Solution   We cannot apply Rule 4 of Section 3.1, since $f(x) = (\sqrt{x^2 + 16} - 4)/x^2$ is not defined for $x = 0$. (If we plug in 0, we get the expression 0/0.) We use a trick that will allow us to find the limit: We rationalize the numerator. For $x \neq 0$, we find that

$$\frac{\sqrt{x^2 + 16} - 4}{x^2} = \frac{(\sqrt{x^2 + 16} - 4)}{x^2} \frac{(\sqrt{x^2 + 16} + 4)}{(\sqrt{x^2 + 16} + 4)}$$

$$= \frac{x^2 + 16 - 16}{x^2(\sqrt{x^2 + 16} + 4)} = \frac{x^2}{x^2(\sqrt{x^2 + 16} + 4)}$$

$$= \frac{1}{\sqrt{x^2 + 16} + 4}$$

Note that we are allowed to divide by x^2 in the last step, since we are assuming that $x \neq 0$. We can now apply Rule 4 to $1/(\sqrt{x^2 + 16} + 4)$. When we do, we obtain

$$\lim_{x \to 0} \frac{\sqrt{x^2 + 16} - 4}{x^2} = \lim_{x \to 0} \frac{1}{\sqrt{x^2 + 16} + 4} = \frac{1}{8} = 0.125$$

as we saw in Example 9 of Section 3.1. In Chapter 5, we will learn another method for finding the limit of expressions of the form 0/0.   ■

Section 3.2 Problems

■ **3.2.1**

In Problems 1–4, show that each function is continuous at the given value.

1. $f(x) = 2x, c = 1/2$

2. $f(x) = -x, c = 1$

3. $f(x) = x^3 - 2x + 1, c = 2$

4. $f(x) = x^2 + 1, c = -1$

5. Show that

$$f(x) = \begin{cases} \dfrac{x^2 - x - 2}{x - 2} & \text{if } x \neq 2 \\ 3 & \text{if } x = 2 \end{cases}$$

is continuous at $x = 2$.

6. Show that

$$f(x) = \begin{cases} \dfrac{2x^2 + x - 6}{x + 2} & \text{if } x \neq -2 \\ -7 & \text{if } x = -2 \end{cases}$$

is continuous at $x = -2$.

7. Let

$$f(x) = \begin{cases} \dfrac{x^2 - 9}{x - 3} & \text{if } x \neq 3 \end{cases}$$

Which value must you assign to a so that $f(x)$ is continuous at $x = 3$?

8. Let

$$f(x) = \begin{cases} \dfrac{x^2 + x - 2}{x - 1} & \text{if } x \neq 1 \\ a & \text{if } x = 1 \end{cases}$$

Which value must you assign to a so that $f(x)$ is continuous at $x = 1$?

In Problems 9–12, determine at which points $f(x)$ is discontinuous.

9. $f(x) = \dfrac{1}{x - 3}$

10. $f(x) = \dfrac{1}{x^2 - 1}$

11. $f(x) = \begin{cases} \dfrac{x^2 - 3x + 2}{x - 2} & \text{if } x \neq 1 \\ 1 & \text{if } x = 1 \end{cases}$

12. $f(x) = \begin{cases} x^2 - 1 & \text{if } x \leq 0 \\ x & \text{if } x > 0 \end{cases}$

13. Show that the floor function $f(x) = \lfloor x \rfloor$ is continuous at $x = 5/2$ but discontinuous at $x = 3$.

14. Show that the floor function $f(x) = \lfloor x \rfloor$ is continuous from the right at $x = 2$.

■ **3.2.2**

In Problems 15–24, find the values of $x \in \mathbb{R}$ for which the given functions are continuous.

15. $f(x) = 3x^4 - x^2 + 4$

16. $f(x) = \sqrt{x^2 - 1}$

17. $f(x) = \dfrac{x^2 + 1}{x - 1}$

18. $f(x) = \cos(2x)$

19. $f(x) = e^{-|x|}$

20. $f(x) = \ln(x - 2)$

21. $f(x) = \ln \dfrac{x}{x + 1}$

22. $f(x) = \exp[-\sqrt{x - 1}]$

23. $f(x) = \tan(2\pi x)$

24. $f(x) = \sin\left(\dfrac{2x}{3 + x}\right)$

25. Let

$$f(x) = \begin{cases} x^2 + 2 & \text{for } x \leq 0 \\ x + c & \text{for } x > 0 \end{cases}$$

(a) Graph $f(x)$ when $c = 1$, and determine whether $f(x)$ is continuous for this choice of c.

(b) How must you choose c so that $f(x)$ is continuous for all $x \in (-\infty, \infty)$?

26. Let

$$f(x) = \begin{cases} \dfrac{1}{x} & \text{for } x \geq 1 \\ 2x + c & \text{for } x < 1 \end{cases}$$

(a) Graph $f(x)$ when $c = 0$, and determine whether $f(x)$ is continuous for this choice of c.

(b) How must you choose c so that $f(x)$ is continuous for all $x \in (-\infty, \infty)$?

27. (a) Show that

$$f(x) = \sqrt{x - 1}, \quad x \geq 1$$

is continuous from the right at $x = 1$.

(b) Graph $f(x)$.

(c) Does it make sense to look at continuity from the left at $x = 1$?

28. (a) Show that

$$f(x) = \sqrt{x^2 - 4}, \quad |x| \geq 2$$

is continuous from the right at $x = 2$ and continuous from the left at $x = -2$.

(b) Graph $f(x)$.

(c) Does it make sense to look at continuity from the left at $x = 2$ and at continuity from the right at $x = -2$?

In Problems 29–48, find the limits.

29. $\displaystyle\lim_{x \to \pi/3} \sin\left(\dfrac{x}{2}\right)$

30. $\displaystyle\lim_{x \to -\pi/2} \cos(2x)$

31. $\displaystyle\lim_{x \to \pi/2} \dfrac{\cos^2 x}{1 - \sin^2 x}$

32. $\displaystyle\lim_{x \to -\pi/2} \dfrac{1 + \tan^2 x}{\sec^2 x}$

33. $\displaystyle\lim_{x \to -1} \sqrt{4 + 5x^4}$

34. $\displaystyle\lim_{x \to -2} \sqrt{6 + x}$

35. $\displaystyle\lim_{x \to -1} \sqrt{x^2 + 2x + 2}$

36. $\displaystyle\lim_{x \to 1} \sqrt{x^3 + 4x - 1}$

37. $\displaystyle\lim_{x \to 0} e^{-x^2/3}$

38. $\displaystyle\lim_{x \to 0} e^{3x+2}$

39. $\displaystyle\lim_{x \to 3} e^{x^2 - 9}$

40. $\displaystyle\lim_{x \to -1} e^{x^2/2 - 1}$

41. $\displaystyle\lim_{x \to 0} \dfrac{e^{2x} - 1}{e^x - 1}$

42. $\displaystyle\lim_{x \to 0} \dfrac{e^{-x} - e^x}{e^{-x} + 1}$

43. $\displaystyle\lim_{x \to -2} \dfrac{1}{\sqrt{5x^2 - 4}}$

44. $\displaystyle\lim_{x \to 1} \dfrac{1}{\sqrt{3 - 2x^2}}$

45. $\displaystyle\lim_{x \to 0} \dfrac{\sqrt{x^2 + 9} - 3}{x^2}$

46. $\displaystyle\lim_{x \to 0} \dfrac{5 - \sqrt{25 + x^2}}{2x^2}$

47. $\displaystyle\lim_{x \to 0} \ln(1 - x)$

48. $\displaystyle\lim_{x \to 1} \ln[e^x \cos(x - 1)]$

■ 3.3 Limits at Infinity

The limit laws discussed in Subsection 3.1.2 also hold as x tends to ∞ (or $-\infty$).

 EXAMPLE 1   Find

$$\lim_{x \to \infty} \dfrac{x}{x + 1}$$

Solution

We set $f(x) = x$ and $g(x) = x+1$. Obviously, neither $\lim_{x \to \infty} f(x)$ nor $\lim_{x \to \infty} g(x)$ exists. Thus, we cannot use Rule 4 from Section 3.1. But we can divide both numerator and denominator by x. When we do, we find that

$$\lim_{x \to \infty} \frac{x}{x+1} = \lim_{x \to \infty} \frac{1}{1 + \frac{1}{x}}$$

Since $\lim_{x \to \infty} 1 = 1$ and $\lim_{x \to \infty}(1 + \frac{1}{x}) = 1$, both limits exist. Furthermore, $\lim_{x \to \infty}(1 + \frac{1}{x}) \neq 0$. We can now apply Rule 4 of Section 3.1 after having done the algebraic manipulation:

$$\lim_{x \to \infty} \frac{x}{x+1} = \lim_{x \to \infty} \frac{1}{1 + \frac{1}{x}} = \frac{\lim_{x \to \infty} 1}{\lim_{x \to \infty}(1 + \frac{1}{x})} = \frac{1}{1} = 1 \qquad ■$$

In Example 1, we computed the limit of a rational function as x tended to infinity. Rational functions are ratios of polynomials. To find out how the limit of a rational function behaves as x tends to infinity, we will first compare the relative growth of functions of the form $y = x^n$: If $n > m$, then x^n dominates x^m for large x, in the sense that

$$\lim_{x \to \infty} \frac{x^n}{x^m} = \infty \qquad \text{and} \qquad \lim_{x \to \infty} \frac{x^m}{x^n} = 0$$

The preceding statement follows immediately if we simplify the fractions

$$\frac{x^n}{x^m} = x^{n-m} \qquad \text{with } n - m > 0$$

and

$$\frac{x^m}{x^n} = \frac{1}{x^{n-m}} \qquad \text{with } n - m > 0$$

This limiting behavior is important when we compute limits of rational functions as $x \to \infty$. We compare the following three limits:

(a) $\lim_{x \to \infty} \dfrac{x^2 + 2x - 1}{x^3 - 3x + 1}$

(b) $\lim_{x \to \infty} \dfrac{2x^3 - 4x + 7}{3x^3 + 7x^2 - 1}$

(c) $\lim_{x \to \infty} \dfrac{x^4 + 2x - 5}{x^2 - x + 2}$

To determine whether the numerator or the denominator dominates, we look at each of their leading terms. (The leading term is the term with the largest exponent.) The leading term of a polynomial tells us how quickly the polynomial increases as x increases.

(a) The leading term in the numerator is x^2, and the leading term in the denominator is x^3. As $x \to \infty$, the denominator grows much faster than the numerator. We therefore expect the limit to be equal to 0. We can show this by dividing both numerator and denominator by the higher of the two powers, namely, x^3. We get

$$\lim_{x \to \infty} \frac{x^2 + 2x - 1}{x^3 - 3x + 1} = \lim_{x \to \infty} \frac{\frac{1}{x} + \frac{2}{x^2} - \frac{1}{x^3}}{1 - \frac{3}{x^2} + \frac{1}{x^3}}$$

Since $\lim_{x \to \infty}(\frac{1}{x} + \frac{2}{x^2} - \frac{1}{x^3})$ exists (it is equal to 0), and $\lim_{x \to \infty}(1 - \frac{3}{x^2} + \frac{1}{x^3})$ exists and is not equal to 0 (it is equal to 1), we can apply Rule 4 to find that

$$\lim_{x \to \infty} \frac{\frac{1}{x} + \frac{2}{x^2} - \frac{1}{x^3}}{1 - \frac{3}{x^2} + \frac{1}{x^3}} = \frac{\lim_{x \to \infty}(\frac{1}{x} + \frac{2}{x^2} - \frac{1}{x^3})}{\lim_{x \to \infty}\left(1 - \frac{3}{x^2} + \frac{1}{x^3}\right)} = \frac{0}{1} = 0$$

(b) The leading term in both the numerator and the denominator is x^3, so we divide numerator and denominator by x^3 and obtain

$$\lim_{x \to \infty} \frac{2x^3 - 4x + 7}{3x^3 + 7x^2 - 1} = \lim_{x \to \infty} \frac{2 - \frac{4}{x^2} + \frac{7}{x^3}}{3 + \frac{7}{x} - \frac{1}{x^3}} = \frac{2}{3}$$

In the last step, we used the facts that the limits in both the numerator and the denominator exist and that the limit in the denominator is not equal to 0. Applying Rule 4 yields the limiting value. Note that the limiting value is equal to the ratio of the coefficients of the leading terms in the numerator and the denominator.

(c) The leading term in the numerator is x^4 and the leading term in the denominator is x^2. Since the leading term in the numerator grows much more quickly than the leading term in the denominator, we expect the limit to be undefined. This is indeed the case and can be seen if we divide the numerator by the denominator. We find that

$$\lim_{x \to \infty} \frac{x^4 + 2x - 5}{x^2 - x + 2} = \lim_{x \to \infty} \left(x^2 + x - 1 - \frac{x + 3}{x^2 - x + 2} \right) \quad \text{does not exist}$$

It is often useful to determine whether the limit tends to $+\infty$ or $-\infty$. Since $x^2 + x - 1$ tends to $+\infty$ as $x \to \infty$ and the ratio $\frac{x+3}{x^2-x+2}$ tends to 0 as $x \to \infty$, the limit of $\frac{x^4+2x-5}{x^2-x+2}$ tends to $+\infty$ as $x \to +\infty$.

Let's summarize our findings: If $f(x)$ is a rational function of the form $f(x) = p(x)/q(x)$, where $p(x)$ is a polynomial of degree $\deg(p)$ and $q(x)$ is a polynomial of degree $\deg(q)$, then

$$\lim_{x \to \infty} f(x) = \lim_{x \to \infty} \frac{p(x)}{q(x)} = \begin{cases} 0 & \text{if } \deg(p) < \deg(q) \\ L \neq 0 & \text{if } \deg(p) = \deg(q) \\ \text{does not exist} & \text{if } \deg(p) > \deg(q) \end{cases}$$

Here, L is a real number that is the ratio of the coefficients of the leading terms in the numerator and denominator. The same behavior holds as $x \to -\infty$.

EXAMPLE 2   Compute

(a) $\displaystyle \lim_{x \to -\infty} \frac{1 - x + 2x^2}{3x - 5x^2}$

(b) $\displaystyle \lim_{x \to \infty} \frac{1 - x^3}{1 + x^5}$

(c) $\displaystyle \lim_{x \to \infty} \frac{2 - x^2}{1 + 2x}$

(d) $\displaystyle \lim_{x \to -\infty} \frac{4 + 3x^2}{1 - 7x}$

Solution   **(a)** Since the degree of the numerator is equal to the degree of the denominator,

$$\lim_{x \to -\infty} \frac{1 - x + 2x^2}{3x - 5x^2} = \frac{2}{-5} = -\frac{2}{5}$$

(b) Since the degree of the numerator is less than the degree of the denominator,

$$\lim_{x \to \infty} \frac{1 - x^3}{1 + x^5} = 0$$

(c) Since the degree of the numerator is greater than the degree of the denominator, the limit does not exist. When x is very large, the expression $\frac{2-x^2}{1+2x}$ behaves like $\frac{-x^2}{2x} = -\frac{x}{2}$, which tends to $-\infty$ as $x \to \infty$. Hence,

$$\lim_{x \to \infty} \frac{2 - x^2}{1 + 2x} = -\infty \quad \text{(limit does not exist)}$$

(d) The degree of the numerator is greater than the degree of the denominator, so

$$\lim_{x \to -\infty} \frac{4 + 3x^2}{1 - 7x} = \infty \qquad \text{(limit does not exist)}$$

since $\frac{4+3x^2}{1-7x}$ behaves like $\frac{3x^2}{-7x} = -\frac{3}{7}x$ for x large, which tends to $+\infty$ as $x \to -\infty$. ■

Rational functions are not the only functions that involve limits as $x \to \infty$ (or $x \to -\infty$). Many important applications in biology involve exponential functions. We will use the following result repeatedly—it is one of the most important limits:

$$\lim_{x \to \infty} e^{-x} = 0$$

The graph of $f(x) = e^{-x}$ is given in Figure 3.15. You should familiarize yourself with the basic shape of the function $f(x) = e^{-x}$ and its behavior as $x \to \infty$.

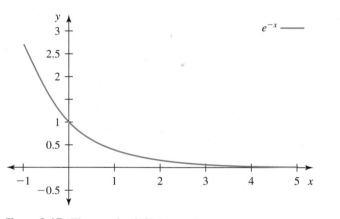

Figure 3.15 The graph of $f(x) = e^{-x}$.

EXAMPLE 3

Logistic Growth The logistic curve describes the density of a population over time, where the rate of growth depends on the population size. We will discuss this function in more detail in coming chapters. It suffices here to say that the per capita rate of growth decreases with increasing population size. If $N(t)$ denotes the size of the population at time t, then the logistic curve is given by

$$N(t) = \frac{K}{1 + \left(\frac{K}{N(0)} - 1\right)e^{-rt}} \qquad \text{for } t \geq 0$$

The parameters K and r are positive numbers that describe the population dynamics. We can check that $N(0)$ on the right-hand side is indeed the population size at time 0 [evaluate $N(t)$ at $t = 0$], and we assume that $N(0)$ is positive. The graph of $N(t)$ is shown in Figure 3.16. We will interpret K now; the interpretation of r must wait until the next chapter.

If we are interested in the long-term behavior of the population as it evolves in accordance with the logistic growth curve, we need to investigate what happens to $N(t)$ as $t \to \infty$. We find that

$$\lim_{t \to \infty} \frac{K}{1 + \left(\frac{K}{N(0)} - 1\right)e^{-rt}} = K$$

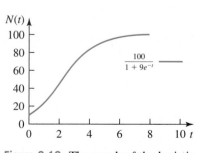

Figure 3.16 The graph of the logistic curve with $K = 100$, $N_0 = 10$, and $r = 1$.

since $\lim_{t \to \infty} e^{-rt} = 0$ for $r > 0$. That is, as $t \to \infty$, the population size approaches K, which is called the **carrying capacity** of the population. You will encounter logistic growth repeatedly in this text; it is one of the most fundamental equations describing population growth. ■

Evaluate the limits in Problems 1–24.

1. $\lim_{x\to\infty} \dfrac{2x^2 - 3x + 5}{x^4 - 2x + 1}$

2. $\lim_{x\to\infty} \dfrac{x^2 + 3}{5x^2 - 2x + 1}$

3. $\lim_{x\to-\infty} \dfrac{x^3 + 3}{x - 2}$

4. $\lim_{x\to-\infty} \dfrac{2x - 1}{3 - 4x}$

5. $\lim_{x\to\infty} \dfrac{1 - x^3 + 2x^4}{2x^2 + x^4}$

6. $\lim_{x\to\infty} \dfrac{1 - 5x^3}{1 + 3x^4}$

7. $\lim_{x\to\infty} \dfrac{x^2 - 2}{2x + 1}$

8. $\lim_{x\to-\infty} \dfrac{3 - x^2}{1 - 2x^2}$

9. $\lim_{x\to-\infty} \dfrac{x^2 - 3x + 1}{4 - x}$

10. $\lim_{x\to-\infty} \dfrac{1 - x^3}{2 + x}$

11. $\lim_{x\to-\infty} \dfrac{2 + x^2}{1 - x^2}$

12. $\lim_{x\to-\infty} \dfrac{2x + x^2}{3x + 1}$

13. $\lim_{x\to\infty} \dfrac{4}{1 + e^{-2x}}$

14. $\lim_{x\to\infty} \dfrac{e^{-x}}{1 - e^{-x}}$

15. $\lim_{x\to\infty} \dfrac{2e^x}{e^x + 3}$

16. $\lim_{x\to\infty} \dfrac{e^x}{2 - e^x}$

17. $\lim_{x\to-\infty} \exp[x]$

18. $\lim_{x\to\infty} \exp[-\ln x]$

19. $\lim_{x\to\infty} \dfrac{3e^{2x}}{2e^{2x} - e^x}$

20. $\lim_{x\to\infty} \dfrac{3e^{2x}}{2e^{2x} - e^{3x}}$

21. $\lim_{x\to\infty} \dfrac{3}{2 + e^{-x}}$

22. $\lim_{x\to-\infty} \dfrac{4}{1 + e^{-x}}$

23. $\lim_{x\to-\infty} \dfrac{e^x}{1 + x}$

24. $\lim_{x\to\infty} \dfrac{2}{e^x(1 + x)}$

25. In Section 1.2.3, Example 6, we introduced the Monod growth function

$$r(N) = a\frac{N}{k + N}, \quad N \ge 0$$

Find $\lim_{N\to\infty} r(N)$.

26. In Problem 86 of Section 1.3, we discussed the Michaelis–Menten equation, which describes the initial velocity of an enzymatic reaction (v_0) as a function of the substrate concentration (s_0). The equation was given by

$$v_0 = \frac{v_{\max} s_0}{s_0 + K_m}$$

Find $\lim_{s_0\to\infty} v_0$.

27. Suppose the size of a population at time t is given by

$$N(t) = \frac{500t}{3 + t}, \quad t \ge 0$$

(a) Use a graphing calculator to sketch the graph of $N(t)$.

(b) Determine the size of the population as $t \to \infty$. We call this the **limiting population size**.

(c) Show that, at time $t = 3$, the size of the population is half its limiting size.

28. Logistic Growth Suppose that the size of a population at time t is given by

$$N(t) = \frac{100}{1 + 9e^{-t}}$$

for $t \ge 0$.
(a) Use a graphing calculator to sketch the graph of $N(t)$.

(b) Determine the size of the population as $t \to \infty$, using the basic rules for limits. Compare your answer with the graph that you sketched in (a).

29. Logistic Growth Suppose that the size of a population at time t is given by

$$N(t) = \frac{50}{1 + 3e^{-t}}$$

for $t \ge 0$.
(a) Use a graphing calculator to sketch the graph of $N(t)$.

(b) Determine the size of the population as $t \to \infty$, using the basic rules for limits. Compare your answer with the graph that you sketched in (a).

■ 3.4 The Sandwich Theorem and Some Trigonometric Limits

What happens during bungee jumping? The jumper is tied to an elastic rope, jumps off a bridge, and experiences damped oscillations until she comes to rest and will be hauled in to safety. The trajectory over time might resemble the function (Figure 3.17)

$$g(x) = e^{-x}\cos(10x), \quad x \ge 0$$

We suspect from the graph that

$$\lim_{x\to\infty} e^{-x}\cos(10x) = 0$$

If we wanted to calculate this limit, we would quickly see that none of the rules we have learned so far apply. Although $\lim_{x\to\infty} e^{-x} = 0$, we find that $\lim_{x\to\infty}\cos(10x)$ does not exist: The function $\cos(10x)$ oscillates between -1 and 1. Still, this property allows us to sandwich the function $g(x) = e^{-x}\cos(10x)$ between $f(x) = -e^{-x}$ and $h(x) = e^{-x}$. To do so, we note that from

$$-1 \le \cos(10x) \le 1$$

it follows that

$$-e^{-x} \le e^{-x}\cos(10x) \le e^{-x}$$

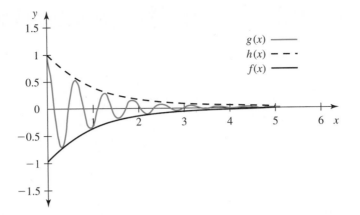

Figure 3.17 The graph of $f(x) = e^{-x} \cos x$, together with the two functions $g(x) = e^{-x}$ and $h(x) = -e^{-x}$.

Then, since

$$\lim_{x \to \infty} (-e^{-x}) = \lim_{x \to \infty} e^{-x} = 0$$

our function $g(x) = e^{-x} \cos(10x)$ gets squeezed in between the two functions $f(x) = -e^{-x}$ and $h(x) = e^{-x}$, which both go to 0 as x tends to infinity. Therefore,

$$\lim_{x \to \infty} e^{-x} \cos(10x) = 0$$

This useful method is known as the sandwich theorem. We will not prove it.

Sandwich Theorem If $f(x) \leq g(x) \leq h(x)$ for all x in an open interval that contains c (except possibly at c) and

$$\lim_{x \to c} f(x) = \lim_{x \to c} h(x) = L$$

then

$$\lim_{x \to c} g(x) = L$$

The theorem is called the sandwich theorem because we "sandwich" the function $g(x)$ between the two functions $f(x)$ and $h(x)$. Since $f(x)$ and $h(x)$ converge to the same value as $x \to c$, $g(x)$ also must converge to that value as $x \to c$, because it is squeezed in between $f(x)$ and $h(x)$. The sandwich theorem also applies to one-sided limits. We demonstrate how to use the sandwich theorem in the next example.

EXAMPLE 1 Show that

$$\lim_{x \to 0} x^2 \sin \frac{1}{x} = 0$$

Solution First, note that we cannot use Rule 3—which says that the limit of a product is equal to the product of the limits—because it requires that the limits of both factors exist. The limit of $\sin(1/x)$ as $x \to 0$ does not exist; instead, it diverges by oscillating. (See Example 8 of Section 3.1 for a similar limit.) However, we know that

$$-1 \leq \sin \frac{1}{x} \leq 1$$

for all $x \neq 0$. To go from this set of inequalities to one that involves $x^2 \sin \frac{1}{x}$, we need to multiply all three parts by x^2. Performing the multiplication, we find that

$$-x^2 \leq x^2 \sin \frac{1}{x} \leq x^2$$

Since $\lim_{x\to 0}(-x^2) = \lim_{x\to 0} x^2 = 0$, we can apply the sandwich theorem to obtain

$$\lim_{x\to 0} x^2 \sin \frac{1}{x} = 0$$

This limit is illustrated in Figure 3.18.

■

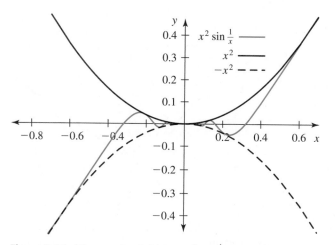

Figure 3.18 The graph of $f(x) = x^2 \sin \frac{1}{x}$.

EXAMPLE 2   Show that

$$\lim_{x\to 0} x \sin \frac{1}{x} = 0$$

Solution   As in Example 1, note that we cannot use Rule 3—which, again, says that the limit of a product is equal to the product of the limits—because it requires that the limits of both factors exist. The limit of $\sin(1/x)$ as $x \to 0$ does not exist; instead, it diverges by oscillating. (See Example 8 of Section 3.1 for a similar limit.) However, we know that

$$-1 \le \sin \frac{1}{x} \le 1$$

for all $x \neq 0$. To go from this set of inequalities to one that involves $x \sin \frac{1}{x}$, we need to multiply all three parts by x. Since multiplying an inequality by x reverses inequality signs when $x < 0$, we need to split the discussion into two cases, one involving $x > 0$, the other $x < 0$.

Multiplying all three parts by $x > 0$, we find that

$$-x \le x \sin \frac{1}{x} \le x$$

Since $\lim_{x\to 0^+}(-x) = \lim_{x\to 0^+} x = 0$, we can apply the sandwich theorem to obtain

$$\lim_{x\to 0^+} x \sin \frac{1}{x} = 0$$

We can repeat the same steps when we multiply by $x < 0$, except we now need to reverse the inequality signs. That is, for $x < 0$,

$$-x \ge x \sin \frac{1}{x} \ge x$$

Because $\lim_{x\to 0^-}(-x) = \lim_{x\to 0^-} x = 0$, we can again apply the sandwich theorem and get

$$\lim_{x\to 0^-} x \sin \frac{1}{x} = 0$$

The left-hand and right-hand limits are the same. Therefore, combining the two results, we find that

$$\lim_{x \to 0} x \sin \frac{1}{x} = 0$$

This limit is illustrated in Figure 3.19. ■

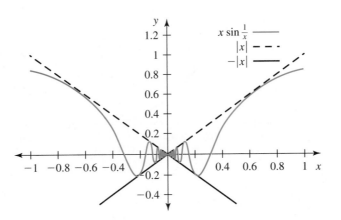

Figure 3.19 The sandwich theorem illustrated on $\lim_{x \to 0} x \sin(1/x)$.

The following two trigonometric limits are important for developing the differential calculus for trigonometric functions:

$$\lim_{x \to 0} \frac{\sin x}{x} = 1 \quad \text{and} \quad \lim_{x \to 0} \frac{1 - \cos x}{x} = 0$$

You should memorize these two limits, noting that the angle x is measured in radians. We will prove both statements. The proof of the first statement uses a nice geometric argument and the sandwich theorem; the second statement follows from the first.

Proof that $\lim_{x \to 0} \frac{\sin x}{x} = 1$ We will need to divide an inequality by $\sin x$. Since dividing an inequality by a negative number reverses the inequality sign (see Example 1), we will split the proof into two cases, one in which $0 < x < \pi/2$, the other in which $-\pi/2 < x < 0$. In the former case, both x and $\sin x$ are positive; in the latter, both x and $\sin x$ are negative. (Since we are interested in the limit as $x \to 0$, we can restrict the values of x to values close to 0.) We start with the case $0 < x < \pi/2$. In Figure 3.20, we draw the unit circle together with the triangles OAD and OBC. The angle x is measured in radians. Since $\overline{OB} = 1$, we find that

$$\text{arc length of } BD = x$$
$$\overline{OA} = \cos x$$
$$\overline{AD} = \sin x$$
$$\overline{BC} = \tan x$$

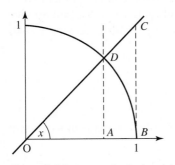

Figure 3.20 The unit circle with the triangles OAD and OBC.

Furthermore, using the symbol $\triangle$ to denote a triangle, we obtain

$$\text{area of } \triangle OAD \leq \text{area of sector } OBD \leq \text{area of } \triangle OBC$$

The area of a sector of central angle x (measured in radians) and radius r is $\frac{1}{2}r^2 x$. Therefore,

$$\frac{1}{2}\overline{OA} \cdot \overline{AD} \leq \frac{1}{2}\overline{OB}^2 \cdot x \leq \frac{1}{2}\overline{OB} \cdot \overline{BC}$$

or

$$\frac{1}{2}\cos x \sin x \le \frac{1}{2} \cdot 1^2 \cdot x \le \frac{1}{2} \cdot 1 \cdot \tan x$$

Dividing this pair of inequalities by $\frac{1}{2}\sin x$ (and noting that $\frac{1}{2}\sin x > 0$ for $0 < x < \pi/2$) yields

$$\cos x \le \frac{x}{\sin x} \le \frac{1}{\cos x}$$

On the rightmost part, we used the fact that $\tan x = \frac{\sin x}{\cos x}$. Taking reciprocals and reversing the inequality signs gives

$$\frac{1}{\cos x} \ge \frac{\sin x}{x} \ge \cos x$$

We can now take the limit as $x \to 0^+$. (Remember, we assumed that $0 < x < \pi/2$, so we can approach 0 only from the right.) We find that

$$\lim_{x \to 0^+} \cos x = 1 \qquad \text{and} \qquad \lim_{x \to 0^+} \frac{1}{\cos x} = \frac{1}{\lim_{x \to 0^+} \cos x} = 1$$

We now apply the sandwich theorem, which yields

$$\lim_{x \to 0^+} \frac{\sin x}{x} = 1$$

We have shown only that $\lim_{x \to 0^+} \frac{\sin x}{x} = 1$, but a similar argument can be carried out when $-\frac{\pi}{2} < x < 0$. In this case, $\lim_{x \to 0^-} \frac{\sin x}{x} = 1$. The left-hand and the right-hand limits are the same, and we conclude that

$$\lim_{x \to 0} \frac{\sin x}{x} = 1 \qquad\qquad ■$$

Proof that $\lim_{x \to 0} \frac{1 - \cos x}{x} = 0$ Multiplying both numerator and denominator of $f(x) = (1 - \cos x)/x$ by $1 + \cos x$, we can reduce the second statement to the first:

$$\lim_{x \to 0} \frac{1 - \cos x}{x} = \lim_{x \to 0} \frac{1 - \cos x}{x} \frac{1 + \cos x}{1 + \cos x}$$

$$= \lim_{x \to 0} \frac{1 - \cos^2 x}{x(1 + \cos x)}$$

Using the identity $\sin^2 x + \cos^2 x = 1$, we write this as

$$\lim_{x \to 0} \frac{\sin^2 x}{x(1 + \cos x)}$$

Rewriting again, we obtain

$$\lim_{x \to 0} \frac{\sin x}{x} \frac{\sin x}{1 + \cos x}$$

and we can determine the limit. First, we note that $\lim_{x \to 0} \frac{\sin x}{x}$ exists by the first statement, and $\lim_{x \to 0} \frac{\sin x}{1 + \cos x}$ exists because $1 + \cos x \ne 0$ for x close to 0. Then, by Rule 3 of the limit laws, the limit of a product is the product of the limits. We therefore find that

$$\lim_{x \to 0} \frac{\sin x}{x} \frac{\sin x}{1 + \cos x} = \lim_{x \to 0} \frac{\sin x}{x} \lim_{x \to 0} \frac{\sin x}{1 + \cos x} = 1 \cdot 0 = 0 \qquad ■$$

EXAMPLE 3 Find the following limits.

$$\textbf{(a)}\ \lim_{x\to0}\frac{\sin 3x}{5x}\qquad\textbf{(b)}\ \lim_{x\to0}\frac{\sin^2 x}{x^2}\qquad\textbf{(c)}\ \lim_{x\to0}\frac{\sec x-1}{x\sec x}$$

Solution **(a)** We cannot apply the first trigonometric limit directly. The trick is to substitute $z=3x$ and observe that $z\to0$ as $x\to0$. Then

$$\lim_{x\to0}\frac{\sin 3x}{5x}=\lim_{z\to0}\frac{\sin z}{5z/3}=\frac{3}{5}\lim_{z\to0}\frac{\sin z}{z}=\frac{3}{5}$$

(b) We note that

$$\lim_{x\to0}\frac{\sin^2 x}{x^2}=\lim_{x\to0}\left(\frac{\sin x}{x}\right)^2=\left(\lim_{x\to0}\frac{\sin x}{x}\right)^2=1$$

Here, we used the fact that the limit of a product is the product of the limits, provided that the individual limits exist.

(c) We first write $\sec x=1/\cos x$ and then multiply both numerator and denominator by $\cos x$:

$$\lim_{x\to0}\frac{\sec x-1}{x\sec x}=\lim_{x\to0}\frac{\frac{1}{\cos x}-1}{\frac{x}{\cos x}}$$

$$=\lim_{x\to0}\frac{\left(\frac{1}{\cos x}-1\right)\cos x}{\frac{x}{\cos x}\cos x}=\lim_{x\to0}\frac{1-\cos x}{x}=0 \qquad ■$$

Section 3.4 Problems

1. Let

$$f(x)=x^2\cos\frac{1}{x},\quad x\neq0$$

(a) Use a graphing calculator to sketch the graph of $y=f(x)$.

(b) Show that

$$-x^2\leq x^2\cos\frac{1}{x}\leq x^2$$

holds for $x\neq0$.

(c) Use your result in (b) and the sandwich theorem to show that

$$\lim_{x\to0}x^2\cos\frac{1}{x}=0$$

2. Let

$$f(x)=x\cos\frac{1}{x},\quad x\neq0$$

(a) Use a graphing calculator to sketch the graph of $y=f(x)$.

(b) Use the sandwich theorem to show that

$$\lim_{x\to0}x\cos\frac{1}{x}=0$$

3. Let

$$f(x)=\frac{\ln x}{x},\quad x>0$$

(a) Use a graphing calculator to graph $y=f(x)$.

(b) Use a graphing calculator to investigate the values of x for which

$$\frac{1}{x}\leq\frac{\ln x}{x}\leq\frac{1}{\sqrt{x}}$$

holds.

(c) Use your result in (b) to explain why the following is true:

$$\lim_{x\to\infty}\frac{\ln x}{x}=0$$

4. Let

$$f(x)=\frac{\sin x}{x},\quad x>0$$

(a) Use a graphing calculator to graph $y=f(x)$.

(b) Explain why you cannot use the basic rules for finding limits to compute

$$\lim_{x\to\infty}\frac{\sin x}{x}$$

(c) Show that

$$-\frac{1}{x}\leq\frac{\sin x}{x}\leq\frac{1}{x}$$

holds for $x>0$, and use the sandwich theorem to compute

$$\lim_{x\to\infty}\frac{\sin x}{x}$$

In Problems 5–20, evaluate the trigonometric limits.

5. $\lim_{x \to 0} \dfrac{\sin(2x)}{2x}$

6. $\lim_{x \to 0} \dfrac{\sin(2x)}{3x}$

7. $\lim_{x \to 0} \dfrac{\sin(5x)}{x}$

8. $\lim_{x \to 0} \dfrac{\sin x}{-x}$

9. $\lim_{x \to 0} \dfrac{\sin(\pi x)}{x}$

10. $\lim_{x \to 0} \dfrac{\sin(-\pi x/2)}{2x}$

11. $\lim_{x \to 0} \dfrac{\sin(\pi x)}{\sqrt{x}}$

12. $\lim_{x \to 0} \dfrac{\sin^2 x}{x}$

13. $\lim_{x \to 0} \dfrac{\sin x \cos x}{x(1-x)}$

14. $\lim_{x \to 0} \dfrac{1 - \cos^2 x}{x^2}$

15. $\lim_{x \to 0} \dfrac{1 - \cos x}{2x}$

16. $\lim_{x \to 0} \dfrac{1 - \cos(2x)}{3x}$

17. $\lim_{x \to 0} \dfrac{1 - \cos(5x)}{2x}$

18. $\lim_{x \to 0} \dfrac{1 - \cos(x/2)}{x}$

19. $\lim_{x \to 0} \dfrac{\sin x(1 - \cos x)}{x^2}$

20. $\lim_{x \to 0} \dfrac{\csc x - \cot x}{x \csc x}$

21. (a) Use a graphing calculator to sketch the graph of

$$f(x) = e^{ax} \sin x, \quad x \ge 0$$

for $a = -0.1, -0.01, 0, 0.01,$ and 0.1.

(b) Which part of the function $f(x)$ produces the oscillations that you see in the graphs sketched in (a)?

(c) Describe in words the effect that the value of a has on the shape of the graph of $f(x)$.

(d) Graph $f(x) = e^{ax} \sin x$, $g(x) = -e^{ax}$, and $h(x) = e^{ax}$ together in one coordinate system for **(i)** $a = 0.1$ and **(ii)** $a = -0.1$. [Use separate coordinate systems for (i) and (ii).] Explain what you see in each case. Show that

$$-e^{ax} \le e^{ax} \sin x \le e^{ax}$$

Use this pair of inequalities to determine the values of a for which

$$\lim_{x \to \infty} f(x)$$

exists, and find the limiting value.

3.5 Properties of Continuous Functions

3.5.1 The Intermediate-Value Theorem

As you hike up a mountain, the temperature decreases with increasing elevation. Suppose the temperature at the bottom of the mountain is 70°F and the temperature at the top of the mountain is 40°F. How do you know that at some time during your hike you must have crossed a point where the temperature was exactly 50°F? Your answer will probably be something like the following: "To go from 70°F to 40°F, I must have passed through 50°F, since 50°F is between 40°F and 70°F and the temperature changed continuously as I hiked up the mountain." This statement represents the content of the intermediate-value theorem.

> **The Intermediate-Value Theorem** Suppose that f is continuous on the closed interval $[a, b]$. If L is any real number with $f(a) < L < f(b)$ or $f(b) < L < f(a)$, then there exists at least one number c on the open interval (a, b) such that $f(c) = L$.

We will not prove this theorem, but Figure 3.21 should convince you that it is true. In the figure, f is continuous and defined on the closed interval $[a, b]$ with $f(a) < L < f(b)$. Therefore, the graph of $f(x)$ must intersect the line $y = L$ at least once on the open interval (a, b).

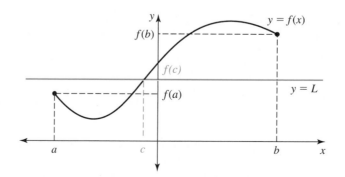

Figure 3.21 The intermediate-value theorem.

EXAMPLE 1   Let

$$f(x) = 3 + \sin x \quad \text{for } 0 \le x \le \frac{3\pi}{2}$$

Show that there exists at least one point c in $(0, 3\pi/2)$ such that $f(c) = 5/2$.

Solution   The graph of $f(x)$ is shown in Figure 3.22. First, note that $f(x)$ is defined on a closed interval and is continuous on $[0, 3\pi/2]$. Furthermore, we find that

$$f(0) = 3 + \sin 0 = 3 + 0 = 3$$
$$f\left(\frac{3\pi}{2}\right) = 3 + \sin\frac{3\pi}{2} = 3 + (-1) = 2$$

Given that

$$2 < \frac{5}{2} < 3$$

we conclude from the intermediate-value theorem that there exists a number c such that $f(c) = 5/2$. Note that the theorem does not tell us where c is or whether there is more than one such number.   ■

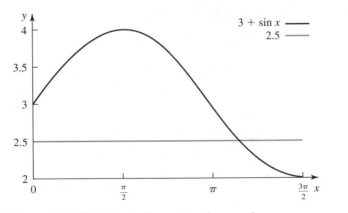

Figure 3.22 The intermediate-value theorem for $f(x) = 3 + \sin x, 0 \le x \le 3\pi/2$, and $L = 2.5$.

In applying the intermediate-value theorem, it is important to check that f is continuous. Discontinuous functions can easily miss values; for example, the floor function in Example 3 of Section 3.2 misses all numbers that are not integers.

As mentioned in Example 1, the intermediate-value theorem gives us only the existence of a number c; it does not tell us how many such points there are or where they are located.

You might wonder how such a result can be of any use. One important application is that the theorem can be used to find approximate roots (or solutions) of equations of the form $f(x) = 0$. We show how in the next example.

EXAMPLE 2   Find a root of the equation $x^5 - 7x^2 + 3 = 0$.

Solution   Let $f(x) = x^5 - 7x^2 + 3 = 0$. Because $f(x)$ is a polynomial, it is continuous for all $x \in \mathbf{R}$. Furthermore,

$$\lim_{x \to -\infty} f(x) = -\infty \quad \text{and} \quad \lim_{x \to \infty} f(x) = \infty$$

That is, if we choose a large enough interval $[a, b]$, then $f(a) < 0$ and $f(b) > 0$ and, therefore, there must be a number $c \in (a, b)$ such that $f(c) = 0$. This number c is a root of the equation $f(x) = 0$. The existence of c is guaranteed by the intermediate-value theorem.

To find a number c for which $f(c) = 0$, we use the **bisection method**. We start by finding a and b such that $f(a) < 0$ and $f(b) > 0$. For example,

$$f(-1) = -5 \quad \text{and} \quad f(2) = 7$$

The intermediate-value theorem then tells us that there must be a number in $(-1, 2)$ for which $f(c) = 0$. To locate this root with more precision, we take the midpoint of $(-1, 2)$, which is 0.5, and evaluate the function at $x = 0.5$. [The midpoint of the interval (a, b) is $(a + b)/2$.] Now, $f(0.5) \approx 1.28$ (rounded to two decimals). We thus have

$$f(-1) = -5 \quad f(0.5) \approx 1.28 \quad f(2) = 7$$

Using the intermediate-value theorem again, we can now guarantee a root in $(-1, 0.5)$, since $f(-1) < 0$ and $f(0.5) > 0$. Bisecting the new interval and computing the respective values of $f(x)$, we find that

$$f(-1) = -5 \quad f(-0.25) \approx 2.562 \quad f(0.5) \approx 1.28$$

Using the intermediate-value theorem yet again, we can guarantee a root in $(-1, -0.25)$, since $f(-1) < 0$ and $f(-0.25) > 0$. Repeating this procedure of bisecting and selecting a new (smaller) interval will eventually produce an interval that is small enough that we can locate the root to any desired accuracy. The first several steps are summarized in Table 3-1.

TABLE 3-1 Bisection Method

a	$\frac{a+b}{2}$	b	$f(a)$	$f(\frac{a+b}{2})$	$f(b)$
-1	0.5	2	-5	1.28	7
-1	0.5	2	-5	1.28	7
-1	-0.25	0.5	-5	2.562	1.28
-1	-0.625	-0.25	-5	0.170	2.562
-1	-0.8125	-0.625	-5	-1.975	0.170
-0.8125	-0.71875	-0.625	-1.975	-0.808	0.170
-0.71875	-0.671875	-0.625	-0.808	-0.297	0.170
-0.671875	-0.6484375	-0.625	-0.297	-0.0579	0.170
-0.6484375	-0.63671875	-0.625	-0.0579	0.0575	0.170
-0.6484375	-0.642578125	-0.63671875	-0.0579	9.9×10^{-5}	0.0575
-0.6484375		-0.642578125			

After nine steps, we find that there exists a root in

$$(-0.6484375, -0.642578125)$$

The length of this interval is 0.005859375. If we are satisfied with that level of precision, we can stop here and choose, for instance, the midpoint of the last interval as an approximate value for a root of the equation $x^5 - 7x^2 + 3 = 0$. The midpoint is

$$\frac{-0.642578125 + (-0.6484375)}{2} = -0.6455078125$$

$$\approx -0.646$$

(rounded to three decimals).

Note that the length of the interval decreases by a factor of $1/2$ at each step. That is, after nine steps, the length of the interval is $(1/2)^9$ of the length of the original interval. In this example, the length of the original interval was 3; hence, the length of the interval after nine steps is

$$3 \cdot \left(\frac{1}{2}\right)^9 = \frac{3}{512} = 0.005859375$$

as we saw. The bisection method is fairly slow when we need high accuracy. For instance, to reduce the length of the interval to 10^{-6}, we would need at least 22 steps, since

$$3 \cdot \left(\frac{1}{2}\right)^{21} > 10^{-6} > 3 \cdot \left(\frac{1}{2}\right)^{22}$$

In Section 5.7, we will learn a faster method.

Figure 3.23 shows the graph of $f(x) = x^5 - 7x^2 + 3$. We see that the graph intersects the x-axis three times. We found an approximation of the leftmost root of the equation $x^5 - 7x^2 + 3$. If we had used another starting interval—say, $[1, 2]$—we would have located an approximation of the rightmost root of the equation. ■

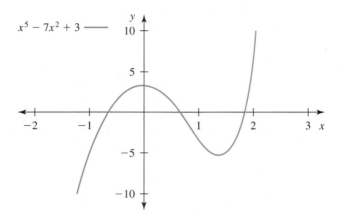

Figure 3.23 The graph of $f(x) = x^5 - 7x^2 + 3$.

■ 3.5.2 A Final Remark on Continuous Functions

Many functions in biology are in fact discontinuous. For example, if we measure the size of a population over time, we find that it takes on discrete values only (namely, nonnegative integers) and therefore changes discontinuously. However, if the population size is sufficiently large, an increase or decrease by 1 changes the population size so slightly that it might be justified to approximate it by a continuous function. For example, if we measure the number of bacteria, in millions, in a petri dish, then the number 2.1 would correspond to 2,100,000 bacteria. An increase by 1 results in 2,100,001 bacteria, or, if we measure the size in millions, in 2.100001, an increase of 10^{-6}.

Section 3.5 Problems

■ 3.5.1, 3.5.2

1. Let

$$f(x) = x^2 - 1, \quad 0 \le x \le 2$$

(a) Graph $y = f(x)$ for $0 \le x \le 2$.

(b) Show that

$$f(0) < 0 < f(2)$$

and use the intermediate-value theorem to conclude that there exists a number $c \in (0, 2)$ such that $f(c) = 0$.

2. Let

$$f(x) = x^3 - 2x + 3, \quad -3 \le x \le -1$$

(a) Graph $y = f(x)$ for $-3 \le x \le -1$.

(b) Use the intermediate-value theorem to conclude that

$$x^3 - 2x + 3 = 0$$

has a solution in $(-3, -1)$.

3. Let

$$f(x) = \sqrt{x^2 + 2}, \quad 1 \le x \le 2$$

(a) Graph $y = f(x)$ for $1 \le x \le 2$.

(b) Use the intermediate-value theorem to conclude that

$$\sqrt{x^2 + 2} = 2$$

has a solution in $(1, 2)$.

4. Let

$$f(x) = \sin x - x, \quad -1 \le x \le 1$$

(a) Graph $y = f(x)$ for $-1 \le x \le 1$.

(b) Use the intermediate-value theorem to conclude that

$$\sin x = x$$

has a solution in $(-1, 1)$.

5. Use the intermediate-value theorem to show that

$$e^{-x} = x$$

has a solution in $(0, 1)$.

6. Use the intermediate-value theorem to show that

$$\cos x = x$$

has a solution in $(0, 1)$.

7. Use the bisection method to find a solution of

$$e^{-x} = x$$

that is accurate to two decimal places.

8. Use the bisection method to find a solution of

$$\cos x = x$$

that is accurate to two decimal places.

9. (a) Use the bisection method to find a solution of $3x^3 - 4x^2 - x + 2 = 0$ that is accurate to two decimal places.

(b) Graph the function $f(x) = 3x^3 - 4x^2 - x + 2$.

(c) Which solution did you locate in (a)? Is it possible in this case to find the other solution by using the bisection method together with the intermediate-value theorem?

10. In Example 2, how many steps are required to guarantee that the approximate root is within 0.0001 of the true value of the root?

11. Suppose that the number of individuals in a population at time t is given by

$$N(t) = \frac{54t}{13 + t}, \quad t \geq 0 \qquad (3.7)$$

(a) Use a calculator to confirm that $N(10)$ is approximately 23.47826. Considering that the number of individuals in a population is an integer, how should you report your answer?

(b) Now suppose that $N(t)$ is given by the same function (3.7), but that the size of the population is measured in millions. How should you report the population size at time $t = 10$? Make some reasonable assumptions about the accuracy of a measurement for the size of such a large population.

(c) Discuss the use of continuous functions in both (a) and (b).

12. Suppose that the biomass of a population at time t is given by

$$B(t) = \frac{32.00t}{17.00 + t}, \quad t \geq 0 \qquad (3.8)$$

(a) Use a calculator to confirm that $B(10)$ is approximately 1.185185. Considering the function $B(t)$, how many significant figures should you report in your answer?

(b) Discuss the use of continuous functions in this problem.

13. Explain why a polynomial of degree 3 has at least one root.

14. Explain why a polynomial of degree n, where n is an odd number, has at least one root.

15. Explain why $y = x^2 - 4$ has at least two roots.

16. On the basis of the intermediate-value theorem, what can you say about the number of roots of a polynomial of even degree?

■ 3.6 A Formal Definition of Limits (Optional)

The ancient Greeks used limiting procedures to compute areas, such as the area of a circle, by the "method of exhaustion." In this method, a region was covered (or "exhausted") as closely as possible by triangles. Adding the areas of the triangles then yielded an approximation of the area of the region of interest. Newton and Leibniz, the inventors of calculus, were aware of the importance of taking limits in their development of the subject; however, they did not give a rigorous definition of the procedure. The French mathematician Augustin-Louis Cauchy (1789–1857) was the first to develop a rigorous definition of limits; the definition we will use goes back to the German mathematician Karl Weierstrass (1815–1897).

Before we write the formal definition, let's return to the informal one. In that definition, we stated that $\lim_{x \to c} f(x) = L$ means that the value of $f(x)$ can be made arbitrarily close to L whenever x is sufficiently close to c. But just how close is sufficient? Take Example 1 from Section 3.1: Suppose we wish to show that

$$\lim_{x \to 2} x^2 = 4$$

without using the continuity of $y = x^2$, which itself was based on $\lim_{x \to c} x = c$ [Equation (3.3)]. What would we have to do? We would need to show that x^2 can be made arbitrarily close to 4 for all values of x sufficiently close, but not equal, to 2. (In what follows, we will always exclude $x = 2$ from the discussion, since the value of x^2 at $x = 2$ is irrelevant in finding the limit.) Suppose we wish to make x^2 within 0.01 of 4; that is, we want $|x^2 - 4| < 0.01$. Does this inequality hold for all x sufficiently close, but not equal, to 2? We begin with

$$|x^2 - 4| < 0.01$$

which is equivalent to

$$-0.01 < x^2 - 4 < 0.01$$
$$3.99 < x^2 < 4.01$$
$$\sqrt{3.99} < |x| < \sqrt{4.01}$$

Now, $\sqrt{3.99} = 1.997498\ldots$ and $\sqrt{4.01} = 2.002498\ldots$. We therefore find that values of $x \neq 2$ in the interval $(1.998, 2.002)$ satisfy $|x^2 - 4| < 0.01$. (We chose a somewhat smaller interval than indicated, to get an interval that is symmetric about 2.) That is, for all values of x within 0.002 of 2 but not equal to 2 (i.e., $0 < |x - 2| < 0.002$), x^2 is within the prescribed precision—that is, within 0.01 of 4.

You might think about this example in the following way: Suppose that you wish to stake out a square of area 4 m². Each side of your square is 2 m long. You bring along a stick, which you cut to a length of 2 m. We can then ask: How accurately do we need to cut the stick so that the area will be within a prescribed precision? Our prescribed precision was 0.01, and we found that if we cut the stick within 0.002 of 2 m, we would be able to obtain the prescribed precision.

There is nothing special about 0.01; we could have chosen any other degree of precision and would have found a corresponding interval of x-values. We translate this procedure into a formal definition of limits. (See Figure 3.24.)

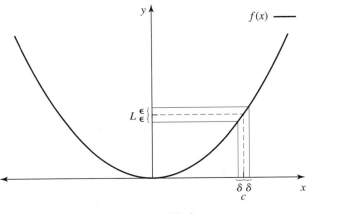

Figure 3.24 The ϵ–δ definition of limits.

Definition The statement
$$\lim_{x \to c} f(x) = L$$
means that, for every $\epsilon > 0$, there exists a number $\delta > 0$ such that

$$|f(x) - L| < \epsilon \quad \text{whenever} \quad 0 < |x - c| < \delta$$

Note that, as in the informal definition of limits, we exclude the value $x = c$ from the statement. (This is done in the inequality $0 < |x - c|$.) To apply the formal definition, we first need to guess the limiting value L. We then choose an $\epsilon > 0$, the prescribed precision, and try to find a $\delta > 0$ such that $f(x)$ is within ϵ of L whenever x is within δ of c but not equal to c. [In our example, $f(x) = x^2$, $c = 2$, $L = 4$, $\epsilon = 0.01$, and $\delta = 0.002$.]

EXAMPLE 1   Show that

$$\lim_{x \to 1} (2x - 3) = -1$$

Solution   We let $f(x) = 2x - 3$. Our guess for the limiting value is $L = -1$. Then

$$|f(x) - L| = |2x - 3 - (-1)|$$
$$= |2x - 2|$$
$$= 2|x - 1|$$

We now choose $\epsilon > 0$. (ϵ is arbitrary, and we do not specify it because our statement needs to hold for all $\epsilon > 0$.) Our goal is to find a $\delta > 0$ such that $2|x - 1| < \epsilon$ whenever x is within δ of 1 but not equal to 1; that is, $0 < |x - 1| < \delta$. The value of δ will typically depend on our choice of ϵ. Since $|x - 1| < \delta$ implies that $2|x - 1| < 2\delta$, we should try $2\delta = \epsilon$. If we choose $\delta = \epsilon/2$, then, indeed,

$$|f(x) - L| = 2|x - 1| < 2\delta = 2\frac{\epsilon}{2} = \epsilon$$

This means that, for every $\epsilon > 0$, we can find a number $\delta > 0$ (namely, $\delta = \epsilon/2$) such that

$$|f(x) - (-1)| < \epsilon \qquad \text{whenever} \qquad 0 < |x - 1| < \delta$$

But this is exactly the definition of

$$\lim_{x \to 1} (2x - 3) = -1 \qquad\qquad ■$$

EXAMPLE 2   We promised in Section 3.2 that we would show that

$$\lim_{x \to c} x = c$$

Solution   Let $f(x) = x$. We need to show that, for every $\epsilon > 0$, there corresponds a number $\delta > 0$ such that

$$|x - c| < \epsilon \qquad \text{whenever} \qquad 0 < |x - c| < \delta \qquad (3.9)$$

This immediately suggests that we should choose $\delta = \epsilon$, and, indeed, if $\delta = \epsilon$, then (3.9) holds. ■

Let's look at an example in which $f(x)$ is not linear.

EXAMPLE 3   Use the formal definition of limits to show that

$$\lim_{x \to 0} x^3 = 0$$

Solution   We need to show that, for every $\epsilon > 0$, there corresponds a number $\delta > 0$ such that

$$|x^3| < \epsilon \quad \text{whenever} \quad 0 < |x| < \delta \qquad (3.10)$$

Now, $|x^3| < \epsilon$ is equivalent to

$$-\epsilon < x^3 < \epsilon$$
$$-\epsilon^{1/3} < x < \epsilon^{1/3}$$

This pair of inequalities suggests that we set $\delta = \epsilon^{1/3}$. Accordingly, if $0 < |x| < \epsilon^{1/3}$, then

$$-\epsilon^{1/3} < x < \epsilon^{1/3}$$

or

$$-\epsilon < x^3 < \epsilon$$

which is the same as $|x^3| < \epsilon$. ■

We can also use the formal definition to show that a limit does not exist.

EXAMPLE 4   Show that

$$\lim_{x \to 0} \frac{|x|}{x}$$

does not exist.

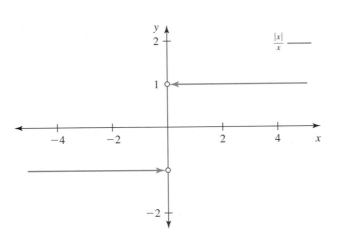

Figure 3.25 The graph of $f(x) = \frac{|x|}{x}$ in Example 4: The limit of $\frac{|x|}{x}$ as x tends to 0 does not exist.

Solution   Showing that this limit does not exist is tricky. (See Figure 3.25.) The approach is as follows: First, we set $f(x) = |x|/x, x \neq 0$. Then we assume that the limit exists and try to find a contradiction.[1] Suppose, then, that there exists an L such that

$$\lim_{x \to 0} \frac{|x|}{x} = L$$

If we look at Figure 3.25, we see that if, for instance, we choose $L = 1$, then we cannot get close to L when x is less than 0. Similarly, we see that, for any value of L, either the distance to $+1$ exceeds 1 or the distance to -1 exceeds 1. That is, regardless of the value of L, if $\epsilon < 1$, we will not be able to find a value of δ such that if $0 < |x| < \delta$, then $|f(x) - L| < \epsilon$, since $f(x)$ takes on both the values $+1$ and -1 for $0 < |x| < \delta$. Therefore, $\lim_{x \to 0} \frac{|x|}{x}$ does not exist.   ■

In the previous section, we considered an example in which $\lim_{x \to c} f(x) = \infty$. This statement can be made precise as well.

Definition The statement
$$\lim_{x \to c} f(x) = \infty$$
means that, for every $M > 0$, there exists a $\delta > 0$ such that

$$f(x) > M \quad \text{whenever} \quad 0 < |x - c| < \delta$$

Similar definitions hold for the case when $f(x)$ decreases without bound as $x \to c$ and for one-sided limits. We will not give definitions for all possible cases; rather, we illustrate how we would use such a definition.

---

[1] This approach is called "indirect proof" or "*reductio ad absurdum.*" We assume the opposite of what we wish to prove, and then we show that assuming the opposite leads to a contradiction. Therefore, what we originally sought to prove must be true.

EXAMPLE 5 Show that

$$\lim_{x \to 0} \frac{1}{x^2} = \infty$$

Figure 3.26 The function $f(x) = \frac{1}{x^2}$ in Example 5: The limit of $\frac{1}{x^2}$ as x tends to 0 does not exist.

Solution The graph of $f(x) = 1/x^2$, $x \neq 0$, is shown in Figure 3.26. We fix $M > 0$. (Again, M is arbitrary, because our solution must hold for all $M > 0$.) We need to find a $\delta > 0$ such that $f(x) > M$ whenever $0 < |x| < \delta$. (Note that $c = 0$.) We start with the inequality $f(x) > M$ and try to determine how to choose δ. We have

$$\frac{1}{x^2} > M \qquad \text{is the same as} \qquad x^2 < \frac{1}{M}$$

Taking square roots on both sides, we find that

$$|x| < \frac{1}{\sqrt{M}}$$

This suggests that we should choose $\delta = 1/\sqrt{M}$. Let's try that value: Given $M > 0$, we choose $\delta = 1/\sqrt{M}$. If $0 < |x| < \delta$, then

$$x^2 < \delta^2, \qquad \text{or} \qquad \frac{1}{x^2} > \frac{1}{\delta^2} = M$$

That is, $1/x^2 > M$ whenever $0 < |x| < \delta = 1/\sqrt{M}$. ■

There is also a formal definition when $x \to \infty$ (and a similar one for $x \to -\infty$). This definition is analogous to that in Chapter 2.

Definition The statement

$$\lim_{x \to \infty} f(x) = L$$

means that, for every $\epsilon > 0$, there exists an $x_0 > 0$ such that

$$|f(x) - L| < \epsilon \quad \text{whenever } x > x_0$$

Note that, in the definition, x_0 is a real number.

EXAMPLE 6 Show that

$$\lim_{x \to \infty} \frac{x}{0.5 + x} = 1$$

Solution　This limit is illustrated in Figure 3.27. You can see that $f(x) = x/(0.5 + x)$, $x \geq 0$, is in the strip of width 2ϵ and centered at the limiting value $L = 1$ for all values of x greater than x_0. (We assume that $0 < \epsilon < 1$, since, when $\epsilon \geq 1$, the choice $x_0 = 1$ works.) We now determine x_0 when $\epsilon < 1$. To do this, we try to solve

$$\left| \frac{x}{0.5 + x} - 1 \right| < \epsilon$$

for $\epsilon > 0$. This inequality is equivalent to the pair of inequalities

$$-\epsilon < \frac{x}{0.5 + x} - 1 < \epsilon$$

or, after adding 1 to all three parts,

$$1 - \epsilon < \frac{x}{0.5 + x} < 1 + \epsilon$$

Since $\frac{x}{0.5+x} < 1$ for $x > 0$, the right-hand inequality always holds. We therefore need only consider

$$1 - \epsilon < \frac{x}{0.5 + x}$$

Because we are interested in the behavior of $f(x)$ as $x \to \infty$, we need only look at large values of x. Multiplying by $0.5 + x$ (and noticing that we can assume that $0.5 + x > 0$, because we let $x \to \infty$), we obtain

$$(1 - \epsilon)(0.5 + x) < x$$

Solving for x yields

$$(1 - \epsilon)(0.5) < x - x(1 - \epsilon)$$
$$(1 - \epsilon)(0.5) < \epsilon x$$
$$\frac{1 - \epsilon}{2\epsilon} < x$$

For instance, if $\epsilon = 0.1$ (as in Figure 3.27), then

$$x > \frac{0.9}{0.2} = 4.5$$

That is, we would set $x_0 = 4.5$ and conclude that, for $x > 4.5$, $|f(x) - 1| < 0.1$.

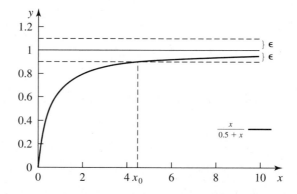

Figure 3.27 The function $f(x) = \frac{x}{0.5+x}$ in Example 6: The limit of $f(x)$ as x tends to infinity is 1.

More generally, we find that, for every $0 < \epsilon < 1$, there exists an

$$x_0 = \frac{1 - \epsilon}{2\epsilon}$$

such that

$$|f(x) - 1| < \epsilon \quad \text{whenever} \quad x > x_0 \qquad ■$$

Section 3.6 Problems

1. Find the values of x such that

$$|2x - 1| < 0.01$$

2. Find the values of x such that

$$|3x - 9| < 0.01$$

3. Find the values of x such that

$$|x^2 - 9| < 0.1$$

4. Find the values of x such that

$$|2\sqrt{x} - 5| < 0.1$$

5. Let

$$f(x) = 2x - 1, \ x \in \mathbf{R}$$

(a) Graph $y = f(x)$ for $-3 \le x \le 5$.
(b) For which values of x is $y = f(x)$ within 0.1 of 3? [*Hint*: Find values of x such that $|(2x - 1) - 3| < 0.1$.]
(c) Illustrate your result in (b) on the graph that you obtained in (a).

6. Let

$$f(x) = \sqrt{x}, \quad x \ge 0$$

(a) Graph $y = f(x)$ for $0 \le x \le 6$.
(b) For which values of x is $y = f(x)$ within 0.2 of 1? (*Hint*: Find values of x such that $|\sqrt{x} - 1| < 0.2$.)
(c) Illustrate your result in (b) on the graph that you obtained in (a).

7. Let

$$f(x) = \frac{1}{x}, \quad x > 0$$

(a) Graph $y = f(x)$ for $0 < x \le 4$.
(b) For which values of x is $y = f(x)$ greater than 4?
(c) Illustrate your result in (b) on the graph that you obtained in (a).

8. Let

$$f(x) = e^{-x}, \quad x \ge 0$$

(a) Graph $y = f(x)$ for $0 \le x \le 6$.
(b) For which values of x is $y = f(x)$ less than 0.1?
(c) Illustrate your result in (b) on the graph that you obtained in (a).

In Problems 9–22, use the formal definition of limits to prove each statement.

9. $\lim\limits_{x \to 2} (2x - 1) = 3$

10. $\lim\limits_{x \to 0} x^2 = 0$

11. $\lim\limits_{x \to 0} x^5 = 0$

12. $\lim\limits_{x \to 1} \dfrac{1}{x} = 1$

13. $\lim\limits_{x \to 0} \dfrac{4}{x^2} = \infty$

14. $\lim\limits_{x \to 0} \dfrac{-2}{x^2} = -\infty$

15. $\lim\limits_{x \to 0} \dfrac{1}{x^4} = \infty$

16. $\lim\limits_{x \to 3} \dfrac{1}{(x - 3)^2} = \infty$

17. $\lim\limits_{x \to \infty} \dfrac{2}{x^2} = 0$

18. $\lim\limits_{x \to \infty} e^{-x} = 0$

19. $\lim\limits_{x \to \infty} \dfrac{x}{x + 1} = 1$

20. $\lim\limits_{x \to -\infty} \dfrac{x}{x + 1} = 1$

21. $\lim\limits_{x \to c} (mx) = mc$, where m is a constant

22. $\lim\limits_{x \to c} (mx + b) = mc + b$, where m and b are constants

Chapter 3 Key Terms

Discuss the following definitions and concepts:

1. Limit of $f(x)$ as x approaches c
2. One-sided limits
3. Infinite limits
4. Divergence by oscillations
5. Convergence
6. Divergence
7. Limit laws
8. Continuity
9. One-sided continuity
10. Continuous function
11. Removable discontinuity
12. Sandwich theorem
13. Trigonometric limits
14. Intermediate-value theorem
15. Bisection method
16. ϵ–δ definition of limits

Chapter 3 Review Problems

In Problems 1–4, determine where each function is continuous. Investigate the behavior as $x \to \pm\infty$. Use a graphing calculator to sketch the corresponding graphs.

1. $f(x) = e^{-|x|}$

2. $f(x) = \begin{cases} \dfrac{\sin x}{x} & \text{if } x \neq 0 \\ 1 & \text{if } x = 0 \end{cases}$

3. $f(x) = \dfrac{2}{e^x + e^{-x}}$

4. $f(x) = \dfrac{1}{\sqrt{x^2 - 1}}$

5. Sketch the graph of a function that is discontinuous from the left and continuous from the right at $x = 1$.

6. Sketch the graph of a function $f(x)$ that is continuous on $[0, 2]$, except at $x = 1$, where $f(1) = 4$, $\lim_{x \to 1^-} f(x) = 2$, and $\lim_{x \to 1^+} f(x) = 3$.

7. Sketch the graph of a continuous function on $[0, \infty)$ with $f(0) = 0$ and $\lim_{x \to \infty} f(x) = 1$.

8. Sketch the graph of a continuous function on $(-\infty, \infty)$ with $f(0) = 1$, $f(x) \ge 0$ for all $x \in \mathbf{R}$, and $\lim_{x \to \pm\infty} f(x) = 0$.

9. Show that the floor function

$$f(x) = \lfloor x \rfloor$$

is continuous from the right, but discontinuous from the left at $x = -2$.

10. Suppose $f(x)$ is continuous on the interval $[1, 3]$. If $f(1) = 0$ and $f(3) = 2$, explain why there must be a number $c \in (1, 3)$ such that $f(c) = 1$.

11. Population Size Assume that the size of a population at time t is

$$N(t) = \frac{at}{k + t}, \quad t \ge 0$$

where a and k are positive constants. Suppose that the limiting population size is

$$\lim_{t \to \infty} N(t) = 1.24 \times 10^6$$

and that, at time $t = 5$, the population size is half the limiting population size. Use the preceding information to determine the constants a and k.

12. Population Size Suppose that

$$N(t) = 10 + 2e^{-0.3t} \sin t, \quad t \geq 0$$

describes the size of a population (in millions) at time t (measured in weeks).

(a) Use a graphing calculator to sketch the graph of $N(t)$, and describe in words what you see.

(b) Give lower and upper bounds on the size of the population; that is, find N_1 and N_2 such that, for all $t \geq 0$,

$$N_1 \leq N(t) \leq N_2$$

(c) Find $\lim_{t \to \infty} N(t)$. Interpret this expression.

13. Physiology Suppose that an organism reacts to a stimulus only when the stimulus exceeds a certain threshold. Assume that the stimulus is a function of time t and that it is given by

$$s(t) = \sin(\pi t), \quad t \geq 0$$

The organism reacts to the stimulus and shows a certain reaction when $s(t) \geq 1/2$. Define a function $g(t)$ such that $g(t) = 0$ when the organism shows no reaction at time t and $g(t) = 1$ when the organism shows the reaction.

(a) Plot $s(t)$ and $g(t)$ in the same coordinate system.

(b) Is $s(t)$ continuous? Is $g(t)$ continuous?

14. Tree Height The following function describes the height of a tree as a function of age:

$$f(x) = 132e^{-20/x}, \quad x \geq 0$$

Find $\lim_{x \to \infty} f(x)$.

15. Predator–Prey Model There are a number of mathematical models that describe predator–prey interactions. Typically, they share the feature that the number of prey eaten per predator increases with the density of the prey. In the simplest version, the number of encounters with prey per predator is proportional to the product of the total number of prey and the period over which the predators search for prey. That is, if we let N be the number of prey, P be the number of predators, T be the period available for searching, and N_e be the number of encounters with prey, then

$$\frac{N_e}{P} = aTN \qquad (3.11)$$

where a is a positive constant. The quantity N_e/P is the number of prey encountered per predator.

(a) Set $f(N) = aTN$, and sketch the graph of $f(N)$ when $a = 0.1$ and $T = 2$ for $N \geq 0$.

(b) Predators usually spend some time eating the prey that they find. Therefore, not all of the time T can be used for searching. The actual searching time is reduced by the per-prey handling time T_h and can be written as

$$T - T_h \frac{N_e}{P}$$

Show that if $T - T_h \frac{N_e}{P}$ is substituted for T in (3.11), then

$$\frac{N_e}{P} = \frac{aTN}{1 + aT_hN} \qquad (3.12)$$

Define

$$g(N) = \frac{aTN}{1 + aT_hN}$$

and graph $g(N)$ for $N \geq 0$ when $a = 0.1$, $T = 2$, and $T_h = 0.1$.

(c) Show that (3.12) reduces to (3.11) when $T_h = 0$.

(d) Find

$$\lim_{N \to \infty} \frac{N_e}{P}$$

in the cases when $T_h = 0$ and when $T_h > 0$. Explain, in words, the difference between the two cases.

16. Community Respiration Duarte and Agustí (1998) investigated the CO_2 balance of aquatic ecosystems. They related the community respiration rates (R) to the gross primary production rates (P) of aquatic ecosystems. (Both quantities were measured in the same units.) They made the following statement:

> Our results confirm the generality of earlier reports that the relation between community respiration rate and gross production is not linear. Community respiration is scaled as the approximate two-thirds power of gross production.

(a) Use the preceding quote to explain why

$$R = aP^b$$

can be used to describe the relationship between the community respiration rates (R) and the gross primary production rates (P). What value would you assign to b on the basis of their quote?

(b) Suppose that you obtained data on the gross production and respiration rates of a number of freshwater lakes. How would you display your data graphically to quickly convince an audience that the exponent b in the power equation relating R and P is indeed approximately 2/3? (*Hint*: Use an appropriate log transformation.)

(c) The ratio R/P for an ecosystem is important in assessing the global CO_2 budget. If respiration exceeds production (i.e., $R > P$), then the ecosystem acts as a carbon dioxide source, whereas if production exceeds respiration (i.e., $P > R$), then the ecosystem acts as a carbon dioxide sink. Assume now that the exponent in the power equation relating R and P is 2/3. Show that the ratio R/P, as a function of P, is continuous for $P > 0$. Furthermore, show that

$$\lim_{P \to 0^+} \frac{R}{P} = \infty$$

and

$$\lim_{P \to \infty} \frac{R}{P} = 0$$

Use a graphing calculator to sketch the graph of the ratio R/P as a function of P for $P > 0$. (Experiment with the graphing calculator to see how the value of a affects the graph.)

(d) Use your results in (c) and the intermediate-value theorem to conclude that there exists a value P^* such that the ratio R/P at P^* is equal to 1. On the basis of your graph in (c), is there more than one such value P^*?

(e) Use your results in (d) to identify production rates P where the ratio $R/P > 1$ (i.e., where respiration exceeds production).

(f) Use your results in (a)–(e) to explain the following quote from Duarte and Agustí:

> Unproductive aquatic ecosystems … tend to be heterotrophic ($R > P$), and act as carbon dioxide sources.

17. Hyperbolic functions are used in the sciences. We take a look at the following three examples: the hyperbolic sine, $\sinh x$; the hyperbolic cosine, $\cosh x$; and the hyperbolic tangent, $\tanh x$, defined respectively as

$$\sinh x = \frac{e^x - e^{-x}}{2}, \quad x \in \mathbf{R}$$

$$\cosh x = \frac{e^x + e^{-x}}{2}, \quad x \in \mathbf{R}$$

$$\tanh x = \frac{e^x - e^{-x}}{e^x + e^{-x}}, \quad x \in \mathbf{R}$$

(a) Show that these three hyperbolic functions are continuous for all $x \in \mathbf{R}$. Use a graphing calculator to sketch the graphs of all three functions.

(b) Find

$$\lim_{x\to\infty} \sinh x \qquad \lim_{x\to-\infty} \sinh x$$
$$\lim_{x\to\infty} \cosh x \qquad \lim_{x\to-\infty} \cosh x$$
$$\lim_{x\to\infty} \tanh x \qquad \lim_{x\to-\infty} \tanh x$$

(c) Show that the two identities

$$\cosh^2 x - \sinh^2 x = 1$$

and

$$\tanh x = \frac{\sinh x}{\cosh x}$$

are valid.

(d) Show that $\sinh x$ and $\tanh x$ are odd functions and that $\cosh x$ is even.

(*Note*: It can be shown that if a flexible cable is suspended between two points at equal heights, the shape of the resulting curve is given by the hyperbolic cosine function. This curve is called a *catenary*.)

Differentiation

Differential calculus allows us to solve two of the basic problems that we mentioned in Chapter 1: constructing a tangent line to a curve (Figure 4.1) and finding maxima and minima of a curve (Figure 4.2). The solutions to these two problems, by themselves, cannot explain the impact calculus has had on the sciences. Calculus is one of the most important analytical tools for investigating dynamic problems. Applications of differential calculus in the life sciences include simple growth models, interactions between organisms, invasions of organisms, the working of neurons, enzymatic reactions, harvesting models in fishery, epidemiological modeling, changes of gene frequencies under random mating, evolutionary strategies, and many others.

Figure 4.1 Tangent line to a curve at a point.

Figure 4.2 Maxima and minima of a curve.

Growth models will be of particular interest to us. Let's revisit the example at the beginning of Chapter 3, in which we looked at a population whose size at time t is given by $N(t)$. The average growth rate during the time interval $[t, t + h]$ is equal to

$$[\text{average growth rate}] = \frac{[\text{change in population size}]}{[\text{length of time interval}]} = \frac{\Delta N}{\Delta t}$$

where

$$\Delta N = N(t + h) - N(t) \quad \text{and} \quad \Delta t = (t + h) - t = h$$

Thus,

$$\frac{\Delta N}{\Delta t} = \frac{N(t + h) - N(t)}{h}$$

The instantaneous rate of growth is defined as the limit of $\Delta N/\Delta t$ as $\Delta t \to 0$ (or $h \to 0$), or

$$\lim_{\Delta t \to 0} \frac{\Delta N}{\Delta t} = \lim_{h \to 0} \frac{N(t+h) - N(t)}{h}$$

provided that this limit exists.

We are interested in the geometric interpretation of the limit when it exists. When we draw a straight line through the points $(t, N(t))$ and $(t+h, N(t+h))$, we obtain the **secant line**. The slope of this line is given by the quantity $\Delta N/\Delta t$ (Figure 4.3). In the limit as $\Delta t \to 0$, the secant line converges to the line that touches the graph at the point $(t, N(t))$. This line is called the **tangent line** at the point $(t, N(t))$ (Figure 4.4). The limit of $\Delta N/\Delta t$ as the length of the time interval $[t, t+h]$ goes to 0 (i.e., $\Delta t \to 0$ or $h \to 0$) will therefore be equal to the slope of the tangent line at $(t, N(t))$. We denote the limiting value of $\Delta N/\Delta t$ as $\Delta t \to 0$ by $N'(t)$ (read "N prime of t") and call this quantity the **derivative** of $N(t)$. That is,

$$N'(t) = \lim_{\Delta t \to 0} \frac{\Delta N}{\Delta t} = \lim_{h \to 0} \frac{N(t+h) - N(t)}{h}$$

provided that this limit exists.

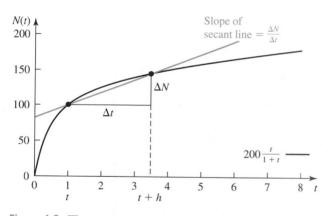

Figure 4.3 The average growth rate $\Delta N/\Delta t$ is equal to the slope of the secant line.

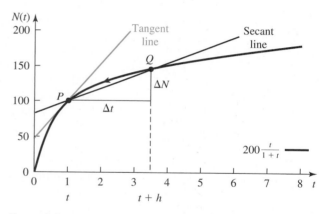

Figure 4.4 The instantaneous growth rate is the limit $\Delta N/\Delta t$ as $\Delta t \to 0$. Geometrically, the point Q moves toward the point P on the graph of $N(t)$, and the secant line through P and Q becomes the tangent line at P. The instantaneous growth rate is then equal to the slope of the tangent line.

Finding derivatives is the topic of this chapter.

4.1 Formal Definition of the Derivative

> **Definition** The derivative of a function f at x, denoted by $f'(x)$, is
>
> $$f'(x) = \lim_{h \to 0} \frac{f(x+h) - f(x)}{h}$$
>
> provided that the limit exists.

If the limit exists, then we say that f is differentiable at x. The phrase "provided that the limit exists" is crucial: If we take an arbitrary function f, the limit may not exist. In fact, we saw many examples in the previous chapter in which limits did not exist. The geometric interpretation will help us to understand when the limit exists and under which conditions we cannot expect the limit to exist. Notice that $\lim_{h \to 0}$ is a two-sided limit (i.e., we approach 0 from both the negative and the positive side). The quotient

$$\frac{f(x+h) - f(x)}{h}$$

is called the **difference quotient**, and we denote it by $\frac{\Delta f}{\Delta x}$. (See Figure 4.5.)

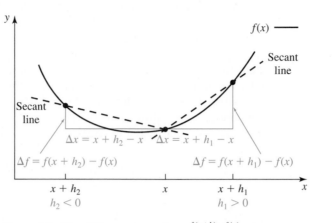

Figure 4.5 The difference quotient $\frac{f(x+h)-f(x)}{h}$ when $h = h_1 > 0$ and $h = h_2 < 0$.

We say that f is differentiable on (a, b) if f is differentiable at every $x \in (a, b)$. (Since the limit in the definition is two sided, we exclude the endpoints of the interval. At endpoints, only one-sided limits can be computed, which yield one-sided derivatives.)

If we want to compute the derivative at $x = c$, we can also write

$$f'(c) = \lim_{x \to c} \frac{f(x) - f(c)}{x - c}$$

which emphasizes that the point $(x, f(x))$ converges to the point $(c, f(c))$ as we take the limit as $x \to c$ (Figure 4.6). This approach will be important when we discuss the geometric interpretation of the derivative in the next subsection.

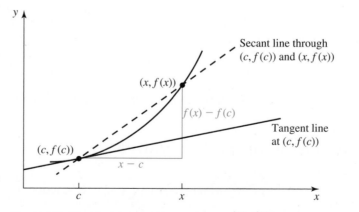

Figure 4.6 The derivative $f'(c) = \lim_{x \to c} \frac{f(x)-f(c)}{x-c}$ is the slope of the tangent line at $(c, f(c))$.

There is more than one way to write the derivative of a function $y = f(x)$. The following expressions are equivalent:

$$y' = \frac{dy}{dx} = f'(x) = \frac{df}{dx} = \frac{d}{dx} f(x)$$

The notation $\frac{df}{dx}$ goes back to Leibniz and is called **Leibniz notation**. (Leibniz had a real talent for finding good notation.) It should remind you that we take the limit of $\Delta f/\Delta x$ as Δx approaches 0.

If we wish to emphasize that we evaluate the derivative of $f(x)$ at $x = c$, we write

$$f'(c) = \frac{df}{dx}\Big|_{x=c}$$

Newton used different notation to denote the derivative of a function. He wrote $\dot{y}$ (read "y dot") for the derivative of y. This notation is still common in physics when derivatives are taken with respect to a variable that denotes time. We will use either Leibniz notation or the notation $f'(x)$.

■ 4.1.1 Geometric Interpretation and Using the Definition

Let's look at $f(x) = x^2$, $x \in \mathbf{R}$. (Refer to Figures 4.7 and 4.8 as we go along.) To compute the derivative of f at, say, $x = 1$ from the definition, we first compute the difference quotient at $x = 1$. For $h \neq 0$,

$$\frac{\Delta f}{\Delta x} = \frac{f(1+h) - f(1)}{h} = \frac{(1+h)^2 - 1^2}{h}$$

$$= \frac{1 + 2h + h^2 - 1}{h} = \frac{2h + h^2}{h} = \frac{h(2+h)}{h}$$

$$= 2 + h$$

The difference quotient $\Delta f/\Delta x$ is the slope of the secant line through the points $(1, 1)$ and $(1 + h, (1 + h)^2)$ (Figure 4.7).

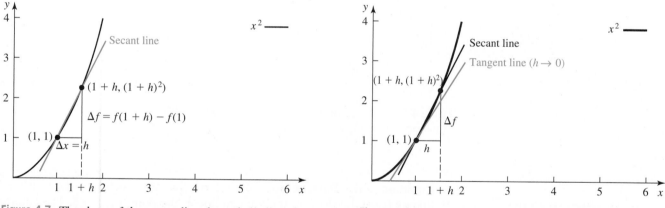

Figure 4.7 The slope of the secant line through $(1, 1)$ and $(1 + h, (1 + h)^2)$ is $\Delta f/\Delta x$.

Figure 4.8 Taking the limit $h \to 0$, the secant line converges to the tangent line at $(1, 1)$.

To find the derivative $f'(1)$, we need to take the limit as $h \to 0$ (Figure 4.8):

$$f'(1) = \lim_{h \to 0} \frac{f(1+h) - f(1)}{h} = \lim_{h \to 0} (2 + h) = 2$$

Taking the limit as $h \to 0$ means that the point $(1 + h, (1 + h)^2)$ approaches the point $(1, 1)$. [The limit as $h \to 0$ is a two-sided limit; in Figure 4.8, we only drew one point $(1 + h, (1 + h)^2)$ for some $h > 0$.] As $h \to 0$, the secant lines through the points $(1, 1)$ and $(1 + h, (1 + h)^2)$ converge to the line that touches the graph at $(1, 1)$. As mentioned earlier, the limiting line is called the tangent line. Since $f'(1)$ is the limiting value of the slope of the secant line as the point $(1 + h, (1 + h)^2)$ approaches $(1, 1)$, we find that $f'(1) = 2$ is the slope of the tangent line at the point $(1, 1)$.

Motivated by this example, we define the tangent line formally:

Definition of the Tangent Line If the derivative of a function f exists at $x = c$, then the tangent line at $x = c$ is the line going through the point $(c, f(c))$ with slope

$$f'(c) = \lim_{h \to 0} \frac{f(c + h) - f(c)}{h}$$

Knowing the derivative at a point (which is the slope of the tangent line at that point) and the coordinates of that point allows us to find the equation of the tangent line at the point by using the point–slope form of a straight line, namely,

$$y - y_0 = m(x - x_0)$$

where (x_0, y_0) is the point and m is the slope. Going back to the function $y = x^2$, we see that the point at $c = x_0 = 1$ has coordinates $(x_0, y_0) = (1, 1)$ and its derivative at $c = 1$ is $m = 2$. The equation of the tangent line is then given by

$$y - 1 = 2(x - 1), \quad \text{or} \quad y = 2x - 1$$

(Figure 4.9).

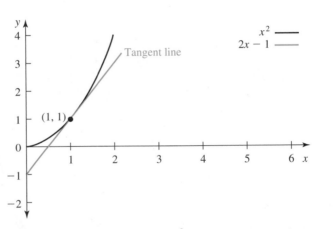

Figure 4.9 The slope of $f(x) = x^2$ at $(1, 1)$ is $m = 2$. The equation of the tangent line at $(1, 1)$ is $y = 2x - 1$.

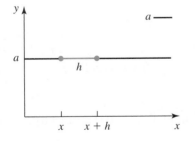

Figure 4.10 The slope of a horizontal line is $m = 0$.

Equation of the Tangent Line If the derivative of a function f exists at $x = c$, then $f'(c)$ is the slope of the tangent line at the point $(c, f(c))$. The equation of the tangent line is given by

$$y - f(c) = f'(c)(x - c)$$

The geometric interpretation will help us to compute derivatives in the next two examples.

EXAMPLE 1

The Derivative of a Constant Function The graph of $f(x) = a$ is a horizontal line that intersects the y-axis at $(0, a)$ (Figure 4.10). Since the graph is a straight line, the tangent line at x coincides with the graph of $f(x)$ and, therefore, the slope of the tangent line at x is equal to the slope of the straight line. The slope of a horizontal line is 0; we thus expect that $f'(x) = 0$. Using the formal definition with $f(x) = a$ and $f(x + h) = a$, we find that

$$f'(x) = \lim_{h \to 0} \frac{f(x + h) - f(x)}{h} = \lim_{h \to 0} \frac{a - a}{h} = \lim_{h \to 0} \frac{0}{h} = \lim_{h \to 0} 0 = 0$$

Here again, it is important to remember that when we take the limit as $h \to 0$, h approaches 0 (from both sides) but is not equal to 0. Since $h \neq 0$, the expression $0/h = 0$. This property was used in going from $\lim_{h\to 0} \frac{0}{h}$ to $\lim_{h\to 0} 0$. ■

EXAMPLE 2

The Derivative of a Linear Function The graph of $f(x) = mx + b$ is a straight line with slope m and y-intercept b (Figure 4.11). The derivative of $f(x)$ is the slope of the tangent line at x. Since the graph is a straight line, the tangent line at x coincides with the graph of $f(x)$ and, therefore, the slope of the tangent line at x is equal to the slope of the straight line. We thus expect that $f'(x) = m$. Using the formal definition, we can confirm this expectation:

$$f'(x) = \lim_{h\to 0} \frac{f(x+h) - f(x)}{h} = \lim_{h\to 0} \frac{m(x+h) + b - (mx+b)}{h}$$

$$= \lim_{h\to 0} \frac{mx + mh + b - mx - b}{h} = \lim_{h\to 0} \frac{mh}{h} = \lim_{h\to 0} m = m$$

In going from $\lim_{h\to 0} \frac{mh}{h}$ to $\lim_{h\to 0} m$, we were able to cancel h because $h \neq 0$.

The preceding reasoning yields the following: If $f(x) = mx + b$, then $f'(x) = m$. This includes the special case of a constant function, for which $m = 0$ (Example 1). ■

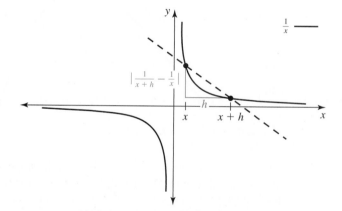

EXAMPLE 3

Using the Definition Find the derivative of

$$f(x) = \frac{1}{x} \quad \text{for } x \neq 0$$

Solution

We will use the formal definition of the derivative to compute $f'(x)$ (Figure 4.12).

Figure 4.11 The slope of the line $y = mx + b$ is m.

Figure 4.12 The graph of $f(x) = 1/x$ for Example 3.

The main algebraic step is the computation of $f(x+h) - f(x)$; we will do this first. With $f(x+h) = \frac{1}{x+h}$, we find that

$$f(x+h) - f(x) = \frac{1}{x+h} - \frac{1}{x}$$
$$= \frac{x - (x+h)}{x(x+h)} = \frac{-h}{x(x+h)}$$

To compute $f'(x)$, we need to divide both sides of this equation by h and take the limit as $h \to 0$:

$$f'(x) = \lim_{h\to 0} \frac{f(x+h) - f(x)}{h} = \lim_{h\to 0} \frac{\frac{-h}{x(x+h)}}{h}$$

$$= \lim_{h\to 0} \left(-\frac{h}{x(x+h)} \frac{1}{h} \right) = \lim_{h\to 0} \left(-\frac{1}{x(x+h)} \right) = -\frac{1}{x^2}$$

That is, if $f(x) = \frac{1}{x}, x \neq 0$, then

$$f'(x) = -\frac{1}{x^2}, \quad x \neq 0 \qquad \blacksquare$$

Looking back at the first three examples, we see that in order to compute $f'(x)$ from the formal definition of the derivative, we evaluate the limit

$$\lim_{h \to 0} \frac{f(x+h) - f(x)}{h}$$

Since both $\lim_{h \to 0}[f(x+h) - f(x)]$ and $\lim_{h \to 0} h$ are equal to 0, we cannot simply evaluate the limits in the numerator and the denominator separately, because this would result in the undefined expression 0/0. It is important to simplify the difference quotient before we take the limit.

■ 4.1.2 The Derivative as an Instantaneous Rate of Change: A First Look at Differential Equations

Velocity Suppose that you ride your bike on a straight road. Your position (in miles) at time t (in hours) is given by (Figure 4.13)

$$s(t) = -t^3 + 6t^2 \quad \text{for } 0 \leq t \leq 6$$

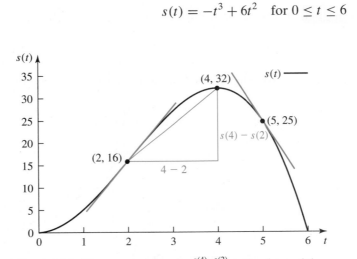

Figure 4.13 The average velocity $\frac{s(4)-s(2)}{4-2}$ is the slope of the secant line through (2, 16) and (4, 32). The velocity at time t is the slope of the tangent line at t: At $t = 2$, the velocity is positive; at $t = 5$, the velocity is negative.

You might ask what the average velocity during the interval (say) [2, 4] is. This velocity is defined as the net change in position during the interval, divided by the length of the interval. To compute the average velocity, find the position at time 2 and at time 4, and take the difference of these two quantities. Then divide this difference by the time that it took you to travel that distance. At time $t = 2$, $s(2) = -8 + 24 = 16$, and at time $t = 4$, $s(4) = -64 + 96 = 32$. Hence, the average velocity is

$$\frac{s(4) - s(2)}{4 - 2} = \frac{32 - 16}{4 - 2} = 8 \text{ mph}$$

We recognize this ratio as the difference quotient

$$\frac{\Delta s}{\Delta t} = \frac{s(t+h) - s(t)}{h}$$

and call the difference quotient $\Delta s / \Delta t$ the **average velocity**, which is an average rate of change.

The **instantaneous velocity** at time t is defined as the limit of $\frac{\Delta s}{\Delta t}$ as $\Delta t \to 0$, or

$$\lim_{\Delta t \to 0} \frac{\Delta s}{\Delta t} = \lim_{h \to 0} \frac{s(t+h) - s(t)}{h}$$

provided that the limit exists. This quantity is the derivative of $s(t)$ at time t, which we denote by

$$\frac{ds}{dt} = \lim_{\Delta t \to 0} \frac{\Delta s}{\Delta t}$$

Note that $\frac{ds}{dt}$ is an instantaneous rate of change. The instantaneous velocity (or, simply, the velocity) is, then, the slope of the tangent line at a given point of the position function $s(t)$, provided that the derivative at this point exists.

Let's look at two points on the graph of $s(t)$, namely, $(2, 16)$ and $(5, 25)$. We find that the slope of the tangent line is positive at $(2, 16)$ and negative at $(5, 25)$. The velocity is therefore positive at time $t = 2$ and negative at time $t = 5$. At $t = 2$ we move away from our starting point, whereas at $t = 5$ we move toward our starting point. At these two times, we move in opposite directions.

There is a difference between *velocity* and *speed*. If you had a speedometer on your bike, it would tell you the speed and not the velocity. Speed is the absolute value of velocity; it ignores direction.

Interpreting the derivative as an instantaneous rate of change will turn out to be extremely important to us. In fact, when you encounter derivatives in your science courses, this will be the interpretation most often used. This interpretation will allow us to describe a quantity in terms of how quickly it changes with respect to another quantity. To illustrate the point, we revisit two previous examples and introduce one new application.

Population Growth At the beginning of this chapter, we described the growth of a population at time t by the continuous function $N(t)$. If the derivative of $N(t)$ exists at time t, we can define the instantaneous growth rate of the population by

$$N'(t) = \frac{dN}{dt} = [\text{instantaneous population growth rate at time } t]$$

We are frequently interested in the **instantaneous per capita growth rate**. This is the growth rate per individual, and it can be obtained by dividing the instantaneous growth rate of the population by the population size at that time. That is,

$$\frac{1}{N(t)} \frac{dN}{dt} = [\text{instantaneous per capita growth rate at time } t]$$

In biology textbooks (and in this book), the dependence on t is often not explicitly spelled out, and we write

$$\frac{1}{N} \frac{dN}{dt} \quad \text{instead of} \quad \frac{1}{N(t)} \frac{dN}{dt}$$

The Rate of a Chemical Reaction Another illustration of the use of the derivative as an instantaneous rate of change is in Example 5 of Subsection 1.2.2, where we discussed the reaction rate of the irreversible chemical reaction

$$\text{A} + \text{B} \to \text{AB}$$

which is proportional to the concentrations of A and B. If the concentration of the product AB is denoted by x, then the reaction rate is equal to

$$k(a - x)(b - x)$$

where $a = [\text{A}]$ is the initial concentration of A and $b = [\text{B}]$ is the initial concentration of B (Figure 4.14). The reaction rate tells us how quickly the concentration of x

Figure 4.14 The reaction rate for $a \leq b$.

changes with time as the reaction proceeds. The concentration x is thus a function of time t: $x = x(t)$. The reaction rate is an instantaneous rate of change, namely,

$$\lim_{\Delta t \to 0} \frac{x(t + \Delta t) - x(t)}{\Delta t}$$

We can identify the limit as $\Delta t \to 0$ as the derivative of the function $x(t)$ with respect to t and therefore write

$$\frac{dx}{dt} = k(a - x)(b - x) \tag{4.1}$$

Equation (4.1) is an example of a **differential equation**—an equation that contains the derivative of a function. We will discuss such equations extensively in later chapters.

From this point on, when we say "rate of change," we will always mean "instantaneous rate of change." When we are interested in the average rate of change, we will always state this explicitly.

The rate of change in a chemical reaction is described by a differential equation. It is sometimes possible to solve such differential equations—that is, to state explicitly a function whose derivative satisfies the given equation. We will discuss this situation in detail later. More often, it is not possible (or not necessary) to explicitly find a solution. Without solving the differential equation, we can still obtain useful information about its behavior. We illustrate this property in the next application.

Tilman's Model for Resource Competition   David Tilman (1982) of the University of Minnesota developed a theoretical framework to describe the outcome of competition for limited resources. To test the predictions of his theory, he conducted many experiments on the grassland habitat at Cedar Creek Natural History Area in Minnesota. For this grassland habitat, nitrogen is a limiting resource; that is, adding nitrogen to the soil will result in an increase in biomass. We will discuss the case where one species competes for a single limited resource. We assume that the rate of change of biomass has two components: rate of growth and rate of loss. We write

[rate of biomass change] = [rate of growth] − [rate of loss]

We denote the biomass of the plant population at time t by $B(t)$ and assume that the rate of growth depends on a single resource whose concentration is denoted by R. We will write an equation for the **specific rate of change** of biomass, which is defined as the change of biomass per unit of biomass, or $\frac{1}{B}\frac{dB}{dt}$. We assume that the per-unit rate of loss of biomass is constant and denote this quantity by m. A simple model for how the biomass changes over time is then

$$\frac{1}{B}\frac{dB}{dt} = f(R) - m \tag{4.2}$$

where the function $f(R)$ describes the specific growth rate as a function of resource concentration. A common choice for $f(R)$ is the Monod growth function (or Michaelis–Menten equation) that we considered in Example 6 of Subsection 1.2.3, or

$$f(R) = a\frac{R}{k + R} \tag{4.3}$$

where a and k are positive constants. Let's graph $f(R)$ and m together in Figure 4.15. Doing so yields the following observations: When $0 < m < a$, the graphs of the functions $y = f(R)$ and $y = m$ intersect at $R = R^*$ (read "R star"). Consequently, at the resource level R^*, $f(R) = m$, and thus the specific rate of change $\frac{1}{B}\frac{dB}{dt}$ is equal to 0. That is, growth balances loss, and the biomass of the species no longer changes. We say that the biomass is at **equilibrium**. If the resource level R were held at a value less than R^*, then $f(R) - m < 0$, and the specific rate of growth would be negative; that is, biomass would decrease. If $R > R^*$, then $f(R) - m > 0$, and biomass would increase.

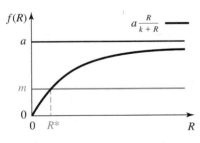

Figure 4.15 Growth balances loss when $R = R^*$.

We can compute R^* in the case when $f(R)$ is given by (4.3). Since R^* satisfies $f(R^*) = m$, we obtain

$$a\frac{R^*}{k + R^*} = m, \quad \text{or} \quad R^* = \frac{mk}{a - m}$$

■ 4.1.3 Differentiability and Continuity

Using the geometric interpretation, we can find situations in which $f'(c)$ does not exist at one or more values of c.

EXAMPLE 4   **A Function with a "Corner"** Let

$$f(x) = |x| = \begin{cases} x & \text{for } x \geq 0 \\ -x & \text{for } x < 0 \end{cases}$$

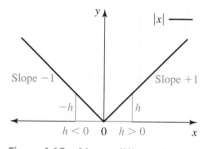

Figure 4.16 f is not differentiable at $x = 0$.

The graph of $f(x)$ is shown in Figure 4.16. Looking at the graph, we realize that there is no tangent line at $x = 0$ and therefore we do not expect that $f'(0)$ exists. We can define the slope of the secant line when we approach 0 from the right and also when we approach 0 from the left; however, the slopes converge to different values in the limit. The former is $+1$, the latter is -1. In this example, we can read off the slopes from the graph. But we can also find the slopes formally by taking appropriate limits. When $h > 0$, $f(h) = |h| = h$ and

$$\lim_{h \to 0+} \frac{f(0 + h) - f(0)}{h} = \frac{h - 0}{h} = 1$$

When $h < 0$, $f(h) = |h| = -h$ and

$$\lim_{h \to 0-} \frac{f(0 + h) - f(0)}{h} = \frac{-h - 0}{h} = -1$$

Since $1 \neq -1$, it follows that

$$\lim_{h \to 0} \frac{f(0 + h) - f(0)}{h}$$

and thus $f'(0)$, do not exist.

At all other points, the derivative exists. We can find the derivative by simply looking at the graph. We see that

$$f'(x) = \begin{cases} +1 & \text{for } x > 0 \\ -1 & \text{for } x < 0 \end{cases}$$ ■

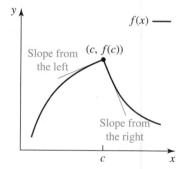

Figure 4.17 $f(x)$ is continuous at $x = c$ but not differentiable at $x = c$: The derivatives from the left and the right are not equal.

Example 4 shows that continuity alone is not enough for a function to be differentiable: The function $f(x) = |x|$ is continuous at all values of x, but it is not differentiable at $x = 0$. To draw the graph of a continuous function that is not differentiable at a point, put in a "corner" at that point (Figure 4.17).

However, if a function is differentiable, it is also continuous. We say that continuity is a necessary, but not a sufficient, condition for differentiability. This result is important enough that we will formulate it as a theorem and prove it:

Theorem If f is differentiable at $x = c$, then f is also continuous at $x = c$.

Proof Since f is differentiable at $x = c$, we know that the limit

$$\lim_{x \to c} \frac{f(x) - f(c)}{x - c} \tag{4.4}$$

exists and is equal to $f'(c)$. To show that f is continuous at $x = c$, we must show that

$$\lim_{x \to c} f(x) = f(c), \quad \text{or} \quad \lim_{x \to c}[f(x) - f(c)] = 0$$

First, note that f is defined at $x = c$. [Otherwise, we could not have computed the difference quotient $\frac{f(x)-f(c)}{x-c}$.] Now,

$$\lim_{x \to c}[f(x) - f(c)] = \lim_{x \to c} \frac{f(x) - f(c)}{x - c}(x - c)$$

Given that

$$\lim_{x \to c} \frac{f(x) - f(c)}{x - c}$$

[this is Equation (4.4)] exists and is equal to $f'(c)$, and that

$$\lim_{x \to c}(x - c)$$

exists (it is equal to 0), we can apply the product rule for limits and find that

$$\lim_{x \to c} \frac{f(x) - f(c)}{x - c}(x - c) = \lim_{x \to c} \frac{f(x) - f(c)}{x - c} \lim_{x \to c}(x - c) = f'(c) \cdot 0 = 0$$

This set of equations shows that

$$\lim_{x \to c}[f(x) - f(c)] = 0$$

and consequently that f is continuous at $x = c$.   ■

It follows from the preceding theorem that if a function f is not continuous at $x = c$, then f is not differentiable at $x = c$. The function $y = f(x)$ in Figure 4.18 is discontinuous at $x = c$; we cannot draw a tangent line there.

Functions can have vertical tangent lines, but since the slope of a vertical line is not defined, the function would not be differentiable at any point where the tangent line is vertical. This situation is illustrated in the next example.

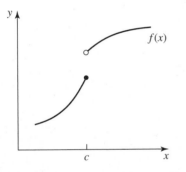

Figure 4.18 The function $y = f(x)$ is not differentiable at $x = c$.

EXAMPLE 5

Vertical Tangent Line Show that

$$f(x) = x^{1/3}$$

is not differentiable at $x = 0$.

Solution

We see from the graph of $f(x)$ in Figure 4.19 that $f(x)$ is continuous at $x = 0$. Using the formal definition, we find that

$$f'(0) = \lim_{h \to 0} \frac{f(h) - f(0)}{h} = \lim_{h \to 0} \frac{h^{1/3} - 0}{h}$$

$$= \lim_{h \to 0} \frac{1}{h^{2/3}} = \infty \qquad \text{does not exist}$$

Since the limit does not exist, $f(x)$ is not differentiable at $x = 0$. We see from the graph that the tangent line at $x = 0$ is vertical.   ■

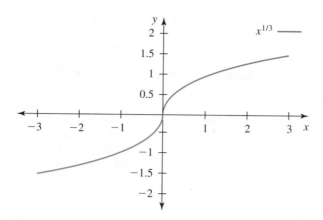

Figure 4.19 The function $f(x) = x^{1/3}$ has a vertical tangent line at $x = 0$. It is therefore not differentiable at $x = 0$.

Section 4.1 Problems

■ **4.1.1**

In Problems 1–8, find the derivative at the indicated point from the graph of each function.

1. $f(x) = 5; x = 1$

2. $f(x) = -3x; x = -2$

3. $f(x) = 4x - 3; x = -1$

4. $f(x) = -5x + 1; x = 0$

5. $f(x) = 2x^2; x = 0$

6. $f(x) = (x + 2)^2; x = 1$

7. $f(x) = \cos x; x = 0$

8. $f(x) = \sin x; x = \dfrac{\pi}{2}$

In Problems 9–16, find c so that $f'(c) = 0$.

9. $f(x) = -3x^2 + 1$

10. $f(x) = -x^2 + 4$

11. $f(x) = (x - 2)^2$

12. $f(x) = (x + 3)^2$

13. $f(x) = x^2 - 6x + 9$

14. $f(x) = x^2 + 4x + 4$

15. $f(x) = \sin\left(\dfrac{\pi}{2}x\right)$

16. $\cos(\pi - x)$

In Problems 17–20, compute $f(c+h) - f(c)$ at the indicated point.

17. $f(x) = -2x + 1; c = 2$

18. $f(x) = 3x^2; c = 1$

19. $f(x) = \sqrt{x}; c = 4$

20. $f(x) = \dfrac{1}{x}; c = -2$

21. (a) Use the formal definition of the derivative to find the derivative of $y = 5x^2$ at $x = -1$.

(b) Show that the point $(-1, 5)$ is on the graph of $y = 5x^2$, and find the equation of the tangent line at the point $(-1, 5)$.

(c) Graph $y = 5x^2$ and the tangent line at the point $(-1, 5)$ in the same coordinate system.

22. (a) Use the formal definition to find the derivative of $y = -2x^2$ at $x = 1$.

(b) Show that the point $(1, -2)$ is on the graph of $y = -2x^2$, and find the equation of the tangent line at the point $(1, -2)$.

(c) Graph $y = -2x^2$ and the tangent line at the point $(1, -2)$ in the same coordinate system.

23. (a) Use the formal definition to find the derivative of $y = 1 - x^3$ at $x = 2$.

(b) Show that the point $(2, -7)$ is on the graph of $y = 1 - x^3$, and find the equation of the normal line at the point $(2, -7)$.

(c) Graph $y = 1 - x^3$ and the tangent line at the point $(2, -7)$ in the same coordinate system.

24. (a) Use the formal definition to find the derivative of $y = \dfrac{1}{x}$ at $x = 2$.

(b) Show that the point $(2, \frac{1}{2})$ is on the graph of $y = \frac{1}{x}$, and find the equation of the normal line at the point $(2, \frac{1}{2})$.

(c) Graph $y = \frac{1}{x}$ and the tangent line at the point $(2, \frac{1}{2})$ in the same coordinate system.

25. Use the formal definition to find the derivative of

$$y = \sqrt{x}$$

for $x > 0$.

26. Use the formal definition to find the derivative of

$$f(x) = \frac{1}{x + 1}$$

for $x \neq -1$.

27. Find the equation of the tangent line to the curve $y = 3x^2$ at the point $(1, 3)$.

28. Find the equation of the tangent line to the curve $y = 2/x$ at the point $(2, 1)$.

29. Find the equation of the tangent line to the curve $y = \sqrt{x}$ at the point $(4, 2)$.

30. Find the equation of the tangent line to the curve $y = x^2 - 3x + 1$ at the point $(2, -1)$.

31. Find the equation of the normal line to the curve $y = -3x^2$ at the point $(-1, -3)$.

32. Find the equation of the normal line to the curve $y = 4/x$ at the point $(-1, -4)$.

33. Find the equation of the normal line to the curve $y = 2x^2 - 1$ at the point $(1, 1)$.

34. Find the equation of the normal line to the curve $y = \sqrt{x - 1}$ at the point $(5, 2)$.

35. The following limit represents the derivative of a function f at the point $(a, f(a))$:

$$\lim_{h \to 0} \frac{2(a + h)^2 - 2a^2}{h}$$

Find $f(x)$.

36. The following limit represents the derivative of a function f at the point $(a, f(a))$:

$$\lim_{h \to 0} \frac{4(a+h)^3 - 4a^3}{h}$$

Find $f(x)$.

37. The following limit represents the derivative of a function f at the point $(a, f(a))$:

$$\lim_{h \to 0} \frac{\frac{1}{(2+h)^2+1} - \frac{1}{5}}{h}$$

Find f and a.

38. The following limit represents the derivative of a function f at the point $(a, f(a))$:

$$\lim_{h \to 0} \frac{\sin(\frac{\pi}{6}+h) - \sin\frac{\pi}{6}}{h}$$

Find f and a.

■ **4.1.2**

39. Velocity A car moves along a straight road. Its location at time t is given by

$$s(t) = 20t^2, \ 0 \le t \le 2$$

where t is measured in hours and $s(t)$ is measured in kilometers.
(a) Graph $s(t)$ for $0 \le t \le 2$.
(b) Find the average velocity of the car between $t = 0$ and $t = 2$. Illustrate the average velocity on the graph of $s(t)$.
(c) Use calculus to find the instantaneous velocity of the car at $t = 1$. Illustrate the instantaneous velocity on the graph of $s(t)$.

40. Velocity A train moves along a straight line. Its location at time t is given by

$$s(t) = \frac{100}{t}, \quad 1 \le t \le 5$$

where t is measured in hours and $s(t)$ is measured in kilometers.
(a) Graph $s(t)$ for $1 \le t \le 5$.
(b) Find the average velocity of the train between $t = 1$ and $t = 5$. Where on the graph of $s(t)$ can you find the average velocity?
(c) Use calculus to find the instantaneous velocity of the train at $t = 2$. Where on the graph of $s(t)$ can you find the instantaneous velocity? What is the speed of the train at $t = 2$?

41. Velocity If $s(t)$ denotes the position of an object that moves along a straight line, then $\Delta s / \Delta t$, called the average velocity, is the average rate of change of $s(t)$, and $v(t) = ds/dt$, called the (instantaneous) velocity, is the instantaneous rate of change of $s(t)$. The speed of the object is the absolute value of the velocity, $|v(t)|$.

Suppose now that a car moves along a straight road. The location at time t is given by

$$s(t) = \frac{160}{3}t^2, \quad 0 \le t \le 1$$

where t is measured in hours and $s(t)$ is measured in kilometers.
(a) Where is the car at $t = 3/4$, and where is it at $t = 1$?
(b) Find the average velocity of the car between $t = 3/4$ and $t = 1$.

(c) Find the velocity and the speed of the car at $t = 3/4$.
42. Velocity Suppose a particle moves along a straight line. The position at time t is given by

$$s(t) = 3t - t^2, \quad t \ge 0$$

where t is measured in seconds and $s(t)$ is measured in meters.
(a) Graph $s(t)$ for $t \ge 0$.
(b) Use the graph in (a) to answer the following questions:
(i) Where is the particle at time 0?
(ii) Is there another time at which the particle visits the location where it was at time 0?
(iii) How far to the right on the straight line does the particle travel?
(iv) How far to the left on the straight line does the particle travel?
(v) Where is the velocity positive? where negative? equal to 0?
(c) Find the velocity of the particle.
(d) When is the velocity of the particle equal to 1 m/s?

43. Tilman's Resource Model In Subsection 4.1.2, we considered Tilman's resource model. Denote the biomass at time t by $B(t)$, and assume that

$$\frac{1}{B}\frac{dB}{dt} = f(R) - m$$

where R denotes the resource level,

$$f(R) = 200\frac{R}{5+R}$$

and $m = 40$. Use the graphical approach to find the value R^* at which $\frac{1}{B}\frac{dB}{dt} = 0$. Then compute R^* by solving $\frac{1}{B}\frac{dB}{dt} = 0$.
44. Exponential Growth Assume that $N(t)$ denotes the size of a population at time t and that $N(t)$ satisfies the differential equation

$$\frac{dN}{dt} = rN$$

where r is a constant.
(a) Find the per capita growth rate.
(b) Assume that $r < 0$ and that $N(0) = 20$. Is the population size at time 1 greater than 20 or less than 20? Explain your answer.
45. Logistic Growth Assume that $N(t)$ denotes the size of a population at time t and that $N(t)$ satisfies the differential equation

$$\frac{dN}{dt} = 3N\left(1 - \frac{N}{20}\right)$$

Let $f(N) = 3N(1 - \frac{N}{20})$ for $N \ge 0$. Graph $f(N)$ as a function of N and identify all equilibria (i.e., all points where $\frac{dN}{dt} = 0$).
46. Island Model Assume that a species lives in a habitat that consists of many islands close to a mainland. The species occupies both the mainland and the islands, but, although it is present on the mainland at all times, it frequently goes extinct on the islands. Islands can be recolonized by migrants from the mainland. The following model keeps track of the fraction of islands occupied: Denote the fraction of islands occupied at time t by $p(t)$. Assume that each island experiences a constant risk of extinction and that vacant islands (the fraction $1 - p$) are colonized from the mainland at a constant rate. Then

$$\frac{dp}{dt} = c(1 - p) - ep$$

where c and e are positive constants.

(a) The gain from colonization is $f(p) = c(1 - p)$ and the loss from extinction is $g(p) = ep$. Graph $f(p)$ and $g(p)$ for $0 \leq p \leq 1$ in the same coordinate system. Explain why the two graphs intersect whenever e and c are both positive. Compute the point of intersection and interpret its biological meaning.

(b) The parameter c measures how quickly a vacant island becomes colonized from the mainland. The closer the islands, the larger is the value of c. Use your graph in (a) to explain what happens to the point of intersection of the two lines as c increases. Interpret your result in biological terms.

47. Chemical Reaction Consider the chemical reaction

$$A + B \longrightarrow AB$$

If $x(t)$ denotes the concentration of AB at time t, then

$$\frac{dx}{dt} = k(a - x)(b - x)$$

where k is a positive constant and a and b denote the concentrations of A and B, respectively, at time 0. Assume that $k = 3$, $a = 7$, and $b = 4$. For what values of x is $dx/dt = 0$? Interpret the meaning of $dx/dt = 0$.

48. Chemical Reaction Consider the autocatalytic reaction

$$A + X \longrightarrow X$$

which was introduced in Problem 30 of Section 1.2. Find a differential equation that describes the rate of change of the concentration of the product X.

49. Logistic Growth Suppose that the rate of change of the size of a population is given by

$$\frac{dN}{dt} = rN\left(1 - \frac{N}{K}\right)$$

where $N = N(t)$ denotes the size of the population at time t and r and K are positive constants. Find the equilibrium size of the population—that is, the size at which the rate of change is equal to 0. Use your answer to explain why K is called the carrying capacity.

50. Biotic Diversity (Adapted from Valentine, 1985.) Walker and Valentine (1984) suggested a model for species diversity which assumes that species extinction rates are independent of diversity but speciation rates are regulated by competition. Denoting the number of species at time t by $N(t)$, the speciation rate by b, and the extinction rate by a, they used the model

$$\frac{dN}{dt} = N\left[b\left(1 - \frac{N}{K}\right) - a\right]$$

where K denotes the number of "niches," or potential places for species in the ecosystem.
(a) Find possible equilibria under the condition $a < b$.

(b) Use your result in (a) to explain the following statement by Valentine (1985):

In this situation, ecosystems are never "full," with all potential niches occupied by species so long as the extinction rate is above zero.

(c) What happens when $a \geq b$?

■ **4.1.3**

51. Which of the following statements is true?
(A) If $f(x)$ is continuous, then $f(x)$ is differentiable.
(B) If $f(x)$ is differentiable, then $f(x)$ is continuous.
52. Explain the relationship between continuity and differentiability.
53. Sketch the graph of a function that is continuous at all points in its domain and differentiable in the domain except at one point.
54. Sketch the graph of a periodic function defined on **R** that is continuous at all points in its domain and differentiable in the domain except at $c = k, k \in \mathbf{Z}$.
55. If $f(x)$ is differentiable for all $x \in \mathbf{R}$ except at $x = c$, is it true that $f(x)$ must be continuous at $x = c$? Justify your answer.

In Problems 56–69, graph each function and, on the basis of the graph, guess where the function is not differentiable. (Assume the largest possible domain.)

56. $y = |x - 2|$ **57.** $y = -|x + 5|$
58. $y = 2 - |x - 3|$ **59.** $y = |x + 2| - 1$
60. $y = \dfrac{1}{2 + x}$ **61.** $y = \dfrac{1}{x - 3}$
62. $y = \dfrac{3 - x}{3 + x}$ **63.** $y = \dfrac{x - 1}{x + 1}$
64. $y = |x^2 - 3|$ **65.** $y = |2x^2 - 1|$

66. $f(x) = \begin{cases} x & \text{for } x \leq 0 \\ x + 1 & \text{for } x > 0 \end{cases}$

67. $f(x) = \begin{cases} 2x & \text{for } x \leq 1 \\ x + 2 & \text{for } x > 1 \end{cases}$

68. $f(x) = \begin{cases} x^2 & \text{for } x \leq -1 \\ 2 - x^2 & \text{for } x > -1 \end{cases}$

69. $f(x) = \begin{cases} x^2 + 1 & \text{for } x \leq 0 \\ e^{-x} & \text{for } x > 0 \end{cases}$

70. Suppose the function $f(x)$ is piecewise defined; that is, $f(x) = f_1(x)$ for $x \leq a$ and $f(x) = f_2(x)$ for $x > a$. Assume that $f_1(x)$ is continuous and differentiable for $x < a$ and that $f_2(x)$ is continuous and differentiable for $x > a$. Sketch graphs of $f(x)$ for the following three cases:
(a) $f(x)$ is continuous and differentiable at $x = a$.
(b) $f(x)$ is continuous, but not differentiable, at $x = a$.
(c) $f(x)$ is neither continuous nor differentiable at $x = a$.

■ 4.2 The Power Rule, the Basic Rules of Differentiation, and the Derivatives of Polynomials

In this section, we will begin a systematic treatment of the computation of derivatives. Knowing how to differentiate is fundamental to your understanding of the rest of the course. Although computer software is now available to compute derivatives of many functions (such as $y = cx^n$ or $y = e^{\sin x}$), it is nonetheless important that you master the techniques of differentiation.

The power rule is the simplest of the differentiation rules. It allows us to compute the derivative of a function of the form $y = x^n$, where n is a positive integer.

> **Power Rule** Let n be a positive integer; then
>
> $$\frac{d}{dx}(x^n) = nx^{n-1}$$

We found the rule for the constant function $f(x) = a$ in the previous section.

> If $f(x)$ is the constant function $f(x) = a$, then
>
> $$\frac{d}{dx}f(x) = 0$$

We prove the power rule first for $n = 2$—that is, for $f(x) = x^2$ (Figure 4.20). In Subsection 4.1.1, we computed the derivative of $y = x^2$ at $x = 1$. In this section, we compute the difference quotient $\frac{\Delta f}{\Delta x}$ at any arbitrary x:

$$\frac{\Delta f}{\Delta x} = \frac{f(x+h) - f(x)}{h} = \frac{(x+h)^2 - x^2}{h}$$

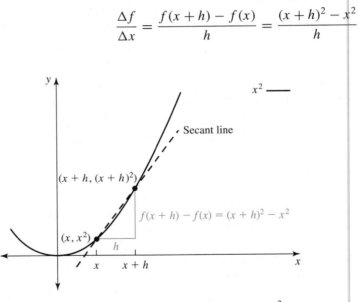

Figure 4.20 The slope of the secant line through (x, x^2) and $(x+h, (x+h)^2)$ is $\frac{(x+h)^2 - x^2}{h}$.

Using the expansion $(x + h)^2 = x^2 + 2xh + h^2$, we find that

$$\frac{\Delta f}{\Delta x} = \frac{x^2 + 2xh + h^2 - x^2}{h} = \frac{2xh + h^2}{h} = 2x + h \tag{4.5}$$

after canceling h in both the numerator and the denominator. To find the derivative, we need to let $h \to 0$:

$$f'(x) = \lim_{\Delta x \to 0} \frac{\Delta f}{\Delta x} = \lim_{h \to 0}(2x + h) = 2x$$

This sequence of steps proves the power rule for $n = 2$. The proof of the rule for other positive integers of n is conceptually no different from the case $n = 2$, but it gets algebraically much more involved. For general n, we need the expansion of $(x + h)^n$, given by the **binomial theorem**, which we will not prove.

> **Binomial Theorem** If n is a positive integer, then
> $$(x + y)^n = x^n + nx^{n-1}y + \frac{n(n-1)}{2 \cdot 1}x^{n-2}y^2$$
> $$+ \frac{n(n-1)(n-2)}{3 \cdot 2 \cdot 1}x^{n-3}y^3$$
> $$+ \cdots + \frac{n(n-1)\cdots(n-k+1)}{k(k-1)\cdots 2 \cdot 1}x^{n-k}y^k$$
> $$+ \cdots + nxy^{n-1} + y^n$$

The expansion of $(x + y)^n$ is thus a sum of terms of the form

$$C_{n,k}x^{n-k}y^k, \quad k = 0, 1, \ldots, n$$

where $C_{n,k}$ is a coefficient that depends on n and k. The exact form of the coefficients $C_{n,k}$ will not be important in the proof of the power rule, except for the two terms $C_{n,0} = 1$ and $C_{n,1} = n$, which are the coefficients for x^n and $x^{n-1}y$, respectively.

Proof of the Power Rule We use the binomial theorem to expand and then compute the difference in the numerator of the difference quotient:

$$\Delta f = f(x + h) - f(x) = (x + h)^n - x^n$$
$$= (C_{n,0}x^n + C_{n,1}x^{n-1}h + C_{n,2}x^{n-2}h^2 + C_{n,3}x^{n-3}h^3$$
$$+ \cdots + C_{n,n-1}xh^{n-1} + C_{n,n}h^n) - x^n$$

Since $C_{n,0} = 1$, the x^n terms cancel. We can then factor h out of the remaining terms and find that

$$f(x + h) - f(x) = h\left[C_{n,1}x^{n-1} + C_{n,2}x^{n-2}h + C_{n,3}x^{n-3}h^2 + \cdots + C_{n,n}h^{n-1}\right]$$

When we divide by h and let $h \to 0$, we obtain

$$f'(x) = \lim_{h \to 0} \frac{f(x + h) - f(x)}{h}$$
$$= \lim_{h \to 0}\left[C_{n,1}x^{n-1} + C_{n,2}x^{n-2}h + C_{n,3}x^{n-3}h^2 + \cdots + C_{n,n}h^{n-1}\right]$$

All terms except for the first have h as a factor and thus tend to 0 as $h \to 0$. (The first term does not depend on h.) We find that

$$f'(x) = C_{n,1}x^{n-1}$$

With $C_{n,1} = n$, this is then

$$f'(x) = nx^{n-1}$$

which proves the power rule. ∎

EXAMPLE 1 We apply the power rule to various functions and take the opportunity to practice the different notations.

(a) If $f(x) = x^6$, then $f'(x) = 6x^5$.

(b) If $f(x) = x^{300}$, then $f'(x) = 300x^{299}$.

(c) If $g(t) = t^5$, then $\frac{d}{dt}g(t) = 5t^4$.

(d) If $z = s^3$, then $\frac{dz}{ds} = 3s^2$.

(e) If $x = y^4$, then $\frac{dx}{dy} = 4y^3$. ∎

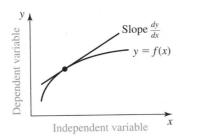

Figure 4.21 If $y = f(x)$, then x is the independent variable and y is the dependent variable.

Example 1 illustrates the importance of knowing how the variables depend on each other (Figure 4.21). If $y = f(x)$, we call x the **independent variable** and y the **dependent variable**, because y depends on the variable x. For instance, in (a), y is a function of x; thus, x is the independent, and y is the dependent, variable. In (e), by contrast, x is a function of y; thus, y is now the independent, and x the dependent, variable. The Leibniz notation $\frac{dy}{dx}$ emphasizes this dependence. When we write $\frac{dy}{dx}$, we consider y to be a function of x (i.e., y is the dependent, and x is the independent, variable) and differentiate y with respect to x.

Since polynomials and rational functions are built up by the basic operations of addition, subtraction, multiplication, and division operating on power functions of the form $y = x^n$, $n = 0, 1, 2, \ldots$, we need differentiation rules for such operations. We begin with the following rules:

Theorem Suppose a is a constant and $f(x)$ and $g(x)$ are differentiable at x. Then the following relationships hold:

1. $\dfrac{d}{dx}[af(x)] = a\dfrac{d}{dx}f(x)$

2. $\dfrac{d}{dx}[f(x) + g(x)] = \dfrac{d}{dx}f(x) + \dfrac{d}{dx}g(x)$

Rule 1 says that a constant factor can be pulled out of the derivative expression; Rule 2 says that the derivative of a sum of two functions is equal to the sum of the derivatives of the functions. Similarly, since $f(x) - g(x) = f(x) + (-1)g(x)$, the derivative of a difference of functions is the difference of the derivatives:

$$\frac{d}{dx}[f(x) - g(x)] = \frac{d}{dx}[f(x) + (-1)g(x)] = \frac{d}{dx}f(x) + \frac{d}{dx}[(-1)g(x)]$$

Using Rule 1 on the rightmost term, we find that $\frac{d}{dx}[(-1)g(x)] = (-1)\frac{d}{dx}g(x)$. Therefore,

$$\frac{d}{dx}[f(x) - g(x)] = \frac{d}{dx}f(x) - \frac{d}{dx}g(x)$$

Rules 1 and 2 allow us to differentiate polynomials, as illustrated in the next three examples.

EXAMPLE 2 Differentiate $y = 2x^4 - 3x^3 + x - 7$.

Solution

$$\frac{d}{dx}(2x^4 - 3x^3 + x - 7) = \frac{d}{dx}(2x^4) - \frac{d}{dx}(3x^3) + \frac{d}{dx}x - \frac{d}{dx}7$$
$$= 2\frac{d}{dx}x^4 - 3\frac{d}{dx}x^3 + \frac{d}{dx}x - \frac{d}{dx}7$$
$$= 2(4x^3) - 3(3x^2) + 1 - 0 = 8x^3 - 9x^2 + 1 \qquad ■$$

EXAMPLE 3

(a) $\frac{d}{dx}(-5x^7 + 2x^3 - 10) = -35x^6 + 6x^2$

(b) $\frac{d}{dt}(t^3 - 8t^2 + 3t) = 3t^2 - 16t + 3$

(c) Suppose that n is a positive integer and a is a constant. Then $\frac{d}{ds}(as^n) = ans^{n-1}$.

(d) $\frac{d}{dN}(\ln 5 + N \ln 7) = \ln 7$

(e) $\frac{d}{dr}(r^2 \sin \frac{\pi}{4} - r^3 \cos \frac{\pi}{12} + \sin \frac{\pi}{6}) = 2r \sin \frac{\pi}{4} - 3r^2 \cos \frac{\pi}{12}$ ■

In the previous section, we related the derivative to the slope of the tangent line; the next example uses this interpretation.

EXAMPLE 4

Tangent and Normal Lines If $f(x) = 2x^3 - 3x + 1$, find the tangent and normal lines at $(-1, 2)$.

Solution

The slope of the tangent line at $(-1, 2)$ is $f'(-1)$. We begin calculating this derivative as follows:

$$f'(x) = 6x^2 - 3$$

Evaluating $f'(x)$ at $x = -1$, we get

$$f'(-1) = 6(-1)^2 - 3 = 3$$

Therefore, the equation of the tangent line at $(-1, 2)$ is

$$y - 2 = 3(x - (-1)), \qquad \text{or} \qquad y = 3x + 5$$

To find the equation of the normal line, recall that the normal line is perpendicular to the tangent line; hence, the slope m of the normal line is given by

$$m = -\frac{1}{f'(-1)} = -\frac{1}{3}$$

The normal line goes through the point $(-1, 2)$ as well. The equation of the normal line is therefore

$$y - 2 = -\frac{1}{3}(x - (-1)), \qquad \text{or} \qquad y = -\frac{1}{3}x + \frac{5}{3}$$

The graph of $f(x)$, including the tangent and normal lines at $(-1, 2)$, is shown in Figure 4.22. ■

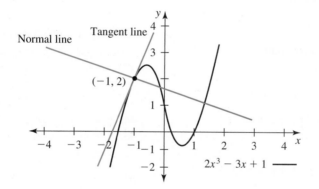

Figure 4.22 The graph of $f(x) = 2x^3 - 3x + 1$, together with the tangent and normal lines at $(-1, 2)$.

Look again at the last example: When we computed $f'(-1)$, we *first* computed $f'(x)$; the *second* step was to evaluate $f'(x)$ at $x = -1$. It makes no sense to plug -1 into $f(x)$ and then differentiate the result. Since $f(-1) = 2$ is a constant, the derivative would be 0, which is obviously not $f'(-1)$. Just look at Figure 4.22 to convince yourself. The notation $f'(-1)$ means that we evaluate the function $f'(x)$ at $x = -1$.

Section 4.2 Problems

Differentiate the functions given in Problems 1–22 with respect to the independent variable.

1. $f(x) = 4x^3 - 7x + 1$

2. $f(x) = -3x^4 + 5x^2$

3. $f(x) = -2x^5 + 7x - 4$

4. $f(x) = -3x^4 + 6x^2 - 2$

5. $f(x) = 3 - 4x - 5x^2$

6. $f(x) = -1 + 3x^2 - 2x^4$

7. $g(s) = 5s^7 + 2s^3 - 5s$

8. $g(s) = 3 - 4s^2 - 4s^3$

9. $h(t) = -\frac{1}{3}t^4 + 4t$

10. $h(t) = \frac{1}{2}t^2 - 3t + 2$

11. $f(x) = x^2 \sin\frac{\pi}{3} + \tan\frac{\pi}{4}$

12. $f(x) = 2x^3 \cos\frac{\pi}{3} + \cos\frac{\pi}{6}$

13. $f(x) = -3x^4 \tan \dfrac{\pi}{6} - \cot \dfrac{\pi}{6}$

14. $f(x) = x^2 \sec \dfrac{\pi}{6} + 3x \sec \dfrac{\pi}{4}$

15. $f(t) = t^3 e^{-2} + t + e^{-1}$

16. $f(x) = \dfrac{1}{2} x^2 e^3 - x^4$

17. $f(s) = s^3 e^3 + 3e$

18. $f(x) = \dfrac{x}{e} + e^2 x + e$

19. $f(x) = 20x^3 - 4x^6 + 9x^8$

20. $f(x) = \dfrac{x^3}{15} - \dfrac{x^4}{20} + \dfrac{2}{15}$

21. $f(x) = \pi x^3 - \dfrac{1}{\pi} + \dfrac{x}{\pi}$

22. $f(x) = \pi x e^2 - \dfrac{x^2 \pi}{e}$

23. Differentiate
$$f(x) = ax^3$$
with respect to x. Assume that a is a constant.

24. Differentiate
$$f(x) = x^3 + a$$
with respect to x. Assume that a is a constant.

25. Differentiate
$$f(x) = ax^2 - 2a$$
with respect to x. Assume that a is a constant.

26. Differentiate
$$f(x) = a^2 x^4 - 2ax^2$$
with respect to x. Assume that a is a constant.

27. Differentiate
$$h(s) = rs^2 - r$$
with respect to s. Assume that r is a constant.

28. Differentiate
$$f(r) = rs^2 - r$$
with respect to r. Assume that s is a constant.

29. Differentiate
$$f(x) = rs^2 x^3 - rx + s$$
with respect to x. Assume that r and s are constants.

30. Differentiate
$$f(x) = \dfrac{r+x}{rs^2} - rsx + (r+s)x - rs$$
with respect to x. Assume that r and s are nonzero constants.

31. Differentiate
$$f(N) = (b-1)N^4 - \dfrac{N^2}{b}$$
with respect to N. Assume that b is a nonzero constant.

32. Differentiate
$$f(N) = \dfrac{bN^2 + N}{K+b}$$
with respect to N. Assume that b and K are positive constants.

33. Differentiate
$$g(t) = a^3 t - at^3$$
with respect to t. Assume that a is a constant.

34. Differentiate
$$h(s) = a^4 s^2 - as^4 + \dfrac{s^2}{a^4}$$
with respect to s. Assume that a is a positive constant.

35. Differentiate
$$V(t) = V_0(1 + \gamma t)$$
with respect to t. Assume that V_0 and γ are positive constants.

36. Differentiate
$$p(T) = \dfrac{NkT}{V}$$
with respect to T. Assume that N, k, and V are positive constants.

37. Differentiate
$$g(N) = N\left(1 - \dfrac{N}{K}\right)$$
with respect to N. Assume that K is a positive constant.

38. Differentiate
$$g(N) = rN\left(1 - \dfrac{N}{K}\right)$$
with respect to N. Assume that K and r are positive constants.

39. Differentiate
$$g(N) = rN^2\left(1 - \dfrac{N}{K}\right)$$
with respect to N. Assume that K and r are positive constants.

40. Differentiate
$$g(N) = rN(a - N)\left(1 - \dfrac{N}{K}\right)$$
with respect to N. Assume that r, a, and K are positive constants.

41. Differentiate
$$R(T) = \dfrac{2\pi^5}{15}\dfrac{k^4}{c^2 h^3}T^4$$
with respect to T. Assume that k, c, and h are positive constants.

In Problems 42–48, find the tangent line, in standard form, to $y = f(x)$ at the indicated point.

42. $y = 3x^2 - 4x + 7$, at $x = 2$

43. $y = 7x^3 + 2x - 1$, at $x = -3$

44. $y = -2x^3 - 3x + 1$, at $x = 1$

45. $y = 2x^4 - 5x$, at $x = 1$

46. $y = -x^3 - 2x^2$, at $x = 0$

47. $y = \dfrac{1}{\sqrt{2}}x^2 - \sqrt{2}$, at $x = 4$

48. $y = 3\pi x^5 - \dfrac{\pi}{2}x^3$, at $x = -1$

In Problems 49–54, find the normal line, in standard form, to $y = f(x)$ at the indicated point.

49. $y = 2 + x^2$, at $x = -1$

50. $y = 1 - 3x^2$, at $x = -2$

51. $y = \sqrt{3}x^4 - 2\sqrt{3}x^2$, at $x = -\sqrt{3}$

52. $y = -2x^2 - x$, at $x = 0$

53. $y = x^3 - 3$, at $x = 1$

54. $y = 1 - \pi x^2$, at $x = -1$

55. Find the tangent line to
$$f(x) = ax^2$$
at $x = 1$. Assume that a is a positive constant.

56. Find the tangent line to
$$f(x) = ax^3 - 2ax$$
at $x = -1$. Assume that a is a positive constant.

57. Find the tangent line to

$$f(x) = \frac{ax^2}{a^2 + 2}$$

at $x = 2$. Assume that a is a positive constant.

58. Find the tangent line to

$$f(x) = \frac{x^2}{a + 1}$$

at $x = a$. Assume that a is a positive constant.

59. Find the normal line to

$$f(x) = ax^3$$

at $x = -1$. Assume that a is a positive constant.

60. Find the normal line to

$$f(x) = ax^2 - 3ax$$

at $x = 2$. Assume that a is a positive constant.

61. Find the normal line to

$$f(x) = \frac{ax^2}{a + 1}$$

at $x = 2$. Assume that a is a positive constant.

62. Find the normal line to

$$f(x) = \frac{x^3}{a + 1}$$

at $x = 2a$. Assume that a is a positive constant.

In Problems 63–70, find the coordinates of all of the points of the graph of $y = f(x)$ that have horizontal tangents.

63. $f(x) = x^2$

64. $f(x) = 2 - x^2$

65. $f(x) = 3x - x^2$

66. $f(x) = 4x + 2x^2$

67. $f(x) = 3x^3 - x^2$

68. $f(x) = -4x^4 + x^3$

69. $f(x) = \frac{1}{2}x^4 - \frac{7}{3}x^3 - 2x^2$

70. $f(x) = 3x^5 - \frac{3}{2}x^4$

71. Find a point on the curve

$$y = 4 - x^2$$

whose tangent line is parallel to the line $y = 2$. Is there more than one such point? If so, find all other points with this property.

72. Find a point on the curve

$$y = (4 - x)^2$$

whose tangent line is parallel to the line $y = -3$. Is there more than one such point? If so, find all other points with this property.

73. Find a point on the curve

$$y = 2x^2 - \frac{1}{2}$$

whose tangent line is parallel to the line $y = x$. Is there more than one such point? If so, find all other points with this property.

74. Find a point on the curve

$$y = 1 - 3x^3$$

whose tangent line is parallel to the line $y = -x$. Is there more than one such point? If so, find all other points with this property.

75. Find a point on the curve

$$y = x^3 + 2x + 2$$

whose tangent line is parallel to the line $3x - y = 2$. Is there more than one such point? If so, find all other points with this property.

76. Find a point on the curve

$$y = 2x^3 - 4x + 1$$

whose tangent line is parallel to the line $y - 2x = 1$. Is there more than one such point? If so, find all other points with this property.

77. Show that the tangent line to the curve

$$y = x^2$$

at the point $(1, 1)$ passes through the point $(0, -1)$.

78. Find all tangent lines to the curve

$$y = x^2$$

that pass through the point $(0, -1)$.

79. Find all tangent lines to the curve

$$y = x^2$$

that pass through the point $(0, -a^2)$, where a is a positive number.

80. How many tangent lines to the curve

$$y = x^2 + 2x$$

pass through the point $(-\frac{1}{2}, -3)$?

81. Suppose that $P(x)$ is a polynomial of degree 4. Is $P'(x)$ a polynomial as well? If yes, what is its degree?

82. Suppose that $P(x)$ is a polynomial of degree k. Is $P'(x)$ a polynomial as well? If yes, what is its degree?

■ 4.3 The Product and Quotient Rules, and the Derivatives of Rational and Power Functions

■ 4.3.1 The Product Rule

The derivative of a sum of differentiable functions is the sum of the derivatives of the functions. The rule for products is not so simple, as can be seen from the following example: Consider $y = x^5 = (x^3)(x^2)$. We know that

$$\frac{d}{dx}x^5 = 5x^4$$

$$\frac{d}{dx}x^3 = 3x^2$$

$$\frac{d}{dx}x^2 = 2x$$

This chain of reasoning shows that

$$\frac{d}{dx}x^5 \qquad \text{is not equal to} \qquad \left(\frac{d}{dx}x^3\right)\left(\frac{d}{dx}x^2\right)$$

(Leibniz first thought that the multiplication rule was as simple as that, but he quickly realized his mistake and found the correct formula for differentiating products of functions.)

The Product Rule If $h(x) = f(x)g(x)$ and both $f(x)$ and $g(x)$ are differentiable at x, then

$$h'(x) = f'(x)g(x) + f(x)g'(x)$$

If we set $u = f(x)$ and $v = g(x)$, then

$$(uv)' = u'v + uv'$$

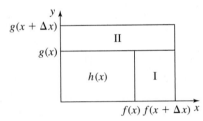

Figure 4.23 The product rule.

Proof Since $h(x) = f(x)g(x)$ is a product of two functions, we can visualize $h(x)$ as the area of a rectangle with sides $f(x)$ and $g(x)$. To compute the derivative, we need $h(x + \Delta x)$; this is given by

$$h(x + \Delta x) = f(x + \Delta x)g(x + \Delta x)$$

To compute $h'(x)$, we must compute $h(x + \Delta x) - h(x)$ (Figure 4.23). We find that

$$h(x + \Delta x) - h(x) = \text{area of I} + \text{area of II}$$
$$= [f(x + \Delta x) - f(x)]g(x)$$
$$\quad + [g(x + \Delta x) - g(x)]f(x + \Delta x)$$

Dividing this result by Δx and taking the limit as $\Delta x \to 0$, we obtain

$$h'(x) = \lim_{\Delta x \to 0} \frac{h(x + \Delta x) - h(x)}{\Delta x}$$

$$= \lim_{\Delta x \to 0} \frac{[f(x + \Delta x) - f(x)]g(x) + [g(x + \Delta x) - g(x)]f(x + \Delta x)}{\Delta x}$$

$$= \lim_{\Delta x \to 0} \left[\frac{f(x + \Delta x) - f(x)}{\Delta x}g(x) + \frac{g(x + \Delta x) - g(x)}{\Delta x}f(x + \Delta x) \right]$$

Now, we need the assumption that $f'(x)$ and $g'(x)$ exist and that $f(x)$ is continuous at x [which follows from the fact that $f(x)$ is differentiable at x]. These assumptions allow us to use the basic rules for limits, and we write the last expression as

$$\left(\lim_{\Delta x \to 0} \frac{f(x + \Delta x) - f(x)}{\Delta x} \right) g(x) + \left(\lim_{\Delta x \to 0} \frac{g(x + \Delta x) - g(x)}{\Delta x} \right)\left(\lim_{\Delta x \to 0} f(x + \Delta x) \right)$$

The limits of the difference quotients are the respective derivatives. Using the fact that $f(x)$ is continuous at x, we find that $\lim_{\Delta x \to 0} f(x + \Delta x) = f(x)$. Therefore,

$$h'(x) = f'(x)g(x) + g'(x)f(x)$$

as claimed.   ■

EXAMPLE 1

Differentiate $f(x) = (3x + 1)(2x^2 - 5)$.

Solution

We write $u = 3x + 1$ and $v = 2x^2 - 5$. The product rule says that $(uv)' = u'v + uv'$. That is, we need the derivatives of both u and v:

$$u = 3x + 1 \qquad v = 2x^2 - 5$$
$$u' = 3 \qquad\qquad v' = 4x$$

Then

$$(uv)' = u'v + uv'$$
$$= 3(2x^2 - 5) + (3x + 1)(4x)$$
$$= 6x^2 - 15 + 12x^2 + 4x = 18x^2 + 4x - 15$$

Of course, we could have gotten this result by first multiplying out $(3x+1)(2x^2 - 5) = 6x^3 - 15x + 2x^2 - 5$, which is simply a polynomial function. We then would have found that

$$\frac{d}{dx}(6x^3 - 15x + 2x^2 - 5) = 18x^2 - 15 + 4x$$

which is the same answer. ■

EXAMPLE 2

Differentiate $f(x) = (3x^3 - 2x)^2$.

Solution

Again, we could simply expand the square and then differentiate the resulting polynomial—but we can also use the product rule. To do so, we write $u = v = 3x^3 - 2x$. Then $f(x) = uv$ and $(uv)' = u'v + uv'$. Since $u = v$, it follows that $u' = v'$, and the formula simplifies to $(uv)' = (u^2)' = u'u + uu' = 2uu'$. Because $u' = 9x^2 - 2$, we have

$$f'(x) = 2(3x^3 - 2x)(9x^2 - 2)$$ ■

EXAMPLE 3

Population Growth In many population models, the population growth rate depends only on the current population size. We can express this quantity by

$$\frac{dN}{dt} = f(N)$$

where $N(t)$ denotes the size of the population at time t and $f(N)$ is the population growth rate, which depends only on the current population size $N = N(t)$. The per capita growth rate $\frac{1}{N}\frac{dN}{dt}$ is then also just a function of N, namely,

$$\frac{1}{N}\frac{dN}{dt} = g(N)$$

with

$$f(N) = Ng(N)$$

Assume that $g(N)$ is differentiable and that $\lim_{N \to 0^+} g(N)$ and $\lim_{N \to 0^+} g'(N)$ exist. Show that

$$g(0) = \lim_{N \to 0^+} \frac{d}{dN} f(N)$$

Solution

Using the product rule, we compute the derivative of the population growth rate $f(N) = Ng(N)$:

$$\frac{d}{dN}(Ng(N)) = g(N) + Ng'(N)$$

Then we take the limit as $N \to 0^+$:

$$\lim_{N \to 0^+} \frac{d}{dN}(Ng(N)) = \lim_{N \to 0^+} [g(N) + Ng'(N)] = g(0)$$

(Note that we can take only one-sided limits here, since $N \geq 0$ for biological reasons.) ■

EXAMPLE 4

Apply the product rule repeatedly to find the derivative of

$$y = (2x + 1)(x + 1)(3x - 4)$$

Solution

Since the product rule is formulated for products of two factors, we group the terms in our function as follows:

$$u = (2x + 1)(x + 1)$$

Then $y = uv$, with $v = 3x - 4$. Note that any other grouping into two factors would work as well. Now, to differentiate u, we need to use the product rule:

$$w = 2x + 1 \qquad z = x + 1$$
$$w' = 2 \qquad z' = 1$$

Therefore,

$$u' = 2(x + 1) + (2x + 1)(1) = 2x + 2 + 2x + 1 = 4x + 3$$

With $v' = 3$, we find that

$$y' = (4x + 3)(3x - 4) + 3(2x + 1)(x + 1)$$
$$= 12x^2 - 16x + 9x - 12 + 6x^2 + 6x + 3x + 3$$
$$= 18x^2 + 2x - 9$$ ■

■ 4.3.2 The Quotient Rule

The quotient rule will allow us to compute the derivative of a quotient of two functions. In particular, the rule will allow us to compute the derivative of a rational function, because a rational function is the quotient of two polynomial functions.

> **The Quotient Rule** If $h(x) = \frac{f(x)}{g(x)}$, $g(x) \neq 0$, and both $f'(x)$ and $g'(x)$ exist, then
>
> $$h'(x) = \frac{f'(x)g(x) - f(x)g'(x)}{[g(x)]^2}$$
>
> In short, with $u = f(x)$ and $v = g(x)$,
>
> $$\left(\frac{u}{v}\right)' = \frac{u'v - uv'}{v^2}$$

We could prove the quotient rule much as we did the product rule, by using the formal definition of derivatives, but that would not be very exciting. Instead, we will give a different proof of the quotient rule in the next subsection.

Note carefully the exact forms of the product and quotient rules. In the product rule we add $f'g$ and fg', whereas in the quotient rule we subtract fg' from $f'g$. As mentioned, we can use the quotient rule to find the derivative of rational functions. We illustrate this application in the next two examples.

EXAMPLE 5

Differentiate $y = \frac{x^3 - 3x + 2}{x^2 + 1}$. (This function is defined for all $x \in \mathbf{R}$, since $x^2 + 1 \neq 0$.)

Solution

We set $u = x^3 - 3x + 2$ and $v = x^2 + 1$. Both u and v are polynomials, which we know how to differentiate. We find that

$$u = x^3 - 3x + 2 \qquad v = x^2 + 1$$
$$u' = 3x^2 - 3 \qquad v' = 2x$$

Using the quotient rule, we compute y':

$$y' = \frac{u'v - uv'}{v^2} = \frac{(3x^2 - 3)(x^2 + 1) - (x^3 - 3x + 2)2x}{(x^2 + 1)^2}$$

$$= \frac{3x^4 + 3x^2 - 3x^2 - 3 - 2x^4 + 6x^2 - 4x}{(x^2 + 1)^2}$$

$$= \frac{x^4 + 6x^2 - 4x - 3}{(x^2 + 1)^2}$$ ■

EXAMPLE 6

Monod Growth Function Differentiate the Monod growth function

$$f(R) = \frac{aR}{k + R}, \quad R \geq 0$$

where a and k are positive constants.

Solution

Since a and k are positive constants, $f(R)$ is defined for all $R \geq 0$. We write $u = aR$ and $v = k + R$ and obtain

$$u = aR \qquad v = k + R$$
$$u' = a \qquad v' = 1$$

Hence,

$$\frac{d}{dR} f(R) = \frac{u'v - uv'}{v^2} = \frac{a(k + R) - aR \cdot 1}{(k + R)^2} = \frac{ak}{(k + R)^2}$$

In Figure 4.24, we graph both $f(R)$ and $f'(R)$. We see that the slope of the tangent line at $(R, f(R))$ is positive for all $R \geq 0$. We can also draw this conclusion from the graph of $f'(R)$, since it is positive for all $R \geq 0$. Furthermore, we see that $f(R)$ becomes less steep as R increases, which is reflected in the fact that $f'(R)$ becomes smaller as R increases. ■

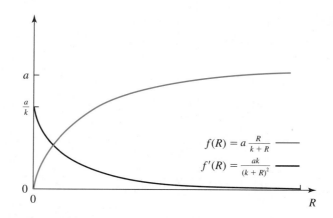

$$f(R) = a\frac{R}{k + R} \quad \underline{\qquad}$$

$$f'(R) = \frac{ak}{(k + R)^2} \quad \underline{\qquad}$$

Figure 4.24 The graph of $f(R)$ and $f'(R)$ in Example 6.

The quotient rule allows us to extend the power rule to the case where the exponent is a negative integer:

> **Power Rule (Negative Integer Exponents)** If $f(x) = x^{-n}$, where n is a positive integer, then
> $$f'(x) = -nx^{-n-1}$$

Note that the power rule for negative integer exponents works the same way as the power rule for positive integer exponents: We write the exponent of the original function x^n in front and decrease the exponent by 1. We will now prove the power rule for negative integer exponents.

Proof We write $f(x) = \frac{1}{x^n}$ and set $u = 1$ and $v = x^n$. Then

$$u = 1 \qquad v = x^n$$
$$u' = 0 \qquad v' = nx^{n-1}$$

and, therefore,

$$f'(x) = \frac{u'v - uv'}{v^2} = \frac{0 \cdot x^n - 1 \cdot nx^{n-1}}{(x^n)^2} = -\frac{nx^{n-1}}{x^{2n}} = -nx^{-n-1} \qquad ■$$

EXAMPLE 7

(a) If $y = \frac{1}{x}$, then

$$y' = \frac{d}{dx}(x^{-1}) = (-1)x^{-1-1} = -\frac{1}{x^2}$$

(b) If $g(x) = \frac{3}{x^4}$, then

$$g'(x) = \frac{d}{dx}(3x^{-4}) = 3\frac{d}{dx}x^{-4} = 3(-4)x^{-4-1} = -12x^{-5} = -\frac{12}{x^5} \qquad ■$$

There is a general form of the power rule in which the exponent can be any real number. In the next section, we give the proof for the case when the exponent is rational; we prove the general case in Section 4.7.

Power Rule (General Form) Let $f(x) = x^r$, where r is any real number. Then

$$f'(x) = rx^{r-1}$$

EXAMPLE 8

(a) If $y = \sqrt{x}$, then

$$y' = \frac{d}{dx}\left(x^{1/2}\right) = \frac{1}{2}x^{\frac{1}{2}-1} = \frac{1}{2}x^{-1/2} = \frac{1}{2\sqrt{x}}$$

(b) If $y = \sqrt[5]{x}$, then

$$y' = \frac{d}{dx}\left(x^{1/5}\right) = \frac{1}{5}x^{(1/5)-1} = \frac{1}{5}x^{-4/5} = \frac{1}{5x^{4/5}}$$

(c) If $g(t) = \frac{1}{\sqrt[3]{t}}$, then

$$g'(t) = \frac{d}{dt}\left(t^{-1/3}\right) = \left(-\frac{1}{3}\right)t^{(-1/3)-1} = \left(-\frac{1}{3}\right)t^{-4/3} = -\frac{1}{3t^{4/3}}$$

(d) If $h(s) = s^\pi$, then $h'(s) = \pi s^{\pi-1}$. ■

The function $f(x) = \sqrt{x}$, $x \geq 0$, appears quite frequently. It is therefore worthwhile to memorize its derivative, which is defined only for $x > 0$:

$$\frac{d}{dx}\sqrt{x} = \frac{1}{2\sqrt{x}}$$

EXAMPLE 9 **Combining the Rules** Differentiate $f(x) = \sqrt{x}(x^2 - 1)$.

Solution I

We can consider $f(x)$ to be a product of two functions. Let

$$u = \sqrt{x} \qquad v = x^2 - 1$$

$$u' = \frac{1}{2\sqrt{x}} \qquad v' = 2x$$

Hence,

$$f'(x) = u'v + uv' = \frac{1}{2\sqrt{x}}(x^2 - 1) + \sqrt{x}(2x)$$

$$= \frac{x^2 - 1 + \sqrt{x}(2x)2\sqrt{x}}{2\sqrt{x}} = \frac{x^2 - 1 + 4x^2}{2\sqrt{x}} = \frac{5x^2 - 1}{2\sqrt{x}}$$

Solution 2

Since $f(x) = \sqrt{x}(x^2 - 1) = x^{5/2} - x^{1/2}$, we can also use the general version of the power rule. We find that

$$f'(x) = \frac{5}{2}x^{(5/2)-1} - \frac{1}{2}x^{(1/2)-1} = \frac{5}{2}x^{3/2} - \frac{1}{2\sqrt{x}} = \frac{5x^{3/2}\sqrt{x} - 1}{2\sqrt{x}} = \frac{5x^2 - 1}{2\sqrt{x}} \qquad ■$$

EXAMPLE 10

A Function That Contains a Constant Differentiate $h(t) = (at)^{1/3}(a+1) - a$, where a is a positive constant.

Solution

Since $h(t)$ is a function of t, we need to differentiate with respect to t, keeping in mind that a is a constant. Rewriting $h(t)$ will make this easier:

$$h(t) = a^{1/3}(a + 1)t^{1/3} - a$$

The factor $a^{1/3}(a + 1)$ in front of $t^{1/3}$ is a constant. Thus,

$$h'(t) = a^{1/3}(a + 1)\frac{1}{3}t^{-2/3} - 0 = \frac{a^{1/3}(a + 1)}{3t^{2/3}} \qquad ■$$

EXAMPLE 11

Differentiating a Function That Is Not Specified Suppose $f(2) = 3$ and $f'(2) = 1/4$. Find

$$\frac{d}{dx}[xf(x)]$$

at $x = 2$.

Solution

Since $xf(x)$ is a product, we can use the product rule

$$\frac{d}{dx}[xf(x)] = f(x) + xf'(x)$$

Hence,

$$\frac{d}{dx}[xf(x)]\Big|_{x=2} = f(2) + 2f'(2) = 3 + \frac{1}{2} = \frac{7}{2} \qquad ■$$

EXAMPLE 12

Differentiating a Function That Is Not Specified Suppose that $f(x)$ is differentiable. Find an expression for the derivative of

$$y = \frac{f(x)}{x^2}$$

Solution

We set

$$u = f(x) \qquad v = x^2$$

$$u' = f'(x) \qquad v' = 2x$$

and use the quotient rule. We find that

$$y' = \frac{f'(x)x^2 - f(x)2x}{x^4} = \frac{xf'(x) - 2f(x)}{x^3} \qquad ■$$

Section 4.3 Problems

■ **4.3.1**

In Problems 1–16, use the product rule to find the derivative with respect to the independent variable.

1. $f(x) = (x + 5)(x^2 - 3)$

2. $f(x) = (2x^3 - 1)(3 + 2x^2)$

3. $f(x) = (3x^4 - 5)(2x - 5x^3)$

4. $f(x) = (3x^4 - x^2 + 1)(2x^2 - 5x^3)$

5. $f(x) = \left(\frac{1}{2}x^2 - 1\right)(2x + 3x^2)$

6. $f(x) = 2(3x^2 - 2x^3)(1 - 5x^2)$

7. $f(x) = \frac{1}{5}(x^2 - 1)(x^2 + 1)$

8. $f(x) = 3(x^2 + 2)(4x^2 - 5x^4) - 3$

9. $f(x) = (3x - 1)^2$ **10.** $f(x) = (4 - 2x^2)^2$

11. $f(x) = 3(1 - 2x)^2$ **12.** $f(x) = \frac{(2x^2 - 3x + 1)^2}{4} + 2$

13. $g(s) = (2s^2 - 5s)^2$ **14.** $h(t) = 4(3t^2 - 1)(2t + 1)$

15. $g(t) = 3(2t^2 - 5t^4)^2$ **16.** $h(s) = (4 - 3s^2 + 4s^3)^2$

In Problems 17–20, apply the product rule to find the tangent line, in slope–intercept form, of $y = f(x)$ at the specified point.

17. $f(x) = (3x^2 - 2)(x - 1)$, at $x = 1$

18. $f(x) = (1 - 2x)(1 + 2x)$, at $x = 2$

19. $f(x) = 4(2x^4 + 3x)(4 - 2x^2)$, at $x = -1$

20. $f(x) = (3x^3 - 3)(2 - 2x^2)$, at $x = 0$

In Problems 21–24, apply the product rule to find the normal line, in slope–intercept form, of $y = f(x)$ at the specified point.

21. $f(x) = (1 - x)(2 - x^2)$, at $x = 2$

22. $f(x) = (2x + 1)(3x^2 - 1)$, at $x = 1$

23. $f(x) = 5(1 - 2x)(x + 1) - 3$, at $x = 0$

24. $f(x) = \frac{(2 - x)(3 - x)}{4}$, at $x = -1$

In Problems 25–28, apply the product rule repeatedly to find the derivative of $y = f(x)$.

25. $f(x) = (2x - 1)(3x + 4)(1 - x)$

26. $f(x) = (x - 3)(2 - 3x)(5 - x)$

27. $f(x) = (x - 3)(2x^2 + 1)(1 - x^2)$

28. $f(x) = (2x + 1)(4 - x^2)(1 + x^2)$

29. Differentiate

$$f(x) = a(x - 1)(2x - 1)$$

with respect to x. Assume that a is a positive constant.

30. Differentiate

$$f(x) = (a - x)(a + x)$$

with respect to x. Assume that a is a positive constant.

31. Differentiate

$$f(x) = 2a(x^2 - a)^2 + a$$

with respect to x. Assume that a is a positive constant.

32. Differentiate

$$f(x) = \frac{3(x - 1)^2}{2 + a}$$

with respect to x. Assume that a is a positive constant.

33. Differentiate

$$g(t) = (at + 1)^2$$

with respect to t. Assume that a is a positive constant.

34. Differentiate

$$h(t) = \sqrt{a}(t - a) + a$$

with respect to t. Assume that a is a positive constant.

35. Suppose that $f(2) = -4$, $g(2) = 3$, $f'(2) = 1$, and $g'(2) = -2$. Find

$$(fg)'(2)$$

36. Suppose that $f(2) = -4$, $g(2) = 3$, $f'(2) = 1$, and $g'(2) = -2$. Find

$$(f^2 + g^2)'(2)$$

In Problems 37–40, assume that $f(x)$ is differentiable. Find an expression for the derivative of y at $x = 1$, assuming that $f(1) = 2$ and $f'(1) = -1$.

37. $y = 2xf(x)$ **38.** $y = 3x^2 f(x)$

39. $y = -5x^3 f(x) - 2x$ **40.** $y = \frac{xf(x)}{2}$

In Problems 41–44, assume that $f(x)$ and $g(x)$ are differentiable at x. Find an expression for the derivative of y.

41. $y = 3f(x)g(x)$ **42.** $y = [f(x) - 3]g(x)$

43. $y = [f(x) + 2g(x)]g(x)$

44. $y = [-2f(x) - 3g(x)]g(x) + \frac{2g(x)}{3}$

45. Let $B(t)$ denote the biomass at time t with specific growth rate $g(B)$. Show that the specific growth rate at $B = 0$ is given by the slope of the tangent line on the graph of the growth rate at $B = 0$.

46. Let $N(t)$ denote the size of a population at time t. Differentiate

$$f(N) = rN\left(1 - \frac{N}{K}\right)$$

with respect to N, where r and K are positive constants.

47. Let $N(t)$ denote the size of a population at time t. Differentiate

$$f(N) = r\left(aN - N^2\right)\left(1 - \frac{N}{K}\right)$$

with respect to N, where r, K, and a are positive constants.

48. Consider the chemical reaction

$$A + B \longrightarrow AB$$

If x denotes the concentration of AB at time t, then the reaction rate $R(x)$ is given by

$$R(x) = k(a - x)(b - x)$$

where k, a, and b are positive constants. Differentiate $R(x)$.

■ **4.3.2**

In Problems 49–70, differentiate with respect to the independent variable.

49. $f(x) = \frac{3x - 1}{x + 1}$ **50.** $f(x) = \frac{1 - 4x^3}{1 - x}$

51. $f(x) = \frac{3x^2 - 2x + 1}{2x + 1}$ **52.** $f(x) = \frac{x^4 + 2x - 1}{5x^2 - 2x + 1}$

53. $f(x) = \frac{3 - x^3}{1 - x}$ **54.** $f(x) = \frac{1 + 2x^2 - 4x^4}{3x^3 - 5x^5}$

55. $h(t) = \dfrac{t^2 - 3t + 1}{t + 1}$

56. $h(t) = \dfrac{3 - t^2}{(t + 1)^2}$

57. $f(s) = \dfrac{4 - 2s^2}{1 - s}$

58. $f(s) = \dfrac{2s^3 - 4s^2 + 5s - 7}{(s^2 - 3)^2}$

59. $f(x) = \sqrt{x}(x - 1)$

60. $f(x) = \sqrt{x}(x^4 - 5x^2)$

61. $f(x) = \sqrt{3x}(x^2 - 1)$

62. $f(x) = \dfrac{\sqrt{5x}(1 + x^2)}{\sqrt{2}}$

63. $f(x) = x^3 - \dfrac{1}{x^3}$

64. $f(x) = x^5 - \dfrac{1}{x^5}$

65. $f(x) = 2x^2 - \dfrac{3x - 1}{x^3}$

66. $f(x) = -x^3 + \dfrac{2x^2 - 3}{4x^4}$

67. $g(s) = \dfrac{s^{1/3} - 1}{s^{2/3} - 1}$

68. $g(s) = \dfrac{s^{1/7} - s^{2/7}}{s^{3/7} + s^{4/7}}$

69. $f(x) = (1 - 2x)\left(\sqrt{2x} + \dfrac{2}{\sqrt{x}}\right)$

70. $f(x) = (x^3 - 3x^2 + 2)\left(\sqrt{x} + \dfrac{1}{\sqrt{x}} - 1\right)$

In Problems 71–74, find the tangent line, in slope–intercept form, of $y = f(x)$ at the specified point.

71. $f(x) = \dfrac{x^2 + 3}{x^3 + 5}$, at $x = -2$

72. $f(x) = \dfrac{3}{x} - \dfrac{4}{\sqrt{x}} + \dfrac{2}{x^2}$, at $x = 1$

73. $f(x) = \dfrac{2x - 5}{x^3}$, at $x = 2$ **74.** $f(x) = \sqrt{x}(x^3 - 1)$, at $x = 1$

75. Differentiate
$$f(x) = \dfrac{ax}{3 + x}$$
with respect to x. Assume that a is a positive constant.

76. Differentiate
$$f(x) = \dfrac{ax}{k + x}$$
with respect to x. Assume that a and k are positive constants.

77. Differentiate
$$f(x) = \dfrac{ax^2}{4 + x^2}$$
with respect to x. Assume that a is a positive constant.

78. Differentiate
$$f(x) = \dfrac{ax^2}{k^2 + x^2}$$
with respect to x. Assume that a and k are positive constants.

79. Differentiate
$$f(R) = \dfrac{R^n}{k^n + R^n}$$
with respect to R. Assume that k is a positive constant and n is a positive integer.

80. Differentiate
$$h(t) = \sqrt{at}(1 - a) + a$$
with respect to t. Assume that a is a positive constant.

81. Differentiate
$$h(t) = \sqrt{at}(t - a) + at$$
with respect to t. Assume that a is a positive constant.

82. Suppose that $f(2) = -4$, $g(2) = 3$, $f'(2) = 1$, and $g'(2) = -2$. Find
$$\left(\dfrac{1}{f}\right)'(2)$$

83. Suppose that $f(2) = -4$, $g(2) = 3$, $f'(2) = 1$, and $g'(2) = -2$. Find
$$\left(\dfrac{f}{2g}\right)'(2)$$

In Problems 84–87, assume that $f(x)$ is differentiable. Find an expression for the derivative of y at $x = 2$, assuming that $f(2) = -1$ and $f'(2) = 1$.

84. $y = \dfrac{f(x)}{x^2 + 1}$

85. $y = \dfrac{x^2 + 4f(x)}{f(x)}$

86. $y = [f(x)]^2 - \dfrac{x}{f(x)}$

87. $y = \dfrac{f(x)}{f(x) + x}$

In Problems 88–91, assume that $f(x)$ and $g(x)$ are differentiable at x. Find an expression for the derivative of y.

88. $y = \dfrac{2f(x) + 1}{3g(x)}$

89. $y = \dfrac{f(x)}{[g(x)]^2}$

90. $y = \dfrac{x^2}{f(x) - g(x)}$

91. $y = \sqrt{x}f(x)g(x)$

92. Assume that $f(x)$ is a differentiable function. Find the derivative of the reciprocal function $g(x) = 1/f(x)$ at those points x where $f(x) \neq 0$.

93. Find the tangent line to the hyperbola $yx = c$, where c is a positive constant, at the point (x_1, y_1) with $x_1 > 0$. Show that the tangent line intersects the x-axis at a point that does not depend on c.

94. *(Adapted from Roff, 1992)* The males in the frog species *Eleutherodactylus coqui* (found in Puerto Rico) take care of their brood. On the other hand, while they protect the eggs, they cannot find other mates and therefore cannot increase their number of offspring. On the other hand, if they do not spend enough time with their brood, then the offspring might not survive. The proportion $w(t)$ of offspring hatching per unit time is given as a function of (1) the probability $f(t)$ of hatching if time t is spent brooding, and (2) the cost C associated with the time spent searching for other mates:
$$w(t) = \dfrac{f(t)}{C + t}$$
Find the derivative of $w(t)$.

■ 4.4 The Chain Rule and Higher Derivatives

■ 4.4.1 The Chain Rule

In Section 1.2, we defined the composition of functions. To find the derivative of composite functions, we need the **chain rule**, the proof of which is given at the end of this section.

> **Chain Rule** If g is differentiable at x and f is differentiable at $y = g(x)$, then the composite function $(f \circ g)(x) = f[g(x)]$ is differentiable at x, and the derivative is given by
>
> $$(f \circ g)'(x) = f'[g(x)]g'(x)$$

This formula looks complicated. Let's take a moment to see what we need to do to find the derivative of the composite function $(f \circ g)(x)$. The function g is the inner function; the function f is the outer function. The expression $f'[g(x)]g'(x)$ thus means that we need to find the derivative of the outer function, evaluated at $g(x)$, and the derivative of the inner function, evaluated at x, and then multiply the two together.

EXAMPLE 1   **A Polynomial** Find the derivative of

$$h(x) = (3x^2 - 1)^2$$

Solution   The inner function is $g(x) = 3x^2 - 1$; the outer function is $f(u) = u^2$. Then

$$g'(x) = 6x \quad \text{and} \quad f'(u) = 2u$$

Evaluating $f'(u)$ at $u = g(x)$ yields

$$f'[g(x)] = 2g(x) = 2(3x^2 - 1)$$

Thus,

$$h'(x) = (f \circ g)'(x) = f'[g(x)]g'(x)$$
$$= 2(3x^2 - 1)6x = 12x(3x^2 - 1) \qquad ■$$

The derivative of $f \circ g$ can be written in Leibniz notation. If we set $u = g(x)$, then

$$\frac{d}{dx}[(f \circ g)(x)] = \frac{df}{du}\frac{du}{dx}$$

This form of the chain rule emphasizes that, in order to differentiate $f \circ g$, we multiply the derivative of the outer function and the derivative of the inner function, the former evaluated at u, the latter at x.

EXAMPLE 2   **A Polynomial** Find the derivative of

$$h(x) = (2x + 1)^3$$

Solution   If we set $u = g(x) = 2x + 1$ and $f(u) = u^3$, then $h(x) = (f \circ g)(x)$. We need to find both $f'[g(x)]$ and $g'(x)$ to compute $h'(x)$. Now,

$$g'(x) = 2 \quad \text{and} \quad f'(u) = 3u^2$$

Hence, since $f'[g(x)] = 3(g(x))^2 = 3(2x + 1)^2$, it follows that

$$h'(x) = f'[g(x)]g'(x) = 3(2x + 1)^2 \cdot 2$$
$$= 6(2x + 1)^2$$

If we use Leibniz notation, this becomes

$$h'(x) = \frac{df}{du}\frac{du}{dx} = 3u^2 \cdot 2 = 3(2x + 1)^2 \cdot 2$$
$$= 6(2x + 1)^2 \qquad ■$$

EXAMPLE 3

A Radical Find the derivative of $h(x) = \sqrt{x^2 + 1}$.

Solution   If we set $u = g(x) = x^2 + 1$ and $f(u) = \sqrt{u}$, then $h(x) = (f \circ g)(x)$. We find that

$$g'(x) = 2x \quad \text{and} \quad f'(u) = \frac{1}{2\sqrt{u}}$$

We need to evaluate f' at $g(x)$—that is,

$$f'[g(x)] = \frac{1}{2\sqrt{g(x)}} = \frac{1}{2\sqrt{x^2 + 1}}$$

Therefore,

$$h'(x) = f'[g(x)]g'(x) = \frac{1}{2\sqrt{x^2 + 1}} 2x = \frac{x}{\sqrt{x^2 + 1}}$$ ■

EXAMPLE 4

A Radical Find the derivative of

$$h(x) = \sqrt[7]{2x^2 + 3x}$$

Solution   We write

$$h(x) = (2x^2 + 3x)^{1/7}$$

The inner function is $u = g(x) = 2x^2 + 3x$ and the outer function is $f(u) = u^{1/7}$. Thus, we find that

$$h'(x) = \frac{df}{du}\frac{du}{dx} = \frac{1}{7}u^{1/7-1}(4x + 3)$$

$$= \frac{1}{7}(2x^2 + 3x)^{-6/7}(4x + 3)$$

$$= \frac{4x + 3}{7(2x^2 + 3x)^{6/7}}$$ ■

EXAMPLE 5

A Rational Function Find the derivative of $h(x) = \left(\frac{x}{x+1}\right)^2$.

Solution   If we set $u = g(x) = \frac{x}{x+1}$ and $f(u) = u^2$, then $h(x) = (f \circ g)(x)$. We use the quotient rule to compute the derivative of $g(x)$:

$$g'(x) = \frac{1 \cdot (x + 1) - x \cdot 1}{(x + 1)^2} = \frac{1}{(x + 1)^2}$$

Since $f'(u) = 2u$, we obtain

$$h'(x) = f'[g(x)]g'(x) = 2\frac{x}{x + 1}\frac{1}{(x + 1)^2} = \frac{2x}{(x + 1)^3}$$ ■

The Proof of the Quotient Rule   We can use the chain rule to prove the quotient rule. Assume that $g(x) \neq 0$ for all x in the domain of g. If we define $h(x) = \frac{1}{x}$, then

$$(h \circ g)(x) = h[g(x)] = \frac{1}{g(x)}$$

We used the formal definition of the derivative in Example 3 in Section 4.1 to show that $h'(x) = -\frac{1}{x^2}$. This, together with the chain rule, yields

$$(h \circ g)'(x) = -\frac{1}{[g(x)]^2}g'(x), \quad \text{or} \quad \left(\frac{1}{g}\right)' = -\frac{g'}{g^2}$$

Since $\frac{f}{g} = f\frac{1}{g}$, we can use the product rule to find the derivative of $\frac{f}{g}$:

$$\left(\frac{f}{g}\right)' = f'\frac{1}{g} + f\left(\frac{1}{g}\right)' = f'\frac{1}{g} + f\left(-\frac{g'}{g^2}\right)$$
$$= \frac{f'g - fg'}{g^2}$$

Note that we did not use the power rule for negative integer exponents (Subsection 4.3.2) to compute $h'(x)$, but instead used the formal definition of derivatives to compute the derivative of $1/x$. Using the power rule for negative integer exponents would have been circular reasoning: We used the quotient rule to prove the power rule for negative integer exponents, so we cannot use the power rule for negative integer exponents to prove the quotient rule.

EXAMPLE 6   **A Function with Parameters** Find the derivative of

$$h(x) = (ax^2 - 2)^n$$

where $a > 0$ and n is a positive integer.

Solution   If we set $u = g(x) = ax^2 - 2$ and $f(u) = u^n$, then $h(x) = (f \circ g)(x)$. Since

$$g'(x) = 2ax \qquad \text{and} \qquad f'(u) = nu^{n-1}$$

it follows that

$$h'(x) = f'[g(x)]g'(x) = n(ax^2 - 2)^{n-1}2ax$$
$$= 2anx(ax^2 - 2)^{n-1}$$

Looking at $h'(x) = n(ax^2 - 2)^{n-1} \cdot 2ax$, we see that we first differentiated the outer function f, which yielded $n(ax^2 - 2)^{n-1}$ via the power rule, and then multiplied the result by $2ax$, the derivative of the inner function g. ∎

EXAMPLE 7   **Differentiating a Function That Is Not Specified** Suppose $f(x)$ is differentiable. Find

$$\frac{d}{dx}\frac{1}{\sqrt{f(x)}}$$

Solution   We set

$$h(x) = \frac{1}{\sqrt{f(x)}} = [f(x)]^{-1/2}$$

Now, $u = f(x)$ is the inner function and $h(u) = u^{-1/2}$ is the outer function; hence,

$$\frac{d}{dx}h(x) = \frac{dh}{du}\frac{du}{dx} = -\frac{1}{2}u^{-3/2}f'(x)$$
$$= -\frac{1}{2u^{3/2}}f'(x) = -\frac{f'(x)}{2[f(x)]^{3/2}}$$ ∎

EXAMPLE 8   **Generalized Power Rule** Suppose $f(x)$ is differentiable and r is a real number. Find

$$\frac{d}{dx}[f(x)]^r$$

Solution   Using the general form of the power rule and the chain rule, we find that

$$\frac{d}{dx}[f(x)]^r = r[f(x)]^{r-1}f'(x)$$ ∎

EXAMPLE 9   **Differentiating a Function That Is Not Specified** Suppose that $f'(x) = 3x - 1$. Find

$$\frac{d}{dx} f(x^2) \quad \text{at } x = 3$$

Solution   The inner function is $u = x^2$, the outer function is $f(u)$, and we find that

$$\frac{d}{dx} f(x^2) = 2x f'(x^2)$$

If we substitute $x = 3$ into $f'(x^2)$, we obtain $f'(3^2) = f'(9) = (3)(9) - 1 = 26$. Thus,

$$\frac{d}{dx} f(x^2) \bigg|_{x=3} = (2)(3) f'(9) = (6)(26) = 156 \qquad ■$$

The chain rule can be applied repeatedly, as shown in the next two examples.

EXAMPLE 10   **Nested Chain Rule** Find the derivative of

$$h(x) = \left(\sqrt{x^2 + 1} + 1 \right)^2$$

Solution   If we set $h(x) = (f \circ g)(x)$, then $g(x) = \sqrt{x^2 + 1} + 1$ and $f(u) = u^2$. We see that $g(x)$ is itself a composition of two functions, with inner function $v = x^2 + 1$ and outer function $\sqrt{v} + 1$. To differentiate $h(x)$, we proceed stepwise. First,

$$h'(x) = \frac{d}{dx} \left(\sqrt{x^2 + 1} + 1 \right)^2 = 2 \left(\sqrt{x^2 + 1} + 1 \right) \frac{d}{dx} \left(\sqrt{x^2 + 1} + 1 \right)$$

Then, since

$$\frac{d}{dx} \left(\sqrt{x^2 + 1} + 1 \right) = \frac{2x}{2\sqrt{x^2 + 1}} = \frac{x}{\sqrt{x^2 + 1}}$$

(where we used the chain rule to differentiate $\sqrt{x^2 + 1}$), we get

$$h'(x) = 2 \left(\sqrt{x^2 + 1} + 1 \right) \frac{x}{\sqrt{x^2 + 1}} \qquad ■$$

EXAMPLE 11   **Nested Chain Rule** Find the derivative of

$$h(x) = \left(2x^3 - \sqrt{3x^4 - 2} \right)^3$$

Solution   As in the previous example, we proceed stepwise:

$$h'(x) = 3 \left(2x^3 - \sqrt{3x^4 - 2} \right)^2 \frac{d}{dx} \left(2x^3 - \sqrt{3x^4 - 2} \right)$$

$$= 3 \left(2x^3 - \sqrt{3x^4 - 2} \right)^2 \left(6x^2 - \frac{12x^3}{2\sqrt{3x^4 - 2}} \right)$$

$$= 18x^2 \left(2x^3 - \sqrt{3x^4 - 2} \right)^2 \left(1 - \frac{x}{\sqrt{3x^4 - 2}} \right) \qquad ■$$

We conclude this subsection with the proof of the chain rule. The first part of the proof follows along the lines of the argument we sketched out at the beginning of the section, but the second part is much more technical and deals with the problem that Δu could be zero.

Proof of the Chain Rule We will use the definition of the derivative to prove the chain rule. Formally,

$$(f \circ g)'(x) = \lim_{x \to c} \frac{f[g(x)] - f[g(c)]}{x - c}$$

We need to show that the right-hand side is equal to $f'[g(c)]g'(c)$. As long as $g(x) \neq g(c)$, we can write

$$\lim_{x \to c} \frac{f[g(x)] - f[g(c)]}{x - c} = \lim_{x \to c} \frac{\frac{f[g(x)] - f[g(c)]}{g(x) - g(c)}[g(x) - g(c)]}{x - c}$$

$$= \lim_{x \to c} \frac{f[g(x)] - f[g(c)]}{g(x) - g(c)} \frac{g(x) - g(c)}{x - c}$$

Since $g(x)$ is continuous at $x = c$, it follows that $\lim_{x \to c} g(x) = g(c)$, and hence,

$$\lim_{x \to c} \frac{f[g(x)] - f[g(c)]}{g(x) - g(c)} = f'[g(c)]$$

Furthermore,

$$\lim_{x \to c} \frac{g(x) - g(c)}{x - c} = g'(c)$$

Since these limits exist, we can use the fact that the limit of a product is the product of the limits. We find that

$$\lim_{x \to c} \frac{f[g(x)] - f[g(c)]}{x - c} = \lim_{x \to c} \frac{f[g(x)] - f[g(c)]}{g(x) - g(c)} \lim_{x \to c} \frac{g(x) - g(c)}{x - c}$$

$$= f'[g(c)]g'(c)$$

In the preceding calculation, we needed to assume that $g(x) - g(c) \neq 0$. Of course, when we take the limit as $x \to c$, there might be x-values such that $g(x) = g(c)$, and we must deal with this possibility.

We set $y = g(x)$ and $d = g(c)$. The expression

$$f^*(y) \equiv \frac{f(y) - f(d)}{y - d}$$

is defined only for $y \neq d$. Since

$$\lim_{y \to d} \frac{f(y) - f(d)}{y - d} = f'(d)$$

we can extend $f^*[g(x)]$ by defining $f^*[g(x)] = f'[g(x)]$ to make $f^*[g(x)]$ a continuous function:

$$f^*[g(x)] = \begin{cases} \dfrac{f[g(x)] - f[g(c)]}{g(x) - g(c)} & \text{for } g(x) \neq g(c) \\ f'[g(c)] & \text{for } g(x) = g(c) \end{cases}$$

This means that, for all x,

$$f[g(x)] - f[g(c)] = f^*[g(x)][g(x) - g(c)]$$

With this equivalence, we can repeat our calculations to obtain

$$\lim_{x \to c} \frac{f[g(x)] - f[g(c)]}{x - c} = \lim_{x \to c} \frac{f^*[g(x)][g(x) - g(c)]}{x - c}$$

$$\lim_{x \to c} f^*[g(x)] \lim_{x \to c} \frac{g(x) - g(c)}{x - c} = f'[g(c)] \cdot g'(c)$$

Note that in the last step we used the fact that $f^*[g(x)]$ is continuous at $x = c$. ■

■ 4.4.2 Implicit Functions and Implicit Differentiation

So far, we have considered only functions of the form $y = f(x)$, which define y *explicitly* as a function of x. It is also possible to define y *implicitly* as a function of x, as in the following equation:

$$y^5x^2 - yx + 2y^2 = \sqrt{x}$$

Here, y is still given as a function of x (i.e., y is the dependent variable), but there is no obvious way to solve for y. Fortunately, there is a very useful technique, based on the chain rule, that will allow us to find dy/dx for implicitly defined functions. This technique is called **implicit differentiation**. We explain the procedure in the next example.

EXAMPLE 12 Find $\frac{dy}{dx}$ if $x^2 + y^2 = 1$.

Solution Remembering that y is a function of x, we differentiate both sides of the equation $x^2 + y^2 = 1$ with respect to x:

$$\frac{d}{dx}\left(x^2 + y^2\right) = \frac{d}{dx}(1)$$

Since the derivative of a sum is the sum of the derivatives, we find that

$$\frac{d}{dx}(x^2) + \frac{d}{dx}(y^2) = \frac{d}{dx}(1)$$

Starting with the left-hand side and using the power rule, we have $\frac{d}{dx}(x^2) = 2x$. To differentiate y^2 with respect to x, we apply the chain rule to get $\frac{d}{dx}(y^2) = 2y\frac{dy}{dx}$. On the right-hand side, we obtain $\frac{d}{dx}(1) = 0$. We therefore have

$$2x + 2y\frac{dy}{dx} = 0$$

We can now solve for $\frac{dy}{dx}$:

$$\frac{dy}{dx} = -\frac{2x}{2y} = -\frac{x}{y}$$

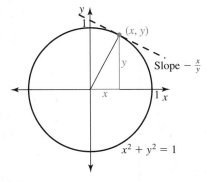

Figure 4.25 The slope of the tangent line at the unit circle $x^2 + y^2 = 1$ at (x, y) is $m = -\frac{x}{y}$.

Since $x^2 + y^2 = 1$ is the equation for the unit circle centered at the origin (Figure 4.25), we can use a geometric argument to convince ourselves that we have indeed obtained the correct derivative. The line that connects $(0, 0)$ and (x, y) has slope y/x and is perpendicular to the tangent line at (x, y). Since the slopes of perpendicular lines are negative reciprocals of each other, the slope of the tangent line at (x, y) must be $-x/y$.

We could have solved $x^2 + y^2 = 1$ for y and then differentiated with respect to x; this would have yielded the same answer but would have been more complicated. ■

We summarize the steps we take to find dy/dx when an equation defines y implicitly as a differentiable function of x:

STEP 1. Differentiate both sides of the equation with respect to x, keeping in mind that y is a function of x.

STEP 2. Solve the resulting equation for dy/dx.

Note that differentiating terms involving y typically requires the chain rule. Here is another example; this time, we can neither use a geometric argument nor easily solve for y.

EXAMPLE 13 Find $\frac{dy}{dx}$ when $y^3x^2 - yx + 2y^2 = x$.

Solution   We differentiate both sides of the equation with respect to x:

$$\frac{d}{dx}(y^3x^2) - \frac{d}{dx}(yx) + \frac{d}{dx}(2y^2) = \frac{d}{dx}(x)$$

To differentiate y^3x^2 and yx with respect to x, we use the product rule:

$$\frac{d}{dx}(y^3x^2) = \left(\frac{d}{dx}y^3\right)x^2 + y^3\left(\frac{d}{dx}x^2\right) = \left(\frac{d}{dx}y^3\right)x^2 + (y^3)(2x)$$

$$\frac{d}{dx}(yx) = \left(\frac{d}{dx}y\right)(x) + y\left(\frac{d}{dx}x\right) = \left(\frac{dy}{dx}\right)x + y$$

To find $\frac{d}{dx}y^3$, we use the chain rule to get

$$\frac{d}{dx}y^3 = 3y^2\frac{dy}{dx}$$

Furthermore,

$$\frac{d}{dx}(2y^2) = 4y\frac{dy}{dx}$$

and

$$\frac{d}{dx}(x) = 1$$

Putting the pieces together, we obtain

$$3y^2\frac{dy}{dx}(x^2) + (y^3)(2x) - \left[\left(\frac{dy}{dx}\right)x + y\right] + 4y\frac{dy}{dx} = 1$$

Factoring $\frac{dy}{dx}$ yields

$$\frac{dy}{dx}[3y^2x^2 - x + 4y] + 2xy^3 - y = 1$$

Solving for $\frac{dy}{dx}$ gives

$$\frac{dy}{dx} = \frac{y + 1 - 2xy^3}{3y^2x^2 - x + 4y}$$

The next example prepares us for the power rule for rational exponents.

EXAMPLE 14   Find $\frac{dy}{dx}$ when $y^2 = x^3$. Assume that $x > 0$ and $y > 0$.

Solution   We differentiate both sides with respect to x:

$$\frac{d}{dx}(y^2) = \frac{d}{dx}(x^3)$$

$$2y\frac{dy}{dx} = 3x^2$$

Therefore,

$$\frac{dy}{dx} = \frac{3}{2}\frac{x^2}{y}$$

Since $y = x^{3/2}$, it follows that

$$\frac{dy}{dx} = \frac{3}{2}\frac{x^2}{y} = \frac{3}{2}\frac{x^2}{x^{3/2}} = \frac{3}{2}x^{1/2}$$

This is the answer we expect from the general version of the power rule:

$$\frac{dy}{dx} = \frac{d}{dx}x^{3/2} = \frac{3}{2}x^{1/2}$$

Power Rule for Rational Exponents   We can generalize Example 14 to functions of the form $y = x^r$, where r is a rational number. This will provide a proof of the generalized form of the power rule when the exponent is a rational number, something we promised in the previous section. We write $r = p/q$, where p and q are integers and are in lowest terms. (If q is even, we require x and y to be positive.) Then

$$y = x^{p/q} \iff y^q = x^p$$

Differentiating both sides of $y^q = x^p$ with respect to x, we find that

$$q y^{q-1} \frac{dy}{dx} = p x^{p-1}$$

Hence,

$$\frac{dy}{dx} = \frac{p}{q} \frac{x^{p-1}}{y^{q-1}} = \frac{p}{q} \frac{x^{p-1}}{(x^{p/q})^{q-1}}$$

$$= \frac{p}{q} \frac{x^{p-1}}{x^{p-p/q}} = \frac{p}{q} x^{p-1-p+p/q}$$

$$= \frac{p}{q} x^{p/q-1} = r x^{r-1}$$

We summarize the preceding result:

If r is a rational number, then

$$\frac{d}{dx}(x^r) = r x^{r-1}$$

■ 4.4.3 Related Rates

An important application of implicit differentiation is related-rates problems. We begin with a motivating example.

Consider a parcel of air rising quickly in the atmosphere. The parcel expands without exchanging heat with the surrounding air. Laws of physics tell us that the volume (V) and the temperature (T) of the parcel of air are related via the formula

$$T V^{\gamma-1} = C$$

where γ (lowercase Greek gamma) is approximately 1.4 for sufficiently dry air and C is a constant. The temperature is measured in kelvin,[1] a scale chosen so that the temperature is always positive. (The Kelvin scale is the absolute temperature scale.) Since rising air expands, the volume of the parcel of air increases with time; we express this relationship mathematically as $dV/dt > 0$, where t denotes time.

To determine how the temperature of the air parcel changes as it rises, we implicitly differentiate $T V^{\gamma-1} = C$ with respect to t:

$$\frac{dT}{dt} V^{\gamma-1} + T(\gamma - 1) V^{\gamma-2} \frac{dV}{dt} = 0$$

or

$$\frac{dT}{dt} = -\frac{T(\gamma - 1) V^{\gamma-2} \frac{dV}{dt}}{V^{\gamma-1}} = -T(\gamma - 1) \frac{1}{V} \frac{dV}{dt}$$

---

[1] To compare the Celsius and the Kelvin scales, note that a temperature difference of 1°C is equal to a temperature difference of 1 K, and that 0°C = 273.15 K and 100°C = 373.15 K.

If we use $\gamma = 1.4$, then

$$\frac{dT}{dt} = -T(0.4)\frac{1}{V}\frac{dV}{dt}$$

implying that if air expands (i.e., $dV/dt > 0$), then temperature decreases (i.e., $dT/dt < 0$), since both T and V are positive: The temperature of a parcel of air decreases as the parcel rises, and the temperature of a falling air parcel increases. These phenomena can be observed close to high mountains.

In a typical related-rates problem, one quantity is expressed in terms of another and both quantities change with time. We usually know how one of the quantities changes with time and are interested in finding out how the other quantity changes. For instance, suppose that y is a function of x and both y and x depend on time. If we know how x changes with time (i.e., if we know dx/dt), then we might want to know how y changes with time (i.e, dy/dt). We illustrate this situation in the next example.

EXAMPLE 15   Find $\frac{dy}{dt}$ when $x^2 + y^3 = 1$ and $\frac{dx}{dt} = 2$ for $x = \sqrt{7/8}$.

Solution   In this example, both x and y are functions of t. Implicit differentiation with respect to t yields

$$\frac{d}{dt}(x^2 + y^3) = \frac{d}{dt}(1)$$

Hence,

$$2x\frac{dx}{dt} + 3y^2\frac{dy}{dt} = 0$$

Solving for $\frac{dy}{dt}$ gives

$$\frac{dy}{dt} = -\frac{2}{3}\frac{x}{y^2}\frac{dx}{dt}$$

When $x = \sqrt{7/8}$,

$$y^3 = 1 - x^2 = 1 - \frac{7}{8} = \frac{1}{8}$$

Thus, $y = 1/2$. Therefore,

$$\frac{dy}{dt} = -\frac{2}{3}\frac{\sqrt{7/8}}{1/4} \cdot 2 = -\frac{16}{3}\sqrt{\frac{7}{8}} = -\frac{4}{3}\sqrt{14}$$ ■

We present two applications of related rates.

EXAMPLE 16   **Changing Volume**   A spherical balloon is being filled with air. When the radius $r = 6$ cm, the radius is increasing at a rate of 2 cm/s. How fast is the volume changing at this time?

Solution   The volume V of a sphere of radius r is given by

$$V = \frac{4}{3}\pi r^3 \tag{4.6}$$

(See Figure 4.26.) Note that V is a function of r. Since r is increasing at a certain rate, we think of r as a function of time t; that is, $r = r(t)$. Because the volume V depends on r, it changes with time t as well. We therefore consider V also as a function of time t. Differentiating both sides of (4.6) with respect to t, we find that

$$\frac{dV}{dt} = \frac{4}{3}\pi 3r^2\frac{dr}{dt} = 4\pi r^2\frac{dr}{dt}$$

When $r = 6$ cm and $dr/dt = 2$ cm/s,

$$\frac{dV}{dt} = 4\pi 6^2\text{cm}^2 2\frac{\text{cm}}{\text{s}} = 288\pi\frac{\text{cm}^3}{\text{s}}$$

Figure 4.26 The volume of a sphere with radius r is $V = \frac{4}{3}\pi r^3$.

Note that the unit of dV/dt is cm³/s, which is what you should expect, because the unit of the volume is cm³ and time is measured in seconds. ■

EXAMPLE 17

Allometric Growth (Adapted from Benton and Harper, 1997) Ichthyosaurs are a group of marine reptiles that were fish shaped and comparable in size to dolphins. They became extinct during the Cretaceous.[2] On the basis of a study of 20 fossil skeletons, it was found that the skull length (in cm) and backbone length (in cm) of an individual ichthyosaur were related through the allometric equation. (We introduced allometric equations in Example 7 of Section 1.2.)

$$[\text{skull length}] = 1.162[\text{backbone length}]^{0.933}$$

How is the growth rate of the backbone related to the growth rate of the skull?

Solution

Let x denote the age of the ichthyosaur, and set

$$S = S(x) = \text{skull length at age } x$$
$$B = B(x) = \text{backbone length at age } x$$

so that

$$S(x) = (1.162)[B(x)]^{0.933}$$

We are interested in the relationship between dS/dx and dB/dx, the growth rates of the skull and the backbone, respectively. Differentiating the equation for $S(x)$ with respect to x, we find that

$$\frac{dS}{dx} = (1.162)(0.933)[B(x)]^{0.933-1}\frac{dB}{dx}$$

Rearranging terms on the right-hand side, we write this as

$$\frac{dS}{dx} = \underbrace{(1.162)[B(x)]^{0.933}}_{S(x)}(0.933)\frac{1}{B(x)}\frac{dB}{dx}$$

Hence,

$$\frac{1}{S(x)}\frac{dS}{dx} = 0.933\frac{1}{B(x)}\frac{dB}{dx}$$

This equation relates the relative growth rates $\frac{1}{S}\frac{dS}{dx}$ and $\frac{1}{B}\frac{dB}{dx}$. The factor 0.933 is less than 1, which indicates that skulls grow less quickly than backbones. This finding should be familiar to us: Relative to their body sizes, juvenile vertebrates often have larger heads than adults. ■

■ 4.4.4 Higher Derivatives

The derivative of a function f is itself a function. We refer to this derivative as the **first derivative**, denoted f'. If the first derivative exists, we say that the function is once differentiable. Given that the first derivative is a function, we can define its derivative (where it exists). This derivative is called the **second derivative** and is denoted f''. If the second derivative exists, we say that the original function is twice differentiable. This second derivative is again a function; hence, we can define its derivative (where it exists). The result is the **third derivative**, denoted f'''. If the third derivative exists, we say that the original function is three times differentiable. We can continue in this manner; from the fourth derivative on, we denote the derivatives by $f^{(4)}$, $f^{(5)}$, and so on. If the nth derivative exists, we say that the original function is n times differentiable.

Polynomials are functions that can be differentiated as many times as desired. The reason is that the first derivative of a polynomial of degree n is a polynomial of degree $n - 1$. Since the derivative is a polynomial as well, we can find its derivative, and so on. Eventually, the derivative will be equal to 0, as is illustrated in the next example.

---

(2) The Cretaceous period began about 144 million years ago and ended about 65 million years ago.

EXAMPLE 18   Find the *n*th derivative of $f(x) = x^5$ for $n = 1, 2, \ldots$.

Solution   Differentiating $f(x)$, we find the first derivative to be

$$f'(x) = 5x^4$$

Differentiating $f'(x)$, we find the second derivative to be

$$f''(x) = 5(4x^3) = 20x^3$$

Differentiating $f''(x)$, we obtain the third derivative:

$$f'''(x) = 20(3x^2) = 60x^2$$

Differentiating $f'''(x)$, we find the fourth derivative:

$$f^{(4)}(x) = 60(2x) = 120x$$

Differentiating $f^{(4)}(x)$, we get the fifth derivative:

$$f^{(5)}(x) = 120$$

Differentiating $f^{(5)}(x)$, we find the sixth derivative:

$$f^{(6)}(x) = 0$$

All higher-order derivatives—that is, $f^{(7)}, f^{(8)}, \ldots$—are equal to 0 as well.   ■

We can write higher-order derivatives in Leibniz notation: The *n*th derivative of $f(x)$ is denoted by

$$\frac{d^n f}{dx^n}$$

EXAMPLE 19   Find the second derivative of $f(x) = \sqrt{x}, x \geq 0$.

Solution   First, we find the first derivative:

$$\frac{d}{dx}\sqrt{x} = \frac{d}{dx}x^{1/2} = \frac{1}{2}x^{-1/2} \quad \text{for } x > 0$$

To find the second derivative, we differentiate the first derivative

$$\frac{d^2}{dx^2}\sqrt{x} = \frac{d}{dx}\left(\frac{d}{dx}\sqrt{x}\right) = \frac{d}{dx}\left(\frac{1}{2}x^{-1/2}\right) = \frac{1}{2}\left(-\frac{1}{2}\right)x^{(-1/2)-1}$$

$$= -\frac{1}{4}x^{-3/2}$$ ■

When functions are implicitly defined, we can use the technique of implicit differentiation to find higher derivatives.

EXAMPLE 20   Find $\frac{d^2y}{dx^2}$ when $x^2 + y^2 = 1$.

Solution   We found

$$\frac{dy}{dx} = -\frac{x}{y}$$

in Example 12. Differentiating both sides of this equation with respect to x, we get

$$\frac{d}{dx}\left[\frac{dy}{dx}\right] = \frac{d}{dx}\left[-\frac{x}{y}\right]$$

The left-hand side can be written as

$$\frac{d^2y}{dx^2}$$

On the right-hand side, we use the quotient rule. Hence,

$$\frac{d^2 y}{dx^2} = -\frac{1 \cdot y - x \frac{dy}{dx}}{y^2}$$

Substituting $-\frac{x}{y}$ for $\frac{dy}{dx}$, we obtain

$$\frac{d^2 y}{dx^2} = -\frac{y - x(-\frac{x}{y})}{y^2}$$

$$= -\frac{y + \frac{x^2}{y}}{y^2} = -\frac{y^2 + x^2}{y^3}$$

Since $x^2 + y^2 = 1$, we can simplify the rightmost expression further and obtain

$$\frac{d^2 y}{dx^2} = -\frac{1}{y^3}$$ ■

We introduced the velocity of an object that moves on a straight line as the derivative of the object's position. The derivative of the velocity is the **acceleration**. If $s(t)$ denotes the position of an object moving on a straight line, $v(t)$ its velocity, and $a(t)$ its acceleration, then the three quantities are related as follows:

$$v(t) = \frac{ds}{dt} \quad \text{and} \quad a(t) = \frac{dv}{dt} = \frac{d^2 s}{dt^2}$$

EXAMPLE 21   **Acceleration** Assume that the position of a car moving along a straight line is given by

$$s(t) = 3t^3 - 2t + 1$$

Find the car's velocity and acceleration.

Solution   To find the velocity, we need to differentiate the position:

$$v(t) = \frac{ds}{dt} = 9t^2 - 2$$

To find the acceleration, we differentiate the velocity:

$$a(t) = \frac{dv}{dt} = \frac{d^2 s}{dt^2} = 18t$$ ■

EXAMPLE 22   Neglecting air resistance, we find that the distance (in meters) an object falls when dropped from rest from a height is

$$s(t) = \frac{1}{2} g t^2$$

where $g = 9.81 \text{m/s}^2$ is the earth's gravitational constant and t is the time (in seconds) elapsed since the object was released.

(a) Find the object's velocity and acceleration.

(b) If the height is 30 m, how long will it take until the object hits the ground, and what is its velocity at the time of impact?

Solution   (a) The velocity is

$$v(t) = \frac{ds}{dt} = gt$$

and the acceleration is

$$a(t) = \frac{dv}{dt} = g$$

Note that the acceleration is constant.

(b) To find the time it takes the object to hit the ground, we set $s(t) = 30$ m and solve for t:

$$30 \, \text{m} = \frac{1}{2}(9.81)\frac{\text{m}}{\text{s}^2}t^2$$

This yields

$$t^2 = \frac{60}{9.81}\text{s}^2, \quad \text{or} \quad t = \sqrt{\frac{60}{9.81}} \, \text{s} \approx 2.47 \, \text{s}$$

(We need consider only the positive solution.) The velocity at the time of impact is then

$$v(t) = gt = (9.81)\frac{\text{m}}{\text{s}^2}\sqrt{\frac{60}{9.81}} \, \text{s} \approx 24.3\frac{\text{m}}{\text{s}} \qquad ■$$

Section 4.4 Problems

■ 4.4.1

In Problems 1–28, differentiate the functions with respect to the independent variable.

1. $f(x) = (x - 3)^2$

2. $f(x) = (4x + 5)^3$

3. $f(x) = (1 - 3x^2)^4$

4. $f(x) = (5x^2 - 3x)^3$

5. $f(x) = \sqrt{x^2 + 3}$

6. $f(x) = \sqrt{2x + 7}$

7. $f(x) = \sqrt{3 - x^3}$

8. $f(x) = \sqrt{5x + 3x^4}$

9. $f(x) = \dfrac{1}{(x^3 - 2)^4}$

10. $f(x) = \dfrac{2}{(1 - 5x^2)^3}$

11. $f(x) = \dfrac{3x - 1}{\sqrt{2x^2 - 1}}$

12. $f(x) = \dfrac{(1 - 2x^2)^3}{(3 - x^2)^2}$

13. $f(x) = \dfrac{\sqrt{2x - 1}}{(x - 1)^2}$

14. $f(x) = \dfrac{\sqrt{x^2 - 1}}{2 + \sqrt{x^2 + 1}}$

15. $f(s) = \sqrt{s + \sqrt{s}}$

16. $g(t) = \sqrt{t^2 + \sqrt{t + 1}}$

17. $g(t) = \left(\dfrac{t}{t - 3}\right)^3$

18. $h(s) = \left(\dfrac{2s^2}{s + 1}\right)^4$

19. $f(r) = (r^2 - r)^3(r + 3r^3)^{-4}$

20. $h(s) = \dfrac{2(3 - s)^2}{s^2 + (7s - 1)^2}$

21. $h(x) = \sqrt[5]{3 - x^4}$

22. $h(x) = \sqrt[3]{1 - 2x}$

23. $f(x) = \sqrt[7]{x^2 - 2x + 1}$

24. $f(x) = \sqrt[4]{2 - 4x^2}$

25. $g(s) = (3s^7 - 7s)^{3/2}$

26. $h(t) = (t^4 - 5t)^{5/2}$

27. $h(t) = \left(3t + \dfrac{3}{t}\right)^{2/5}$

28. $h(t) = \left(4t^4 + \dfrac{4}{t^4}\right)^{1/4}$

29. Differentiate

$$f(x) = (ax + 1)^3$$

with respect to x. Assume that a is a positive constant.

30. Differentiate

$$f(x) = \sqrt{ax^2 - 2}$$

with respect to x. Assume that a is a positive constant.

31. Differentiate

$$g(N) = \frac{bN}{(k + N)^2}$$

with respect to N. Assume that b and k are positive constants.

32. Differentiate

$$g(N) = \frac{N}{(k + bN)^3}$$

with respect to N. Assume that b and k are positive constants.

33. Differentiate

$$g(T) = a(T_0 - T)^3 - b$$

with respect to T. Assume that a, b, and T_0 are positive constants.

34. Suppose that $f'(x) = 2x + 1$. Find the following:

(a) $\dfrac{d}{dx}f(x^2)$ at $x = -1$

(b) $\dfrac{d}{dx}f(\sqrt{x})$ at $x = 4$

35. Suppose that $f'(x) = \frac{1}{x}$. Find the following:

(a) $\dfrac{d}{dx}f(x^2 + 3)$

(b) $\dfrac{d}{dx}f(\sqrt{x - 1})$

In Problems 36–39, assume that $f(x)$ and $g(x)$ are differentiable.

36. Find $\dfrac{d}{dx}\sqrt{f(x) + g(x)}$.

37. Find $\dfrac{d}{dx}\left(\dfrac{f(x)}{g(x)} + 1\right)^2$.

38. Find $\dfrac{d}{dx}f\left[\dfrac{1}{g(x)}\right]$.

39. Find $\dfrac{d}{dx}\dfrac{[f(x)]^2}{g(2x) + 2x}$.

In Problems 40–46, find $\frac{dy}{dx}$ by applying the chain rule repeatedly.

40. $y = (\sqrt{1 - 2x^2} + 1)^2$

41. $y = (\sqrt{x^3 - 3x} + 3x)^4$

42. $y = \left(1 + 2(x + 3)^4\right)^2$

43. $y = \left(1 + (3x^2 - 1)^3\right)^2$

44. $y = \left(\dfrac{x}{2(x^2 - 1)^2 - 1}\right)^2$

45. $y = \left(\dfrac{2x + 1}{3(x^3 - 1)^3 - 1}\right)^3$

46. $y = \left(\dfrac{(2x + 1)^2 - x}{(3x^3 + 1)^3 - x}\right)^2$

■ 4.4.2

In Problems 47–54, find $\frac{dy}{dx}$ by implicit differentiation.

47. $x^2 + y^2 = 4$

48. $y = x^2 + 3yx$

49. $x^{3/4} + y^{3/4} = 1$

50. $xy - y^3 = 1$

51. $\sqrt{xy} = x^2 + 1$

52. $\dfrac{1}{2xy} - y^3 = 4$

53. $\dfrac{x}{y} = \dfrac{y}{x}$

54. $\dfrac{x}{xy + 1} = 2xy$

In Problems 55–57, find the lines that are (a) tangential and (b) normal to each curve at the given point.

55. $x^2 + y^2 = 25$, $(4, -3)$ (circle)

56. $\dfrac{x^2}{4} + \dfrac{y^2}{9} = 1$, $(1, \frac{3}{2}\sqrt{3})$ (ellipse)

57. $\dfrac{x^2}{25} - \dfrac{y^2}{9} = 1$, $(\frac{25}{3}, 4)$ (hyperbola)

58. Lemniscate

(a) The curve with equation $y^2 = x^2 - x^4$ is shaped like the numeral eight. Find $\frac{dy}{dx}$ at $(\frac{1}{2}, \frac{1}{4}\sqrt{3})$.

(b) Use a graphing calculator to graph the curve in (a). If the calculator cannot graph implicit functions, graph the upper and the lower halves of the curve separately; that is, graph

$$y_1 = \sqrt{x^2 - x^4}$$
$$y_2 = -\sqrt{x^2 - x^4}$$

Choose the viewing rectangle $-2 \leq x \leq 2, -1 \leq y \leq 1$.

59. Astroid

(a) Consider the curve with equation $x^{2/3} + y^{2/3} = 4$. Find $\frac{dy}{dx}$ at $(-1, 3\sqrt{3})$.

(b) Use a graphing calculator to graph the curve in (a). If the calculator cannot graph implicit functions, graph the upper and the lower halves of the curve separately. To get the left half of the graph, make sure that your calculator evaluates $x^{2/3}$ in the order $(x^2)^{1/3}$. Choose the viewing rectangle $-10 \leq x \leq 10$, $-10 \leq y \leq 10$.

60. Kampyle of Eudoxus

(a) Consider the curve with equation $y^2 = 10x^4 - x^2$. Find $\frac{dy}{dx}$ at $(1, 3)$.

(b) Use a graphing calculator to graph the curve in (a). If the calculator cannot graph implicit functions, graph the upper and the lower halves of the curve separately. Choose the viewing rectangle $-3 \leq x \leq 3, -10 \leq y \leq 10$.

■ **4.4.3**

61. Assume that x and y are differentiable functions of t. Find $\frac{dy}{dt}$ when $x^2 + y^2 = 1$, $\frac{dx}{dt} = 2$ for $x = \frac{1}{2}$, and $y > 0$.

62. Assume that x and y are differentiable functions of t. Find $\frac{dy}{dt}$ when $y^2 = x^2 - x^4$, $\frac{dx}{dt} = 1$ for $x = \frac{1}{2}$, and $y > 0$.

63. Assume that x and y are differentiable functions of t. Find $\frac{dy}{dt}$ when $x^2 y = 1$ and $\frac{dx}{dt} = 3$ for $x = 2$.

64. Assume that u and v are differentiable functions of t. Find $\frac{du}{dt}$ when $u^2 + v^3 = 12$, $\frac{dv}{dt} = 2$ for $v = 2$, and $u > 0$.

65. Assume that the side length x and the volume $V = x^3$ of a cube are differentiable functions of t. Express dV/dt in terms of dx/dt.

66. Assume that the radius r and the area $A = \pi r^2$ of a circle are differentiable functions of t. Express dA/dt in terms of dr/dt.

67. Assume that the radius r and the surface area $S = 4\pi r^2$ of a sphere are differentiable functions of t. Express dS/dt in terms of dr/dt.

68. Assume that the radius r and the volume $V = \frac{4}{3}\pi r^3$ of a sphere are differentiable functions of t. Express dV/dt in terms of dr/dt.

69. Suppose that water is stored in a cylindrical tank of radius 5 m. If the height of the water in the tank is h, then the volume of the water is $V = \pi r^2 h = (25 \text{m}^2)\pi h = 25\pi h \text{ m}^2$. If we drain the water at a rate of 250 liters per minute, what is the rate at which the water level inside the tank drops? (Note that 1 cubic meter contains 1000 liters.)

70. Suppose that we pump water into an inverted right circular conical tank at the rate of 5 cubic feet per minute (i.e., the tank stands with its point facing downward). The tank has a height of 6

ft and the radius on top is 3 ft. What is the rate at which the water level is rising when the water is 2 ft deep? (Note that the volume of a right circular cone of radius r and height h is $V = \frac{1}{3}\pi r^2 h$.)

71. Two people start biking from the same point. One bikes east at 15 mph, the other south at 18 mph. What is the rate at which the distance between the two people is changing after 20 minutes and after 40 minutes?

72. Allometric equations describe the scaling relationship between two measurements, such as skull length versus body length. In vertebrates, we typically find that

$$[\text{skull length}] \propto [\text{body length}]^a$$

for $0 < a < 1$. Express the growth rate of the skull length in terms of the growth rate of the body length.

■ **4.4.4**

In Problems 73–82, find the first and the second derivatives of each function.

73. $f(x) = x^3 - 3x^2 + 1$

74. $f(x) = (2x^2 + 4)^3$

75. $g(x) = \dfrac{x - 1}{x + 1}$

76. $h(s) = \dfrac{1}{s^2 + 2}$

77. $g(t) = \sqrt{3t^3 + 2t}$

78. $f(x) = \dfrac{1}{x^2} + x - x^3$

79. $f(s) = \sqrt{s^{3/2} - 1}$

80. $f(x) = \dfrac{2x}{x^2 + 1}$

81. $g(t) = t^{-5/2} - t^{1/2}$

82. $f(x) = x^3 - \dfrac{1}{x^3}$

83. Find the first 10 derivatives of $y = x^5$.

84. Find $f^{(n)}(x)$ and $f^{(n+1)}(x)$ of $f(x) = x^n$.

85. Find a second-degree polynomial $p(x) = ax^2 + bx + c$ with $p(0) = 3$, $p'(0) = 2$, and $p''(0) = 6$.

86. The position at time t of a particle that moves along a straight line is given by the function $s(t)$. The first derivative of $s(t)$ is called the velocity, denoted by $v(t)$; that is, the velocity is the rate of change of the position. The rate of change of the velocity is called **acceleration**, denoted by $a(t)$; that is,

$$\frac{d}{dt}v(t) = a(t)$$

Given that $v(t) = s'(t)$, it follows that

$$\frac{d^2}{dt^2}s(t) = a(t)$$

Find the velocity and the acceleration at time $t = 1$ for the following position functions:

(a) $s(t) = t^2 - 3t$ **(b)** $s(t) = \sqrt{t^2 + 1}$ **(c)** $s(t) = t^4 - 2t$

87. Neglecting air resistance, the height h (in meters) of an object thrown vertically from the ground with initial velocity v_0 is given by

$$h(t) = v_0 t - \frac{1}{2}gt^2$$

where $g = 9.81 \text{ m/s}^2$ is the earth's gravitational constant and t is the time (in seconds) elapsed since the object was released.

(a) Find the velocity and the acceleration of the object.

(b) Find the time when the velocity is equal to 0. In which direction is the object traveling right before this time? in which direction right after this time?

■ 4.5 Derivatives of Trigonometric Functions

We will need the trigonometric limits from Section 3.4 to compute the derivatives of the sine and cosine functions. Note that all angles are measured in radians.

> **Theorem** The functions $\sin x$ and $\cos x$ are differentiable for all x, and
>
> $$\frac{d}{dx}\sin x = \cos x \qquad \text{and} \qquad \frac{d}{dx}\cos x = -\sin x$$

Graphs of the derivatives of each of the trigonometric functions, based on the geometric interpretation of a derivative as the slope of the tangent line, confirm these rules. (See Figures 4.27 and 4.28.) Pay particular attention to the points on the graph of $f(x)$ with horizontal tangent lines. These correspond to the points of intersection of the graph of $f'(x)$ with the x-axis.

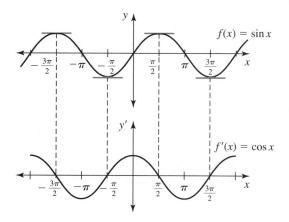

Figure 4.27 The function $f(x) = \sin x$ and its derivative $f'(x) = \cos x$. The derivative $f'(x) = 0$ where $f(x)$ has a horizontal tangent line.

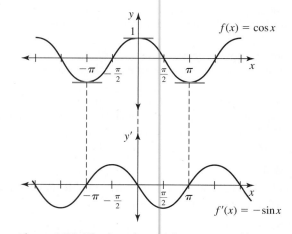

Figure 4.28 The function $f(x) = \cos x$ and its derivative $f'(x) = -\sin x$. The derivative $f'(x) = 0$ where $f(x)$ has a horizontal tangent line.

Proof We prove the first formula; a similar proof of the second formula is discussed in Problem 61. We need the trigonometric identity

$$\sin(\alpha + \beta) = \sin\alpha\cos\beta + \cos\alpha\sin\beta$$

Using the formal definition of derivatives, we find that

$$\frac{d}{dx}\sin x = \lim_{h\to 0}\frac{\sin(x+h) - \sin x}{h}$$

$$= \lim_{h\to 0}\frac{\sin x\cos h + \cos x\sin h - \sin x}{h}$$

$$= \lim_{h\to 0}\left[\sin x\frac{\cos h - 1}{h} + \cos x\frac{\sin h}{h}\right]$$

In Section 3.4, we showed that

$$\lim_{h\to 0}\frac{\cos h - 1}{h} = 0 \qquad \text{and} \qquad \lim_{h\to 0}\frac{\sin h}{h} = 1$$

We can therefore apply the basic rules for limits to obtain

$$\frac{d}{dx}\sin x = \sin x\lim_{h\to 0}\frac{\cos h - 1}{h} + \cos x\lim_{h\to 0}\frac{\sin h}{h}$$

$$= (\sin x)(0) + (\cos x)(1) = \cos x \qquad\blacksquare$$

EXAMPLE 1   Find the derivative of $f(x) = -4 \sin x + \cos \frac{\pi}{6}$.

Solution

$$f'(x) = \frac{d}{dx}(-4 \sin x + \cos \frac{\pi}{6})$$

$$= -4 \frac{d}{dx} \sin x + \frac{d}{dx} \cos \frac{\pi}{6}$$

The first term is a trigonometric function and the second term is a constant (namely, $\frac{1}{2}\sqrt{3}$). Differentiating each term, we find that

$$f'(x) = -4(\cos x) + 0 = -4 \cos x$$ ■

EXAMPLE 2   Find the derivative of $y = \cos(x^2 + 1)$.

Solution   We set $u = g(x) = x^2 + 1$ and $f(u) = \cos u$; then $y = f[g(x)]$. Using the chain rule, we then obtain

$$y' = \frac{df}{du}\frac{du}{dx} = \frac{d}{du}(\cos u)\frac{d}{dx}(x^2 + 1) = (-\sin u)(2x)$$

$$= -[\sin(x^2 + 1)]2x = -2x \sin(x^2 + 1)$$ ■

EXAMPLE 3   Find the derivative of $y = x^2 \sin(3x) - \cos(5x)$.

Solution   We will use the product rule for the first term; in addition, we will need the chain rule for both $\sin(3x)$ and $\cos(5x)$:

$$y' = \frac{d}{dx}[x^2 \sin(3x) - \cos(5x)]$$

$$= \frac{d}{dx}[x^2 \sin(3x)] - \frac{d}{dx}\cos(5x)$$

$$= \left(\frac{d}{dx}x^2\right)\sin(3x) + x^2\frac{d}{dx}\sin(3x) - \frac{d}{dx}\cos(5x)$$

$$= 2x \sin(3x) + x^2 3 \cos(3x) - 5(-\sin(5x))$$

$$= 2x \sin(3x) + 3x^2 \cos(3x) + 5 \sin(5x)$$ ■

The derivatives of the other trigonometric functions can be found using the following identities:

$$\tan x = \frac{\sin x}{\cos x} \qquad\qquad \cot x = \frac{\cos x}{\sin x}$$

$$\sec x = \frac{1}{\cos x} \qquad\qquad \csc x = \frac{1}{\sin x}$$

For instance, to find the derivative of the tangent, we use the quotient rule:

$$\frac{d}{dx}\tan x = \frac{d}{dx}\frac{\sin x}{\cos x}$$

$$= \frac{(\frac{d}{dx}\sin x)\cos x - \sin x(\frac{d}{dx}\cos x)}{\cos^2 x}$$

$$= \frac{(\cos x)(\cos x) - (\sin x)(-\sin x)}{\cos^2 x}$$

$$= \frac{\cos^2 x + \sin^2 x}{\cos^2 x} = \frac{1}{\cos^2 x} = \sec^2 x$$

In the penultimate step, we used the identity $\cos^2 x + \sin^2 x = 1$.

The other derivatives can be found in a similar fashion, as explained in Problems 62–64.

We summarize the derivatives of the six fundamental trigonometric functions in the following box:

$$\frac{d}{dx} \sin x = \cos x \qquad\qquad \frac{d}{dx} \cos x = -\sin x$$

$$\frac{d}{dx} \tan x = \sec^2 x \qquad\qquad \frac{d}{dx} \cot x = -\csc^2 x$$

$$\frac{d}{dx} \sec x = \sec x \tan x \qquad\qquad \frac{d}{dx} \csc x = -\csc x \cot x$$

EXAMPLE 4 Compare the derivatives of

(a) $\tan x^2$ (b) $\tan^2 x$

Solution (a) If $y = \tan x^2 = \tan(x^2)$, then, using the chain rule, we find that

$$\frac{dy}{dx} = \frac{d}{dx} \tan(x^2) = (\sec^2(x^2))(2x) = 2x \sec^2(x^2)$$

(b) If $y = \tan^2 x = (\tan x)^2$, then, using the chain rule, we obtain

$$\frac{dy}{dx} = \frac{d}{dx}(\tan x)^2 = 2(\tan x)\frac{d}{dx} \tan x = 2 \tan x \sec^2 x$$

The two derivatives are clearly different, and you should look again at $\tan x^2$ and $\tan^2 x$ to make sure that you understand which is the inner and which the outer function. ■

EXAMPLE 5 **Repeated Application of the Chain Rule** Find the derivative of $f(x) = \sec \sqrt{x^2 + 1}$.

Solution This is a composite function; the inner function is $\sqrt{x^2 + 1}$ and the outer function is $\sec x$. Applying the chain rule once, we find that

$$\frac{df}{dx} = \frac{d}{dx} \sec \sqrt{x^2 + 1} = \sec \sqrt{x^2 + 1} \tan \sqrt{x^2 + 1} \frac{d}{dx} \sqrt{x^2 + 1}$$

To evaluate $\frac{d}{dx}\sqrt{x^2 + 1}$, we need to apply the chain rule a second time:

$$\frac{d}{dx}\sqrt{x^2 + 1} = \frac{1}{2\sqrt{x^2 + 1}} 2x = \frac{x}{\sqrt{x^2 + 1}}$$

Combining the two steps, we obtain

$$\frac{df}{dx} = \left(\sec \sqrt{x^2 + 1}\right)\left(\tan \sqrt{x^2 + 1}\right)\frac{x}{\sqrt{x^2 + 1}}$$

The function $f(x)$ can be thought of as a composition of three functions. The innermost function is $u = g(x) = x^2 + 1$, the middle function is $v = h(u) = \sqrt{u}$, and the outermost function is $f(v) = \sec v$. When we computed the derivative, we applied the chain rule twice in the form

$$\frac{df}{dx} = \frac{df}{dv}\frac{dv}{du}\frac{du}{dx}$$

■

Section 4.5 Problems

In Problems 1–58, find the derivative with respect to the independent variable.

1. $f(x) = 2\sin x - \cos x$ **2.** $f(x) = 3\cos x - 2\sin x$
3. $f(x) = 3\sin x + 5\cos x - 2\sec x$
4. $f(x) = -\sin x + \cos x - 3\csc x$
5. $f(x) = \tan x - \cot x$ **6.** $f(x) = \sec x - \csc x$
7. $f(x) = \sin(3x)$ **8.** $f(x) = \cos(-5x)$
9. $f(x) = 2\sin(3x + 1)$ **10.** $f(x) = -3\cos(1 - 2x)$
11. $f(x) = \tan(4x)$ **12.** $f(x) = \cot(2 - 3x)$
13. $f(x) = 2\sec(1 + 2x)$ **14.** $f(x) = -3\csc(3 - 5x)$
15. $f(x) = 3\sin(x^2)$ **16.** $f(x) = 2\cos(x^3 - 3x)$
17. $f(x) = \sin^3(x^2 - 3)$ **18.** $f(x) = \cos^2(x^2 - 1)$
19. $f(x) = 3\sin^2 x^2$ **20.** $f(x) = -\sin^2(2x^3 - 1)$
21. $f(x) = 4\cos x^2 - 2\cos^2 x$
22. $f(x) = -5\cos(2 - x^3) + 2\cos^3(x - 4)$
23. $f(x) = 4\cos^2 x + 2\cos x^4$ **24.** $f(x) = -3\cos^2(3x^2 - 4)$
25. $f(x) = 2\tan(1 - x^2)$ **26.** $f(x) = -\cot(3x^3 - 4x)$
27. $f(x) = -2\tan^3(3x - 1)$ **28.** $f(x) = \sqrt{\sin x} + \sin\sqrt{x}$
29. $f(x) = \sqrt{\sin(2x^2 - 1)}$ **30.** $g(s) = (\cos^2 s - 3s^2)^2$
31. $g(s) = \sqrt{\cos s} - \cos\sqrt{x}$ **32.** $g(t) = \dfrac{\sin(3t)}{\cos(5t)}$
33. $g(t) = \dfrac{\sin(2t) + 1}{\cos(6t) - 1}$ **34.** $f(x) = \dfrac{\cot(2x)}{\tan(4x)}$
35. $f(x) = \dfrac{\sec(x^2 - 1)}{\csc(x^2 + 1)}$ **36.** $f(x) = \sin x \cos x$
37. $f(x) = \sin(2x - 1)\cos(3x + 1)$
38. $f(x) = \tan x \cot x$
39. $f(x) = \tan(3x^2 - 1)\cot(3x^2 + 1)$
40. $f(x) = \sec x \cos x$
41. $f(x) = \sin x \sec x$ **42.** $f(x) = \dfrac{1}{\sin^2 x + \cos^2 x}$
43. $f(x) = \dfrac{1}{\tan^2 x - \sec^2 x}$ **44.** $g(x) = \dfrac{1}{\sin(3x)}$
45. $g(x) = \dfrac{1}{\sin(3x^2 - 1)}$ **46.** $g(x) = \dfrac{1}{\csc^2(5x)}$
47. $g(x) = \dfrac{1}{\csc^3(1 - 5x^2)}$ **48.** $h(x) = \cot(3x)\csc(3x)$
49. $h(x) = \dfrac{3}{\tan(2x) - x}$ **50.** $g(t) = \left(\dfrac{1}{\sin t^2}\right)^{3/2}$
51. $h(s) = \sin^3 s + \cos^3 s$ **52.** $f(x) = (2x^3 - x)\cos(1 - x^2)$
53. $f(x) = \dfrac{\sin(2x)}{1 + x^2}$ **54.** $f(x) = \dfrac{1 + \cos(3x)}{2x^3 - x}$
55. $f(x) = \tan\dfrac{1}{x}$ **56.** $f(x) = \sec\dfrac{1}{1 + x^2}$
57. $f(x) = \dfrac{\sec x^2}{\sec^2 x}$ **58.** $f(x) = \dfrac{\csc(3 - x^2)}{1 - x^2}$

59. Find the points on the curve $y = \sin(\frac{\pi}{3}x)$ that have a horizontal tangent.
60. Find the points on the curve $y = \cos^2 x$ that have a horizontal tangent.
61. Use the identity
$$\cos(\alpha + \beta) = \cos\alpha\cos\beta - \sin\alpha\sin\beta$$
and the definition of the derivative to show that
$$\frac{d}{dx}\cos x = -\sin x$$

62. Use the quotient rule to show that
$$\frac{d}{dx}\cot x = -\csc^2 x$$
(*Hint*: Write $\cot x = \frac{\cos x}{\sin x}$.)
63. Use the quotient rule to show that
$$\frac{d}{dx}\sec x = \sec x \tan x$$
[*Hint*: Write $\sec x = (\cos x)^{-1}$.]
64. Use the quotient rule to show that
$$\frac{d}{dx}\csc x = -\csc x \cot x$$
[*Hint*: Write $\csc x = (\sin x)^{-1}$.]
65. Find the derivative of
$$f(x) = \sin\sqrt{x^2 + 1}$$
66. Find the derivative of
$$f(x) = \cos\sqrt{x^2 + 1}$$
67. Find the derivative of
$$f(x) = \sin\sqrt{3x^3 + 3x}$$
68. Find the derivative of
$$f(x) = \cos\sqrt{1 - 4x^4}$$
69. Find the derivative of
$$f(x) = \sin^2(x^2 - 1)$$
70. Find the derivative of
$$f(x) = \cos^2(2x^2 + 3)$$
71. Find the derivative of
$$f(x) = \tan^3(3x^3 - 3)$$
72. Find the derivative of
$$f(x) = \sec^2(2x^2 - 2)$$

73. Suppose that the concentration of nitrogen in a lake exhibits periodic behavior. That is, if we denote the concentration of nitrogen at time t by $c(t)$, then we assume that
$$c(t) = 2 + \sin\left(\frac{\pi}{2}t\right)$$
(a) Find
$$\frac{dc}{dt}$$
(b) Use a graphing calculator to graph both $c(t)$ and $\frac{dc}{dt}$ in the same coordinate system.
(c) By inspecting the graph in (b), answer the following questions:
(i) When $c(t)$ reaches a maximum, what is the value of dc/dt?
(ii) When dc/dt is positive, is $c(t)$ increasing or decreasing?
(iii) What can you say about $c(t)$ when $dc/dt = 0$?

■ 4.6 Derivatives of Exponential Functions

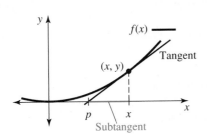

Figure 4.29 The subtangent problem.

Pierre de Fermat (1601–1665) devised a method of finding the tangent line at a given point (x, y) on a curve by constructing the **subtangent**, defined as the line segment between the point where the tangent line intersects the x-axis and the point $(x, 0)$. (See Figure 4.29.)

Fermat's procedure essentially amounted to finding the slope of the tangent line by considering the secant line through two points: (x, y) and a nearby point of the graph of $y = f(x)$. After computing the slope of the secant line, he set the two points he used to compute the slope equal to each other, thus obtaining the slope of the tangent line. This sounds very much like the definition of the derivative that we use today, and, in fact, it is the same idea. Fermat did not, however, develop and formalize a general framework for the differential calculus; that was done by Leibniz and Newton.

Using the definition of derivatives, we can relate the subtangent to the slope of the tangent at the corresponding point of the graph of $y = f(x)$. Suppose the tangent line at (x, y), where $y = f(x)$, intersects the x-axis at $(p(x), 0)$; the location of $p(x)$ depends on x. We set $c(x) = x - p(x)$. This is the equation of the subtangent. We see from Figure 4.29 that the slope of the tangent line at (x, y) is given by $y/c(x)$. Since the slope of the tangent line at (x, y) is the derivative of the function of the curve, evaluated at x [i.e., $f'(x)$], we find that

$$\frac{dy}{dx} = \frac{y}{c(x)}$$

A natural problem (which was posed to Descartes by Debaune in 1639) is to find a curve whose subtangent is a given constant. That is, we wish to find the function $y = f(x)$ that satisfies

$$\frac{dy}{dx} = \frac{y}{c}$$

where c is a constant other than 0. (This problem was solved by Leibniz in 1684, when he published his differential calculus for the first time.) In words, we are looking for a function $y = f(x)$ whose derivative is proportional to the function itself. As we will see next, exponential functions are the solutions to this problem.

Recall from Section 1.2 that the function f is an exponential function with base a if

$$f(x) = a^x, \quad x \in \mathbf{R}$$

where a is a positive constant other than 1. (See Figure 4.30.) We can use the formal definition of the derivative to compute $f'(x)$:

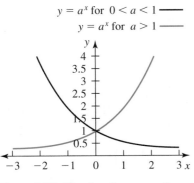

$y = a^x$ for $0 < a < 1$ ———
$y = a^x$ for $a > 1$ ———

Figure 4.30 The function $y = a^x$.

$$f'(x) = \frac{d}{dx}a^x = \lim_{h \to 0} \frac{f(x+h) - f(x)}{h} = \lim_{h \to 0} \frac{a^{x+h} - a^x}{h}$$

$$= \lim_{h \to 0} \frac{a^x(a^h - 1)}{h} = a^x \lim_{h \to 0} \frac{a^h - 1}{h}$$

In the final step, we were able to write the term a^x in front of the limit because a^x does not depend on h. Thus, we are left with investigating

$$\lim_{h \to 0} \frac{a^h - 1}{h}$$

We first note that this limit does not depend on x. If we assume that the limit exists, then it follows that it is equal to $f'(0)$. To see why, use the formal definition of the derivative to compute $f'(0)$. (It can be formally shown that the limit exists, but doing so is beyond the scope of this course.)

It follows from the preceding calculation that if $f'(0)$ exists, then $f'(x)$ exists and

$$f'(x) = \underbrace{a^x}_{f(x)} f'(0)$$

This equation shows that the exponential function is a function whose derivative is proportional to the function itself, provided that that $f'(0)$ exists. [The constant of proportionality is $f'(0)$.] That is, exponential functions solve the subtangent problems just mentioned. We single out the case where the value of the base a is such that

$$\lim_{h \to 0} \frac{a^h - 1}{h}$$

is equal to 1. We denote this base by e. The number e is thus defined by

$$\lim_{h \to 0} \frac{e^h - 1}{h} = 1 \tag{4.7}$$

We find, for the derivative of $f(x) = e^x$, that

$$\frac{d}{dx} e^x = e^x \tag{4.8}$$

Figure 4.31 The function $y = e^x$. The slope of the tangent line at $x = 0$ is $m = 1$.

A graph of $f(x) = e^x$ is shown in Figure 4.31. The domain of this function is **R** and its range is the open interval $(0, \infty)$. (In particular, $e^x > 0$ for all $x \in$ **R**.) Denoting by e the base of the exponential function for which (4.7) and (4.8) hold is no accident; it is indeed the natural exponential base that we introduced in Section 1.2. Although we cannot prove this here, a table should convince you: With $e = 2.71828\ldots$, we find that

h	0.1	0.01	0.001	0.0001
$\frac{e^h - 1}{h}$	1.0517	1.0050	1.00050	1.000050

Now recall that there is an alternative notation for e^x, namely, $\exp[x]$. Using the identity

$$a^x = \exp[\ln a^x]$$

and the fact that $\ln a^x = x \ln a$, we can find the derivative of a^x with the help of the chain rule:

$$\frac{d}{dx} a^x = \frac{d}{dx} \exp[\ln a^x] = \frac{d}{dx} \exp[x \ln a]$$
$$= \exp[x \ln a] \ln a = (\ln a) a^x$$

That is, we have

$$\frac{d}{dx} a^x = (\ln a) a^x \tag{4.9}$$

which allows us to obtain the following identity:

$$\lim_{h \to 0} \frac{a^h - 1}{h} = \ln a \tag{4.10}$$

EXAMPLE 1

Find the derivative of $f(x) = e^{-x^2/2}$.

Solution   We use the chain rule:

$$f'(x) = e^{-x^2/2} \left(-\frac{2x}{2} \right) = -x e^{-x^2/2}$$

EXAMPLE 2   Find the derivative of $f(x) = 3^{\sqrt{x}}$.

Solution   We can use (4.9) and the chain rule to get

$$\frac{d}{dx}3^{\sqrt{x}} = (\ln 3)3^{\sqrt{x}}\frac{1}{2\sqrt{x}}$$

However, since every exponential function can be written in terms of the base e, and the differentiation rule for e^x is particularly simple ($\frac{d}{dx}e^x = e^x$), it is often easier to rewrite the exponential function in terms of e and then differentiate. That is, we write

$$3^{\sqrt{x}} = \exp[\ln 3^{\sqrt{x}}] = \exp[\sqrt{x}\ln 3]$$

Then, using the chain rule, we obtain

$$\frac{d}{dx}\exp[\sqrt{x}\ln 3] = \frac{\ln 3}{2\sqrt{x}}\exp[\sqrt{x}\ln 3] = \frac{\ln 3}{2\sqrt{x}}3^{\sqrt{x}} \qquad ■$$

Since we must frequently differentiate functions of the form $y = e^{g(x)}$, we state this differentiation in a separate rule. Using the chain rule, we have

$$\frac{d}{dx}e^{g(x)} = g'(x)e^{g(x)} \qquad (4.11)$$

EXAMPLE 3   Find the derivative of $f(x) = \exp[\sin \sqrt{x}]$.

Solution   We set $g(x) = \sin \sqrt{x}$. To differentiate $g(x)$, we must apply the chain rule:

$$\frac{d}{dx}g(x) = (\cos \sqrt{x})\frac{1}{2\sqrt{x}}$$

Using Equation (4.11), we can now differentiate $f(x)$:

$$\frac{d}{dx}f(x) = (\cos \sqrt{x})\frac{1}{2\sqrt{x}}\exp[\sin \sqrt{x}] \qquad ■$$

Here is an example that shows how (4.10) is used:

EXAMPLE 4   Find

$$\lim_{h \to 0}\frac{3^{2h} - 1}{h}$$

Solution   We make the substitution $l = 2h$ and note that $l \to 0$ as $h \to 0$. Then

$$\lim_{h \to 0}\frac{3^{2h} - 1}{h} = \lim_{l \to 0}\frac{3^l - 1}{l/2}$$

$$= 2\lim_{l \to 0}\frac{3^l - 1}{l} = 2\ln 3 \qquad ■$$

The exponential function with base e appears in many scientific problems; the next example involves radioactive decay.

EXAMPLE 5   **Radioactive Decay** Find the derivative of the radioactive decay function, which describes the amount of material left after t units of time. (See Example 10 in Subsection 1.2.5.) The function is

$$W(t) = W_0 e^{-\lambda t}, \quad t \geq 0$$

where W_0 is the amount of material at time 0 and λ is called the *radioactive decay rate*. Show that $W(t)$ satisfies the differential equation

$$\frac{dW}{dt} = -\lambda W(t)$$

Solution   We use the chain rule to find the derivative of $W(t)$:

$$\frac{d}{dt} W(t) = \underbrace{W_0 e^{-\lambda t}}_{W(t)}(-\lambda)$$

That is,

$$\frac{dW}{dt} = -\lambda W(t)$$

In words, the rate of decay is proportional to the amount of material left. This equation should remind you of the subtangent problem; there, we wanted to find a function whose derivative is proportional to the function itself. That is exactly the situation we have in this example: The derivative of $W(t)$ is proportional to $W(t)$. ■

EXAMPLE 6   **Exponential Growth** Find the per capita growth rate of a population whose size $N(t)$ at time t follows the exponential growth function

$$N(t) = N(0)e^{rt}$$

where $N(0)$ is the population size at time 0 and r is a constant.

Solution   We first find the derivative of $N(t)$:

$$\frac{dN}{dt} = N(0)re^{rt}$$

Since $N(0)e^{rt} = N(t)$, we can write

$$\frac{dN}{dt} = rN(t)$$

Thus, the per capita growth rate of an exponentially growing population is constant; that is,

$$\frac{1}{N}\frac{dN}{dt} = r$$

■

Section 4.6 Problems

Differentiate the functions in Problems 1–52 with respect to the independent variable.

1. $f(x) = e^{3x}$

2. $f(x) = e^{-2x}$

3. $f(x) = 4e^{1-3x}$

4. $f(x) = 3e^{2-5x}$

5. $f(x) = e^{-2x^2+3x-1}$

6. $f(x) = e^{4x^2-2x+1}$

7. $f(x) = e^{7(x^2+1)^2}$

8. $f(x) = e^{-3(x^3-1)^4}$

9. $f(x) = xe^x$

10. $f(x) = 2xe^{-3x}$

11. $f(x) = x^2e^{-x}$

12. $f(x) = (3x^2-1)e^{1-x^2}$

13. $f(x) = \frac{1+e^x}{1+x^2}$

14. $f(x) = \frac{x-e^{-x}}{1+xe^{-x}}$

15. $f(x) = \frac{e^x+e^{-x}}{2+e^x}$

16. $f(x) = \frac{x}{e^x+e^{-x}}$

17. $f(x) = e^{\sin(3x)}$

18. $f(x) = e^{\cos(4x)}$

19. $f(x) = e^{\sin(x^2-1)}$

20. $f(x) = e^{\cos(1-2x^3)}$

21. $f(x) = \sin(e^x)$

22. $f(x) = \cos(e^x)$

23. $f(x) = \sin(e^{2x} + x)$

24. $f(x) = \cos(3x - e^{x^2-1})$

25. $f(x) = \exp[x - \sin x]$

26. $f(x) = \exp[x^2 - 2\cos x]$

27. $g(s) = \exp[\sec s^2]$

28. $g(s) = \exp[\tan s^3]$

29. $f(x) = e^{x \sin x}$

30. $f(x) = e^{1-x\cos x}$

31. $f(x) = -3e^{x^2+\tan x}$

32. $f(x) = 2e^{-x\sec(3x)}$

33. $f(x) = 2^x$

34. $f(x) = 3^x$

35. $f(x) = 2^{x+1}$

36. $f(x) = 3^{x-1}$

37. $f(x) = 5^{\sqrt{2x-1}}$

38. $f(x) = 3^{\sqrt{1-3x}}$

39. $f(x) = 2^{x^2+1}$

40. $f(x) = 3^{x^3-1}$

41. $h(t) = 2^{t^2-1}$

42. $h(t) = 4^{2t^3-t}$

43. $f(x) = 2^{\sqrt{x}}$

44. $f(x) = 3^{\sqrt{x+1}}$

45. $f(x) = 2^{\sqrt{x^2-1}}$

46. $f(x) = 4^{\sqrt{1-2x^3}}$

47. $h(t) = 5^{\sqrt{t}}$

48. $h(t) = 6^{\sqrt{6t^6-6}}$

49. $g(x) = 2^{2\cos x}$

50. $g(r) = 2^{-3\sin r}$

51. $g(r) = 3^{r^{1/5}}$

52. $g(r) = 4^{r^{1/4}}$

Compute the limits in Problems 53–56.

53. $\lim\limits_{h \to 0} \dfrac{e^{2h} - 1}{h}$

54. $\lim\limits_{h \to 0} \dfrac{e^{5h} - 1}{3h}$

55. $\lim\limits_{h \to 0} \dfrac{e^h - 1}{\sqrt{h}}$

56. $\lim\limits_{h \to 0} \dfrac{2^h - 1}{h}$

57. Find the length of the subtangent to the curve $y = 2^x$ at the point $(1, 2)$.

58. Find the length of the subtangent to the curve $y = \exp[x^2]$ at the point $(2, e^4)$.

59. Population Growth Suppose that the population size at time t is

$$N(t) = e^{2t}, \quad t \geq 0$$

(a) What is the population size at time 0?

(b) Show that

$$\frac{dN}{dt} = 2N$$

60. Population Growth Suppose that the population size at time t is

$$N(t) = N_0 e^{rt}, \quad t \geq 0$$

where N_0 is a positive constant and r is a real number.

(a) What is the population size at time 0?

(b) Show that

$$\frac{dN}{dt} = rN$$

61. Bacterial Growth Suppose that a bacterial colony grows in such a way that at time t the population size is

$$N(t) = N(0)2^t$$

where $N(0)$ is the population size at time 0. Find the rate of growth dN/dt. Express your solution in terms of $N(t)$. Show that the growth rate of the population is proportional to the population size.

62. Bacterial Growth Suppose that a bacterial colony grows in such a way that at time t the population size is

$$N(t) = N(0)2^t$$

where $N(0)$ is the population size at time 0. Find the per capita growth rate.

63. Logistic Growth

(a) Find the derivative of the logistic growth curve (see Example 3 in Section 3.3)

$$N(t) = \frac{K}{1 + \left(\frac{K}{N(0)} - 1\right)e^{-rt}}$$

where r and K are positive constants and $N(0)$ is the population size at time 0.

(b) Show that $N(t)$ satisfies the equation

$$\frac{dN}{dt} = rN\left(1 - \frac{N}{K}\right)$$

[*Hint*: Use the function $N(t)$ given in (a) for the right-hand side, and simplify until you obtain the derivative of $N(t)$ that you computed in (a).]

(c) Plot the per capita rate of growth $\frac{1}{N}\frac{dN}{dt}$ as a function of N, and note that it decreases with increasing population size.

64. Fish Recruitment Model The following model is used in the fisheries literature to describe the recruitment of fish as a function of the size of the parent stock: If we denote the number of recruits by R and the size of the parent stock by P, then

$$R(P) = \alpha P e^{-\beta P}, \quad P \geq 0$$

where α and β are positive constants.

(a) Sketch the graph of the function $R(P)$ when $\beta = 1$ and $\alpha = 2$.

(b) Differentiate $R(P)$ with respect to P.

(c) Find all the points on the curve that have a horizontal tangent.

65. Von Bertalanffy Growth Model The growth of fish can be described by the von Bertalanffy growth function

$$L(x) = L_\infty - (L_\infty - L_0)e^{-kx}$$

where x denotes the age of the fish and k, L_∞, and L_0 are positive constants.

(a) Set $L_0 = 1$ and $L_\infty = 10$. Graph $L(x)$ for $k = 1.0$ and $k = 0.1$.

(b) Interpret L_∞ and L_0.

(c) Compare the graphs for $k = 0.1$ and $k = 1.0$. According to which graph do fish reach $L = 5$ more quickly?

(d) Show that

$$\frac{d}{dx}L(x) = k(L_\infty - L(x))$$

That is, $dL/dx \propto L_\infty - L$. What does this proportionality say about how the rate of growth changes with age?

(e) The constant k is the proportionality constant in (d). What does the value of k tell you about how quickly a fish grows?

66. Radioactive Decay Suppose $W(t)$ denotes the amount of a radioactive material left after time t (measured in days). Assume that the radioactive decay rate of the material is 0.2/day. Find the differential equation for the radioactive decay function $W(t)$.

67. Radioactive Decay Suppose $W(t)$ denotes the amount of a radioactive material left after time t (measured in days). Assume that the radioactive decay rate of the material is 4/day. Find the differential equation for the radioactive decay function $W(t)$.

68. Radioactive Decay Suppose $W(t)$ denotes the amount of a radioactive material left after time t (measured in days). Assume that the half-life of the material is 3 days. Find the differential equation for the radioactive decay function $W(t)$.

69. Radioactive Decay Suppose $W(t)$ denotes the amount of a radioactive material left after time t (measured in days). Assume that the half-life of the material is 5 days. Find the differential equation for the radioactive decay function $W(t)$.

70. Radioactive Decay Suppose $W(t)$ denotes the amount of a radioactive material left after time t. Assume that $W(0) = 15$ and that

$$\frac{dW}{dt} = -2W(t)$$

(a) How much material is left at time $t = 2$?

(b) What is the half-life of this material?

71. Radioactive Decay Suppose $W(t)$ denotes the amount of a radioactive material left after time t. Assume that $W(0) = 6$ and that

$$\frac{dW}{dt} = -3W(t)$$

(a) How much material is left at time $t = 4$?

(b) What is the half-life of the material?

72. Radioactive Decay Suppose $W(t)$ denotes the amount of a radioactive material left after time t. Assume that $W(0) = 10$ and $W(1) = 8$.

(a) Find the differential equation that describes this situation.

(b) How much material is left at time $t = 5$?

(c) What is the half-life of the material?

73. Radioactive Decay Suppose $W(t)$ denotes the amount of a radioactive material left after time t. Assume that $W(0) = 5$ and $W(1) = 2$.

(a) Find the differential equation that describes this situation.

(b) How much material is left at time $t = 3$?

(c) What is the half-life of the material?

■ 4.7 Derivatives of Inverse Functions, Logarithmic Functions, and the Inverse Tangent Function

Recall that the logarithmic function is the inverse of the exponential function. To find the derivative of the logarithmic function, we must therefore learn how to compute the derivative of an inverse function.

■ 4.7.1 Derivatives of Inverse Functions

We begin with an example (Figure 4.32). Let $f(x) = x^2$, $x \geq 0$. We computed the inverse function of f in Subsection 1.2.6. First note that $f(x) = x^2$, $x \geq 0$, is one to one (use the horizontal line test from Subsection 1.2.6); hence, we can define its inverse. We repeat the steps from Subsection 1.2.6 to find an inverse function. [Recall that we obtain the graph of the inverse function by reflecting $y = f(x)$ about the line $y = x$.]

1. Write $y = f(x)$:

$$y = x^2$$

2. Solve for x:

$$x = \sqrt{y}$$

3. Interchange x and y:

$$y = \sqrt{x}$$

Since the range of $f(x)$, which is the interval $[0, \infty)$, becomes the domain for the inverse function, it follows that

$$f^{-1}(x) = \sqrt{x} \quad \text{for } x \geq 0$$

We already know the derivative of $\sqrt{x}$, namely, $1/(2\sqrt{x})$. But we will try to find the derivative in a different way that we can generalize to get a formula for finding the derivative of any inverse function. Let $g(x) = f^{-1}(x)$. Then

$$(f \circ g)(x) = f[g(x)] = (\sqrt{x})^2 = x, \quad x \geq 0$$

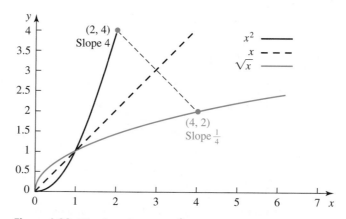

Figure 4.32 The function $y = x^2$, $x \geq 0$, and its inverse function $y = \sqrt{x}$, $x \geq 0$.

Therefore, the derivatives of $(\sqrt{x})^2$ and x must be equal. Applying the chain rule, we find that

$$\frac{d}{dx}(\sqrt{x})^2 = 2\sqrt{x}\frac{d}{dx}\sqrt{x}$$

Since $\frac{d}{dx}x = 1$, we obtain

$$2\sqrt{x}\frac{d}{dx}\sqrt{x} = 1$$

or, for $x > 0$,

$$\frac{d}{dx}\sqrt{x} = \frac{1}{2\sqrt{x}}$$

To prepare for how the derivatives of a function and its inverse function are related geometrically, look at Figure 4.32, where the slope at the point $(2, 4)$ of $f(x) = x^2$ is $m = 4$ and the slope at the point $(4, 2)$ of $f^{-1}(x) = \sqrt{x}$ is $m = 1/4$. We will also find this reciprocal relationship of slopes at related points in the general case.

The steps that led us to the derivative of $\sqrt{x}$ can be used to find a general formula for the derivatives of inverse functions. We assume that $f(x)$ is one to one in its domain. If $g(x)$ is the inverse function of $f(x)$, then $f[g(x)] = x$. Applying the chain rule, we find that

$$\frac{d}{dx}f[g(x)] = f'[g(x)]g'(x)$$

Since $\frac{d}{dx}x = 1$, we obtain

$$f'[g(x)]g'(x) = 1$$

If $f'[g(x)] \neq 0$, we can divide by $f'[g(x)]$ to get

$$g'(x) = \frac{1}{f'[g(x)]}$$

Because $g(x) = f^{-1}(x)$ and $g'(x) = \frac{d}{dx}g(x) = \frac{d}{dx}f^{-1}(x)$, we obtain the following rule:

Derivative of an Inverse Function If $f(x)$ is one to one and differentiable with inverse function $f^{-1}(x)$ and $f'[f^{-1}(x)] \neq 0$, then $f^{-1}(x)$ is differentiable and

$$\frac{d}{dx}f^{-1}(x) = \frac{1}{f'[f^{-1}(x)]} \qquad (4.12)$$

This reciprocal relationship is illustrated in Figure 4.33.

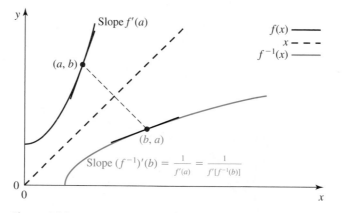

Figure 4.33 The graphs of $y = f(x)$ and its inverse function $y = f^{-1}(x)$ have reciprocal slopes at the points (a, b) and (b, a).

We return to the example of $f(x) = x^2$, $x \geq 0$, where $f^{-1}(x) = \sqrt{x}$, $x \geq 0$, to illustrate how to use (4.12). Now, $f'(x) = 2x$ and we need to evaluate

$$f'[f^{-1}(x)] = f'[\sqrt{x}] = 2\sqrt{x}$$

To apply the formula, we assume that $f'[f^{-1}(x)] \neq 0$. Then

$$\frac{d}{dx}\sqrt{x} = \frac{1}{2\sqrt{x}} \quad \text{for } x > 0$$

Looking at the graphs of $y = x^2$ and $y = \sqrt{x}$, $x \geq 0$, we easily see why $f^{-1}(x)$ is not differentiable at $x = 0$. [Recall that we obtain the graph of the inverse function by reflecting $y = f(x)$ about the line $y = x$.] If we draw the tangent line to the curve $y = x^2$ at $x = 0$, we find that the tangent line is horizontal; that is, its slope is 0. Reflecting a horizontal line about $y = x$ results in a vertical line, for which the slope is not defined (Figure 4.32).

The formula for finding derivatives of inverse functions takes on a particularly easy to remember form when we use Leibniz notation. To see this, note that (without interchanging x and y)

$$y = f(x) \quad \Longleftrightarrow \quad x = f^{-1}(y)$$

and hence

$$\frac{dx}{dy} = \frac{1}{\frac{dy}{dx}}$$

This formula again emphasizes the reciprocal relationship. We illustrate the formula with the example

$$y = x^2 \quad \Longleftrightarrow \quad x = \sqrt{y}$$

for $x > 0$. Since $\frac{dy}{dx} = 2x$, we have

$$\frac{dx}{dy} = \frac{1}{\frac{dy}{dx}} = \frac{1}{2x} = \frac{1}{2\sqrt{y}}$$

That is,

$$\frac{d}{dy}\sqrt{y} = \frac{1}{2\sqrt{y}}$$

The answer is now in terms of y, because we did not interchange x and y when we computed the inverse function. If we now do so, we again find that

$$\frac{d}{dx}\sqrt{x} = \frac{1}{2\sqrt{x}}$$

EXAMPLE 1

Let

$$f(x) = \frac{x}{1+x} \quad \text{for } x \geq 0$$

Find $\frac{d}{dx}f^{-1}(x)\big|_{x=\frac{1}{3}}$.

Solution

To show that $f^{-1}(x)$ exists, we use the horizontal line test and conclude that $f(x)$ is one to one on its domain, since each horizontal line intersects the graph of $f(x)$ at most once. (See Figure 4.34.)

We can actually compute the inverse of $f(x)$. This will give us two different ways to compute the derivative of the inverse of $f(x)$: We can compute the inverse function explicitly and then differentiate the result, or we can use the formula for finding derivatives of inverse functions. We begin with the latter way.

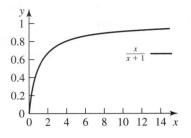

Figure 4.34 The graph of $f(x) = \frac{x}{x+1}$ for $x \geq 0$.

1. To use the formula (4.12), we need to find $f^{-1}(\frac{1}{3})$; this means that we need to find x so that $f(x) = 1/3$. Now,

$$\frac{x}{1+x} = \frac{1}{3} \quad \text{implies that} \quad 2x = 1, \quad \text{or} \quad x = \frac{1}{2}$$

Therefore, $f^{-1}(\frac{1}{3}) = \frac{1}{2}$. Formula (4.12) thus becomes

$$\frac{d}{dx} f^{-1}(x) \Big|_{x=\frac{1}{3}} = \frac{1}{f'[f^{-1}(\frac{1}{3})]} = \frac{1}{f'(\frac{1}{2})}$$

We use the quotient rule to find the derivative of $f(x)$:

$$f'(x) = \frac{(1)(1+x) - (x)(1)}{(1+x)^2} = \frac{1}{(1+x)^2}$$

At $x = 1/2$, $f'(\frac{1}{2}) = \frac{1}{(1+1/2)^2} = \frac{4}{9}$. Therefore,

$$\frac{d}{dx} f^{-1}(x) \Big|_{x=\frac{1}{3}} = \frac{1}{\frac{4}{9}} = \frac{9}{4}$$

2. We can compute the inverse function: Set $y = \frac{x}{1+x}$. Then, solving for x yields

$$x = \frac{y}{1-y}$$

Interchanging x and y, we find that

$$y = \frac{x}{1-x}$$

Since the domain of $f(x)$ is $[0, \infty)$, the range of f is $[0, 1)$. Now, the range of f becomes the domain of the inverse; therefore, the inverse function is

$$f^{-1}(x) = \frac{x}{1-x} \quad \text{for } 0 \leq x < 1$$

We use the quotient rule to find $\frac{d}{dx} f^{-1}(x)$:

$$\frac{d}{dx} f^{-1}(x) = \frac{(1)(1-x) - (x)(-1)}{(1-x)^2} = \frac{1}{(1-x)^2}$$

Therefore,

$$\frac{d}{dx} f^{-1}(x) \Big|_{x=\frac{1}{3}} = \frac{1}{(1-1/3)^2} = \frac{9}{4}$$

which agrees with the answer in part (1). ■

The inverse cannot always be computed explicitly, as the next example shows.

EXAMPLE 2 Let

$$f(x) = 2x + e^x \quad \text{for } x \in \mathbf{R}$$

Find $\frac{d}{dx} f^{-1}(x) \big|_{x=1}$.

Solution In this case, it is not possible to solve $y = 2x + e^x$ for x. Therefore, we must use (4.12) if we wish to compute the derivative of the inverse function at a particular point. Equation (4.12) becomes

$$\frac{d}{dx} f^{-1}(x) \Big|_{x=1} = \frac{1}{f'[f^{-1}(1)]}$$

We need to find $f'(x)$:

$$f'(x) = 2 + e^x$$

Since $f(0) = 1$, it follows that $f^{-1}(1) = 0$, and, hence,

$$\frac{1}{f'[f^{-1}(1)]} = \frac{1}{f'(0)} = \frac{1}{2+1} = \frac{1}{3}$$

■

The next example, in which we again need to use (4.12), involves finding the derivative of the inverse of a trigonometric function.

EXAMPLE 3   Let $f(x) = \tan x$, $-\frac{\pi}{2} < x < \frac{\pi}{2}$. Find $\frac{d}{dx} f^{-1}(x)\big|_{x=1}$.

Solution   Since $f(\frac{\pi}{4}) = \tan \frac{\pi}{4} = 1$, it follows that $f^{-1}(1) = \frac{\pi}{4}$. Recall that $f'(x) = \sec^2 x$. We therefore have

$$\frac{d}{dx} f^{-1}(x)\bigg|_{x=1} = \frac{1}{f'[f^{-1}(1)]} = \frac{1}{f'(\frac{\pi}{4})} = \frac{1}{\sec^2(\frac{\pi}{4})}$$

$$= \cos^2\left(\frac{\pi}{4}\right) = \left(\frac{1}{2}\sqrt{2}\right)^2 = \frac{1}{2}$$

■

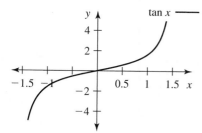

Figure 4.35   The function $f(x) = \tan x$, $(-\frac{\pi}{2}, \frac{\pi}{2})$, is one to one on its domain.

If we define $f(x) = \tan x$ on the domain $(-\frac{\pi}{2}, \frac{\pi}{2})$, then $f(x)$ is one to one, as can be seen from Figure 4.35. (Use the horizontal line test.) The range of $f(x)$ is $(-\infty, \infty)$. We cannot use algebra to solve $y = \tan x$ for x; instead, the inverse of the tangent function gets its own name. It is called $y = \arctan x$ (or $y = \tan^{-1} x$) and its domain is $(-\infty, \infty)$. In the next example, we will find the derivative of $y = \arctan x$, which will turn out to have a surprisingly simple form.

EXAMPLE 4   Let $f(x) = \tan x$, $-\frac{\pi}{2} < x < \frac{\pi}{2}$. Find $\frac{d}{dx} f^{-1}(x)$.

Solution   As mentioned previously, $f^{-1}(x)$ exists, since $f(x)$ is one to one on its domain. Recall that

$$\frac{d}{dx} \tan x = \sec^2 x$$

The inverse of the tangent function is denoted by $\tan^{-1} x$ or $\arctan x$. (Note that $\tan^{-1} x$ is different from $\frac{1}{\tan x}$. The superscript "-1" refers to the function being an inverse function.) We set $y = \arctan x$ (and hence $x = \tan y$). Then

$$\frac{dy}{dx} = \frac{d}{dx} \arctan x = \frac{1}{\frac{dx}{dy}} = \frac{1}{\frac{d}{dy}\tan y} = \frac{1}{\sec^2 y} = \frac{1}{1 + \tan^2 y}$$

where we used the trigonometric identity $\sec^2 y = 1 + \tan^2 y$ to get the denominator in the rightmost term. Since $x = \tan y$, it follows that $x^2 = \tan^2 y$, and, hence,

$$\frac{1}{1 + \tan^2 y} = \frac{1}{1 + x^2}$$

Therefore,

$$\frac{d}{dx} \arctan x = \frac{1}{1 + x^2}$$

■

The result in the preceding example is important, and we summarize it in the following box:

$$\frac{d}{dx} \arctan x = \frac{d}{dx} \tan^{-1} x = \frac{1}{1 + x^2}$$

The derivative of the inverse sine function, $y = \arcsin x$, is discussed in Problem 22 of this section. We list it here:

$$\frac{d}{dx}\arcsin x = \frac{d}{dx}\sin^{-1}x = \frac{1}{\sqrt{1-x^2}}$$

The derivatives of the remaining inverse trigonometric functions are listed in the table of derivatives on the inside back cover of the book.

■ 4.7.2 The Derivative of the Logarithmic Function

We introduced the logarithmic function to the base a, $\log_a x$, as the inverse function of the exponential function a^x (Figures 4.36 and 4.37). We can therefore use the formula for derivatives of inverse functions to find the derivative of $y = \log_a x$. Since

$$\log_a x = \frac{\ln x}{\ln a}$$

and $\ln a$ is a constant, it is enough to find the derivative of $\ln x$ (Figure 4.38). We set $f(x) = e^x$; then $f'(x) = e^x$ and $f^{-1}(x) = \ln x$. Therefore,

$$\frac{d}{dx}\ln x = \frac{d}{dx}[f^{-1}(x)] = \frac{1}{f'[f^{-1}(x)]} = \frac{1}{\exp[\ln x]} = \frac{1}{x}$$

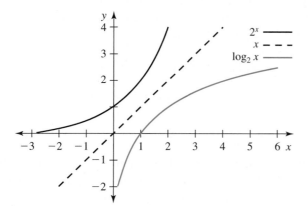

Figure 4.36 The function $y = \log_2 x$ as the inverse function of $y = 2^x$.

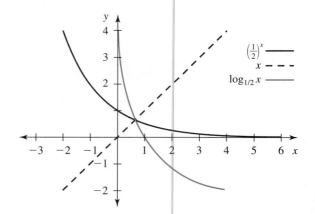

Figure 4.37 The function $y = \log_{1/2} x$ as the inverse function of $y = \left(\frac{1}{2}\right)^x$.

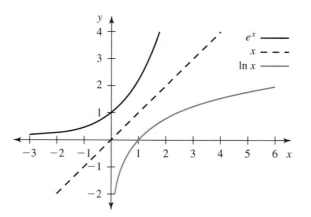

Figure 4.38 The natural logarithm $y = \ln x$ as the inverse function of the natural exponential function $y = e^x$.

We summarize this differentiation rule in the following box:

$$\frac{d}{dx}\ln x = \frac{1}{x}$$

$$\frac{d}{dx}\log_a x = \frac{1}{(\ln a)x}$$

EXAMPLE 5   Find the derivative of $y = \ln(3x)$.

Solution   We use the chain rule with $u = g(x) = 3x$ and $f(u) = \ln u$:

$$\frac{dy}{dx} = \frac{dy}{du}\frac{du}{dx} = \frac{1}{u}3 = \frac{3}{3x} = \frac{1}{x}$$

If you are surprised that the factor 3 disappeared, note that

$$y = \ln(3x) = \ln 3 + \ln x$$

Since $\ln 3$ is a constant, its derivative is 0. Hence,

$$\frac{d}{dx}(\ln 3 + \ln x) = \frac{d}{dx}\ln 3 + \frac{d}{dx}\ln x = 0 + \frac{1}{x} = \frac{1}{x}$$ ■

EXAMPLE 6   Find the derivative of $y = \ln(x^2 + 1)$.

Solution   We can use the chain rule with $u = g(x) = x^2 + 1$ and $f(u) = \ln u$. We obtain

$$y' = \frac{df}{du}\frac{du}{dx} = \frac{1}{u}2x = \frac{1}{x^2+1}2x = \frac{2x}{x^2+1}$$ ■

The preceding example is of the form $y = \ln f(x)$. We will frequently encounter such functions; to find their derivatives, we need to use the chain rule, as shown in the following box:

$$\frac{d}{dx}\ln f(x) = \frac{f'(x)}{f(x)}$$

EXAMPLE 7   Differentiate $y = \ln(\sin x)$.

Solution   This function is also of the form $y = \ln f(x)$, with $f(x) = \sin x$. Since

$$\frac{d}{dx}\sin x = \cos x$$

it follows that

$$\frac{dy}{dx} = \frac{\cos x}{\sin x} = \cot x$$ ■

EXAMPLE 8   Differentiate

$$y = \ln(\tan x + x)$$

Solution   This function is of the form $y = \ln f(x)$ with $f(x) = \tan x + x$. Thus,

$$y' = \frac{\sec^2 x + 1}{\tan x + x}$$ ■

EXAMPLE 9   Differentiate
$$y = \log(2x^3 - 1)$$

Solution   This function is of the form $y = \log f(x)$ with $f(x) = 2x^3 - 1$. The logarithm is to base 10. Hence,
$$y' = \frac{1}{\ln 10} \frac{6x^2}{2x^3 - 1}$$ ■

■ 4.7.3 Logarithmic Differentiation

In 1695, Leibniz introduced logarithmic differentiation, following Johann Bernoulli's suggestion to find derivatives of functions of the form $y = [f(x)]^x$. Bernoulli generalized this method and published his results two years later. The basic idea is to take logarithms on both sides and then to use implicit differentiation.

EXAMPLE 10   Find $\frac{dy}{dx}$ when $y = x^x$.

Solution   We take logarithms on both sides of the equation $y = x^x$:
$$\ln y = \ln x^x$$

Applying properties of the logarithm, we can simplify the right-hand side to $\ln x^x = x \ln x$. We can now differentiate both sides with respect to x. Since y is a function of x, we need to use the chain rule to differentiate $\ln y$ (as we learned in the section on implicit differentiation):

$$\frac{d}{dx}[\ln y] = \frac{d}{dx}[x \ln x]$$
$$\frac{1}{y}\frac{dy}{dx} = 1 \cdot \ln x + x\frac{1}{x}$$
$$\frac{dy}{dx} = y[\ln x + 1]$$
$$\frac{dy}{dx} = (\ln x + 1)x^x$$ ■

If the function $y = x^x$ looks strange, write it as
$$y = x^x = \exp[\ln x^x] = \exp[x \ln x]$$

That is, $y = e^{x \ln x}$. We can differentiate this function without using logarithmic differentiation; that is,

$$\frac{dy}{dx} = e^{x \ln x}\frac{d}{dx}(x \ln x)$$
$$= e^{x \ln x}\left(1 \cdot \ln x + x\frac{1}{x}\right)$$
$$= e^{x \ln x}(\ln x + 1)$$

Either approach will give you the correct answer.

EXAMPLE 11   Find the derivative of $y = (\sin x)^x$.

Solution   We take logarithms on both sides of the equation and simplify:
$$\ln y = x \ln(\sin x)$$

Differentiating with respect to x yields

$$\frac{d}{dx}[\ln y] = \frac{d}{dx}[x \ln(\sin x)]$$

$$\frac{1}{y}\frac{dy}{dx} = 1 \cdot \ln(\sin x) + x\frac{d}{dx}[\ln(\sin x)]$$

$$= \ln(\sin x) + x\frac{\cos x}{\sin x}$$

$$= \ln(\sin x) + x \cot x$$

Hence, after multiplying by y and substituting $(\sin x)^x$ for y, we obtain

$$\frac{dy}{dx} = [\ln(\sin x) + x \cot x](\sin x)^x \qquad ■$$

The next example should convince you that logarithmic differentiation can simplify finding the derivatives of complicated expressions.

EXAMPLE 12   Differentiate

$$y = \frac{e^x x^{3/2}\sqrt{1+x}}{(x^2+3)^4(3x-2)^3}$$

Solution   Without logarithmic differentiation, differentiating y would be rather difficult. Taking logarithms on both sides, however, we can simplify the right-hand side. Note that it is very important that we apply the properties of the logarithm before differentiating, as this will simplify the expressions that we must differentiate. We have

$$\ln y = \ln \frac{e^x x^{3/2}\sqrt{1+x}}{(x^2+3)^4(3x-2)^3}$$

$$= \ln e^x + \ln x^{3/2} + \ln \sqrt{1+x} - \ln(x^2+3)^4 - \ln(3x-2)^3$$

$$= x + \frac{3}{2}\ln x + \frac{1}{2}\ln(1+x) - 4\ln(x^2+3) - 3\ln(3x-2)$$

This no longer looks so daunting, and we can differentiate both sides:

$$\frac{d}{dx}[\ln y] = \frac{d}{dx}\left[x + \frac{3}{2}\ln x + \frac{1}{2}\ln(1+x) - 4\ln(x^2+3) - 3\ln(3x-2)\right]$$

$$\frac{1}{y}\frac{dy}{dx} = 1 + \frac{3}{2}\frac{1}{x} + \frac{1}{2}\frac{1}{1+x} - 4\frac{2x}{x^2+3} - 3\frac{3}{3x-2}$$

Finally, solving for dy/dx yields

$$\frac{dy}{dx} = \left(1 + \frac{3}{2x} + \frac{1}{2(1+x)} - \frac{8x}{x^2+3} - \frac{9}{3x-2}\right)\frac{e^x x^{3/2}\sqrt{1+x}}{(x^2+3)^4(3x-2)^3} \qquad ■$$

We can also use this method to prove the general power rule (as promised in Section 4.3).

Power Rule (General Form)   Let $f(x) = x^r$, where r is any real number. Then

$$\frac{d}{dx}(x^r) = rx^{r-1}$$

Proof We set $y = x^r$ and use logarithmic differentiation to obtain

$$\frac{d}{dx}[\ln y] = \frac{d}{dx}[\ln x^r]$$

$$\frac{1}{y}\frac{dy}{dx} = \frac{d}{dx}[r \ln x]$$

$$\frac{1}{y}\frac{dy}{dx} = r\frac{1}{x}$$

Solving for dy/dx yields

$$\frac{dy}{dx} = r\frac{1}{x}y = r\frac{1}{x}x^r = rx^{r-1}$$

■

Section 4.7 Problems

■ 4.7.1

In Problems 1–6, find the inverse of each function and differentiate each inverse in two ways: (i) Differentiate the inverse function directly, and (ii) use (4.12) to find the derivative of the inverse.

1. $f(x) = \sqrt{2x+1}, x \geq -\frac{1}{2}$ **2.** $f(x) = \sqrt{x-1}, x \geq 1$

3. $f(x) = 2x^2 - 1, x \geq 0$ **4.** $f(x) = 3x^2 + 2, x \geq 0$

5. $f(x) = 3 - 2x^3, x \geq 0$ **6.** $f(x) = \dfrac{2x^2 - 1}{x^2 - 1}, x > 1$

In Problems 7–22, use (4.12) to find the derivative of the inverse at the indicated point.

7. Let
$$f(x) = 2x^2 - 2, \quad x \geq 0$$
Find $\frac{d}{dx}f^{-1}(x)\big|_{x=0}$. [Note that $f(1) = 0$.]

8. Let
$$f(x) = -x^3 + 7, \quad x \geq 0$$
Find $\frac{d}{dx}f^{-1}(x)\big|_{x=-1}$. [Note that $f(2) = -1$.]

9. Let
$$f(x) = \sqrt{x+1}, \quad x \geq 0$$
Find $\frac{d}{dx}f^{-1}(x)\big|_{x=2}$. [Note that $f(3) = 2$.]

10. Let
$$f(x) = \sqrt{2+x^2}, \quad x \geq 0$$
Find $\frac{d}{dx}f^{-1}(x)\big|_{x=\sqrt{3}}$. [Note that $f(1) = \sqrt{3}$.]

11. Let
$$f(x) = x + e^x, \quad x \in \mathbb{R}$$
Find $\frac{d}{dx}f^{-1}(x)\big|_{x=1}$. [Note that $f(0) = 1$.]

12. Let
$$f(x) = x + \ln(x+1), \quad x > -1$$
Find $\frac{d}{dx}f^{-1}(x)\big|_{x=0}$. [Note that $f(0) = 0$.]

13. Let
$$f(x) = x - \sin x, \quad x \in \mathbb{R}$$
Find $\frac{d}{dx}f^{-1}(x)\big|_{x=\pi}$. [Note that $f(\pi) = \pi$.]

14. Let
$$f(x) = x - \cos x, \quad x \in \mathbb{R}$$
Find $\frac{d}{dx}f^{-1}(x)\big|_{x=-1}$. [Note that $f(0) = -1$.]

15. Let
$$f(x) = x^2 + \tan x, \quad x \in \left(-\frac{\pi}{2}, \frac{\pi}{2}\right)$$
Find $\frac{d}{dx}f^{-1}(x)\big|_{x=0}$. [Note that $f(0) = 0$.]

16. Let
$$f(x) = x^2 + \tan x, \quad x \in \left(-\frac{\pi}{2}, \frac{\pi}{2}\right)$$
Find $\frac{d}{dx}f^{-1}(x)\big|_{x=\frac{\pi^2}{16}+1}$. [Note that $f(\frac{\pi}{4}) = \frac{\pi^2}{16} + 1$.]

17. Let $f(x) = \ln(\sin x), 0 < x < \pi/2$. Find $\frac{d}{dx}f^{-1}(x)$ at $x = -\ln 2$.

18. Let $f(x) = \ln(\tan x), 0 < x < \pi/2$. Find $\frac{d}{dx}f^{-1}(x)$ at $x = \frac{\ln 3}{2}$.

19. Let $f(x) = x^5 + x + 1, -1 < x < 1$. Find $\frac{d}{dx}f^{-1}(x)$ at $x = 1$.

20. Let $f(x) = e^{-x^2} + x$. Find $\frac{d}{dx}f^{-1}(x)$ at $x = 1$.

21. Let $f(x) = e^{-x^2/2} + 2x$. Find $\frac{d}{dx}f^{-1}(x)$ at $x = 1$.

22. Denote the inverse of $y = \sin x, -\frac{\pi}{2} \leq x \leq \frac{\pi}{2}$, by $y = \arcsin x, -1 \leq x \leq 1$. Show that

$$\frac{d}{dx}\arcsin x = \frac{1}{\sqrt{1-x^2}}, \quad -1 < x < 1$$

■ 4.7.2

In Problems 23–60, differentiate the functions with respect to the independent variable. (Note that log denotes the logarithm to base 10.)

23. $f(x) = \ln(x+1)$ **24.** $f(x) = \ln(3x+4)$

25. $f(x) = \ln(1-2x)$ **26.** $f(x) = \ln(4-3x)$

27. $f(x) = \ln x^2$ **28.** $f(x) = \ln(1-x^2)$

29. $f(x) = \ln(2x^3 - x)$ **30.** $f(x) = \ln(1-x^3)$

31. $f(x) = (\ln x)^2$ **32.** $f(x) = (\ln x)^3$

33. $f(x) = (\ln x^2)^2$ **34.** $f(x) = (\ln(1-x^2))^3$

35. $f(x) = \ln\sqrt{x^2+1}$ **36.** $f(x) = \ln\sqrt{2x^2-x}$

37. $f(x) = \ln\dfrac{x}{x+1}$ **38.** $f(x) = \ln\dfrac{2x}{1+x^2}$

39. $f(x) = \ln\dfrac{1-x}{1+2x}$ **40.** $f(x) = \ln\dfrac{x^2-1}{x^3-1}$

41. $f(x) = \exp[x - \ln x]$ **42.** $g(s) = \exp[s^2 + \ln s]$

43. $f(x) = \ln(\sin x)$ **44.** $f(x) = \ln(\cos(1-x))$

45. $f(x) = \ln(\tan x^2)$ **46.** $g(s) = \ln(\sin^2(3s))$

47. $f(x) = x \ln x$ **48.** $f(x) = x^2 \ln x^2$

49. $f(x) = \dfrac{\ln x}{x}$ **50.** $h(t) = \dfrac{\ln t}{1+t^2}$

51. $h(t) = \sin(\ln(3t))$ **52.** $h(s) = \ln(\ln s)$

53. $f(x) = \ln|x^2 - 3|$

54. $f(x) = \log(2x^2 - 1)$

55. $f(x) = \log(1 - x^2)$

56. $f(x) = \log(3x^3 - x + 2)$

57. $f(x) = \log(x^3 - 3x)$

58. $f(x) = \log(\sqrt[3]{\tan x^2})$

59. $f(u) = \log_3(3 + u^4)$

60. $g(s) = \log_5(3^s - 2)$

61. Let $f(x) = \ln x$. We know that $f'(x) = \frac{1}{x}$. We will use this fact and the definition of derivatives to show that

$$\lim_{n \to \infty} \left(1 + \frac{1}{n}\right)^n = e$$

(a) Use the definition of the derivative to show that

$$f'(1) = \lim_{h \to 0} \frac{\ln(1 + h)}{h}$$

(b) Show that (a) implies that

$$\ln[\lim_{h \to 0}(1 + h)^{1/h}] = 1$$

(c) Set $h = \frac{1}{n}$ in (b) and let $n \to \infty$. Show that this implies that

$$\lim_{n \to \infty} \left(1 + \frac{1}{n}\right)^n = e$$

62. Assume that $f(x)$ is differentiable with respect to x. Show that

$$\frac{d}{dx} \ln\left[\frac{f(x)}{x}\right] = \frac{f'(x)}{f(x)} - \frac{1}{x}$$

■ **4.7.3**

In Problems 63–74, use logarithmic differentiation to find the first derivative of the given functions.

63. $f(x) = 2x^x$

64. $f(x) = (2x)^{2x}$

65. $f(x) = (\ln x)^x$

66. $f(x) = (\ln x)^{3x}$

67. $f(x) = x^{\ln x}$

68. $f(x) = x^{2\ln x}$

69. $f(x) = x^{1/x}$

70. $f(x) = x^{3/x}$

71. $y = x^{x^x}$

72. $y = (x^x)^x$

73. $y = x^{\cos x}$

74. $y = (\cos x)^x$

75. Differentiate

$$y = \frac{e^{2x}(9x - 2)^3}{\sqrt[4]{(x^2 + 1)(3x^3 - 7)}}$$

76. Differentiate

$$y = \frac{e^{x-1} \sin^2 x}{(x^2 + 5)^{2x}}$$

■ 4.8 Linear Approximation and Error Propagation

Suppose we want to find an approximation to $\ln(1.05)$ without using a calculator. The method for solving this problem will be useful in many other applications. Let's look at the graph of $f(x) = \ln x$ (Figure 4.39). We know that $\ln 1 = 0$, and we see that 1.05 is quite close to 1—so close, in fact, that the curve connecting $(1, 0)$ to $(1.05, \ln 1.05)$ is close to a straight line. This suggests that we should approximate the curve by a straight line—but not just any straight line: We choose the tangent line to the graph of $f(x) = \ln x$ at $x = 1$ (Figure 4.39). We can find the equation of the tangent line without a calculator. We note that the slope of $f(x) = \ln x$ at $x = 1$ is $f'(1) = \frac{1}{x}\big|_{x=1} = 1$. This, together with the point $(1, 0)$, allows us to find the tangent line at $x = 1$:

$$L(x) = f(1) + f'(1)(x - 1) = 0 + (1)(x - 1) = x - 1$$

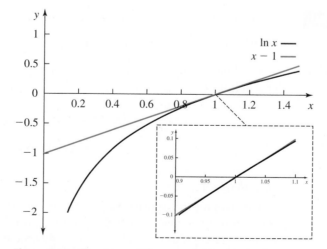

Figure 4.39 The tangent line approximation for $\ln x$ at $x = 1$ to approximate $\ln(1.05)$. When x is close to 1, the tangent line and the graph of $y = \ln x$ are close (see inset).

We call $L(x)$ the **tangent line approximation**, or **the linearization**, of $f(x)$ at $x = 1$. If we evaluate $L(x)$ at $x = 1.05$, we find that $L(1.05) = 1.05 - 1 = 0.05$, which is a good approximation to $\ln 1.05 = 0.048790\ldots$ (Here, we used the calculator to see how close the approximation is to the exact value.)

The Tangent Line Approximation

Assume that $y = f(x)$ is differentiable at $x = a$; then

$$L(x) = f(a) + f'(a)(x - a)$$

is the **tangent line approximation**, or the **linearization**, of f at $x = a$.

Geometrically, the linearization of f at $x = a$, $L(x) = f(a) + f'(a)(x - a)$, is the equation of the tangent line to the graph of $f(x)$ at the point $(a, f(a))$. (See Figure 4.40.)

If $|x - a|$ is sufficiently small, then $f(x)$ can be linearly approximated by $L(x)$; that is,

$$f(x) \approx f(a) + f'(a)(x - a)$$

This approximation is illustrated in Figure 4.41.

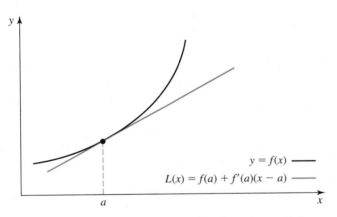

Figure 4.40 The tangent line approximation of $y = f(x)$ at $x = a$.

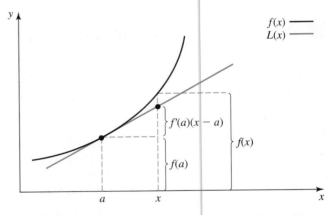

Figure 4.41 The linearization of f at $x = a$ can be used to approximate $f(x)$ for x close to a.

EXAMPLE 1

(a) Find the linear approximation of $f(x) = \sqrt{x}$ at $x = a$, and

(b) use your answer in (a) to find an approximate value of $\sqrt{50}$.

Solution

(a) Since $f(x) = \sqrt{x}$, it follows that $f'(x) = \frac{1}{2\sqrt{x}}$, and the linear approximation at $x = a$ is

$$L(x) = f(a) + f'(a)(x - a)$$
$$= \sqrt{a} + \frac{1}{2\sqrt{a}}(x - a)$$

(See Figure 4.42.)

(b) To find the approximate value of $f(50) = \sqrt{50}$, we need to choose a value for a close to 50 and for which we know $\sqrt{a}$ exactly. Our choice is $a = 49$. We thus approximate $f(50)$ by $L(50)$ with $a = 49$ and find that

$$\sqrt{50} \approx \sqrt{49} + \frac{50 - 49}{2\sqrt{49}} = 7 + \frac{1}{14} \approx 7.0714$$

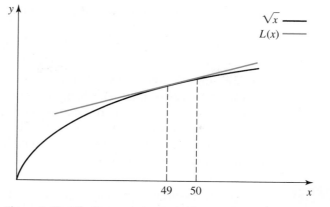

Figure 4.42 The linear approximation of $f(x) = \sqrt{x}$ at $x = 49$ is the line $y = L(x)$.

Using a calculator to compute $\sqrt{50} = 7.0711\ldots$, we see that the error in the linear approximation is quite small. ■

EXAMPLE 2

Find the linear approximation of $f(x) = \sin x$ at $x = 0$.

Solution

Since $f'(x) = \cos x$, it follows that

$$L(x) = f(0) + f'(0)(x - 0)$$
$$= \sin 0 + (\cos 0)x = x$$

(Figure 4.43). That is, for small values of x, we can approximate $\sin x$ by x. This approximation is often used in physics. (Note that x is measured in radians.) ■

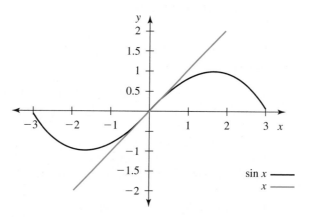

Figure 4.43 The linear approximation of $y = \sin x$ is the line $y = x$.

EXAMPLE 3

Let $N(t)$ be the size of a population at time t, and assume that the growth rate dN/dt of the population is given by

$$\frac{dN}{dt} = f(N)$$

where $f(N)$ is a differentiable function with $f(0) = 0$. Find the linearization of the growth rate at $N = 0$.

Solution

We need to find the tangent line approximation of $f(N)$ at $N = 0$. If we denote the linearization of $f(N)$ by $L(N)$, we obtain

$$L(N) = f(0) + f'(0)N$$

Now, $f(0) = 0$. If we set $r = f'(0)$, we find that, for N close to 0,

$$\frac{dN}{dt} \approx rN$$

The preceding formula shows that the population changes approximately exponentially when its size is small. This behavior is observed, for instance, when bacteria are grown in a nutrient-rich environment at a low population density.

We choose $f(0) = 0$ for biological reasons: When the population size is 0, the growth rate should be 0; otherwise, we would have spontaneous creation if $f(0) > 0$, or the population size would become negative if $f(0) < 0$. ■

EXAMPLE 4   Let $N(t)$ (measured in millions) be the size of a bacterial population at time t, and assume that the per capita growth rate is equal to 2%. We can express this statement in a differential equation, namely,

$$\frac{1}{N}\frac{dN}{dt} = 0.02$$

Suppose we know that at time $t = 10$ the size of the population is 250,000,000; that is, $N(10) = 250$ (since we measure the population size in millions). Use a linear approximation to predict the approximate population size at time $t = 10.1$.

Solution   To predict the population size at time 10.1, we use the following linearization of $N(t)$ at $t = 10$:

$$L(t) = N(10) + N'(10)(t - 10)$$

To evaluate $L(t)$ at $t = 10.1$, we need to find $N'(10)$.

Using the differential equation, we find that

$$N'(t) = (0.02)N(t)$$

When $t = 10$, we obtain

$$N'(10) = (0.02)N(10) = (0.02)(250) = 5$$

Hence,

$$L(10.1) = N(10) + N'(10)(10.1 - 10)$$
$$= 250 + (5)(0.1) = 250.5$$

Thus, we predict that the population size at time 10.1 is approximately 250,500,000. Note that this approximation is good only if the time at which we want to predict the population size is very close to the time at which we know the population size. ■

Error Propagation   Linear approximations are used in problems of **error propagation**. Suppose that you wish to determine the surface area of a spherical cell. Since the surface area S of a sphere with radius r is given by

$$S = 4\pi r^2$$

it suffices to measure the radius r of the cell. If your measurement of the radius is accurate within 3%, how does this affect the accuracy of the surface area?

First we must discuss what it means for a measurement to be accurate within a certain percentage. Suppose that x_0 is the true value of an observation and x is the measured value. Then $|\Delta x| = |x - x_0|$ is the **absolute error**, or tolerance, in measurement. The **relative error** is defined as $|\Delta x / x_0|$ and the **percentage error** as $100|\Delta x / x_0|$.

Returning to our example, let's find the error that arises in computing the surface area. We start with the absolute error of the surface area,

$$|\Delta S| = |S(r_0 + \Delta r) - S(r_0)|$$

where r_0 is the true radius and $|\Delta r|$ is the absolute error in the measurement of the radius. We approximate $S(r_0 + \Delta r) - S(r_0)$ by its linear approximation $S'(r_0)\Delta r$; that is,

$$\Delta S \approx S'(r_0)\Delta r$$

Since $S'(r) = 8\pi r$, the percentage error in the measurement of the surface area is

$$100 \left| \frac{\Delta S}{S(r_0)} \right| \approx 100 \left| \frac{S'(r_0)\Delta r}{S(r_0)} \right| = 100 \left| \frac{\Delta r}{r_0} \right| \left| \frac{S'(r_0)r_0}{S(r_0)} \right|$$

$$= 100 \underbrace{\left| \frac{\Delta r}{r_0} \right|}_{=3} \underbrace{\left| \frac{8\pi r_0^2}{4\pi r_0^2} \right|}_{=2} = 6$$

because $100|\Delta r / r_0| = 3$. In other words, the surface area is (approximately) accurate within 6% if the radius is accurate within 3%. Where the doubling of the percentage error comes from can be seen in the next example.

EXAMPLE 5   Suppose that you wish to determine $f(x)$ from a measurement of x. If $f(x)$ is given by a power function, namely, $f(x) = cx^s$, how does an error in the measurement of x propagate?

Solution   Since $f'(x) = csx^{s-1}$, we have

$$\Delta f \approx f'(x)\Delta x = csx^{s-1}\Delta x$$

The percentage error $100 \left| \frac{\Delta f}{f} \right|$ is therefore related to the percentage error $100 \left| \frac{\Delta x}{x} \right|$ as follows:

$$100 \left| \frac{\Delta f}{f} \right| \approx 100 \left| \frac{f'(x)\Delta x}{f(x)} \right| = 100 \left| \frac{\Delta x}{x} \right| \left| \frac{f'(x)x}{f(x)} \right|$$

$$= 100 \left| \frac{\Delta x}{x} \right| \left| \frac{csx^s}{cx^s} \right| = \left(100 \left| \frac{\Delta x}{x} \right| \right) |s|$$

In our previous example, $s = 2$; hence, the percentage error in the surface area measurement is twice the percentage error in the radius measurement. ■

EXAMPLE 6   **Allometric Growth**   Suppose that you wish to estimate the total leaf area of a tree in a certain plot. Experimental data obtained from the plot you are studying (Niklas, 1994) indicate that

$$\text{[leaf area]} \propto \text{[stem diameter]}^{1.84}$$

Instead of trying to measure the total leaf area directly, you measure the stem diameter and then use the scaling relationship to estimate the total leaf area. How accurately must you measure the stem diameter if you want to estimate the leaf area within an error of 10%?

Solution   We denote the leaf area by A and stem diameter by d. Then

$$A(d) = cd^{1.84}$$

where c is the constant of proportionality. An error in measurement of d is propagated as

$$\Delta A \approx A'(d)\Delta d = c(1.84)d^{0.84}\Delta d$$

The percentage error $100 \left| \frac{\Delta A}{A} \right|$ is related to the percentage error $100 \left| \frac{\Delta d}{d} \right|$ as

$$100 \left| \frac{\Delta A}{A} \right| \approx 100 \left| \frac{A'(d)\Delta d}{A(d)} \right| = 100 \left| \frac{\Delta d}{d} \right| \left| \frac{A'(d)d}{A(d)} \right|$$

$$= 100 \left| \frac{\Delta d}{d} \right| \left| \frac{c(1.84)d^{0.84}d}{cd^{1.84}} \right| = \left(100 \left| \frac{\Delta d}{d} \right| \right) (1.84)$$

We require that $100 \left| \frac{\Delta A}{A} \right| = 10$. Hence,

$$10 = (1.84) \left(100 \left| \frac{\Delta d}{d} \right| \right)$$

or

$$100 \left| \frac{\Delta d}{d} \right| = \frac{10}{1.84} = 5.4$$

That is, we must measure the stem diameter to within an error of 5.4%.

Using the result of Example 5, we could have found the same error immediately. Since

$$A(d) = cd^{1.84}$$

we get $s = 1.84$, where s is the exponent defined in Example 5. Using

$$100 \left| \frac{\Delta A}{A} \right| = |s| \left(100 \left| \frac{\Delta d}{d} \right| \right)$$

we obtain

$$100 \left| \frac{\Delta d}{d} \right| = \frac{1}{|s|} \left(100 \left| \frac{\Delta A}{A} \right| \right) = \frac{10}{1.84} = 5.4$$

as before. ■

EXAMPLE 7   Suppose that you wish to determine the percentage error of $f(x)$ from a measurement of x, where $f(x) = \ln x$, $x = 10$, and the percentage error for x is equal to 2%. Find the percentage error of $f(x)$.

Solution   The function $f(x)$ is not a power function, so there is no simple rule. We find that

$$100 \frac{\Delta f}{f} \approx 100 \frac{f'(x) \Delta x}{f(x)}$$

Since we know $100 \left| \frac{\Delta x}{x} \right|$, we multiply and divide the right-hand side by x and rearrange terms to get

$$100 \frac{f'(x) \Delta x}{f(x)} = 100 \frac{\Delta x}{x} \frac{x f'(x)}{f(x)}$$

Since $f'(x) = 1/x$, at $x = 10$ we obtain

$$100 \left| \frac{\Delta x}{x} \right| \left| \frac{x f'(x)}{f(x)} \right| \bigg|_{x=10} = 2 \frac{(10)(1/10)}{\ln 10} = \frac{2}{\ln 10} \approx 0.869$$

Thus, the percentage error of f is approximately 0.9%. ■

Section 4.8 Problems

In Problems 1–10, use the formula

$$f(x) \approx f(a) + f'(a)(x - a)$$

to approximate the value of the given function. Then compare your result with the value you get from a calculator.

1. $\sqrt{65}$; let $f(x) = \sqrt{x}$, $a = 64$, and $x = 65$

2. $\sqrt{35}$; let $f(x) = \sqrt{x}$, $a = 36$, and $x = 35$

3. $\sqrt[3]{124}$ **4.** $(7.9)^3$

5. $(0.99)^{25}$ **6.** $\tan(0.01)$

7. $\sin \left(\frac{\pi}{2} + 0.02 \right)$ **8.** $\cos \left(\frac{\pi}{4} - 0.01 \right)$

9. $\ln(1.01)$ **10.** $e^{0.1}$

In Problems 11–30, approximate $f(x)$ at a by the linear approximation

$$L(x) = f(a) + f'(a)(x - a)$$

11. $f(x) = \dfrac{1}{1 + x}$ at $a = 0$ **12.** $f(x) = \dfrac{1}{1 - x}$ at $a = 0$

13. $f(x) = \dfrac{2}{1 + x}$ at $a = 1$ **14.** $f(x) = \dfrac{1}{3 - 2x}$ at $a = 2$

15. $f(x) = \dfrac{1}{(1+x)^2}$ at $a = 0$ **16.** $f(x) = \dfrac{1}{(1-x)^2}$ at $a = 0$

17. $f(x) = \ln(1+x)$ at $a = 0$ **18.** $f(x) = \ln(1+2x)$ at $a = 0$

19. $f(x) = \log x$ at $a = 1$ **20.** $f(x) = \log(1+x^2)$ at $a = 0$

21. $f(x) = e^x$ at $a = 0$ **22.** $f(x) = e^{2x}$ at $a = 0$

23. $f(x) = e^{-x}$ at $a = 0$ **24.** $f(x) = e^{-3x}$ at $a = 0$

25. $f(x) = e^{x-1}$ at $a = 1$ **26.** $f(x) = e^{2x+1}$ at $a = -1/2$

27. $f(x) = (1+x)^{-n}$ at $a = 0$. (Assume that n is a positive integer.)

28. $f(x) = (1-x)^{-n}$ at $a = 0$. (Assume that n is a positive integer.)

29. $f(x) = \sqrt{1+x^2}$ at $a = 0$

30. $f(x) = \left(1 + \dfrac{1}{x}\right)^{1/4}$ at $a = 1$

31. Population Growth Suppose that the per capita growth rate of a population is 3%; that is, if $N(t)$ denotes the population size at time t, then

$$\frac{1}{N}\frac{dN}{dt} = 0.03$$

Suppose also that the population size at time $t = 4$ is equal to 100. Use a linear approximation to compute the population size at time $t = 4.1$.

32. Population Growth Suppose that the per capita growth rate of a population is 2%; that is, if $N(t)$ denotes the population size at time t, then

$$\frac{1}{N}\frac{dN}{dt} = 0.02$$

Suppose also that the population size at time $t = 2$ is equal to 50. Use a linear approximation to compute the population size at time $t = 2.1$.

33. Plant Biomass Suppose that the specific growth rate of a plant is 1%; that is, if $B(t)$ denotes the biomass at time t, then

$$\frac{1}{B(t)}\frac{dB}{dt} = 0.01$$

Suppose that the biomass at time $t = 1$ is equal to 5 grams. Use a linear approximation to compute the biomass at time $t = 1.1$.

34. Plant Biomass Suppose that a certain plant is grown along a gradient ranging from nitrogen-poor to nitrogen-rich soil. Experimental data show that the average mass per plant grown in a soil with a total nitrogen content of 1000 mg nitrogen per kg of soil is 2.7 g and the rate of change of the average mass per plant at this nitrogen level is 1.05×10^{-3} g per mg change in total nitrogen per kg soil. Use a linear approximation to predict the average mass per plant grown in a soil with a total nitrogen content of 1100 mg nitrogen per kg of soil.

In Problems 35–40, a measurement error in x affects the accuracy of the value f(x). In each case, determine an interval of the form

$$[f(x) - \Delta f, f(x) + \Delta f]$$

that reflects the measurement error Δx. In each problem, the quantities given are f(x) and x = true value of x $\pm |\Delta x|$.

35. $f(x) = 2x, x = 1 \pm 0.1$

36. $f(x) = 1 - 3x, x = -2 \pm 0.3$

37. $f(x) = 3x^2, x = 2 \pm 0.1$

38. $f(x) = \sqrt{x}, x = 10 \pm 0.5$

39. $f(x) = e^x, x = 2 \pm 0.2$

40. $f(x) = \sin x, x = -1 \pm 0.05$

In Problems 41–44, assume that the measurement of x is accurate within 2%. In each case, determine the error Δf in the calculation of f and find the percentage error $100\frac{\Delta f}{f}$. The quantities f(x) and the true value of x are given.

41. $f(x) = 4x^3, x = 1.5$ **42.** $f(x) = x^{1/4}, x = 10$

43. $f(x) = \ln x, x = 20$ **44.** $f(x) = \dfrac{1}{1+x}, x = 4$

45. The volume V of a spherical cell of radius r is given by

$$V(r) = \frac{4}{3}\pi r^3$$

If you can determine the radius to within an accuracy of 3%, how accurate is your calculation of the volume?

46. Poiseuille's Law The speed v of blood flowing along the central axis of an artery of radius R is given by Poiseuille's law,

$$v(R) = cR^2$$

where c is a constant. If you can determine the radius of the artery to within an accuracy of 5%, how accurate is your calculation of the speed?

47. Allometric Growth Suppose that you are studying reproduction in moss. The scaling relation

$$N \propto L^{2.11}$$

has been found (Niklas, 1994) between the number of moss spores (N) and the capsule length (L). This relation is not very accurate, but it turns out that it suffices for your purpose. To estimate the number of moss spores, you measure the capsule length. If you wish to estimate the number of moss spores within an error of 5%, how accurately must you measure the capsule length?

48. Tilman's Resource Model Suppose that the rate of growth of a plant in a certain habitat depends on a single resource—for instance, nitrogen. Assume that the growth rate $f(R)$ depends on the resource level R in accordance with the formula

$$f(R) = a\frac{R}{k+R}$$

where a and k are constants. Express the percentage error of the growth rate, $100\frac{\Delta f}{f}$, as a function of the percentage error of the resource level, $100\frac{\Delta R}{R}$.

49. Chemical Reaction The reaction rate $R(x)$ of the irreversible reaction

$$A + B \rightarrow AB$$

is a function of the concentration x of the product AB and is given by

$$R(x) = k(a - x)(b - x)$$

where k is a constant, a is the concentration of A at the beginning of the reaction, and b is the concentration of B at the beginning of the reaction. Express the percentage error of the reaction rate, $100\frac{\Delta R}{R}$, as a function of the percentage error of the concentration x, $100\frac{\Delta x}{x}$.

Chapter 4 Key Terms

Discuss the following definitions and concepts:

1. Derivative, formal definition
2. Difference quotient
3. Secant line and tangent line
4. Instantaneous rate of change
5. Average rate of change
6. Differential equation
7. Differentiability and continuity
8. Power rule

9. Basic rules of differentiation
10. Product rule
11. Quotient rule
12. Chain rule
13. Implicit function
14. Implicit differentiation
15. Related rates
16. Higher derivatives
17. Derivatives of trigonometric functions

18. Derivatives of exponential functions
19. Derivatives of inverse and logarithmic functions
20. Logarithmic differentiation
21. Tangent line approximation
22. Error propagation
23. Absolute error, relative error, percentage error

Chapter 4 Review Problems

In Problems 1–8, differentiate with respect to the independent variable.

1. $f(x) = -3x^4 + \dfrac{2}{\sqrt{x}} + 1$ **2.** $g(x) = \dfrac{1}{\sqrt{x^3 + 4}}$

3. $h(t) = \left(\dfrac{1-t}{1+t}\right)^{1/3}$ **4.** $f(x) = (x^2 + 1)e^{-x}$

5. $f(x) = e^{2x} \sin\left(\dfrac{\pi}{2}x\right)$ **6.** $g(s) = \dfrac{\sin(3s+1)}{\cos(3s)}$

7. $f(x) = 2\dfrac{\ln(x+1)}{\ln x^2}$ **8.** $g(x) = e^{-x}\ln(x+1)$

In Problems 9–12, find the first and second derivatives of the given functions.

9. $f(x) = e^{-x^2/2}$ **10.** $g(x) = \tan(x^2 + 1)$

11. $h(x) = \dfrac{x}{x+1}$ **12.** $f(x) = \dfrac{e^{-x}}{e^{-x} + 1}$

In Problems 13–16, find dy/dx.

13. $x^2 y - y^2 x = \sin x$ **14.** $e^{x^2 + y^2} = 2x$

15. $\ln(x - y) = 2x$ **16.** $\tan(x - y) = x^2$

In Problems 17–19, find dy/dx and d^2y/dx^2.

17. $x^2 + y^2 = 16$ **18.** $x = \tan y$ **19.** $e^y = \ln x$

20. Assume that x is a function of t. Find $\frac{dy}{dt}$ when $y = \cos x$ and $\frac{dx}{dt} = \sqrt{3}$ for $x = \frac{\pi}{3}$.

21. Velocity A flock of birds passes directly overhead, flying horizontally at an altitude of 100 feet and a speed of 6 feet per second. How quickly is the distance between you and the birds increasing when the distance is 320 feet? (You are on the ground and are not moving.)

22. Find the derivative of

$$y = \ln|\cos x|$$

23. Suppose that $f(x)$ is differentiable. Find an expression for the derivative of each of the following functions:
(a) $y = e^{f(x)}$ **(b)** $y = \ln f(x)$ **(c)** $y = [f(x)]^2$

24. Find the tangent line and the normal line to $y = \ln(x + 1)$ at $x = 1$.

25. Suppose that

$$f(x) = \dfrac{x^2}{1 + x^2}, \quad x \geq 0$$

(a) Use a graphing calculator to graph $f(x)$ for $x \geq 0$. Note that the graph is S shaped.

(b) Find a line through the origin that touches the graph of $f(x)$ at some point $(c, f(c))$ with $c > 0$. This is the tangent line at $(c, f(c))$ that goes through the origin. Graph the tangent line in the same coordinate system that you used in (a).

In Problems 26–29, find an equation for the tangent line to the curve at the specified point.

26. $y = (\sin x)^{\cos x}$ at $x = \dfrac{\pi}{2}$ **27.** $y = e^{-x^2}\cos x$ at $x = \dfrac{\pi}{3}$

28. $x^2 + y = e^y$ at $x = \sqrt{e - 1}$
29. $x \ln y = y \ln x$ at $x = 1$

30. In Review Problem 17 of Chapter 2, we introduced the following hyperbolic functions:

$$\sinh x = \dfrac{e^x - e^{-x}}{2}, \quad x \in \mathbf{R}$$

$$\cosh x = \dfrac{e^x + e^{-x}}{2}, \quad x \in \mathbf{R}$$

$$\tanh x = \dfrac{e^x - e^{-x}}{e^x + e^{-x}}, \quad x \in \mathbf{R}$$

(a) Show that

$$\dfrac{d}{dx}\sinh x = \cosh x$$

and

$$\dfrac{d}{dx}\cosh x = \sinh x$$

(b) Use the facts that

$$\tanh x = \dfrac{\sinh x}{\cosh x}$$

and

$$\cosh^2 x - \sinh^2 x = 1$$

together with your results in (a) to show that

$$\dfrac{d}{dx}\tanh x = \dfrac{1}{\cosh^2 x}$$

31. Find a second-degree polynomial

$$p(x) = ax^2 + bx + c$$

with $p(-1) = 6$, $p'(1) = 8$, and $p''(0) = 4$.

32. Use the geometric interpretation of the derivative to find the equations of the tangent lines to the curve

$$x^2 + y^2 = 1$$

at the following points:
(a) $(1, 0)$ **(b)** $\left(\frac{1}{2}, \frac{1}{2}\sqrt{3}\right)$

(c) $(-\frac{1}{2}\sqrt{2}, -\frac{1}{2}\sqrt{2})$ **(d)** $(0, -1)$

33. Distance and Velocity Geradedorf[3] and Straightville are connected by a very straight, but rather hilly, road. Biking from Geradedorf to Straightville, your position at time t (measured in hours) is given by the function

$$s(t) = 3\pi t + 3(1 - \cos(\pi t))$$

for $0 \le t \le 5.5$, where $s(t)$ is measured in miles.

(a) Use a graphing calculator to convince yourself that you didn't backtrack during your trip. How can you check this? Assuming that your trip takes 5.5 hours, find the distance between Geradedorf and Straightville.

(b) Find the velocity $v(t)$ and the acceleration $a(t)$.

(c) Use a graphing calculator to graph $s(t)$, $v(t)$, and $a(t)$. In (a), you used the function $s(t)$ to conclude that you didn't backtrack during your trip. Can you use any of the other two functions to answer the question of backtracking? Explain your answer.

(d) Assuming that you slow down going uphill and speed up going downhill, how many peaks and valleys does this road have?

34. Distance and Velocity Suppose your position at time t on a straight road is given by

$$s(t) = \cos(\pi t)$$

for $0 \le t \le 2$, where t is measured in hours.

(a) What is your position at the beginning and end of your trip?

(b) Use a graphing calculator to help describe your trip in words.

(c) What is the total distance you have traveled?

(d) Determine your velocity and your acceleration during the trip. When is your velocity equal to 0? Relate this velocity to your position, and explain what it means.

35. Population Growth In one very simple population model, the growth rate at time t depends on the number of individuals at time $t - T$, where T is a positive constant. (That is, the model incorporates a time delay into the birthrate.) This assumption is useful, for instance, if one wishes to take into account the fact that individuals must mature before reproducing.

---

(3) Those who are curious may look up the words *gerade* and *Dorf* in a German–English dictionary.

Denote the size of the population at time t by $N(t)$, and assume that

$$\frac{dN}{dt} = \frac{\pi}{2T}(K - N(t - T)) \qquad (4.13)$$

where K and T are positive constants.

(a) Show that

$$N(t) = K + A\cos\frac{\pi t}{2T}$$

is a solution of (4.13).

(b) Graph $N(t)$ for $K = 100$, $A = 50$, and $T = 1$.

(c) Explain in words how the size of the population changes over time.

36. Radioactive Decay We denote by $W(t)$ the amount of a radioactive material left at time t if the initial amount present was $W(0) = W_0$.

(a) Show that

$$W(t) = W_0 e^{-\lambda t}$$

solves the differential equation

$$\frac{dW}{dt} = -\lambda W(t)$$

(b) Show that if you graph $W(t)$ on semilog paper, then the result is a straight line.

(c) Use your result in (b) to explain why

$$\frac{d\ln W(t)}{dt} = \text{constant}$$

Determine the constant, and relate it to the graph in (b).

(d) Show that

$$\frac{d\ln W(t)}{dt} = \text{constant}$$

implies that

$$\frac{dW}{dt} \propto W(t)$$

37. Allometric Growth In Example 17 of Subsection 4.4.3, we introduced an allometric relationship between skull length (in cm) and backbone length (in cm) of ichthyosaurs, a group of extinct marine reptiles. The relationship is

$$S = (1.162)B^{0.933}$$

where S and B denote skull length and backbone length, respectively. Suppose that you found only the skull of an individual and that, on the basis of the skull length, you wish to estimate the backbone length of this specimen. How accurately must you measure skull length if you want to estimate backbone length to within an error of 10%?

5 | Applications of Differentiation

■ 5.1 Extrema and the Mean-Value Theorem

The primary focus of this chapter is how calculus can help us to understand the behavior of functions. Points where a function is smallest or largest, called extrema, are of particular importance. This section defines extrema, gives conditions that guarantee extrema (via the extreme-value theorem), and provides a characterization of extrema (Fermat's theorem). Fermat's theorem will be crucial in establishing the mean-value theorem, a result that can be understood graphically. The mean-value theorem has far-reaching consequences that arise in later sections, where we learn methods for characterizing the behavior of functions.

■ 5.1.1 The Extreme-Value Theorem

Suppose that you measure the depth of a creek along a transect between two points A and B (see Figure 5.1). Looking at the profile of the creek, you see that there is a location of maximum depth and a location of minimum depth. The existence of such locations is the content of the extreme-value theorem. To formalize the theorem, we must introduce some terminology.

Figure 5.1 A transect of a creek between the points A and B.

> **Definition** Let f be a function defined on the set D that contains the number c. Then
>
> f has a **global** (or **absolute**) **maximum** at $x = c$ if
> $$f(c) \geq f(x) \quad \text{for all } x \in D$$
>
> and
>
> f has a **global** (or **absolute**) **minimum** at $x = c$ if
> $$f(c) \leq f(x) \quad \text{for all } x \in D$$

The following result gives conditions under which global maxima and global minima, collectively called **global** (or **absolute**) **extrema**, exist:

> **The Extreme-Value Theorem** If f is continuous on a closed interval $[a, b]$, $-\infty < a < b < \infty$, then f has a global maximum and a global minimum on $[a, b]$.

The proof of the extreme-value theorem is beyond the scope of this text and will be omitted. However, the result is quite intuitive, and we illustrate it in Figures 5.2 and 5.3. Figure 5.2 shows that a function may attain its extreme values at the endpoints of the interval $[a, b]$, whereas in Figure 5.3 the extreme values are attained in the interior of the interval $[a, b]$. The function must be continuous and defined on a closed interval in order for it to have global maxima and global minima. But note that the extreme-value theorem tells us only that global extrema exist, not where they are. Furthermore, they need not be unique, meaning that a function can have more than one global maximum or global minimum.

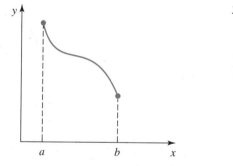

Figure 5.2 Extreme values at the endpoint.

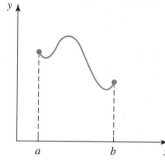

Figure 5.3 Extreme values in the interior.

EXAMPLE 1

Optimal Strategy Suppose a plant has two reproductive strategies, one asexual by clonal reproduction and the other sexual by seed production. The plant's fitness depends on how it allocates its resources to the two strategies. Suppose that the plant allocates a fixed amount of resources to reproduction, a fraction p of which is allocated to clonal reproduction ($0 \leq p \leq 1$) and a fraction $1 - p$ to sexual reproduction. Denote by $f(p)$ the plant's fitness as a function of p. Assuming that $f(p)$ is a continuous function, why is there a strategy of resource allocation (called an optimal strategy) that maximizes the plant's fitness?

Solution

According to the extreme-value theorem, since $f(p)$ is continuous on the closed interval $[0, 1]$, $f(p)$ has a global maximum (and a global minimum) on $[0, 1]$. The global maximum represents the optimal strategy. Note that the theorem guarantees only the existence of an optimal strategy; it does not tell us *which* strategy is optimal. Furthermore, there could be more than one global maximum, meaning that there could be more than one optimal strategy of resource allocation. ■

Figure 5.2 illustrates the importance of one of the assumptions in the extreme-value theorem, namely, that the interval is *closed*. If the interval from a to b in Figure 5.2 did not include the endpoints, we would have neither a global maximum nor a global minimum.

The next two examples illustrate that the theorem cannot be used if either f is discontinuous or the interval is not closed.

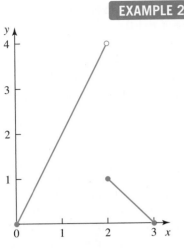

Figure 5.4 The graph of $f(x)$ in Example 2.

Let

$$f(x) = \begin{cases} 2x & \text{if } 0 \leq x < 2 \\ 3 - x & \text{if } 2 \leq x \leq 3 \end{cases}$$

Note that $f(x)$ is defined on a *closed* interval, namely, [0, 3]. However, f is discontinuous at $x = 2$, as can be seen in Figure 5.4. The graph of $f(x)$ shows that there is no value $c \in [0, 3]$ where $f(c)$ attains a global maximum. Do not be tempted to say that there should be a global maximum close to $x = 2$: For any candidate for a global maximum that you might come up with, you will be able to find a point whose y-coordinate exceeds the y-coordinate of your previous candidate. Try it! The reason for this is that

$$\lim_{x \to 2^-} f(x) = 4$$

but the function f takes on the value 1 at $x = 2$. We conclude that the function does not have a global maximum. It does, however, have global minima, at $x = 0$ and $x = 3$, where $f(x)$ takes on the value 0. (This is a function that has more than one global minimum.)
■

EXAMPLE 3   Let

$$f(x) = x \qquad \text{for } 0 < x < 1$$

Note that $f(x)$ is continuous on its domain, (0, 1), but is *not* defined on a closed interval (Figure 5.5). The function $f(x)$ attains neither a global maximum nor a global minimum. Although

$$\lim_{x \to 0^+} f(x) = 0 \qquad \text{and} \qquad \lim_{x \to 1^-} f(x) = 1$$

and $0 < f(x) < 1$ for all $x \in (0, 1)$, there is no number c in the open interval (0, 1) where $f(c) = 0$ or $f(c) = 1$.
■

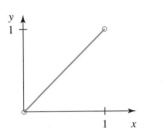

Figure 5.5 The graph of $f(x)$ in Example 3.

■ 5.1.2 Local Extrema

We will now discuss local (or relative) extrema, which are points where a graph is higher or lower than all *nearby* points. This discussion will allow us to identify the peaks and the valleys of the graph of a function. (See Figure 5.6.) The graph of the function in Figure 5.6 has three peaks—at $x = a, c,$ and e—and two valleys—at $x = b$ and d. A peak (or local maximum) has the property that the graph is lower nearby; a valley (or local minimum) has the property that the graph is higher nearby.

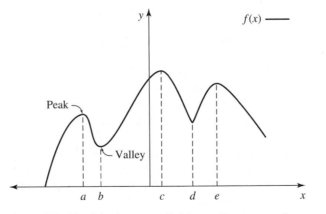

Figure 5.6 The function $y = f(x)$ has valleys at $x = b$ and d and peaks at $x = a, c,$ and e.

The formal definitions follow (see Figures 5.7 and 5.8):

Local maximum

$f(x)$

$c - \delta \; c \; c + \delta$ x

Figure 5.7 The function $y = f(x)$ has a local maximum at $x = c$.

> A function f defined on a set D has a **local** (or **relative**) **maximum** at a point c if there exists a $\delta > 0$ such that
>
> $$f(c) \geq f(x) \qquad \text{for all } x \in (c - \delta, c + \delta) \cap D$$
>
> A function f defined on a set D has a **local** (or **relative**) **minimum** at a point c if there exists a $\delta > 0$ such that
>
> $$f(c) \leq f(x) \qquad \text{for all } x \in (c - \delta, c + \delta) \cap D$$

Local maxima and local minima are collectively called local (or relative) extrema. If D is an interval and c is in the interior of D (i.e., not a boundary point), then the definitions simplify: The function f has a local maximum at c if there exists an open interval I such that $f(c) \geq f(x)$ for all $x \in I$; likewise, the function f has a local minimum at c if there exists an open interval I such that $f(c) \leq f(x)$ for all $x \in I$. In the definitions in the preceding box, we wrote $(c - \delta, c + \delta) \cap D$. If c is an interior point of D, δ can be chosen small enough so that $(c - \delta, c + \delta)$ is contained in D and we can set $I = (c - \delta, c + \delta)$. Intersecting the interval $(c - \delta, c + \delta)$ with D becomes important when c is a boundary point, as we will see in Example 4.

Local minimum

$f(x)$

$c - \delta \;\; c \;\; c + \delta$ x

Figure 5.8 The function $y = f(x)$ has a local minimum at $x = c$.

We examine local and global extrema in the next two examples; the discussion is based on looking at the graphs of functions. In the first example, we consider a function that is defined on a closed interval; this allows us to compute the value of the function at both endpoints of its domain. In the second example, we consider a function that is defined on a half-open interval; thus, the value of the function can be computed at one endpoint of its domain, but not at the other endpoint.

EXAMPLE 4   Let

$$f(x) = (x - 1)^2(x + 2) \quad \text{for } -2 \leq x \leq 3$$

(a) Use the graph of $f(x)$ to find all local extrema.

(b) Find the global extrema.

Solution   **(a)** The graph of $f(x)$ is illustrated in Figure 5.9. The function f is defined on the closed interval $[-2, 3]$. We begin with local extrema that occur at interior points of the domain $D = [-2, 3]$; looking at the figure, we see that a local maximum occurs at $x = -1$, as there are no greater values of f nearby. That is, we can find a small interval I about $x = -1$ so that $f(-1) \geq f(x)$ for all $x \in I$. For instance, we can choose $\delta = 0.1$ in the preceding definition and obtain $I = (-1.1, -0.9)$ (Figure 5.10).

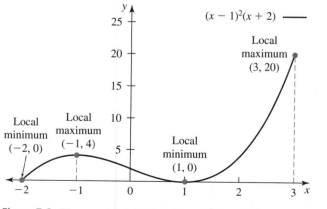

Figure 5.9 The graph of $f(x) = (x - 1)^2(x + 2)$ for $-2 \leq x \leq 3$ in Example 4.

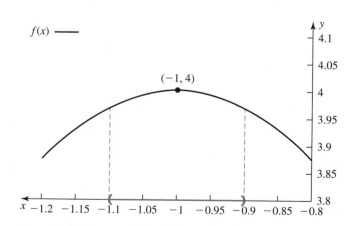

Figure 5.10 The graph of $f(x) = (x - 1)^2(x + 2)$ near $x = -1$. The point $(-1, 4)$ is a local maximum: $f(-1) \geq f(x)$ for all nearby x in the domain of f.

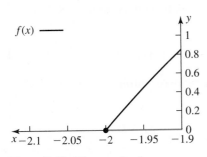

Figure 5.11 The graph of $f(x) = (x-1)^2(x+2)$ near $x = -2$. The point $(-2, 0)$ is a local minimum: $f(-2) \leq f(x)$ for all nearby x in the domain of f.

There is a local minimum at $x = 1$, since there are no smaller values of f nearby. This time, we need to find a small interval about $x = 1$ so that $f(1) \leq f(x)$ for all $x \in I$; for example, $I = (0.9, 1.1)$.

A local minimum also occurs at $x = -2$, one of the endpoints of the domain of f. As discussed in the definition of a local minimum, we require an interval I about c such that $f(c) \leq f(x)$ for all $x \in I \cap D$, where D is the domain of the function. If c is an interior point, we can always choose I small enough so that $I \subset D$ and, therefore, $I \cap D = I$, but this is not possible at an endpoint. To show that there is a local minimum at $x = -2$, we must find $\delta > 0$ such that $f(-2) \leq f(x)$ for all $x \in (-2 - \delta, -2 + \delta) \cap D = [-2, -2 + \delta)$. We can again choose $\delta = 0.1$ and see that $f(-2) \leq f(x)$ for all $x \in [-2, -1.9)$ (Figure 5.11).

Similarly, we see that there is a local maximum at $x = 3$, since $f(3) \geq f(x)$ for all $x \in (2.9, 3]$; that is, there is no larger value of f nearby.

(b) Global extrema are points at which a function is either largest or smallest. Since f is defined on a closed interval, it follows from the extreme-value theorem that both a global maximum and a global minimum exist. These global extrema may occur either in the interior or at the endpoints of the domain $D = [-2, 3]$.

To find the global minimum, we compare the local minima. Since $f(-2) = 0$ and $f(1) = 0$, it follows that the global minima occur at $x = -2$ and $x = 1$ (Figure 5.9). To find the global maximum, we compare the local maxima. Since $f(3) = 20$ and $f(-1) = 4$, it follows that $f(3) > f(-1)$; therefore, the global maximum occurs at the endpoint $x = 3$ (Figure 5.9).   ■

EXAMPLE 5   Let

$$f(x) = |x^2 - 4| \quad \text{for } -2.5 \leq x < 3$$

Find all local and global extrema.

Solution   The graph of $f(x)$, illustrated in Figure 5.12, reveals that local minima occur at $x = -2$ and $x = 2$ and local maxima occur at $x = -2.5$ and $x = 0$. Note that $f(x)$ is not defined at $x = 3$; thus, $x = 3$ cannot be a local maximum. To find the global extrema, we need to look at the function values close to the boundary $x = 3$. Candidates for global extrema are all the local extrema, which must be compared against the value of the function near the boundary $x = 3$. We discuss the global maximum first. Since

$$f(-2.5) = 2.25 \qquad f(0) = 4 \qquad \lim_{x \to 3^-} f(x) = 5$$

the function is largest near the point $x = 3$. But because $f(x)$ is not defined at $x = 3$, the function has no global maximum. (This does not contradict the extreme-value theorem, as the function is not defined on a closed interval, which is an assumption of the theorem.) To find the global minimum, we need only compare $f(-2)$ and $f(2)$.

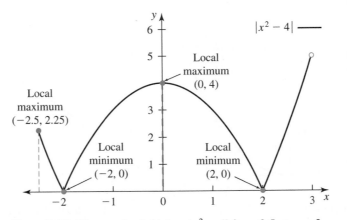

Figure 5.12 The graph of $f(x) = |x^2 - 4|$ for $-2.5 \leq x < 3$ in Example 5.

We find that $f(-2) = 0$ and $f(2) = 0$; therefore, global minima occur at $x = -2$ and $x = 2$. ■

Looking at Figures 5.9 and 5.12, we see that if the function f is differentiable at an interior point where f has a local extremum, then there is a horizontal tangent line at that point. This statement is known as Fermat's theorem (see Figure 5.13).

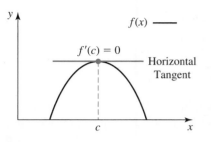

Figure 5.13 Fermat's theorem.

> **Fermat's Theorem** If f has a local extremum at an interior point c and $f'(c)$ exists, then $f'(c) = 0$.

Proof We prove Fermat's theorem for the case where the local extremum is a maximum; the proof where the local extremum is a minimum is similar. We need to show that $f'(c) = 0$. To do so, we use the formal definition of the derivative to compute $f'(c)$:

$$f'(c) = \lim_{x \to c} \frac{f(x) - f(c)}{x - c}$$

To compute the limit, we will separately compute the left-hand limit ($x \to c^-$) and the right-hand limit ($x \to c^+$). We begin with the following observation (Figure 5.14): Suppose that f has a local maximum at an interior point c. Then there exists a $\delta > 0$ such that

$$f(x) \le f(c) \quad \text{for all } x \in (c - \delta, c + \delta)$$

Since $f(x) - f(c) \le 0$ and $x - c < 0$ if $x < c$, we find that the left-hand limit is

$$\lim_{x \to c^-} \frac{f(x) - f(c)}{x - c} \ge 0 \tag{5.1}$$

and since $f(x) - f(c) \le 0$ and $x - c > 0$ if $x > c$, we find that the right-hand limit is

$$\lim_{x \to c^+} \frac{f(x) - f(c)}{x - c} \le 0 \tag{5.2}$$

Now, because f is differentiable at c, it follows that

$$f'(c) = \lim_{x \to c^-} \frac{f(x) - f(c)}{x - c} = \lim_{x \to c^+} \frac{f(x) - f(c)}{x - c}$$

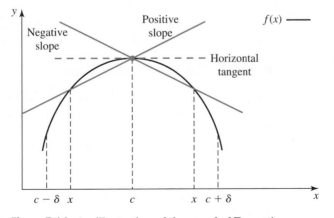

Figure 5.14 An illustration of the proof of Fermat's theorem: For $x < c$, the slope of the secant line is positive; for $x > c$, the slope of the secant line is negative. In the limit $x \to c$, the secant lines converge to the horizontal tangent line.

This, together with Equation (5.1), shows that $f'(c) \geq 0$ and, together with Equation (5.2), that $f'(c) \leq 0$. Now, if you have a number that is simultaneously nonnegative and nonpositive, the number must be 0. Therefore, $f'(c) = 0$. ∎

EXAMPLE 6

Explain why $y = \tan x$ does not have a local extremum at $x = 0$.

Solution

$y = \tan x$ is differentiable at $x = 0$, with

$$\frac{d}{dx} \tan x = \sec^2 x$$

Hence, the derivative of $y = \tan x$ at $x = 0$ is equal to 1. Since the derivative is not equal to 0, Fermat's theorem (or, more precisely, its contrapositive) implies that $x = 0$ is not a local extremum. ∎

Caution!

1. The condition that $f'(c) = 0$ is a necessary, but not sufficient, condition for the existence of local extrema at interior points where $f'(c)$ exists. In particular, the fact that f is differentiable at c with $f'(c) = 0$ tells us nothing about whether f has a local extremum at $x = c$. For instance, $f(x) = x^3$, $x \in \mathbf{R}$, is differentiable at $x = 0$ and $f'(0) = 0$, but there is no local extremum at $x = 0$. The graph of $y = x^3$ is shown in Figure 5.15. Although there is a horizontal tangent at $x = 0$, there is no local extremum at $x = 0$. Fermat's theorem does tell you, however, that if $x = c$ is an interior point with $f'(c) \neq 0$, then $x = c$ cannot be a local extremum (Example 6). Interior points with horizontal tangents are *candidates* for local extrema.

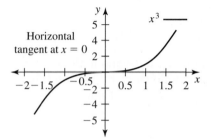

Figure 5.15 The graph of $y = x^3$ has a horizontal tangent at $x = 0$, but $(0, 0)$ is not an extremum.

2. The function f may not be differentiable at a local extremum. For instance, in Example 5, the function $f(x)$ is not differentiable at $x = -2$ and $x = 2$, but both points turned out to be local extrema. This means that, in order to identify candidates for local extrema, it will not be enough simply to look at points with horizontal tangents; you also must look at points where the function $f(x)$ is not differentiable.

3. Local extrema may occur at endpoints of the domain. Since Fermat's theorem says nothing about what happens at endpoints, you will have to look at endpoints separately.

To summarize our discussion, here are some guidelines for finding local extrema:

---

1. Don't assume that points where $f'(x) = 0$ give you local extrema; these are just candidates.

2. Check points where the derivative is not defined.

3. Check endpoints of the domain.

We will return to local extrema in Section 5.3, where we will learn methods for deciding whether candidates for local extrema are indeed local extrema.

■ 5.1.3 The Mean-Value Theorem

The mean-value theorem (MVT) is a very important, yet easily understood, result in calculus. Its consequences are far reaching, and we will use it in every section in this chapter to derive important results that will help us to analyze functions.

Here is an example that explains the MVT: Consider the function

$$f(x) = x^2 \quad \text{for } 0 \leq x \leq 1$$

The secant line connecting the endpoints $(0, 0)$ and $(1, 1)$ of the graph of $f(x)$ has slope

$$m = \frac{f(1) - f(0)}{1 - 0} = \frac{1 - 0}{1 - 0} = 1$$

The graph of $f(x)$ and the secant line are shown in Figure 5.16. Note that $f(x)$ is differentiable in $(0, 1)$; that is, you can draw a tangent line at every point of the graph

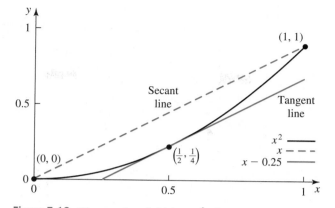

Figure 5.16 The graphs of $f(x) = x^2$, the secant line through $(0, 0)$ and $(1, 1)$, and the tangent line parallel to the secant line.

in the open interval $(0, 1)$. If you look at the graph of $f(x)$, you see that there exists a number $c \in (0, 1)$ such that the slope of the tangent line at $(c, f(c))$ is the same as the slope of the secant line through the points $(0, 0)$ and $(1, 1)$. That is, we claim that there exists a number $c \in (0, 1)$ such that

$$\frac{f(1) - f(0)}{1 - 0} = f'(c)$$

Proving that there exists such a value c is the thrust of the mean-value theorem.

We can compute the value of c in this example. Since $f'(x) = 2x$ and the slope of the secant line is $m = 1$, we must solve

$$1 = 2c, \quad \text{or} \quad c = \tfrac{1}{2}$$

Using the point–slope form $[y - y_0 = m(x - x_0)]$, we can find the equation of the tangent line at $(\tfrac{1}{2}, f(\tfrac{1}{2})) = (\tfrac{1}{2}, \tfrac{1}{4})$, namely,

$$y - \tfrac{1}{4} = 1 \left(x - \tfrac{1}{2} \right), \quad \text{or} \quad y = x - \tfrac{1}{4}$$

(This tangent line is shown in Figure 5.16.)

The Mean-Value Theorem (MVT) If f is continuous on the closed interval $[a, b]$ and differentiable on the open interval (a, b), then there exists at least one number $c \in (a, b)$ such that

$$\frac{f(b) - f(a)}{b - a} = f'(c)$$

The fraction on the left-hand side of the equation in the theorem is the slope of the secant line connecting the points $(a, f(a))$ and $(b, f(b))$, and the quantity on the right-hand side is the slope of the tangent line at $(c, f(c))$ (see Figure 5.17).

Geometrically, the MVT is indeed easily understood: It states that there exists a point on the graph between $(a, f(a))$ and $(b, f(b))$ where the tangent line at this point is parallel to the secant line through $(a, f(a))$ and $(b, f(b))$. [We denoted the point in question by $(c, f(c))$.] The MVT is an "existence" result: It tells us neither how many such points there are nor where they are in the interval (a, b).

Going back to the example $f(x) = x^2, 0 \le x \le 1$, we see that $f(x)$ satisfies the assumptions of the MVT, namely, that $f(x)$ is continuous on the closed interval $[0, 1]$ and differentiable on the open interval $(0, 1)$. The MVT then guarantees the

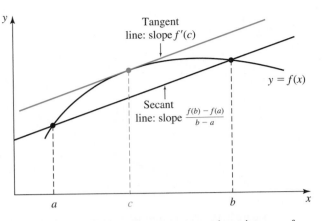

Figure 5.17 The mean value guarantees the existence of a number $c \in (a, b)$ such that the tangent line at $(c, f(c))$ has the same slope as the secant line through $(a, f(a))$ and $(b, f(b))$.

existence of at least one number $c \in (0, 1)$ such that the slope of the secant line through the points $(0, 0)$ and $(1, 1)$ is equal to the slope of the tangent line at $(c, f(c))$.

At this point, you might be wondering how such a seemingly simple theorem can be so important. In the sections that follow, you will encounter the theorem mostly in proofs of other important results that will enable us to understand properties of functions by using calculus. The next example, however, is an application that gives physical meaning to the theorem.

EXAMPLE 7   **Velocity**   A car moves in a straight line. At time t (measured in seconds), its position (measured in meters) is

$$s(t) = \frac{1}{25}t^3, \ 0 \le t \le 10$$

Show that there is a time $t \in (0, 10)$ when the velocity is equal to the average velocity between $t = 0$ and $t = 10$.

Solution   The average velocity between $t = 0$ and $t = 10$ is

$$\frac{s(10) - s(0)}{10 - 0} = \frac{\frac{1}{25} \cdot 1000 \,\text{m}}{10 \,\text{s}} = 4\frac{\text{m}}{\text{s}}$$

This is the slope of the secant line connecting the points $(0, 0)$ and $(10, 40)$. Since $s(t)$ is continuous on $[0, 10]$ and differentiable on $(0, 10)$, the MVT tells us that there must exist a number $c \in (0, 10)$ such that $s'(c) = 4\frac{\text{m}}{\text{s}}$. Now, $s'(t)$ is the (instantaneous) velocity. So, at some point during this short trip, the speedometer must have read $4\frac{\text{m}}{\text{s}}$.   ■

The rest of this section is devoted to the proof of the MVT, which is typically proved by first showing a special case of the theorem called Rolle's theorem.

> **Rolle's Theorem**   If f is continuous on the closed interval $[a, b]$ and differentiable on the open interval (a, b), and if $f(a) = f(b)$, then there exists a number $c \in (a, b)$ such that $f'(c) = 0$.

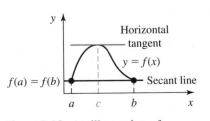

Figure 5.18 An illustration of Rolle's theorem.

Figure 5.18 illustrates Rolle's theorem. The function in the graph is defined on the closed interval $[a, b]$ and takes on the same values at the two endpoints of $[a, b]$ [namely, $f(a) = f(b)$]. Thus, the secant line connecting the two endpoints is a horizontal line. We see, then, that there is a point in (a, b) with a horizontal tangent line.

Before we prove Rolle's theorem, we check why it is a special case of the MVT. If we compare the assumptions in the two theorems, we find that Rolle's theorem has an additional requirement, $f(a) = f(b)$; that is, the function values must agree at the endpoints of the interval on which f is defined. If we apply the MVT to such a function, it says that there exists a number $c \in (a, b)$ such that

$$f'(c) = \frac{f(b) - f(a)}{b - a}$$

But since $f(b) = f(a)$, it follows that the expression on the right-hand side is equal to 0; that is, $f'(c) = 0$, which is the conclusion of Rolle's theorem.

Proof of Rolle's Theorem If f is the constant function, then $f'(x) = 0$ for all $x \in (a, b)$ and the theorem is true in this particular case. For the more general case, we assume that f is not constant. Since $f(x)$ is continuous on the closed interval $[a, b]$, it follows from the extreme-value theorem that the function has a global maximum and a global minimum in that interval. To see that the function must have a global extremum inside the open interval (a, b), we observe that if f is not constant, then there exists an $x_0 \in (a, b)$ such that either $f(x_0) > f(a) = f(b)$ or $f(x_0) < f(a) = f(b)$. This global extremum is also a local extremum. Suppose that the local extremum is at $c \in (a, b)$; then it follows from Fermat's theorem that $f'(c) = 0$. (In Figure 5.18, the global minima occur at the endpoints of the interval $[a, b]$, but the global maximum occurs in the open interval (a, b), and that's where the horizontal tangent is.) ■

The MVT follows from Rolle's theorem and can be thought of as a "tilted" version of that theorem. (The secant and tangent lines in the MVT are no longer necessarily horizontal [Figure 5.17], as in Rolle's theorem [Figure 5.18], but are "tilted"; they are still parallel, though.)

Proof of the MVT We define the following function:

$$F(x) = f(x) - \frac{f(b) - f(a)}{b - a}(x - a)$$

The function F is continuous on $[a, b]$ and differentiable on (a, b). Furthermore,

$$F(a) = f(a) - \frac{f(b) - f(a)}{b - a}(a - a) = f(a)$$

$$F(b) = f(b) - \frac{f(b) - f(a)}{b - a}(b - a) = f(a)$$

Therefore, $F(a) = F(b)$. We can apply Rolle's theorem to the function $F(x)$: There exists a $c \in (a, b)$ with $F'(c) = 0$. Since

$$F'(x) = f'(x) - \frac{f(b) - f(a)}{b - a}$$

it follows that, for this value of c,

$$0 = F'(c) = f'(c) - \frac{f(b) - f(a)}{b - a}$$

and hence

$$f'(c) = \frac{f(b) - f(a)}{b - a}$$ ■

We next discuss two consequences of the MVT.

Corollary 1 If f is continuous on the closed interval $[a, b]$ and differentiable on the open interval (a, b) such that

$$m \leq f'(x) \leq M \qquad \text{for all } x \in (a, b)$$

then

$$m(b - a) \leq f(b) - f(a) \leq M(b - a)$$

This corollary is useful in obtaining information about a function on the basis of its derivative.

EXAMPLE 8

Population Growth Denote the population size at time t by $N(t)$, and assume that $N(t)$ is continuous on the interval $[0, 10]$ and differentiable on the interval $(0, 10)$ with $N(0) = 100$ and $|dN/dt| \leq 3$ for all $t \in (0, 10)$. What can you say about $N(10)$?

Solution Since $|dN/dt| \leq 3$ implies that $-3 \leq dN/dt \leq 3$, we can set $m = -3$ and $M = 3$ in Corollary 1. With $a = 0$ and $b = 10$, Corollary 1 yields the following estimate:

$$(-3)(10 - 0) \leq N(10) - N(0) \leq (3)(10 - 0)$$

Simplifying and solving for $N(10)$ gives

$$-30 + N(0) \leq N(10) \leq 30 + N(0)$$

Since $N(0) = 100$, we have

$$70 \leq N(10) \leq 130$$

That is, the population size at time $t = 10$ is bounded between 70 and 130. ∎

EXAMPLE 9

Show that

$$|\sin b - \sin a| \leq |b - a|$$

Solution If $a = b$, then, trivially, $|\sin a - \sin a| \leq |a - a|$. We therefore assume that $a < b$. (The case $a > b$ is similar.) Let $f(x) = \sin x$, $a \leq x \leq b$. Then $f(x)$ is continuous on $[a, b]$ and differentiable on (a, b). Since $f'(x) = \cos x$, it follows that

$$-1 \leq f'(x) \leq 1$$

for all $x \in (a, b)$. Applying Corollary 1, with $m = -1$ and $M = 1$, to $f(x) = \sin x$, $a < x < b$, we find that

$$-(b - a) \leq \sin b - \sin a \leq (b - a)$$

which is the same as

$$|\sin b - \sin a| \leq |b - a|$$ ∎

The next corollary is important, and we will see it again in Section 5.8.

Figure 5.19 An illustration of Corollary 2.

Corollary 2 If f is continuous on the closed interval $[a, b]$ and differentiable on the open interval (a, b), with $f'(x) = 0$ for all $x \in (a, b)$, then f is constant on $[a, b]$.

Figure 5.19 explains why Corollary 2 is true: Each point on the graph has a horizontal tangent, so the function must be constant.

EXAMPLE 10 Assume that f is continuous on $[-1, 1]$ and differentiable on $(-1, 1)$, with $f(0) = 2$ and $f'(x) = 0$ for all $x \in (-1, 1)$. Find $f(x)$.

Solution Corollary 2 tells us that $f(x)$ is a constant. Since we know that $f(0) = 2$, we have $f(x) = 2$ for all $x \in [-1, 1]$. ■

Proof of Corollary 2 Let $x_1, x_2 \in (a, b), x_1 < x_2$. Then f satisfies the assumptions of the MVT on the closed interval $[x_1, x_2]$. Therefore, there exists a number $c \in (x_1, x_2)$ such that

$$\frac{f(x_2) - f(x_1)}{x_2 - x_1} = f'(c)$$

Since $f'(c) = 0$, it follows that $f(x_2) = f(x_1)$. Finally, because x_1, x_2 are arbitrary numbers from the interval (a, b), we conclude that f is constant. ■

EXAMPLE 11 Show that

$$\sin^2 x + \cos^2 x = 1 \quad \text{for all } x \in [0, 2\pi]$$

Solution This identity can be shown without calculus, but let's see what we get if we use Corollary 2. We define $f(x) = \sin^2 x + \cos^2 x, 0 \le x \le 2\pi$. Then $f(x)$ is continuous on $[0, 2\pi]$ and differentiable on $(0, 2\pi)$, with

$$f'(x) = 2 \sin x \cos x - 2 \cos x \sin x = 0$$

Using Corollary 2 now, we conclude that $f(x)$ is equal to a constant on $[0, 2\pi]$. To find the constant, we need only evaluate $f(x)$ at one point in the interval, say, $x = 0$. We find that

$$f(0) = \sin^2 0 + \cos^2 0 = 1$$

This proves the identity. ■

Section 5.1 Problems

■ **5.1.1**

In Problems 1–8, each function is continuous and defined on a closed interval. It therefore satisfies the assumptions of the extreme-value theorem. With the help of a graphing calculator, graph each function and locate its global extrema. (Note that a function may assume a global extremum at more than one point.)

1. $f(x) = 2x - 1, 0 \le x \le 1$

2. $f(x) = -x^2 + 1, -1 \le x \le 1$

3. $f(x) = \sin(2x), 0 \le x \le \pi$

4. $f(x) = \cos\frac{x}{2}, 0 \le x \le 2\pi$

5. $f(x) = |x|, -1 \le x \le 1$

6. $f(x) = (x - 1)^2(x + 2), -2 \le x \le 2$

7. $f(x) = e^{-|x|}, -1 \le x \le 1$

8. $f(x) = \ln(x + 1), 0 \le x \le 2$

9. Sketch the graph of a function that is continuous on the closed interval $[0, 3]$ and has a global maximum at the left endpoint and a global minimum at the right endpoint.

10. Sketch the graph of a function that is continuous on the closed interval $[-2, 1]$ and has a global maximum and a global minimum in the interior of the domain of the function.

11. Sketch the graph of a function that is continuous on the open interval $(0, 2)$ and has neither a global maximum nor a global minimum in its domain.

12. Sketch the graph of a function that is continuous on the closed interval $[1, 4]$, except at $x = 2$, and has neither a global maximum nor a global minimum in its domain.

■ **5.1.2**

In Problems 13–18, use a graphing calculator to determine all local and global extrema of the functions on their respective domains.

13. $f(x) = 3 - x, x \in [-1, 3)$

14. $f(x) = 5 + 2x, x \in (-2, 1)$

15. $f(x) = x^2 - 2, x \in [-1, 1]$

16. $f(x) = (x - 2)^2, x \in [0, 3]$

17. $f(x) = -x^2 + 1, x \in [-2, 1]$

18. $f(x) = x^2 - x, x \in [0, 1]$

In Problems 19–26, find c such that $f'(c) = 0$ and determine whether $f(x)$ has a local extremum at $x = c$.

19. $f(x) = x^2$

20. $f(x) = (x - 4)^2$

21. $f(x) = -x^2$

22. $f(x) = -(x + 3)^2$

23. $f(x) = x^3$

24. $f(x) = x^5$

25. $f(x) = (x + 1)^3$

26. $f(x) = -(x - 3)^5$

27. Show that $f(x) = |x|$ has a local minimum at $x = 0$ but $f(x)$ is not differentiable at $x = 0$.

28. Show that $f(x) = |x - 1|$ has a local minimum at $x = 1$ but $f(x)$ is not differentiable at $x = 1$.

29. Show that $f(x) = |x^2 - 1|$ has local minima at $x = 1$ and $x = -1$ but $f(x)$ is not differentiable at $x = 1$ or $x = -1$.

30. Show that $f(x) = -|x^2 - 4|$ has local maxima at $x = 2$ and $x = -2$ but $f(x)$ is not differentiable at $x = 2$ or $x = -2$.

31. Graph
$$f(x) = |1 - |x||, \quad -1 \le x \le 2$$
and determine all local and global extrema on $[-1, 2]$.

32. Graph
$$f(x) = -||x| - 2|, \quad -3 \le x \le 3$$
and determine all local and global extrema on $[-3, 3]$.

33. Suppose the size of a population at time t is $N(t)$ and its growth rate is given by the logistic growth function
$$\frac{dN}{dt} = rN\left(1 - \frac{N}{K}\right), \quad t \ge 0$$
where r and K are positive constants.
(a) Graph the growth rate $\frac{dN}{dt}$ as a function of N for $r = 2$ and $K = 100$, and find the population size for which the growth rate is maximal.
(b) Show that $f(N) = rN(1 - N/K)$, $N \ge 0$, is differentiable for $N > 0$, and compute $f'(N)$.
(c) Show that $f'(N) = 0$ for the value of N that you determined in (a) when $r = 2$ and $K = 100$.

34. Suppose that the size of a population at time t is $N(t)$ and its growth rate is given by the logistic growth function
$$\frac{dN}{dt} = rN\left(1 - \frac{N}{K}\right), \quad t \ge 0$$
where r and K are positive constants. The per capita growth rate is defined by
$$g(N) = \frac{1}{N}\frac{dN}{dt}$$
(a) Show that
$$g(N) = r\left(1 - \frac{N}{K}\right)$$
(b) Graph $g(N)$ as a function of N for $N \ge 0$ when $r = 2$ and $K = 100$, and find the population size for which the per capita growth rate is maximal.

■ **5.1.3**

35. Suppose $f(x) = x^2, x \in [0, 2]$.
(a) Find the slope of the secant line connecting the points $(0, 0)$ and $(2, 4)$.
(b) Find a number $c \in (0, 2)$ such that $f'(c)$ is equal to the slope of the secant line you computed in (a), and explain why such a number must exist in $(0, 2)$.

36. Suppose $f(x) = 1/x, x \in [1, 2]$.
(a) Find the slope of the secant line connecting the points $(1, 1)$ and $(2, 1/2)$.
(b) Find a number $c \in (1, 2)$ such that $f'(c)$ is equal to the slope of the secant line you computed in (a), and explain why such a number must exist in $(1, 2)$.

37. Suppose that $f(x) = x^2, x \in [-1, 1]$.
(a) Find the slope of the secant line connecting the points $(-1, 1)$ and $(1, 1)$.
(b) Find a number $c \in (-1, 1)$ such that $f'(c)$ is equal to the slope of the secant line you computed in (a), and explain why such a number must exist in $(-1, 1)$.

38. Suppose that $f(x) = x^2 - x - 2, x \in [-1, 2]$.
(a) Find the slope of the secant line connecting the points $(-1, 0)$ and $(2, 0)$.

(b) Find a number $c \in (-1, 2)$ such that $f'(c)$ is equal to the slope of the secant line you computed in (a), and explain why such a number must exist in $(-1, 2)$.

39. Let $f(x) = x(1 - x)$. Use the MVT to find an interval that contains a number c such that $f'(c) = 0$.

40. Let $f(x) = 1/(1 + x^2)$. Use the MVT to find an interval that contains a number c such that $f'(c) = 0$.

41. Suppose that $f(x) = -x^2 + 2$. Explain why there exists a point c in the interval $(-1, 2)$ such that $f'(c) = -1$.

42. Suppose that $f(x) = x^3$. Explain why there exists a point c in the interval $(-1, 1)$ such that $f'(c) = 1$.

43. Sketch the graph of a function $f(x)$ that is continuous on the closed interval $[0, 1]$ and differentiable on the open interval $(0, 1)$ such that there exists exactly one point $(c, f(c))$ on the graph at which the slope of the tangent line is equal to the slope of the secant line connecting the points $(0, f(0))$ and $(1, f(1))$. Why can you be sure that there is such a point?

44. Sketch the graph of a function $f(x)$ that is continuous on the closed interval $[0, 1]$ and differentiable on the open interval $(0, 1)$ such that there exist exactly two points $(c_1, f(c_1))$ and $(c_2, f(c_2))$ on the graph at which the slope of the tangent lines is equal to the slope of the secant line connecting the points $(0, f(0))$ and $(1, f(1))$. Why can you be sure that there is at least one such point?

45. Suppose that $f(x) = x^2, x \in [a, b]$.
(a) Compute the slope of the secant line through the points $(a, f(a))$ and $(b, f(b))$.
(b) Find the point $c \in (a, b)$ such that the slope of the tangent line to the graph of f at $(c, f(c))$ is equal to the slope of the secant line determined in (a). How do you know that such a point exists? Show that c is the midpoint of the interval (a, b); that is, show that $c = (a + b)/2$.

46. Assume that f is continuous on $[a, b]$ and differentiable on (a, b). Show that if $f(a) < f(b)$, then f' is positive at some point between a and b.

47. Assume that f is continuous on $[a, b]$ and differentiable on (a, b). Assume further that $f(a) = f(b) = 0$ but f is not constant on $[a, b]$. Explain why there must be a point $c_1 \in (a, b)$ with $f'(c_1) > 0$ and a point $c_2 \in (a, b)$ with $f'(c_2) < 0$.

48. A car moves in a straight line. At time t (measured in seconds), its position (measured in meters) is
$$s(t) = \frac{1}{10}t^2, \ 0 \le t \le 10$$
(a) Find its average velocity between $t = 0$ and $t = 10$.
(b) Find its instantaneous velocity for $t \in (0, 10)$.
(c) At what time is the instantaneous velocity of the car equal to its average velocity?

49. A car moves in a straight line. At time t (measured in seconds), its position (measured in meters) is
$$s(t) = \frac{1}{100}t^3, \ 0 \le t \le 5$$
(a) Find its average velocity between $t = 0$ and $t = 5$.
(b) Find its instantaneous velocity for $t \in (0, 5)$.
(c) At what time is the instantaneous velocity of the car equal to its average velocity?

50. Denote the population size at time t by $N(t)$, and assume that $N(0) = 50$ and $|dN/dt| \le 2$ for all $t \in [0, 5]$. What can you say about $N(5)$?

51. Denote the biomass at time t by $B(t)$, and assume that $B(0) = 3$ and $|dB/dt| \leq 1$ for all $t \in [0, 3]$. What can you say about $B(3)$?

52. Suppose that f is differentiable for all $x \in \mathbf{R}$ and, furthermore, that f satisfies $f(0) = 0$ and $1 \leq f'(x) \leq 2$ for all $x > 0$.

(a) Use Corollary 1 of the MVT to show that

$$x \leq f(x) \leq 2x$$

for all $x \geq 0$.

(b) Use your result in (a) to explain why $f(1)$ cannot be equal to 3.

(c) Find an upper and a lower bound for the value of $f(1)$.

53. Suppose that f is differentiable for all $x \in \mathbf{R}$ with $f(2) = 3$ and $f'(x) = 0$ for all $x \in \mathbf{R}$. Find $f(x)$.

54. Suppose that $f(x) = e^{-|x|}$, $x \in [-2, 2]$.

(a) Show that $f(-2) = f(2)$.

(b) Compute $f'(x)$, where defined.

(c) Show that there is no number $c \in (-2, 2)$ such that $f'(c) = 0$.

(d) Explain why your results in (a) and (c) do not contradict Rolle's theorem.

(e) Use a graphing calculator to sketch the graph of $f(x)$.

55. Use Corollary 2 of the MVT to show that if $f(x)$ is differentiable for all $x \in \mathbf{R}$ and satisfies

$$|f(x) - f(y)| \leq |x - y|^2 \tag{5.3}$$

for all $x, y \in \mathbf{R}$, then $f(x)$ is constant. [*Hint:* Show that (5.3) implies that

$$\lim_{x \to y} \frac{f(x) - f(y)}{x - y} = 0 \tag{5.4}$$

and use the definition of the derivative to interpret the left-hand side of (5.4).]

56. We have seen that

$$f(x) = f_0 e^{rx}$$

satisfies the differential equation

$$\frac{df}{dx} = rf(x)$$

with $f(0) = f_0$. This exercise will show that $f(x)$ is in fact the only solution. Suppose that r is a constant and f is a differentiable function,

$$\frac{df}{dx} = rf(x) \tag{5.5}$$

for all $x \in \mathbf{R}$, and $f(0) = f_0$. The following steps will show that $f(x) = f_0 e^{rx}$, $x \in \mathbf{R}$, is the only solution of (5.5).

(a) Define the function

$$F(x) = f(x)e^{-rx}, \quad x \in \mathbf{R}$$

Use the product rule to show that

$$F'(x) = e^{-rx}[f'(x) - rf(x)]$$

(b) Use (a) and (5.5) to show that $F'(x) = 0$ for all $x \in \mathbf{R}$.

(c) Use Corollary 2 to show that $F(x)$ is a constant and, hence, $F(x) = F(0) = f_0$.

(d) Show that (c) implies that

$$f_0 = f(x)e^{-rx}$$

and therefore,

$$f(x) = f_0 e^{rx}$$

■ 5.2 Monotonicity and Concavity

Fish are indeterminate growers; they increase in body size throughout their life. However, as they become older, they grow proportionately more slowly. Their growth is often described mathematically by the von Bertalanffy equation, which fits a large number of both freshwater and marine fishes. This equation is given by

$$L(x) = L_\infty - (L_\infty - L_0)e^{-Kx}$$

where $L(x)$ denotes the length of the fish at age x, L_0 the length at age 0, and L_∞ the asymptotic maximum attainable length. We assume that $L_\infty > L_0$. K is related to how quickly the fish grows. Figure 5.20 shows examples for two different values of K; L_∞ and L_0 are the same in both cases. We see from the graphs that for larger K, the asymptotic length L_∞ is approached more quickly.

The fact that fish increase their body size throughout their life can be expressed mathematically by the first derivative of the function $L(x)$. Looking at the graph, we see that $L(x)$ is an increasing function of x: The tangent line at any point of the graph has a positive slope, or, equivalently, $L'(x) > 0$. We can compute

$$L'(x) = K(L_\infty - L_0)e^{-Kx}$$

Since $L_\infty > L_0$ (by assumption) and $e^{-Kx} > 0$ (this holds for all x, regardless of K), we see that, indeed, $L'(x) > 0$. The graph of $L'(x)$ is shown in Figure 5.21.

The graph of $L'(x)$ shows that $L'(x)$ is a decreasing function of x: Although fish increase their body size throughout their life, they do so at a rate that decreases with age. Mathematically, this relationship can be expressed with the second derivative of

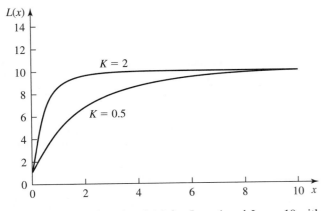

Figure 5.20 The function $L(x)$ for $L_0 = 1$ and $L_\infty = 10$ with $K = 0.5$ and $K = 2$.

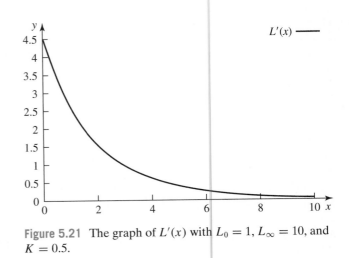

Figure 5.21 The graph of $L'(x)$ with $L_0 = 1$, $L_\infty = 10$, and $K = 0.5$.

$L(x)$—the derivative of the first derivative. The tangent line at any point on the graph of $L'(x)$ has a negative slope; that is, the derivative of $L'(x)$ is negative: $L''(x) < 0$. The fact that the rate of growth decreases with age can also be seen directly from the graph of $L(x)$: It bends downward. The second derivative thus tells us something about which way the graph of $L(x)$ bends.

This section discusses the important concepts of monotonicity—whether a function is decreasing or increasing—and concavity—whether a function bends upward or downward.

■ 5.2.1 Monotonicity

We saw in the motivating example that the first derivative tells us something about whether a function increases or decreases. Not every function is differentiable, however, so we phrase the definitions of increasing and decreasing in terms of the function f alone. (See Figures 5.22 and 5.23.)

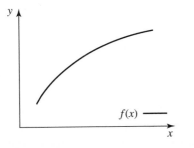

Figure 5.22 An increasing function.

> **Definition** A function f defined on an interval I is called **(strictly) increasing** on I if
> $$f(x_1) < f(x_2) \quad \text{whenever } x_1 < x_2 \text{ in } I$$
> and is called **(strictly) decreasing** on I if
> $$f(x_1) > f(x_2) \quad \text{whenever } x_1 < x_2 \text{ in } I$$

An increasing or decreasing function is called **monotonic**. The word *strictly* in the preceding definition refers to having a strict inequality ($f(x_1) < f(x_2)$ and $f(x_1) > f(x_2)$). We will frequently drop *strictly*. If, instead of the strict inequality $f(x_1) < f(x_2)$, we have the inequality $f(x_1) \le f(x_2)$, whenever $x_1 < x_2$ in I, we call f *nondecreasing*. If $f(x_1) \ge f(x_2)$ whenever $x_1 < x_2$ in I, then f is called *nonincreasing*. (See Figures 5.24 and 5.25.)

When the function f is differentiable, there is a useful test to determine whether f is increasing or decreasing. This criterion is a consequence of the MVT.

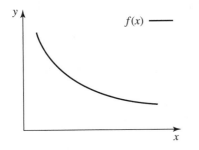

Figure 5.23 A decreasing function.

> **First-Derivative Test for Monotonicity** Suppose f is continuous on $[a, b]$ and differentiable on (a, b).
>
> **(a)** If $f'(x) > 0$ for all $x \in (a, b)$, then f is increasing on $[a, b]$.
> **(b)** If $f'(x) < 0$ for all $x \in (a, b)$, then f is decreasing on $[a, b]$.

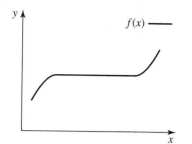

Figure 5.24 A nondecreasing function may have regions where the function is constant.

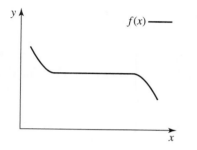

Figure 5.25 A nonincreasing function may have regions where the function is constant.

Proof (See Figure 5.26.) We choose two numbers x_1 and x_2 in $[a, b]$, $x_1 < x_2$. Then f is continuous on $[x_1, x_2]$ and differentiable on (x_1, x_2). We can therefore apply the MVT to f defined on $[x_1, x_2]$: There exists a number $c \in (x_1, x_2)$ such that

$$\frac{f(x_2) - f(x_1)}{x_2 - x_1} = f'(c)$$

In part (a) of the theorem, we assume that $f'(x) > 0$ for all $x \in (a, b)$. Since $c \in (x_1, x_2) \subset (a, b)$, it follows that $f'(c) > 0$. Since, in addition, $x_2 > x_1$, it follows that

$$f(x_2) - f(x_1) = f'(c)(x_2 - x_1) > 0$$

which implies that $f(x_2) > f(x_1)$. Because x_1 and x_2 are arbitrary numbers in $[a, b]$ satisfying $x_1 < x_2$, it follows that f is increasing. The proof of part (b) is similar and relegated to Problem 24. ■

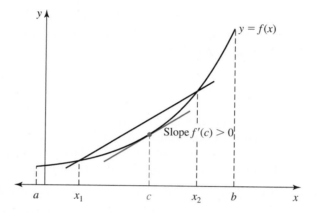

Figure 5.26 An illustration of the proof of "$f'(x) > 0$ for all $x \in (a, b)$ implies that $f(x)$ is increasing on $[a, b]$."

EXAMPLE 1

Determine where the function

$$f(x) = x^3 - \frac{3}{2}x^2 - 6x + 3, \quad x \in \mathbf{R}$$

is increasing and where it is decreasing.

Solution Since $f(x)$ is continuous and differentiable for all $x \in \mathbf{R}$, we can use the first-derivative test for monotonic functions. We differentiate $f(x)$ and obtain

$$f'(x) = 3x^2 - 3x - 6 = 3(x - 2)(x + 1), \quad x \in \mathbf{R}$$

The graphs of $f(x)$ and $f'(x)$ are shown in Figure 5.27. The graph of $f'(x)$ is a parabola that intersects the x-axis at $x = 2$ and $x = -1$. The function $f'(x)$ therefore changes sign at $x = -1$ and $x = 2$. We find that

$$f'(x) \begin{cases} > 0 & \text{if } x < -1 \text{ or } x > 2 \\ < 0 & \text{if } -1 < x < 2 \end{cases}$$

Thus, $f(x)$ is increasing for $x < -1$ or $x > 2$ and decreasing for $-1 < x < 2$. A look at the graph of $f(x)$ in Figure 5.27 confirms this conclusion. ■

EXAMPLE 2

Host–Parasitoid Interactions Parasitoids are insects whose larvae develop inside other, host insects. The larvae eventually kill the host. An example is the parasitoid *Macrocentrus grandii*, a wasp, which parasitizes *Ostrinia nubilis*, the European corn

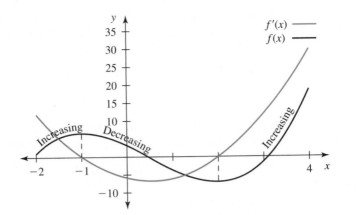

Figure 5.27 The graph of $f(x) = x^3 - \frac{3}{2}x^2 - 6x + 3$ and $f'(x) = 3x^2 - 3x - 6$.

borer. To understand host–parasitoid interactions, a large number of models have been developed. The function

$$f(N) = \left(1 + \frac{a\beta P}{k(\beta + aN)}\right)^{-k}$$

is one example that describes the likelihood of a host escaping parasitism as a function of host density N, where a, β, and k are positive parameters and P is the density of the parasitoid. On the basis of $f(N)$, what is the effect of an increase in host density on the likelihood of escaping parasitism?

Solution   To find the effect of an increase in host density, we compute the derivative of $f(N)$:

$$\frac{df}{dN} = (-k)\left[1 + \frac{a\beta P}{k(\beta + aN)}\right]^{-k-1} \frac{-a^2\beta P}{k(\beta + aN)^2}$$

$$= \left[1 + \frac{a\beta P}{k(\beta + aN)}\right]^{-k-1} \frac{a^2\beta P}{(\beta + aN)^2}$$

Both factors in the final expression are positive; hence, $df/dN > 0$. In words, if the density of the parasitoid is fixed and the host density increases, a host is more likely to escape parasitism in cases where the interaction is described by $f(N)$.   ■

■ 5.2.2 Concavity

We saw in the motivating example at the beginning of this section that the second derivative tells us something about whether a function bends upward or downward. We arrived at this conclusion by checking whether the first derivative was increasing or decreasing.

A function is called **concave up** if it bends upward, and **concave down** if it bends downward. Before stating a precise definition of concavity for differentiable functions, we give two examples, in Figure 5.28.

First, look at the graph of the differentiable function $y = x^2$: It bends upward, so we call it concave up. Bending upward means that the slopes of the tangent lines are increasing as x increases. We can check this hypothesis by computing the slope of the tangent line at x, which is given by the first derivative, $y' = 2x$. Since $y' = 2x$ is an increasing function, the slopes of the tangent lines are increasing as x increases.

Looking at the graph of the differentiable function $y = -x^2 + 4$, we see that it bends downward, so we call it concave down. Bending downward means that the slopes of the tangent lines are decreasing as x increases. We can check this hypothesis by computing the first derivative of y, which is $y' = -2x$, a decreasing function.

The following definition pertaining to differentiable functions is based on the preceding discussion (see Figures 5.29 and 5.30):

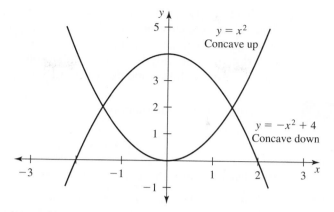

Figure 5.28 The graphs of $f(x) = x^2$ and $g(x) = -x^2 + 4$.

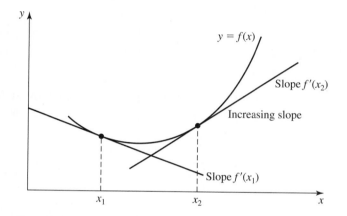

Figure 5.29 A function is concave up if its derivative is increasing.

Figure 5.30 A function is concave down if its derivative is decreasing.

> **Definition** A differentiable function $f(x)$ is **concave up** on an interval I if the first derivative $f'(x)$ is an increasing function on I. $f(x)$ is **concave down** on an interval I if the first derivative $f'(x)$ is a decreasing function on I.

Note that the definition assumes that $f(x)$ is differentiable. There is a more general definition that does not require differentiability. (After all, not all functions are differentiable.) The more general definition is more difficult to use, however. The definition given here suffices for our purposes and has the added advantage that it provides the following criterion, which we can use to determine whether a twice-differentiable function is concave up or concave down:

> **Second-Derivative Test for Concavity** Suppose that f is twice differentiable on an open interval I.
>
> **(a)** If $f''(x) > 0$ for all $x \in I$, then f is concave up on I.
> **(b)** If $f''(x) < 0$ for all $x \in I$, then f is concave down on I.

Proof Since f is twice differentiable, we can apply the first-derivative criterion to the function $f'(x)$. The proof of part (a) proceeds, then, as follows: If $f''(x) > 0$ on I, then $f'(x)$ is an increasing function on I. From the definition of *concave up*, it follows that f is concave up on I. The proof of part (b) is similar and relegated to Problem 25. ■

You can use the function $y = x^2$ to remember which functions are concave up: The "u" in "concave up" should remind you of the U-shaped form of the graph of $y = x^2$. You can also use the function $y = x^2$ to remember the second-derivative criterion. You already know that the graph of $y = x^2$ is concave up, and you can easily compute the second derivative of $y = x^2$, namely, $y'' = 2 > 0$.

EXAMPLE 3 Determine where the function

$$f(x) = x^3 - \frac{3}{2}x^2 - 6x + 3, \quad x \in \mathbf{R}$$

is concave up and where it is concave down.

Solution This is the same function as in Example 1 (redrawn in Figure 5.31). Since $f(x)$ is a polynomial, it is twice differentiable. In Example 1, we found that $f'(x) = 3x^2 - 3x - 6$; differentiating $f'(x)$, we get the second derivative of f:

$$f''(x) = 6x - 3$$

We find that

$$f''(x) \begin{cases} > 0 & \text{if } x > \frac{1}{2} \\ < 0 & \text{if } x < \frac{1}{2} \end{cases}$$

Thus, $f(x)$ is concave up for $x > 1/2$ and concave down for $x < 1/2$. A look at Figure 5.31 confirms this result. ■

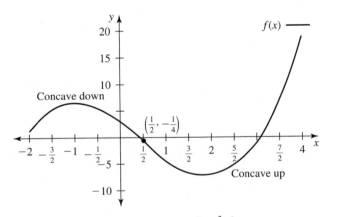

Figure 5.31 The graph of $f(x) = x^3 - \frac{3}{2}x^2 - 6x + 3$.

A very common mistake is to associate monotonicity and concavity. One has nothing to do with the other. For instance, an increasing function can bend downward or upward. (This possibility will be discussed in Problem 21.)

There are many biological examples of increasing functions that have a decreasing derivative and are therefore concave down.

EXAMPLE 4 **Crop Yield** The response of crop yield Y to soil nitrogen level N can often be described by a function of the form

$$Y(N) = Y_{\max}\frac{N}{K + N}, \quad N \geq 0$$

where $Y_{\max}$ is the maximum attainable yield and K is a positive constant. The graph of $Y(N)$ is shown in Figure 5.32. We see from the graph that $Y(N)$ is an increasing function of N. The graph bends downward and hence is concave down. Before continuing, we will check this conclusion against the results we obtained in this section. Using the quotient rule, we obtain

$$Y'(N) = Y_{\max}\frac{K + N - N}{(K + N)^2} = Y_{\max}\frac{K}{(K + N)^2}$$

Figure 5.32 The graph of $Y(N)$ in Example 4.

and using the chain rule, we get

$$Y''(N) = \frac{d}{dN}\left(Y_{max}K(K+N)^{-2}\right)$$

$$= Y_{max}K(-2)(K+N)^{-3}(1) = -Y_{max}\frac{2K}{(K+N)^3}$$

Since Y_{max} and K are positive constants and $N \geq 0$, it follows that

$$Y'(N) > 0$$

which implies that $Y(N)$ is an increasing function. Furthermore,

$$Y''(N) < 0$$

which implies that $Y(N)$ is concave down. That is, $Y(N)$ is an increasing function, but the rate of increase is decreasing. We say that Y is *increasing at a decelerating rate*. What does this mean? It means that as we increase fertilizer levels, the yield will increase, but at a proportionally lesser rate. This type of curve is called a *diminishing return*. To be concrete, we choose values for Y_{max} and K:

$$Y(N) = 50\frac{N}{5+N}, \quad N \geq 0$$

The graph of this function is shown in Figure 5.33.

Suppose that initially $N = 5$. If we increase N by 5 (i.e., from 5 to 10), then $Y(N)$ changes from $Y(5) = 25$ to $Y(10) = 33.3$, an increase of 8.3. If we increase N by double the original amount, namely 10 (i.e., from 5 to 15), then $Y(15) = 37.5$ and the increase in yield is only 12.5, less than twice 8.3. Diminishing return can also be understood by comparing successive increments. If we start with $N = 5$ and increase by 5 to $N = 10$, then the change in Y (the Y-increment) is 8.3. Increasing N by the same amount, but starting at 10, we see that the Y-increment changes by $Y(15) - Y(10) = 4.2$. In general, changing N by equal increments has less of an effect for larger values of N; thus, we say that the return is diminishing.

You should compare a function representing a diminishing return with a linear function, say, $f(x) = 2x$, which is neither concave up nor concave down. (See Figure 5.34.) With a linear function, if we increase x from 5 to 10, $f(x)$ changes from 10 to 20. That is, $f(x)$ increases by 10. Then, if we increase x from 10 to 15, $f(x)$ changes from 20 to 30, again an increase of 10. That is, for linear functions, the increase is proportional. ■

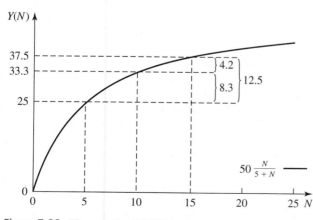

Figure 5.33 The graph of $Y(N)$ in Example 4 for $Y_{max} = 50$ and $K = 5$.

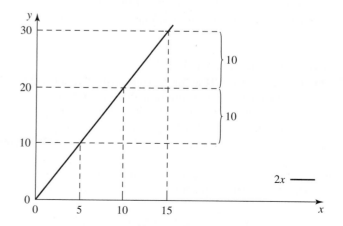

Figure 5.34 The graph of a linear function: Increases are proportional.

Section 5.2 Problems

■ 5.2.1 and 5.2.2

In Problems 1–20, determine where each function is increasing, decreasing, concave up, and concave down. With the help of a graphing calculator, sketch the graph of each function and label the intervals where it is increasing, decreasing, concave up, and concave down. Make sure that your graphs and your calculations agree.

1. $y = 3x - x^2$, $x \in \mathbf{R}$
2. $y = x^2 + 5x$, $x \in \mathbf{R}$
3. $y = x^2 + x - 4$, $x \in \mathbf{R}$
4. $y = x^2 - x + 3$, $x \in \mathbf{R}$
5. $y = -\frac{2}{3}x^3 + \frac{7}{2}x^2 - 3x + 4$, $x \in \mathbf{R}$
6. $y = (x - 2)^3 + 3$, $x \in \mathbf{R}$
7. $y = \sqrt{x + 1}$, $x \geq -1$
8. $y = (3x - 1)^{1/3}$, $x \in \mathbf{R}$
9. $y = \frac{1}{x}$, $x \neq 0$
10. $y = \frac{-2}{x^2 + 3}$
11. $(x^2 + 1)^{1/3}$, $x \in \mathbf{R}$
12. $y = \frac{5}{x - 2}$, $x \neq 2$
13. $y = \frac{1}{(1 + x)^2}$, $x \neq -1$
14. $y = \frac{x^2}{x^2 + 1}$, $x \geq 0$
15. $y = \sin x$, $0 \leq x \leq 2\pi$
16. $y = \cos[\pi(x^2 - 1)]$, $2 \leq x \leq 3$
17. $y = e^x$, $x \in \mathbf{R}$
18. $y = \ln x$, $x > 0$
19. $y = e^{-x^2/2}$, $x \in \mathbf{R}$
20. $y = \frac{1}{1 + e^{-x}}$, $x \in \mathbf{R}$

21. Sketch the graph of
(a) a function that is increasing at an accelerating rate; and
(b) a function that is increasing at a decelerating rate.
(c) Assume that your functions in (a) and (b) are twice differentiable. Explain in each case how you could check the respective properties by using the first and the second derivatives. Which of the functions is concave up, and which is concave down?

22. Show that if $f(x)$ is the linear function $y = mx + b$, then increases in $f(x)$ are proportional to increases in x. That is, if we increase x by Δx, then $f(x)$ increases by the same amount Δy, regardless of the value of x. Compute Δy as a function of Δx.

23. We frequently must solve equations of the form $f(x) = 0$. When f is a continuous function on $[a, b]$ and $f(a)$ and $f(b)$ have opposite signs, the intermediate-value theorem guarantees that there exists at least one solution of the equation $f(x) = 0$ in $[a, b]$.
(a) Explain in words why there exists exactly one solution in (a, b) if, in addition, f is differentiable in (a, b) and $f'(x)$ is either strictly positive or strictly negative throughout (a, b).
(b) Use the result in (a) to show that

$$x^3 - 4x + 1 = 0$$

has exactly one solution in $[-1, 1]$.

24. **First-Derivative Test for Monotonicity** Suppose that f is continuous on $[a, b]$ and differentiable on (a, b). Show that if $f'(x) < 0$ for all $x \in (a, b)$, then f is decreasing on $[a, b]$.

25. **Second-Derivative Test for Concavity** Suppose that f is twice differentiable on an open interval I. Show that if $f''(x) < 0$, then f is concave down.

26. Suppose the size of a population at time t is $N(t)$, and the growth rate of the population is given by the logistic growth function

$$\frac{dN}{dt} = rN\left(1 - \frac{N}{K}\right), \quad t \geq 0$$

where r and K are positive constants.

(a) Graph the growth rate $\frac{dN}{dt}$ as a function of N for $r = 3$ and $K = 10$.
(b) The function $f(N) = rN(1 - N/K)$, $N \geq 0$, is differentiable for $N > 0$. Compute $f'(N)$, and determine where the function $f(N)$ is increasing and where it is decreasing.

27. **Logistic Growth** Suppose that the size of a population at time t is $N(t)$ and the growth rate of the population is given by the logistic growth function

$$\frac{dN}{dt} = rN\left(1 - \frac{N}{K}\right), \quad t \geq 0$$

where r and K are positive constants. The per capita growth rate is defined by

$$g(N) = \frac{1}{N}\frac{dN}{dt} = r\left(1 - \frac{N}{K}\right)$$

(a) Graph $g(N)$ as a function of N for $N \geq 0$ when $r = 3$ and $K = 10$.
(b) The function $g(N) = r(1 - N/K)$, $N \geq 0$, is differentiable for $N > 0$. Compute $g'(N)$, and determine where the function $g(N)$ is increasing and where it is decreasing.

28. **Resource-Dependent Growth** The growth rate of a plant depends on the amount of resources available. A simple and frequently used model for resource-dependent growth is the Monod model, according to which the growth rate is equal to

$$f(R) = \frac{aR}{k + R}, \quad R \geq 0$$

where R denotes the resource level and a and k are positive constants. When is the growth rate increasing? When is it decreasing?

29. **Population Growth** Suppose that the growth rate of a population is given by

$$f(N) = N\left(1 - \left(\frac{N}{K}\right)^\theta\right)$$

where N is the size of the population, K is a positive constant denoting the carrying capacity, and θ is a parameter greater than 1. Find $f'(N)$, and determine where the growth rate is increasing and where it is decreasing.

30. **Predation** Spruce budworms are a major pest that defoliates balsam fir. They are preyed upon by birds. A model for the per capita predation rate is given by

$$f(N) = \frac{aN}{k^2 + N^2}$$

where N denotes the density of spruce budworms and a and k are positive constants. Find $f'(N)$, and determine where the predation rate is increasing and where it is decreasing.

31. **Host–Parasitoid Interactions** Parasitoids are insects that lay their eggs in, on, or close to other (host) insects. Parasitoid larvae then devour the host insect. The likelihood of escaping parasitism may depend on parasitoid density. One model expressing this dependence sets the probability of escaping parasitism equal to

$$f(P) = e^{-aP}$$

where P is the parasitoid density and a is a positive constant. Determine whether the probability of escaping parasitism increases or decreases with parasitoid density.

32. Host–Parasitoid Interactions As an alternative to the model set forth in Problem 31, another model sets the probability of escaping parasitism equal to

$$f(P) = \left(1 + \frac{aP}{k}\right)^{-k}$$

where P is the parasitoid density and a and k are positive constants. Determine whether the probability of escaping parasitism increases or decreases with parasitoid density.

33. Tree Growth Suppose that the height y in feet of a tree as a function of the age x in years of the tree is given by

$$y = 117e^{-10/x}, \quad x > 0$$

(a) Show that the height of the tree increases with age. What is the maximum attainable height?

(b) Where is the graph of height versus age concave up, and where is it concave down?

(c) Use a graphing calculator to sketch the graph of height versus age.

(d) Use a graphing calculator to verify that the rate of growth is greatest at the point where the graph in (c) changes concavity.

34. Reproduction Plants employ two basic reproductive strategies: *polycarpy*, in which reproduction occurs repeatedly during the lifetime of the organism, and *monocarpy*, in which reproduction occurs only once during the lifetime of the organism. (Bamboo, for instance, is a monocarpic plant.) The following quote is taken from Iwasa et al. (1995):

> The optimal strategy is polycarpy (repeated reproduction) if reproductive success increases with the investment at a decreasing rate, [or] monocarpy ("big bang" reproduction) or intermittent reproduction if the reproductive success increases at an increasing rate.

(a) Sketch the graph of reproductive success as a function of reproductive investment for the cases of **(i)** polycarpy and **(ii)** monocarpy.

(b) Given that the second derivative describes whether a curve bends upward or downward, explain the preceding quote in terms of the second derivative of the reproductive success function.

35. Pollinator Visits Assume that the formula (Iwasa et al., 1995)

$$X(F) = cF^\gamma$$

where c is a positive constant, expresses the relationship between the number of flowers on a plant, F, and the average number of pollinator visits, $X(F)$. Find the range of values for the parameter γ such that the average number of pollinator visits to a plant increases with the number of flowers F but the rate of increase decreases with F. Explain your answer in terms of appropriate derivatives of the function $X(F)$.

36. Pollinator Visits Assume that the dependence of the average number of pollinator visits to a plant, X, on the number of flowers, F, is given by

$$X(F) = cF^\gamma$$

where γ is a positive constant less than 1 and c is a positive constant (Iwasa et al., 1995). How does the average number of pollen grains exported per flower, $E(F)$, change with the number of flowers on the plant, F, if $E(F)$ is proportional to

$$1 - \exp\left[-k\frac{X(F)}{F}\right]$$

where k is a positive constant?

37. Population Size Denote the size of a population by $N(t)$, and assume that $N(t)$ satisfies

$$\frac{dN}{dt} = Ne^{-aN} - N^2$$

where a is a positive constant.
(a) Show that the nontrivial equilibrium N^* satisfies

$$e^{-aN^*} = N^*$$

(b) Assume now that the nontrivial equilibrium N^* is a function of the parameter a. Use implicit differentiation to show that N^* is a decreasing function of a.

38. Population Size Denote the size of a population by $N(t)$, and assume that $N(t)$ satisfies

$$\frac{dN}{dt} = N\left(1 - \frac{N}{K}\right) - N \ln N$$

where K is a positive constant.
(a) Show that if $K > 1$, then there exists a nontrivial equilibrium $N^* > 0$ that satisfies

$$1 - \frac{N^*}{K} = \ln N^*$$

(b) Assume now that the nontrivial equilibrium N^* is a function of the parameter K. Use implicit differentiation to show that N^* is an increasing function of K.

39. Intraspecific Competition *(Adapted from Bellows, 1981)* Suppose that a study plot contains N annual plants, each of which produces S seeds that are sown within the same plot. The number of surviving plants in the next year is given by

$$A(N) = \frac{NS}{1 + (aN)^b} \tag{5.6}$$

for some positive constants a and b. This mathematical model incorporates density-dependent mortality: The greater the number of plants in the plot, the lower is the number of surviving offspring per plant, which is given by $A(N)/N$ and is called the *net reproductive rate*.

(a) Use calculus to show that $A(N)/N$ is a decreasing function of N.

(b) The following quantity, called the *k-value*, can be used to quantify the effects of intraspecific competition (i.e., competition between individuals of the same species):

$$k = \log[\text{initial density}] - \log[\text{final density}]$$

Here, "log" denotes the logarithm to base 10. The initial density is the product of the number of plants (N) and the number of seeds each plant produces (S). The final density is given by (5.6). Use the expression for k and (5.6) to show that

$$k = \log[NS] - \log\left[\frac{NS}{1 + (aN)^b}\right]$$
$$= \log\left[1 + (aN)^b\right]$$

We typically plot k versus $\log N$; the slope of the resulting curve is then used to quantify the effects of competition.
(i) Show that

$$\frac{d \log N}{dN} = \frac{1}{N \ln 10}$$

where ln denotes the natural logarithm.

(ii) Show that

$$\frac{dk}{d \log N} = (\ln 10) N \frac{dk}{dN} = \frac{b}{1 + (aN)^{-b}}$$

(iii) Find

$$\lim_{N \to \infty} \frac{dk}{d \log N}$$

(iv) Show that if

$$\frac{dk}{d \log N} < 1$$

then $A(N)$ is increasing, whereas if

$$\frac{dk}{d \log N} > 1$$

then $A(N)$ is decreasing. [*Hint*: Compute $A'(N)$.] Explain in words what the two inequalities mean with respect to varying the initial density of seeds and observing the number of surviving plants the next year. (*Hint*: The first case is called *undercompensation* and the second case is called *overcompensation*.)

(v) The case

$$\frac{dk}{d \log N} = 1$$

is referred to as *exact compensation*. Suppose that you plot k versus $\log N$ and observe that, over a certain range of values of N, the slope of the resulting curve is equal to 1. Explain what this means.

40. *(Adapted from Reiss, 1989)* Suppose that the rate at which body weight W changes with age x is

$$\frac{dW}{dx} \propto W^a \tag{5.7}$$

where a is some species-specific positive constant.

(a) The relative growth rate (percentage weight gained per unit of time) is defined as

$$\frac{1}{W} \frac{dW}{dx}$$

What is the relationship between the relative growth rate and body weight? For which values of a is the relative growth rate increasing, and for which values is it decreasing?

(b) As fish grow larger, their weight increases each day but the relative growth rate decreases. If the rate of growth is described by (5.7), what values of a can you exclude on the basis of your results in (a)? Explain how the increase in percentage weight (relative to the current body weight) differs for juvenile fish and for adult fish.

41. Allometric Growth Allometric equations describe the scaling relationship between two measurements, such as tree height versus tree diameter or skull length versus backbone length. These equations are often of the form

$$Y = bX^a \tag{5.8}$$

where b is some positive constant and a is a constant that can be positive, negative, or zero.

(a) Assume that X and Y are body measurements (and therefore positive) and that their relationship is described by an allometric equation of the form (5.8). For what values of a is Y an increasing function of X, but one such that the ratio Y/X decreases with increasing X? Is Y concave up or concave down in this case?

(b) In vertebrates, we typically find

$$[\text{skull length}] \propto [\text{body length}]^a$$

for some $a \in (0, 1)$. Use your answer in (a) to explain what this means for skull length versus body length in juveniles versus adults; that is, at which developmental stage do vertebrates have larger skulls relative to their body length?

42. pH The pH value of a solution measures the concentration of hydrogen ions, denoted by $[H^+]$, and is defined as

$$pH = -\log[H^+]$$

Use calculus to decide whether the pH value of a solution increases or decreases as the concentration of H^+ increases.

43. Allometric Growth The differential equation

$$\frac{dy}{dx} = k \frac{y}{x}$$

describes allometric growth, where k is a positive constant. Assume that x and y are both positive variables and that $y = f(x)$ is twice differentiable. Use implicit differentiation to determine for which values of k the function $y = f(x)$ is concave up.

44. Population Size Let $N(t)$ denote the population size at time t, and assume that $N(t)$ is twice differentiable and satisfies the differential equation

$$\frac{dN}{dt} = rN$$

where r is a real number. Differentiate the differential equation with respect to t, and state whether $N(t)$ is concave up or down.

■ 5.3 Extrema, Inflection Points, and Graphing

■ 5.3.1 Extrema

If f is a continuous function on the closed interval $[a, b]$, then f has a global maximum and a global minimum in $[a, b]$. This is the content of the extreme-value theorem, which is an existence result: It tells us only that global extrema exist under certain conditions, but it does not tell us how to find them.

Our strategy for finding global extrema in the case where f is a continuous function defined on a closed interval will be, first, to identify all local extrema of the function and, then, to select the global extrema from the set of local extrema. If f is a continuous function defined on an open interval or half-open interval, the existence of global extrema is no longer guaranteed, and we must compare the local extrema with the behavior of the function near the open boundaries of the domain. (See Example 5 in Section 5.1.) In particular, if $f(x)$ is defined on **R**, we need to investigate the behavior of $f(x)$ as $x \to \pm\infty$. For if the function $f(x)$ goes to $+\infty$ (or $-\infty$) as $x \to +\infty$ or $-\infty$, it cannot have a global maximum (global minimum). We discuss this in Example 1 of this section.

Local extrema can be found in a systematic way, using a straightforward recipe to identify candidates. We showed in Section 5.1 that if f has a local extremum at an interior point c and $f'(c)$ exists, then $f'(c) = 0$ (Fermat's theorem). That is, points where f is differentiable and where the first derivative is equal to 0 are certainly candidates for local extrema in the interior of the domain. Of course, these are only candidates, as explained in Section 5.1. (Recall that $y = x^3$ has a horizontal tangent at $x = 0$, but $y = x^3$ does not have a local extremum at $x = 0$.) In addition to points where the first derivative is equal to 0, we must check all points where the function is not differentiable. (For instance, $y = |x|$ has a local minimum at $x = 0$, although it is not differentiable at 0.) Points where the first derivative is equal to 0 or does not exist are called **critical points**. In addition to checking the critical points, we must always check the endpoints of the interval on which f is defined (provided that there are such endpoints).

There are no other points where local extrema can occur. We are thus equipped with a systematic way of searching for *candidates* for local extrema:

---

1. Find all numbers c where $f'(c) = 0$.
2. Find all numbers c where $f'(c)$ does not exist.
3. Find the endpoints of the domain of f.

We illustrate this procedure in the following example: We wish to find all local and global extrema of the function

$$f(x) = |x^2 - 4|, \quad -3 \leq x < 2.5$$

We know from Example 5 of Subsection 5.1.2 what the graph of the function looks like (the domain is different here). We plot it again (Figure 5.35), which will make it easier to understand the procedure for finding relative extrema. But note that very often we do not know what a graph looks like, and we find relative extrema in order to gain a better understanding of the graph!

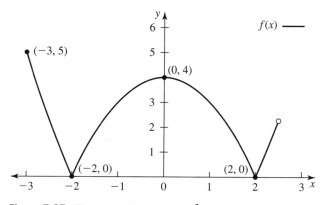

Figure 5.35 The graph of $f(x) = |x^2 - 4|$ for $-3 \leq x < 2.5$.

We will first rewrite $f(x)$ as a piecewise-defined function in order to get rid of the absolute-value sign:

$$f(x) = \begin{cases} x^2 - 4 & \text{for } -3 \leq x \leq -2 \text{ or } 2 \leq x < 2.5 \\ -x^2 + 4 & \text{for } -2 \leq x \leq 2 \end{cases}$$

This piecewise-defined function is differentiable on the open intervals $(-3, -2)$,

$(-2, 2)$, and $(2, 2.5)$. We find that

$$f'(x) = \begin{cases} 2x & \text{for } -3 < x < -2 \text{ or } 2 < x < 2.5 \\ -2x & \text{for } -2 < x < 2 \end{cases}$$

Since $f'(x) = 0$ for $x = 0$ and $0 \in (-2, 2)$, it follows that $(0, f(0))$ is a critical point and is our first candidate for a local extremum. There are no other points where $f'(x) = 0$.

The second step is to identify interior points where the function is not differentiable. Since the function is differentiable on the open intervals $(-3, -2)$, $(-2, 2)$, and $(2, 2.5)$, we must look at the points where the function is pieced together, namely at $x = -2$ and at $x = 2$. We obtain

$$\lim_{x \to -2^-} f'(x) = -4 \qquad \text{and} \qquad \lim_{x \to -2^+} f'(x) = 4$$

and

$$\lim_{x \to 2^-} f'(x) = -4 \qquad \text{and} \qquad \lim_{x \to 2^+} f'(x) = 4$$

These limits show that the function is not differentiable at $x = -2$ and $x = 2$. Therefore, there are critical points at $x = -2$ and $x = 2$, and those points are also candidates for local extrema.

The third step is to identify endpoints of the domain. Since f is defined on $[-3, 2.5)$, there is an endpoint at $x = -3$. The fourth candidate is thus at $x = -3$. The interval $[-3, 2.5)$ is open at $x = 2.5$; hence, 2.5 is not in the domain of the function. The point $(2.5, f(2.5))$ is therefore not a candidate for an extremum.

Our systematic procedure has provided us with four candidates for local extrema, at $x = -3, -2, 0$, and 2. In each case, we must decide whether the associated point is in fact a local extremum and, if so, whether it is a local maximum or minimum. The following observation, although rather obvious, is the key (see Figures 5.36 and 5.37):

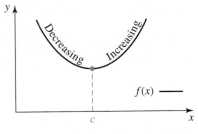

Figure 5.36 The function $y = f(x)$ has a local minimum at $x = c$.

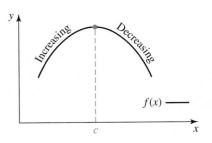

Figure 5.37 The function $y = f(x)$ has a local maximum at $x = c$.

A continuous function has a local minimum at c if the function is decreasing to the left of c and increasing to the right of c. A continuous function has a local maximum at c if the function is increasing to the left of c and decreasing to the right of c.

If the function is differentiable, as in our example, we can use the first-derivative test to identify regions where the function is increasing and regions where it is decreasing.

Since $f'(x) = 2x$ for $-3 < x < -2$ and $2 < x < 2.5$, it follows that $f'(x) > 0$ for $2 < x < 2.5$ and $f'(x) < 0$ for $-3 < x < -2$. Also, since $f'(x) = -2x$ for $-2 < x < 2$, it follows that $f'(x) > 0$ for $x \in (-2, 0)$ and $f'(x) < 0$ for $x \in (0, 2)$. We illustrate these regions on the following number line for x [the plus (minus) signs show where $f'(x)$ is positive (negative)]:

We start with the interior points. At $x = -2$, the function changes from decreasing to increasing; that is, $f(x)$ has a local minimum at $x = -2$. At $x = 0$, the function changes from increasing to decreasing; that is, $f(x)$ has a local maximum at $x = 0$. At $x = 2$, the function changes from decreasing to increasing; that is, $f(x)$ has a local minimum at $x = 2$.

We still need to analyze the endpoint at $x = -3$. We see that the function is decreasing to the right of $x = -3$; that is, $f(x)$ has a local maximum at $x = -3$. You should compare all of our findings with the graph of $f(x)$.

The last step is to select the global extrema from the local extrema, but since the domain of f is not a closed interval, we must compare the values of the local extrema against the value at the boundary $x = 2.5$. We have

$$f(-3) = 5 \qquad f(-2) = 0 \qquad f(0) = 4 \qquad f(2) = 0 \qquad \lim_{x \to 2.5^-} f(x) = 2.25$$

Since 5 is the maximum value and 0 the minimum, the absolute maximum occurs at $x = -3$ and the global minima (there are two) occur at $x = -2$ and $x = 2$.

When a function is twice differentiable at the point where the first derivative is equal to 0, there is a shortcut for determining whether a local maximum or a local minimum exists. (See Figure 5.38.) We assume that the function $f(x)$ is twice differentiable. The graph of $f(x)$ in Figure 5.38 has a local maximum at $x = c$, since the function is increasing to the left of $x = c$ and decreasing to the right of $x = c$. If we look at how the slopes of the tangent lines change as we cross $x = c$ from the left, we see that the slopes are decreasing; that is, $f''(c) < 0$. In other words, the function is concave down at $x = c$ (which is immediately apparent when you look at the graph, but remember that typically you don't have the graph in front of you). There is an analogous result where f has a local minimum at $x = c$. (See Figure 5.39.) This discussion yields the following test:

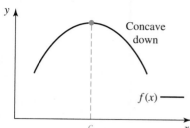

Figure 5.38 The function $y = f(x)$ has a local maximum at $x = c$.

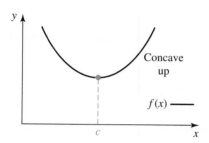

Figure 5.39 The function $y = f(x)$ has a local minimum at $x = c$.

> **The Second-Derivative Test for Local Extrema** Suppose that f is twice differentiable on an open interval containing c.
>
> If $f'(c) = 0$ and $f''(c) < 0$, then f has a local maximum at $x = c$.
> If $f'(c) = 0$ and $f''(c) > 0$, then f has a local minimum at $x = c$.

Note that finding the point c where $f'(c) = 0$ gives us a candidate for a local extremum. If the second derivative $f''(c) \neq 0$, the local extremum is established: Not only does this information tell us whether there is a local extremum, it identifies it. The test is easy to apply, as we have only to check the sign of the second derivative at $x = c$; we do not have to check the behavior of the function in a neighborhood of $x = c$. Still, the second-derivative test will not always work. For instance, if $f(x) = x^4$, then $f'(x) = 4x^3$ and $f''(x) = 12x^2$. On the basis of the graph of $y = f(x)$, we know that $f(x)$ has a local minimum at $x = 0$. We find that $f'(0) = 0$ and $f''(0) = 0$. Thus, the theorem cannot be used to draw any conclusions about $y = f(x)$ at $x = 0$.

We next look at two examples in which the second-derivative test can be applied.

EXAMPLE 1

Find all local and global extrema of

$$f(x) = \frac{3}{2}x^4 - 2x^3 - 6x^2 + 2, \quad x \in \mathbf{R}$$

Solution

Figure 5.40 shows the graph of $f(x)$. Since $f(x)$ is twice differentiable for all $x \in \mathbf{R}$, we begin by finding the first two derivatives of f. The first derivative is

$$f'(x) = 6x^3 - 6x^2 - 12x = 6x(x - 2)(x + 1)$$

Factoring $f'(x)$ will make it easier to find its zeros. The second derivative is

$$f''(x) = 18x^2 - 12x - 12$$

Since $f'(x)$ exists for all $x \in \mathbf{R}$ and the domain has no endpoints, the only candidates for local extrema are points where $f'(x) = 0$:

$$6x(x - 2)(x + 1) = 0$$

We thus find that $x = 0$, $x = 2$, and $x = -1$. Since $f''(x)$ exists, we can use the second-derivative test to determine whether $(0, f(0))$, $(2, f(2))$, and $(-1, f(-1))$

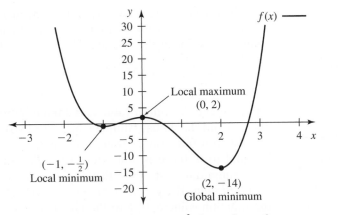

Figure 5.40 The graph of $f(x) = \frac{3}{2}x^4 - 2x^3 - 6x^2 + 2$ in Example 1.

are local extrema and, if so, of what type they are. We need to evaluate the second derivative at each x-coordinate:

$$f''(0) = -12 < 0 \implies \text{local maximum at } x = 0$$
$$f''(2) = 36 > 0 \implies \text{local minimum at } x = 2$$
$$f''(-1) = 18 > 0 \implies \text{local minimum at } x = -1$$

The function $f(x)$ is defined on **R**. As mentioned at the beginning of this section, in order to find global extrema, we must check the local extrema and compare their values against each other and against the function values as $x \to \infty$ and $x \to -\infty$. Thus, we have

$$f(0) = 2 \qquad f(2) = -14 \qquad f(-1) = -\frac{1}{2}$$

and

$$\lim_{x \to \infty} f(x) = \infty \qquad \lim_{x \to -\infty} f(x) = \infty$$

Although the local maximum is 2 at $x = 0$, the point $(0, 2)$ is not a global maximum: Since the function goes to ∞ as $|x| \to \infty$, it certainly exceeds the value 2. In fact, there is no global maximum; there is, however, a global minimum at $x = 2$. ■

EXAMPLE 2

Find all local and global extrema of

$$f(x) = x(1 - x)^{2/3}, \quad x \in \mathbf{R}$$

Solution

The graph of $f(x)$ is shown in Figure 5.41. We differentiate $f(x)$ by the product rule:

$$f'(x) = (1 - x)^{2/3} + x\frac{2}{3}(1 - x)^{-1/3}(-1)$$

$$= (1 - x)^{2/3} - \frac{2x}{3(1 - x)^{1/3}} \quad \text{for } x \neq 1$$

$$f''(x) = \frac{2}{3}(1 - x)^{-1/3}(-1) - \frac{2}{3}\left[(1 - x)^{-1/3} + x\left(-\frac{1}{3}\right)(1 - x)^{-4/3}(-1)\right]$$

$$= -\frac{2}{3(1 - x)^{1/3}} - \frac{2}{3(1 - x)^{1/3}} - \frac{2x}{9(1 - x)^{4/3}}$$

$$= -\frac{4}{3(1 - x)^{1/3}} - \frac{2x}{9(1 - x)^{4/3}} \quad \text{for } x \neq 1$$

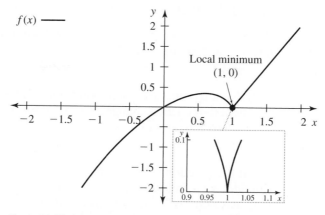

Figure 5.41 The graph of $f(x) = x(1 - x)^{2/3}$.

We first find points where $f'(x) = 0$. If $x \neq 1$, then $f'(x) = 0$ when

$$(1 - x)^{2/3} = \frac{2x}{3(1 - x)^{1/3}}$$

$$3(1 - x) = 2x$$

$$3 = 5x$$

$$x = \frac{3}{5}$$

Next, we check the second derivative at $x = \frac{3}{5}$:

$$f''\left(\frac{3}{5}\right) = -\frac{4}{3(\frac{2}{5})^{1/3}} - \frac{(2)(\frac{3}{5})}{9(\frac{2}{5})^{4/3}} < 0$$

That is, $f(x)$ has a local maximum at $x = \frac{3}{5}$.

The first derivative is not defined at $x = 1$. We therefore must investigate the function in the neighborhood of $x = 1$. Note that $f(x)$ is continuous at $x = 1$. We investigate where the function is increasing and where it is decreasing. The only x-values at which the derivative of the function can change sign are $x = \frac{3}{5}$ and $x = 1$. To determine whether $f'(x)$ is positive or negative in each of the intervals associated with these values, we simply need to evaluate $f'(x)$ at one value in each interval. For instance, $f'(0) = 1 > 0$; that is, $f'(x) > 0$ for $x < 3/5$. A similar computation can be carried out for the other subintervals. We find that $f'(x)$ is positive to the left of $\frac{3}{5}$, negative between $\frac{3}{5}$ and 1, and positive for $x > 1$. The situation is summarized in the following number line:

$$\underbrace{+++++++}_{} \quad \underbrace{-----}_{} \quad \underbrace{+++++++}_{}$$
$$\qquad\qquad \frac{3}{5} \qquad\qquad 1$$

We conclude that $f(x)$ has a local minimum at $x = 1$. We can say a bit more about what is happening at $x = 1$ (see inset in Figure 5.41):

$$\lim_{x \to 1^-} f'(x) = \lim_{x \to 1^-} \left[(1 - x)^{2/3} - \frac{2x}{3(1 - x)^{1/3}}\right] = -\infty$$

$$\lim_{x \to 1^+} f'(x) = \lim_{x \to 1^+} \left[(1 - x)^{2/3} - \frac{2x}{3(1 - x)^{1/3}}\right] = \infty$$

We see that there is a vertical tangent at $x = 1$; such points are called **cusps**.

Since

$$\lim_{x \to \infty} f(x) = \infty \qquad \text{and} \qquad \lim_{x \to -\infty} f(x) = -\infty$$

there are no absolute extrema.   ■

■ 5.3.2 Inflection Points

We begin with a verbal definition of **inflection points** and then give a procedure for locating such points. (See Figure 5.42.)

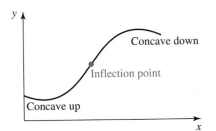

> Inflection points are points where the concavity of a function changes—that is, where the function changes from concave up to concave down or from concave down to concave up.

Figure 5.42 Inflection point.

If the function is twice differentiable, there is an algebraic condition for finding candidates for inflection points. Recall that if a function f is twice differentiable, it is concave up if $f'' > 0$ and concave down if $f'' < 0$. At an inflection point, f'' must therefore change sign; that is, the second derivative must be 0 at an inflection point. More formally,

> If $f(x)$ is twice differentiable and has an inflection point at $x = c$, then $f''(c) = 0$.

Note that $f''(c) = 0$ is a necessary, but not sufficient, condition for the existence of an inflection point of a twice-differentiable function. For instance, $f(x) = x^4$ has $f''(0) = 0$, but $f(x)$ does not have an inflection point at $x = 0$. The function $f(x) = x^4$ is concave up and has a local minimum at $x = 0$. (See Figure 5.43; we encountered a similar situation in Fermat's theorem, where we had a necessary, but not sufficient, condition for the existence of local extrema.) We can therefore use this test only for finding *candidates* for inflection points. To determine whether a candidate *is* an inflection point, we must check whether the second derivative changes sign.

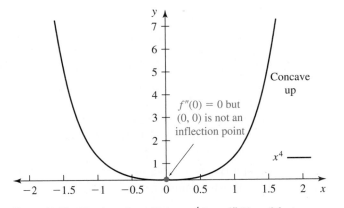

Figure 5.43 The function $f(x) = x^4$ has $f''(0) = 0$ but no inflection point at $x = 0$.

EXAMPLE 3

Show that the function

$$f(x) = \frac{1}{2}x^3 - \frac{3}{2}x^2 + 2x + 1, \quad x \in \mathbf{R}$$

has an inflection point at $x = 1$.

Solution The graph of $f(x)$ is shown in Figure 5.44. We compute the first two derivatives:

$$f'(x) = \frac{3}{2}x^2 - 3x + 2$$

$$f''(x) = 3x - 3$$

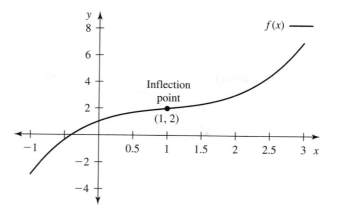

Figure 5.44 The graph of $f(x) = \frac{1}{2}x^3 - \frac{3}{2}x^2 + 2x + 1$ has an inflection point at $x = 1$.

Now, $f''(x) = 0$ if $x = 1$. Therefore, $(1, f(1))$ is a candidate for an inflection point. Since $f''(x)$ is positive for $x > 1$ and negative for $x < 1$, $f''(x)$ changes sign at $x = 1$. We therefore conclude that $f(x)$ has an inflection point at $x = 1$. ■

■ 5.3.3 Graphing and Asymptotes

Using the first and the second derivatives of a twice-differentiable function, we can obtain a fair amount of information about the function. We can determine intervals on which the function is increasing, decreasing, concave up, and concave down. We can identify local and global extrema and find inflection points. To graph the function, we also need to know how the function behaves in the neighborhood of points where either the function or its derivative is not defined, and we need to know how the function behaves at the endpoints of its domain (or, if the function is defined for all $x \in \mathbf{R}$, how the function behaves for $x \to \pm\infty$).

We will need limits again—this time to determine the behavior of a function at points where it is not defined and when $x \to \pm\infty$. We illustrate as follows: Consider

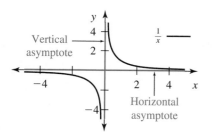

Figure 5.45 The graph of $f(x) = \frac{1}{x}$ with horizontal asymptote $y = 0$ and vertical asymptote $x = 0$.

$$f(x) = \frac{1}{x}, \quad x \neq 0$$

You are familiar with the graph of $f(x)$. (See Figure 5.45.) You can see that the graph of $f(x)$ approaches the line $y = 0$ when $x \to \infty$ and also when $x \to -\infty$. Such a line is called an **asymptote**, and we say that $f(x)$ approaches the line $y = 0$ asymptotically as $x \to \infty$ and also as $x \to -\infty$. Since $y = 0$ is a horizontal line, it is called a **horizontal asymptote**. We can represent horizontal asymptotes mathematically by the following limits:

$$\lim_{x \to \infty} f(x) = \lim_{x \to \infty} \frac{1}{x} = 0 \quad \text{and} \quad \lim_{x \to -\infty} f(x) = \lim_{x \to -\infty} \frac{1}{x} = 0$$

The function $f(x) = \frac{1}{x}$ is not defined at $x = 0$. Looking at the graph of the function $f(x) = \frac{1}{x}$, we see that it approaches the line $x = 0$ asymptotically. Since $x = 0$ is a vertical line, it is called a **vertical asymptote**. We can also represent vertical asymptotes mathematically by limits:

$$\lim_{x \to 0^-} f(x) = \lim_{x \to 0^-} \frac{1}{x} = -\infty \quad \text{and} \quad \lim_{x \to 0^+} f(x) = \lim_{x \to 0^+} \frac{1}{x} = +\infty$$

Definition A line $y = b$ is a **horizontal** asymptote if either

$$\lim_{x \to -\infty} f(x) = b \qquad \text{or} \qquad \lim_{x \to \infty} f(x) = b$$

A line $x = c$ is a **vertical** asymptote if

$$\lim_{x \to c^+} f(x) = +\infty \qquad \text{or} \qquad \lim_{x \to c^+} f(x) = -\infty$$

or

$$\lim_{x \to c^-} f(x) = +\infty \qquad \text{or} \qquad \lim_{x \to c^-} f(x) = -\infty$$

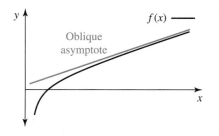

Figure 5.46 The function $y = f(x)$ has an oblique asymptote.

In addition to horizontal and vertical asymptotes, there are **oblique** asymptotes. (See Figure 5.46.) These are straight lines that are neither horizontal nor vertical and are such that the graph of the function approaches them as either $x \to +\infty$ or $x \to -\infty$. Mathematically, we express oblique asymptotes in the following way: If

$$\lim_{x \to +\infty} [f(x) - (mx + b)] = 0 \qquad \text{or} \qquad \lim_{x \to -\infty} [f(x) - (mx + b)] = 0$$

then the line $y = mx + b$ is an oblique asymptote. The simplest case of an oblique asymptote occurs with a rational function in which the degree of the numerator is one higher than the degree of the denominator.

EXAMPLE 4   **Oblique Asymptote** Let

$$f(x) = \frac{x^2 - 3}{x - 2}, \quad x \neq 2$$

To determine whether $f(x)$ has an oblique asymptote, we use long division. We find that

$$\frac{x^2 - 3}{x - 2} = x + 2 + \frac{1}{x - 2}$$

We see that $f(x)$ is the sum of a linear term $x + 2$ and a remainder term $\frac{1}{x-2}$, the latter of which goes to 0 as $x \to \pm\infty$. To check that $y = x + 2$ is indeed an oblique asymptote, we carry out the following computation:

$$\lim_{x \to \pm\infty} [f(x) - (x + 2)] = \lim_{x \to \pm\infty} \left[\frac{x^2 - 3}{x - 2} - (x + 2) \right]$$

$$= \lim_{x \to \pm\infty} \left[x + 2 + \frac{1}{x - 2} - (x + 2) \right]$$

$$= \lim_{x \to \pm\infty} \frac{1}{x - 2} = 0$$

Hence, $y = x + 2$ is an oblique asymptote. The graph of $f(x) = \frac{x^2-3}{x-2}$, together with its oblique asymptote, is shown in Figure 5.47. [The graph of $f(x)$ also has a vertical asymptote at $x = 2$.] ■

We can now combine the results that we have obtained so far to produce the graph of a given function. We illustrate the steps leading to the graph in the next two examples.

EXAMPLE 5   Sketch the graph of the function

$$f(x) = \frac{2}{3}x^3 - 2x + 1, \quad x \in \mathbf{R}$$

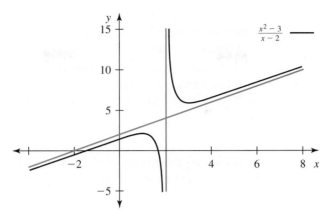

Figure 5.47 The graph of $f(x) = \frac{x^2-3}{x-2}$ with oblique asymptote $g(x) = x + 2$ (and a vertical asymptote at $x = 2$).

Solution

STEP 1. Find $f'(x)$ and $f''(x)$:

$$f'(x) = 2x^2 - 2 = 2(x - 1)(x + 1), \quad x \in \mathbf{R}$$
$$f''(x) = 4x, \quad x \in \mathbf{R}$$

STEP 2. Find the places where $f'(x)$ is positive, negative, zero, or undefined. Identify the intervals where the function is increasing and where it is decreasing. Find local extrema.

We begin by setting $f'(x) = 0$; that is,

$$2(x - 1)(x + 1) = 0$$

We find two solutions, namely $x = 1$ and $x = -1$. The following number line shows where $f'(x)$ is positive and where it is negative. Since $f'(x)$ is a polynomial, it is differentiable for all $x \in \mathbf{R}$; it can change sign only at $x = 1$ and at $x = -1$. These x-values break the number line into three intervals: $x < -1, -1 < x < 1$, and $x > 1$. To determine whether $f'(x)$ is positive or negative in each of these intervals, we simply need to evaluate $f'(x)$ at one value in each interval. For instance, $f'(-2) = 2(-3)(-1) = 6 > 0$; that is, $f'(x) > 0$ for $x < -1$. A similar calculation can be carried out for the other subintervals. When we finish calculating, we have the following number line:

$$\underset{\underset{-1}{\big|}\qquad\quad\underset{1}{\big|}}{+++++++\quad-----\quad+++++++}$$

That is, $f(x)$ is increasing for $x < -1$ and $x > 1$ and decreasing for $-1 < x < 1$. The number line also shows that the function has a local maximum at $x = -1$, namely, $(-1, \frac{7}{3})$, and a local minimum at $x = 1$, namely, $(1, -\frac{1}{3})$. (Alternatively, we could have used the second-derivative test.)

STEP 3. Find the places where $f''(x)$ is positive, negative, zero, or undefined. Find inflection points.

We set $f''(x) = 0$; that is,

$$4x = 0$$

We find one solution: $x = 0$. Since $f''(x)$ exists for all $x \in \mathbf{R}$, it can change sign only at $x = 0$. Evaluating $f''(x)$ at a value to the left of 0, say, $x = -1$, shows that $f''(x) < 0$ for $x < 0$. Similarly, evaluating $f''(x)$ at a value to the right of 0, say, $x = 1$, shows that $f''(x) > 0$ for $x > 0$. The situation is illustrated on the following number line:

$$\underset{\underset{0}{\big|}}{----------\quad++++++++++}$$

Since $f''(x)$ changes sign at $x = 0$, we conclude that the function has an inflection point at $x = 0$, namely, $(0, 1)$. The function is concave down for $x < 0$ and concave up for $x > 0$.

STEP 4. Determine the behavior at endpoints of the domain.

The domain is $(-\infty, +\infty)$. We therefore need to check the behavior of $f(x)$ as $x \to +\infty$ and as $x \to -\infty$. We find that

$$\lim_{x \to -\infty} f(x) = -\infty \qquad \text{and} \qquad \lim_{x \to +\infty} f(x) = +\infty$$

Combining the results of these four steps allows us to sketch the graph of the function, as illustrated in Figure 5.48. You should label all extrema and inflection points. ■

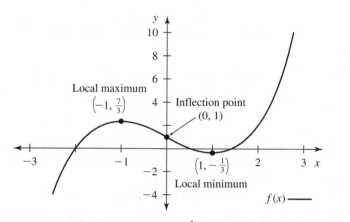

Figure 5.48 The graph of $f(x) = \frac{2}{3}x^3 - 2x + 1$.

EXAMPLE 6

Sketch the graph of the function

$$f(x) = e^{-x^2/2}, \quad x \in \mathbf{R}$$

Solution

STEP 1.

$$f'(x) = -xe^{-x^2/2}, \quad x \in \mathbf{R}$$

and

$$f''(x) = (-1)e^{-x^2/2} + (-x)(-x)e^{-x^2/2}$$
$$= e^{-x^2/2}(x^2 - 1), \quad x \in \mathbf{R}$$

STEP 2. Since $f'(x)$ is defined for all $x \in \mathbf{R}$, we need to identify only those points where $f'(x) = 0$:

$$f'(x) = 0 \quad \text{for } x = 0$$

The sign of $f'(x)$ is illustrated on the following number line for x:

$$\begin{array}{c} + + + + + + + + + + \quad - - - - - - - - - \\ \hline 0 \end{array}$$

We find that $f(x)$ is increasing for $x < 0$ and decreasing for $x > 0$. Hence, $f(x)$ has a local maximum at $x = 0$, namely, $(0, 1)$.

STEP 3. We have

$$f''(x) = 0 \quad \text{for } x = 1 \text{ and } x = -1$$

The sign of $f''(x)$ is illustrated on the following number line for x:

$$\begin{array}{c} + + + + + + + \quad - - - - - \quad + + + + + + + + \\ \hline -1 1 \end{array}$$

We find that $f(x)$ is concave up for $x < -1$ and $x > 1$ and is concave down for $-1 < x < 1$. There are two inflection points, one at $x = -1$, namely, $(-1, e^{-1/2})$, and the other at $x = 1$, namely, $(1, e^{-1/2})$. There are no other inflection points, since $f''(x)$ is defined for all $x \in \mathbf{R}$.

STEP 4. We have

$$\lim_{x \to -\infty} f(x) = 0 \qquad \text{and} \qquad \lim_{x \to +\infty} f(x) = 0$$

This shows that $y = 0$ is a horizontal asymptote.
The graph of $f(x)$ is shown in Figure 5.49.　■

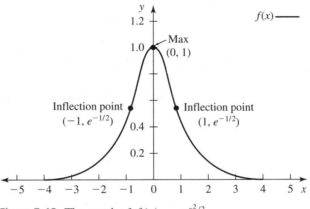

Figure 5.49 The graph of $f(x) = e^{-x^2/2}$.

Section 5.3 Problems

■ 5.3.1

Find the local maxima and minima of each of the functions in Problems 1–16. Determine whether each function has absolute maxima and minima and find their coordinates. For each function, find the intervals on which it is increasing and the intervals on which it is decreasing.

1. $y = (2 - x)^2, -2 \le x \le 3$
2. $y = \sqrt{x - 1}, 1 \le x \le 2$

3. $y = \ln(2x - 1), 1 \le x \le 2$
4. $y = \ln \frac{x}{x+1}, x > 0$

5. $y = xe^{-x}, 0 \le x \le 1$
6. $y = |16 - x^2|, -5 \le x \le 8$

7. $y = (x - 1)^3 + 1, x \in \mathbf{R}$
8. $y = x^3 - 3x + 1, x \in \mathbf{R}$

9. $y = \cos(\pi x^2), -1 \le x \le 1$

10. $y = \sin[2\pi(x - 3)], 2 \le x \le 3$

11. $y = e^{-|x|}, x \in \mathbf{R}$
12. $y = e^{-x^2/4}, x \in \mathbf{R}$

13. $y = \frac{1}{3}x^3 + \frac{1}{2}x^2 - 6x + 2, x \in \mathbf{R}$

14. $y = x^2(1 - x), x \in \mathbf{R}$

15. $y = (x - 1)^{1/3}, x \in \mathbf{R}$
16. $y = \sqrt{1 + x^2}, x \in \mathbf{R}$

17. [This problem illustrates the fact that $f'(c) = 0$ is not a sufficient condition for the existence of a local extremum of a differentiable function.] Show that the function $f(x) = x^3$ has a horizontal tangent at $x = 0$; that is, show that $f'(0) = 0$, but $f'(x)$ does not change sign at $x = 0$ and, hence, $f(x)$ does not have a local extremum at $x = 0$.

18. Suppose that $f(x)$ is twice differentiable on $\mathbf{R}$, with $f(x) > 0$ for $x \in \mathbf{R}$. Show that if $f(x)$ has a local maximum at $x = c$, then $g(x) = \ln f(x)$ also has a local maximum at $x = c$.

■ 5.3.2

In Problems 19–24, determine all inflection points.

19. $f(x) = x^3 - 2, x \in \mathbf{R}$
20. $f(x) = (x - 3)^5, 0 \in \mathbf{R}$

21. $f(x) = e^{-x^2}, x \ge 0$
22. $f(x) = xe^{-x}, x \ge 0$

23. $f(x) = \tan x, -\dfrac{\pi}{2} < x < \dfrac{\pi}{2}$

24. $f(x) = \ln x + \dfrac{1}{x}, x > 0$

25. [This problem illustrates the fact that $f''(c) = 0$ is not a sufficient condition for an inflection point of a twice-differentiable function.] Show that the function $f(x) = x^4$ has $f''(0) = 0$ but that $f''(x)$ does not change sign at $x = 0$ and, hence, $f(x)$ does not have an inflection point at $x = 0$.

26. Logistic Equation Suppose that the size of a population at time t is denoted by $N(t)$ and satisfies

$$N(t) = \frac{100}{1 + 3e^{-2t}}$$

for $t \ge 0$.
(a) Show that $N(0) = 25$.
(b) Show that $N(t)$ is strictly increasing.
(c) Show that

$$\lim_{t \to \infty} N(t) = 100$$

(d) Show that $N(t)$ has an inflection point when $N(t) = 50$—that is, when the size of the population is at half its limiting value.
(e) Use your results in (a)–(d) to sketch the graph of $N(t)$.

■ **5.3.3**

Find the local maxima and minima of the functions in Problems 27–34. Determine whether the functions have absolute maxima and minima, and, if so, find their coordinates. Find inflection points. Find the intervals on which the function is increasing, on which it is decreasing, on which it is concave up, and on which it is concave down. Sketch the graph of each function.

27. $y = \frac{2}{3}x^3 - 2x^2 - 6x + 2$ for $-2 \leq x \leq 5$

28. $y = x^4 - 2x^2, x \in \mathbf{R}$

29. $y = |x^2 - 9|, -4 \leq x \leq 5$

30. $y = \sqrt{|x|}, x \in \mathbf{R}$

31. $y = x + \cos x, x \in \mathbf{R}$

32. $y = \tan x - x, x \in \left(-\frac{\pi}{2}, \frac{\pi}{2}\right)$

33. $y = \frac{x^2 - 1}{x^2 + 1}, x \in \mathbf{R}$

34. $y = \ln(x^2 + 1), x \in \mathbf{R}$

35. Let
$$f(x) = \frac{x}{x-1}, \quad x \neq 1$$

(a) Show that
$$\lim_{x \to -\infty} f(x) = 1$$

and
$$\lim_{x \to +\infty} f(x) = 1$$

That is, show that $y = 1$ is a horizontal asymptote of the curve $y = \frac{x}{x-1}$.

(b) Show that
$$\lim_{x \to 1^-} f(x) = -\infty$$

and
$$\lim_{x \to 1^+} f(x) = +\infty$$

That is, show that $x = 1$ is a vertical asymptote of the curve $y = \frac{x}{x-1}$.

(c) Determine where $f(x)$ is increasing and where it is decreasing. Does $f(x)$ have local extrema?

(d) Determine where $f(x)$ is concave up and where it is concave down. Does $f(x)$ have inflection points?

(e) Sketch the graph of $f(x)$ together with its asymptotes.

36. Let
$$f(x) = -\frac{2}{x^2 - 1}, \quad x \neq -1, 1$$

(a) Show that
$$\lim_{x \to +\infty} f(x) = 0$$

and
$$\lim_{x \to -\infty} f(x) = 0$$

That is, show that $y = 0$ is a horizontal asymptote of $f(x)$.

(b) Show that
$$\lim_{x \to -1^-} f(x) = -\infty$$

and
$$\lim_{x \to -1^+} f(x) = +\infty$$

and that
$$\lim_{x \to 1^-} f(x) = +\infty$$

and
$$\lim_{x \to 1^+} f(x) = -\infty$$

That is, show that $x = -1$ and $x = 1$ are vertical asymptotes of $f(x)$.

(c) Determine where $f(x)$ is increasing and where it is decreasing. Does $f(x)$ have local extrema?

(d) Determine where $f(x)$ is concave up and where it is concave down. Does $f(x)$ have inflection points?

(e) Sketch the graph of $f(x)$ together with its asymptotes.

37. Let
$$f(x) = \frac{2x^2 - 5}{x + 2}, \quad x \neq -2$$

(a) Show that $x = -2$ is a vertical asymptote.

(b) Determine where $f(x)$ is increasing and where it is decreasing. Does $f(x)$ have local extrema?

(c) Determine where $f(x)$ is concave up and where it is concave down. Does $f(x)$ have inflection points?

(d) Since the degree of the numerator is one higher than the degree of the denominator, $f(x)$ has an oblique asymptote. Find it.

(e) Sketch the graph of $f(x)$ together with its asymptotes.

38. Let
$$f(x) = \frac{\sin x}{x}, \quad x \neq 0$$

(a) Show that $y = 0$ is a horizontal asymptote.

(b) Since $f(x)$ is not defined at $x = 0$, does this mean that $f(x)$ has a vertical asymptote at $x = 0$? Find $\lim_{x \to 0^+} f(x)$ and $\lim_{x \to 0^-} f(x)$.

(c) Use a graphing calculator to sketch the graph of $f(x)$.

39. Let
$$f(x) = \frac{x^2}{1 + x^2}, \quad x \in \mathbf{R}$$

(a) Determine where $f(x)$ is increasing and where it is decreasing.

(b) Where is the function concave up and where is it concave down? Find all inflection points of $f(x)$.

(c) Find $\lim_{x \to \pm\infty} f(x)$ and decide whether $f(x)$ has a horizontal asymptote.

(d) Sketch the graph of $f(x)$ together with its asymptotes and inflection points (if they exist).

40. Let
$$f(x) = \frac{x^k}{1 + x^k}, \quad x \geq 0$$

where k is a positive integer greater than 1.

(a) Determine where $f(x)$ is increasing and where it is decreasing.

(b) Where is the function concave up and where is it concave down? Find all inflection points of $f(x)$.

(c) Find $\lim_{x \to \infty} f(x)$ and decide whether $f(x)$ has a horizontal asymptote.

(d) Sketch the graph of $f(x)$ together with its asymptotes and inflection points (if they exist).

41. Let
$$f(x) = \frac{x}{a + x}, \quad x \geq 0$$

where a is a positive constant.

(a) Determine where $f(x)$ is increasing and where it is decreasing.

(b) Where is the function concave up and where is it concave down? Find all inflection points of $f(x)$.

(c) Find $\lim_{x \to \infty} f(x)$ and decide whether $f(x)$ has a horizontal asymptote.

(d) Sketch the graph of $f(x)$ together with its asymptotes and inflection points (if they exist).

42. Let

$$f(x) = \frac{2}{1 + e^{-x}}, \quad x \in \mathbf{R}$$

(a) Determine where $f(x)$ is increasing and where it is decreasing.

(b) Where is the function concave up and where is it concave down? Find all inflection points of $f(x)$.

(c) Find $\lim_{x \to \infty} f(x)$ and decide whether $f(x)$ has a horizontal asymptote.

(d) Find $\lim_{x \to -\infty} f(x)$ and decide whether $f(x)$ has a horizontal asymptote.

(e) Sketch the graph of $f(x)$ together with its asymptotes and inflection points (if they exist).

43. Population Growth Suppose that the growth rate of a population is given by

$$f(N) = N\left(1 - \left(\frac{N}{K}\right)^{\theta}\right) \quad N \geq 0$$

where N is the size of the population, K is a positive constant denoting the carrying capacity, and θ is a parameter greater than 1. Find the population size for which the growth rate is maximal.

44. Predation Rate Spruce budworms are a major pest that defoliate balsam fir. They are preyed upon by birds. A model for the per capita predation rate is given by

$$f(N) = \frac{aN}{k^2 + N^2}$$

where N denotes the density of spruce budworm and a and k are positive constants. For which density of spruce budworms is the per capita predation rate maximal?

■ 5.4 Optimization

There are many situations in which we wish to maximize or minimize certain quantities. For instance, in a chemical reaction, you might wish to know under which conditions the reaction rate is maximized. In an agricultural setting, you might be interested in finding the amount of fertilizer that would maximize the yield of some crops. In a medical setting, you might wish to optimize the dosage of a drug for maximum benefit. Optimization problems also arise in the study of the evolution of life histories and involve questions such as when an organism should begin reproduction in order to maximize the number of surviving offspring. In each case, we are interested in finding global extrema.

EXAMPLE 1   **Chemical Reaction** Consider the chemical reaction

$$A + B \to AB$$

In Example 5 of Subsection 1.2.2, we found that the reaction rate is given by the function

$$R(x) = k(a - x)(b - x), \quad 0 \leq x \leq \min(a, b)$$

where x is the concentration of the product AB and $\min(a, b)$ denotes the minimum of the two values of a and b. The constants a and b are the concentrations of the reactants A and B at the beginning of the reaction. To be concrete, we choose $k = 2$, $a = 2$, and $b = 5$. Then

$$R(x) = 2(2 - x)(5 - x) \quad \text{for } 0 \leq x \leq 2$$

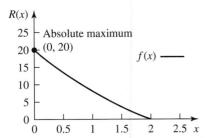

Figure 5.50 The chemical reaction rate $R(x)$ in Example 1. The graph of $R(x) = 2(2 - x)(5 - x), 0 \leq x \leq 2$, has an absolute maximum at $(0, 20)$.

(See Figure 5.50.)

We are interested in finding the concentration x that maximizes the reaction rate; this is the absolute maximum of $R(x)$. Since $R(x)$ is differentiable on $(0, 2)$, we can find all local extrema on $(0, 2)$ by investigating the first derivative. To compute the first derivative of $R(x)$, we multiply $R(x)$ out:

$$R(x) = 20 - 14x + 2x^2 \quad \text{for } 0 \leq x \leq 2$$

Differentiating with respect to x yields

$$R'(x) = -14 + 4x \quad \text{for } 0 < x < 2$$

To find candidates for local extrema, we set $R'(x) = 0$:

$$-14 + 4x = 0 \qquad \text{or} \qquad x = \frac{7}{2}$$

Since $\frac{7}{2} \notin (0, 2)$, there are no points in the interval $(0, 2)$ with horizontal tangents. Given that

$$R'(x) = -14 + 4x < 0 \quad \text{for } x \in (0, 2)$$

we conclude that $R(x)$ is decreasing in $(0, 2)$. The absolute maximum is therefore attained at the left endpoint of the interval $[0, 2]$, namely, $x = 0$. Thus, the reaction rate is maximal when the concentration of the product AB is equal to 0. You should compare this result with the graph of $R(x)$ in Figure 5.50. Since the reaction rate is proportional to the product of the concentrations of A and B, and since A and B react to form the product AB, their concentrations decrease during the reaction. Hence, we expect the reaction rate to be highest at the beginning of the reaction, when the concentrations of A and B are highest. ■

EXAMPLE 2

Crop Yield Let $Y(N)$ be the yield of an agricultural crop as a function of nitrogen level N in the soil. A model that is used for this relationship is

$$Y(N) = \frac{N}{1 + N^2} \quad \text{for } N \geq 0$$

(where N is measured in appropriate units). Find the nitrogen level that maximizes yield.

Solution

The function $Y(N)$, shown in Figure 5.51, is differentiable for $N > 0$. We find that

$$Y'(N) = \frac{(1 + N^2) - N \cdot 2N}{(1 + N^2)^2} = \frac{1 - N^2}{(1 + N^2)^2}$$

Setting $Y'(N) = 0$, we obtain the x-coordinates of candidates for local extrema:

$$Y'(N) = 0 \quad \text{if } 1 - N^2 = 0 \text{ or } N = \pm 1$$

Since $N = -1$ is not in the domain of $Y(N)$, we can discard it. The other x-coordinate, $N = 1$, is in the domain, and we see that

$$Y'(N) \begin{cases} > 0 & \text{for } 0 < N < 1 \\ < 0 & \text{for } N > 1 \end{cases}$$

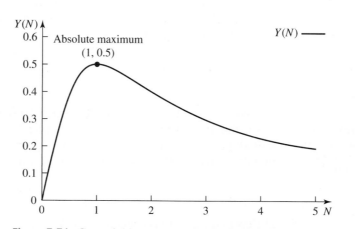

Figure 5.51 Crop yield $Y(N)$ in Example 2. The graph of $Y(N) = \frac{N}{1+N^2}$, $N \geq 0$, has an absolute maximum at $(1, \frac{1}{2})$.

Since $Y(N)$ changes from increasing to decreasing at $N = 1$, $(1, Y(1))$ is a local maximum. To find the global maximum, we still need to check the endpoints of the domain. We obtain

$$Y(0) = 0 \qquad \text{and} \qquad \lim_{N \to \infty} Y(N) = \lim_{N \to \infty} \frac{N}{1 + N^2} = 0$$

Since $Y(1) = \frac{1}{2}$, we conclude that the global maximum is $(1, \frac{1}{2})$. That is, $N = 1$ is the nitrogen level that maximizes yield. ■

EXAMPLE 3

Maximizing Area A field biologist wants to enclose a rectangular study plot. She has 1600 ft of fencing. Using this fencing, determine the dimensions of the study plot that will have the largest area.

Solution

Figure 5.52 illustrates the situation. The area A of this study plot is given by

$$A = xy \tag{5.9}$$

and the perimeter of the study plot is given by

$$1600 = 2x + 2y \tag{5.10}$$

Solving (5.10) for y yields

$$y = 800 - x$$

We can substitute this value for y in equation (5.9) and obtain

$$A(x) = x(800 - x) = 800x - x^2 \quad \text{for } 0 \le x \le 800$$

Figure 5.52 The rectangular study plot in Example 3.

It is important to state the domain of the function. Clearly, the smallest value of x is 0, in which case the enclosed area is also 0, since $A(0) = 0$. The largest possible value for x is 800, which will also produce a rectangle with one side of length 0; the corresponding area is $A(800) = 0$.

We wish to maximize the enclosed area $A(x)$. The function $A(x)$ is differentiable for $x \in (0, 800)$:

$$A'(x) = 800 - 2x \quad \text{for } 0 < x < 800$$
$$A''(x) = -2 \qquad\quad \text{for } 0 < x < 800$$

To find candidates for local extrema, we set $A'(x) = 0$ and solve for x:

$$800 - 2x = 0, \qquad \text{or} \qquad x = 400$$

Since $A''(400) < 0$, the point $(400, 160{,}000)$ is a local maximum. To find the global maximum, we need to check the function $A(x)$ at the endpoints of the interval $[0, 800]$. We have

$$A(0) = A(800) = 0$$

Because $A(400) = 400^2 = 160{,}000$, the area is maximized when $x = 400$, which implies that the study plot is a square. This relationship is true in general: For a rectangle with fixed perimeter, the maximum area occurs when the rectangle is a square. (See Problem 2 in this section.) ■

EXAMPLE 4

Minimizing Material Aluminum soda cans are shaped like a right circular cylinder and hold about 12 ounces of liquid. The production of aluminum requires a lot of energy, so it is desirable to design soda cans that use the least amount of material. What dimensions would such an optimal soda can have?

Solution

We approximate the soda can by a right circular cylinder in which the cylinder wall and both ends are made of aluminum. We denote the height of the cylinder by h and the radius by r. (See Figure 5.53.)

Figure 5.53 A right circular cylinder with height h and radius r.

If we measure h and r in centimeters, we must convert the 12 ounces into a volume measured in cubic centimeters (cm^3). For this, we need to know that 12 ounces are about 0.355 liter and that 1 liter equals 1000 cm^3.

We are now ready to set up the problem. The can with the least amount of material is the cylinder whose surface area is minimal for a given volume. Using formulas from geometry, we find that the surface area A of a right circular cylinder with top and bottom closed is given by

$$A = \underbrace{2\pi rh}_{\text{cylinder wall}} + \underbrace{2(\pi r^2)}_{\text{cylinder ends}}$$

The volume V of a right circular cylinder is given by

$$V = \pi r^2 h$$

We therefore need to minimize A when the volume $V = 12$ ounces $= 355$ cm^3. Solving $\pi r^2 h = 355$ for h yields

$$h = \frac{355}{\pi r^2}$$

Substituting this value for h in the formula for A, we find that

$$A(r) = 2\pi r \frac{355}{\pi r^2} + 2\pi r^2 = \frac{710}{r} + 2\pi r^2 \quad \text{for } r > 0$$

The graph of $A(r)$ is shown in Figure 5.54.

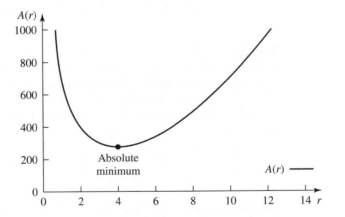

Figure 5.54 The surface area A of a right circular cylinder with given volume as a function of radius r.

To find the global minimum, we differentiate $A = A(r)$ and set the derivative equal to 0:

$$A'(r) = -\frac{710}{r^2} + 4\pi r = 0$$

Solving for r then yields

$$4\pi r = \frac{710}{r^2}, \quad \text{or} \quad r^3 = \frac{710}{4\pi} = \frac{355}{2\pi}$$

We thus find that

$$r = \left(\frac{355}{2\pi}\right)^{1/3} \approx 3.84 \text{ cm} \tag{5.11}$$

To check whether this value of r is indeed where a minimum occurs, we compute the second derivative of $A(r)$:

$$A''(r) = 2\frac{710}{r^3} + 4\pi > 0 \quad \text{for } r > 0$$

Since $A''(r) > 0, r = (\frac{355}{2\pi})^{1/3}$ is where a local minimum occurs. To determine whether this value of r is also where the global minimum occurs, we need to compute the surface area at the boundaries of the domain. Given that

$$\lim_{r \to 0+} A(r) = \infty \qquad \text{and} \qquad \lim_{r \to \infty} A(r) = \infty$$

it follows that the global minimum is achieved at $r = (\frac{355}{2\pi})^{1/3}$.

To find h, we use

$$h = \frac{355}{\pi r^2} = \frac{355}{\pi (\frac{355}{2\pi})^{2/3}} = \frac{\frac{355}{\pi}}{(\frac{355}{2\pi})^{2/3}}$$

$$= 2 \left(\frac{355}{2\pi}\right)^{1/3} = 2r \approx 7.67 \, \text{cm}$$

In the penultimate step, we used (5.11) to simplify the expression. We thus find that the can which uses the least amount of material is one whose height is equal to its diameter.

A real soda can has $h = 12.5$ cm and $r = 3.1$ cm. In our computation, we assumed that the material used to manufacture the can is of equal thickness throughout. The top of a real soda can, however, is of thicker material than the rest of the can, which could explain why soda cans don't have the dimensions we computed in this example.

■

The previous four examples illustrate the basic types of optimization problems. In the first two examples, the functions that we wished to optimize were given; in the third and fourth examples, we needed to set up equations for the functions that we wished to optimize. Once you obtain a function that you wish to optimize, the results from the previous sections in this chapter will help you to find the global extremum. It is important to state the domain of the function, since global extrema may be found at the endpoints of the domain as well as within it.

Life histories of organisms are thought to evolve to some optimal state, within given constraints. Theoretical models can help to find such optimal states. As an example, we will look at clutch size. Suppose an organism can produce more than one offspring at a time. What is the optimal clutch size? The clutch size is determined by the amount of resources the parents can provide to each offspring. On the other hand, if resources are limited (as they usually are), the more offspring per clutch, the less is available to each individual offspring. On the other hand, if the number of offspring is too small, then the chances of having any offspring that survive to reproductive age might be quite small for reasons other than insufficient resources. This trade-off between not enough resources per offspring if there are too many offspring and the chance of losing the entire clutch if there are too few offspring suggests that an intermediate number of offspring might be optimal. We discuss a simple model that addresses the trade-off in the next example.

EXAMPLE 5

(Adapted from Roff, 1992) Lloyd (1987) proposed the following model to determine the optimal clutch size: If the clutch size is equal to N and the total amount of resources allocated is R, then the amount of resources allocated to each offspring is $x = R/N$. The chance $f(x)$ of survival for an individual is related to the investment x per individual. Lloyd proposed an S-shaped, or sigmoidal, curve; that is, survival chances are very low when the investment is low, and the curve shows a saturation effect for large investments. With $N = N(x) = R/x$, the success of a clutch, or its **fitness**, can be measured as

$$w(x) = [\text{number of offspring}] \times [\text{probability of survival of offspring}]$$

$$= N(x)f(x) = \frac{R}{x} f(x) \quad \text{for } x > 0$$

where R is a positive constant. We wish to find the value of x that maximizes the fitness $w(x)$. If we assume that $f(x)$ is differentiable for $x > 0$, then differentiating $w(x)$ with respect to x yields

$$\frac{dw}{dx} = R\frac{d}{dx}\left(\frac{f(x)}{x}\right)$$

$$= R\frac{f'(x)x - f(x)}{x^2} = \frac{R}{x}\left(f'(x) - \frac{f(x)}{x}\right) \quad \text{for } x > 0$$

Setting $w'(x) = 0$ gives

$$f'(x) = \frac{f(x)}{x} \tag{5.12}$$

We denote the solution of (5.12) by $\hat{x}$ (read "x hat"). If $f(x)$ is twice differentiable at $\hat{x}$, we can use the second-derivative test to learn whether $\hat{x}$ is a local maximum or minimum. We find that

$$\frac{d^2w}{dx^2} = -\frac{R}{x^2}\left(f'(x) - \frac{f(x)}{x}\right) + \frac{R}{x}\left(f''(x) - \frac{d}{dx}\frac{f(x)}{x}\right)$$

Since $f'(\hat{x}) = f(\hat{x})/\hat{x}$, the first term on the right-hand side is equal to 0 when $x = \hat{x}$. Furthermore, because of (5.12), $\frac{d}{dx}[f(x)/x] = \frac{xf'(x)-f(x)}{x^2} = 0$ when $x = \hat{x}$. Hence,

$$\left.\frac{d^2w}{dx^2}\right|_{x=\hat{x}} = \frac{R}{\hat{x}}f''(\hat{x})$$

If we choose a function $f(x)$ that is concave down at $\hat{x}$, then $f''(\hat{x}) < 0$ and it follows that $w(x)$ has a local maximum at $\hat{x}$.

A common choice for $f(x)$ is

$$f(x) = \frac{x^2}{k^2 + x^2} \quad \text{for } x \geq 0$$

where k is a positive constant. The graph of $f(x)$ is shown in Figure 5.55; the curve is sigmoidal. Differentiating $f(x)$ with respect to x yields

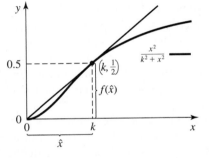

$$f'(x) = \frac{2x(k^2 + x^2) - x^2(2x)}{(k^2 + x^2)^2} = \frac{2k^2x}{(k^2 + x^2)^2}$$

Figure 5.55 The graph of $f(x) = \frac{x^2}{k^2+x^2}$ together with the tangent line at $(k, 1/2)$.

Since

$$\frac{dw}{dx} = 0 \quad \text{for } f'(x) = \frac{f(x)}{x}$$

we obtain

$$\frac{2k^2x}{(k^2 + x^2)^2} = \frac{1}{x}\frac{x^2}{k^2 + x^2}$$

which yields

$$2k^2 = k^2 + x^2 \quad \text{or} \quad k^2 = x^2$$

Because k is a positive constant and $x \geq 0$, we can discard the solution $x = -k$ and find $\hat{x} = k$. To see whether $\hat{x} = k$ is a local maximum for the function $w(x)$, we evaluate $w''(k)$; however, we need to find $f''(x)$ first:

$$f''(x) = \frac{2k^2(k^2 + x^2)^2 - (2k^2x)2(k^2 + x^2)(2x)}{(k^2 + x^2)^4}$$

$$= \frac{2k^2(k^2 + x^2) - 8k^2x^2}{(k^2 + x^2)^3} = \frac{2k^4 - 6k^2x^2}{(k^2 + x^2)^3}$$

Since

$$\left.\frac{d^2w}{dx^2}\right|_{x=k} = \frac{R}{k}f''(k) = \frac{R}{k}\frac{-4k^4}{(2k^2)^3} < 0$$

we conclude that there is a local maximum at $\hat{x} = k$. To see whether it is a global maximum, we compare $w(k)$ with $w(0)$ and $\lim_{x \to \infty} w(x)$. We have

$$w(x) = \frac{R}{x} f(x) = \frac{R}{x} \frac{x^2}{k^2 + x^2} = R \frac{x}{k^2 + x^2}$$

so

$$w(0) = 0 \qquad w(k) = \frac{R}{2k} \qquad \lim_{x \to \infty} w(x) = 0$$

Hence, $\hat{x} = k$ is where the absolute maximum occurs; for our choice of $f(x) = \frac{x^2}{k^2 + x^2}$, the optimal clutch size N_{opt} satisfies $N_{opt} = R/k$. [Other choices of $f(x)$ would give a different result.]

There is a geometric way of finding $\hat{x}$. Since

$$f'(\hat{x}) = \frac{f(\hat{x})}{\hat{x}}$$

it follows that the tangent line at $(\hat{x}, f(\hat{x}))$ has slope $\frac{f(\hat{x})}{\hat{x}}$. This line can be obtained by drawing a straight line through the origin that just touches the graph of $y = f(x)$, as illustrated in Figure 5.55. ■

Section 5.4 Problems

1. Find the smallest perimeter possible for a rectangle whose area is 25 in.^{2}.

2. Show that, among all rectangles with a given perimeter, the square has the largest area.

3. A rectangle has its base on the x-axis and its upper two vertices on the parabola $y = 3 - x^2$, as shown in Figure 5.56. What is the largest area the rectangle can have?

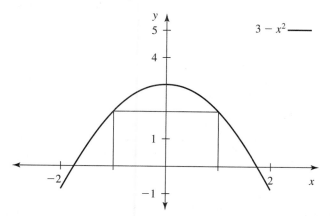

Figure 5.56 The graph of $y = 3 - x^2$ together with the inscribed rectangle in Problem 3.

4. A rectangular study area is to be enclosed by a fence and divided into two equal parts, with the fence running along the division parallel to one of the sides. If the total area is 384 ft^2, find the dimensions of the study area that will minimize the total length of the fence. How much fencing will be required?

5. A rectangular field is bounded on one side by a river and on the other three sides by a fence. Find the dimensions of the field that will maximize the enclosed area if the fence has a total length of 320 ft.

6. Find the largest possible area of a right triangle whose hypotenuse is 4 cm long.

7. Suppose that a and b are the side lengths in a right triangle whose hypotenuse is 5 cm long. What is the largest perimeter possible?

8. Suppose that a and b are the side lengths in a right triangle whose hypotenuse is 10 cm long. Show that the area of the triangle is largest when $a = b$.

9. A rectangle has its base on the x-axis, its lower left corner at $(0, 0)$, and its upper right corner on the curve $y = 1/x$. What is the smallest perimeter the rectangle can have?

10. A rectangle has its base on the x-axis and its upper left and right corners on the curve $y = \sqrt{4 - x^2}$, as shown in Figure 5.57. The left and the right corners are equidistant from the vertical axis. What is the largest area the rectangle can have?

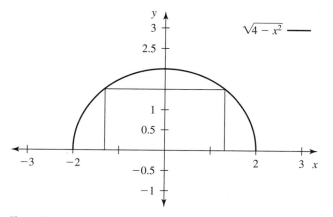

Figure 5.57 The graph of $y = (4 - x^2)^{1/2}$ together with the inscribed rectangle in Problem 10.

11. Denote by (x, y) a point on the straight line $y = 4 - 3x$. (See Figure 5.58.)

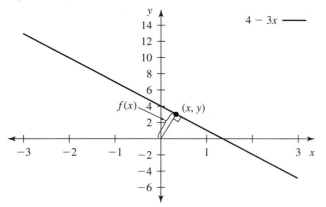

Figure 5.58 The graph of $y = 4 - 3x$ in Problem 11.

(a) Show that the distance from (x, y) to the origin is given by

$$f(x) = \sqrt{x^2 + (4 - 3x)^2}$$

(b) Give the coordinates of the point on the line $y = 4 - 3x$ that is closest to the origin. (*Hint*: Find x so that the distance you computed in (a) is minimized.)

(c) Show that the *square* of the distance between the point (x, y) on the line and the origin is given by

$$g(x) = [f(x)]^2 = x^2 + (4 - 3x)^2$$

and find the minimum of $g(x)$. Show that this minimum agrees with your answer in (b).

12. How close does the line $y = 1 + 2x$ come to the origin?

13. How close does the curve $y = 1/x$ come to the origin? (*Hint*: Find the point on the curve that minimizes the *square* of the distance between the origin and the point on the curve. If you use the square of the distance instead of the distance, you avoid dealing with square roots.)

14. How close does the circle with radius $\sqrt{2}$ and center $(2, 2)$ come to the origin.

15. Show that if $f(x)$ is a positive twice-differentiable function that has a local minimum at $x = c$, then $g(x) = [f(x)]^2$ has a local minimum at $x = c$ as well.

16. Show that if $f(x)$ is a differentiable function with $f(x) < 0$ for all $x \in \mathbf{R}$ and with a local maximum at $x = c$, then $g(x) = [f(x)]^2$ has a local minimum at $x = c$.

17. Find the dimensions of a right circular cylindrical can (with bottom and top closed) that has a volume of 1 liter and that minimizes the amount of material used. (*Note*: One liter corresponds to 1000 cm³.)

18. Find the dimensions of a right circular cylinder that is open on the top, is closed on the bottom, holds 1 liter, and uses the least amount of material.

19. A circular sector with radius r and angle θ has area A. Find r and θ so that the perimeter is smallest when (a) $A = 2$ and (b) $A = 10$. (*Note*: $A = \frac{1}{2}r^2\theta$, and the length of the arc $s = r\theta$, when θ is measured in radians; see Figure 5.59.)

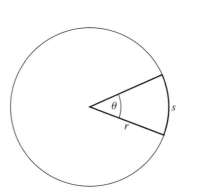

Figure 5.59 The circular sector in Problems 19 and 20.

20. A circular sector with radius r and angle θ has area A. Find r and θ so that the perimeter is smallest for a given area A. (*Note*: $A = \frac{1}{2}r^2\theta$, and the length of the arc $s = r\theta$, when θ is measured in radians; see Figure 5.59.)

21. Repeat Example 4 under the assumption that the top of the can is made out of aluminum that is three times as thick as the aluminum used for the wall and the bottom.

22. Find two positive numbers a and b such that $a + b = 20$ and ab is a maximum.

23. Find two numbers a and b such that $a - b = 4$ and ab is a minimum.

24. Classical Model of Viability Selection Consider a population of diploid organisms (i.e., each individual carries two copies of each chromosome). Genes reside on chromosomes, and we call the location of a gene on a chromosome a **locus**. Different versions of the same gene are called **alleles**. Let us examine the case of one locus with two possible alleles, A_1 and A_2. Since the individuals are diploid, the following types, called **genotypes**, may occur: A_1A_1, A_1A_2, and A_2A_2 (where A_1A_2 and A_2A_1 are considered to be equivalent). If two parents mate and produce an offspring, the offspring receives one gene from each parent. If mating is random, then we can imagine all genes being put into one big gene pool from which we choose two genes at random. If we assume that the frequency of A_1 in the population is p and the frequency of A_2 is $q = 1 - p$, then the combination A_1A_1 is picked with probability p^2, the combination A_1A_2 with probability $2pq$ (the factor 2 appears because A_1 can come from either the father or the mother), and the combination A_2A_2 with probability q^2.

We assume that the survival chances of offspring depend on their genotypes. We define the quantities w_{11}, w_{12}, and w_{22} to describe the differential survival chances of the types A_1A_1, A_1A_2, and A_2A_2, respectively. The ratio $A_1A_1:A_1A_2:A_2A_2$ among adults is given by

$$p^2w_{11}:2pqw_{12}:q^2w_{22}$$

The average fitness of this population is defined as

$$\overline{w} = p^2w_{11} + 2pqw_{12} + q^2w_{22}$$

We will investigate the preceding function. Since $q = 1 - p$, $\overline{w}$ is a function of p only; specifically,

$$\overline{w}(p) = p^2w_{11} + 2p(1 - p)w_{12} + (1 - p)^2w_{22}$$

for $0 \leq p \leq 1$. We consider the following three cases:

(i) Directional selection: $w_{11} > w_{12} > w_{22}$

(ii) Overdominance: $w_{12} > w_{11}, w_{22}$

(iii) Underdominance: $w_{12} < w_{11}, w_{22}$

(a) Show that

$$\overline{w}(p) = p^2(w_{11} - 2w_{12} + w_{22}) + 2p(w_{12} - w_{22}) + w_{22}$$

and graph $\overline{w}(p)$ for each of the three cases, where we choose the parameters as follows:

(i) $w_{11} = 1$, $w_{12} = 0.7$, $w_{22} = 0.3$
(ii) $w_{11} = 0.7$, $w_{12} = 1$, $w_{22} = 0.3$
(iii) $w_{11} = 1$, $w_{12} = 0.3$, $w_{22} = 0.7$

(b) Show that

$$\frac{d\overline{w}}{dp} = 2p(w_{11} - 2w_{12} + w_{22}) + 2(w_{12} - w_{22})$$

(c) Find the global maximum of $\overline{w}(p)$ in each of the three cases considered in (a). (Note that the global maximum may occur at the boundary of the domain of $\overline{w}$.)

(d) We can show that under a certain mating scheme the gene frequencies change until $\overline{w}$ reaches its global maximum. Assume that this is the case, and state what the equilibrium frequency will be for each of the three cases considered in (a).

25. Continuation of Problem 94 from Section 4.3 We discussed the properties of hatching offspring per unit time, $w(t)$, in the species *Eleutherodactylus coqui*. The function $w(t)$ was given by

$$w(t) = \frac{f(t)}{C + t}$$

where $f(t)$ is the proportion of offspring that survive if t is the time spent brooding and where C is the cost associated with the time spent searching for other mates.

We assume now that $f(t)$, $t \geq 0$, is twice differentiable and concave down with $f(0) = 0$ and $0 \leq f \leq 1$. The optimal brooding time is defined as the time that maximizes $w(t)$.

(a) Show that the optimal brooding time can be obtained by finding the point on the curve $f(t)$ where the line through $(-C, 0)$ is tangential to the curve $f(t)$.

(b) Use the procedure in (a) to find the optimal brooding time for $f(t) = \frac{t}{1+t}$ and $C = 2$. Determine the equation of the line through $(-2, 0)$ that is tangential to the curve $f(t) = \frac{t}{1+t}$, and graph both $f(t)$ and the tangent together.

26. Optimal Age of Reproduction *(from Roff, 1992)*
Semelparous organisms breed only once during their lifetime. Examples of this type of reproduction can be found in Pacific salmon and bamboo. The per capita rate of increase, r, can be thought of as a measure of reproductive fitness. The greater the value of r, the more offspring an individual produces. The intrinsic rate of increase is typically a function of age x. Models for age-structured populations of semelparous organisms predict that the intrinsic rate of increase as a function of x is given by

$$r(x) = \frac{\ln[l(x)m(x)]}{x}$$

where $l(x)$ is the probability of surviving to age x and $m(x)$ is the number of female offspring at age x. The optimal age of reproduction is the age x that maximizes $r(x)$.

(a) Find the optimal age of reproduction for

$$l(x) = e^{-ax}$$

and

$$m(x) = bx^c$$

where a, b, and c are positive constants.

(b) Use a graphing calculator to sketch the graph of $r(x)$ when $a = 0.1$, $b = 4$, and $c = 0.9$.

27. Optimal Age at First Reproduction *(from Lloyd, 1987)*
Iteroparous organisms breed more than once during their lifetime. Consider a model in which the intrinsic rate of increase, r, depends on the age of first reproduction, denoted by x, and satisfies the equation

$$\frac{e^{-x(r(x)+L)}(1 - e^{-kx})^3 c}{1 - e^{-(r(x)+L)}} = 1 \qquad (5.13)$$

where k, L, and c are positive constants describing the life history of the organism. The optimal age of first reproduction is the age x for which $r(x)$ is maximized. Since we cannot separate $r(x)$ in the preceding equation, we must use implicit differentiation to find a candidate for the optimal age of reproduction.

(a) Find an equation for $\frac{dr}{dx}$. [*Hint*: Take logarithms of both sides of (5.13) before differentiating with respect to x.]

(b) Set $\frac{dr}{dx} = 0$ and show that this gives

$$r(x) = \frac{3ke^{-kx}}{1 - e^{-kx}} - L$$

[To find the candidate for the optimal age x, you would need to substitute for $r(x)$ in (5.13) and solve the equation numerically. Then you would still need to check that this solution actually gives you the absolute maximum. It can, in fact, be done.]

■ 5.5 L'Hospital's Rule

Guillaume François l'Hospital was born in France in 1661. He became interested in calculus around 1690, when articles on the new calculus by Leibniz and the Bernoulli brothers began to appear. Johann Bernoulli was in Paris in 1691, and l'Hospital asked Bernoulli to teach him some calculus. Bernoulli left Paris a year later, but continued to provide l'Hospital with new material on calculus. Bernoulli received a monthly salary for his service and agreed that he would not give anyone else access to the material. Once l'Hospital thought he understood the material well enough, he decided to write a book on the subject, which was published under his name and met with great success. Bernoulli was not particularly happy about this development, as his contributions were hardly acknowledged in the book; l'Hospital perhaps felt that because he had paid for the course material, he had a right to publish it.

Today, l'Hospital is most famous for his treatment of the limits of fractions in which both the numerator and the denominator tend to 0 in the limit. The rule that bears his name was discovered by Johann Bernoulli but was published in l'Hospital's book. The rule also works when both the numerator and the denominator tend to infinity. We have encountered such examples before—for instance,

$$\lim_{x \to 3} \frac{x^2 - 9}{x - 3} = \lim_{x \to 3} (x + 3) = 6$$

and

$$\lim_{x \to \infty} \frac{kx}{1 + x} = \lim_{x \to \infty} \frac{k}{\frac{1}{x} + 1} = k$$

In both examples, we were able to find the limit by algebraic manipulations.

Using algebraic manipulations, however, is not always possible, as in the following example:

$$\lim_{x \to 0} \frac{e^x - 1}{x}$$

Here, both numerator and denominator tend to 0 as $x \to 0$. (See Figure 5.60.) There is no way of algebraically simplifying the ratio. Instead, we linearize both numerator and denominator at $a = 0$, and the linearization serves as an approximation. Recall from Section 4.8 that the linear approximation of a function $f(x)$ at $x = a$ is defined as

$$L(x) = f(a) + f'(a)(x - a)$$

If $f(x) = e^x - 1$, then $f'(x) = e^x$, $f(0) = 0$, and $f'(0) = 1$. That is, the linear approximation of the numerator at $a = 0$ is

$$L(x) = 0 + (1)(x - 0) = x$$

The denominator is already a linear function, namely, $g(x) = x$. We use linearization to approximate $\frac{e^x - 1}{x}$, obtaining the ratio $\frac{x}{x} = 1$. We might then expect that the limiting value of $\frac{e^x - 1}{x}$ as $x \to 0$ is 1, and indeed that can be shown.

To see clearly what we have just done, we look at the general case

$$\lim_{x \to a} \frac{f(x)}{g(x)}$$

and assume that both

$$\lim_{x \to a} f(x) = 0 \quad \text{and} \quad \lim_{x \to a} g(x) = 0$$

(See Figure 5.61.) Using a linear approximation as before, we find that, for x close to a,

$$\frac{f(x)}{g(x)} \approx \frac{f(a) + f'(a)(x - a)}{g(a) + g'(a)(x - a)}$$

Since $f(a) = g(a) = 0$ and $x \neq a$, the right-hand side is equal to

$$\frac{f'(a)(x - a)}{g'(a)(x - a)} = \frac{f'(a)}{g'(a)}$$

provided that $\frac{f'(a)}{g'(a)}$ is defined. We therefore hope that something like

$$\lim_{x \to a} \frac{f(x)}{g(x)} = \frac{f'(a)}{g'(a)}$$

holds when $\frac{f(a)}{g(a)}$ is of the form $\frac{0}{0}$ and $\frac{f'(a)}{g'(a)}$ is defined. In fact, something like this does hold; it is called l'Hospital's rule. The rule is more general than what we just did, and its proof would require a generalized version of the mean-value theorem, which is beyond the scope of this book.

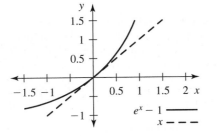

Figure 5.60 The graphs of $y = e^x - 1$ and $y = x$.

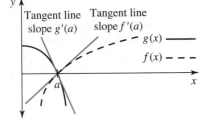

Figure 5.61 L'Hospital's rule.

L'Hospital's Rule Suppose that f and g are differentiable functions and that

$$\lim_{x \to a} f(x) = \lim_{x \to a} g(x) = 0$$

or

$$\lim_{x \to a} f(x) = \lim_{x \to a} g(x) = \infty$$

If

$$\lim_{x \to a} \frac{f'(x)}{g'(x)} = L$$

then

$$\lim_{x \to a} \frac{f(x)}{g(x)} = L$$

L'Hospital's rule works for $a = +\infty$ or $-\infty$ as well, and it also applies to one-sided limits.

Using l'Hospital's rule, we can redo the three examples we just presented. In each case, the ratio $\frac{f(x)}{g(x)}$ is an **indeterminate expression**: When a is substituted for x, the ratio is of the form $\frac{0}{0}$ or $\frac{\infty}{\infty}$. We apply l'Hospital's rule in each case: We differentiate both numerator and denominator and then take the limit of $\frac{f'(x)}{g'(x)}$ as $x \to a$. If this limit exists, it is equal to the limit of $\frac{f(x)}{g(x)}$ as $x \to a$. L'Hospital's rule then gives

$$\lim_{x \to 3} \frac{x^2 - 9}{x - 3} = \lim_{x \to 3} \frac{2x}{1} = 6$$

$$\lim_{x \to \infty} \frac{kx}{1 + x} = \lim_{x \to \infty} \frac{k}{1} = k$$

$$\lim_{x \to 0} \frac{e^x - 1}{x} = \lim_{x \to 0} \frac{e^x}{1} = 1$$

As we just said, if the limit $\frac{f'(x)}{g'(x)}$ exists as $x \to a$, then we can conclude that the limit of $\frac{f(x)}{g(x)}$ exists as $x \to a$ and the two limits are equal. Thus, each of the first equalities in the three examples are true *because* the second equality holds. This should be kept in mind in the examples that follow. In each one, you might wish to use a graphing calculator to graph the function. The first three examples are straightforward applications of l'Hospital's rule.

EXAMPLE 1   **Indeterminate Expression 0/0** Evaluate

$$\lim_{x \to 2} \frac{x^6 - 64}{x^2 - 4}$$

Solution   This limit is of the form $\frac{0}{0}$, since $2^6 - 64 = 0$ and $2^2 - 4 = 0$. Applying l'Hospital's rule yields

$$\lim_{x \to 2} \frac{x^6 - 64}{x^2 - 4} = \lim_{x \to 2} \frac{6x^5}{2x} = \frac{(6)(2^5)}{(2)(2)} = (6)(2^3) = 48 \qquad ■$$

EXAMPLE 2   **Indeterminate Expression 0/0** Evaluate

$$\lim_{x \to 0} \frac{1 - \cos^2 x}{\sin x}$$

Solution This limit is of the form $\frac{0}{0}$, since $1 - \cos^2 0 = 0$ and $\sin 0 = 0$. Applying l'Hospital's rule yields

$$\lim_{x \to 0} \frac{1 - \cos^2 x}{\sin x} = \lim_{x \to 0} \frac{2 \cos x \sin x}{\cos x} = \lim_{x \to 0} 2 \sin x = 0 \qquad ■$$

EXAMPLE 3 **Indeterminate Expression ∞/∞** Evaluate

$$\lim_{x \to \infty} \frac{\ln x}{x}$$

Solution This limit is of the form $\frac{\infty}{\infty}$. We can again apply l'Hospital's rule:

$$\lim_{x \to \infty} \frac{\frac{1}{x}}{1} = \lim_{x \to \infty} \frac{1}{x} = 0 \qquad ■$$

EXAMPLE 4 **Indeterminate Expression ∞/∞, One-Sided Limit** Evaluate

$$\lim_{x \to (\pi/2)^-} \frac{\tan x}{1 + \tan x}$$

Solution This limit is of the form $\frac{\infty}{\infty}$. We apply l'Hospital's rule and find that

$$\lim_{x \to (\pi/2)^-} \frac{\tan x}{1 + \tan x} = \lim_{x \to (\pi/2)^-} \frac{\sec^2 x}{\sec^2 x} = \lim_{x \to (\pi/2)^-} 1 = 1 \qquad ■$$

EXAMPLE 5 **Applying L'Hospital's Rule More Than Once** Evaluate

$$\lim_{x \to \infty} \frac{x^3 - 3x + 1}{3x^3 - 2x^2}$$

Solution This limit is of the form $\frac{\infty}{\infty}$. Applying l'Hospital's rule yields

$$\lim_{x \to \infty} \frac{x^3 - 3x + 1}{3x^3 - 2x^2} = \lim_{x \to \infty} \frac{3x^2 - 3}{9x^2 - 4x}$$

which is still of the form $\frac{\infty}{\infty}$. Applying l'Hospital's rule again, we obtain

$$\lim_{x \to \infty} \frac{3x^2 - 3}{9x^2 - 4x} = \lim_{x \to \infty} \frac{6x}{18x - 4}$$

Since this is still of the form $\frac{\infty}{\infty}$, we can apply l'Hospital's rule yet again. We now find that

$$\lim_{x \to \infty} \frac{6x}{18x - 4} = \lim_{x \to \infty} \frac{6}{18} = \frac{1}{3}$$

In this example, we could have found the answer without using l'Hospital's rule: We just divide both numerator and denominator by x^3 to get

$$\lim_{x \to \infty} \frac{x^3 - 3x + 1}{3x^3 - 2x^2} = \lim_{x \to \infty} \frac{x^3(1 - \frac{3}{x^2} + \frac{1}{x^3})}{x^3(3 - \frac{2}{x})}$$

$$= \lim_{x \to \infty} \frac{1 - \frac{3}{x^2} + \frac{1}{x^3}}{3 - \frac{2}{x}} = \frac{1}{3} \qquad ■$$

When you apply l'Hospital's rule, it can certainly happen that the limit is infinite, as in the next example.

EXAMPLE 6 **Infinite Limit** Evaluate

$$\lim_{x \to \infty} \frac{e^x}{x}$$

Solution This limit is of the form $\frac{\infty}{\infty}$. We apply l'Hospital's rule.

$$\lim_{x \to \infty} \frac{e^x}{x} = \lim_{x \to \infty} \frac{e^x}{1} = \infty \qquad \text{(limit does not exist)}$$ ■

L'Hospital's rule can sometimes be applied to limits of the form

$$\lim_{x \to a} f(x)g(x)$$

where

$$\lim_{x \to a} f(x) = 0 \qquad \text{and} \qquad \lim_{x \to a} g(x) = \infty$$

since we can write the limit in the form

$$\lim_{x \to a} f(x)g(x) = \lim_{x \to a} \frac{f(x)}{\frac{1}{g(x)}} = \lim_{x \to a} \frac{g(x)}{\frac{1}{f(x)}}$$

which is again of the form $\frac{0}{0}$ or $\frac{\infty}{\infty}$.

EXAMPLE 7 **Indeterminate Expression $0 \cdot \infty$** Evaluate

$$\lim_{x \to 0^+} x \ln x$$

Solution This limit is of the form $(0)(-\infty)$. We apply l'Hospital's rule after rewriting it in the form $\frac{\infty}{\infty}$:

$$\lim_{x \to 0^+} x \ln x = \lim_{x \to 0^+} \frac{\ln x}{\frac{1}{x}} = \lim_{x \to 0^+} \frac{\frac{1}{x}}{-\frac{1}{x^2}}$$

$$= \lim_{x \to 0^+} \frac{1}{x}\left(-\frac{x^2}{1}\right) = \lim_{x \to 0^+} (-x) = 0$$

We could have written the limit in the form

$$\lim_{x \to 0^+} \frac{x}{\frac{1}{\ln x}}$$

and then applied l'Hospital's rule. In this case,

$$\lim_{x \to 0^+} \frac{x}{\frac{1}{\ln x}} = \lim_{x \to 0^+} \frac{1}{(-1)(\ln x)\frac{1}{x}} = \lim_{x \to 0^+} \frac{x}{-\ln x} = 0$$

You will probably agree that the first way was easier, because it is more difficult to differentiate $1/\ln x$ than $\ln x$. Before you apply l'Hospital's rule to expressions of the form $0 \cdot \infty$, you should always determine which way will be easier to evaluate. ■

EXAMPLE 8 **Indeterminate Expression $0 \cdot \infty$** Evaluate

$$\lim_{x \to 0^+} x \cot x$$

Solution This limit is of the form $0 \cdot \infty$. We have two choices: We can evaluate

$$\lim_{x \to 0^+} x \cot x = \lim_{x \to 0^+} \frac{x}{\frac{1}{\cot x}} = \lim_{x \to 0^+} \frac{x}{\tan x}$$

where we use the fact that $\cot x = \frac{1}{\tan x}$, or we can use

$$\lim_{x \to 0^+} x \cot x = \lim_{x \to 0^+} \frac{\cot x}{\frac{1}{x}}$$

It certainly seems easier to apply l'Hospital's rule to the first form:

$$\lim_{x \to 0^+} x \cot x = \lim_{x \to 0^+} \frac{x}{\tan x} = \lim_{x \to 0^+} \frac{1}{\sec^2 x} = \lim_{x \to 0^+} \cos^2 x = 1$$

In the penultimate step, we used the fact that $\sec x = \frac{1}{\cos x}$. ■

Limits of the form $\infty - \infty$ sometimes can be evaluated with l'Hospital's rule if we can algebraically transform such limits into the form $\frac{0}{0}$ or $\frac{\infty}{\infty}$.

EXAMPLE 9 **Indeterminate Expression $\infty - \infty$** Evaluate

$$\lim_{x \to (\frac{\pi}{2})^-} (\tan x - \sec x)$$

Solution This limit is of the form $\infty - \infty$. Note that

$$\tan x = \frac{\sin x}{\cos x} \qquad \text{and} \qquad \sec x = \frac{1}{\cos x}$$

Using these two identities, we can write the limit as

$$\lim_{x \to (\frac{\pi}{2})^-} (\tan x - \sec x) = \lim_{x \to (\frac{\pi}{2})^-} \left(\frac{\sin x}{\cos x} - \frac{1}{\cos x} \right)$$

$$= \lim_{x \to (\frac{\pi}{2})^-} \frac{\sin x - 1}{\cos x}$$

This is now of the form $\frac{0}{0}$, and we can apply l'Hospital's rule to obtain

$$\lim_{x \to (\frac{\pi}{2})^-} \frac{\sin x - 1}{\cos x} = \lim_{x \to (\frac{\pi}{2})^-} \frac{\cos x}{-\sin x} = \frac{0}{-1} = 0 \qquad ■$$

EXAMPLE 10 **Indeterminate Expression $\infty - \infty$** Evaluate

$$\lim_{x \to \infty} (x - \sqrt{x^2 + x})$$

Solution This limit is of the form $\infty - \infty$. We need to get a product or a ratio. To do so, we can factor x and get

$$\lim_{x \to \infty} (x - \sqrt{x^2 + x}) = \lim_{x \to \infty} x \left(1 - \sqrt{1 + \frac{1}{x}} \right)$$

This is of the form $\infty \cdot 0$. We can transform it to the form $\frac{0}{0}$ and then apply l'Hospital's rule:

$$\lim_{x \to \infty} x \left(1 - \sqrt{1 + \frac{1}{x}} \right) = \lim_{x \to \infty} \frac{1 - \sqrt{1 + \frac{1}{x}}}{\frac{1}{x}}$$

$$= \lim_{x \to \infty} \frac{-\frac{1}{2}(1 + \frac{1}{x})^{-1/2}(-\frac{1}{x^2})}{-\frac{1}{x^2}}$$

$$= \lim_{x \to \infty} \frac{-1}{2\sqrt{1 + \frac{1}{x}}} = -\frac{1}{2} \qquad ■$$

Finally, we consider expressions of the form

$$\lim_{x \to a} [f(x)]^{g(x)}$$

when they are of the type 0^0, ∞^0, or 1^∞. The key to solving such limits is to rewrite them as

$$\lim_{x \to a} [f(x)]^{g(x)} = \lim_{x \to a} \exp \left\{ \ln [f(x)]^{g(x)} \right\}$$

$$= \lim_{x \to a} \exp [g(x) \cdot \ln f(x)]$$

$$= \exp \left[\lim_{x \to a} (g(x) \cdot \ln f(x)) \right]$$

The last step, in which we interchanged lim and exp, uses the fact that the exponential function is continuous. Rewriting the limit in this way transforms

$$0^0 \quad \text{into} \quad \exp[0 \cdot (-\infty)]$$

$$\infty^0 \quad \text{into} \quad \exp[0 \cdot (\infty)]$$

$$1^\infty \quad \text{into} \quad \exp[\infty \cdot \ln 1] = \exp[\infty \cdot 0]$$

Since we know how to deal with limits of the form $0 \cdot \infty$, we are in good shape again. We present a couple of examples.

EXAMPLE 11   Indeterminate Expression 0^0 Evaluate

$$\lim_{x \to 0^+} x^x$$

Solution   This limit is of the form 0^0; we rewrite the limit first:

$$\lim_{x \to 0^+} x^x = \lim_{x \to 0^+} \exp[\ln x^x] = \lim_{x \to 0^+} \exp[x \ln x]$$

$$= \exp \left[\lim_{x \to 0^+} (x \ln x) \right]$$

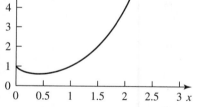

We evaluated this limit in Example 7 and obtained

$$\lim_{x \to 0^+} (x \ln x) = 0$$

Hence,

$$\lim_{x \to 0^+} x^x = \exp \left[\lim_{x \to 0^+} (x \ln x) \right] = \exp[0] = 1$$

Figure 5.62 The graph of $y = x^x$.

(See Figure 5.62.)   ■

EXAMPLE 12   Indeterminate Expression 1^∞ Evaluate

$$\lim_{x \to (\pi/4)^-} (\tan x)^{\tan(2x)}$$

Solution   Since $\tan \frac{\pi}{4} = 1$ and $\tan \left(2 \frac{\pi}{4} \right) = \infty$, this limit is of the form 1^∞. We rewrite it as

$$\lim_{x \to (\pi/4)^-} (\tan x)^{\tan(2x)} = \lim_{x \to (\pi/4)^-} \exp[\tan(2x) \cdot \ln \tan x]$$

$$= \exp \left[\lim_{x \to (\pi/4)^-} (\tan(2x) \cdot \ln \tan x) \right]$$

The limit is now of the form $\infty \cdot 0$ (since $\ln \tan \frac{\pi}{4} = \ln 1 = 0$). We evaluate the limit by writing it in the form $\frac{0}{0}$ and then applying l'Hospital's rule:

$$\lim_{x \to (\pi/4)^-} (\tan(2x) \cdot \ln \tan x) = \lim_{x \to (\pi/4)^-} \frac{\ln \tan x}{\frac{1}{\tan(2x)}} = \lim_{x \to (\pi/4)^-} \frac{\ln \tan x}{\cot(2x)}$$

Since

$$\frac{d}{dx} \ln \tan x = \frac{\sec^2 x}{\tan x} = \frac{\cos x}{\cos^2 x \sin x} = \frac{1}{\sin x \cos x}$$

and

$$\frac{d}{dx} \cot(2x) = -(\csc^2(2x)) \cdot 2 = \frac{-2}{\sin^2(2x)}$$

it follows that

$$\lim_{x \to (\pi/4)^-} \frac{\ln \tan x}{\cot(2x)} = \lim_{x \to (\pi/4)^-} \frac{\frac{1}{\sin x \cos x}}{\frac{-2}{\sin^2(2x)}} = \lim_{x \to (\pi/4)^-} \frac{\sin^2(2x)}{-2 \sin x \cos x}$$

$$= \frac{1}{(-2)(\frac{1}{2}\sqrt{2})(\frac{1}{2}\sqrt{2})} = -1$$

Therefore,

$$\lim_{x \to (\pi/4)^-} (\tan x)^{\tan(2x)} = \exp \left[\lim_{x \to (\pi/4)^-} (\tan(2x) \ln \tan x) \right]$$

$$= \exp[-1] = e^{-1}$$

Figure 5.63 The graph of $y = (\tan x)^{\tan(2x)}$.

The graph of $f(x) = (\tan x)^{\tan(2x)}$ is shown in Figure 5.63.

Section 5.5 Problems

Use l'Hospital's rule to find the limits in Problems 1–50.

1. $\lim\limits_{x \to 5} \dfrac{x^2 - 25}{x - 5}$

2. $\lim\limits_{x \to 2} \dfrac{x - 2}{x^2 - 4}$

3. $\lim\limits_{x \to -2} \dfrac{3x^2 + 5x - 2}{x + 2}$

4. $\lim\limits_{x \to -3} \dfrac{x + 3}{x^2 + 2x - 3}$

5. $\lim\limits_{x \to 0} \dfrac{\sqrt{2x + 4} - 2}{x}$

6. $\lim\limits_{x \to 0} \dfrac{3 - \sqrt{2x + 9}}{2x}$

7. $\lim\limits_{x \to 0} \dfrac{\sin x}{x \cos x}$

8. $\lim\limits_{x \to 0} \dfrac{x \sin x}{1 - \cos x}$

9. $\lim\limits_{x \to 0} \dfrac{1 - \cos x}{x \tan x}$

10. $\lim\limits_{x \to \pi/2} \dfrac{\sin(\frac{\pi}{2} - x)}{\cos x}$

11. $\lim\limits_{x \to 0^+} \dfrac{\sqrt{x}}{\ln(x + 1)}$

12. $\lim\limits_{x \to \infty} \dfrac{\ln x}{\sqrt{x}}$

13. $\lim\limits_{x \to \infty} \dfrac{\ln(\ln x)}{x}$

14. $\lim\limits_{x \to \infty} \dfrac{\ln(\ln x)}{\ln x}$

15. $\lim\limits_{x \to 0} \dfrac{2^x - 1}{3^x - 1}$

16. $\lim\limits_{x \to 0} \dfrac{5^x - 1}{7^x - 1}$

17. $\lim\limits_{x \to 0} \dfrac{3^{-x} - 1}{2^x - 1}$

18. $\lim\limits_{x \to 0} \dfrac{2^{-x} - 1}{5^x - 1}$

19. $\lim\limits_{x \to 0} \dfrac{e^x - 1 - x}{x^2}$

20. $\lim\limits_{x \to 0} \dfrac{e^x - 1 - x - \dfrac{x^2}{2}}{x^3}$

21. $\lim\limits_{x \to \infty} \dfrac{(\ln x)^2}{x^2}$

22. $\lim\limits_{x \to \infty} \dfrac{x^7}{e^x}$

23. $\lim\limits_{x \to (\pi/2)^-} \dfrac{\tan x}{\sec^2 x}$

24. $\lim\limits_{x \to 0} \dfrac{e^x - 1}{\sin x}$

25. $\lim\limits_{x \to \infty} x e^{-x}$

26. $\lim\limits_{x \to \infty} x^2 e^{-x}$

27. $\lim\limits_{x \to \infty} x^5 e^{-x}$

28. $\lim\limits_{x \to \infty} x^n e^{-x}, n \in \mathbf{N}$

29. $\lim\limits_{x \to 0^+} \sqrt{x} \ln x$

30. $\lim\limits_{x \to 0^+} x^2 \ln x$

31. $\lim\limits_{x \to 0^+} x^5 \ln x$

32. $\lim\limits_{x \to 0^+} x^n \ln x, n \in \mathbf{N}$

33. $\lim\limits_{x \to (\pi/2)^-} \left(\dfrac{\pi}{2} - x \right) \sec x$

34. $\lim\limits_{x \to 1^-} (1 - x) \tan \left(\dfrac{\pi}{2} x \right)$

35. $\lim\limits_{x \to \infty} \sqrt{x} \sin \dfrac{1}{x}$

36. $\lim\limits_{x \to \infty} x^2 \sin \dfrac{1}{x^2}$

37. $\lim\limits_{x \to 0^+} (\cot x - \csc x)$

38. $\lim\limits_{x \to \infty} (x - \sqrt{x^2 - 1})$

39. $\lim\limits_{x \to 0^+} \left(\dfrac{1}{\sin x} - \dfrac{1}{x} \right)$

40. $\lim\limits_{x \to 0^+} \left(\dfrac{1}{\sin^2 x} - \dfrac{1}{x} \right)$

41. $\lim_{x \to 0^+} x^{2x}$

42. $\lim_{x \to 0^+} x^{\sin x}$

43. $\lim_{x \to \infty} x^{1/x}$

44. $\lim_{x \to \infty} (1 + e^x)^{1/x}$

45. $\lim_{x \to \infty} \left(1 + \dfrac{3}{x}\right)^x$

46. $\lim_{x \to \infty} \left(1 + \dfrac{5}{x}\right)^x$

47. $\lim_{x \to \infty} \left(1 - \dfrac{2}{x}\right)^x$

48. $\lim_{x \to \infty} \left(1 + \dfrac{3}{x^2}\right)^x$

49. $\lim_{x \to \infty} \left(\dfrac{x}{1 + x}\right)^x$

50. $\lim_{x \to 0^+} (\cos(2x))^{3/x}$

Find the limits in Problems 51–60. Be sure to check whether you can apply l'Hospital's rule before you evaluate the limit.

51. $\lim_{x \to 0} xe^x$

52. $\lim_{x \to 0^+} \dfrac{e^x}{x}$

53. $\lim_{x \to (\pi/2)^-} (\tan x + \sec x)$

54. $\lim_{x \to (\pi/2)^-} \dfrac{\tan x}{1 + \sec x}$

55. $\lim_{x \to 1} \dfrac{x^2 - 1}{x + 1}$

56. $\lim_{x \to 0} \dfrac{1 - \cos x}{\sec x}$

57. $\lim_{x \to -\infty} xe^x$

58. $\lim_{x \to 0^+} \left(\dfrac{1}{x} - \dfrac{1}{\sqrt{x}}\right)$

59. $\lim_{x \to 0^+} x^{3x}$

60. $\lim_{x \to \infty} \left(\dfrac{x + 1}{x + 2}\right)^x$

61. Use l'Hospital's rule to find

$$\lim_{x \to 0} \frac{a^x - 1}{b^x - 1}$$

where $a, b > 0$.

62. Use l'Hospital's rule to find

$$\lim_{x \to \infty} \left(1 + \frac{c}{x}\right)^x$$

where c is a constant.

63. For $p > 0$, determine the values of p for which the following limit is either 1 or ∞ or a constant that is neither 1 nor ∞:

$$\lim_{x \to \infty} \left(1 + \frac{c}{x^p}\right)^x$$

64. Show that

$$\lim_{x \to \infty} x^p e^{-x} = 0$$

for any positive number p. Graph $f(x) = x^p e^{-x}$, $x > 0$, for $p = 1/2, 1$, and 2. Since $f(x) = x^p e^{-x} = x^p/e^x$, the limiting behavior ($\lim_{x \to \infty} \frac{x^p}{e^x} = 0$) shows that the exponential function grows faster than any power of x as $x \to \infty$.

65. Show that

$$\lim_{x \to \infty} \frac{\ln x}{x^p} = 0$$

for any number $p > 0$. This shows that the logarithmic function grows more slowly than any positive power of x as $x \to \infty$.

66. When l'Hospital introduced indeterminate limits in his textbook, his *first* example was

$$\lim_{x \to a} \frac{\sqrt{2a^3 x - x^4} - a\sqrt[3]{a^2 x}}{a - \sqrt[4]{ax^3}}$$

where a is a positive constant. (This example was communicated to him by Bernoulli.) Show that this limit is equal to $(16/9)a$.

67. The height y in feet of a tree as a function of the tree's age x in years is given by

$$y = 121e^{-17/x} \quad \text{for } x > 0$$

(a) Determine (1) the rate of growth when $x \to 0^+$ and (2) the limit of the height as $x \to \infty$.

(b) Find the age at which the growth rate is maximal.

(c) Show that the height of the tree is an increasing function of age. At what age is the height increasing at an accelerating rate and at what age at a decelerating rate?

(d) Sketch the graph of both the height and the rate of growth of the tree as functions of age.

■ 5.6 Difference Equations: Stability (Optional)

In Chapter 2, we introduced difference equations and saw that first-order difference equations can be described by recursions of the form

$$x_{t+1} = f(x_t), \quad t = 0, 1, 2, \ldots \tag{5.14}$$

where $f(x)$ is a function. There, we were able to analyze difference equations only numerically (except for equations describing exponential growth, which we were able to solve). We saw that fixed points (or equilibria) played a special role. A fixed point x^* of (5.14) satisfies the equation

$$x^* = f(x^*) \tag{5.15}$$

and has the property that if $x_0 = x^*$, then $x_t = x^*$ for $t = 1, 2, 3, \ldots$. We also saw in a number of applications that, under certain conditions, x_t converged to the fixed point as $t \to \infty$ even if $x_0 \neq x^*$. However, back then, we were not able to predict when this behavior would occur.

In this section, we will return to fixed points and use calculus to come up with a condition that allows us to check whether convergence to a fixed point occurs. We start with the simplest example: exponential growth.

■ 5.6.1 Exponential Growth

Exponential growth in discrete time is given by the recursion

$$N_{t+1} = RN_t, \quad t = 0, 1, 2, \ldots \tag{5.16}$$

where N_t is the population size at time t and $R > 0$ is the growth parameter. We assume throughout that $N_0 \geq 0$, which implies that $N_t \geq 0$.

The fixed point of (5.16) can be found by solving $N = RN$. The only solution of this equation is $N^* = 0$, unless $R = 1$. If $R = 1$, then the population size never changes, regardless of N_0. A consequence of being a fixed point is that if $N_0 = N^*$, then $N_t = N^*$ for $t = 1, 2, 3, \ldots$. That is, with $N^* = 0$, if $N_0 = 0$, then $N_t = 0$ for $t = 1, 2, 3, \ldots$. But what happens if we start with a value that is different from 0? In Chapter 2, we found that $N_t = N_0 R^t$ is a solution of (5.16) with initial condition N_0. Using this fact, we concluded that if $N_0 > 0$ and $0 < R < 1$, then $N_t \to 0$ as $t \to \infty$, whereas if $N_0 > 0$ and $R > 1$, then $N_t \to \infty$ as $t \to \infty$. If $R = 1$, then $N_t = N_0$ for $t = 1, 2, 3, \ldots$.

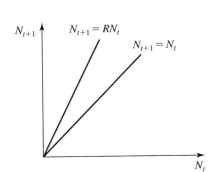

Figure 5.64 The graphs of $N_{t+1} = N_t$ and $N_{t+1} = RN_t$ intersect at $N = 0$ only when $R \neq 1$.

We can interpret the behavior of N_t as follows: If $0 < R < 1$ and $N_0 > 0$, then N_t will return to the equilibrium $N^* = 0$; if $R \geq 1$ and $N_0 > 0$, then N_t will not return to the equilibrium $N^* = 0$; if $R = 1$, N_t will stay at N_0; if $R > 1$, N_t will go to infinity. We say that $N^* = 0$ is **stable** if $0 < R < 1$ and **unstable** if $R > 1$. The case $R = 1$ is called **neutral**, since, no matter what the value of N_0 is, $N_t = N_0$ for $t = 1, 2, 3, \ldots$.

Cobwebbing There is a graphical approach to determining whether a fixed point is stable or unstable. The fixed points of (5.16) are found graphically where the graphs of $N_{t+1} = RN_t$ and $N_{t+1} = N_t$ intersect, as already pointed out in Chapter 2. We see (Figure 5.64) that the two graphs intersect where $N_t = 0$ only when $R \neq 1$, confirming what we found earlier.

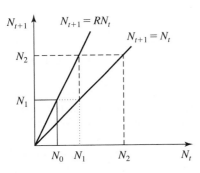

Figure 5.65 The graphs of $N_{t+1} = N_t$ and $N_{t+1} = RN_t$ can be used to determine successive population sizes.

We can use the two graphs in Figure 5.65 to follow successive population sizes ($R > 1$ in the figure). Start at N_0 on the horizontal axis. Since $N_1 = RN_0$, we find N_1 on the vertical axis as shown by the solid horizontal and vertical line segments in Figure 5.65. Using the line $N_{t+1} = N_t$, we can locate N_1 on the horizontal axis, shown in the figure by the dotted horizontal and vertical line segments. Using the line $N_{t+1} = RN_t$ again, we can find N_2 on the vertical axis, shown in the figure by the broken horizontal and vertical line segments. Using the line $N_{t+1} = N_t$ once more, we can locate N_2 on the horizontal axis and then repeat the preceding steps to find N_3 on the vertical axis, and so on (Figure 5.66). This procedure is called **cobwebbing**.

In Figures 5.64–5.66, $R > 1$, and we see that if $N_0 > 0$, then N_t will not converge to the fixed point $N^* = 0$, but instead will move away from 0 (and, in fact, go to infinity as t tends to infinity).

In Figure 5.67, we use the cobwebbing procedure when $0 < R < 1$. We see that if $N_0 > 0$, then N_t will return to the fixed point $N^* = 0$.

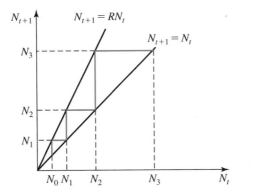

Figure 5.66 The cobwebbing procedure when $R > 1$.

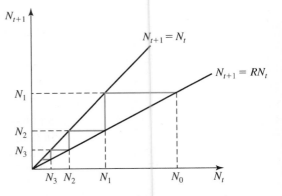

Figure 5.67 The cobwebbing procedure when $0 < R < 1$.

Figure 5.68 Multiple equilibria.

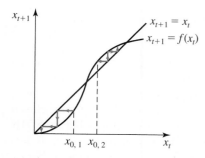

Figure 5.69 Depending on the initial value, the dynamical system converges to different limiting values.

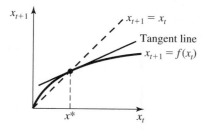

Figure 5.70 Linearizing about the equilibrium.

When we discuss the general case, we will find that the slope of the function $N_{t+1} = f(N_t)$ at the fixed point [i.e., $f'(N^*)$] determines whether the solution moves away from the fixed point or converges to it. In the example of exponential growth, $N^* = 0$ and $f'(0) = R$. For $0 < R < 1$, N^* is stable; if $R > 1$, N^* is unstable.

■ 5.6.2 Stability: General Case

The general form of a first-order recursion is

$$x_{t+1} = f(x_t), \ t = 0, 1, 2, \ldots \tag{5.17}$$

We assume that the function f is differentiable in its domain. To find fixed points algebraically, we solve $x = f(x)$. To find them graphically, we look for points of intersection of the graphs of $x_{t+1} = f(x_t)$ and $x_{t+1} = x_t$ (Figure 5.68). The graphs in Figure 5.68 intersect more than once, which means that there are multiple equilibria or fixed points.

We can use the cobwebbing procedure from the previous subsection to graphically investigate the behavior of the difference equation for different initial values. Two cases are shown in Figure 5.69, one starting at $x_{0,1}$ and the other at $x_{0,2}$. We see that x_t converges to different values, depending on the initial value. This is important to keep in mind in the discussion that follows.

Stability To determine the **stability** of an equilibrium—that is, whether it is stable or unstable—we will proceed as in the previous subsection: We will start at a value that is different from the equilibrium and check whether the solution will return to the equilibrium. There is one important difference, however: We will *not* allow just any initial value that is different from the equilibrium; rather, we allow only initial values that are "close" to the equilibrium. We think of starting at a different value as a **perturbation** of the equilibrium, and since the initial value is close to the equilibrium, we call it a **small perturbation**. The reason for looking only at small perturbations is that if there are multiple equilibria and if we start too far away from the equilibrium of interest, we might end up at a different equilibrium, not because the equilibrium of interest is unstable, but simply because we are drawn to another equilibrium (as in Figure 5.69).

If we are concerned only with small perturbations, we can approximate the function $f(x)$ by its tangent-line approximation at the equilibrium x^* (Figure 5.70). We will therefore first look at graphs in which we replace $f(x)$ by its tangent-line approximation at x^*.

There are four different cases, which can be divided according to whether the slope of the tangent line at x^* is between 0 and 1 (Figure 5.71a), greater than 1 (Figure 5.71b), between −1 and 0 (Figure 5.72a), and less than −1 (Figure 5.72b).

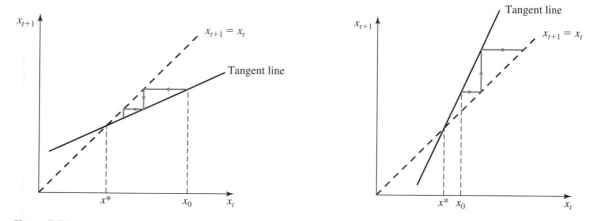

Figure 5.71a A locally stable node.

Figure 5.71b An unstable node.

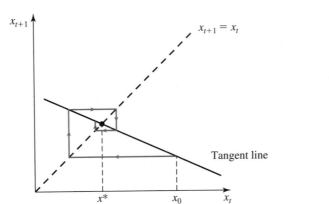

Figure 5.72a A locally stable spiral.

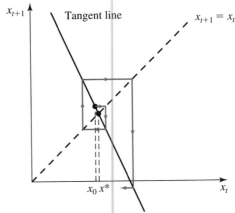

Figure 5.72b An unstable spiral.

We see that when the slope of the tangent line is between -1 and 1, x_t converges to the equilibrium (Figures 5.71a and 5.72a). The difference between Figures 5.71a and 5.72a is that in Figure 5.72a the solution x_t approaches the equilibrium in a spiral (thus exhibiting oscillatory behavior), whereas in Figure 5.71a it approaches it in one direction (thus exhibiting nonoscillatory behavior). Looking at Figure 5.71b, in which the slope is greater than 1, and Figure 5.72b, in which the slope is less than -1, we see that the solution x_t does not return to the equilibrium. In Figure 5.72b the solution moves away from the equilibrium in a spiral (thus exhibiting oscillatory behavior), whereas in Figure 5.71b it moves away in one direction (thus exhibiting nonoscillatory behavior). We call the equilibria in Figures 5.71a and 5.72a **locally stable**, and in Figures 5.71b and 5.72b **unstable**. Note that we added the word *locally* to *stable* to emphasize that this is a local property since we consider only perturbations close to the equilibrium.

Since the slope of the tangent-line approximation of $f(x)$ at x^* is given by $f'(x^*)$, we are led to the following criterion, which we will prove by calculus:

Criterion An equilibrium x^* of $x_{t+1} = f(x_t)$ is locally stable if

$$|f'(x^*)| < 1$$

Proof In Figures 5.71 and 5.72, we looked at the linearization of $f(x)$ about the equilibrium x^* and investigated how a small perturbation affects the future of the solution. Translating this approach into equations, we denote a small perturbation at time t by z_t and write

$$x_t = x^* + z_t$$

Then

$$x_{t+1} = f(x_t) = f(x^* + z_t)$$

Recall that the linear approximation for $f(x)$ at $x = a$ is $L(x) = f(a) + f'(a)(x - a)$. With $x = x^* + z_t$ and $a = x^*$, the linear approximation for $f(x^* + z_t)$ at x^* is

$$L(x^* + z_t) = f(x^*) + f'(x^*)z_t$$

We can approximate $x_{t+1} = x^* + z_{t+1}$ by

$$x^* + z_{t+1} \approx f(x^*) + f'(x^*)z_t$$

Since $f(x^*) = x^*$ (x^* is an equilibrium), we find that

$$z_{t+1} \approx f'(x^*)z_t \qquad (5.18)$$

This approximation should remind you of the equation $y_{t+1} = Ry_t$ for exponential growth, where we can identify y_t with z_t and R with $f'(x^*)$. Since the solution of $y_{t+1} = Ry_t$ is equal to $y_t = y_0 R^t$ and $R^t \to 0$ as $t \to \infty$ for $|R| < 1$, we obtain the criterion $|f'(x^*)| < 1$ for local stability. That is, if $|f'(x^*)| < 1$, then the perturbation z_t will converge to $z^* = 0$ or, equivalently, $x_t \to x^*$ as $t \to \infty$. ■

Looking back at Figures 5.71 and 5.72, we can see, in addition, that the equilibrium is approached without oscillations if $f'(x^*) > 0$ and with oscillations if $f'(x^*) < 0$.

EXAMPLE 1

Use the stability criterion to characterize the stability of the equilibria of

$$x_{t+1} = \frac{1}{4} - \frac{5}{4}x_t^2, \quad t = 0, 1, 2, \ldots$$

Solution

To find the equilibria, we need to solve

$$x = \frac{1}{4} - \frac{5}{4}x^2$$

$$\frac{5}{4}x^2 + x - \frac{1}{4} = 0$$

$$5x^2 + 4x - 1 = 0$$

The left-hand side can be factored into $(5x - 1)(x + 1)$, and we find that

$$(5x - 1)(x + 1) = 0 \quad \text{if} \quad x = \frac{1}{5} \quad \text{or} \quad x = -1$$

To determine stability, we need to evaluate the derivative of $f(x) = \frac{1}{4} - \frac{5}{4}x^2$ at the equilibria. Now,

$$f'(x) = -\frac{5}{2}x$$

so if $x = \frac{1}{5}$, then $|f'(\frac{1}{5})| = |-\frac{1}{2}| = \frac{1}{2} < 1$ and if $x = -1$, then $|f'(-1)| = |\frac{5}{2}| = \frac{5}{2} > 1$. Thus, $x = \frac{1}{5}$ is locally stable and $x = -1$ is unstable.

We can say a bit more, namely, that since $f'(\frac{1}{5}) = -\frac{1}{2} < 0$, the equilibrium $x^* = 1/5$ is approached with oscillations. ■

EXAMPLE 2

Use the stability criterion to characterize the stability of the equilibria of

$$x_{t+1} = \frac{x_t}{0.1 + x_t}, \quad t = 0, 1, 2, \ldots$$

Solution

To find the equilibria, we need to solve

$$x = \frac{x}{0.1 + x}$$

This immediately yields $x = 0$ as a solution. If $x \neq 0$, then after dividing by x, we have

$$1 = \frac{1}{0.1 + x} \quad \text{or} \quad 0.1 + x = 1 \quad \text{or} \quad x = 0.9$$

With $f(x) = \frac{x}{0.1+x}$, we find that

$$f'(x) = \frac{0.1 + x - x}{(0.1 + x)^2} = \frac{0.1}{(0.1 + x)^2}$$

Since $f'(0) = \frac{1}{0.1} = 10 > 1$, we conclude that $x^* = 0$ is unstable. Because $f'(0.9) = 0.1 \in (0, 1)$, we conclude that $x^* = 0.9$ is stable and is approached without oscillations. ■

■ **5.6.3 Examples**

In the remaining examples in this section, we will revisit three of the growth models we discussed in Chapter 2. There, we analyzed these models by simulations, and you simply had to believe for which parameters a nontrivial locally stable solution existed. We are now in the position to determine stability analytically, using the criterion from the previous subsection.

EXAMPLE 3

Beverton–Holt Recruitment Curve Denote by N_t the size of a population at time $t, t = 0, 1, 2, \ldots$. Find all equilibria and determine their stability for the Beverton–Holt recruitment curve

$$N_{t+1} = \frac{RN_t}{1 + \frac{R-1}{K} N_t}$$

where we assume that the parameters R and K satisfy $R > 1$ and $K > 0$.

Solution

To find the equilibria, we set

$$N = \frac{RN}{1 + \frac{R-1}{K} N}$$

and solve for N. This gives immediately the trivial solution $N = 0$ and, after division by $N \neq 0$,

$$1 = \frac{R}{1 + \frac{R-1}{K} N}$$

Solving the latter expression for N yields the nontrivial solution

$$1 + \frac{R-1}{K} N = R, \qquad \text{or} \qquad N = K$$

To determine stability of the two equilibria, we need to differentiate

$$f(N) = \frac{RN}{1 + \frac{R-1}{K} N}$$

Using the quotient rule, we find that

$$f'(N) = \frac{R\left(1 + \frac{R-1}{K} N\right) - RN \frac{R-1}{K}}{\left(1 + \frac{R-1}{K} N\right)^2}$$

$$= \frac{R}{\left(1 + \frac{R-1}{K} N\right)^2}$$

To determine the stability of the trivial equilibrium $N^* = 0$, we compute

$$f'(0) = R > 1$$

(since we assumed that $R > 1$). Thus, $N^* = 0$ is unstable. The stability of the nontrivial equilibrium $N^* = K$ can be determined by computing

$$f'(K) = \frac{R}{\left(1 + \frac{R-1}{K} K\right)^2} = \frac{1}{R}$$

Hence, $|f'(K)| < 1$ because $R > 1$. Consequently, $N^* = K$ is locally stable when $R > 1$, as we found in Chapter 2. Since $f'(K) > 0$, the equilibrium is approached without oscillations. ■

EXAMPLE 4

Logistic Growth Denote by N_t the size of a population at time $t, t = 0, 1, 2, \ldots$. Find all equilibria and determine their stability for the discrete logistic growth equation

$$N_{t+1} = N_t \left[1 + R \left(1 - \frac{N_t}{K} \right) \right]$$

where we assume that the parameters R and K are both positive.

Solution   To find the equilibria, we set

$$N = N \left[1 + R \left(1 - \frac{N}{K} \right) \right]$$

This yields the trivial solution $N = 0$ and the nontrivial solution $N = K$.
To determine stability, we need to differentiate

$$f(N) = N \left[1 + R \left(1 - \frac{N}{K} \right) \right]$$

Using the product rule, we find that

$$f'(N) = 1 + R \left(1 - \frac{N}{K} \right) + N \left(-\frac{R}{K} \right)$$

$$= 1 + R - \frac{2NR}{K}$$

Since $f'(0) = 1 + R > 1$, we conclude that $N^* = 0$ is unstable. Now,

$$f'(K) = 1 + R - 2R = 1 - R$$

Because $|f'(K)| = |1 - R| < 1$ if $-1 < 1 - R < 1$ or $2 > R > 0$, we conclude that $N^* = K$ is locally stable if $0 < R < 2$, as we saw in Chapter 2. We can say a bit more now: If $0 < R < 1$, then $N^* = K$ is approached without oscillations, since $f'(K) > 0$; if $1 < R < 2$, $N^* = K$ is approached *with* oscillations, since $f'(K) < 0$. ■

EXAMPLE 5

Ricker's Curve Denote by N_t the size of a population at time $t, t = 0, 1, 2, \ldots$. Find all equilibria and determine their stability for Ricker's curve,

$$N_{t+1} = N_t \exp \left[R \left(1 - \frac{N_t}{K} \right) \right]$$

where we assume that the parameter R is positive.

Solution   To find the equilibria, we set

$$N = N \exp \left[R \left(1 - \frac{N}{K} \right) \right]$$

This gives the trivial equilibrium $N = 0$ and

$$1 = \exp \left[R \left(1 - \frac{N}{K} \right) \right]$$

Solving the latter equation for N yields

$$R \left(1 - \frac{N}{K} \right) = 0, \quad \text{or} \quad N = K$$

To determine stability, we need to differentiate

$$f(N) = N \exp \left[R \left(1 - \frac{N}{K} \right) \right]$$

Using the product rule and the chain rule, we find that

$$f'(N) = \exp\left[R\left(1 - \frac{N}{K}\right)\right] + N \exp\left[R\left(1 - \frac{N}{K}\right)\right]\left(-\frac{R}{K}\right)$$

$$= \exp\left[R\left(1 - \frac{N}{K}\right)\right]\left(1 - \frac{NR}{K}\right)$$

Now,

$$f'(0) = e^R > 1$$

for $R > 0$, so $N^* = 0$ is unstable. Since

$$f'(K) = 1 - R$$

and $|f'(K)| = |1 - R| < 1$ if $-1 < 1 - R < 1$ or $0 < R < 2$, we conclude that $N^* = K$ is locally stable if $0 < R < 2$. We can say a bit more now: If $0 < R < 1$, then $N^* = K$ is approached without oscillations, since $f'(K) > 0$; if $1 < R < 2$, $N^* = K$ is approached *with* oscillations, since $f'(K) < 0$. ■

Section 5.6 Problems

■ 5.6.1

1. Assume a discrete-time population whose size at generation $t + 1$ is related to the size of the population at generation t by

$$N_{t+1} = (1.03)N_t, \quad t = 0, 1, 2, \ldots$$

(a) If $N_0 = 10$, how large will the population be at generation $t = 5$?

(b) How many generations will it take for the population size to reach double the size at generation 0?

2. Suppose a discrete-time population evolves according to

$$N_{t+1} = (0.9)N_t, \quad t = 0, 1, 2 \ldots$$

(a) If $N_0 = 50$, how large will the population be at generation $t = 6$?

(b) After how many generations will the size of the population be one-quarter of its original size?

(c) What will happen to the population in the long run—that is, as $t \to \infty$?

3. Assume the discrete-time population model

$$N_{t+1} = bN_t, \quad t = 0, 1, 2 \ldots$$

Assume also that the population increases by 2% each generation.
(a) Determine b.

(b) Find the size of the population at generation 10 when $N_0 = 20$.

(c) After how many generations will the population size have doubled?

4. Assume the discrete-time population model

$$N_{t+1} = bN_t, \quad t = 0, 1, 2, \ldots$$

Assume also that the population decreases by 3% each generation.
(a) Determine b.

(b) Find the size of the population at generation 10 when $N_0 = 50$.

(c) How long will it take until the population is one-half its original size?

5. Assume the discrete-time population model

$$N_{t+1} = bN_t, \quad t = 0, 1, 2, \ldots$$

Assume that the population increases by $x\%$ each generation.
(a) Determine b.

(b) After how many generations will the population size have doubled? Compute the doubling time for $x = 0.1, 0.5, 1, 2, 5$, and 10.

6. (a) Find all equilibria of

$$N_{t+1} = 1.3N_t, \quad t = 0, 1, 2, \ldots$$

(b) Use cobwebbing to determine the stability of the equilibria you found in (a).

7. (a) Find all equilibria of

$$N_{t+1} = 0.9N_t, \quad t = 0, 1, 2, \ldots$$

(b) Use cobwebbing to determine the stability of the equilibria you found in (a).

8. (a) Find all equilibria of

$$N_{t+1} = N_t, \quad t = 0, 1, 2, \ldots$$

(b) How will the population size N_t change over time, starting at time 0 with N_0?

■ 5.6.2

9. Use the stability criterion to characterize the stability of the equilibria of

$$x_{t+1} = \frac{2}{3} - \frac{2}{3}x_t^2, \quad t = 0, 1, 2, \ldots$$

10. Use the stability criterion to characterize the stability of the equilibria of

$$x_{t+1} = \frac{3}{5}x_t^2 - \frac{2}{5}, \quad t = 0, 1, 2, \ldots$$

11. Use the stability criterion to characterize the stability of the equilibria of

$$x_{t+1} = \frac{x_t}{0.5 + x_t}, \quad t = 0, 1, 2, \ldots$$

12. Use the stability criterion to characterize the stability of the equilibria of

$$x_{t+1} = \frac{x_t}{0.3 + x_t}, \quad t = 0, 1, 2, \ldots$$

13. (a) Use the stability criterion to characterize the stability of the equilibria of

$$x_{t+1} = \frac{5x_t^2}{4 + x_t^2}, \quad t = 0, 1, 2, \ldots$$

(b) Use cobwebbing to decide to which value x_t converges as $t \to \infty$ if **(i)** $x_0 = 0.5$ and **(ii)** $x_0 = 2$.

14. (a) Use the stability criterion to characterize the stability of the equilibria of

$$x_{t+1} = \frac{10x_t^2}{9 + x_t^2}, \quad t = 0, 1, 2, \ldots$$

(b) Use cobwebbing to decide to which value x_t converges as $t \to \infty$ if (i) $x_0 = 0.5$ and (ii) $x_0 = 3$.

■ **5.6.3**

15. Ricker's curve is given by

$$R(P) = \alpha P e^{-\beta P}$$

for $P \geq 0$, where P denotes the size of the parental stock and $R(P)$ the number of recruits. The parameters α and β are positive constants.

(a) Show that $R(0) = 0$ and $R(P) > 0$ for $P > 0$.

(b) Find

$$\lim_{P \to \infty} R(P)$$

(c) For what size of the parental stock is the number of recruits maximal?

(d) Does $R(P)$ have inflection points? If so, find them.

(e) Sketch the graph of $f(x)$ when $\alpha = 2$ and $\beta = 1/2$.

16. Suppose that the size of a fish population at generation t is given by

$$N_{t+1} = 1.5N_t e^{-0.001N_t}$$

for $t = 0, 1, 2, \ldots$.

(a) Assume that $N_0 = 100$. Find the size of the fish population at generation t for $t = 1, 2, \ldots, 20$.

(b) Assume that $N_0 = 800$. Find the size of the fish population at generation t for $t = 1, 2, \ldots, 20$.

(c) Determine all fixed points. On the basis of your computations in (a) and (b), make a guess as to what will happen to the population in the long run, starting from **(i)** $N_0 = 100$ and **(ii)** $N_0 = 800$.

(d) Use the cobwebbing method to illustrate your answer in (a).

(e) Explain why the dynamical system converges to the nontrivial fixed point.

17. Suppose that the size of a fish population at generation t is given by

$$N_{t+1} = 10N_t e^{-0.01N_t}$$

for $t = 0, 1, 2, \ldots$.

(a) Assume that $N_0 = 100$. Find the size of the fish population at generation t for $t = 1, 2, \ldots, 20$.

(b) Show that if $N_0 = 100 \ln 10$, then $N_t = 100 \ln 10$ for $t = 1, 2, 3, \ldots$; that is, show that $N^* = 100 \ln 10$ is a nontrivial fixed point, or equilibrium. How would you find N^*? Are there any other equilibria?

(c) On the basis of your computations in (a), make a prediction about the long-term behavior of the fish population when $N_0 = 100$. How does your answer compare with that in (b)?

(d) Use the cobwebbing method to illustrate your answer in (c).

In Problems 18–20, consider the following discrete-time dynamical system, which is called the discrete logistic model and which models the size of a population over time:

$$N_{t+1} = N_t \left[1 + R \left(1 - \frac{N_t}{100} \right) \right]$$

for $t = 0, 1, 2, \ldots$.

18. (a) Find all equilibria when $R = 0.5$.

(b) Investigate the system when $N_0 = 10$ and describe what you see.

19. (a) Find all equilibria when $R = 1.5$.

(b) Investigate the system when $N_0 = 10$ and describe what you see.

20. (a) Find all equilibria when $R = 2.5$.

(b) Investigate the system when $N_0 = 10$ and describe what you see.

In Problems 21–22, we investigate the canonical discrete-time logistic growth model

$$x_{t+1} = rx_t(1 - x_t)$$

for $t = 0, 1, 2, \ldots$.

21. Show that for $r > 1$, there are two fixed points. For which values of r is the nonzero fixed point locally stable?

22. Use a calculator or a spreadsheet to simulate the canonical discrete-time logistic growth model with $x_0 = 0.1$ for $t = 0, 1, 2, \ldots, 100$, and describe the behavior when

(a) $r = 3.20$ **(b)** $r = 3.52$ **(c)** $r = 3.80$

(d) $r = 3.83$ **(e)** $r = 3.828$

In Problems 23–25, we consider density-dependent population growth models of the form

$$N_{t+1} = R(N_t)N_t$$

The function $R(N)$ describes the per capita growth. Various forms have been considered. For each function $R(N)$, find all nontrivial fixed points N^* (i.e., $N^* > 0$) and determine the stability as a function of the parameter values. We assume that the function parameters are $r > 0$, $K > 0$, and $\gamma > 1$.

23. $R(N) = rN^{1-\gamma}$ **24.** $R(N) = \dfrac{r}{1 + N/K}$

25. $R(N) = e^{r(1-N/K)}$

■ 5.7 Numerical Methods: The Newton–Raphson Method (Optional)

Numerical methods are very important in the sciences, where we frequently encounter situations in which exact solutions are impossible. In Section 3.5, we encountered one method, the bisection method, for solving equations of the form $f(x) = 0$. Here we examine another method that is often much more efficient than the bisection method.

The Newton–Raphson method allows us to find solutions of equations of the form

$$f(x) = 0$$

The idea behind the method can be best explained graphically. (See Figure 5.73.)

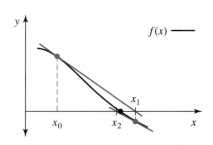

Figure 5.73 The graph of $y = f(x)$ and the first two iterations in the Newton–Raphson method.

Suppose that we wish to find the roots of $f(x) = 0$. We begin by choosing a value x_0 as our initial guess; then we replace the graph of $f(x)$ by its tangent line $y = L(x)$ to our initial guess x_0. Since the slope of the tangent line $y = L(x)$ is equal to $f'(x_0)$, and since $(x_0, f(x_0))$ is a point on the tangent line, we can use the point–slope form by

$$y - f(x_0) = f'(x_0)(x - x_0)$$

to find the equation of the tangent line. This line will intersect the x-axis at some point $x = x_1$, provided that $f'(x_0) \neq 0$. To find x_1, we set $x = x_1$ and $y = 0$ and solve for x_1:

$$0 - f(x_0) = f'(x_0)(x_1 - x_0) \qquad \text{or} \qquad x_1 = x_0 - \frac{f(x_0)}{f'(x_0)}$$

We use x_1 as our next guess and repeat the procedure, which will result in x_2, and so on. The value of x_{n+1} is then given by

$$x_{n+1} = x_n - \frac{f(x_n)}{f'(x_n)} \quad \text{for } n = 0, 1, 2, \ldots \tag{5.19}$$

This equation produces a sequence of numbers $x_1, x_2, \ldots$. If these values converge to the root of $f(x) = 0$, denoted by r, as $n \to \infty$, or

$$\lim_{n \to \infty} x_n = r$$

then computing $x_1, x_2, \ldots$ provides a numerical way to approximate r. The method will not always converge, but before we discuss situations in which the method fails, we give two examples in which it works.

EXAMPLE 1

Use the Newton–Raphson method to find a numerical approximation to a solution of the equation

$$x^2 - 3 = 0$$

Solution

The equation $x^2 - 3 = 0$ has roots $r = \sqrt{3}$ and $-\sqrt{3}$. Finding a numerical approximation to a root of this equation is therefore the same as finding a numerical approximation to $\sqrt{3}$ or $-\sqrt{3}$. You might ask why we don't just use our calculators and get the numerical value of $\sqrt{3}$. In fact, that is what you would do if you had to solve an equation such as $x^2 - 3 = 0$, but we will use this simple example to illustrate the method.

We will find a numerical approximation to $\sqrt{3}$ by the Newton–Raphson method. As our initial value, we choose a number close to $\sqrt{3}$, say, $x_0 = 2$. The function $f(x)$ is

$$f(x) = x^2 - 3$$

and its derivative is

$$f'(x) = 2x$$

Using (5.19), we find that

$$x_{n+1} = x_n - \frac{x_n^2 - 3}{2x_n}$$

$$= x_n - \frac{x_n}{2} + \frac{3}{2x_n} = \frac{x_n}{2} + \frac{3}{2x_n} \quad \text{for } n = 0, 1, 2, \ldots$$

(See Figure 5.74 for the first step of the approximation.)
The following table shows the results of the procedure:

n	x_n	$x_{n+1} = \dfrac{x_n}{2} + \dfrac{3}{2x_n}$	$\left\lvert \sqrt{3} - x_{n+1} \right\rvert$
0	2	1.75	0.0179
1	1.75	1.7321429	9.2×10^{-5}
2	1.7321429	1.7320508	2.45×10^{-9}

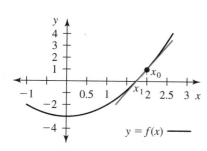

Figure 5.74 The graph of $f(x) = x^2 - 3$ and the first step in the Newton–Raphson method.

With the starting value $x_0 = 2$, after three steps we obtain the approximation 1.732050. Since a calculator yields the approximation 1.732050080757 for $\sqrt{3}$, our approximation by the Newton–Raphson method is correct to six decimal places. As you can see, the method can converge very quickly. In fact, it is used in many calculators to calculate roots. ■

In the next example, we cannot simply solve for x by algebraic manipulations. To find a solution of the given equation, we must resort to a numerical method.

EXAMPLE 2   Solve the equation

$$e^x + 1 + x = 0$$

Solution   It turns out that this equation can be solved only numerically. We will use the Newton–Raphson method to determine the root of $e^x + 1 + x = 0$. Since $e^x + 1 + x$ is positive for $x = 0$ and negative for $x = -2$, we conclude that there is a root of $e^x + 1 + x = 0$ in the interval $(-2, 0)$. We therefore choose a starting value in this interval, say, $x_0 = -1$. We find that

$$f(x) = e^x + x + 1$$
$$f'(x) = e^x + 1$$

Therefore,

$$x_{n+1} = x_n - \frac{f(x_n)}{f'(x_n)}$$

$$= x_n - \frac{e^{x_n} + x_n + 1}{e^{x_n} + 1} = x_n - 1 - \frac{x_n}{e^{x_n} + 1} \quad \text{for } n = 0, 1, 2, \ldots$$

The first step in the approximation is shown in Figure 5.75.
We summarize the first few steps of the iteration in the following table, where $x_0 = -1$:

n	x_n	$x_{n+1} = x_n - 1 - \dfrac{x_n}{e^{x_n} + 1}$
0	-1	-1.26894142
1	-1.26894142	-1.27845462
2	-1.27845462	-1.27846454
3	-1.27846454	-1.27846454

Note that x_2 and x_3 agree to four decimal places and that x_3 and x_4 agree to eight decimal places. We therefore suspect that $x = -1.278464$ is a root of $e^x + x + 1 = 0$, accurate to six decimal places. This can be confirmed with, for instance, a graphing calculator. ■

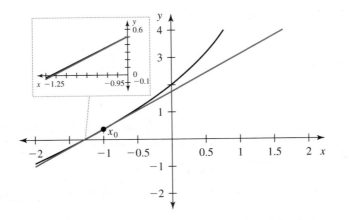

Figure 5.75 The graph of $f(x) = e^x + x + 1$ and the first step of the approximations in the Newton–Raphson method.

It is not our goal here to provide a complete description of the Newton–Raphson method (when it works, how quickly it converges, etc.). Instead, we give a few examples that illustrate some of the problems we can encounter when using the method.

EXAMPLE 3 This example illustrates a situation in which the Newton–Raphson method does not work. Let

$$f(x) = \begin{cases} \sqrt{x - 2.5} & \text{for } x \geq 2.5 \\ -\sqrt{2.5 - x} & \text{for } x \leq 2.5 \end{cases}$$

We wish to solve the equation $f(x) = 0$. We see immediately that $x = 2.5$ is a solution. Given that

$$f'(x) = \begin{cases} \dfrac{1}{2\sqrt{x - 2.5}} & \text{for } x > 2.5 \\ \dfrac{1}{2\sqrt{2.5 - x}} & \text{for } x < 2.5 \end{cases}$$

we apply the Newton–Raphson method and find that

$$x_{n+1} = x_n - \frac{f(x_n)}{f'(x_n)}$$

$$= \begin{cases} x_n - \dfrac{\sqrt{x_n - 2.5}}{\dfrac{1}{2\sqrt{x_n - 2.5}}} & \text{for } x_n > 2.5 \\ x_n - \dfrac{-\sqrt{2.5 - x_n}}{\dfrac{1}{2\sqrt{2.5 - x_n}}} & \text{for } x_n < 2.5 \end{cases}$$

$$= \begin{cases} x_n - 2(x_n - 2.5) = -x_n + 5 & \text{for } x_n > 2.5 \\ x_n + 2(2.5 - x_n) = -x_n + 5 & \text{for } x_n < 2.5 \end{cases}$$

$$= -x_n + 5 \quad \text{for } x_n \neq 2.5$$

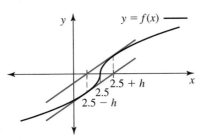

Figure 5.76 The graph of $f(x)$ in Example 3 where the Newton–Raphson method does not converge.

If $x_0 = 2.5 + h$, then $x_1 = 2.5 - h$, $x_2 = 2.5 + h$, $x_3 = 2.5 - h$, and so on; that is, successive approximations oscillate between $2.5 + h$ and $2.5 - h$ and never approach the root $r = 2.5$. A graph of $f(x)$, together with the iterations, is shown in Figure 5.76. Note that the two tangent lines are parallel and are used alternately in the approximation. ■

EXAMPLE 4

This example shows that an initial approximation can get worse. Use the Newton–Raphson method to find the root of

$$x^{1/3} = 0$$

when the starting value is $x_0 = 1$.

Solution   We set $f(x) = x^{1/3}$. Then $f'(x) = \frac{1}{3}x^{-2/3}$, and

$$x_{n+1} = x_n - \frac{f(x_n)}{f'(x_n)} = x_n - \frac{x_n^{1/3}}{\frac{1}{3}x_n^{-2/3}} = x_n - 3x_n = -2x_n$$

The following table shows successive values of x_n:

n	x_n	$x_{n+1} = -2x_n$
0	1	−2
1	−2	4
2	4	−8
3	−8	16

The successive values do not converge to the root $r = 0$. The situation is graphically illustrated in Figure 5.77.

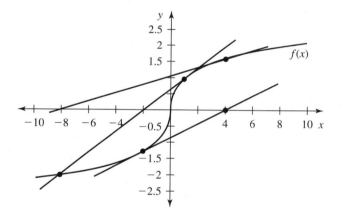

Figure 5.77 The graph of $f(x)$ in Example 4 where the Newton–Raphson method does not converge.

Graphing calculators have an option in the graphing menu that allows you to find roots of equations $f(x) = 0$ if you graph $f(x)$ and put the cursor close to the root. For this example, the calculator cannot produce an answer.   ■

EXAMPLE 5

This example shows that our successive approximations do not necessarily converge to the closest root. We wish to find a root of the equation

$$x^4 - x^2 = 0$$

If we set $f(x) = x^4 - x^2$, then $f'(x) = 4x^3 - 2x$ and

$$x_{n+1} = x_n - \frac{f(x_n)}{f'(x_n)}$$

$$= x_n - \frac{x_n^4 - x_n^2}{4x_n^3 - 2x_n} = x_n - \frac{x_n^3 - x_n}{4x_n^2 - 2}$$

We can solve for the roots of $x^4 - x^2 = 0$. Since $x^4 - x^2 = x^2(x^2 - 1)$, there are three roots: $r = 0, -1,$ and 1. Suppose that we set $x_0 = -0.7$. The closest root is -1. If we compute x_1, we find that

$$x_1 = (-0.7) - \frac{(-0.7)^3 - (-0.7)}{4(-0.7)^2 - 2} = 8.225$$

Successive values are collected in the following list:

$x_2 = 6.184$	$x_7 = 1.613$
$x_3 = 4.659$	$x_8 = 1.306$
$x_4 = 3.521$	$x_9 = 1.115$
$x_5 = 2.678$	$x_{10} = 1.024$
$x_6 = 2.059$	$x_{11} = 1.001$

We conclude that the method converges to the root $r = 1$. The situation is illustrated in Figure 5.78. ■

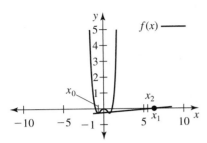

Figure 5.78 The graph of $f(x)$ in Example 5 together with the first two approximations.

Section 5.7 Problems

1. Use the Newton–Raphson method to find a numerical approximation to the solution of

$$x^2 - 7 = 0$$

that is correct to six decimal places.

2. Use the Newton–Raphson method to find a numerical approximation to the solution of

$$e^{-x} = x$$

that is correct to six decimal places.

3. Use the Newton–Raphson method to find a numerical approximation to the solution of

$$x^2 + \ln x = 0$$

that is correct to six decimal places.

4. The equation

$$x^2 - 5 = 0$$

has two solutions. Use the Newton–Raphson method to approximate the two solutions.

5. Use the Newton–Raphson method to solve the equation

$$\sin x = \frac{1}{2}x$$

in the interval $(0, \pi)$.

6. Let

$$f(x) = \begin{cases} \sqrt{x-1} & \text{for } x \geq 1 \\ -\sqrt{1-x} & \text{for } x \leq 1 \end{cases}$$

(a) Show that if you use the Newton–Raphson method to solve $f(x) = 0$, then the following statement holds: If $x_0 = 1 + h$, then $x_1 = 1 - h$, and if $x_0 = 1 - h$, then $x_1 = 1 + h$.

(b) Does the Newton–Raphson method converge? Use a graph to explain what happens.

7. In Example 4, we discussed the case of finding the root of $x^{1/3} = 0$.

(a) Given x_0, find a formula for $|x_n|$.

(b) Find

$$\lim_{n \to \infty} |x_n|$$

(c) Graph $f(x) = x^{1/3}$ and illustrate what happens when you apply the Newton–Raphson method.

8. In Example 5, we considered the equation

$$x^4 - x^2 = 0$$

(a) What happens if you choose

$$x_0 = -\frac{1}{2}\sqrt{2}$$

in the Newton–Raphson method? Give a graphical illustration.

(b) Repeat the procedure in (a) for $x_0 = -0.71$, and compare your result with the result we obtained in Example 5 when $x_0 = -0.70$. Give a graphical illustration and explain it in words. What happens when $x_0 = -0.6$? (This is an example in which small changes in the initial value can drastically change the outcome.)

9. Use the Newton–Raphson method to find a numerical approximation to the solution of

$$x^2 - 16 = 0$$

when your initial guess is (a) $x_0 = 3$ and (b) $x_0 = 4$.

10. Suppose that you wish to use the Newton–Raphson method to solve

$$f(x) = 0$$

numerically. It just so happens that your initial guess x_0 satisfies $f(x_0) = 0$. What happens to subsequent iterations? Give a graphical illustration of your results. [Assume that $f'(x_0) \neq 0$.]

■ 5.8 Antiderivatives

Throughout this and the previous chapter, we have repeatedly encountered differential equations. Occasionally, we showed that a certain function would solve a given differential equation. In this section, we will discuss a particular type of differential equation and address two important general questions: First, given a differential equation, how can we find its solutions? Second, given a solution of a differential equation, how do we know if it is the only one?

We will consider differential equations of the form

$$\frac{dy}{dx} = f(x)$$

That is, the rate of change of y with respect to x depends *only* on x. Our goal is to find functions y that satisfy $y' = f(x)$. We will see that if we can find one such function, then there is a whole family of functions with this property, all related by vertical translations. If we want to pick out one of these functions, we need to specify an **initial condition**—a point (x_0, y_0) on the graph of the function. Such a function is called a **solution** of the **initial-value problem**

$$\frac{dy}{dx} = f(x) \quad \text{with } y = y_0 \text{ when } x = x_0 \tag{5.20}$$

Let's look at an example before we begin a systematic treatment of the solution of differential equations of the form (5.20). Consider a population whose size at time t is denoted by $N(t)$, and assume that the growth rate is given by

$$\frac{dN}{dt} = \frac{1}{2\sqrt{t}} \quad \text{for } t > 0 \tag{5.21}$$

and that the size of the population at time 0 is $N(0) = 20$ [i.e., the initial condition is $N(0) = 20$]. Then

$$N(t) = \sqrt{t} + 20 \quad \text{for } t \geq 0 \tag{5.22}$$

is a solution of the differential equation (5.21) that satisfies the initial condition $N(0) = 20$. This is easy to check: First, note that $N(0) = \sqrt{0} + 20 = 20$. Second, differentiating $N(t)$, we find that

$$\frac{dN}{dt} = \frac{d}{dt}\left(\sqrt{t} + 20\right) = \frac{1}{2\sqrt{t}} \quad \text{for } t > 0$$

That is, $N(t) = \sqrt{t} + 20$ satisfies (5.21) with $N(0) = 20$.

This example shows that if we have a solution of a differential equation, we can verify that the solution indeed satisfies the differential equation by differentiating the solution. The method suggests that, in order to find solutions, we reverse the process of differentiation. This leads us to what is called an antiderivative, which is defined as follows:

Definition A function F is called an **antiderivative** of f on an interval I if $F'(x) = f(x)$ for all $x \in I$.

How can we find antiderivatives? Let

$$f(x) = 3x^2 \quad \text{for } x \in \mathbf{R}$$

To find the antiderivative of $f(x) = 3x^2$, we need to find a function whose derivative is $3x^2$. We can guess an answer, namely,

$$F(x) = x^3 \quad \text{for } x \in \mathbf{R}$$

which certainly satisfies $F'(x) = 3x^2$. But this is not the only answer. For example, take $F(x) = x^3 + 4$. Then $F'(x) = 3x^2$; hence, $x^3 + 4$ is also an antiderivative of $3x^2$. In fact, $F(x) = x^3 + C$, $x \in \mathbf{R}$, where C is any constant, is an antiderivative of $3x^2$. [We will soon show that there are no other functions $F(x)$ such that $F'(x) = 3x^2$.] The function $f(x)$ and some of its antiderivatives are shown in Figure 5.79. All of the antiderivatives are related through vertical shifts, since they all have the same derivative.

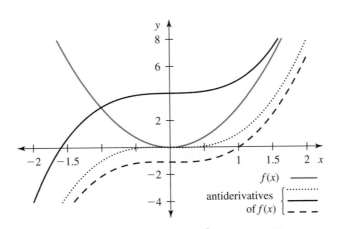

Figure 5.79 The function $f(x) = 3x^2$ and some of its antiderivatives.

Although we will learn rules that allow us to compute antiderivatives, this process is typically much more difficult than finding derivatives, and sometimes it takes ingenuity to come up with the correct answer; in addition, there are even cases where it is impossible to find an expression for an antiderivative.

We begin with two corollaries of the mean-value theorem that will help us in finding antiderivatives. The first of these is Corollary 2 from Section 5.1:

Corollary 2 If f is continuous on the closed interval $[a, b]$ and differentiable on the open interval (a, b), with $f'(x) = 0$ for all $x \in (a, b)$, then f is constant on $[a, b]$.

Note that Corollary 2 is the converse of the rule which states that $f'(x) = 0$ when $f(x) = c$, where c is a constant. It tells us that *all* antiderivatives of a function that is identically 0 are constant functions.

The next corollary tells us that functions with identical derivatives differ only by a constant; that is, to find all antiderivatives of a given function, we need only find one.

Corollary 3 If $F(x)$ and $G(x)$ are antiderivatives of the continuous function $f(x)$ on an interval I, then there exists a constant C such that

$$G(x) = F(x) + C \quad \text{for all } x \in I$$

Proof Since $F(x)$ and $G(x)$ are both antiderivatives of $f(x)$, it follows that $F'(x) = f(x)$ and $G'(x) = f(x)$; that is, $F'(x) = G'(x)$, or $F'(x) - G'(x) = 0$. Also, since

$$0 = F'(x) - G'(x) = \frac{d}{dx}[F(x) - G(x)]$$

it follows from Corollary 2, applied to the function $F - G$, that $F(x) - G(x) = C$, where C is a constant. ■

EXAMPLE 1

Find general antiderivatives for the given functions. Assume that all functions are defined for $x \in \mathbf{R}$.

(a) $f(x) = 3x^2$ (b) $f(x) = \cos x$ (c) $f(x) = e^x$

Solution

(a) If $F(x) = x^3$, then $F'(x) = 3x^2$; that is, $F(x) = x^3$ is a particular antiderivative of $3x^2$. Using Corollary 3, we find the general antiderivative simply by adding a constant; that is, the general antiderivative of $f(x) = 3x^2$ is the function $G(x) = x^3 + C$, where C is a constant.

(b) If $F(x) = \sin x$, then $F'(x) = \cos x$. Hence, the general antiderivative of $f(x) = \cos x$ is the function $G(x) = \sin x + C$, where C is a constant.

(c) If $F(x) = e^x$, then $F'(x) = e^x$ and the general antiderivative of $f(x) = e^x$ is the function $G(x) = e^x + C$, where C is a constant.   ■

EXAMPLE 2

Find general antiderivatives for the given functions. (Assume the largest possible domain.)

(a) $f(x) = 3x^5$ (b) $f(x) = x^2 + 2x - 1$
(c) $f(x) = e^{2x}$ (d) $f(x) = \sec^2(3x)$

Solution

(a) Since $\frac{d}{dx}(\frac{1}{2}x^6) = 3x^5$, $F(x) = \frac{1}{2}x^6 + C$ is the general antiderivative of $f(x) = 3x^5$.

(b) Since $\frac{d}{dx}(\frac{1}{3}x^3 + x^2 - x) = x^2 + 2x - 1$, $F(x) = \frac{1}{3}x^3 + x^2 - x + C$ is the general antiderivative of $f(x) = x^2 + 2x - 1$.

(c) Since $\frac{d}{dx}(\frac{1}{2}e^{2x}) = \frac{1}{2}e^{2x}(2) = e^{2x}$, $F(x) = \frac{1}{2}e^{2x} + C$ is the general antiderivative of $f(x) = e^{2x}$.

(d) Since $\frac{d}{dx}(\frac{1}{3}\tan(3x)) = \frac{1}{3}(\sec^2(3x))(3) = \sec^2(3x)$, $F(x) = \frac{1}{3}\tan(3x) + C$ is the general antiderivative of $f(x) = \sec^2(3x)$.   ■

Table 5-1 summarizes some of the rules for finding antiderivatives. We denote functions by $f(x)$ and $g(x)$ and their particular antiderivatives by $F(x)$ and $G(x)$, respectively. The general antiderivative is then obtained simply by adding a constant. The quantities a and k denote nonzero constants.

TABLE 5-1 A Collection of Antiderivatives

Function	Particular Antiderivative
$kf(x)$	$kF(x)$
$f(x) + g(x)$	$F(x) + G(x)$
$x^n, n \neq -1$	$\frac{1}{n+1}x^{n+1}$
$\frac{1}{x}$	$\ln|x|$
e^{ax}	$\frac{1}{a}e^{ax}$
$\sin(ax)$	$-\frac{1}{a}\cos(ax)$
$\cos(ax)$	$\frac{1}{a}\sin(ax)$
$\sec^2(ax)$	$\frac{1}{a}\tan(ax)$

We can now return to our initial question, namely, How do we solve differential equations of the form (5.20)?

EXAMPLE 3   Find the general solution of

$$\frac{dy}{dx} = \frac{3}{x^2} - 2x^2, \quad x \neq 0$$

Solution   Finding the general solution of this differential equation means finding the antiderivative of the function $f(x) = \frac{3}{x^2} - 2x^2$. Using Table 5-1, we obtain

$$\frac{3}{-1}x^{-1} - \frac{2}{3}x^3 = -\frac{3}{x} - \frac{2}{3}x^3$$

as a particular antiderivative. That is, the general solution is

$$y = -\frac{3}{x} - \frac{2}{3}x^3 + C, \quad x \neq 0 \qquad\qquad ■$$

In Example 3, we found the general solution of the given differential equation. Often, we wish to select a particular solution; for instance, we may know that the solution has to pass through a specific point (x_0, y_0). Such a problem is called an *initial-value problem*, as explained at the beginning of this section. We consider the initial-value problem posed in (5.21) again now.

EXAMPLE 4   Solve the initial-value problem

$$\frac{dN}{dt} = \frac{1}{2\sqrt{t}} \quad \text{for } t > 0 \text{ with } N(0) = 20$$

Solution   The general antiderivative of $f(t) = \frac{1}{2\sqrt{t}}$ is $F(t) = \sqrt{t} + C$. Since $N(0) = 20$, we have

$$N(0) = \sqrt{0} + C = 20, \qquad \text{or} \qquad C = 20$$

That is, the function

$$N(t) = \sqrt{t} + 20, \quad t \geq 0$$

solves the initial-value problem, and it is the only solution thereof. (See Figure 5.80.)
■

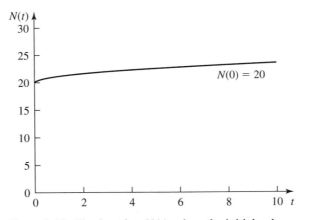

Figure 5.80 The function $N(t)$ solves the initial-value problem with $N(0) = 20$.

EXAMPLE 5   Solve the initial-value problem

$$\frac{dy}{dx} = -2x^2 + 3 \quad \text{for } x \in \mathbf{R} \text{ and } y = 10 \text{ when } x = 3$$

Solution   The general antiderivative of $f(x) = -2x^2 + 3$ is $F(x) = -\frac{2}{3}x^3 + 3x + C$. Since

$$F(3) = -\frac{2}{3}3^3 + (3)(3) + C = -9 + C = 10$$

it follows that $C = 19$. That is,

$$y = -\frac{2}{3}x^3 + 3x + 19, \quad x \in \mathbf{R}$$

solves the initial-value problem, and it is the only solution thereof.   ■

EXAMPLE 6   An object that falls freely in a vacuum, close to the surface of the earth, has a constant acceleration of

$$g = 9.81\frac{\text{m}}{\text{s}^2}$$

If the object is dropped from rest, find its velocity and the distance it has traveled t seconds after it was released.

Solution   If the distance function is $s(t)$, then the velocity $v(t)$ is given by

$$v(t) = \frac{d}{dt}s(t)$$

and the acceleration is given by

$$a(t) = \frac{d}{dt}v(t) = \frac{d^2}{dt^2}s(t)$$

We wish to solve the initial-value problem

$$\frac{d}{dt}v(t) = 9.81\frac{\text{m}}{\text{s}^2} \quad \text{when } v(0) = 0$$

A general solution is

$$v(t) = \left(9.81\frac{\text{m}}{\text{s}^2}\right)t + C$$

Since $v(0) = 0$, it follows that $C = 0$. Hence,

$$v(t) = \left(9.81\frac{\text{m}}{\text{s}^2}\right)t, \quad t \geq 0$$

To find the distance traveled, note that $s(0) = 0$ and

$$\frac{d}{dt}s(t) = v(t) = \left(9.81\frac{\text{m}}{\text{s}^2}\right)t, \quad t \geq 0$$

A general solution is

$$s(t) = \frac{1}{2}\left(9.81\frac{\text{m}}{\text{s}^2}\right)t^2 + C$$

Since $s(0) = 0$, it follows that $C = 0$. Thus,

$$s(t) = \frac{1}{2}\left(9.81\frac{\text{m}}{\text{s}^2}\right)t^2, \quad t \geq 0$$

Note that if t is measured in seconds, the unit of $v(t)$ is $\frac{\text{m}}{\text{s}}$ and the unit of $s(t)$ is m.

■

Section 5.8 Problems

In Problems 1–40, find the general antiderivative of the given function.

1. $f(x) = 4x^2 - x$

2. $f(x) = 2 - 5x^2$

3. $f(x) = x^2 + 3x - 4$

4. $f(x) = 3x^2 - x^4$

5. $f(x) = x^4 - 3x^2 + 1$

6. $f(x) = 2x^3 + x^2 - 5x$

7. $f(x) = 4x^3 - 2x + 3$

8. $f(x) = x - 2x^2 - 3x^3 - 4x^4$

9. $f(x) = 1 + \dfrac{1}{x} + \dfrac{1}{x^2}$

10. $f(x) = x^2 - \dfrac{2}{x^2} + \dfrac{3}{x^3}$

11. $f(x) = 1 - \dfrac{1}{x^2}$

12. $f(x) = x^3 - \dfrac{1}{x^3}$

13. $f(x) = \dfrac{1}{1+x}$

14. $f(x) = \dfrac{x}{1+x}$

15. $f(x) = 5x^4 + \dfrac{5}{x^4}$

16. $f(x) = x^7 + \dfrac{1}{x^7}$

17. $f(x) = \dfrac{1}{1+2x}$

18. $f(x) = \dfrac{1}{1+3x}$

19. $f(x) = e^{-3x}$

20. $f(x) = e^{x/2} + e^{-x/2}$

21. $f(x) = 2e^{2x}$

22. $f(x) = -3e^{-4x}$

23. $f(x) = \dfrac{1}{e^{2x}}$

24. $f(x) = \dfrac{3}{e^{-x}}$

25. $f(x) = \sin(2x)$

26. $f(x) = \cos(3x)$

27. $f(x) = \sin\left(\dfrac{x}{3}\right) + \cos\left(\dfrac{x}{3}\right)$

28. $f(x) = \cos\left(\dfrac{x}{5}\right) - \sin\left(\dfrac{x}{5}\right)$

29. $f(x) = 2\sin\left(\dfrac{\pi}{2}x\right) - 3\cos\left(\dfrac{\pi}{2}x\right)$

30. $f(x) = -3\sin\left(\dfrac{\pi}{3}x\right) + 4\cos\left(-\dfrac{\pi}{4}x\right)$

31. $f(x) = \sec^2(2x)$

32. $f(x) = \sec^2(-4x)$

33. $f(x) = \sec^2\left(\dfrac{x}{3}\right)$

34. $f(x) = \sec^2\left(-\dfrac{x}{4}\right)$

35. $f(x) = \dfrac{\sec x + \cos x}{\cos x}$

36. $f(x) = \sin^2 x + \cos^2 x$

37. $f(x) = x^{-7} + 3x^5 + \sin(2x)$

38. $f(x) = 2e^{-3x} + \sec^2\left(-\dfrac{x}{2}\right)$

39. $f(x) = \sec^2(3x - 1) + \dfrac{x^2 - 3}{x}$

40. $f(x) = 5e^{3x} - \sec^2(x - 3)$

In Problems 41–46, assume that a is a positive constant. Find the general antiderivative of the given function.

41. $f(x) = \dfrac{e^{(a+1)x}}{a}$

42. $f(x) = \sin^2(a^2x + 1)$

43. $f(x) = \dfrac{1}{ax + 3}$

44. $f(x) = \dfrac{a}{a + x}$

45. $f(x) = x^{a+2} - a^{x+2}$

46. $f(x) = \dfrac{e^{-ax} + e^{ax}}{2a}$

In Problems 47–58, find the general solution of the differential equation.

47. $\dfrac{dy}{dx} = \dfrac{2}{x} - x,\ x > 0$

48. $\dfrac{dy}{dx} = \dfrac{2}{x^3} - x^3,\ x > 0$

49. $\dfrac{dy}{dx} = x(1 + x),\ x > 0$

50. $\dfrac{dy}{dx} = e^{-4x},\ x > 0$

51. $\dfrac{dy}{dt} = t(1 - t),\ t \geq 0$

52. $\dfrac{dy}{dt} = t^2(1 - t^2),\ t \geq 0$

53. $\dfrac{dy}{dt} = e^{-t/2},\ t \geq 0$

54. $\dfrac{dy}{dt} = 1 - e^{-3t},\ t \geq 0$

55. $\dfrac{dy}{ds} = \sin(\pi s),\ 0 \leq s \leq 1$

56. $\dfrac{dy}{ds} = \cos(2\pi s),\ 0 \leq s \leq 1$

57. $\dfrac{dy}{dx} = \sec^2\left(\dfrac{x}{2}\right),\ -1 < x < 1$

58. $\dfrac{dy}{dx} = 1 + \sec^2\left(\dfrac{x}{4}\right),\ -1 < x < 1$

In Problems 59–72, solve the initial-value problem.

59. $\dfrac{dy}{dx} = 3x^2$, for $x \geq 0$ with $y = 1$ when $x = 0$

60. $\dfrac{dy}{dx} = \dfrac{x^2}{3}$, for $x \geq 0$ with $y = 2$ when $x = 0$

61. $\dfrac{dy}{dx} = 2\sqrt{x}$, for $x \geq 0$ with $y = 2$ when $x = 1$

62. $\dfrac{dy}{dx} = \dfrac{1}{2\sqrt{x}}$, for $x \geq 1$ with $y = 3$ when $x = 4$

63. $\dfrac{dN}{dt} = \dfrac{1}{t}$, for $t \geq 1$ with $N(1) = 10$

64. $\dfrac{dN}{dt} = \dfrac{t}{t + 2}$, for $t \geq 0$ with $N(0) = 2$

65. $\dfrac{dW}{dt} = e^t$, for $t \geq 0$ with $W(0) = 1$

66. $\dfrac{dW}{dt} = e^{-3t}$, for $t \geq 0$ with $W(0) = 2$

67. $\dfrac{dW}{dt} = e^{-3t}$, for $t \geq 0$ with $W(0) = 2/3$

68. $\dfrac{dW}{dt} = e^{-5t}$, for $t \geq 0$ with $W(0) = 1$

69. $\dfrac{dT}{dt} = \sin(\pi t)$, for $t \geq 0$ with $T(0) = 3$

70. $\dfrac{dT}{dt} = \cos(\pi t)$, for $t \geq 0$ with $T(0) = 3$

71. $\dfrac{dy}{dx} = \dfrac{e^{-x} + e^x}{2}$, for $x \geq 0$ with $y = 0$ when $x = 0$

72. $\dfrac{dN}{dt} = t^{-1/3}$, for $t > 0$ with $N(0) = 60$

73. Suppose that the length of a certain organism at age x is given by $L(x)$, which satisfies the differential equation

$$\frac{dL}{dx} = e^{-0.1x}, \quad x \geq 0$$

Find $L(x)$ if the limiting length L_∞ is given by

$$L_\infty = \lim_{x \to \infty} L(x) = 25$$

How big is the organism at age $x = 0$?

74. Fish are indeterminate growers; that is, their length $L(x)$ increases with age x throughout their lifetime. If we plot the growth rate dL/dx versus age x on semilog paper, a straight line with negative slope results. Set up a differential equation that relates growth rate and age. Solve this equation under the assumption that $L(0) = 5$, $L(1) = 10$, and

$$\lim_{x \to \infty} L(x) = 20$$

Graph the solution $L(x)$ as a function of x.

75. An object is dropped from a height of 100 ft. Its acceleration is 32 ft/s^2. When will the object hit the ground, and what will its speed be at impact?

76. Suppose that the growth rate of a population at time t undergoes seasonal fluctuations according to

$$\frac{dN}{dt} = 3\sin(2\pi t)$$

where t is measured in years and $N(t)$ denotes the size of the population at time t. If $N(0) = 10$ (measured in thousands), find an expression for $N(t)$. How are the seasonal fluctuations in the growth rate reflected in the population size?

77. Suppose that the amount of water contained in a plant at time t is denoted by $V(t)$. Due to evaporation, $V(t)$ changes over time. Suppose that the change in volume at time t, measured over a 24-hour period, is proportional to $t(24 - t)$, measured in grams per hour. To offset the water loss, you water the plant at a constant rate of 4 grams of water per hour.
(a) Explain why

$$\frac{dV}{dt} = -at(24 - t) + 4$$

$0 \le t \le 24$, for some positive constant a, describes this situation.
(b) Determine the constant a for which the net water loss over a 24-hour period is equal to 0.

Chapter 5 Key Terms

Discuss the following definitions and concepts:

1. Global or absolute extrema
2. Local or relative extrema: local minimum and local maximum
3. The extreme-value theorem
4. Fermat's theorem
5. Mean-value theorem
6. Rolle's theorem
7. Increasing and decreasing function
8. Monotonicity and the first derivative

9. Concavity: concave up and concave down
10. Concavity and the second derivative
11. Diminishing return
12. Candidates for local extrema
13. Monotonicity and local extrema
14. The second-derivative test for local extrema
15. Inflection points
16. Inflection points and the second derivative

17. Asymptotes: horizontal, vertical, and oblique
18. Using calculus to graph functions
19. L'Hospital's rule
20. Dynamical systems: cobwebbing
21. Stability of equilibria
22. Newton–Raphson method for finding roots
23. Antiderivative

Chapter 5 Review Problems

1. Suppose that

$$f(x) = xe^{-x}, \quad x \ge 0$$

(a) Show that $f(0) = 0$, $f(x) > 0$ for $x > 0$, and

$$\lim_{x \to \infty} f(x) = 0$$

(b) Find local and absolute extrema.
(c) Find inflection points.
(d) Use the foregoing information to graph $f(x)$.

2. Suppose that

$$f(x) = x \ln x, \quad x > 0$$

(a) Define $f(x)$ at $x = 0$ so that $f(x)$ is continuous for all $x \ge 0$.
(b) Find extrema and inflection points.
(c) Graph $f(x)$.

3. In Review Problem 17 of Chapter 2 we introduced the hyperbolic functions

$$\sinh x = \frac{e^x - e^{-x}}{2}, \quad x \in \mathbf{R}$$

$$\cosh x = \frac{e^x + e^{-x}}{2}, \quad x \in \mathbf{R}$$

$$\tanh x = \frac{e^x - e^{-x}}{e^x + e^{-x}}, \quad x \in \mathbf{R}$$

(a) Show that $f(x) = \tanh x$, $x \in \mathbf{R}$, is a strictly increasing function on $\mathbf{R}$. Evaluate

$$\lim_{x \to -\infty} \tanh x$$

and

$$\lim_{x \to \infty} \tanh x$$

(b) Use your results in (a) to explain why $f(x) = \tanh x$, $x \in \mathbf{R}$, is invertible, and show that its inverse function $f^{-1}(x) = \tanh^{-1} x$ is given by

$$f^{-1}(x) = \frac{1}{2} \ln \frac{1 + x}{1 - x}$$

What is the domain of $f^{-1}(x)$?
(c) Show that

$$\frac{d}{dx} f^{-1}(x) = \frac{1}{1 - x^2}$$

(d) Use your result in (c) and the facts that

$$\tanh x = \frac{\sinh x}{\cosh x}$$

and

$$\cosh^2 x - \sinh^2 x = 1$$

to show that

$$\frac{d}{dx} \tanh x = \frac{1}{\cosh^2 x}$$

4. Let

$$f(x) = \frac{x}{1 + e^{-x}}, \quad x \in \mathbf{R}$$

(a) Show that $y = 0$ is a horizontal asymptote as $x \to -\infty$.

(b) Show that $y = x$ is an oblique asymptote as $x \to +\infty$.

(c) Show that

$$f'(x) = \frac{1 + e^{-x}(1 + x)}{(1 + e^{-x})^2}$$

(d) Use your result in (c) to show that $f(x)$ has exactly one local extremum at $x = c$, where c satisfies the equation

$$1 + c + e^c = 0$$

[*Hint*: Use your result in (c) to show that $f'(x) = 0$ if and only if $1 + e^{-x}(1 + x) = 0$. Let $g(x) = 1 + e^{-x}(1 + x)$. Show that $g(x)$ is strictly increasing for $x < 0$, that $g(0) > 0$, and $g(-2) < 0$. This implies that $g(x) = 0$ has exactly one solution on $(-2, 0)$. Since $g(-2) < 0$ and $g(x)$ is strictly increasing for $x < 0$, there are no solutions of $g(x) = 0$ for $x < -2$. Furthermore, $g(x) > 0$ for $x > 0$; hence, there are no solutions of $g(x) = 0$ for $x > 0$.]

(e) The equation $1 + c + e^c = 0$ can be solved for c only numerically . With the help of a calculator, find a numerical approximation to c. [*Hint*: From (d), you know that $c \in (-2, 0)$.]

(f) Show that $f(x) < 0$ for $x < 0$. [This implies that, for $x < 0$, the graph of $f(x)$ is below the horizontal asymptote $y = 0$.]

(g) Show that $x - f(x) > 0$ for $x > 0$. [This implies that, for $x > 0$, the graph of $f(x)$ is below the oblique asymptote $y = x$.]

(h) Use your results in (a)–(g) and the fact that $f(0) = 0$ and $f'(0) = \frac{1}{2}$ to sketch the graph of $f(x)$.

5. Recruitment Model Ricker's curve describes the relationship between the size of the parental stock of some fish and the number of recruits. If we denote the size of the parental stock by P and the number of recruits by R, then Ricker's curve is given by

$$R(P) = \alpha P e^{-\beta P} \quad \text{for } P \geq 0$$

where α and β are positive constants. [Note that $R(0) = 0$; that is, without parents there are no offspring. Furthermore, $R(P) > 0$ when $P > 0$.]

We are interested in the size P of the parental stock that maximizes the number $R(P)$ of recruits. Since $R(P)$ is differentiable, we can use its first derivative to solve this problem.

(a) Use the product rule to show that, for $P > 0$,

$$R'(P) = \alpha e^{-\beta P}(1 - \beta P)$$
$$R''(P) = -\alpha \beta e^{-\beta P}(2 - \beta P)$$

(b) Show that $R'(P) = 0$ if $P = 1/\beta$ and that $R''(1/\beta) < 0$. This shows that $R(P)$ has a *local* maximum at $P = \frac{1}{\beta}$. Show that $R(1/\beta) = \frac{\alpha}{\beta}e^{-1} > 0$.

(c) To find the global maximum, you need to check $R(0)$ and $\lim_{P \to \infty} R(P)$. Show that

$$R(0) = 0 \quad \text{and} \quad \lim_{P \to \infty} R(P) = 0$$

and that this implies that there is a global maximum at $P = 1/\beta$.

(d) Show that $R(P)$ has an inflection point at $P = 2/\beta$.

(e) Sketch the graph of $R(P)$ for $\alpha = 2$ and $\beta = 1$.

6. Gompertz Growth Model The *Gompertz growth curve* is sometimes used to study the growth of populations. Its properties are quite similar to the properties of the logistic growth curve. The Gompertz growth curve is given by

$$N(t) = K \exp[-ae^{-bt}]$$

for $t \geq 0$, where K and b are positive constants.

(a) Show that $N(0) = Ke^{-a}$ and, hence,

$$a = \ln \frac{K}{N_0}$$

if $N_0 = N(0)$.

(b) Show that $y = K$ is a horizontal asymptote and that $N(t) < K$ if $N_0 < K$, $N(t) = K$ if $N_0 = K$, and $N(t) > K$ if $N_0 > K$.

(c) Show that

$$\frac{dN}{dt} = bN(\ln K - \ln N)$$

and

$$\frac{d^2N}{dt^2} = b\frac{dN}{dt}[\ln K - \ln N - 1]$$

(d) Use your results in (b) and (c) to show that $N(t)$ is strictly increasing if $N_0 < K$ and strictly decreasing if $N_0 > K$.

(e) When does $N(t), t \geq 0$, have an inflection point? Discuss its concavity.

(f) Graph $N(t)$ when $K = 100$ and $b = 1$ if **(i)** $N_0 = 20$, **(ii)** $N_0 = 70$, and **(iii)** $N_0 = 150$, and compare your graphs with your answers in (b)–(e).

7. Monod Growth Model The Monod growth curve is given by

$$f(x) = \frac{cx}{k + x}$$

for $x \geq 0$, where c and k are positive constants. The equation can be used to describe the specific growth rate of a species as a function of a resource level x.

(a) Show that $y = c$ is a horizontal asymptote for $x \to \infty$. The constant c is called the *saturation value*.

(b) Show that $f(x)$, $x \geq 0$, is strictly increasing and concave down. Explain why this implies that the saturation value is equal to the maximal specific growth rate.

(c) Show that if $x = k$, then $f(x)$ is equal to half the saturation value. (For this reason, the constant k is called the *half-saturation constant*.)

(d) Sketch a graph of $f(x)$ for $k = 2$ and $c = 5$, clearly marking the saturation value and the half-saturation constant. Compare this graph with one where $k = 3$ and $c = 5$.

(e) Without graphing the three curves, explain how you can use the saturation value and the half-saturation constant to decide quickly that

$$\frac{10x}{3 + x} > \frac{10x}{5 + x} > \frac{8x}{5 + x}$$

for $x \geq 0$.

8. Logistic Growth The logistic growth curve is given by

$$N(t) = \frac{K}{1 + (\frac{K}{N_0} - 1)e^{-rt}}$$

for $t \geq 0$, where K, N_0, and r are positive constants and $N(t)$ denotes the population size at time t.

(a) Show that $N(0) = N_0$ and that $y = K$ is a horizontal asymptote as $t \to \infty$.

(b) Show that $N(t) < K$ if $N_0 < K$, $N(t) = K$ if $N_0 = K$, and $N(t) > K$ if $N_0 > K$.

(c) Show that

$$\frac{dN}{dt} = rN\left(1 - \frac{N}{K}\right)$$

and

$$\frac{d^2N}{dt^2} = r\frac{dN}{dt}\left(1 - \frac{2N}{K}\right)$$

(d) Use your results in (b) and (c) to show that $N(t)$ is strictly increasing if $N_0 < K$ and strictly decreasing if $N_0 > K$.

(e) Show that if $N_0 < K/2$, then $N(t)$, $t \geq 0$, has exactly one inflection point $(t^*, N(t^*))$, with $t^* > 0$ and

$$N(t^*) = \frac{K}{2}$$

(i.e., half the carrying capacity). What happens if $K/2 < N_0 < K$? What if $N_0 > K$? Where is the function $N(t)$, $t \geq 0$, concave up, and where is it concave down?

(f) Sketch the graphs of $N(t)$ for $t \geq 0$ when
(i) $K = 100$, $N_0 = 10$, $r = 1$
(ii) $K = 100$, $N_0 = 70$, $r = 1$
(iii) $K = 100$, $N_0 = 150$, $r = 1$

Sketch the respective horizontal asymptotes. Mark the inflection point clearly if it exists.

9. **Genetics** A population is said to be in Hardy–Weinberg equilibrium, with respect to a single gene with two alleles A and a, if the three genotypes AA, Aa, and aa have respective frequencies $p_{AA} = \theta^2$, $p_{Aa} = 2\theta(1-\theta)$, and $p_{aa} = (1-\theta)^2$ for some $\theta \in [0, 1]$. Suppose that we take a random sample of size n from a population. We can show that the probability of observing n_1 individuals of type AA, n_2 individuals of type Aa, and n_3 individuals of type aa is given by

$$\frac{n!}{n_1!\,n_2!\,n_3!} p_{AA}^{n_1} p_{Aa}^{n_2} p_{aa}^{n_3}$$

where $n! = n(n-1)(n-2)\cdots 3 \cdot 2 \cdot 1$ (read "n factorial"). Here, $n_1 + n_2 + n_3 = n$. This probability depends on θ. There is a method, called the *maximum likelihood method*, that can be used to estimate θ. The principle is simple: We find the value of θ that maximizes the probability of the observed data. Since the coefficient

$$\frac{n!}{n_1!\,n_2!\,n_3!}$$

does not depend on θ, we need only maximize

$$L(\theta) = p_{AA}^{n_1} p_{Aa}^{n_2} p_{aa}^{n_3}$$

(a) Suppose $n_1 = 8$, $n_2 = 6$, and $n_3 = 3$. Compute $L(\theta)$.

(b) Show that if $L(\theta)$ is maximal for $\theta = \hat{\theta}$ (read "theta hat"), then $\ln L(\theta)$ is also maximal for $\theta = \hat{\theta}$.

(c) Use your result in (b) to find the value $\hat{\theta}$ that maximizes $L(\theta)$ for the data given in (a). The number $\hat{\theta}$ is the maximum likelihood estimate.

10. **Cell Volume** Suppose the volume of a cell is increasing at a constant rate of 10^{-12} cm^3/s.
(a) If $V(t)$ denotes the cell volume at time t, set up an initial-value problem that describes this situation if the initial volume is 10^{-10} cm^3.

(b) Solve the initial-value problem given in (a), and determine the volume of the cell after 10 seconds.

11. **Drug Concentration** Suppose the concentration $c(t)$ of a drug in the bloodstream at time t satisfies

$$\frac{dc}{dt} = -0.1e^{-0.3t}$$

for $t \geq 0$.
(a) Solve the differential equation under the assumption that there will eventually be no trace of the drug in the blood.

(b) How long does it take until the concentration reaches half its initial value?

12. **Resource-Limited Growth** Sterner (1997) investigated the effect of food quality on zooplankton dynamics. In his model, zooplankton may be limited by either carbon (C) or phosphorus (P). He argued that when food quantity is low, demand for carbon increases relative to demand for phosphorus in order to to satisfy basic metabolic requirements and that there should be a curve separating C- and P-limited growth when food quantity C_F (measured in amount of carbon per liter) is graphed as a function of the C:P ratio of the food, $f = C_F : P_F$. He derived the following equation for the curve separating the two regions:

$$C_F = \frac{m}{a_C g - \frac{C_Z a_P g}{P_Z f}}$$

Here, m denotes the respiration rate, g the ingestion rate, and a_C (a_P) the assimilation rate of carbon (phosphorus). C_Z and P_Z are, respectively, the carbon and the phosphorus content of the zooplankton.
(a) Show that the graph of $y = C_F(f)$ approaches the horizontal line $y = \frac{m}{a_C g}$ as $f \to \infty$.

(b) The graph of $C_F(f)$ has a vertical asymptote. Let $f = C_F : P_F$ (the C:P ratio of the food). Show that the vertical asymptote is at

$$\frac{C_F}{P_F} = \frac{C_Z}{P_Z}\frac{a_P}{a_C}$$

(c) Sketch a graph of $C_F(f)$ as a function of f.

(d) The graph of $C_F(f)$ separates C-limited (below the curve) from P-limited (above the curve) growth. Explain why this graph indicates that when food quantity is low, the demand for carbon relative to phosphorus increases.

13. **Velocity and Distance** Neglecting air resistance, the height (in meters) of an object thrown vertically from the ground with initial velocity v_0 is given by

$$h(t) = v_0 t - \frac{1}{2}gt^2$$

where $g = 9.81$m/s^2 is the earth's gravitational constant and t is the time (in seconds) elapsed since the object was released
(a) Find the time at which the object reaches its maximum height.
(b) Find the maximum height.
(c) Find the velocity of the object at the time it reaches its maximum height.
(d) At what time $t > 0$ will the object reach the initial height again?

6 | Integration

LEARNING OBJECTIVES

In this chapter, we introduce integration. Specifically, we will learn how to

- calculate areas, volumes, and lengths of curves;
- relate integration and differention; and
- use integration to calculate cumulative rates of change and average values.

■ 6.1 The Definite Integral

Computing the area of a region bounded by curves is an ancient problem that was solved in certain cases by Greek mathematicians. Foremost among them was Archimedes (circa 287–212 B.C.), who lived more than 2000 years ago. The Greeks used a method called *exhaustion*, which goes back to the Greek mathematician Eudoxus (circa 408–355 B.C.). The basic idea is to divide an area into very small regions consisting mostly of rectilinear figures of known area (such as triangles) so that the total area of the rectilinear figures is close to the area of the region of interest.

Interest in this problem resurfaced in the 17th century, when many new curves had been defined and an attempt was made to determine the areas of regions bounded by such curves. The first curves that were considered were of the form $y = x^n$, where n is a positive integer. In a letter to Roberval on September 22, 1636, Fermat wrote that he had succeeded in computing the area under the curve $y = x^n$. He noted that his method was different from the method employed by the Greeks, notably Archimedes. Whereas Archimedes used triangles to exhaust the area bounded by curved lines, Fermat used rectangles. This sounds like a minor difference, but it enabled Fermat to compute areas bounded by other curves that previously could not have been computed. He found that the area under the curve $y = x^n$ inscribed in a rectangle of width b and height b^n is $1/(n + 1)$ times the area of the rectangle; that is,

$$\frac{1}{n+1} b \cdot b^n = \frac{1}{n+1} b^{n+1}$$

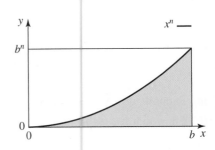

Figure 6.1 The area of the shaded region under the curve $y = x^n$ inscribed in the rectangle is $1/(n + 1)$ times the area of the rectangle.

(See Figure 6.1.) (This problem was independently solved around 1640 by Cavalieri, Pascal, Roberval, and Torricelli as well.)

Augustin-Louis Cauchy (1789–1857) was the first to define areas on the basis of the limit of the sum of areas of approximating rectangles. The definition we will use goes back to Georg Bernhard Riemann (1826–1866). His definition is more general than Cauchy's and allows for a larger class of functions to be used as boundary curves for areas.

■ 6.1.1 The Area Problem

We wish to find the surface area of the lake shown in Figure 6.2; to do so, we overlay a grid and count the number of squares that have a nonempty intersection with the lake. The sum of the areas of these squares will then approximate the area of the lake. The finer the grid, the closer our approximation will be to the true area of the lake.

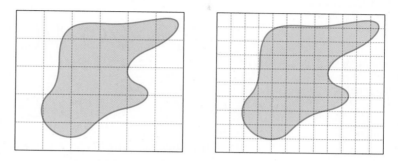

Figure 6.2 The outline of a lake with superimposed grid. The finer the grid, the more accurately the area of the lake can be determined.

Dividing a region into smaller regions of known area is the basic principle we will employ in this section to find the area of a region bounded by curves of continuous functions.

EXAMPLE 1   We will try to find the area of the region below the parabola $f(x) = x^2$ and above the x-axis between 0 and 1. (See Figure 6.3.) To do this, we divide the interval $[0, 1]$ into n subintervals of equal length and approximate the area of interest by a sum of the areas of rectangles, the widths of whose bases are equal to the lengths of the subintervals and whose heights are the values of the function at the left endpoints of these subintervals. This technique is illustrated in Figure 6.4 with $n = 5$.

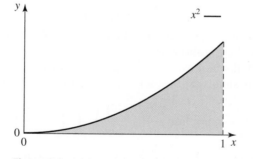

Figure 6.3 The region under the curve $y = x^2$ in Example 1.

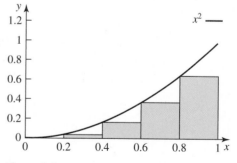

Figure 6.4 Approximation of the area under the curve $y = x^2$ from $x = 0$ to $x = 1$ by five rectangles.

In the figure, the base of each rectangle has width $1/5 = 0.2$. The height of the first rectangle is $f(0) = 0$, the height of the second rectangle is $f(0.2) = (0.2)^2$, the height of the third rectangle is $f(0.4) = (0.4)^2$, and so on. The area of a rectangle is the product of its width and height; adding up the areas of the approximating rectangles in Figure 6.4 yields

$$(0.2)(0)^2 + (0.2)(0.2)^2 + (0.2)(0.4)^2 + (0.2)(0.6)^2 + (0.2)(0.8)^2$$
$$= (0.2)\left[0^2 + (0.2)^2 + (0.4)^2 + (0.6)^2 + (0.8)^2\right] = 0.24$$

Thus, an approximation of the area between 0 and 1 is 0.24 when we use five subintervals.

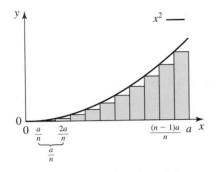

Figure 6.5 Approximation of the area under the curve $y = x^2$ from $x = 0$ to $x = a$ by n rectangles.

We turn now to the general case (illustrated in Figure 6.5), where the interval is $[0, a]$ and the number of subintervals is n. Since the interval $[0, a]$ has length a and the number of subintervals is n, each subinterval has length a/n. The left endpoints of successive subintervals are therefore $0, a/n, 2a/n, 3a/n, \ldots, (n-1)a/n$. The heights of the successive rectangles are then $f(0) = 0$, $f(a/n) = (a/n)^2$, $f(2a/n) = (2a/n)^2, \ldots, f((n-1)a/n) = ((n-1)a/n)^2$. We denote the sum of the areas of the n rectangles by S_n, where S stands for "sum" and the subscript n denotes the number of subintervals. We find that

$$S_n = \frac{a}{n} f(0) + \frac{a}{n} f\left(\frac{a}{n}\right) + \frac{a}{n} f\left(\frac{2a}{n}\right) + \cdots + \frac{a}{n} f\left(\frac{(n-1)a}{n}\right)$$

$$= \frac{a}{n} 0^2 + \frac{a}{n} \frac{a^2}{n^2} + \frac{a}{n} \frac{2^2 a^2}{n^2} + \cdots + \frac{a}{n} \frac{(n-1)^2 a^2}{n^2}$$

$$= \frac{a^3}{n^3}[1^2 + 2^2 + \cdots + (n-1)^2]$$

The sum of the squares of the first k integers can be computed:

$$1^2 + 2^2 + 3^2 + \cdots + k^2 = \frac{k(k+1)(2k+1)}{6} \qquad (6.1)$$

(For a proof of this formula, see Problem 31.) Using the preceding formula for $k = n - 1$, we obtain

$$S_n = \frac{a^3}{n^3} \frac{(n-1)(n-1+1)(2(n-1)+1)}{6}$$

$$= \frac{a^3}{n^3} \frac{(n-1)n(2n-1)}{6} = \frac{a^3}{6} \frac{n-1}{n} \frac{n}{n} \frac{2n-1}{n}$$

$$= \frac{a^3}{6}\left(1 - \frac{1}{n}\right) \cdot 1 \cdot \left(2 - \frac{1}{n}\right)$$

The finer the subdivision of $[0, a]$ (i.e., the larger n), the more accurate is the approximation, as illustrated in Figure 6.6, in which we see that the area of the region below the parabola and above the x-axis between 0 and a is more accurately approximated when we use a larger number of rectangles. Choosing finer and finer subdivisions means that we let n go to infinity. We find that

$$\lim_{n\to\infty} S_n = \lim_{n\to\infty} \frac{a^3}{6}\left(1 - \frac{1}{n}\right) \cdot 1 \cdot \left(2 - \frac{1}{n}\right) = \frac{a^3}{6}(1)(1)(2) = \frac{a^3}{3}$$

That is, the area under the parabola $y = x^2$ from 0 to a is equal to $a^3/3$. ■

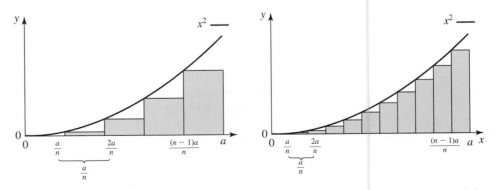

Figure 6.6 Increasing the number of approximating rectangles improves the accuracy of the approximation.

We see from Example 1 that computing areas entails summing a large number of terms. Therefore, before we continue our discussion of the computation of areas, we will spend some time on sums of the type we encountered in Example 1.

It will be convenient to have a shorthand notation for sums that involve a large number of terms:

> **Sigma Notation for Finite Sums** Let $a_1, a_2, \cdots, a_n$ be real numbers and n be a positive integer. Then
>
> $$\sum_{k=1}^{n} a_k = a_1 + a_2 + \cdots + a_n$$

The letter Σ is the capital Greek letter sigma, and the symbol $\sum_{k=1}^{n}$ means that we sum from $k = 1$ to $k = n$, where k is called the *index of summation*, the number 1 is the lower limit of summation, and the number n is the upper limit of summation. In text, instead of $\sum_{k=1}^{n}$ we will write $\sum_{k=1}^{n}$.

EXAMPLE 2

(a) Write each sum in expanded form:

(i) $\displaystyle\sum_{k=1}^{4} k = 1 + 2 + 3 + 4$

(ii) $\displaystyle\sum_{k=3}^{6} k^2 = 3^2 + 4^2 + 5^2 + 6^2$

(iii) $\displaystyle\sum_{k=1}^{n} \frac{1}{k} = 1 + \frac{1}{2} + \frac{1}{3} + \cdots + \frac{1}{n}$

(iv) $\displaystyle\sum_{k=1}^{5} 1 = 1 + 1 + 1 + 1 + 1$

(b) Write each sum in sigma notation:

(i) $2 + 3 + 4 + 5 = \displaystyle\sum_{k=2}^{5} k$

(ii) $1^3 + 2^3 + 3^3 + 4^3 = \displaystyle\sum_{k=1}^{4} k^3$

(iii) $1 + 3 + 5 + 7 + \cdots + (2n + 1) = \displaystyle\sum_{k=0}^{n} (2k + 1)$

(iv) $x + 2x^2 + 3x^3 + \cdots + nx^n = \displaystyle\sum_{k=1}^{n} kx^k$ ■

Occasionally, we will need a formula to sum the first n integers. The next example shows how this is done.

EXAMPLE 3

Show that

$$S_n = \sum_{k=1}^{n} k = 1 + 2 + 3 + \cdots + n = \frac{n(n + 1)}{2}$$

Solution   The following "trick" will enable us to compute the sum S_n: We write the sum in the usual order and in reverse order:

$$S_n = 1 + 2 + 3 + \cdots + n$$
$$S_n = n + (n - 1) + (n - 2) + \cdots + 2 + 1$$

Adding vertically, we find that

$$2S_n = (1 + n) + (2 + n - 1) + (3 + n - 2) + \cdots + (n + 1)$$
$$= (n + 1) + (n + 1) + (n + 1) + \cdots + (n + 1)$$

There are n terms (each $n + 1$) on the right-hand side. Hence,

$$2S_n = n(n + 1), \qquad \text{or} \qquad S_n = \frac{n(n + 1)}{2}$$

This method was used by Karl Friedrich Gauss (1777–1855) when he was 10 years old. One day, to keep the students busy, his teacher asked them to add all the numbers from 1 to 100. To his teacher's astonishment, Karl Friedrich quickly gave the correct answer, 5050. To find the answer, he did not add the numbers in their numerical order, but rather added $1 + 100, 2 + 99, 3 + 98, \ldots, 50 + 51$ (just as we did in the preceding derivation). Each term is equal to 101 and there are 50 such terms; thus, the answer is $50 \cdot 101 = 5050$. (Gauss went on to become one of the greatest mathematicians in history, contributing to geometry, number theory, astronomy, and other areas.) ■

The following rules are useful in evaluating finite sums:

Algebraic Rules

1. Constant-value rule: $\displaystyle\sum_{k=1}^{n} 1 = n$

2. Constant-multiple rule: $\displaystyle\sum_{k=1}^{n} c \cdot a_k = c \sum_{k=1}^{n} a_k$, where c is a constant that does not depend on k

3. Sum rule: $\displaystyle\sum_{k=1}^{n} (a_k + b_k) = \sum_{k=1}^{n} a_k + \sum_{k=1}^{n} b_k$

EXAMPLE 4   Use the algebraic rules to simplify the following sums:

(a) $\displaystyle\sum_{k=2}^{4} (3 + k) = \sum_{k=2}^{4} 3 + \sum_{k=2}^{4} k = (3 + 3 + 3) + (2 + 3 + 4) = 18$

(b) $\displaystyle\sum_{k=1}^{n} (k^2 - 2k) = \sum_{k=1}^{n} k^2 - 2 \sum_{k=1}^{n} k$

$$= \frac{n(n + 1)(2n + 1)}{6} - 2\frac{n(n + 1)}{2} = \frac{n(n + 1)(2n - 5)}{6}$$

[We used (6.1) to evaluate the first sum and Example 3 to evaluate the second sum.] ■

■ 6.1.2 Riemann Integrals

We will now develop a more systematic solution to the area problem. Although our approach will be similar to that in the previous subsection, we will look at a more general situation. We will now allow the function whose graph makes up the boundary of the region of interest to take on negative values as well as positive ones.

Furthermore, we will allow the rectangles that we use to approximate the area to vary in width and the points that we choose to compute the heights of the rectangles to be anywhere in their respective subintervals (which form the bases of the rectangles).

Accordingly, let f be a continuous function on the interval $[a, b]$. (See Figure 6.7.) We partition $[a, b]$ into n subintervals by choosing $n-1$ numbers $x_1, x_2, \ldots, x_{n-1}$ in (a, b) such that

$$a = x_0 < x_1 < x_2 < \cdots < x_{n-1} < x_n = b$$

The n subintervals

$$[x_0, x_1], [x_1, x_2], \ldots, [x_{n-1}, x_n]$$

form a **partition** of $[a, b]$, which we denote by $P = [x_0, x_1, x_2, \ldots, x_n]$. The partition P depends on the number of subintervals n and on the choice of points $x_0, x_1, \ldots, x_n$. For notational convenience, however, we will simply call a partition P. The length of the kth subinterval $[x_{k-1}, x_k]$ is denoted by Δx_k. The length of the longest subinterval is called the **norm** of P and is denoted by $\|P\|$ (read "norm of P"); thus,

$$\|P\| = \max\{\Delta x_1, \Delta x_2, \ldots, \Delta x_n\}$$

where $\max\{\Delta x_1, \Delta x_2, \ldots, \Delta x_n\}$ denotes the largest element of the set $\{\Delta x_1, \Delta x_2, \ldots, \Delta x_n\}$. In each subinterval $[x_{k-1}, x_k]$, we choose a point c_k and construct a rectangle with base Δx_k and height $|f(c_k)|$, as shown in Figure 6.7. If $f(c_k)$ is positive, then $f(c_k)\Delta x_k$ is the area of the rectangle. If $f(c_k)$ is negative, then $f(c_k)\Delta x_k$ is the negative of the rectangle's area. The sum of these products is denoted by S_P; that is,

$$S_P = \sum_{k=1}^{n} f(c_k)\, \Delta x_k$$

The value of the sum depends on the choice of the partition P (hence the subscript P on S) and the choice of the points $c_k \in [x_{k-1}, x_k]$ and is called a **Riemann sum** for f on $[a, b]$.

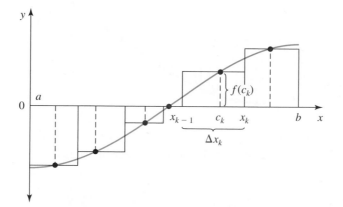

Figure 6.7 An illustration of a Riemann sum.

EXAMPLE 5

Use five equal subintervals with **(a)** left endpoints, **(b)** midpoints, and **(c)** right endpoints to find the Riemann sum for $f(x) = x^2$ on $[0, 1]$.

Solution

We partition $[0, 1]$ into five equal subintervals, each of length 0.2: $[0, 0.2]$, $[0.2, 0.4]$, $[0.4, 0.6]$, $[0.6, 0.8]$, and $[0.8, 1.0]$. The Riemann sum is given by

$$S_P = \sum_{k=1}^{5} f(c_k)\, \Delta x_k$$

where $\Delta x_k = 0.2$ in all three cases (a)–(c).

(a) We use left endpoints; thus, $c_1 = 0$, $c_2 = 0.2$, $c_3 = 0.4$, $c_4 = 0.6$, and $c_5 = 0.8$. We find that

$$S_P = (0.2)[0^2 + (0.2)^2 + (0.4)^2 + (0.6)^2 + (0.8)^2] = 0.24$$

(b) We use midpoints; hence, $c_1 = 0.1$, $c_2 = 0.3$, $c_3 = 0.5$, $c_4 = 0.7$, and $c_5 = 0.9$. We get

$$S_P = (0.2)[(0.1)^2 + (0.3)^2 + (0.5)^2 + (0.7)^2 + (0.9)^2] = 0.33$$

(c) We use right endpoints; therefore, $c_1 = 0.2$, $c_2 = 0.4$, $c_3 = 0.6$, $c_4 = 0.8$, and $c_5 = 1.0$. We obtain

$$S_P = (0.2)[(0.2)^2 + (0.4)^2 + (0.6)^2 + (0.8)^2 + (1.0)^2] = 0.44$$

Comparing S_P in (a)–(c) shows that the Riemann sum depends on the choice of the points $c_k \in [x_{k-1}, x_k]$. ■

To obtain a better approximation, we need to choose finer and finer partitions of $[a, b]$ so that the rectangles fill out the region between the curve and the x-axis more and more accurately. (See Figure 6.8.) A finer partition means that both the number of subintervals becomes larger and the length of the longest subinterval becomes smaller, so the norm of the partition P becomes smaller. One way to do this is to choose a sequence of partitions $P = P(n) = [x_0, x_1, x_2, \ldots, x_n]$, $n = 1, 2, 3, \ldots$, in such a way that $\|P(n)\| > \|P(n+1)\|$ for $n = 1, 2, 3, \ldots$. In fact, we will take the limit $\|P(n)\| \to 0$ as $n \to \infty$. For notational convenience, we will omit n and simply write $\|P\| \to 0$, but keep in mind that this means that, simultaneously, the number of subintervals goes to infinity and the length of the longest subinterval goes to 0. The limit of S_P as $\|P\| \to 0$ (if it exists) is called the *definite integral* of f from a to b.

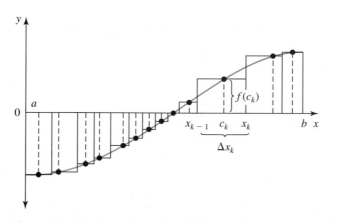

Figure 6.8 A finer partition than in Figure 6.7.

Definite Integral Let $P = [x_0, x_1, x_2, \ldots, x_n]$, $n = 1, 2, \ldots$, be a sequence of partitions of $[a, b]$ with $\|P\| \to 0$. Set $\Delta x_k = x_k - x_{k-1}$ and $c_k \in [x_{k-1}, x_k]$. The **definite integral** of f from a to b is

$$\int_a^b f(x)\,dx = \lim_{\|P\| \to 0} \sum_{k=1}^n f(c_k)\,\Delta x_k$$

if the limit exists, in which case f is said to be (Riemann) **integrable** on the interval $[a, b]$.

The symbol $\int$ is an elongated S (as in "sum") and was introduced by Leibniz. It is called the **integral sign**. In the notation $\int_a^b f(x)\,dx$ [read "the integral from a

to b of $f(x)\,dx$"], $f(x)$ is called the **integrand**, the number a is the **lower limit of integration**, and b is the **upper limit of integration**. Although the symbol dx by itself has no meaning, it should remind you that, as we take the limit, the widths of the subintervals become ever smaller. The x in dx indicates that x is the independent variable and that we integrate with respect to x.

The phrase "if the limit exists" means, in particular, that the value of $\lim_{\|P\|\to 0} S_P$ does not depend on how we choose the partitions and the points $c_k \in [x_{k-1}, x_k]$ as we take the limit. An important result tells us that if f is continuous on $[a, b]$, the definite integral of f on $[a, b]$ exists.

> **Theorem** All continuous functions are Riemann integrable; that is, if $f(x)$ is continuous on $[a, b]$, then
>
> $$\int_a^b f(x)\,dx$$
>
> exists.

The class of functions that are Riemann integrable is quite a bit larger than the set of continuous functions; for instance, functions that are both bounded (functions for which there exists an $M < \infty$ such that $|f(x)| < M$ for all x over which we wish to integrate) and piecewise continuous (continuous except for a finite number of discontinuities) are integrable. (See Figure 6.9.) We will be concerned primarily with continuous functions in this text; knowing that continuous functions are Riemann integrable will therefore suffice for the most part.

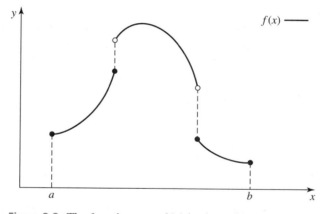

Figure 6.9 The function $y = f(x)$ is piecewise continuous and bounded on $[a, b]$.

Note that $\int_a^b f(x)\,dx$ is a number that does not depend on x. We could have written $\int_a^b f(u)\,du$ (or any other letter in place of x) and meant the same thing.

EXAMPLE 6 Express the definite integral

$$\int_3^7 (x^2 - 1)\,dx$$

as a limit of Riemann sums.

Solution We have

$$\int_3^7 (x^2 - 1)\,dx = \lim_{\|P\|\to 0} \sum_{k=1}^n (c_k^2 - 1)\,\Delta x_k$$

where $x_0 = 3 < x_1 < x_2 < \cdots < x_n = 7$, $n = 1, 2, \ldots$, is a sequence of partitions of $[3, 7]$, $c_k \in [x_{k-1}, x_k]$, $\Delta x_k = x_k - x_{k-1}$, and the limit as $\|P\| \to 0$ means that the norm of the partition tends to 0 (and, simultaneously, the number of subintervals goes to infinity). ■

EXAMPLE 7

Express the limit

$$\lim_{\|P\|\to 0} \sum_{k=1}^{n} \sqrt{c_k - 1}\, \Delta x_k$$

as a definite integral, where $P = [x_0, x_1, \ldots, x_n]$, $n = 1, 2, \ldots$, is a sequence of partitions of $[2, 4]$ into n subintervals, $\Delta x_k = x_k - x_{k-1}$, and $c_k \in [x_{k-1}, x_k]$.

Solution

$$\lim_{\|P\|\to 0} \sum_{k=1}^{n} \sqrt{c_k - 1}\, \Delta x_k = \int_{2}^{4} \sqrt{x - 1}\, dx$$ ∎

EXAMPLE 8

Evaluate

$$\int_{0}^{2} x^2\, dx$$

Solution

We evaluated the Riemann sum and its limit for $y = x^2$ from 0 to a in Example 1 and found that

$$\lim_{n\to\infty} S_n = \int_{0}^{a} x^2\, dx = \frac{a^3}{3}$$

With $a = 2$, we therefore have

$$\int_{0}^{2} x^2\, dx = \frac{8}{3}$$ ∎

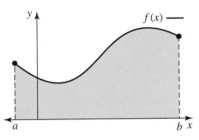

Figure 6.10 The area of a region under the curve of a positive function is given by the definite integral $\int_a^b f(x)\, dx$.

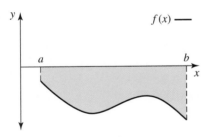

Figure 6.11 The area of a region under the curve of a negative function is given by $-\int_a^b f(x)\, dx$.

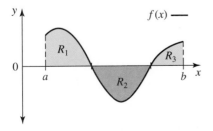

Figure 6.12 A_+ is the combined area of R_1 and R_3, and A_- is the area of R_2. Then $\int_a^b f(x)\, dx = A_+ - A_-$.

Geometric Interpretation of Definite Integrals In Example 1, we computed the area of the region below the parabola $y = x^2$ and above the x-axis between 0 and a by approximating the region with n rectangles of equal width and then taking the limit as $n \to \infty$. More generally, we can now define the **area** of a region A above the x-axis, as shown in Figure 6.10, as the limiting value (if it exists) of the Riemann sum of approximating rectangles. (Note that an area is always a positive number.) This definition allows us to interpret the definite integral of a nonnegative function as an area.

If $f(x) \leq 0$ on $[a, b]$, then the definite integral $\int_a^b f(x)\, dx$ is less than or equal to 0 and its value is the negative of the area of the region above the graph of f and below the x-axis between a and b. (See Figure 6.11.) We refer to the latter region as a "signed area." (A signed area may be either positive or negative.)

In general, a definite integral can thus be interpreted as a difference of areas, as illustrated in Figure 6.12. If A_+ denotes the total area of the region above the x-axis and below the graph of f (where $f \geq 0$) and A_- denotes the total area of the region below the x-axis and above the graph of f (where $f \leq 0$), then

$$\int_{a}^{b} f(x)\, dx = A_+ - A_-$$

---

1. If f is integrable on $[a, b]$ and $f(x) \geq 0$ on $[a, b]$, then

$$\int_{a}^{b} f(x)\, dx = \left[\begin{array}{l} \text{the area of the region between the} \\ \text{graph of } f \text{ and the } x\text{-axis from } a \text{ to } b \end{array} \right]$$

2. If f is integrable on $[a, b]$, then

$$\int_{a}^{b} f(x)\, dx = [\text{area above } x\text{-axis}] - [\text{area below } x\text{-axis}]$$

EXAMPLE 9 Find the value of

$$\int_{-2}^{3} (2x + 1)\, dx$$

by interpreting it as the signed area of an appropriately chosen region.

Solution We graph $y = 2x + 1$ between -2 and 3. (See Figure 6.13.) The line intersects the x-axis at $x = -1/2$. The area of the region to the left of $-1/2$ between the graph of $y = 2x + 1$ and the x-axis is denoted by A_-; the area of the region to the right of $-1/2$ between the graph of $y = 2x + 1$ and the x-axis is denoted by A_+. Both regions are triangles whose areas can be computed with the formula $A = \frac{1}{2}bh$ from geometry:

$$A_- = \frac{1}{2} \cdot \frac{3}{2} \cdot 3 = \frac{9}{4}$$

$$A_+ = \frac{1}{2} \cdot \frac{7}{2} \cdot 7 = \frac{49}{4}$$

Therefore,

$$\int_{-2}^{3} (2x + 1)\, dx = A_+ - A_- = \frac{49}{4} - \frac{9}{4} = \frac{40}{4} = 10$$

■

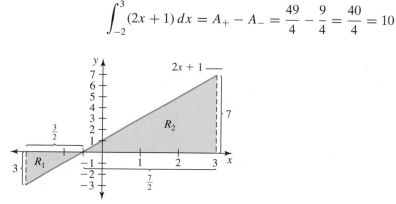

Figure 6.13 The area of R_1 is A_-; the area of R_2 is A_+.

EXAMPLE 10 Find the value of

$$\int_{0}^{2\pi} \sin x\, dx$$

by interpreting it as the signed area of an appropriately chosen region.

Solution We graph $y = \sin x$ from 0 to 2π. (See Figure 6.14.) The function $f(x) = \sin x$ is symmetric about $x = \pi$. It follows from this symmetry that the area of the region below the graph of f and above the x-axis between 0 and π (denoted by A_+) is the

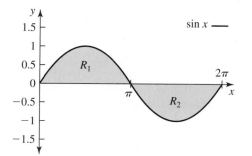

Figure 6.14 The graph of $f(x) = \sin x$, $0 \le x \le 2\pi$, in Example 10. The area of R_1 is A_+; the area of R_2 is A_-.

same as the area of the region above the graph of f and below the x-axis between π and 2π (denoted by A_-). Therefore, $A_+ = A_-$ and

$$\int_0^{2\pi} \sin x \, dx = A_+ - A_- = 0$$ ■

EXAMPLE 11   Find the value of

$$\int_0^2 \sqrt{4 - x^2} \, dx$$

by interpreting it as the signed area of an appropriately chosen region.

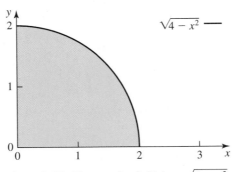

Figure 6.15   The graph of $f(x) = \sqrt{4 - x^2}$ is a quarter-circle. The area of the shaded region is equal to $\int_0^2 \sqrt{4 - x^2} \, dx$.

Solution   The graph of $y = \sqrt{4 - x^2}, 0 \le x \le 2$, is the quarter-circle with center at $(0, 0)$ and radius 2 in the first quadrant. (See Figure 6.15.) Since the area of a circle with radius 2 is $\pi(2)^2 = 4\pi$, the area of a quarter-circle is $4\pi/4 = \pi$. Hence,

$$\int_0^2 \sqrt{4 - x^2} \, dx = \pi$$ ■

■ 6.1.3 Properties of the Riemann Integral

In this subsection, we collect important properties that will help us to evaluate definite integrals.

Properties   Assume that f is integrable over $[a, b]$. Then

1. $\displaystyle\int_a^a f(x) \, dx = 0$ and

2. $\displaystyle\int_b^a f(x) \, dx = -\int_a^b f(x) \, dx$

The first integral says that the signed area between a and a is equal to 0; given that the width of the area is equal to 0, we expect the area to be 0 as well. The second property gives an orientation to the integral; for instance, if $f(x)$ is nonnegative on $[a, b]$, then $\int_a^b f(x) \, dx$ is nonnegative and can be interpreted as the area of the region between the graph of $f(x)$ and the x-axis from a to b. If we reverse the direction of the integration—that is, compute $\int_b^a f(x) \, dx$—we want the integral to be negative.

The next three properties follow immediately from the definition of the definite integral as the limit of a sum of areas of approximating rectangles.

Properties Assume that f and g are integrable over $[a, b]$.

3. If k is a constant, then

$$\int_a^b kf(x)\,dx = k \int_a^b f(x)\,dx$$

4. $\displaystyle\int_a^b [f(x) + g(x)]\,dx = \int_a^b f(x)\,dx + \int_a^b g(x)\,dx$

5. If f is integrable over an interval containing the three numbers a, b, and c, then

$$\int_a^b f(x)\,dx = \int_a^c f(x)\,dx + \int_c^b f(x)\,dx$$

We prove Property (4) to illustrate how the definition of the definite integral can be used to prove its properties.

Proof of (4)   We choose a sequence of partitions $P = [x_0, x_1, \ldots, x_n]$ of $[a, b]$ into n subintervals, $n = 1, 2, \ldots$, with $\|P\| \to 0$, $\Delta x_k = x_k - x_{k-1}$, and $c_k \in [x_{k-1}, x_k]$, and we then use the definition of a definite integral:

$$\int_a^b [f(x) + g(x)]\,dx = \lim_{\|P\| \to 0} \sum_{k=1}^n [f(c_k) + g(c_k)]\,\Delta x_k$$

$$= \lim_{\|P\| \to 0} \sum_{k=1}^n [f(c_k)\,\Delta x_k + g(c_k)\,\Delta x_k]$$

Applying the sum rule for finite sums, we find that

$$= \lim_{\|P\| \to 0} \left[\sum_{k=1}^n f(c_k)\,\Delta x_k + \sum_{k=1}^n g(c_k)\,\Delta x_k \right]$$

Since f and g are integrable, the individual limits exist, and we get

$$= \lim_{\|P\| \to 0} \sum_{k=1}^n f(c_k)\,\Delta x_k + \lim_{\|P\| \to 0} \sum_{k=1}^n g(c_k)\,\Delta x_k$$

Using the definition of definite integrals again, we obtain

$$= \int_a^b f(x)\,dx + \int_a^b g(x)\,dx \qquad ■$$

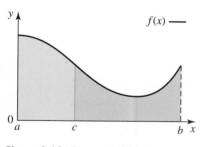

Figure 6.16 Property (5) when $a < c < b$.

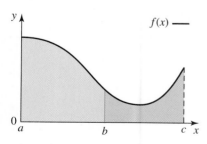

Figure 6.17 Property (5) when $a < b < c$.

Property (5) is an addition property. Rather than proving that it holds, we give two special cases (illustrated in Figures 6.16 and 6.17). In the first case (Figure 6.16), $a < c < b$ and $f(x) \geq 0$ for $x \in [a, b]$. The definite integral $\int_a^b f(x)\,dx$ can then be interpreted as the area between the graph of $f(x)$ and the x-axis from a to b. We see from the figure that this area is composed of two areas: the area between the graph of $f(x)$ and the x-axis from a to c and the area between the graph of $f(x)$ and the x-axis from c to b. We can express this relationship mathematically as

$$\int_a^b f(x)\,dx = \int_a^c f(x)\,dx + \int_c^b f(x)\,dx$$

which is Property (5) in this special case.

In the second case we wish to discuss, $a < b < c$ and $f(x) \geq 0$ for $x \in [a, c]$ (Figure 6.17). From the figure, we see that

$$\int_a^b f(x)\,dx = \int_a^c f(x)\,dx - \int_b^c f(x)\,dx$$

But because of Property (2),

$$\int_b^c f(x)\,dx = -\int_c^b f(x)\,dx$$

Therefore,

$$\int_a^b f(x)\,dx = \int_a^c f(x)\,dx + \int_c^b f(x)\,dx$$

as stated in Property (5).

Property (5) is much more general: The function f need not be positive as in Figures 6.16 and 6.17 (it merely needs to be integrable), and the numbers a, b, and c can be arranged in any order on the number line (not just $a < c < b$, as in Figure 6.16, or $a < b < c$, as in Figure 6.17). The next example shows how to use this property.

EXAMPLE 12   Given that $\int_0^a x^2 = a^3/3$, evaluate

$$\int_1^4 (3x^2 + 2)\,dx$$

Solution

$$\int_1^4 (3x^2 + 2)\,dx = 3\int_1^4 x^2\,dx + \int_1^4 2\,dx$$

To evaluate $\int_1^4 x^2\,dx$, we use the addition property (5) and write

$$\int_1^4 x^2\,dx = \int_1^0 x^2\,dx + \int_0^4 x^2\,dx$$

Since

$$\int_1^0 x^2\,dx = -\int_0^1 x^2\,dx$$

it follows that

$$\int_1^4 x^2\,dx = -\int_0^1 x^2\,dx + \int_0^4 x^2\,dx$$

which can be evaluated with the use of $\int_0^a x^2 = a^3/3$. To evaluate $\int_1^4 2\,dx$, we note that $y = 2$ is a horizontal line that intersects the y-axis at $y = 2$. The region under $y = 2$ from 1 to 4 is therefore a rectangle with base $4 - 1 = 3$ and height 2. Hence,

$$\int_1^4 (3x^2 + 2)\,dx = 3\left[\int_0^4 x^2\,dx - \int_0^1 x^2\,dx\right] + \int_1^4 2\,dx$$

$$= 3\left(\frac{4^3}{3} - \frac{1^3}{3}\right) + (2)(3)$$

$$= 64 - 1 + 6 = 69 \qquad\blacksquare$$

The next three properties are called order properties. They allow us either to compare definite integrals or say something about how big or small a particular definite integral can be. We first state the properties and then explain what they mean geometrically.

Properties Assume that f and g are integrable over $[a, b]$.

6. If $f(x) \geq 0$ on $[a, b]$, then $\int_a^b f(x)\,dx \geq 0$.

7. If $f(x) \leq g(x)$ on $[a, b]$, then $\int_a^b f(x)\,dx \leq \int_a^b g(x)\,dx$.

8. If $m \leq f(x) \leq M$ on $[a, b]$, then

$$m(b - a) \leq \int_a^b f(x)\,dx \leq M(b - a)$$

Property (6), illustrated in Figure 6.18, says that if f is nonnegative over the interval $[a, b]$, then the definite integral over that interval is also nonnegative. We can understand this statement from its geometric interpretation: If $f(x) \geq 0$ for $x \in [a, b]$, then $\int_a^b f(x)\,dx$ is the area between the curve and the x-axis between a and b. But an area must be a nonnegative number.

Property (7) is explained in Figure 6.19 when both f and g are positive functions on $[a, b]$. We use the fact that in this case the definite integral can be interpreted as an area. Looking at the figure, we see that the function f has a smaller area than g has. Property (7) holds without the assumption that both f and g are positive, and we can draw an analogous figure for the general case as well. The definite integral then needs to be interpreted as a signed area.

Property (8) is explained in Figure 6.20 for $f(x) \geq 0$ in $[a, b]$. We see that the rectangle with height m is contained in the area between the graph of f and the x-axis, which in turn is contained in the rectangle with height M. Since (1) $m(b - a)$ is the area of the small rectangle, (2) $\int_a^b f(x)\,dx$ is the area between the graph of f and the x-axis for nonnegative f, and (3) $M(b - a)$ is the area of the big rectangle, the inequalities in (8) follow. Note that the statement does not require that f be nonnegative; you can draw an analogous figure when f is negative on parts or all of $[a, b]$.

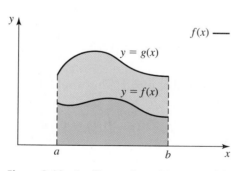

Figure 6.18 An illustration of Property (6).

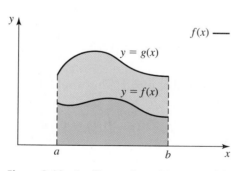

Figure 6.19 An illustration of Property (7).

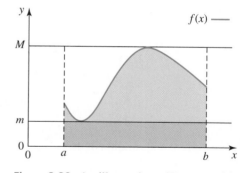

Figure 6.20 An illustration of Property (8).

The next example illustrates how the order properties (6)–(8) are used.

EXAMPLE 13 Show that

$$0 \leq \int_0^\pi \sin x\,dx \leq \pi$$

Solution   Note that $0 \leq \sin x \leq 1$ for $x \in [0, \pi]$. Using Property (6), we find that

$$\int_0^\pi \sin x\,dx \geq 0$$

Using Property (8), we obtain

$$\int_0^\pi \sin x \, dx \le (1)(\pi) = \pi$$

(See Figure 6.21.) ■

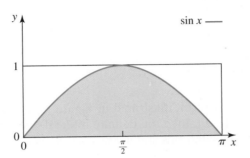

Figure 6.21 An illustration of the integral in Example 13. The shaded area is nonnegative and less than the area of the rectangle with base length π and height 1.

The next example will help us to deepen our understanding of signed areas; it also uses the order properties.

EXAMPLE 14   Find the value of $a \ge 0$ that maximizes

$$\int_0^a (1 - x^2) \, dx$$

Solution   We graph the integrand $f(x) = 1 - x^2$ for $x \ge 0$ in Figure 6.22. Using the interpretation of the definite integral as the signed area, we see from the graph of $f(x)$ that $a = 1$ maximizes the integral, since the graph of $f(x)$ is positive for $x < 1$ and negative for $x > 1$.

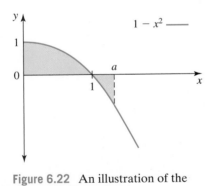

Figure 6.22 An illustration of the integral in Example 14.

We also wish to give a rigorous argument; our goal is to show that

$$\int_0^1 f(x) \, dx > \int_0^a f(x) \, dx$$

for all $a \ge 0$, provided that $a \ne 1$. This would then imply that $a = 1$ maximizes the integral

$$\int_0^a (1 - x^2) \, dx$$

First, note that $f(x)$ is continuous for $x \ge 0$ and that

$$f(x) \begin{cases} > 0 & \text{for } 0 \le x < 1 \\ < 0 & \text{for } x > 1 \end{cases}$$

which implies that, for $0 \le a < 1$,

$$\int_a^1 f(x) \, dx > 0$$

and, therefore, for $0 \le a < 1$,

$$\int_0^1 f(x) \, dx = \int_0^a f(x) \, dx + \underbrace{\int_a^1 f(x) \, dx}_{>0} > \int_0^a f(x) \, dx \qquad (6.2)$$

Now, for $a > 1$,

$$\int_1^a f(x)\,dx < 0$$

Therefore,

$$\int_0^a f(x)\,dx = \int_0^1 f(x)\,dx + \underbrace{\int_1^a f(x)\,dx}_{<0} < \int_0^1 f(x)\,dx \qquad (6.3)$$

Combining (6.2) and (6.3) shows that

$$\int_0^a f(x)\,dx < \int_0^1 f(x)\,dx$$

for all $a \geq 0$ and $a \neq 1$. Hence, $a = 1$ maximizes the integral $\int_0^a (1 - x^2)\,dx$. ■

Section 6.1 Problems

■ **6.1.1**

1. Approximate the area under the parabola $y = x^2$ from 0 to 1, using four equal subintervals with left endpoints.

2. Approximate the area under the parabola $y = x^2$ from 0 to 1, using five equal subintervals with midpoints.

3. Approximate the area under the parabola $y = x^2$ from 0 to 1, using four equal subintervals with right endpoints.

4. Approximate the area under the parabola $y = 1 - x^2$ from 0 to 1, using five equal subintervals with **(a)** left endpoints and **(b)** right endpoints.

In Problems 5–14, write each sum in expanded form.

5. $\displaystyle\sum_{k=1}^{4} \sqrt{k}$

6. $\displaystyle\sum_{k=3}^{5} (k-1)^2$

7. $\displaystyle\sum_{k=2}^{6} 3^k$

8. $\displaystyle\sum_{k=1}^{3} \frac{k^2}{k^2+1}$

9. $\displaystyle\sum_{k=0}^{3} (x+1)^k$

10. $\displaystyle\sum_{k=0}^{4} k^x$

11. $\displaystyle\sum_{k=0}^{3} (-1)^{k+1}$

12. $\displaystyle\sum_{k=1}^{n} f(c_k)\Delta x_k$

13. $\displaystyle\sum_{k=1}^{n} \left(\frac{k}{n}\right)^2 \frac{1}{n}$

14. $\displaystyle\sum_{k=1}^{n} \cos\left(k\frac{\pi}{n}\right)\frac{\pi}{n}$

In Problems 15–22, write each sum in sigma notation.

15. $2 + 4 + 6 + 8 + \cdots + 2n$

16. $\dfrac{1}{\sqrt{1}} + \dfrac{1}{\sqrt{2}} + \dfrac{1}{\sqrt{3}} + \dfrac{1}{\sqrt{4}}$

17. $\ln 2 + \ln 3 + \ln 4 + \ln 5$

18. $\dfrac{3}{5} + \dfrac{4}{6} + \dfrac{5}{7} + \dfrac{6}{8} + \dfrac{7}{9}$

19. $-\dfrac{1}{4} + \dfrac{1}{6} + \dfrac{2}{7} + \dfrac{3}{8}$

20. $\dfrac{1}{1} + \dfrac{1}{2} + \dfrac{1}{4} + \dfrac{1}{8} + \dfrac{1}{16} + \cdots + \dfrac{1}{2^n}$

21. $1 + q + q^2 + q^3 + q^4 + \cdots + q^{n-1}$

22. $1 - a + a^2 - a^3 + a^4 - a^5 + \cdots + (-1)^n a^n$

In Problems 23–30, use the algebraic rules for sums to evaluate each sum. Recall that

$$\sum_{k=1}^{n} k = \frac{n(n+1)}{2}$$

and

$$\sum_{k=1}^{n} k^2 = \frac{n(n+1)(2n+1)}{6}$$

23. $\displaystyle\sum_{k=1}^{15} (2k+3)$

24. $\displaystyle\sum_{k=1}^{5} (4 - k^2)$

25. $\displaystyle\sum_{k=0}^{6} k(k+1)$

26. $\displaystyle\sum_{k=1}^{n} 4k$

27. $\displaystyle\sum_{k=1}^{n} 4(k-1)^2$

28. $\displaystyle\sum_{k=1}^{n} (k+2)(k-2)$

29. $\displaystyle\sum_{k=1}^{10} (-1)^k$

30. $\displaystyle\sum_{k=0}^{10} (-1)^k$

31. The steps that follow will show that

$$\sum_{k=1}^{n} k^2 = \frac{n(n+1)(2n+1)}{6}$$

(a) Show that

$$\sum_{k=1}^{n}[(1+k)^3 - k^3] = (2^3 - 1^3) + (3^3 - 2^3) + (4^3 - 3^3)$$
$$+ \cdots + [(1+n)^3 - n^3]$$
$$= (1+n)^3 - 1^3$$

(Sums that "collapse" like this due to cancellation of terms are called *telescoping* or *collapsing* sums.)

(b) Use Example 3 and the algebraic rules for sums to show that

$$\sum_{k=1}^{n}[(1+k)^3 - k^3] = 3\sum_{k=1}^{n} k^2 + 3\frac{n(n+1)}{2} + n$$

(c) In (a) and (b), we found two expressions for the sum

$$\sum_{k=1}^{n}[(1+k)^3 - k^3]$$

Those two expressions are therefore equal; that is,

$$(1+n)^3 - 1^3 = 3\sum_{k=1}^{n}k^2 + 3\frac{n(n+1)}{2} + n$$

Solve this equation for $\sum_{k=1}^{n}k^2$, and show that

$$\sum_{k=1}^{n}k^2 = \frac{n(n+1)(2n+1)}{6}$$

■ **6.1.2**

32. Approximate

$$\int_{-1}^{1}(1-x^2)\,dx$$

using five equal subintervals and left endpoints.

33. Approximate

$$\int_{-1}^{1}(1-x^2)\,dx$$

using five equal subintervals and midpoints.

34. Approximate

$$\int_{-1}^{1}(2+x^2)\,dx$$

using five equal subintervals and right endpoints.

35. Approximate

$$\int_{-2}^{2}(2+x^2)\,dx$$

using four equal subintervals and left endpoints.

36. Approximate

$$\int_{-1}^{2}e^{-x}\,dx$$

using three equal subintervals and midpoints.

37. Approximate

$$\int_{0}^{3\pi/2}\sin x\,dx$$

using three equal subintervals and right endpoints.

38. (a) Assume that $a > 0$. Evaluate $\int_0^a x\,dx$, using the fact that the region bounded by $y = x$ and the x-axis between 0 to a is a triangle. (See Figure 6.23.)

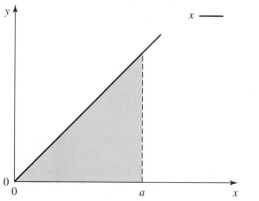

Figure 6.23 The region for Problem 38.

(b) Assume that $a > 0$. Evaluate $\int_0^a x\,dx$ by approximating the region bounded by $y = x$ and the x-axis from 0 to a with rectangles. Use equal subintervals and take right endpoints. (*Hint*: Use the result in Example 3 to evaluate the sum of the areas of the rectangles.)

39. Assume that $0 < a < b < \infty$. Use a geometric argument to show that

$$\int_a^b x\,dx = \frac{b^2 - a^2}{2}$$

40. Assume that $0 < a < b < \infty$. Use a geometric argument and Example 1 to show that

$$\int_a^b x^2\,dx = \frac{b^3 - a^3}{3}$$

Express the limits in Problems 41–47 as definite integrals. Note that (1) $P = [x_0, x_1, \ldots, x_n]$ is a partition of the indicated interval, (2) $c_k \in [x_{k-1}, x_k]$, and (3) $\Delta x_k = x_k - x_{k-1}$.

41. $\displaystyle\lim_{\|P\|\to 0}\sum_{k=1}^{n}2c_k^3\Delta x_k$, where P is a partition of $[1, 2]$

42. $\displaystyle\lim_{\|P\|\to 0}\sum_{k=1}^{n}\sqrt{c_k}\,\Delta x_k$, where P is a partition of $[1, 4]$

43. $\displaystyle\lim_{\|P\|\to 0}\sum_{k=1}^{n}(2c_k - 1)\Delta x_k$, where P is a partition of $[-3, 2]$

44. $\displaystyle\lim_{\|P\|\to 0}\sum_{k=1}^{n}\frac{1}{c_k + 1}\Delta x_k$, where P is a partition of $[1, 2]$

45. $\displaystyle\lim_{\|P\|\to 0}\sum_{k=1}^{n}\frac{c_k - 1}{c_k + 2}\Delta x_k$, where P is a partition of $[2, 3]$

46. $\displaystyle\lim_{\|P\|\to 0}\sum_{k=1}^{n}(\sin c_k)\Delta x_k$, where P is a partition of $[0, \pi]$

47. $\displaystyle\lim_{\|P\|\to 0}\sum_{k=1}^{n}e^{c_k}\Delta x_k$, where P is a partition of $[-5, 2]$

In Problems 48–53, express the definite integrals as limits of Riemann sums.

48. $\displaystyle\int_{-2}^{-1}\frac{x^2}{1+x^2}\,dx$ **49.** $\displaystyle\int_{2}^{6}(x+1)^{1/3}\,dx$

50. $\displaystyle\int_{1}^{3}e^{-2x}\,dx$ **51.** $\displaystyle\int_{1}^{e}\ln x\,dx$ **52.** $\displaystyle\int_{0}^{\pi}\cos\frac{2x}{\pi}\,dx$

53. $\displaystyle\int_{0}^{5}g(x)\,dx$, where $g(x)$ is a continuous function on $[0, 5]$

In Problems 54–60, use a graph to interpret the definite integral in terms of areas. Do not compute the integrals.

54. $\displaystyle\int_{0}^{3}(2x+1)\,dx$ **55.** $\displaystyle\int_{-1}^{2}(x^2 - 1)\,dx$

56. $\displaystyle\int_{-2}^{2}\frac{1}{2}x^3\,dx$ **57.** $\displaystyle\int_{0}^{5}e^{-x}\,dx$

58. $\displaystyle\int_{-\pi}^{\pi}\cos x\,dx$ **59.** $\displaystyle\int_{1/2}^{4}\ln x\,dx$

60. $\displaystyle\int_{-3}^{2}\left(1 - \frac{1}{2}x\right)dx$

In Problems 61–67, use an area formula from geometry to find the value of each integral by interpreting it as the (signed) area under the graph of an appropriately chosen function.

61. $\displaystyle\int_{-2}^{3} |x|\, dx$

62. $\displaystyle\int_{-3}^{3} \sqrt{9 - x^2}\, dx$

63. $\displaystyle\int_{2}^{5} \left(\frac{1}{2}x - 4\right) dx$

64. $\displaystyle\int_{1/2}^{1} \sqrt{1 - x^2}\, dx$

65. $\displaystyle\int_{-2}^{2} (\sqrt{4 - x^2} - 2)\, dx$

66. $\displaystyle\int_{0}^{1} \sqrt{2 - x^2}\, dx$

67. $\displaystyle\int_{-3}^{0} (4 - \sqrt{9 - x^2})\, dx$

■ **6.1.3**

68. Given that
$$\int_{0}^{a} x^2\, dx = \frac{1}{3}a^3$$
evaluate the following:

(a) $\displaystyle\int_{0}^{2} \frac{1}{2}x^2\, dx$

(b) $\displaystyle\int_{-3}^{-2} 3x^2\, dx$

(c) $\displaystyle\int_{-1}^{3} \frac{1}{3}x^2\, dx$

(d) $\displaystyle\int_{1}^{1} 3x^2\, dx$

(e) $\displaystyle\int_{-2}^{3} (x + 1)^2\, dx$

(f) $\displaystyle\int_{2}^{4} (x - 2)^2\, dx$

69. Find $\displaystyle\int_{2}^{2} \cos(3x^2)\, dx$.

70. Find $\displaystyle\int_{-3}^{-3} e^{-x^2/2}\, dx$.

71. Find $\displaystyle\int_{-2}^{2} \frac{x^3}{3}\, dx$.

72. Find $\displaystyle\int_{-5}^{5} 2x^5\, dx$.

73. Find $\displaystyle\int_{-1}^{1} \tan x\, dx$.

74. Explain geometrically why
$$\int_{1}^{2} x^2\, dx = \int_{0}^{2} x^2\, dx - \int_{0}^{1} x^2\, dx \qquad (6.4)$$
and show that (6.4) can be written as
$$\int_{1}^{2} x^2\, dx = \int_{1}^{0} x^2\, dx + \int_{0}^{2} x^2\, dx \qquad (6.5)$$
Relate (6.5) to addition property (5).

In Problems 75–79, verify each inequality without evaluating the integrals.

75. $\displaystyle\int_{0}^{1} x\, dx \geq \int_{0}^{1} x^2\, dx$

76. $\displaystyle\int_{1}^{2} x\, dx \leq \int_{1}^{2} x^2\, dx$

77. $\displaystyle 0 \leq \int_{0}^{4} \sqrt{x}\, dx \leq 8$

78. $\displaystyle\frac{1}{2} \leq \int_{0}^{1} \sqrt{1 - x^2}\, dx \leq 1$

79. $\displaystyle\frac{\pi}{3} \leq \int_{\pi/6}^{5\pi/6} \sin x\, dx \leq \frac{2\pi}{3}$

80. Find the value of $a \geq 0$ that maximizes $\int_{0}^{a} (4 - x^2)\, dx$.

81. Find the value of $a \in [0, 2\pi]$ that maximizes $\int_{0}^{a} \cos x\, dx$.

82. Find $a \in (0, 2\pi]$ such that
$$\int_{0}^{a} \sin x\, dx = 0$$

83. Find $a > 1$ such that
$$\int_{1}^{a} (x - 2)^3\, dx = 0$$

84. Find $a > 0$ such that
$$\int_{-a}^{a} (1 - |x|)\, dx = 0$$

85. To determine age-specific mortality, a group of individuals, all born at the same time, is followed over time. If $N(t)$ denotes the number still alive at time t, then $N(t)/N(0)$ is the fraction surviving at time t. The quantity $r(t)$, called the *hazard rate function*, measures the rate at which individuals die at time t; that is, $r(t)\, dt$ is the probability that an individual who is alive at time t dies during the infinitesimal time interval $(t, t + dt)$. The cumulative hazard during the time interval $[0, t]$, $\int_{0}^{t} r(s)\, ds$, can be estimated as $-\ln \frac{N(t)}{N(0)}$. Show that the cumulative hazard during the time interval $[t, t+1]$, $\int_{t}^{t+1} r(s)\, ds$, can be estimated as $-\ln \frac{N(t+1)}{N(t)}$.

■ **6.2 The Fundamental Theorem of Calculus**

In Section 6.1, we used the definition of definite integrals to compute $\int_{0}^{a} x^2\, dx$. This required the summation of a large number of terms, which was facilitated by the explicit summation formula for $\sum_{k=1}^{n} k^2$. Fermat and others were able to carry out similar calculations for the area under curves of the form $y = x^r$, where r was a rational number different from -1. The solution to the case $r = -1$ was found by the Belgian mathematician Gregory of St. Vincent (1584–1667) and published in 1647. At that time, it seemed that methods specific to a given function needed to be developed to compute the area under the curve of that function. Such methods would not have been practical.

Fortunately, it turns out that the area problem is related to the tangent problem. This relationship is not at all obvious; among the first to notice it were Isaac Barrow (1630–1677) and James Gregory (1638–1675). Each presented the relationship in geometrical terms, without realizing the importance of his discovery.

Both Newton and Leibniz are to be credited with systematically developing the connection between the tangent and area problems, which ultimately resulted in a method for computing areas and for solving problems that can be translated into area problems. The result is known as the *fundamental theorem of calculus*, which says that the tangent and area problems are inversely related.

The fundamental theorem of calculus has two parts: The first part links antiderivatives and integrals, and the second part provides a method for computing definite integrals.

■ 6.2.1 The Fundamental Theorem of Calculus (Part I)

Let $f(x)$ be a continuous function on $[a, b]$, and let

$$F(x) = \int_a^x f(u)\, du$$

Geometrically, $F(x) = \int_a^x f(u)\, du$ represents the signed area between the graph of $f(u)$ and the horizontal axis between a and x. (See Figure 6.24.) Note that the independent variable x appears as the upper limit of integration. We can now ask how the signed area $F(x)$ changes as x varies. To answer this question, we compute $\frac{dF}{dx}$, using the definition of the derivative. That is,

$$\frac{d}{dx} F(x) = \lim_{h \to 0} \frac{F(x+h) - F(x)}{h}$$

$$= \lim_{h \to 0} \frac{1}{h} \left[\int_a^{x+h} f(u)\, du - \int_a^x f(u)\, du \right] \tag{6.6}$$

$$= \lim_{h \to 0} \frac{1}{h} \int_x^{x+h} f(u)\, du$$

[In the last step, we used property (5) of Subsection 6.1.3.] To evaluate

$$\lim_{h \to 0} \frac{1}{h} \int_x^{x+h} f(u)\, du$$

we will resort to the geometric interpretation of definite integrals. The following argument is illustrated in Figure 6.25: Note that

$$\int_x^{x+h} f(u)\, du$$

is the signed area of the region bounded by the graph of $f(u)$ and the horizontal axis between x and $x + h$. If h is small, then this area is closely approximated by the area of the inscribed rectangle with height $|f(x)|$. The signed area of this rectangle is $f(x)h$. Hence,

$$\int_x^{x+h} f(u)\, du \approx f(x)h$$

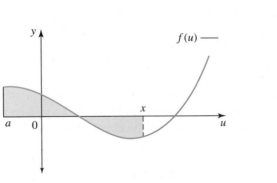

Figure 6.24 The shaded signed area is $F(x)$.

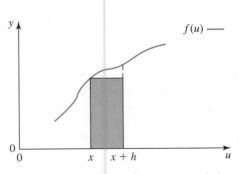

Figure 6.25 The approximate rectangle in the fundamental theorem of calculus.

If we divide both sides by h and let h tend to 0, then

$$\lim_{h \to 0} \frac{1}{h} \int_x^{x+h} f(u)\,du = f(x) \tag{6.7}$$

(as will be shown rigorously at the end of this subsection). Combining (6.6) and (6.7), we arrive at the remarkable result

$$\frac{d}{dx} F(x) = f(x)$$

In addition, we see that $F(x)$ is continuous since it is differentiable.

The preceding argument relies on geometric intuition. To show how we can make the argument mathematically rigorous, we give the complete proof at the end of this subsection. The result is summarized in the following theorem:

The Fundamental Theorem of Calculus (FTC) (Part I) If f is continuous on $[a, b]$, then the function F defined by

$$F(x) = \int_a^x f(u)\,du, \quad a \le x \le b$$

is continuous on $[a, b]$ and differentiable on (a, b), with

$$\frac{d}{dx} F(x) = f(x)$$

Simply stated, the FTC (part I) says that if we first integrate $f(x)$ and then differentiate the result, we get $f(x)$ again. In this sense, it shows that integration and differentiation are inverse operations.

We begin with an example that can be immediately solved by using the FTC.

EXAMPLE 1 Compute

$$\frac{d}{dx} \int_0^x (\sin u - e^{-u})\,du$$

for $x > 0$.

Solution First, note that $f(x) = \sin x - e^{-x}$ is continuous for $x \ge 0$. If we set $F(x) = \int_0^x (\sin u - e^{-u})\,du$ and apply the FTC, then

$$\frac{d}{dx} F(x) = \frac{d}{dx} \int_0^x (\sin u - e^{-u})\,du = \sin x - e^{-x} \quad ■$$

EXAMPLE 2 Compute

$$\frac{d}{dx} \int_3^x \frac{1}{1+u^2}\,du$$

for $x > 3$.

Solution First, note that $f(x) = \frac{1}{1+x^2}$ is continuous for $x \ge 3$. If we set $F(x) = \int_3^x \frac{1}{1+u^2}\,du$ and apply the FTC, then

$$\frac{d}{dx} F(x) = \frac{d}{dx} \int_3^x \frac{1}{1+u^2}\,du = \frac{1}{1+x^2} \quad ■$$

The remainder of this subsection can be omitted.

Leibniz's Rule (Optional) Combining the chain rule and the FTC (part I), we can differentiate integrals with respect to x when the upper and/or lower limits of integration are functions of x.

In the first example, the upper limit of integration is a function of x.

EXAMPLE 3   Compute

$$\frac{d}{dx} \int_0^{x^2} (u^3 - 2) \, du, \quad x > 0$$

Solution   Note that $f(u) = u^3 - 2$ is continuous for all $u \in \mathbf{R}$. We set $F(v) = \int_0^v (u^3 - 2) \, du$, $v > 0$. Then, for $x > 0$,

$$F(x^2) = \int_0^{x^2} (u^3 - 2) \, du$$

We wish to compute $\frac{d}{dx} F(x^2)$. To do so, we need to apply the chain rule. We set $v(x) = x^2$. Then

$$\frac{d}{dx} F(x^2) = \frac{dF(v)}{dv} \frac{dv}{dx}$$

To evaluate $\frac{d}{dv} F(v) = \frac{d}{dv} \int_0^v (u^3 - 2) \, du$, we use the FTC:

$$\frac{d}{dv} \int_0^v (u^3 - 2) \, du = v^3 - 2$$

Since $\frac{dv}{dx} = \frac{d}{dx}(x^2) = 2x$, it follows that

$$\frac{d}{dx} F(x^2) = (v^3 - 2)2x = [(x^2)^3 - 2]2x$$
$$= (x^6 - 2)2x$$

■

Thus far, we have dealt only with the case where the upper limit of integration depends on x. The next example shows what we must do when the lower limit of integration depends on x.

EXAMPLE 4   Compute

$$\frac{d}{dx} \int_{\sin x}^1 u^2 \, du$$

Solution   Note that $f(u) = u^2$ is continuous for all $u \in \mathbf{R}$. We use the fact that

$$\int_{\sin x}^1 u^2 \, du = - \int_1^{\sin x} u^2 \, du$$

The upper limit now depends on x, but we introduced a minus sign. Hence

$$\frac{d}{dx} \int_{\sin x}^1 u^2 \, du = -\frac{d}{dx} \int_1^{\sin x} u^2 \, du$$
$$= -(\sin x)^2 \cos x$$

where, as in Example 3, we used the chain rule in the last step.   ■

The preceding example makes an important point: We need to be careful about whether the upper or the lower limit of integration depends on x. In the next example, we show what we must do when both limits of integration depend on x.

EXAMPLE 5   For $x \in \mathbf{R}$, compute

$$\frac{d}{dx} \int_{x^2}^{x^3} e^u \, du$$

Solution
Note that $f(u) = e^u$ is continuous for all $u \in \mathbf{R}$. The given integral is therefore defined for all $x \in \mathbf{R}$, and we can split it into two integrals at any $a \in \mathbf{R}$. We choose $a = 0$, which yields

$$\int_{x^2}^{x^3} e^u \, du = \int_{x^2}^{0} e^u \, du + \int_{0}^{x^3} e^u \, du = -\int_{0}^{x^2} e^u \, du + \int_{0}^{x^3} e^u \, du$$

The right-hand side is now written in a form that we know how to differentiate, and we find that

$$\frac{d}{dx} \int_{x^2}^{x^3} e^u \, du = -\frac{d}{dx} \int_{0}^{x^2} e^u \, du + \frac{d}{dx} \int_{0}^{x^3} e^u \, du$$

$$= -\left[e^{x^2} \frac{d}{dx} x^2 \right] + \left[e^{x^3} \frac{d}{dx} x^3 \right]$$

$$= -e^{x^2} 2x + e^{x^3} 3x^2$$ ■

The preceding example illustrates the most general case that we can encounter, namely, when both limits of integration are functions of x. We summarize this case in the following box, in a property known as Leibniz's rule:

Leibniz's Rule If $g(x)$ and $h(x)$ are differentiable functions and $f(u)$ is continuous for u between $g(x)$ and $h(x)$, then

$$\frac{d}{dx} \int_{g(x)}^{h(x)} f(u) \, du = f[h(x)]h'(x) - f[g(x)]g'(x)$$

We can check that Examples 3–5 can be solved with the preceding formula; for instance, in Example 4, we have $f(u) = u^2$, $g(x) = \sin x$, and $h(x) = 1$. Then $g'(x) = \cos x$ and $h'(x) = 0$. We therefore find that

$$f[h(x)]h'(x) - f[g(x)]g'(x) = 0 - (\sin x)^2 \cos x$$

which is the answer we obtained in Example 4.

Proof of the Fundamental Theorem of Calculus (Part I) (Optional)   At the beginning of this subsection, we found that if

$$F(x) = \int_{a}^{x} f(u) \, du$$

then

$$\frac{d}{dx} F(x) = \lim_{h \to 0} \frac{1}{h} \int_{x}^{x+h} f(u) \, du \qquad (6.8)$$

We now give a mathematically rigorous argument which will show that

$$\frac{d}{dx} F(x) = f(x)$$

We begin with the observation that, according to the extreme-value theorem, the continuous function $f(u)$ defined on the closed interval $[x, x+h]$ attains an absolute minimum and an absolute maximum on $[x, x+h]$. That is, there exist m and M such that m is the minimum of f on $[x, x+h]$ and M is the maximum of f on $[x, x+h]$, which implies that

$$m \le f(u) \le M \quad \text{for all } u \in [x, x+h] \qquad (6.9)$$

Of course, m and M depend on both x and h. Applying property (8) from Subsection 6.1.3 to (6.9), we find that

$$\int_x^{x+h} m\, du \le \int_x^{x+h} f(u)\, du \le \int_x^{x+h} M\, du$$

and hence

$$mh \le \int_x^{x+h} f(u)\, du \le Mh$$

Dividing by h, we obtain

$$m \le \frac{1}{h}\int_x^{x+h} f(u)\, du \le M \qquad (6.10)$$

We set

$$I = \frac{1}{h}\int_x^{x+h} f(u)\, du$$

Then (6.10) becomes $m \le I \le M$; that is, I is a number between m and M. We compare this inequality with (6.9), which says that $f(u)$ also lies between m, the minimum of f on $[x, x+h]$, and M, the maximum of f on $[x, x+h]$, for all $u \in [x, x+h]$. The intermediate-value theorem applied to $f(u)$ tells us that any value between m and M is attained by $f(u)$ for some number on the interval $[x, x+h]$. Specifically, since I lies between m and M, there must exist a number $c_h \in [x, x+h]$ such that $f(c_h) = I$; that is,

$$f(c_h) = \frac{1}{h}\int_x^{x+h} f(u)\, du \qquad (6.11)$$

Because $x \le c_h \le x + h$, it follows that

$$\lim_{h \to 0} c_h = x$$

Since f is continuous,

$$\lim_{h \to 0} f(c_h) = f\left(\lim_{h \to 0} c_h\right) = f(x) \qquad (6.12)$$

Combining (6.8), (6.11), and (6.12) yields the result we seek:

$$\frac{d}{dx}F(x) = f(x) \qquad\qquad ■$$

■ 6.2.2 Antiderivatives and Indefinite Integrals

The first part of the fundamental theorem of calculus tells us that if

$$F(x) = \int_a^x f(u)\, du$$

then $F'(x) = f(x)$ [provided that $f(x)$ is continuous over the range of integration]. This statement says that $F(x)$ is an *antiderivative* of $f(x)$. (We introduced antiderivatives in Section 5.8.) Now, if we let

$$F(x) = \int_a^x f(u)\, du \qquad \text{and} \qquad G(x) = \int_b^x f(u)\, du$$

where a and b are two numbers, then both integrals have the same derivative, namely, $F'(x) = G'(x) = f(x)$ [again, provided that $f(x)$ is continuous over the range of integration]. That is, both $F(x)$ and $G(x)$ are antiderivatives of $f(x)$. We saw in

Section 5.8 that antiderivatives of a given function differ only by a constant. We can identify the constant, namely

$$F(x) = \int_a^x f(u)\, du = \int_a^b f(u)\, du + \int_b^x f(u)\, du = C + G(x)$$

where C is a constant denoting the number $\int_a^b f(u)\, du$.

The general antiderivative of a function $f(x)$ is $F(x) + C$, where $F'(x) = f(x)$ and C is a constant. It follows that $C + \int_a^x f(u)\, du$ is the general antiderivative of $f(x)$. We will use the notation $\int f(x)\, dx$ to denote both the general antiderivative of $f(x)$ and the function $C + \int_a^x f(u)\, du$; that is,

$$\int f(x)\, dx = C + \int_a^x f(u)\, du \tag{6.13}$$

We call $\int f(x)\, dx$ an **indefinite integral**. Thus, the first part of the FTC says that indefinite integrals and antiderivatives are the same.

When we write $\int_a^x f(u)\, du$, we use a letter other than x in the integrand because x already appears as the upper limit of integration. However, in the symbolic notation $\int f(x)\, dx$ we write x. This notation is to be interpreted as in (6.13); it is a convenient shorthand for $C + \int_a^x f(u)\, du$. The choice of value of a for the lower limit of integration on the right side of (6.13) is not important, because different indefinite integrals of the same function $f(x)$ differ only by an additive constant that can be absorbed into the constant C.

Examples 6–8 show how to compute indefinite integrals.

EXAMPLE 6   Compute $\int x^4\, dx$.

Solution   We need to find a function $F(x)$ such that $F'(x) = x^4$. The solution is

$$\int x^4\, dx = \frac{1}{5}x^5 + C$$

where C is a constant. We check that, indeed,

$$\frac{d}{dx}\left(\frac{1}{5}x^5 + C\right) = x^4$$ ■

EXAMPLE 7   Compute $\int (e^x + \sin x)\, dx$.

Solution   We need to find an antiderivative of $f(x) = e^x + \sin x$. Since

$$\frac{d}{dx}(e^x - \cos x) = e^x - (-\sin x) = e^x + \sin x$$

it follows that

$$\int (e^x + \sin x)\, dx = e^x - \cos x + C$$ ■

When we compute the indefinite integral $\int f(x)\, dx$, we want to know the general antiderivative of $f(x)$; this is why we added the constant C in the previous two examples.

EXAMPLE 8   Show that

$$\int \frac{1}{x}\, dx = \ln |x| + C \quad \text{for } x \neq 0$$

Solution   Since the absolute value of x appears on the right-hand side, we split our discussion into two parts, according to whether $x \geq 0$ or $x \leq 0$. Recall that

$$|x| = \begin{cases} x & \text{for } x \geq 0 \\ -x & \text{for } x < 0 \end{cases}$$

Since $\ln x$ is not defined at $x = 0$, we consider the two cases $x > 0$ and $x < 0$.

(i) $x > 0$: Since $\ln|x| = \ln x$ when $x > 0$, we have

$$\frac{d}{dx}\ln x = \frac{1}{x}$$

Hence,

$$\int \frac{1}{x}\,dx = \ln x + C \quad \text{for } x > 0$$

(ii) $x < 0$: Since $\ln|x| = \ln(-x)$ when $x < 0$, it follows that

$$\frac{d}{dx}\ln(-x) = \frac{1}{-x}(-1) = \frac{1}{x}$$

Hence,

$$\int \frac{1}{x}\,dx = \ln(-x) + C \quad \text{for } x < 0$$

Combining (i) and (ii), we obtain

$$\int \frac{1}{x}\,dx = \ln|x| + C \quad \text{for } x \neq 0 \qquad\blacksquare$$

When we introduced logarithmic and exponential functions in Sections 4.6 and 4.7, we had to resort to the calculator to convince ourselves that e^x and $\ln x$ were indeed the functions we knew from precalculus. To give a mathematically rigorous definition of these functions, we typically start by *defining* $\ln x$ as $\int_1^x \frac{1}{u}\,du$ and then derive the algebraic rules for $\ln x$ from this integral representation. The exponential function e^x is defined as the inverse function of $\ln x$, and the number e is defined so that $\ln e = 1$. This definition is then consistent with the definition in Section 4.6.

We have seen that in order to evaluate indefinite integrals, we must find antiderivatives. Table 6-1 gives a list of indefinite integrals. (The table is a slightly expanded form of the table of antiderivatives from Section 5.8.) Examples 9 and 10 show how to use this list to compute indefinite integrals.

TABLE 6-1 A Collection of Indefinite Integrals

$$\int x^n\,dx = \frac{x^{n+1}}{n+1} + C \quad (n \neq -1) \qquad \int \frac{1}{x}\,dx = \ln|x| + C$$

$$\int e^x\,dx = e^x + C \qquad \int a^x\,dx = \frac{a^x}{\ln a} + C$$

$$\int \cos x\,dx = \sin x + C \qquad \int \sin x\,dx = -\cos x + C$$

$$\int \sec^2 x\,dx = \tan x + C \qquad \int \csc^2 x\,dx = -\cot x + C$$

$$\int \sec x \tan x\,dx = \sec x + C \qquad \int \csc x \cot x\,dx = -\csc x + C$$

$$\int \tan x\,dx = \ln|\sec x| + C \qquad \int \cot x\,dx = -\ln|\csc x| + C$$

$$\int \frac{1}{1+x^2}\,dx = \tan^{-1} x + C \qquad \int \frac{1}{\sqrt{1-x^2}}\,dx = \sin^{-1} x + C$$

EXAMPLE 9   Evaluate

$$\int \frac{1}{\sin^2 x - 1}\,dx$$

Solution   We first work on the integrand. Using the fact that $\sin^2 x + \cos^2 x = 1$, we find that

$$\frac{1}{\sin^2 x - 1} = -\frac{1}{\cos^2 x} = -\sec^2 x$$

Hence,

$$\int \frac{1}{\sin^2 x - 1}\, dx = -\int \sec^2 x\, dx = -\tan x + C \qquad ■$$

EXAMPLE 10   Evaluate

$$\int \frac{x^2}{x^2 + 1}\, dx$$

Solution   We first rewrite the integrand:

$$\frac{x^2}{x^2 + 1} = \frac{x^2 + 1 - 1}{x^2 + 1} = \frac{x^2 + 1}{x^2 + 1} - \frac{1}{x^2 + 1} = 1 - \frac{1}{x^2 + 1}$$

Then, from Table 6-1,

$$\int \frac{x^2}{x^2 + 1}\, dx = \int \left(1 - \frac{1}{x^2 + 1}\right) dx = x - \tan^{-1} x + C \qquad ■$$

■ 6.2.3 The Fundamental Theorem of Calculus (Part II)

The first part of the FTC allows us to compute integrals of the form $\int_a^x f(u)\, du$ only up to an additive constant; for instance,

$$F(x) = \int_1^x u^2\, du = \frac{1}{3}x^3 + C$$

To evaluate the definite integral $F(2) = \int_1^2 u^2\, du$, which represents the area under the graph of $f(x) = x^2$ between $x = 1$ and $x = 2$, we would need to know the value of the constant. This value is provided by the second part of the FTC, which allows us to evaluate definite integrals. The calculation that follows shows us how this constant is determined.

We saw in the last subsection that if we set

$$G(x) = \int_a^x f(u)\, du$$

then $G(x)$ is an antiderivative of $f(x)$. Furthermore, if $F(x)$ is another antiderivative of $f(x)$, then $G(x)$ and $F(x)$ differ only by an additive constant. That is,

$$G(x) = F(x) + C$$

where C is a constant. Now,

$$G(a) = \int_a^a f(u)\, du = 0$$

Hence,

$$0 = G(a) = F(a) + C$$

which implies that $C = -F(a)$ and, therefore, $G(x) = F(x) - F(a)$, or, in the integral representation of $G(x)$,

$$\int_a^x f(u)\, du = F(x) - F(a)$$

If we set $x = b$, then

$$\int_a^b f(u)\, du = F(b) - F(a)$$

This formula allows us to evaluate definite integrals and is the content of the second part of the FTC.

The Fundamental Theorem of Calculus (Part II) Assume that f is continuous on $[a, b]$; then

$$\int_a^b f(x)\, dx = F(b) - F(a)$$

where $F(x)$ is an antiderivative of $f(x)$; that is, $F'(x) = f(x)$.

So how do we use this part of the FTC? To compute the definite integral $\int_a^b f(x)\, dx$ when f is continuous on $[a, b]$, we first need to find an antiderivative $F(x)$ of $f(x)$ (any antiderivative will do) and then compute $F(b) - F(a)$. This number is then equal to $\int_a^b f(x)\, dx$. Table 6-1 will help us find the required antiderivative. (Note that an indefinite integral is a function, whereas a definite integral is simply a number.)

Using the FTC (Part II) to Evaluate Definite Integrals

EXAMPLE 11

Evaluate $\int_{-1}^2 (x^2 - 3x)\, dx$.

Solution

Note that $f(x) = x^2 - 3x$ is continuous on $[-1, 2]$. We need to find an antiderivative of $f(x) = x^2 - 3x$; for instance, $F(x) = \frac{1}{3}x^3 - \frac{3}{2}x^2$ is an antiderivative of $f(x)$ since $F'(x) = f(x)$. We then must evaluate $F(2) - F(-1)$:

$$F(2) = \frac{1}{3}2^3 - \frac{3}{2}2^2 = \frac{8}{3} - 6 = -\frac{10}{3}$$

$$F(-1) = \frac{1}{3}(-1)^3 - \frac{3}{2}(-1)^2 = -\frac{1}{3} - \frac{3}{2} = -\frac{11}{6}$$

We find that $F(2) - F(-1) = -\frac{10}{3} - (-\frac{11}{6}) = -\frac{9}{6} = -\frac{3}{2}$. Therefore,

$$\int_{-1}^2 (x^2 - 3x)\, dx = F(2) - F(-1) = -\frac{3}{2}$$

In the preceding calculation, we chose the simplest antiderivative—that is, the one in which the constant C is equal to 0. We could have chosen any $C \neq 0$, and the answer would have been the same. Let's see why. The general antiderivative of $f(x)$ is $G(x) = \frac{1}{3}x^3 - \frac{3}{2}x^2 + C$. We can write this as $G(x) = F(x) + C$, where $F(x)$ is the antiderivative we used previously. Then, using $G(x)$ to evaluate the integral, we find that

$$\int_{-1}^2 (x^2 - 3)\, dx = G(2) - G(-1)$$

$$= [F(2) + C] - [F(-1) + C] = F(2) - F(-1)$$

which is the same answer as before, since the constant C cancels out. We thus see that we can use the simplest antiderivative (the one in which $C = 0$), and we will do so from now on. ■

EXAMPLE 12   Evaluate $\int_0^\pi \sin x \, dx$.

Solution   Note that $\sin x$ is continuous on $[0, \pi]$. Since $F(x) = -\cos x$ is an antiderivative of $\sin x$, we have

$$\int_0^\pi \sin x \, dx = F(\pi) - F(0) = -\cos \pi - (-\cos 0) = -(-1) + 1 = 2 \qquad ■$$

EXAMPLE 13   Evaluate

$$\int_{-5}^{-1} \frac{1}{x} \, dx$$

Solution   Note that $\frac{1}{x}$ is continuous on $[-5, -1]$. Now $\ln |x|$ is an antiderivative of $\frac{1}{x}$. We use this antiderivative to evaluate the integral and obtain

$$\int_{-5}^{-1} \frac{1}{x} \, dx = (\ln |-1|) - (\ln |-5|) = -\ln 5$$

since $\ln |-1| = \ln 1 = 0$ and $|-5| = 5$.   ■

We next introduce additional notation. If $F(x)$ is an antiderivative of $f(x)$, then we write

$$\int_a^b f(x) \, dx = F(x)\Big]_a^b = F(b) - F(a)$$

For instance,

$$\int_{-5}^{-1} \frac{1}{x} \, dx = \ln |x|\Big]_{-5}^{-1} = \ln |-1| - \ln |-5|$$

The notation $F(x)\Big]_a^b$ indicates that we evaluate the antiderivative $F(x)$ at b and a, respectively, and compute the difference $F(b) - F(a)$.

EXAMPLE 14   Evaluate

$$\int_0^3 2x e^{x^2} \, dx$$

Solution   Observe that $2x e^{x^2}$ is continuous on $[0, 3]$ and that $F(x) = e^{x^2}$ is an antiderivative of $f(x) = 2x e^{x^2}$, since, applying the chain rule, we find that

$$F'(x) = e^{x^2}\left(\frac{d}{dx} x^2\right) = e^{x^2} 2x$$

Therefore,

$$\int_0^3 2x e^{x^2} \, dx = e^{x^2}\Big]_0^3 = e^9 - e^0 = e^9 - 1 \qquad ■$$

EXAMPLE 15   Evaluate

$$\int_1^4 \frac{2x^2 - 3x + \sqrt{x}}{\sqrt{x}} \, dx$$

Solution   The integrand is continuous on $[1, 4]$. We first simplify the integrand:

$$f(x) = \frac{2x^2 - 3x + \sqrt{x}}{\sqrt{x}} = 2x^{3/2} - 3\sqrt{x} + 1$$

An antiderivative of $f(x)$ is, therefore,

$$F(x) = 2 \cdot \frac{2}{5} x^{5/2} - 3 \cdot \frac{2}{3} x^{3/2} + x = \frac{4}{5} x^{5/2} - 2x^{3/2} + x$$

which can be checked by differentiating $F(x)$. We can now evaluate the integral:

$$\int_1^4 \frac{2x^2 - 3x + \sqrt{x}}{\sqrt{x}}\,dx = \frac{4}{5}x^{5/2} - 2x^{3/2} + x \Big]_1^4$$

$$= \left(\frac{4}{5} \cdot 4^{5/2} - 2 \cdot 4^{3/2} + 4\right) - \left(\frac{4}{5} \cdot 1^{5/2} - 2 \cdot 1^{3/2} + 1\right)$$

$$= \left(\frac{4}{5} \cdot 32 - (2)(8) + 4\right) - \left(\frac{4}{5} - 2 + 1\right)$$

$$= \frac{68}{5} - \left(-\frac{1}{5}\right) = \frac{69}{5} \qquad\qquad ■$$

Finding an Integrand

EXAMPLE 16   Suppose that

$$\int_0^x f(t)\,dt = \cos(2x) + a$$

where a is a constant. Find $f(x)$ and a.

Solution   We solve this problem in two steps. First, we use the FTC, part I, to conclude that

$$\frac{d}{dx}\int_0^x f(t)\,dt = f(x)$$

Hence,

$$f(x) = \frac{d}{dx}[\cos(2x) + a] = -2\sin(2x)$$

In the second step, we use the FTC, part II, to determine a:

$$\int_0^x (-2\sin(2t))\,dt = \cos(2t)\Big]_0^x$$

$$= \cos(2x) - \cos(0) = \cos(2x) - 1$$

We conclude that $a = -1$. ■

Discontinuous Integrand   You might wonder why we always check that the integrand is continuous on the interval between the lower and the upper limit of integration. The next example shows what can go wrong when the integrand is discontinuous.

EXAMPLE 17   Evaluate

$$\int_{-2}^1 \frac{1}{x^2}\,dx$$

Solution   An antiderivative of $f(x) = 1/x^2$ is $F(x) = -\frac{1}{x}$. We find that $F(1) = -1$ and $F(-2) = \frac{1}{2}$. When we compute $F(1) - F(-2)$, we get $-\frac{3}{2}$. This is obviously not equal to $\int_{-2}^1 \frac{1}{x^2}\,dx$, since $f(x) = \frac{1}{x^2}$ is positive on $[-2, 1]$ (see Figure 6.26) and, therefore, the integral of $f(x)$ between -2 and 1 should not be negative. The function $f(x)$ is discontinuous at $x = 0$ (where it has a vertical asymptote). Hence, the second part of the FTC therefore cannot be applied. We will learn how to deal with such discontinuities in Section 7.4. In any case, before you evaluate an integral, always check whether the integrand is continuous between the limits of integration. ■

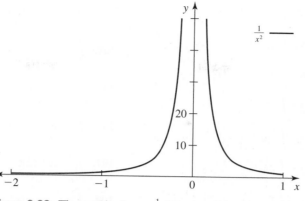

Figure 6.26 The graph of $y = \frac{1}{x^2}$ between $x = -2$ and $x = 1$. The function is discontinuous at $x = 0$.

Section 6.2 Problems

■ 6.2.1

In Problems 1–14, find $\frac{dy}{dx}$.

1. $y = \int_0^x 2u^3 \, du$

2. $y = \int_0^x (1 - \frac{u^4}{2}) \, du$

3. $y = \int_0^x (4u^2 - 3) \, du$

4. $y = \int_0^x (3 + u^4) \, du$

5. $y = \int_0^x \sqrt{1 + 2u} \, du, \; x > 0$

6. $y = \int_0^x \sqrt{1 + u^2} \, du, \; x > 0$

7. $y = \int_0^x \sqrt{1 + \sin^2 u} \, du, \; x > 0$

8. $y = \int_0^x \sqrt{2 + \csc^2 u} \, du, \; x > 0$

9. $y = \int_3^x u e^{4u} \, du$

10. $y = \int_1^x u e^{-u^2} \, du$

11. $y = \int_{-2}^x \frac{1}{u + 3} \, du, \; x > -2$

12. $y = \int_{-1}^x \frac{2}{2 + u^2} \, du$

13. $y = \int_{\pi/2}^x \sin(u^2 + 1) \, du$

14. $y = \int_{\pi/4}^x \cos^2(u - 3) \, du$

In Problems 15–38, use Leibniz's rule to find $\frac{dy}{dx}$.

15. $y = \int_0^{3x} (1 + t^2) \, dt$

16. $y = \int_0^{2x-1} (t^3 - 2) \, dt$

17. $y = \int_0^{1-4x} (2t^2 + 1) \, dt$

18. $y = \int_0^{3x+2} (1 + t^3) \, dt$

19. $y = \int_4^{x^2+1} \sqrt{t} \, dt, \; x > 0$

20. $y = \int_2^{x^2-2} \sqrt{3 + u} \, du, \; x > 0$

21. $y = \int_0^{3x} (1 + e^t) \, dt$

22. $y = \int_0^{2x^2-1} (e^{-2t} + e^{2t}) \, dt$

23. $y = \int_1^{3x^2+x} (1 + t e^t) \, dt$

24. $y = \int_2^{\ln x} e^{-t} \, dt, \; x > 0$

25. $y = \int_x^3 (1 + t) \, dt$

26. $y = \int_x^5 (1 + e^t) \, dt$

27. $y = \int_{2x}^3 (1 + \sin t) \, dt$

28. $y = \int_{2x^2}^6 (1 + \tan t) \, dt$

29. $y = \int_x^5 \frac{1}{u^2} \, du, \; x > 0$

30. $y = \int_{x^2}^3 \frac{1}{1 + t} \, dt, \; x > 0$

31. $y = \int_{x^2}^1 \sec t \, dt, \; -1 < x < 1$

32. $y = \int_{2+x^2}^2 \cot t \, dt$

33. $y = \int_x^{2x} (1 + t^2) \, dt$

34. $y = \int_{-x}^x \tan u \, du, \; 0 < x < \frac{\pi}{4}$

35. $y = \int_{x^2}^{x^3} \ln(t - 3) \, dt, \; x > 0$

36. $y = \int_{x^3}^{x^4} \ln(1 + t^2) \, dt, \; x > 0$

37. $y = \int_{2-x^2}^{x+x^3} \sin t \, dt$

38. $y = \int_{1+x^2}^{x^3-2x} \cos t \, dt$

■ 6.2.2

In Problems 39–96, compute the indefinite integrals.

39. $\int (1 + 3x^2) \, dx$

40. $\int (x^3 - 4) \, dx$

41. $\int (\frac{1}{3}x^2 - \frac{1}{2}x) \, dx$

42. $\int (4x^3 + 5x^2) \, dx$

43. $\int \left(\frac{1}{2}x^2 + 3x - \frac{1}{3} \right) dx$

44. $\int \left(\frac{1}{2}x^5 + 2x^3 - 1 \right) dx$

45. $\int \frac{2x^2 - x}{\sqrt{x}} \, dx$

46. $\int \frac{x^3 + 3x}{2\sqrt{x}} \, dx$

47. $\int x^2 \sqrt{x} \, dx$

48. $\int (1 + x^3) \sqrt{x} \, dx$

49. $\int (x^{7/2} + x^{2/7}) \, dx$

50. $\int (x^{3/5} + x^{5/3}) \, dx$

51. $\int \left(\sqrt{x} + \frac{1}{\sqrt{x}} \right) dx$

52. $\int \left(3x^{1/3} + \frac{1}{3x^{1/3}} \right) dx$

53. $\int (x - 1)(x + 1) \, dx$

54. $\int (x - 1)^2 \, dx$

55. $\int (x-2)(3-x)\,dx$ **56.** $\int (2x+3)^2\,dx$

57. $\int e^{2x}\,dx$ **58.** $\int 2e^{3x}\,dx$

59. $\int 3e^{-x}\,dx$ **60.** $\int 2e^{-x/3}\,dx$

61. $\int xe^{-x^2/2}\,dx$ **62.** $\int e^x(1-e^{-x})\,dx$

63. $\int \sin(2x)\,dx$ **64.** $\int \sin\frac{1-x}{3}\,dx$

65. $\int \cos(3x)\,dx$ **66.** $\int \cos\frac{2-4x}{5}\,dx$

67. $\int \sec^2(3x)\,dx$ **68.** $\int \csc^2(2x)\,dx$

69. $\int \frac{\sin x}{1-\sin^2 x}\,dx$ **70.** $\int \frac{\cos x}{1-\cos^2 x}\,dx$

71. $\int \tan(2x)\,dx$ **72.** $\int \cot(3x)\,dx$

73. $\int (\sec^2 x + \tan x)\,dx$ **74.** $\int (\cot x - \csc^2 x)\,dx$

75. $\int \frac{4}{1+x^2}\,dx$ **76.** $\int \left(1-\frac{x^2}{1+x^2}\right)\,dx$

77. $\int \frac{1}{\sqrt{1-x^2}}\,dx$ **78.** $\int \frac{5}{\sqrt{1-x^2}}\,dx$

79. $\int \frac{1}{x+2}\,dx$ **80.** $\int \frac{1}{x-3}\,dx$

81. $\int \frac{2x-1}{3x}\,dx$ **82.** $\int \frac{2x+5}{x}\,dx$

83. $\int \frac{x+3}{x^2-9}\,dx$ **84.** $\int \frac{x+4}{x^2-16}\,dx$

85. $\int \frac{3-x}{x^2-9}\,dx$ **86.** $\int \frac{4-x}{x^2-16}\,dx$

87. $\int \frac{5x^2}{x^2+1}\,dx$ **88.** $\int \frac{2x^2}{1+x^2}\,dx$

89. $\int 3^x\,dx$ **90.** $\int 2^x\,dx$

91. $\int 3^{-2x}\,dx$ **92.** $\int 4^{-x}\,dx$

93. $\int (x^2+2^x)\,dx$ **94.** $\int (x^{-3}+3^{-x})\,dx$

95. $\int (\sqrt{x}+\sqrt{e^x})\,dx$ **96.** $\int \left(\frac{1}{\sqrt{x}}+\frac{1}{\sqrt{e^x}}\right)\,dx$

■ **6.2.3**

In Problems 97–122, evaluate the definite integrals.

97. $\int_2^4 (3-2x)\,dx$ **98.** $\int_{-1}^3 (2x^2-1)\,dx$

99. $\int_0^1 (x^3-x^{1/3})\,dx$ **100.** $\int_1^2 x^{5/2}\,dx$

101. $\int_1^8 x^{-2/3}\,dx$ **102.** $\int_4^9 \frac{1+\sqrt{x}}{\sqrt{x}}\,dx$

103. $\int_0^2 (2t-1)(t+3)\,dt$ **104.** $\int_{-1}^2 (2+3t)^2\,dt$

105. $\int_0^{\pi/4} \sin(2x)\,dx$ **106.** $\int_{-\pi/3}^{\pi/3} 2\cos\left(\frac{x}{2}\right)\,dx$

107. $\int_0^{\pi/8} \sec^2(2x)\,dx$ **108.** $\int_{-\pi/4}^{\pi/4} \tan x\,dx$

109. $\int_0^1 \frac{1}{1+x^2}\,dx$ **110.** $\int_{-\sqrt{3}}^{-1} \frac{4}{1+x^2}\,dx$

111. $\int_0^{1/2} \frac{1}{\sqrt{1-x^2}}\,dx$ **112.** $\int_{-1/2}^{1/2} \frac{2}{\sqrt{1-x^2}}\,dx$

113. $\int_0^{\pi/6} \tan(2x)\,dx$ **114.** $\int_{\pi/20}^{\pi/15} \sec(5x)\tan(5x)\,dx$

115. $\int_{-1}^0 e^{3x}\,dx$ **116.** $\int_0^2 2te^{t^2}\,dt$

117. $\int_{-1}^1 |x|\,dx$ **118.** $\int_{-1}^1 e^{-|s|}\,ds$

119. $\int_1^e \frac{1}{x}\,dx$ **120.** $\int_2^3 \frac{1}{z+1}\,dz$

121. $\int_{-2}^{-1} \frac{1}{1-u}\,du$ **122.** $\int_2^3 \frac{2}{t-1}\,dt$

123. Use l'Hospital's rule to compute

$$\lim_{x\to 0}\frac{1}{x^2}\int_0^x \sin t\,dt$$

124. Use l'Hospital's rule to compute

$$\lim_{h\to 0}\frac{1}{h}\int_0^h e^x\,dx$$

125. Suppose that

$$\int_0^x f(t)\,dt = 2x^2$$

Find $f(x)$.

126. Suppose that

$$\int_0^x f(t)\,dt = \frac{1}{2}\tan(2x)$$

Find $f(x)$.

■ **6.3 Applications of Integration**

In this section, we will discuss a number of applications of integrals. In the first application, we will revisit the interpretation of integrals as areas; the second application interprets integrals as cumulative (or net) change; the third will allow us to compute averages using integrals; and, finally, we will use integrals to compute volumes. In each application, you will see that integrals can be interpreted as "sums of many small increments."

■ 6.3.1 Areas

The first application is already familiar to us. If f is a nonnegative, continuous function on $[a, b]$, then

$$A = \int_a^b f(x)\,dx$$

represents the area of the region bounded by the graph of $f(x)$ between a and b, the vertical lines $x = a$ and $x = b$, and the x-axis between a and b. In all of the examples we have presented thus far, one of the boundaries of the region whose area we wanted to know has been the x-axis. We will now discuss how to find the geometric area between two arbitrary curves. We emphasize that we want to compute geometric areas; that is, the areas we compute in this subsection will always be positive.

Suppose that $f(x)$ and $g(x)$ are continuous functions on $[a, b]$. We wish to find the area between the graphs of f and g. We start with a simple example. (See Figure 6.27.) We assume for the moment that both f and g are nonnegative on $[a, b]$ and that $f(x) \geq g(x)$ on $[a, b]$. From the figure, we see that

$$A = \left[\begin{array}{c} \text{area between} \\ f \text{ and } x\text{-axis} \end{array} \right] - \left[\begin{array}{c} \text{area between} \\ g \text{ and } x\text{-axis} \end{array} \right]$$

$$= \int_a^b f(x)\,dx - \int_a^b g(x)\,dx$$

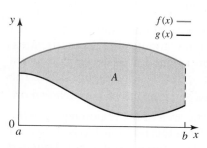

Figure 6.27 Computing the area between the two curves.

Using Property (4) of Subsection 6.1.3, we can write this equation as

$$A = \int_a^b [f(x) - g(x)]\,dx$$

We obtained this formula under the assumption that both f and g are nonnegative on $[a, b]$; we now show that this assumption is not necessary. To do so, we use Riemann sums to derive a formula for the area between two curves.

We assume that f and g are continuous on $[a, b]$ and that $f(x) \geq g(x)$ for all $x \in [a, b]$. We approximate the area between the two curves by rectangles: We divide $[a, b]$ into n equal subintervals, each of length Δx; that is, we set $a = x_0 < x_1 < x_2 < \cdots < x_n = b$ with $\Delta x = x_k - x_{k-1} = (b - a)/n$. The kth subinterval is thus between x_{k-1} and x_k. We choose left endpoints to compute the heights of the approximating rectangles. From Figure 6.28, we see that the height of the kth rectangle is equal to $f(x_{k-1}) - g(x_{k-1})$. We therefore obtain

$$A = \lim_{n \to \infty} \sum_{k=1}^n [f(x_{k-1}) - g(x_{k-1})]\,\Delta x$$

$$= \int_a^b [f(x) - g(x)]\,dx$$

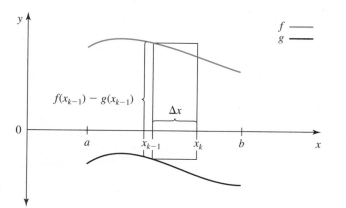

Figure 6.28 The kth rectangle between x_{k-1} and x_k.

The following box expresses this result:

> If f and g are continuous on $[a, b]$, with $f(x) \geq g(x)$ for all $x \in [a, b]$, then the area of the region between the curves $y = f(x)$ and $y = g(x)$ from a to b is equal to
>
> $$\text{Area} = \int_a^b [f(x) - g(x)]\, dx$$

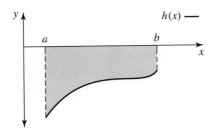

Figure 6.29 The definite integral $\int_a^b h(x)\, dx$ represents a signed area; it is negative here.

Before looking at a number of examples, we point out once more that this area formula always yields a nonnegative number since it computes the geometric area. To contrast this kind of area with the concept of a signed area, let us consider a function $h(x)$ on $[a, b]$, with $h(x) \leq 0$ for all $x \in [a, b]$. (See Figure 6.29.) The definite integral $\int_a^b h(x)\, dx$ represents a signed area and is negative in this case; more precisely, $\int_a^b h(x)\, dx$ is the negative of the geometric area of the region between the x-axis and the graph of $h(x)$ from $x = a$ to $x = b$. That this relationship is consistent with our definition of area can be seen as follows: The region of interest is bounded by the two curves $y = 0$ and $y = h(x)$. Since $h(x) \leq 0$ for $x \in [a, b]$, the area formula yields

$$\text{Area} = \int_a^b [0 - h(x)]\, dx = -\int_a^b h(x)\, dx$$

which is a positive number.

When you compute the area between two curves, you should always graph the bounding curves. This will show you how to set up the appropriate integral(s).

EXAMPLE 1 Find the area between the curves $y = \sec^2 x$ and $y = \cos x$ from $x = 0$ to $x = \pi/4$.

Solution We first graph the bounding curves, as shown in Figure 6.30. We see that both $\sec^2 x$ and $\cos x$ are continuous on $[0, \pi/4]$ and that $\sec^2 x \geq \cos x$ for $x \in [0, \frac{\pi}{4}]$. Therefore,

$$\begin{aligned}
\text{Area} &= \int_0^{\pi/4} [\sec^2 x - \cos x]\, dx \\
&= \tan x - \sin x\big]_0^{\pi/4} \\
&= \left(\tan \frac{\pi}{4} - \sin \frac{\pi}{4}\right) - (\tan 0 - \sin 0) \\
&= \left(1 - \frac{1}{2}\sqrt{2}\right) - (0 - 0) = 1 - \frac{1}{2}\sqrt{2}
\end{aligned}$$

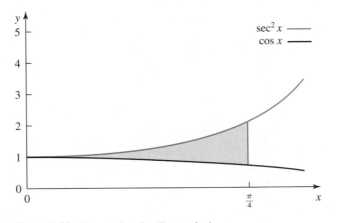

Figure 6.30 The region for Example 1.

EXAMPLE 2   Find the area of the region enclosed by $y = (x - 1)^2 - 1$ and $y = -x + 2$.

Solution   The bounding curves are graphed in Figure 6.31. To find the points where the two curves intersect, we solve

$$(x - 1)^2 - 1 = -x + 2$$
$$x^2 - 2x + 1 - 1 = -x + 2$$
$$x^2 - x - 2 = 0$$
$$(x + 1)(x - 2) = 0$$

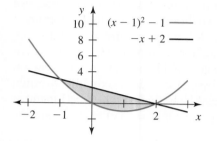

Figure 6.31 The region for Example 2.

Therefore,

$$x = -1 \quad \text{and} \quad x = 2$$

are the x-coordinates of the points of intersection. Note that both $y = (x - 1)^2$ and $y = -x + 2$ are continuous on $[-1, 2]$. Since $-x + 2 \geq (x - 1)^2 - 1$ for $x \in [-1, 2]$, the area of the enclosed region is

$$\text{Area} = \int_{-1}^{2} [(-x + 2) - ((x - 1)^2 - 1)]\,dx$$

$$= \int_{-1}^{2} [-x + 2 - x^2 + 2x - 1 + 1]\,dx$$

$$= \int_{-1}^{2} [-x^2 + x + 2]\,dx = -\frac{1}{3}x^3 + \frac{1}{2}x^2 + 2x \Big]_{-1}^{2}$$

$$= \left(-\frac{1}{3}(2)^3 + \frac{1}{2}(2)^2 + (2)(2)\right) - \left(-\frac{1}{3}(-1)^3 + \frac{1}{2}(-1)^2 + (2)(-1)\right)$$

$$= -\frac{8}{3} + 2 + 4 - \frac{1}{3} - \frac{1}{2} + 2 = \frac{9}{2} \qquad ■$$

EXAMPLE 3   Find the area of the region bounded by $y = \sqrt{x}$, $y = x - 2$, and the x-axis.

Solution   We first graph the bounding curves in Figure 6.32. We see that $y = \sqrt{x}$ and $y = x - 2$ intersect. To find the point of intersection, we solve

$$x - 2 = \sqrt{x}$$

Squaring both sides, we find that

$$(x - 2)^2 = (\sqrt{x})^2 \quad \text{and thus} \quad x^2 - 4x + 4 = x$$

or

$$x^2 - 5x + 4 = 0$$

Factoring the preceding equation, we obtain

$$(x - 4)(x - 1) = 0$$

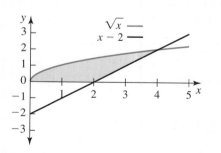

Figure 6.32 The region for Example 3.

which yields the solutions $x = 4$ and $x = 1$. Since squaring an equation can introduce extraneous solutions, we need to check whether the solutions satisfy $x - 2 = \sqrt{x}$. When $x = 4$, we find that $4 - 2 = \sqrt{4}$, which is a true statement; when $x = 1$, we find that $1 - 2 = \sqrt{1}$, which is a false statement. Hence, $x = 4$ is the only solution.

The graph of $y = x - 2$ intersects the x-axis at $x = 2$. To compute the area, we need to split the integral into two parts, because the lower bounding curve is composed of two parts: the x-axis from $x = 0$ to $x = 2$ and the line $y = x - 2$ from $x = 2$ to $x = 4$. We see from the graph that all bounding curves are continuous on their respective intervals. We get

$$\text{Area} = \int_0^2 \sqrt{x}\,dx + \int_2^4 [\sqrt{x} - (x-2)]\,dx$$

$$= \left[\frac{2}{3}x^{3/2}\right]_0^2 + \left[\frac{2}{3}x^{3/2} - \frac{1}{2}x^2 + 2x\right]_2^4$$

$$= \frac{2}{3}\cdot 2^{3/2} + \frac{2}{3}\cdot 4^{3/2} - \frac{1}{2}\cdot 16 + 8 - \frac{2}{3}\cdot 2^{3/2} + 2 - 4 = \frac{10}{3} \quad\blacksquare$$

In the preceding example, we needed to split the integral into two parts because the lower boundary of the area was composed of two different curves that determined the heights of the approximating rectangles in the Riemann integral. Recall that the rectangles are obtained by partitioning the x-axis. It is sometimes more convenient to partition the y-axis. We illustrate this approach in Figure 6.33 for the region of Example 3, where we partition the interval $[0, 2]$ on the y-axis into n equal subintervals, each of length Δy. We need to express the boundary curves as functions of y. In the case of Example 3, the right boundary curve is $x = f(y) = y + 2$ and the left boundary curve is $x = g(y) = y^2$.

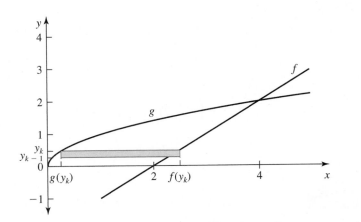

Figure 6.33 The area between $f(y)$ and $g(y)$ in Example 3 together with an approximating rectangle.

We set $y_0 = 0 < y_1 < y_2 < \cdots < y_n = 2$, where $y_k - y_{k-1} = \Delta y$ for $k = 1, 2, \ldots, n$. As shown in Figure 6.33, the kth rectangle has width $y_k - y_{k-1}$ and height $f(y_k) - g(y_k)$. Therefore, the area of the kth rectangle is

$$[f(y_k) - g(y_k)]\,\Delta y$$

If we sum this quantity from $k = 1$ to $k = n$ and let $\Delta y \to 0$, we find that

$$\lim_{\Delta y \to 0} \sum_{k=1}^n [f(y_k) - g(y_k)]\,\Delta y = \int_0^2 [f(y) - g(y)]\,dy$$

Since each rectangle is bounded by the same two curves, there is no need to split the integral. Using $f(y) = y + 2$ and $g(y) = y^2$, we obtain, for the total area,

$$\text{Area} = \int_0^2 (y + 2 - y^2)\,dy$$

$$= \frac{1}{2}y^2 + 2y - \frac{1}{3}y^3 \Big]_0^2 = 2 + 4 - \frac{8}{3} = \frac{10}{3}$$

which is the same as the result in Example 3.

To summarize the general case, as illustrated in Figure 6.34, suppose that a region is bounded by $x = f(y)$ and $x = g(y)$, with $g(y) \leq f(y)$ for $c \leq y \leq d$; that is, $f(y)$ is to the right of $g(y)$ for all $y \in [c, d]$. Then the area of the shaded region is given by the following formula:

$$\text{Area} = \int_c^d [f(y) - g(y)]\, dy$$

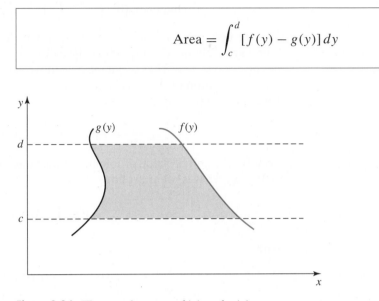

Figure 6.34 The area between $f(y)$ and $g(y)$.

■ 6.3.2 Cumulative Change

Consider a population whose size at time t, $t \geq 0$, is $N(t)$ and whose dynamics are given by the initial-value problem

$$\frac{dN}{dt} = f(t) \quad \text{with } N(0) = N_0$$

for some $f(t)$ that is continuous for $t \geq 0$. To solve this initial-value problem, we must find an antiderivative of $f(t)$ and determine the constant so that $N(0) = N_0$. Using (6.13) (i.e., the FTC, part I), we find that

$$N(t) = \int_0^t f(u)\, du + C$$

is the general antiderivative of $N(t)$. We choose 0 as the lower limit of integration for convenience, because it gives a simple expression for the constant when we take the initial condition into account. That is, with

$$N(0) = \int_0^0 f(u)\, du + C = C$$

it follows that $C = N(0) = N_0$. Therefore,

$$N(t) = N_0 + \int_0^t f(u)\, du$$

or

$$N(t) - N_0 = \int_0^t f(u)\, du \tag{6.14}$$

Since $f(u) = dN/du$ and $N(0) = N_0$, we can write

$$N(t) - N(0) = \int_0^t \frac{dN}{du}\, du$$

which allows us to interpret the definite integral $\int_0^t \frac{dN}{du}\, du$ as the **net**, or **cumulative**, **change** in population size between times 0 and t, since it is a "sum" of instantaneous changes accumulated over time. That is,

$$\begin{bmatrix} \text{cumulative} \\ \text{change in } [0, t] \end{bmatrix} = \int_0^t \begin{bmatrix} \text{instantaneous rate of} \\ \text{change at time } u \end{bmatrix} du$$

We now present another example in which we can interpret an integral as the cumulative change in a quantity. Recall that velocity is the instantaneous rate of change of distance. That is, if a particle moves along a straight line, and we denote by $s(t)$ the location of the particle at time t, with $s(0) = s_0$, and by $v(t)$ the velocity at time t, then $s(t)$ and $v(t)$ are related via

$$\frac{ds}{dt} = v(t) \quad \text{for } t > 0 \text{ with } s(0) = s_0$$

To solve this initial-value problem, we must find an antiderivative of $v(t)$ that satisfies $s(0) = s_0$. That is, we must have

$$s(t) = \int_0^t v(u)\, du + C$$

with

$$s_0 = s(0) = \int_0^0 v(u)\, du + C = 0 + C$$

which implies that $C = s_0$. Hence,

$$s(t) = s_0 + \int_0^t v(u)\, du$$

or, with $s(0) = s_0$,

$$s(t) - s(0) = \int_0^t v(u)\, du$$

Again, the cumulative change in distance, $s(t) - s(0)$, can be represented as a "sum" of instantaneous changes.

■ 6.3.3 Average Values

The concentration of soil nitrogen in g/m^3 was measured every meter along a transect in moist tundra and yielded the following data:

Distance from Origin (m)	Concentration (g/m^3)
1	589.3
2	602.7
3	618.5
4	667.2
5	641.2
6	658.3
7	672.8
8	661.2
9	652.3
10	669.8

If we denote the concentration at distance x from the origin by $c(x)$, then the average concentration, denoted by $\bar{c}$ (read "c bar"), is the arithmetic average

$$\bar{c} = \frac{1}{10} \sum_{k=1}^{10} c(k) = 643.3 \, g/m^3$$

More generally, to find the average concentration between two points a and b along a transect, we measure the concentration at equal distances. To formulate this notation mathematically, we divide $[a, b]$ into n subintervals of equal lengths $\Delta x = \frac{b-a}{n}$ and measure the concentration at, say, the right endpoint of each subinterval. If the concentration at location x_k is denoted by $c(x_k)$, then the average concentration $\bar{c}$ is

$$\bar{c} = \frac{1}{n} \sum_{k=1}^{n} c(x_k)$$

Since $\Delta x = \frac{b-a}{n}$, we can write $n = \frac{b-a}{\Delta x}$. Hence,

$$\bar{c} = \frac{1}{b-a} \sum_{k=1}^{n} c(x_k)\, \Delta x$$

If we let the number of subintervals grow ($n \to \infty$), then the length of each subinterval goes to 0 ($\Delta x \to 0$) and

$$\bar{c} = \frac{1}{b-a} \lim_{n \to \infty} \sum_{k=1}^{n} c(x_k)\, \Delta x = \frac{1}{b-a} \int_{a}^{b} c(x)\, dx$$

That is, the average concentration can be expressed as an integral over $c(x)$ between a and b, divided by the length of the interval $[a, b]$:

> Assume that $f(x)$ is a continuous function on $[a, b]$. The average value of f on the interval $[a, b]$ is
>
> $$f_{avg} = \frac{1}{b-a} \int_{a}^{b} f(x)\, dx$$

EXAMPLE 4   Find the average value of $f(x) = 4 - x^2$ on the interval $[-2, 2]$.

Solution   We use the formula for computing average values. Note that $f(x) = 4 - x^2$ is continuous on $[-2, 2]$. Then the average value of f on the interval $[-2, 2]$ is

$$f_{avg} = \frac{1}{2-(-2)} \int_{-2}^{2} (4 - x^2)\, dx = \frac{1}{4}\left[4x - \frac{1}{3}x^3 \right]_{-2}^{2}$$

$$= \frac{1}{4}\left[8 - \frac{8}{3} + 8 - \frac{8}{3} \right] = \frac{1}{4} \cdot \frac{32}{3} = \frac{8}{3} \qquad \blacksquare$$

The following theorem says a bit more about the value of f_{avg}.

> **The Mean-Value Theorem for Definite Integrals**   Assume that $f(x)$ is a continuous function on $[a, b]$. Then there exists a number $c \in [a, b]$ such that
>
> $$f(c)(b-a) = \int_{a}^{b} f(x)\, dx$$

That is, when we compute the average value of a function that is continuous on $[a, b]$, we find that there exists a number c such that $f(c) = f_{avg}$. We can understand this concept graphically when we look at the graph of a function f and at f_{avg}. For simplicity, let's assume that $f(x) \geq 0$. Since (1) $\int_{a}^{b} f(x)\, dx$ is then equal to the area between the graph of $f(x)$ and the x-axis, (2) $f_{avg}(b-a)$ is equal to the area of the rectangle with height f_{avg} and width $b - a$, and (3) the two areas are equal, the horizontal line $y = f_{avg}$ must intersect the graph of $f(x)$ at some point on the interval $[a, b]$. (See Figure 6.35.) The x-coordinate of this point of intersection is then the value c in the MVT for definite integrals. (Note that there could be more than one such number.) A similar argument can be made when we do not assume that $f(x)$ is positive. In this case, "area" is replaced by "signed area." The proof of this theorem is short, and we supply it for completeness at the end of this subsection.

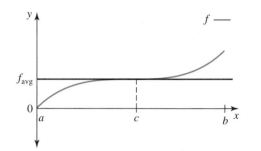

Figure 6.35 An illustration of the average value of a function:
$\int_a^b f(x)\, dx = f_{avg}(b - a)$.

EXAMPLE 5

Find the average value of $f(x) = x^3$ on the interval $[-1, 1]$, and determine $x \in [-1, 1]$ such that $f(x)$ equals the average value.

Solution

The function $f(x) = x^3$ is continuous on $[-1, 1]$. Then

$$f_{avg} = \frac{1}{2} \int_{-1}^{1} x^3\, dx = \frac{1}{2} \left[\frac{1}{4} x^4 \right]_{-1}^{1} = \frac{1}{8}(1^4 - (-1)^4) = 0$$

The graph of $y = x^3$ (Figure 6.36) is symmetric about the origin, which explains why the average value is 0. Since $x^3 = 0$ for $x = 0$, the function $f(x)$ takes on its average value at $x = 0$. ■

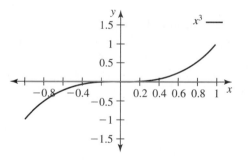

Figure 6.36 The graph of $y = x^3$, $-1 \le x \le 1$. The average value is 0.

EXAMPLE 6

The speed of water in a channel varies considerably with depth. Because of friction, the velocity reaches zero at the bottom and along the sides of the channel; the velocity is greatest near the surface of the water. The average velocity of a stream is of interest in characterizing rivers. One way to obtain this average value would be to measure a stream's velocity at various depths along a vertical transect and then average the values obtained. In practice, however, a much simpler method is employed: The speed is measured at 60% of the depth from the surface, because the speed at that depth is very close to the average speed. Explain why it is possible that the measurement at just one depth would yield the average stream velocity.

Solution

Assuming that the velocity profile of the stream along a vertical transect is a continuous function of depth, the MVT for definite integrals guarantees that there exists a depth h at which the velocity is equal to the average stream velocity.

The MVT only gives the existence of such a depth; it does not tell us where the velocity is equal to the average velocity. It is surprising and fortunate that the depth where the velocity reaches its average is quite universal; that is, it does not depend much on the specifics of the river. (The 60%-depth rule is derived in Problems 3–5 of Chapter 6 Review Problems.)   ■

Proof of the Mean-Value Theorem for Definite Integrals   Since $f(x)$ is continuous on $[a, b]$, we can apply the extreme-value theorem to conclude that f attains an absolute maximum and an absolute minimum in $[a, b]$. If we denote the absolute maximum by M and the absolute minimum by m, then

$$m \le f(x) \le M \quad \text{for all } x \in [a, b]$$

and f takes on both m and M for some values in $[a, b]$. We therefore find that

$$m(b - a) \le \int_a^b f(x)\, dx \le M(b - a)$$

or

$$m \le \frac{1}{b - a} \int_a^b f(x)\, dx \le M$$

We set $I = \frac{1}{b-a} \int_a^b f(x)\, dx$; then $m \le I \le M$.

Using the facts that $f(x)$ takes on all values between m and M in the interval $[a, b]$ (this follows from the intermediate-value theorem) and that I is a number between m and M, it follows (also from the intermediate-value theorem) that there must be a number $c \in [a, b]$ such that $f(c) = I$; that is,

$$f(c) = \frac{1}{b - a} \int_a^b f(x)\, dx \qquad ■$$

■ 6.3.4 The Volume of a Solid (Optional)

From geometry, we know various formulas for computing the volumes of certain regular solids, such as a right circular cylinder. To compute the volume of a less regularly shaped solid, we will use an approach that is similar to that for computing areas of irregularly shaped regions; there, we approximated the areas of such regions by rectangles whose areas were easy to compute with simple formulas from geometry.

We begin with the volume of a generalized cylinder; the volume is the base area times the height. The base can be any arbitrarily shaped region. (See Figure 6.37, for example.)

If we denote the base area by A and the height of the cylinder by h, then the volume of the generalized cylinder is

$$V = Ah$$

As an example, consider the circular cylinder whose base is a disk. If the disk has radius r and the cylinder has height h, then the volume of the circular cylinder is $\pi r^2 h$. We will use cylinders to approximate volumes of more complicated solids.

Suppose that we wish to compute the volume of the solid shown in Figure 6.38. We can slice the solid into small slabs by cutting it perpendicular to the x-axis at points $x_0 = a < x_1 < x_2 < \cdots < x_n = b$ that partition the interval $[a, b]$ and then slice the solid into planes. The intersection of such a plane and the solid is called a *cross section*. We denote the area of the cross section at x_k by $A(x_k)$. By cutting the solid along these planes, we obtain slices, just as we do when we cut bread. We will approximate the volume of a slice between x_{k-1} and x_k by the volume of a cylinder with base area equal to that of the slice at x_k and height $\Delta x_k = x_k - x_{k-1}$. The volume of the slice between x_{k-1} and x_k is then approximately

$$A(x_k)\, \Delta x_k$$

Figure 6.37 A right cylinder with an irregularly shaped base.

Figure 6.38 The volume of an irregularly shaped solid, found by the disk method.

Adding the volumes of all the slices gives us an approximation for the total volume of the solid:

$$V \approx \sum_{k=1}^{n} A(x_k)\, \Delta x_k$$

By making the partition of $[a, b]$ finer, we can improve the approximation.

Recall that we used $\|P\|$ (the norm of P) as a measure of how fine the partition is; that is, $\|P\| = \max_{k=1,2,\dots,n} \Delta x_k$. This approach suggests that we should define the volume of the solid as the limit of our approximation as $\|P\| \to 0$. We summarize this concept in the following definition:

Definition The **volume of a solid** of integrable cross-sectional area $A(x)$ between a and b is

$$\int_a^b A(x)\, dx$$

EXAMPLE 7

Find the volume of the sphere of radius r centered at the origin.

Solution

The cross section at x is perpendicular to the x-axis. (See Figure 6.39.) It is a disk of radius $y = \sqrt{r^2 - x^2}$ and whose area is

$$A(x) = \pi y^2 = \pi (r^2 - x^2)$$

Since the solid is between $-r$ and r, it follows that

$$\text{Volume} = \int_{-r}^{r} \pi (r^2 - x^2)\, dx$$

The integrand is continuous on $[-r, r]$. Evaluating the integral yields

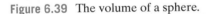

$$= \pi \left[r^2 x - \frac{1}{3} x^3 \right]_{-r}^{r}$$

$$= \pi \left[(r^3 - \frac{1}{3} r^3) - (-r^3 + \frac{1}{3} r^3) \right] = \pi \left(\frac{2}{3} r^3 + \frac{2}{3} r^3 \right) = \frac{4}{3} \pi r^3$$

Figure 6.39 The volume of a sphere.

This result agrees with the formula for a sphere that we know from geometry. ■

The cross sections of the sphere in the last example were of a particularly simple form, namely, disks. However, we can think of a sphere as a **solid of revolution**—that is, a solid obtained by revolving a curve about the x-axis (or the y-axis). In this case, we rotate the curve $y = \sqrt{r^2 - x^2}$, $-r \le x \le r$, about the x-axis, which creates circular cross sections.

We can use other curves $y = f(x)$, rotate them about the x-axis, and obtain solids in the same way. We illustrate in Figure 6.40, in which we rotate the graph of $y = f(x), a \le x \le b$, about the x-axis. A cross section through x perpendicular to the x-axis is then a disk with radius $f(x)$; hence, its cross-sectional area is $A(x) = \pi[f(x)]^2$. If we use the formula $\int_a^b A(x)\, dx$ to compute the volume of the solid, we find that the volume of the solid of revolution is

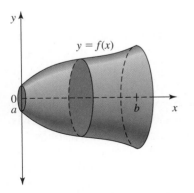

Figure 6.40 The solid of rotation when rotating $f(x)$ about the x-axis.

$$V = \int_a^b \pi [f(x)]^2\, dx \tag{6.15}$$

Computing volumes by using (6.15) is called the **disk method**.

EXAMPLE 8

Compute the volume of the solid obtained by rotating $y = x^2, 0 \leq x \leq 2$, about the x-axis.

Solution

We illustrate the solid in Figure 6.41. When we rotate the graph of $y = x^2$ about the x-axis, we find that the cross section at x is a disk with radius $y = f(x) = x^2$. The cross-sectional area at x, $A(x)$, is then $\pi(x^2)^2 = \pi x^4$, which is integrable on $[0, 2]$. Thus, the volume is

$$V = \int_0^2 \pi [f(x)]^2 \, dx = \int_0^2 \pi x^4 \, dx$$

$$= \frac{\pi}{5} x^5 \Big]_0^2 = \frac{32}{5} \pi$$

■

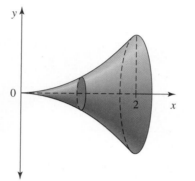

Figure 6.41 The solid of rotation for Example 8.

EXAMPLE 9

Rotate the area bounded by the curves $y = \sqrt{x}$ and $y = x/2$ about the x-axis, and compute the volume of the solid of rotation.

Solution

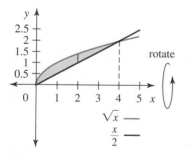

Figure 6.42 The plane region for Example 9.

The curves $y = \sqrt{x}$ and $y = x/2$ are graphed in Figure 6.42, together with a vertical bar to indicate the cross section. We see from the graph that the curves intersect at $x = 0$ and $x = 4$. To find the points of intersection algebraically, we need to equate $\sqrt{x}$ and $x/2$ and solve for x:

$$\sqrt{x} = \frac{x}{2}$$

This equation immediately yields the solution $x = 0$. If $x > 0$, we can divide by $\sqrt{x}$ and find

$$1 = \frac{\sqrt{x}}{2}, \quad \text{or} \quad 2 = \sqrt{x}$$

Squaring yields $x = 4$. Thus, the curves intersect at $x = 0$ and $x = 4$. We can compute the volume of this solid of rotation by first rotating $y = \sqrt{x}, 0 \leq x \leq 4$, about the x-axis and computing the volume of this solid, and then subtracting the volume of the solid obtained by rotating $y = x/2, 0 \leq x \leq 4$. When we do so, we get

$$V = \int_0^4 \pi (\sqrt{x})^2 \, dx - \int_0^4 \pi \left(\frac{1}{2} x \right)^2 \, dx$$

Both integrands are continuous on $[0, 4]$ and we find that

$$V = \pi \frac{1}{2} x^2 \Big]_0^4 - \pi \frac{1}{12} x^3 \Big]_0^4$$

$$= 8\pi - \frac{16}{3} \pi = \frac{8}{3} \pi$$

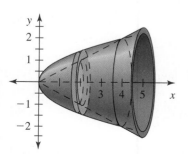

Figure 6.43 The solid of rotation for Example 9.

Looking at Figure 6.43, we see that the cross-sectional area is that of a washer. In this case, the disk method is also referred to as the washer method.

■

In the next example, we rotate a curve about the y-axis.

EXAMPLE 10

Rotate the region bounded by $y = 2$, $x = 0$, $y = 0$, and $y = \ln x$ about the y-axis, and compute the volume of the resulting solid.

Solution

When we rotate about the y-axis, the cross sections are perpendicular to the y-axis. At $y = \ln x$, the radius of the cross-sectional disk is x. (See Figure 6.44.) The solid is shown in Figure 6.45. Since we "sum" the slices along the y-axis, we must integrate with respect to y; because $y = \ln x$, we get $x = e^y$. Therefore, the cross-sectional area at y is $A(y) = \pi(e^y)^2$, which is integrable on $[0, 2]$, and the volume is

$$V = \int_0^2 \pi(e^y)^2 \, dy = \int_0^2 \pi e^{2y} \, dy = \pi \frac{1}{2} e^{2y} \Big]_0^2 = \frac{\pi}{2}(e^4 - 1) \qquad ■$$

Figure 6.44 The plane region for Example 10.

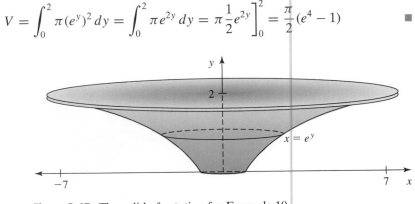

Figure 6.45 The solid of rotation for Example 10.

■ 6.3.5 Rectification of Curves (Optional)

In this subsection, we will study how to compute the lengths of curves in the plane. This is yet another example in which integrals appear as we add up a large number of small increments. The first curve whose length was determined was the semicubical parabola $y^2 = x^3$; this was done by William Neile (1637–1670) in 1657. Interestingly enough, only about 20 years earlier Descartes asserted that there was no rigorous way to determine the exact length of a curve. It turns out that the exact formula is, in essence, an infinitesimal version of the Pythagorean theorem.

How would we *rectify* a curve (i.e., determine its length) with a ruler? We could approximate the curve by short line segments, measure the length of each segment, and add up the measurements, as illustrated in Figure 6.46. By choosing smaller line segments, our approximation would improve. This is precisely the method that we will employ to find an exact formula.

Figure 6.46 The length of a curve in the plane.

To find the exact formula, assume that the curve whose length we want to find is given by a function $y = f(x)$, $a \leq x \leq b$, which has a continuous first derivative on (a, b). We partition the interval $[a, b]$ into subintervals by using the partition $P = [x_0, x_1, x_2, \ldots, x_n]$, where $a = x_0 < x_1 < x_2 < \cdots < x_n = b$, and approximate the curve by a polygon that consists of the straight-line segments connecting neighboring points on the curve, as shown in Figure 6.46. A typical line segment connecting the points P_{k-1} and P_k is shown in Figure 6.47. Using the Pythagorean theorem, we can find its length. We set $\Delta x_k = x_k - x_{k-1}$ and $\Delta y_k = y_k - y_{k-1}$. Then the length of the line segment is given by

$$\sqrt{(\Delta x_k)^2 + (\Delta y_k)^2}$$

Figure 6.47 A typical line segment.

Using the partition P, we then find that the length of the polygon is

$$L_P = \sum_{k=1}^{n} \sqrt{(\Delta x_k)^2 + (\Delta y_k)^2} \tag{6.16}$$

With finer and finer partitions, the length of the polygon will become a better and better approximation of the length of the corresponding curve. However, before we can take this limit, we need to work on the sum.

The difference Δy_k is equal to $f(x_k) - f(x_{k-1})$. The MVT guarantees that there is a number $c_k \in [x_{k-1}, x_k]$ such that

$$f'(c_k) = \frac{f(x_k) - f(x_{k-1})}{x_k - x_{k-1}}$$

Since $\Delta x_k = x_k - x_{k-1}$, it follows that

$$\Delta y_k = f'(c_k)(x_k - x_{k-1}) = f'(c_k)\,\Delta x_k \tag{6.17}$$

Replacing Δy_k in (6.16) by (6.17), we find that the length of the polygon is given by

$$L_P = \sum_{k=1}^{n} \sqrt{(\Delta x_k)^2 + [f'(c_k)\,\Delta x_k]^2}$$

$$= \sum_{k=1}^{n} \sqrt{1 + [f'(c_k)]^2}\,\Delta x_k$$

This form allows us to take the limit as $\|P\| \to 0$ and obtain

$$L = \lim_{\|P\| \to 0} \sum_{k=1}^{n} \sqrt{1 + [f'(c_k)]^2}\,\Delta x_k$$

$$= \int_a^b \sqrt{1 + [f'(x)]^2}\,dx$$

Thus,

If $f(x)$ is differentiable on (a, b) and $f'(x)$ is continuous on $[a, b]$, then the length of the curve $y = f(x)$ from a to b is given by

$$L = \int_a^b \sqrt{1 + [f'(x)]^2}\,dx$$

As mentioned, the first curve that was rectified was $y = f(x) = x^{3/2}$. We will choose this function for our first example of rectification.

EXAMPLE 11

Determine the length of the curve given by the graph of $y = f(x) = x^{3/2}$ between $a = 5/9$ and $b = 21/9$.

Solution

To determine the length of the curve, we need to find $f'(x)$ first. We have

$$f'(x) = \frac{3}{2}x^{1/2}$$

Then

$$L = \int_{5/9}^{21/9} \sqrt{1 + \left[\frac{3}{2}x^{1/2}\right]^2}\,dx = \int_{5/9}^{21/9} \sqrt{1 + \frac{9}{4}x}\,dx$$

An antiderivative of $\sqrt{1 + \frac{9}{4}x}$ is $\frac{4}{9} \cdot \frac{2}{3}\left(1 + \frac{9}{4}x\right)^{3/2}$, as can be checked by differentiating the latter with respect to x. Thus, the length is

$$L = \left[\frac{4}{9} \cdot \frac{2}{3}\left(1 + \frac{9}{4}x\right)^{3/2}\right]_{5/9}^{21/9} = \frac{8}{27}\left[\left(1 + \frac{9}{4} \cdot \frac{21}{9}\right)^{3/2} - \left(1 + \frac{9}{4} \cdot \frac{5}{9}\right)^{3/2}\right]$$

$$= \frac{8}{27}\left[\left(\frac{5}{2}\right)^3 - \left(\frac{3}{2}\right)^3\right] = \frac{8}{27}\left(\frac{125}{8} - \frac{27}{8}\right) = \frac{98}{27}$$

The length of the curve is therefore 98/27.   ■

Before we present another example, we discuss the formula in more detail. Using

$$\frac{dy}{dx} = f'(x)$$

we can write

$$L = \int_a^b \sqrt{1 + \left(\frac{dy}{dx}\right)^2}\, dx$$

for the length. Treating dy and dx as if they were numbers, we can rewrite this equation as

$$L = \int_a^b \sqrt{(dx)^2 + (dy)^2}$$

We call the expression $\sqrt{(dx)^2 + (dy)^2}$ the **arc length differential** and denote it by ds. We can think of ds as a typical infinitesimal line segment. The Pythagorean theorem in this infinitesimal form then becomes $(ds)^2 = (dx)^2 + (dy)^2$. "Adding up" these line segments (i.e., computing $\int_a^b ds$) then yields the length of the curve.

EXAMPLE 12   Determine the length of

$$f(x) = \frac{1}{4}x^2 - \frac{1}{2}\ln x \quad \text{from } x = 1 \text{ to } x = e$$

Solution   Differentiating $f(x)$, we find that

$$f'(x) = \frac{x}{2} - \frac{1}{2x}$$

The length of the curve is then given by

$$L = \int_1^e \sqrt{1 + \left(\frac{x}{2} - \frac{1}{2x}\right)^2}\, dx = \int_1^e \sqrt{1 + \left(\frac{x^2}{4} - \frac{1}{2} + \frac{1}{4x^2}\right)}\, dx$$

$$= \int_1^e \sqrt{\frac{x^2}{4} + \frac{1}{2} + \frac{1}{4x^2}}\, dx$$

We notice that the expression under the square root is a perfect square, namely,

$$\frac{x^2}{4} + \frac{1}{2} + \frac{1}{4x^2} = \left(\frac{x}{2} + \frac{1}{2x}\right)^2$$

Hence, the integral for the length simplifies to

$$L = \int_1^e \sqrt{\left(\frac{x}{2} + \frac{1}{2x}\right)^2}\, dx$$

$$= \int_1^e \left(\frac{x}{2} + \frac{1}{2x}\right) dx = \frac{1}{4}x^2 + \frac{1}{2}\ln|x| \Big]_1^e$$

$$= \frac{1}{4}e^2 + \frac{1}{2} - \frac{1}{4} = \frac{1}{4}(e^2 + 1) \qquad\blacksquare$$

Because of the somewhat complicated form of the integrand in the computation of the length, we quickly run into problems when we actually try to compute the integral. In practice, the integrand rarely simplifies enough for easy computation, as it does in Example 12, where it turned out to be a perfect square.

Even seemingly simple looking functions, such as $y = 1/x$, quickly turn into complicated integrals when we compute the length of the curve.

EXAMPLE 13 Set up, but do not evaluate, the length of the curve of the hyperbola $f(x) = \frac{1}{x}$ between $a = 1$ and $b = 2$.

Solution To determine the length of the curve, we need to find $f'(x)$ first.

$$f'(x) = -\frac{1}{x^2}$$

Then the length of the curve is given by the integral

$$L = \int_1^2 \sqrt{1 + \left(-\frac{1}{x^2}\right)^2}\, dx = \int_1^2 \sqrt{1 + \frac{1}{x^4}}\, dx \qquad ■$$

The antiderivative of the integrand in Example 13 is quite complicated, and we will not be able to find it with the techniques available in this text. In Section 7.5, we will learn numerical methods for evaluating integrals, and some of these methods can be used to evaluate the integral in Example 13. There are also computer software packages that can numerically evaluate such integrals. Using either of these approaches on the integral in Example 13, we would find that the length L is approximately 1.13.

Section 6.3 Problems

■ 6.3.1

Find the areas of the regions bounded by the lines and curves in Problems 1–12.

1. $y = x^2 - 4$, $y = x + 2$
2. $y = 2x^2 - 1$, $y = 2 - x^4$
3. $y = e^{x/2}$, $y = -x$, $x = 0$, $x = 2$
4. $y = \cos x$, $y = 1$, $x = 0$, $x = \frac{\pi}{2}$
5. $y = x^2 + 1$, $y = 4x - 2$ (in the first quadrant)
6. $y = x^2$, $y = 2 - x$, $y = 0$ (in the first quadrant)
7. $y = x^2$, $y = \frac{1}{x}$, $y = 4$ (in the first quadrant)
8. $y = \sin x$, $y = \cos x$ from $x = 0$ to $x = \frac{\pi}{4}$
9. $y = \sin x$, $y = 1$ from $x = 0$ to $x = \frac{\pi}{2}$
10. $y = x^2$, $y = (x - 2)^2$, $y = 0$ from $x = 0$ to $x = 2$
11. $y = x^2$, $y = x^3$ from $x = 0$ to $x = 2$
12. $y = e^{-x}$, $y = x + 1$ from $x = -1$ to $x = 1$

In Problems 13–16, find the areas of the regions bounded by the lines and curves by expressing x as a function of y and integrating with respect to y.

13. $y = x^2$, $y = (x - 2)^2$, $y = 0$ from $x = 0$ to $x = 2$
14. $y = x$, $yx = 1$, $y = \frac{1}{2}$ (in the first quadrant)
15. $x = (y - 1)^2 + 3$, $x = 1 - (y - 1)^2$ from $y = 0$ to $y = 2$ (in the first quadrant)
16. $x = (y - 1)^2 - 1$, $x = (y - 1)^2 + 1$ from $y = 0$ to $y = 2$

■ 6.3.2

17. Consider a population whose size at time t is $N(t)$ and whose dynamics are given by the initial-value problem

$$\frac{dN}{dt} = e^{-t}$$

with $N(0) = 100$.

(a) Find $N(t)$ by solving the initial-value problem.

(b) Compute the cumulative change in population size between $t = 0$ and $t = 5$.

(c) Express the cumulative change in population size between time 0 and time t as an integral. Give a geometric interpretation of this quantity.

18. Suppose that a change in biomass $B(t)$ at time t during the interval $[0, 12]$ follows the equation

$$\frac{d}{dt} B(t) = \cos\left(\frac{\pi}{6} t\right)$$

for $0 \le t \le 12$.

(a) Graph $\frac{dB}{dt}$ as a function of t.

(b) Suppose that $B(0) = B_0$. Express the cumulative change in biomass during the interval $[0, t]$ as an integral. Give a geometric interpretation. What is the value of the biomass at the end of the interval $[0, 12]$ compared with the value at time 0? How are these two quantities related to the cumulative change in the biomass during the interval $[0, 12]$?

19. A particle moves along the x-axis with velocity

$$v(t) = -(t - 2)^2 + 1$$

for $0 \le t \le 5$. Assume that the particle is at the origin at time 0.

(a) Graph $v(t)$ as a function of t.

(b) Use the graph of $v(t)$ to determine when the particle moves to the left and when it moves to the right.

(c) Find the location $s(t)$ of the particle at time t for $0 \le t \le 5$. Give a geometric interpretation of $s(t)$ in terms of the graph of $v(t)$.

(d) Graph $s(t)$ and find the leftmost and rightmost positions of the particle.

20. Recall that the acceleration $a(t)$ of a particle moving along a straight line is the instantaneous rate of change of the velocity $v(t)$; that is,

$$a(t) = \frac{d}{dt}v(t)$$

Assume that $a(t) = 32$ ft/s^2. Express the cumulative change in velocity during the interval $[0, t]$ as a definite integral, and compute the integral.

21. If $\frac{dl}{dt}$ represents the growth rate of an organism at time t (measured in months), explain what

$$\int_2^7 \frac{dl}{dt} \, dt$$

represents.

22. If $\frac{dw}{dx}$ represents the rate of change of the weight of an organism of age x, explain what

$$\int_3^5 \frac{dw}{dx} \, dx$$

means.

23. If $\frac{dB}{dt}$ represents the rate of change of biomass at time t, explain what

$$\int_1^6 \frac{dB}{dt} \, dt$$

means.

24. Let $N(t)$ denote the size of a population at time t, and assume that

$$\frac{dN}{dt} = f(t)$$

Express the cumulative change of the population size in the interval $[0, 3]$ as an integral.

■ 6.3.3

25. Let $f(x) = x^2 - 2$. Compute the average value of $f(x)$ over the interval $[0, 2]$.

26. Let $g(t) = \sin(\pi t)$. Compute the average value of $g(t)$ over the interval $[-1, 1]$.

27. Suppose that the temperature T (measured in degrees Fahrenheit) in a growing chamber varies over a 24-hour period according to

$$T(t) = 68 + \sin\left(\frac{\pi}{12}t\right)$$

for $0 \le t \le 24$.
(a) Graph the temperature T as a function of time t.
(b) Find the average temperature and explain your answer graphically.

28. Suppose that the concentration (measured in gm^{-3}) of nitrogen in the soil along a transect in moist tundra yields data points that follow a straight line with equation

$$y = 673.8 - 34.7x$$

for $0 \le x \le 10$, where x is the distance to the beginning of the transect. What is the average concentration of nitrogen in the soil along this transect?

29. Let $f(x) = \tan x$. Give a geometric argument to explain why the average value of $f(x)$ over $[-1, 1]$ is equal to 0.

30. Suppose that you drive from St. Paul to Duluth and you average 50 mph. Explain why there must be a time during your trip at which your speed is exactly 50 mph.

31. Let $f(x) = 2x, 0 \le x \le 2$. Use a geometric argument to find the average value of f over the interval $[0, 2]$, and find x such that $f(x)$ is equal to this average value.

32. A particle moves along the x-axis with velocity

$$v(t) = -(t - 3)^2 + 5$$

for $0 \le t \le 6$.
(a) Graph $v(t)$ as a function of t for $0 \le t \le 6$.
(b) Find the average velocity of this particle during the interval $[0, 6]$.
(c) Find a time $t^* \in [0, 6]$ such that the velocity at time t^* is equal to the average velocity during the interval $[0, 6]$. Is it clear that such a point exists? Is there more than one such point in this case? Use your graph in (a) to explain how you would find t^* graphically.

■ 6.3.4

33. Find the volume of a right circular cone with base radius r and height h.

34. Find the volume of a pyramid with square base of side length a and height h.

In Problems 35–40, find the volumes of the solids obtained by rotating the region bounded by the given curves about the x-axis. In each case, sketch the region and a typical disk element.

35. $y = 4 - x^2, y = 0, x = 0$ (in the first quadrant)
36. $y = \sqrt{2x}, y = 0, x = 2$
37. $y = \sqrt{\sin x}, 0 \le x \le \pi, y = 0$
38. $y = e^x, y = 0, x = 0, x = \ln 2$
39. $y = \sec x, -\frac{\pi}{3} \le x \le \frac{\pi}{3}, y = 0$
40. $y = \sqrt{1 - x^2}, 0 \le x \le 1, y = 0$

In Problems 41–46, find the volumes of the solids obtained by rotating the region bounded by the given curves about the x-axis. In each case, sketch the region together with a typical disk element.

41. $y = x^2, y = x, 0 \le x \le 1$
42. $y = 2 - x^3, y = 2 + x^3, 0 \le x \le 1$
43. $y = e^x, y = e^{-x}, 0 \le x \le 2$
44. $y = \sqrt{1 - x^2}, y = 1, x = 1$ (in the first quadrant)
45. $y = \sqrt{\cos x}, y = 1, x = \frac{\pi}{2}$
46. $y = \frac{1}{x}, x = 0, y = 1, y = 2$ (in the first quadrant)

In Problems 47–52, find the volumes of the solids obtained by rotating the region bounded by the given curves about the y-axis. In each case, sketch the region together with a typical disk element.

47. $y = \sqrt{x}, y = 2, x = 0$
48. $y = x^2, y = 4, x = 0$ (in the first quadrant)
49. $y = \ln(x + 1), y = \ln 3, x = 0$
50. $y = \sqrt{x}, y = x, 0 \le x \le 1$
51. $y = x^2, y = \sqrt{x}, 0 \le x \le 1$
52. $y = \frac{1}{x}, x = 0, y = \frac{1}{2}, y = 1$

■ 6.3.5

53. Find the length of the straight line

$$y = 2x$$

from $x = 0$ to $x = 2$ by each of the following methods:
(a) planar geometry

(b) the integral formula for the lengths of curves, derived in Subsection 6.3.5

54. Find the length of the straight line

$$y = mx$$

from $x = 0$ to $x = a$, where m and a are positive constants, by each of the following methods:
(a) planar geometry
(b) the integral formula for the lengths of curves, derived in Subsection 6.3.5

55. Find the length of the curve

$$y^2 = x^3$$

from $x = 1$ to $x = 4$.

56. Find the length of the curve

$$2y^2 = 3x^3$$

from $x = 0$ to $x = 1$.

57. Find the length of the curve

$$y = \frac{x^3}{6} + \frac{1}{2x}$$

from $x = 1$ to $x = 3$.

58. Find the length of the curve

$$y = \frac{x^4}{4} + \frac{1}{8x^2}$$

from $x = 2$ to $x = 4$.

In Problems 59–62, set up, but do not evaluate, the integrals for the lengths of the following curves:

59. $y = x^2, -1 \le x \le 1$

60. $y = \sin x, 0 \le x \le \dfrac{\pi}{2}$

61. $y = e^{-x}, 0 \le x \le 1$

62. $y = \ln x, 1 \le x \le e$

63. Find the length of the quarter-circle

$$y = \sqrt{1 - x^2}$$

for $0 \le x \le 1$, by each of the following methods:
(a) a formula from geometry
(b) the integral formula from Subsection 6.3.5

64. A cable that hangs between two poles at $x = -M$ and $x = M$ takes the shape of a catenary, with equation

$$y = \frac{1}{2a}(e^{ax} + e^{-ax})$$

where a is a positive constant. Compute the length of the cable when $a = 1$ and $M = \ln 2$.

65. Show that if

$$f(x) = \frac{e^x + e^{-x}}{2}$$

then the length of the curve $f(x)$ between $x = 0$ and $x = a$ for any $a > 0$ is given by $f'(a)$.

Chapter 6 Key Terms

Discuss the following definitions and concepts:

1. Area

2. Summation notation

3. Algebraic rules for sums

4. A partition of an interval and the norm of a partition

5. Riemann sum

6. Definite integral

7. Riemann integrable

8. Geometric interpretation of definite integrals

9. The constant-value and constant-multiple rules for integrals

10. The definite integral over a union of intervals

11. Comparison rules for definite integrals

12. The fundamental theorem of calculus, part I

13. Leibniz's rule

14. Antiderivatives

15. The fundamental theorem of calculus, part II

16. Evaluating definite integrals by using the FTC, part II

17. Computing the area between curves by using definite integrals

18. Cumulative change and definite integrals

19. The mean-value theorem for definite integrals

20. The volume of a solid and definite integrals

21. Rectification of curves

22. Length of a curve

23. Arc length differential

Chapter 6 Review Problems

1. Discharge of a River In studying the flow of water in an open channel, such as a river in its bed, the amount of water passing through a cross section per second—the discharge (Q)—is of interest. The following formula is used to compute the discharge:

$$Q = \int_0^B \overline{v}(b)h(b)\,db \qquad (6.18)$$

In this formula, b is the distance from one bank of the river to the point where the depth $h(b)$ of the river and the average velocity $\overline{v}(b)$ of the vertical velocity profile of the river at b were measured. The total width of the cross section is B. (See Figure 6.48.)

Figure 6.48 The river for Problem 1.

To evaluate the integral in (6.18), we would need to know $\overline{v}(b)$ and $h(b)$ at every location b along the cross section. In practice, the cross section is divided into a finite number of subintervals and measurements of $\overline{v}$ and h are taken at, say, the right endpoints of each subinterval. The following table contains an example of such measurements:

Location	h	$\bar{v}$
0	0	0
1	0.28	0.172
3	0.76	0.213
5	1.34	0.230
7	1.57	0.256
9	1.42	0.241
11	1.21	0.206
13	0.83	0.187
15	0.42	0.116
16	0	0

The location 0 corresponds to the left bank, and the location $B = 16$ to the right bank, of the river. The units of the location and of h are meters, and of $\bar{v}$, meters per second. Approximate the integral in (6.18) by a Riemann sum, using the locations in the table, and find the approximate discharge, using the data from the table.

2. Biomass Growth Suppose that you grow plants in several study plots and wish to measure the response of total biomass to the treatment in each plot. One way to measure this response would be to determine the average specific growth rate of the biomass for each plot over the course of the growing season.

We denote by $B(t)$ the biomass in a given plot at time t. Then the specific growth rate of the biomass at time t is given by

$$\frac{1}{B(t)}\frac{dB}{dt}$$

(a) Explain why

$$\frac{1}{t}\int_0^t \frac{1}{B(s)}\frac{dB(s)}{ds}\,ds$$

is a way to express the average specific growth rate over the interval $[0, t]$.

(b) Use the chain rule to show that

$$\frac{1}{B(t)}\frac{dB}{dt} = \frac{d}{dt}(\ln B(t))$$

(c) Use the results in (a) and (b) to show that the average specific growth rate of $B(s)$ over the interval $[0, t]$ is given by

$$\frac{1}{t}\int_0^t \frac{d}{ds}(\ln(B(s)))\,ds = \frac{1}{t}\ln\frac{B(t)}{B(0)}$$

provided that $B(s) > 0$ for $s \in [0, t]$.

(d) Explain the measurements that you would need to take if you wanted to determine the average specific growth rate of biomass in a given plot over the interval $[0, t]$.

Problems 3–6 discuss stream speed profiles and provide a justification for the two measurement methods described next.

(Adapted from Herschy, 1995) The speed of water in a channel varies considerably with depth. Due to friction, the speed reaches zero at the bottom and along the sides of the channel. The speed is greatest near the surface of the stream. To find the average speed for the vertical speed profile, two methods are frequently employed in practice:

1. The 0.6 depth method: The speed is measured at 0.6 of the depth from the surface, and this value is taken as the average speed.

2. The 0.2 and 0.8 depth method: The speed is measured at 0.2 and 0.8 of the depth from the surface, and the average of the two readings is taken as the average speed.

The theoretical speed distribution of water flowing in an open channel is given approximately by

$$v(d) = \left(\frac{D-d}{a}\right)^{1/c} \tag{6.19}$$

where $v(d)$ is the speed at depth d below the water surface, c is a constant varying from 5 for coarse beds to 7 for smooth beds, D is the total depth of the channel, and a is a constant that is equal to the distance above the bottom of the channel at which the speed has unit value.

3. (a) Sketch the graph of $v(d)$ as a function of d for $D = 3$ m and $a = 1$ m for (i) $c = 5$ and (ii) $c = 7$.

(b) Show that the speed is equal to 0 at the bottom ($d = D$) and is maximal at the surface ($d = 0$).

4. (a) Show by integration that the average speed $\bar{v}$ in the vertical profile is given by

$$\bar{v} = \frac{c}{c+1}\left(\frac{D}{a}\right)^{1/c} \tag{6.20}$$

(b) What fraction of the maximum speed is the average speed $\bar{v}$?

(c) If you knew that the maximum speed occurred at the surface of the river [as predicted in the approximate formula for $v(d)$], how could you find $\bar{v}$? (In practice, the maximum speed may occur quite a bit below the surface due to friction between the water on the surface and the atmosphere. Therefore, the speed at the surface would not be an accurate measure of the maximum speed.)

5. Explain why the depth d_1, at which $v = \bar{v}$, is given by the equation

$$\bar{v} = \left(\frac{D-d_1}{a}\right)^{1/c} \tag{6.21}$$

We can find d_1 by equating (6.20) and (6.21). Show that

$$\frac{d_1}{D} = 1 - \left(\frac{c}{c+1}\right)^c$$

and that d_1/D is approximately 0.6 for values of c between 5 and 7, thus resulting in the rule

$$\bar{v} \approx v_{0.6}$$

where $v_{0.6}$ is the speed at depth $0.6D$. (*Hint*: Graph $1-(c/(c+1))^c$ as a function of c for $c \in [5, 7]$, and investigate the range of this function.)

6. We denote by $v_{0.2}$ the speed at depth $0.2D$. We will now find the depth d_2 such that

$$\bar{v} = \tfrac{1}{2}(v_{0.2} + v_{d_2})$$

(a) Show that d_2 satisfies

$$\frac{1}{2}\left[\left(\frac{D-0.2D}{a}\right)^{1/c} + \left(\frac{D-d_2}{a}\right)^{1/c}\right] = \frac{c}{c+1}\left(\frac{D}{a}\right)^{1/c}$$

[*Hint*: Use (6.19) and (6.20).]

(b) Show that

$$\frac{d_2}{D} = 1 - \left[\frac{2c}{c+1} - (0.8)^{1/c}\right]^c$$

and confirm that d_2/D is approximately 0.8 for values of c between 5 and 7, thus resulting in the rule

$$\bar{v} \approx \tfrac{1}{2}(v_{0.2} + v_{0.8})$$

Integration Techniques and Computational Methods

LEARNING OBJECTIVES

The primary focus of this chapter is on integration techniques. Specifically, we will learn how to

- integrate by using the substitution rule and integration by parts;
- integrate rational functions;
- integrate when either the integrand is discontinuous or the limits of integration are infinite;
- integrate numerically; and
- approximate functions by polynomials.

In the first two sections of the chapter, we will learn two important integration techniques that are essentially differentiation rules applied backward. (Because of the connection between differentiation and integration, it should not come as a surprise that some integration techniques are closely related to differentiation rules.) The first technique, called the substitution rule, is the chain rule applied backward; the second, called integration by parts, is the product rule applied backward. The chapter's second section is devoted to these integration techniques. An additional technique called the method of partial fractions is introduced in the third section. The fourth section deals with improper integrals, which are integrals for which the integrand goes to infinity somewhere over the interval of integration or for which the interval of integration is unbounded. Finally, we devote a section to numerical integration, another to the approximation of functions by polynomials, and a last section to the use of tables to evaluate more complicated integrals.

■ 7.1 The Substitution Rule

■ 7.1.1 Indefinite Integrals

The substitution rule is the chain rule in integral form. We therefore begin by recalling the chain rule. Suppose that we wish to differentiate

$$f(x) = \sin(3x^2 + 1)$$

This is clearly a situation in which we need to use the chain rule. We set

$$f(u) = \sin u \qquad \text{and} \qquad u = 3x^2 + 1$$

and find that $f'(u) = \cos u$. To differentiate the inner function $u = 3x^2 + 1$, we use Leibniz notation to get

$$\frac{du}{dx} = 6x$$

or, if we treat du and dx like any other variables,

$$du = 6x\,dx$$

The latter form will be particularly convenient when we reverse the chain rule. But let's first use the chain rule. We obtain

$$\frac{d}{dx}\sin(3x^2 + 1) = \frac{df}{du}\frac{du}{dx} = (\cos u)(6x) = \cos(3x^2 + 1) \cdot 6x$$

Reversing these steps and integrating along the way, we get

$$\int \underbrace{\cos(3x^2 + 1)}_{\cos u} \cdot \underbrace{6x\,dx}_{du} = \int \cos u\,du = \underbrace{\sin u}_{\sin(3x^2+1)} + C = \sin(3x^2 + 1) + C$$

In the first step, we substituted u for $3x^2 + 1$ and used $du = 6x\,dx$. This substitution simplified the integrand. At the end, we substitute back $3x^2 + 1$ for u to get the final answer in terms of x.

To see the general principle behind this technique, we write $u = g(x)$ [and hence $du = g'(x)\,dx$]. Our integral is then of the form

$$\int f[g(x)]g'(x)\,dx$$

If we denote by $F(x)$ an antiderivative of $f(x)$ [i.e., $F'(x) = f(x)$], then, using the chain rule to differentiate $F[g(x)]$, we find that

$$\frac{d}{dx}F[g(x)] = F'[g(x)]g'(x) = f[g(x)]g'(x)$$

which shows that $F[g(x)]$ is an antiderivative of $f[g(x)]g'(x)$. We can therefore write

$$\int f[g(x)]g'(x)\,dx = F[g(x)] + C \tag{7.1}$$

If we set $u = g(x)$, then we can write the right-hand side of (7.1) as $F(u) + C$. But since $F(u)$ is an antiderivative of $f(u)$, it also has the representation

$$\int f(u)\,du = F(u) + C \tag{7.2}$$

which shows that the left-hand side of (7.1) is the same as the left-hand side of (7.2). Equating the left-hand sides of (7.1) and (7.2) results in the substitution rule.

Substitution Rule for Indefinite Integrals If $u = g(x)$, then

$$\int f[g(x)]g'(x)\,dx = \int f(u)\,du$$

We present a number of examples that will illustrate how to use this rule and when the rule can be successfully applied. In the first two examples, we can apply the substitution rule immediately.

EXAMPLE 1   **Using Substitution** Evaluate

$$\int (2x + 1)e^{x^2+x}\,dx$$

Solution   The expression $2x + 1$ is the derivative of $x^2 + x$, which is the inner function of e^{x^2+x}. This suggests the following substitution:

$$u = x^2 + x \qquad \text{with} \quad \frac{du}{dx} = 2x + 1 \text{ or } du = (2x + 1)\,dx$$

Hence,

$$\int \underbrace{e^{x^2+x}}_{e^u}\underbrace{(2x + 1)\,dx}_{du} = \int e^u\,du = e^u + C = e^{x^2+x} + C$$

In the last step, we substituted $x^2 + x$ back for u, since we want the final result in terms of x.   ■

EXAMPLE 2   **Using Substitution** Evaluate

$$\int \frac{1}{x \ln x} \, dx$$

Solution   We see that $1/x$ is the derivative of $\ln x$. We try

$$u = \ln x \qquad \text{with } \frac{du}{dx} = \frac{1}{x} \text{ or } du = \frac{1}{x} \, dx$$

Then

$$\int \underbrace{\frac{1}{\ln x}}_{\frac{1}{u}} \underbrace{\frac{1}{x} \, dx}_{du} = \int \frac{1}{u} \, du = \ln |u| + C = \ln |\ln x| + C \qquad \blacksquare$$

Examples 1 and 2 illustrate types of integrals that are frequently encountered, and we display those integrals as follows for ease of reference:

$$\int g'(x) e^{g(x)} \, dx = e^{g(x)} + C$$

$$\int \frac{g'(x)}{g(x)} \, dx = \ln |g(x)| + C$$

In the previous two examples, the derivative of the inner function appeared exactly in the integrand. This will not always be the case, as shown in the next example, in which the derivative of the inner function appears in the integrand only up to a multiplicative constant.

EXAMPLE 3   **Multiplicative Constant** Evaluate

$$\int 4x \sqrt{x^2 + 1} \, dx$$

Solution   If we set $u = x^2 + 1$, then

$$\frac{du}{dx} = 2x \qquad \text{or} \qquad \frac{du}{2} = x \, dx$$

The integrand contains the derivative of the inner function up to a multiplicative constant. But this is good enough, so we write

$$\int 4x \sqrt{x^2 + 1} \, dx = \int 4 \underbrace{\sqrt{x^2 + 1}}_{\sqrt{u}} \underbrace{x \, dx}_{\frac{du}{2}} = \int 4\sqrt{u} \, \frac{du}{2}$$

$$= \int 2\sqrt{u} \, du = 2 \cdot \frac{2}{3} u^{3/2} + C = \frac{4}{3} (x^2 + 1)^{3/2} + C \qquad \blacksquare$$

EXAMPLE 4   **Rewriting the Integrand** Evaluate

$$\int \tan x \, dx$$

Solution   The trick here is to rewrite $\tan x$ as $\frac{\sin x}{\cos x}$ and realize that the derivative of the function in the denominator is the function in the numerator (up to a minus sign). The integral is therefore of the type $\int [g'(x)/g(x)] \, dx$, which we discussed before. We use the substitution

$$u = \cos x \qquad \text{with } \frac{du}{dx} = -\sin x \text{ or } -du = \sin x \, dx$$

Then

$$\int \tan x \, dx = \int \frac{\sin x}{\cos x} \, dx = \int \underbrace{\frac{1}{\cos x}}_{\frac{1}{u}} \underbrace{\sin x \, dx}_{-du}$$

$$= -\int \frac{1}{u} \, du = -\ln|u| + C = -\ln|\cos x| + C$$

In Table 6-1 of Section 6.2, we listed $\int \tan x \, dx = \ln|\sec x| + C$. This is the same result that we have just obtained, since $-\ln|\cos x| = \ln|\cos x|^{-1} = \ln|\sec x|$. ■

In Problem 59, you will evaluate $\int \cot x \, dx$, where the same trick as that in Example 4 is used. That is, both $\int \tan x \, dx$ and $\int \cot x \, dx$ are special cases of $\int [g'(x)/g(x)] \, dx$. We collect both integrals as follows:

$$\int \tan x \, dx = -\ln|\cos x| + C$$

$$\int \cot x \, dx = \ln|\sin x| + C$$

It is not always obvious that substitution will be successful, as in the next example.

EXAMPLE 5 **Substitution and Square Roots** Evaluate

$$\int x\sqrt{2x - 1} \, dx$$

Solution Obviously, x is not the derivative of $2x - 1$, so this integral does not seem to fit our scheme. But watch what we do: Set

$$u = 2x - 1 \quad \text{with } \frac{du}{dx} = 2 \text{ or } dx = \frac{du}{2}$$

Since $u = 2x - 1$, we have $x = \frac{1}{2}(u + 1)$. Making all the substitutions, we find that

$$\int x\sqrt{2x - 1} \, dx = \int \frac{1}{2}(u + 1)\sqrt{u} \, \frac{du}{2} = \frac{1}{4} \int (u^{3/2} + u^{1/2}) \, du$$

$$= \frac{1}{4} \left(\frac{2}{5} u^{5/2} + \frac{2}{3} u^{3/2} \right) + C$$

$$= \frac{1}{10}(2x - 1)^{5/2} + \frac{1}{6}(2x - 1)^{3/2} + C \qquad ■$$

Functions in the integrand are not always explicitly given, as in the next example.

EXAMPLE 6 Assume that $g(x)$ is a differentiable function whose derivative $g'(x)$ is continuous. Evaluate

$$\int g'(x) \cos[g(x)] \, dx$$

Solution If we set

$$u = g(x) \quad \text{with } \frac{du}{dx} = g'(x) \text{ or } du = g'(x) \, dx$$

then

$$\int g'(x) \cos[g(x)] \, dx = \int \cos u \, du = \sin u + C = \sin[g(x)] + C \qquad ■$$

■ 7.1.2 Definite Integrals

Part II of the FTC says that when we evaluate a definite integral, we must find an antiderivative of the integrand and then evaluate the antiderivative at the limits of integration. When we use the substitution $u = g(x)$ to find an antiderivative of an integrand, the antiderivative will be given in terms of u at first. To complete the calculation, we can proceed in either of two ways: (1) We can leave the antiderivative in terms of u and change the limits of integration according to $u = g(x)$, or (2) we can substitute $g(x)$ for u in the antiderivative and then evaluate the antiderivative at the limits of integration in terms of x. We illustrate these two ways by evaluating

$$\int_0^4 2x\sqrt{x^2+1}\,dx$$

Recall that when we compute definite integrals, we need to check whether the integrand is continuous over the interval of integration . This is the case here.

First Way We change the limits of integration along with the substitution. That is, we set

$$u = x^2 + 1 \quad \text{with } \frac{du}{dx} = 2x \text{ or } du = 2x\,dx$$

as before, and note that

$$\text{if } x = 0, \quad \text{then } u = 1$$
$$\text{if } x = 4, \quad \text{then } u = 17$$

Hence,

$$\int_0^4 2x\sqrt{x^2+1}\,dx = \int_1^{17} \sqrt{u}\,du = \frac{2}{3}u^{3/2}\Big]_1^{17} = \frac{2}{3}[(17)^{3/2}-1]$$

After substitution, the integrand is $\sqrt{u}$, and the limits of integration are $u = 1$ and $u = 17$. The region corresponding to the definite integral after substitution is shown in Figure 7.1.

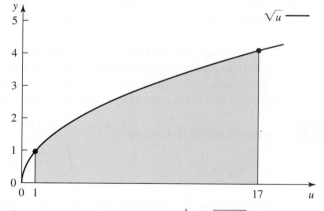

Figure 7.1 The definite integral $\int_0^4 2x\sqrt{x^2+1}\,dx$ becomes $\int_1^{17}\sqrt{u}\,du$ after the substitution $u = x^2 + 1$. The region corresponding to $\int_1^{17}\sqrt{u}\,du$ has an area of $\frac{2}{3}[(17)^{3/2}-1]$.

This way is the more common one, and we summarize the procedure as follows:

Substitution Rule for Definite Integrals If $u = g(x)$, then

$$\int_a^b f[g(x)]g'(x)\,dx = \int_{g(a)}^{g(b)} f(u)\,du$$

Second Way   We can also find the antiderivative of $f(x) = 2x\sqrt{x^2+1}$ first and then use part II of the FTC. To find an antiderivative of $f(x)$, we choose the substitution

$$u = x^2 + 1 \quad \text{with} \quad \frac{du}{dx} = 2x \text{ or } du = 2x\,dx$$

Then

$$\int 2x\sqrt{x^2+1}\,dx = \int \sqrt{u}\,du = \frac{2}{3}u^{3/2} + C = \frac{2}{3}(x^2+1)^{3/2} + C$$

and $F(x) = \frac{2}{3}(x^2+1)^{3/2}$ is an antiderivative of $2x\sqrt{x^2+1}$. Using part II of the FTC, we can now compute the definite integral:

$$\int_0^4 2x\sqrt{x^2+1}\,dx = F(4) - F(0)$$

$$= \frac{2}{3}(17)^{3/2} - \frac{2}{3}(1)^{3/2} = \frac{2}{3}[(17)^{3/2} - 1]$$

The region corresponding to the definite integral before substitution is shown in Figure 7.2.

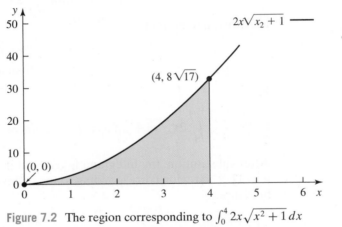

Figure 7.2   The region corresponding to $\int_0^4 2x\sqrt{x^2+1}\,dx$ before substitution has an area of $\frac{2}{3}[(17)^{3/2} - 1]$.

   Although we will not use the second way in this section, since the first way turns out to be easier to use when we apply the substitution rule, the second way of first finding an antiderivative will be more convenient in the next section, when we discuss another integration technique (integration by parts).

EXAMPLE 7   ***Definite Integral***   Compute

$$\int_1^2 \frac{3x^2+1}{x^3+x}\,dx$$

Solution   The region corresponding to the definite integral is shown in Figure 7.3. The integrand is continuous on $[1, 2]$ and is of the form $\frac{g'(x)}{g(x)}$. We set

$$u = x^3 + x \quad \text{with} \quad \frac{du}{dx} = 3x^2 + 1 \quad \text{or} \quad du = (3x^2+1)\,dx$$

and change the limits of integration:

$$\text{if } x = 1, \quad \text{then } u = 2$$
$$\text{if } x = 2, \quad \text{then } u = 10$$

Therefore,

$$\int_1^2 \frac{3x^2+1}{x^3+x}\,dx = \int_2^{10} \frac{1}{u}\,du = \ln|u|]_2^{10} = \ln 10 - \ln 2 = \ln\frac{10}{2} = \ln 5 \qquad ■$$

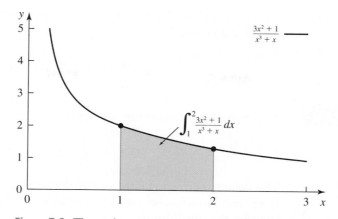

Figure 7.3 The region corresponding to the definite integral in Example 7.

EXAMPLE 8

Substitution Function Is Decreasing Compute

$$\int_{1/2}^{1} \frac{1}{x^2} e^{1/x} \, dx$$

Solution

The region corresponding to the definite integral is shown in Figure 7.4. The integrand is continuous on $[1/2, 1]$. Since $-1/x^2$ is the derivative of $1/x$, we set

$$u = \frac{1}{x} \quad \text{with} \quad \frac{du}{dx} = -\frac{1}{x^2} \quad \text{or} \quad -du = \frac{1}{x^2} \, dx$$

and change the limits of integration:

$$\text{if } x = \frac{1}{2}, \quad \text{then } u = 2$$

$$\text{if } x = 1, \quad \text{then } u = 1$$

Therefore,

$$\int_{1/2}^{1} \frac{1}{x^2} e^{1/x} \, dx = -\int_{2}^{1} e^u \, du = \int_{1}^{2} e^u \, du = e^u \big]_1^2 = e^2 - e$$

Note that because $\frac{1}{x}$ is a decreasing function for $x > 0$, the lower limit is greater than the upper limit of integration after the substitution. When we reversed the order of integration in the second step, we removed the negative sign.

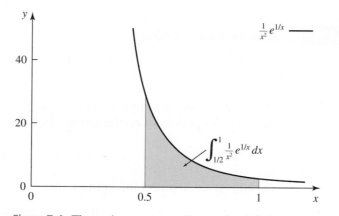

Figure 7.4 The region corresponding to the definite integral in Example 8.

However, you need not reverse the order of integration. Instead, you can compute directly:

$$-\int_2^1 e^u \, du = -e^u]_2^1 = -(e^1 - e^2) = e^2 - e \qquad ■$$

EXAMPLE 9 **Trigonometric Substitution** Compute

$$\int_0^{\pi/6} \cos x \, e^{\sin x} \, dx$$

Solution The region corresponding to the definite integral is shown in Figure 7.5. The integrand is continuous on the interval $[0, \pi/6]$ and is of the form $g'(x)e^{g(x)}$, which suggests the substitution

$$u = \sin x \quad \text{with} \quad \frac{du}{dx} = \cos x \quad \text{or} \quad du = \cos x \, dx$$

Now we change the limits of integration:

$$\text{if } x = 0, \quad \text{then } u = \sin 0 = 0$$

$$\text{if } x = \frac{\pi}{6}, \quad \text{then } u = \sin \frac{\pi}{6} = \frac{1}{2}$$

Therefore,

$$\int_0^{\pi/6} \cos x \, e^{\sin x} \, dx = \int_0^{1/2} e^u \, du = e^u]_0^{1/2} = e^{1/2} - 1 \qquad ■$$

Figure 7.5 The region corresponding to the definite integral in Example 9.

EXAMPLE 10 **Rational Function** Compute

$$\int_4^9 \frac{2}{x-3} \, dx$$

Solution The region corresponding to the definite integral is shown in Figure 7.6. The integrand is continuous on the interval $[4, 9]$. We set

$$u = x - 3 \quad \text{with} \quad \frac{du}{dx} = 1 \text{ or } du = dx$$

and change the limits of integration:

$$\text{if } x = 4, \quad \text{then } u = 1$$

$$\text{if } x = 9, \quad \text{then } u = 6$$

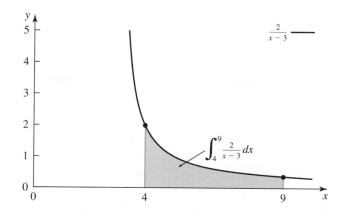

Figure 7.6 The region corresponding to the definite integral in Example 10.

Therefore,

$$\int_4^9 \frac{2}{x-3}\,dx = \int_1^6 \frac{2}{u}\,du = 2\ln|u|\Big]_1^6 = 2(\ln 6 - \ln 1) = 2\ln 6 \qquad ■$$

We can easily spend a great deal of time on integration techniques. The problems can get very involved, and to solve them all, we would need a big bag full of tricks. There are excellent software programs (such as *Mathematica* and MATLAB®) that can integrate symbolically. These programs do not render integration techniques useless; in fact, they use them. Understanding the basic techniques conceptually and being able to apply them in simple situations makes such software packages less of a "black box." Nevertheless, their availability has made it less important to acquire a large number of tricks.

So far, we have learned only one technique: substitution. Unless you can immediately recognize an antiderivative, substitution is the only method you can try at this point.

As we proceed, you will learn other techniques. An additional complication will then be to recognize which technique to use. If you don't see right away what to do, just try something. Don't always expect the first attempt to succeed. With practice, you will see much more quickly whether or not your approach will succeed. If your attempt does not seem to work, try to determine the reason. That way, failed attempts can be quite useful for gaining experience in integration.

Section 7.1 Problems

■ **7.1.1**

In Problems 1–16, evaluate the indefinite integral by making the given substitution.

1. $\int 2x\sqrt{x^2+3}\,dx$, with $u = x^2+3$

2. $\int 3x^2\sqrt{x^3+1}\,dx$, with $u = x^3+1$

3. $\int 3x(1-x^2)^{1/4}\,dx$, with $u = 1-x^2$

4. $\int 4x^3(4+x^4)^{1/3}\,dx$, with $u = 4+x^4$

5. $\int 5\cos(3x)\,dx$, with $u = 3x$

6. $\int 5\sin(1-2x)\,dx$, with $u = 1-2x$

7. $\int 7x^2\sin(4x^3)\,dx$, with $u = 4x^3$

8. $\int x\cos(x^2-1)\,dx$, with $u = x^2-1$

9. $\int e^{2x+3}\,dx$, with $u = 2x+3$

10. $\int 3e^{1-x}\,dx$, with $u = 1-x$

11. $\int xe^{-x^2/2}\,dx$, with $u = -x^2/2$

12. $\int xe^{1-3x^2}\,dx$, with $u = 1-3x^2$

13. $\int \dfrac{x+2}{x^2+4x}\,dx$, with $u = x^2 + 4x$

14. $\int \dfrac{2x}{3-x^2}\,dx$, with $u = 3 - x^2$

15. $\int \dfrac{3x}{x+4}\,dx$, with $u = x + 4$

16. $\int \dfrac{x}{5-x}\,dx$, with $u = 5 - x$

In Problems 17–36, use substitution to evaluate the indefinite integrals.

17. $\int \sqrt{x+3}\,dx$

18. $\int (4-x)^{1/7}\,dx$

19. $\int (4x-3)\sqrt{2x^2 - 3x + 2}\,dx$

20. $\int (x^2 - 2x)(x^3 - 3x^2 + 3)^{2/3}\,dx$

21. $\int \dfrac{x-1}{1+4x-2x^2}\,dx$

22. $\int \dfrac{x^2-1}{x^3 - 3x + 1}\,dx$

23. $\int \dfrac{2x}{1+2x^2}\,dx$

24. $\int \dfrac{x^3-1}{x^4 - 4x}\,dx$

25. $\int 3xe^{x^2}\,dx$

26. $\int \cos x\, e^{\sin x}\,dx$

27. $\int \dfrac{1}{x}\csc^2(\ln x)\,dx$

28. $\int \sec^2 x\, e^{\tan x}\,dx$

29. $\int \sin\left(\dfrac{3\pi}{2}x + \dfrac{\pi}{4}\right)\,dx$

30. $\int \cos(2x - 1)\,dx$

31. $\int \tan x \sec^2 x\,dx$

32. $\int \sin^3 x \cos x\,dx$

33. $\int \dfrac{(\ln x)^2}{x}\,dx$

34. $\int \dfrac{dx}{(x-3)\ln(x-3)}$

35. $\int x^3\sqrt{5+x^2}\,dx$

36. $\int \sqrt{1+\ln x}\,\dfrac{\ln x}{x}\,dx$

In Problems 37–42, a, b, and c are constants and g(x) is a continuous function whose derivative g'(x) is also continuous. Use substitution to evaluate the indefinite integrals.

37. $\int \dfrac{2ax+b}{ax^2 + bx + c}\,dx$

38. $\int \dfrac{1}{ax+b}\,dx$

39. $\int g'(x)[g(x)]^n\,dx$

40. $\int g'(x)\sin[g(x)]\,dx$

41. $\int g'(x)e^{-g(x)}\,dx$

42. $\int \dfrac{g'(x)}{[g(x)]^2 + 1}\,dx$

■ **7.1.2**

In Problems 43–58, use substitution to evaluate the definite integrals.

43. $\int_0^3 x\sqrt{x^2+1}\,dx$

44. $\int_1^2 x^5\sqrt{x^3+2}\,dx$

45. $\int_2^3 \dfrac{2x+3}{(x^2+3x)^3}\,dx$

46. $\int_0^2 \dfrac{2x}{(4x^2+3)^{1/3}}\,dx$

47. $\int_2^5 (x-2)e^{-(x-2)^2/2}\,dx$

48. $\int_{\ln 4}^{\ln 7} \dfrac{e^x}{(e^x-3)^2}\,dx$

49. $\int_0^{\pi/3} \sin x \cos x\,dx$

50. $\int_{-\pi/6}^{\pi/6} \sin^2 x \cos x\,dx$

51. $\int_0^{\pi/4} \tan x \sec^2 x\,dx$

52. $\int_0^{\pi/3} \dfrac{\sin x}{\cos^2 x}\,dx$

53. $\int_5^9 \dfrac{x}{x-3}\,dx$

54. $\int_0^2 \dfrac{x}{x+2}\,dx$

55. $\int_e^{e^2} \dfrac{dx}{x(\ln x)^2}$

56. $\int_1^2 \dfrac{x\,dx}{(x^2+1)\ln(x^2+1)}$

57. $\int_1^9 \dfrac{1}{\sqrt{x}}e^{-\sqrt{x}}\,dx$

58. $\int_0^2 x\sqrt{4-x^2}\,dx$

59. Use the fact that

$$\cot x = \frac{\cos x}{\sin x}$$

to evaluate

$$\int \cot x\,dx$$

■ 7.2 Integration by Parts and Practicing Integration

■ 7.2.1 Integration by Parts

As mentioned at the beginning of this chapter, integration by parts is the product rule in integral form. Let $u = u(x)$ and $v = v(x)$ be differentiable functions. Then, differentiating with respect to x yields

$$(uv)' = u'v + uv'$$

or, after rearranging,

$$uv' = (uv)' - u'v$$

Integrating both sides with respect to x, we find that

$$\int uv'\,dx = \int (uv)'\,dx - \int u'v\,dx$$

Since uv is an antiderivative of $(uv)'$, it follows that

$$\int (uv)'\,dx = uv + C$$

Therefore,

$$\int uv'\,dx = uv - \int u'v\,dx$$

(Note that the constant C can be absorbed into the indefinite integral on the right-hand side.) Because $u' = du/dx$ and $v' = dv/dx$, we can write the preceding equation in the short form

$$\int u\,dv = uv - \int v\,du$$

We summarize this result as follows:

Rule for Integration by Parts If $u(x)$ and $v(x)$ are differentiable functions, then

$$\int u(x)v'(x)\,dx = u(x)v(x) - \int u'(x)v(x)\,dx$$

or, in short form,

$$\int u\,dv = uv - \int v\,du$$

You are probably wondering how this technique will help, given that we traded one integral for another one. Here is a first example.

EXAMPLE 1 **Integration by Parts** Evaluate

$$\int x\sin x\,dx$$

Solution   The integrand $x\sin x$ is a product of two functions, one of which will be designated as u, the other as v'. Since integration by parts will result in another integral of the form $\int u'v\,dx$, we must choose u and v' so that $u'v$ is of a simpler form. This suggests the following choices:

$$u = x \qquad \text{and} \qquad v' = \sin x$$

Because $v = -\cos x$ and $u' = 1$, the integral $\int u'v\,dx$ is of the form $-\int \cos x\,dx$, which is indeed simpler. We obtain

$$\int x\sin x\,dx = (-\cos x)(x) - \int(-\cos x)(1)\,dx$$

$$= -x\cos x + \int \cos x\,dx$$

$$= -x\cos x + \sin x + C$$

If we had chosen $v' = x$ and $u = \sin x$, then we would have had $v = \frac{1}{2}x^2$ and $u' = \cos x$. The integral $\int u'v\,dx$ would have been of the form $\int \frac{1}{2}x^2\cos x\,dx$, which is even more complicated than $\int x\sin x\,dx$.

If we use the short form $\int u\,dv = uv - \int v\,du$, we would write

$$u = x \qquad \text{and} \qquad dv = \sin x\,dx$$

Then

$$du = dx \qquad \text{and} \qquad v = -\cos x$$

and

$$\int \underbrace{x}_{u}\,\underbrace{\sin x\,dx}_{dv} = \underbrace{x}_{u}\,\underbrace{(-\cos x)}_{v} - \int \underbrace{(-\cos x)}_{v}\,\underbrace{dx}_{du}$$

$$= -x\cos x + \sin x + C \qquad \blacksquare$$

EXAMPLE 2   **Integration by Parts** Evaluate

$$\int x \ln x \, dx$$

Solution   Since we do not know an antiderivative of $\ln x$, we try

$$u = \ln x \quad \text{and} \quad v' = x$$

Then

$$u' = \frac{1}{x} \quad \text{and} \quad v = \frac{1}{2}x^2$$

and

$$\int x \ln x \, dx = \frac{1}{2}x^2 \ln x - \int \frac{1}{2}x^2 \cdot \frac{1}{x} \, dx$$
$$= \frac{1}{2}x^2 \ln x - \int \frac{1}{2}x \, dx = \frac{1}{2}x^2 \ln x - \frac{1}{4}x^2 + C \quad ■$$

Before we present a few more useful "tricks," we show how to evaluate definite integrals with this method.

EXAMPLE 3   **Definite Integral** Compute

$$\int_0^1 xe^{-x} \, dx$$

Solution   The region representing the definite integral is shown in Figure 7.7. The integrand is continuous on $[0, 1]$. We set

$$u = x \quad \text{and} \quad dv = e^{-x} \, dx$$

Then

$$du = dx \quad \text{and} \quad v = -e^{-x}$$

Therefore,

$$\int_0^1 xe^{-x} \, dx = -xe^{-x}\Big]_0^1 - \int_0^1 (-e^{-x}) \, dx$$

$$= -1e^{-1} - (-0e^{-0}) + \int_0^1 e^{-x} \, dx$$

$$= -e^{-1} + \left[-e^{-x}\right]_0^1 = -e^{-1} + (-e^{-1} - (-e^{-0}))$$

$$= -e^{-1} - e^{-1} + 1 = 1 - 2e^{-1} \quad ■$$

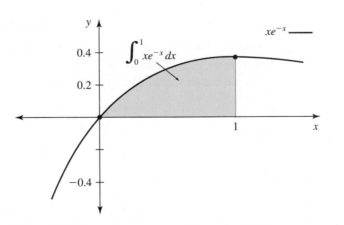

Figure 7.7 The region corresponding to the definite integral in Example 3.

In the next two examples, we demonstrate a "trick" that is sometimes useful in integration by parts. The technique is called "multiplying by 1."

EXAMPLE 4 **Multiplying by 1** Evaluate

$$\int \ln x \, dx$$

Solution The integrand $\ln x$ is not a product of two functions, but we can write it as $(1)(\ln x)$ and set

$$u = \ln x \quad \text{and} \quad v' = 1$$

Then

$$u' = \frac{1}{x} \quad \text{and} \quad v = x$$

We find that

$$\int \ln x \, dx = \int (1)(\ln x) \, dx = x \ln x - \int x \frac{1}{x} \, dx$$

$$= x \ln x - \int 1 \, dx = x \ln x - x + C$$

Our choices for u and v' might surprise you, as we said that our goal was to make the integral look simpler, which often means that we try to reduce the power of functions of the form x^n. In this case, however, integrating 1 and differentiating $\ln x$ yielded a simpler integral. In fact, if we had chosen $u' = \ln x$ and $v = 1$, we would not have been able to carry out the integration by parts, since we would have needed the antiderivative of $\ln x$ to compute uv and $\int uv' \, dx$.

If you prefer the short-form notation, you don't need to multiply by 1, because

$$u = \ln x \quad \text{and} \quad dv = dx$$

together with

$$du = \frac{1}{x} dx \quad \text{and} \quad v = x$$

immediately produces

$$\int \ln x \, dx = x \ln x - \int x \frac{1}{x} \, dx = x \ln x - \int dx$$

$$= x \ln x - x + C$$ ■

EXAMPLE 5 **Multiplying by 1** Evaluate

$$\int \tan^{-1} x \, dx$$

Solution We write $\tan^{-1} x = (1)(\tan^{-1} x)$ and

$$u = \tan^{-1} x \quad \text{and} \quad v' = 1$$

Then

$$u' = \frac{1}{x^2 + 1} \quad \text{and} \quad v = x$$

We find that

$$\int \tan^{-1} x \, dx = x \tan^{-1} x - \int \frac{x}{x^2 + 1} \, dx$$

We need to use substitution to evaluate the integral on the right-hand side. With

$$w = x^2 + 1 \quad \text{and} \quad \frac{dw}{dx} = 2x$$

we obtain

$$\int \frac{x}{x^2+1}\,dx = \frac{1}{2}\int \frac{dw}{w} = \frac{1}{2}\ln|w| + C_1 = \frac{1}{2}\ln(x^2+1) + C_1$$

where C_1 is the constant of integration. Hence,

$$\int \tan^{-1} x\,dx = x\tan^{-1} x - \frac{1}{2}\ln(x^2+1) - C_1$$

We write the final answer as

$$\int \tan^{-1} x\,dx = x\tan^{-1} x - \frac{1}{2}\ln(x^2+1) + C$$

where C is the constant of integration with $C = -C_1$. Replacing $-C_1$ by C is done for purely aesthetical reasons. ∎

EXAMPLE 6 **Using Integration by Parts Repeatedly** Compute

$$\int_0^1 x^2 e^x\,dx$$

Solution When you integrate a definite integral by parts, it is often easier to integrate the indefinite integral first and then to use part II of the FTC to evaluate the definite integral. To evaluate $\int x^2 e^x\,dx$, we set

$$u = x^2 \qquad \text{and} \qquad v' = e^x$$

Then

$$u' = 2x \qquad \text{and} \qquad v = e^x$$

Therefore,

$$\int x^2 e^x\,dx = x^2 e^x - \int 2x e^x\,dx \tag{7.3}$$

To evaluate the integral $\int x e^x\,dx$, we must use integration by parts a second time. We set

$$u = x \qquad \text{and} \qquad v' = e^x$$

Then

$$u' = 1 \qquad \text{and} \qquad v = e^x$$

Therefore,

$$\int x e^x\,dx = x e^x - \int e^x\,dx = x e^x - e^x + C \tag{7.4}$$

Combining (7.3) and (7.4), we find that

$$\int x^2 e^x\,dx = x^2 e^x - 2[x e^x - e^x + C]$$

$$= x^2 e^x - 2x e^x + 2e^x - 2C$$

After evaluating the indefinite integral, we can compute the definite integral. Note that the integrand is continuous on $[0, 1]$. We set $F(x) = x^2 e^x - 2x e^x + 2e^x$. Then

$$\int_0^1 x^2 e^x\,dx = F(1) - F(0)$$

$$= (e - 2e + 2e) - (0 - 0 + 2) = e - 2 \qquad ∎$$

EXAMPLE 7   **Using Integration by Parts Repeatedly** Evaluate

$$\int e^x \cos x \, dx$$

Solution   You can check that it does not matter which of the functions you call u and which v'. We set

$$u = \cos x \qquad \text{and} \qquad v' = e^x$$

Then

$$u' = -\sin x \qquad \text{and} \qquad v = e^x$$

Therefore,

$$\int e^x \cos x \, dx = e^x \cos x + \int e^x \sin x \, dx \tag{7.5}$$

We now integrate by parts a second time. This time, the choice matters. We need to set

$$u = \sin x \qquad \text{and} \qquad v' = e^x$$

Then

$$u' = \cos x \qquad \text{and} \qquad v = e^x$$

Therefore,

$$\int e^x \sin x \, dx = e^x \sin x - \int e^x \cos x \, dx \tag{7.6}$$

Combining (7.5) and (7.6) yields

$$\int e^x \cos x \, dx = e^x \cos x + e^x \sin x - \int e^x \cos x \, dx$$

We see that the integral $\int e^x \cos x \, dx$ appears on both sides. Rearranging the equation, we obtain

$$2 \int e^x \cos x \, dx = e^x \cos x + e^x \sin x + C_1$$

or

$$\int e^x \cos x \, dx = \frac{1}{2} e^x (\cos x + \sin x) + C$$

with $C = C_1/2$. [Note that we introduced the constant C_1 (and C) in the final answer.]

We said that the choices for u and v' in the second integration by parts matter. If we had designated $u = e^x$ and $v' = \sin x$, then $u' = e^x$ and $v = -\cos x$, yielding

$$\int e^x \sin x \, dx = -e^x \cos x + \int e^x \cos x \, dx$$

Combining this equation with (7.5), we then obtain

$$\int e^x \cos x \, dx = e^x \cos x - e^x \cos x + \int e^x \cos x \, dx$$

which is a correct, but useless, statement.   ∎

We conclude this subsection with a piece of practical advice: In integrals of the form $\int P(x) \sin(ax) \, dx$, $\int P(x) \cos(ax) \, dx$, and $\int P(x) e^{ax} \, dx$, where $P(x)$ is a polynomial and a is a constant, the polynomial $P(x)$ should be considered as u and the expressions $\sin(ax)$, $\cos(ax)$, and e^{ax} as v'. If an integral contains the function $\ln x$, $\tan^{-1} x$, or $\sin^{-1} x$, the function is usually treated as u. After practicing the problems at the end of the section, you can confirm this advice.

■ 7.2.2 Practicing Integration

Thus far in this chapter, we have learned the two main integration techniques: substitution and integration by parts. One of the major difficulties in integration is deciding which rule to use. This subsection is devoted to practicing integration, because such practice aids in making the right decision.

EXAMPLE 8   Find

$$\int \tan x \sec^2 x e^{\tan x}\, dx$$

Solution   Since $\frac{d}{dx}\tan x = \sec^2 x$, we try the substitution $w = \tan x$. Then $dw = \sec^2 x\, dx$ and

$$\int \tan x \sec^2 x e^{\tan x}\, dx = \int w e^w\, dw$$

To continue, we need to use integration by parts, with

$$u = w \qquad \text{and} \qquad v' = e^w$$

Then

$$u' = 1 \qquad \text{and} \qquad v = e^w$$

and

$$\int w e^w\, dw = w e^w - \int e^w\, dw = w e^w - e^w + C$$

With $w = \tan x$, we therefore find that

$$\int \tan x \sec^2 x e^{\tan x}\, dx = e^{\tan x}(\tan x - 1) + C \qquad\blacksquare$$

Frequently, we must perform algebraic manipulations of the integrand before we can integrate.

EXAMPLE 9   Find

$$\int_0^{\sqrt{3}} \frac{1}{9 + x^2}\, dx$$

Solution   The region corresponding to the definite integral is shown in Figure 7.8. The integrand is continuous on $[0, \sqrt{3}]$. The integrand should remind you of the function $\frac{1}{1+u^2}$, whose antiderivative is $\tan^{-1} u$. However, we have a 9 in the denominator. To get a 1 there, we factor 9 in the denominator to obtain

$$\frac{1}{9 + x^2} = \frac{1}{9\left(1 + \frac{x^2}{9}\right)} = \frac{1}{9\left(1 + \left(\frac{x}{3}\right)^2\right)}$$

The last expression now suggests that we should try the substitution

$$u = \frac{x}{3} \qquad \text{with} \qquad dx = 3\, du$$

Since we wish to evaluate a definite integral, we must change the limits of integration as well. We find that $x = 0$ corresponds to $u = 0$ and $x = \sqrt{3}$ corresponds to $u = \frac{1}{3}\sqrt{3}$. We end up with

$$\int_0^{\sqrt{3}} \frac{1}{9+x^2}\, dx = \frac{1}{9}\int_0^{\sqrt{3}} \frac{1}{1+\left(\frac{x}{3}\right)^2}\, dx = \frac{1}{9}\int_0^{\frac{1}{3}\sqrt{3}} \frac{3}{1+u^2}\, du$$

$$= \frac{1}{3}\int_0^{\frac{1}{3}\sqrt{3}} \frac{1}{1+u^2}\, du = \frac{1}{3}\tan^{-1} u \Big]_0^{\frac{1}{3}\sqrt{3}}$$

$$= \frac{1}{3}\left[\tan^{-1}\left(\frac{1}{3}\sqrt{3}\right) - \tan^{-1} 0\right] = \frac{1}{3}\left(\frac{\pi}{6} - 0\right) = \frac{\pi}{18} \qquad\blacksquare$$

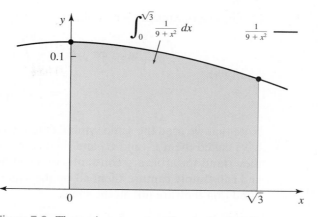

Figure 7.8 The region corresponding to the definite integral in Example 9.

The next example shows that simplifying the integrand first can help greatly.

EXAMPLE 10 Find

$$\int x^{1/2} \ln(x^{1/2} e^x) \, dx$$

Solution Before we try any of our techniques, let's simplify the logarithm first. We find that

$$\ln(x^{1/2} e^x) = \ln x^{1/2} + \ln e^x = \frac{1}{2} \ln x + x$$

The integral now becomes

$$\int x^{1/2} \ln(x^{1/2} e^x) \, dx = \int x^{1/2} \left(\frac{1}{2} \ln x + x \right) dx$$

$$= \frac{1}{2} \int x^{1/2} \ln x \, dx + \int x^{3/2} \, dx$$

We can integrate the first integral by parts, with

$$u = \ln x \qquad \text{and} \qquad v' = x^{1/2}$$

Then

$$u' = \frac{1}{x} \qquad \text{and} \qquad v = \frac{2}{3} x^{3/2}$$

For the first integral, we obtain

$$\int x^{1/2} \ln x \, dx = \frac{2}{3} x^{3/2} \ln x - \int \frac{1}{x} \cdot \frac{2}{3} x^{3/2} \, dx$$

$$= \frac{2}{3} x^{3/2} \ln x - \frac{2}{3} \int x^{1/2} \, dx$$

$$= \frac{2}{3} x^{3/2} \ln x - \frac{2}{3} \cdot \frac{2}{3} x^{3/2} + C$$

$$= \frac{2}{3} x^{3/2} \left(\ln x - \frac{2}{3} \right) + C$$

The other integral is straightforward:

$$\int x^{3/2} \, dx = \frac{2}{5} x^{5/2} + C$$

Combining our results, we find that

$$\int x^{1/2} \ln(x^{1/2} e^x) \, dx = \frac{1}{2} \int x^{1/2} \ln x \, dx + \int x^{3/2} dx$$

$$= \frac{1}{3} x^{3/2} \left(\ln x - \frac{2}{3} \right) + \frac{2}{5} x^{5/2} + C$$

Note that we used the same symbol C to denote the integration constants. We could have called them C_1 and C_2 and then combined them into $C = C_1 + C_2$, but since they stand for arbitrary constants, we need not keep track of how they are related and can simply capture them all by the same symbol. However, we should keep in mind that they are not all the same. ■

Section 7.2 Problems

■ **7.2.1**

In Problems 1–30, use integration by parts to evaluate the integrals.

1. $\int x \cos x \, dx$

2. $\int 3x \cos x \, dx$

3. $\int 2x \cos(3x - 1) \, dx$

4. $\int 3x \cos(4 - x) \, dx$

5. $\int 2x \sin(x - 1) \, dx$

6. $\int x \sin(1 - 2x) \, dx$

7. $\int x e^x \, dx$

8. $\int 3x e^{-x/2} \, dx$

9. $\int x^2 e^x \, dx$

10. $\int 2x^2 e^{-x} \, dx$

11. $\int x \ln x \, dx$

12. $\int x^2 \ln x \, dx$

13. $\int x \ln(3x) \, dx$

14. $\int x^2 \ln x^2 \, dx$

15. $\int x \sec^2 x \, dx$

16. $\int x \csc^2 x \, dx$

17. $\int_0^{\pi/3} x \sin x \, dx$

18. $\int_0^{\pi/4} 2x \cos x \, dx$

19. $\int_1^2 \ln x \, dx$

20. $\int_1^e \ln x^2 \, dx$

21. $\int_1^4 \ln \sqrt{x} \, dx$

22. $\int_1^4 \sqrt{x} \ln \sqrt{x} \, dx$

23. $\int_0^1 x e^{-x} \, dx$

24. $\int_0^3 x^2 e^{-x} \, dx$

25. $\int_0^{\pi/3} e^x \sin x \, dx$

26. $\int_0^{\pi/6} e^x \cos x \, dx$

27. $\int e^{-3x} \cos \left(\frac{\pi}{2} x \right) dx$

28. $\int e^{-2x} \sin \left(\frac{x}{2} \right) dx$

29. $\int \sin(\ln x) \, dx$

30. $\int \cos(\ln x) \, dx$

31. Evaluating the integral

$$\int \cos^2 x \, dx$$

requires two steps.

First, write

$$\cos^2 x = (\cos x)(\cos x)$$

and integrate by parts to show that

$$\int \cos^2 x \, dx = \sin x \cos x + \int \sin^2 x \, dx$$

Then, use $\sin^2 x + \cos^2 x = 1$ to replace $\sin^2 x$ in the integral on the right-hand side, and complete the integration of $\int \cos^2 x \, dx$.

32. Evaluating the integral

$$\int \sin^2 x \, dx$$

requires two steps.

First, write

$$\sin^2 x = (\sin x)(\sin x)$$

and integrate by parts to show that

$$\int \sin^2 x \, dx = -\sin x \cos x + \int \cos^2 x \, dx$$

Then, use $\sin^2 x + \cos^2 x = 1$ to replace $\cos^2 x$ in the integral on the right-hand side, and complete the integration of $\int \sin^2 x \, dx$.

33. Evaluating the integral

$$\int \arcsin x \, dx$$

requires two steps.

(a) Write

$$\arcsin x = 1 \cdot \arcsin x$$

and integrate by parts once to show that

$$\int \arcsin x \, dx = x \arcsin x - \int \frac{x}{\sqrt{1 - x^2}} \, dx$$

(b) Use substitution to compute

$$\int \frac{x}{\sqrt{1 - x^2}} \, dx \qquad (7.7)$$

and combine your result in (a) with (7.7) to complete the computation of $\int \arcsin x \, dx$.

34. Evaluating the integral

$$\int \arccos x \, dx$$

requires two steps.

(a) Write

$$\arccos x = 1 \cdot \arccos x$$

and integrate by parts once to show that

$$\int \arccos x \, dx = x \arccos x + \int \frac{x}{\sqrt{1 - x^2}} \, dx$$

(b) Use substitution to compute

$$\int \frac{x}{\sqrt{1 - x^2}} \, dx \qquad (7.8)$$

and combine your result in (a) with (7.8) to complete the computation of $\int \arccos x \, dx$.

35. (a) Use integration by parts to show that, for $x > 0$,

$$\int \frac{1}{x} \ln x \, dx = (\ln x)^2 - \int \frac{1}{x} \ln x \, dx$$

(b) Use your result in (a) to evaluate

$$\int \frac{1}{x} \ln x \, dx$$

36. (a) Use integration by parts to show that

$$\int x^n e^x \, dx = x^n e^x - n \int x^{n-1} e^x \, dx$$

Such formulas are called **reduction formulas**, since they reduce the exponent of x by 1 each time they are applied.

(b) Apply the reduction formula in (a) repeatedly to compute

$$\int x^3 e^x \, dx$$

37. (a) Use integration by parts to verify the validity of the reduction formula

$$\int x^n e^{ax} \, dx = \frac{1}{a} x^n e^{ax} - \frac{n}{a} \int x^{n-1} e^{ax} \, dx$$

where a is a constant not equal to 0.

(b) Apply the reduction formula in (a) to compute

$$\int x^2 e^{-3x} \, dx$$

38. (a) Use integration by parts to verify the validity of the reduction formula

$$\int (\ln x)^n \, dx = x (\ln x)^n - n \int (\ln x)^{n-1} \, dx$$

(b) Apply the reduction formula in (a) repeatedly to compute

$$\int (\ln x)^3 \, dx$$

In Problems 39–48, first make an appropriate substitution and then use integration by parts to evaluate the indefinite integrals.

39. $\displaystyle \int \cos \sqrt{x} \, dx$

40. $\displaystyle \int \sin \sqrt{x} \, dx$

41. $\displaystyle \int x^3 e^{-x^2/2} \, dx$

42. $\displaystyle \int x^5 e^{x^2} \, dx$

43. $\displaystyle \int \sin x \cos x e^{\sin x} \, dx$

44. $\displaystyle \int \sin x \cos^3 x e^{1-\sin^2 x} \, dx$

45. $\displaystyle \int_0^1 e^{\sqrt{x}} \, dx$

46. $\displaystyle \int_1^2 e^{\sqrt{x+1}} \, dx$

47. $\displaystyle \int_1^4 \ln(\sqrt{x} + 1) \, dx$

48. $\displaystyle \int_0^1 x^3 \ln(x^2 + 1) \, dx$

■ 7.2.2

In Problems 49–60, use either substitution or integration by parts to evaluate each integral.

49. $\displaystyle \int x e^{-2x} \, dx$

50. $\displaystyle \int x e^{-2x^2} \, dx$

51. $\displaystyle \int \frac{1}{\tan x} \, dx$

52. $\displaystyle \int \frac{1}{\csc x \sec x} \, dx$

53. $\displaystyle \int 2x \sin(x^2) \, dx$

54. $\displaystyle \int 2x^2 \sin x \, dx$

55. $\displaystyle \int \frac{1}{16 + x^2} \, dx$

56. $\displaystyle \int \frac{1}{x^2 + 5} \, dx$

57. $\displaystyle \int \frac{x}{x + 3} \, dx$

58. $\displaystyle \int \frac{1}{x^2 + 3} \, dx$

59. $\displaystyle \int \frac{x}{x^2 + 3} \, dx$

60. $\displaystyle \int \frac{x + 2}{x^2 + 2} \, dx$

61. The integral

$$\int \ln x \, dx$$

can be evaluated in two ways.

(a) Write $\ln x = 1 \cdot \ln x$ and use integration by parts to evaluate the integral.

(b) Use the substitution $u = \ln x$ and integration by parts to evaluate the integral.

62. Use an appropriate substitution followed by integration by parts to evaluate

$$\int x^3 e^{-x^2/2} \, dx$$

63. Use an appropriate substitution to evaluate

$$\int x(x - 2)^{1/4} \, dx$$

64. Simplify the integrand and then use an appropriate substitution to evaluate

$$\int \frac{\sin^2 x - \cos^2 x}{(\sin x - \cos x)^2} \, dx$$

In Problems 65–70, evaluate each definite integral.

65. $\displaystyle \int_1^4 e^{\sqrt{x}} \, dx$

66. $\displaystyle \int_1^2 \ln(x^2 e^x) \, dx$

67. $\displaystyle \int_{-1}^0 \frac{2}{1 + x^2} \, dx$

68. $\displaystyle \int_1^2 x^2 \ln x \, dx$

69. $\displaystyle \int_0^{\pi/4} e^x \sin x \, dx$

70. $\displaystyle \int_0^{\pi/6} (1 + \tan^2 x) \, dx$

■ 7.3 Rational Functions and Partial Fractions

A rational function f is the quotient of two polynomials. That is,

$$f(x) = \frac{P(x)}{Q(x)} \tag{7.9}$$

where $P(x)$ and $Q(x)$ are polynomials. To integrate rational functions, we use an algebraic technique, called the method of **partial fractions**, to write $f(x)$ as a sum of a polynomial and simpler rational functions. Such a sum is called a **partial-fraction decomposition**. These simpler rational functions, which can be integrated with the methods we have learned, are of the form

$$\frac{A}{(ax+b)^n} \quad \text{or} \quad \frac{Bx+C}{(ax^2+bx+c)^n} \tag{7.10}$$

where A, B, C, a, b, and c are constants and n is a positive integer. In this form, the quadratic polynomial $ax^2 + bx + c$ can no longer be factored into a product of two linear functions with real coefficients. Such polynomials are called **irreducible**. (In other words, $ax^2 + bx + c$ is irreducible if and only if $ax^2 + bx + c = 0$ has no real roots.)

■ 7.3.1 Proper Rational Functions

If the degree of $P(x)$ in (7.9) is greater than or equal to the degree of $Q(x)$, then the first step in the partial-fraction decomposition is to use long division to write $f(x)$ as a sum of a polynomial and a rational function, where the rational function is such that the degree of the polynomial in the numerator is less than the degree of the polynomial in the denominator. (Such rational functions are called **proper**.) We illustrate this step in the next two examples.

EXAMPLE 1　**Long Division before Integration** Find

$$\int \frac{x}{x+2}\, dx$$

Solution　The degree of the numerator is equal to the degree of the denominator; using long division or writing the integrand in the form

$$\frac{x}{x+2} = \frac{x+2-2}{x+2} = 1 - \frac{2}{x+2}$$

results in a polynomial of degree 0 and a proper rational function. We can integrate the integrand in this new form:

$$\int \frac{x}{x+2}\, dx = \int \left(1 - \frac{2}{x+2}\right) dx = x - 2\ln|x+2| + C \qquad \blacksquare$$

EXAMPLE 2　**Long Division before Integration** Find

$$\int \frac{3x^3 - 7x^2 + 17x - 3}{x^2 - 2x + 5}\, dx$$

Solution　Since the degree of the numerator is higher than the degree of the denominator, we use long division to simplify the integrand:

$$
\begin{array}{r}
3x - 1 \\
x^2 - 2x + 5 \,\overline{)\, 3x^3 - 7x^2 + 17x - 3} \\
\underline{3x^3 - 6x^2 + 15x } \\
-x^2 + 2x - 3 \\
\underline{-x^2 + 2x - 5} \\
2
\end{array}
$$

That is,

$$\frac{3x^3 - 7x^2 + 17x - 3}{x^2 - 2x + 5} = 3x - 1 + \frac{2}{x^2 - 2x + 5}$$

Now, by the quadratic formula, $x^2 - 2x + 5 = 0$ does not have real solutions. This means that it is irreducible, so we will complete the square instead and obtain

$$x^2 - 2x + 5 = (x^2 - 2x + 1) + 4 = (x - 1)^2 + 4$$

The integral we wish to evaluate is therefore

$$\int \left(3x - 1 + \frac{2}{(x-1)^2 + 4} \right) dx = \int (3x - 1) \, dx + 2 \int \frac{1}{(x-1)^2 + 4} \, dx \quad (7.11)$$

The first integral on the right-hand side is straightforward since the integrand is a polynomial; we find that

$$\int (3x - 1) \, dx = \frac{3}{2}x^2 - x + C$$

To evaluate the second integral, we use a "trick" similar to the one we used in Example 9 of the previous section—namely, we factor 4 in the denominator:

$$\int \frac{1}{(x-1)^2 + 4} \, dx = \frac{1}{4} \int \frac{1}{1 + (\frac{x-1}{2})^2} \, dx$$

Setting

$$u = \frac{x - 1}{2} \qquad \text{with} \qquad dx = 2 \, du$$

yields

$$\frac{1}{4} \int \frac{1}{1 + (\frac{x-1}{2})^2} \, dx = \frac{1}{4} \int \frac{2 \, du}{1 + u^2} = \frac{1}{2} \tan^{-1} u + C = \frac{1}{2} \tan^{-1} \left(\frac{x-1}{2} \right) + C$$

Putting the pieces together [and remembering that there was a factor 2 in front of the second integral in (7.11)], we find that

$$\int \frac{3x^3 - 7x^2 + 17x - 3}{x^2 - 2x + 5} \, dx = \frac{3}{2}x^2 - x + \tan^{-1} \left(\frac{x-1}{2} \right) + C \qquad ■$$

Assume now that the rational function is proper. Then, unless the integrand is already of one of the types in (7.10), we need to decompose it further. A result in algebra tells us that every polynomial can be written as the product of linear and irreducible quadratic factors. Factoring the denominator into a product of linear and irreducible quadratic factors is the key to partial-fraction decomposition.

■ 7.3.2 Partial-Fraction Decomposition

Linear Factors   We discuss the cases of distinct and repeated linear factors.

If the linear factor $ax + b$ is contained n times in the factorization of the denominator of a proper rational function, then the partial-fraction decomposition contains terms of the form

$$\frac{A_1}{ax + b} + \frac{A_2}{(ax + b)^2} + \cdots + \frac{A_n}{(ax + b)^n}$$

where $A_1, A_2, \ldots, A_n$ are constants.

We discuss the case of distinct linear factors first.

Case 1a: $Q(x)$ is a product of m distinct linear factors. $Q(x)$ is thus of the form

$$Q(x) = a(x - x_1)(x - x_2)\ldots(x - x_m)$$

where $x_1, x_2, \ldots, x_m$ are the m distinct roots of $Q(x)$. The rational function can then be written as

$$\frac{P(x)}{Q(x)} = \frac{1}{a}\left[\frac{A_1}{x - x_1} + \frac{A_2}{x - x_2} + \cdots + \frac{A_m}{x - x_m}\right]$$

We will see in the next example how the constants $A_1, A_2, \ldots, A_m$ are determined.

EXAMPLE 3

Distinct Linear Factors Find

$$\int \frac{1}{x(x - 1)}\, dx$$

Solution
The integrand is a proper rational function whose denominator is a product of two distinct linear functions. We claim that the integrand can be written in the form

$$\frac{1}{x(x - 1)} = \frac{A}{x} + \frac{B}{x - 1} \tag{7.12}$$

where A and B are constants that we need to determine. To find A and B, we write the right-hand side of (7.12) with a common denominator. That is,

$$\frac{A}{x} + \frac{B}{x - 1} = \frac{A(x - 1) + Bx}{x(x - 1)} = \frac{(A + B)x - A}{x(x - 1)}$$

Since this must be equal to $\frac{1}{x(x-1)}$, we conclude that

$$(A + B)x - A = 1$$

Therefore,

$$A + B = 0 \qquad \text{and} \qquad -A = 1$$

This yields $A = -1$ and $B = -A = 1$. We thus find that

$$\frac{1}{x(x - 1)} = -\frac{1}{x} + \frac{1}{x - 1}$$

The integrand can now be written as a sum of two rational functions, which can be integrated immediately:

$$\int \frac{1}{x(x - 1)}\, dx = \int \left[\frac{1}{x - 1} - \frac{1}{x}\right] dx = \int \frac{1}{x - 1}\, dx - \int \frac{1}{x}\, dx$$

$$= \ln|x - 1| - \ln|x| + C = \ln\left|\frac{x - 1}{x}\right| + C \qquad ■$$

Case 1b: $Q(x)$ is a product of repeated linear factors. We illustrate how to proceed in the next two examples.

EXAMPLE 4

Repeated Linear Factors Evaluate

$$\int \frac{x}{(x + 1)^2}\, dx$$

Solution
The integrand is a proper rational function whose denominator is a linear factor that is repeated. We therefore write the integrand in the form

$$\frac{x}{(x + 1)^2} = \frac{A}{x + 1} + \frac{B}{(x + 1)^2} \tag{7.13}$$

where A and B are constants. Writing the right-hand side of (7.13) with a common denominator yields

$$\frac{A}{x+1} + \frac{B}{(x+1)^2} = \frac{A(x+1)+B}{(x+1)^2} = \frac{Ax+(A+B)}{(x+1)^2}$$

Equating coefficients on the left-hand side of (7.13), we find that

$$A = 1 \quad \text{and} \quad A + B = 0$$

This implies that $B = -1$. Therefore,

$$\frac{x}{(x+1)^2} = \frac{1}{x+1} - \frac{1}{(x+1)^2}$$

and

$$\int \frac{x}{(x+1)^2}\, dx = \int \left(\frac{1}{x+1} - \frac{1}{(x+1)^2} \right) dx$$

$$= \ln|x+1| + \frac{1}{x+1} + C \qquad \blacksquare$$

EXAMPLE 5   **Repeated Linear Factors**   Evaluate

$$\int \frac{dx}{x^2(x+1)}$$

Solution   The integrand is a proper rational function whose denominator is a product of three linear functions: x, x (again), and $x+1$. The factor x is repeated, whereas $x+1$ only occurs once. We write the integrand in the form

$$\frac{1}{x^2(x+1)} = \frac{A}{x} + \frac{B}{x^2} + \frac{C}{x+1} \qquad (7.14)$$

where A, B, and C are constants. As in the previous example, we find A, B, and C by writing the right-hand side of (7.14) with a common denominator. This yields

$$\frac{A}{x} + \frac{B}{x^2} + \frac{C}{x+1} = \frac{Ax(x+1)+B(x+1)+Cx^2}{x^2(x+1)}$$

$$= \frac{Ax^2+Ax+Bx+B+Cx^2}{x^2(x+1)}$$

$$= \frac{(A+C)x^2+(A+B)x+B}{x^2(x+1)}$$

Comparing the last expression with the left-hand side of (7.14), we conclude that

$$A + C = 0, \quad A + B = 0, \quad \text{and} \quad B = 1$$

This implies that $A = -1$ and $C = 1$. Therefore,

$$\frac{1}{x^2(x+1)} = -\frac{1}{x} + \frac{1}{x^2} + \frac{1}{x+1}$$

and

$$\int \frac{1}{x^2(x+1)}\, dx = \int \left(-\frac{1}{x} + \frac{1}{x^2} + \frac{1}{x+1} \right) dx$$

$$= -\ln|x| - \frac{1}{x} + \ln|x+1| + C \qquad \blacksquare$$

Irreducible Quadratic Factors Irreducible quadratic factors in the denominator of a proper rational functions are dealt with in the partial-fraction decomposition as follows:

If the irreducible quadratic factor $ax^2 + bx + c$ is contained n times in the factorization of the denominator of a proper rational function, then the partial-fraction decomposition contains terms of the form

$$\frac{B_1 x + C_1}{ax^2 + bx + c} + \frac{B_2 x + C_2}{(ax^2 + bx + c)^2} + \cdots + \frac{B_n x + C_n}{(ax^2 + bx + c)^n}$$

EXAMPLE 6   **Distinct Irreducible Quadratic Factors** Evaluate

$$\int \frac{2x^3 - x^2 + 2x - 2}{(x^2 + 2)(x^2 + 1)}\, dx$$

Solution   The rational function in the integrand is proper. The denominator is already factored, each factor is an irreducible quadratic polynomial, and the two factors are distinct. We can therefore write the integrand as

$$\frac{2x^3 - x^2 + 2x - 2}{(x^2 + 2)(x^2 + 1)} = \frac{Ax + B}{x^2 + 2} + \frac{Cx + D}{x^2 + 1}$$

$$= \frac{(Ax + B)(x^2 + 1) + (Cx + D)(x^2 + 2)}{(x^2 + 2)(x^2 + 1)}$$

$$= \frac{Ax^3 + Ax + Bx^2 + B + Cx^3 + 2Cx + Dx^2 + 2D}{(x^2 + 2)(x^2 + 1)}$$

$$= \frac{(A + C)x^3 + (B + D)x^2 + (A + 2C)x + B + 2D}{(x^2 + 2)(x^2 + 1)}$$

Comparing the last expression with the integrand, we find that

$$A + C = 2, \quad B + D = -1, \quad A + 2C = 2, \quad \text{and} \quad B + 2D = -2$$

which yields $C = 0$ (write $A + 2C = 2$ as $A + C + C = 2$ and use $A + C = 2$) and $D = -1$ (write $B + 2D = -2$ as $B + D + D = -2$ and use $B + D = -1$). Then $A = 2$ and $B = 0$. Therefore,

$$\frac{2x^3 - x^2 + 2x - 2}{(x^2 + 2)(x^2 + 1)} = \frac{2x}{x^2 + 2} - \frac{1}{x^2 + 1}$$

and

$$\int \frac{2x^3 - x^2 + 2x - 2}{(x^2 + 2)(x^2 + 1)}\, dx = \int \frac{2x}{x^2 + 2}\, dx - \int \frac{1}{x^2 + 1}\, dx$$

$$= \int \frac{du}{u} - \tan^{-1} x + C = \ln |u| - \tan^{-1} x + C$$

$$= \ln |x^2 + 2| - \tan^{-1} x + C$$

where we used the substitution $u = x^2 + 2$.   ■

EXAMPLE 7   **Repeated Irreducible Quadratic Factors** Evaluate

$$\int \frac{x^2 + x + 1}{(x^2 + 1)^2}\, dx$$

Solution   The rational function $\frac{x^2+x+1}{(x^2+1)^2}$ is proper, since the numerator is a polynomial of degree 2 and the denominator is of degree 4 [because $(x^2 + 1)^2 = x^4 + 2x^2 + 1$]. The denominator contains the irreducible quadratic factor $x^2 + 1$ twice. We can therefore write the integrand as

$$\frac{x^2 + x + 1}{(x^2 + 1)^2} = \frac{Ax + B}{x^2 + 1} + \frac{Cx + D}{(x^2 + 1)^2} = \frac{(Ax + B)(x^2 + 1) + Cx + D}{(x^2 + 1)^2}$$

$$= \frac{Ax^3 + Ax + Bx^2 + B + Cx + D}{(x^2 + 1)^2}$$

$$= \frac{Ax^3 + Bx^2 + (A + C)x + (B + D)}{(x^2 + 1)^2}$$

Comparing the last expression with the integrand, we conclude that

$$A = 0, \quad B = 1, \quad A + C = 1, \quad \text{and} \quad B + D = 1$$

which implies that $C = 1$ and $D = 0$. Therefore,

$$\frac{x^2 + x + 1}{(x^2 + 1)^2} = \frac{1}{x^2 + 1} + \frac{x}{(x^2 + 1)^2}$$

and

$$\int \frac{x^2 + x + 1}{(x^2 + 1)^2}\, dx = \int \frac{1}{x^2 + 1}\, dx + \int \frac{x}{(x^2 + 1)^2}\, dx$$

The first integral on the right-hand side is $\tan^{-1} x + C$. To evaluate the second integral on the right-hand side, we use substitution: $u = x^2 + 1$ with $du/2 = x\, dx$. This yields

$$\int \frac{x}{(x^2 + 1)^2}\, dx = \frac{1}{2} \int \frac{du}{u^2} = -\frac{1}{2u} + C = -\frac{1}{2(x^2 + 1)} + C$$

Combining the two results, we find that

$$\int \frac{x^2 + x + 1}{(x^2 + 1)^2}\, dx = \tan^{-1} x - \frac{1}{2(x^2 + 1)} + C \qquad ■$$

EXAMPLE 8   (*Dalzell 1944, 1971*) Show that

$$\int_0^1 \frac{x^4 (1 - x)^4}{1 + x^2}\, dx = \frac{22}{7} - \pi$$

In Problem 53, we will show that this equation allows us to find lower and upper bounds on π, namely,

$$3.140 \le \pi \le 3.142$$

Solution   The integrand is an improper rational function. Long division (see Problem 53) yields

$$\frac{x^4 (1 - x)^4}{1 + x^2} = x^6 - 4x^5 + 5x^4 - 4x^2 + 4 - \frac{4}{1 + x^2}$$

Therefore,

$$\int_0^1 \frac{x^4 (1 - x)^4}{1 + x^2}\, dx = \int_0^1 \left(x^6 - 4x^5 - 4x^2 + 4 - \frac{4}{1 + x^2} \right) dx$$

$$= \frac{x^7}{7} - \frac{2}{3}x^6 + x^5 - \frac{4}{3}x^3 + 4x - 4\tan^{-1} x \Big]_0^1$$

$$= \left(\frac{1}{7} - \frac{2}{3} + 1 - \frac{4}{3} + 4 - 4\tan^{-1} 1 \right) - \left(0 - 4\tan^{-1} 0 \right)$$

$$= \left(\frac{22}{7} - \pi \right) - 0 = \frac{22}{7} - \pi \qquad ■$$

We conclude this section by providing a summary of the two most important cases: when the integrand is a rational function for which the denominator is a polynomial of degree 2 and is either (1) a product of two not necssarily distinct linear factors or (2) an irreducible quadratic polynomial.

The first step is to make sure that the degree of the numerator is less than the degree of the denominator. If not, then we use long division to simplify the integrand.

We will now assume that the degree of the numerator is strictly less than the degree of the denominator (i.e., the integrand is a proper rational function). We write the rational function $f(x)$ as

$$f(x) = \frac{P(x)}{Q(x)}$$

with $Q(x) = ax^2 + bx + c, a \neq 0$, and $P(x) = rx + s$. Either $Q(x)$ can be factored into two linear factors, or it is irreducible (i.e., does not have real roots).

Case 1a: $Q(x)$ is a product of two distinct linear factors. In this case, we write

$$Q(x) = a(x - x_1)(x - x_2)$$

where x_1 and x_2 are the two distinct roots of $Q(x)$. We then use the method of partial fractions to simplify the rational function:

$$\frac{P(x)}{Q(x)} = \frac{rx + s}{ax^2 + bx + c} = \frac{1}{a}\left[\frac{A}{x - x_1} + \frac{B}{x - x_2}\right]$$

The constants A and B must now be determined as in Example 3.

Case 1b: $Q(x)$ is a product of two identical linear factors. In this case, we write

$$Q(x) = a(x - x_1)^2$$

where x_1 is the root of $Q(x)$. We then use the method of partial fractions to simplify the rational function:

$$\frac{P(x)}{Q(x)} = \frac{rx + s}{ax^2 + bx + c} = \frac{1}{a}\left[\frac{A}{x - x_1} + \frac{B}{(x - x_1)^2}\right]$$

The constants A and B must now be determined as in Example 5.

Case 2: $Q(x)$ is an irreducible quadratic polynomial. In this case,

$$Q(x) = ax^2 + bx + c \quad \text{with } b^2 - 4ac < 0$$

and we must complete the square as in Example 2. Doing so then leads to integrals of the form

$$\int \frac{dx}{x^2 + 1} \qquad \text{or} \qquad \int \frac{x}{x^2 + 1}\, dx$$

The first integral is $\tan^{-1} x + C$, whereas the second integral can be evaluated by substitution. (See Examples 6 and 7.)

Section 7.3 Problems

■ 7.3.1, 7.3.2

In Problems 1–4, use long division to write $f(x)$ as a sum of a polynomial and a proper rational function.

1. $f(x) = \dfrac{2x^2 + 5x - 1}{x + 2}$

2. $f(x) = -\dfrac{x^2 - 4x - 1}{x - 1}$

3. $f(x) = \dfrac{3x^3 + 5x - 2x^2 - 2}{x^2 + 1}$

4. $f(x) = \dfrac{x^3 - 3x^2 - 15}{x^2 + x + 3}$

In Problems 5–8, write out the partial-fraction decomposition of the function $f(x)$.

5. $f(x) = \dfrac{2x - 3}{x(x + 1)}$

6. $f(x) = -\dfrac{x + 1}{(2x + 1)(x - 1)}$

7. $f(x) = \dfrac{4x^2 - 14x - 6}{x(x - 3)(x + 1)}$

8. $f(x) = \dfrac{16x - 6}{(2x - 5)(3x + 1)}$

In Problems 9–12, write out the partial-fraction decomposition of the function $f(x)$.

9. $f(x) = \dfrac{5x - 1}{x^2 - 1}$

10. $f(x) = \dfrac{9x - 7}{2x^2 - 7x + 3}$

11. $f(x) = \dfrac{4x + 1}{x^2 - 3x - 10}$

12. $f(x) = -\dfrac{10}{3x^2 + 8x - 3}$

In Problems 13–18, use partial-fraction decomposition to evaluate the integrals.

13. $\displaystyle\int \dfrac{1}{x(x - 2)}\,dx$

14. $\displaystyle\int \dfrac{1}{x(2x + 1)}\,dx$

15. $\displaystyle\int \dfrac{1}{(x + 1)(x - 3)}\,dx$

16. $\displaystyle\int \dfrac{1}{(x - 1)(x + 2)}\,dx$

17. $\displaystyle\int \dfrac{x^2 - 2x - 2}{x^2(x + 2)}\,dx$

18. $\displaystyle\int \dfrac{4x^2 - x - 1}{(x + 1)^2(x - 3)}\,dx$

In Problems 19–22, use partial-fraction decompositon to evaluate each integral.

19. $\displaystyle\int \dfrac{x^3 - x^2 + x - 4}{(x^2 + 1)(x^2 + 4)}\,dx$

20. $\displaystyle\int \dfrac{x^3 - 3x^2 + x - 6}{(x^2 + 2)(x^2 + 1)}\,dx$

21. $\displaystyle\int \dfrac{2x^2 - 3x + 2}{(x^2 + 1)^2}\,dx$

22. $\displaystyle\int \dfrac{3x^2 + 4x + 3}{(x^2 + 1)^2}\,dx$

In Problems 23–26, complete the square in the denominator and evaluate the integral.

23. $\displaystyle\int \dfrac{1}{x^2 - 2x + 2}\,dx$

24. $\displaystyle\int \dfrac{1}{x^2 + 4x + 5}\,dx$

25. $\displaystyle\int \dfrac{1}{x^2 - 4x + 13}\,dx$

26. $\displaystyle\int \dfrac{1}{x^2 + 2x + 5}\,dx$

In Problems 27–36, evaluate each integral.

27. $\displaystyle\int \dfrac{1}{(x - 3)(x + 2)}\,dx$

28. $\displaystyle\int \dfrac{2x - 1}{(x + 4)(x + 1)}\,dx$

29. $\displaystyle\int \dfrac{1}{x^2 - 9}\,dx$

30. $\displaystyle\int \dfrac{1}{x^2 + 9}\,dx$

31. $\displaystyle\int \dfrac{1}{x^2 - x - 2}\,dx$

32. $\displaystyle\int \dfrac{1}{x^2 - x + 2}\,dx$

33. $\displaystyle\int \dfrac{x^2 + 1}{x^2 + 3x + 2}\,dx$

34. $\displaystyle\int \dfrac{x^3 + 1}{x^2 + 3}\,dx$

35. $\displaystyle\int \dfrac{x^2 + 4}{x^2 - 4}\,dx$

36. $\displaystyle\int \dfrac{x^4 + 3}{x^2 - 4x + 3}\,dx$

In Problems 37–44, evaluate each definite integral.

37. $\displaystyle\int_3^5 \dfrac{x - 1}{x}\,dx$

38. $\displaystyle\int_3^5 \dfrac{x}{x - 1}\,dx$

39. $\displaystyle\int_0^1 \dfrac{x}{x^2 + 1}\,dx$

40. $\displaystyle\int_1^2 \dfrac{x^2 + 1}{x}\,dx$

41. $\displaystyle\int_2^3 \dfrac{1}{1 - x}\,dx$

42. $\displaystyle\int_2^3 \dfrac{1}{1 - x^2}\,dx$

43. $\displaystyle\int_0^1 \tan^{-1} x\,dx$

44. $\displaystyle\int_0^1 x \tan^{-1} x\,dx$

In Problems 45–52, evaluate each integral.

45. $\displaystyle\int \dfrac{1}{(x + 1)^2 x}\,dx$

46. $\displaystyle\int \dfrac{1}{x^2(x - 1)^2}\,dx$

47. $\displaystyle\int \dfrac{4}{(1 - x)(1 + x)^2}\,dx$

48. $\displaystyle\int \dfrac{2x^2 + 2x - 1}{x^3(x - 3)}\,dx$

49. $\displaystyle\int \dfrac{1}{(x^2 - 9)^2}\,dx$

50. $\displaystyle\int \dfrac{1}{(x^2 - x - 2)^2}\,dx$

51. $\displaystyle\int \dfrac{1}{x^2(x^2 + 1)}\,dx$

52. $\displaystyle\int \dfrac{1}{(x + 1)^2(x^2 + 1)}\,dx$

53. (a) To complete Example 8, show that

$$\dfrac{x^4(1 - x)^4}{1 + x^2} = x^6 - 4x^5 + 5x^4 - 4x^2 + 4 - \dfrac{4}{1 + x^2}$$

(b) Show that

$$\int_0^1 \dfrac{x^4(1 - x)^4}{2}\,dx \leq \int_0^1 \dfrac{x^4(1 - x)^4}{1 + x^2}\,dx \leq \int_0^1 x^4(1 - x)^4\,dx$$

and conclude that

$$\dfrac{1}{1260} \leq \dfrac{22}{7} - \pi \leq \dfrac{1}{630}$$

Use this result to show that

$$3.140 \leq \pi \leq 3.142$$

■ 7.4 Improper Integrals

In this section, we discuss definite integrals of two types with the following characteristics:

1. One or both limits of integration are infinite; that is, the integration interval is unbounded; or
2. The integrand becomes infinite at one or more points of the interval of integration.

We call such integrals **improper integrals**.

■ 7.4.1 Type 1: Unbounded Intervals

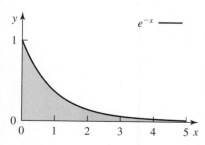

Figure 7.9 The unbounded region between the graph of $y = e^{-x}$ and the x-axis for $x \geq 0$.

Suppose that we wanted to compute the area of the unbounded region below the graph of $f(x) = e^{-x}$ and above the x-axis for $x \geq 0$. (See Figure 7.9.) How would we proceed? We know how to find the area of a region bounded by the graph of a continuous function [here, $f(x) = e^{-x}$] and the x-axis between 0 and z, namely,

$$A(z) = \int_0^z e^{-x}\,dx = -e^{-x}\Big]_0^z = 1 - e^{-z}$$

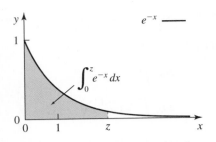

Figure 7.10 The region between 0 and z.

This is the shaded area in Figure 7.10. If we now let z tend to infinity, we may regard the limiting value (if it exists) as the area of the unbounded region below the graph of $f(x) = e^{-x}$ and above the x-axis for $x \geq 0$ (see Figure 7.9):

$$A = \lim_{z \to \infty} A(z) = \lim_{z \to \infty} (1 - e^{-z}) = 1$$

We write

$$\int_0^\infty e^{-x}\, dx = 1$$

Therefore, for functions that are continuous on unbounded intervals (see Figures 7.11 and 7.12), we define

$$\int_a^\infty f(x)\, dx = \lim_{z \to \infty} \int_a^z f(x)\, dx$$

and

$$\int_{-\infty}^a f(x)\, dx = \lim_{z \to -\infty} \int_z^a f(x)\, dx$$

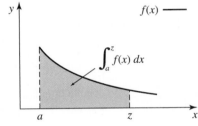

Figure 7.11 The definition of the improper integral $\int_a^\infty f(x)\, dx$ as the limit of $\int_a^z f(x)\, dx$ as $z \to \infty$.

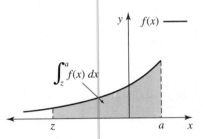

Figure 7.12 The definition of the improper integral $\int_{-\infty}^a f(x)\, dx$ as the limit of $\int_z^a f(x)\, dx$ as $z \to -\infty$.

You might be surprised that the area of an unbounded region can be finite. This need not be the case, and it happens only if the graph of $f(x)$ approaches the x-axis sufficiently fast. We illustrate this property in the next two examples.

EXAMPLE 1 **Finite Area** Compute

$$\int_1^\infty \frac{1}{x^2}\, dx$$

Solution The function $y = 1/x^2$ is continuous on $[1, \infty)$. We first compute

$$A(z) = \int_1^z \frac{1}{x^2}\, dx = -\frac{1}{x}\Big]_1^z = 1 - \frac{1}{z}$$

(see Figure 7.13) and then let $z \to \infty$. We find that

$$\lim_{z \to \infty} A(z) = \lim_{z \to \infty} \left(1 - \frac{1}{z}\right) = 1$$

Hence,

$$\int_1^\infty \frac{1}{x^2}\, dx = 1$$

∎

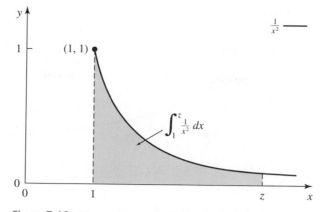

Figure 7.13 The region corresponding to $A(z)$ in Example 1.

EXAMPLE 2

Infinite Area Compute

$$\int_1^\infty \frac{1}{\sqrt{x}}\,dx$$

Solution   The function $f(x) = 1/\sqrt{x}$ is continuous on $[1, \infty)$. We first compute

$$A(z) = \int_1^z \frac{1}{\sqrt{x}}\,dx = 2\sqrt{x}\,\Big]_1^z = 2(\sqrt{z} - 1)$$

(see Figure 7.14) and then let $z \to \infty$. We find that

$$\lim_{z\to\infty} A(z) = \lim_{z\to\infty} 2(\sqrt{z} - 1) = \infty$$

Hence,

$$\int_1^\infty \frac{1}{\sqrt{x}}\,dx$$

does not exist.   ■

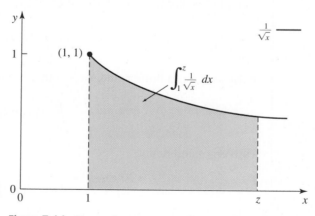

Figure 7.14 The region corresponding to $A(z)$ in Example 2.

Looking back at Examples 1 and 2, we see that, in both cases, the respective integrands approached the x-axis as $x \to \infty$; that is, both

$$\lim_{x\to\infty} \frac{1}{x^2} = 0 \qquad \text{and} \qquad \lim_{x\to\infty} \frac{1}{\sqrt{x}} = 0$$

However, $\frac{1}{x^2}$ approaches the x-axis much faster than $\frac{1}{\sqrt{x}}$, as can be seen from the graphs in Figure 7.15. The exponent of x in the denominator determines how fast

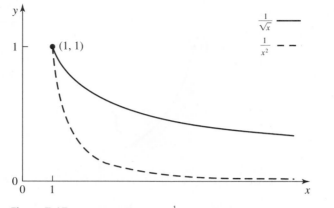

Figure 7.15 The function $y = \frac{1}{x^2}$ approaches the x-axis much faster than the function $y = \frac{1}{\sqrt{x}}$.

the function approaches the x-axis. The area between the graph and the x-axis from $x = 1$ to infinity is finite only if the graph approaches the x-axis fast enough. Indeed, if we tried to compute

$$\int_1^\infty \frac{1}{x^p}\, dx$$

for $0 < p < \infty$, we would find that

$$\int_1^\infty \frac{1}{x^p}\, dx = \begin{cases} \frac{1}{p-1} & \text{for } p > 1 \\ \infty & \text{for } 0 < p \le 1 \end{cases}$$

(Note that $y = 1/x^p$ is continuous on $[1, \infty)$.) For $p > 1$, the function $\frac{1}{x^p}$ approaches the x-axis fast enough as $x \to \infty$ for the area under the graph to be finite. (We investigate this integral further in Problem 33.)

We will use the following terminology to indicate whether an improper integral is finite or infinite:

Let $f(x)$ be continuous on the interval $[a, \infty)$. If

$$\lim_{z \to \infty} \int_a^z f(x)\, dx$$

exists and has a finite value, we say that the improper integral

$$\int_a^\infty f(x)\, dx$$

converges and define

$$\int_a^\infty f(x)\, dx = \lim_{z \to \infty} \int_a^z f(x)\, dx$$

Otherwise, we say that the improper integral **diverges**.

Analogous definitions can be given when the lower limit of integration is infinite.

EXAMPLE 3   **Infinite Lower Limit** Show that the improper integral

$$\int_{-\infty}^0 \frac{1}{(x-1)^2}\, dx$$

converges.

Solution

Note that $y = 1/(x-1)^2$ is continuous on $(-\infty, 0]$. To show that the integral converges, we compute its value. We need to find

$$A(z) = \int_z^0 \frac{1}{(x-1)^2} \, dx \quad \text{for } z < 0$$

and then let $z \to -\infty$. (See Figure 7.16.) We find that

$$\int_z^0 (x-1)^{-2} \, dx = -(x-1)^{-1}\Big]_z^0$$

$$= -\frac{1}{x-1}\Big]_z^0 = -\frac{1}{-1} + \frac{1}{z-1} = 1 + \frac{1}{z-1}$$

and

$$\lim_{z \to -\infty} \left(1 + \frac{1}{z-1}\right) = 1$$

Therefore,

$$\int_{-\infty}^0 \frac{1}{(x-1)^2} \, dx = 1 \qquad \blacksquare$$

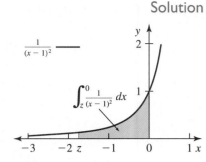

Figure 7.16 The region corresponding to $A(z)$ in Example 3.

We next discuss the case when both limits of integration are infinite.

Assume that $f(x)$ is continuous on $(-\infty, \infty)$. Then

$$\int_{-\infty}^{\infty} f(x) \, dx = \int_{-\infty}^{a} f(x) \, dx + \int_{a}^{\infty} f(x) \, dx \qquad (7.15)$$

where a is a real number. If *both* improper integrals on the right-hand side of (7.15) are convergent, then the value of the improper integral on the left-hand side of (7.15) is the sum of the two limiting values on the right-hand side.

Suppose that we wish to compute

$$\int_{-\infty}^{\infty} x^3 \, dx$$

We choose a value $a \in (-\infty, \infty)$—for instance, $a = 0$. Then

$$\int_{-\infty}^{\infty} x^3 \, dx = \int_{-\infty}^{0} x^3 \, dx + \int_{0}^{\infty} x^3 \, dx$$

Looking at Figure 7.17, you can see that both improper integrals on the right-hand side are divergent. We check this assertion for the second one: We have

$$\int_0^{\infty} x^3 \, dx = \lim_{z \to \infty} \int_0^z x^3 \, dx = \frac{1}{4}x^4\Big]_0^z = \frac{1}{4} \lim_{z \to \infty} (z^4 - 0)$$

which does not exist. Hence,

$$\int_{-\infty}^{\infty} x^3 \, dx$$

is divergent.

It is important to realize that the definition of $\int_{-\infty}^{\infty} f(x) \, dx$ is different from that of

$$\lim_{b \to \infty} \int_{-b}^{b} f(x) \, dx$$

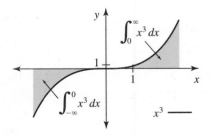

Figure 7.17 The integral $\int_{-\infty}^{\infty} x^3 \, dx$ is divergent.

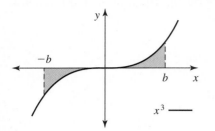

Figure 7.18 Because of symmetry, $\int_{-b}^{b} x^3 \, dx = 0$.

We use $f(x) = x^3$ again to illustrate this difference. For any $b > 0$, we find that

$$\int_{-b}^{b} x^3 \, dx = \frac{1}{4} x^4 \Big]_{-b}^{b} = \frac{1}{4}(b^4 - (-b)^4) = 0$$

(See Figure 7.18.) Therefore,

$$\lim_{b \to \infty} \int_{-b}^{b} x^3 \, dx = 0$$

This limit is not the same as $\int_{-\infty}^{\infty} x^3 \, dx$.

Looking at (7.15), we see that, in order to evaluate $\int_{-\infty}^{\infty} f(x) \, dx$, we need to split up the integral at some $a \in \mathbf{R}$. There are often natural choices for a; we illustrate this in the next example.

EXAMPLE 4 **Infinite Upper and Lower Limit** Compute

$$\int_{-\infty}^{\infty} \frac{1}{1 + x^2} \, dx$$

Solution The graph of $f(x) = \frac{1}{1+x^2}$ is shown in Figure 7.19. The function $f(x) = 1/(1 + x^2)$ is continuous for all $x \in \mathbf{R}$. It is symmetric about $x = 0$; a good choice for splitting up the integral is therefore $a = 0$. We write

$$\int_{-\infty}^{\infty} \frac{1}{1 + x^2} \, dx = \int_{-\infty}^{0} \frac{1}{1 + x^2} \, dx + \int_{0}^{\infty} \frac{1}{1 + x^2} \, dx$$

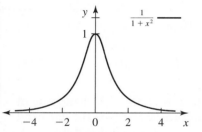

Figure 7.19 The graph of $f(x) = \frac{1}{1+x^2}$ in Example 4.

Now,

$$\lim_{z \to \infty} \int_{0}^{z} \frac{1}{1 + x^2} \, dx = \lim_{z \to \infty} \left[\tan^{-1} x \right]_{0}^{z}$$
$$= \lim_{z \to \infty} (\tan^{-1} z - \tan^{-1} 0) = \frac{\pi}{2}$$

and

$$\lim_{z \to -\infty} \int_{z}^{0} \frac{1}{1 + x^2} \, dx = \lim_{z \to -\infty} \left[\tan^{-1} x \right]_{z}^{0}$$
$$= \lim_{z \to -\infty} (\tan^{-1} 0 - \tan^{-1} z) = \frac{\pi}{2}$$

That $\int_{-\infty}^{0} \frac{1}{1+x^2} \, dx = \frac{\pi}{2}$ is expected because of symmetry. The area of the region to the left of the y-axis is equal to the area of the region to the right of the y-axis. Putting things together, we find that

$$\int_{-\infty}^{\infty} \frac{1}{1 + x^2} \, dx = \frac{\pi}{2} + \frac{\pi}{2} = \pi$$ ■

EXAMPLE 5 **Infinite Upper and Lower Limit** Compute

$$\int_{-\infty}^{\infty} \frac{x}{1 + x^2} \, dx$$

Solution The graph of $f(x) = \frac{x}{1+x^2}$ is shown in Figure 7.20. The function $f(x) = x/(1 + x^2)$ is continuous for all $x \in \mathbf{R}$. Because of the symmetry about the origin, you might be tempted to say that the signed area to the left of 0 is the negative of the area to the

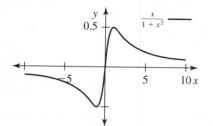

Figure 7.20 The graph of $f(x) = \frac{x}{1+x^2}$ in Example 5.

right of 0 and, therefore, the value of the improper integral should be 0. But this is wrong! We choose $a = 0$, and write

$$\int_{-\infty}^{\infty} \frac{x}{1+x^2}\, dx = \int_{-\infty}^{0} \frac{x}{1+x^2}\, dx + \int_{0}^{\infty} \frac{x}{1+x^2}\, dx$$

We begin by computing

$$\int_{0}^{z} \frac{x}{1+x^2}\, dx$$

Using the substitution $u = 1 + x^2$ and $du = 2x\, dx$, we find that

$$\int_{0}^{z} \frac{x}{1+x^2}\, dx = \int_{1}^{1+z^2} \frac{1}{2u}\, du = \frac{1}{2} \ln|u| \Big]_{1}^{1+z^2}$$

$$= \frac{1}{2}[\ln(1+z^2) - \ln 1] = \frac{1}{2}\ln(1+z^2)$$

Taking the limit as $z \to \infty$, we obtain

$$\int_{0}^{\infty} \frac{x}{1+x^2}\, dx = \lim_{z \to \infty} \frac{1}{2}\ln(1+z^2) = \infty$$

Since one of the integrals is already divergent, we conclude that

$$\int_{-\infty}^{\infty} \frac{x}{1+x^2}\, dx$$

is divergent and therefore cannot be equal to 0. This example has an important take-home message: Before we can use symmetry to compute an improper integral, we need to make sure that the integral exists. ■

■ 7.4.2 Type 2: Unbounded Integrand

Figure 7.21 The graph of $y = \frac{1}{\sqrt{x}}$.

So far, when we computed a definite integral, we made sure that the integrand was continuous over the interval of integration. We will now explain what to do when the integrand becomes infinite at one or more points of the interval. Suppose we wish to integrate

$$\int_{0}^{1} \frac{dx}{\sqrt{x}}$$

The graph of $f(x) = \frac{1}{\sqrt{x}}$ is shown in Figure 7.21. We see immediately that $f(x)$ is continuous on $(0, 1]$ and undefined at $x = 0$, and that

$$\lim_{x \to 0^+} \frac{1}{\sqrt{x}} = \infty$$

Let's choose a number $c \in (0, 1)$ and compute

$$\int_{c}^{1} \frac{dx}{\sqrt{x}} = 2\sqrt{x}\, \Big]_{c}^{1} = 2(1 - \sqrt{c})$$

(See Figure 7.22.) If we now let $c \to 0^+$, we may regard the limiting value (if it exists) as the definite integral $\int_{0}^{1} \frac{1}{\sqrt{x}}\, dx$. That is,

Figure 7.22 The area of the shaded region is $\int_{c}^{1} \frac{1}{\sqrt{x}}\, dx = 2(1 - \sqrt{c})$.

$$\int_{0}^{1} \frac{dx}{\sqrt{x}} = \lim_{c \to 0^+} \int_{c}^{1} \frac{dx}{\sqrt{x}} = \lim_{c \to 0^+} 2(1 - \sqrt{c}) = 2$$

If f is continuous on $(a, b]$ and $\lim_{x \to a^+} f(x) = \pm\infty$ (see Figure 7.23), we define

$$\int_a^b f(x)\,dx = \lim_{c \to a^+} \int_c^b f(x)\,dx$$

provided that this limit exists. If the limit exists, we say that the improper integral on the left-hand side **converges**; if the limit does not exist, we say that the integral **diverges**.

Similarly, if f is continuous on $[a, b)$ and $\lim_{x \to b^-} f(x) = \pm\infty$ (see Figure 7.24), we define

$$\int_a^b f(x)\,dx = \lim_{c \to b^-} \int_a^c f(x)\,dx$$

provided that this limit exists.

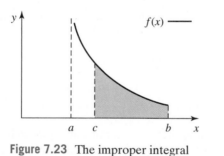

Figure 7.23 The improper integral $\int_a^b f(x)\,dx$ is defined as the limit of $\int_c^b f(x)\,dx$ as $c \to a^+$.

Figure 7.24 The improper integral $\int_a^b f(x)\,dx$ is defined as the limit of $\int_a^c f(x)\,dx$ as $c \to b^-$.

EXAMPLE 6

Integrand Undefined at Right Endpoint Compute

$$\int_0^1 \frac{dx}{(x-1)^{2/3}}$$

Solution

The graph of $f(x) = \frac{1}{(x-1)^{2/3}}$ is shown in Figure 7.25. We see immediately that $f(x)$ is continuous on $[0, 1)$ and undefined at $x = 1$, and that

$$\lim_{x \to 1^-} f(x) = \infty$$

Figure 7.25 The graph of $f(x) = \frac{1}{(x-1)^{2/3}}, 0 \le x < 1$, in Example 6.

To compute the integral, we choose a number $c \in (0, 1)$ and compute

$$\int_0^c \frac{dx}{(x-1)^{2/3}}$$

(See Figure 7.26.) Letting $c \to 1^-$ will then produce the desired integral. That is,

$$\int_0^1 \frac{dx}{(x-1)^{2/3}} = \lim_{c \to 1^-} \int_0^c \frac{dx}{(x-1)^{2/3}}$$

We first compute the indefinite integral

$$\int \frac{dx}{(x-1)^{2/3}} = 3(x-1)^{1/3} + C$$

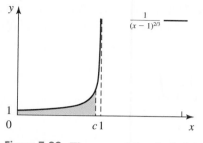

Figure 7.26 The area of the shaded region is $\int_0^c \frac{1}{(x-1)^{2/3}}\, dx$.

If we set $F(x) = 3(x-1)^{1/3}$, then

$$\lim_{c\to 1^-} \int_0^c \frac{dx}{(x-1)^{2/3}} = \lim_{c\to 1^-} [F(c) - F(0)]$$

$$= \lim_{c\to 1^-} \left[3(c-1)^{1/3} - 3(-1)^{1/3}\right] = 3 \qquad (7.16)$$

We therefore find that

$$\int_0^1 \frac{dx}{(x-1)^{2/3}} = 3 \qquad\blacksquare$$

EXAMPLE 7

Integrand Undefined at Left Endpoint Compute

$$\int_0^1 \ln x\, dx$$

Solution   The graph of $f(x) = \ln x, 0 < x \le 1$, is shown in Figure 7.27. We immediately see that $f(x)$ is continuous on $(0, 1]$ and not defined at $x = 0$, and that

$$\lim_{x\to 0^+} f(x) = -\infty$$

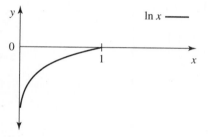

Figure 7.27 The graph of $f(x) = \ln x, 0 < x \le 1$, in Example 7.

To determine the definite integral, we need to compute

$$\lim_{c\to 0^+} \int_c^1 \ln x\, dx$$

Since $F(x) = x \ln x - x$ is an antiderivative of $f(x) = \ln x$, we find that

$$\lim_{c\to 0^+} \int_c^1 \ln x\, dx = \lim_{c\to 0^+} [F(1) - F(c)]$$

$$= \lim_{c\to 0^+} [1 \ln 1 - 1 - c \ln c + c]$$

We need to find $\lim_{c\to 0^+} c \ln c$. The limit is of the form $0 \cdot \infty$. L'Hospital's rule yields

$$\lim_{c\to 0^+} c \ln c = \lim_{c\to 0^+} \frac{\ln c}{\frac{1}{c}} = \lim_{c\to 0^+} \frac{\frac{1}{c}}{-\frac{1}{c^2}}$$

$$= \lim_{c\to 0^+} \left(-\frac{1}{c} \cdot \frac{c^2}{1}\right) = -\lim_{c\to 0^+} c = 0$$

Together with $\lim_{c\to 0^+} c = 0$, we therefore find that

$$\int_0^1 \ln x\, dx = \lim_{c\to 0^+} \int_c^1 \ln x\, dx = -1 \qquad\blacksquare$$

EXAMPLE 8

Integrand Discontinuous in Interval Compute

$$\int_{-1}^1 \frac{1}{x^2}\, dx$$

Solution   The function $f(x) = \frac{1}{x^2}$ is not defined at $x = 0$. In fact, it has a vertical asymptote at $x = 0$, since

$$\lim_{x\to 0^-} \frac{1}{x^2} = \infty \qquad \text{and} \qquad \lim_{x\to 0^+} \frac{1}{x^2} = \infty$$

The graph of $f(x) = \frac{1}{x^2}$, $x \neq 0$, is shown in Figure 7.28. We see that $f(x) = 1/x^2$ is continuous, except at $x = 0$. To deal with this discontinuity, we split the integral at $x = 0$. We write

$$\int_{-1}^{1} \frac{1}{x^2}\,dx = \lim_{c \to 0^-} \int_{-1}^{c} \frac{1}{x^2}\,dx + \lim_{c \to 0^+} \int_{c}^{1} \frac{1}{x^2}\,dx$$

(See Figure 7.29.) The function

$$F(x) = -\frac{1}{x}$$

Figure 7.28 The graph of $f(x) = \frac{1}{x^2}$.

is an antiderivative of $\frac{1}{x^2}$. Therefore,

$$\lim_{c \to 0^-} \int_{-1}^{c} \frac{1}{x^2}\,dx = \lim_{c \to 0^-} [F(c) - F(-1)]$$

$$= \lim_{c \to 0^-} \left[-\frac{1}{c} - 1 \right] = \infty$$

We can already conclude that the integral is divergent. But to see what the other limit looks like, we will compute it. That is,

$$\lim_{c \to 0^+} \int_{c}^{1} \frac{1}{x^2}\,dx = \lim_{c \to 0^+} [F(1) - F(c)]$$

$$= \lim_{c \to 0^+} \left[-1 + \frac{1}{c} \right] = \infty$$

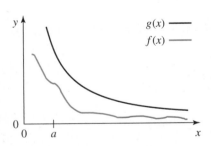

Figure 7.29 The improper integral $\int_{-1}^{1} \frac{1}{x^2}\,dx$.

Therefore,

$$\int_{-1}^{1} \frac{1}{x^2}\,dx$$

is divergent. ■

■ 7.4.3 A Comparison Result for Improper Integrals

Figure 7.30 The function $g(x)$ lies above the function $f(x)$.

In many cases, it is impossible to evaluate an integral exactly. In dealing with improper integrals, we frequently must know whether the integral converges. Instead of computing the value of the improper integral exactly, we can then resort to simpler integrals that either dominate or are dominated by the improper integral of interest. We will explain this idea graphically.

We assume that $f(x) \geq 0$ for $x \geq a$. Suppose we wish to show that $\int_{a}^{\infty} f(x)\,dx$ is convergent. Then it is enough to find a function $g(x)$ such that $g(x) \geq f(x)$ for all $x \geq a$ and $\int_{a}^{\infty} g(x)\,dx$ is convergent. This is illustrated in Figure 7.30. It is clear from the graph that

$$0 \leq \int_{a}^{\infty} f(x)\,dx \leq \int_{a}^{\infty} g(x)\,dx$$

If $\int_{a}^{\infty} g(x)\,dx < \infty$, it follows that $\int_{a}^{\infty} f(x)\,dx$ is convergent, since $\int_{a}^{\infty} f(x)\,dx$ must take on a value between 0 and a finite number, given by $\int_{a}^{\infty} g(x)\,dx$.

We again assume that $f(x) \geq 0$ for all $x \geq a$. Suppose we now wish to show that $\int_{a}^{\infty} f(x)\,dx$ is divergent. It is then enough to find a function $g(x)$ such that $0 \leq g(x) \leq f(x)$ for all $x \geq a$ and $\int_{a}^{\infty} g(x)\,dx$ is divergent. This is illustrated in Figure 7.31. It is clear from the graph that

$$\int_{a}^{\infty} f(x)\,dx \geq \int_{a}^{\infty} g(x)\,dx \geq 0$$

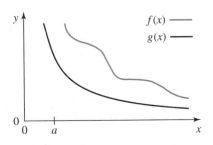

Figure 7.31 The graph of $g(x)$ is below the graph of $f(x)$.

If $\int_{a}^{\infty} g(x)\,dx$ is divergent, it follows that $\int_{a}^{\infty} f(x)\,dx$ is divergent.

You can see from the preceding discussion that in one case we selected a function that dominated $f(x)$, whereas in the other case we selected a function that was dominated by $f(x)$. This indicates that, before you find a comparison function, you must first guess whether the integral is likely to converge. (With practice, you get better at guessing whether an integral converges or diverges.) Sketching the functions involved can help you convince yourself that you are making the comparison in the right direction. Your comparison function, of course, should be simple enough so that you can integrate it without any problems. We present two examples that illustrate both cases.

EXAMPLE 9 **Convergence** Show that

$$\int_0^\infty e^{-x^2}\,dx$$

is convergent.

Solution The function $f(x) = e^{-x^2}$ is continuous and positive for $x \in [0, \infty)$. We cannot compute the antiderivative of $f(x) = e^{-x^2}$ with any of the techniques we have learned in this text. In fact, there is no simple way to express the value of $\int_0^z e^{-x^2}\,dx$ for $z > 0$. (It can be expressed as a sum of infinitely many terms.) But we can still determine whether the integral is convergent. To do so, we write $\int_0^\infty e^{-x^2}\,dx$ as a sum of two integrals and then show that each one is finite. We have

$$\int_0^\infty e^{-x^2}\,dx = \int_0^1 e^{-x^2}\,dx + \int_1^\infty e^{-x^2}\,dx$$

Since $0 < e^{-x^2} \le 1$, it follows that

$$0 < \int_0^1 e^{-x^2}\,dx \le \int_0^1 dx = 1 < \infty$$

To show that $\int_1^\infty e^{-x^2}\,dx$ is convergent, we use the fact that e^{-x} is a decreasing function and that if $x \ge 1$, then $x \le x^2$. It then follows that

$$0 \le e^{-x^2} \le e^{-x} \quad \text{for } x \ge 1$$

Therefore,

$$0 \le \int_1^\infty e^{-x^2}\,dx \le \int_1^\infty e^{-x}\,dx = \lim_{c \to \infty}\left[-e^{-x}\right]_1^c = e^{-1} < \infty$$

Since both integrals are convergent, $\int_0^\infty e^{-x^2}\,dx$ is convergent.

Although for $0 < z < \infty$, $\int_0^z e^{-x^2}\,dx$ can be computed only approximately (e.g., using numerical methods of the sort we will discuss in Section 7.5), we can show with very different tools (which we do not cover in this text) that

$$\int_0^\infty e^{-x^2}\,dx = \frac{\sqrt{\pi}}{2}$$

■

EXAMPLE 10 **Divergence** Show that

$$\int_1^\infty \frac{1}{\sqrt{x + \sqrt{x}}}\,dx$$

is divergent.

Solution   The function $f(x) = 1/\sqrt{x + \sqrt{x}}$ is continuous on $[1, \infty)$. The integrand looks rather complicated, but since $x + \sqrt{x} \leq x + x$ for $x \geq 1$, it follows that

$$\frac{1}{\sqrt{x + \sqrt{x}}} \geq \frac{1}{\sqrt{2x}} \quad \text{for } x \geq 1$$

Hence,

$$\int_1^\infty \frac{1}{\sqrt{x + \sqrt{x}}} \, dx \geq \int_1^\infty \frac{1}{\sqrt{2x}} \, dx = \frac{1}{\sqrt{2}} \int_1^\infty \frac{1}{\sqrt{x}} \, dx = \infty$$

as shown in Example 2. Therefore,

$$\int_1^\infty \frac{dx}{\sqrt{x + \sqrt{x}}}$$

is divergent.

Section 7.4 Problems

■ 7.4.1, 7.4.2

All the integrals in Problems 1–16 are improper and converge. Explain in each case why the integral is improper, and evaluate each integral.

1. $\int_0^\infty 3e^{-6x} \, dx$

2. $\int_0^\infty xe^{-x} \, dx$

3. $\int_0^\infty \frac{2}{1 + x^2} \, dx$

4. $\int_e^\infty \frac{dx}{x(\ln x)^2}$

5. $\int_1^\infty \frac{1}{x^{3/2}} \, dx$

6. $\int_{-\infty}^{-1} \frac{1}{1 + x^2} \, dx$

7. $\int_{-\infty}^\infty e^{-|x|} \, dx$

8. $\int_{-\infty}^\infty xe^{-x^2/2} \, dx$

9. $\int_{-\infty}^\infty \frac{x}{(1 + x^2)^2} \, dx$

10. $\int_{-\infty}^\infty x^3 e^{-x^4} \, dx$

11. $\int_0^9 \frac{dx}{\sqrt{9 - x}}$

12. $\int_1^e \frac{dx}{x\sqrt{\ln x}}$

13. $\int_0^{\pi/2} \frac{\cos x}{\sqrt{\sin x}} \, dx$

14. $\int_{-2}^0 \frac{dx}{(x + 1)^{1/3}}$

15. $\int_{-1}^1 \ln |x| \, dx$

16. $\int_0^2 \frac{dx}{(x - 1)^{2/5}}$

In Problems 17–28, determine whether each integral is convergent. If the integral is convergent, compute its value.

17. $\int_1^\infty \frac{1}{x^3} \, dx$

18. $\int_1^\infty \frac{1}{x^{1/3}} \, dx$

19. $\int_0^4 \frac{1}{x^4} \, dx$

20. $\int_0^4 \frac{1}{x^{1/4}} \, dx$

21. $\int_0^2 \frac{1}{(x - 1)^{1/3}} \, dx$

22. $\int_0^2 \frac{1}{(x - 1)^4} \, dx$

23. $\int_0^\infty \frac{1}{\sqrt{x + 1}} \, dx$

24. $\int_{-1}^0 \frac{1}{\sqrt{x + 1}} \, dx$

25. $\int_e^\infty \frac{dx}{x \ln x}$

26. $\int_1^e \frac{dx}{x \ln x}$

27. $\int_{-2}^2 \frac{2x \, dx}{(x^2 - 1)^{1/3}}$

28. $\int_{-\infty}^1 \frac{3}{1 + x^2} \, dx$

29. Determine whether

$$\int_{-\infty}^\infty \frac{1}{x^2 - 1} \, dx$$

is convergent. *Hint*: Use the partial-fraction decomposition

$$\frac{1}{x^2 - 1} = \frac{1}{2}\left(\frac{1}{x - 1} - \frac{1}{x + 1}\right)$$

30. Although we cannot compute the antiderivative of $f(x) = e^{-x^2/2}$, it is known that

$$\int_{-\infty}^\infty e^{-x^2/2} \, dx = \sqrt{2\pi}$$

Use this fact to show that

$$\int_{-\infty}^\infty x^2 e^{-x^2/2} \, dx = \sqrt{2\pi}$$

Hint: Write the integrand as

$$x \cdot (xe^{-x^2/2})$$

and use integration by parts.

31. Determine the constant c so that

$$\int_0^\infty ce^{-3x} \, dx = 1$$

32. Determine the constant c so that

$$\int_{-\infty}^\infty \frac{c}{1 + x^2} \, dx = 1$$

33. In this problem, we investigate the integral

$$\int_1^\infty \frac{1}{x^p} \, dx$$

for $0 < p < \infty$.

(a) For $z > 1$, set

$$A(z) = \int_1^z \frac{1}{x^p} \, dx$$

and show that

$$A(z) = \frac{1}{1-p}(z^{-p+1} - 1)$$

for $p \neq 1$ and

$$A(z) = \ln z$$

for $p = 1$.

(b) Use your results in (a) to show that, for $0 < p \leq 1$,

$$\lim_{z \to \infty} A(z) = \infty$$

(c) Use your results in (a) to show that, for $p > 1$,

$$\lim_{z \to \infty} A(z) = \frac{1}{p-1}$$

34. In this problem, we investigate the integral

$$\int_0^1 \frac{1}{x^p} \, dx$$

for $0 < p < \infty$.

(a) Compute

$$\int \frac{1}{x^p} \, dx$$

for $0 < p < \infty$. (*Hint:* Treat the case where $p = 1$ separately.)

(b) Use your result in (a) to compute

$$\int_c^1 \frac{1}{x^p} \, dx$$

for $0 < c < 1$.

(c) Use your result in (b) to show that

$$\int_0^1 \frac{1}{x^p} \, dx = \frac{1}{1-p}$$

for $0 < p < 1$.

(d) Show that

$$\int_0^1 \frac{1}{x^p} \, dx$$

is divergent for $p \geq 1$.

■ **7.4.3**

35. (a) Show that

$$0 \leq e^{-x^2} \leq e^{-x}$$

for $x \geq 1$.

(b) Use your result in (a) to show that

$$\int_1^\infty e^{-x^2} \, dx$$

is convergent.

36. (a) Show that

$$0 \leq \frac{1}{\sqrt{1+x^4}} \leq \frac{1}{x^2}$$

for $x > 0$.

(b) Use your result in (a) to show that

$$\int_1^\infty \frac{1}{\sqrt{1+x^4}} \, dx$$

is convergent.

37. (a) Show that

$$\frac{1}{\sqrt{1+x^2}} \geq \frac{1}{2x} > 0$$

for $x \geq 1$.

(b) Use your result in (a) to show that

$$\int_1^\infty \frac{1}{\sqrt{1+x^2}} \, dx$$

is divergent.

38. (a) Show that

$$\frac{1}{\sqrt{x + \ln x}} \geq \frac{1}{\sqrt{2x}} > 0$$

for $x \geq 1$.

(b) Use your result in (a) to show that

$$\int_1^\infty \frac{1}{\sqrt{x + \ln x}} \, dx$$

is divergent.

In Problems 39–42, find a comparison function for each integrand and determine whether the integral is convergent.

39. $\displaystyle\int_{-\infty}^\infty e^{-x^2/2} \, dx$ **40.** $\displaystyle\int_1^\infty \frac{1}{\sqrt{1+x^6}} \, dx$

41. $\displaystyle\int_1^\infty \frac{1}{\sqrt{1+x}} \, dx$ **42.** $\displaystyle\int_{-\infty}^\infty \frac{1}{e^x + e^{-x}} \, dx$

43. (a) Show that

$$\lim_{x \to \infty} \frac{\ln x}{\sqrt{x}} = 0$$

(b) Use your result in (a) to show that

$$2 \ln x \leq \sqrt{x} \qquad (7.17)$$

for sufficiently large x. Use a graphing calculator to determine just how large x must be for (7.17) to hold.

(c) Use your result in (b) to show that

$$\int_0^\infty e^{-\sqrt{x}} \, dx \qquad (7.18)$$

converges. Use a graphing calculator to sketch the function $f(x) = e^{-\sqrt{x}}$ together with its comparison function(s), and use your graph to explain how you showed that the integral in (7.18) is convergent.

44. (a) Show that

$$\lim_{x \to \infty} \frac{\ln x}{x} = 0$$

(b) Use your result in (a) to show that, for any $c > 0$,

$$cx \geq \ln x$$

for sufficiently large x.

(c) Use your result in (b) to show that, for any $p > 0$,

$$x^p e^{-x} \leq e^{-x/2}$$

provided that x is sufficiently large.

(d) Use your result in (c) to show that, for any $p > 0$,

$$\int_0^\infty x^p e^{-x} \, dx$$

is convergent.

■ 7.5 Numerical Integration

Some integrals, such as

$$\int_0^4 e^{-x^2}\, dx$$

are impossible to evaluate exactly. In such situations, numerical approximations are needed.

One way to approximate an integral numerically should be obvious from our initial approach to the area problem. To solve that problem, we approximated areas by rectangles; that is, we used the Riemann sum approximation. Recall that for f continuous,

$$\int_a^b f(x)\, dx = \lim_{\|P\| \to 0} \sum_{k=1}^n f(c_k)\, \Delta x_k$$

where $P = [x_0, x_1, \ldots, x_n]$, $n = 1, 2, \ldots$, is a sequence of partitions of $[a, b]$ with $x_0 = a$ and $x_n = b$ and $\|P\| \to 0$ as $n \to \infty$. The number c_k is in $[x_{k-1}, x_k]$, and $\Delta x_k = x_k - x_{k-1}$ for $1 \le k \le n$.

In what follows, we will assume that we partition the interval $[a, b]$ into n equal subintervals; that is, each subinterval is of length

$$\Delta x = \frac{b - a}{n}$$

We assume that the function f is continuous on $[a, b]$. We will discuss two methods: the midpoint rule and the trapezoidal rule.

■ 7.5.1 The Midpoint Rule

This is the Riemann sum approximation, where we choose the midpoint of each subinterval for the point c_k. The midpoint of the interval $[x_{k-1}, x_k]$ is

$$c_k = \frac{x_{k-1} + x_k}{2}$$

The rule is defined as follows (see also Figure 7.32):

Midpoint Rule Suppose that $f(x)$ is continuous on $[a, b]$ and that $[x_0, x_1, \ldots, x_n]$ is a partition of $[a, b]$ into n subintervals of equal length. We approximate

$$\int_a^b f(x)\, dx$$

by

$$M_n = \frac{b - a}{n} \sum_{k=1}^n f(c_k)$$

where $c_k = \frac{x_{k-1} + x_k}{2}$ is the midpoint of $[x_{k-1}, x_k]$.

In the next example, we choose an integral that we can evaluate exactly, so that we can see how close the approximation is.

EXAMPLE 1   **Midpoint Rule** Use the midpoint rule with $n = 4$ to approximate

$$\int_0^1 x^2\, dx$$

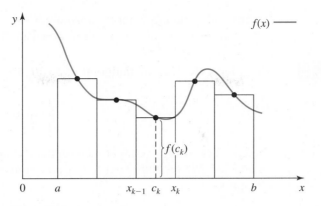

Figure 7.32 The midpoint rule.

Solution The function $f(x) = x^2$ is continuous on [0, 1]. For $n = 4$, we find $\Delta x = \frac{b-a}{4} = \frac{1}{4}$ and we obtain four subintervals, each of length $\frac{1}{4}$ (see Figure 7.33), as given in the following table:

Subinterval $[x_{k-1}, x_k]$	Midpoint c_k	$f(c_k)$	$f(c_k)\,\Delta x$
$\left[0, \dfrac{1}{4}\right]$	$\dfrac{1}{8}$	$\dfrac{1}{64}$	$\dfrac{1}{64}\dfrac{1}{4}$
$\left[\dfrac{1}{4}, \dfrac{1}{2}\right]$	$\dfrac{3}{8}$	$\dfrac{9}{64}$	$\dfrac{9}{64}\dfrac{1}{4}$
$\left[\dfrac{1}{2}, \dfrac{3}{4}\right]$	$\dfrac{5}{8}$	$\dfrac{25}{64}$	$\dfrac{25}{64}\dfrac{1}{4}$
$\left[\dfrac{3}{4}, 1\right]$	$\dfrac{7}{8}$	$\dfrac{49}{64}$	$\dfrac{49}{64}\dfrac{1}{4}$

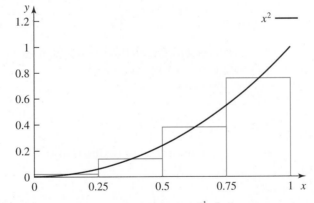

Figure 7.33 The midpoint rule for $\int_0^1 x^2\,dx$ with $n = 4$.

We find the approximation

$$M_4 = \frac{b-a}{4}\sum_{k=1}^{4} f(c_k) = \frac{1}{4}\left(\frac{1}{64} + \frac{9}{64} + \frac{25}{64} + \frac{49}{64}\right) = \frac{1}{4}\frac{84}{64} = \frac{21}{64} \approx 0.3281$$

We know that $\int_0^1 x^2\,dx = \frac{1}{3}$. Hence, the error is

$$\left|\int_0^1 x^2\,dx - M_4\right| \approx 0.0052$$

Larger values of n improve the approximation. ■

Instead of memorizing the formula for the midpoint rule, it is easier to keep a picture in mind. We illustrate this heuristic in the next example.

EXAMPLE 2 **Midpoint Rule** Use the midpoint rule with $n = 5$ to approximate

$$\int_1^2 \frac{1}{x}\, dx$$

Solution The graph of $f(x) = \frac{1}{x}$, together with the five approximating rectangles, is shown in Figure 7.34. We see that $f(x) = 1/x$ is continuous on $[1, 2]$.

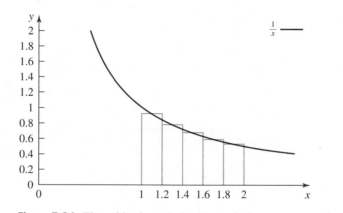

Figure 7.34 The midpoint rule for Example 2.

With $n = 5$, the partition of $[1, 2]$ is given by $P = [1, 1.2, 1.4, 1.6, 1.8, 2]$ and the midpoints are 1.1, 1.3, 1.5, 1.7, and 1.9. Since the width of each rectangle is 0.2 and $f(x) = \frac{1}{x}$, the area of the first rectangle is $(0.2)\frac{1}{1.1}$, the area of the second rectangle is $(0.2)\frac{1}{1.3}$, and so on. We thus find that

$$M_5 = (0.2)\left[\frac{1}{1.1} + \frac{1}{1.3} + \frac{1}{1.5} + \frac{1}{1.7} + \frac{1}{1.9}\right] = 0.6919$$

Note that we factored out 0.2, the width of each rectangle, since it is a common factor of the areas of the five rectangles.

We know that

$$\int_1^2 \frac{1}{x}\, dx = \ln x\Big]_1^2 = \ln 2 - \ln 1 = \ln 2$$

Hence, the error in the approximation is

$$\left|\int_1^2 \frac{1}{x}\, dx - M_5\right| = |\ln 2 - 0.6919| = 0.0012 \qquad \blacksquare$$

We typically use an approximation when we cannot evaluate the integral exactly. Thus, we cannot use the exact value to determine how close the approximation is. Fortunately, there are results that allow us to obtain upper bounds for the error:

Error Bound for the Midpoint Rule Suppose that $|f''(x)| \le K$ for all $x \in [a, b]$. Then the error in the midpoint rule is at most

$$\left|\int_a^b f(x)\, dx - M_n\right| \le K\frac{(b - a)^3}{24n^2}$$

Let's check this bound for our two examples. In the first example, $f(x) = x^2$, and therefore $f''(x) = 2$. Hence, with $n = 4$, the error is at most

$$2\frac{(1-0)^3}{24(4^2)} \approx 0.0052$$

This is in fact the error that we obtained.

In the second example, $f(x) = \frac{1}{x}$. Since $f'(x) = -\frac{1}{x^2}$ and $f''(x) = \frac{2}{x^3}$, it follows that

$$|f''(x)| = \left|\frac{2}{x^3}\right| \leq 2 \quad \text{for } 1 \leq x \leq 2$$

Hence, with $n = 5$, the error is at most

$$2\frac{(2-1)^3}{24(5^2)} = 0.0033$$

The actual error was in fact smaller, only 0.0012.

The actual error can be quite a bit smaller than the theoretical error bound, which is the worst-case scenario, but it will never be larger. The advantage of such an error bound is that it allows us to find the number of subintervals required to obtain a certain accuracy. For instance, if we want to numerically approximate $\int_0^1 x^2\, dx$ so that the error is at most 10^{-4}, then we must choose n so that

$$K\frac{(b-a)^3}{24n^2} \leq 10^{-4}$$

$$2\frac{1}{24n^2} \leq 10^{-4}$$

$$\frac{1}{12}10^4 \leq n^2$$

$$28.9 \leq n$$

That is, $n = 29$ would suffice to produce an error of at most 10^{-4}.

Finding a value for K in the estimate is not always easy. A graph of $f''(x)$ over the interval of interest can facilitate finding a bound on the second derivative. We need not find the best possible bound. For instance, if we wanted to integrate $f(x) = e^x$ over the interval $[1, 2]$, we would need to find a bound on $f''(x) = e^x$ over the interval $[1, 2]$. Since $|e^x| \leq e^2$ over that interval, we could use, for instance, $K = 9$. (See Figure 7.35.) This is not the best possible bound, but it is a number that we can find without using a calculator. (The best possible bound would be $K = e^2$.)

Figure 7.35 An upper bound on $|e^x|$ over [1, 2] is 9.

■ 7.5.2 The Trapezoidal Rule

In this method, we use trapezoids instead of rectangles to approximate integrals, as illustrated in Figure 7.36. We assume again that f is a continuous function on $[a, b]$ and divide $[a, b]$ into n equal subintervals. But this time we approximate the function $f(x)$ by a polygon $P(x)$. To obtain the polygon $P(x)$, we connect the points $(x_k, f(x_k))$, $k = 0, 1, 2, \ldots, n$, by straight lines, as shown in the figure. The integral $\int_a^b P(x)\, dx$ is then the approximation to $\int_a^b f(x)\, dx$.

We see from Figure 7.36 that all this amounts to adding up (signed) areas of trapezoids. Recall from planar geometry that the area of the trapezoid in Figure 7.37 is

$$A = d\frac{h_1 + h_2}{2}$$

The width of each trapezoid in Figure 7.36 is $d = \frac{b-a}{n}$. Adding up the areas of the

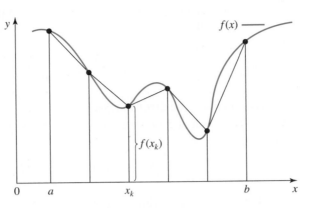

Figure 7.36 The trapezoidal rule.

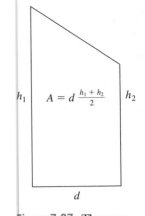

Figure 7.37 The area of a trapezoid.

trapezoids then yields

$$T_n = \frac{b-a}{n}\left[\frac{f(x_0) + f(x_1)}{2} + \frac{f(x_1) + f(x_2)}{2} + \frac{f(x_2) + f(x_3)}{2}\right.$$

$$\left. + \cdots + \frac{f(x_{n-2}) + f(x_{n-1})}{2} + \frac{f(x_{n-1}) + f(x_n)}{2}\right]$$

$$= \frac{b-a}{n}\left[\frac{f(x_0)}{2} + f(x_1) + f(x_2) + \cdots + f(x_{n-1}) + \frac{f(x_n)}{2}\right]$$

> **Trapezoidal Rule** Suppose that $f(x)$ is continuous on $[a, b]$ and that $P = [x_0, x_1, x_2, \ldots, x_n]$ is a partition of $[a, b]$ into n subintervals of equal length. Then we can approximate
>
> $$\int_a^b f(x)\, dx$$
>
> by
>
> $$T_n = \frac{b-a}{n}\left[\frac{f(x_0)}{2} + f(x_1) + f(x_2) + \cdots + f(x_{n-1}) + \frac{f(x_n)}{2}\right]$$

EXAMPLE 3 **Trapezoidal Rule** Use the trapezoidal rule with $n = 4$ to approximate

$$\int_0^1 x^2\, dx$$

Solution The function $f(x) = x^2$ is continuous on $[0, 1]$. As in Example 1, there are four subintervals, each of length $\frac{1}{4}$. (See Figure 7.38.) We find the following:

k	x_k	$f(x_k)$
0	0	0
1	$\frac{1}{4}$	$\frac{1}{16}$
2	$\frac{1}{2}$	$\frac{1}{4}$
3	$\frac{3}{4}$	$\frac{9}{16}$
4	1	1

The approximation is

$$T_4 = \frac{1}{4}\left[\frac{0}{2} + \frac{1}{16} + \frac{1}{4} + \frac{9}{16} + \frac{1}{2}\right] = 0.34375$$

Since we know that $\int_0^1 x^2\,dx = \frac{1}{3}$, we can compute the error:

$$\left|\int_0^1 x^2\,dx - T_4\right| = 0.0104$$

Note that because $y = x^2$ is concave up, the line segments of the polygon are above the curve $y = x^2$. The area of the trapezoid therefore exceeds the area under the curve $y = x^2$ in each subinterval, and the trapezoidal approximation overestimates the integral. ∎

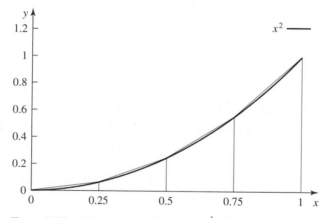

Figure 7.38 The trapezoidal rule for $\int_0^1 x^2\,dx$ with $n = 4$.

EXAMPLE 4   **Trapezoidal Rule** Use the trapezoidal rule with $n = 5$ to approximate

$$\int_1^2 \frac{1}{x}\,dx$$

Solution   The situation is illustrated in Figure 7.39. The function $1/x$ is continuous on $[1, 2]$. With $n = 5$, the partition of $[1, 2]$ is given by $P = [1.0, 1.2, 1.4, 1.6, 1.8, 2.0]$. The base of each trapezoid has length 0.2. Hence,

$$T_5 = (0.2)\left[\frac{1}{2}\cdot\frac{1}{1.0} + \frac{1}{1.2} + \frac{1}{1.4} + \frac{1}{1.6} + \frac{1}{1.8} + \frac{1}{2}\cdot\frac{1}{2.0}\right] = 0.69563$$

Since we know from Example 2 that $\int_1^2 \frac{1}{x}\,dx = \ln 2$, we can compute the error:

$$\left|\int_1^2 \frac{1}{x}\,dx - T_5\right| = 0.00249$$ ∎

There is also a theoretical error bound for the trapezoidal rule:

Error Bound for the Trapezoidal Rule Suppose that $|f''(x)| \le K$ for all $x \in [a, b]$. Then the error in the trapezoidal rule is at most

$$\left|\int_a^b f(x)\,dx - T_n\right| \le K\frac{(b-a)^3}{12n^2}$$

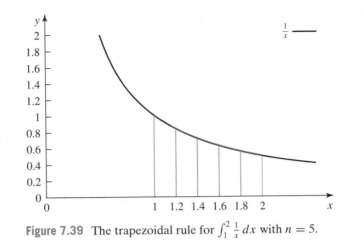

Figure 7.39 The trapezoidal rule for $\int_1^2 \frac{1}{x}\,dx$ with $n = 5$.

In Example 3, since $f(x) = x^2$, it follows that $f''(x) = 2$ and hence $K = 2$. The error is therefore bounded by

$$2\frac{1}{12(4^2)} = 0.0104$$

which is the same as the actual error.

In Example 4, since $f(x) = 1/x$, we have $|f''(x)| = 2/x^3 \le 2$ for $1 \le x \le 2$ (as in Example 2). Hence, with $n = 5$, the error bound is at most

$$2\frac{(2-1)^3}{12(5^2)} = 0.0067$$

The actual error was in fact smaller, only 0.00249. As with the midpoint rule, the theoretical error can be quite a bit larger than the actual error.

Section 7.5 Problems

■ **7.5.1, 7.5.2**

In Problems 1–4, use the midpoint rule to approximate each integral with the specified value of n.

1. $\int_1^2 x^2\,dx, n = 4$ **2.** $\int_{-1}^0 (x+1)^3\,dx, n = 5$

3. $\int_0^1 e^{-x}\,dx, n = 3$ **4.** $\int_0^{\pi/2} \sin x\,dx, n = 4$

In Problems 5–8, use the midpoint rule to approximate each integral with the specified value of n. Compare your approximation with the exact value.

5. $\int_2^4 \frac{1}{x}\,dx, n = 4$ **6.** $\int_{-1}^1 (e^{2x} - 1)\,dx, n = 4$

7. $\int_0^4 \sqrt{x}\,dx, n = 4$ **8.** $\int_2^4 \frac{2}{\sqrt{x}}\,dx, n = 5$

In Problems 9–12, use the trapezoidal rule to approximate each integral with the specified value of n.

9. $\int_1^2 x^2\,dx, n = 4$ **10.** $\int_{-1}^0 x^3\,dx, n = 5$

11. $\int_0^1 e^{-x}\,dx, n = 3$ **12.** $\int_0^{\pi/2} \sin x\,dx, n = 4$

In Problems 13–16, use the trapezoidal rule to approximate each integral with the specified value of n. Compare your approximation with the exact value.

13. $\int_1^3 x^3\,dx, n = 5$ **14.** $\int_{-1}^1 (1 - e^{-x})\,dx, n = 4$

15. $\int_0^2 \sqrt{x}\,dx, n = 4$ **16.** $\int_1^2 \frac{1}{x}\,dx, n = 5$

17. How large should n be so that the midpoint rule approximation of

$$\int_0^2 x^2\,dx$$

is accurate to within 10^{-4}?

In Problems 18–24, use the theoretical error bound to determine how large n should be. [Hint: In each case, find the second derivative of the integrand, graph it, and use a graphing calculator to find an upper bound on $|f''(x)|$.]

18. How large should n be so that the midpoint rule approximation of

$$\int_1^2 \frac{1}{x}\,dx$$

is accurate to within 10^{-3}?

19. How large should n be so that the midpoint rule approximation of

$$\int_0^2 e^{-x^2/2}\,dx$$

is accurate to within 10^{-4}?

20. How large should n be so that the midpoint rule approximation of

$$\int_2^8 \frac{1}{\ln t}\,dt$$

is accurate to within 10^{-3}?

21. How large should n be so that the trapezoidal rule approximation of

$$\int_0^1 e^{-x}\,dx$$

is accurate to within 10^{-5}?

22. How large should n be so that the trapezoidal rule approximation of

$$\int_0^2 \sin x\,dx$$

is accurate to within 10^{-4}?

23. How large should n be so that the trapezoidal rule approximation of

$$\int_1^2 \frac{e^t}{t}\,dt$$

is accurate to within 10^{-4}?

24. How large should n be so that the trapezoidal rule approximation of

$$\int_1^2 \frac{\cos x}{x}\,dx$$

is accurate to within 10^{-3}?

25. (a) Show graphically that, for $n = 5$, the trapezoidal rule overestimates, and the midpoint rule underestimates,

$$\int_0^1 x^3\,dx$$

In each case, compute the approximate value of the integral and compare it with the exact value.

(b) The result in (a) has to do with the fact that $y = x^3$ is concave up on $[0, 1]$. To generalize that result to functions with this concavity property, we assume that the function $f(x)$ is continuous, nonnegative, and concave up on the interval $[a, b]$. Denote by M_n the midpoint rule approximation, and by T_n the trapezoidal rule approximation, of $\int_a^b f(x)\,dx$. Explain in words why

$$M_n \le \int_a^b f(x)\,dx \le T_n$$

(c) If we assume that $f(x)$ is continuous, nonnegative, and concave *down* on $[a, b]$, then

$$M_n \ge \int_a^b f(x)\,dx \ge T_n$$

Explain why this is so. Use this result to give an upper and a lower bound on

$$\int_0^1 \sqrt{x}\,dx$$

when $n = 4$ in the approximation.

■ 7.6 The Taylor Approximation

In many ways, polynomials are the easiest functions to work with. Therefore, in this section we will learn how to approximate functions by polynomials. We will see that the approximation typically improves when we use higher-degree polynomials.

■ 7.6.1 Taylor Polynomials

In Section 4.8, we discussed how to linearize a function about a given point. This discussion led to the linear, or tangent, approximation. We found the following:

The linear approximation of $f(x)$ at $x = a$ is

$$L(x) = f(a) + f'(a)(x - a)$$

As an example, we look at

$$f(x) = e^x$$

and approximate this function by its linearization at $x = 0$. We find that

$$L(x) = f(0) + f'(0)x = 1 + x \qquad (7.19)$$

since $f'(x) = e^x$ and $f(0) = f'(0) = 1$. To see how close the approximation is, we graph both $f(x)$ and $L(x)$ in the same coordinate system. (The result is shown in Figure 7.40.) The approximation is quite good as long as x is close to 0. The figure suggests that it gets gradually worse as we move away from 0. In the approximation, we required only that $f(x)$ and $L(x)$ have in common $f(0) = L(0)$ and $f'(0) = L'(0)$.

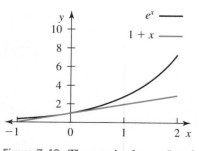

Figure 7.40 The graph of $y = e^x$ and its linear approximation at 0.

To improve the approximation, we may wish to use an approximating function whose higher-order derivatives also agree with those of $f(x)$ at $x = 0$. The function $L(x)$ is a polynomial of degree 1. To improve the approximation, we will continue to work with polynomials, but require that the function and its first n derivatives at $x = 0$ agree with those of the polynomial. To be able to match up the first n derivatives, the polynomial must be of degree n. (If the degree of the polynomial is $k < n$, then all derivatives of degree $k + 1$ or higher are equal to 0.) A polynomial of degree n can be written as

$$P_n(x) = a_0 + a_1 x + a_2 x^2 + \cdots + a_n x^n \tag{7.20}$$

If we want to approximate $f(x)$ at $x = 0$, then we require that

$$
\begin{aligned}
f(0) &= P_n(0) \\
f'(0) &= P_n'(0) \\
f''(0) &= P_n''(0) \\
&\vdots \\
f^{(n)}(0) &= P_n^{(n)}(0)
\end{aligned}
\tag{7.21}
$$

Now,

$$
\begin{aligned}
P_n(0) &= a_0 + a_1 x + \cdots + a_n x^n \big|_{x=0} = a_0 \\
P_n'(0) &= a_1 + 2a_2 x + 3a_3 x^2 + \cdots + na_n x^{n-1} \big|_{x=0} = a_1 \\
P_n''(0) &= 2a_2 + (3)(2)a_3 x + \cdots + n(n-1)a_n x^{n-2} \big|_{x=0} = 2a_2 \\
P_n'''(0) &= (3)(2)a_3 + (4)(3)(2)a_4 x + \cdots + n(n-1)(n-2)a_n x^{n-3} \big|_{x=0} \\
&= (3)(2)a_3 \\
&\vdots \\
P_n^{(n)}(0) &= n(n-1)(n-2) \cdots (3)(2)(1)a_n \big|_{x=0} \\
&= n(n-1)(n-2) \cdots (3)(2)(1)a_n
\end{aligned}
$$

We introduce the notation

$$k! = k(k-1)(k-2) \cdots (3)(2)(1)$$

where $k!$ is read "k factorial." Solving these equations for a_k, $k = 0, 1, 2, \ldots, n$, and using $f^{(k)}(0) = P^{(k)}(0)$, $k = 0, 1, 2, \ldots, n$, we find that

$$
\begin{aligned}
a_0 &= P_n(0) = f(0) \\
a_1 &= P_n'(0) = f'(0) \\
a_2 &= \frac{1}{2} P_n''(0) = \frac{1}{2!} f''(0) \\
a_3 &= \frac{1}{3 \cdot 2} P_n'''(0) = \frac{1}{3!} f'''(0) \\
&\vdots \\
a_n &= \frac{1}{n(n-1) \cdots 2 \cdot 1} P_n^{(n)}(0) = \frac{1}{n!} f^{(n)}(0)
\end{aligned}
\tag{7.22}
$$

A polynomial of degree n of the form (7.20) and whose coefficients satisfy (7.22) is called a **Taylor polynomial** of degree n. We summarize this definition as follows:

> **Definition** The Taylor polynomial of degree n about $x = 0$ for the function $f(x)$ is given by
>
> $$P_n(x) = f(0) + f'(0)x + \frac{f''(0)}{2!}x^2 + \frac{f'''(0)}{3!}x^3$$
> $$+ \frac{f^{(4)}(0)}{4!}x^4 + \cdots + \frac{f^{(n)}(0)}{n!}x^n$$

EXAMPLE 1

Compute the Taylor polynomial of degree 3 about $x = 0$ for the function $f(x) = e^x$.

Solution

To find the Taylor polynomial of degree 3, we need the first three derivatives of $f(x)$ at $x = 0$. We have

$$f(x) = e^x, \qquad \text{so } f(0) = 1$$
$$f'(x) = e^x, \qquad \text{so } f'(0) = 1$$
$$f''(x) = e^x, \qquad \text{so } f''(0) = 1$$
$$f'''(x) = e^x, \qquad \text{so } f'''(0) = 1$$

Therefore,

$$P_3(x) = 1 + x + \frac{x^2}{2!} + \frac{x^3}{3!} = 1 + x + \frac{x^2}{2} + \frac{x^3}{6}$$

since $2! = (2)(1) = 2$ and $3! = (3)(2)(1) = 6$.

Our claim was that this polynomial would provide a better approximation to e^x than the linearization $1 + x$. We check this claim by evaluating e^x, $L(x)$, and $P_3(x)$ at a few values. The results are summarized in the following table:

x	e^x	$1 + x$	$1 + x + \frac{x^2}{2} + \frac{x^3}{6}$
-1	0.36788	0	0.3333
-0.1	0.90484	0.9	0.9048
0	1	1	1.0000
0.1	1.1052	1.1	1.1052
1	2.7183	2	2.6667

We see from the table that the third-degree Taylor polynomial provides a better approximation. Indeed, for x sufficiently close to 0, the values of $f(x)$ and $P_3(x)$ are very close. For instance,

$$f(0.1) = 1.105170918 \qquad \text{and} \qquad P_3(0.1) = 1.105166667$$

That is, their first five digits are identical. The error of approximation is

$$|f(0.1) - P_3(0.1)| = 4.25 \times 10^{-6}$$

which is quite small.

In Figure 7.41, we display the graphs of $f(x)$ and the Taylor polynomials $P_1(x)$, $P_2(x)$, and $P_3(x)$. We see from the graphs that the approximation is good only as long as x is close to 0. We also see that increasing the degree of the Taylor polynomial improves the approximation. ■

When we look at Example 1, we find that the successive Taylor polynomials for $f(x) = e^x$ about $x = 0$ are

$$P_0(x) = 1$$
$$P_1(x) = 1 + x$$
$$P_2(x) = 1 + x + \frac{x^2}{2!}$$
$$P_3(x) = 1 + x + \frac{x^2}{2!} + \frac{x^3}{3!}$$

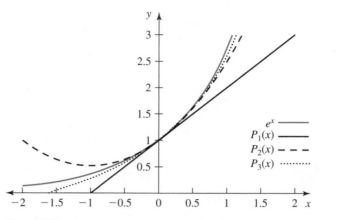

Figure 7.41 The graph of $y = e^x$ and the first three Taylor polynomials.

The first thing we notice is that $P_1(x)$ is the linear approximation $L(x)$ that we found in (7.19).

The next thing we notice is that there is a pattern, and we might be tempted to guess the form of $P_n(x)$ for an arbitrary n. Our guess would be

$$P_n(x) = 1 + x + \frac{x^2}{2!} + \frac{x^3}{3!} + \cdots + \frac{x^n}{n!}$$

and this is indeed the case.

You might wonder why we bother to find an approximation for the function $f(x) = e^x$. To compute $f(1) = e$, for instance, it seems a lot easier simply to use a calculator. However, since $f(x) = e^x$ is not an algebraic function, the values of $f(x)$ cannot be found exactly with only the basic algebraic operations. Taylor polynomials are one way to evaluate such functions on a computer.

We now give additional functions for which we can find the Taylor polynomial of degree n.

EXAMPLE 2

Compute the Taylor polynomial of degree n about $x = 0$ for the function $f(x) = \sin x$.

Solution

We begin by computing successive derivatives of $f(x) = \sin x$ at $x = 0$:

$$
\begin{aligned}
f(x) &= \sin x && \text{and} && f(0) = 0 \\
f'(x) &= \cos x && \text{and} && f'(0) = 1 \\
f''(x) &= -\sin x && \text{and} && f''(0) = 0 \\
f'''(x) &= -\cos x && \text{and} && f'''(0) = -1 \\
f^{(4)}(x) &= \sin x && \text{and} && f^{(4)}(0) = 0
\end{aligned}
$$

Since $f^{(4)}(x) = f(x)$, we find that $f^{(5)}(x) = f'(x)$, $f^{(6)}(x) = f''(x)$, and so on. We also conclude that all even derivatives are equal to 0 at $x = 0$ and that the odd derivatives alternate between 1 and -1 at $x = 0$. We find that

$$P_1(x) = P_2(x) = x$$

$$P_3(x) = P_4(x) = x - \frac{x^3}{3!}$$

$$P_5(x) = P_6(x) = x - \frac{x^3}{3!} + \frac{x^5}{5!}$$

$$P_7(x) = P_8(x) = x - \frac{x^3}{3!} + \frac{x^5}{5!} - \frac{x^7}{7!}$$

and so on. To find the Taylor polynomial of degree n, we must find out how to write the last term. Note that the sign in front of successive terms alternates between plus and minus. To account for this alternating sign, we introduce the factor

$$(-1)^n = \begin{cases} 1 & \text{if } n \text{ is even} \\ -1 & \text{if } n \text{ is odd} \end{cases}$$

An odd number can be written as $2n + 1$ for any integer n. For a term of the form $\pm \frac{x^k}{k!}$ with k odd, we write

$$(-1)^n \frac{x^{2n+1}}{(2n+1)!} \tag{7.23}$$

where n is an integer. Inserting successive values of n into (7.23), we find the following:

n	$(-1)^n \dfrac{x^{2n+1}}{(2n+1)!}$
0	x
1	$-\dfrac{x^3}{3!}$
2	$\dfrac{x^5}{5!}$
3	$-\dfrac{x^7}{7!}$

We see from the table that the term (7.23) produces successive terms in the Taylor polynomial for $f(x) = \sin x$. The Taylor polynomial of degree $2n + 1$ is thus

$$P_{2n+1}(x) = x - \frac{x^3}{3!} + \frac{x^5}{5!} - \frac{x^7}{7!} + \cdots + (-1)^n \frac{x^{2n+1}}{(2n+1)!} \qquad ■$$

EXAMPLE 3

Compute the Taylor polynomial of degree n about $x = 0$ for the function $f(x) = \frac{1}{1-x}, x \neq 1$.

Solution

We begin by computing successive derivatives of $f(x) = \frac{1}{1-x}$ at $x = 0$.

$$f(x) = \frac{1}{1-x}, \qquad \text{so } f(0) = 1$$

$$f'(x) = \frac{1}{(1-x)^2}, \qquad \text{so } f'(0) = 1$$

$$f''(x) = \frac{2}{(1-x)^3}, \qquad \text{so } f''(0) = 2 = 2!$$

$$f'''(x) = \frac{(2)(3)}{(1-x)^4}, \qquad \text{so } f'''(0) = (2)(3) = 3!$$

$$f^{(4)}(x) = \frac{(2)(3)(4)}{(1-x)^5}, \qquad \text{so } f^{(4)}(0) = (2)(3)(4) = 4!$$

and so on. Continuing in this way, we find that

$$f^{(k)}(x) = \frac{(2)(3)(4)\cdots(k)}{(1-x)^{k+1}} \quad \text{so } f^{(k)}(0) = k!$$

For the Taylor polynomial of degree n about $x = 0$, we obtain

$$P_n(x) = 1 + x + \frac{2!}{2!}x^2 + \frac{3!}{3!}x^3 + \frac{4!}{4!}x^4 + \frac{5!}{5!}x^5 + \cdots + \frac{n!}{n!}x^n$$

$$= 1 + x + x^2 + x^3 + \cdots + x^n \qquad ■$$

Taylor approximations are widely used in biology. Here is an example that is already familiar to us.

EXAMPLE 4

Denote the size of a population at time t by $N(t)$. A general model that describes the dynamics of this population is given by

$$\frac{dN}{dt} = f(N) \qquad \text{with } f(0) = 0$$

Find the linear and the quadratic approximation of $f(N)$ about $N = 0$.

Solution

The linear approximation of $f(N)$ about $N = 0$ is the Taylor polynomial of degree 1:

$$P_1(N) = \underbrace{f(0)}_{=0} + f'(0)N$$

If we set $r = f'(0)$, then the first-order approximation of this growth model is

$$\frac{dN}{dt} = rN$$

which is the equation that describes exponential growth.

The quadratic approximation of $f(N)$ about $N = 0$ is the Taylor polynomial of degree 2:

$$P_2(N) = \underbrace{f(0)}_{=0} + f'(0)N + \frac{f''(0)}{2}N^2$$

Factoring $f'(0)N$ yields

$$P_2(N) = f'(0)N \left[1 + \frac{f''(0)}{2f'(0)}N \right]$$

If we set $r = f'(0)$ and $K = -\frac{2f'(0)}{f''(0)}$, then the second-order approximation of the growth model is

$$\frac{dN}{dt} = rN \left(1 - \frac{N}{K} \right)$$

which is the equation that describes logistic growth if K and r are positive. In either approximation, $r = f'(0)$ is the intrinsic rate of growth. ■

■ **7.6.2 The Taylor Polynomial about $x = a$**

Thus far, we have considered Taylor polynomials about $x = 0$. Because Taylor polynomials typically are good approximations only close to the point of approximation, it is useful to have approximations about points other than $x = 0$. We have already done this for linear approximations. For instance, the tangent-line approximation of $f(x)$ at $x = a$ is

$$L(x) = f(a) + f'(a)(x - a) \qquad (7.24)$$

Note that $L(a) = f(a)$ and $L'(a) = f'(a)$. That is, the linear approximation and the original function, together with their first derivatives, agree at $x = a$. If we want to approximate $f(x)$ at $x = a$ by a polynomial of degree n, we then require that the polynomial and the original function, together with their first n derivatives, agree at $x = a$. This leads us to a polynomial of the form

$$P_n(x) = c_0 + c_1(x - a) + c_2(x - a)^2 + \cdots + c_n(x - a)^n \qquad (7.25)$$

Comparing (7.25) and (7.24), we conclude that $c_0 = f(a)$ and $c_1 = f'(a)$. To find the remaining coefficients, we proceed as in the case $a = 0$. That is, we differentiate $f(x)$ and $P_n(x)$ and require that their first n derivatives agree at $x = a$. We then arrive at the following formula:

The Taylor polynomial of degree n about $x = a$ for the function $f(x)$ is given by

$$P_n(x) = f(a) + f'(a)(x - a) + \frac{f''(a)}{2!}(x - a)^2$$
$$+ \frac{f'''(a)}{3!}(x - a)^3 + \cdots + \frac{f^{(n)}(a)}{n!}(x - a)^n$$

EXAMPLE 5   Find the Taylor polynomial of degree 3 for

$$f(x) = \ln x$$

at $x = 1$.

Solution   We need to evaluate $f(x)$ and its first three derivatives at $x = 1$. We find that

$$f(x) = \ln x, \qquad \text{so } f(1) = 0$$
$$f'(x) = \frac{1}{x}, \qquad \text{so } f'(1) = 1$$
$$f''(x) = -\frac{1}{x^2}, \qquad \text{so } f''(1) = -1$$
$$f'''(x) = \frac{2}{x^3}, \qquad \text{so } f'''(1) = 2$$

Using the definition of the Taylor polynomial, we get

$$P_3(x) = 0 + (1)(x - 1) + \frac{(-1)}{2!}(x - 1)^2 + \frac{2}{3!}(x - 1)^3$$
$$= (x - 1) - \frac{1}{2}(x - 1)^2 + \frac{1}{3}(x - 1)^3$$

Figure 7.42 shows $f(x)$, the linear approximation $P_1(x) = x - 1$, and $P_3(x)$. We see that the approximation is good when x is close to 1 and that the approximation $P_3(x)$ is better than the linear approximation.   ■

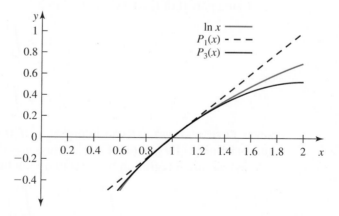

Figure 7.42 The graph of $y = \ln x$, the linear approximation, and the Taylor polynomial of degree 3.

■ 7.6.3 How Accurate Is the Approximation? (Optional)

We saw in Example 1 that the approximation improved when the degree of the polynomial was higher. We will now investigate how accurate the Taylor approximation is. We can assess the accuracy of the approximation directly for the function in Example 3.

In Example 3, we showed that the Taylor polynomial of degree n about $x = 0$ for $f(x) = \frac{1}{1-x}, x \neq 1$, is

$$P_n(x) = 1 + x + x^2 + x^3 + \cdots + x^{n-1} + x^n \tag{7.26}$$

There is a nice "trick" we may use to find an expression for the sum (7.26). Note that

$$x P_n(x) = x + x^2 + x^3 + \cdots + x^n + x^{n+1} \tag{7.27}$$

Subtracting (7.27) from (7.26), we find that

$$P_n(x) - x P_n(x) = 1 + x + x^2 + \cdots + x^n - x - x^2 - \cdots - x^n - x^{n+1}$$
$$= 1 - x^{n+1}$$

That is,

$$(1 - x) P_n(x) = 1 - x^{n+1}$$

or

$$P_n(x) = \frac{1 - x^{n+1}}{1 - x} = \frac{1}{1 - x} - \frac{x^{n+1}}{1 - x}$$

provided that $x \neq 1$. We therefore conclude that

$$|f(x) - P_n(x)| = \left| \frac{x^{n+1}}{1 - x} \right|$$

We can interpret the term $x^{n+1}/(1-x)$ as the error of approximation. Since

$$\lim_{n \to \infty} \left| \frac{x^{n+1}}{1-x} \right| = \begin{cases} \infty & \text{if } |x| > 1 \\ 0 & \text{if } |x| < 1 \end{cases}$$

it follows that the error of approximation can be made small only when $|x| < 1$. For $|x| > 1$, the error of approximation increases with increasing n. [When $x = 1$, the function $f(x)$ is not defined.]

In general, it is not straightforward to obtain error estimates. In its general form, the error is given as an integral. Let's first look at the error terms $P_0(x)$ and $P_1(x)$ before stating the error term for arbitrary n.

Using part II of the FTC, we find that

$$f(x) - f(a) = \int_a^x f'(t)\, dt$$

or

$$f(x) = f(a) + \int_a^x f'(t)\, dt$$

Since $f(a) = P_0(x)$, we can interpret $\int_a^x f'(t)\, dt$ as the error term in the Taylor approximation of $f(x)$ about $x = a$ when $n = 0$.

We can use integration by parts to obtain the next-higher approximation:

$$\int_a^x f'(t)\, dt = \int_a^x 1 \cdot f'(t)\, dt$$

We set $u' = 1$ with $u = -(x-t)$ and $v = f'(t)$ with $v' = f''(t)$. [Writing $u = -(x-t)$ turns out to be a more convenient antiderivative of $u' = 1$ than $u = t$, as you will see shortly.] We obtain

$$\int_a^x f'(t)\, dt = -(x-t)f'(t)\Big]_a^x + \int_a^x (x-t)f''(t)\, dt$$
$$= (x-a)f'(a) + \int_a^x (x-t)f''(t)\, dt$$

That is,

$$f(x) = f(a) + f'(a)(x-a) + \int_a^x (x-t)f''(t)\,dt$$

The expression $f(a) + f'(a)(x-a)$ is the linear approximation $P_1(x)$; the integral can then be considered as the error term.

Continuing in this way, we find the general formula:

Taylor's Formula Suppose that $f : I \to \mathbf{R}$, where I is an interval, $a \in I$, and f and its first $n+1$ derivatives are continuous at $a \in I$. Then, for $x, a \in I$,

$$f(x) = f(a) + \frac{f'(a)}{1!}(x-a) + \frac{f''(a)}{2!}(x-a)^2$$
$$+ \cdots + \frac{f^{(n)}(a)}{n!}(x-a)^n + R_{n+1}(x)$$

where

$$R_{n+1}(x) = \frac{1}{n!}\int_a^x (x-t)^n f^{(n+1)}(t)\,dt$$

We will now examine the error term in Taylor's formula more closely. The error term is given in integral form, and it is often difficult (or impossible) to evaluate the integral. We will first look at the case $n = 0$; that is, we approximate $f(x)$ by the constant function $f(a)$. The error term is then $R_{n+1}(x)$ when $n = 0$; that is,

$$R_1(x) = \int_a^x f'(t)\,dt$$

Using the MVT for integrals, we can find a value c in the interval between a and x such that

$$\int_a^x f'(t)\,dt = f'(c)(x-a)$$

That is, we find

$$R_1(x) = f'(c)(x-a)$$

for some number c between a and x. Although we don't know the value of c, this form is quite useful, for if we set

$$K = \left[\begin{array}{c} \text{largest value of } |f'(t)| \\ \text{for } t \text{ between } a \text{ and } x \end{array} \right]$$

then

$$|R_1(x)| \le K|x-a|$$

Before we give the corresponding results for $R_{n+1}(x)$, we look at one example that illustrates how to find K.

EXAMPLE 6

Estimate the error in the approximation of $f(x) = e^x$ by $P_0(x)$ about $x = 0$ on the interval $[0, 1]$.

Solution

Since $f(0) = 1$, it follows that

$$P_0(x) = 1$$

and

$$f(x) = 1 + R_1(x)$$

with

$$R_1(x) = \int_0^x f'(t)\,dt = xf'(c)$$

for some c between 0 and x. Because $f'(t) = e^t$, the largest value of $|f'(t)|$ in the interval $[0, 1]$, namely, $|f'(1)| = e$, occurs when $t = 1$. Since we want to find an approximation of $f(x) = e^x$ for $x \in [0, 1]$, we should not use e in our error estimate, as e is one of the values we want to estimate. Instead, we use $|f'(t)| \leq 3$ for $t \in [0, 1]$. We thus have

$$|R_1(x)| \leq 3x \quad \text{for } x \in [0, 1]$$

■

The error term for general n can be dealt with in a similar fashion, so that we find the following:

There exists a c between a and x such that the error term in Taylor's formula is of the form

$$R_{n+1}(x) = \frac{f^{(n+1)}(c)}{(n + 1)!}(x - a)^{n+1}$$

As in the case where $n = 0$, this form of the error term is quite useful. Although we don't know what c is, we can try to estimate $f^{(n+1)}(c)$ between a and x as before, when $n = 0$. Let

$$K = \begin{bmatrix} \text{largest value of } |f^{(n+1)}(t)| \\ \text{for } t \text{ between } a \text{ and } x \end{bmatrix}$$

Then

$$|R_{n+1}(x)| \leq \frac{K|x - a|^{n+1}}{(n + 1)!}$$

We will use this inequality in the next example to determine in advance what degree of Taylor polynomial will allow us to achieve a given accuracy.

EXAMPLE 7   Suppose that $f(x) = e^x$. What degree of Taylor polynomial about $x = 0$ will allow us to approximate $f(1)$ so that the error is less than 10^{-5}?

Solution   In Example 1, we found that, for any $n \geq 1$,

$$f^{(n+1)}(t) = e^t$$

We need to find out how large $f^{(n+1)}(t)$ can get for $t \in [0, 1]$. We obtain

$$|f^{(n+1)}(t)| = e^t \leq e \quad \text{for } 0 \leq t \leq 1$$

As in Example 6, instead of using e as a bound, we use a slightly larger value, namely, 3. Therefore,

$$|R_{n+1}(1)| \leq \frac{3|1|^{n+1}}{(n + 1)!} = \frac{3}{(n + 1)!} \tag{7.28}$$

We want the error to be less than 10^{-5}; that is, we want

$$|R_{n+1}(1)| < 10^{-5}$$

Inserting different values of n shows that

$$\frac{3}{8!} = 7.44 \times 10^{-5} \quad \text{and} \quad \frac{3}{9!} = 8.27 \times 10^{-6}$$

That is, when $n = 8$,

$$|R_{n+1}(1)| = |R_9(1)| \leq 8.27 \times 10^{-6} < 10^{-5}$$

Because the estimate of the error is greater than 10^{-5} when $n = 7$, we conclude that a polynomial of degree 8 would certainly give us the desired accuracy, whereas a polynomial of degree 7 might not. We can easily check this; we find that

$$1 + 1 + \frac{1}{2!} + \frac{1}{3!} + \frac{1}{4!} + \cdots + \frac{1}{7!} = 2.71825396825$$

$$1 + 1 + \frac{1}{2!} + \frac{1}{3!} + \frac{1}{4!} + \cdots + \frac{1}{8!} = 2.71827876984$$

Comparing these with $e = 2.71828182846\ldots$, we see that the error is equal to 2.79×10^{-5} when $n = 7$ and 3.06×10^{-6} when $n = 8$. The error that we computed with (7.28) is a worst-case scenario; that is, the true error can be (and typically is) smaller than the error bound. ■

We have already seen one example in which a Taylor polynomial was useful only for values close to the point at which we approximated the function, regardless of n, the degree of the polynomial. In some situations, the error in the approximation cannot be made small for *any* value close to the point of approximation, regardless of n. One such example is the continuous function

$$f(x) = \begin{cases} e^{-1/x} & \text{for } x > 0 \\ 0 & \text{for } x \le 0 \end{cases}$$

which is used, for instance, to describe the height of a tree as a function of age. We can show that $f^{(k)}(0) = 0$ for *all* $k \ge 1$, which implies that a Taylor polynomial of degree n about $x = 0$ is

$$P_n(x) = 0$$

for all n. This example clearly shows that it will not help to increase n; the approximation just will not improve.

When we use Taylor polynomials to approximate functions, it is important to know for which values of x the approximation can be made arbitrarily close by choosing n large.

Following are a few of the most important functions, together with their Taylor polynomials about $x = 0$ and the range of x values over which the approximation can be made arbitrarily close by choosing n large enough:

$$e^x = 1 + x + \frac{x^2}{2!} + \frac{x^3}{3!} + \cdots + \frac{x^n}{n!} + R_{n+1}(x), \quad -\infty < x < \infty$$

$$\sin x = x - \frac{x^3}{3!} + \frac{x^5}{5!} - \frac{x^7}{7!} + \frac{x^9}{9!} - \cdots + (-1)^n \frac{x^{2n+1}}{(2n+1)!} + R_{n+1}(x), \quad -\infty < x < \infty$$

$$\cos x = 1 - \frac{x^2}{2!} + \frac{x^4}{4!} - \frac{x^6}{6!} + \frac{x^8}{8!} - \cdots + (-1)^n \frac{x^{2n}}{(2n)!} + R_{n+1}(x), \quad -\infty < x < \infty$$

$$\ln(1+x) = x - \frac{x^2}{2} + \frac{x^3}{3} - \frac{x^4}{4} + \frac{x^5}{5} - \cdots + (-1)^{n+1} \frac{x^n}{n} + R_{n+1}(x), \quad -1 < x \le 1$$

$$\frac{1}{1-x} = 1 + x + x^2 + x^3 + x^4 + \cdots + x^n + R_{n+1}(x), \quad -1 < x < 1$$

Section 7.6 Problems

■ 7.6.1

In Problems 1–5, find the linear approximation of $f(x)$ at $x = 0$.

1. $f(x) = e^{2x}$

2. $f(x) = \sin(3x)$

3. $f(x) = \dfrac{1}{1-x}$

4. $f(x) = x^4$

5. $f(x) = \ln(2 + x^2)$

In Problems 6–10, compute the Taylor polynomial of degree n about $a = 0$ for the indicated functions.

6. $f(x) = \dfrac{1}{1+x}, n = 4$

7. $f(x) = \cos x, n = 5$

8. $f(x) = e^{3x}, n = 3$

9. $f(x) = x^5, n = 6$

10. $f(x) = \sqrt{1+x}, n = 3$

In Problems 11–16, compute the Taylor polynomial of degree n about a = 0 for the indicated functions and compare the value of the functions at the indicated point with the value of the corresponding Taylor polynomial.

11. $f(x) = \sqrt{2 + x}, n = 3, x = 0.1$

12. $f(x) = \dfrac{1}{1 - x}, n = 3, x = 0.1$

13. $f(x) = \sin x, n = 5, x = 1$

14. $f(x) = e^{-x}, n = 4, x = 0.3$

15. $f(x) = \tan x, n = 2, x = 0.1$

16. $f(x) = \ln(1 + x), n = 3, x = 0.1$

17. (a) Find the Taylor polynomial of degree 3 about $a = 0$ for $f(x) = \sin x$.

(b) Use your result in (a) to give an intuitive explanation why

$$\lim_{x \to 0} \frac{\sin x}{x} = 1$$

18. (a) Find the Taylor polynomial of degree 2 about $a = 0$ for $f(x) = \cos x$.

(b) Use your result in (a) to give an intuitive explanation why

$$\lim_{x \to 0} \frac{\cos x - 1}{x} = 0$$

■ **7.6.2**

In Problems 19–23, compute the Taylor polynomial of degree n about a and compare the value of the approximation with the value of the function at the given point x.

19. $f(x) = \sqrt{x}, a = 1, n = 3; x = 2$

20. $f(x) = \ln x, a = 1, n = 3; x = 2$

21. $f(x) = \cos x, a = \frac{\pi}{6}, n = 3; x = \frac{\pi}{7}$

22. $f(x) = x^{1/5}, a = -1, n = 3; x = -0.9$

23. $f(x) = e^x, a = 2, n = 3; x = 2.1$

24. Show that

$$T^4 \approx T_a^4 + 4T_a^3(T - T_a)$$

for T close to T_a.

25. Show that, for positive constants r and k,

$$rN\left(1 - \frac{N}{K}\right) \approx rN$$

for N close to 0.

26. (a) Show that, for positive constants a and k,

$$f(R) = \frac{aR}{k + R} \approx \frac{a}{k}R$$

for R close to 0.

(b) Show that, for positive constants a and k,

$$f(R) = \frac{aR}{k + R} \approx \frac{a}{2} + \frac{a}{4k}(R - k)$$

for R close to k.

■ **7.6.3**

In Problems 27–30, use the following form of the error term

$$R_{n+1}(x) = \frac{f^{(n+1)}(c)}{(n + 1)!}x^{n+1}$$

where c is between 0 and x, to determine in advance the degree of Taylor polynomial at a = 0 that would achieve the indicated accuracy in the interval [0, x]. (Do not compute the Taylor polynomial.)

27. $f(x) = e^x, x = 2, \text{error} < 10^{-3}$

28. $f(x) = \cos x, x = 1, \text{error} < 10^{-2}$

29. $f(x) = 1/(1 + x), x = 0.2, \text{error} < 10^{-2}$

30. $f(x) = \ln(1 + x), x = 0.1, \text{error} < 10^{-2}$

31. Let $f(x) = e^{-1/x}$ for $x > 0$ and $f(x) = 0$ for $x = 0$. Compute a Taylor polynomial of degree 2 at $x = 0$, and determine how large the error is.

32. We can show that the Taylor polynomial for $f(x) = (1 + x)^\alpha$ about $x = 0$, with α a positive constant, converges for $x \in (-1, 1)$. Show that

$$(1 + x)^\alpha = 1 + \alpha x + \frac{\alpha(\alpha - 1)}{2!}x^2$$
$$+ \frac{\alpha(\alpha - 1)(\alpha - 2)}{3!}x^3 + \cdots + R_{n+1}(x)$$

33. We can show that the Taylor polynomial for $f(x) = \tan^{-1} x$ about $x = 0$ converges for $|x| \leq 1$.

(a) Show that the following is true:

$$\tan^{-1} x = x - \frac{x^3}{3} + \frac{x^5}{5} - \frac{x^7}{7} + \cdots + R_{n+1}(x)$$

(b) Explain why the following holds:

$$\frac{\pi}{4} = 1 - \frac{1}{3} + \frac{1}{5} - \frac{1}{7} + \cdots$$

(This series converges very slowly, as you would see if you used it to approximate π.)

■ 7.7 Tables of Integrals (Optional)

Before the advent of software that could integrate functions, tables of indefinite integrals were useful aids for evaluating integrals. In using a table of integrals, it is still necessary to bring the integrand of interest into a form that is listed in the table—and there are many integrals that simply cannot be evaluated exactly and must be evaluated numerically. We will give a very brief list of indefinite integrals and explain how to use such tables.

I. Basic Functions

1. $\displaystyle\int x^n dx = \frac{1}{n+1}x^{n+1} + C \,, n \neq -1$

2. $\displaystyle\int \frac{1}{x}\, dx = \ln|x| + C$

3. $\displaystyle\int e^x\, dx = e^x + C$

4. $\displaystyle\int a^x\, dx = \frac{a^x}{\ln a} + C \text{ with } a > 0, a \neq 1$

5. $\displaystyle\int \ln x\, dx = x\ln x - x + C$

6. $\displaystyle\int \sin x\, dx = -\cos x + C$

7. $\displaystyle\int \cos x\, dx = \sin x + C$

8. $\displaystyle\int \tan x\, dx = -\ln|\cos x| + C$

II. Rational Functions

9. $\displaystyle\int \frac{1}{ax+b}\, dx = \frac{1}{a}\ln|ax+b| + C$

10. $\displaystyle\int \frac{x}{ax+b}\, dx = \frac{x}{a} - \frac{b}{a^2}\ln|ax+b| + C$

11. $\displaystyle\int \frac{x}{(ax+b)^2}\, dx = \frac{b}{a^2(ax+b)} + \frac{1}{a^2}\ln|ax+b| + C$

12. $\displaystyle\int \frac{x}{ax^2+bx+c}\, dx = \frac{1}{2a}\ln|ax^2+bx+c| - \frac{b}{2a}\int \frac{1}{ax^2+bx+c}\, dx$

13. $\displaystyle\int \frac{1}{a^2+x^2}\, dx = \frac{1}{a}\arctan\frac{x}{a} + C$

14. $\displaystyle\int \frac{1}{a^2-x^2}\, dx = \frac{1}{2a}\ln\left|\frac{x+a}{x-a}\right| + C$

III. Integrands Involving $\sqrt{a^2+x^2}$, $\sqrt{a^2-x^2}$, or $\sqrt{x^2-a^2}$

15. $\displaystyle\int \frac{1}{\sqrt{a^2-x^2}}\, dx = \arcsin\frac{x}{a} + C$

16. $\displaystyle\int \frac{1}{\sqrt{x^2\pm a^2}}\, dx = \ln|x + \sqrt{x^2\pm a^2}| + C$

17. $\displaystyle\int \sqrt{a^2\pm x^2}\, dx = \frac{1}{2}\left(x\sqrt{a^2\pm x^2} + a^2\int \frac{1}{\sqrt{a^2\pm x^2}}\, dx\right)$

18. $\displaystyle\int \sqrt{x^2-a^2}\, dx = \frac{1}{2}\left(x\sqrt{x^2-a^2} - a^2\int \frac{1}{\sqrt{x^2-a^2}}\, dx\right)$

IV. Integrands Involving Trigonometric Functions

19. $\displaystyle\int \sin(ax)\, dx = -\frac{1}{a}\cos(ax) + C$

20. $\displaystyle\int \sin^2(ax)\, dx = \frac{1}{2}x - \frac{1}{4a}\sin(2ax) + C$

21. $\displaystyle\int \sin(ax)\sin(bx)\, dx = \frac{\sin(a-b)x}{2(a-b)} - \frac{\sin(a+b)x}{2(a+b)} + C, \text{ for } a^2 \neq b^2$

22. $\displaystyle\int \cos(ax)\, dx = \frac{1}{a}\sin(ax) + C$

23. $\displaystyle\int \cos^2(ax)\,dx = \frac{1}{2}x + \frac{1}{4a}\sin(2ax) + C$

24. $\displaystyle\int \cos(ax)\cos(bx)\,dx = \frac{\sin(a-b)x}{2(a-b)} + \frac{\sin(a+b)x}{2(a+b)} + C,\ \text{for } a^2 \neq b^2$

25. $\displaystyle\int \sin(ax)\cos(ax)\,dx = \frac{1}{2a}\sin^2(ax) + C$

26. $\displaystyle\int \sin(ax)\cos(bx)\,dx = -\frac{\cos(a+b)x}{2(a+b)} - \frac{\cos(a-b)x}{2(a-b)} + C,\ \text{for } a^2 \neq b^2$

V. Integrands Involving Exponential Functions

27. $\displaystyle\int e^{ax}\,dx = \frac{1}{a}e^{ax} + C$

28. $\displaystyle\int xe^{ax}\,dx = \frac{e^{ax}}{a^2}(ax - 1) + C$

29. $\displaystyle\int x^n e^{ax}\,dx = \frac{1}{a}x^n e^{ax} - \frac{n}{a}\int x^{n-1}e^{ax}\,dx$

30. $\displaystyle\int e^{ax}\sin(bx)\,dx = \frac{e^{ax}}{a^2 + b^2}(a\sin(bx) - b\cos(bx)) + C$

31. $\displaystyle\int e^{ax}\cos(bx)\,dx = \frac{e^{ax}}{a^2 + b^2}(a\cos(bx) + b\sin(bx)) + C$

VI. Integrands Involving Logarithmic Functions

32. $\displaystyle\int \ln x\,dx = x\ln x - x + C$

33. $\displaystyle\int (\ln x)^2\,dx = x(\ln x)^2 - 2x\ln x + 2x + C$

34. $\displaystyle\int x^m \ln x\,dx = x^{m+1}\left[\frac{\ln x}{m+1} - \frac{1}{(m+1)^2}\right] + C,\ m \neq -1$

35. $\displaystyle\int \frac{\ln x}{x}\,dx = \frac{(\ln x)^2}{2} + C$

36. $\displaystyle\int \frac{1}{x\ln x}\,dx = \ln(\ln x) + C$

37. $\displaystyle\int \sin(\ln x)\,dx = \frac{x}{2}(\sin(\ln x) - \cos(\ln x)) + C$

38. $\displaystyle\int \cos(\ln x)\,dx = \frac{x}{2}(\sin(\ln x) + \cos(\ln x)) + C$

We will now illustrate how to use the preceding table. We begin with examples that fit one of the listed integrals exactly.

EXAMPLE 1   **Square Root** Find

$$\int \sqrt{3 - x^2}\,dx$$

Solution   The integrand involves $\sqrt{a^2 - x^2}$ and is of the form III.17 with $a^2 = 3$. Hence,

$$\int \sqrt{3 - x^2}\,dx = \frac{1}{2}\left(x\sqrt{3 - x^2} + 3\int \frac{1}{\sqrt{3 - x^2}}\,dx\right)$$

To evaluate $\int \frac{1}{\sqrt{3-x^2}}\,dx$, we use III.15 with $a^2 = 3$ and find that

$$\int \frac{1}{\sqrt{3 - x^2}}\,dx = \arcsin\frac{x}{\sqrt{3}} + C$$

Thus,

$$\int \sqrt{3 - x^2}\, dx = \frac{1}{2}\left(x\sqrt{3 - x^2} + 3\arcsin \frac{x}{\sqrt{3}} \right) + C \qquad ■$$

EXAMPLE 2   **Trigonometric Function** Find

$$\int \sin(3x)\cos(4x)\, dx$$

Solution   The integrand involves trigonometric functions, and we can find it in IV.26 with $a = 3$ and $b = 4$. Hence,

$$\int \sin(3x)\cos(4x)\, dx = -\frac{\cos(7x)}{14} - \frac{\cos(-x)}{(2)(-1)} + C$$

Since $\cos(-x) = \cos x$, this simplifies to

$$\int \sin(3x)\cos(4x)\, dx = -\frac{\cos(7x)}{14} + \frac{\cos x}{2} + C \qquad ■$$

EXAMPLE 3   **Exponential Function** Find

$$\int x^2 e^{3x}\, dx$$

Solution   This integrand is of the form V.29 with $n = 2$ and $a = 3$. Hence,

$$\int x^2 e^{3x}\, dx = \frac{1}{3}x^2 e^{3x} - \frac{2}{3}\int x e^{3x}\, dx$$

We now use V.28 to continue the evaluation of the integral and find that

$$\int x e^{3x}\, dx = \frac{e^{3x}}{9}(3x - 1) + C$$

Thus,

$$\int x^2 e^{3x}\, dx = \frac{1}{3}x^2 e^{3x} - \frac{2}{3}\left[\frac{e^{3x}}{9}(3x - 1) \right] + C \qquad ■$$

Thus far, each of our examples exactly matched one of the integrals in our table. Often, this will not be the case, and the integrand must be manipulated until it matches one of the integrals in the table. Among the manipulations that are used are expansions, long division, completion of the square, and substitution. We give a few examples to illustrate.

EXAMPLE 4   **Exponential Function** Find

$$\int e^{2x}\sin(3x - 1)\, dx$$

Solution   This integrand looks similar to V.30. If we use the substitution

$$u = 3x - 1 \quad \text{with } dx = \frac{1}{3}\, du \text{ and } 2x = \frac{2}{3}(u + 1)$$

then the integrand can be transformed so that it matches V.30 exactly, and we have

$$\int e^{2x} \sin(3x - 1)\, dx = \int e^{2(u+1)/3} (\sin u) \frac{1}{3}\, du$$

$$= \frac{e^{2/3}}{3} \int e^{2u/3} \sin u\, du$$

$$= \frac{e^{2/3}}{3} \frac{e^{2u/3}}{\frac{4}{9} + 1} \left[\frac{2}{3} \sin u - \cos u \right] + C$$

$$= \frac{e^{2/3}}{3 \cdot \frac{13}{9}} e^{2(3x-1)/3} \left[\frac{2}{3} \sin(3x - 1) - \cos(3x - 1) \right] + C$$

$$= \frac{3}{13} e^{2x} \left[\frac{2}{3} \sin(3x - 1) - \cos(3x - 1) \right] + C \qquad ■$$

EXAMPLE 5   **Rational Function** Find

$$\int \frac{x^2}{9 + x^2}\, dx$$

Solution   The integrand is a rational function; we can use long division to simplify it:

$$\frac{x^2}{9 + x^2} = 1 - \frac{9}{9 + x^2}$$

Then, using II.13 with $a = 3$, we obtain

$$\int \frac{x^2}{9 + x^2}\, dx = \int dx - 9 \int \frac{1}{9 + x^2}\, dx$$

$$= x - 9 \left(\frac{1}{3} \arctan \frac{x}{3} \right) + C$$

$$= x - 3 \arctan \frac{x}{3} + C \qquad ■$$

EXAMPLE 6   **Rational Function** Find

$$\int \frac{1}{x^2 - 2x - 3}\, dx$$

Solution   The first step is to complete the square in the denominator:

$$\frac{1}{x^2 - 2x - 3} = \frac{1}{(x^2 - 2x + 1) - 1 - 3}$$

$$= \frac{1}{(x - 1)^2 - 4}$$

Then, using the substitution $u = x - 1$ with $du = dx$, we find that

$$\int \frac{dx}{(x - 1)^2 - 4} = \int \frac{du}{u^2 - 4} = -\int \frac{du}{4 - u^2}$$

which is of the form II.14 with $a = 2$. Therefore,

$$\int \frac{1}{x^2 - 2x - 3}\, dx = -\int \frac{du}{4 - u^2} = -\frac{1}{4} \ln \left| \frac{u + 2}{u - 2} \right| + C$$

$$= -\frac{1}{4} \ln \left| \frac{x + 1}{x - 3} \right| + C \qquad ■$$

■ 7.7.1 A Note on Software Packages That Can Integrate

Mathematicians and scientists use software packages to integrate functions. They are not difficult to use with some practice. Although they will not give you insight into what technique could be used to solve an integration problem, they quickly give you the correct answer. For instance, if we used MATLAB, one of the common software packages, to calculate the integral in Example 8 of Section 7.3, we would enter the following string of commands into our computer:

```
syms x;
f=x^4*(1-x)^4/(1+x^2);
int(f,0,1)
```

MATLAB then returns

```
ans = 22/7-pi
```

Section 7.7 Problems

In Problems 1–8, use the table on pages 383–384 to compute each integral.

1. $\displaystyle\int \frac{x}{2x-3}\,dx$

2. $\displaystyle\int \frac{dx}{16+x^2}$

3. $\displaystyle\int \sqrt{x^2-16}\,dx$

4. $\displaystyle\int \sin(2x)\cos(2x)\,dx$

5. $\displaystyle\int_0^1 x^3 e^{-x}\,dx$

6. $\displaystyle\int_0^{\pi/4} e^{-x}\cos(2x)\,dx$

7. $\displaystyle\int_1^e x^2 \ln x\,dx$

8. $\displaystyle\int_e^{e^2} \frac{dx}{x\ln x}$

In Problems 9–22, use the table on pages 383–384 to compute each integral after manipulating the integrand in a suitable way.

9. $\displaystyle\int_0^{\pi/6} e^x \cos\left(x-\frac{\pi}{6}\right)dx$

10. $\displaystyle\int_1^2 x\ln(2x-1)\,dx$

11. $\displaystyle\int (x^2-1)e^{-x/2}\,dx$

12. $\displaystyle\int (x+1)^2 e^{-2x}\,dx$

13. $\displaystyle\int \cos^2(5x-3)\,dx$

14. $\displaystyle\int \frac{x^2}{4x^2+4x+1}\,dx$

15. $\displaystyle\int \sqrt{9+4x^2}\,dx$

16. $\displaystyle\int \frac{1}{\sqrt{16-9x^2}}\,dx$

17. $\displaystyle\int e^{2x+1}\sin\left(\frac{\pi}{2}x\right)dx$

18. $\displaystyle\int (x-1)^2 e^{2x}\,dx$

19. $\displaystyle\int_2^4 \frac{1}{x\ln\sqrt{x}}\,dx$

20. $\displaystyle\int_1^e (x+2)^2 \ln x\,dx$

21. $\displaystyle\int \cos(\ln(3x))\,dx$

22. $\displaystyle\int \frac{3}{x^2-4x+8}\,dx$

Chapter 7 Key Terms

Discuss the following definitions and concepts:

1. The substitution rule for indefinite integrals

2. The substitution rule for definite integrals

3. Integration by parts

4. The "trick" of "multiplying by 1"

5. Partial-fraction decomposition

6. Partial-fraction method

7. Proper rational function

8. Irreducible quadratic factor

9. Improper integral

10. Integration when the interval is unbounded

11. Integration when the integrand is discontinuous

12. Convergence and divergence of improper integrals

13. Comparison results for improper integrals

14. Numerical integration: midpoint and trapezoidal rule

15. Error bounds for the midpoint and the trapezoidal rule

16. Using tables of integrals for integration

17. Linear approximation

18. Taylor polynomial of degree n

19. Taylor's formula

Chapter 7 Review Problems

In Problems 1–30, evaluate the given indefinite integrals.

1. $\displaystyle\int x^2(1-x^3)^2\,dx$

2. $\displaystyle\int \frac{\cos x}{1+\sin^2 x}\,dx$

3. $\displaystyle\int 4xe^{-x^2}\,dx$

4. $\displaystyle\int \frac{x\ln(1+x^2)}{1+x^2}\,dx$

5. $\displaystyle\int (1+\sqrt{x})^{1/3}\,dx$

6. $\displaystyle\int x\sqrt{3x+1}\,dx$

7. $\displaystyle\int x\sec^2(3x^2)\,dx$

8. $\displaystyle\int \tan x\sec^2 x\,dx$

9. $\displaystyle\int x\ln x\,dx$

10. $\displaystyle\int x^3\ln x^2\,dx$

11. $\displaystyle\int \sec^2 x \ln(\tan x)\,dx$

12. $\displaystyle\int \sqrt{x}\ln\sqrt{x}\,dx$

13. $\displaystyle\int \frac{1}{4+x^2}\,dx$

14. $\displaystyle\int \frac{1}{4-x^2}\,dx$

15. $\displaystyle\int \tan x\,dx$

16. $\displaystyle\int \tan^{-1}x\,dx$

17. $\displaystyle\int e^{2x}\sin x\,dx$

18. $\displaystyle\int x\sin x\,dx$

19. $\displaystyle\int \sqrt{e^x}\,dx$

20. $\displaystyle\int \ln\sqrt{x}\,dx$

21. $\displaystyle\int \sin^2 x\,dx$

22. $\displaystyle\int \sin x\cos x\,e^{\sin x}\,dx$

23. $\displaystyle\int \frac{1}{x(x-1)}\,dx$

24. $\displaystyle\int \frac{1}{(x+1)(x-2)}\,dx$

25. $\displaystyle\int \frac{x}{x+5}\,dx$

26. $\displaystyle\int \frac{x}{x^2+5}\,dx$

27. $\displaystyle\int \frac{1}{x+5}\,dx$

28. $\displaystyle\int \frac{1}{x^2+5}\,dx$

29. $\displaystyle\int \frac{(x+1)^2}{x-1}\,dx$

30. $\displaystyle\int \frac{2x+1}{\sqrt{1-x^2}}\,dx$

In Problems 31–50, evaluate the given definite integrals.

31. $\displaystyle\int_1^3 \frac{x^2+1}{x}\,dx$

32. $\displaystyle\int_0^{\pi/2} x\sin x\,dx$

33. $\displaystyle\int_0^1 xe^{-x^2/2}\,dx$

34. $\displaystyle\int_1^2 \ln x\,dx$

35. $\displaystyle\int_0^2 \frac{1}{4+x^2}\,dx$

36. $\displaystyle\int_0^{1/2} \frac{2}{\sqrt{1-x^2}}\,dx$

37. $\displaystyle\int_2^6 \frac{1}{\sqrt{x-2}}\,dx$

38. $\displaystyle\int_0^2 \frac{1}{x-2}\,dx$

39. $\displaystyle\int_0^\infty \frac{1}{9+x^2}\,dx$

40. $\displaystyle\int_0^\infty \frac{1}{x^2+3}\,dx$

41. $\displaystyle\int_0^\infty \frac{1}{x+3}\,dx$

42. $\displaystyle\int_0^\infty \frac{1}{(x+3)^2}\,dx$

43. $\displaystyle\int_0^1 \frac{1}{x^2}\,dx$

44. $\displaystyle\int_1^\infty \frac{1}{x^2}\,dx$

45. $\displaystyle\int_0^1 \frac{1}{\sqrt{x}}\,dx$

46. $\displaystyle\int_1^\infty \frac{1}{\sqrt{x}}\,dx$

47. $\displaystyle\int_0^1 x\ln x\,dx$

48. $\displaystyle\int_0^1 x2^x\,dx$

49. $\displaystyle\int_0^{\pi/4} e^{\cos x}\sin x\,dx$

50. $\displaystyle\int_0^{\pi/4} x\sin(2x)\,dx$

In Problems 51–54, use (a) the midpoint rule and (b) the trapezoidal rule to approximate each integral with the specified value of n.

51. $\displaystyle\int_0^2 (x^2-1)\,dx,\, n=4$

52. $\displaystyle\int_{-1}^1 (x^3-1)\,dx,\, n=4$

53. $\displaystyle\int_0^1 e^{-x}\,dx,\, n=5$

54. $\displaystyle\int_0^{\pi/4} \sin(4x)\,dx,\, n=4$

In Problems 55–58, find the Taylor polynomial of degree n about x = a for each function.

55. $f(x)=\sin(2x), a=0, n=3$

56. $f(x)=e^{-x^2/2}, a=0, n=3$

57. $f(x)=\ln x, a=1, n=3$

58. $f(x)=\dfrac{1}{x-3}, a=4, n=4$

59. Cost of Gene Substitution (*Adapted from Roughgarden, 1996*) Suppose that an advantageous mutation arises in a population. Initially, the gene carrying this mutation is at a low frequency. As the gene spreads through the population, the average fitness of the population increases. We denote by $f_{\text{avg}}(t)$ the average fitness of the population at time t, by $f_{\text{avg}}(0)$ the average fitness of the population at time 0 (when the mutation arose), and by K the final value of the average fitness after the mutation has spread through the population. Haldane (1957) suggested measuring the *cost of evolution* (now known as the *cost of gene substitution*) by the cumulative difference between the current and the final fitness—that is, by

$$\int_0^\infty (K - f_{\text{avg}}(t))\,dt$$

In Figure 7.43, shade the region whose area is equal to the cost of gene substitution.

Figure 7.43 The cost of gene substitution. See Problem 59.

Differential Equations

8

In Section 4.6, Example 6, we looked at the exponential growth of a population, given by

$$N(t) = N(0)e^{rt}, \ t \geq 0$$

where $N(t)$ denotes the size of the population at time t and r is a parameter. Differentiating $N(t)$ with respect to t, we found that exponential growth satisfies the differential equation

$$\frac{dN}{dt} = rN(t), \ t \geq 0 \tag{8.1}$$

We conclude from (8.1) that if a population grows exponentially, then its per capita growth rate $\frac{1}{N}\frac{dN}{dt}$ is a constant—namely, the parameter r. Equation (8.1) contains the derivative of a function, and such equations are therefore called **differential equations**.

While the function $N(t) = N(0)e^{rt}, t \geq 0$, reveals the behavior of the population size over time, it tells us little about what processes lead to exponential growth. In contrast, the differential equation $\frac{1}{N}\frac{dN}{dt} = r$ reveals the process—a constant per capita growth rate—but does not tell us much about the population size trajectory over time. It is therefore important to be able to go back and forth between the two descriptions.

We saw in Example 6 of Section 4.6 that differentiating the function $N(t)$ results in a differential equation describing the rate of change of $N(t)$. On the basis of what we learned in Chapters 6 and 7, we now know that going from the derivative of a function to the function involves integration. In Section 8.1, we will learn how to use integration to go from a differential equation to a function that satisfies the differential equation.

Differential equations can contain derivatives of any order; for example,

$$\frac{d^2 y}{dx^2} + \frac{dy}{dx} = xy$$

is a differential equation that contains the first and second derivative of the function $y = y(x)$. If a differential equation, such as (8.1), contains only the first derivative, it is called a **first-order** differential equation.

Throughout this chapter, we will restrict ourselves to first-order differential equations of the form

$$\frac{dy}{dx} = f(x)g(y) \tag{8.2}$$

The right-hand side of (8.2) is the product of two functions, one depending only on x, the other only on y. Such equations are called **separable differential equations**. (The reason for this name will become clear shortly.) This type of differential equation includes two special cases:

$$\frac{dy}{dx} = f(x) \tag{8.3}$$

and

$$\frac{dy}{dx} = g(y) \tag{8.4}$$

We discussed differential equations of the form (8.3) in Section 5.8. Differential equations of the form (8.4) include (8.1) and are frequently used in biological models.

■ 8.1 Solving Differential Equations

Let's return to the growth model in (8.1), and, to be concrete, let's choose $r = 2$. This results in the equation

$$\frac{dN}{dt} = 2N(t), \ t \geq 0 \tag{8.5}$$

We are interested in finding a function $N(t)$ that satisfies (8.5). Such a function is called a **solution** of the differential equation. We already know that, with $N_0 = N(0)$,

$$N(t) = N_0 e^{2t}, \ t \geq 0$$

is a solution of (8.5). To check whether the function $N(t)$ is indeed a solution, we differentiate $N(t)$:

$$\frac{dN}{dt} = 2\underbrace{N_0 e^{2t}}_{N(t)} = 2N(t)$$

Figure 8.1 The per capita growth rate for the differential equation (8.6) shows oscillations and is described by the function $g(t) = 2(1 + \sin(2\pi t))$.

Modeling biological situations frequently leads to differential equations. A description of the instantaneous rate of change is often a good starting point for building models. As an example, let's look again at (8.5) and let's write it in the form

$$\frac{1}{N}\frac{dN}{dt} = 2$$

In this model, the per capita growth rate is constant. Suppose, however, we observed that the per capita growth rate shows oscillations, as illustrated in Figure 8.1. Then we can immediately modify our original differential equation (8.5) to reflect this observation and obtain

$$\frac{dN}{dt} = 2(1 + \sin(2\pi t)) N(t), \ t \geq 0 \tag{8.6}$$

In this section, we will learn how to solve differential equations like (8.6). (See Problem 57 at the end of the section.)

A first-order differential equation tells us what the derivative of a function is. Therefore, in order to find a solution, we must integrate. (Since it is not always possible to integrate a function, it is not always possible to write the solution of a differential equation in explicit form.)

We begin with a general method for solving separable differential equations of the form (8.2), or

$$\frac{dy}{dx} = f(x)g(y) \tag{8.7}$$

We divide both sides of (8.7) by $g(y)$ [assuming that $g(y) \neq 0$]:

$$\frac{1}{g(y)}\frac{dy}{dx} = f(x) \tag{8.8}$$

Now, if $y = u(x)$ is a solution of (8.8), then $u(x)$ satisfies $\frac{dy}{dx} = u'(x)$, and hence

$$\frac{1}{g[u(x)]}u'(x) = f(x)$$

If we integrate both sides with respect to x, we find that

$$\int \frac{1}{g[u(x)]}u'(x)\,dx = \int f(x)\,dx$$

or

$$\int \frac{1}{g(y)}\,dy = \int f(x)\,dx \tag{8.9}$$

since $g[u(x)] = g(y)$ and $u'(x)\,dx = dy$.

The preceding analysis suggests the following procedure for solving separable differential equations: We separate the variables x and y so that one side of the equation depends only on y and the other side only on x. To do so, we treat dy and dx as if they were regular numbers. Specifically, separating variables in (8.7) yields

$$\frac{dy}{g(y)} = f(x)\,dx$$

and integrating both sides results in equation (8.9):

$$\int \frac{dy}{g(y)} = \int f(x)\,dx$$

The method of separating the variables x and y works because the right-hand side of (8.7) is of the special form $f(x)g(y)$, which gave this type of differential equation its name. Note that when we divided (8.7) by $g(y)$, we had to be careful, since $g(y)$ might be 0 for some values of y, in which case we cannot divide. We will address this problem in Subsection 8.1.2. In the next two subsections, we discuss how to solve differential equations of the forms (8.3) and (8.4); we give an example of the general type (8.2) in Subsection 8.1.3.

■ 8.1.1 Pure-Time Differential Equations

In many applications, the independent variable represents time. If the rate of change of a function depends only on time, we call the resulting differential equation a **pure-time differential equation**. Such a differential equation is of the form

$$\frac{dy}{dx} = f(x), \quad x \in I \tag{8.10}$$

where I is an interval and x represents time. We discussed such equations in Section 5.8, where we found that their solution is of the form

$$y(x) = \int_{x_0}^{x} f(u)\,du + C \tag{8.11}$$

The constant C comes from finding the general antiderivative of $f(x)$; the number x_0 is in the interval I. To determine C, we must phrase the problem as an initial-value problem (see Section 5.8); if we assume that $y(x_0) = y_0$, then plugging x_0 into (8.11) yields $y(x_0) = C$, and thus $C = y_0$. The solution can then be written as

$$y(x) = y_0 + \int_{x_0}^{x} f(u)\,du$$

To solve (8.10) formally, we separate variables; we write the differential equation in the form

$$dy = f(x)\,dx$$

and then integrate both sides:

$$\int dy = \int f(x)\,dx$$

or

$$y(x) = \int f(x)\,dx$$

which is the same as (8.11).

EXAMPLE 1   Suppose that the volume $V(t)$ of a cell at time t changes according to

$$\frac{dV}{dt} = \sin t \quad \text{with } V(0) = 3$$

Find $V(t)$.

Solution   Since

$$V(t) = V(0) + \int_0^t \sin u\,du$$

it follows that

$$\begin{aligned} V(t) &= 3 + [-\cos u]_0^t \\ &= 3 + (-\cos t + \cos 0) \\ &= 4 - \cos t \end{aligned}$$

because $\cos 0 = 1$. (See Figure 8.2.)   ■

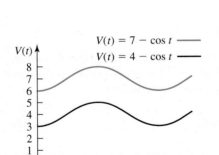

Figure 8.2 The solution $V(t) = 4 - \cos t$ in Example 1.

If we changed the initial condition in Example 1, the graph of the new solution would be obtained from the old solution by shifting the solution vertically so that it would satisfy the new initial condition. (See Figure 8.3; we discussed the situation in Section 5.8.)

■ 8.1.2 Autonomous Differential Equations

Many of the differential equations, such as (8.1), that model biological situations are of the form

$$\frac{dy}{dx} = g(y) \tag{8.12}$$

where the right-hand side does not explicitly depend on x. These equations are called **autonomous differential equations**.

To interpret the biological meaning of *autonomous*, let's return to the growth model

$$\frac{dN}{dt} = 2N(t) \tag{8.13}$$

We will see shortly that the general solution of (8.13) is

$$N(t) = Ce^{2t} \tag{8.14}$$

where C is a constant that can be determined if the population size at one time is known. Suppose we conduct an experiment in which we follow a population over time, and suppose the population satisfies (8.13) with $N(0) = 20$. Using (8.14), we find that $N(0) = C = 20$. Then the size of the population at time t is given by

$$N(t) = 20e^{2t}$$

Figure 8.3 The curve $V(t) = 7 - \cos t$ is obtained from $V(t) = 4 - \cos t$ by a vertical shift. The function $V(t) = 7 - \cos t$ solves the differential equation in Example 1 with $V(0) = 6$.

If we repeat the experiment at, say, time $t = 10$ with the exact same initial population size, then, everything else being equal, the population evolves in exactly the same way as the one starting at $t = 0$. Using (8.14) now with $N(10) = 20$, we find that $N(10) = Ce^{20} = 20$, or $C = 20e^{-20}$. The size of the population is then given by

$$N(t) = 20e^{-20}e^{2t} = 20e^{2(t-10)}$$

The graph of this solution can be obtained from the previous graph, where $N(0) = 20$, by shifting the old graph 10 units to the right to the new starting point $(10, 20)$. (See Figure 8.4.)

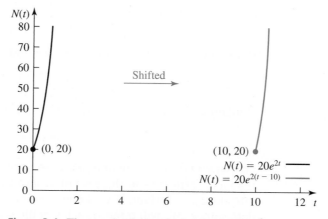

Figure 8.4 The graph of the solution $N(t) = 20e^{2t}$ is shifted to the new starting point $(10, 20)$.

This means that a population starting with $N = 20$ follows the same trajectory, regardless of when we start the experiment. This statement makes sense biologically: If the growth conditions do not depend explicitly on time, the experiment should yield the same outcome, regardless of when the experiment is performed. If the growth conditions for the population do change over time, we would not be able to use an autonomous differential equation to describe the growth of the population; we would need to include the time dependence explicitly in the dynamics.

Formally, we can solve (8.12) by separation of variables. We begin by dividing both sides of (8.12) by $g(y)$ and multiplying both sides of (8.12) by dx, to obtain

$$\frac{dy}{g(y)} = dx$$

Integrating both sides then gives

$$\int \frac{dy}{g(y)} = \int dx$$

We will discuss two cases: $g(y) = k(y - a)$ and $g(y) = k(y - a)(y - b)$. The growth model (8.1) is an example of the first case; the logistic equation, which we saw in Section 3.3 for the first time, is an example of the second case.

Case 1: $g(y) = k(y - a)$. We wish to solve

$$\frac{dy}{dx} = k(y - a) \tag{8.15}$$

where k and a are constants. We assume that $k \neq 0$. Separating variables then yields

$$\frac{dy}{y - a} = k\,dx \tag{8.16}$$

where we need to assume that $y \neq a$ to divide by $y - a$. It is somewhat arbitrary whether we leave k on the right-hand side or move it to the left-hand side; leaving it on the right is more convenient. Integrating both sides of (8.16) results in

$$\int \frac{dy}{y-a} = \int k\,dx$$

or

$$\ln|y - a| = kx + C_1$$

Exponentiating both sides then yields

$$|y - a| = e^{kx + C_1}$$

or

$$|y - a| = e^{C_1} e^{kx}$$

Removing the absolute-value signs, we find

$$y - a = \pm e^{C_1} e^{kx}$$

Renaming the constant by setting $C = \pm e^{C_1}$, we can write the solution as

$$y = Ce^{kx} + a \qquad\qquad (8.17)$$

(Renaming the constant only serves the purpose of getting the equation in a more readable form.) As in the case of a pure-time differential equation, integration introduces a constant. If we know a point (x_0, y_0) of the solution, then C can be determined. We will refer to such a point as an **initial condition**.

　　To obtain (8.16), we divided by $y - a$. We are allowed to do this only as long as $y \neq a$. If $y = a$, then $dy/dx = 0$ and the constant function $y = a$ is a solution of (8.15). We *lost* this solution when we divided (8.15) by $y - a$. Note that the constant C in (8.17) is different from 0, since $C = \pm e^{C_1}$ and $e^{C_1} \neq 0$ for any real number C_1. But we can combine the constant solution $y = a$ and the solution in (8.17) by allowing C to be equal to 0 in (8.17).

　　Before we turn to biological applications, we give an example in which we see how to solve a differential equation such as (8.15) and how to determine the value of C.

EXAMPLE 2　　Solve

$$\frac{dy}{dx} = 2 - 3y, \quad \text{where } y_0 = 1 \text{ when } x_0 = 1$$

Solution　　Instead of trying to identify the constants C, k, and a in equation (8.17), it is easier to solve the equation directly. We separate variables and then integrate, which results in

$$\int \frac{dy}{2 - 3y} = \int dx$$

Since an antiderivative of $\frac{1}{2-3y}$ is $-\frac{1}{3}\ln|2 - 3y|$, we find that

$$-\frac{1}{3}\ln|2 - 3y| = x + C_1$$

Solving for y yields

$$\ln|2 - 3y| = -3x - 3C_1$$
$$|2 - 3y| = e^{-3x - 3C_1}$$
$$2 - 3y = \pm e^{-3C_1} e^{-3x}$$

Setting $C = \pm e^{-3C_1}$, we obtain

$$2 - 3y = Ce^{-3x}$$

$$y = \frac{2}{3} + \frac{1}{3}e^{3-3x}$$ ———

(1, 1)

Figure 8.5 The solution to Example 2.

To determine C, we use the initial condition $y_0 = 1$ when $x_0 = 1$. That is,

$$2 - 3 = Ce^{-3}, \qquad \text{or} \qquad C = -e^3$$

Hence,

$$2 - 3y = -e^3 e^{-3x}$$

or

$$y = \frac{2}{3} + \frac{1}{3}e^{3-3x}$$

(See Figure 8.5.)   ∎

We now turn to two biological applications that are covered by Case 1.

EXAMPLE 3

Exponential Population Growth This model is given by (8.1) and was introduced in Section 4.6. We denote the population size at time $t \geq 0$ by $N(t)$ and assume that $N(0) = N_0 > 0$. The change in population size is described by the initial-value problem

$$\frac{dN}{dt} = rN, \quad \text{where } N(0) = N_0 \tag{8.18}$$

The parameter r is called the *intrinsic rate of growth* and is the per capita rate of growth, since

$$r = \frac{1}{N}\frac{dN}{dt}$$

When $r > 0$, this model represents a growing population. When $r < 0$, the size of the population decreases. (Note that the per capita growth rate r is independent of the population size.)

We can either solve (8.18) directly, as in Example 2, or use (8.15). Let's use (8.15). If we compare (8.18) with (8.15), we find that $k = r$ and $a = 0$. Hence, using (8.17), we find the solution

$$N(t) = Ce^{rt}$$

Since $N(0) = N_0 = C$, we write the solution as

$$N(t) = N_0 e^{rt} \tag{8.19}$$

Equation (8.19) shows that the population size grows exponentially when $r > 0$. When $r < 0$, the population size decreases exponentially. When $r = 0$, the population size stays constant.

We show solution curves of $N(t) = N_0 e^{rt}$ for $r > 0$, $r = 0$, and $r < 0$ in Figure 8.6. Exponential growth (or decay) is one of the most important growth phenomena in biology. You should therefore memorize both the differential equation (8.18) and its solution (8.19), together with the graphs in Figure 8.6, and know that (8.18) describes a situation in which the per capita growth rate (or intrinsic rate of growth) is a constant.

When $r > 0$, the population size grows without bound ($\lim_{t \to \infty} N(t) = \infty$). This kind of growth can be found when individuals are not limited by the availability of food or by competition. If we start a bacterial colony on a nutrient-rich substrate by inoculating the substrate with a few bacteria, then the bacteria initially can grow and divide unrestrictedly. Subsequently, when the substrate becomes more crowded and the food source depleted, the growth will be restricted and a different differential equation will be required to describe this situation. (We will discuss this scenario in Case 2.) Exponential decay in a population can be seen when the death rate exceeds the birth rate (for instance, in cases of starvation).   ∎

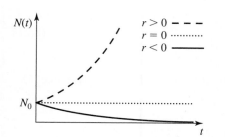

$r > 0$ – – –
$r = 0$ ··········
$r < 0$ ———

N_0

Figure 8.6 Solution curves for $dN/dt = rN$.

The type of growth in Example 3 is referred to as *Malthusian growth*, named after Thomas Malthus (1766–1834), a British clergyman and economist. Malthus wrote about the consequences of unrestricted growth on the welfare of humans. He claimed that populations and food show two fundamentally different growth patterns: Populations grow exponentially and food grows linearly. He concluded that, since exponential growth ultimately overtakes linear growth, populations would eventually experience starvation. (See Problem 56.)

Recall from Section 4.6 that when $r < 0$, (8.18) has the same form as the differential equation that describes radioactive decay. $N(t)$ would then denote the amount of radioactive material left at time t. We will revisit this application in Problem 20.

EXAMPLE 4

Restricted Growth: von Bertalanffy Equation This example describes the simplest form of restricted growth and can be used to describe the growth of fish. We denote by $L(t)$ the length of the fish at age t and assume that $L(0) = L_0$. Then

$$\frac{dL}{dt} = k(A - L) \tag{8.20}$$

where A is a positive constant that we will interpret shortly. We assume that $L_0 < A$ and explain this restriction subsequently as well. The constant k is also positive; the equation says that the growth rate dL/dt is proportional to $A - L$, so k should be interpreted as a constant of proportionality. We see that the growth rate dL/dt is positive and decreases linearly with length as long as $L < A$ and that the growth stops (i.e., $dL/dt = 0$) when $L = A$. To solve (8.20), we separate variables and integrate, yielding

$$\int \frac{dL}{A - L} = \int k \, dt$$

Hence,

$$-\ln|A - L| = kt + C_1$$

After multiplying this equation by -1 and exponentiating, we obtain

$$|A - L| = e^{-C_1} e^{-kt}$$

or

$$A - L = Ce^{-kt}$$

with $C = \pm e^{-C_1}$. Since $L(0) = L_0$, it follows that

$$A - L_0 = C$$

The solution is then given by

$$A - L(t) = (A - L_0)e^{-kt}$$

or

$$L(t) = A \left[1 - \left(1 - \frac{L_0}{A} \right) e^{-kt} \right] \tag{8.21}$$

Figure 8.7 The graph of the von Bertalanffy equation.

(See Figure 8.7.)

This is the von Bertalanffy equation that we encountered previously. Since

$$\lim_{t \to \infty} L(t) = A$$

the parameter A denotes the **asymptotic length** of the fish. Mathematically, there are no restrictions on L_0; biologically, however, we require that $0 < L_0 < A$. Otherwise, the growth rate would be negative, meaning that the fish shrinks in size. Note that A is an asymptotic length that is never reached, since there is no finite age T with $L(T) = A$ if $L(0) < A$.

We can now interpret the differential equation (8.16): The growth rate is proportional to the difference of the current length and the asymptotic length, with k representing the constant of proportionality. Since this difference is decreasing over time, (8.16) also shows that the growth rate decreases over time, implying that juveniles grow at a faster rate than adults. Moreover, the growth rate is always positive. This means that fish grow throughout their lives, which is indeed the case. ■

Case 2: $g(y) = k(y - a)(y - b)$. We now turn to differential equations of the form

$$\frac{dy}{dx} = k(y - a)(y - b) \tag{8.22}$$

where k, a, and b are constant. We assume that $k \neq 0$. Separating variables and integrating both sides yields

$$\int \frac{dy}{(y - a)(y - b)} = \int k \, dx \tag{8.23}$$

provided that $y \neq a$ and $y \neq b$. When $a = b$, we must find an antiderivative of $\frac{1}{(y-a)^2}$, which is $-\frac{1}{y-a}$. In this case,

$$-\frac{1}{y - a} = kx + C$$

or

$$y = a - \frac{1}{kx + C}$$

The constant C can then be determined from the initial condition. When $y = a$, $dy/dx = 0$ and, consequently, y is equal to a constant (specifically, $y = a$).

To find the solution when $a \neq b$, we must use the partial-fraction method that we introduced in Section 7.3. Let's first do an example.

EXAMPLE 5   Solve

$$\frac{dy}{dx} = 2(y - 1)(y + 2) \quad \text{with } y_0 = 2 \text{ when } x_0 = 0$$

Solution   Separation of variables yields

$$\int \frac{dy}{(y - 1)(y + 2)} = \int 2 \, dx$$

We use partial fractions to integrate the left-hand side:

$$\frac{1}{(y - 1)(y + 2)} = \frac{A}{y - 1} + \frac{B}{y + 2}$$

$$= \frac{A(y + 2) + B(y - 1)}{(y - 1)(y + 2)}$$

$$= \frac{(A + B)y + 2A - B}{(y - 1)(y + 2)}$$

Comparing the last term with the integrand, we find that

$$A + B = 0 \quad \text{and} \quad 2A - B = 1$$

which yields

$$A = -B \quad \text{and} \quad 2A - B = 3A = 1$$

Thus,

$$A = \frac{1}{3} \quad \text{and} \quad B = -\frac{1}{3}$$

Using the partial-fraction decomposition, we must integrate

$$\frac{1}{3} \int \left(\frac{1}{y - 1} - \frac{1}{y + 2} \right) dy = \int 2 \, dx$$

This produces

$$\frac{1}{3}[\ln|y - 1| - \ln|y + 2|] = 2x + C_1$$

Simplifying the latter equation results in

$$\ln\left|\frac{y-1}{y+2}\right| = 6x + 3C_1$$

$$\left|\frac{y-1}{y+2}\right| = e^{3C_1}e^{6x}$$

$$\frac{y-1}{y+2} = \pm e^{3C_1}e^{6x}$$

$$\frac{y-1}{y+2} = Ce^{6x}$$

Using the initial condition $y_0 = 2$ when $x_0 = 0$, we find that

$$\frac{1}{4} = C$$

The solution is therefore

$$\frac{y-1}{y+2} = \frac{1}{4}e^{6x}$$

If we want the solution in the form $y = f(x)$, we must solve for y:

$$y - 1 = (y+2)\frac{1}{4}e^{6x}$$

$$y\left(1 - \frac{1}{4}e^{6x}\right) = \frac{1}{2}e^{6x} + 1$$

$$y = \frac{\frac{1}{2}e^{6x} + 1}{1 - \frac{1}{4}e^{6x}}$$

$$y = \frac{2e^{6x} + 4}{4 - e^{6x}}$$

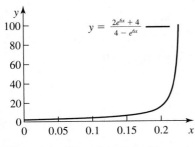

$$y = \frac{2e^{6x} + 4}{4 - e^{6x}}$$

Figure 8.8 The solution for Example 5.

(See Figure 8.8.) ■

We return now to (8.23) when $a \neq b$ and $y \neq a$ and $y \neq b$. We use the partial-fraction method to simplify the integral on the left-hand side. We decompose the integrand into a sum of simpler rational functions that we know how to integrate; that is, we write the integrand in the form

$$\frac{1}{(y-a)(y-b)} = \frac{A}{y-a} + \frac{B}{y-b} \qquad (8.24)$$

where A and B are constants that we must find. Next, we do the following algebraic manipulation on the right-hand side of (8.24):

$$\frac{A}{y-a} + \frac{B}{y-b} = \frac{A(y-b) + B(y-a)}{(y-a)(y-b)}$$

$$= \frac{y(A+B) - (Ab + Ba)}{(y-a)(y-b)} \qquad (8.25)$$

Comparing the last expression in (8.25) with the left-hand side of (8.24), we conclude that the constants A and B must satisfy

$$A + B = 0 \qquad \text{and} \qquad Ab + Ba = -1$$

Substituting $B = -A$ from the first equation into the second, we find that

$$Ab - Aa = -1, \qquad \text{or} \qquad A(b-a) = -1, \qquad \text{or} \qquad A = \frac{1}{a-b}$$

Therefore,

$$B = -\frac{1}{a - b}$$

That is,

$$\frac{1}{(y - a)(y - b)} = \frac{1}{a - b}\left[\frac{1}{y - a} - \frac{1}{y - b}\right] \tag{8.26}$$

Equation (8.26) allows us to evaluate

$$\int \frac{dy}{(y - a)(y - b)}$$

which is the left-hand side of (8.23). Evaluating yields

$$\int \frac{dy}{(y - a)(y - b)} = \frac{1}{a - b}\int \left(\frac{1}{y - a} - \frac{1}{y - b}\right) dy$$

$$= \frac{1}{a - b}\left[\ln|y - a| - \ln|y - b|\right] + C_1$$

Integrating the right-hand side of (8.23) as well, and combining the constants of integration into a new constant, C_2, we find that

$$\frac{1}{a - b}\left[\ln|y - a| - \ln|y - b|\right] = kx + C_2$$

or

$$\ln\left|\frac{y - a}{y - b}\right| = k(a - b)x + C_2(a - b)$$

Exponentiating the latter equation yields

$$\left|\frac{y - a}{y - b}\right| = e^{C_2(a-b)}e^{k(a-b)x}$$

or

$$\frac{y - a}{y - b} = \pm e^{C_2(a-b)}e^{k(a-b)x}$$

Defining $C = \pm e^{C_2(a-b)}$, we obtain the solution of (8.22):

$$\frac{y - a}{y - b} = Ce^{k(a-b)x}$$

We solve this for y as follows:

$$y = a + (y - b)Ce^{k(a-b)x}$$

$$y(1 - Ce^{k(a-b)x}) = a - bCe^{k(a-b)x}$$

or

$$y = \frac{a - bCe^{k(a-b)x}}{1 - Ce^{k(a-b)x}} \tag{8.27}$$

The constant C can then be determined from the initial condition. When $y = a$ or $y = b$, $dy/dx = 0$ and, consequently, y is equal to a constant (specifically, $y = a$ or $y = b$).

The next example, of density-dependent growth, is one of the most important applications of this case. In Example 3, we considered unrestricted (or density-*in*dependent) growth, which has the unrealistic feature that if the intrinsic rate of growth is positive, the size of a population grows without bound. Frequently, however, the per capita growth rate decreases as the population size increases. The growth rate thus depends on the population density. The simplest such model for which the growth rate is density dependent is the logistic equation, in which the size of the population cannot grow without bound.

The logistic equation was originally developed around 1835 by Pierre-François Verhulst, who used the term *logistic* to describe the equation. His work was completely forgotten until 1920, when Raymond Pearl and Lowell J. Reed published a series of papers on population growth. They used the same equation as Verhulst. After discovering Verhulst's work, Pearl and Reed adopted the name *logistic* for the equation. Pearl and Reed (1920) used the logistic equation to predict future growth of the U.S. population on the basis of census data from 1790 to 1920. Their equation would have predicted about 185 million people in the United States in the year 2000, which is a gross underestimate of the actual population size (over 260 million people). Although the logistic equation does not seem to fit actual populations very well, it is a useful model for analyzing growth under limiting resources.

EXAMPLE 6

The Logistic Equation The logistic equation describes the change in size of a population for which per capita growth is density dependent. If we denote the population size at time t by $N(t)$, then the change in growth is given by the initial-value problem

$$\frac{dN}{dt} = rN\left(1 - \frac{N}{K}\right) \quad \text{with } N(0) = N_0 \tag{8.28}$$

where r and K are positive constants. This is the simplest way of incorporating density dependence in the per capita growth rate, namely, having it decrease linearly with population size (see Figure 8.9):

$$\frac{1}{N}\frac{dN}{dt} = r\left(1 - \frac{N}{K}\right)$$

We can interpret the parameter r as follows: The per capita growth rate is equal to r when $N = 0$. Therefore, one way to measure r is to grow the organism at a very low density (i.e., when N is much smaller than K), so that the per capita growth rate is close to r. (In Problem 42, we discuss a different way to find r.)

The quantity K is called the **carrying capacity**. Looking at Figure 8.9, we see that the per capita growth rate is 0 when the population size is at the carrying capacity. Since the per capita growth rate is positive below K and negative above K, the size of the population will increase below K and decrease above K. The number K thus determines the population size that can be supported by the environment. To show this claim mathematically, we need to solve (8.28), from whose solution we will then see that, starting from any positive initial population size, the size of the population will eventually reach K [i.e., $\lim_{t \to \infty} N(t) = K$ if $N(0) > 0$].

Let's solve (8.28). We write the right-hand side as

$$g(N) = -\frac{r}{K}(N - 0)(N - K) \tag{8.29}$$

Comparing (8.29) with the right-hand side of (8.22), we find that

$$k = -\frac{r}{K} \qquad a = 0 \qquad b = K \tag{8.30}$$

The solution of (8.28) is given in (8.27) and is, with the use of (8.30),

$$N(t) = \frac{0 - KCe^{-(r/K)(0-K)t}}{1 - Ce^{-(r/K)(0-K)t}}$$

or

$$N(t) = \frac{-CKe^{rt}}{1 - Ce^{rt}} = \frac{CK}{C - e^{-rt}}$$

$$= \frac{K}{1 - e^{-rt}/C} \tag{8.31}$$

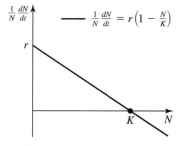

Figure 8.9 The per capita growth rate in the logistic equation is a linearly decreasing function of population size.

Since $N(0) = N_0$, we have

$$N_0 = \frac{CK}{C-1}$$

Solving this equation for C yields

$$N_0(C-1) = CK \qquad \text{or} \qquad C(N_0 - K) = N_0$$

Thus,

$$C = \frac{N_0}{N_0 - K} \qquad (8.32)$$

Substituting (8.32) into (8.31), we find that

$$N(t) = \frac{K}{1 - \frac{N_0 - K}{N_0}e^{-rt}}$$

or

$$N(t) = \frac{K}{1 + (\frac{K}{N_0} - 1)e^{-rt}} \qquad (8.33)$$

It follows from (8.33) that

$$\lim_{t \to \infty} N(t) = K$$

We see from Figure 8.10, where we plot (8.33) for $N_0 > K$ and $N_0 < K$, that the solution $N(t)$ approaches K, the carrying capacity, as $t \to \infty$. When $N_0 > K$, $N(t)$ approaches K from above; when $0 < N_0 < K$, $N(t)$ approaches K from below. To use separation of variables to solve (8.28), we need to exclude $N = 0$ and $N = K$. When $N(0) = 0$, $dN/dt = 0$ and, consequently, $N(t) = \text{constant} = N(0) = 0$ for all $t \geq 0$. When $N(0) = K$, $dN/dt = 0$, and consequently, $N(t) = \text{constant} = K$ for all $t \geq 0$. [The constant solution $N(t) = K$ is contained in (8.33) if we choose $N_0 = K$.]

The constant solutions K and 0 are called *equilibria*. The constant solution $N(t) = 0$ is not a very interesting one. We call it the *trivial equilibrium*. If $N(0) = 0$, then nothing happens; that is, $N(t)$ stays equal to 0 for all later times. That makes sense: If there aren't any individuals to begin with, then there won't be any later on.

We can show that if $0 < N_0 < \frac{K}{2}$, then the solution curve is S shaped, as seen in Figure 8.10. An S-shaped curve is characteristic of populations that show this type of density-dependent growth. At low densities, the growth is almost like unrestricted growth. At higher densities, the growth is restricted and the curve bends around and eventually levels off at the carrying capacity. If the population size is initially greater than the carrying capacity K, the population size decreases and becomes asymptotically (i.e., when $t \to \infty$) equal to K. ■

Figure 8.10 Solution curves for different initial values N_0.

■ 8.1.3 Allometric Growth

Solving differential equations of the form (8.2) gets complicated quickly. Before we show the role such differential equations play in biology, we will give an example in which we solve a differential equation of that type.

EXAMPLE 7   Solve

$$\frac{dy}{dx} = \frac{y+1}{x} \qquad \text{with } y_0 = 0 \text{ when } x_0 = 1$$

Solution   To solve this differential equation, we separate variables first and then integrate. We find that

$$\int \frac{dy}{y+1} = \int \frac{dx}{x}$$

Carrying out the integration on both sides, we obtain

$$\ln|y+1| = \ln|x| + C_1 \qquad (8.34)$$

Solving for y, we get

$$|y + 1| = e^{C_1}|x|$$

$$y + 1 = \pm e^{C_1}x$$

Setting $C = \pm e^{C_1}$ yields

$$y = Cx - 1$$

Using the initial condition $y_0 = 0$ when $x_0 = 1$ allows us to determine C:

$$0 = C - 1, \quad \text{or} \quad C = 1$$

The solution is therefore

$$y = x - 1$$

■

We now turn to a biological application of equation (8.2): allometric growth. We have seen a number of allometric relationships throughout the earlier chapters of this book—typically, relationships between sizes of parts of an organism (e.g., skull length and body length, or leaf area and stem diameter). We denote by $L_1(t)$ and $L_2(t)$ the respective sizes of two organs of an individual of age t. We say that L_1 and L_2 are related through an allometric law if their specific growth rates are proportional—that is, if

$$\frac{1}{L_1}\frac{dL_1}{dt} = k\frac{1}{L_2}\frac{dL_2}{dt} \tag{8.35}$$

If the constant k is equal to 1, then the growth is called *isometric*; otherwise it is called *allometric*. Cancelling dt on both sides of (8.35) and integrating, we find that

$$\int \frac{dL_1}{L_1} = k \int \frac{dL_2}{L_2}$$

or

$$\ln |L_1| = k \ln |L_2| + C_1 \tag{8.36}$$

Solving for L_1, we obtain

$$L_1 = CL_2^k \tag{8.37}$$

where $C = \pm e^{C_1}$. (Since L_1 and L_2 are typically positive, the constant C will typically be positive.)

EXAMPLE 8

In a study of 45 species of unicellular algae, the relationship between cell volume V and cell biomass B was found to be

$$B \propto V^{0.794}$$

Find a differential equation that relates the relative growth rates of cell biomass and volume.

Solution

The relationship between cell biomass and volume can be expressed as

$$B(V) = CV^{0.794} \tag{8.38}$$

where C is the constant of proportionality. Comparing (8.38) with (8.37), we see that $k = 0.794$. It therefore follows from (8.35) that

$$\frac{1}{B}\frac{dB}{dt} = (0.794)\frac{1}{V}\frac{dV}{dt} \tag{8.39}$$

We can also get (8.39) by differentiating (8.38) with respect to V; that is,

$$\frac{dB}{dV} = C(0.794)V^{0.794-1}$$

Equation (8.38) allows us to eliminate C. Solving (8.38) for C, we find that $C = BV^{-0.794}$ and, therefore,

$$\frac{dB}{dV} = BV^{-0.794}(0.794)V^{0.794-1} = (0.794)BV^{-1}$$

Rearranging terms yields

$$\frac{dB}{B} = (0.794)\frac{dV}{V}$$

Dividing both sides by dt, we get

$$\frac{1}{B}\frac{dB}{dt} = (0.794)\frac{1}{V}\frac{dV}{dt}$$

which is the same as (8.39). ■

EXAMPLE 9

Homeostasis The nutrient content of a consumer (e.g., the percent nitrogen of the consumer's biomass) can range from reflecting the nutrient content of its food to being constant. The former is referred to as *absence of homeostasis*, the latter as *strict homeostasis*. A model for homeostatic regulation is provided in Sterner and Elser (2002). The model relates a consumer's nutrient content (denoted by y) to its food's nutrient content (denoted by x) as

$$\frac{dy}{dx} = \frac{1}{\theta}\frac{y}{x} \tag{8.40}$$

where $\theta \geq 1$ is a constant. Solve the differential equation and relate θ to absence of homeostasis and strict homeostasis.

Solution   We can solve (8.40) by separation of variables:

$$\int \frac{dy}{y} = \frac{1}{\theta}\int \frac{dx}{x}$$

Integrating and simplifying yields

$$\ln|y| = \frac{1}{\theta}\ln|x| + C_1$$
$$|y| = e^{(1/\theta)\ln|x|+C_1}$$
$$|y| = |x|^{1/\theta}e^{C_1}$$
$$y = \pm e^{C_1}x^{1/\theta}$$

Since x and y are positive (they denote nutrient contents), it follows that

$$y = Cx^{1/\theta}$$

where C is a positive constant.

Absence of homeostasis means that the consumer reflects the food's nutrient content. This occurs when $y = Cx$ and thus when $\theta = 1$. *Strict homeostasis* means that the nutrient content of the consumer is independent of the nutrient content of the food; that is, $y = C$; this occurs in the limit as $\theta \to \infty$. ■

Section 8.1 Problems

■ **8.1.1**

In Problems 1–8, solve each pure-time differential equation.

1. $\frac{dy}{dx} = x + \sin x$, where $y_0 = 0$ for $x_0 = 0$

2. $\frac{dy}{dx} = e^{-3x}$, where $y_0 = 10$ for $x_0 = 0$

3. $\frac{dy}{dx} = \frac{1}{x}$, where $y_0 = 0$ when $x_0 = 1$

4. $\dfrac{dy}{dx} = \dfrac{1}{1+x^2}$, where $y_0 = 1$ when $x_0 = 0$

5. $\dfrac{dx}{dt} = \dfrac{1}{1-t}$, where $x(0) = 2$

6. $\dfrac{dx}{dt} = \cos(2\pi(t-3))$, where $x(3) = 1$

7. $\dfrac{ds}{dt} = \sqrt{3t+1}$, where $s(0) = 1$

8. $\dfrac{dh}{dt} = 5 - 16t^2$, where $h(3) = -11$

9. Suppose that the volume $V(t)$ of a cell at time t changes according to

$$\frac{dV}{dt} = 1 + \cos t \quad \text{with } V(0) = 5$$

Find $V(t)$.

10. Suppose that the amount of phosphorus in a lake at time t, denoted by $P(t)$, follows the equation

$$\frac{dP}{dt} = 3t + 1 \quad \text{with } P(0) = 0$$

Find the amount of phosphorus at time $t = 10$.

■ 8.1.2

In Problems 11–16, solve the given autonomous differential equations.

11. $\dfrac{dy}{dx} = 3y$, where $y_0 = 2$ for $x_0 = 0$

12. $\dfrac{dy}{dx} = 2(1-y)$, where $y_0 = 2$ for $x_0 = 0$

13. $\dfrac{dx}{dt} = -2x$, where $x(1) = 5$

14. $\dfrac{dx}{dt} = 1 - 3x$, where $x(-1) = -2$

15. $\dfrac{dh}{ds} = 2h + 1$, where $h(0) = 4$

16. $\dfrac{dN}{dt} = 5 - N$, where $N(2) = 3$

17. Suppose that a population, whose size at time t is denoted by $N(t)$, grows according to

$$\frac{dN}{dt} = 0.3N(t) \quad \text{with } N(0) = 20$$

Solve this differential equation, and find the size of the population at time $t = 5$.

18. Suppose that you follow the size of a population over time. When you plot the size of the population versus time on a semilog plot (i.e., the horizontal axis, representing time, is on a linear scale, whereas the vertical axis, representing the size of the population, is on a logarithmic scale), you find that your data fit a straight line which intercepts the vertical axis at 1 (on the log scale) and has slope -0.43. Find a differential equation that relates the growth rate of the population at time t to the size of the population at time t.

19. Suppose that a population, whose size at time t is denoted by $N(t)$, grows according to

$$\frac{1}{N}\frac{dN}{dt} = r \tag{8.41}$$

where r is a constant.

(a) Solve (8.41).

(b) Transform your solution in (a) appropriately so that the resulting graph is a straight line. How can you determine the constant r from your graph?

(c) Suppose now that, over time, you followed a population which evolved according to (8.41). Describe how you would determine r from your data.

20. Assume that $W(t)$ denotes the amount of radioactive material in a substance at time t. Radioactive decay is then described by the differential equation

$$\frac{dW}{dt} = -\lambda W(t) \quad \text{with } W(0) = W_0 \tag{8.42}$$

where λ is a positive constant called the *decay constant.*

(a) Solve (8.42).

(b) Assume that $W(0) = 123\,\text{gr}$ and $W(5) = 20\,\text{gr}$ and that time is measured in minutes. Find the decay constant λ and determine the half-life of the radioactive substance.

21. Suppose that a population, whose size at time t is given by $N(t)$, grows according to

$$\frac{dN}{dt} = \frac{1}{100}N^2, \quad \text{with } N(0) = 10 \tag{8.43}$$

(a) Solve (8.43).

(b) Graph $N(t)$ as a function of t for $0 \le t < 10$. What happens as $t \to 10$? Explain in words what this means.

22. Denote by $L(t)$ the length of a fish at time t, and assume that the fish grows according to the von Bertalanffy equation

$$\frac{dL}{dt} = k(34 - L(t)) \quad \text{with } L(0) = 2 \tag{8.44}$$

(a) Solve (8.44).

(b) Use your solution in (a) to determine k under the assumption that $L(4) = 10$. Sketch the graph of $L(t)$ for this value of k.

(c) Find the length of the fish when $t = 10$.

(d) Find the asymptotic length of the fish; that is, find $\lim_{t\to\infty} L(t)$.

23. Denote by $L(t)$ the length of a certain fish at time t, and assume that this fish grows according to the von Bertalanffy equation

$$\frac{dL}{dt} = k(L_\infty - L(t)) \quad \text{with } L(0) = 1 \tag{8.45}$$

where k and L_∞ are positive constants. A study showed that the asymptotic length is equal to 123 in and that it takes this fish 27 months to reach half its asymptotic length.

(a) Use this information to determine the constants k and L_∞ in (8.45). [*Hint*: Solve (8.45).]

(b) Determine the length of the fish after 10 months.

(c) How long will it take until the fish reaches 90% of its asymptotic length?

24. Let $N(t)$ denote the size of a population at time t. Assume that the population exhibits exponential growth.

(a) If you plot $\log N(t)$ versus t, what kind of graph do you get?

(b) Find a differential equation that describes the growth of this population and sketch possible solution curves.

25. Use the partial-fraction method to solve

$$\frac{dy}{dx} = y(1+y)$$

where $y_0 = 2$ for $x_0 = 0$.

26. Use the partial-fraction method to solve

$$\frac{dy}{dx} = y(1-y)$$

where $y_0 = 2$ for $x_0 = 0$.

27. Use the partial-fraction method to solve

$$\frac{dy}{dx} = y(y-5)$$

where $y_0 = 1$ for $x_0 = 0$.

28. Use the partial-fraction method to solve

$$\frac{dy}{dx} = (y-1)(y-2)$$

where $y_0 = 0$ for $x_0 = 0$.

29. Use the partial-fraction method to solve

$$\frac{dy}{dx} = 2y(3-y)$$

where $y_0 = 5$ for $x_0 = 1$.

30. Use the partial-fraction method to solve

$$\frac{dy}{dt} = \frac{1}{2}y^2 - 2y$$

where $y_0 = -3$ for $t_0 = 0$.

In Problems 31–34, solve the given differential equations.

31. $\dfrac{dy}{dx} = y(1+y)$ **32.** $\dfrac{dy}{dx} = (1+y)^2$

33. $\dfrac{dy}{dx} = (1+y)^3$ **34.** $\dfrac{dy}{dx} = (3-y)(2+y)$

35. (a) Use partial fractions to show that

$$\int \frac{du}{u^2 - a^2} = \frac{1}{2a} \ln \left| \frac{u-a}{u+a} \right| + C$$

(b) Use your result in (a) to find a solution of

$$\frac{dy}{dx} = y^2 - 4$$

that passes through (i) $(0,0)$, (ii) $(0,2)$, and (iii) $(0,4)$.

36. Find a solution of

$$\frac{dy}{dx} = y^2 + 4$$

that passes through $(0,2)$.

37. Suppose that the size of a population at time t is denoted by $N(t)$ and that $N(t)$ satisfies the differential equation

$$\frac{dN}{dt} = 0.34N \left(1 - \frac{N}{200} \right) \quad \text{with } N(0) = 50$$

Solve this differential equation, and determine the size of the population in the long run; that is, find $\lim_{t \to \infty} N(t)$.

38. Assume that the size of a population, denoted by $N(t)$, evolves according to the logistic equation. Find the intrinsic rate of growth if the carrying capacity is 100, $N(0) = 10$, and $N(1) = 20$.

39. Suppose that $N(t)$ denotes the size of a population at time t and that

$$\frac{dN}{dt} = 1.5N \left(1 - \frac{N}{50} \right)$$

(a) Solve this differential equation when $N(0) = 10$.
(b) Solve this differential equation when $N(0) = 90$.
(c) Graph your solutions in (a) and (b) in the same coordinate system.
(d) Find $\lim_{t \to \infty} N(t)$ for your solutions in (a) and (b).
40. Suppose that the size of a population, denoted by $N(t)$, satisfies

$$\frac{dN}{dt} = 0.7N \left(1 - \frac{N}{35} \right) \tag{8.46}$$

(a) Determine all equilibria by solving $dN/dt = 0$.
(b) Solve (8.46) for **(i)** $N(0) = 10$, **(ii)** $N(0) = 35$, **(iii)** $N(0) = 50$, and **(iv)** $N(0) = 0$. Find $\lim_{t \to \infty} N(t)$ for each of the four initial conditions.
(c) Compare your answer in (a) with the limiting values you found in (b).
41. Let $N(t)$ denote the size of a population at time t. Assume that the population evolves according to the logistic equation. Assume also that the intrinsic growth rate is 5 and that the carrying capacity is 30.
(a) Find a differential equation that describes the growth of this population.
(b) Without solving the differential equation in (a), sketch solution curves of $N(t)$ as a function of t when **(i)** $N(0) = 10$, **(ii)** $N(0) = 20$, and **(iii)** $N(0) = 40$.
42. Logistic growth is described by the differential equation

$$\frac{dN}{dt} = rN \left(1 - \frac{N}{K} \right)$$

The solution of this differential equation with initial condition $N(0) = N_0$ is given by

$$N(t) = \frac{K}{1 + (\frac{K}{N_0} - 1)e^{-rt}} \tag{8.47}$$

(a) Show that

$$r = \frac{1}{t} \ln \left(\frac{K - N_0}{N_0} \right) + \frac{1}{t} \ln \left(\frac{N(t)}{K - N(t)} \right) \tag{8.48}$$

by solving (8.47) for r.
(b) Equation (8.48) can be used to estimate r. Suppose we follow a population that grows according to the logistic equation and find that $N(0) = 10$, $N(5) = 22$, $N(100) = 30$, and $N(200) = 30$. Estimate r.
43. Selection at a Single Locus We consider one locus with two alleles, A_1 and A_2, in a randomly mating diploid population. That is, each individual in the population is either of type $A_1 A_1$, $A_1 A_2$, or $A_2 A_2$. We denote by $p(t)$ the frequency of the A_1 allele and by $q(t)$ the frequency of the A_2 allele in the population at time t. Note that $p(t) + q(t) = 1$. We denote the fitness of the $A_i A_j$ type by w_{ij} and assume that $w_{11} = 1$, $w_{12} = 1 - s/2$, and $w_{22} = 1 - s$, where s is a nonnegative constant less than or equal to 1. That is, the fitness of the heterozygote $A_1 A_2$ is halfway between the fitness of the two homozygotes, and the type $A_1 A_1$ is the fittest. If s is small, we can show that, approximately,

$$\frac{dp}{dt} = \frac{1}{2} sp(1-p) \quad \text{with } p(0) = p_0 \tag{8.49}$$

(a) Use separation of variables and partial fractions to find the solution of (8.49).
(b) Suppose $p_0 = 0.1$ and $s = 0.01$; how long will take until $p(t) = 0.5$?
(c) Find $\lim_{t \to \infty} p(t)$. Explain in words what this limit means.

■ **8.1.3**

In Problems 44–52, solve each differential equation with the given initial condition.

44. $\dfrac{dy}{dx} = 2\dfrac{y}{x}$, with $y_0 = 1$ if $x_0 = 1$

45. $\dfrac{dy}{dx} = \dfrac{x+1}{y}$, with $y_0 = 2$ if $x_0 = 0$

46. $\dfrac{dy}{dx} = \dfrac{y}{x+1}$, with $y_0 = 1$ if $x_0 = 0$

47. $\dfrac{dy}{dx} = (y+1)e^{-x}$, with $y_0 = 2$ if $x_0 = 0$

48. $\dfrac{dy}{dx} = x^2 y^2$, with $y_0 = 1$ if $x_0 = 1$

49. $\dfrac{dy}{dx} = \dfrac{y+1}{x-1}$, with $y_0 = 5$ if $x_0 = 2$

50. $\dfrac{du}{dt} = \dfrac{\sin t}{u^2 + 1}$, with $u_0 = 3$ if $t_0 = 0$

51. $\dfrac{dr}{dt} = re^{-t}$, with $r_0 = 1$ if $t_0 = 0$

52. $\dfrac{dx}{dy} = \dfrac{1}{2}\dfrac{x}{y}$, with $x_0 = 2$ if $y_0 = 3$

53. *(Adapted from Reiss, 1989)* In a case study by Taylor et al. (1980) in which the maximal rate of oxygen consumption (in ml s^{-1}) for nine species of wild African mammals was plotted against body mass (in kg) on a log–log plot, it was found that the data points fall on a straight line with slope approximately equal to 0.8. Find a differential equation that relates maximal oxygen consumption to body mass.

54. Consider the following differential equation, which is important in population genetics:

$$a(x)g(x) - \frac{1}{2}\frac{d}{dx}\left[b(x)g(x)\right] = 0$$

Here, $b(x) > 0$.

(a) Define $y = b(x)g(x)$, and show that y satisfies

$$\frac{a(x)}{b(x)}y - \frac{1}{2}\frac{dy}{dx} = 0 \qquad (8.50)$$

(b) Separate variables in (8.50), and show that if $y > 0$, then

$$y = C\exp\left[2\int \frac{a(x)}{b(x)}\,dx\right]$$

55. When phosphorus content in *Daphnia* was plotted against phosphorus content of its algal food on a log–log plot, a straight line with slope $1/7.7$ resulted. (See Sterner and Elser, 2002; data from DeMott et al., 1998.) Find a differential equation that relates the phosphorus content of *Daphnia* to the phosphorus content of its algal food.

56. This problem addresses Malthus's concerns. Assume that a population size grows exponentially according to

$$N(t) = 1000e^t$$

and the food supply grows linearly according to

$$F(t) = 3t$$

(a) Write a differential equation for each of $N(t)$ and $F(t)$.

(b) What assumptions do you need to make to be able to compare whether and, if so, when food supply will be insufficient? Does exponential growth eventually overtake linear growth? Explain.

(c) Do a Web search to determine whether food supply has grown linearly, as claimed by Malthus.

57. At the beginning of this section, we modified the exponential-growth equation to include oscillations in the per capita growth rate. Solve the differential equation we obtained, namely,

$$\frac{dN}{dt} = 2\left(1 + \sin(2\pi t)\right)N(t)$$

with $N(0) = 5$.

■ 8.2 Equilibria and Their Stability

In Subsection 8.1.2, we learned how to solve autonomous differential equations and graphed their solutions as functions of the independent variable for given initial conditions. For instance, logistic growth

$$\frac{dN}{dt} = rN\left(1 - \frac{N}{K}\right) \qquad (8.51)$$

with initial condition $N(0) = N_0$ has the solution given in (8.33) and graphed in Figure 8.10 for different initial values.

The solution of a differential equation can inform us about long-term behavior, as we saw in the case of logistic growth. In particular, if $N_0 > 0$, then $N(t) \to K$, the carrying capacity, as $t \to \infty$, and if $N_0 = 0$, then $N(t) = 0$ for all $t > 0$. Also, if $N_0 = K$, then $N(t) = K$ for all $t > 0$. What is so special about $N_0 = K$ or $N_0 = 0$? We see from Equation (8.51) that if $N = K$ or $N = 0$, then $dN/dt = 0$, implying that $N(t)$ is constant.

Constant solutions form a very special class of solutions of autonomous differential equations. These solutions are called **point equilibria** or, simply, equilibria. The constant solutions $N = K$ and $N = 0$ are point equilibria of the logistic equation.

In this section, we will consider autonomous differential equations of the form

$$\frac{dy}{dx} = g(y) \tag{8.52}$$

where we will typically think of x as time. We will learn how to find point equilibria, and we will discuss what they can tell us about the long-term behavior of the solution $y = y(x)$—that is, the behavior of $y(x)$ as $x \to \infty$. If we can solve (8.52), we can study the solution directly to obtain information about its long-term behavior. But what should we do if we cannot solve (8.52)?

Candidates for describing long-term behavior are the constant solutions or equilibria $y = \hat{y}$ (read "y hat") that satisfy $g(\hat{y}) = 0$. Such solutions, of course, need not exist. The following holds, however, if they do exist:

If $\hat{y}$ satisfies
$$g(\hat{y}) = 0$$
then $\hat{y}$ is an equilibrium of
$$\frac{dy}{dx} = g(y)$$

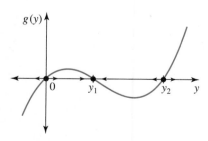

Figure 8.11 Stability illustrated with a ball on a hill and in a valley.

Let's look at equilibria in more detail before we discuss specific examples. The basic property of equilibria is that if, initially (say, at $x = 0$), $y(0) = \hat{y}$ and $\hat{y}$ is an equilibrium, then $y(x) = \hat{y}$ for all $x > 0$.

A physical analogue of equilibria is provided in Figure 8.11. On the left side, a ball rests on top of a hill; on the right side, a ball rests at the bottom of a valley. In either case, the ball is in equilibrium because it does not move.

Of great interest is the **stability** of equilibria. What we mean by this is best explained by our example of a ball on a hill versus a ball in the valley. If we perturb the ball by a small amount—that is, if we move it out of its equilibrium slightly—the ball on the left side will roll down the hill and not return to the top, whereas the ball on the right side will return to the bottom of the valley. We call the situation on the left side *unstable* and the situation on the right side *stable*.

The analogue of stability for equilibria of differential equations is as follows: Suppose that $\hat{y}$ is an equilibrium of $\frac{dy}{dx} = g(y)$; that is, $g(\hat{y}) = 0$. We say that $\hat{y}$ is **locally stable** if the solution returns to $\hat{y}$ after a small perturbation; this means that we look at what happens to the solution when we start close to the equilibrium (i.e., the solution moves away from the equilibrium by a small amount, called a *small perturbation*). If the solution does not return to the equilibrium after a small perturbation, we say that $\hat{y}$ is **unstable**. These concepts will be developed in the next subsection, in which we will discuss a graphical and an analytical method for analyzing stability of equilibria. A number of applications then follow in the subsequent subsections.

■ 8.2.1 A First Look at Stability

Figure 8.12 The function $g(y)$. The arrows close to the equilibria indicate the type of stability.

Graphical Approach Suppose that $g(y)$ is of the form given in Figure 8.12. To find the equilibria of (8.52), we set $g(y) = 0$. Graphically, this means that if we graph $g(y)$ (i.e., the *derivative* of y with respect to x) as a function of y, then the equilibria are the points of intersection of $g(y)$ with the horizontal axis, which is the y-axis in this case, since y is the independent variable. (Look at the labels of the axes in the figure to see what is graphed there.) We see that, for our choice of $g(y)$, the equilibria are at $y = 0$, y_1, and y_2.

Why does this work? Remember, we are discussing *autonomous* differential equations. This means that the derivative of y (dy/dx) is a function of y; it does *not* depend explicitly on x. This fact allows us to graph the derivative of y as a function of y. Since $dy/dx = g(y)$, we can graph $g(y)$ as a function of y. We can then use the graph of $g(y)$ to say the following about the fate of a solution on the basis of its current value: If the current value y is such that $g(y) > 0$ (i.e., $dy/dx > 0$), then y

will increase as a function of x; if, however, y is such that $g(y) < 0$ (i.e., $dy/dx < 0$), then y will decrease as a function of x. The points y where $g(y) = 0$ are the points where y will not change as a function of x [since $g(y) = dy/dx = 0$]. These points are the equilibria.

Equilibria are characterized by the property that a system in an equilibrium state stays there for all later times (unless some external force disturbs the system). Note that this implies *neither* that the system will necessarily reach a particular equilibrium when starting from some initial value that is different from the equilibrium nor that the system will return to the equilibrium after a small perturbation.

Whether or not the system will return to an equilibrium after a small perturbation depends on the local stability of the equilibrium. By this statement, we mean the following: Suppose that the system is in equilibrium, which we denote by $\hat{y}$. We apply a small perturbation to the system, so that after the perturbation the new state of the system is

$$y = \hat{y} + z \tag{8.53}$$

where z is small and may be positive or negative. We explain what can happen to the system with the use of the function $g(y)$ in Figure 8.12.

Suppose that $\hat{y} = 0$ and we subject this equilibrium to a small perturbation. If the new value $y = \hat{y} + z = z > 0$ for z small, then $dy/dx > 0$; that is, y will increase. If the new value $y = \hat{y} + z = z < 0$ for z small, then $dy/dx < 0$ and y will decrease. In either case, the system will not return to 0. We say that $\hat{y} = 0$ is *unstable*.

We now turn to the equilibrium $\hat{y} = y_1$. We perturb the equilibrium to $y = y_1 + z$ for z small. From Figure 8.12, if $z > 0$, then $dy/dx < 0$ and hence y decreases; if $z < 0$, then $dy/dx > 0$ and y increases. The system will therefore return to the equilibrium y_1 after a small perturbation. We say that y_1 is *locally stable*. The attribute *locally* refers to the fact that the system will return to y_1 if the perturbation is sufficiently small. It does not say anything about what happens when the perturbation is large. For instance, if the perturbation z is large and the new value is less than 0, then $dy/dx < 0$ and the system will not return to y_1.

Analyzing the equilibrium y_2 in the same way shows that it is an unstable equilibrium, just like the equilibrium $\hat{y} = 0$.

The preceding discussion illustrates the fact that it is not necessarily the case that the system will reach an equilibrium value. The only equilibrium in Figure 8.12 that can be reached is y_1: If, initially, $y(0) \in (0, y_2)$, then $y(x)$ will approach y_1; if, however, $y(0) < 0$, then $y(x) \to -\infty$, and if $y > y_2$, then $y(x) \to \infty$.

The preceding discussion also illustrates that we can subject an equilibrium only to a small perturbation if we want to learn something about its stability: If we perturb y_1 too much, so that the value after the perturbation is either less than 0 or greater than y_2, the solution will not return to the equilibrium value y_1.

EXAMPLE 1

Let $N(t)$ denote the size of a population at time t such that the population evolves according to the logistic equation

$$\frac{dN}{dt} = 2N\left(1 - \frac{N}{100}\right) \quad \text{for } N \geq 0$$

Find the equilibria and analyze their stability.

Solution

To find the equilibria, we set $\frac{dN}{dt} = 0$; that is,

$$2N\left(1 - \frac{N}{100}\right) = 0$$

We see that either

$$N_1 = 0 \quad \text{or} \quad N_2 = 100$$

To analyze the stability, we draw the graph of dN/dt versus N in Figure 8.13. Note that $N \geq 0$, since N represents the size of a population. To perturb the trivial equilibrium $N_1 = 0$, we therefore need choose only values that are slightly bigger than 0.

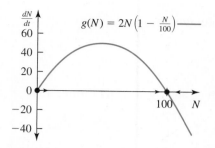

Figure 8.13 The graph of dN/dt versus N in Example 1.

We see from the graph that such small perturbations result in $dN/dt > 0$, and hence N increases. This implies that $N_1 = 0$ is an unstable equilibrium.

To perturb the nontrivial equilibrium $N_2 = 100$, we can either increase or decrease the population size a bit. If we decrease it, then $dN/dt > 0$ and the population size will increase. If we increase it, then $dN/dt < 0$ and the population size will decrease. That is, if the system is subjected to a small perturbation about the nontrivial equilibrium $N_2 = 100$, the population size will return to the equilibrium value of $N_2 = 100$. This implies that $N_2 = 100$ is a locally stable equilibrium. ■

Analytical Approach   We assume that $\hat{y}$ is an equilibrium of

$$\frac{dy}{dx} = g(y)$$

[Thus, $\hat{y}$ satisfies $g(\hat{y}) = 0$.] We consider a small perturbation about the equilibrium $\hat{y}$; we express this perturbation as

$$y = \hat{y} + z$$

where z is small and may be either positive or negative. Then

$$\frac{dy}{dx} = \frac{d}{dx}(\hat{y} + z) = \frac{dz}{dx}$$

since $d\hat{y}/dx = 0$ ($\hat{y}$ is a constant). We find that

$$\frac{dz}{dx} = g(\hat{y} + z)$$

If z is sufficiently small, we can approximate $g(\hat{y} + z)$ by its linear approximation. The linear approximation of $g(y)$ about $\hat{y}$ is given by

$$L(y) = g(\hat{y}) + (y - \hat{y})g'(\hat{y})$$

Since $g(\hat{y}) = 0$,

$$L(y) = (y - \hat{y})g'(\hat{y})$$

Therefore, the linear approximation of $g(\hat{y} + z)$ is given by

$$L(\hat{y} + z) = (\hat{y} + z - \hat{y})g'(\hat{y}) = zg'(\hat{y})$$

If we set

$$\lambda = g'(\hat{y})$$

then

$$\frac{dz}{dx} = \lambda z$$

is the first-order approximation of the perturbation. This equation has the solution

$$z(x) = z(0)e^{\lambda x} \tag{8.54}$$

which has the property that

$$\lim_{x \to \infty} z(x) = 0 \quad \text{if } \lambda < 0$$

Now, on the one hand, since $y(x) = \hat{y} + z(x)$, it follows that if $\lambda < 0$, the system returns to the equilibrium $\hat{y}$ after a small perturbation $z(0)$. This means that $\hat{y}$ is locally stable if $\lambda < 0$. On the other hand, if $\lambda > 0$, then $z(x)$ does not go to 0 as $x \to \infty$, implying that the system will *not* return to the equilibrium $\hat{y}$ after a small perturbation, and $\hat{y}$ is unstable. The value $\lambda = g'(\hat{y})$ is called an **eigenvalue** and is the slope of the tangent line of $g(y)$ at $\hat{y}$. This is summarized in the following box:

Stability Criterion Consider the differential equation

$$\frac{dy}{dx} = g(y)$$

where $g(y)$ is a differentiable function. Assume that $\hat{y}$ is an equilibrium; that is, $g(\hat{y}) = 0$. Then

$$\hat{y} \text{ is locally stable if } g'(\hat{y}) < 0$$
$$\hat{y} \text{ is unstable if } g'(\hat{y}) > 0$$

When the eigenvalue $\lambda = 0$, the first-order approximation $\frac{dz}{dx} = \lambda z$ does not allow us to draw any conclusions about the behavior of $z(x)$, since higher-order terms then become important.

We wish to tie this analytical approach in with the more informal graphical analysis presented at the beginning of this section. Looking at Figure 8.12, we find that the slope of the tangent line at $y = 0$ is positive [i.e., $g'(0) > 0$]; similarly, we find that $g'(y_1) < 0$ and $g'(y_2) > 0$. Hence, the equilibria 0 and y_2 are unstable and the equilibrium y_1 is locally stable, as found in the graphical analysis.

Let's try out the analytical approach on Example 1. There,

$$g(N) = 2N\left(1 - \frac{N}{100}\right)$$

To differentiate $g(N)$, we multiply out:

$$g(N) = 2N - \frac{N^2}{50}$$

Differentiating $g(N)$ yields

$$g'(N) = 2 - \frac{2N}{50}$$

If $N = 0$, then $g'(0) = 2 > 0$; thus, $N = 0$ is an unstable equilibrium. If $N = 100$, then $g'(100) = 2 - 200/50 = -2 < 0$; hence, $N = 100$ is a locally stable equilibrium.

The analytical approach is more powerful than the graphical approach: In addition to determining whether an equilibrium is locally stable or unstable, the analytical approach allows us to say something about how quickly a solution returns to an equilibrium after a small perturbation. This property follows from (8.54). There, we found that the perturbation has the approximate solution

$$z(x) = z(0)e^{\lambda x}$$

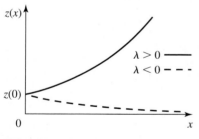

Figure 8.14 The graph of $z(x) = z(0)e^{\lambda x}$ for $\lambda > 0$ and $\lambda < 0$.

If $\lambda > 0$, then the larger λ, the faster the solution moves away from the equilibrium. If $\lambda < 0$, then the more negative λ, the faster the solution will return to the equilibrium after a small perturbation. (See Figure 8.14.)

The preceding derivation of the analytical stability criterion was based on linearizing $g(y)$ about the equilibrium $\hat{y}$. Since the linearization is close only for values close to $\hat{y}$ [unless $g(y)$ is linear], the perturbations about the equilibrium must be small, and hence the stability analysis is always local (i.e., within close vicinity of the equilibrium). When $g(y)$ is linear, the analysis is exact; in particular, we can compute exactly how quickly a solution returns to a locally stable equilibrium (or moves away from an unstable equilibrium) after a perturbation.

EXAMPLE 2 **$g(y)$ Is Linear** Show that the differential equation

$$\frac{dy}{dx} = 1 - y \tag{8.55}$$

has a locally stable equilibrium at $\hat{y} = 1$, and determine how quickly a solution starting at $y(0) = y_0 \neq 1$ will reach $\hat{y} = 1$.

Solution

Since $g(y) = 0$ for $y = 1$, $\hat{y} = 1$ is an equilibrium. Differentiating $g(y) = 1 - y$, we find that $g'(y) = -1$, which is negative regardless of y. Therefore, $\hat{y} = 1$ is a locally stable equilibrium.

We can solve (8.55) exactly by separation of variables:

$$\int \frac{dy}{1 - y} = \int dx$$
$$-\ln|1 - y| = x + C_1$$
$$1 - y = \pm e^{-C_1} e^{-x}$$
$$1 - y = Ce^{-x}$$

Note that we set $\pm e^{-C_1} = C$. Solving for y yields

$$y(x) = 1 - Ce^{-x}$$

If we set $y(0) = y_0$, then $y_0 = 1 - C$, and the solution is

$$y(x) = 1 - (1 - y_0)e^{-x}$$

We see that, for any initial value y_0,

$$\lim_{x \to \infty} y(x) = 1$$

and it takes an infinite amount of time to reach the equilibrium $\hat{y} = 1$ if $y_0 \neq 1$.

Instead of asking how long it takes until the equilibrium is reached (which yielded the uninformative answer "an infinite amount of time"), we compute the time it takes to reduce the initial deviation from the equilibrium, $y_0 - 1$, to a fraction e^{-1}; that is, we wish to determine the number x_R such that

$$\underbrace{y(x_R) - 1}_{\text{deviation from } x = x_R} = \underbrace{e^{-1}(y_0 - 1)}_{\text{initial deviation}}$$

Since $y(x_R) - 1 = -(1 - y_0)e^{-x_R}$, it follows that

$$-(1 - y_0)e^{-x_R} = e^{-1}(y_0 - 1)$$
$$e^{-x_R} = e^{-1}$$
$$x_R = 1$$

(See Figure 8.15.) It takes one unit of time to reduce the initial deviation to a fraction e^{-1}; this does not depend on how large the initial deviation is. ■

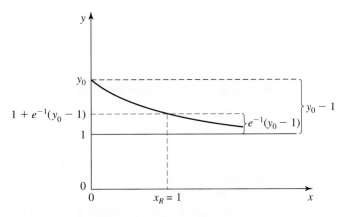

Figure 8.15 An illustration of the time x_R in Example 2.

Using Both Approaches

EXAMPLE 3 Suppose that

$$\frac{dy}{dx} = y(4 - y)$$

Find the equilibria of this differential equation and discuss their stability, using both the graphical and the analytical approach.

Solution To find the equilibria, we set $dy/dx = 0$. That is,

$$y(4 - y) = 0$$

which yields

$$y_1 = 0 \quad \text{and} \quad y_2 = 4$$

On the one hand, we set $g(y) = y(4 - y)$ and graph $g(y)$ (see Figure 8.16); we see that $y_1 = 0$ is unstable, since, if we perturb $y_1 = 0$ to $y = z > 0$, where z is small, then $g(y) = dy/dx > 0$, and if we perturb $y_1 = 0$ to $y = z < 0$, where z is small, then $g(y) = dy/dx < 0$. That is, in either case, y will not return to 0. On the other hand, y_2 is a locally stable equilibrium: A small perturbation to the right of $y = y_2$ results in $dy/dx < 0$, whereas a small perturbation to the left results in $dy/dx > 0$. In either case, the solution y will return to $y_2 = 4$.

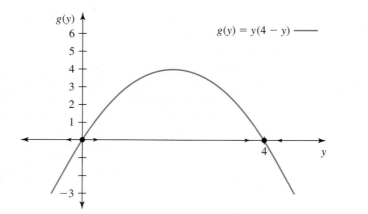

Figure 8.16 The graph of $g(y)$ in Example 3.

If we use the analytical approach, we need to compute the eigenvalues. The eigenvalue associated with $y_1 = 0$ is

$$\lambda_1 = g'(0) = 4 - 2y|_{y=0} = 4 > 0$$

which implies that y_1 is unstable. The eigenvalue associated with $y_2 = 4$ is

$$\lambda_2 = g'(4) = 4 - 2y|_{y=4} = -4 < 0$$

which implies that $y_2 = 4$ is locally stable. ■

■ 8.2.2 Single Compartment or Pool

Figure 8.17 Input and output rates are the same: The water in the tank remains at the same level.

This example is adapted from DeAngelis (1992). Compartment models are frequently used to model the flow of matter (nutrients, energy, and so forth). The simplest such model consists of one compartment—for instance, a fixed volume V of water (such as a tank or lake) containing a solute (such as phosphorus). Assume that water enters the compartment at a constant rate q and leaves the compartment at the same rate. (See Figure 8.17.) (Having the same input and output rate keeps the volume of the pool constant.) We will investigate the effects of different input concentrations of the solution on the concentration of the solution in the pool.

We denote by $C(t)$ the concentration of the solution in the compartment at time t. Then the total mass of the solute is $C(t)V$, where V is the volume of the compartment. For instance, if the concentration of the solution is 2 grams per liter and the volume of the compartment is 10 liters, then the total mass of the solute in the compartment is 2 g liter^{-1} times 10 liters, which is equal to 20 g.

If C_I is the concentration of the incoming solution and q is the rate at which water enters, then qC_I, the **input loading**, is the rate at which mass enters. For instance, if the concentration of the incoming solution is 5 g liter^{-1} and the rate at which the solution enters is 0.1 liter s^{-1}, then the input loading—that is, the rate at which mass enters—is 5 g liter^{-1} times 0.1 liter s^{-1}, which is equal to 0.5 g s^{-1}.

If we assume that the solution in the compartment is well mixed, so that the outflowing solution has the same concentration as the solution in the compartment— namely, $C(t)$ at time t—then $qC(t)$ is the rate at which mass leaves the compartment at time t.

Figure 8.18 Flow diagram for the single-compartment model.

These different processes can be schematically illustrated in a flow diagram, as shown in Figure 8.18. Because mass is conserved in the system, we can use the **law of conservation of mass** to derive an equation that describes the flow of matter in this system:

$$\begin{bmatrix} \text{rate of change} \\ \text{of mass of} \\ \text{solute in pool} \end{bmatrix} = \begin{bmatrix} \text{rate at which} \\ \text{mass enters} \end{bmatrix} - \begin{bmatrix} \text{rate at which} \\ \text{mass leaves} \end{bmatrix}$$

Writing C for $C(t)$ (and being careful not to confuse C with a constant), we obtain

$$\frac{d}{dt}(CV) = qC_I - qC \tag{8.56}$$

Since V is constant, we can write (8.56) as

$$\frac{dC}{dt} = \frac{q}{V}(C_I - C) \tag{8.57}$$

This is a linear differential equation of the type discussed in Subsection 8.1.2, Case 1. It can be solved by separation of variables. We skip the steps and instead concentrate on the discussion of the system. If $C(0) = C_0$, then the solution of the differential equation is

$$C(t) = C_I\left[1 - \left(1 - \frac{C_0}{C_I}\right)e^{-(q/V)t}\right] \tag{8.58}$$

Solution curves for different values of C_0 are shown in Figure 8.19. From (8.57), we conclude that C_I is the only equilibrium. Looking at the solution $C(t)$ in (8.58), we see that

$$\lim_{t \to \infty} C(t) = C_I$$

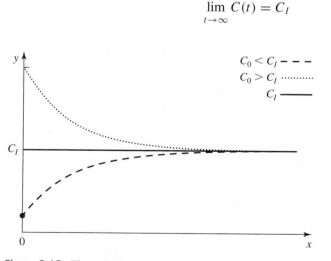

Figure 8.19 The solution curves for the single-compartment model for different values of C_0.

regardless of the initial concentration C_0 in the compartment. This shows that C_I is **globally stable**, which implies that no matter how much we perturb the equilibrium, the system will return to it.

We can also obtain this result by using the eigenvalue method. We write

$$C(t) = C_I + z(t)$$

Then, as before,

$$\frac{dC(t)}{dt} = \frac{d}{dt}[C_I + z(t)] = \frac{dz(t)}{dt}$$

That is,

$$\frac{dz}{dt} = -\frac{q}{V}z(t) \qquad (8.59)$$

since $C_I - C(t) = -z(t)$. Note that (8.59) is exact; that is, we need not use a linear approximation for the right-hand side of (8.57), since it is already linear. We see from (8.59) that $-q/V$ is the eigenvalue associated with the equilibrium C_I. Since q and V are both positive, the eigenvalue is negative, and it follows that C_I is locally stable. We can obtain more information here, since (8.59) is exact with solution

$$z(t) = z(0)e^{-(q/V)t} \qquad (8.60)$$

which shows that, for *any* perturbation $z(0)$, $z(t) \to 0$ as $t \to \infty$. That is, the system returns to the equilibrium C_I [$C(t) \to C_I$ as $t \to \infty$] regardless of how much we perturb the system. The reason that the eigenvalue method allows us to show global stability lies in the fact that the differential equation (8.56) is *linear*. In other cases, the eigenvalue method allows us to obtain only local stability, because we must first linearize and, consequently, the linearized differential equation is only an approximation.

Equation (8.60) shows that the system returns to the equilibrium C_I exponentially fast and that the eigenvalue $-(q/V)$ determines the time scale—that is, how quickly the system reaches the equilibrium. The larger q/V, the faster the system recovers from a perturbation. In this context, we can define a return time to equilibrium (just as in Example 2), denoted by T_R. By convention, the return time to equilibrium, T_R, is defined as the amount of time it takes to reduce the initial difference $C_0 - C_I$ to a fraction e^{-1}, or

$$C(T_R) - C_I = e^{-1}(C_0 - C_I) \qquad (8.61)$$

Using (8.58), we see that

$$C(T_R) = C_I \left[1 - \left(1 - \frac{C_0}{C_I}\right)e^{-(q/V)T_R}\right]$$

which we can rewrite as

$$C(T_R) - C_I = (C_0 - C_I)e^{-(q/V)T_R} \qquad (8.62)$$

Equating (8.61) and (8.62), we find that

$$e^{-1}(C_0 - C_I) = (C_0 - C_I)e^{-(q/V)T_R}$$

which yields

$$1 = \frac{q}{V}T_R, \qquad \text{or} \qquad T_R = \frac{V}{q}$$

The last equation shows that the return time increases with the volume V and decreases with the flow rate q. This relationship can be understood intuitively: The larger the volume and the smaller the input rate, the longer the system takes to return to equilibrium. It can be shown that T_R is the mean residence time of a molecule of the solute; that is, when the system is in equilibrium, T_R is the average time a molecule of the solute spends in the compartment before leaving.

■ 8.2.3 The Levins Model

The ecological importance of spatial structure to the maintenance of populations was pointed out by Andrewartha and Birch (1954) on the basis of studies of insect populations. They observed that, although local populations frequently become extinct, their patches of habitat subsequently become recolonized by migrants from other patches occupied by individuals from the same kind of population, thus allowing the population to persist globally. Fifteen years later, Richard Levins introduced the concept of *metapopulations* (Levins, 1969). A major theoretical advance, the concept provided a framework for studying spatially structured populations.

A metapopulation is a collection of subpopulations. Each subpopulation occupies a patch, and different patches are linked via the migration of individuals between patches. (See Figure 8.20.) In this setting, we keep track only of what proportion of patches is occupied by subpopulations. Subpopulations go extinct at a constant rate, denoted by m, which stands for *mortality*. Vacant patches can be colonized at a rate that is proportional to the fraction of occupied patches; the constant of proportionality is denoted by c, which stands for *colonization rate*. If we denote by $p(t)$ the fraction of patches that are occupied at time t, then writing $p = p(t)$, we have

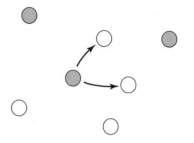

Figure 8.20 A schematic description of a metapopulation model. The shaded patches are occupied; arrows indicate migration events.

$$\frac{dp}{dt} = cp(1 - p) - mp \qquad (8.63)$$

The first term on the right-hand side describes the colonization process. Note that an increase in the fraction of occupied patches occurs only if a vacant patch becomes occupied—hence the product $p(1 - p)$ in the first term on the right-hand side. The minus sign in front of m shows that an extinction decreases the fraction of occupied patches.

We will not solve (8.63); instead, we focus on its equilibria. We set

$$cp(1 - p) - mp = 0$$

Isolating the factor cp, we obtain

$$cp\left(1 - \frac{m}{c} - p\right) = 0$$

which has the two solutions

$$p_1 = 0 \qquad \text{and} \qquad p_2 = 1 - \frac{m}{c}$$

We call the solution $p_1 = 0$ a trivial solution, because it corresponds to the situation in which all patches are vacant. Since individuals are not created spontaneously, a vacant patch can be recolonized only through migration from other, occupied patches. Therefore, once a metapopulation is extinct, it stays extinct. The other equilibrium, $p_2 = 1 - m/c$, is relevant only when $p_2 \in (0, 1]$, because p represents a fraction that is a number between 0 and 1. Since m and c are both positive, it follows immediately that $p_2 < 1$ for all choices of m and c. To see when $p_2 > 0$, we check

$$1 - \frac{m}{c} > 0$$

which holds when

$$m < c$$

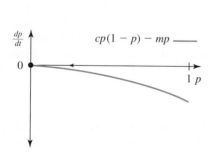

Figure 8.21a The case $m > c$.

That is, the nontrivial equilibrium $p_2 = 1 - m/c$ is in $(0, 1]$ if the extinction rate m is less than the colonization rate c. If $m \geq c$, then there is only one equilibrium in $[0, 1]$, namely, $p_1 = 0$. We illustrate this scenario in Figures 8.21a and 8.22a; looking at the graphs, we can analyze the stability of the equilibria.

Case 1: $m > c$ There is only the trivial equilibrium $p_1 = 0$. For any $p \in (0, 1]$, $dp/dt < 0$; hence, the fraction of occupied patches declines. The equilibrium is locally and globally stable.

Figure 8.22a The case $m < c$.

Case 2: $m < c$ There are two equilibria: 0 and $1 - m/c$. The trivial equilibrium $p_1 = 0$ is now unstable, since, if we perturb $p_1 = 0$ to some value in $(0, 1 - m/c)$, then $dp/dt > 0$, which implies that $p(t)$ increases. The system will therefore not return to 0.

The other equilibrium, $p_2 = 1 - m/c$, is locally stable. After a small perturbation of this equilibrium to the right of p_2, $dp/dt < 0$; a small perturbation to the left of p_2 gives $dp/dt > 0$. Therefore, the system will return to p_2.

We can also use the eigenvalue approach to analyze the stability of the equilibria. In addition, this approach will allow us to obtain information on how quickly the system returns to the stable equilibrium. We set

$$g(p) = cp(1 - p) - mp$$

To linearize this function about the equilibrium values, we must find

$$g'(p) = c - 2cp - m$$

Now, if $p_1 = 0$, then

$$g'(0) = c - m$$

whereas if $p_2 = 1 - m/c$, then

$$g'\left(1 - \frac{m}{c}\right) = c - 2c\left(1 - \frac{m}{c}\right) - m = c - 2c + 2m - m = m - c$$

From these equations, we see that $c - m$ is the eigenvalue corresponding to $p_1 = 0$ and $m - c$ is the eigenvalue corresponding to $p_2 = 1 - m/c$. We find that

if $\quad c - m < 0, \quad$ then $\quad p_1 = 0$ is locally stable

if $\quad m - c < 0, \quad$ then $\quad p_1 = 0$ is unstable and $p_2 = 1 - \dfrac{m}{c}$ is locally stable

To summarize our results, if $m > c$, then $p_1 = 0$ is the only equilibrium in $[0, 1]$ and $p_1 = 0$ is locally stable. (In fact, the graphical analysis showed that $p_1 = 0$ is *globally* stable.) Figure 8.21b shows a solution curve when $m = 2$ and $c = 1$, starting at $p(0) = 2$. We see that the solution approaches $p_1 = 0$. If $m < c$, then there are two equilibria in $[0, 1]$. The equilibrium $p_1 = 0$ is now unstable, and the equilibrium $p_2 = 1 - m/c$ is locally stable. Figure 8.22b shows solution curves when $m = 1$ and $c = 2$, starting from different initial conditions: $p(0) = 0.1, 0.5,$ and 1.5. All solution curves eventually approach $p_2 = 1 - m/c = 0.5$. When $p(0) = p_2 = 0.5$, the solution curve stays at $p_2 = 0.5$.

Figure 8.21b Solution curve when $m > c$: $m = 2, c = 1$, and $p(0) = 2$. The solution approaches the locally stable equilibrium $p_1 = 0$.

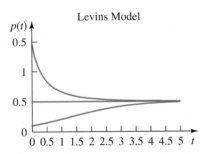

Figure 8.22b Solution curves when $m < c$: $m = 1, c = 2$, and $p(0) = 0.1$, $0.5, 1.5$. When $p(0) = 0.1$, the solution approaches the locally stable equilibrium $p_2 = 1 - m/c$ = 0.5. When $p(0) = 0.5$, the solution remains at the locally stable equilibrium $p_2 = 1 - m/c = 0.5$. When $p(0) = 1.5$, the solution approaches the locally stable equilibrium $p_2 = 1 - m/c = 0.5$.

■ 8.2.4 The Allee Effect

A sexually reproducing species may experience a disproportionately low recruitment rate when the population density falls below a certain level, due to lack of suitable mates. This phenomenon is called an *Allee effect* (Allee, 1931). A simple extension of the logistic equation incorporates the effect. We denote the size of a population at time t by $N(t)$; then, writing $N = N(t)$, we have

$$\frac{dN}{dt} = rN(N - a)\left(1 - \frac{N}{K}\right) \tag{8.64}$$

where r, a, and K are positive constants. We assume that $0 < a < K$. We will see that, as in the logistic equation, K denotes the carrying capacity. The constant a is a threshold population size below which the recruitment rate is negative, meaning that the population will shrink and ultimately go to extinction.

The equilibria of (8.64) are given by $\hat{N} = 0, a,$ and K. We set

$$g(N) = rN(N - a)\left(1 - \frac{N}{K}\right) = r\left(N^2 + \frac{a}{K}N^2 - \frac{N^3}{K} - aN\right)$$

A graph of $g(N)$ is shown in Figure 8.23a. Differentiating $g(N)$ yields

$$g'(N) = r\left(2N + \frac{2a}{K}N - \frac{3N^2}{K} - a\right) = \frac{r}{K}(2NK + 2aN - 3N^2 - aK)$$

We can compute the eigenvalue $g'(\hat{N})$ associated with the equilibrium $\hat{N}$:

$$\text{if} \quad \hat{N} = 0, \quad \text{then} \quad g'(0) = \frac{r}{K}(-aK) < 0$$

$$\text{if} \quad \hat{N} = a, \quad \text{then} \quad g'(a) = \frac{r}{K}a(K - a) > 0$$

$$\text{if} \quad \hat{N} = K, \quad \text{then} \quad g'(K) = \frac{r}{K}K(a - K) < 0$$

As we continue, you should compare the results from the eigenvalue method with the graph of $g(N)$.

Since $g'(0) < 0$, it follows that $\hat{N} = 0$ is locally stable. Likewise, since $g'(K) < 0$, it follows that $\hat{N} = K$ is locally stable. The equilibrium $\hat{N} = a$ is unstable, because $g'(a) > 0$. This instability is also evident from Figure 8.23a. The Allee effect is an example in which both stable equilibria are locally, but not globally, stable.

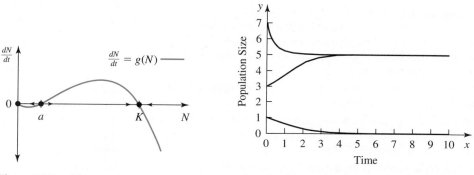

Figure 8.23a The graph of $g(N)$ illustrating the Allee effect.

Figure 8.23b Solution curves when $r = 0.5$, $a = 2$, and $K = 5$. When the initial condition $N(0)$ is between 0 and 2, the solution curve approaches the locally stable equilibrium $\hat{N} = 0$. When the initial condition $N(0)$ is greater than 2, the solution curve approaches the locally stable equilibrium $\hat{N} = K = 5$. The approach is from below when $2 < N(0) < 5$ and from above when $N(0) > 5$.

We see from Figures 8.23a and 8.23b that if $0 \leq N(0) < a$, then $N(t) \to 0$ as $t \to \infty$. If $a < N(0) \leq K$ or $N(0) \geq K$, then $N(t) \to K$ as $t \to \infty$. To interpret our results, we observe that if the initial population $N(0)$ is too small [i.e., $N(0) < a$], then the population goes extinct, and if the initial population is large enough [i.e., $N(0) > a$], then the population persists. That is, the parameter a is a threshold level. The recruitment rate is large enough only when the population size exceeds this level.

Section 8.2 Problems

■ 8.2.1

1. Suppose that

$$\frac{dy}{dx} = y(2 - y)$$

(a) Find the equilibria of this differential equation.

(b) Graph dy/dx as a function of y, and use your graph to discuss the stability of the equilibria.

(c) Compute the eigenvalues associated with each equilibrium, and discuss the stability of the equilibria.

2. Suppose that

$$\frac{dy}{dx} = (4 - y)(5 - y)$$

(a) Find the equilibria of this differential equation.

(b) Graph dy/dx as a function of y, and use your graph to discuss the stability of the equilibria.

(c) Compute the eigenvalues associated with each equilibrium, and discuss the stability of the equilibria.

3. Suppose that

$$\frac{dy}{dx} = y(y-1)(y-2)$$

(a) Find the equilibria of this differential equation.

(b) Graph dy/dx as a function of y, and use your graph to discuss the stability of the equilibria.

(c) Compute the eigenvalues associated with each equilibrium, and discuss the stability of the equilibria.

4. Suppose that

$$\frac{dy}{dx} = y(2-y)(y-3)$$

(a) Find the equilibria of this differential equation.

(b) Graph dy/dx as a function of y, and use your graph to discuss the stability of the equilibria.

(c) Compute the eigenvalues associated with each equilibrium, and discuss the stability of the equilibria.

5. Logistic Equation Assume that the size of a population evolves according to the logistic equation with intrinsic rate of growth $r = 1.5$. Assume that the carrying capacity $K = 100$.

(a) Find the differential equation that describes the rate of growth of this population.

(b) Find all equilibria, and, using the graphical approach, discuss the stability of the equilibria.

(c) Find the eigenvalues associated with the equilibria, and use the eigenvalues to determine the stability of the equilibria. Compare your answers with your results in (b).

6. A Simple Model of Predation Suppose that $N(t)$ denotes the size of a population at time t. The population evolves according to the logistic equation, but, in addition, predation reduces the size of the population so that the rate of change is given by

$$\frac{dN}{dt} = N\left(1 - \frac{N}{50}\right) - \frac{9N}{5+N} \qquad (8.65)$$

The first term on the right-hand side describes the logistic growth; the second term describes the effect of predation.

(a) Set

$$g(N) = N\left(1 - \frac{N}{50}\right) - \frac{9N}{5+N}$$

and graph $g(N)$.

(b) Find all equilibria of (8.65).

(c) Use your graph in (a) to determine the stability of the equilibria you found in (b).

(d) Use the method of eigenvalues to determine the stability of the equilibria you found in (b).

7. Logistic Equation Assume that the size of a population evolves according to the logistic equation with intrinsic rate of growth $r = 2$. Assume that $N(0) = 10$.

(a) Determine the carrying capacity K if the population grows fastest when the population size is 1000. (*Hint:* Show that the graph of dN/dt as a function of N has a maximum at $K/2$.)

(b) If $N(0) = 10$, how long will it take the population size to reach 1000?

(c) Find $\lim_{t\to\infty} N(t)$.

8. Logistic Equation The logistic curve $N(t)$ is an S-shaped curve that satisfies

$$\frac{dN}{dt} = rN\left(1 - \frac{N}{K}\right) \quad \text{with } N(0) = N_0 \qquad (8.66)$$

when $N_0 < K$.

(a) Use the differential equation (8.66) to show that the inflection point of the logistic curve is at exactly half the saturation value of the curve. [*Hint:* Do not solve (8.66); instead, differentiate the right-hand side with respect to t.]

(b) The solution $N(t)$ of (8.66) can be defined for all $t \in \mathbf{R}$. Show that $N(t)$ is symmetric about the inflection point and that $N(0) = N_0$. That is, first use the solution of (8.66) that is given in (8.33), and find the time t_0 so that $N(t_0) = K/2$ (i.e., the inflection point) is at $t = t_0$. Compute $N(t_0 + h)$ and $N(t_0 - h)$ for $h > 0$, and show that

$$N(t_0 + h) - \frac{K}{2} = \frac{K}{2} - N(t_0 - h)$$

Use a sketch of the graph of $N(t)$ to explain why the preceding equation shows that $N(t)$ is symmetric about the inflection point $(t_0, N(t_0))$.

9. Suppose that a fish population evolves according to the logistic equation and that a fixed number of fish per unit time are removed. That is,

$$\frac{dN}{dt} = rN\left(1 - \frac{N}{K}\right) - H$$

Assume that $r = 2$ and $K = 1000$.

(a) Find possible equilibria, and discuss their stability when $H = 100$.

(b) What is the maximal harvesting rate that maintains a positive population size?

10. Suppose that a fish population evolves according to a logistic equation and that fish are harvested at a rate proportional to the population size. If $N(t)$ denotes the population size at time t, then

$$\frac{dN}{dt} = rN\left(1 - \frac{N}{K}\right) - hN$$

Assume that $r = 2$ and $K = 1000$.

(a) Find possible equilibria, use the graphical approach to discuss their stability when $h = 0.1$, and find the maximal harvesting rate that maintains a positive population size.

(b) Show that if $h < r = 2$, then there is a nontrivial equilibrium. Find the equilibrium.

(c) Use **(i)** the eigenvalue approach and **(ii)** the graphical approach to analyze the stability of the equilibrium found in (b).

■ **8.2.2**

11. Assume the single-compartment model defined in Subsection 8.2.2: If $C(t)$ is the concentration of the solute at time t, then dC/dt is given by (8.57); that is,

$$\frac{dC}{dt} = \frac{q}{V}(C_I - C)$$

where q, V, and C_I are defined as in Subsection 8.2.2. Use the graphical approach to discuss the stability of the equilibrium $\hat{C} = C_I$.

12. Assume the single-compartment model defined in Subsection 8.2.2; that is, denote the concentration of the solute at time t by $C(t)$, and assume that

$$\frac{dC}{dt} = 3(20 - C(t)) \quad \text{for } t \geq 0 \qquad (8.67)$$

(a) Solve (8.67) when $C(0) = 5$.

(b) Find $\lim_{t \to \infty} C(t)$.

(c) Use your answer in (a) to determine t so that $C(t) = 10$.

13. Assume the single-compartment model defined in Subsection 8.2.2; that is, denote the concentration of the solution at time t by $C(t)$, and assume that the concentration of the incoming solution is 3 g liter^{-1} and the rate at which mass enters is 0.2 liter s^{-1}. Assume, further, that the volume of the compartment $V = 400$ liters.

(a) Find the differential equation for the rate of change of the concentration at time t.

(b) Solve the differential equation in (a) when $C(0) = 0$, and find $\lim_{t \to \infty} C(t)$.

(c) Find all equilibria of the differential equation and discuss their stability.

14. Suppose that a tank holds 1000 liters of water, and 2 kg of salt is poured into the tank.

(a) Compute the concentration of salt in g liter^{-1}.

(b) Assume now that you want to reduce the salt concentration. One method would be to remove a certain amount of the salt water from the tank and then replace it by pure water. How much salt water do you have to replace by pure water to obtain a salt concentration of 1 g liter^{-1}?

(c) Another method for reducing the salt concentration would be to hook up an overflow pipe and pump pure water into the tank. That way, the salt concentration would be gradually reduced. Assume that you have two pumps, one that pumps water at a rate of 1 liter s^{-1}, the other at a rate of 2 liter s^{-1}. For each pump, find out how long it would take to reduce the salt concentration from the original concentration to 1 g liter^{-1} and how much pure water is needed in each case. (Note that the rate at which water enters the tank is equal to the rate at which water leaves the tank.) Compare the amount of water needed using the pumps with the amount of water needed in part (b).

15. Assume the single-compartment model introduced in Subsection 8.2.2. Denote the concentration at time t by $C(t)$, measured in mg/L, and assume that

$$\frac{dC}{dt} = 0.37(254 \text{ mg/L} - C(t)) \quad \text{for } t \geq 0$$

(a) Find the equilibrium concentration.

(b) Assume that the concentration is suddenly increased from the equilibrium concentration to 400 mg/L. Find the return time to equilibrium, denoted by T_R, which is the amount of time until the initial difference is reduced to a fraction e^{-1}.

(c) Repeat (b) for the case when the concentration is suddenly increased from the equilibrium concentration to 800 mg/L.

(d) Are the values for T_R computed in (b) and (c) different?

16. Assume the compartment model as in Subsection 8.2.2. Suppose that the equilibrium concentration is C_I and the initial concentration is C_0. Express the time it takes until the initial deviation $C_0 - C_I$ is reduced to a fraction p in terms of T_R.

17. Assume the compartment model as in Subsection 8.2.2. Suppose that the equilibrium concentration is C_I. The time T_R has an integral representation that can be generalized to systems with more than one compartment. Show that

$$T_R = \int_0^\infty \frac{C(t) - C_I}{C(0) - C_I} \, dt$$

[*Hint*: Use (8.58) to show that

$$\frac{C(t) - C_I}{C(0) - C_I} = e^{-(q/V)t}$$

and integrate both sides with respect to t from 0 to ∞.]

18. Use the compartment model defined in Subsection 8.2.2 to investigate how the size of a lake influences nutrient dynamics in the lake after a perturbation. Mary Lake and Elizabeth Lake are two fictitious lakes in the North Woods that are used as experimental lakes to study nutrient dynamics. Mary Lake has a volume of 6.8×10^3 m^3, and Elizabeth Lake has twice that volume, or 13.6×10^3 m^3. Both lakes have the same inflow/outflow rate $q = 170$ liter s^{-1}. Because both lakes share the same drainage area, the concentration C_I of the incoming solute is the same for both lakes, namely, $C_I = 0.7$ mg liter^{-1}. Assume that at the beginning of the experiment both lakes are in equilibrium; that is, the concentration of the solution in both lakes is 0.7 mg liter^{-1}. Your experiment consists of increasing the concentration of the solution by 10% in each lake at time 0 and then watching how the concentration of the solution in each lake changes with time. Assume the single-compartment model to make predictions about how the concentration of the solution will evolve. (Note that 1 m^3 of water corresponds to 1000 liters of water.)

(a) Find the initial concentration C_0 of the solution in each lake at time 0 (i.e., immediately after the 10% increase in concentration of the solution).

(b) Use Equation (8.58) to determine how the concentration of the solution changes over time in each lake. Graph your results.

(c) Which lake returns to equilibrium faster? Compute the return time to equilibrium, T_R, for each lake, and explain how it is related to the eigenvalues corresponding to the equilibrium concentration C_I for each lake.

19. Use the single-compartment model defined in Subsection 8.2.2 to investigate the effect of an increase in the input concentration C_I on the nutrient concentration in a lake. Suppose a lake in a pristine environment has an equilibrium phosphorus concentration of 0.3 mg^{-1}. The volume V of the lake is 12.3×10^6 m^3, and the inflow/outflow rate q is equal to 220 liter s^{-1}. Conversion of land in the drainage area of the lake to agricultural use has increased the input concentration from 0.3 mg liter^{-1} to 1.1 mg liter^{-1}. Assume that this increase happened instantaneously. Compute the return time to the new equilibrium, denoted by T_R, in days, and find the nutrient concentration in the lake T_R units of time after the change in input concentration. (Note that 1 m^3 of water corresponds to about 1000 liters of water.)

■ **8.2.3**

20. Levins Model Denote by $p = p(t)$ the fraction of occupied patches in a metapopulation model, and assume that

$$\frac{dp}{dt} = 2p(1 - p) - p \quad \text{for } t \geq 0 \tag{8.68}$$

(a) Set $g(p) = 2p(1 - p) - p$. Graph $g(p)$ for $p \in [0, 1]$.

(b) Find all equilibria in (8.68) that are in $[0, 1]$. Use your graph in (a) to determine their stability.

(c) Use the eigenvalue approach to analyze the stability of the equilibria that you found in (b).

21. Levins Model Denote by $p = p(t)$ the fraction of occupied patches in a metapopulation model, and assume that

$$\frac{dp}{dt} = 0.5p(1 - p) - 1.5p \quad \text{for } t \geq 0 \tag{8.69}$$

(a) Set $g(p) = 0.5p(1 - p) - 1.5p$. Graph $g(p)$ for $p \in [0, 1]$.

(b) Find all equilibria of (8.69) that are in $[0, 1]$. Use your graph in (a) to determine their stability.

(c) Use the eigenvalue approach to analyze the stability of the equilibria that you found in (b).

22. A Metapopulation Model with Density-Dependent Extinction Denote by $p = p(t)$ the fraction of occupied patches in a metapopulation model, and assume that

$$\frac{dp}{dt} = cp(1 - p) - p^2 \quad \text{for } t \geq 0 \tag{8.70}$$

where $c > 0$. The term p^2 describes the density-dependent extinction of patches; that is, the per-patch extinction rate is p, and a fraction p of patches are occupied, resulting in an extinction rate of p^2. The colonization of vacant patches is the same as in the Levins model.

(a) Set $g(p) = cp(1 - p) - p^2$ and sketch the graph of $g(p)$.

(b) Find all equilibria of (8.70) in $[0, 1]$, and determine their stability.

(c) Is there a nontrivial equilibrium when $c > 0$? Contrast your findings with the corresponding results in the Levins model.

23. Habitat Destruction In Subsection 8.2.3, we introduced the Levins model. To study the effects of habitat destruction on a single species, we modify equation (8.63) in the following way: We assume that a fraction D of patches is permanently destroyed. Consequently, only patches that are vacant and undestroyed can be successfully colonized. These patches have frequency $1 - p(t) - D$ if $p(t)$ denotes the fraction of occupied patches at time t. Then

$$\frac{dp}{dt} = cp(1 - p - D) - mp \tag{8.71}$$

(a) Explain in words the meaning of the different terms in (8.71).

(b) Show that there are two possible equilibria: the trivial equilibrium $p_1 = 0$ and the nontrivial equilibrium $p_2 = 1 - D - \frac{m}{c}$. Sketch the graph of p_2 as a function of D.

(c) Assume that $m < c$ such that the nontrivial equilibrium is stable when $D = 0$. Find a condition for D such that the nontrivial equilibrium is between 0 and 1, and investigate the stability of both the nontrivial equilibrium and the trivial equilibrium under that condition.

(d) Assume the condition that you derived in (c); that is, the nontrivial equilibrium is between 0 and 1. Show that when the system is in equilibrium, the fraction of patches that are

vacant and undestroyed—that is, the sites that are *available* for colonization—is independent of D. Show that the **effective colonization rate** in equilibrium—that is, c times the fraction of available patches—is equal to the extinction rate. This equality shows that the effective birth rate of new colonies balances their extinction rate at equilibrium.

■ **8.2.4**

24. Allee Effect Denote the size of a population at time t by $N(t)$, and assume that

$$\frac{dN}{dt} = 2N(N - 10)\left(1 - \frac{N}{100}\right) \quad \text{for } t \geq 0 \tag{8.72}$$

(a) Find all equilibria of (8.72).

(b) Use the eigenvalue approach to determine the stability of the equilibria you found in (a).

(c) Set

$$g(N) = 2N(N - 10)\left(1 - \frac{N}{100}\right)$$

for $N \geq 0$, and graph $g(N)$. Identify the equilibria of (8.72) on your graph, and use the graph to determine the stability of the equilibria. Compare your results with your findings in (b). Use your graph to give a graphical interpretation of the eigenvalues associated with the equilibria.

25. Allee Effect Denote the size of a population at time t by $N(t)$, and assume that

$$\frac{dN}{dt} = 0.3N(N - 17)\left(1 - \frac{N}{200}\right) \quad \text{for } t \geq 0 \tag{8.73}$$

(a) Find all equilibria of (8.73).

(b) Use the eigenvalue approach to determine the stability of the equilibria you found in (a).

(c) Set

$$g(N) = 0.3N(N - 17)\left(1 - \frac{N}{200}\right)$$

for $N \geq 0$, and graph $g(N)$. Identify the equilibria of (8.73) on your graph, and use the graph to determine the stability of the equilibria. Compare your results with your findings in (b). Use your graph to give a graphical interpretation of the eigenvalues associated with the equilibria.

■ 8.3 Systems of Autonomous Equations (Optional)

In the preceding two sections, we discussed models that could be described by a single differential equation. If we wish to describe models in which several quantities interact, such as a competition model in which various species interact, more than one differential equation is needed. We call this model a *system of differential equations*. We will restrict ourselves again to autonomous systems—that is, systems whose dynamics do not depend explicitly on the independent variable (which typically is time).

This section is a preview of Chapter 11, in which we will discuss systems of differential equations in detail. A thorough analysis of such systems requires a fair amount of theory, which we will develop in Chapters 9 and 10. Since we are not yet equipped with the right tools to analyze systems of differential equations, this section will be rather informal. As with movie previews, you will not know the full story after you finish reading the section, but reading it will (hopefully) convince you that systems of differential equations provide a rich tool for modeling biological systems.

■ 8.3.1 A Simple Model of Epidemics

We begin our discussion of systems of autonomous differential equations with a classical model of an infectious disease: the Kermack–McKendrick model (Kermack & McKendrick, 1927, 1932, 1933). We consider a population of fixed size N that, at time t, can be divided into three classes: the susceptibles, $S(t)$, which can get infected; the infectives, $I(t)$, which are infected and can transmit the disease; and the removed class, $R(t)$, which are immune to the disease. The flow among these classes can be described by

$$S \longrightarrow I \longrightarrow R$$

We assume that the infection spreads according to the mass action law that we encountered in the discussion of chemical reactions. Each susceptible becomes infected at a rate that is proportional to the number of infectives I. Each infected individual recovers at a constant rate. A gain in the class of infectives is a simultaneous loss in the class of susceptibles. Likewise, a gain in the class of recovered individuals is a loss in the class of infectives. We can therefore describe the dynamics by

$$\frac{dS}{dt} = -bSI \tag{8.74}$$

$$\frac{dI}{dt} = bSI - aI \tag{8.75}$$

$$\frac{dR}{dt} = aI \tag{8.76}$$

Note that

$$\frac{dS}{dt} + \frac{dI}{dt} + \frac{dR}{dt} = -bSI + bSI - aI + aI = 0$$

Since

$$\frac{dS}{dt} + \frac{dI}{dt} + \frac{dR}{dt} = \frac{d}{dt}(S + I + R)$$

it follows that $S(t) + I(t) + R(t)$ is a constant that we can identify as the population size N. To analyze the system, we assume that, at time 0,

$$S(0) > 0 \qquad I(0) > 0 \qquad R(0) = 0$$

A question of interest is whether the infection will spread. We say that the infection spreads if

$$I(t) > I(0) \quad \text{for some } t > 0$$

Equation (8.75) allows us to answer the question: If

$$\frac{dI}{dt} = I(bS - a) > 0 \quad \text{at } t = 0$$

then $I(t)$ increases at the beginning, and hence the infection can spread. This condition can be written as

$$\frac{bS(0)}{a} > 1$$

The value $bS(0)/a$ is called the **basic reproductive rate** of the infection and is typically denoted by R_0 [not to be confused with the number $R(0)$ of recovered individuals at time 0]. The quantity R_0 is of great importance in epidemiology, because it tells us whether an infection can spread. R_0 is the key to understanding why vaccination programs work. It explains why it is not necessary to vaccinate everyone against an infectious disease: As long as the number of susceptibles is reduced below a certain threshold, the infection will not spread. The theoretical threshold, based on the Kermack–McKendrick model, is a/b. In practice, there are additional factors (such as the spatial proximity of infected individuals) that influence whether or not an infection will spread. But the basic conclusion is the same: As long as the number of susceptibles is below a certain threshold, the infection will not spread.

To find out how the infection progresses over time, we divide (8.74) by (8.76). For $I > 0$, we obtain

$$\frac{dS/dt}{dR/dt} = \frac{dS}{dR} = -\frac{bSI}{aI} = -\frac{b}{a}S$$

That is,

$$\frac{dS}{dR} = -\frac{b}{a}S$$

Separating variables and integrating yields

$$\int \frac{dS}{S} = -\frac{b}{a} \int dR$$

Since $S(t) > 0$, we have

$$\ln S(t) = -\frac{b}{a}R(t) + C$$

With $R(0) = 0$, we obtain

$$\ln S(0) = C$$

Hence,

$$S(t) = S(0)e^{-(b/a)R(t)} \tag{8.77}$$

Figure 8.24 The solution $S(t)$ of the Kermack–McKendrick model.

Since $R(t)$ is nondecreasing (no one can leave the removed class), it follows from (8.77) that $S(t)$ is nonincreasing. (See Figure 8.24.)

Letting $t \to \infty$, we find that

$$S(\infty) = \lim_{t \to \infty} S(t) = \lim_{t \to \infty} S(0)e^{-(b/a)R(t)}$$
$$= S(0)e^{-(b/a)R(\infty)}$$

Since $R(\infty) \le N$, it follows that

$$S(\infty) \ge S(0)e^{-(b/a)N} > 0$$

That is, not everyone becomes infected. As long as the number of susceptibles is greater than a/b, $dI/dt > 0$ and the number of infected individuals increases. Because the population size is constant, the infection "uses up" susceptibles, and there will be a time when the number of susceptibles falls below a/b. (See Figure 8.25.) From then on, dI/dt is negative, and the number of infected individuals begins to decline. The infection will eventually cease; that is,

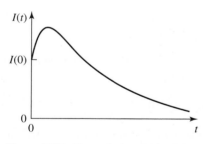

Figure 8.25 The solution $I(t)$ of the Kermack–McKendrick model.

$$\lim_{t \to \infty} I(t) = 0$$

Since not everyone becomes infected $[S(\infty) > 0]$, when the infection finally comes to an end, it is because the population runs out of infectives, not because of a lack of susceptibles.

■ 8.3.2 A Compartment Model

In Subsection 8.2.2, we introduced a single-compartment model that led to an autonomous differential equation with one dependent variable. In this subsection, we introduce a model with two compartments that describes the interaction of an *autotroph*[1] and its nutrient pool. (The discussion that follows is partially adapted from DeAngelis, 1992.) A schematic description of the interactions can be found in Figure 8.26.

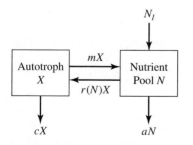

Figure 8.26 A schematic description of the interaction between autotrophs and the nutrient pool.

---

(1) An autotroph is an organism that can manufacture organic compounds entirely from inorganic components. Examples include most chlorophyll-containing plants and blue-green algae.

We assume that the nutrient pool has an external source and denote the input rate of nutrients by N_I. That is, N_I denotes the total mass per unit time that flows into the system. Nutrients may be washed out, and we assume that the output rate is proportional to the total mass of nutrients, N, in the compartment, with proportionality constant a.

The autotroph feeds on the nutrients in the nutrient pool, and its growth rate is proportional to the quantity of autotroph biomass, X. The specific rate of growth, $r(N)$, depends on the amount of available nutrients.

Autotrophs can leave the autotroph compartment in two ways: (1) At rate c, they get completely lost (e.g., through harvesting or by being washed out of their habitat); (2) at rate m, the biomass of the autotroph gets recycled back into the nutrient pool (e.g., after the death of an organism).

This system is then described by the following set of differential equations:

$$\frac{dN}{dt} = N_I - aN - r(N)X + mX \tag{8.78}$$

$$\frac{dX}{dt} = r(N)X - (m + c)X \tag{8.79}$$

You should compare the set of differential equations with the schematic description of the model in Figure 8.26. In particular, you should pay attention to the direction of the arrows. For instance, the term $r(N)X$ shows up in both equations: Because this term corresponds to the nutrient uptake by the autotrophs, it appears as a loss to the nutrient pool, which is indicated by the minus sign in front of the term in equation (8.78); and the same term appears with a plus sign in equation (8.79), because the uptake of nutrients by the autotroph results in an increase in autotroph biomass. Note that the arrow labeled $r(N)X$ goes from the nutrient pool to the autotroph pool. Hence, it represents a loss in the nutrient pool [a minus sign in (8.78) in front of $r(N)X$] and a gain in the autotroph pool [a plus sign in (8.79) in front of $r(N)X$].

It is important to learn how to go from the schematic description to the set of differential equations and back. A schematic description quickly summarizes the flow of matter (such as nutrients), whereas the set of differential equations is indispensable if we want to analyze the system.

To keep the discussion concrete, we assume that the function $r(N)$ is linear; that is,

$$r(N) = bN$$

for some constant $b > 0$.

As in Section 8.2, we can introduce the concept of equilibria. An equilibrium for the system given by (8.78) and (8.79) is characterized by simultaneously requiring that

$$\frac{dN}{dt} = 0 \quad \text{and} \quad \frac{dX}{dt} = 0 \tag{8.80}$$

since, when the rates of change of both quantities are equal to 0, the values for N and X no longer change.

There is a graphical method for finding equilibria: plotting the **zero isoclines**. These curves are obtained by setting $dN/dt = 0$ and $dX/dt = 0$. The curves for which $dX/dt = 0$ are obtained by setting the right-hand side of (8.79) equal to 0; that is, $X(bN - (m + c)) = 0$, yielding

$$N = \frac{m + c}{b} \quad \text{or} \quad X = 0$$

The curve for which $dN/dt = 0$ is obtained by setting the right-hand side of (8.78) equal to 0, yielding

$$X = \frac{N_I - aN}{bN - m}$$

We plot the three curves in the N–X plane, as illustrated in Figure 8.27.

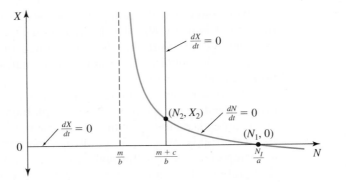

Figure 8.27 The zero isoclines in the N–X plane.

The zero isocline for N intersects the horizontal line $X = 0$ at $(N_1, 0)$, where N_1 satisfies

$$0 = \frac{N_I - aN_1}{bN_1 - m}$$

and thus $N_1 = N_I/a$. We call this a *trivial equilibrium*, since it corresponds to the case in which there is no autotroph in the system. (The system reduces to the single-compartment model we discussed in Section 8.2.)

Depending on where the vertical zero isocline for X is located, there can be another equilibrium (N_1, N_2) for which both N_2 and X_2 are positive. Looking at Figure 8.27, we see that when

$$\frac{m}{b} < \frac{m+c}{b} < \frac{N_I}{a} \tag{8.81}$$

the two zero isoclines $N = (m + c)/b$ and $X = (N_I - aN)/(bN - m)$ intersect in the first quadrant. We call this equilibrium a *nontrivial equilibrium*.

Since $(m + c)/b > m/b$ when $c > 0$, the vertical zero isocline $dX/dt = 0$ is always to the right of the vertical asymptote $N = m/b$. The first inequality in (8.81) therefore always holds. If we solve the second inequality,

$$\frac{m+c}{b} < \frac{N_I}{a}$$

for b, we find that

$$b > \frac{a}{N_I}(m + c)$$

That is, the growth parameter b must exceed a certain threshold in order for the autotroph to survive. This should be intuitively clear, since m and c are the rates at which the autotroph pool is depleted. The depletion must be balanced by an increase in biomass of the autotroph.

Suppose that the ratio m/b is fixed. Then the smaller c (the rate at which autotrophs get lost) is, the farther to the left is the vertical zero isocline $dX/dt = 0$ and, consequently, the larger is the equilibrium value X_2. This should also be intuitively clear, since losing fewer autotrophs should result in a higher equilibrium value.

To find the nontrivial equilibrium (N_2, X_2), we need to solve the system of equations

$$0 = N_I - aN_2 - bN_2X_2 + mX_2 \tag{8.82}$$
$$0 = bN_2X_2 - (m + c)X_2 \tag{8.83}$$

where we now assume that $X_2 \neq 0$. Equation (8.83) yields

$$N_2 = \frac{m+c}{b}$$

Using this relationship in (8.82), we find that

$$0 = N_I - a\frac{m+c}{b} - (m+c)X_2 + mX_2$$

or

$$X_2 = \frac{1}{c}\left(N_I - a\frac{m+c}{b}\right)$$

It can be shown that if the nontrivial equilibrium exists, it is locally stable; that is, the system will return to this equilibrium after a small perturbation. In Chapter 11, we will learn two methods for analyzing the stability of a system of differential equations: a graphical and an analytical method; these are extensions of the methods we developed in Section 8.2.

■ 8.3.3 A Hierarchical Competition Model

We now extend the metapopulation (Levins) model, introduced in Subsection 8.2.3, to a multispecies setting. We recall Levins model but use a somewhat different interpretation. In the model, the dynamics of subpopulations was described. We will now view the Levins model as an occupancy model of sites for single individuals; that is, we suppose that the habitat is divided into patches that are now so small that at most one individual occupies a patch. Using this interpretation, we denote by $p(t)$ the fraction of sites that are occupied by a single individual at time t. Then

$$\frac{dp}{dt} = cp(1-p) - mp$$

where m is the extinction rate (the death rate of an individual) and c is the colonization rate (the rate at which offspring migrate to new sites). Tilman (1994) extended this model to a system in which species are ranked according to their competitiveness. The fraction of sites that are occupied by species i at time t is denoted by $p_i(t)$. We assume that species 1 is the best competitor, species 2 the next best, and so on. A superior competitor can invade a site that is occupied by an inferior competitor. The inferior competitor is displaced upon invasion of the superior competitor. The dynamics are then given by

$$\frac{dp_1}{dt} = c_1 p_1(1-p_1) - m_1 p_1$$

$$\frac{dp_2}{dt} = c_2 p_2(1-p_1-p_2) - m_2 p_2 - c_1 p_1 p_2$$

$$\frac{dp_3}{dt} = c_3 p_3(1-p_1-p_2-p_3) - m_3 p_3 - c_1 p_1 p_3 - c_2 p_2 p_3$$

$$\vdots$$

$$\frac{dp_i}{dt} = c_i\left(1-\sum_{j=1}^{i} p_j\right) - m_i p_i - \sum_{j=1}^{i-1} c_j p_j p_i$$

$$\vdots$$

The first term describes the colonization by species i of sites that are either occupied by an inferior competitor or vacant, the second term describes the extinction of sites that are occupied by species i, and the remaining terms describe the competitive displacement by superior competitors.

To simplify the discussion, we assume that there are only two species and that $m_1 = m_2 = 1$. The equations are then given by

$$\frac{dp_1}{dt} = c_1 p_1(1-p_1) - p_1$$

$$\frac{dp_2}{dt} = c_2 p_2(1-p_1-p_2) - p_2 - c_1 p_1 p_2$$

The hierarchical structure of the model makes it easy to find possible equilibria. Setting $dp_1/dt = 0$, we find that

$$0 = \hat{p}_1[c_1(1 - \hat{p}_1) - 1]$$

which, aside from the trivial equilibrium 0, gives

$$\hat{p}_1 = 1 - \frac{1}{c_1}$$

The equation for species 1 is identical to the Levins model in Subsection 8.2.3, and we can use the results from that subsection. Thus, $c_1 > 1$, $\hat{p}_1 \in (0, 1)$, and the nontrivial equilibrium is stable. Because we are interested in the coexistence of species, we assume that $c_1 > 1$ in what follows, so that species 1 can survive.

Setting $dp_2/dt = 0$ with $\hat{p}_1 = 1 - 1/c_1$ allows us to find $\hat{p}_2$:

$$0 = \hat{p}_2[c_2(1 - \hat{p}_1 - \hat{p}_2) - 1 - c_1\hat{p}_1]$$

This equation yields, aside from the trivial equilibrium 0,

$$\hat{p}_2 = (1 - \hat{p}_1) - \frac{1}{c_2} - \frac{c_1}{c_2}\hat{p}_1$$

$$= \frac{1}{c_1} - \frac{1}{c_2} - \frac{c_1}{c_2} + \frac{1}{c_2}$$

$$= \frac{1}{c_1} - \frac{c_1}{c_2}$$

Coexistence of the two species means that both $\hat{p}_1$ and $\hat{p}_2$ are positive and that their sum $\hat{p}_1 + \hat{p}_2$, which denotes the total fraction of occupied sites, is less than 1. Now, the sum of the two nontrivial equilibria, $\hat{p}_1 + \hat{p}_2 = 1 - c_1/c_2$, is automatically less than 1. Furthermore, since we assumed that $c_1 > 1$, we have $\hat{p}_1 > 0$. Therefore, we need only to find out when $\hat{p}_2 > 0$—that is, when

$$\frac{1}{c_1} - \frac{c_1}{c_2} > 0$$

To satisfy this inequality, we need

$$c_2 > c_1^2$$

So far, we know that there is a nontrivial equilibrium if $c_1 > 1$ and $c_2 > c_1^2$, but this does not tell us anything about stability. Still, even though we cannot yet analyze stability directly, we can take an approach that is very common in the ecological literature: We determine whether species 2 can invade the *monoculture equilibrium* of species 1—that is, the positive equilibrium of species 1 in the absence of species 2. Why does this help us? First, note that species 1 is unaffected by the presence of species 2: As long as species 1 can survive in the absence of species 2, it can also survive in its presence. Therefore, we need only worry about species 2. If species 2 can invade the monoculture equilibrium of species 1, then if species 2 is at a low density, it will be able to increase its density, and it will therefore be able to coexist with species 1. The invasion criterion is then

$$\left.\frac{dp_2}{dt}\right|_{p_1 = \hat{p}_1} > 0 \quad \text{when } p_2 \text{ is small}$$

This works as follows:

$$\frac{dp_2}{dt} = p_2[c_2(1 - p_1 - p_2) - 1 - c_1 p_1]$$

We assume that $p_1 = \hat{p}_1 = 1 - 1/c_1$ and that p_2 is very small. Then $1 - \hat{p}_1 - p_2 \approx 1 - \hat{p}_1$. Hence,

$$\frac{dp_2}{dt} \approx p_2\left[c_2\frac{1}{c_1} - 1 - c_1 + 1\right] = p_2\left[\frac{c_2}{c_1} - c_1\right] > 0$$

if

$$\frac{c_2}{c_1} - c_1 > 0, \qquad \text{or} \qquad c_2 > c_1^2$$

Since $dp_2/dt > 0$ when species 1 is in equilibrium and species 2 has a low abundance, it follows that species 2 can invade. We conclude that species 1 and 2 can coexist when $c_2 > c_1^2$.

This mechanism of coexistence is referred to as the **competition–colonization trade-off**. That is, the weaker competitor (species 2) can compensate for its inferior competitiveness by being a superior colonizer ($c_2 > c_1^2$).

Section 8.3 Problems

■ 8.3.1

In Problems 1–4, we will investigate the classical Kermack–McKendrick model for the spread of an infectious disease in a population of fixed size N. (This model was introduced in Subsection 8.3.1, and you should refer to that subsection when working out the problems.) If $S(t)$ denotes the number of susceptibles at time t, $I(t)$ the number of infectives at time t, and $R(t)$ the number of immune individuals at time t, then

$$\frac{dS}{dt} = -bSI$$

$$\frac{dI}{dt} = bSI - aI$$

and $R(t) = N - S(t) - I(t)$.

1. Determine, in each of the following cases, whether or not the disease can spread (*Hint*: Compute R_0.):
(a) $S(0) = 1000, a = 200, b = 0.3$
(b) $S(0) = 1000, a = 200, b = 0.1$

2. Assume that $a = 100$ and $b = 0.2$. The **critical number** of susceptibles $S_c(0)$ at time 0 for the spread of a disease that is introduced into a population at time 0 is defined as the minimum number of susceptibles for which the disease can spread. Find $S_c(0)$.

3. Suppose that $a = 100$, $b = 0.01$, and $N = 10,000$. Can the disease spread if, at time 0, there is one infected individual?

4. Refer to the simple model of epidemics in Subsection 8.3.1.
(a) Divide (8.75) by (8.74) to show that when $I > 0$,

$$\frac{dI}{dS} = \frac{a}{b}\frac{1}{S} - 1 \qquad (8.84)$$

Also, show that when $R(0) = 0, I(0) = I_0$, and $S(0) = S_0$, the solution of (8.84) satisfies

$$I(t) = N - S(t) + \frac{a}{b}\ln\frac{S(t)}{S_0}$$

where $I(t)$ denotes the number of infectives, N the total number of individuals in the population, and $S(t)$ the number of susceptibles at time t.

(b) Since $I(t)$ gives the number of infectives at time t and $dI/dt = bSI - aI$, if $S(0) > a/b$, then $dI/dt > 0$ at time $t = 0$. Also, since $\lim_{t\to\infty} I(t) = 0$, there is a time $t > 0$ at which $I(t)$ is maximal. Show that the number of susceptibles when $I(t)$ is maximal is given by $S = a/b$. [*Hint*: When $I(t)$ attains a maximum, the derivative of $I(t)$ with respect to t, dI/dt, is equal to 0.]

(c) In (a), you expressed $I(t)$ as a function of $S(t)$. Use your result in (b) to show that the maximal number of infectives is given by

$$I_{max} = N - \frac{a}{b} + \frac{a}{b}\ln\left(\frac{a/b}{S_0}\right)$$

(d) Use your result in (c) to show that I_{max} is a decreasing function of the parameter a/b for $a/b < S_0$ (i.e., in the case in which the infection can spread). Use the latter statement to explain how a and b determine the severity (as measured by I_{max}) of the disease. Does this make sense?

■ 8.3.2

5. Assume the compartment model of Subsection 8.3.2, with $a = 5, b = 0.02, m = 1$, and $c = 1$.
(a) Find the system of differential equations that corresponds to these values.
(b) Determine which values of N_I result in a nontrivial equilibrium, and find the equilibrium values for both the autotroph and the nutrient pool.

6. Assume the compartment model of Subsection 8.2.3, with $a = 1, b = 0.01, m = 2, c = 1$, and $N_I = 500$.
(a) Find the system of differential equations that corresponds to these values.
(b) Plot the zero isoclines corresponding to this system.
(c) Use your graph in (b) to determine whether the system has a nontrivial equilibrium.

7. Assume the compartment model of Subsection 8.3.2, with $a = 1, b = 0.01, m = 2, c = 1$ and $N_I = 200$.
(a) Find the system of differential equations that corresponds to these values.
(b) Plot the zero isoclines corresponding to this system.
(c) Use your graph in (b) to determine whether the system has a nontrivial equilibrium.

■ **8.3.3**

8. Assume the hierarchical competition model introduced in Subsection 8.3.3, and assume that the model describes two species. Specifically, assume that

$$\frac{dp_1}{dt} = 2p_1(1 - p_1) - p_1$$

$$\frac{dp_2}{dt} = 5p_2(1 - p_1 - p_2) - p_2 - 2p_1p_2$$

(a) Find all equilibria.

(b) Determine whether species 2 can invade a monoculture of species 1. (Assume that species 1 is in equilibrium.)

9. Assume the hierarchical competition model introduced in Subsection 8.3.3, and assume that the model describes two species. Specifically, assume that

$$\frac{dp_1}{dt} = 2p_1(1 - p_1) - p_1$$

$$\frac{dp_2}{dt} = 3p_2(1 - p_1 - p_2) - p_2 - 2p_1p_2$$

(a) Find all equilibria.

(b) Determine whether species 2 can invade a monoculture equilibrium of species 1.

10. Assume the hierarchical competition model introduced in Subsection 8.3.3, and assume that the model describes two species. Specifically, assume that

$$\frac{dp_1}{dt} = 2p_1(1 - p_1) - p_1$$

$$\frac{dp_2}{dt} = 6p_2(1 - p_1 - p_2) - p_2 - 2p_1p_2$$

(a) Use the zero-isocline approach to find all equilibria graphically.

(b) Determine the numerical values of all equilibria.

11. Assume the hierarchical competition model introduced in Subsection 8.3.3, and assume that the model describes two species. Specifically, assume

$$\frac{dp_1}{dt} = 3p_1(1 - p_1) - p_1$$

$$\frac{dp_2}{dt} = 5p_2(1 - p_1 - p_2) - p_2 - 3p_1p_2$$

(a) Use the zero-isocline approach to find all equilibria graphically.

(b) Determine the numerical values of all equilibria.

12. *(Adapted from Crawley, 1997)* Denote plant biomass by V, and herbivore number by N. The plant–herbivore interaction is modeled as

$$\frac{dV}{dt} = aV(1 - \frac{V}{K}) - bVN$$

$$\frac{dN}{dt} = cVN - dN$$

(a) Suppose the herbivore number is equal to 0. What differential equation describes the dynamics of the plant biomass? Can you explain the resulting equation? Determine the plant biomass equilibrium in the absence of herbivores.

(b) Now assume that herbivores are present. Describe the effect of herbivores on plant biomass; that is, explain the term $-bVN$ in the first equation. Describe the dynamics of the herbivores— that is, how their population size increases and what contributes to decreases in their population size.

(c) Determine the equilibria (1) by solving

$$\frac{dV}{dt} = 0 \quad \text{and} \quad \frac{dN}{dt} = 0$$

and (2) graphically. Explain why this model implies that "plant abundance is determined solely by attributes of the herbivore," as stated in Crawley (1997).

Chapter 8 Key Terms

Discuss the following definitions and concepts:

1. Differential equation
2. Separable differential equation
3. Solution of a differential equation
4. Pure-time differential equation
5. Autonomous differential equation
6. Exponential growth
7. Von Bertalanffy equation
8. Logistic equation
9. Allometric growth
10. Equilibrium
11. Stability
12. Eigenvalue
13. Single-compartment model
14. Levins model
15. Allee effect
16. Kermack–McKendrick model
17. Zero isocline
18. Hierarchical competition model

Chapter 8 Review Problems

1. Newton's Law of Cooling Suppose that an object has temperature T and is brought into a room that is kept at a constant temperature T_a. Newton's law of cooling states that the rate of temperature change of the object is proportional to the difference between the temperature of the object and the surrounding medium.

(a) Denote the temperature at time t by $T(t)$, and explain why

$$\frac{dT}{dt} = k(T - T_a)$$

is the differential equation that expresses Newton's law of cooling.

(b) Suppose that it takes the object 20 min to cool from 30°C to 28°C in a room whose temperature is 21°C. How long will it take the object to cool to 25°C if it is at 30°C when it is brought into the room? [*Hint*: Solve the differential equation in (a) with the initial condition $T(0) = 30°C$ and with $T_a = 21°C$. Use $T(20) = 28°C$ to determine the constant k.]

2. *(Adapted from Cain et al., 1995)* In this problem, we discuss a model for clonal growth in the white clover *Trifolium repens*. *T. repens* is a widespread perennial clonal plant species that spreads through stolon growth. (A *stolon* is a horizontal stem.) By mapping the shape of a clone over time, Cain et al. estimated stolon elongation and dieback rates as follows. Denote by $S(t)$ the stolon length of the clone at time t. Cain et al. observed that the change in stolon length was proportional to the stolon length; that is,

$$\frac{dS}{dt} \propto S$$

Introducing the proportionality constant r, called the *net growth rate*, we find that

$$\frac{dS}{dt} = rS \qquad (8.85)$$

(a) Suppose that S_f and S_0 are the final and the initial stolon lengths, respectively, and that T denotes the period of observation. Use (8.85) to show that r, the net growth rate, can be estimated from

$$r = \frac{1}{T} \ln \frac{S_f}{S_0}$$

[*Hint*: Solve the differential equation (8.85) with initial condition $S(0) = S_0$, and use the fact that $S(T) = S_f$.]

(b) The net growth rate r is the difference between the stolon elongation rate b and the stolon dieback rate m; that is,

$$r = b - m$$

Let B be the total amount of stolon elongation and D be the total amount of stolon dieback over the observation period of length T. Show that

$$B = \int_0^T bS(t)\, dt = \frac{bS_0}{r}(e^{rT} - 1)$$

$$D = \int_0^T mS(t)\, dt = \frac{mS_0}{r}(e^{rT} - 1)$$

(c) Show that $B - D = S_f - S_0$, and rearrange the equations for B and D in (b) so that you can estimate b and m from r, B, and D; that is, show that

$$b = \frac{rB}{S_f - S_0} = \frac{rB}{B - D}$$

$$m = \frac{rD}{S_f - S_0} = \frac{rD}{B - D}$$

(d) Explain how B and r can be estimated if S_f, S_0, and D are known from field measurements. Use your result in (c) to explain how you would then find estimates for b and m.

3. Diversification of Life (*Adapted from Benton, 1997, and Walker, 1985*) Several models have been proposed to explain the diversification of life during geological periods. According to Benton (1997),

> The diversification of marine families in the past 600 million years (Myr) appears to have followed two or three logistic curves, with equilibrium levels that lasted for up to 200 Myr. In contrast, continental organisms clearly show an exponential pattern of diversification, and although it is not clear whether the empirical diversification patterns are real or are artifacts of a poor fossil record, the latter explanation seems unlikely.

In this problem, we will investigate three models for diversification. They are analogous to models for population growth; however, the quantities involved have a different interpretation. We denote by $N(t)$ the diversification function, which counts the number of taxa as a function of time, and by r the intrinsic rate of diversification.

(a) (*Exponential Model*) This model is described by

$$\frac{dN}{dt} = r_e N \qquad (8.86)$$

Solve (8.86) with the initial condition $N(0)$ at time 0, and show that r_e can be estimated from

$$r_e = \frac{1}{t} \ln \left[\frac{N(t)}{N(0)} \right] \qquad (8.87)$$

[*Hint*: To find (8.87), solve for r in the solution of (8.86).]

(b) (*Logistic Growth*) This model is described by

$$\frac{dN}{dt} = r_l N \left(1 - \frac{N}{K} \right) \qquad (8.88)$$

where K is the equilibrium value. Solve (8.88) with the initial condition $N(0)$ at time 0, and show that r_l can be estimated from

$$r_l = \frac{1}{t} \ln \left[\frac{K - N(0)}{N(0)} \right] + \frac{1}{t} \ln \left[\frac{N(t)}{K - N(t)} \right] \qquad (8.89)$$

for $N(t) < K$.

(c) Assume that $N(0) = 1$ and $N(10) = 1000$. Estimate r_e and r_l for both $K = 1001$ and $K = 10,000$.

(d) Use your answer in (c) to explain the following quote from Stanley (1979):

> There must be a general tendency for calculated values of [r] to represent underestimates of exponential rates, because some radiation will have followed distinctly sigmoid paths during the interval evaluated.

(e) Explain why the exponential model is a good approximation to the logistic model when N/K is small compared with 1.

4. A Simple Model for Photosynthesis of Individual Leaves (*Adapted from Horn, 1971*) Photosynthesis is a complex mechanism; the following model is a very simplified caricature: Suppose that a leaf contains a number of traps that can capture light. If a trap captures light, the trap becomes energized. The energy in the trap can then be used to produce sugar, which causes the energized trap to become unenergized. The number of traps that can become energized is proportional to the number of unenergized traps and the intensity of the light. Denote by T the total number of traps (unenergized and energized) in a leaf, by I the light intensity, and by x the number of energized traps. Then the following differential equation describes how the number of energized traps changes over time:

$$\frac{dx}{dt} = k_1(T - x)I - k_2 x$$

Here, k_1 and k_2 are positive constants. Find all equilibria, and use the eigenvalue approach to study their stability.

5. Gompertz Growth Model This model is sometimes used to study the growth of a population for which the per capita growth rate is density dependent. Denote the size of a population at time t by $N(t)$; then, for $N \geq 0$,

$$\frac{dN}{dt} = kN(\ln K - \ln N) \quad \text{with } N(0) = N_0 \qquad (8.90)$$

(a) Show that

$$N(t) = K \exp \left[-\left(\ln \frac{K}{N_0} \right) e^{-kt} \right]$$

is a solution of (8.90). To do this, differentiate $N(t)$ with respect to t and show that the derivaive can be written in the form (8.90). Don't forget to show that $N(0) = N_0$. Use a graphing calculator to sketch the graph of $N(t)$ for $N_0 = 100$, $k = 2$, and $K = 1000$. The function $N(t)$ is called the *Gompertz growth curve*.

(b) Use l'Hospital's rule to show that

$$\lim_{N \to 0} N \ln N = 0$$

and use this equation to show that $\lim_{N \to 0} dN/dt = 0$. Are there any other values of N where $dN/dt = 0$?

(c) Sketch the graph of dN/dt as a function of N for $k = 2$ and $K = 1000$. Find the equilibria, and use your graph to and discuss their stability. Explain the meaning of K.

6. Island Biogeography Preston (1962) and MacArthur and Wilson (1963) investigated the effect of area on species diversity in oceanic islands. It is assumed that species can immigrate to an island from a species pool of size P and that species on the island can go extinct. We denote the immigration rate by $I(S)$ and the extinction rate by $E(S)$, where S is the number of species on the island. Then the change in species diversity over time is

$$\frac{dS}{dt} = I(S) - E(S) \tag{8.91}$$

For a fixed island, the simplest functional forms for $I(S)$ and $E(S)$ are

$$I(S) = c \left(1 - \frac{S}{P} \right) \tag{8.92}$$

$$E(S) = m \frac{S}{P} \tag{8.93}$$

where c, m, and P are positive constants.

(a) Find the equilibrium species diversity $\hat{S}$ of (8.91) with $I(S)$ and $E(S)$ given in (8.92) and (8.93).

(b) It is reasonable to assume that the extinction rate is a decreasing function of island size. That is, we assume that if A denotes the area of the island, then m is a function of the island size A, with $dm/dA < 0$. Furthermore, we assume that the immigration rate I does not depend on the size of the island. Use these assumptions to investigate how the equilibrium species diversity changes with island size.

(c) Assume that $S(0) = S_0$. Solve (8.91) with $I(S)$ and $E(S)$ as given in (8.92) and (8.93), respectively.

(d) Assume that $S_0 = 0$. That is, the island is initially void of species. The time constant T for the system is defined as

$$S(T) = (1 - e^{-1}) \hat{S}$$

Show that, under the assumption $S_0 = 0$,

$$T = \frac{P}{c + m}$$

(e) Use the assumptions in (b) and your answer in (d) to investigate the effect of island size on the time constant T; that is, determine whether $T(A)$ is an increasing or decreasing function of A.

7. Chemostat A chemostat is an apparatus for growing bacteria in a medium in which all nutrients but one are available in excess. One nutrient, whose concentration can be controlled, is held at a concentration that limits the growth of bacteria. The growth chamber of the chemostat is continually flushed by adding nutrients dissolved in liquid at a constant rate and allowing the liquid in the growth chamber, which contains bacteria, to leave the chamber at the same rate. If X denotes the number of bacteria in the growth chamber, then the growth dynamics of the bacteria are given by

$$\frac{dX}{dt} = r(N)X - qX \tag{8.94}$$

where $r(N)$ is the growth rate depending on the nutrient concentration N and q is the input and output flow rate. The equation for the nutrient flow is given by

$$\frac{dN}{dt} = qN_0 - qN - r(N)X \tag{8.95}$$

Note that (8.94) is (8.79) with $m = 0$, $N_I = qN_0$, and $a = e = q$ and that (8.95) is (8.78) with $m = 0$.

(a) Explain in words the meaning of the terms in (8.94) and (8.95).

(b) Assume that $r(N)$ is given by the Monod growth function

$$r(N) = b \frac{N}{k + N}$$

where k and b are positive constants. Draw the zero isoclines in the N–X plane, and explain how to find the equilibria $(\hat{N}, \hat{X})$ graphically.

(c) Show that a nontrivial equilibrium (an equilibrium for which $\hat{N}$ and $\hat{X}$ are both positive) satisfies

$$r(\hat{N}) - q = 0 \tag{8.96}$$

$$qN_0 - q\hat{N} - r(\hat{N})\hat{X} = 0 \tag{8.97}$$

Show also that (8.96) has a positive solution $\hat{N}$ if $q < b$, and find an expression for $\hat{N}$. Use this expression and (8.97) to find $\hat{X}$.

(d) Assume that $q < b$. Use your results in (c) to show that $\hat{X} > 0$ if $\hat{N} < N_0$ and $\hat{N} < N_0$ if $q < bN_0/(k + N_0)$. Furthermore, show that $\hat{N}$ is an increasing function of q for $q < b$.

(e) Use your results in (d) to explain why the following is true: As we increase the flow rate q from 0 to $bN_0/(k + N_0)$, the nutrient concentration $\hat{N}$ increases until it reaches the value N_0 and the number of bacteria decreases to 0.

8. *(Adapted from Nee and May, 1992, and Tilman, 1994)* In Subsection 8.3.3, we introduced a hierarchical competition model. We will use this model to investigate the effects of habitat destruction on coexistence. We assume that a fraction D of the sites is permanently destroyed. Furthermore, we restrict our discussion to two species and assume that species 1 is the superior and species 2 the inferior competitor. In the case in which both species have the same mortality ($m_1 = m_2$), which we set equal to 1, the dynamics are described by

$$\frac{dp_1}{dt} = c_1 p_1 (1 - p_1 - D) - p_1 \tag{8.98}$$

$$\frac{dp_2}{dt} = c_2 p_2 (1 - p_1 - p_2 - D) - p_2 - c_1 p_1 p_2 \tag{8.99}$$

where p_i, $i = 1, 2$, is the fraction of sites occupied by species i.

(a) Explain in words the meanings of the different terms in (8.98) and (8.99).

(b) Show that

$$\hat{p}_1 = 1 - \frac{1}{c_1} - D$$

is an equilibrium for species 1, which is in $(0, 1)$, and is stable if $D < 1 - 1/c_1$ and $c_1 > 1$.

(c) Assume that $c_1 > 1$ and $D < 1 - 1/c_1$. Show that species 2 can invade the nontrivial equilibrium of species 1 [computed in (b)] if

$$c_2 > c_1^2 (1 - D)$$

(d) Assume that $c_1 = 2$ and $c_2 = 5$. Then species 1 can survive as long as $D < 1/2$. Show that the fraction of sites that are occupied by species 1 is then

$$\hat{p}_1 = \begin{cases} \dfrac{1}{2} - D & \text{for } 0 \leq D \leq \dfrac{1}{2} \\[2ex] 0 & \text{for } \dfrac{1}{2} \leq D \leq 1 \end{cases}$$

Show also that

$$\hat{p}_2 = \frac{1}{10} + \frac{2}{5}D \qquad \text{for} \quad 0 \leq D \leq \frac{1}{2}$$

For $D > 1/2$, species 1 can no longer persist. Explain why the dynamics for species 2 reduce to

$$\frac{dp_2}{dt} = 5p_2(1 - p_2 - D) - p_2$$

in this case. Show, in addition, that the nontrivial equilibrium is of the form

$$\hat{p}_2 = 1 - \frac{1}{5} - D \quad \text{for } \frac{1}{2} \leq D \leq 1 - \frac{1}{5}$$

Plot $\hat{p}_1$ and $\hat{p}_2$ as functions of D in the same coordinate system. What happens for $D > 1 - 1/5$? Use the plot to explain in words how each species is affected by habitat destruction.

(e) Repeat (d) for $c_1 = 2$ and $c_2 = 3$.

9 Linear Algebra and Analytic Geometry

LEARNING OBJECTIVES

The primary focus of this chapter is on tools from linear algebra that are needed to develop multidimensional differential calculus in Chapter 10 and to analyze systems of differential equations in Chapter 11. Specifically, we will learn how to

- solve systems of linear equations;
- define matrices and perform algebraic operations on matrices;
- define and analyze linear maps; and
- define lines and planes in more than two dimensions.

■ 9.1 Linear Systems

Two different species of insects are reared together in a laboratory cage. They are supplied with two different types of food each day. Each individual of species 1 consumes 5 units of food A and 3 units of food B, whereas each individual of species 2 consumes 2 units of food A and 4 units of food B, on average, per day. Each day, a lab technician supplies 900 units of food A and 960 units of food B. How many of each species are reared together?

To solve such a problem, we will set up a system of equations. If

$$x = \text{number of individuals of species 1}$$
$$y = \text{number of individuals of species 2}$$

then the following two equations must be satisfied:

$$\text{food } A: \quad 5x + 2y = 900$$
$$\text{food } B: \quad 3x + 4y = 960$$

We refer to these two equations as a *system of two linear equations in two variables*. This section is devoted to finding solutions of such systems.

■ 9.1.1 Graphical Solution

In this subsection, we restrict ourselves to systems of two linear equations in two variables. Recall that the standard form of a linear equation in two variables is

$$Ax + By = C$$

Figure 9.1 The graph of a linear equation in standard form.

where A, B, and C are constants, A and B are not both equal to 0, and x and y are the two variables; the graph of this equation is a straight line. (See Figure 9.1.) Any point (x, y) on this straight line satisfies (or solves) the equation $Ax + By = C$. If we extend this situation to more than one equation, we call the resulting set of equations a **system** of linear equations. In the case of two linear equations in two variables, the system is of the form

$$
\begin{aligned}
Ax + By &= C \\
Dx + Ey &= F
\end{aligned}
\qquad (9.1)
$$

where A, B, C, D, E, and F are constants and x and y are the two variables. (We require that A and B and that D and E are not both equal to 0.) When we say that we "solve" (9.1) for x and y, we mean that we find an ordered pair (x, y) that satisfies *each* equation of the system (9.1). Because each equation in (9.1) describes a straight line, we are therefore asking for the point of intersection of these two lines. The following three cases are possible:

1. The two lines have exactly one point of intersection. In this case, the system (9.1) has exactly one solution, as illustrated in Figure 9.2.
2. The two lines are parallel and do not intersect. In this case, the system (9.1) has no solution, as illustrated in Figure 9.3.
3. The two lines are parallel and intersect (i.e., they are identical). In this case, the system (9.1) has infinitely many solutions—namely, each point on the line, as illustrated in Figure 9.4.

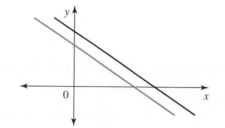

Figure 9.2 The two lines have exactly one point of intersection.

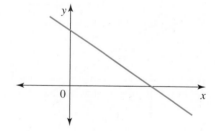

Figure 9.3 The two lines are parallel but do not intersect.

Figure 9.4 The two lines are identical.

Exactly One Solution

EXAMPLE 1   Find the solution of

$$
\begin{aligned}
2x + 3y &= 6 \\
2x + y &= 4
\end{aligned}
\qquad (9.2)
$$

Solution   The line corresponding to $2x + 3y = 6$ has y-intercept $(0, 2)$ and x-intercept $(3, 0)$; the line corresponding to $2x + y = 4$ has y-intercept $(0, 4)$ and x-intercept $(2, 0)$. (See Figure 9.5.) To find the solution of the linear system (9.2), we need to find the point of intersection of the two lines. Solving each equation for y produces the new set of equations

$$
\begin{aligned}
y &= 2 - \frac{2}{3}x \\
y &= 4 - 2x
\end{aligned}
$$

Setting the right-hand sides equal to each other, we obtain

$$2 - \frac{2}{3}x = 4 - 2x$$

$$\frac{4}{3}x = 2$$

$$x = \frac{3}{2}$$

To find y, we substitute the value of x back into one of the two original equations (say, $2x + y = 4$; it does not matter which one you choose). Then

$$y = 4 - 2x = 4 - (2)\left(\frac{3}{2}\right) = 1$$

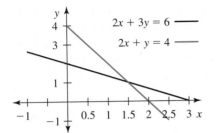

and the solution is the point $(3/2, 1)$.

Looking at Figure 9.5, we see that the two lines have exactly one point of intersection. This is so because the lines have different slopes: $-2/3$ and -2, respectively. The point of intersection is $(3/2, 1)$ and corresponds to the solution of (9.2), which is the only solution. ■

Figure 9.5 The two lines have exactly one point of intersection.

No Solution

EXAMPLE 2

Solve

$$\begin{align} 2x + y &= 4 \\ 4x + 2y &= 6 \end{align} \tag{9.3}$$

Solution

The line corresponding to $2x + y = 4$ has y-intercept $(0, 4)$ and x-intercept $(2, 0)$; the line corresponding to $4x + 2y = 6$ has y-intercept $(0, 3)$ and x-intercept $(3/2, 0)$. (See Figure 9.6.) Since

$$\begin{align} 2x + y = 4 &\iff y = 4 - 2x \\ 4x + 2y = 6 &\iff y = 3 - 2x \end{align}$$

it follows immediately that both lines have the same slope, namely, -2, but different y-intercepts: 4 and 3, respectively. This implies that the two lines are parallel and do not intersect.

Let's see what happens when we solve the system (9.3). We equate the two equations $y = 4 - 2x$ and $y = 3 - 2x$:

$$4 - 2x = 3 - 2x$$

This implies that

$$4 = 3$$

Figure 9.6 The two lines are parallel but do not intersect.

The last expression is obviously wrong. We conclude that there is no point (x, y) that satisfies both equations in (9.3) simultaneously.

Looking at Figure 9.6, we see that the two lines are parallel. Since parallel lines that are not identical do not intersect, (9.3) has no solution. (Just look at the graph.) In this case, we write the solution as ∅, the symbol for the empty set. ■

Infinitely Many Solutions

EXAMPLE 3

Solve

$$\begin{align} 2x + y &= 4 \\ 4x + 2y &= 8 \end{align} \tag{9.4}$$

Solution

If we divide the second equation by 2, we find that both equations are identical, namely, $2x + y = 4$. That is, both equations describe the same line with x-intercept

(2, 0) and y-intercept (0, 4), as shown in Figure 9.7. Every point (x, y) on this line is therefore a solution of (9.4). To find the solution algebraically, we use the same procedure as in Examples 1 and 2. We first solve each equation for y:

$$2x + y = 4 \implies y = 4 - 2x$$
$$4x + 2y = 8 \implies y = 4 - 2x$$

Equating the two equations $y = 4 - 2x$ and $y = 4 - 2x$ yields

$$4 - 2x = 4 - 2x$$

which simplifies to

$$0 = 0$$

This is a true statement, which implies that any value of x is a solution. A convenient way to write the solution is to introduce a new variable, say, t, to denote the x-coordinate. The new variable t can take on any real number. To find the corresponding y-coordinate, we substitute $t \in \mathbf{R}$ for x:

$$y = 4 - 2x = 4 - 2t$$

The solution can then be written as the set of points

$$\{(t, 4 - 2t) : t \in \mathbf{R}\}$$

reflecting the fact that the system (9.4) has infinitely many solutions, as expected from the graphical considerations. [Figure 9.7 shows that the two lines representing (9.4) are identical; hence, only one line is visible.] ■

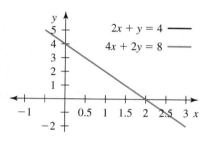

Figure 9.7 The two lines are identical.

In Example 3, we introduced the variable t to describe the set of solutions. We call t a **dummy variable**; it stands for any real number. Introducing a dummy variable is a convenient way to describe the set of solutions when there are infinitely many.

A Solution Method The graphical and algebraic way of solving systems of linear equations we have employed so far works only for systems in two variables. To solve systems of m linear equations in n variables, we will need to develop a method that will work for a system of any size . The basic strategy will be to transform the system of linear equations into a new system of equations that has the same solutions as the original. The new system is called an **equivalent system**. It will be of a simpler form, so that we can solve for the unknown variables one by one and thus arrive at a solution. We illustrate this approach in the next example. We tag all equations with labels of the form (R_i); R_i stands for "ith row." The labels will allow us to keep track of our computations.

EXAMPLE 4   Solve

$$3x + 2y = 8 \quad (R_1)$$
$$2x + 4y = 5 \quad (R_2)$$

Solution   There are two basic operations that transform a system of linear equations into an equivalent system: (1) We can multiply an equation by a nonzero number, and (2) we can add one equation to the other.

Our goal will be to eliminate x in the second equation. If we multiply the first equation by 2 and the second equation by -3, we obtain

$$\begin{aligned} 2(R_1) \quad & 6x + 4y = 16 \\ -3(R_2) \quad & -6x - 12y = -15 \end{aligned}$$

If we add the two equations, we find that

$$-8y = 1$$

Through algebraic manipulations, we eliminated x from the second equation. We can replace the original system of equations by a new (equivalent) system by leaving the first equation unchanged and replacing the second equation by $-8y = 1$. We then obtain the equivalent system of equations [labeled (R_3) and (R_4)]

$$\begin{array}{lll} (R_1) & 3x + 2y = 8 & (R_3) \\ 2(R_1) - 3(R_2) & -8y = 1 & (R_4) \end{array}$$

We can now successively solve the system. It follows from equation (R_4) that

$$y = -\frac{1}{8}$$

Substituting this value into equation (R_3) and solving for x, we find that

$$3x + (2)\left(-\frac{1}{8}\right) = 8$$
$$3x = 8 + \frac{1}{4}$$
$$x = \frac{11}{4}$$

The solution is therefore $(11/4, -1/8)$. ■

When we look at equations (R_3) and (R_4), we see that the left-hand side has the shape of a triangle:

$$*x + *y = *$$
$$*y = *$$

We therefore call a system that is written in this form **upper triangular**. In the preceding example, we reduced the system of linear equations to upper triangular form and then solved it by back-substitution. As we will see in the next subsection, this method can be generalized to larger systems of equations.

■ 9.1.2 Solving Systems of Linear Equations

In this subsection, we will extend the solution method of Example 4 to systems of m equations in n variables, which we write in the form

$$\begin{aligned} a_{11}x_1 + a_{12}x_2 + \cdots + a_{1n}x_n &= b_1 \\ a_{21}x_1 + a_{22}x_2 + \cdots + a_{2n}x_n &= b_2 \\ &\cdots\cdots\cdots\cdots\cdots\cdots\cdots\cdots \\ a_{m1}x_1 + a_{m2}x_2 + \cdots + a_{mn}x_n &= b_m \end{aligned} \qquad (9.5)$$

The variables are now $x_1, x_2, \ldots, x_n$. The coefficients a_{ij} on the left-hand side have two subscripts. The first subscript indicates the equation, and the second subscript indicates to which x_j a_{ij} belongs. For instance, you would find a_{21} in the second equation as the coefficient of x_1; a_{43} would be in the fourth equation as the coefficient of x_3. Using double subscripts is a convenient way of labeling the coefficients. The subscripts on the numbers b_i on the right-hand side of (9.5) indicate the equation.

We will transform this system into an equivalent system in upper triangular form. (Recall that *equivalent* means that the new system has the same solutions as the old system.) To do so, we will use the following three basic operations:

1. Multiplying an equation by a nonzero constant
2. Adding one equation to another
3. Rearranging the order of the equations

This method is also called **Gaussian elimination**.

As in the case of a linear system with two equations in two variables, the general system (9.5) may have

1. Exactly one solution
2. No solutions
3. Infinitely many solutions

When a system has no solutions, we say that the system is **inconsistent**.

Exactly One Solution

EXAMPLE 5 Solve

$$3x + 5y - z = 10 \quad (R_1)$$
$$2x - y + 3z = 9 \quad (R_2)$$
$$4x + 2y - 3z = -1 \quad (R_3)$$

Solution   Our goal is to reduce this system to upper triangular form. The first step is to eliminate $2x$ from the second equation and $4x$ from the third equation. We leave the first equation unchanged. Multiplying the first equation by 2 and the second equation by -3 and then adding the two equations yields a new equation that replaces the second equation:

$$2(R_1) \quad 6x + 10y - 2z = 20$$
$$-3(R_2) \quad -6x + 3y - 9z = -27$$

This system yields

$$13y - 11z = -7$$

In the new equation, the term involving x has been eliminated.

We transform the third equation by multiplying the second equation by 2 and the third equation by -1.

$$2(R_2) \quad 4x - 2y + 6z = 18$$
$$-(R_3) \quad -4x - 2y + 3z = 1$$

Adding the two equations gives

$$-4y + 9z = 19$$

The term involving x has been eliminated.

These two steps transform the original set of equations into the following equivalent set of equations, labeled (R_4)–(R_6):

$$(R_1) \quad 3x + 5y - z = 10 \quad (R_4)$$
$$2(R_1) - 3(R_2) \quad 13y - 11z = -7 \quad (R_5)$$
$$2(R_2) - (R_3) \quad - 4y + 9z = 19 \quad (R_6)$$

Our next step is to eliminate y from (R_6). To do this, we multiply (R_5) by 4 and (R_6) by 13:

$$4(R_5) \quad 52y - 44z = -28$$
$$13(R_6) \quad -52y + 117z = 247$$

Adding the two equations then yields

$$73z = 219$$

We leave the first two equations unchanged. The new (equivalent) system of equations is thus

$$(R_4) \quad 3x + 5y - z = 10 \quad (R_7)$$
$$(R_5) \quad 13y - 11z = -7 \quad (R_8) \qquad (9.6)$$
$$4(R_5) + 13(R_6) \quad 73z = 219 \quad (R_9)$$

The system of equations is now in upper triangular form, as can be seen from its shape. We use back-substitution to find the solution. Solving equation (R_9) for z yields

$$z = \frac{219}{73} = 3$$

Solving (R_8) for y and substituting the value of z gives

$$y = \frac{1}{13}(-7 + 11z) = \frac{1}{13}(-7 + (11)(3)) = \frac{26}{13} = 2$$

Solving (R_7) for x and substituting the values of y and z, we find that

$$x = \frac{1}{3}(10 - 5y + z)$$

$$= \frac{1}{3}(10 - (5)(2) + 3) = 1$$

Hence, the solution is $x = 1$, $y = 2$, and $z = 3$. ■

In Example 5, when we eliminate variables, we have a choice in how to do this. We used equations (R_1) and (R_2) to replace (R_2) and used (R_2) and (R_3) to replace (R_3). We could instead have used (R_2) and (R_3) to replace (R_2) by $(R_3) - 2(R_2)$, resulting in $4y - 9z = -19$. However, in this case, we could not have used (R_2) and (R_3) again to replace (R_3). If we used (R_2) and (R_3) again, we would need to replace (R_3) by $(R_3) - 2(R_2)$, which is $4y - 9z = -19$. We would replace both (R_2) and (R_3) by the same equation, thus effectively losing one equation. To replace (R_3), we would need to choose (R_1) and (R_3) and replace (R_3) by $4(R_1) - 3(R_3)$, which is $14y + 5z = 43$.

A comment about eliminating y from (R_5) and (R_6) is in order: When we replaced (R_6), we had to use (R_5) and (R_6). If we had used (R_6) and (R_4), say, to eliminate y, we would have computed $(R_6) + 2(R_4)$, which results in the equation $4x + z = 8$. This would have introduced the variable x into the resulting equation, and the system would not have been reduced to upper triangular form.

No Solution

EXAMPLE 6 Solve

$$\begin{array}{rrrl} 2x - & y + & z = 3 & (R_1) \\ 4x - & 4y + & 3z = 2 & (R_2) \\ 2x - & 3y + & 2z = 1 & (R_3) \end{array}$$

Solution As in Example 5, we try to reduce the system to upper triangular form. We begin by eliminating terms involving x from the second and third equations. We leave the first equation unchanged. We replace the second equation by the result of subtracting the second equation from the first, after multiplying the first equation by 2. We replace the third equation by the result of subtracting the third equation from the first. The result of all these operations is

$$\begin{array}{lrrl} (R_1) & 2x - & y + z = 3 & (R_4) \\ 2(R_1) - (R_2) & & 2y - z = 4 & (R_5) \\ (R_1) - (R_3) & & 2y - z = 2 & (R_6) \end{array}$$

To obtain (R_5), we also could have computed $(R_2) - 2(R_1)$; we then would have found that $-2y + z = -4$. Similarly, for (R_6), we could have computed $(R_3) - (R_1)$. As long as we do not use the same pair of equations twice, we have some freedom in how we reduce the system to upper triangular form.

Now, let's continue with our calculations. To eliminate terms involving y from Equation (R_6), we replace (R_6) by the difference of (R_5) and (R_6). We keep the first two equations. Then

$$(R_4) \quad 2x - y + z = 3$$
$$(R_5) \quad\quad\quad 2y - z = 4$$
$$(R_5) - (R_6) \quad\quad\quad\quad 0 = 2$$

The last equation, $0 = 2$, is a false statement, which means that this system does not have a solution. We therefore write the solution as $\emptyset$, the symbol denoting the empty set. ■

Infinitely Many Solutions

EXAMPLE 7   Solve

$$x - 3y + z = 4 \quad (R_1)$$
$$x - 2y + 3z = 6 \quad (R_2)$$
$$2x - 6y + 2z = 8 \quad (R_3)$$

Solution   We proceed as in the preceding examples. The first step is to eliminate terms involving x from the second and third equations. We leave the first equation unchanged. We replace the second equation by the difference of the first and second equations. We replace the third equation by the difference of the first and the third equations after dividing the third equation by 2. We find that

$$(R_1) \quad x - 3y + z = 4 \quad (R_4)$$
$$(R_2) - (R_1) \quad\quad y + 2z = 2 \quad (R_5)$$
$$(R_1) - \tfrac{1}{2}(R_3) \quad\quad\quad 0z = 0 \quad (R_6)$$

The third equation, $0z = 0$, is a correct statement. It means that we can substitute any number for z and still obtain a correct result. We introduce the dummy variable t and set

$$z = t \quad \text{for } t \in \mathbf{R}$$

If we solve (R_5) for y and substitute $z = t$ into the resulting equation, we get

$$y = 2 - 2z = 2 - 2t$$

If we solve (R_4) for x and substitute $y = 2 - 2t$ and $z = t$ into the resulting equation, we find that

$$x = 4 + 3y - z = 4 + 3(2 - 2t) - t$$
$$= 4 + 6 - 6t - t = 10 - 7t$$

The solution is therefore the set

$$\{(x, y, z) : x = 10 - 7t, y = 2 - 2t, z = t, \text{for } t \in \mathbf{R}\}$$ ■

A Shorthand Notation   When we transform a system of linear equations, we make changes only to the coefficients of the variables. It is therefore convenient to introduce notation that will simply keep track of all the coefficients. This motivates the following definition:

Definition A **matrix** is a rectangular array of numbers:

$$A = \begin{bmatrix} a_{11} & a_{12} & \cdots & a_{1n} \\ a_{21} & a_{22} & \cdots & a_{2n} \\ \cdots\cdots\cdots\cdots\cdots\cdots \\ a_{m1} & a_{m2} & \cdots & a_{mn} \end{bmatrix}$$

The elements a_{ij} of the matrix A are called **entries**. If the matrix has m rows and n columns, it is called an $m \times n$ matrix.

We can think of the matrix A in the definition as the **coefficient matrix** of the linear system we introduced in (9.5), namely,

$$
\begin{aligned}
a_{11}x_1 + a_{12}x_2 + \cdots + a_{1n}x_n &= b_1 \\
a_{21}x_1 + a_{22}x_2 + \cdots + a_{2n}x_n &= b_2 \\
&\cdots\cdots\cdots\cdots\cdots\cdots\cdots \\
a_{m1}x_1 + a_{m2}x_2 + \cdots + a_{mn}x_n &= b_m
\end{aligned}
\tag{9.7}
$$

The entries a_{ij} of the matrix have two subscripts: The entry a_{ij} is located in the ith row and the jth column. You can think of this numbering system as the street address of the entries, using streets and avenues. Suppose that avenues go north–south and streets go east–west; then finding the corner of Second Street and Third Avenue would be like finding the element of A that is in the second row and third column, which is a_{23} in the matrix.

If a matrix has the same number of rows as columns, it is called a **square matrix**. An $m \times 1$ matrix is called a **column vector** and a $1 \times n$ matrix is called a **row vector**. Following are examples of a 3×3 square matrix, a 3×1 column vector, and a 1×4 row vector (the shape of each of the matrices explains their names):

$$
\begin{bmatrix} 1 & 3 & 0 \\ 0 & 1 & -1 \\ 5 & 4 & 3 \end{bmatrix}, \qquad \begin{bmatrix} 2 \\ 7 \\ 4 \end{bmatrix}, \qquad \begin{bmatrix} 1 & 3 & 0 & 5 \end{bmatrix}
$$

If A is a square matrix, then the **diagonal line** of A consists of the elements $a_{11}, a_{22}, \ldots, a_{nn}$. In the preceding 3×3 square matrix, the diagonal line thus consists of the elements 1, 1, and 3.

To solve systems of linear equations with the use of matrices, we introduce the **augmented matrix**—the coefficient matrix of the linear system (9.7), augmented by an additional column representing the right-hand side of (9.7). The augmented matrix representing the linear system (9.7) is therefore

$$
\left[\begin{array}{cccc|c} a_{11} & a_{12} & \cdots & a_{1n} & b_1 \\ a_{21} & a_{22} & \cdots & a_{2n} & b_2 \\ \multicolumn{5}{c}{\cdots\cdots\cdots\cdots\cdots\cdots} \\ a_{m1} & a_{m2} & \cdots & a_{mn} & b_m \end{array} \right]
$$

So how can we use augmented matrices to solve systems of linear equations?

EXAMPLE 8

Find the augmented matrix for the system given in Example 5, and solve the system by using the augmented matrix.

Solution

The augmented matrix is given by

$$
\left[\begin{array}{ccc|c} 3 & 5 & -1 & 10 \\ 2 & -1 & 3 & 9 \\ 4 & 2 & -3 & -1 \end{array} \right] \begin{array}{l} (R_1) \\ (R_2) \\ (R_3) \end{array}
$$

Our goal is to transform the augmented matrix into the simpler, upper triangular form, in which all the entries below the diagonal line are zero. We use the same basic transformations as before, and you should compare the steps that follow with the steps in Example 5. We indicate on the left side of the matrix the transformation that we used, and we relabel the rows on the right side of the matrix:

$$
\begin{array}{r}
(R_1) \\
2(R_1) - 3(R_2) \\
2(R_2) - (R_3)
\end{array}
\left[\begin{array}{ccc|c} 3 & 5 & -1 & 10 \\ 0 & 13 & -11 & -7 \\ 0 & -4 & 9 & 19 \end{array} \right]
\begin{array}{l} (R_4) \\ (R_5) \\ (R_6) \end{array}
$$

That is, we copied row 1, which we now call row 4; to get the second row, we multiplied the first row by 2 and the second row by 3 and then took the difference ($2(R_1) - 3(R_2)$), calling the resulting row (R_5), and so on.

The next step is to eliminate the -4 in (R_6). We find that

$$\begin{array}{c} (R_4) \\ (R_5) \\ 4(R_5) + 13(R_6) \end{array} \left[\begin{array}{ccc|c} 3 & 5 & -1 & 10 \\ 0 & 13 & -11 & -7 \\ 0 & 0 & 73 & 219 \end{array} \right]$$

This is now the augmented matrix for the system of linear equations in (9.6), and we can proceed as in Example 5 to solve the system by back-substitution. ■

So far, all of the systems that we have considered have had the same number of equations as variables. This need not be the case, however. When a system has fewer equations than variables, we say that it is **underdetermined**. Although underdetermined systems can be inconsistent (have no solutions), they frequently have infinitely many solutions. When a system has more equations than variables, we say that it is **overdetermined**. Overdetermined systems are frequently inconsistent.

In Example 9 we will look at an underdetermined system, and in Example 10 an overdetermined system.

EXAMPLE 9   Solve the following underdetermined system:

$$\begin{aligned} 2x + 2y - z &= & (R_1) \\ 2x - y + z &= 2 & (R_2) \end{aligned}$$

Solution   The system has fewer equations than variables and is therefore underdetermined. We use the augmented matrix

$$\left[\begin{array}{ccc|c} 2 & 2 & -1 & 1 \\ 2 & -1 & 1 & 2 \end{array} \right] \begin{array}{c} (R_1) \\ (R_2) \end{array}$$

to solve the system. Transforming the augmented matrix into upper triangular form, we find that

$$\begin{array}{c} (R_1) \\ (R_1) - (R_2) \end{array} \left[\begin{array}{ccc|c} 2 & 2 & -1 & 1 \\ 0 & 3 & -2 & -1 \end{array} \right]$$

Translating this matrix back into a system of equations, we obtain

$$\begin{aligned} 2x + 2y - z &= 1 \\ 3y - 2z &= -1 \end{aligned}$$

It then follows that

$$y = -\frac{1}{3} + \frac{2}{3}z$$

and

$$2x = 1 - 2y + z = 1 - 2\left(-\frac{1}{3} + \frac{2}{3}z \right) + z = \frac{5}{3} - \frac{1}{3}z$$

or

$$x = \frac{5}{6} - \frac{1}{6}z$$

We use a dummy variable again and set $z = t, t \in \mathbf{R}$; therefore, $x = \frac{5}{6} - \frac{1}{6}t$ and $y = -\frac{1}{3} + \frac{2}{3}t$. The solution can then be written as

$$\left\{ (x, y, z) : x = \frac{5}{6} - \frac{1}{6}t, y = -\frac{1}{3} + \frac{2}{3}t, z = t, t \in \mathbf{R} \right\}$$

■

EXAMPLE 10   Solve the following overdetermined system:

$$\begin{aligned} 2x - y &= 1 & (R_1) \\ x + y &= 2 & (R_2) \\ x - y &= 3 & (R_3) \end{aligned}$$

Solution   The system has more equations than variables and is therefore overdetermined. To
solve it, we write

$$
\begin{array}{lrcl}
(R_1) & 2x - y = & 1 & (R_4) \\
2(R_2) - (R_1) & 3y = & 3 & (R_5) \\
(R_2) - (R_3) & 2y = & -1 & (R_6)
\end{array}
$$

It follows from (R_6) that $y = -\frac{1}{2}$ and from (R_5) that $y = 1$. Since $1 \neq -\frac{1}{2}$, there
cannot be a solution. The system is inconsistent, and the solution set is the empty set.
In this example, we can show that the system is inconsistent if we graph the three
straight lines represented by (R_1), (R_2), and (R_3), as in Figure 9.8.

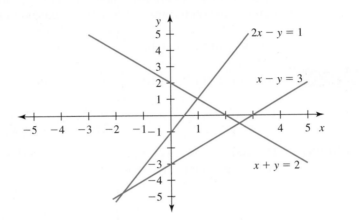

Figure 9.8 The straight lines in Example 10 do not intersect
in one point.

In order for the system to have solutions, the three lines would need to intersect
in one point (or be identical). They don't, so there is no solution.   ■

| EXAMPLE 11 | **More Than One Dummy Variable** Solve

$$
\begin{array}{rcll}
3x - 3y + 6z = & 9 & (R_1) \\
-x + y - 2z = & -3 & (R_2)
\end{array}
$$

Solution   We see that $(R_1) = -3(R_2)$. Thus, eliminating the variable x from (R_2) yields

$$
\begin{array}{lrl}
(R_1) & 3x - 3y + 6z = 9 \\
(R_1) + 3(R_2) & 0z = 0
\end{array}
$$

We can write the solution by introducing the dummy variable t and setting $z = t$,
$t \in \mathbf{R}$. To solve (R_1), we need a second dummy variable s, which satisfies $y = s$,
$s \in \mathbf{R}$:

$$
3x = 9 + 3s - 6t
$$

or

$$
x = 3 + s - 2t
$$

The solution can be written as

$$
\{(x, y, z) : x = 3 + s - 2t, \, y = s, \, z = t; \, s \in \mathbf{R}, t \in \mathbf{R}\}
$$ ■

Section 9.1 Problems

■ 9.1.1

In Problems 1–4, solve each linear system of equations. In addition, for each system, graph the two lines corresponding to the two equations in a single coordinate system and use your graph to explain your solution.

1. $x - y = 1$
$x - 2y = -2$

2. $2x + 3y = 6$
$x - 4y = -4$

3. $x - 3y = 6$
$y = 3 + \frac{1}{3}x$

4. $2x + y = \frac{1}{3}$
$6x + 3y = 1$

5. Determine c such that

$$2x - 3y = 5$$
$$4x - 6y = c$$

has **(a)** infinitely many solutions and **(b)** no solutions. **(c)** Is it possible to choose a number for c so that the system has exactly one solution? Explain your answer.

6. (a) Determine the solution of

$$-2x + 3y = 5$$
$$ax - y = 1$$

in terms of a.

(b) For which values of a are there no solutions, exactly one solution, and infinitely many solutions?

7. Show that the solution of

$$a_{11}x_1 + a_{12}x_2 = b_1$$
$$a_{21}x_1 + a_{22}x_2 = b_2$$

is given by

$$x_1 = \frac{a_{22}b_1 - a_{12}b_2}{a_{11}a_{22} - a_{21}a_{12}}$$

and

$$x_2 = \frac{-a_{21}b_1 + a_{11}b_2}{a_{11}a_{22} - a_{21}a_{12}}$$

8. Assume that the system

$$a_{11}x_1 + a_{12}x_2 = b_1$$
$$a_{21}x_1 + a_{22}x_2 = b_2$$

has infinitely many solutions. Determine the number of solutions if you change
(a) a_{11} **(b)** b_1

In Problems 9–16, reduce the system of linear equations to upper triangular form and solve.

9. $2x - y = 3$
$x - 3y = 7$

10. $5x - 3y = 2$
$2x + 7y = 3$

11. $7x - y = 4$
$3x + 2y = 1$

12. $5x + 2y = 8$
$-x + 3y = 9$

13. $3x - y = 1$
$-3x + y = 4$

14. $2x + 3y = 5$
$-y = -2 + \frac{2}{3}x$

15. $x + 2y = 3$
$4y + 2x = 6$

16. $x - 2y = 2$
$4y - 2x = -4$

17. Zach wants to buy fish and plants for his aquarium. Each fish costs \$2.30; each plant costs \$1.70. He buys a total of 11 items and spends a total of \$21.70. Set up a system of linear equations that will allow you to determine how many fish and how many plants Zach bought, and solve the system.

18. Laboratory mice are fed with a mixture of two foods that contain two essential nutrients. Food 1 contains 3 units of nutrient A and 2 units of nutrient B per ounce; food 2 contains 4 units of nutrient A and 5 units of nutrient B per ounce.
(a) In what proportion should you mix the food if the mice require nutrients A and B in equal amounts?
(b) Assume now that the mice require nutrients A and B in the ratio 1:2. Is it possible to satisfy their dietary needs with the two foods available?

19. Show that if

$$a_{11}a_{22} - a_{21}a_{12} \neq 0$$

then the system

$$a_{11}x_1 + a_{12}x_2 = 0$$
$$a_{21}x_1 + a_{22}x_2 = 0$$

has exactly one solution, namely, $x_1 = 0$ and $x_2 = 0$.

■ 9.1.2

In Problems 20–24, solve each system of linear equations.

20. $2x - 3y + z = -1$
$x + y - 2z = -3$
$3x - 2y + z = 2$

21. $5x - y + 2z = 6$
$x + 2y - z = -1$
$3x + 2y - 2z = 1$

22. $x + 4y - 3z = -13$
$2x - 3y + 5z = 18$
$3x + y - 2z = 1$

23. $-2x + 4y - z = -1$
$x + 7y + 2z = -4$
$3x - 2y + 3z = -3$

24. $2x - y + 3z = 3$
$2x + y + 4z = 4$
$2x - 3y + 2z = 2$

In Problems 25–28, find the augmented matrix and use it to solve the system of linear equations.

25. $-x - 2y + 3z = -9$
$2x + y - z = 5$
$4x - 3y + 5z = -9$

26. $3x - 2y + z = 4$
$4x + y - 2z = -12$
$2x - 3y + z = 7$

27. $y + x = 3$
$z - y = -1$
$x + z = 2$

28. $2x - z = 1$
$y + 3z = x - 1$
$x + z = y - 3$

In Problems 29–34, determine whether each system is overdetermined or underdetermined; then solve each system.

29. $x - 2y + z = 3$
$2x - 3y + z = 8$

30. $x - y = 2$
$x + y + z = 3$

31. $2x - y = 3$
$x - y = 4$
$3x - y = 1$

32. $4y - 3z = 6$
$2y + z = 1$
$y + z = 0$

33. $2x - 7y + z = 2$
$x + y - 2z = 4$

34. $3x + y = 1$
$x - y = 0$
$4x = 1$

35. SplendidLawn sells three types of lawn fertilizer: SL 24–4–8, SL 21–7–12 and SL 17–0–0. The three numbers refer to the percentages of nitrogen, phosphate, and potassium, in that order, of the contents. (For instance, 100 g of SL 24–4–8 contains 24 g of nitrogen.) Suppose that each year your lawn requires 500 g of nitrogen, 100 g of phosphate, and 180 g of potassium per 1000 square feet. How much of each of the three types of fertilizer should you apply per 1000 square feet per year?

36. Three different species of insects are reared together in a laboratory cage. They are supplied with two different types of food each day. Each individual of species 1 consumes 3 units of food A and 5 units of food B, each individual of species 2 consumes 2 units of food A and 3 units of food B, and individual of species 3 consumes 1 unit of food A and 2 units of food B. Each day, 500 units of food A and 900 units of food B are supplied. How many individuals of each species can be reared together? Is there more than one solution? What happens if we add 550 units of a third type of food, called C, and each individual of species 1 consumes 2 units of food C, each individual of species 2 consumes 4 units of food C, and each individual of species 3 consumes 1 unit of food C?

■ 9.2 Matrices

We introduced matrices in the previous section; in this section, we will learn various matrix operations.

■ 9.2.1 Basic Matrix Operations

Recall that an $m \times n$ matrix A is a rectangular array of numbers with m rows and n columns. We write it as

$$A = \begin{bmatrix} a_{11} & a_{12} & \cdots & a_{1n} \\ a_{21} & a_{22} & \cdots & a_{2n} \\ \cdots\cdots\cdots\cdots\cdots\cdots \\ a_{m1} & a_{m2} & \cdots & a_{mn} \end{bmatrix} = \begin{bmatrix} a_{ij} \end{bmatrix}_{\substack{1 \le i \le m \\ 1 \le j \le n}}$$

We will also use the shorthand notation $A = [a_{ij}]$ if the size of the matrix is clear. We list a few simple definitions that do not need much explanation.

> **Definition** Suppose that $A = [a_{ij}]$ and $B = [b_{ij}]$ are two $m \times n$ matrices. Then
> $$A = B$$
> if and only if, for all $1 \le i \le m$ and $1 \le j \le n$,
> $$a_{ij} = b_{ij}$$

This definition says that we can compare matrices of the same size, and they are equal if all their corresponding entries are equal. The next definition shows how to add matrices.

> **Definition** Suppose that $A = [a_{ij}]$ and $B = [b_{ij}]$ are two $m \times n$ matrices. Then
> $$C = A + B$$
> is an $m \times n$ matrix with entries
> $$c_{ij} = a_{ij} + b_{ij} \quad \text{for } 1 \le i \le m, 1 \le j \le n$$

Note that addition is defined only for matrices of equal size. Matrix addition satisfies the following two properties:

1. $A + B = B + A$
2. $(A + B) + C = A + (B + C)$

The matrix with all its entries equal to zero is called the **zero matrix** and is denoted by **0**. The following holds:

$$A + \mathbf{0} = A$$

We can multiply matrices by numbers. Such numbers are called **scalars**.

> **Definition** Suppose that $A = [a_{ij}]$ is an $m \times n$ matrix and c is a scalar. Then cA is an $m \times n$ matrix with entries ca_{ij} for $1 \leq i \leq m$ and $1 \leq j \leq n$.

EXAMPLE 1   Find $A + 2B - 3C$ if

$$A = \begin{bmatrix} 2 & 3 \\ 1 & 0 \end{bmatrix}, \quad B = \begin{bmatrix} 0 & 1 \\ -1 & -3 \end{bmatrix}, \quad C = \begin{bmatrix} 1 & 0 \\ 0 & 3 \end{bmatrix}$$

Solution

$$
\begin{aligned}
A + 2B - 3C &= \begin{bmatrix} 2 & 3 \\ 1 & 0 \end{bmatrix} + 2\begin{bmatrix} 0 & 1 \\ -1 & -3 \end{bmatrix} - 3\begin{bmatrix} 1 & 0 \\ 0 & 3 \end{bmatrix} \\
&= \begin{bmatrix} 2 & 3 \\ 1 & 0 \end{bmatrix} + \begin{bmatrix} 0 & 2 \\ -2 & -6 \end{bmatrix} + \begin{bmatrix} -3 & 0 \\ 0 & -9 \end{bmatrix} \\
&= \begin{bmatrix} 2+0-3 & 3+2+0 \\ 1-2+0 & 0-6-9 \end{bmatrix} = \begin{bmatrix} -1 & 5 \\ -1 & -15 \end{bmatrix}
\end{aligned}
$$

■

EXAMPLE 2   Let

$$A = \begin{bmatrix} 1 & 3 \\ 0 & 4 \end{bmatrix}$$

Find B so that

$$A + B = \begin{bmatrix} 1 & 1 \\ 1 & 1 \end{bmatrix}$$

Solution   If

$$A + B = \begin{bmatrix} 1 & 1 \\ 1 & 1 \end{bmatrix}$$

then

$$B = \begin{bmatrix} 1 & 1 \\ 1 & 1 \end{bmatrix} - A = \begin{bmatrix} 1 & 1 \\ 1 & 1 \end{bmatrix} - \begin{bmatrix} 1 & 3 \\ 0 & 4 \end{bmatrix}$$

Using rules for matrix addition, we see that this simplifies to

$$B = \begin{bmatrix} 1-1 & 1-3 \\ 1-0 & 1-4 \end{bmatrix} = \begin{bmatrix} 0 & -2 \\ 1 & -3 \end{bmatrix}$$

■

The operation that interchanges rows and columns of a matrix is called **transposition**.

> **Definition** Suppose that $A = [a_{ij}]$ is an $m \times n$ matrix. Then the **transpose** of A, denoted by A', is an $n \times m$ matrix with entries
>
> $$a'_{ij} = a_{ji}$$

The next example enables us to understand how this operation works.

EXAMPLE 3   Transpose the following matrices:

$$A = \begin{bmatrix} 1 & 2 & 3 \\ 4 & 5 & 6 \end{bmatrix}, \quad B = \begin{bmatrix} 1 \\ 2 \end{bmatrix}, \quad C = \begin{bmatrix} 3 & 4 \end{bmatrix}$$

Solution   To find the transpose, we need to interchange rows and columns. Since A is a 2×3 matrix, its transpose is the 3×2 matrix

$$A' = \begin{bmatrix} 1 & 4 \\ 2 & 5 \\ 3 & 6 \end{bmatrix}$$

To check this against our definition, note that

$$a'_{11} = a_{11} = 1, \quad a'_{12} = a_{21} = 4, \quad a'_{21} = a_{12} = 2, \ldots$$

The matrix B is a 2×1 matrix, which is a column vector. Its transpose is the 1×2 matrix

$$B' = \begin{bmatrix} 1 & 2 \end{bmatrix}$$

which is a row vector. Similarly, C is a 1×2 row vector; its transpose is then the 2×1 column vector

$$C' = \begin{bmatrix} 3 \\ 4 \end{bmatrix}$$

■

Additional properties of the transpose of a matrix are discussed in Problems 17–20.

In the preceding example, we saw that the transpose of a row vector is a column vector and vice versa. When we need to write a column vector in text, such as $X = \begin{bmatrix} 1 \\ 2 \\ 3 \end{bmatrix}$, we can instead write $X = \begin{bmatrix} 1 & 2 & 3 \end{bmatrix}'$, which, written as the transpose of a row vector, is really a column vector. A large column vector written as the transpose of a row vector is more legible in text.

■ **9.2.2 Matrix Multiplication**

The multiplication of two matrices is more complicated than any of the operations we learned in the previous subsection. We give the definition first.

Definition Suppose that $A = [a_{ij}]$ is an $m \times l$ matrix and $B = [b_{ij}]$ is an $l \times n$ matrix. Then

$$C = AB$$

is an $m \times n$ matrix with

$$c_{ij} = a_{i1}b_{1j} + a_{i2}b_{2j} + \cdots + a_{il}b_{lj} = \sum_{k=1}^{l} a_{ik}b_{kj}$$

for $1 \leq i \leq m$ and $1 \leq j \leq n$.

To explain this mathematical definition in words, note that c_{ij} is the entry in C that is located in the ith row and the jth column. To obtain it, we multiply the entries of the ith row of A with the entries of the jth column of B, as indicated in the definition. For the product AB to be defined, the number of columns in A must equal the number of rows in B. The definition looks rather formidable; let's see how it actually works.

EXAMPLE 4 **Matrix Multiplication** Compute AB when

$$A = \begin{bmatrix} 1 & 2 & 3 \\ -1 & 0 & 4 \end{bmatrix} \quad \text{and} \quad B = \begin{bmatrix} 1 & 2 & 3 & -3 \\ 0 & -1 & 4 & 0 \\ -1 & 0 & -2 & 1 \end{bmatrix}$$

Solution First, note that A is a 2×3 matrix and B is a 3×4 matrix. That is, A has 3 columns and B has 3 rows. Therefore, the product AB is defined and AB is a 2×4 matrix. We write the product as

$$C = AB = \begin{bmatrix} 1 & 2 & 3 \\ -1 & 0 & 4 \end{bmatrix} \cdot \begin{bmatrix} 1 & 2 & 3 & -3 \\ 0 & -1 & 4 & 0 \\ -1 & 0 & -2 & 1 \end{bmatrix}$$

For instance, to find c_{11}, we multiply the first row of A with the first column of B as follows:

$$c_{11} = \begin{bmatrix} 1 & 2 & 3 \end{bmatrix} \begin{bmatrix} 1 \\ 0 \\ -1 \end{bmatrix} = (1)(1) + (2)(0) + (3)(-1) = -2$$

You can see from this calculation that the number of columns of A must be equal to the number of rows of B. Otherwise, we would run out of numbers when multiplying the corresponding entries.

To find c_{23}, we multiply the second row of A by the third column of B:

$$c_{23} = \begin{bmatrix} -1 & 0 & 4 \end{bmatrix} \begin{bmatrix} 3 \\ 4 \\ -2 \end{bmatrix} = (-1)(3) + (0)(4) + (4)(-2) = -11$$

The other entries are obtained in a similar way, and we find that

$$C = \begin{bmatrix} -2 & 0 & 5 & 0 \\ -5 & -2 & -11 & 7 \end{bmatrix} \quad ■$$

In multiplying matrices, the order is important. For instance, BA is not defined for the matrices in Example 4, since the number of columns of B (which is 4) is different from the number of rows of A (which is 2). From this, it should be clear that, typically, $AB \neq BA$. Even if both AB and BA exist, the resulting matrices are usually not the same. They can differ both in the number of rows and columns and in the actual entries. The next two examples illustrate this important point.

EXAMPLE 5 **Order Is Important** Suppose that

$$A = \begin{bmatrix} 2 & 1 & -1 \end{bmatrix} \quad \text{and} \quad B = \begin{bmatrix} 1 \\ -1 \\ 0 \end{bmatrix}$$

Show that both AB and BA are defined but $AB \neq BA$.

Solution Note that A is a 1×3 matrix and B is a 3×1 matrix. Thus, the product AB is defined, since the number of columns of A is the same as the number of rows of B (namely, 3); the product AB is a 1×1 matrix. When we carry out the matrix multiplication, we find that

$$AB = \begin{bmatrix} 2 & 1 & -1 \end{bmatrix} \begin{bmatrix} 1 \\ -1 \\ 0 \end{bmatrix} = \begin{bmatrix} 2 - 1 + 0 \end{bmatrix} = \begin{bmatrix} 1 \end{bmatrix} = 1$$

Note that a 1×1 matrix is simply a number.

The product BA is also defined, since the number of columns of B is the same as the number of rows of A (namely, 1); the product BA is a 3×3 matrix. When we carry out the matrix multiplication, we obtain

$$BA = \begin{bmatrix} 1 \\ -1 \\ 0 \end{bmatrix} \begin{bmatrix} 2 & 1 & -1 \end{bmatrix} = \begin{bmatrix} 2 & 1 & -1 \\ -2 & -1 & 1 \\ 0 & 0 & 0 \end{bmatrix}$$

Comparing AB and BA, we see immediately that $AB \neq BA$, since the two matrices are not even of the same size. ■

EXAMPLE 6

Order Is Important Suppose that

$$A = \begin{bmatrix} 2 & -1 \\ 4 & -2 \end{bmatrix} \quad \text{and} \quad B = \begin{bmatrix} 1 & -1 \\ 2 & -2 \end{bmatrix}$$

Show that both AB and BA are defined but $AB \neq BA$.

Solution A and B are each 2×2 matrices. We conclude that both AB and BA are 2×2 matrices as well. We find that

$$AB = \begin{bmatrix} 0 & 0 \\ 0 & 0 \end{bmatrix} \quad \text{and} \quad BA = \begin{bmatrix} -2 & 1 \\ -4 & 2 \end{bmatrix}$$

Since the entries in AB are not the same as in BA, it follows that $AB \neq BA$. ■

The product AB in Example 6 resulted in a matrix with all entries equal to 0. Earlier, we called this matrix the zero matrix and denoted it by **0**. Example 6 shows that a product of two matrices can be the zero matrix without either matrix being the zero matrix. This is an important fact and is different from multiplying real numbers. When multiplying two real numbers, we know that if the product is equal to 0, at least one of the two factors must have been equal to 0. This rule, however, does not hold for matrix multiplication; if A and B are two matrices whose product AB is defined, then the product AB can be equal to the zero matrix **0** without either A or B being equal to **0**.

The following properties of matrix multiplication assume that all matrices are of appropriate sizes so that all of the matrix multiplications are defined:

1. $(A + B)C = AC + BC$
2. $A(B + C) = AB + AC$
3. $(AB)C = A(BC)$
4. $A0 = 0A = 0$

(We will practice applying these properties in the problems at the end of this section.)

Next, we note that if A is a square matrix (i.e., if A has the same number of rows as columns), we can define powers of A: If k is a positive integer, then

$$A^k = A^{k-1}A = AA^{k-1} = \underbrace{AA \ldots A}_{k \text{ factors}}$$

For instance, $A^2 = AA$, $A^3 = AAA$, and so on.

EXAMPLE 7

Powers of Matrices Find A^3 if

$$A = \begin{bmatrix} 1 & 2 \\ 3 & 4 \end{bmatrix}$$

Solution   To find A^3, we first need to compute A^2. We obtain

$$A^2 = AA = \begin{bmatrix} 1 & 2 \\ 3 & 4 \end{bmatrix}\begin{bmatrix} 1 & 2 \\ 3 & 4 \end{bmatrix} = \begin{bmatrix} 7 & 10 \\ 15 & 22 \end{bmatrix}$$

To compute A^3 now, we compute either A^2A or AA^2. Both will yield A^3. This is a case in which the order of multiplication does not matter, since $A^3 = (AA)A = A(AA) = AAA$. We do it both ways.

$$A^3 = A^2A = \begin{bmatrix} 7 & 10 \\ 15 & 22 \end{bmatrix} \begin{bmatrix} 1 & 2 \\ 3 & 4 \end{bmatrix} = \begin{bmatrix} 37 & 54 \\ 81 & 118 \end{bmatrix}$$

and

$$A^3 = AA^2 = \begin{bmatrix} 1 & 2 \\ 3 & 4 \end{bmatrix} \begin{bmatrix} 7 & 10 \\ 15 & 22 \end{bmatrix} = \begin{bmatrix} 37 & 54 \\ 81 & 118 \end{bmatrix} \qquad ■$$

An important square matrix is the **identity matrix**, denoted by I_n. The identity matrix is an $n \times n$ matrix with 1's on its diagonal line and 0's elsewhere; that is,

$$I_n = \begin{bmatrix} 1 & 0 & 0 & \cdots & 0 \\ 0 & 1 & 0 & \cdots & 0 \\ \multicolumn{5}{c}{\dotfill} \\ 0 & 0 & 0 & \cdots & 1 \end{bmatrix}$$

For instance,

$$I_1 = \begin{bmatrix} 1 \end{bmatrix}, \qquad I_2 = \begin{bmatrix} 1 & 0 \\ 0 & 1 \end{bmatrix}, \qquad I_3 = \begin{bmatrix} 1 & 0 & 0 \\ 0 & 1 & 0 \\ 0 & 0 & 1 \end{bmatrix}$$

(We frequently write I instead of I_n if the size of I_n is clear from the context.) The identity matrix serves the same role as the number 1 in the multiplication of real numbers: If A is an $m \times n$ matrix, then

$$AI_n = I_mA = A$$

It follows that $I^k = I$ for any positive integer k.

If a and x are numbers, then $ax = x$ can be written as $ax - x = 0$. To factor x on the left-hand side, we transform that side as follows: $ax - 1x = (a - 1)x$. To factor x, we needed to introduce the factor 1 in front of x. We obtain $(a - 1)x = 0$. There is a similar procedure for matrices, but we must be more careful, since the order in which we multiply matrices is important.

EXAMPLE 8

Matrix Equation Let A be a 2×2 matrix and X be a 2×1 matrix. Show that $AX = X$ can be written as $(A - I_2)X = \mathbf{0}$.

Solution

We know that
$$AX = X \qquad \text{is equivalent to} \qquad AX - X = \mathbf{0}$$

To factor X, we multiply X from the left by I_2. The matrix I_2 will then allow us to factor X just as the factor 1 allowed us to factor x in $ax - x = 0$ in the case of real numbers. Multiplying X from the left by I_2 gives

$$AX - I_2X = \mathbf{0}$$

We can now factor X to get

$$(A - I_2)X = \mathbf{0}$$

Observe again that we multiplied X from the left by $A - I_2$. We are not allowed to reverse the order. [Indeed, the product $X(A - I_2)$ is not even defined.] ■

There is a close connection between systems of linear equations and matrices: The system of linear equations

$$\begin{bmatrix} a_{11}x_1 & a_{12}x_2 & \cdots & a_{1n}x_n & b_1 \\ a_{21}x_1 & a_{22}x_2 & \cdots & a_{2n}x_n & b_2 \\ \multicolumn{5}{c}{\dotfill} \\ a_{m1}x_1 & a_{m2}x_2 & \cdots & a_{mn}x_n & b_m \end{bmatrix}$$

can be written in the matrix form

$$AX = B$$

with

$$A = \begin{bmatrix} a_{11} & \cdots & a_{1n} \\ a_{21} & \cdots & a_{2n} \\ \cdots\cdots\cdots\cdots \\ a_{m1} & \cdots & a_{mn} \end{bmatrix}, \quad X = \begin{bmatrix} x_1 \\ x_2 \\ \vdots \\ x_n \end{bmatrix}, \quad B = \begin{bmatrix} b_1 \\ b_2 \\ \vdots \\ b_m \end{bmatrix}$$

EXAMPLE 9

Matrix Representation of Linear Systems Write

$$2x_1 + 3x_2 + 4x_3 = 1$$
$$-x_1 + 5x_2 - 6x_3 = 7$$

in matrix form.

Solution In matrix form, we have

$$\begin{bmatrix} 2 & 3 & 4 \\ -1 & 5 & -6 \end{bmatrix} \begin{bmatrix} x_1 \\ x_2 \\ x_3 \end{bmatrix} = \begin{bmatrix} 1 \\ 7 \end{bmatrix}$$

We see that A is a 2×3 matrix and X is a 3×1 matrix. We therefore expect (on the basis of the rules of multiplication) that B is a 2×1 matrix, which it is indeed. ■

We will use matrix representation to solve systems of linear equations in the next subsection.

■ 9.2.3 Inverse Matrices

In this subsection, we will learn how to solve systems of n linear equations in n unknowns when they are written in the matrix form

$$AX = B \tag{9.8}$$

where A is an $n \times n$ square matrix and X and B are $n \times 1$ column vectors. We start with a simple example in which $n = 1$. When we solve

$$5x = 10$$

for x, we simply divide both sides by 5 or, equivalently, multiply both sides by $\frac{1}{5} = 5^{-1}$, and obtain

$$5^{-1} \cdot 5x = 5^{-1} \cdot 10$$

Since $5^{-1} \cdot 5 = 1$ and $5^{-1} \cdot 10 = 2$, we find that $x = 2$. To solve (9.8), we therefore need an operation that is analogous to division, or multiplication by the "reciprocal" of A. We will define a matrix A^{-1} that will serve this function. It is called the **inverse matrix** of A. If this inverse matrix exists, then we can write the solution of (9.8) as

$$A^{-1}AX = A^{-1}B$$

In order for the inverse matrix to have the same effect as multiplying a number by its reciprocal, it should have the property $A^{-1}A = I$, where I is the identity matrix. The solution would then be of the form

$$X = A^{-1}B$$

Definition Suppose that $A = [a_{ij}]$ is an $n \times n$ square matrix. If there exists an $n \times n$ square matrix B such that

$$AB = BA = I_n$$

then B is called the inverse matrix of A and is denoted by A^{-1}.

If A has an inverse matrix, A is called **invertible** or **nonsingular**; if A does not have an inverse matrix, A is called **singular**. If A is invertible, its inverse matrix is unique; that is, if B and C are both inverse matrices of A, then $B = C$. (In other words, if you and your friend compute the inverse of A, you both should get the exact same answer.) To see why, assume that B and C are both inverse matrices of A. Using $BA = I_n$ and $AC = I_n$, we find that

$$B = BI_n = B(AC) = (BA)C = I_nC = C$$

The following properties hold for inverse matrices:

1. If A is an invertible $n \times n$ matrix, then

$$(A^{-1})^{-1} = A$$

2. If A and B are invertible $n \times n$ matrices, then

$$(AB)^{-1} = B^{-1}A^{-1}$$

The first statement says that if we apply the inverse operation twice, we get the original matrix back. This property is familiar from dealing with real numbers: Take the number 2; its inverse is $2^{-1} = 1/2$. If we take the inverse of $1/2$, we get 2 again. To see why the analogous property holds for matrices, assume that A is an $n \times n$ matrix with inverse matrix A^{-1}. Then, by definition,

$$AA^{-1} = A^{-1}A = I_n$$

But this equation also says that A is the inverse matrix of A^{-1}; that is, $A = (A^{-1})^{-1}$.

To see why the second statement is true, we must verify that

$$(AB)(B^{-1}A^{-1}) = (B^{-1}A^{-1})(AB) = I_n$$

That this is indeed the case is seen when we compute

$$(AB)(B^{-1}A^{-1}) = A(BB^{-1})A^{-1} = AI_nA^{-1} = AA^{-1} = I_n$$

and

$$(B^{-1}A^{-1})(AB) = B^{-1}(A^{-1}A)B = B^{-1}I_nB = B^{-1}B = I_n$$

In the next example, we will see how we can check whether two matrices are inverses of each other. The matrices will be of size 3×3. It is somewhat cumbersome to find the inverse of a matrix of size 3×3 or larger (unless you use a calculator or computer software). Here, we start out with the two matrices and simply check whether they are inverse matrices of each other.

EXAMPLE 10 **Checking Whether Matrices Are Inverses of Each Other** Show that the inverse of

$$A = \begin{bmatrix} 3 & 5 & -1 \\ 2 & -1 & 3 \\ 4 & 2 & -3 \end{bmatrix} \quad \text{is} \quad B = \begin{bmatrix} -\frac{3}{73} & \frac{13}{73} & \frac{14}{73} \\ \frac{18}{73} & -\frac{5}{73} & -\frac{11}{73} \\ \frac{8}{73} & \frac{14}{73} & -\frac{13}{73} \end{bmatrix}$$

Solution   Carrying out the matrix multiplications AB and BA, we see that

$$AB = I_3 \quad \text{and} \quad BA = I_3$$

Therefore, $B = A^{-1}$.   ■

In this book, we will need to invert only 2×2 matrices. In the next example, we will show one way to do this.

EXAMPLE 11   **Finding an Inverse**   Find the inverse of

$$A = \begin{bmatrix} 2 & 5 \\ 1 & 3 \end{bmatrix}$$

Solution   We need to find a matrix

$$B = \begin{bmatrix} b_{11} & b_{12} \\ b_{21} & b_{22} \end{bmatrix}$$

such that $AB = I_2$. We will then check that $BA = I_2$ and, hence, that B is the inverse of A. We must solve

$$\begin{bmatrix} 2 & 5 \\ 1 & 3 \end{bmatrix} \begin{bmatrix} b_{11} & b_{12} \\ b_{21} & b_{22} \end{bmatrix} = \begin{bmatrix} 1 & 0 \\ 0 & 1 \end{bmatrix}$$

or

$$\begin{array}{ccc} 2b_{11} + 5b_{21} = 1 & & 2b_{12} + 5b_{22} = 0 \\ b_{11} + 3b_{21} = 0 & \text{and} & b_{12} + 3b_{22} = 1 \end{array} \tag{9.9}$$

We solve both equations by reducing them to upper triangular form. Leaving the first equation of either set of equations and eliminating b_{11} from the first set of equations and b_{12} from the second set of equations by multiplying the second equation by (-2) and adding the first and second equation, we find that

$$\begin{array}{ccc} 2b_{11} + 5b_{21} = 1 & & 2b_{12} + 5b_{22} = 0 \\ -\ b_{21} = 1 & \text{and} & -\ b_{22} = -2 \end{array} \tag{9.10}$$

The left set of equations has the solution

$$b_{21} = -1 \quad \text{and} \quad b_{11} = \frac{1}{2}(1 - 5b_{21}) = \frac{1}{2}(1 + 5) = 3$$

The right set of equations has the solution

$$b_{22} = 2 \quad \text{and} \quad b_{12} = \frac{1}{2}(-5b_{22}) = -5$$

Hence,

$$B = \begin{bmatrix} 3 & -5 \\ -1 & 2 \end{bmatrix}$$

You can verify that both $AB = I_2$ and $BA = I_2$; thus, $B = A^{-1}$.   ■

In the optional Subsection 9.2.4, we will present a method that can be employed for larger matrices. It is essentially the same method as in Example 11, but uses augmented matrices. Since manually inverting larger matrices takes a long time, graphing calculators or computer software should be used to find inverse matrices.

At the end of Subsection 9.2.2, we mentioned the connection between systems of linear equations and matrix multiplication. A system of linear equations can be written as

$$AX = B \tag{9.11}$$

In a linear system of n equations in n unknowns, A is an $n \times n$ matrix. If A is invertible, then multiplying (9.11) by A^{-1} from the left, we find that

$$A^{-1}AX = A^{-1}B$$

Since $A^{-1}A = I_n$ and $I_nX = X$, it follows that

$$X = A^{-1}B$$

We will use this equation to redo Example 5 of Section 9.1.

EXAMPLE 12 **Using Inverse Matrices to Solve Linear Systems** Solve

$$
\begin{aligned}
3x + 5y - z &= 10 \\
2x - y + 3z &= 9 \\
4x + 2y - 3z &= -1
\end{aligned}
$$

Solution The coefficient matrix of A is

$$
A = \begin{bmatrix} 3 & 5 & -1 \\ 2 & -1 & 3 \\ 4 & 2 & -3 \end{bmatrix}
$$

which we encountered in Example 10, where we saw that

$$
A^{-1} = \begin{bmatrix} -\frac{3}{73} & \frac{13}{73} & \frac{14}{73} \\ \frac{18}{73} & -\frac{5}{73} & -\frac{11}{73} \\ \frac{8}{73} & \frac{14}{73} & -\frac{13}{73} \end{bmatrix}
$$

We set

$$
B = \begin{bmatrix} 10 \\ 9 \\ -1 \end{bmatrix}
$$

and compute $A^{-1}B$. We find that

$$
A^{-1} \begin{bmatrix} 10 \\ 9 \\ -1 \end{bmatrix} = \begin{bmatrix} 1 \\ 2 \\ 3 \end{bmatrix}
$$

That is, $x = 1$, $y = 2$, and $z = 3$, as we found in Example 5 of Section 9.1. ■

We now have two methods for solving systems of linear equations when the number of equations is equal to the number of variables: We can reduce the system to upper triangular form and then use back-substitution, or we can write the system in the matrix form $AX = B$, find A^{-1}, and then compute $X = A^{-1}B$. Of course, the second method works only when A^{-1} exists. If A^{-1} does not exist, the system has either no solution or infinitely many solutions. But when A^{-1} exists, $AX = B$ has exactly one solution.

There is a simple and very useful criterion for checking whether a 2×2 matrix is invertible. Deriving this criterion will also provide us with a formula for the inverse of an invertible 2×2 matrix. We set

$$
A = \begin{bmatrix} a_{11} & a_{12} \\ a_{21} & a_{22} \end{bmatrix} \quad \text{and} \quad B = \begin{bmatrix} b_{11} & b_{12} \\ b_{21} & b_{22} \end{bmatrix}
$$

We wish to write the entries of B in terms of the entries of A such that $AB = BA = I$ or $B = A^{-1}$. From

$$
\begin{bmatrix} a_{11} & a_{12} \\ a_{21} & a_{22} \end{bmatrix} \begin{bmatrix} b_{11} & b_{12} \\ b_{21} & b_{22} \end{bmatrix} = \begin{bmatrix} 1 & 0 \\ 0 & 1 \end{bmatrix}
$$

we obtain the following set of equations:

$$
\begin{aligned}
a_{11}b_{11} + a_{12}b_{21} &= 1 \\
a_{21}b_{11} + a_{22}b_{21} &= 0
\end{aligned}
\tag{9.12}
$$

$$
\begin{aligned}
a_{11}b_{12} + a_{12}b_{22} &= 0 \\
a_{21}b_{12} + a_{22}b_{22} &= 1
\end{aligned}
\tag{9.13}
$$

We solve the system of linear equations (9.12) for b_{11} and b_{21}:

$$a_{11}b_{11} + a_{12}b_{21} = 1 \quad (R_1)$$
$$a_{21}b_{11} + a_{22}b_{21} = 0 \quad (R_2)$$

If we compute $a_{11}(R_2) - a_{21}(R_1)$, we find that

$$a_{11}a_{22}b_{21} - a_{12}a_{21}b_{21} = -a_{21}$$

Solving this equation for b_{21}, we first factor b_{21} on the left-hand side. This yields

$$(a_{11}a_{22} - a_{12}a_{21})b_{21} = -a_{21}$$

If $a_{11}a_{22} - a_{12}a_{21} \neq 0$, we can isolate b_{21}:

$$b_{21} = -\frac{a_{21}}{a_{11}a_{22} - a_{12}a_{21}}$$

Substituting b_{21} into (R_1) and solving for b_{11}, we obtain, for $a_{11} \neq 0$,

$$b_{11} = \frac{1}{a_{11}}[1 - a_{12}b_{21}]$$
$$= \frac{1}{a_{11}}\left[1 - \frac{-a_{12}a_{21}}{a_{11}a_{22} - a_{12}a_{21}}\right]$$
$$= \frac{1}{a_{11}}\frac{a_{11}a_{22}}{a_{11}a_{22} - a_{12}a_{21}}$$
$$= \frac{a_{22}}{a_{11}a_{22} - a_{12}a_{21}}$$

(When $a_{11} = 0$, the same final result holds.) When we solve (9.13), we find that

$$b_{12} = -\frac{a_{12}}{a_{11}a_{22} - a_{12}a_{21}}$$
$$b_{22} = \frac{a_{11}}{a_{11}a_{22} - a_{12}a_{21}}$$

Our calculations yield two important results:

Theorem If

$$A = \begin{bmatrix} a_{11} & a_{12} \\ a_{21} & a_{22} \end{bmatrix}$$

and $a_{11}a_{22} - a_{12}a_{21} \neq 0$, then A is invertible and

$$A^{-1} = \frac{1}{a_{11}a_{22} - a_{12}a_{21}} \begin{bmatrix} a_{22} & -a_{12} \\ -a_{21} & a_{11} \end{bmatrix}$$

The expression $a_{11}a_{22} - a_{12}a_{21}$ is called the **determinant** of A and is denoted by det A.

Definition If

$$A = \begin{bmatrix} a_{11} & a_{12} \\ a_{21} & a_{22} \end{bmatrix}$$

then the determinant of A is defined as

$$\det A = a_{11}a_{22} - a_{12}a_{21}$$

Looking back now at the formula for A^{-1}, where A is a 2×2 matrix whose determinant is nonzero, we see that, to find the inverse of A, we divide A by the determinant of A, switch the diagonal elements of A, and change the sign of the off-diagonal elements. If the determinant is equal to 0, then the inverse of A does not exist.

The determinant can be defined for any $n \times n$ matrix. We will not give the general formula, which is computationally complicated for $n \geq 3$. Graphing calculators or more sophisticated computer software can compute determinants. But we mention the following result, which allows us to determine whether or not an $n \times n$ matrix has an inverse:

Theorem Suppose that A is an $n \times n$ matrix. Then A is nonsingular if and only if $\det A \neq 0$.

EXAMPLE 13

Using The Determinant to Find Inverse Matrices Determine which of the following matrices is invertible, and in each case compute the inverse if it exists:

(a) $A = \begin{bmatrix} 3 & 5 \\ 2 & 4 \end{bmatrix}$ (b) $B = \begin{bmatrix} 2 & 6 \\ 1 & 3 \end{bmatrix}$

Solution (a) To check whether A is invertible, we compute the determinant of A:

$$\det A = (3)(4) - (2)(5) = 2 \neq 0$$

Hence, A is invertible and we find that

$$A^{-1} = \frac{1}{2} \begin{bmatrix} 4 & -5 \\ -2 & 3 \end{bmatrix} = \begin{bmatrix} 2 & -\frac{5}{2} \\ -1 & \frac{3}{2} \end{bmatrix}$$

(b) Since

$$\det B = (2)(3) - (1)(6) = 0$$

B is not invertible. ∎

We mentioned that if A is invertible, then

$$AX = B$$

has exactly one solution, namely, $X = A^{-1}B$. Of particular importance is the case when $B = \mathbf{0}$. That is, assume that A is a 2×2 matrix and $B = \begin{bmatrix} 0 \\ 0 \end{bmatrix}$; then $X = \begin{bmatrix} 0 \\ 0 \end{bmatrix}$ is a solution of

$$AX = \begin{bmatrix} 0 \\ 0 \end{bmatrix}$$

We call the solution $\begin{bmatrix} 0 \\ 0 \end{bmatrix}$ a **trivial** solution. It is the only solution of $AX = \mathbf{0}$ when A is invertible. If $AX = \mathbf{0}$ has a solution $X \neq \begin{bmatrix} 0 \\ 0 \end{bmatrix}$, then X is called a **nontrivial** solution. In order to get a nontrivial solution for

$$AX = \mathbf{0},$$

A must be singular. We formulate this condition as a theorem in the more general case when A is an $n \times n$ matrix; we will need this result repeatedly.

Theorem Suppose that A is an $n \times n$ matrix, and X and $\mathbf{0}$ are $n \times 1$ matrices. Then the equation

$$AX = \mathbf{0}$$

has a nontrivial solution if and only if A is singular.

EXAMPLE 14   **Nontrivial Solutions** Let

$$A = \begin{bmatrix} a & 6 \\ 3 & 2 \end{bmatrix}$$

Determine a so that $AX = \mathbf{0}$ has at least one nontrivial solution, and find the nontrivial solution(s).

Solution   To determine when $AX = \mathbf{0}$ has a nontrivial solution, we must find conditions under which A is singular or, equivalently, det $A = 0$.

Since det $A = 2a - 18$, it follows that A is singular if $2a - 18 = 0$, or $a = 9$. Therefore, if $a = 9$, $AX = \mathbf{0}$ has a nontrivial solution. To compute that nontrivial solution, we must solve

$$\begin{bmatrix} 9 & 6 \\ 3 & 2 \end{bmatrix}\begin{bmatrix} x_1 \\ x_2 \end{bmatrix} = \begin{bmatrix} 0 \\ 0 \end{bmatrix}$$

which can be written as a system of two linear equations:

$$9x_1 + 6x_2 = 0$$
$$3x_1 + 2x_2 = 0$$

We see that the first equation is three times the second equation. We can therefore simplify the system and obtain

$$3x_1 + 2x_2 = 0$$
$$0x_2 = 0$$

Together, these equations show that the system has infinitely many solutions, namely,

$$\left\{ (x_1, x_2) : x_2 = t \text{ and } x_1 = -\frac{2}{3}t, \text{ for } t \in \mathbf{R} \right\}$$

In particular, the system has nontrivial solutions; for instance, choosing $t = 3$, we find that $x_1 = -2$ and $x_2 = 3$, and choosing $t = -1$, we find that $x_1 = 2/3$ and $x_2 = -1$. Any value for t that is different from 0 will yield a nontrivial solution. If $t = 0$, the trivial solution $(0, 0)$ results.

If we graphed the two lines corresponding to the two equations when $a = 9$, the lines would be identical. For all other values of a, the two lines would intersect only at the point $(0, 0)$, which corresponds to the trivial solution. (Try it!)   ■

■ 9.2.4 Computing Inverse Matrices (Optional)

In the preceding subsection, we saw how to invert 2×2 matrices by solving two systems of linear equations. Manually inverting larger matrices takes a long time, and algorithms have been implemented into graphing calculators and computers to carry out these calculations. Here is one method: Inverting an $n \times n$ matrix results in n linear systems, each consisting of n equations in n unknowns. By solving these n systems simultaneously, we can speed up the process of finding the inverse matrix.

Recall Equation (9.9) of Subsection 9.2.3:

$$\begin{array}{ccc} 2b_{11} + 5b_{21} = 1 & & 2b_{12} + 5b_{22} = 0 \\ b_{11} + 3b_{21} = 0 & \text{and} & b_{12} + 3b_{22} = 1 \end{array}$$

Let us rewrite both systems as augmented matrices:

$$\begin{bmatrix} 2 & 5 & | & 1 \\ 1 & 3 & | & 0 \end{bmatrix} \quad \text{and} \quad \begin{bmatrix} 2 & 5 & | & 0 \\ 1 & 3 & | & 1 \end{bmatrix}$$

We see that each augmented matrix has the same matrix A on its left side. To solve each system, we must perform row transformations until we can read off the solutions. Reading off the solutions is particularly easy when the matrix on the left is transformed into the identity matrix. We do this for the augmented matrix

$$\begin{bmatrix} 2 & 5 & 1 \\ 1 & 3 & 0 \end{bmatrix} \begin{array}{c} (R_1) \\ (R_2) \end{array}$$

We divide the first row by 2 in order to get a 1 in the first position. In the second row we want a 0 in the first position; to get this, we subtract $2(R_2)$ from (R_1); that is,

$$\begin{matrix} \frac{1}{2}(R_1) \\ (R_1) - 2(R_2) \end{matrix} \begin{bmatrix} 1 & \frac{5}{2} & \bigm| & \frac{1}{2} \\ 0 & -1 & \bigm| & 1 \end{bmatrix} \begin{matrix} (R_3) \\ (R_4) \end{matrix}$$

By adding $\frac{5}{2}(R_4)$ to (R_3), we get a 0 in the second position in the first row. Multiplying (R_4) by -1, we get a 1 in the second position in the second row. That is, we obtain the identity matrix on the left side:

$$\begin{matrix} (R_3) + \frac{5}{2}(R_4) \\ (-1)(R_4) \end{matrix} \begin{bmatrix} 1 & 0 & \bigm| & 3 \\ 0 & 1 & \bigm| & -1 \end{bmatrix}$$

Rewriting this matrix as a system of linear equations, we see that

$$b_{11} = 3$$
$$b_{21} = -1$$

To find b_{12} and b_{22}, we need to transform the augmented matrix

$$\begin{bmatrix} 2 & 5 & \bigm| & 0 \\ 1 & 3 & \bigm| & 1 \end{bmatrix}$$

so that the matrix on the left is the identity matrix. This involves the exact same transformations as before, since the coefficient matrix is the same for both systems of linear equations. Now comes the trick that speeds up the calculation: Instead of doing each augmented matrix separately, we do them simultaneously. That is, we write

$$\begin{bmatrix} 2 & 5 & \bigm| & 1 & 0 \\ 1 & 3 & \bigm| & 0 & 1 \end{bmatrix}$$

and then make row transformations until this matrix is of the form

$$\begin{bmatrix} 1 & 0 & \bigm| & b_{11} & b_{12} \\ 0 & 1 & \bigm| & b_{21} & b_{22} \end{bmatrix}$$

It then follows that

$$A^{-1} = \begin{bmatrix} b_{11} & b_{12} \\ b_{21} & b_{22} \end{bmatrix}$$

Let's do the calculation:

$$\begin{bmatrix} 2 & 5 & \bigm| & 1 & 0 \\ 1 & 3 & \bigm| & 0 & 1 \end{bmatrix} \begin{matrix} (R_1) \\ (R_2) \end{matrix}$$

$$\begin{matrix} \frac{1}{2}(R_1) \\ (R_1) - 2(R_2) \end{matrix} \begin{bmatrix} 1 & \frac{5}{2} & \bigm| & \frac{1}{2} & 0 \\ 0 & -1 & \bigm| & 1 & -2 \end{bmatrix} \begin{matrix} (R_3) \\ (R_4) \end{matrix}$$

$$\begin{matrix} (R_3) + \frac{5}{2}(R_4) \\ (-1)(R_4) \end{matrix} \begin{bmatrix} 1 & 0 & \bigm| & 3 & -5 \\ 0 & 1 & \bigm| & -1 & 2 \end{bmatrix}$$

We recognize the column

$$\begin{bmatrix} 3 \\ -1 \end{bmatrix}$$

which gives

$$b_{11} = 3 \quad \text{and} \quad b_{21} = -1$$

The column on the right,

$$\begin{bmatrix} -5 \\ 2 \end{bmatrix}$$

corresponds to the solution

$$b_{12} = -5 \quad \text{and} \quad b_{22} = 2$$

We conclude that

$$\begin{bmatrix} 3 & -5 \\ -1 & 2 \end{bmatrix}$$

as we saw in Example 11.

This method works for larger matrices as well. It is quite efficient, considering that we must solve a large number of linear equations. We illustrate the technique in the next example on a 3×3 matrix.

EXAMPLE 15

Find the inverse (if it exists) of

$$\begin{bmatrix} 1 & -1 & -1 \\ 2 & -1 & 1 \\ -1 & 1 & -1 \end{bmatrix}$$

Solution   We write

$$\left[\begin{array}{ccc|ccc} 1 & -1 & -1 & 1 & 0 & 0 \\ 2 & -1 & 1 & 0 & 1 & 0 \\ -1 & 1 & -1 & 0 & 0 & 1 \end{array} \right] \begin{array}{c} (R_1) \\ (R_2) \\ (R_3) \end{array}$$

and perform appropriate row transformations in order to get the identity matrix on the left side. The matrix on the right side is then the inverse matrix. The first step is to get 0's below the 1 in the first column:

$$\begin{array}{c} (R_1) \\ 2(R_1) - (R_2) \\ (R_1) + (R_3) \end{array} \left[\begin{array}{ccc|ccc} 1 & -1 & -1 & 1 & 0 & 0 \\ 0 & -1 & -3 & 2 & -1 & 0 \\ 0 & 0 & -2 & 1 & 0 & 1 \end{array} \right] \begin{array}{c} (R_4) \\ (R_5) \\ (R_6) \end{array}$$

Now we turn to the second column. We want a 1 in the middle and 0's on top and bottom of the second column:

$$\begin{array}{c} (R_4) - (R_5) \\ (-1)(R_5) \\ (R_6) \end{array} \left[\begin{array}{ccc|ccc} 1 & 0 & 2 & -1 & 1 & 0 \\ 0 & 1 & 3 & -2 & 1 & 0 \\ 0 & 0 & -2 & 1 & 0 & 1 \end{array} \right] \begin{array}{c} (R_7) \\ (R_8) \\ (R_9) \end{array}$$

Next, we turn to the third column. We need 0's in the first two rows and a 1 in the third row. Note that we can use only (R_9) to do the transformations:

$$\begin{array}{c} (R_7) + (R_9) \\ 2(R_8) + 3(R_9) \\ -\frac{1}{2}(R_9) \end{array} \left[\begin{array}{ccc|ccc} 1 & 0 & 0 & 0 & 1 & 1 \\ 0 & 2 & 0 & -1 & 2 & 3 \\ 0 & 0 & 1 & -\frac{1}{2} & 0 & -\frac{1}{2} \end{array} \right] \begin{array}{c} (R_{10}) \\ (R_{11}) \\ (R_{12}) \end{array}$$

Finally, we need to get a 1 in the second row and second column:

$$\begin{array}{c} (R_{10}) \\ \frac{1}{2}(R_{11}) \\ (R_{12}) \end{array} \left[\begin{array}{ccc|ccc} 1 & 0 & 0 & 0 & 1 & 1 \\ 0 & 1 & 0 & -\frac{1}{2} & 1 & \frac{3}{2} \\ 0 & 0 & 1 & -\frac{1}{2} & 0 & -\frac{1}{2} \end{array} \right]$$

We succeeded in getting the identity matrix on the left side. Therefore,

$$A^{-1} = \begin{bmatrix} 0 & 1 & 1 \\ -\frac{1}{2} & 1 & \frac{3}{2} \\ -\frac{1}{2} & 0 & -\frac{1}{2} \end{bmatrix}$$

■

This method is reasonably quick as long as the matrices are not too big. If you cannot get the identity matrix on the left side, then the matrix does not have an inverse, as is illustrated in the next example.

EXAMPLE 16

Find the inverse (if it exists) of

$$A = \begin{bmatrix} 2 & 6 \\ 1 & 3 \end{bmatrix}$$

Solution

Since $\det A = (2)(3) - (1)(6) = 0$, we already know that A is singular and therefore does not have an inverse. But let's see what happens if we try to find the inverse. First, we have

$$\begin{bmatrix} 2 & 6 & | & 1 & 0 \\ 1 & 3 & | & 0 & 1 \end{bmatrix} \begin{matrix} (R_1) \\ (R_2) \end{matrix}$$

As before, we try to get a 1 in the first row and first column and a 0 in the second row and first column. We find that

$$\begin{matrix} \frac{1}{2}(R_1) \\ (R_1) - 2(R_2) \end{matrix} \begin{bmatrix} 1 & 3 & | & \frac{1}{2} & 0 \\ 0 & 0 & | & 1 & -2 \end{bmatrix}$$

Since the last row in the left half consists only of 0's, we can never obtain the identity matrix on the left side. That is, our method fails to provide an inverse matrix, and we conclude that A does not have an inverse. ■

■ 9.2.5 An Application: The Leslie Matrix

In this subsection, we discuss age-structured populations with discrete breeding seasons, such as perennial plants in the temperate zone, where reproduction is limited to a particular season of the year. Such populations are described by discrete-time models.

We begin with the simplest case. We measure time in units of generations and denote the size of a population at generation t by $N(t)$, where $t = 0, 1, 2, \ldots$. The discrete-time analogue of unrestricted growth was discussed in Chapter 2 and is described by

$$N(t+1) = RN(t) \quad \text{for } t = 0, 1, 2, \ldots \tag{9.14}$$

with $N(0) = N_0$. The constant R can be interpreted as the number of individuals in the next generation per individual in the current generation. We assume that $R \geq 0$. The solution of (9.14) can be found by first computing $N(t)$ for $t = 1, 2$, and 3:

$$N(1) = RN(0)$$
$$N(2) = RN(1) = R[RN(0)] = R^2 N(0)$$
$$N(3) = RN(2) = R[R^2 N(0)] = R^3 N(0)$$

Recognizing the pattern, we conclude (as in Chapter 2) that

$$N(t) = R^t N(0)$$

Unrestricted growth results in exponential growth of the population. Depending on the value of R, we obtain the following long-term behavior:

$$\lim_{t \to \infty} N(t) = \begin{cases} \infty & \text{if } R > 1 \\ N(0) & \text{if } R = 1 \\ 0 & \text{if } 0 \leq R < 1 \end{cases}$$

(See Figure 9.9.)

The population model in (9.14) is the simplest model of discrete-time population growth. The quantity R describes the relative change of the population size from generation to generation.

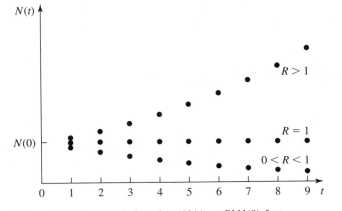

Figure 9.9 The population sizes $N(t) = R^t N(0)$ for successive values of t when $R > 1$, $R = 1$, and $0 < R < 1$.

In many species, reproduction is highly age dependent. For instance, periodical cicadas spend 13 to 17 years in the nymphal stage; they reproduce only once in their life. The purple coneflower, a prairie flower, does not reproduce until it is about three years old, but then continues to produce seeds throughout its life.   To take the life history into account, we will examine discrete-time, age-structured models. Introduced by Patrick Leslie in 1945, these models are formulated as matrix models and are based on demographic data. They are used to calculate important demographic quantities, such as the population size in each age class and the growth rate of the population. Leslie's approach is widely used not only in population biology, but also in the life insurance industry.

We begin with a numerical example to illustrate the method. Since only females produce offspring, we will follow only the females of the population. We assume that breeding occurs once a year and that we take a census of the population at the end of each breeding season. Individuals born during a particular breeding season are of age 0 at the end of that season. If a zero-year-old survives until the end of the next breeding season, it will be age 1 when we take a census at the end of that season. If a one-year-old survives until the end of the next breeding season, it will be age 2 at the end of that breeding season, and so on. In our example, we will assume that no individual lives beyond age 3; that is, there are no individuals age 4 or older in the population. We define

$$N_x(t) = \text{number of females of age } x \text{ at time } t$$

where $t = 0, 1, 2, \ldots$. Then

$$N(t) = \begin{bmatrix} N_0(t) \\ N_1(t) \\ N_2(t) \\ N_3(t) \end{bmatrix}$$

is the vector that describes the number of females in each age class at time t.

We assume that 40% of the females age 0, 30% of the females age 1, and 10% of the females age 2 at time t are alive when we take the census at time $t + 1$. That is,

$$N_1(t + 1) = (0.4)N_0(t)$$
$$N_2(t + 1) = (0.3)N_1(t)$$
$$N_3(t + 1) = (0.1)N_2(t)$$

The number of zero-year-old females at time $t + 1$ is equal to the number of female offspring during the breeding season that survive until the end of the breeding season, when the census is taken. We assume that

$$N_0(t + 1) = 2N_1(t) + 1.5N_2(t)$$

which means that females reach sexual maturity when they are of age 1. The factor 2 in front of $N_1(t)$ should be interpreted as the average number of surviving female offspring of a one-year-old female. [The factor 1.5 in front of $N_2(t)$ is then the average number of surviving female offspring of a two-year-old female.] There is no contribution of three-year-olds to the newborn class.

We can summarize the dynamics in matrix form:

$$
\begin{bmatrix} N_0(t+1) \\ N_1(t+1) \\ N_2(t+1) \\ N_3(t+1) \end{bmatrix} = \begin{bmatrix} 0 & 2 & 1.5 & 0 \\ 0.4 & 0 & 0 & 0 \\ 0 & 0.3 & 0 & 0 \\ 0 & 0 & 0.1 & 0 \end{bmatrix} \begin{bmatrix} N_0(t) \\ N_1(t) \\ N_2(t) \\ N_3(t) \end{bmatrix} \tag{9.15}
$$

The 4×4 matrix in this equation is called the **Leslie matrix**, which we denote by L. We can write the matrix equation (9.15) in short form as

$$
N(t+1) = LN(t) \tag{9.16}
$$

To see how this equation works, let's assume that the population at time t has the following age distribution:

$$
N_0(t) = 1000, \quad N_1(t) = 200, \quad N_2(t) = 100, \quad \text{and} \quad N_3(t) = 10
$$

When we take a census after the next breeding season, we find that 40% of the females age 0 at time t are alive at time $t+1$; that is, $N_1(t+1) = (0.4)(1000) = 400$. Also, 30% of the females age 1 at time t are alive at time $t+1$; that is, $N_2(t+1) = (0.3)(200) = 60$. And 10% of the females age 2 at time t are alive at time $t+1$; that is, $N_3(t+1) = (0.1)(100) = 10$. The number of surviving female offspring at time $t+1$ is

$$
N_0(t+1) = 2N_1(t) + 1.5N_2(t) = (2)(200) + (1.5)(100) = 550
$$

We obtain the same result when we use the Leslie matrix (9.15):

$$
\begin{bmatrix} N_0(t+1) \\ N_1(t+1) \\ N_2(t+1) \\ N_3(t+1) \end{bmatrix} = \begin{bmatrix} 0 & 2 & 1.5 & 0 \\ 0.4 & 0 & 0 & 0 \\ 0 & 0.3 & 0 & 0 \\ 0 & 0 & 0.1 & 0 \end{bmatrix} \begin{bmatrix} 1000 \\ 200 \\ 100 \\ 10 \end{bmatrix}
$$

$$
= \begin{bmatrix} (2)(200) + (1.5)(100) \\ (0.4)(1000) \\ (0.3)(200) \\ (0.1)(100) \end{bmatrix} = \begin{bmatrix} 550 \\ 400 \\ 60 \\ 10 \end{bmatrix}
$$

To obtain the age distribution at time $t+2$, we apply the Leslie matrix to the population vector at time $t+1$; that is,

$$
\begin{bmatrix} N_0(t+2) \\ N_1(t+2) \\ N_2(t+2) \\ N_3(t+2) \end{bmatrix} = \begin{bmatrix} 0 & 2 & 1.5 & 0 \\ 0.4 & 0 & 0 & 0 \\ 0 & 0.3 & 0 & 0 \\ 0 & 0 & 0.1 & 0 \end{bmatrix} \begin{bmatrix} 550 \\ 400 \\ 60 \\ 10 \end{bmatrix} = \begin{bmatrix} 890 \\ 220 \\ 120 \\ 6 \end{bmatrix}
$$

These three breeding seasons are illustrated in Figure 9.10.

From the preceding discussion, we can now find the general form of the Leslie matrix. We assume that the population is divided into $m+1$ age classes. The census is again taken at the end of each breeding season. Then, for the survival of each age class, we find that

$$
N_1(t+1) = P_0 N_0(t)
$$
$$
N_2(t+1) = P_1 N_1(t)
$$
$$
\vdots
$$
$$
N_m(t+1) = P_{m-1} N_{m-1}(t)
$$

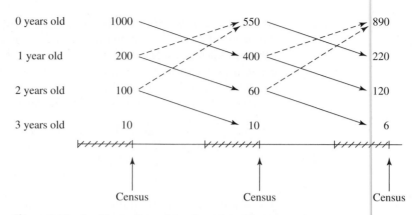

Figure 9.10 An illustration of the three breeding seasons.

where P_i denotes the fraction of females age i at time t that survive to time $t + 1$. Since P_i denotes a fraction, it follows that $0 \le P_i \le 1$ for $i = 0, 1, \ldots, m - 1$. The equation for zero-year-old females is given by

$$N_0(t + 1) = F_0 N_0(t) + F_1 N_1(t) + \cdots + F_m N_m(t)$$

where F_i is the average number of surviving female offspring per female individual age i. Writing this equation in matrix form, we find that the Leslie matrix is given by

$$
\begin{bmatrix}
F_0 & F_1 & F_2 & \cdots & F_{m-1} & F_m \\
P_0 & 0 & \cdots & \cdots & \cdots & 0 \\
0 & P_1 & 0 & \cdots & \cdots & 0 \\
\vdots & \cdots & \cdots & \cdots & \cdots & \vdots \\
0 & \cdots & \cdots & 0 & P_{m-1} & 0
\end{bmatrix}
$$

This is a matrix in which all elements are 0, except possibly those in the first row and in the first subdiagonal below the diagonal. The size of the matrix is $(m+1) \times (m+1)$, reflecting the $m + 1$ age classes.

EXAMPLE 17 Suppose that the Leslie matrix of a population is

$$
\begin{bmatrix}
5 & 7 & 1.5 \\
0.2 & 0 & 0 \\
0 & 0.4 & 0
\end{bmatrix}
$$

Interpret this matrix, and determine what happens if you follow a population for two breeding seasons, starting with 1000 zero-year-old females.

Solution The population is divided into three age classes: zero-year-olds, one-year-olds, and two-year-olds. The elements in the subdiagonal describe the survival probabilities; that is, 20% of the zero-year-olds survive until the next census, and 40% of the one-year-olds survive until the next census. The maximum age is two years. Zero-year-olds produce an average of five surviving females per female; one-year-olds produce an average of seven surviving females per female; two-year-olds produce an average of 1.5 surviving females per female.

A population that starts with 1000 zero-year-old females at time 0 has the population vector

$$
N(0) = \begin{bmatrix} 1000 \\ 0 \\ 0 \end{bmatrix}
$$

Using $N(t + 1) = LN(t)$, we find that

$$
N(1) = \begin{bmatrix}
5 & 7 & 1.5 \\
0.2 & 0 & 0 \\
0 & 0.4 & 0
\end{bmatrix}
\begin{bmatrix} 1000 \\ 0 \\ 0 \end{bmatrix}
=
\begin{bmatrix} 5000 \\ 200 \\ 0 \end{bmatrix}
$$

and

$$N(2) = \begin{bmatrix} 5 & 7 & 1.5 \\ 0.2 & 0 & 0 \\ 0 & 0.4 & 0 \end{bmatrix} \begin{bmatrix} 5000 \\ 200 \\ 0 \end{bmatrix} = \begin{bmatrix} 26,400 \\ 1000 \\ 80 \end{bmatrix}$$ ■

Stable Age Distribution

We will now investigate what happens when we run a model like the preceding one for a long time. Let's assume that the Leslie matrix is given by

$$L = \begin{bmatrix} 1.5 & 2 \\ 0.08 & 0 \end{bmatrix}$$

Using the equation

$$N(t+1) = LN(t)$$

we can compute successive population vectors; that is,

$$N_0(t+1) = 1.5N_0(t) + 2N_1(t)$$
$$N_1(t+1) = 0.08N_0(t)$$

Suppose that we start with $N_0(0) = 100$ and $N_1(0) = 100$. Then

$$\begin{bmatrix} N_0(1) \\ N_1(1) \end{bmatrix} = \begin{bmatrix} 1.5 & 2 \\ 0.08 & 0 \end{bmatrix} \begin{bmatrix} 100 \\ 100 \end{bmatrix} = \begin{bmatrix} 350 \\ 8 \end{bmatrix}$$

Continuing in this way, we find the population vectors at successive times, starting at time 0:

$$\begin{bmatrix} 100 \\ 100 \end{bmatrix}, \begin{bmatrix} 350 \\ 8 \end{bmatrix}, \begin{bmatrix} 541 \\ 28 \end{bmatrix}, \begin{bmatrix} 868 \\ 43 \end{bmatrix}, \begin{bmatrix} 1388 \\ 69 \end{bmatrix},$$
$$\begin{bmatrix} 2221 \\ 111 \end{bmatrix}, \begin{bmatrix} 3553 \\ 178 \end{bmatrix}, \begin{bmatrix} 5685 \\ 284 \end{bmatrix}, \dots$$

(In these calculations, we rounded to the nearest integer.) The first thing you notice is that the total population is growing. [The population size at time t is $N_0(t) + N_1(t)$.] But we can say a lot more. If we look at the successive ratios

$$q_0(t) = \frac{N_0(t)}{N_0(t-1)}$$

we find the following:

t	1	2	3	4	5	6	7
$q_0(t)$	3.5	1.55	1.60	1.5991	1.6001	1.5997	1.6001

That is, $q_0(t)$ seems to approach a limiting value, namely, 1.6. The same happens when we look at the ratio

$$q_1(t) = \frac{N_1(t)}{N_1(t-1)}$$

We find

t	1	2	3	4	5	6	7
$q_0(t)$	0.08	3.5	1.536	1.605	1.609	1.604	1.596

Both ratios seem to approach 1.6. In fact, we can show that

$$\lim_{t \to \infty} \frac{N_0(t)}{N_0(t-1)} = \lim_{t \to \infty} \frac{N_1(t)}{N_1(t-1)} = 1.6$$

(In the next section, we will see how to get this result analytically from the Leslie matrix.) If we look at the fraction of females in age class 0, namely,

$$p(t) = \frac{N_0(t)}{N_0(t) + N_1(t)}$$

we find

t	0	1	2	3	4	5	6	7
$p(t)$	0.5	0.9777	0.9508	0.9528	0.9526	0.9524	0.9523	0.9524

That is, this fraction also seems to converge. It looks as if about 95.2% of the population is in age class 0 when t is sufficiently large.

Although the population is increasing in size, the fraction of females in age class 0 (and hence also in age class 1) seems to converge. This constant fraction is referred to as the **stable age distribution**.

The preceding method of finding the stable age distribution does not always work, and we will give an example in Problem 81 in which the population does not reach a stable age distribution.

If we start the population in the stable age distribution, the fraction of females in age class 0 will remain the same, about 95.2%, and the population will increase by a constant by a factor of 1.6 each generation. Here is a numerical illustration of these two important properties: A stable age distribution for this population is

$$N(0) = \begin{bmatrix} 2000 \\ 100 \end{bmatrix}$$

(We will learn in the next section how to find this vector.) If we start with the foregoing stable age distribution, then we have

$$N(1) = LN(0) = \begin{bmatrix} 1.5 & 2 \\ 0.08 & 0 \end{bmatrix} \begin{bmatrix} 2000 \\ 100 \end{bmatrix} = \begin{bmatrix} 3200 \\ 160 \end{bmatrix}$$

The fraction of females in age class 0 remains the same, namely, $3200/3360 = 2000/2100$. We compare this with

$$(1.6) \begin{bmatrix} 2000 \\ 100 \end{bmatrix} = \begin{bmatrix} 3200 \\ 160 \end{bmatrix}$$

which yields the same result. That is, if N denotes a stable age distribution, then

$$LN = \lambda N \tag{9.17}$$

where $\lambda = 1.6$ and $N = [N_0, N_1]'$, with $N_0/(N_0 + N_1) \approx 95.2\%$. [In the next section, we will see how to compute λ and the ratio $N_0/(N_0 + N_1)$.] Equation (9.17) is used to determine the stable age distribution. The vector N is called an **eigenvector**; the value λ is the corresponding **eigenvalue**. A 2×2 matrix has two eigenvalues (which might be identical). In the case of the Leslie matrix, the largest eigenvalue is interpreted as the growth parameter; that is, it determines how the population grows. Its associated eigenvector is a stable age distribution.

In the next section, we will learn how to compute eigenvalues and eigenvectors. Because we will need them in other contexts as well, we will develop the theory in a more general setting. At the end of the next section, we will return to Leslie matrices and see how to compute the stable age distribution and the growth parameter of the population.

Section 9.2 Problems

■ **9.2.1, 9.2.2**

In Problems 1–6, let

$$A = \begin{bmatrix} -1 & 2 \\ 0 & -3 \end{bmatrix}, \quad B = \begin{bmatrix} 0 & 1 \\ 2 & 4 \end{bmatrix}, \quad C = \begin{bmatrix} 1 & -2 \\ 1 & -1 \end{bmatrix}$$

1. Find $A - B + 2C$. **2.** Find $-2A + 3B$.

3. Determine D so that $A + B = 2A - B + D$.

4. Show that $A + B = B + A$.

5. Show that $(A + B) + C = A + (B + C)$.

6. Show that $2(A + B) = 2A + 2B$.

In Problems 7–12, let

$$A = \begin{bmatrix} 1 & 3 & -1 \\ 2 & 4 & 1 \\ 0 & -2 & 2 \end{bmatrix}, \quad B = \begin{bmatrix} 5 & -1 & 4 \\ 2 & 0 & 1 \\ 1 & -3 & -3 \end{bmatrix},$$

$$C = \begin{bmatrix} -2 & 0 & 4 \\ 1 & -3 & 1 \\ 0 & 0 & 2 \end{bmatrix}$$

7. Find $2A + 3B - C$. **8.** Find $3C - B + \frac{1}{2}A$.

9. Determine D so that $A + B + C + D = \mathbf{0}$.

10. Determine D so that $A + 4B = 2(A + B) + D$.

11. Show that $A + B = B + A$.

12. Show that $(A + B) + C = A + (B + C)$.

13. Show that if $A + B = C$, then $A = C - B$.

14. Find the transpose of

$$A = \begin{bmatrix} 2 \\ -3 \\ 5 \end{bmatrix}$$

15. Find the transpose of

$$A = \begin{bmatrix} -1 & 0 & 3 \\ 2 & 1 & -4 \end{bmatrix}$$

16. Suppose A is a 2×2 matrix. Find conditions on the entries of A such that

$$A + A' = \mathbf{0}$$

17. Suppose that A and B are $m \times n$ matrices. Show that

$$(A + B)' = A' + B'$$

18. Suppose that A is an $m \times n$ matrix. Show that

$$(A')' = A$$

19. Suppose that A is an $m \times n$ matrix and k is a real number. Show that

$$(kA)' = kA'$$

20. Suppose that A is an $m \times k$ matrix and B is a $k \times n$ matrix. Show that

$$(AB)' = B'A'$$

In Problems 21–26, let

$$A = \begin{bmatrix} -1 & 0 \\ 1 & 2 \end{bmatrix}, \quad B = \begin{bmatrix} 2 & 3 \\ -1 & 1 \end{bmatrix}, \quad C = \begin{bmatrix} 1 & 2 \\ 0 & -1 \end{bmatrix}$$

21. Compute the following:

(a) AB **(b)** BA

22. Compute ABC.

23. Show that $AC \neq CA$.

24. Show that $(AB)C = A(BC)$.

25. Show that $(A + B)C = AC + BC$.

26. Show that $A(B + C) = AB + AC$.

27. Suppose that A is a 3×4 matrix and B is a 4×2 matrix. What is the size of the product AB?

28. Suppose A is a 3×4 matrix and B is an $m \times n$ matrix. What are values of m and n such that the following products are defined?

(a) AB **(b)** BA

29. Suppose that A is a 4×3 matrix, B is a 1×3 matrix, C is a 3×1 matrix, and D is a 4×3 matrix. Which of the matrix multiplications that follow are defined? Whenever it is defined, state the size of the resulting matrix.

(a) BD' **(b)** $D'A$ **(c)** ACB

30. Suppose that A is an $l \times p$ matrix, B is an $m \times q$, matrix, and C is an $n \times r$ matrix. What can you say about l, m, n, p, q, and r if the products that follow are defined? State the size of the resulting matrix.

(a) ABC **(b)** $AB'C$ **(c)** BAC' **(d)** $A'CB'$

31. Let

$$A = \begin{bmatrix} 1 & 3 \\ 0 & -2 \end{bmatrix} \quad \text{and} \quad B = \begin{bmatrix} 1 & 2 & 0 & -1 \\ 2 & 1 & 3 & 0 \end{bmatrix}$$

(a) Compute AB. **(b)** Compute $B'A$.

32. Let

$$A = \begin{bmatrix} 1 & 4 & -2 \end{bmatrix} \quad \text{and} \quad B = \begin{bmatrix} -1 \\ 2 \\ 3 \end{bmatrix}$$

(a) Compute AB. **(b)** Compute BA.

33. Let

$$A = \begin{bmatrix} 2 & 1 \\ -1 & -3 \end{bmatrix}$$

Find A^2, A^3, and A^4.

34. Suppose that

$$A = \begin{bmatrix} 1 & -1 \\ 3 & 0 \\ 5 & 2 \end{bmatrix} \quad \text{and} \quad B = \begin{bmatrix} 2 & 4 & 1 \\ 6 & 0 & 0 \end{bmatrix}$$

Show that $(AB)' = B'A'$.

35. Let

$$B = \begin{bmatrix} 0 & 1 \\ 1 & 0 \end{bmatrix}$$

(a) Find B^2, B^3, B^4, and B^5.

(b) What can you say about B^k when **(i)** k is even and **(ii)** k is odd?

36. Let

$$I_3 = \begin{bmatrix} 1 & 0 & 0 \\ 0 & 1 & 0 \\ 0 & 0 & 1 \end{bmatrix}$$

Show that $I_3 = I_3^2 = I_3^3$.

37. Let

$$A = \begin{bmatrix} 1 & 3 \\ 0 & -2 \end{bmatrix} \quad \text{and} \quad I_2 = \begin{bmatrix} 1 & 0 \\ 0 & 1 \end{bmatrix}$$

Show that $AI_2 = I_2A = A$.

38. Let

$$A = \begin{bmatrix} 1 & 3 & 0 \\ 0 & -1 & 2 \\ -1 & -2 & 1 \end{bmatrix} \quad \text{and} \quad I_3 = \begin{bmatrix} 1 & 0 & 0 \\ 0 & 1 & 0 \\ 0 & 0 & 1 \end{bmatrix}$$

Show that $A I_3 = I_3 A = A$.

In Problems 39–42, write each system in matrix form.

39. $2x_1 + 3x_2 - x_3 = 0$
$2x_2 + x_3 = 1$
$x_1 - 2x_3 = 2$

40. $2x_2 - x_1 = x_3$
$4x_1 + x_3 = 7x_2$
$x_2 - x_1 = x_3$

41. $2x_1 - 3x_2 = 4$
$-x_1 + x_2 = 3$
$3x_1 = 4$

42. $x_1 - 2x_2 + x_3 = 1$
$-2x_1 + x_2 - 3x_3 = 0$

■ **9.2.3**

43. Show that the inverse of

$$A = \begin{bmatrix} 2 & 1 \\ 1 & 1 \end{bmatrix}$$

is

$$B = \begin{bmatrix} 1 & -1 \\ -1 & 2 \end{bmatrix}$$

44. Show that the inverse of

$$A = \begin{bmatrix} 2 & 3 & 1 \\ 5 & 2 & 3 \\ 1 & 2 & 0 \end{bmatrix}$$

is

$$B = \begin{bmatrix} -\frac{6}{5} & \frac{2}{5} & \frac{7}{5} \\ \frac{3}{5} & -\frac{1}{5} & -\frac{1}{5} \\ \frac{8}{5} & -\frac{1}{5} & -\frac{11}{5} \end{bmatrix}$$

In Problems 45–48, let

$$A = \begin{bmatrix} -1 & 1 \\ 2 & 3 \end{bmatrix}, \quad B = \begin{bmatrix} 2 & -2 \\ 3 & 2 \end{bmatrix}$$

45. Find the inverse (if it exists) of A.
46. Find the inverse (if it exists) of B.
47. Show that $(A^{-1})^{-1} = A$.
48. Show that $(AB)^{-1} = B^{-1}A^{-1}$.
49. Find the inverse (if it exists) of

$$C = \begin{bmatrix} 2 & 4 \\ 3 & 6 \end{bmatrix}$$

50. Find the inverse (if it exists) of

$$I_3 = \begin{bmatrix} 1 & 0 & 0 \\ 0 & 1 & 0 \\ 0 & 0 & 1 \end{bmatrix}$$

51. Suppose that

$$A = \begin{bmatrix} -1 & 0 \\ 2 & -3 \end{bmatrix} \quad \text{and} \quad D = \begin{bmatrix} -2 \\ -5 \end{bmatrix}$$

Find X such that $AX = D$ by
(a) solving the associated system of linear equations and
(b) using the inverse of A.

52. (a) Show that if $X = AX + D$, then

$$X = (I - A)^{-1} D$$

provided that $I - A$ is invertible.
(b) Suppose that

$$A = \begin{bmatrix} 3 & 2 \\ 0 & -1 \end{bmatrix} \quad \text{and} \quad D = \begin{bmatrix} -2 \\ 2 \end{bmatrix}$$

Compute $(I - A)^{-1}$, and use your result in (a) to compute X.
53. Use the determinant to determine whether the matrix

$$A = \begin{bmatrix} 2 & -1 \\ 1 & 3 \end{bmatrix}$$

is invertible.
54. Use the determinant to determine whether the matrix

$$A = \begin{bmatrix} -1 & 3 \\ 0 & 3 \end{bmatrix}$$

is invertible.
55. Use the determinant to determine whether the matrix

$$A = \begin{bmatrix} 4 & -1 \\ 8 & -2 \end{bmatrix}$$

is invertible.
56. Use the determinant to determine whether the matrix

$$A = \begin{bmatrix} -1 & 2 \\ -1 & 2 \end{bmatrix}$$

is invertible.
57. Suppose that

$$A = \begin{bmatrix} 2 & 4 \\ 3 & 6 \end{bmatrix}$$

(a) Compute det A. Is A invertible?
(b) Suppose that

$$X = \begin{bmatrix} x \\ y \end{bmatrix} \quad \text{and} \quad B = \begin{bmatrix} b_1 \\ b_2 \end{bmatrix}$$

Write $AX = B$ as a system of linear equations.
(c) Show that if

$$B = \begin{bmatrix} 3 \\ \frac{9}{2} \end{bmatrix}$$

then

$$AX = B$$

has infinitely many solutions. Graph the two straight lines associated with the corresponding system of linear equations, and explain why the system has infinitely many solutions.
(d) Find a column vector

$$B = \begin{bmatrix} b_1 \\ b_2 \end{bmatrix}$$

so that

$$AX = B$$

has no solutions.
58. Suppose that

$$A = \begin{bmatrix} a & 8 \\ 2 & 4 \end{bmatrix}, \quad X = \begin{bmatrix} x \\ y \end{bmatrix}, \quad \text{and} \quad B = \begin{bmatrix} b_1 \\ b_2 \end{bmatrix}$$

(a) Show that when $a \neq 4$, $AX = B$ has exactly one solution.

(b) Suppose $a = 4$. Find conditions on b_1 and b_2 such that $AX = B$ has **(i)** infinitely many solutions and **(ii)** no solutions.

(c) Explain your results in (a) and (b) graphically.

In Problems 59–62, use the determinant to find the inverse of A.

59. $A = \begin{bmatrix} 2 & 1 \\ 3 & -1 \end{bmatrix}$ **60.** $A = \begin{bmatrix} 1 & 2 \\ 0 & 3 \end{bmatrix}$

61. $A = \begin{bmatrix} -1 & 4 \\ 5 & 1 \end{bmatrix}$ **62.** $A = \begin{bmatrix} -2 & 1 \\ 3 & 2 \end{bmatrix}$

63. Use the determinant to determine whether

$$A = \begin{bmatrix} 1 & -1 \\ 0 & 2 \end{bmatrix}$$

is invertible. If it is invertible, compute its inverse. In either case, solve $AX = \mathbf{0}$.

64. Use the determinant to determine whether

$$B = \begin{bmatrix} 1 & 1 \\ 2 & 1 \end{bmatrix}$$

is invertible. If it is invertible, compute its inverse. In either case, solve $BX = \mathbf{0}$.

65. Use the determinant to determine whether

$$C = \begin{bmatrix} 1 & 3 \\ 1 & 3 \end{bmatrix}$$

is invertible. If it is invertible, compute its inverse. In either case, solve $CX = \mathbf{0}$.

66. Use the determinant to determine whether

$$D = \begin{bmatrix} -3 & 6 \\ -4 & 8 \end{bmatrix}$$

is invertible. If it is invertible, compute its inverse. In either case, solve $DX = \mathbf{0}$.

■ 9.2.4

In Problems 67–70, find the inverse matrix to each given matrix if the inverse matrix exists.

67. $A = \begin{bmatrix} 2 & -1 & -1 \\ 2 & 1 & 1 \\ -1 & 1 & -1 \end{bmatrix}$ **68.** $A = \begin{bmatrix} -1 & 3 & -1 \\ 2 & -2 & 3 \\ -1 & 1 & 2 \end{bmatrix}$

69. $A = \begin{bmatrix} -1 & 0 & -1 \\ 0 & -2 & 0 \\ -1 & 1 & 2 \end{bmatrix}$ **70.** $A = \begin{bmatrix} -1 & 0 & 2 \\ -1 & -2 & 3 \\ 0 & 2 & -1 \end{bmatrix}$

■ 9.2.5

In Problems 71–74, suppose that breeding occurs once a year and that a census is taken at the end of each breeding season.

71. Assume that a population is divided into three age classes and that 20% of the females age 0 and 70% of the females age 1 survive until the end of the next breeding season. Assume further that females age 1 have an average of 3.2 female offspring and females age 2 have an average of 1.7 female offspring. If, at time 0, the population consists of 2000 females age 0, 800 females age 1, and 200 females age 2, find the Leslie matrix and the age distribution at time 2.

72. Assume that a population is divided into three age classes and that 80% of the females age 0 and 10% of the females age 1 survive until the end of the next breeding season. Assume further that females age 1 have an average of 1.6 female offspring and females age 2 have an average of 3.9 female offspring. If, at time 0, the population consists of 1000 females age 0, 100 females age 1, and 20 females age 2, find the Leslie matrix and the age distribution at time 3.

73. Assume that a population is divided into four age classes and that 70% of the females age 0, 50% of the females age 1, and 10% of the females age 2 survive until the end of the next breeding season. Assume further that females age 2 have an average of 4.6 female offspring and females age 3 have an average of 3.7 female offspring. If, at time 0, the population consists of 1500 females age 0, 500 females age 1, 250 females age 2, and 50 females age 3, find the Leslie matrix and the age distribution at time 2.

74. Assume that a population is divided into four age classes and that 65% of the females age 0, 40% of the females age 1, and 30% of the females age 2 survive until the end of the next breeding season. Assume further that females age 1 have an average of 2.8 female offspring, females age 2 have an average of 7.2 female offspring, and females age 3 have an average of 3.7 female offspring. If, at time 0, the population consists of 1500 females age 0, 500 females age 1, 250 females age 2, and 50 females age 3, find the Leslie matrix and the age distribution at time 3.

In Problems 75–76, assume the given Leslie matrix L. Determine the number of age classes in the population, the fraction of one-year-olds that survive until the end of the next breeding season, and the average number of female offspring of a two-year-old female.

75. $L = \begin{bmatrix} 2 & 3 & 2 & 1 \\ 0.4 & 0 & 0 & 0 \\ 0 & 0.6 & 0 & 0 \\ 0 & 0 & 0.8 & 0 \end{bmatrix}$ **76.** $L = \begin{bmatrix} 0 & 5 & 0 \\ 0.8 & 0 & 0 \\ 0 & 0.3 & 0 \end{bmatrix}$

In Problems 77–78, assume the given Leslie matrix L. Determine the number of age classes in the population. What fraction of two-year-olds survive until the end of the next breeding season? Determine the average number of female offspring of a one-year-old female.

77. $L = \begin{bmatrix} 1 & 2.5 & 3 & 1.5 \\ 0.9 & 0 & 0 & 0 \\ 0 & 0.3 & 0 & 0 \\ 0 & 0 & 0.2 & 0 \end{bmatrix}$ **78.** $L = \begin{bmatrix} 0 & 4.2 & 3.7 \\ 0.7 & 0 & 0 \\ 0 & 0.1 & 0 \end{bmatrix}$

79. Assume that the Leslie matrix is

$$L = \begin{bmatrix} 1.2 & 3.2 \\ 0.8 & 0 \end{bmatrix}$$

Suppose that, at time $t = 0$, $N_0(0) = 100$ and $N_1(0) = 0$. Find the population vectors for $t = 0, 1, 2, \ldots, 10$. Compute the successive ratios

$$q_0(t) = \frac{N_0(t)}{N_0(t-1)} \quad \text{and} \quad q_1(t) = \frac{N_1(t)}{N_1(t-1)}$$

for $t = 1, 2, \ldots, 10$. What value do $q_0(t)$ and $q_1(t)$ approach as $t \to \infty$? (Take a guess.) Compute the fraction of females age 0 for $t = 0, 1, \ldots, 10$. Can you find a stable age distribution?

80. Assume that the Leslie matrix is

$$L = \begin{bmatrix} 0.2 & 3 \\ 0.33 & 0 \end{bmatrix}$$

Suppose that, at time $t = 0$, $N_0(0) = 10$ and $N_1(0) = 5$. Find the population vectors for $t = 0, 1, 2, \ldots, 10$. Compute the successive ratios

$$q_0(t) = \frac{N_0(t)}{N_0(t-1)} \quad \text{and} \quad q_1(t) = \frac{N_1(t)}{N_1(t-1)}$$

for $t = 1, 2, \ldots, 10$. What value do $q_0(t)$ and $q_1(t)$ approach as $t \to \infty$? (Take a guess.) Compute the fraction of females age 0 for $t = 0, 1, \ldots, 10$. Can you find a stable age distribution?

81. Assume that the Leslie matrix is

$$L = \begin{bmatrix} 0 & 2 \\ 0.6 & 0 \end{bmatrix}$$

Suppose that, at time $t = 0$, $N_0(0) = 5$ and $N_1(0) = 1$. Find the population vectors for $t = 0, 1, 2, \ldots, 10$. Compute the successive

ratios

$$q_0(t) = \frac{N_0(t)}{N_0(t-1)} \quad \text{and} \quad q_1(t) = \frac{N_1(t)}{N_1(t-1)}$$

for $t = 1, 2, \ldots, 10$. Do $q_0(t)$ and $q_1(t)$ converge? Compute the fraction of females age 0 for $t = 0, 1, \ldots, 10$. Describe the long-term behavior of $q_0(t)$.

82. Assume that the Leslie matrix is

$$L = \begin{bmatrix} 0 & a \\ b & 0 \end{bmatrix}$$

Suppose that, at time $t = 0$, $N_0(0) = 1$ and $N_1(0) = 1$. Find the population vectors for $t = 0, 1, 2, \ldots, 10$. Compute the successive ratios

$$q_0(t) = \frac{N_0(t)}{N_0(t-1)} \quad \text{and} \quad q_1(t) = \frac{N_1(t)}{N_1(t-1)}$$

for $t = 1, 2, \ldots, 10$. Do $q_0(t)$ and $q_1(t)$ converge? Compute the fraction of females age 0 for $t = 0, 1, \ldots, 10$. Describe the long-term behavior of $q_0(t)$.

■ 9.3 Linear Maps, Eigenvectors, and Eigenvalues

In this section, we will denote vectors by boldface lowercase letters. Consider a map of the form

$$\mathbf{x} \to A\mathbf{x} \tag{9.18}$$

where A is a 2×2 matrix and $\mathbf{x}$ is a 2×1 column vector (or, simply, vector). Since $A\mathbf{x}$ is a 2×1 vector, this map takes a 2×1 vector and maps it into a 2×1 vector. That enables us to apply A repeatedly: We can compute $A(A\mathbf{x}) = A^2\mathbf{x}$, which is again a 2×1 vector, and so on. We will first look at vectors, then at maps $A\mathbf{x}$, and finally at iterates of the map A (i.e., $A^2\mathbf{x}$, $A^3\mathbf{x}$, and so on).

According to the properties of matrix multiplication, the map (9.18) satisfies the following conditions:

1. $A(\mathbf{x} + \mathbf{y}) = A\mathbf{x} + A\mathbf{y}$, and
2. $A(\lambda\mathbf{x}) = \lambda(A\mathbf{x})$, where λ is a scalar.

Because of these two properties, we say that the map $\mathbf{x} \to A\mathbf{x}$ is *linear*.

We saw an example of such a map in the previous section: If A is a 2×2 Leslie matrix and $\mathbf{x}$ is a population vector at time 0, then $A\mathbf{x}$ represents the population vector at time 1.

Linear maps are important in other contexts as well, and we will encounter them in Chapters 10 and 11. Here, we restrict our discussion to 2×2 matrices but point out that we can generalize the discussion that follows to arbitrary $n \times n$ matrices. (These topics are covered in courses on matrix or linear algebra.)

■ 9.3.1 Graphical Representation

Vectors We begin with a graphical representation of vectors. We assume that $\mathbf{x}$ is a 2×1 matrix. We call $\mathbf{x}$ a *column vector* or simply a *vector*. Since a 2×1 matrix has just two components, we can represent a vector in the plane. For instance, to represent the vector

$$\mathbf{x} = \begin{bmatrix} 3 \\ 4 \end{bmatrix}$$

in the x_1–x_2 plane, we draw an arrow from the origin $(0, 0)$ to the point $(3, 4)$, as illustrated in Figure 9.11. We see from Figure 9.11 that a vector has a length and a direction. The length of the vector $\mathbf{x} = \begin{bmatrix} 3 \\ 4 \end{bmatrix}$, denoted by $|\mathbf{x}|$, is the distance from the origin $(0, 0)$ to the point $(3, 4)$; that is,

$$\text{length of } \mathbf{x} = |\mathbf{x}| = \sqrt{9 + 16} = 5$$

We define the direction of $\mathbf{x}$ as the angle α between the positive x_1-axis and the vector $\mathbf{x}$ (measured counterclockwise), as shown in Figure 9.11. The angle α is in the interval $[0, 2\pi)$. In this example, α satisfies $\tan \alpha = 4/3$.

More generally (see Figure 9.11 again), a vector $\begin{bmatrix} x_1 \\ x_2 \end{bmatrix}$ has length

$$|\mathbf{x}| = \sqrt{x_1^2 + x_2^2}$$

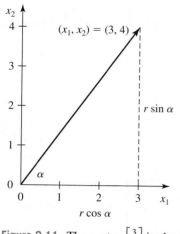

Figure 9.11 The vector $\begin{bmatrix} 3 \\ 4 \end{bmatrix}$ in the x_1–x_2 plane.

and direction α, where $\alpha \in [0, 2\pi)$ satisfies

$$\tan \alpha = \frac{x_2}{x_1}$$

The angle α is always measured counterclockwise from the positive x_1-axis.

If we denote the length of $\mathbf{x}$ by r (i.e., $r = |\mathbf{x}|$), then, as shown in Figure 9.11, since $x_1 = r \cos \alpha$ and $x_2 = r \sin \alpha$, the vector $\mathbf{x}$ can also be written as

$$\mathbf{x} = \begin{bmatrix} r \cos \alpha \\ r \sin \alpha \end{bmatrix}$$

We thus have two distinct ways of representing vectors in the plane: We can use either the endpoint (x_1, x_2) or the length and direction (r, α). The first representation leads to our familiar Cartesian coordinate system. The second representation, using the length and direction of the corresponding vector, leads to the **polar coordinate system**. We will use both representations in what follows.

EXAMPLE 1

If the length of the vector $\begin{bmatrix} x_1 \\ x_2 \end{bmatrix}$ is 4 and its angle with the positive x_1-axis is $120°$ (measured clockwise), what is its representation in Cartesian coordinates?

Solution

An angle of $120°$ measured clockwise from the positive x_1-axis corresponds to an angle of $360° - 120° = 240°$ measured counterclockwise from the positive x_1-axis (Figure 9.12). Since the length of the vector is 4, we obtain

$$x_1 = 4 \cos(240°) = (4)\left(-\frac{1}{2}\right) = -2$$

$$x_2 = 4 \sin(240°) = (4)\left(-\frac{1}{2}\sqrt{3}\right) = -2\sqrt{3}$$

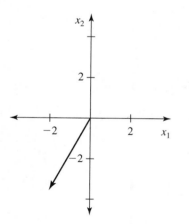

Figure 9.12 The vector in Example 1.

which yields the Cartesian coordinate representation

$$\mathbf{x} = \begin{bmatrix} -2 \\ -2\sqrt{3} \end{bmatrix}$$

■

Because vectors are matrices, we can use matrix addition to add vectors. For instance,

$$\begin{bmatrix} 1 \\ 2 \end{bmatrix} + \begin{bmatrix} 2 \\ 2 \end{bmatrix} = \begin{bmatrix} 3 \\ 4 \end{bmatrix}$$

This vector sum has a graphical representation. (See Figure 9.13.) To add $\begin{bmatrix} 2 \\ 2 \end{bmatrix}$ to $\begin{bmatrix} 1 \\ 2 \end{bmatrix}$, we move the vector $\begin{bmatrix} 2 \\ 2 \end{bmatrix}$ to the tip of the vector $\begin{bmatrix} 1 \\ 2 \end{bmatrix}$ without changing the direction or the length of $\begin{bmatrix} 2 \\ 2 \end{bmatrix}$. The sum of the two vectors then starts at $(0, 0)$ and ends at the point where the tip of the vector that was moved ended. This series of operations can also be described in the following way: The sum is the diagonal in the parallelogram that

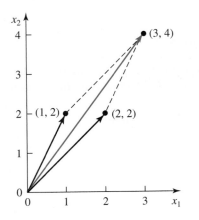

Figure 9.13 Addition of two vectors.

is formed by the two vectors $\begin{bmatrix} 1 \\ 2 \end{bmatrix}$ and $\begin{bmatrix} 2 \\ 2 \end{bmatrix}$. The rules for vector addition are therefore referred to as the **parallelogram law**.

Multiplication of a vector by a scalar is carried out componentwise. For instance, if we multiply $\begin{bmatrix} 1 \\ 2 \end{bmatrix}$ by 2, we get

$$2\begin{bmatrix} 1 \\ 2 \end{bmatrix} = \begin{bmatrix} 2 \\ 4 \end{bmatrix}$$

This operation corresponds to changing the length of the vector. The vector $\begin{bmatrix} 1 \\ 2 \end{bmatrix}$ has length $\sqrt{1+4} = \sqrt{5}$; the vector $2\begin{bmatrix} 1 \\ 2 \end{bmatrix}$ has length $\sqrt{4+16} = \sqrt{20} = 2\sqrt{5}$. That is, multiplying the vector $\begin{bmatrix} 1 \\ 2 \end{bmatrix}$ by 2 increases its length by the factor 2. Since 2 is positive, the resulting vector points in the same direction as the vector $\begin{bmatrix} 1 \\ 2 \end{bmatrix}$. If we multiply $\begin{bmatrix} 1 \\ 2 \end{bmatrix}$ by -1, then the resulting vector is $\begin{bmatrix} -1 \\ -2 \end{bmatrix}$, which has the same length as the original vector, but points in the opposite direction, as illustrated in Figure 9.14.

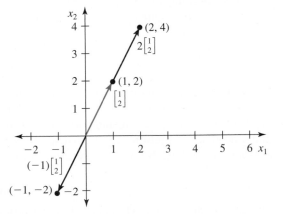

Figure 9.14 The vectors in Example 1.

EXAMPLE 2 Let

$$\mathbf{u} = \begin{bmatrix} 2 \\ 1 \end{bmatrix}, \qquad \mathbf{v} = \begin{bmatrix} -1 \\ -3 \end{bmatrix}, \qquad \mathbf{w} = \begin{bmatrix} -2 \\ 3 \end{bmatrix}$$

Find $-\frac{1}{2}\mathbf{u}$, $\mathbf{u} + \mathbf{v}$, $\mathbf{v} + \mathbf{w}$, and $-\mathbf{v}$, and illustrate the results graphically.

Solution

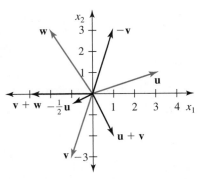

Figure 9.15 The vectors in Example 2.

$$-\frac{1}{2}\mathbf{u} = -\frac{1}{2}\begin{bmatrix} 2 \\ 1 \end{bmatrix} = \begin{bmatrix} -1 \\ -\frac{1}{2} \end{bmatrix}$$

$$\mathbf{u} + \mathbf{v} = \begin{bmatrix} 2 \\ 1 \end{bmatrix} + \begin{bmatrix} -1 \\ -3 \end{bmatrix} = \begin{bmatrix} 1 \\ -2 \end{bmatrix}$$

$$\mathbf{v} + \mathbf{w} = \begin{bmatrix} -1 \\ -3 \end{bmatrix} + \begin{bmatrix} -2 \\ 3 \end{bmatrix} = \begin{bmatrix} -3 \\ 0 \end{bmatrix}$$

$$-\mathbf{v} = -\begin{bmatrix} -1 \\ -3 \end{bmatrix} = \begin{bmatrix} 1 \\ 3 \end{bmatrix}$$

The results are illustrated in Figure 9.15. ■

The following box summarizes vector addition, and multiplication of a vector by a scalar:

Let $\mathbf{x} = \begin{bmatrix} x_1 \\ x_2 \end{bmatrix}$ and $\mathbf{y} = \begin{bmatrix} y_1 \\ y_2 \end{bmatrix}$. Then

$$\mathbf{x} + \mathbf{y} = \begin{bmatrix} x_1 + y_1 \\ x_2 + y_2 \end{bmatrix}$$

If a is a scalar, then

$$a\mathbf{x} = \begin{bmatrix} ax_1 \\ ax_2 \end{bmatrix}$$

If $|\mathbf{x}|$ denotes the length of $\mathbf{x}$, then the length of $a\mathbf{x}$ is the absolute value of a times the length of $\mathbf{x}$—that is, $|a||\mathbf{x}|$.

Linear Maps   We will first use a graphical approach to study maps of the form

$$\mathbf{x} \rightarrow A\mathbf{x}$$

where A is a 2×2 matrix and $\mathbf{x}$ is a 2×1 vector. Since $A\mathbf{x}$ is a 2×1 vector as well, the map A takes the 2×1 vector $\mathbf{x}$ and maps it into a 2×1 vector.

The simplest such map is the **identity map,** represented by the identity matrix I_2:

$$\begin{bmatrix} x_1 \\ x_2 \end{bmatrix} \rightarrow \begin{bmatrix} 1 & 0 \\ 0 & 1 \end{bmatrix} \begin{bmatrix} x_1 \\ x_2 \end{bmatrix}$$

Since

$$I_2\mathbf{x} = \begin{bmatrix} 1 & 0 \\ 0 & 1 \end{bmatrix} \begin{bmatrix} x_1 \\ x_2 \end{bmatrix} = \begin{bmatrix} x_1 \\ x_2 \end{bmatrix} = \mathbf{x}$$

it follows that the identity matrix leaves the vector $\mathbf{x}$ unchanged.

Slightly more complicated is the map

$$\begin{bmatrix} x_1 \\ x_2 \end{bmatrix} \rightarrow \begin{bmatrix} a & 0 \\ 0 & b \end{bmatrix} \begin{bmatrix} x_1 \\ x_2 \end{bmatrix} = \begin{bmatrix} ax_1 \\ bx_2 \end{bmatrix}$$

This map stretches or contracts each coordinate separately. In Figure 9.16, we show the action of the map on $\begin{bmatrix} 1 \\ 2 \end{bmatrix}$ when $a = 2$ and $b = -\frac{1}{2}$. We find that

$$\begin{bmatrix} 2 & 0 \\ 0 & -\frac{1}{2} \end{bmatrix} \begin{bmatrix} 1 \\ 2 \end{bmatrix} = \begin{bmatrix} 2 \\ -1 \end{bmatrix}$$

This map stretches the first coordinate by a factor of 2 and contracts the second coordinate by a factor of 1/2. The minus sign in front of $\frac{1}{2}$ corresponds to reflecting the second coordinate about the x_1-axis.

Another linear map is a **rotation**, which rotates a vector in the x_1–x_2 plane by a fixed angle. The following matrix rotates a vector by an angle θ:

$$R_\theta = \begin{bmatrix} \cos\theta & -\sin\theta \\ \sin\theta & \cos\theta \end{bmatrix}$$

If $\theta > 0$, then the rotation is counterclockwise; if $\theta < 0$, the rotation is clockwise by the angle $|\theta|$.

To check that this is indeed a rotation, we investigate the action of R_θ on a vector with coordinates (x_1, x_2), as illustrated in Figure 9.17. Using the polar coordinate system, we can write this vector as

$$\begin{bmatrix} x_1 \\ x_2 \end{bmatrix} = \begin{bmatrix} r\cos\alpha \\ r\sin\alpha \end{bmatrix}$$

where r is the length of the vector and α is the angle it forms with the positive x_1-axis.

Figure 9.16 The action of the matrix A on the vector $\begin{bmatrix} 1 \\ 2 \end{bmatrix}$.

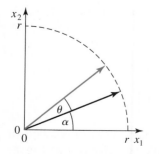

Figure 9.17 The rotation of a vector.

Applying R_θ, we find that

$$\begin{bmatrix} \cos\theta & -\sin\theta \\ \sin\theta & \cos\theta \end{bmatrix} \begin{bmatrix} r\cos\alpha \\ r\sin\alpha \end{bmatrix} = \begin{bmatrix} r(\cos\theta\cos\alpha - \sin\theta\sin\alpha) \\ r(\sin\theta\cos\alpha + \cos\theta\sin\alpha) \end{bmatrix}$$

$$= \begin{bmatrix} r\cos(\theta+\alpha) \\ r\sin(\theta+\alpha) \end{bmatrix}$$

where we used the trigonometric identities

$$\cos(\theta+\alpha) = \cos\theta\cos\alpha - \sin\theta\sin\alpha$$
$$\sin(\theta+\alpha) = \sin\theta\cos\alpha + \cos\theta\sin\alpha$$

We see that the resulting vector still has length r and that the angle with the x_1-axis is $\alpha + \theta$. If $\theta > 0$, as in Figure 9.17, then the vector is rotated counterclockwise by the angle θ.

EXAMPLE 3 Use a rotation matrix to rotate the vector $\begin{bmatrix} 1 \\ 3 \end{bmatrix}$ counterclockwise by the angle $\pi/3$.

Solution The rotation matrix we seek is

$$R_{\pi/3} = \begin{bmatrix} \cos(\pi/3) & -\sin(\pi/3) \\ \sin(\pi/3) & \cos(\pi/3) \end{bmatrix} = \begin{bmatrix} 1/2 & -\sqrt{3}/2 \\ \sqrt{3}/2 & 1/2 \end{bmatrix}$$

Hence, the rotated vector has coordinates

$$\begin{bmatrix} 1/2 & -\sqrt{3}/2 \\ \sqrt{3}/2 & 1/2 \end{bmatrix} \begin{bmatrix} 1 \\ 3 \end{bmatrix} = \begin{bmatrix} \frac{1}{2} - \frac{3}{2}\sqrt{3} \\ \frac{1}{2}\sqrt{3} + \frac{3}{2} \end{bmatrix} = \frac{1}{2} \begin{bmatrix} 1 - 3\sqrt{3} \\ \sqrt{3} + 3 \end{bmatrix} \qquad ■$$

From this brief discussion of linear maps, we see that the map $\mathbf{x} \to A\mathbf{x}$ typically takes the vector $\mathbf{x}$ and rotates, stretches, or contracts it. For an arbitrary matrix A, vectors may be moved in a way that has no simple geometric interpretation.

EXAMPLE 4 Investigate the action of

$$A = \begin{bmatrix} 1 & 2 \\ 3 & 2 \end{bmatrix}$$

on $\begin{bmatrix} 3 \\ -1 \end{bmatrix}$ and $\begin{bmatrix} 2 \\ 3 \end{bmatrix}$.

Solution If $\mathbf{x} = \begin{bmatrix} 3 \\ -1 \end{bmatrix}$, then

$$A\mathbf{x} = \begin{bmatrix} 1 & 2 \\ 3 & 2 \end{bmatrix} \begin{bmatrix} 3 \\ -1 \end{bmatrix} = \begin{bmatrix} 1 \\ 7 \end{bmatrix}$$

On the one hand, we can compute the outcome of this map, but there does not seem to be a straightforward geometric explanation of it. (See Figure 9.18.) On the other hand, if $\mathbf{x} = \begin{bmatrix} 2 \\ 3 \end{bmatrix}$, then

$$A\mathbf{x} = \begin{bmatrix} 1 & 2 \\ 3 & 2 \end{bmatrix} \begin{bmatrix} 2 \\ 3 \end{bmatrix} = \begin{bmatrix} 8 \\ 12 \end{bmatrix} = 4\begin{bmatrix} 2 \\ 3 \end{bmatrix}$$

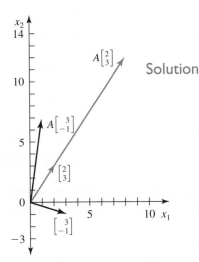

Figure 9.18 The action of the matrix $A = \begin{bmatrix} 1 & 2 \\ 3 & 2 \end{bmatrix}$ on two vectors.

The result of $A\mathbf{x}$ is simply a multiple of $\mathbf{x}$. (See Figure 9.18.) Such a vector is called an *eigenvector*, and the stretching or contracting factor is called an *eigenvalue*. Eigenvectors and eigenvalues are the topic of the next subsection. ■

■ 9.3.2 Eigenvalues and Eigenvectors

In Subsection 9.3.1, we saw that there are matrices and vectors for which the map $\mathbf{x} \to A\mathbf{x}$ takes on a particularly simple form, namely,

$$A\mathbf{x} = \lambda\mathbf{x} \tag{9.19}$$

where λ is a scalar. Now we investigate this relationship. Again restricting our discussion to 2×2 matrices, we begin with the following definition:

> **Definition** Assume that A is a square matrix. A nonzero vector $\mathbf{x}$ that satisfies the equation
>
> $$A\mathbf{x} = \lambda\mathbf{x}$$
>
> is an eigenvector of the matrix A, and the number λ is an eigenvalue of the matrix A.

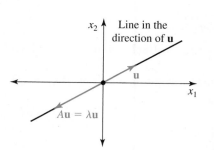

Figure 9.19 Any vector on the line in the direction of the eigenvector will remain on the line under the map A.

Note that we assume that the vector $\mathbf{x}$ is different from the zero vector. (The zero vector $\mathbf{x} = \mathbf{0}$ always satisfies the equation $A\mathbf{x} = \lambda\mathbf{x}$ and thus would not be special.) The eigenvalue λ can be 0, however. We will see that λ can even be a complex number.

The action of A on eigenvectors produces a particularly simple form: If we apply A to an eigenvector $\mathbf{x}$ (i.e., if we compute $A\mathbf{x}$), the result is a constant multiple of $\mathbf{x}$. This property of an eigenvector has an important geometric interpretation when the eigenvalue λ is a real number: If we draw a straight line through the origin in the direction of an eigenvector, then any vector on this straight line will remain on the line after the map A is applied. (See Figure 9.19.)

How can we find eigenvalues and eigenvectors for 2×2 matrices? We will see that a 2×2 matrix has two eigenvalues, which are either distinct or identical. We will discuss only the case in which the two eigenvalues are distinct; the case in which the eigenvalues are identical is more complicated and is covered in courses on linear algebra.

We will show how to find eigenvalues and eigenvectors by way of example. We will use the same matrix as in Example 4.

EXAMPLE 5   Find all eigenvalues and eigenvectors of

$$A = \begin{bmatrix} 1 & 2 \\ 3 & 2 \end{bmatrix}$$

Solution   We are interested in finding $\mathbf{x} \neq \mathbf{0}$ and λ such that

$$A\mathbf{x} = \lambda\mathbf{x}$$

We can rewrite this equation as

$$A\mathbf{x} - \lambda\mathbf{x} = \mathbf{0}$$

In order to factor $\mathbf{x}$, we must multiply $\lambda\mathbf{x}$ by the identity matrix $I = I_2$. (Because we will be dealing only with 2×2 matrices, we simply write I instead of I_2.) Multiplication by I yields

$$A\mathbf{x} - \lambda I\mathbf{x} = \mathbf{0}$$

We can now factor $\mathbf{x}$, resulting in

$$(A - \lambda I)\mathbf{x} = \mathbf{0}$$

In Section 9.2, we showed that in order to obtain a nontrivial solution ($\mathbf{x} \neq \mathbf{0}$), $A - \lambda I$ must be singular; that is,

$$\det(A - \lambda I) = 0$$

This is the key equation that will allow us to find eigenvalues. Now,

$$A - \lambda I = \begin{bmatrix} 1 - \lambda & 2 \\ 3 & 2 - \lambda \end{bmatrix}$$

and

$$\begin{aligned} \det(A - \lambda I) &= (1 - \lambda)(2 - \lambda) - (2)(3) \\ &= 2 - 3\lambda + \lambda^2 - 6 = \lambda^2 - 3\lambda - 4 \\ &= (\lambda + 1)(\lambda - 4) \end{aligned}$$

Solving

$$(\lambda + 1)(\lambda - 4) = 0$$

we find that

$$\lambda_1 = -1 \quad \text{and} \quad \lambda_2 = 4$$

These two numbers are the eigenvalues of the matrix A. Each eigenvalue will have its own eigenvector.

To compute the eigenvectors, we carry out the following calculations: If $\lambda_1 = -1$, we must find a nonzero vector $\mathbf{x} = \begin{bmatrix} x_1 \\ x_2 \end{bmatrix}$ such that

$$A \begin{bmatrix} x_1 \\ x_2 \end{bmatrix} = (-1) \begin{bmatrix} x_1 \\ x_2 \end{bmatrix}$$

Writing this matrix equation as a system of equations gives

$$\begin{aligned} x_1 + 2x_2 &= -x_1 \\ 3x_1 + 2x_2 &= -x_2 \end{aligned}$$

Simplifying, we obtain

$$\begin{aligned} 2x_1 + 2x_2 &= 0 \\ 3x_1 + 3x_2 &= 0 \end{aligned}$$

The two equations are identical and reduce to

$$x_1 + x_2 = 0$$

We are looking for a nonzero vector that satisfies this equation. For instance,

$$\begin{bmatrix} x_1 \\ x_2 \end{bmatrix} = \begin{bmatrix} 1 \\ -1 \end{bmatrix}$$

would be a reasonable choice. To check that $\begin{bmatrix} 1 \\ -1 \end{bmatrix}$ is indeed an eigenvector corresponding to the eigenvalue $\lambda_1 = -1$, we compute

$$\begin{bmatrix} 1 & 2 \\ 3 & 2 \end{bmatrix} \begin{bmatrix} 1 \\ -1 \end{bmatrix} = \begin{bmatrix} -1 \\ 1 \end{bmatrix} = (-1) \begin{bmatrix} 1 \\ -1 \end{bmatrix}$$

That is,

$$A \begin{bmatrix} 1 \\ -1 \end{bmatrix} = (-1) \begin{bmatrix} 1 \\ -1 \end{bmatrix}$$

The vector $\begin{bmatrix} 1 \\ -1 \end{bmatrix}$ is not the only eigenvector associated with $\lambda_1 = -1$. In fact, any vector $a \begin{bmatrix} -1 \\ 1 \end{bmatrix} = \begin{bmatrix} -a \\ a \end{bmatrix}, a \neq 0$, is an eigenvector associated with the eigenvalue -1. For instance, $\begin{bmatrix} -1 \\ 1 \end{bmatrix}$ or $\begin{bmatrix} 2 \\ -2 \end{bmatrix}$ are other choices. (See Figure 9.20.)

To find an eigenvector associated with $\lambda_2 = 4$, we must solve

$$A \begin{bmatrix} x_1 \\ x_2 \end{bmatrix} = 4 \begin{bmatrix} x_1 \\ x_2 \end{bmatrix}$$

which yields

$$x_1 + 2x_2 = 4x_1$$
$$3x_1 + 2x_2 = 4x_2$$

Simplifying, we obtain

$$-3x_1 + 2x_2 = 0$$
$$3x_1 - 2x_2 = 0$$

For instance,

$$\begin{bmatrix} x_1 \\ x_2 \end{bmatrix} = \begin{bmatrix} 2 \\ 3 \end{bmatrix}$$

satisfies the preceding system. We see that A has two eigenvalues: -1 and 4. An eigenvector associated with -1 is $\begin{bmatrix} 1 \\ -1 \end{bmatrix}$, and an eigenvector associated with 4 is $\begin{bmatrix} 2 \\ 3 \end{bmatrix}$. (See Figure 9.21.) As before, any vector $b\begin{bmatrix} 2 \\ 3 \end{bmatrix}$, $b \neq 0$, is an eigenvector associated with the eigenvalue 4. ■

Figure 9.20 The three vectors are all eigenvectors corresponding to the eigenvalue $\lambda_1 = -1$.

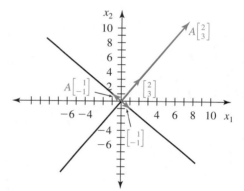

Figure 9.21 The two eigenvectors with their corresponding lines. The images of the eigenvectors under the map A remain on their respective lines.

Eigenvectors are not unique; they are determined only up to a multiplicative constant. When the eigenvalues are real, as in the previous example, all eigenvectors corresponding to a particular eigenvalue lie on the *same* straight line through the origin. The line represented by the vector $\begin{bmatrix} 1 \\ -1 \end{bmatrix}$ is given by

$$l_1 = \{(x_1, x_2) : x_1 + x_2 = 0\}$$

and the line represented by the vector $\begin{bmatrix} 2 \\ 3 \end{bmatrix}$ is given by

$$l_2 = \{(x_1, x_2) : 3x_1 - 2x_2 = 0\}$$

The lines l_1 and l_2 are **invariant** under the map $\mathbf{x} \to A\mathbf{x}$, in the sense that if we choose a point (x_1, x_2) on a line that is represented by an eigenvector, then since

$$A\begin{bmatrix} x_1 \\ x_2 \end{bmatrix} = \lambda \begin{bmatrix} x_1 \\ x_2 \end{bmatrix}$$

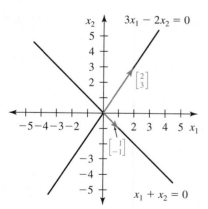

Figure 9.22 Eigenvectors and associated lines of the matrix $A = \begin{bmatrix} 1 & 2 \\ 3 & 2 \end{bmatrix}$.

the result of the map is a point that is on the same line. We check this claim for the line l_1. Assume that $(x_1, x_2) \in l_1$; that is, $x_1 + x_2 = 0$. Then the point $(\lambda x_1, \lambda x_2) \in l_1$ as well, since $\lambda x_1 + \lambda x_2 = \lambda(x_1 + x_2) = 0$. This situation is illustrated in Figure 9.22, where we draw both eigenvectors and the corresponding lines. We will refer to a graph such as this as the *geometric interpretation* of eigenvalues and eigenvectors. Since eigenvectors are often used in further calculations, you should choose values that are easy to work with. Small integers are good choices. How do we choose them? If we look at the straight line $3x_1 - 2x_2 = 0$ representing the eigenvectors corresponding to the eigenvalue $\lambda_2 = 4$ in Example 5, we see that the line goes through the origin and has slope 3/2. The entries of the vector $\begin{bmatrix} 2 \\ 3 \end{bmatrix}$ are thus the smallest integers that produce a vector on this line.

In the next example, we use a different matrix, which will also have real and distinct eigenvalues, to show the procedure for finding eigenvalues and eigenvectors once more.

EXAMPLE 6

Finding Eigenvalues and Eigenvectors Find the eigenvalues and eigenvectors of the matrix

$$A = \begin{bmatrix} 1 & 4 \\ 1 & -2 \end{bmatrix}$$

Solution

To find the eigenvalues of A, we must solve

$$\det(A - \lambda I) = \det \begin{bmatrix} 1 - \lambda & 4 \\ 1 & -2 - \lambda \end{bmatrix}$$

$$= (1 - \lambda)(-2 - \lambda) - 4 = \lambda^2 + \lambda - 6$$

$$= (\lambda - 2)(\lambda + 3) = 0$$

Solving yields

$$\lambda_1 = 2 \quad \text{and} \quad \lambda_2 = -3$$

To find an eigenvector associated with the eigenvalue $\lambda_1 = 2$, we must determine x_1 and x_2 (not both equal to 0) such that

$$\begin{bmatrix} 1 & 4 \\ 1 & -2 \end{bmatrix} \begin{bmatrix} x_1 \\ x_2 \end{bmatrix} = 2 \begin{bmatrix} x_1 \\ x_2 \end{bmatrix}$$

Writing this matrix equation as a system of linear equations, we get

$$x_1 + 4x_2 = 2x_1$$
$$x_1 - 2x_2 = 2x_2$$

We see that both equations are the same. Simplifying yields

$$-x_1 + 4x_2 = 0$$

Setting $x_2 = 1$, we find that $x_1 = 4$; that is, $\begin{bmatrix} 4 \\ 1 \end{bmatrix}$ is an eigenvector associated with the eigenvalue $\lambda_1 = 2$.

To find an eigenvector associated with the eigenvalue $\lambda_2 = -3$, we must determine x_1 and x_2 (not both equal to 0) such that

$$\begin{bmatrix} 1 & 4 \\ 1 & -2 \end{bmatrix} \begin{bmatrix} x_1 \\ x_2 \end{bmatrix} = -3 \begin{bmatrix} x_1 \\ x_2 \end{bmatrix}$$

Writing this matrix equation as a system of linear equations, we obtain

$$x_1 + 4x_2 = -3x_1$$
$$x_1 - 2x_2 = -3x_2$$

We see that both equations are the same. Simplifying yields

$$x_1 + x_2 = 0$$

Setting $x_1 = 1$, we find that $x_2 = -1$; that is, $\begin{bmatrix} 1 \\ -1 \end{bmatrix}$ is an eigenvector associated with the eigenvalue $\lambda_2 = -3$.

Both eigenvectors and the corresponding lines are illustrated in Figure 9.23. ■

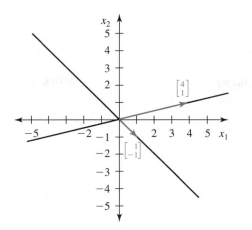

Figure 9.23 The eigenvectors and their corresponding lines in Example 6.

EXAMPLE 7   **Eigenvalues May Be 0** Let

$$A = \begin{bmatrix} 1 & 1 \\ 1 & 1 \end{bmatrix}$$

Find the eigenvalues and corresponding eigenvectors of A.

Solution   To find the eigenvalues of A, we must solve

$$\begin{aligned} \det(A - \lambda I) &= \det \begin{bmatrix} 1 - \lambda & 1 \\ 1 & 1 - \lambda \end{bmatrix} \\ &= (1 - \lambda)^2 - (1)(1) \\ &= 1 - 2\lambda + \lambda^2 - 1 = \lambda^2 - 2\lambda \\ &= \lambda(\lambda - 2) = 0 \end{aligned}$$

Solving yields

$$\lambda_1 = 0 \quad \text{and} \quad \lambda_2 = 2$$

The eigenvector corresponding to $\lambda_1 = 0$ satsifies

$$\begin{bmatrix} 1 & 1 \\ 1 & 1 \end{bmatrix} \begin{bmatrix} x_1 \\ x_2 \end{bmatrix} = 0 \begin{bmatrix} x_1 \\ x_2 \end{bmatrix}$$

Writing this matrix equation as a system of linear equations, we get

$$x_1 + x_2 = 0$$
$$x_1 + x_2 = 0$$

The two equations are the same, and we can choose $\begin{bmatrix} -1 \\ 1 \end{bmatrix}$ as an eigenvector corresponding to $\lambda_1 = 0$.

The eigenvector corresponding to $\lambda_2 = 2$ satisfies

$$\begin{bmatrix} 1 & 1 \\ 1 & 1 \end{bmatrix} \begin{bmatrix} x_1 \\ x_2 \end{bmatrix} = 2 \begin{bmatrix} x_1 \\ x_2 \end{bmatrix}$$

which results in

$$x_1 + x_2 = 2x_1$$
$$x_1 + x_2 = 2x_2$$

We can choose $\begin{bmatrix} 1 \\ 1 \end{bmatrix}$ as an eigenvector corresponding to $\lambda_2 = 2$.   ■

Let's look at one last example of finding eigenvalues and eigenvectors. This time, we choose a matrix that illustrates a case when we can read off eigenvalues from the matrix directly.

EXAMPLE 8

Reading Off Eigenvalues from a Matrix Find the eigenvalues and eigenvectors of the matrix

$$A = \begin{bmatrix} -2 & 1 \\ 0 & -1 \end{bmatrix}$$

Solution

To find the eigenvalues of A, we must solve

$$\det(A - \lambda I) = \det \begin{bmatrix} -2 - \lambda & 1 \\ 0 & -1 - \lambda \end{bmatrix}$$
$$= (-2 - \lambda)(-1 - \lambda) - (0)(1)$$
$$= (-2 - \lambda)(-1 - \lambda) = 0$$

Solving gives

$$\lambda_1 = -2 \quad \text{and} \quad \lambda_2 = -1$$

Let's pause for a moment and look back at the matrix A. The eigenvalues we found are identical to the diagonal elements of A! That is because one of the off-diagonal elements of A is equal to 0. Knowing this simplifies finding eigenvalues: If one (or both) off-diagonal elements of the 2×2 matrix A are equal to 0, then the eigenvalues of A are equal to the diagonal elements.

Now, let's find the associated eigenvectors. To find an eigenvector associated with the eigenvalue $\lambda_1 = -2$, we must determine x_1 and x_2 (not both equal to 0) such that

$$\begin{bmatrix} -2 & 1 \\ 0 & -1 \end{bmatrix} \begin{bmatrix} x_1 \\ x_2 \end{bmatrix} = -2 \begin{bmatrix} x_1 \\ x_2 \end{bmatrix}$$

Writing this matrix equation as a system of linear equations, we obtain

$$-2x_1 + x_2 = -2x_1$$
$$- x_2 = -2x_2$$

Simplifying either equation yields

$$x_2 = 0$$

This time, we cannot choose a value for x_2 (since $x_2 = 0$). But looking at the first equation, $-2x_1 + x_2 = -2x_1$, tells us that any value for x_1 will yield an identity, provided that $x_2 = 0$. Since $x_2 = 0$, we cannot choose $x_1 = 0$ (because $\begin{bmatrix} 0 \\ 0 \end{bmatrix}$ is not an eigenvector); any value of $x_1 \neq 0$ will do, however, so let's choose $x_1 = 1$. Then $\begin{bmatrix} 1 \\ 0 \end{bmatrix}$ is an eigenvector associated with $\lambda_1 = -2$.

To find an eigenvector associated with the eigenvalue $\lambda_2 = -1$, we must determine x_1 and x_2 (not both equal to 0) such that

$$\begin{bmatrix} -2 & 1 \\ 0 & -1 \end{bmatrix} \begin{bmatrix} x_1 \\ x_2 \end{bmatrix} = (-1) \begin{bmatrix} x_1 \\ x_2 \end{bmatrix}$$

Writing this matrix equation as a system of linear equations, we get

$$-2x_1 + x_2 = -x_1$$
$$- x_2 = -x_2$$

which simplifies to

$$-x_1 + x_2 = 0$$
$$0x_2 = 0$$

The first equation tells us that $x_1 = x_2$; the second equation tells us that we can choose any value for x_2. Choosing $x_2 = 1$, we need to set $x_1 = 1$. When we do, we find that $\begin{bmatrix} 1 \\ 1 \end{bmatrix}$ is an eigenvector associated with the eigenvalue $\lambda_2 = -1$. ■

So far, we have seen only examples in which the eigenvalues were real and distinct. In the next example, we will see that eigenvalues can be complex. When they are, we will not compute the corresponding eigenvectors, because the eigenvectors will be complex as well and hence beyond the scope of this book.

EXAMPLE 9   **Complex Eigenvalues** Let

$$A = \begin{bmatrix} \cos 30° & -\sin 30° \\ \sin 30° & \cos 30° \end{bmatrix}$$

Describe the action of A on the vector $\begin{bmatrix} 1 \\ 0 \end{bmatrix}$. Compute the eigenvalues of A.

Solution   We recognize that A is a matrix that describes a counterclockwise rotation by 30°. (The matrix A is the matrix R_θ for $\theta = 30°$, as defined in Subsection 9.3.1.) The vector $\begin{bmatrix} 1 \\ 0 \end{bmatrix}$ is thus rotated counterclockwise by 30°, and $A\begin{bmatrix} 1 \\ 0 \end{bmatrix} = \begin{bmatrix} \cos 30° \\ \sin 30° \end{bmatrix} = \begin{bmatrix} \sqrt{3}/2 \\ 1/2 \end{bmatrix}$. (See Figure 9.24.)

We will now compute the eigenvalues associated with this matrix. We set

$$\det(A - \lambda I) = 0$$

Using $\cos 30° = \frac{1}{2}\sqrt{3}$ and $\sin 30° = \frac{1}{2}$, we find that

$$\det \begin{bmatrix} \frac{1}{2}\sqrt{3} - \lambda & -\frac{1}{2} \\ \frac{1}{2} & \frac{1}{2}\sqrt{3} - \lambda \end{bmatrix} = 0$$

or

$$\left(\frac{1}{2}\sqrt{3} - \lambda \right)^2 + \frac{1}{4} = 0$$

$$\frac{3}{4} - \lambda\sqrt{3} + \lambda^2 + \frac{1}{4} = 0$$

$$\lambda^2 - \sqrt{3}\lambda + 1 = 0$$

Solving this quadratic equation, we obtain

$$\lambda_{1,2} = \frac{\sqrt{3} \pm \sqrt{3-4}}{2} = \frac{1}{2}\left(\sqrt{3} \pm i \right)$$

where $i^2 = -1$. This solution shows that eigenvalues can be complex numbers. ■

In Chapter 11, we will examine the stability of equilibria in systems of ordinary differential equations. This examination will lead us to investigate the eigenvalues of certain linear maps. It will be important to determine whether the real parts of the eigenvalues are positive or negative. In the case where the linear map is given by a 2×2 matrix, there is a useful criterion. Consider

$$A = \begin{bmatrix} a & b \\ c & d \end{bmatrix}$$

The determinant of A is

$$\det A = ad - bc$$

and the **trace** of A (denoted by tr A) is defined as

$$\text{tr } A = a + d$$

The trace is the sum of the diagonal elements of A. The trace and the determinant of a matrix are related to its eigenvalues.

Figure 9.24 The vector $\begin{bmatrix} 1 \\ 0 \end{bmatrix}$ is rotated counterclockwise by 30° in Example 9.

The eigenvalues of A satisfy

$$\det(A - \lambda I) = 0$$

or

$$\det \begin{bmatrix} a - \lambda & b \\ c & d - \lambda \end{bmatrix} = 0$$

which yields

$$(a - \lambda)(d - \lambda) - bc = 0$$
$$\lambda^2 - (a + d)\lambda + ad - bc = 0$$

Since $\det A = ad - bc$ and $\operatorname{tr} A = a + d$, we can write the last equation for λ as the quadratic equation

$$\lambda^2 - (\operatorname{tr} A)\lambda + \det A = 0 \qquad (9.20)$$

If λ_1 and λ_2 are the two solutions of (9.20), then λ_1 and λ_2 must satisfy

$$(\lambda - \lambda_1)(\lambda - \lambda_2) = 0$$

Multiplying the left-hand side out, we find that λ_1 and λ_2 satisfy

$$\lambda^2 - (\lambda_1 + \lambda_2)\lambda + \lambda_1\lambda_2 = 0$$

Comparing this equation with (9.20), we find the following important result:

If A is a 2×2 matrix with eigenvalues λ_1 and λ_2, then

$$\operatorname{tr} A = \lambda_1 + \lambda_2 \qquad \text{and} \qquad \det A = \lambda_1\lambda_2$$

To prepare for the next theorem, we make the following observations: Assume that λ_1 and λ_2 are both real and negative. Then the trace of A, which is the sum of the two eigenvalues, is negative, and the determinant of A, which is the product of the two eigenvalues, is positive.

In the case when the eigenvalues are complex conjugates, we have the same result. Assume that λ_1 and λ_2 are complex conjugates and that their real parts are negative. Then we can show that the trace of A is negative and the determinant is positive. (We will see an example of this fact later.)

That is, whenever the real parts of the two eigenvalues λ_1 and λ_2 are negative, it follows that $\operatorname{tr} A < 0$ and $\det A > 0$. The converse of this result is also true: If $\operatorname{tr} A < 0$ and $\det A > 0$, then both eigenvalues have negative real parts. This will be a useful criterion in Chapter 11, since it will enable us to determine whether or not both eigenvalues have negative real parts without computing the eigenvalues.

Theorem Let A be a 2×2 matrix with eigenvalues λ_1 and λ_2. Then the real parts of λ_1 and λ_2 are negative if and only if

$$\operatorname{tr} A < 0 \qquad \text{and} \qquad \det A > 0$$

EXAMPLE 10

Trace and Determinant Use the preceding theorem to show that both of the eigenvalues of

$$A = \begin{bmatrix} -1 & -3 \\ 1 & -2 \end{bmatrix}$$

have negative real parts. Then compute the eigenvalues, and use them to recompute the trace and the determinant of A.

Solution   Since

$$\operatorname{tr} A = -1 - 2 = -3 < 0 \quad \text{and} \quad \det A = (-1)(-2) - (1)(-3) = 5 > 0$$

it follows from the theorem that both eigenvalues have negative real parts. To compute the eigenvalues, we solve

$$\det(A - \lambda I) = 0$$

or

$$\det \begin{bmatrix} -1 - \lambda & -3 \\ 1 & -2 - \lambda \end{bmatrix} = 0$$

Evaluating the preceding equation, we find that

$$(-1 - \lambda)(-2 - \lambda) - (1)(-3) = 0$$
$$\lambda^2 + 3\lambda + 5 = 0$$

That is,

$$\lambda_{1,2} = \frac{-3 \pm \sqrt{9 - 20}}{2} = -\frac{3}{2} \pm \frac{1}{2} i \sqrt{11}$$

This equation also shows that the real parts of both eigenvalues are negative.

We can now use the eigenvalues to recompute the trace and the determinant of A. For the trace, we have

$$\operatorname{tr} A = \lambda_1 + \lambda_2 = \left(-\frac{3}{2} + \frac{1}{2} i \sqrt{11} \right) + \left(-\frac{3}{2} - \frac{1}{2} i \sqrt{11} \right) = -3$$

This is the same result that we obtained previously. Note that since λ_1 and λ_2 are complex conjugates, the imaginary parts cancel when we add λ_1 and λ_2. The determinant of A is

$$\det A = \lambda_1 \lambda_2 = \left(-\frac{3}{2} + \frac{1}{2} i \sqrt{11} \right) \left(-\frac{3}{2} - \frac{1}{2} i \sqrt{11} \right)$$
$$= \frac{9}{4} - \frac{1}{4} i^2 (11) = \frac{9}{4} + \frac{11}{4} = \frac{20}{4} = 5$$

which also is the same result that we obtained previously. Note that because λ_1 and λ_2 are complex conjugates, we computed a product of the form $(x + iy)(x - iy) = x^2 - (iy)^2 = x^2 + y^2$, which is a real number (and positive).   ■

■ 9.3.3 Iterated Maps (Needed for Section 10.7)

We restrict ourselves to the case in which A is a 2×2 matrix with real eigenvalues. We saw that in this case the eigenvectors define lines through the origin that are invariant under the map A. If the two invariant lines are *not* identical, we say that the two eigenvectors are **linearly independent**. (See Figure 9.25.) This notion can be formulated as follows in terms of eigenvectors: If we denote the two eigenvectors by $\mathbf{u}_1$ and $\mathbf{u}_2$, then $\mathbf{u}_1$ and $\mathbf{u}_2$ are linearly independent if there does not exist a number a such that $\mathbf{u}_1 = a\mathbf{u}_2$. (Linear independence is defined not just for eigenvectors: *Any* two nonzero vectors $\mathbf{x}_1$ and $\mathbf{x}_2$ are linearly independent if there does not exist a number a such that $\mathbf{x}_1 = a\mathbf{x}_2$.)

The following criterion is useful when we want to determine whether eigenvalues are linearly independent:

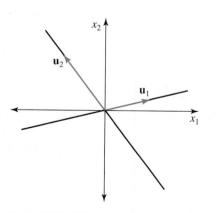

Figure 9.25 The two eigenvectors $\mathbf{u}_1$ and $\mathbf{u}_2$ are linearly independent.

Let A be a 2×2 matrix with eigenvalues λ_1 and λ_2. Denote by $\mathbf{u}_1$ the eigenvector associated with λ_1 and by $\mathbf{u}_2$ the eigenvector associated with λ_2. If $\lambda_1 \neq \lambda_2$, then $\mathbf{u}_1$ and $\mathbf{u}_2$ are linearly independent.

There are also cases in which $\mathbf{u}_1$ and $\mathbf{u}_2$ are linearly independent even though $\lambda_1 = \lambda_2$. We will, however, be concerned primarily with cases in which $\lambda_1 \neq \lambda_2$. Hence, the preceding criterion will suffice for our purposes. (The other cases are covered in courses on linear algebra.)

A consequence of linear independence is that we can write any vector uniquely as a **linear combination** of two eigenvectors. Suppose that $\mathbf{u}_1$ and $\mathbf{u}_2$ are linearly independent eigenvectors of a 2×2 matrix; then any 2×1 vector $\mathbf{x}$ can be written as

$$\mathbf{x} = a_1\mathbf{u}_1 + a_2\mathbf{u}_2$$

where a_1 and a_2 are uniquely determined. We will not prove this statement but will examine what we can do with it.

If we apply A to $\mathbf{x}$ (written as a linear combination of the two eigenvectors of A), then, using the linearity of the map A, we find that

$$A\mathbf{x} = A(a_1\mathbf{u}_1 + a_2\mathbf{u}_2) = a_1A\mathbf{u}_1 + a_2A\mathbf{u}_2$$

Now, $\mathbf{u}_1$ and $\mathbf{u}_2$ are both eigenvectors corresponding to A. Hence, $A\mathbf{u}_1 = \lambda_1\mathbf{u}_1$ and $A\mathbf{u}_2 = \lambda_2\mathbf{u}_2$. We thus obtain

$$A\mathbf{x} = a_1\lambda_1\mathbf{u}_1 + a_2\lambda_2\mathbf{u}_2$$

This representation of $\mathbf{x}$ is particularly useful if we apply A repeatedly to $\mathbf{x}$. Applying A to $A\mathbf{x}$, we find that

$$A^2\mathbf{x} = A(A\mathbf{x}) = A(a_1\lambda_1\mathbf{u}_1 + a_2\lambda_2\mathbf{u}_2) = a_1\lambda_1A\mathbf{u}_1 + a_2\lambda_2A\mathbf{u}_2$$
$$= a_1\lambda_1^2\mathbf{u}_1 + a_2\lambda_2^2\mathbf{u}_2$$

where we again used the fact that $\mathbf{u}_1$ and $\mathbf{u}_2$ are eigenvectors of the matrix A. Continuing in this way yields

$$A^n\mathbf{x} = a_1\lambda_1^n\mathbf{u}_1 + a_2\lambda_2^n\mathbf{u}_2 \tag{9.21}$$

Thus, instead of multiplying A with itself n times (which is rather time consuming), we can use the right-hand side of (9.21), which just amounts to adding two vectors (a much faster task). Are you convinced that this formula might be useful? If not, the next two examples should convince you.

EXAMPLE 11   Let

$$A = \begin{bmatrix} 1 & 2 \\ 3 & 2 \end{bmatrix}$$

Find $A^{10}\mathbf{x}$ for $\mathbf{x} = \begin{bmatrix} 4 \\ 1 \end{bmatrix}$.

Solution   We computed eigenvalues and eigenvectors for the matrix A earlier, and we found that

$$\lambda_1 = -1 \quad \text{with} \quad \mathbf{u}_1 = \begin{bmatrix} 1 \\ -1 \end{bmatrix}$$

and

$$\lambda_2 = 4 \quad \text{with} \quad \mathbf{u}_2 = \begin{bmatrix} 2 \\ 3 \end{bmatrix}$$

We first represent $\mathbf{x} = \begin{bmatrix} 4 \\ 1 \end{bmatrix}$ as a linear combination of $\mathbf{u}_1$ and $\mathbf{u}_2$. For this, we must find a_1 and a_2 so that

$$\begin{bmatrix} 4 \\ 1 \end{bmatrix} = a_1 \begin{bmatrix} 1 \\ -1 \end{bmatrix} + a_2 \begin{bmatrix} 2 \\ 3 \end{bmatrix}$$

Writing this matrix equation as a system of linear equations yields

$$a_1 + 2a_2 = 4$$
$$-a_1 + 3a_2 = 1$$

Using the method of elimination, we obtain

$$a_1 + 2a_2 = 4$$
$$5a_2 = 5$$

Hence, $a_2 = 1$ and $a_1 = 4 - 2a_2 = 4 - 2 = 2$. We now find that

$$\begin{bmatrix} 4 \\ 1 \end{bmatrix} = 2 \begin{bmatrix} 1 \\ -1 \end{bmatrix} + \begin{bmatrix} 2 \\ 3 \end{bmatrix}$$

To compute $A^{10}\mathbf{x}$, we use the right-hand side of (9.21):

$$\begin{aligned} A^{10}\mathbf{x} = A^{10} \begin{bmatrix} 4 \\ 1 \end{bmatrix} &= A^{10} \left(2 \begin{bmatrix} 1 \\ -1 \end{bmatrix} + \begin{bmatrix} 2 \\ 3 \end{bmatrix} \right) \\ &= 2A^{10} \begin{bmatrix} 1 \\ -1 \end{bmatrix} + A^{10} \begin{bmatrix} 2 \\ 3 \end{bmatrix} \\ &= 2(-1)^{10} \begin{bmatrix} 1 \\ -1 \end{bmatrix} + 4^{10} \begin{bmatrix} 2 \\ 3 \end{bmatrix} \\ &= 2 \begin{bmatrix} 1 \\ -1 \end{bmatrix} + 4^{10} \begin{bmatrix} 2 \\ 3 \end{bmatrix} = \begin{bmatrix} 2,097,154 \\ 3,145,726 \end{bmatrix} \end{aligned}$$

To compute A^{10} directly would have taken a much longer time.   ■

The Leslie Matrix Revisited

EXAMPLE 12 In Subsection 9.2.5, we investigated an age-structured population with Leslie matrix

$$L = \begin{bmatrix} 1.5 & 2 \\ 0.08 & 0 \end{bmatrix}$$

Find both eigenvalues and eigenvectors.

Solution To find the eigenvalues, we must solve

$$\det(L - \lambda I) = 0$$

Since

$$L - \lambda I = \begin{bmatrix} 1.5 - \lambda & 2 \\ 0.08 & -\lambda \end{bmatrix}$$

it follows that

$$\begin{aligned} \det(L - \lambda I) &= (1.5 - \lambda)(-\lambda) - (2)(0.08) \\ &= -1.5\lambda + \lambda^2 - 0.16 \end{aligned}$$

Also, since

$$\lambda^2 - 1.5\lambda - 0.16 = (\lambda - 1.6)(\lambda + 0.1)$$

the eigenvalues are

$$\lambda_1 = 1.6 \quad \text{and} \quad \lambda_2 = -0.1$$

To compute the corresponding eigenvectors, we start with the larger eigenvalue, $\lambda_1 = 1.6$. We need to solve

$$\begin{bmatrix} 1.5 & 2 \\ 0.08 & 0 \end{bmatrix} \begin{bmatrix} x_1 \\ x_2 \end{bmatrix} = (1.6) \begin{bmatrix} x_1 \\ x_2 \end{bmatrix}$$

or

$$1.5x_1 + 2x_2 = 1.6x_1$$
$$0.08x_1 \qquad = 1.6x_2$$

Simplifying, we obtain

$$-0.1x_1 + 2x_2 = 0$$
$$0.08x_1 - 1.6x_2 = 0$$

or

$$x_1 - 20x_2 = 0$$
$$x_1 - 20x_2 = 0$$

From this system of equations, it follows that

$$x_1 = 20x_2$$

For instance, $x_2 = 1$ and $x_1 = 20$ satisfies the preceding equation. Therefore, an eigenvector corresponding to $\lambda_1 = 1.6$ is

$$\mathbf{u}_1 = \begin{bmatrix} 20 \\ 1 \end{bmatrix}$$

The eigenvector corresponding to $\lambda_2 = -0.1$ satisfies

$$\begin{bmatrix} 1.5 & 2 \\ 0.08 & 0 \end{bmatrix} \begin{bmatrix} x_1 \\ x_2 \end{bmatrix} = -0.1 \begin{bmatrix} x_1 \\ x_2 \end{bmatrix}$$

or

$$1.5x_1 + 2x_2 = -0.1x_1$$
$$0.08x_1 \qquad = -0.1x_2$$

Simplifying, we find that

$$1.6x_1 + 2x_2 = 0$$
$$0.08x_1 + 0.1x_2 = 0$$

or

$$0.8x_1 + x_2 = 0$$
$$0.8x_1 + x_2 = 0$$

From this system, it follows that

$$0.8x_1 = -x_2$$

For instance, $x_1 = 5$ and $x_2 = -4$ satisfies the preceding equation. Therefore, an eigenvector corresponding to the eigenvalue $\lambda_2 = -0.1$ is

$$\mathbf{u}_2 = \begin{bmatrix} 5 \\ -4 \end{bmatrix}$$ ■

We will now show that the larger eigenvalue determines the growth rate of the population and the eigenvector corresponding to the larger eigenvalue is a stable age distribution. We assume that the population vector at time 0 is

$$N(0) = \begin{bmatrix} 105 \\ 1 \end{bmatrix}$$

We can compute $N(1)$ by evaluating

$$N(1) = LN(0) = \begin{bmatrix} 1.5 & 2 \\ 0.08 & 0 \end{bmatrix} \begin{bmatrix} 105 \\ 1 \end{bmatrix} = \begin{bmatrix} 159.5 \\ 8.4 \end{bmatrix}$$

If we wanted to compute $N(t)$ for some integer value t, we would need to find

$$N(t) = L^t N(0)$$

$N(t)$ can be computed from (9.21). We just need to write $N(0)$ as a linear combination of the two eigenvectors and then apply L^t to that combination.

Looking at the eigenvalues, we see that $N(0)$, as a linear combination of the two eigenvectors, is

$$\begin{bmatrix} 105 \\ 1 \end{bmatrix} = 5 \begin{bmatrix} 20 \\ 1 \end{bmatrix} + \begin{bmatrix} 5 \\ -4 \end{bmatrix}$$

Now, if we want to compute $N(1)$, we must find

$$N(1) = LN(0) = L \begin{bmatrix} 105 \\ 1 \end{bmatrix} = L \left(5 \begin{bmatrix} 20 \\ 1 \end{bmatrix} + \begin{bmatrix} 5 \\ -4 \end{bmatrix} \right)$$

$$= 5L \begin{bmatrix} 20 \\ 1 \end{bmatrix} - L \begin{bmatrix} 5 \\ -4 \end{bmatrix}$$

Since $\begin{bmatrix} 20 \\ 1 \end{bmatrix}$ is an eigenvector corresponding to $\lambda_1 = 1.6$, it follows that $L\begin{bmatrix} 20 \\ 1 \end{bmatrix} = 1.6\begin{bmatrix} 20 \\ 1 \end{bmatrix}$. Likewise, since $\begin{bmatrix} 5 \\ -4 \end{bmatrix}$ is an eigenvector corresponding to $\lambda_2 = -0.1$, it follows that $L\begin{bmatrix} 5 \\ -4 \end{bmatrix} = -0.1\begin{bmatrix} 5 \\ -4 \end{bmatrix}$. Hence,

$$N(1) = (5)(1.6) \begin{bmatrix} 20 \\ 1 \end{bmatrix} + (-0.1) \begin{bmatrix} 5 \\ -4 \end{bmatrix}$$

$$= \begin{bmatrix} 159.5 \\ 8.4 \end{bmatrix}$$

which, of course, is the same answer as before. Now, let's find $N(t)$. For this, we need to compute

$$N(t) = L^t N(0) = L^t \begin{bmatrix} 105 \\ 1 \end{bmatrix} = L^t \left(5 \begin{bmatrix} 20 \\ 1 \end{bmatrix} + \begin{bmatrix} 5 \\ -4 \end{bmatrix} \right)$$

$$= 5(1.6)^t \begin{bmatrix} 20 \\ 1 \end{bmatrix} + (-0.1)^t \begin{bmatrix} 5 \\ -4 \end{bmatrix}$$

We see from this equation that $(1.6)^t$ grows much faster than $(-0.1)^t$. In fact, $(-0.1)^t$ tends to 0 as $t \to \infty$. We conclude that the larger eigenvalue determines the growth rate of the population.

It turns out that the larger eigenvalue of a Leslie matrix is always real and positive, provided that a positive fraction of zero-year-old females survive and either zero-year-old or one-year-old females have offspring. Furthermore, $\lambda_1 \geq |\lambda_2|$. The strict inequality $\lambda_1 > |\lambda_2|$ holds if zero-year-olds can give birth. We thus find the following:

> If L is a 2×2 Leslie matrix with eigenvalues λ_1 and λ_2, then the larger eigenvalue determines the growth parameter of the population.

If λ_1 is the larger eigenvalue and $0 < \lambda_1 < 1$, then the population size decreases over time. If $\lambda_1 > 1$, then the population size increases over time.

We will now show that the eigenvector corresponding to the larger eigenvalue is a stable age distribution. Let λ_1 be the larger eigenvalue and $\mathbf{u}_1$ its corresponding eigenvector; then, if $N(0) = \mathbf{u}_1$, it follows that

$$N(t) = L^t N(0) = L^t \mathbf{u}_1 = \lambda_1^t \mathbf{u}_1$$

To show that $\mathbf{u}_1$ is a stable age distribution, observe that if $\mathbf{u}_1 = \begin{bmatrix} x \\ y \end{bmatrix}$, then the fraction of zero-year-olds at time t is

$$\frac{\lambda^t x}{\lambda^t x + \lambda^t y} = \frac{x}{x + y}$$

which is the same fraction of zero-year-olds at time 0. Furthermore, since $\lambda_1 > 0$, we can choose $\mathbf{u}_1$ so that both entries are positive (a condition that is needed if the entries represent population sizes). We summarize this demonstration as follows:

> If L is a 2×2 Leslie matrix with eigenvalues λ_1 and λ_2, then the eigenvector corresponding to the larger eigenvalue is a stable age distribution.

For the matrix

$$L = \begin{bmatrix} 1.5 & 2 \\ 0.08 & 0 \end{bmatrix}$$

the vector

$$\begin{bmatrix} 20 \\ 1 \end{bmatrix}$$

is an eigenvector corresponding to the larger eigenvalue; thus, it is a stable age distribution. In Section 9.2.5, we claimed that $\begin{bmatrix} 2000 \\ 100 \end{bmatrix}$ was a stable age distribution. In both cases, the fraction of zero-year-olds is the same, namely, $20/21 = 2000/2100$. That is, both vectors represent the same proportion of zero-year-olds in the population. We can check that $\begin{bmatrix} 2000 \\ 100 \end{bmatrix} = 100 \begin{bmatrix} 20 \\ 1 \end{bmatrix}$, which shows that both vectors are eigenvectors. (Recall that if $\mathbf{u}$ is an eigenvector, then any vector $a\mathbf{u}$, $a \neq 0$, is also an eigenvector.) When we list a stable age distribution, we make sure that both entries are positive since they represent numbers of individuals in each age class.

If $\lambda_1 > |\lambda_2|$, then the population vector $N(t)$ will converge to a stable age distribution as $t \to \infty$, provided that $N(0) \neq \mathbf{u}_2$. This follows from writing $N(0)$ as a linear combination of the two eigenvectors $\mathbf{u}_1$ and $\mathbf{u}_2$ and applying L^t to the result; that is,

$$L^t N(0) = L^t(a_1 \mathbf{u}_1 + a_2 \mathbf{u}_2) = a_1 \lambda_1^t \mathbf{u}_1 + a_2 \lambda_2^t \mathbf{u}_2$$

$$= a_1 \lambda_1^t \begin{bmatrix} x_1 \\ y_1 \end{bmatrix} + a_2 \lambda_2^t \begin{bmatrix} x_2 \\ y_2 \end{bmatrix}$$

where $\mathbf{u}_1 = \begin{bmatrix} x_1 \\ y_1 \end{bmatrix}$ and $\mathbf{u}_2 = \begin{bmatrix} x_2 \\ y_2 \end{bmatrix}$. Here, $a_1 \neq 0$, since $N(0) \neq \mathbf{u}_2$. The fraction of zero-year-olds at time t is

$$\frac{a_1 \lambda_1^t x_1 + a_2 \lambda_2^t x_2}{a_1 \lambda_1^t x_1 + a_2 \lambda_2^t x_2 + a_1 \lambda_1^t y_1 + a_2 \lambda_2^t y_2} \to \frac{x_1}{x_1 + y_1}$$

as $t \to \infty$.

Section 9.3 Problems

■ 9.3.1

1. Let

$$A = \begin{bmatrix} 2 & 1 \\ 3 & 4 \end{bmatrix}, \quad \mathbf{x} = \begin{bmatrix} x_1 \\ x_2 \end{bmatrix}, \quad \text{and} \quad \mathbf{y} = \begin{bmatrix} y_1 \\ y_2 \end{bmatrix}$$

(a) Show by direct calculation that $A(\mathbf{x} + \mathbf{y}) = A\mathbf{x} + A\mathbf{y}$.

(b) Show by direct calculation that $A(\lambda \mathbf{x}) = \lambda(A\mathbf{x})$.

2. Let

$$A = \begin{bmatrix} a_{11} & a_{12} \\ a_{21} & a_{22} \end{bmatrix}, \quad \mathbf{x} = \begin{bmatrix} x_1 \\ x_2 \end{bmatrix}, \quad \text{and} \quad \mathbf{y} = \begin{bmatrix} y_1 \\ y_2 \end{bmatrix}$$

(a) Show by direct calculation that $A(\mathbf{x} + \mathbf{y}) = A\mathbf{x} + A\mathbf{y}$.

(b) Show by direct calculation that $A(\lambda \mathbf{x}) = \lambda(A\mathbf{x})$.

In Problems 3–8, represent each given vector $\mathbf{x} = \begin{bmatrix} x_1 \\ x_2 \end{bmatrix}$ in the x_1–x_2 plane, and determine its length and the angle that it forms with the positive x_1-axis (measured counterclockwise).

3. $\mathbf{x} = \begin{bmatrix} 2 \\ 2 \end{bmatrix}$ **4.** $\mathbf{x} = \begin{bmatrix} -2 \\ 0 \end{bmatrix}$ **5.** $\mathbf{x} = \begin{bmatrix} 0 \\ 3 \end{bmatrix}$

6. $\mathbf{x} = \begin{bmatrix} -1 \\ -1 \end{bmatrix}$ **7.** $\mathbf{x} = \begin{bmatrix} -\sqrt{3} \\ 1 \end{bmatrix}$ **8.** $\mathbf{x} = \begin{bmatrix} 1 \\ -\sqrt{3} \end{bmatrix}$

In Problems 9–12, vectors are given in their polar coordinate representation (length r, and angle α measured counterclockwise from the positive x_1-axis). Find the representation of the vector $\begin{bmatrix} x_1 \\ x_2 \end{bmatrix}$

in Cartesian coordinates.

9. $r = 2, \alpha = 30°$ **10.** $r = 3, \alpha = 150°$

11. $r = 1, \alpha = 120°$ **12.** $r = 5, \alpha = 240°$

13. Suppose a vector $\begin{bmatrix} x_1 \\ x_2 \end{bmatrix}$ has length 3 and is 15° clockwise from the positive x_1-axis. Find x_1 and x_2.

14. Suppose a vector $\begin{bmatrix} x_1 \\ x_2 \end{bmatrix}$ has length 2 and is 140° clockwise from the positive x_1-axis. Find x_1 and x_2.

15. Suppose a vector $\begin{bmatrix} x_1 \\ x_2 \end{bmatrix}$ has length 5 and is 25° counterclockwise from the positive x_2-axis. Find x_1 and x_2.

16. Suppose a vector $\begin{bmatrix} x_1 \\ x_2 \end{bmatrix}$ has length 4 and is 70° counterclockwise from the negative x_2-axis. Find x_1 and x_2.

In Problems 17–22, find $\mathbf{x} + \mathbf{y}$ for the given vectors $\mathbf{x}$ and $\mathbf{y}$. Represent $\mathbf{x}$, $\mathbf{y}$, and $\mathbf{x} + \mathbf{y}$ in the plane, and explain graphically how you add $\mathbf{x}$ and $\mathbf{y}$.

17. $\mathbf{x} = \begin{bmatrix} 1 \\ 2 \end{bmatrix}$ and $\mathbf{y} = \begin{bmatrix} 3 \\ 2 \end{bmatrix}$ **18.** $\mathbf{x} = \begin{bmatrix} -1 \\ 0 \end{bmatrix}$ and $\mathbf{y} = \begin{bmatrix} 3 \\ 2 \end{bmatrix}$

19. $\mathbf{x} = \begin{bmatrix} 0 \\ -2 \end{bmatrix}$ and $\mathbf{y} = \begin{bmatrix} 1 \\ -1 \end{bmatrix}$ **20.** $\mathbf{x} = \begin{bmatrix} -1 \\ -1 \end{bmatrix}$ and $\mathbf{y} = \begin{bmatrix} 1 \\ 2 \end{bmatrix}$

21. $\mathbf{x} = \begin{bmatrix} 1 \\ 0 \end{bmatrix}$ and $\mathbf{y} = \begin{bmatrix} -1 \\ 0 \end{bmatrix}$ **22.** $\mathbf{x} = \begin{bmatrix} -3 \\ -1 \end{bmatrix}$ and $\mathbf{y} = \begin{bmatrix} -2 \\ 3 \end{bmatrix}$

In Problems 23–28, compute $a\mathbf{x}$ for the given vector $\mathbf{x}$ and scalar a. Represent $\mathbf{x}$ and $a\mathbf{x}$ in the plane, and explain graphically how you obtain $a\mathbf{x}$.

23. $\mathbf{x} = \begin{bmatrix} -2 \\ 1 \end{bmatrix}$ and $a = 2$ **24.** $\mathbf{x} = \begin{bmatrix} 3 \\ -1 \end{bmatrix}$ and $a = -1$

25. $\mathbf{x} = \begin{bmatrix} 0 \\ -2 \end{bmatrix}$ and $a = 0.5$ **26.** $\mathbf{x} = \begin{bmatrix} 3 \\ -9 \end{bmatrix}$ and $a = -1/3$

27. $\mathbf{x} = \begin{bmatrix} -4 \\ 1 \end{bmatrix}$ and $a = 1/4$ **28.** $\mathbf{x} = \begin{bmatrix} 0.5 \\ 0.25 \end{bmatrix}$ and $a = 5$

In Problems 29–34, let

$$\mathbf{u} = \begin{bmatrix} 3 \\ 4 \end{bmatrix}, \quad \mathbf{v} = \begin{bmatrix} -1 \\ -2 \end{bmatrix}, \quad and \quad \mathbf{w} = \begin{bmatrix} 1 \\ -2 \end{bmatrix}$$

29. Compute $\mathbf{u} + \mathbf{v}$ and illustrate the result graphically.

30. Compute $\mathbf{u} - \mathbf{v}$ and illustrate the result graphically.

31. Compute $\mathbf{w} - \mathbf{u}$ and illustrate the result graphically.

32. Compute $\mathbf{v} - \frac{1}{2}\mathbf{u}$ and illustrate the result graphically.

33. Compute $\mathbf{u} + \mathbf{v} + \mathbf{w}$ and illustrate the result graphically.

34. Compute $2\mathbf{v} - \mathbf{w}$ and illustrate the result graphically.

In Problems 35–40, give a geometric interpretation of the map $\mathbf{x} \rightarrow A\mathbf{x}$ for each given map A.

35. $A = \begin{bmatrix} 1 & 0 \\ 0 & 1 \end{bmatrix}$ **36.** $A = \begin{bmatrix} 2 & 0 \\ 0 & -1 \end{bmatrix}$

37. $A = \begin{bmatrix} 0 & -1 \\ 1 & 0 \end{bmatrix}$ **38.** $A = \begin{bmatrix} 0 & 1 \\ -1 & 0 \end{bmatrix}$

39. $A = \frac{1}{2}\begin{bmatrix} \sqrt{3} & -1 \\ 1 & \sqrt{3} \end{bmatrix}$ **40.** $A = \frac{1}{2}\begin{bmatrix} \sqrt{2} & -\sqrt{2} \\ \sqrt{2} & \sqrt{2} \end{bmatrix}$

41. Use a rotation matrix to rotate the vector $\begin{bmatrix} -1 \\ 2 \end{bmatrix}$ counterclockwise by the angle $\pi/6$.

42. Use a rotation matrix to rotate the vector $\begin{bmatrix} 4 \\ -1 \end{bmatrix}$ counterclockwise by the angle $\pi/3$.

43. Use a rotation matrix to rotate the vector $\begin{bmatrix} 5 \\ 2 \end{bmatrix}$ counterclockwise by the angle $\pi/12$.

44. Use a rotation matrix to rotate the vector $\begin{bmatrix} -2 \\ -3 \end{bmatrix}$ counterclockwise by the angle $\pi/9$.

45. Use a rotation matrix to rotate the vector $\begin{bmatrix} 2 \\ 1 \end{bmatrix}$ clockwise by the angle $\pi/4$.

46. Use a rotation matrix to rotate the vector $\begin{bmatrix} 1 \\ -2 \end{bmatrix}$ clockwise by the angle $\pi/3$.

47. Use a rotation matrix to rotate the vector $\begin{bmatrix} 5 \\ -3 \end{bmatrix}$ clockwise by the angle $\pi/7$.

48. Use a rotation matrix to rotate the vector $\begin{bmatrix} -2 \\ -3 \end{bmatrix}$ clockwise by the angle $\pi/8$.

■ **9.3.2**

In Problems 49–56, find the eigenvalues λ_1 and λ_2 and corresponding eigenvectors $\mathbf{v}_1$ and $\mathbf{v}_2$ for each matrix A. Determine the equations of the lines through the origin in the direction of the eigenvectors $\mathbf{v}_1$ and $\mathbf{v}_2$, and graph the lines together with the eigenvectors $\mathbf{v}_1$ and $\mathbf{v}_2$ and the vectors $A\mathbf{v}_1$ and $A\mathbf{v}_2$.

49. $A = \begin{bmatrix} 2 & 3 \\ 0 & -1 \end{bmatrix}$ **50.** $A = \begin{bmatrix} 0 & 0 \\ 1 & -3 \end{bmatrix}$

51. $A = \begin{bmatrix} 1 & 0 \\ 0 & -1 \end{bmatrix}$ **52.** $A = \begin{bmatrix} -1 & 0 \\ 0 & 2 \end{bmatrix}$

53. $A = \begin{bmatrix} -4 & 2 \\ -3 & 1 \end{bmatrix}$ **54.** $A = \begin{bmatrix} 3 & 6 \\ -1 & -4 \end{bmatrix}$

55. $A = \begin{bmatrix} 2 & 1 \\ 2 & 3 \end{bmatrix}$ **56.** $A = \begin{bmatrix} -3 & -0.5 \\ 7 & 1.5 \end{bmatrix}$

In Problems 57–60, find the eigenvalues λ_1 and λ_2 for each matrix A.

57. $A = \begin{bmatrix} 4 & 0 \\ 0 & 3 \end{bmatrix}$ **58.** $A = \begin{bmatrix} -7 & 0 \\ 0 & 6 \end{bmatrix}$

59. $A = \begin{bmatrix} 1 & -3 \\ 0 & 2 \end{bmatrix}$ **60.** $A = \begin{bmatrix} -1 & 4 \\ 0 & -2 \end{bmatrix}$

61. Find the eigenvalues λ_1 and λ_2 for

$$A = \begin{bmatrix} a & 0 \\ c & b \end{bmatrix}$$

62. Find the eigenvalues λ_1 and λ_2 for

$$A = \begin{bmatrix} a & c \\ 0 & b \end{bmatrix}$$

63. Let

$$A = \begin{bmatrix} 2 & 4 \\ -2 & -3 \end{bmatrix}$$

Without explicitly computing the eigenvalues of A, decide whether the real parts of both eigenvalues are negative.

64. Let

$$A = \begin{bmatrix} -2 & 3 \\ -1 & 1 \end{bmatrix}$$

Without explicitly computing the eigenvalues of A, decide whether the real parts of both eigenvalues are negative.

65. Let

$$A = \begin{bmatrix} 4 & 4 \\ -4 & -3 \end{bmatrix}$$

Without explicitly computing the eigenvalues of A, decide whether the real parts of both eigenvalues are negative.

66. Let

$$A = \begin{bmatrix} 0 & 1 \\ 2 & -1 \end{bmatrix}$$

Without explicitly computing the eigenvalues of A, decide whether the real parts of both eigenvalues are negative.

67. Let

$$A = \begin{bmatrix} 2 & -5 \\ 2 & -3 \end{bmatrix}$$

Without explicitly computing the eigenvalues of A, decide whether the real parts of both eigenvalues are negative.

68. Let

$$A = \begin{bmatrix} -2 & 5 \\ 2 & 3 \end{bmatrix}$$

Without explicitly computing the eigenvalues of A, decide whether the real parts of both eigenvalues are negative.

■ **9.3.3**

69. Let

$$A = \begin{bmatrix} -1 & 1 \\ 0 & 2 \end{bmatrix}$$

(a) Show that

$$\mathbf{u}_1 = \begin{bmatrix} 1 \\ 0 \end{bmatrix} \quad \text{and} \quad \mathbf{u}_2 = \begin{bmatrix} 1 \\ 3 \end{bmatrix}$$

are eigenvectors of A and that $\mathbf{u}_1$ and $\mathbf{u}_2$ are linearly independent.

(b) Represent

$$\mathbf{x} = \begin{bmatrix} 1 \\ -3 \end{bmatrix}$$

as a linear combination of $\mathbf{u}_1$ and $\mathbf{u}_2$.

(c) Use your results in (a) and (b) to compute $A^{20}\mathbf{x}$.

70. Let

$$A = \begin{bmatrix} -2 & 1 \\ -4 & 3 \end{bmatrix}$$

(a) Show that

$$\mathbf{u}_1 = \begin{bmatrix} 1 \\ 4 \end{bmatrix} \quad \text{and} \quad \mathbf{u}_2 = \begin{bmatrix} 1 \\ 1 \end{bmatrix}$$

are eigenvectors of A and that $\mathbf{u}_1$ and $\mathbf{u}_2$ are linearly independent.

(b) Represent

$$\mathbf{x} = \begin{bmatrix} -1 \\ 2 \end{bmatrix}$$

as a linear combination of $\mathbf{u}_1$ and $\mathbf{u}_2$.

(c) Use your results in (a) and (b) to compute $A^{10}\mathbf{x}$.

71. Let

$$A = \begin{bmatrix} -1 & 0 \\ 3 & 1 \end{bmatrix}$$

Find

$$A^{15} \begin{bmatrix} 2 \\ 0 \end{bmatrix}$$

without using a calculator.

72. Let

$$A = \begin{bmatrix} 4 & -3 \\ 2 & -1 \end{bmatrix}$$

Find

$$A^{30} \begin{bmatrix} -4 \\ -2 \end{bmatrix}$$

without using a calculator.

73. Let

$$A = \begin{bmatrix} 5 & 7 \\ -2 & -4 \end{bmatrix}$$

Find

$$A^{20} \begin{bmatrix} -3 \\ -2 \end{bmatrix}$$

without using a calculator.

74. Let

$$A = \begin{bmatrix} 1 & -1/4 \\ 1/2 & 1/4 \end{bmatrix}$$

Find

$$A^{30} \begin{bmatrix} 1/2 \\ 3/2 \end{bmatrix}$$

without using a calculator.

75. Suppose that

$$L = \begin{bmatrix} 2 & 4 \\ 0.3 & 0 \end{bmatrix}$$

is the Leslie matrix for a population with two age classes.
(a) Determine both eigenvalues.
(b) Give a biological interpretation of the larger eigenvalue.
(c) Find the stable age distribution.

76. Suppose that

$$L = \begin{bmatrix} 1 & 3 \\ 0.7 & 0 \end{bmatrix}$$

is the Leslie matrix for a population with two age classes.
(a) Determine both eigenvalues.
(b) Give a biological interpretation of the larger eigenvalue.
(c) Find the stable age distribution.

77. Suppose that

$$L = \begin{bmatrix} 7 & 3 \\ 0.1 & 0 \end{bmatrix}$$

is the Leslie matrix for a population with two age classes.
(a) Determine both eigenvalues.
(b) Give a biological interpretation of the larger eigenvalue.
(c) Find the stable age distribution.

78. Suppose that

$$L = \begin{bmatrix} 0 & 5 \\ 0.9 & 0 \end{bmatrix}$$

is the Leslie matrix for a population with two age classes.
(a) Determine both eigenvalues.
(b) Give a biological interpretation of the larger eigenvalue.
(c) Find the stable age distribution.

79. Suppose that

$$L = \begin{bmatrix} 0 & 5 \\ 0.09 & 0 \end{bmatrix}$$

is the Leslie matrix for a population with two age classes.
(a) Determine both eigenvalues.
(b) Give a biological interpretation of the larger eigenvalue.
(c) Find the stable age distribution.

■ 9.4 Analytic Geometry

René Descartes (1596–1650) and Pierre de Fermat (1601–1665) are credited with the invention of analytic geometry, which combines techniques from algebra and geometry and provides important tools for multidimensional calculus. In Chapter 10, we will discuss some aspects of multidimensional calculus; we will therefore need a few results from analytic geometry. Our first task will be to generalize points and vectors in the plane to higher dimensions. We will then introduce a product between vectors that will allow us to determine the length of a vector and the angle between two vectors. Finally, we will give a vector representation of lines and planes in three-dimensional space.

■ 9.4.1 Points and Vectors in Higher Dimensions

We are familiar with points and vectors in the plane. To represent points, we use a Cartesian coordinate system that consists of two axes—the x_1-axis and the x_2-axis—that are perpendicular to each other. Any point in the plane can be represented by an ordered pair (a_1, a_2) of real numbers, where a_1 is the x_1-coordinate and a_2 is the x_2-coordinate. Since we need two numbers to locate such a point, we call the plane "two dimensional."

The plane can thus be thought of as the set of all points (x_1, x_2) with $x_1 \in \mathbf{R}$ and $x_2 \in \mathbf{R}$. We introduce the notation $\mathbf{R}^2$ to denote the set of these points. The two-dimensional plane can be described as

$$\mathbf{R}^2 = \{(x_1, x_2) : x_1 \in \mathbf{R}, x_2 \in \mathbf{R}\}$$

To generalize this equation to n dimensions, we set

$$\mathbf{R}^n = \{(x_1, x_2, \ldots, x_n) : x_1 \in \mathbf{R}, x_2 \in \mathbf{R}, \ldots, x_n \in \mathbf{R}\}$$

For instance, $\mathbf{R}^3$ is three-dimensional space; it consists of all points (x_1, x_2, x_3) with $x_i \in \mathbf{R}$ for $i = 1, 2,$ and 3. To represent points and vectors in $\mathbf{R}^3$, we use a coordinate system that consists of three mutually perpendicular axes oriented in a "right-handed" manner. That is, the axes are perpendicular to each other and oriented so that the thumb of your right hand points along the positive x_1-axis, the index finger of your right hand points along the positive x_2-axis, and the middle finger of your right hand points along the positive x_3-axis. This coordinate system is shown in Figure 9.26.

In four and higher dimensions, we can no longer draw a coordinate system, although we can still represent such systems algebraically and work with them.

We introduced vectors in two dimensions in Section 9.3. A vector is a quantity that has a direction and a magnitude. A vector in two dimensions is an ordered pair that can be represented by a directed segment, as illustrated in Figure 9.27. In the figure, the vector $\mathbf{x} = \begin{bmatrix} x_1 \\ x_2 \end{bmatrix}$ is represented by a directed segment with initial point $(0, 0)$ and endpoint (x_1, x_2). The arrow at the tip indicates the direction of the vector. We will now generalize this representation to n dimensions.

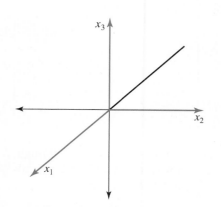

Figure 9.26 A right-handed three-dimensional coordinate system. The axes are perpendicular. The x_1-axis points in the direction of the thumb, the x_2-axis in the direction of the index finger, and the x_3-axis in the direction of the middle finger of the right hand.

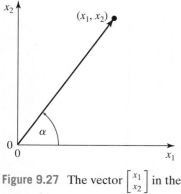

Figure 9.27 The vector $\begin{bmatrix} x_1 \\ x_2 \end{bmatrix}$ in the

x_1–x_2 plane.

> **Definition** A vector in n-dimensional space is an ordered n-tuple
>
> $$\mathbf{x} = \begin{bmatrix} x_1 \\ x_2 \\ \vdots \\ x_n \end{bmatrix}$$
>
> of real numbers. The numbers $x_1, x_2, \ldots, x_n$ are called the **components** of $\mathbf{x}$.

Vectors in n-dimensional space also may be represented by directed segments with initial point $(0, 0, \ldots, 0)$ and endpoint $(x_1, x_2, \ldots, x_n)$. In three dimensions, we

Figure 9.28 A vector in n-dimensional space.

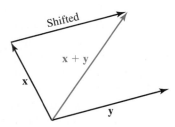

Figure 9.29 The parallelogram for vector addition.

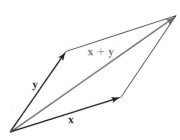

Figure 9.30 Adding two vectors.

can use the three-dimensional right-handed Cartesian coordinate system to visualize vectors. In four and higher dimensions, we can no longer draw the coordinate system. Instead, to represent a vector **x** whose endpoint has coordinates $(x_1, x_2, \ldots, x_n)$, we draw a directed arrow from the origin $(0, 0, \ldots, 0)$ to the point $P = (x_1, x_2, \ldots, x_n)$ (Figure 9.28).

Next, we generalize vector addition to n dimensions. In the previous section, we saw how to add vectors in two-dimensional space. In n dimensions, vector addition is defined in a similar way.

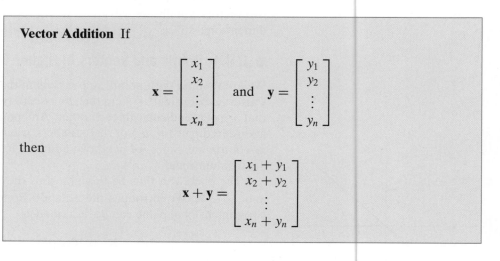

Vector Addition If

$$\mathbf{x} = \begin{bmatrix} x_1 \\ x_2 \\ \vdots \\ x_n \end{bmatrix} \quad \text{and} \quad \mathbf{y} = \begin{bmatrix} y_1 \\ y_2 \\ \vdots \\ y_n \end{bmatrix}$$

then

$$\mathbf{x} + \mathbf{y} = \begin{bmatrix} x_1 + y_1 \\ x_2 + y_2 \\ \vdots \\ x_n + y_n \end{bmatrix}$$

The geometric interpretation is the same as in two-dimensional space (although, again, visualization is impossible). This law of addition is similarly called the parallelogram law, for reasons that can be seen from Figure 9.29, where the addition of two vectors is illustrated.

We see from Figure 9.29 that vector addition can also be interpreted in the following way: To obtain $\mathbf{x} + \mathbf{y}$, we place the vector **y** at the tip of the vector **x**. The sum $\mathbf{x} + \mathbf{y}$ is then the vector that starts at the same point as **x** and ends at the point where the moved vector **y** ends, as illustrated in Figure 9.30.

The multiplication of a vector by a scalar generalizes to higher dimensions as well and has the same geometric interpretation.

Multiplication of a Vector by a Scalar If a is a scalar and **x** is a vector in n-dimensional space, then

$$a\mathbf{x} = a \begin{bmatrix} x_1 \\ x_2 \\ \vdots \\ x_n \end{bmatrix} = \begin{bmatrix} ax_1 \\ ax_2 \\ \vdots \\ ax_n \end{bmatrix}$$

Vector Representation A vector **x** is a directed line segment $\overrightarrow{AB}$ from the initial point A to the terminal point B. A particular representation of **x** has the origin at its initial point and the same direction and length as the directed segment $\overrightarrow{AB}$. We can use the terminal point of **x** as the representation of **x** when its initial point is the origin. We will call this particular representation the **vector representation** of $\overrightarrow{AB}$.

We can use Figure 9.31 to find the vector representation of a directed segment $\overrightarrow{AB}$ from point A with coordinates $(a_1, a_2, \ldots, a_n)$ to point B with coordinates $(b_1, b_2, \ldots, b_n)$. Using the parallelogram law, we see that

$$\overrightarrow{OA} + \overrightarrow{AB} = \overrightarrow{OB}$$

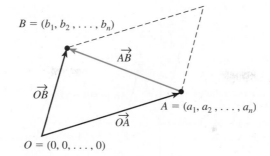

Figure 9.31 The vector representation of a directed segment.

Solving for $\overrightarrow{AB}$ yields

$$\overrightarrow{AB} = \overrightarrow{OB} - \overrightarrow{OA}$$

Now, $\overrightarrow{OB}$ has the vector representation $[b_1, b_2, \ldots, b_n]'$ and $\overrightarrow{OA}$ has the vector representation $[a_1, a_2, \ldots, a_n]'$. The difference $\overrightarrow{OB} - \overrightarrow{OA}$ is then the mathematical difference of the two vectors; that is,

$$\overrightarrow{OB} - \overrightarrow{OA} = \begin{bmatrix} b_1 - a_1 \\ b_2 - a_2 \\ \vdots \\ b_n - a_n \end{bmatrix}$$

The vector representing the line segment $\overrightarrow{AB}$ is thus given by

$$\overrightarrow{AB} = \begin{bmatrix} b_1 - a_1 \\ b_2 - a_2 \\ \vdots \\ b_n - a_n \end{bmatrix}$$

EXAMPLE 1   Find the vector representation of $\overrightarrow{AB}$ when $A = (2, -1)$ and $B = (1, 3)$.

Solution   The vector representation of $\overrightarrow{AB}$ is given by

$$\overrightarrow{AB} = \begin{bmatrix} b_1 - a_1 \\ b_2 - a_2 \end{bmatrix} = \begin{bmatrix} 1 - 2 \\ 3 - (-1) \end{bmatrix} = \begin{bmatrix} -1 \\ 4 \end{bmatrix}$$

We illustrate this equation graphically in Figure 9.32.

We see from Figure 9.32 that if we shift the vector $\overrightarrow{AB}$ to the origin, its tip ends at $(-1, 4)$, which confirms that the vector representation of $\overrightarrow{AB}$ is $\begin{bmatrix} -1 \\ 4 \end{bmatrix}$.   ■

Length of a Vector   The length of a vector in two dimensions is computed from the Pythagorean theorem. If $\mathbf{x} = \begin{bmatrix} x_1 \\ x_2 \end{bmatrix}$, then the length of $\mathbf{x}$ is denoted by $|\mathbf{x}|$, and we find that

$$|\mathbf{x}| = \sqrt{x_1^2 + x_2^2}$$

as illustrated in Figure 9.33. We can generalize this equation to n dimensions as follows: Recall that the transpose of a vector is denoted with the prime symbol; that is,

$$\begin{bmatrix} x_1, x_2, \ldots, x_n \end{bmatrix}' = \begin{bmatrix} x_1 \\ x_2 \\ \vdots \\ x_n \end{bmatrix}$$

Figure 9.32 The vector representation of the vector $\overrightarrow{AB}$.

Figure 9.33 The length of the vector $\begin{bmatrix} x_1 \\ x_2 \end{bmatrix}$.

This representation allows us to denote column vectors as the transposes of the corresponding row vectors—a convenient notation when one is writing large column vectors.

The length of a vector $\mathbf{x} = \begin{bmatrix} x_1, x_2, \ldots, x_n \end{bmatrix}'$ is

$$|\mathbf{x}| = \sqrt{x_1^2 + x_2^2 + \cdots + x_n^2}$$

EXAMPLE 2 Find the length of

$$\mathbf{x} = \begin{bmatrix} 1 \\ -3 \\ 4 \end{bmatrix}$$

Solution The length of $\mathbf{x}$ is given by

$$|\mathbf{x}| = \sqrt{x_1^2 + x_2^2 + x_3^2} = \sqrt{(1)^2 + (-3)^2 + (4)^2} = \sqrt{26}$$ ■

Figure 9.34 The normalized vector $\hat{\mathbf{x}}$ has the same direction as $\mathbf{x}$; its length is 1.

If we know the length of a vector $\mathbf{x}$, we can **normalize x** to obtain a vector of length 1 in the same direction as $\mathbf{x}$. (See Figure 9.34.) We call such a vector a **unit vector** in the direction of $\mathbf{x}$ and denote it by $\hat{\mathbf{x}}$; that is,

$$\hat{\mathbf{x}} = \frac{\mathbf{x}}{|\mathbf{x}|}$$

Of course, $|\hat{\mathbf{x}}| = 1$. We summarize all this as follows:

$\dfrac{\mathbf{x}}{|\mathbf{x}|}$ is a vector of length 1 in the direction of $\mathbf{x}$.

EXAMPLE 3 Normalize the vector

$$\mathbf{x} = \begin{bmatrix} 3 \\ -6 \\ 6 \end{bmatrix}$$

Solution   We must first find the length of **x**:

$$|\mathbf{x}| = \sqrt{x_1^2 + x_2^2 + x_3^2} = \sqrt{(3)^2 + (-6)^2 + (6)^2} = \sqrt{81} = 9$$

The unit vector $\hat{\mathbf{x}}$ is then given by

$$\hat{\mathbf{x}} = \frac{\mathbf{x}}{|\mathbf{x}|} = \frac{1}{9}\begin{bmatrix} 3 \\ -6 \\ 6 \end{bmatrix} = \begin{bmatrix} 1/3 \\ -2/3 \\ 2/3 \end{bmatrix}$$

We check that $\hat{\mathbf{x}}$ is indeed a vector of length 1:

$$\hat{\mathbf{x}} = \sqrt{(1/3)^2 + (-2/3)^2 + (2/3)^2} = 1$$

■

■ 9.4.2 The Dot Product

The dot product of two vectors allows us to determine the angle between the vectors.

> **Definition** The **scalar product**, or **dot product**, of two vectors $\mathbf{x} = [x_1, x_2, \ldots, x_n]'$ and $\mathbf{y} = [y_1, y_2, \ldots, y_n]'$ is the number
>
> $$\mathbf{x} \cdot \mathbf{y} = \mathbf{x}'\mathbf{y} = \sum_{i=1}^{n} x_i y_i$$

Note that the dot product is a scalar (hence the name "scalar product"). It is also called "dot product" because the notation uses a dot between **x** and **y**.

EXAMPLE 4   Find the dot product of

$$\mathbf{x} = \begin{bmatrix} 2 \\ 3 \\ 1 \end{bmatrix} \quad \text{and} \quad \mathbf{y} = \begin{bmatrix} -1 \\ 2 \\ 0 \end{bmatrix}$$

Solution   Using the definition of the dot product, we find that

$$\mathbf{x} \cdot \mathbf{y} = \begin{bmatrix} 2 & 3 & 1 \end{bmatrix} \begin{bmatrix} -1 \\ 2 \\ 0 \end{bmatrix} = -2 + 6 + 0 = 4$$

■

The dot product satisfies the following two properties:

1. $\mathbf{x} \cdot \mathbf{y} = \mathbf{y} \cdot \mathbf{x}$
2. $\mathbf{x} \cdot (\mathbf{y} + \mathbf{z}) = \mathbf{x} \cdot \mathbf{y} + \mathbf{x} \cdot \mathbf{z}$

We can use the dot product to express the length of a vector. Recall that in n dimensions the length of a vector $\mathbf{x} = [x_1, x_2, \ldots, x_n]'$ is defined as

$$|\mathbf{x}| = \sqrt{x_1^2 + x_2^2 + \cdots + x_n^2}$$

If we compute the dot product of **x** with itself, we obtain

$$\mathbf{x} \cdot \mathbf{x} = \sum_{i=1}^{n} x_i^2$$

Comparing $|\mathbf{x}|$ and $\mathbf{x} \cdot \mathbf{x}$, we see that the following relationship holds:

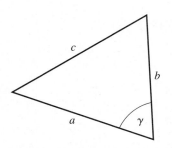

Figure 9.35 The law of cosines: $c^2 = a^2 + b^2 - 2ab\cos\gamma$.

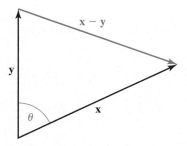

Figure 9.36 The angle between two vectors.

$$|\mathbf{x}|^2 = \mathbf{x} \cdot \mathbf{x}$$

The length of a vector is therefore $|\mathbf{x}| = \sqrt{\mathbf{x} \cdot \mathbf{x}}$.

The Angle between Two Vectors The other important application of the dot product is that it allows us to find the angle between two vectors. To derive this result, we need the trigonometric law of cosines, as illustrated in Figure 9.35.

Let $\mathbf{x}$ and $\mathbf{y}$ be two nonzero vectors whose initial points coincide. Then $\mathbf{x} - \mathbf{y}$ is a vector that connects the endpoint of $\mathbf{y}$ to the endpoint of $\mathbf{x}$. This follows from the parallelogram law for adding two vectors, namely, $\mathbf{y} + \mathbf{x} - \mathbf{y} = \mathbf{x}$, as illustrated in Figure 9.36.

Using the law of cosines, we find that

$$|\mathbf{x} - \mathbf{y}|^2 = |\mathbf{x}|^2 + |\mathbf{y}|^2 - 2|\mathbf{x}||\mathbf{y}|\cos\theta$$

where θ is the angle between $\mathbf{x}$ and $\mathbf{y}$. (See Figure 9.36.) Alternatively, the length of the vector $\mathbf{x} - \mathbf{y}$, denoted $|\mathbf{x} - \mathbf{y}|$, can be computed with the dot product:

$$\begin{aligned}
|\mathbf{x} - \mathbf{y}|^2 &= (\mathbf{x} - \mathbf{y}) \cdot (\mathbf{x} - \mathbf{y}) \\
&= \mathbf{x} \cdot \mathbf{x} - \mathbf{x} \cdot \mathbf{y} - \mathbf{y} \cdot \mathbf{x} + \mathbf{y} \cdot \mathbf{y} \\
&= |\mathbf{x}|^2 - 2\mathbf{x} \cdot \mathbf{y} + |\mathbf{y}|^2
\end{aligned}$$

Setting the two expressions for $|\mathbf{x} - \mathbf{y}|^2$ equal to each other gives

$$|\mathbf{x}|^2 + |\mathbf{y}|^2 - 2|\mathbf{x}||\mathbf{y}|\cos\theta = |\mathbf{x}|^2 - 2\mathbf{x} \cdot \mathbf{y} + |\mathbf{y}|^2$$

Solving this equation for $\mathbf{x} \cdot \mathbf{y}$, we find the following:

$$\mathbf{x} \cdot \mathbf{y} = |\mathbf{x}||\mathbf{y}|\cos\theta \qquad (9.22)$$

The significance of equation (9.22) is that it allows us to find the angle between the two nonzero vectors $\mathbf{x}$ and $\mathbf{y}$, because

$$\theta = \cos^{-1}\left(\frac{\mathbf{x} \cdot \mathbf{y}}{|\mathbf{x}||\mathbf{y}|}\right)$$

Note that $\theta \in [0, \pi)$, since the interval $[0, \pi)$ is used to find the inverse of the cosine function.

EXAMPLE 5 Find the angle between

$$\mathbf{x} = \begin{bmatrix} 2 \\ 1 \end{bmatrix} \quad \text{and} \quad \mathbf{y} = \begin{bmatrix} 1 \\ 1 \end{bmatrix}$$

Solution To determine the angle θ between $\mathbf{x}$ and $\mathbf{y}$, we use

$$\mathbf{x} \cdot \mathbf{y} = |\mathbf{x}||\mathbf{y}|\cos\theta$$

We find that

$$\mathbf{x} \cdot \mathbf{y} = \begin{bmatrix} 2 \\ 1 \end{bmatrix} \cdot \begin{bmatrix} 1 \\ 1 \end{bmatrix} = 2 + 1 = 3$$

$$|\mathbf{x}| = \sqrt{(2)^2 + (1)^2} = \sqrt{5}$$

$$|\mathbf{y}| = \sqrt{(1)^2 + (1)^2} = \sqrt{2}$$

Hence,

$$\cos\theta = \frac{\mathbf{x}\cdot\mathbf{y}}{|\mathbf{x}||\mathbf{y}|} = \frac{3}{\sqrt{5}\sqrt{2}} = \frac{3}{\sqrt{10}}$$

and therefore,

$$\theta = \cos^{-1}\frac{3}{\sqrt{10}} \approx 18.4° \quad \text{or} \quad 0.3218 \qquad ■$$

Figure 9.37 The vectors **x** and **y** are perpendicular.

We wish to single out the case in which $\theta = \pi/2$. We say that two vectors are **perpendicular** to each other if the angle between them is $\pi/2$; this situation is illustrated in Figure 9.37.

An important consequence of (9.22) is that it gives us a criterion with which to determine whether two vectors are perpendicular. Since $\cos(\pi/2) = 0$, we have the following theorem:

Theorem

x and **y** are perpendicular if $\mathbf{x}\cdot\mathbf{y} = 0$.

We will now give two examples that illustrate how to use this theorem.

EXAMPLE 6   Let $\mathbf{x} = \begin{bmatrix} 1 \\ 2 \end{bmatrix}$. Find $\mathbf{y} = \begin{bmatrix} y_1 \\ y_2 \end{bmatrix}$ so that **x** and **y** are perpendicular.

Solution   The vectors **x** and **y** are perpendicular if $\mathbf{x}\cdot\mathbf{y} = 0$. We find that

$$\mathbf{x}\cdot\mathbf{y} = \begin{bmatrix} 1 \\ 2 \end{bmatrix} \cdot \begin{bmatrix} y_1 \\ y_2 \end{bmatrix} = y_1 + 2y_2$$

We set the right-hand side equal to 0:

$$y_1 + 2y_2 = 0$$

Any choice of numbers (y_1, y_2) that satisfies this equation would thus give us a vector that is perpendicular to **x**. For instance, if we choose $y_2 = 1$ and $y_1 = -2$, then

$$\mathbf{y} = \begin{bmatrix} -2 \\ 1 \end{bmatrix}$$

is perpendicular to **x**.   ■

EXAMPLE 7   Show that the coordinate axes in a two-dimensional Cartesian coordinate system are perpendicular .

Solution   A two-dimensional Cartesian coordinate system is illustrated in Figure 9.38. We see that the x-axis can be represented by the vector $\begin{bmatrix} 1 \\ 0 \end{bmatrix}$ and the y-axis by the vector $\begin{bmatrix} 0 \\ 1 \end{bmatrix}$. Computing the dot product between these two vectors, we find that

$$\begin{bmatrix} 1 \\ 0 \end{bmatrix} \cdot \begin{bmatrix} 0 \\ 1 \end{bmatrix} = (1)(0) + (0)(1) = 0$$

and we conclude that the two vectors are perpendicular. Therefore, the x-axis and the y-axis are perpendicular.   ■

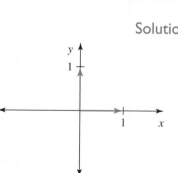

Figure 9.38 The coordinate axes in a two-dimensional Cartesian coordinate system are perpendicular.

We will now use the dot product and the result that the dot product between perpendicular vectors is zero to obtain the equation of a line in two-dimensional space and the equation of a plane in three-dimensional space.

Lines in the Plane   We will use the dot product to write equations of lines in the plane. Suppose that we wish to find the solution of the equation of a line through the point (x_0, y_0) and perpendicular to the vector $\mathbf{n} = \begin{bmatrix} a \\ b \end{bmatrix}$. If (x, y) is another point on the line, then the vector $\mathbf{r} - \mathbf{r}_0$ is perpendicular to $\mathbf{n}$, as illustrated in Figure 9.39.

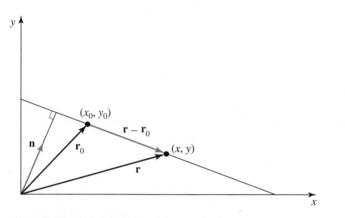

Figure 9.39   The vector $\mathbf{r} - \mathbf{r}_0$ is perpendicular to $\mathbf{n}$.

Therefore,

$$\mathbf{n} \cdot (\mathbf{r} - \mathbf{r}_0) = 0$$

This equation is called the **vector equation** of a line in the plane. Note that, written in vector form, the equation looks identical to the vector equation of a line in the plane. The difference is that the vectors are now in $\mathbf{R}^3$ and not in $\mathbf{R}^2$.

To obtain the **scalar equation** of this line, we set

$$\mathbf{n} = \begin{bmatrix} a \\ b \end{bmatrix}, \quad \mathbf{r} = \begin{bmatrix} x \\ y \end{bmatrix}, \quad \text{and} \quad \mathbf{r}_0 = \begin{bmatrix} x_0 \\ y_0 \end{bmatrix}$$

Then

$$\mathbf{n} \cdot (\mathbf{r} - \mathbf{r}_0) = \begin{bmatrix} a \\ b \end{bmatrix} \cdot \begin{bmatrix} x - x_0 \\ y - y_0 \end{bmatrix}$$
$$= a(x - x_0) + b(y - y_0) = 0$$

That is, we obtain the following result:

The line through (x_0, y_0) and perpendicular to $\begin{bmatrix} a \\ b \end{bmatrix}$ has the equation

$$a(x - x_0) + b(y - y_0) = 0 \tag{9.23}$$

EXAMPLE 8   Find an equation of the line through $(4, 3)$ and perpendicular to $\begin{bmatrix} 1 \\ 2 \end{bmatrix}$.

Solution   Using (9.23), we find that

$$(1)(x - 4) + (2)(y - 3) = 0$$

Simplifying yields

$$x + 2y = 10$$

■

Planes in Space   We can characterize a plane by a point P in the plane, together with the vector with initial point P and that is perpendicular to all vectors in the plane whose initial points coincide with P, as illustrated in Figure 9.40. If P is the endpoint of $\mathbf{r}_0$, $\mathbf{r} - \mathbf{r}_0$ is a vector in the plane with initial point P, and if $\mathbf{n}$ is a vector that is perpendicular to the plane, then

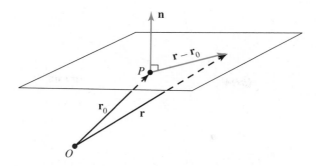

Figure 9.40 The vector $\mathbf{r} - \mathbf{r}_0$ is perpendicular to $\mathbf{n}$.

$$\mathbf{n} \cdot (\mathbf{r} - \mathbf{r}_0) = 0$$

This equation is called the *vector equation of a plane*. To obtain a scalar equation of a plane in three-dimensional space, we set

$$\mathbf{n} = \begin{bmatrix} a \\ b \\ c \end{bmatrix}, \quad \mathbf{r} = \begin{bmatrix} x \\ y \\ z \end{bmatrix}, \quad \text{and} \quad \mathbf{r}_0 = \begin{bmatrix} x_0 \\ y_0 \\ z_0 \end{bmatrix}$$

Then setting $\mathbf{n} \cdot (\mathbf{r} - \mathbf{r}_0) = 0$ results in the scalar equation of the plane through the point (x_0, y_0, z_0) with normal vector $\mathbf{n} = \begin{bmatrix} a, & b, & c \end{bmatrix}'$.

Evaluating the dot product yields

$$a(x - x_0) + b(y - y_0) + c(z - z_0) = 0$$

Summarizing the preceding discussion, we obtain the following result:

The plane through (x_0, y_0, z_0) and perpendicular to $\begin{bmatrix} a, & b, & c \end{bmatrix}'$ has the equation

$$a(x - x_0) + b(y - y_0) + c(z - z_0) = 0 \qquad (9.24)$$

EXAMPLE 9

Find the equation of a plane in three-dimensional space through $(2, 0, 3)$ and perpendicular to $\begin{bmatrix} -1, & 4, & 1 \end{bmatrix}'$.

Solution

Using (9.24), we find that

$$(-1)(x - 2) + (4)(y - 0) + (1)(z - 3) = 0$$

Simplifying yields

$$-x + 4y + z = 1$$

■ 9.4.3 Parametric Equations of Lines

There is another way to write lines in $\mathbf{R}^2$. Looking at Figure 9.41, we can describe the line shown there by the vector connecting the points O and P_0, denoted by $\overrightarrow{OP_0}$, and the vector $\mathbf{u}$: Any point P on the line can be thought of as the endpoint of the sum of the vector $\overrightarrow{OP_0}$ and the vector $\overrightarrow{P_0P}$. Now, $\overrightarrow{P_0P}$ is a multiple of the vector $\mathbf{u}$. Thus, for the equation of the line in vector form, we have

$$\overrightarrow{OP} = \overrightarrow{OP_0} + \overrightarrow{P_0P} = \overrightarrow{OP_0} + t\mathbf{u} \qquad (9.25)$$

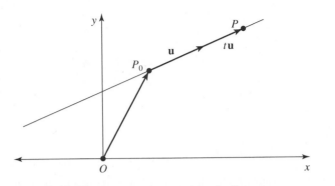

Figure 9.41 The parametric equation of a line.

for some $t \in \mathbf{R}$. If P has coordinates (x, y), P_0 has coordinates (x_0, y_0), and $\mathbf{u} = \begin{bmatrix} u_1 \\ u_2 \end{bmatrix}$, then

$$\begin{bmatrix} x \\ y \end{bmatrix} = \begin{bmatrix} x_0 \\ y_0 \end{bmatrix} + t \begin{bmatrix} u_1 \\ u_2 \end{bmatrix} \tag{9.26}$$

for some $t \in \mathbf{R}$. By varying t, we can reach *any* point on the line. Equation (9.25) [or (9.26)] is called a **parametric equation**, in vector form, of a line and t is called a **parameter**. We can also write the equation of the line in parametric form for each coordinate separately:

$$x = x_0 + tu_1$$
$$y = y_0 + tu_2$$

for $t \in \mathbf{R}$.

EXAMPLE 10

Find the parametric equation of the line in the x–y plane that goes through the point $(2, 1)$ in the direction of $\begin{bmatrix} -1 \\ -3 \end{bmatrix}$.

Solution

We find that

$$\begin{bmatrix} x \\ y \end{bmatrix} = \begin{bmatrix} 2 \\ 1 \end{bmatrix} + t \begin{bmatrix} -1 \\ -3 \end{bmatrix}, \; t \in \mathbf{R}$$

or $x = 2 - t$ and $y = 1 - 3t$, for $t \in \mathbf{R}$. By eliminating t, we can write the equation in the familiar standard form of a line in the x–y plane, namely, $t = 2 - x$. Therefore,

$$y = 1 - 3(2 - x) = 1 - 6 + 3x$$

and

$$3x - y - 5 = 0$$

is the standard form of the equation of this line. ■

EXAMPLE 11

Find the parametric equation of the line in the x–y plane that goes through the points $(-1, 2)$ and $(3, 5)$.

Solution

We designate one of the two points, say, $(-1, 2)$, as the point P_0 and let $\mathbf{u}$ be the vector that starts at $(-1, 2)$ and ends at $(3, 5)$. Then

$$\mathbf{u} = \begin{bmatrix} 3 - (-1) \\ 5 - 2 \end{bmatrix} = \begin{bmatrix} 4 \\ 3 \end{bmatrix}$$

and the parametric equation of this line is

$$\begin{bmatrix} x \\ y \end{bmatrix} = \begin{bmatrix} -1 \\ 2 \end{bmatrix} + t \begin{bmatrix} 4 \\ 3 \end{bmatrix}, \; t \in \mathbf{R}$$ ■

In Example 10, we saw that, by eliminating t, we can obtain the standard form of a linear equation. We can also go from the standard form to the parametric form by introducing a parameter t.

EXAMPLE 12   Find a parametric form of the line in standard form

$$2x - 3y + 1 = 0$$

Solution   The easiest way to parameterize this equation is to set $x = t$, solve the equation for y, and substitute t for x:

$$3y = 2x + 1 \quad \text{is equivalent to} \quad y = \frac{2}{3}x + \frac{1}{3}$$

With $x = t$, we can write the parametric equation as

$$x = t$$
$$y = \frac{2}{3}t + \frac{1}{3}$$

for $t \in \mathbf{R}$.

This is by no means the only way to parameterize the line. If we had chosen $t = \frac{1}{3}(x - 1)$, we would have found that

$$x = 3t + 1$$
$$y = \frac{2}{3}(3t + 1) + \frac{1}{3} = 2t + 1$$ ■

The vector representation of the line can be used in higher dimensions. For instance, a line in $\mathbf{R}^3$ that goes through a point $P_0 = (x_0, y_0, z_0)$ in the direction of $\mathbf{u} = \begin{bmatrix} u_1 \\ u_2 \\ u_3 \end{bmatrix}$ would have the form

$$\begin{bmatrix} x \\ y \\ z \end{bmatrix} = \begin{bmatrix} x_0 \\ y_0 \\ z_0 \end{bmatrix} + t \begin{bmatrix} u_1 \\ u_2 \\ u_3 \end{bmatrix}, \quad t \in \mathbf{R}$$

or, if we write out the coordinates separately,

$$x = x_0 + tu_1$$
$$y = y_0 + tu_2$$
$$z = z_0 + tu_3$$

for $t \in \mathbf{R}$.

EXAMPLE 13   Find the parametric equation of the line in x–y–z space that goes through the points $(1, -1, 3)$ and $(2, 4, -1)$.

Solution   We designate one point as P_0, say, $(2, 4, -1)$, and let $\mathbf{u}$ be the vector connecting the two points (it does not matter which of the two points we select as the starting point):

$$\mathbf{u} = \begin{bmatrix} 2 - 1 \\ 4 - (-1) \\ -1 - 3 \end{bmatrix} = \begin{bmatrix} 1 \\ 5 \\ -4 \end{bmatrix}$$

Then the parametric equation of this line is

$$\begin{bmatrix} x \\ y \\ z \end{bmatrix} = \begin{bmatrix} 2 \\ 4 \\ -1 \end{bmatrix} + t \begin{bmatrix} 1 \\ 5 \\ -4 \end{bmatrix}, \quad t \in \mathbf{R}$$

or

$$x = 2 + t$$
$$y = 4 + 5t$$
$$z = -1 - 4t$$

for $t \in \mathbf{R}$.   ■

Eliminating t in the parametric equation for a line in $\mathbf{R}^3$ is not very useful, since doing so does not yield just one equation as in the case of a line in $\mathbf{R}^2$. We will therefore forgo eliminating t in $\mathbf{R}^3$.

Section 9.4 Problems

■ 9.4.1

1. Let $\mathbf{x} = [1, 4, -1]'$ and $\mathbf{y} = [-2, 1, 0]'$.
(a) Find $\mathbf{x} + \mathbf{y}$. **(b)** Find $2\mathbf{x}$. **(c)** Find $-3\mathbf{y}$.

2. Let $\mathbf{x} = [-4, 3, 1]'$ and $\mathbf{y} = [0, -2, 3]'$.
(a) Find $\mathbf{x} - \mathbf{y}$. **(b)** Find $2\mathbf{x} + 3\mathbf{y}$. **(c)** Find $-\mathbf{x} - 2\mathbf{y}$.

3. Let $A = (2, 3)$ and $B = (4, 1)$. Find the vector representation of $\overrightarrow{AB}$.

4. Let $A = (-1, 0)$ and $B = (2, -4)$. Find the vector representation of $\overrightarrow{AB}$.

5. Let $A = (0, 1, -3)$ and $B = (-1, -1, 2)$. Find the vector representation of $\overrightarrow{AB}$.

6. Let $A = (1, 3, -2)$ and $B = (0, -1, 0)$. Find the vector representation of $\overrightarrow{AB}$.

7. Find the length of $\mathbf{x} = [1, 3]'$.

8. Find the length of $\mathbf{x} = [-2, 7]'$.

9. Find the length of $\mathbf{x} = [0, 1, 5]'$.

10. Find the length of $\mathbf{x} = [-2, 1, -3]'$.

11. Normalize $[1, 3, -1]'$. **12.** Normalize $[2, 0, -4]'$.

13. Normalize $[6, 0, 0]'$. **14.** Normalize $[0, -3, 1, 3]'$.

■ 9.4.2

15. Find the dot product of $\mathbf{x} = [1, 2]'$ and $\mathbf{y} = [3, -1]'$.

16. Find the dot product of $\mathbf{x} = [-1, 2]'$ and $\mathbf{y} = [-3, -4]'$.

17. Find the dot product of $\mathbf{x} = [0, -1, 3]'$ and $\mathbf{y} = [-3, 1, 1]'$.

18. Find the dot product of $\mathbf{x} = [2, -3, 1]'$ and $\mathbf{y} = [3, 1, -2]'$.

19. Use the dot product to compute the length of $[0, -1, 2]'$.

20. Use the dot product to compute the length of $[-1, 4, 3]'$.

21. Use the dot product to compute the length of $[1, 2, 3, 4]'$.

22. Use the dot product to compute the length of $[-1, -2, -3, -4]'$.

23. Find the angle between $\mathbf{x} = [1, 2]'$ and $\mathbf{y} = [3, -1]'$.

24. Find the angle between $\mathbf{x} = [-1, 2]'$ and $\mathbf{y} = [-2, -4]'$.

25. Find the angle between $\mathbf{x} = [0, -1, 3]'$ and $\mathbf{y} = [-3, 1, 1]'$.

26. Find the angle between $\mathbf{x} = [1, -3, 2]'$ and $\mathbf{y} = [3, 1, -4]'$.

27. Let $\mathbf{x} = [1, -1]'$. Find $\mathbf{y}$ so that $\mathbf{x}$ and $\mathbf{y}$ are perpendicular.

28. Let $\mathbf{x} = [-2, 1]'$. Find $\mathbf{y}$ so that $\mathbf{x}$ and $\mathbf{y}$ are perpendicular.

29. Let $\mathbf{x} = [1, -2, 4]'$. Find $\mathbf{y}$ so that $\mathbf{x}$ and $\mathbf{y}$ are perpendicular.

30. Let $\mathbf{x} = [2, 0, -1]'$. Find $\mathbf{y}$ so that $\mathbf{x}$ and $\mathbf{y}$ are perpendicular.

31. A triangle has vertices at coordinates $P = (0, 0)$, $Q = (4, 0)$, and $R = (4, 3)$.
(a) Use basic trigonometry to compute the lengths of all three sides and the measures of all three angles.
(b) Use the results of this section to repeat (a).

32. A triangle has vertices at coordinates $P = (0, 0)$, $Q = (0, 3)$, and $R = (5, 0)$.
(a) Use basic trigonometry to compute the lengths of all three sides and the measures of all three angles.
(b) Use the results of this section to repeat (a).

33. A triangle has vertices at coordinates $P = (1, 2, 3)$, $Q = (1, 5, 2)$, and $R = (2, 4, 2)$.
(a) Compute the lengths of all three sides.
(b) Compute all three angles in both radians and degrees.

34. A triangle has vertices at coordinates $P = (2, 1, 5)$, $Q = (-1, -3, 7)$, and $R = (2, -4, 1)$.
(a) Compute the lengths of all three sides.
(b) Compute all three angles in both radians and degrees.

35. Find the equation of the line through $(2, 1)$ and perpendicular to $[1, 2]'$.

36. Find the equation of the line through $(3, 2)$ and perpendicular to $[-1, 1]'$.

37. Find the equation of the line through $(1, -2)$ and perpendicular to $[4, 1]'$.

38. Find the equation of the line through $(0, 1)$ and perpendicular to $[1, 0]'$.

39. Find the equation of the plane through $(1, 2, 3)$ and perpendicular to $[0, -1, 1]'$.

40. Find the equation of the plane through $(1, 0, -3)$ and perpendicular to $[1, -2, -1]'$.

41. Find the equation of the plane through $(0, 0, 0)$ and perpendicular to $[1, 0, 0]'$.

42. Find the equation of the plane through $(3, -1, 2)$ and perpendicular to $[-1, 1, 2]'$.

■ 9.4.3

In Problems 43–46, find the parametric equation of the line in the x–y plane that goes through the indicated point in the direction of the indicated vector.

43. $(1, -1)$, $\begin{bmatrix} 2 \\ 1 \end{bmatrix}$ **44.** $(3, -4)$, $\begin{bmatrix} -1 \\ 2 \end{bmatrix}$

45. $(-1, -2)$, $\begin{bmatrix} 1 \\ -3 \end{bmatrix}$ **46.** $(-1, 4)$, $\begin{bmatrix} 2 \\ 3 \end{bmatrix}$

In Problems 47–50, find the parametric equation of the line in the x–y plane that goes through the given points. Then eliminate the parameter to find the equation of the line in standard form.

47. $(-1, 2)$ and $(3, 4)$ **48.** $(2, 1)$ and $(3, 5)$

49. $(1, -3)$ and $(4, 0)$ **50.** $(2, 3)$ and $(-1, -4)$

In Problems 51–54, parameterize the equation of the line given in standard form.

51. $3x + 4y - 1 = 0$ **52.** $x - 2y + 5 = 0$

53. $2x + y - 3 = 0$ **54.** $x - 5y + 7 = 0$

In Problems 55–58, find the parametric equation of the line in x–y–z space that goes through the indicated point in the direction of the indicated vector.

55. $(1, -1, 2)$, $\begin{bmatrix} 1 \\ -2 \\ 1 \end{bmatrix}$ **56.** $(2, 0, 4)$, $\begin{bmatrix} 1 \\ 2 \\ 3 \end{bmatrix}$

57. $(-1, 3, -2)$, $\begin{bmatrix} -1 \\ -2 \\ 4 \end{bmatrix}$ **58.** $(2, 1, -3)$, $\begin{bmatrix} 3 \\ -1 \\ 2 \end{bmatrix}$

In Problems 59–62, find the parametric equation of the line in x–y–z space that goes through the given points.

59. $(5, 4, -1)$ and $(2, 0, 3)$ **60.** $(2, 0, -3)$ and $(4, 1, 0)$

61. $(2, -3, 1)$ and $(-5, 2, 1)$ **62.** $(1, 0, 4)$ and $(3, 2, 0)$

63. Given are (1) a plane through $(1, -1, 2)$ and perpendicular to $\begin{bmatrix} 1 \\ 2 \\ 1 \end{bmatrix}$ and (2) a line through the points $(0, -3, 2)$ and $(-1, -2, 3)$. Where do the plane and the line intersect?

64. Given are (1) a plane through $(2, 0, -1)$ and perpendicular to

$\begin{bmatrix} -1 \\ 1 \\ 3 \end{bmatrix}$ and (2) a line through the points $(1, 0, -2)$ and $(-1, -1, 1)$. Where do the plane and the line intersect?

65. Given is a plane through $(0, -2, 1)$ and perpendicular to $\begin{bmatrix} -1 \\ 1 \\ -1 \end{bmatrix}$. Find a line through $(5, -1, 0)$ and that is parallel to the plane.

66. Given is the plane $x + 2y - z + 1 = 0$. Find a line in parametric form that is perpendicular to the plane.

Chapter 9 Key Terms

Discuss the following definitions and concepts:

1. Linear system of equations

2. Solving a linear system of equations

3. Upper triangular form

4. Gaussian elimination

5. Matrix

6. Basic matrix operations; addition, multiplication by a scalar

7. Transposition

8. Matrix multiplication

9. Identity matrix

10. Inverse matrix

11. Determinant

12. Leslie matrix

13. Stable age distribution

14. Vector

15. Parallelogram law

16. Linear map

17. Eigenvalues and eigenvectors

18. Vector addition

19. Multiplication of a vector by a scalar

20. Length of a vector

21. Dot product

22. Angle between two vectors

23. Perpendicular vectors

24. Line in the plane and in space

25. Equation of a plane

26. Parametric equation of a line

Chapter 9 Review Problems

1. Let

$$A = \begin{bmatrix} -1 & 1 \\ 0 & 2 \end{bmatrix}$$

(a) Find $A\mathbf{x}$ when $\mathbf{x} = \begin{bmatrix} 1 \\ 1 \end{bmatrix}$. Graph both $\mathbf{x}$ and $A\mathbf{x}$ in the same coordinate system.

(b) Find the eigenvalues λ_1 and λ_2, and the corresponding eigenvectors $\mathbf{u}_1$ and $\mathbf{u}_2$, of A.

(c) If $\mathbf{u}_i$ is the eigenvector corresponding to λ_i, find $A\mathbf{u}_i$ and explain graphically what happens when you apply A to $\mathbf{u}_i$.

(d) Write $\mathbf{x}$ as a linear combination of $\mathbf{u}_1$ and $\mathbf{u}_2$; that is, find a_1 and a_2 so that

$$\mathbf{x} = a_1\mathbf{u}_1 + a_2\mathbf{u}_2$$

Show that

$$A\mathbf{x} = a_1\lambda_1\mathbf{u}_1 + a_2\lambda_2\mathbf{u}_2$$

and illustrate this equation graphically.

2. Let

$$A = \begin{bmatrix} 3 & 1/2 \\ -5 & -1/2 \end{bmatrix}$$

(a) Find $A\mathbf{x}$ when $\mathbf{x} = \begin{bmatrix} 2 \\ 1 \end{bmatrix}$. Graph both $\mathbf{x}$ and $A\mathbf{x}$ in the same coordinate system.

(b) Find the eigenvalues λ_1 and λ_2, and the corresponding eigenvectors $\mathbf{u}_1$ and $\mathbf{u}_2$, of A.

(c) If $\mathbf{u}_i$ is the eigenvector corresponding to λ_i, find $A\mathbf{u}_i$, and explain graphically what happens when you apply A to $\mathbf{u}_i$.

(d) Write $\mathbf{x}$ as a linear combination of $\mathbf{u}_1$ and $\mathbf{u}_2$; that is, find a_1 and a_2 so that

$$\mathbf{x} = a_1\mathbf{u}_1 + a_2\mathbf{u}_2$$

Show that

$$A\mathbf{x} = a_1\lambda_1\mathbf{u}_1 + a_2\lambda_2\mathbf{u}_2$$

and illustrate this equation graphically.

3. Given the Leslie matrix

$$L = \begin{bmatrix} 1.5 & 0.875 \\ 0.5 & 0 \end{bmatrix}$$

find the growth rate of the population and determine its stable age distribution.

4. Given the Leslie matrix

$$L = \begin{bmatrix} 0.5 & 2.99 \\ 0.25 & 0 \end{bmatrix}$$

find the growth rate of the population and determine its stable age distribution.

5. Let

$$AB = \begin{bmatrix} 0 & 1 \\ 2 & -1 \end{bmatrix} \quad \text{and} \quad A^{-1} = \begin{bmatrix} 4 & -1 \\ 8 & -1 \end{bmatrix}$$

Find B.

6. Let

$$(AB)^{-1} = \begin{bmatrix} -1 & 3 \\ 0 & 2 \end{bmatrix} \quad \text{and} \quad B = \begin{bmatrix} 0 & -2 \\ 1 & 3 \end{bmatrix}$$

Find A.

7. Explain two different ways to solve a system of the form

$$a_{11}x_1 + a_{12}x_2 = b_1$$
$$a_{21}x_1 + a_{22}x_2 = b_2$$

when $a_{11}a_{22} - a_{12}a_{21} \neq 0$.

8. Suppose that

$$a_{11}x_1 + a_{12}x_2 = b_1$$
$$a_{21}x_1 + a_{22}x_2 = b_2$$

has infinitely many solutions. If you wrote this system in matrix form $AX = B$, could you find X by computing $A^{-1}B$?

9. Let

$$ax + 3y = 0$$
$$x - y = 0$$

How do you need to choose a so that this system has infinitely many solutions?

10. Let A be a 2×2 matrix and X and B be 2×1 matrices. Assume that $\det A = 0$. Explain how the choice of B affects the number of solutions of $AX = B$.

11. Suppose that

$$L = \begin{bmatrix} 0.5 & 2.3 \\ a & 0 \end{bmatrix}$$

is the Leslie matrix of a population with two age classes. For which values of a does this population grow?

12. Suppose that

$$L = \begin{bmatrix} 0.5 & 2.0 \\ 0.1 & 0 \end{bmatrix}$$

is the Leslie matrix of a population with two age classes.

(a) If you were to manage this population, would you need to be concerned about its long-term survival?

(b) Suppose that you can improve either the fecundity or the survival of the zero-year-olds, but due to physiological and environmental constraints, the fecundity of zero-year-olds will not exceed 1.5 and the survival of zero-year-olds will not exceed 0.4. Investigate how the growth rate of the population is affected by changing either the survival or the fecundity of zero-year-olds, or both. What would be the maximum achievable growth rate?

(c) In real situations, what other factors might you need to consider when you decide on management strategies?

13. Show that the eigenvalues of

$$\begin{bmatrix} a & c \\ 0 & b \end{bmatrix}$$

are equal to a and b.

14. Show that the eigenvalues of

$$\begin{bmatrix} a & 0 \\ c & b \end{bmatrix}$$

are equal to a and b.

Multivariable Calculus

LEARNING OBJECTIVES

The primary focus of this chapter is differential calculus for functions that depend on more than one variable. This discipline is needed to analyze systems of differential equations in Chapter 11. Specifically, we will learn how to

- find limits;
- calculate partial derivatives;
- linearize functions of two variables; and
- calculate and graphically interpret the gradient and find extrema.

To survive in cold temperatures, humans must either maintain a sufficiently high metabolic rate or regulate heat loss by covering their skin with an insulating material. There is a functional relationship that gives the lowest temperature for survival (T_e) as a function of metabolic heat production (M) and whole-body thermal conductance (g_{Hb}). The metabolic heat production depends on the type of activity; some values for humans are summarized in the following table:

Activity	M in Wm^{-2}
Sleeping	50
Working at a desk	95
Level walking at 2.5 mph	180
Level walking at 3.5 mph with a 40-lb pack	350

Data adapted from Landsberg (1969).

The whole-body thermal conductance g_{Hb} describes how quickly heat is lost. The value of g_{Hb} depends on the type of protection; for instance, $g_{Hb} = 0.45 \, \text{mol m}^{-2} \, \text{s}^{-1}$ without clothing, $g_{Hb} = 0.14 \, \text{mol m}^{-2} \, \text{s}^{-1}$ for a wool suit, and $g_{Hb} = 0.04 \, \text{mol m}^{-2} \, \text{s}^{-1}$ for a warm sleeping bag. That is, the smaller g_{Hb}, the better protection from the cold the material provides. The relationship among T_e, M, and g_{Hb} is given by

$$T_e = 36 - \frac{(0.9M - 12)(g_{Hb} + 0.95)}{27.8 g_{Hb}}$$

where M is measured in $W \, m^{-2}$, g_{Hb} is measured in $\text{mol m}^{-2} \, \text{s}^{-1}$, and T_e is measured in degree Celsius (Campbell, 1986). The temperature T_e is a function of two variables: M and g_{Hb}; to meet the required temperature, we can change either M (by starting to move when we get cold) or g_{Hb} (by putting on more clothes when we get cold). We can plot T_e as a function of M for different values of g_{Hb} or plot T_e as a function of g_{Hb} for different values of M, as shown in Figures 10.1 and 10.2.

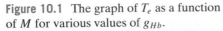

Figure 10.1 The graph of T_e as a function of M for various values of g_{Hb}.

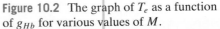

Figure 10.2 The graph of T_e as a function of g_{Hb} for various values of M.

Looking at Figure 10.1, we see that, for a given thermal conductance (say, wearing a wool suit), we need to increase metabolic heat production as it gets colder in order to stay above the minimum temperature for survival. Looking at Figure 10.2, we see that, for a given activity (say, level walking at 2.5 mph), we need to decrease thermal conductance (i.e., dress warmer) as it gets colder in order to stay above the minimum temperature for survival.

Since we can vary the two variables M and g_{Hb} independently, we can think of T_e as a function of two independent variables. In this chapter, we will discuss functions of two or more independent variables, develop the theory of differential calculus for such functions, and discuss a number of applications.

■ 10.1 Functions of Two or More Independent Variables

Figure 10.3 The domain and range of a function.

To recall the notation and terminology that we used when we considered functions of one variable, let

$$f : [0, 4] \longrightarrow \mathbf{R}$$
$$x \longrightarrow \sqrt{x}$$

The function $y = f(x)$ depends on one variable: x. Its domain is the set of numbers that we can use to evaluate $f(x)$, namely, the interval $[0, 4]$. Its range is the set of all possible values $y = f(x)$ for x in the domain of f. We see from Figure 10.3 that the range of $f(x)$ is the interval $[0, 2]$.

We now consider functions for which the domain consists of pairs of real numbers (x, y) with $x, y \in \mathbf{R}$ or, more generally, of n-tuples of real numbers $(x_1, x_2, \ldots, x_n)$ with $x_1, x_2, \ldots, x_n \in \mathbf{R}$. We also call n-tuples *points*. We use the notation $\mathbf{R}^n$ to denote the set of all n-tuples $(x_1, x_2, \ldots, x_n)$ with $x_1, x_2, \ldots, x_n \in \mathbf{R}$; that is,

$$\mathbf{R}^n = \{(x_1, x_2, \ldots, x_n) : x_1 \in \mathbf{R}, x_2 \in \mathbf{R}, \ldots, x_n \in \mathbf{R}\}$$

For $n = 1$, $\mathbf{R}^1 = \mathbf{R}$, which is the set of all real numbers. For $n = 2$, $\mathbf{R}^2$ is the set of all points in the plane, and so on. Note that n-tuples are *ordered*; for instance, $(2, 3) \neq (3, 2)$. We consider functions whose ranges are subsets of the real numbers; such functions are called *real valued*.

> **Definition** Suppose $D \subset \mathbf{R}^n$. Then a **real-valued function** f on D assigns a real number to each element in D, and we write
>
> $$f : D \longrightarrow \mathbf{R}$$
> $$(x_1, x_2, \ldots, x_n) \longrightarrow f(x_1, x_2, \ldots x_n)$$
>
> The set D is the domain of the function f, and the set
>
> $$\{w \in \mathbf{R} : f(x_1, x_2, \ldots, x_n) = w \text{ for some } (x_1, x_2, \ldots, x_n) \in D\}$$
>
> is the range of the function f.

If a function f depends on just two independent variables, we will often denote the independent variables by x and y, and write $f(x, y)$. In the case of three variables, we will often write $f(x, y, z)$. If f is a function of more than three independent variables, it is more convenient to use subscripts to label the variables—for example, $f(x_1, x_2, x_3, x_4)$.

The first example of this chapter shows how a function of more than one independent variable is evaluated at given points. You should pay attention to the fact that the domain consists of *ordered n-tuples*.

EXAMPLE 1

Evaluate the function

$$f(x, y, z) = \frac{xy}{z^2}$$

at the points $(2, 3, -1)$ and $(-1, 2, 3)$.

Solution

Since $f(x, y, z)$ lists the independent variables in the order x, y, and z, to evaluate the function at $(2, 3, -1)$, we need to subsititute 2 for x, 3 for y, and -1 for z:

$$f(2, 3, -1) = \frac{(2)(3)}{(-1)^2} = 6$$

Similarly,

$$f(-1, 2, 3) = \frac{(-1)(2)}{(3)^2} = -\frac{2}{9}$$

■

The next example shows how to determine the range of a real-valued function for a given domain.

EXAMPLE 2

Let $D = \{(x, y) : 0 \le x \le 1, \ 0 \le y \le 1\}$ and

$$f : D \longrightarrow \mathbf{R}$$
$$(x, y) \longrightarrow x + y$$

Graph the domain of f in the x–y plane and determine the range of f.

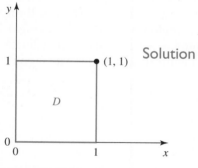

Solution

The domain of f is the set D, which consists of all points (x, y) whose x- and y-coordinates are between 0 and 1. This is the square shown in Figure 10.4.

To find the range of f, we need to determine what values f can take when we plug in points (x, y) from the domain D. The function $z = f(x, y)$ is smallest when $(x, y) = (0, 0)$; that is, $f(0, 0) = 0$. The function $z = f(x, y)$ is largest when $(x, y) = (1, 1)$; that is, $f(1, 1) = 2$. The function takes on all values in between. Hence, the range of f is the set $\{z : 0 \le z \le 2\}$. ■

Figure 10.4 The domain of the function in Example 2.

As in the case of functions of a single variable, the domain sometimes needs to be restricted. The next example illustrates how to find the largest possible domain.

EXAMPLE 3

Find the largest possible domain of the function

$$f(x, y) = \sqrt{y^2 - x}$$

Solution

The largest possible domain of f is the set of all points (x, y) such that $y^2 - x \ge 0$. We can illustrate this domain as a set in the x–y plane, where the boundary is given by the graph of $y^2 - x = 0$, which is a parabola that opens to the right. The set of points (x, y) that satisfies $y^2 - x \ge 0$ is then the shaded area in Figure 10.5, including the boundary curve.

The easiest way to see that this is the domain is to use a test point. The boundary curve $y^2 - x = 0$ divides the plane into two regions: the set of points that lie inside the parabola and the set of points that lie outside the parabola. In one of the regions, $y^2 - x > 0$; in the other, $y^2 - x < 0$. If we use a test point from the inside of the parabola—say, the point $(1, 0)$—then $y^2 - x = 0 - 1 < 0$; if we use a test point from the outside of the parabola—say, the point $(-1, 0)$—then $y^2 - x = 0 - (-1) > 0$. We therefore conclude that the set of points for which $y^2 - x \ge 0$ is the outside of the parabola, including the boundary curve. ■

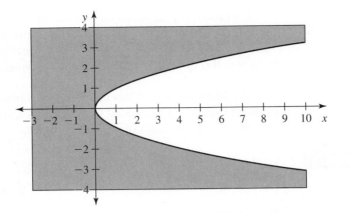

Figure 10.5   The domain of $f(x, y)$ in Example 3.

The Graph of a Function of Two Independent Variables   It is possible to graph a real-valued function of two variables. To do so, we write $z = f(x, y)$ and then locate the point (x, y, z) in three-dimensional space. The following definition makes this idea precise.

> **Definition** If f is a function of two independent variables with domain D, then the graph of f is the set of all points (x, y, z) such that $z = f(x, y)$ for $(x, y) \in D$. That is, the graph of f is the set
>
> $$S = \{(x, y, z) : z = f(x, y), \ (x, y) \in D\}$$

To locate a point (x, y, z) in three-dimensional space, we use the Cartesian coordinate system, which we introduced in Chapter 9. This system consists of three mutually perpendicular axes that emanate from a common point, called the *origin*, which has coordinates $(0, 0, 0)$. The axes are oriented in a right-handed coordinate frame, as explained in Chapter 9. This is shown in Figure 10.6, together with a point P that has coordinates (x_0, y_0, z_0).

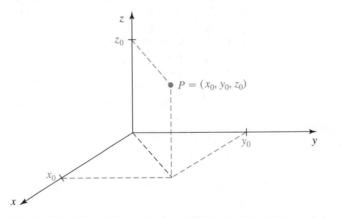

Figure 10.6   The right-handed coordinate system with a point.

Figure 10.6 suggests that we can visualize the graph of $f(x, y)$, $(x, y) \in D$, as lying directly above (or below) the domain D in the x–y plane. The graph of $f(x, y)$ is therefore a **surface** in three-dimensional space, as illustrated in Figure 10.7 for $f(x, y) = 2x^2 - y^2$.

Graphing a surface in three-dimensional space is difficult. Fortunately, good computer software is now available that facilitates this task. We show a few such surfaces, generated by computer software, in Figures 10.7 through 10.10. We will make no effort to learn how to graph such functions, but will provide problems at the end of this section in which surfaces and functions must be matched.

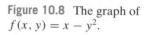

Figure 10.7 The graph of $z = f(x, y)$ is a surface in three-dimensional space [here, $f(x, y) = 2x^2 - y^2$].

Figure 10.8 The graph of $f(x, y) = x - y^2$.

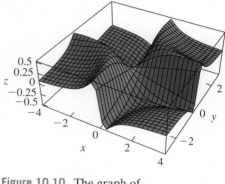

Figure 10.9 The graph of $f(x, y) = \sin x + \cos y$.

Figure 10.10 The graph of $f(x, y) = \frac{xy}{1+x^2y^2}$.

To understand the shape of the graph of $f(x, y) = 2x^2 - y^2$ in Figure 10.7, we can fix a value for x and then "walk" on the surface in the y-direction. Since the function is of the form "a constant minus y^2" for fixed x, we expect a curve that has the shape of an upside-down parabola. Looking at the surface in Figure 10.7, we see that this is indeed the case. If, now, we instead fix y and walk in the direction of x, then we walk along a curve of the form "$2x^2$ minus a constant," which is a parabola. Checking Figure 10.7 confirms this statement.

Figure 10.8 can be analyzed in a similar way. In Figure 10.9, we recognize the waves created by the sine and cosine functions. Figure 10.10 is included to show that functions of two variables can result in interesting (and complicated) shapes.

Another way to visualize functions is with **level curves** or **contour lines**. This approach is used, for instance, in topographical maps. (See Figure 10.11.) There is a subtle distinction between level curves and contour lines, in that level curves are drawn in the function domain whereas contour lines are drawn on the surface. This distinction is not always made, and often the two terms are used interchangeably. In this text, we will almost exclusively use level curves, for which we now give the precise definition:

Definition Suppose that $f : D \to \mathbf{R}$, $D \subset \mathbf{R}^2$. Then the level curves of f comprise the set of points (x, y) in the x–y plane where the function f has a constant value; that is, $f(x, y) = c$.

In Figure 10.12 we graph the surface of $f(x, y) = (2x^2 + y^2)e^{-(x+y)^2}$; in Figure 10.13 its level curves. To get an informative picture from the graph of the level curves, you should choose equidistant values for c—for instance, $c = 0, 1, 2, \dots$ or $c = 0, -0.1, -0.2, \dots$—so that you can infer the steepness of the curve from how close

Figure 10.11 A topographical map with level curves.

together the level curves are. In Figure 10.13, the level curves are equidistant, with $c = 0.5, 1, 1.5, \ldots$.

Figure 10.12 The graph of $f(x, y) = (2x^2 + y^2)e^{-(x+y)^2}$.

Figure 10.13 Level curves for $f(x, y) = (2x^2 + y^2)e^{-(x+y)^2}$.

If we project a level curve $f(x, y) = c$ up to its height c, the curve is called a contour line. The curve lies on the surface of $f(x, y)$ and traces the graph of f in a horizontal plane at height c (Figure 10.14).

EXAMPLE 4 Set $D = \{(x, y) : x^2 + y^2 \le 4\}$. Compare the level curves of

$$f(x, y) = 4 - x^2 - y^2 \quad \text{for } (x, y) \in D$$

and

$$g(x, y) = \sqrt{4 - x^2 - y^2} \quad \text{for } (x, y) \in D$$

Solution The level curve $f(x, y) = c$ is the set of all points (x, y) that satisfy

$$4 - x^2 - y^2 = c \quad \text{or} \quad x^2 + y^2 = 4 - c$$

We recognize the latter equation as that of a circle with center at the origin and radius $\sqrt{4 - c}$, illustrated in Figure 10.15 for $c = 0, 0.5, 1, 1.5,$ and 2.

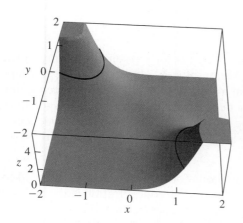

Figure 10.14 A contour line for $f(x, y) = (2x^2 + y^2)e^{-(x+y)^2}$.

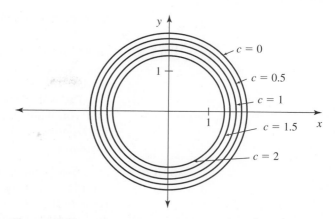

Figure 10.15 Level curves for $f(x, y) = 4 - x^2 - y^2$.

The level curve $g(x, y) = c$ satisfies

$$\sqrt{4 - x^2 - y^2} = c, \qquad \text{or} \qquad x^2 + y^2 = 4 - c^2$$

This is also the equation of a circle with center at the origin, but the radius is $\sqrt{4 - c^2}$; the circle is illustrated in Figure 10.16 for $c = 0, 0.5, 1, 1.5,$ and 2.

To compare the level curves of two functions, we choose $c = 0, 0.5, 1, 1.5,$ and 2 for both $f(x, y)$ and $g(x, y)$. The level curves for $c = 0$ and 1 are the same for the two functions, but the contour lines of $f(x, y)$ for $c = 1.5$ and 2 are a lot closer to the contour line for $c = 1$ than are those of $g(x, y)$. We can understand why when we look at the surfaces. The graph of $f(x, y)$ is obtained by rotating the parabola $z = 4 - x^2$ in the x–z plane about the z-axis; the surface thereby generated is called a *paraboloid*. (See Figure 10.17.)

The graph of $g(x, y)$ is obtained by rotating the half circle $z = \sqrt{4 - x^2}$ in the x–z plane about the z-axis; the surface generated is the top half of a sphere. (See Figure 10.18.)

Figure 10.16 Level curves for $g(x, y) = \sqrt{4 - x^2 - y^2}$.

Figure 10.17 The surface of $f(x, y) = 4 - x^2 - y^2$ is a paraboloid.

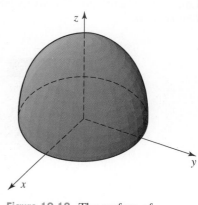

Figure 10.18 The surface of $g(x, y) = \sqrt{4 - x^2 - y^2}$ is the top half of a sphere.

Because the paraboloid is steeper than the sphere for c between 1 and 2, but the reverse holds for c between 0 and 1, the contour lines of the paraboloid are closer together for c between 1 and 2, whereas the contour lines of the sphere are closer together for c between 0 and 1. ■

Biological Applications

EXAMPLE 5

In Figure 10.19, we show level curves for oxygen concentration (in mg/l) in Long Lake, Clear Water County (Minnesota), as a function of date and depth. For instance, on day 140 (May 20, 1998) at a 10-m depth, the oxygen concentration was 12 mg/l.

The water flea *Daphnia* needs a minimum of 3 mg/l oxygen to survive. Suppose that you went out to Long Lake on day 200 (July 19, 1998) and wanted to look for *Daphnia* in the lake. Below what depth could you be fairly sure not to find any *Daphnia*?

Solution

Because *Daphnia* needs a minimum of 3 mg/l of oxygen, we need to find the depth on July 19, 1998 below which the oxygen concentration is always less than 3 mg/l. We see that the 3-mg/l oxygen-level curve goes through the point (200, −17.5). Thus, on July 19, 1998, *Daphnia* needed to stay above 17.5 m; that is, we could have been fairly sure not to have found *Daphnia* below 17.5 m. ■

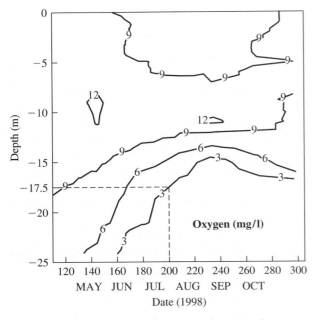

Figure 10.19 Level curves for oxygen concentration on Long Lake, Clear Water County (Minnesota). Courtesy of Leif Hembre.

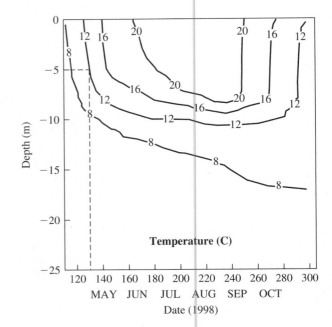

Figure 10.20 Isotherms for Long Lake, Clear Water County (Minnesota). Courtesy of Leif Hembre.

EXAMPLE 6

In Figure 10.20, we show temperature **isoclines** (also called **isotherms**) for Long Lake, Clear Water County (Minnesota). Temperature isoclines are lines of equal temperature—that is, level curves. The isoclines shown are functions of date and depth. For instance, on day 130 (May 10, 1998), the temperature at 5 m was approximately 12°C. If we are interested in how temperature changes over time at a fixed depth, we would draw a horizontal line through the appropriate depth and then read off the days and the temperature where the horizontal line intersects the isoclines: If we look at 5-meter depths, we find that the temperature climbs from roughly 8°C in the spring to more than 20°C in the summer and then falls back to below 12°C later in the fall. ■

Section 10.1 Problems

1. Cardiac output (CO) is a physiological quantity that is calculated as the product of heart rate (HR) and stroke volume (SV). Write cardiac output as a function of heart rate and stroke volume. If heart rate is measured in beats per minute and stroke volume in liters per beat, what is the unit for cardiac output? Determine the domain and range of the function describing cardiac output.

2. Mean arterial blood pressure (MAP) is a function of systolic blood pressure (SP) and diastolic blood pressure (DP). At a resting heart rate,

$$MAP \approx DP + \frac{1}{3}(SP - DP)$$

If systolic pressure is greater than diastolic pressure and both are nonnegative, what is the range of the function describing mean arterial pressure?

3. Locate the following points in a three-dimensional Cartesian coordinate system:
(a) $(1, 3, 2)$ **(b)** $(-1, -2, 1)$
(c) $(0, 1, 2)$ **(d)** $(2, 0, 3)$

4. Describe in words the set of all points in $\mathbf{R}^3$ that satisfy the following expressions:
(a) $x = 0$ **(b)** $y = 0$ **(c)** $z = 0$
(d) $z \geq 0$ **(e)** $y \leq 0$

In Problems 5–12, evaluate each function at the given point.

5. $f(x, y) = \dfrac{2x}{x^2 + y^2}$ at $(2, 3)$

6. $f(x, y, z) = \sqrt{x^2 - 3y + z}$ at $(3, -1, 1)$

7. **(a)** $f_1(x, y) = 2x - 3y^2$ at $(-1, 2)$
(b) $f_2(y, x) = 2x - 3y^2$ at $(-1, 2)$

8. **(a)** $f_1(x, y) = \dfrac{x}{y}$ at $(3, 2)$ **(b)** $f_2(y, x) = \dfrac{x}{y}$ at $(3, 2)$

(c) $f_3(y, x) = \dfrac{y}{x}$ at $(3, 2)$

9. $h(x, t) = \exp\left[-\dfrac{(x - 2)^2}{2t}\right]$ at $(1, 5)$

10. $g(n, p) - np(1 - p)^{n-1}$ at $(5, 0.1)$

11. $h(x_1, x_2) = x_2 e^{-x_1/x_2}$ at $(2, -1)$

12. $g(x_1, x_2, x_3, x_4) = x_1 x_4 \sqrt{x_2 x_3}$ at $(1, 8, 2, -1)$

In Problems 13–18, find the largest possible domain and the corresponding range of each function. Determine the equation of the level curves $f(x, y) = c$, together with the possible values of c.

13. $f(x, y) = x^2 + y^2$ **14.** $f(x, y) = \sqrt{9 - x^2 - y^2}$

15. $f(x, y) = \ln(y - x^2)$ **16.** $f(x, y) = \exp[-(x^2 + y^2)]$

17. $f(x, y) = \dfrac{x - y}{x + y}$ **18.** $f(x, y) = \dfrac{x + y}{x - y}$

In Problems 19–22, match each function with the appropriate graph in Figures 10.21–10.24.

19. $f(x, y) = 1 + x^2 + y^2$ **20.** $f(x, y) = \sin(x)\sin(y)$

21. $f(x, y) = y^2 - x^2$ **22.** $f(x, y) = 4 - x^2$

Figure 10.21

Figure 10.22

Figure 10.23

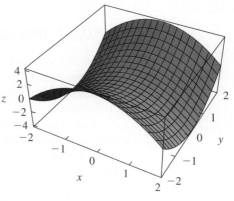

Figure 10.24

23. Let

$$f_a(x, y) = ax^2 + y^2$$

for $(x, y) \in \mathbf{R}$, where a is a positive constant.

(a) Assume that $a = 1$ and describe the level curves of f_1. The graph of $f_1(x, y)$ intersects both the x–z and the y–z planes; show that these two curves of intersection are parabolas.

(b) Assume that $a = 4$. Then

$$f_4(x, y) = 4x^2 + y^2$$

and the level curves satisfy

$$4x^2 + y^2 = c$$

Use a graphing calculator to sketch the level curves for $c = 0, 1, 2, 3$, and 4. These curves are ellipses. Find the curves of intersection of $f_4(x, y)$ with the x–z and the y–z planes.

(c) Repeat (b) for $a = 1/4$.

(d) Explain in words how the surfaces of $f_a(x, y)$ change when a changes.

Figure 10.25 Isotherms for a typical lake in the Northern Hemisphere.

24. The graph in Figure 10.25 shows isotherms of a lake in the temperate climate of the Northern Hemisphere.

(a) Use this plot to sketch the temperature profiles in March and June. That is, plot the temperature as a function of depth for a day in March and for a day in June.

(b) Explain how it follows from your temperature plots that the lake is **homeothermic**—that is, has the same temperature from the surface to the bottom—in March.

(c) Explain how it follows from your temperature plots that the lake is **stratified**—that is, has a warm layer on top (called the **epilimnion**), followed by a region where the temperature changes quickly (called the **metalimnion**), followed by a cold layer deeper down (called the **hypolimnion**)—in June.

25. Figure 10.26 shows the oxygen concentration for Long Lake, Clear Water County (Minnesota). The water flea *Daphnia* can survive only if the oxygen concentration is higher than 3 mg/l. Suppose that you wanted to sample the *Daphnia* population in 1997 on days 180, 200, and 220. Below which depths can you be fairly sure not to find any *Daphnia*?

Figure 10.26 Level curves for oxygen concentration on Long Lake, Clear Water County (Minnesota). Courtesy of Leif Hembre.

26. At the beginning of this chapter, we discussed the minimum temperature required for survival as a function of metabolic heat production and whole-body thermal conductance. Suppose that you wish to go winter camping in Northern Minnesota and the predicted low temperature for the night is $-15°$F. Use the information provided at the beginning of the chapter to find the maximum value of g_{Hb} for your sleeping bag that would allow you to sleep safely.

■ 10.2 Limits and Continuity

■ 10.2.1 Informal Definition of Limits

We need to extend the notion of limits and continuity to the multivariable setting. The ideas are the same as in the one-dimensional case. We will discuss only the two-dimensional case, but note that everything in this section can be generalized to higher dimensions.

Let's start with an informal definition of limits. We say that the "limit of $f(x, y)$ as (x, y) approaches (x_0, y_0) is equal to L" if $f(x, y)$ can be made arbitrarily close to L whenever the point (x, y) is sufficiently close (but not equal) to the point (x_0, y_0). We denote this concept by

$$\lim_{(x,y) \to (x_0, y_0)} f(x, y) = L$$

As in the one-dimensional case in Chapter 3, there is a formal definition of limits, which is difficult to use. Fortunately, laws similar to those in the one-dimensional case allow us to compute limits in the two-dimensional case. We thus extend the limit laws from Chapter 3 to the two-dimensional case.

Limit Laws for the Two-Dimensional Case If a is a constant and if

$$\lim_{(x,y)\to(x_0,y_0)} f(x, y) = L_1 \quad \text{and} \quad \lim_{(x,y)\to(x_0,y_0)} g(x, y) = L_2$$

where L_1 and L_2 are real numbers, then the following hold:

1. (Addition Rule)

$$\lim_{(x,y)\to(x_0,y_0)} [f(x, y)+g(x, y)] = \lim_{(x,y)\to(x_0,y_0)} f(x, y)+ \lim_{(x,y)\to(x_0,y_0)} g(x, y)$$

2. (Constant-Factor Rule)

$$\lim_{(x,y)\to(x_0,y_0)} af(x, y) = a \lim_{(x,y)\to(x_0,y_0)} f(x, y)$$

3. (Multiplication Rule)

$$\lim_{(x,y)\to(x_0,y_0)} f(x, y)g(x, y) = \left[\lim_{(x,y)\to(x_0,y_0)} f(x, y) \right] \left[\lim_{(x,y)\to(x_0,y_0)} g(x, y) \right]$$

4. (Quotient Rule)

$$\lim_{(x,y)\to(x_0,y_0)} \frac{f(x, y)}{g(x, y)} = \frac{\lim_{(x,y)\to(x_0,y_0)} f(x, y)}{\lim_{(x,y)\to(x_0,y_0)} g(x, y)}$$

provided that $L_2 \neq 0$.

The next two examples show how to compute limits by using the limit laws.

Limits of Polynomials when the Limits Exist

EXAMPLE 1

(a) $\displaystyle\lim_{(x,y)\to(0,0)} (x^2 + y^2) = \lim_{(x,y)\to(0,0)} x^2 + \lim_{(x,y)\to(0,0)} y^2 = 0^2 + 0^2 = 0$

(b) $\displaystyle\lim_{(x,y)\to(4,-3)} (x^2 + y^2) = \lim_{(x,y)\to(4,-3)} x^2 + \lim_{(x,y)\to(4,-3)} y^2$
$$= 4^2 + (-3)^2 = 25$$

(c) $\displaystyle\lim_{(x,y)\to(-1,2)} x^2 y = \left(\lim_{(x,y)\to(-1,2)} x^2 \right)\left(\lim_{(x,y)\to(-1,2)} y \right) = (-1)^2(2) = 2$

(d) $\displaystyle\lim_{(x,y)\to(1,2)} (x^2 y + 3x) = \left(\lim_{(x,y)\to(1,2)} x^2 y \right) + \left(3 \lim_{(x,y)\to(1,2)} x \right)$

$$= \left(\lim_{(x,y)\to(1,2)} x^2 \right)\left(\lim_{(x,y)\to(1,2)} y \right) + \left(3 \lim_{(x,y)\to(1,2)} x \right)$$

$$= (1)^2(2) + (3)(1) = 5$$ ■

We see from Example 1 that limits of polynomials are calculated by evaluating the functions at the respective points, provided that the functions are defined at those points. This is the case for rational functions as well.

Limits of Rational Functions when the Limits Exist

EXAMPLE 2

(a) $\displaystyle\lim_{(x,y)\to(-1,3)} \frac{3x}{y} = \frac{(3)(-1)}{3} = -1$

(b) $\displaystyle\lim_{(x,y)\to(2,0)} \frac{4y+2x}{x^2+2xy-3} = \frac{(4)(0)+(2)(2)}{(2)^2+(2)(2)(0)-3} = \frac{4}{4-3} = 4$ ■

Limits That Do Not Exist In the one-dimensional case, there were only two ways in which we could approach a number: from the left or from the right. If the two limits were different, we said that the limit did not exist. In two dimensions, there are many more ways that we can approach the point (x_0, y_0), namely, by any curve in the x–y plane that ends up at the point (x_0, y_0). We call such curves **paths**.

> If $f(x, y)$ approaches L_1 as $(x, y) \to (x_0, y_0)$ along path C_1 and $f(x, y)$ approaches L_2 as $(x, y) \to (x_0, y_0)$ along path C_2, and if $L_1 \neq L_2$, then $\lim_{(x,y)\to(x_0,y_0)} f(x, y)$ does not exist.

EXAMPLE 3 Show that

$$\lim_{(x,y)\to(0,0)} \frac{x^2 - y^2}{x^2 + y^2}$$

does not exist.

Solution We first let $(x, y) \to (0, 0)$ along the positive x-axis; this is the curve C_1 in Figure 10.27. On C_1, $y = 0$ and $x > 0$. Then

$$\lim_{x\to 0+} \frac{x^2}{x^2} = 1$$

Next, we let $(x, y) \to (0, 0)$ along the positive y-axis; this is the curve C_2 in Figure 10.27. On C_2, $x = 0$ and $y > 0$. Then

$$\lim_{y\to 0+} \frac{-y^2}{y^2} = -1$$

Since $1 \neq -1$, we conclude that the limit does not exist. ■

Figure 10.27 The paths C_1 and C_2 for Example 3.

Unless we have a lot of experience, it is not easy to find paths for which limits differ. Therefore, in the Problems section, in case a limit does not exist, we will always provide the paths along which you should check the limits.

EXAMPLE 4 Show that

$$\lim_{(x,y)\to(0,0)} \frac{4xy}{xy + y^3}$$

does not exist.

Solution A natural choice for paths to $(0, 0)$ are straight lines of the form $y = mx$. We assume that $m \neq 0$. If we substitute $y = mx$ in the preceding limit, then $(x, y) \to (0, 0)$ reduces to $x \to 0$ and we find that

$$\lim_{x\to 0} \frac{4mx^2}{mx^2 + (mx)^3} = \lim_{x\to 0} \frac{4}{1 + m^2 x} = 4$$

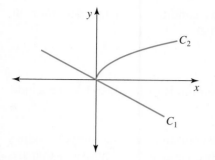

Figure 10.28 The paths C_1 and C_2 for Example 4.

That is, as long as we approach $(0, 0)$ along the straight line $y = mx, m \neq 0$, the limit is always 4, irrespective of m. Such a path, labeled C_1, is shown in Figure 10.28.

You might be tempted to say that the limit exists. But let's approach $(0, 0)$ along the parabola $x = y^2$. This is the curve C_2 in Figure 10.28. Substituting y^2 for x then yields

$$\lim_{y \to 0} \frac{4y^3}{y^3 + y^3} = 2 \neq 4$$

Thus, we have found paths along which the limits differ. Therefore, the limit does not exist. ■

To show that a limit does not exist, we must identify two paths along which the limits differ. To show that a limit exists, we cannot use paths, since the limits along *all* possible paths must be the same and there is no way to check all possible paths. Accordingly, to show that a limit exists, we proceed as in the one-dimensional case: We combine the formal definition of limits and the limit laws to compute limits.

■ 10.2.2 Continuity

The definition of continuity is also analogous to that in the one-dimensional case:

A function $f(x, y)$ is continuous at (x_0, y_0) if the following hold:

1. $f(x, y)$ is defined at (x_0, y_0).

2. $\lim\limits_{(x,y) \to (x_0,y_0)} f(x, y)$ exists.

3. $\lim\limits_{(x,y) \to (x_0,y_0)} f(x, y) = f(x_0, y_0)$.

EXAMPLE 5 Show that

$$f(x, y) = 2 + x^2 + y^2$$

is continuous at $(0, 0)$.

1. $f(x, y)$ is defined at $(0, 0)$; specifically,

$$f(0, 0) = 2$$

2. To show that the limit exists, we refer to Example 1(a), where we claimed that

$$\lim_{(x,y) \to (0,0)} (x^2 + y^2) = 0$$

which indicates that the limit exists. Using the limit laws, we conclude that

$$\lim_{(x,y) \to (0,0)} f(x, y) = 2 + \lim_{(x,y) \to (0,0)} (x^2 + y^2)$$

exists.

3. Using the fact that

$$\lim_{(x,y) \to (0,0)} (x^2 + y^2) = 0$$

we find that

$$\lim_{(x,y) \to (0,0)} (2 + x^2 + y^2) = 2 + \lim_{(x,y) \to (0,0)} (x^2 + y^2) = 2 + 0 = 2$$

Since $f(0, 0) = 2$, $f(x, y)$ is continuous at $(0, 0)$. ■

EXAMPLE 6 Show that

$$f(x, y) = \begin{cases} \frac{x^2 - y^2}{x^2 + y^2} & \text{for } (x, y) \neq (0, 0) \\ 0 & \text{for } (x, y) = (0, 0) \end{cases}$$

is discontinuous at $(0, 0)$.

Solution   The function $f(x, y)$ is defined at $(0,0)$. Hence, condition 1 of the definition of continuity holds. But in Example 3, we showed that

$$\lim_{(x,y)\to(0,0)} f(x, y)$$

does not exist. Therefore, condition 2 of the definition is violated, and we conclude that $f(x, y)$ is discontinuous at $(x, y) = (0, 0)$.   ■

Composition of Functions   Using the definition of continuity and the rules for finding limits, we can show that polynomial functions of two variables (i.e., functions that are sums of terms of the form $ax^n y^m$, where a is a constant and n and m are nonnegative integers) are continuous.

We can obtain a much larger class of continuous functions when we allow the composition of functions. For instance, the function $h(x, y) = e^{x^2 + y^2}$ can be written as a composition of two functions. To see this, set $z = f(x, y) = x^2 + y^2$ and $g(z) = e^z$. Then

$$h(x, y) = (g \circ f)(x, y) = g[f(x, y)] = e^{x^2 + y^2}$$

Another example is $h(x, y) = \sqrt{y^2 - x}$. Here, $z = f(x, y) = y^2 - x$ and $g(z) = \sqrt{z}$. Then $h(x, y) = (g \circ f)(x, y)$. More generally, if

$$f : D \to \mathbf{R}, \quad D \subset \mathbf{R}^2$$

and

$$g : I \to \mathbf{R}, \quad I \subset \mathbf{R}$$

then the composition $(g \circ f)(x, y)$ is defined as

$$h(x, y) = (g \circ f)(x, y) = g[f(x, y)]$$

We can show that if f is continuous at (x_0, y_0) and g is continuous at $z = f(x_0, y_0)$, then

$$h(x, y) = (g \circ f)(x, y) = g[f(x, y)]$$

is continuous at (x_0, y_0). As an example, let's return to the function

$$h(x, y) = e^{x^2 + y^2}$$

With $z = f(x, y) = x^2 + y^2$ and $g(z) = e^z$, $h(x, y) = g[f(x, y)]$ is continuous, since $f(x, y)$ is continuous for all $(x, y) \in \mathbf{R}^2$ and $g(z)$ is continuous for all z in the range of $f(x, y)$. Likewise, $h(x, y) = \sqrt{y^2 - x}$ is continuous for all $(x, y) \in \{(x, y) : y^2 - x \geq 0\}$, since $z = f(x, y) = y^2 - x$ is continuous for all $(x, y) \in \mathbf{R}^2$ and $g(z) = \sqrt{z}$ is continuous for all $z \geq 0$.

■ 10.2.3 Formal Definition of Limits (Optional)

We recall the formal definition of limits in the one-dimensional case. To show that

$$\lim_{x \to 2} x^2 = 4$$

we must show that whenever x is close (but not equal) to 2, then x^2 is close to 4. Formally, this means that we must show that, for every $\epsilon > 0$, there exists a $\delta > 0$ such that

$$|x^2 - 4| < \epsilon \qquad \text{whenever} \qquad 0 < |x - 2| < \delta$$

To generalize this idea to higher dimensions, we need a notion of proximity. In the one-dimensional case, we chose an interval centered at a particular point that contained all points within a distance δ of the center of the interval, except for the center of the interval. In two dimensions, we replace the interval by a disk. A disk of radius δ centered at the point (x_0, y_0) is the set of all points that are within a distance δ of (x_0, y_0).

As with open and closed intervals, there are open and closed disks. We have the following definition:

Definition An **open disk** with radius r centered at $(x_0, y_0) \in \mathbf{R}^2$ is the set

$$B_r(x_0, y_0) = \left\{ (x, y) \in \mathbf{R}^2 : \sqrt{(x - x_0)^2 + (y - y_0)^2} < r \right\}$$

A **closed disk** with radius r centered at $(x_0, y_0) \in \mathbf{R}^2$ is the set

$$\overline{B}_r(x_0, y_0) = \left\{ (x, y) \in \mathbf{R}^2 : \sqrt{(x - x_0)^2 + (y - y_0)^2} \leq r \right\}$$

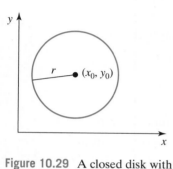

Figure 10.29 A closed disk with radius r centered at (x_0, y_0).

An open disk is thus a disk in which the boundary line is not part of the disk, whereas a closed disk contains the boundary line. This concept is analogous to that of intervals: An open interval does not contain the two endpoints (which are the boundary of the interval), whereas a closed interval does contain them. A closed disk is shown in Figure 10.29.

When we defined $\lim_{x \to x_0} f(x)$, we emphasized that the value of f at x_0 is not important. This idea is expressed in the formal ϵ–δ definition by excluding x_0 from the interval $(x_0 - \delta, x_0 + \delta)$, stated as $0 < |x - x_0| < \delta$. We will need to do the same when we generalize the limit to two dimensions. We write $B_\delta(x_0, y_0) - \{(x_0, y_0)\}$ to denote the open disk with radius δ centered at (x_0, y_0), where the center (x_0, y_0) is removed. We can now generalize the notion of limits to two dimensions:

Definition The limit of $f(x, y)$ as (x, y) approaches (x_0, y_0), denoted by

$$\lim_{(x,y) \to (x_0,y_0)} f(x, y)$$

is the number L such that, for every $\epsilon > 0$, there exists a $\delta > 0$ so that

$$|f(x, y) - L| < \epsilon \qquad \text{whenever} \qquad (x, y) \in B_\delta(x_0, y_0) - \{(x_0, y_0)\}$$

This definition is similar to the one in one dimension: We require that whenever (x, y) is close (but not equal) to (x_0, y_0), it follows that $f(x, y)$ is close to L.

We provide one example in which we use the formal definition of limits. After that, we will give limit laws, which are extensions of the laws in the one-dimensional case. Using the formal definition is difficult, and we will not need to do so subsequently. But seeing an example might help you to understand the definition.

EXAMPLE 7   Let

$$f(x, y) = x^2 + y^2$$

Show that

$$\lim_{(x,y) \to (0,0)} f(x, y) = 0$$

Solution   We must show that, for every $\epsilon > 0$, there exists a $\delta > 0$ such that

$$|x^2 + y^2 - 0| < \epsilon \qquad \text{whenever} \qquad (x, y) \in B_\delta(0, 0) - \{(0, 0)\}$$

Now, $(x, y) \in B_\delta(0, 0)$ if $\sqrt{x^2 + y^2} < \delta$, or $x^2 + y^2 < \delta^2$. We need to show $|x^2 + y^2| < \epsilon$. This suggests that we should choose δ so that $\delta^2 = \epsilon$. Let's try this. We set $\delta = \sqrt{\epsilon}$ for $\epsilon > 0$. Then

$$\sqrt{x^2 + y^2} < \delta = \sqrt{\epsilon}$$

implies that

$$|x^2 + y^2| < \epsilon$$

But this is exactly what we need to show. In other words, we have shown that, for every $\epsilon > 0$, we can find a $\delta > 0$ (namely $\delta = \sqrt{\epsilon}$) such that whenever (x, y) is close to $(0, 0)$, it follows that $x^2 + y^2$ is close to 0. ■

Section 10.2 Problems

■ 10.2.1

In Problems 1–14, use the properties of limits to calculate the following limits:

1. $\lim_{(x,y)\to(1,0)}(x^2 - 3y^2)$

2. $\lim_{(x,y)\to(-1,1)}(2xy + 3x^2)$

3. $\lim_{(x,y)\to(2,-1)}(x^2y^3 - 3xy)$

4. $\lim_{(x,y)\to(1,-2)}(2x^3 - 3y)(xy - 2)$

5. $\lim_{(x,y)\to(-1,3)}x^2(y^2 - 3xy)$

6. $\lim_{(x,y)\to(-5,1)}y(xy + x^2y^2)$

7. $\lim_{(x,y)\to(0,2)}\left(4xy^2 - \frac{x+1}{y}\right)$ **8.** $\lim_{(x,y)\to(1,1)}\frac{xy}{x^2+y^2}$

9. $\lim_{(x,y)\to(1,0)}\frac{x^2+y^2}{x^2-y^2}$ **10.** $\lim_{(x,y)\to(-1,3)}\frac{x^2-xy}{2x+y}$

11. $\lim_{(x,y)\to(0,1)}\frac{2xy-3}{x^2+y^2+1}$ **12.** $\lim_{(x,y)\to(-1,-2)}\frac{x^2-y^2}{2xy+2}$

13. $\lim_{(x,y)\to(2,0)}\frac{2x+4y^2}{y^2+3x}$ **14.** $\lim_{(x,y)\to(1,-2)}\frac{2x^2+y}{2xy+3}$

15. Show that

$$\lim_{(x,y)\to(0,0)}\frac{x^2 - 2y^2}{x^2 + y^2}$$

does not exist by computing the limit along the positive x-axis and the positive y-axis.

16. Show that

$$\lim_{(x,y)\to(0,0)}\frac{3x^2 - y^2}{x^2 + y^2}$$

does not exist by computing the limit along the positive x-axis and the positive y-axis.

17. Compute

$$\lim_{(x,y)\to(0,0)}\frac{4xy}{x^2 + y^2}$$

along the x-axis, the y-axis, and the line $y = x$. What can you conclude?

18. Compute

$$\lim_{(x,y)\to(0,0)}\frac{3xy}{x^2 + y^3}$$

along lines of the form $y = mx$, for $m \neq 0$. What can you conclude?

19. Compute

$$\lim_{(x,y)\to(0,0)}\frac{2xy}{x^3 + yx}$$

along lines of the form $y = mx$, for $m \neq 0$, and along the parabola $y = x^2$. What can you conclude?

20. Compute

$$\lim_{(x,y)\to(0,0)}\frac{3x^2y^2}{x^3 + y^6}$$

along lines of the form $y = mx$, for $m \neq 0$, and along the parabola $x = y^2$. What can you conclude?

■ 10.2.2

21. Use the definition of continuity to show that

$$f(x, y) = x^2 + y^2$$

is continuous at $(0, 0)$.

22. Use the definition of continuity to show that

$$f(x, y) = \sqrt{9 + x^2 + y^2}$$

is continuous at $(0, 0)$.

23. Show that

$$f(x, y) = \begin{cases} \frac{4xy}{x^2+y^2} & \text{for } (x, y) \neq (0, 0) \\ 0 & \text{for } (x, y) = (0, 0) \end{cases}$$

is discontinuous at $(0, 0)$. (*Hint*: Use Problem 17.)

24. Show that

$$f(x, y) = \begin{cases} \frac{3xy}{x^2+y^3} & \text{for } (x, y) \neq (0, 0) \\ 0 & \text{for } (x, y) = (0, 0) \end{cases}$$

is discontinuous at $(0, 0)$. (*Hint*: Use Problem 18.)

25. Show that

$$f(x, y) = \begin{cases} \frac{2xy}{x^3+yx} & \text{for } (x, y) \neq (0, 0) \\ 0 & \text{for } (x, y) = (0, 0) \end{cases}$$

is discontinuous at $(0, 0)$. (*Hint*: Use Problem 19.)

26. Show that

$$f(x, y) = \begin{cases} \frac{3x^2y^2}{x^3+y^6} & \text{for } (x, y) \neq (0, 0) \\ 0 & \text{for } (x, y) = (0, 0) \end{cases}$$

is discontinuous at $(0, 0)$. (*Hint*: Use Problem 20.)

27. (a) Write

$$h(x, y) = \sin(x^2 + y^2)$$

as a composition of two functions.

(b) For which values of (x, y) is $h(x, y)$ continuous?

28. (a) Write

$$h(x, y) = \sqrt{x + y}$$

as a composition of two functions.

(b) For which values of (x, y) is $h(x, y)$ continuous?

29. (a) Write

$$h(x, y) = e^{xy}$$

as a composition of two functions.

(b) For which values of (x, y) is $h(x, y)$ continuous?

30. (a) Write

$$h(x, y) = \cos(y - x)$$

as a composition of two functions.

(b) For which values of (x, y) is $h(x, y)$ continuous?

■ 10.2.3

31. Draw an open disk with radius 2 centered at $(1, -1)$ in the x–y plane, and give a mathematical description of this set.

32. Draw a closed disk with radius 3 centered at $(2, 0)$ in the x–y plane, and give a mathematical description of this set.

33. Give a geometric interpretation of the set

$$A = \left\{(x, y) \in \mathbf{R}^2 : \sqrt{x^2 + y^2 - 4y + 4} < 3\right\}$$

34. Give a geometric interpretation of the set

$$A = \left\{(x, y) \in \mathbf{R}^2 : \sqrt{x^2 + 6x + y^2 - 2y + 10} < 2\right\}$$

35. Let

$$f(x, y) = 2x^2 + y^2$$

Use the ϵ–δ definition of limits to show that

$$\lim_{(x,y)\to(0,0)} f(x, y) = 0$$

36. Let

$$f(x, y) = x^2 + 3y^2$$

Use the ϵ–δ definition of limits to show that

$$\lim_{(x,y)\to(0,0)} f(x, y) = 0$$

■ 10.3 Partial Derivatives

■ 10.3.1 Functions of Two Variables

Suppose that the response of an organism depends on a number of independent variables. To investigate this dependency, a common experimental design is to measure the response when one variable is changed while all other variables are kept fixed. As an example, Pisek et al. (1969) measured the net assimiliation of CO_2 of *Ranunculus glacialis*, a member of the buttercup family, as a function of temperature and light intensity. They varied the temperature while keeping the light intensity constant. Repeating this experiment at different light intensities, they were able to determine how the net assimiliation of CO_2 changes as a function of both temperature and light intensity.

This experimental design illustrates the idea behind partial derivatives. Suppose that we want to know how the function $f(x, y)$ changes when x and y change. Instead of changing both variables simultaneously, we might get an idea of how $f(x, y)$ depends on x and y when we change one variable while keeping the other variable fixed.

To illustrate, we look at

$$f(x, y) = x^2 y$$

We want to know how $f(x, y)$ changes if we change, say, x and keep y fixed. So we fix $y = y_0$. Then the change in f with respect to x is simply the derivative of f with respect to x when $y = y_0$. That is,

$$\frac{d}{dx} f(x, y_0) = \frac{d}{dx} x^2 y_0 = 2x y_0$$

Such a derivative is called a *partial derivative*.

Definition Suppose that f is a function of two independent variables x and y. The **partial derivative** of f with respect to x is defined by

$$\frac{\partial f(x, y)}{\partial x} = \lim_{h \to 0} \frac{f(x + h, y) - f(x, y)}{h}$$

The partial derivative of f with respect to y is defined by

$$\frac{\partial f(x, y)}{\partial y} = \lim_{h \to 0} \frac{f(x, y + h) - f(x, y)}{h}$$

To denote partial derivatives, we use "∂" instead of "d." We will also use the notation

$$f_x(x, y) = \frac{\partial f(x, y)}{\partial x} \quad \text{and} \quad f_y(x, y) = \frac{\partial f(x, y)}{\partial y}$$

In the definition of partial derivatives, you should recognize the formal definition of derivatives of Chapter 4. That is, to compute $\partial f / \partial x$, we look at the ratio of the difference in the f-values, $f(x + h, y) - f(x, y)$, and the difference in the x-values, $x + h - x$. The other variable, y, is not changed. We then let h tend to 0.

To compute $\partial f(x, y) / \partial x$, we differentiate f with respect to x while treating y as a constant. When we read $\partial f(x, y) / \partial x$, we can say "the partial derivative of f of x and y with respect to x." To read $f_x(x, y)$, we say "f sub x of x and y."

Finding partial derivatives is no different from finding derivatives of functions of one variable, since, by keeping all but one variable fixed, computing a partial derivative is reduced to computing a derivative of a function of one variable. You just need to keep straight which of the variables you have fixed and which one you will vary.

EXAMPLE 1

Find $\partial f / \partial x$ and $\partial f / \partial y$ when

$$f(x, y) = ye^{xy}$$

Solution

To compute $\partial f / \partial x$, we treat y as a constant and use the chain rule to differentiate f with respect to x:

$$\frac{\partial f(x, y)}{\partial x} = \frac{\partial}{\partial x}(ye^{xy}) = ye^{xy}y = y^2 e^{xy}$$

To compute $\partial f / \partial y$, we treat x as a constant and use the product rule combined with the chain rule to differentiate f with respect to y:

$$\frac{\partial f(x, y)}{\partial y} = \frac{\partial}{\partial y}(ye^{xy})$$
$$= 1 \cdot e^{xy} + ye^{xy}x$$
$$= e^{xy}(1 + xy)$$

EXAMPLE 2

Find $\partial f / \partial x$ when

$$f(x, y) = \frac{\sin(xy)}{x^2 + \cos y}$$

Solution

We treat y as a constant and use the quotient rule:

$$u = \sin(xy) \qquad\qquad v = x^2 + \cos y$$
$$\frac{\partial u}{\partial x} = y\cos(xy) \qquad\qquad \frac{\partial v}{\partial x} = 2x$$

Hence,

$$\frac{\partial f(x, y)}{\partial x} = \frac{y(x^2 + \cos y)\cos(xy) - 2x\sin(xy)}{(x^2 + \cos y)^2}$$

Geometric Interpretation As with ordinary derivatives, partial derivatives are slopes of lines that are tangent to certain curves. We can find these curves on the surface $z = f(x, y)$.

Let's start with the interpretation of $\partial f / \partial x$. We fix $y = y_0$; then $f(x, y_0)$ as a function of x is obtained by intersecting the surface $z = f(x, y)$ with a vertical plane that is parallel to the x–z plane and goes through $y = y_0$. The curve of intersection is the graph of $z = f(x, y_0)$, as illustrated in Figure 10.30.

We can now project this curve onto the x–z plane, as illustrated in Figure 10.31. The curve is the graph of a function that depends only on x. Consequently, we can find the slope of the tangent line at any point $P(x_0, z_0)$, where $z_0 = f(x_0, y_0)$. The slope of the tangent line is given by the derivative of the function f in the x-direction—that is, by $\partial f / \partial x$. This is the geometric meaning of $\partial f / \partial x$. We summarize it as follows:

> The partial derivative $\partial f / \partial x$ evaluated at (x_0, y_0) is the slope of the tangent line to the curve $z = f(x, y_0)$ at the point (x_0, y_0, z_0), with $z_0 = f(x_0, y_0)$.

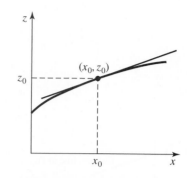

Figure 10.30 The surface of $f(x, y)$ intersected with the plane $y = y_0$.

Figure 10.31 The projection of the curve $z = f(x, y_0)$ onto the x–z plane.

The derivative $\partial f / \partial y$ has a similar meaning. This time, we fix $x = x_0$ and intersect the surface $z = f(x, y)$ with a vertical plane that is parallel to the y–z plane and goes through $x = x_0$. The curve of intersection is the graph of $z = f(x_0, y)$, as illustrated in Figure 10.32.

Figure 10.32 The surface of $f(x, y)$ intersected with the plane $x = x_0$.

Figure 10.33 The projection of the curve $z = f(x_0, y)$ onto the y–z plane.

The projection of this curve onto the y–z plane, together with the tangent line at (y_0, z_0), is illustrated in Figure 10.33. Analogous to the interpretation of $\partial f / \partial x$, we then have the following geometric meaning of $\partial f / \partial y$:

> The partial derivative $\partial f / \partial y$ evaluated at (x_0, y_0) is the slope of the tangent line to the curve $z = f(x_0, y)$ at the point (x_0, y_0, z_0), with $z_0 = f(x_0, y_0)$.

EXAMPLE 3   Let

$$f(x, y) = 3 - x^3 - y^2$$

Find $f_x(1, 1)$ and $f_y(1, 1)$, and interpret your results geometrically.

Solution   We have

$$f_x(x, y) = -3x^2 \qquad \text{and} \qquad f_y(x, y) = -2y$$

Hence,

$$f_x(1, 1) = -3 \qquad \text{and} \qquad f_y(1, 1) = -2$$

To interpret $f_x(1, 1)$ geometrically, we fix $y = 1$. The vertical plane $y = 1$ intersects the graph of $f(x, y)$. The curve of intersection has slope -3 when $x = 1$. The projection of the curve of intersection is shown in Figure 10.34.

Similarly, to interpret $f_y(1, 1)$, we intersect the graph of $f(x, y)$ with the vertical plane $x = 1$. The tangent line at the curve of intersection has slope -2 when $y = 1$. The projection of the curve of intersection is shown in Figure 10.35.   ■

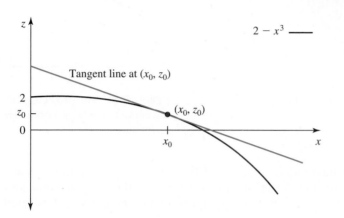

Figure 10.34 The curve of intersection of the graph $z = f(x, y)$ with the plane $y = y_0$. Drawn is the curve in the plane of intersection, together with its tangent line at (x_0, z_0).

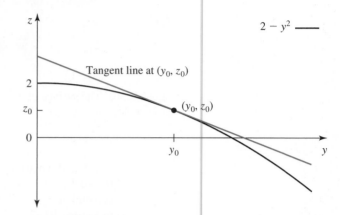

Figure 10.35 The curve of intersection of the graph $z = f(x, y)$ with the plane $x = x_0$. Drawn is the curve in the plane of intersection, together with its tangent line at (y_0, z_0).

A Biological Application

EXAMPLE 4

Holling (1959) derived an expression for the number of prey items P_e eaten by a predator during an interval T as a function of prey density N and the handling time T_h of each prey item:

$$P_e = \frac{aNT}{1 + aT_hN} \tag{10.1}$$

Here, a is a positive constant called the *predator attack rate*. Equation (10.1) is called *Holling's disk equation*. [Holling came up with the equation when he measured how many sandpaper disks (representing prey) a blindfolded assistant (representing the predator) could pick up during a certain interval.] We can consider P_e as a function of N and T_h.

To determine how handling time influences the number of prey eaten, we compute

$$\frac{\partial P_e(N, T_h)}{\partial T_h} = aNT(-1)(1 + aT_hN)^{-2}aN$$

$$= -\frac{a^2N^2T}{(1 + aT_hN)^2} < 0$$

since $a^2N^2T > 0$ and $(1 + aT_hN)^2 > 0$. Because $\partial P_e/\partial T_h$ is negative, the number of prey items eaten decreases with increasing handling time, as expected.

To determine how P_e changes with N, we compute

$$\frac{\partial P_e(N, T_h)}{\partial N} = \frac{aT(1 + aT_hN) - aNTaT_h}{(1 + aT_hN)^2}$$

$$= \frac{aT}{(1 + aT_hN)^2} > 0$$

since $aT > 0$ and $(1 + aT_hN)^2 > 0$. Because $\partial P_e/\partial N$ is positive, the number of prey items eaten increases with increasing prey density, again as expected.   ■

■ 10.3.2 Functions of More Than Two Variables

The definition of partial derivatives extends in a straightforward way to functions of more than two variables. These are ordinary derivatives with respect to one variable while all other variables are treated as constants.

EXAMPLE 5

Let f be a function of three independent variables x, y, and z:

$$f(x, y, z) = e^{yz}(x^2 + z^3)$$

Find $\partial f/\partial x$, $\partial f/\partial y$, and $\partial f/\partial z$.

Solution

To compute $\partial f/\partial x$, we treat y and z as constants and differentiate f with respect to x.

$$\frac{\partial f}{\partial x} = e^{yz}2x$$

Likewise,

$$\frac{\partial f}{\partial y} = ze^{yz}(x^2 + z^3)$$

$$\frac{\partial f}{\partial z} = ye^{yz}(x^2 + z^3) + e^{yz}3z^2$$

■

■ 10.3.3 Higher-Order Partial Derivatives

As in the case of functions of one variable, we can define higher-order partial derivatives for functions of more than one variable. For instance, to find the second partial derivative of $f(x, y)$ with respect to x, denoted by $\partial^2 f/\partial x^2$, we compute

$$\frac{\partial^2 f}{\partial x^2} = \frac{\partial}{\partial x}\left(\frac{\partial f}{\partial x}\right)$$

We can write $\partial^2 f/\partial x^2$ as f_{xx}.

We can also compute **mixed derivatives**. For instance,

$$f_{yx} = \frac{\partial^2 f}{\partial x \partial y} = \frac{\partial}{\partial x}\left(\frac{\partial f}{\partial y}\right)$$

Note the order of yx in the subscript of f and the order of $\partial x \, \partial y$ in the denominator: Either notation means that we differentiate with respect to y first.

EXAMPLE 6

Set

$$f(x, y) = \sin x + xe^y$$

Find $f_{xx} = \dfrac{\partial^2 f}{\partial x^2}$, $f_{yx} = \dfrac{\partial^2 f}{\partial x \partial y}$, and $f_{xy} = \dfrac{\partial^2 f}{\partial y \partial x}$.

Solution

$$f_{xx} = \frac{\partial^2 f}{\partial x^2} = \frac{\partial}{\partial x}\left(\frac{\partial f}{\partial x}\right) = \frac{\partial}{\partial x}\left[\frac{\partial}{\partial x}(\sin x + xe^y)\right] = \frac{\partial}{\partial x}[\cos x + e^y] = -\sin x$$

$$f_{yx} = \frac{\partial^2 f}{\partial x \partial y} = \frac{\partial}{\partial x}\left(\frac{\partial f}{\partial y}\right) = \frac{\partial}{\partial x}\left[\frac{\partial}{\partial y}(\sin x + xe^y)\right] = \frac{\partial}{\partial x}[0 + xe^y] = e^y$$

$$f_{xy} = \frac{\partial^2 f}{\partial y \partial x} = \frac{\partial}{\partial y}\frac{\partial f}{\partial x} = \frac{\partial}{\partial y}\left[\frac{\partial}{\partial x}(\sin x + xe^y)\right] = \frac{\partial}{\partial y}[\cos x + e^y] = e^y$$

■

In the preceding example, $f_{xy} = f_{yx}$, implying that the order of differentiation did not matter. Although this is not always the case, there are conditions under which the order of differentiation in mixed partial derivatives does not matter. To state this theorem, we need the notion of an open and a closed disk. We have the following definition:

Definition An **open disk** with radius r centered at $(x_0, y_0) \in \mathbf{R}^2$ is the set

$$B_r(x_0, y_0) = \left\{ (x, y) \in \mathbf{R}^2 : \sqrt{(x - x_0)^2 + (y - y_0)^2} < r \right\}$$

A **closed disk** with radius r centered at $(x_0, y_0) \in \mathbf{R}^2$ is the set

$$\overline{B}_r(x_0, y_0) = \left\{ (x, y) \in \mathbf{R}^2 : \sqrt{(x - x_0)^2 + (y - y_0)^2} \leq r \right\}$$

An open disk is thus a disk in which the boundary line is not part of the disk, whereas a closed disk contains the boundary line. This concept is analogous to that of intervals: An open interval does not contain the two endpoints (which are the boundary of the interval), whereas a closed interval does contain them.

We can now state the mixed-derivative theorem:

The Mixed-Derivative Theorem If $f(x, y)$ and its partial derivatives f_x, f_y, f_{xy}, and f_{yx} are continuous on an open disk centered at the point (x_0, y_0), then

$$f_{xy}(x_0, y_0) = f_{yx}(x_0, y_0)$$

It is straightforward to define partial derivatives of even higher order. For instance, if f is a function of two independent variables x and y, then

$$\frac{\partial^3 f}{\partial x^2 \, \partial y} = \frac{\partial}{\partial x} \frac{\partial^2 f}{\partial x \, \partial y} = \frac{\partial}{\partial x} \frac{\partial}{\partial x} \frac{\partial f}{\partial y}$$

We illustrate this definition with

$$f(x, y) = y^2 \sin x$$

for which

$$\frac{\partial^3 f}{\partial x^2 \, \partial y} = \frac{\partial}{\partial x} \frac{\partial}{\partial x} \frac{\partial}{\partial y} (y^2 \sin x)$$

$$= \frac{\partial}{\partial x} \frac{\partial}{\partial x} (2y \sin x)$$

$$= \frac{\partial}{\partial x} (2y \cos x) = -2y \sin x$$

and

$$\frac{\partial^3 f}{\partial x \, \partial y^2} = \frac{\partial}{\partial x} \frac{\partial}{\partial y} \frac{\partial}{\partial y} (y^2 \sin x)$$

$$= \frac{\partial}{\partial x} \frac{\partial}{\partial y} (2y \sin x) = \frac{\partial}{\partial x} (2 \sin x) = 2 \cos x$$

and so on.

The mixed-derivative theorem can be extended to higher-order derivatives. The order of differentiation does not matter, as long as the function and all of its derivatives through the order in question are continuous on an open disk centered at the point at which we want to compute the derivative. For instance,

$$\frac{\partial^3 f}{\partial y^2 \, \partial x} = \frac{\partial}{\partial y} \frac{\partial}{\partial y} \frac{\partial}{\partial x} (y^2 \sin x)$$

$$= \frac{\partial}{\partial y} \frac{\partial}{\partial y} (y^2 \cos x)$$

$$= \frac{\partial}{\partial y} (2y \cos x) = 2 \cos x$$

which is the same as $\partial^3 f / (\partial x \, \partial y^2)$.

Section 10.3 Problems

■ 10.3.1

In Problems 1–16, find $\partial f / \partial x$ and $\partial f / \partial y$ for the given functions.

1. $f(x, y) = x^2 y + xy^2$

2. $f(x, y) = 2x\sqrt{y} - \frac{3}{xy^2}$

3. $f(x, y) = (xy)^{3/2} - (xy)^{2/3}$

4. $f(x, y) = \frac{y^4}{x^3} - \frac{1}{x^3 y^4}$

5. $f(x, y) = \sin(x + y)$

6. $f(x, y) = \tan(x - 2y)$

7. $f(x, y) = \cos^2(x^2 - 2y)$

8. $f(x, y) = \sec(y^2 x - x^3)$

9. $f(x, y) = e^{\sqrt{x+y}}$

10. $f(x, y) = x^2 e^{-xy/2}$

11. $f(x, y) = e^x \sin(xy)$

12. $f(x, y) = e^{-y^2} \cos(x^2 - y^2)$

13. $f(x, y) = \ln(2x + y)$

14. $f(x, y) = \ln(3x^2 - xy)$

15. $f(x, y) = \log_3(y^2 - x^2)$

16. $f(x, y) = \log_5(3xy)$

In Problems 17–24, find the indicated partial derivatives.

17. $f(x, y) = 3x^2 - y - 2y^2; \ f_x(1, 0)$

18. $f(x, y) = x^{1/3} y - xy^{1/3}; \ f_y(1, 1)$

19. $g(x, y) = e^{x+3y}; \ g_y(2, 1)$

20. $h(u, v) = e^u \sin(u + v); \ h_u(1, -1)$

21. $f(x, z) = \ln(xz); \ f_z(e, 1)$

22. $g(v, w) = \frac{w^2}{v+w}; \ g_v(1, 1)$

23. $f(x, y) = \frac{xy}{x^2+2}; \ f_x(-1, 2)$

24. $f(u, v) = e^{u^2/2} \ln(u + v); \ f_u(2, 1)$

25. Let

$$f(x, y) = 4 - x^2 - y^2$$

Compute $f_x(1, 1)$ and $f_y(1, 1)$, and interpret these partial derivatives geometrically.

26. Let

$$f(x, y) = \sqrt{4 - x^2 - y^2}$$

Compute $f_x(1, 1)$ and $f_y(1, 1)$, and interpret these partial derivatives geometrically.

27. Let

$$f(x, y) = 1 + x^2 y$$

Compute $f_x(-2, 1)$ and $f_y(-2, 1)$, and interpret these partial derivatives geometrically.

28. Let

$$f(x, y) = 2x^3 - 3yx$$

Compute $f_x(1, 2)$ and $f_y(1, 2)$, and interpret these partial derivatives geometrically.

29. In Example 4, we investigated Holling's disk equation

$$P_e = \frac{aNT}{1 + aT_h N}$$

(See Example 4 for the meaning of this equation.) We will now consider P_e as a function of the predator attack rate a and the length T of the interval during which the predator searches for food.

(a) Determine how the predator attack rate a influences the number of prey eaten per predator.

(b) Determine how the length T of the interval influences the number of prey eaten per predator.

30. Suppose that the per capita growth rate of some prey at time t depends on both the prey density $H(t)$ at t and the predator density $P(t)$ at t. Assume the relationship

$$\frac{1}{H} \frac{dH}{dt} = r \left(1 - \frac{H}{K}\right) - aP \qquad (10.2)$$

where r, K, and a are positive constants. The right-hand side of (10.2) is a function of both prey density and predator density. Investigate how an increase in (a) prey density and (b) predator density affects the per capita growth rate of this prey species.

■ 10.3.2

In Problems 31–38, find $\partial f / \partial x$, $\partial f / \partial y$, and $\partial f / \partial z$ for the given functions.

31. $f(x, y, z) = x^2 z + yz^2 - xy$

32. $f(x, y, z) = xyz$

33. $f(x, y, z) = x^3 y^2 z + \frac{x}{yz}$

34. $f(x, y, z) = \frac{xyz}{x^2+y^2+z^2}$

35. $f(x, y, z) = e^{x+y+z}$

36. $f(x, y, z) = e^{yz} \sin x$

37. $f(x, y, z) = \ln(x + y + z)$

38. $f(x, y, z) = y \tan(x^2 + z)$

■ 10.3.3

In Problems 39–48, find the indicated partial derivatives.

39. $f(x, y) = x^2 y + xy^2; \ \frac{\partial^2 f}{\partial x^2}$

40. $f(x, y) = y^2(x - 3y); \ \frac{\partial^2 f}{\partial y^2}$

41. $f(x, y) = xe^y; \ \frac{\partial^2 f}{\partial x \partial y}$

42. $f(x, y) = \sin(x - y); \ \frac{\partial^2 f}{\partial y \partial x}$

43. $f(u, w) = \tan(u + w); \ \frac{\partial^2 f}{\partial u^2}$

44. $g(s, t) = \ln(s^2 + 3st); \ \frac{\partial^2 g}{\partial t^2}$

45. $f(x, y) = x^3 \cos y; \ \frac{\partial^3 f}{\partial x^2 \partial y}$

46. $f(x, y) = e^{x^2 - y}; \ \frac{\partial^3 f}{\partial y^2 \partial x}$

47. $f(x, y) = \ln(x + y); \ \frac{\partial^3 f}{\partial x^3}$

48. $f(x, y) = \sin(3xy); \ \frac{\partial^3 f}{\partial y^2 \partial x}$

49. The functional responses of some predators are sigmoidal; that is, the number of prey attacked per predator as a function of prey density has a sigmoidal shape. If we denote the prey density by N, the predator density by P, the time available for searching for prey by T, and the handling time of each prey item per predator by T_h, then the number of prey encounters per predator as a function of N, T, and T_h can be expressed as

$$f(N, T, T_h) = \frac{b^2 N^2 T}{1 + cN + bT_h N^2}$$

where b and c are positive constants.

(a) Investigate how an increase in the prey density N affects the function $f(N, T, T_h)$.

(b) Investigate how an increase in the time T available for search affects the function $f(N, T, T_h)$.

(c) Investigate how an increase in the handling time T_h affects the function $f(N, T, T_h)$.

(d) Graph $f(N, T, T_h)$ as a function of N when $T = 2.4$ hours, $T_h = 0.2$ hours, $b = 0.8$, and $c = 0.5$.

50. In this problem, we will investigate how mutual interference of parasitoids affects their searching efficiency for a host. We assume that N is the host density and P is the parasitoid density. A frequently used model for host–parasitoid interactions is the **Nicholson–Bailey model** (Nicholson, 1933; Nicholson and Bailey, 1935), in which it is assumed that the number of parasitized hosts, denoted by N_a, is given by

$$N_a = N[1 - e^{-bP}] \tag{10.3}$$

where b is the searching efficiency.

(a) Show that

$$b = \frac{1}{P} \ln \frac{N}{N - N_a}$$

by solving (10.3) for b.

(b) Consider

$$b = f(P, N, N_a) = \frac{1}{P} \ln \frac{N}{N - N_a}$$

as a function of P, N, and N_a. How is the searching efficiency b affected when the parasitoid density increases?

(c) Assume now that the fraction of parasitized host depends on the host density; that is, assume that

$$N_a = g(N)$$

where $g(N)$ is a nonnegative, differentiable function. The searching efficiency b can then be written as follows as a function of P and N:

$$b = h(P, N) = \frac{1}{P} \ln \frac{N}{N - g(N)}$$

How does the searching efficiency depend on host density when $g(N)$ is a decreasing function of N? (Use the fact that $g(N) < N$.)

51. Leopold and Kriedemann (1975) measured the crop growth rate of sunflowers as a function of leaf area index and percent of full sunlight. (Leaf area index is the ratio of leaf surface area to the ground area the plant covers.) They found that, for a fixed level of sunlight, crop growth rate first increases and then decreases as a function of leaf area index. For a given leaf area index, the crop growth rate increases with the level of sunlight. The leaf area index that maximizes the crop growth rate is an increasing function of sunlight. Sketch the crop growth rate as a function of leaf area index for different values of percent of full sunlight.

■ 10.4 Tangent Planes, Differentiability, and Linearization

■ 10.4.1 Functions of Two Variables

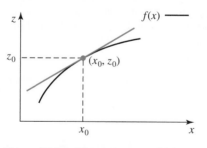

Figure 10.36 The curve $z = f(x)$ and its tangent line at the point (x_0, z_0).

Tangent Planes Suppose that $z = f(x)$ is differentiable at $x = x_0$. Then the equation of the tangent line of $z = f(x)$ at (x_0, z_0) with $z_0 = f(x_0)$ is given by

$$z - z_0 = f'(x_0)(x - x_0) \tag{10.4}$$

The curve z and the tangent line are illustrated in Figure 10.36.

We now generalize this situation to functions of two variables. The analogue of a tangent line is called a **tangent plane**, an example of which is shown in Figure 10.37. Let $z = f(x, y)$ be a function of two variables. We saw in the previous section that the partial derivatives $\partial f/\partial x$ and $\partial f/\partial y$, evaluated at (x_0, y_0), are the slopes of tangent lines at the point (x_0, y_0, z_0), with $z_0 = f(x_0, y_0)$, to certain curves through (x_0, y_0, z_0) on the surface $z = f(x, y)$. These two tangent lines, one in the x-direction, the other in the y-direction, define a unique plane. If, in addition, $f(x, y)$ has partial derivatives that are continuous on an open disk containing (x_0, y_0), then we can show that the tangent line at (x_0, y_0, z_0) to any other smooth curve on the surface $z = f(x, y)$ through (x_0, y_0, z_0) is contained in this plane. The plane is then called the *tangent plane*.

We will use the two original tangent lines to find the equation of the tangent plane at a point (x_0, y_0, z_0) on the surface $z = f(x, y)$. We take the curve that is obtained as the intersection of the surface $z = f(x, y)$ with the plane that is parallel to the y–z plane and contains the point (x_0, y_0, z_0)—that is, the plane $x = x_0$—and we denote this curve by C_1. Its tangent line at (x_0, y_0, z_0) is contained in the tangent plane. (See Figure 10.37.) Likewise, we take the curve of intersection between $z = f(x, y)$ and

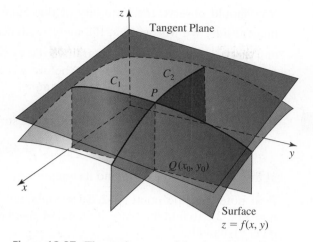

Figure 10.37 The surface $z = f(x, y)$ and its tangent plane at $P = (x_0, y_0, z_0)$.

the plane that is parallel to the x–z plane and contains the point (x_0, y_0, z_0)—that is, the plane $y = y_0$—and we denote this curve by C_2. Its tangent line, too, is contained in the tangent plane. (See Figure 10.37.)

In Section 9.4, we gave the general equation of a plane:

$$z - z_0 = A(x - x_0) + B(y - y_0) \qquad (10.5)$$

We will use curves C_1 and C_2 to determine the constants A and B. C_1 satisfies the equation

$$z = f(x_0, y)$$

The tangent line to C_1 at (x_0, y_0, z_0) therefore satisfies

$$z - z_0 = \frac{\partial f(x_0, y_0)}{\partial y}(y - y_0) \qquad (10.6)$$

The tangent line at (x_0, y_0, z_0) is contained in the tangent plane at (x_0, y_0, z_0). The equation of the tangent line at (x_0, y_0, z_0) can therefore also be obtained by setting $x = x_0$ in (10.5). Doing so yields

$$z - z_0 = (A)(0) + B(y - y_0) = B(y - y_0)$$

Comparing this with (10.6), we find that

$$B = \frac{\partial f(x_0, y_0)}{\partial y}$$

Similarly, using C_2, we find that

$$A = \frac{\partial f(x_0, y_0)}{\partial x}$$

We thus arrive at the following result:

If the tangent plane to the surface $z = f(x, y)$ at the point (x_0, y_0, z_0) exists, then that tangent plane has the equation

$$z - z_0 = \frac{\partial f(x_0, y_0)}{\partial x}(x - x_0) + \frac{\partial f(x_0, y_0)}{\partial y}(y - y_0)$$

You should observe the similarity of this equation to (10.4), the equation of the tangent line. We will see that the mere existence of the partial derivatives $\frac{\partial f(x_0, y_0)}{\partial x}$ and $\frac{\partial f(x_0, y_0)}{\partial y}$ is *not* enough to guarantee the existence of a tangent plane at (x_0, y_0); something stronger is needed. But before we discuss the conditions under which tangent planes exist, let's look at an example.

EXAMPLE 1

It can be shown that the tangent plane to the surface

$$z = f(x, y) = 4x^2 + y^2$$

at the point $(1, 2, 8)$ exists. Find its equation.

Solution

First note that the point $(1, 2, 8)$ is contained in the surface $z = f(x, y)$, since $8 = f(1, 2)$. To find the tangent plane, we need to compute the partial derivatives of $f(x, y)$, namely,

$$\frac{\partial f}{\partial x} = 8x \quad \text{and} \quad \frac{\partial f}{\partial y} = 2y$$

and evaluate them at $(x_0, y_0) = (1, 2)$:

$$\frac{\partial f(1, 2)}{\partial x} = 8 \quad \text{and} \quad \frac{\partial f(1, 2)}{\partial y} = 4$$

Hence, the equation of the tangent plane is

$$z - 8 = 8(x - 1) + 4(y - 2)$$

We can write this equation in the more compact form

$$z - 8 = 8x - 8 + 4y - 8$$

or

$$8x + 4y - z = 8 \qquad\blacksquare$$

Differentiability In discussing the conditions under which tangent planes exist, we need to define what it means for a function of two variables to be *differentiable*. To make the connection with functions of one variable clear, recall that the tangent line is used to linearly approximate $f(x)$ at $x = x_0$. The linear approximation of $f(x)$ at $x = x_0$ is given by

$$L(x) = f(x_0) + f'(x_0)(x - x_0) \tag{10.7}$$

The distance between $f(x)$ and its linear approximation at $x = x_0$ is then

$$|f(x) - L(x)| = |f(x) - f(x_0) - f'(x_0)(x - x_0)| \tag{10.8}$$

From the definition of the derivative, we have

$$f'(x_0) = \lim_{x \to x_0} \frac{f(x) - f(x_0)}{x - x_0} \tag{10.9}$$

If we divide (10.8) by the distance between x and x_0, $|x - x_0|$, we find that

$$\frac{|f(x) - L(x)|}{|x - x_0|} = \left| \frac{f(x) - L(x)}{x - x_0} \right| = \left| \frac{f(x) - f(x_0) - f'(x_0)(x - x_0)}{x - x_0} \right|$$

$$= \left| \frac{f(x) - f(x_0)}{x - x_0} - f'(x_0) \right|$$

Taking the limit $x \to x_0$ and using (10.9), we obtain

$$\lim_{x \to x_0} \left| \frac{f(x) - L(x)}{x - x_0} \right| = 0 \tag{10.10}$$

We say that $f(x)$ is differentiable at $x = x_0$ if (10.10) holds.

For functions of two variables, the definition of differentiability is based on the same idea. Before we can state it, however, we need to address one more point. In the preceding discussion, we divided by the distance between x and x_0. In two dimensions, the distance between two points (x, y) and (x_0, y_0) is $\sqrt{(x - x_0)^2 + (y - y_0)^2}$.

Definition Suppose that $f(x, y)$ is a function of two independent variables and that both $\partial f/\partial x$ and $\partial f/\partial y$ are defined throughout an open disk containing (x_0, y_0). Set

$$L(x, y) = f(x_0, y_0) + \frac{\partial f(x_0, y_0)}{\partial x}(x - x_0) + \frac{\partial f(x_0, y_0)}{\partial y}(y - y_0)$$

Then $f(x, y)$ is differentiable at (x_0, y_0) if

$$\lim_{(x,y) \to (x_0,y_0)} \left| \frac{f(x, y) - L(x, y)}{\sqrt{(x - x_0)^2 + (y - y_0)^2}} \right| = 0$$

Furthermore, if $f(x, y)$ is differentiable at (x_0, y_0), then $z = L(x, y)$ defines the tangent plane to the graph of f at $(x_0, y_0, f(x_0, y_0))$. We say that $f(x, y)$ is differentiable if it is differentiable at every point of its domain.

The key idea in both the one- and the two-dimensional case is to approximate functions by linear functions, so that the error in the approximation vanishes as we approach the point at which we approximated the function [x_0 in the one-dimensional case, (x_0, y_0) in the two-dimensional case].

As in the one-dimensional case, the following theorem holds:

Theorem If $f(x, y)$ is differentiable at (x_0, y_0), then f is continuous at (x_0, y_0).

That $f(x, y)$ is differentiable at (x_0, y_0) means that the function $f(x, y)$ is close to the tangent plane at (x_0, y_0) for all (x, y) close to (x_0, y_0). The mere existence of the partial derivatives $\partial f/\partial x$ and $\partial f/\partial y$ at (x_0, y_0), however, is *not* enough to guarantee differentiability (and, consequently, the existence of a tangent plane at a certain point). The next, very simple, example will show what can go wrong with that assumption.

EXAMPLE 2   Assume that

$$f(x, y) = \begin{cases} 0 & \text{if } xy \neq 0 \\ 1 & \text{if } xy = 0 \end{cases}$$

Show that $\frac{\partial f(0,0)}{\partial x}$ and $\frac{\partial f(0,0)}{\partial y}$ exist, but $f(x, y)$ is not continuous and, consequently, not differentiable, at $(0, 0)$.   ■

Solution   The graph of $f(x, y)$ is shown in Figure 10.38. To compute $\partial f/\partial x$ at $(0, 0)$, we set $y = 0$. Then $f(x, 0) = 1$, and therefore,

$$\frac{\partial f(0, 0)}{\partial x} = 0$$

Likewise, setting $x = 0$, we find that $f(0, y) = 1$ and

$$\frac{\partial f(0, 0)}{\partial y} = 0$$

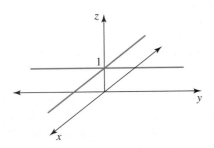

Figure 10.38 The graph of the function

$$f(x, y) = \begin{cases} 0 & \text{if } xy \neq 0 \\ 1 & xy = 0 \end{cases}$$

Even though $\frac{\partial f(0,0)}{\partial x}$ and $\frac{\partial f(0,0)}{\partial y}$ exist, the function is not continuous at $(0, 0)$.

That is, both partial derivatives exist at $(0, 0)$. However, $f(x, y)$ is not continuous at $(0, 0)$. To see this, it is enough to show that $f(x, y)$ has different limits along two different paths as (x, y) approaches $(0, 0)$. For the first path, denoted by C_1, we choose $y = 0$. Then

$$\lim_{\substack{C_1 \\ (x,y) \to (0,0)}} f(x, y) = 1$$

For the second path, denoted by C_2, we choose $y = x$. Then

$$\lim_{\substack{C_2 \\ (x,y) \to (0,0)}} f(x, y) = 0$$

Since $1 \neq 0$, $f(x, y)$ is not continuous at $(0, 0)$, and because differentiability implies continuity, a function that is not continuous cannot be differentiable. ■

The definition of differentiability is not easy to use if we actually want to check whether a function is differentiable at a certain point. Fortunately, there is another criterion, which suffices for all practical purposes, that can be used to check whether $f(x, y)$ is differentiable. We saw in Example 2 that the mere existence of the partial derivatives $\partial f/\partial x$ and $\partial f/\partial y$ at a point (x_0, y_0) is not enough to guarantee that $f(x, y)$ is differentiable. However, if the partial derivatives are continuous on a disk centered at (x_0, y_0), that is enough to guarantee differentiability.

> **Sufficient Condition For Differentiability** Suppose $f(x, y)$ is defined on an open disk centered at (x_0, y_0) and the partial derivatives $\partial f/\partial x$ and $\partial f/\partial y$ are continuous on an open disk centered at (x_0, y_0). Then $f(x, y)$ is differentiable at (x_0, y_0).

EXAMPLE 3   Show that $f(x, y) = 2x^2 y - y^2$ is differentiable for all $(x, y) \in \mathbf{R}^2$.

Solution   We use the sufficient condition for differentiability. First, observe that $f(x, y)$ is defined for all $(x, y) \in \mathbf{R}^2$. The partial derivatives are given by

$$\frac{\partial f}{\partial x} = 4xy \quad \text{and} \quad \frac{\partial f}{\partial y} = 2x^2 - 2y$$

Since both $\partial f/\partial x$ and $\partial f/\partial y$ are polynomials, both are continuous for all $(x, y) \in \mathbf{R}^2$ and, hence, $f(x, y)$ is differentiable for all $(x, y) \in \mathbf{R}^2$. ■

Linearization

> **Definition** Suppose that $f(x, y)$ is differentiable at (x_0, y_0). The linearization of $f(x, y)$ at (x_0, y_0) is the function
>
> $$L(x, y) = f(x_0, y_0) + \frac{\partial f(x_0, y_0)}{\partial x}(x - x_0) + \frac{\partial f(x_0, y_0)}{\partial y}(y - y_0)$$
>
> The approximation
>
> $$f(x, y) \approx L(x, y)$$
>
> is the **standard linear approximation**, or the **tangent plane approximation**, of $f(x, y)$ at (x_0, y_0).

EXAMPLE 4   Find the linear approximation of

$$f(x, y) = x^2 y + 2xe^y$$

at the point $(2, 0)$.

Solution   The linearization of $f(x, y)$ at $(2, 0)$ is given by

$$L(x, y) = f(2, 0) + \frac{\partial f(2, 0)}{\partial x}(x - 2) + \frac{\partial f(2, 0)}{\partial y}(y - 0)$$

Now, $f(2, 0) = 4$,

$$\frac{\partial f}{\partial x} = 2xy + 2e^y, \quad \text{and} \quad \frac{\partial f}{\partial y} = x^2 + 2xe^y$$

Hence,

$$\frac{\partial f(2, 0)}{\partial x} = 2 \quad \text{and} \quad \frac{\partial f(2, 0)}{\partial y} = 4 + 4 = 8$$

and we find that

$$L(x, y) = 4 + 2(x - 2) + 8(y - 0) = 4 + 2x - 4 + 8y$$
$$= 2x + 8y$$

■

EXAMPLE 5   Find the linear approximation of

$$f(x, y) = \ln(x - 2y^2)$$

at the point $(3, 1)$, and use it to find an approximation for $f(3.05, 0.95)$. Use a calculator to compute the value of $f(3.05, 0.95)$ and compare it with the approximation.

Solution   The linearization of $f(x, y)$ at $(3, 1)$ is given by

$$L(x, y) = f(3, 1) + \frac{\partial f(3, 1)}{\partial x}(x - 3) + \frac{\partial f(3, 1)}{\partial y}(y - 1)$$

Now, $f(3, 1) = \ln(3 - 2) = \ln 1 = 0$,

$$\frac{\partial f(x, y)}{\partial x} = \frac{1}{x - 2y^2}, \quad \text{and} \quad \frac{\partial f(x, y)}{\partial y} = \frac{-4y}{x - 2y^2}$$

Hence,

$$\frac{\partial f(3, 1)}{\partial x} = \frac{1}{3 - 2} = 1 \quad \text{and} \quad \frac{\partial f(3, 1)}{\partial y} = \frac{-4}{3 - 2} = -4$$

and we find that

$$L(x, y) = 0 + (1)(x - 3) + (-4)(y - 1)$$
$$= x - 3 - 4y + 4 = x - 4y + 1$$

Using $(3.05, 0.95)$, we get

$$L(3.05, 0.95) = 3.05 - (4)(0.95) + 1 = 0.25$$

Comparing this with $f(3.05, 0.95) \approx 0.2191$, we see that the error of approximation is only about $|0.25 - 0.2191| = 0.031$.

■

■ 10.4.2 Vector-Valued Functions

So far, we have considered only real-valued functions

$$f : \mathbf{R}^n \to \mathbf{R}$$

We will now extend our discussion to functions whose the range is a subset of $\mathbf{R}^m$—that is,

$$f : \mathbf{R}^n \to \mathbf{R}^m$$

Such functions are **vector-valued functions**, since they take on values that are represented by vectors:

$$f : \mathbf{R}^n \to \mathbf{R}^m$$

$$(x_1, x_2, \ldots, x_n) \to \begin{bmatrix} f_1(x_1, x_2, \ldots, x_n) \\ f_2(x_1, x_2, \ldots, x_n) \\ \vdots \\ f_m(x_1, x_2, \ldots, x_n) \end{bmatrix}$$

Here, each function $f_i(x_1, \ldots, x_n)$ is a real-valued function:

$$f_i : \mathbf{R}^n \to \mathbf{R}$$
$$(x_1, x_2, \ldots, x_n) \to f_i(x_1, x_2, \ldots, x_n)$$

We will encounter vector-valued functions where $n = m = 2$ extensively in Chapter 11. As an example, consider a community consisting of two species. Let u and v denote the respective densities of the species and $f(u, v)$ and $g(u, v)$ the per capita growth rates of the species as functions of the densities u and v. We can then write this relationship as a map

$$(u, v) \to \begin{bmatrix} f(u, v) \\ g(u, v) \end{bmatrix}$$

from $\mathbf{R}^2$ to $\mathbf{R}^2$.

Our main task in this subsection will be to define the linearization of vector-valued functions whose domain and range are $\mathbf{R}^2$. We will motivate this definition by analogy with the cases of functions $f : \mathbf{R} \to \mathbf{R}$ and $f : \mathbf{R}^2 \to \mathbf{R}$. At the end of the section, we will mention how to generalize our results to arbitrary vector-valued functions.

We begin with differentiable functions $f : \mathbf{R} \to \mathbf{R}$. The linearization of $f : \mathbf{R} \to \mathbf{R}$ about x_0 is given by

$$L(x) = f(x_0) + f'(x_0)(x - x_0) \tag{10.11}$$

EXAMPLE 6   Find the linearization of

$$f(x) = 2 \ln x$$

at $x_0 = 1$.

Solution   The linearization of $f(x)$ at $x = 1$ is given by

$$L(x) = f(1) + f'(1)(x - 1)$$
$$= 0 + (2)(x - 1) = 2x - 2$$

since $f'(x) = 2/x$.   ■

We found the linearization of a real-valued function $f : \mathbf{R}^2 \to \mathbf{R}$ in the previous subsection, namely,

$$L(x, y) = f(x_0, y_0) + \frac{\partial f(x_0, y_0)}{\partial x}(x - x_0) + \frac{\partial f(x_0, y_0)}{\partial y}(y - y_0)$$

We can write this equation in matrix notation as

$$L(x, y) = f(x_0, y_0) + \begin{bmatrix} \frac{\partial f(x_0, y_0)}{\partial x} & \frac{\partial f(x_0, y_0)}{\partial y} \end{bmatrix} \begin{bmatrix} x - x_0 \\ y - y_0 \end{bmatrix}$$

EXAMPLE 7   Find the linearization of

$$f(x, y) = \ln x + \ln y$$

at $(1, 1)$.

Solution We find $f(1, 1) = 0$. Since $f_x(x, y) = \frac{1}{x}$ and $f_y(x, y) = \frac{1}{y}$, it follows that

$$\frac{\partial f(1, 1)}{\partial x} = 1 \quad \text{and} \quad \frac{\partial f(1, 1)}{\partial y} = 1$$

Hence,

$$L(x, y) = 0 + \begin{bmatrix} 1, & 1 \end{bmatrix} \begin{bmatrix} x - 1 \\ y - 1 \end{bmatrix}$$
$$= x - 1 + y - 1 = x + y - 2 \qquad ■$$

We will denote vector-valued functions by boldface letters. Suppose that

$$\mathbf{h} : \mathbf{R}^2 \to \mathbf{R}^2$$
$$(x, y) \to \begin{bmatrix} f(x, y) \\ g(x, y) \end{bmatrix}$$

and assume that all first partial derivatives are continuous on a disk centered at (x_0, y_0). We can linearize each component of the function $\mathbf{h}$. The linearization of f is

$$\alpha(x, y) = f(x_0, y_0) + \frac{\partial f(x_0, y_0)}{\partial x}(x - x_0) + \frac{\partial f(x_0, y_0)}{\partial y}(y - y_0)$$

and the linearization of g is

$$\beta(x, y) = g(x_0, y_0) + \frac{\partial g(x_0, y_0)}{\partial x}(x - x_0) + \frac{\partial g(x_0, y_0)}{\partial y}(y - y_0)$$

We define the vector-valued function $\mathbf{L}(x, y) = \begin{bmatrix} \alpha(x, y) \\ \beta(x, y) \end{bmatrix}$. The linearization of $\mathbf{h}$ can thus be written in matrix form as

$$\mathbf{L}(x, y) = \begin{bmatrix} \alpha(x, y) \\ \beta(x, y) \end{bmatrix} = \begin{bmatrix} f(x_0, y_0) \\ g(x_0, y_0) \end{bmatrix} + \begin{bmatrix} \frac{\partial f(x_0, y_0)}{\partial x}(x - x_0) + \frac{\partial f(x_0, y_0)}{\partial y}(y - y_0) \\ \frac{\partial g(x_0, y_0)}{\partial x}(x - x_0) + \frac{\partial g(x_0, y_0)}{\partial y}(y - y_0) \end{bmatrix}$$

or

$$\mathbf{L}(x, y) = \begin{bmatrix} f(x_0, y_0) \\ g(x_0, y_0) \end{bmatrix} + \begin{bmatrix} \frac{\partial f(x_0, y_0)}{\partial x} & \frac{\partial f(x_0, y_0)}{\partial y} \\ \frac{\partial g(x_0, y_0)}{\partial x} & \frac{\partial g(x_0, y_0)}{\partial y} \end{bmatrix} \begin{bmatrix} x - x_0 \\ y - y_0 \end{bmatrix}$$

where the matrix

$$\begin{bmatrix} \frac{\partial f(x_0, y_0)}{\partial x} & \frac{\partial f(x_0, y_0)}{\partial y} \\ \frac{\partial g(x_0, y_0)}{\partial x} & \frac{\partial g(x_0, y_0)}{\partial y} \end{bmatrix}$$

is a 2×2 matrix that is called the **Jacobi matrix** or the **derivative matrix**. We denote the Jacobi matrix by $(D\mathbf{h})$; that is,

$$(D\mathbf{h})(x_0, y_0) = \begin{bmatrix} \frac{\partial f(x_0, y_0)}{\partial x} & \frac{\partial f(x_0, y_0)}{\partial y} \\ \frac{\partial g(x_0, y_0)}{\partial x} & \frac{\partial g(x_0, y_0)}{\partial y} \end{bmatrix}$$

EXAMPLE 8 Assume that

$$\mathbf{f} : \mathbf{R}^2 \to \mathbf{R}^2$$
$$(x, y) \to \begin{bmatrix} u \\ v \end{bmatrix}$$

with

$$u(x, y) = x^2 y - y^3 \quad \text{and} \quad v(x, y) = 2x^3 y^2 + y$$

Compute the Jacobi matrix and evaluate it at $(1, 2)$.

Solution The Jacobi matrix is

$$\begin{bmatrix} \frac{\partial u}{\partial x} & \frac{\partial u}{\partial y} \\ \frac{\partial v}{\partial x} & \frac{\partial v}{\partial y} \end{bmatrix} = \begin{bmatrix} 2xy & x^2 - 3y^2 \\ 6x^2y^2 & 4x^3y + 1 \end{bmatrix}$$

At $(1, 2)$, we find that

$$\begin{bmatrix} 4 & -11 \\ 24 & 9 \end{bmatrix}$$ ■

EXAMPLE 9 Let

$$\mathbf{f} : \mathbf{R}^2 \to \mathbf{R}^2$$

with

$$u(x, y) = ye^{-x} \qquad \text{and} \qquad v(x, y) = \sin x + \cos y$$

Find the linear approximation to $\mathbf{f}(x, y)$ at $(0, 0)$. Compare $\mathbf{f}(0.1, -0.1)$ with its linear approximation.

Solution We compute the Jacobi matrix first:

$$(D\mathbf{f})(x, y) = \begin{bmatrix} -ye^{-x} & e^{-x} \\ \cos x & -\sin y \end{bmatrix}$$

The linear approximation of $\mathbf{f}(x, y)$ at $(0, 0)$ is

$$\mathbf{L}(x, y) = \begin{bmatrix} \alpha(x, y) \\ \beta(x, y) \end{bmatrix} = \begin{bmatrix} u(0, 0) \\ v(0, 0) \end{bmatrix} + (D\mathbf{f})(0, 0) \begin{bmatrix} x - 0 \\ y - 0 \end{bmatrix}$$

$$= \begin{bmatrix} 0 \\ 1 \end{bmatrix} + \begin{bmatrix} 0 & 1 \\ 1 & 0 \end{bmatrix} \begin{bmatrix} x \\ y \end{bmatrix}$$

$$= \begin{bmatrix} 0 \\ 1 \end{bmatrix} + \begin{bmatrix} y \\ x \end{bmatrix}$$

Using $(x, y) = (0.1, -0.1)$, we find that

$$\mathbf{L}(0.1, -0.1) = \begin{bmatrix} 0 \\ 1 \end{bmatrix} + \begin{bmatrix} -0.1 \\ 0.1 \end{bmatrix} = \begin{bmatrix} -0.1 \\ 1.1 \end{bmatrix}$$

and

$$\mathbf{f}(0.1, -0.1) = \begin{bmatrix} -0.1e^{-0.1} \\ \sin 0.1 + \cos(-0.1) \end{bmatrix} = \begin{bmatrix} -0.09 \\ 1.09 \end{bmatrix}$$

We see that the linear approximation is close to the actual value.   ■

We can generalize the Jacobi matrix to functions $\mathbf{f} : \mathbf{R}^n \to \mathbf{R}^m$. If

$$\mathbf{f}(x_1, x_2, \ldots, x_n) = \begin{bmatrix} f_1(x_1, x_2, \ldots, x_n) \\ f_2(x_1, x_2, \ldots, x_n) \\ \vdots \\ f_m(x_1, x_2, \ldots, x_n) \end{bmatrix}$$

where $f_i : \mathbf{R}^n \to \mathbf{R}$, $i = 1, 2, \ldots, n$, are real-valued functions of n independent variables, then the Jacobi matrix is an $m \times n$ matrix of the form

$$D\mathbf{f} = \begin{bmatrix} \frac{\partial f_1}{\partial x_1} & \cdots & \frac{\partial f_1}{\partial x_n} \\ \vdots & & \vdots \\ \frac{\partial f_m}{\partial x_1} & \cdots & \frac{\partial f_m}{\partial x_n} \end{bmatrix}$$

The linearization of **f** about the point $(x_1^*, x_2^*, \ldots, x_n^*)$ is then

$$L(x_1^*, \ldots, x_n^*) = \begin{bmatrix} f_1(x_1^*, \ldots, x_n^*) \\ f_2(x_1^*, \ldots, x_n^*) \\ \vdots \\ f_m(x_1^*, \ldots, x_n^*) \end{bmatrix} + D\mathbf{f}(x_1^*, \ldots, x_n^*) \begin{bmatrix} x_1 - x_1^* \\ \vdots \\ x_n - x_n^* \end{bmatrix}$$

Section 10.4 Problems

■ 10.4.1

In Problems 1–10, the tangent plane at the indicated point (x_0, y_0, z_0) exists. Find its equation.

1. $f(x, y) = 2x^3 + y^2;\ (1, 2, 6)$
2. $f(x, y) = x^2 - 3y^2;\ (-1, 1, -2)$
3. $f(x, y) = xy;\ (-1, -2, 2)$
4. $f(x, y) = \sin x + \cos y;\ (0, 0, 1)$
5. $f(x, y) = \sin(xy);\ (1, 0, 0)$
6. $f(x, y) = e^{x-y};\ (1, -1, e^2)$
7. $f(x, y) = e^{x^2+y^2};\ (1, 0, e)$
8. $f(x, y) = e^x \cos y;\ (0, 0, 1)$
9. $f(x, y) = \ln(x + y);\ (2, -1, 0)$
10. $f(x, y) = \ln(x^2 + y^2);\ (1, 1, \ln 2)$

In Problems 11–16, show that $f(x, y)$ is differentiable at the indicated point.

11. $f(x, y) = y^2 x + x^2 y;\ (1, 1)$ 12. $f(x, y) = xy - 3x^2;\ (1, 1)$
13. $f(x, y) = \cos(x + y);\ (0, 0)$ 14. $f(x, y) = e^{x-y};\ (0, 0)$
15. $f(x, y) = x + y^2 - 2xy;\ (-1, 2)$
16. $f(x, y) = \tan(x^2 + y^2);\ \left(\dfrac{\pi}{4}, -\dfrac{\pi}{4}\right)$

In Problems 17–24, find the linearization of $f(x, y)$ at the indicated point (x_0, y_0).

17. $f(x, y) = x - 3y;\ (3, 1)$
18. $f(x, y) = 2xy;\ (1, -1)$
19. $f(x, y) = \sqrt{x} + 2y;\ (1, 0)$
20. $f(x, y) = \cos(x^2 y);\ \left(\dfrac{\pi}{2}, 0\right)$
21. $f(x, y) = \tan(x + y);\ (0, 0)$
22. $f(x, y) = e^{3x+2y};\ (1, 2)$
23. $f(x, y) = \ln(x^2 + y);\ (1, 1)$
24. $f(x, y) = x^2 e^y;\ (1, 0)$
25. Find the linear approximation of

$$f(x, y) = e^{x+y}$$

at $(0, 0)$, and use it to approximate $f(0.1, 0.05)$. Using a calculator, compare the approximation with the exact value of $f(0.1, 0.05)$.

26. Find the linear approximation of

$$f(x, y) = \sin(x + 2y)$$

at $(0, 0)$, and use it to approximate $f(-0.1, 0.2)$. Using a calculator, compare the approximation with the exact value of $f(-0.1, 0.2)$.

27. Find the linear approximation of

$$f(x, y) = \ln(x^2 - 3y)$$

at $(1, 0)$, and use it to approximate $f(1.1, 0.1)$. Using a calculator, compare the approximation with the exact value of $f(1.1, 0.1)$.

28. Find the linear approximation of

$$f(x, y) = \tan(2x - 3y^2)$$

at $(0, 0)$, and use it to approximate $f(0.03, 0.05)$. Using a calculator, compare the approximation with the exact value of $f(0.03, 0.05)$.

■ 10.4.2

In Problems 29–36, find the Jacobi matrix for each given function.

29. $\mathbf{f}(x, y) = \begin{bmatrix} x + y \\ x^2 - y^2 \end{bmatrix}$ 30. $\mathbf{f}(x, y) = \begin{bmatrix} 2x - 3y \\ 4x^2 \end{bmatrix}$

31. $\mathbf{f}(x, y) = \begin{bmatrix} e^{x-y} \\ e^{x+y} \end{bmatrix}$ 32. $\mathbf{f}(x, y) = \begin{bmatrix} (x - y)^2 \\ \sin(x - y) \end{bmatrix}$

33. $\mathbf{f}(x, y) = \begin{bmatrix} \cos(x - y) \\ \cos(x + y) \end{bmatrix}$ 34. $\mathbf{f}(x, y) = \begin{bmatrix} \ln(x + y) \\ e^{x+y} \end{bmatrix}$

35. $\mathbf{f}(x, y) = \begin{bmatrix} 2x^2 y - 3y + x \\ e^x \sin y \end{bmatrix}$

36. $\mathbf{f}(x, y) = \begin{bmatrix} \sqrt{x^2 + y^2} \\ e^{-x^2} \end{bmatrix}$

In Problems 37–42, find a linear approximation to each function $f(x, y)$ at the indicated point.

37. $\mathbf{f}(x, y) = \begin{bmatrix} 2x^2 y \\ \frac{1}{xy} \end{bmatrix}$ at $(1, 1)$

38. $\mathbf{f}(x, y) = \begin{bmatrix} 3x - y^2 \\ 4y \end{bmatrix}$ at $(-1, -2)$

39. $\mathbf{f}(x, y) = \begin{bmatrix} e^{2x-y} \\ \ln(2x - y) \end{bmatrix}$ at $(1, 1)$

40. $\mathbf{f}(x, y) = \begin{bmatrix} e^x \sin y \\ e^{-y} \cos x \end{bmatrix}$ at $(0, 0)$

41. $\mathbf{f}(x, y) = \begin{bmatrix} \frac{x}{y} \\ \frac{y}{x} \end{bmatrix}$ at $(1, 1)$

42. $\mathbf{f}(x, y) = \begin{bmatrix} (x + y)^2 \\ xy \end{bmatrix}$ at $(-1, 1)$

43. Find a linear approximation to

$$\mathbf{f}(x, y) = \begin{bmatrix} x^2 - xy \\ 3y^2 - 1 \end{bmatrix}$$

at $(1, 2)$. Use your result to find an approximation for $f(1.1, 1.9)$, and compare the approximation with the value of $f(1.1, 1.9)$ that you get when you use a calculator.

44. Find a linear approximation to

$$\mathbf{f}(x, y) = \begin{bmatrix} x/y \\ 2xy \end{bmatrix}$$

at $(-1, 1)$. Use your result to find an approximation for $f(-0.9, 1.05)$, and compare the approximation with the value of $f(-0.9, 1.05)$ that you get when you use a calculator.

45. Find a linear approximation to

$$\mathbf{f}(x, y) = \begin{bmatrix} (x - y)^2 \\ 2x^2 y \end{bmatrix}$$

at $(2, -3)$. Use your result to find an approximation for $f(1.9, -3.1)$, and compare the approximation with the value of $f(1.9, -3.1)$ that you get when you use a calculator.

46. Find a linear approximation to

$$\mathbf{f}(x, y) = \begin{bmatrix} \sqrt{2x + y} \\ x - y^2 \end{bmatrix}$$

at $(1, 2)$. Use your result to find an approximation for $f(1.05, 2.05)$, and compare the approximation with the value of $f(1.05, 2.05)$ that you get when you use a calculator.

■ 10.5 More about Derivatives (Optional)

■ 10.5.1 The Chain Rule for Functions of Two Variables

In Section 10.3, we discussed how the net assimilation of CO_2 can change as a function of both temperature and light intensity. If we follow the net assimilation of CO_2 over time, we must take into account the fact that both temperature and light intensity depend on time. If we denote the temperature at time t by $T(t)$, the light intensity at time t by $I(t)$, and the net assimilation of CO_2 at time t by $N(t)$, then $N(t)$ is a function of both $T(t)$ and $I(t)$, and we can write

$$N(t) = f(T(t), I(t))$$

Net assimilation is thus a composite function.

To differentiate composite functions of one variable, we use the chain rule. Suppose that $w = f(x)$ is a function of one variable and that x depends on t. Then, by the chain rule, to differentiate w with respect to t, we have

$$\frac{dw}{dt} = \frac{dw}{dx}\frac{dx}{dt} \tag{10.12}$$

The chain rule can be extended to functions of more than one variable:

Chain Rule for Functions of Two Independent Variables If $w = f(x, y)$ is differentiable and x and y are differentiable functions of t, then w is a differentiable function of t and

$$\frac{dw}{dt} = \frac{\partial w}{\partial x}\frac{dx}{dt} + \frac{\partial w}{\partial y}\frac{dy}{dt}$$

We will not prove this formula, but merely outline the steps that lead to it. We approximate $w = f(x, y)$ at (x_0, y_0) by its linear approximation

$$L(x, y) = f(x_0, y_0) + \frac{\partial f(x_0, y_0)}{\partial x}(x - x_0) + \frac{\partial f(x_0, y_0)}{\partial y}(y - y_0)$$

If we set $\Delta x = x - x_0$, $\Delta y = y - y_0$, and $\Delta w = f(x, y) - f(x_0, y_0)$, we can approximate $\Delta w = f(x_0 + \Delta x, y_0 + \Delta y) - f(x_0, y_0)$ by its linear approximation. We find that

$$\Delta w \approx \frac{\partial f(x_0, y_0)}{\partial x}\Delta x + \frac{\partial f(x_0, y_0)}{\partial y}\Delta y$$

Dividing both sides by Δt, we obtain

$$\frac{\Delta w}{\Delta t} \approx \frac{\partial f(x_0, y_0)}{\partial x}\frac{\Delta x}{\Delta t} + \frac{\partial f(x_0, y_0)}{\partial y}\frac{\Delta y}{\Delta t}$$

If we let $\Delta t \to 0$, then

$$\frac{\Delta w}{\Delta t} \to \frac{dw}{dt}, \qquad \frac{\Delta x}{\Delta t} \to \frac{dx}{dt}, \qquad \frac{\Delta y}{\Delta t} \to \frac{dy}{dt}$$

and we get

$$\frac{dw}{dt} = \frac{\partial f(x_0, y_0)}{\partial x}\frac{dx}{dt} + \frac{\partial f(x_0, y_0)}{\partial y}\frac{dy}{dt}$$

or, in short,

$$\frac{dw}{dt} = \frac{\partial w}{\partial x}\frac{dx}{dt} + \frac{\partial w}{\partial y}\frac{dy}{dt}$$

EXAMPLE 1 Let

$$w = f(x, y) = x^2 y^3$$

with $x(t) = \sin t$ and $y(t) = e^{-t}$. Find the derivative of $w = f(x, y)$ with respect to t when $t = \pi/2$.

Solution Using the chain rule, we find that

$$\begin{aligned}
\frac{dw}{dt} &= \frac{\partial w}{\partial x}\frac{dx}{dt} + \frac{\partial w}{\partial y}\frac{dy}{dt} \\
&= 2xy^3\frac{dx}{dt} + x^2 3y^2\frac{dy}{dt} \\
&= 2xy^3 \cos t + 3x^2 y^2(-1)e^{-t}
\end{aligned}$$

Now, $\cos\frac{\pi}{2} = 0$,

$$x\left(\frac{\pi}{2}\right) = \sin\frac{\pi}{2} = 1, \qquad \text{and} \qquad y\left(\frac{\pi}{2}\right) = e^{-\pi/2}$$

Hence,

$$\frac{dw}{dt}\Big|_{t=\pi/2} = (2)(1)(e^{-3\pi/2})(0) - (3)(1)^2 e^{-\pi}e^{-\pi/2} = -3e^{-3\pi/2}$$

We can check the answer directly:

$$w(t) = (\sin^2 t)(e^{-3t})$$

Therefore,

$$\frac{dw}{dt}\Big|_{t=\pi/2} = 2(\sin t)(\cos t)e^{-3t} - 3(\sin^2 t)e^{-3t}\Big|_{t=\pi/2} = -3e^{-3\pi/2} \qquad ■$$

EXAMPLE 2 Suppose that we wish to predict the abundance of a particular plant species. We suspect that the two major factors influencing the abundance of the plant are nitrogen levels and disturbance levels. Previous studies have shown that an increase in nitrogen in the soil has a negative effect on the abundance of this species; also, an increase in disturbance due to grazing seems to have a negative effect on abundance. If both nitrogen and disturbance due to grazing increase over the next few years, how would the abundance of the species be affected?

Solution We denote the abundance of the plant at time t by $B(t)$. We are interested in how $B(t)$ will change over time; that is, we want to find out whether $B(t)$ will increase or decrease over time. For this, we need to compute the derivative of $B(t)$. If we assume that the abundance of the plant is affected primarily by nitrogen and disturbance levels, we can consider B as a function of both N, the nitrogen level, and D, the disturbance level. N and D change over time and are thus functions of time. The function $B(t)$ is therefore a function of both $N(t)$ and $D(t)$. Using the chain rule for functions of two independent variables, we find that

$$\frac{dB}{dt} = \underbrace{\frac{\partial B}{\partial N}}_{<0} \underbrace{\frac{dN}{dt}}_{>0} + \underbrace{\frac{\partial B}{\partial D}}_{<0} \underbrace{\frac{dD}{dt}}_{>0} < 0$$

since abundance B is a decreasing function of both N and D and since both nitrogen and disturbance levels are assumed to increase over the next few years. We thus find that the abundance of the plant will decrease over the next few years. ■

■ 10.5.2 Implicit Differentiation

We discussed implicit differentiation in Section 4.4. This was a useful technique for differentiating a function $y = f(x)$ when y was given implicitly, as in

$$x^2 y - e^{-y} = 0 \qquad (10.13)$$

To find dy/dx, we differentiate both sides with respect to x, keeping in mind that y is a function of x. We obtain

$$2xy + x^2\frac{dy}{dx} - e^{-y}\left(-\frac{dy}{dx}\right) = 0$$

$$2xy + \frac{dy}{dx}(x^2 + e^{-y}) = 0$$

Solving for dy/dx, we find that

$$\frac{dy}{dx} = -\frac{2xy}{x^2 + e^{-y}} \qquad (10.14)$$

When we look at (10.13), we can define a function $F(x, y)$ as

$$F(x, y) = x^2 y - e^{-y} \quad \text{with } F(x, y) = 0$$

$F(x, y)$ is a function of two variables. We say that the equation $F(x, y) = 0$ defines y implicitly as a function of x.

We will turn to the general case to see why reformulating an implicitly defined function as a function of two variables is a useful way to calculate derivatives. We think of y as a function of x and define a function $F(x, y) = 0$. This defines y implicitly as a function of x. To find the derivative of y with respect to x, we set

$$w = F(u, v) \quad \text{with } u(x) = x \text{ and } v(x) = y$$

This makes w a function of x; that is, $w = w(x)$. We can now use the chain rule to differentiate w with respect to x:

$$\frac{dw}{dx} = \frac{\partial F}{\partial u}\frac{du}{dx} + \frac{\partial F}{\partial v}\frac{dv}{dx}$$

Since $\frac{du}{dx} = 1$ and $\frac{dv}{dx} = \frac{dy}{dx}$, we obtain, with $u = x$ and $v = y$,

$$\frac{dw}{dx} = \frac{\partial F}{\partial x} + \frac{\partial F}{\partial y}\frac{dy}{dx} \qquad (10.15)$$

Because $F(x, y) = 0$, it follows that $w(x) = 0$ for all values of x and, therefore,

$$\frac{dw}{dx} = 0 \qquad (10.16)$$

Equating (10.15) and (10.16), we obtain

$$0 = \frac{\partial F}{\partial x} + \frac{\partial F}{\partial y}\frac{dy}{dx}$$

We can solve this partial differential equation for dy/dx as

$$\frac{dy}{dx} = -\frac{\frac{\partial F}{\partial x}}{\frac{\partial F}{\partial y}} = -\frac{F_x}{F_y}$$

provided that $\partial F/\partial y \neq 0$. We summarize this result as follows:

Suppose that $w = F(x, y)$ is differentiable and $F(x, y) = 0$ defines y implicitly as a differentiable function of x. Then, at any point where $F_y \neq 0$,

$$\frac{dy}{dx} = -\frac{F_x}{F_y}$$

EXAMPLE 3 Find dy/dx if $x^2 y - e^{-y} = 0$.

Solution We set $F(x, y) = x^2 y - e^{-y}$. We need to find F_x and F_y:

$$F_x = \frac{\partial F}{\partial x} = 2xy$$

$$F_y = \frac{\partial F}{\partial y} = x^2 + e^{-y}$$

Then, since $F(x, y) = 0$,

$$\frac{dy}{dx} = -\frac{F_x}{F_y} = -\frac{2xy}{x^2 + e^{-y}}$$

as in (10.14). Since $x^2 + e^{-y} \neq 0$ for all $(x, y) \in \mathbf{R}^2$, dy/dx is defined for all $(x, y) \in \mathbf{R}^2$ with $F(x, y) = 0$. ■

The next example shows that we can use this rule to find the derivatives of inverse trigonometric functions.

EXAMPLE 4 Find dy/dx for

$$y = \arcsin x$$

Solution We already know the answer, since we discussed the derivative of $\arcsin x$ earlier, in the context of inverse functions. But let's see how we can use the results of this section to find the answer. Since

$$x = \sin y \quad \text{for } -\frac{\pi}{2} \leq y \leq \frac{\pi}{2}$$

is equivalent to

$$y = \arcsin x \quad \text{for } -1 \leq x \leq 1$$

we can define a function $F(x, y)$ that satisfies $F(x, y) = 0$ and defines y implicitly as a function of x, namely,

$$F(x, y) = x - \sin y$$

Then

$$\frac{dy}{dx} = -\frac{F_x}{F_y} = -\frac{1}{-\cos y} = \frac{1}{\cos y}$$

Since $\sin^2 y + \cos^2 y = 1$, we have

$$\cos y = \sqrt{1 - \sin^2 y} = \sqrt{1 - x^2}$$

Here, we must use the fact that $x = \sin y$ is defined for $-\pi/2 \leq y \leq \pi/2$ and $\cos y \geq 0$ for y in this interval. Using this representation for $\cos y$, we find that

$$\frac{dy}{dx} = \frac{1}{\sqrt{1 - x^2}}$$

which is defined for $-1 < x < 1$. ■

■ 10.5.3 Directional Derivatives and Gradient Vectors

Suppose that you are on a sloped surface, such as a hillside. Depending on which direction you walk, you must either go uphill, stay at the same level, or go downhill. That is, by choosing a particular direction, you have some control over the steepness of your path. How steep your path is can be described by the slope of the tangent line at your starting point in the direction of your path. This slope is given by the **directional derivative**.

We assume that $z = f(x, y)$ is a differentiable function of two independent variables. We choose a point (x_0, y_0, z_0), with $z_0 = f(x_0, y_0)$ on the surface defined by $z = f(x, y)$. To define the slope of a tangent line at (x_0, y_0, z_0), we must specify a direction in which we wish to go. We know how to deal with this problem when we go in the x- or y-direction. In these cases, the partial derivatives $\partial f/\partial x$ and $\partial f/\partial y$ tell us how $f(x, y)$ changes. We will now explain how we can express the slope when we choose an arbitrary direction.

The first step is to find a way to express what we mean by "going in a certain direction." We start at a point (x_0, y_0) and wish to go in the direction of a unit vector $\mathbf{u} = \begin{bmatrix} u_1 \\ u_2 \end{bmatrix}$. (Recall that a unit vector has length 1.) This is illustrated in Figure 10.39, from which we see that

$$\mathbf{r} = \mathbf{r}_0 + t\mathbf{u} \tag{10.17}$$

where $\mathbf{r}_0 = \begin{bmatrix} x_0 \\ y_0 \end{bmatrix}$, $\mathbf{r} = \begin{bmatrix} x \\ y \end{bmatrix}$, and t is a real number; different values of t get us to different points on the straight line through (x_0, y_0) that points in the direction of $\mathbf{u}$. We can also write (10.17) as

$$\begin{bmatrix} x \\ y \end{bmatrix} = \begin{bmatrix} x_0 \\ y_0 \end{bmatrix} + \begin{bmatrix} tu_1 \\ tu_2 \end{bmatrix} \quad \text{for } t \in \mathbf{R} \tag{10.18}$$

Equation (10.18) is called the *parametric equation of a line*, which we introduced in Section 9.4; t is called the parameter. Since $x = x_0 + tu_1$ and $y = y_0 + tu_2$, it follows that

$$\frac{dx}{dt} = u_1 \quad \text{and} \quad \frac{dy}{dt} = u_2 \tag{10.19}$$

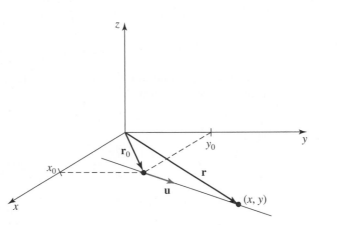

Figure 10.39 Going in the direction of $\mathbf{u}$ from (x_0, y_0).

We can now find out what happens to $f(x, y)$ when we start at (x_0, y_0) and go in the direction of the unit vector **u**, as illustrated in Figure 10.40. To begin, we use the parametric equation (10.18) of the line that passes through (x_0, y_0) and that is oriented in the direction of **u**:

$$x = x_0 + tu_1 \qquad \text{and} \qquad y = y_0 + tu_2$$

Since $z(t) = f(x(t), y(t))$, we can use the chain rule to find out how f changes when we vary t (i.e., as we move along the straight line):

$$\frac{dz}{dt} = \frac{\partial f}{\partial x}\frac{dx}{dt} + \frac{\partial f}{\partial y}\frac{dy}{dt} = \frac{\partial f}{\partial x}u_1 + \frac{\partial f}{\partial y}u_2 \qquad (10.20)$$

[Note that we used (10.19) in the last step.] Equation (10.20) can be written as a dot product:

$$\frac{dz}{dt} = \begin{bmatrix} \frac{\partial f}{\partial x} \\ \frac{\partial f}{\partial y} \end{bmatrix} \cdot \begin{bmatrix} u_1 \\ u_2 \end{bmatrix}$$

The first vector in the dot product is called the **gradient** of f.

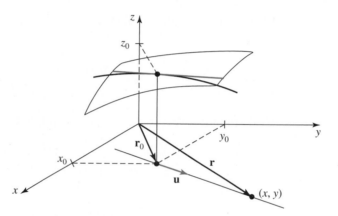

Figure 10.40 An illustration of the directional derivative.

Definition Assume that $z = f(x, y)$ is a function of two independent variables and that $\partial f/\partial x$ and $\partial f/\partial y$ exist. Then the vector

$$\nabla f(x, y) = \begin{bmatrix} \frac{\partial f(x,y)}{\partial x} \\ \frac{\partial f(x,y)}{\partial y} \end{bmatrix}$$

is called the gradient of f at (x, y).

The notation ∇f is read "grad f" or "gradient of f." The symbol ∇ is called "del," so you can also say "del f." An alternative notation is grad f.

We can now define the derivative of f in a particular direction:

Definition The directional derivative of $f(x, y)$ at (x_0, y_0) in the direction of the unit vector $\mathbf{u} = \begin{bmatrix} u_1 \\ u_2 \end{bmatrix}$ is

$$D_\mathbf{u} f(x_0, y_0) = (\nabla f(x_0, y_0)) \cdot \mathbf{u}$$

Note that in the definition of the directional derivative, we assume that **u** is a *unit vector*. Choosing a unit vector (as opposed to a vector of some other length) ensures that the directional derivative of $f(x, y)$ agrees with the partial derivatives when we go along the positive x- or y-axis. That is, if $\mathbf{u} = \begin{bmatrix} 1 \\ 0 \end{bmatrix}$, then

$$D_{\mathbf{u}}f(x_0, y_0) = \begin{bmatrix} \frac{\partial f(x_0, y_0)}{\partial x} \\ \frac{\partial f(x_0, y_0)}{\partial y} \end{bmatrix} \cdot \begin{bmatrix} 1 \\ 0 \end{bmatrix} = \frac{\partial f(x_0, y_0)}{\partial x}$$

and if $\mathbf{u} = \begin{bmatrix} 0 \\ 1 \end{bmatrix}$, then

$$D_{\mathbf{u}}f(x_0, y_0) = \frac{\partial f(x_0, y_0)}{\partial y}$$

EXAMPLE 5

Compute the directional derivative of

$$f(x, y) = \sqrt{x^2 + 2y^2}$$

at the point $(-1, 2)$ in the direction $\begin{bmatrix} -1 \\ 3 \end{bmatrix}$.

Solution

We first compute the gradient vector

$$\nabla f(x, y) = \begin{bmatrix} \frac{x}{\sqrt{x^2+2y^2}} \\ \frac{2y}{\sqrt{x^2+2y^2}} \end{bmatrix}$$

Evaluating this vector at $(-1, 2)$, we find that

$$\nabla f(-1, 2) = \begin{bmatrix} \frac{-1}{\sqrt{1+8}} \\ \frac{4}{\sqrt{1+8}} \end{bmatrix} = \begin{bmatrix} -1/3 \\ 4/3 \end{bmatrix}$$

Since $\begin{bmatrix} -1 \\ 3 \end{bmatrix}$ is not a unit vector, we normalize it first. The vector $\begin{bmatrix} -1 \\ 3 \end{bmatrix}$ has length $\sqrt{1+9} = \sqrt{10}$; hence,

$$\mathbf{u} = \frac{1}{\sqrt{10}} \begin{bmatrix} -1 \\ 3 \end{bmatrix}$$

and

$$D_{\mathbf{u}}f(-1, 2) = (\nabla f(-1, 2)) \cdot \mathbf{u}$$

$$= \begin{bmatrix} -\frac{1}{3} \\ \frac{4}{3} \end{bmatrix} \cdot \begin{bmatrix} -\frac{1}{\sqrt{10}} \\ \frac{3}{\sqrt{10}} \end{bmatrix}$$

$$= \frac{1}{3\sqrt{10}} + \frac{12}{3\sqrt{10}} = \frac{13}{3\sqrt{10}}$$

■

EXAMPLE 6

Compute the directional derivative of

$$f(x, y) = x^2 y - 2y^2$$

at the point $(-3, 2)$ in the direction of $(-1, 1)$.

Solution

We first compute the gradient vector

$$\nabla f(x, y) = \begin{bmatrix} 2xy \\ x^2 - 4y \end{bmatrix}$$

Evaluating this vector at $(-3, 2)$, we find that

$$\nabla f(-3, 2) = \begin{bmatrix} (2)(-3)(2) \\ (-3)^2 - (4)(2) \end{bmatrix} = \begin{bmatrix} -12 \\ 1 \end{bmatrix}$$

The vector that goes from $(-3, 2)$ to $(-1, 1)$ has the form

$$\begin{bmatrix} -1 - (-3) \\ 1 - 2 \end{bmatrix} = \begin{bmatrix} 2 \\ -1 \end{bmatrix}$$

This vector has length $\sqrt{4 + 1} = \sqrt{5}$. Normalizing the vector yields

$$\mathbf{u} = \frac{1}{\sqrt{5}} \begin{bmatrix} 2 \\ -1 \end{bmatrix}$$

and the directional derivative is

$$D_\mathbf{u} f(-3, 2) = (\nabla f(-3, 2)) \cdot \mathbf{u}$$
$$= \frac{1}{\sqrt{5}} \begin{bmatrix} -12 \\ 1 \end{bmatrix} \cdot \begin{bmatrix} 2 \\ -1 \end{bmatrix} = \frac{1}{\sqrt{5}}(-24 - 1) = -\frac{25}{\sqrt{5}} = -5\sqrt{5} \quad ■$$

Properties of the Gradient Vector   The directional derivative is a dot product. We can therefore write

$$D_\mathbf{u} f(x, y) = (\nabla f(x, y)) \cdot \mathbf{u}$$
$$= |\nabla f(x, y)||\mathbf{u}| \cos \theta$$

where θ is the angle between ∇f and $\mathbf{u}$. (See Figure 10.41.) Since $|\mathbf{u}| = 1$ ($\mathbf{u}$ is a unit vector), we have

$$D_\mathbf{u} f(x, y) = |\nabla f(x, y)| \cos \theta$$

The angle θ is in the interval $[0, 2\pi)$, and $\cos \theta$ is maximal when $\theta = 0$. We therefore find that $D_\mathbf{u} f(x, y)$ is maximal when $\mathbf{u}$ is in the direction of ∇f.

We will now show that, geometrically, the gradient vector at a point (x_0, y_0) is perpendicular to the level curve $f(x, y) = c$ that passes through this point. The level curve $f(x, y) = c$ is a curve in the x–y plane. It will be useful to think of traveling on this curve starting at time $t = 0$ at a point labeled $t = 0$. We can then refer to any point on the curve by giving the time t at which we pass through it. We say that we **parameterize** the curve by using the parameter t. We write the curve as the vector $\mathbf{r}(t) = \begin{bmatrix} x(t) \\ y(t) \end{bmatrix}$, just as we did in Subsection 9.4.3, where we parameterized lines. If we differentiate $\mathbf{r}(t)$ with respect to t, we obtain

$$\frac{d}{dt}\mathbf{r}(t) = \begin{bmatrix} \frac{dx(t)}{dt} \\ \frac{dy(t)}{dt} \end{bmatrix}$$

which is the tangent vector at the point $(x(t), y(t))$. Because

$$f(x, y) = c \tag{10.21}$$

and because x and y are both functions of t, we can write $w(t) = f(x(t), y(t)) = c$ and differentiate $w(t)$ with respect to t. We of course find that $dw/dt = 0$, since $w(t)$ is a constant. Using the chain rule to differentiate $f(x(t), y(t))$ yields

$$\frac{d}{dt} f(x(t), y(t)) = \frac{\partial f}{\partial x} \frac{dx}{dt} + \frac{\partial f}{\partial y} \frac{dy}{dt}$$

Hence,

$$\frac{\partial f}{\partial x} \frac{dx}{dt} + \frac{\partial f}{\partial y} \frac{dy}{dt} = 0$$

Figure 10.41 The angle θ between ∇f and the unit vector $\mathbf{u}$.

or

$$\nabla f \cdot \frac{d\mathbf{r}}{dt} = 0$$

This equation shows that the gradient of f at (x_0, y_0) is perpendicular to the level curve at (x_0, y_0), as illustrated in Figure 10.42.

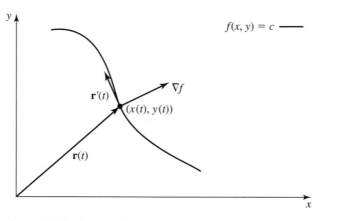

Figure 10.42 The gradient is perpendicular to the level curve.

We summarize these two important properties of the gradient as follows:

Suppose that $f(x, y)$ is a differentiable function. The gradient vector $\nabla f(x, y)$ has the following properties:

1. At each point (x_0, y_0), $f(x, y)$ increases most rapidly in the direction of the gradient vector $\nabla f(x_0, y_0)$.
2. The gradient vector of f at a point (x_0, y_0) is perpendicular to the level curve through (x_0, y_0).

EXAMPLE 7

Let

$$f(x, y) = x^2 y + y^2$$

In what direction does $f(x, y)$ increase most rapidly at $(1, 1)$?

Solution

The function $f(x, y)$ increases most rapidly at $(1, 1)$ in the direction of $\nabla f(1, 1)$. Since

$$\nabla f(x, y) = \begin{bmatrix} \frac{\partial f}{\partial x} \\ \frac{\partial f}{\partial y} \end{bmatrix} = \begin{bmatrix} 2xy \\ x^2 + 2y \end{bmatrix}$$

it follows that

$$\nabla f(1, 1) = \begin{bmatrix} 2 \\ 3 \end{bmatrix}$$

That is, $f(x, y)$ increases most rapidly in the direction $\begin{bmatrix} 2 \\ 3 \end{bmatrix}$ at the point $(1, 1)$. ■

EXAMPLE 8

Find a unit vector that is perpendicular to the level curve of the function

$$f(x, y) = x^2 - y^2$$

at $(1, 2)$.

Solution   The gradient of f at $(1, 2)$ is perpendicular to the level curve at $(1, 2)$. The gradient of f is given as

$$\nabla f(x, y) = \begin{bmatrix} 2x \\ -2y \end{bmatrix}$$

Hence,

$$\nabla f(1, 2) = \begin{bmatrix} 2 \\ -4 \end{bmatrix}$$

To normalize this vector, we divide $\nabla f(1, 2)$ by its length. Since

$$|\nabla f(1, 2)| = \sqrt{(2)^2 + (-4)^2} = \sqrt{4 + 16} = 2\sqrt{5}$$

the unit vector that is perpendicular to the level curve of $f(x, y)$ at $(1, 2)$ is

$$\mathbf{u} = \frac{1}{2\sqrt{5}} \begin{bmatrix} 2 \\ -4 \end{bmatrix} = \frac{1}{\sqrt{5}} \begin{bmatrix} 1 \\ -2 \end{bmatrix} = \begin{bmatrix} \frac{1}{5}\sqrt{5} \\ -\frac{2}{5}\sqrt{5} \end{bmatrix}$$ ■

Section 10.5 Problems

■ 10.5.1

1. Let $f(x, y) = x^2 + y^2$ with $x(t) = 3t$ and $y(t) = e^t$. Find the derivative of $w = f(x, y)$ with respect to t when $t = \ln 2$.

2. Let $f(x, y) = e^x \sin y$ with $x(t) = t$ and $y(t) = t^3$. Find the derivative of $w = f(x, y)$ with respect to t when $t = 1$.

3. Let $f(x, y) = \sqrt{x^2 + y^2}$ with $x(t) = t$ and $y(t) = \sin t$. Find the derivative of $w = f(x, y)$ with respect to t when $t = \pi/3$.

4. Let $f(x, y) = \ln(xy - x^2)$ with $x(t) = t^2$ and $y(t) = t$. Find the derivative of $w = f(x, y)$ with respect to t when $t = 5$.

5. Let $f(x, y) = \frac{1}{x} + \frac{1}{y}$ with $x(t) = \sin t$ and $y(t) = \cos t$. Find the derivative of $w = f(x, y)$ with respect to t when $t = \pi/4$.

6. Let $f(x, y) = xe^y$ with $x(t) = e^t$ and $y(t) = t^2$. Find the derivative of $w = f(x, y)$ with respect to t when $t = 0$.

7. Find $\frac{dz}{dt}$ for $z = f(x, y)$ with $x = u(t)$ and $y = v(t)$.

8. Find $\frac{dw}{dt}$ for $w = e^{f(x,y)}$ with $x = u(t)$ and $y = v(t)$.

■ 10.5.2

9. Find $\frac{dy}{dx}$ if $(x^2 + y^2)e^y = 0$.

10. Find $\frac{dy}{dx}$ if $(\sin x + \cos y)x^2 = 0$.

11. Find $\frac{dy}{dx}$ if $\ln(x^2 + y^2) = 3xy$.

12. Find $\frac{dy}{dx}$ if $\cos(x^2 + y^2) = \sin(x^2 - y^2)$.

13. Find $\frac{dy}{dx}$ if $y = \arccos x$.

14. Find $\frac{dy}{dx}$ if $y = \arctan x$.

15. The growth rate r of a particular organism is affected by both the availability of food and the number of competitors for the food source. Denote the amount of food available at time t by $F(t)$ and the number of competitors at time t by $N(t)$. The growth rate r can then be thought of as a function of the two time-dependent variables $F(t)$ and $N(t)$. Assume that the growth rate is an increasing function of the availability of food and a decreasing function of the number of competitors. How is the growth rate r affected if the availability of food decreases over time while the number of competitors increases?

16. Suppose that you travel along an environmental gradient, along which both temperature and precipitation increase. If the abundance of a particular plant species increases with both temperature and precipitation, would you expect to encounter this species more often or less often during your journey? (Use calculus to answer this question.)

■ 10.5.3

In Problems 17–24, find the gradient of each function.

17. $f(x, y) = x^3 y^2$

18. $f(x, y) = \frac{xy}{x^2 + y^2}$

19. $f(x, y) = \sqrt{x^3 - 3xy}$

20. $f(x, y) = x(x^2 - y^2)^{2/3}$

21. $f(x, y) = \exp\left[\sqrt{x^2 + y^2}\right]$

22. $f(x, y) = \tan\frac{x-y}{x+y}$

23. $f(x, y) = \ln\left(\frac{x}{y} + \frac{y}{x}\right)$

24. $f(x, y) = \cos(3x^2 - 2y^2)$

In Problems 25–30, compute the directional derivative of $f(x, y)$ at the given point in the indicated direction.

25. $f(x, y) = \sqrt{2x^2 + y^2}$ at $(1, 2)$ in the direction $\begin{bmatrix} 1 \\ 1 \end{bmatrix}$

26. $f(x, y) = x^2 \sin y$ at $(-1, 0)$ in the direction $\begin{bmatrix} 2 \\ -1 \end{bmatrix}$

27. $f(x, y) = e^{x+y}$ at $(0, 0)$ in the direction $\begin{bmatrix} -1 \\ -1 \end{bmatrix}$

28. $f(x, y) = x^3 y^2$ at $(2, 3)$ in the direction $\begin{bmatrix} -2 \\ 1 \end{bmatrix}$

29. $f(x, y) = 2xy^3 - 3x^2 y$ at $(1, -1)$ in the direction $\begin{bmatrix} 3 \\ 1 \end{bmatrix}$

30. $f(x, y) = ye^{x^2}$ at $(0, 2)$ in the direction $\begin{bmatrix} 4 \\ -1 \end{bmatrix}$

In Problems 31–34, compute the directional derivative of $f(x, y)$ at the point P in the direction of the point Q.

31. $f(x, y) = 2x^2 y - 3x$, $P = (2, 1)$, $Q = (3, 2)$

32. $f(x, y) = 4xy + y^2$, $P = (-1, 1)$, $Q = (3, 2)$

33. $f(x, y) = \sqrt{xy - 2x^2}$, $P = (1, 6)$, $Q = (3, 1)$

34. $f(x, y) = e^{x-y}$, $P = (2, 2)$, $Q = (1, -1)$

35. In what direction does $f(x, y) = 3xy - x^2$ increase most rapidly at $(-1, 1)$?

36. In what direction does $f(x, y) = e^x \cos y$ increase most rapidly at $(0, \pi/2)$?

37. In what direction does $f(x, y) = \sqrt{x^2 - y^2}$ increase most rapidly at $(5, 3)$?

38. In what direction does $f(x, y) = \ln(x^2 + y^2)$ increase most rapidly at $(1, 1)$?

39. Find a unit vector that is normal to the level curve of the function

$$f(x, y) = 3x + 4y$$

at the point $(-1, 1)$.

40. Find a unit vector that is normal to the level curve of the function

$$f(x, y) = x^2 + \frac{y^2}{9}$$

at the point $(1, 3)$.

41. Find a unit vector that is normal to the level curve of the function

$$f(x, y) = x^2 - y^3$$

at the point $(1, 3)$.

42. Find a unit vector that is normal to the level curve of the function

$$f(x, y) = xy$$

at the point $(2, 3)$.

43. Chemotaxis Chemotaxis is the chemically directed movement of organisms up a concentration gradient—that is, in the direction in which the concentration increases most rapidly. The slime mold *Dictyostelium discoideum* exhibits this phenomenon. Single-celled amoebas of this species move up the concentration gradient of a chemical called cyclic adenosine monophosphate (AMP). Suppose the concentration of cyclic AMP at the point (x, y) in the x–y plane is given by

$$f(x, y) = \frac{4}{|x| + |y| + 1}$$

If you place an amoeba at the point $(3, 1)$ in the x–y plane, determine in which direction the amoeba will move if its movement is directed by chemotaxis.

44. Suppose an organism moves down a sloped surface along the steepest line of descent. If the surface is given by

$$f(x, y) = x^2 - y^2$$

find the direction in which the organism will move at the point $(2, 3)$.

■ 10.6 Applications (Optional)

■ 10.6.1 Maxima and Minima

In Section 5.1, we introduced local extrema for functions of one variable. Local extrema can also be defined for functions of more than one independent variable; here, we will restrict our discussion to functions of two variables. Recall that we denoted by $B_\delta(x_0, y_0)$ the open disk with radius δ centered at (x_0, y_0). The following definition, with which you should compare the corresponding definition in Section 5.1, extends the notion of local extrema to functions of two variables:

Definition A function $f(x, y)$ defined on a set $D \subset \mathbf{R}^2$ has a **local** (or **relative**) **maximum** at a point (x_0, y_0) if there exists a $\delta > 0$ such that

$$f(x, y) \leq f(x_0, y_0) \quad \text{for all } (x, y) \in B_\delta(x_0, y_0) \cap D$$

A function $f(x, y)$ defined on a set $D \subset \mathbf{R}^2$ has a **local** (or **relative**) **minimum** at a point (x_0, y_0) if there exists a $\delta > 0$ such that

$$f(x, y) \geq f(x_0, y_0) \quad \text{for all } (x, y) \in B_\delta(x_0, y_0) \cap D$$

Informally, a local maximum (local minimum) is a point that is higher (lower) than all nearby points. We can define **global** (or **absolute**) extrema as well: If the inequalities in the definition hold for all $(x, y) \in D$, then f has an absolute maximum (minimum) at (x_0, y_0). Figure 10.43 shows an example of a function of two variables with a local maximum at $(0, 0)$.

How can we find local extrema? Recall that in the single-variable case, a horizontal tangent line at a point on the graph of a differentiable function is a necessary condition for the point to be a local extremum (Fermat's theorem). We can generalize this statement to functions of more than one variable: Looking at Figure 10.43, we see that the tangent plane at the local extremum is horizontal. The equation of a horizontal tangent plane on the graph of a differentiable function $f(x, y)$ at (x_0, y_0) is

$$z = f(x_0, y_0)$$

Comparing this equation with the general form of a tangent plane (Section 10.4), we

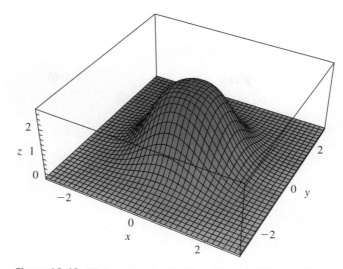

Figure 10.43 The graph of a function $f(x, y)$ with a local maximum at $(0, 0)$.

see that both $\partial f/\partial x$ and $\partial f/\partial y$ are equal to 0 at (x_0, y_0); in other words, $\nabla f(x_0, y_0) = \begin{bmatrix} 0 \\ 0 \end{bmatrix}$. Another way to see this is to argue that if the gradient were nonzero at (x_0, y_0), then the function would increase in the direction of the gradient and decrease in the opposite direction, so we could not be at a local extremum. We thus have the following criterion:

> If $f(x, y)$ has a local extremum at (x_0, y_0) and if f is differentiable at (x_0, y_0), then
> $$\nabla f(x_0, y_0) = \begin{bmatrix} 0 \\ 0 \end{bmatrix} \qquad (10.22)$$

A point (x_0, y_0) that satisfies (10.22) is called a **critical point**; points where $f(x, y)$ is not differentiable are also called critical points. We wish to emphasize that, as in Section 5.1, (10.22) is a *necessary* condition: It identifies only *candidates* for local extrema. Other critical points are also just candidates for local extrema. A further investigation is then needed to determine whether a candidate is indeed a local extremum.

EXAMPLE 1   Figure 10.44 shows the graph of the differentiable function $f(x, y) = x^2 + y^2 + 1$, $(x, y) \in \mathbf{R}^2$. We see that $f(x, y)$ has a local minimum at $(0, 0)$. Show that $\nabla f(0, 0) = \begin{bmatrix} 0 \\ 0 \end{bmatrix}$ and determine the equation of the tangent plane at $(0, 0)$.

Solution   We compute

$$\nabla f(x, y) = \begin{bmatrix} 2x \\ 2y \end{bmatrix}$$

Evaluating $\nabla f(0, 0)$ at $(0, 0)$, we find that

$$\nabla f(0, 0) = \begin{bmatrix} 0 \\ 0 \end{bmatrix}$$

The equation of the tangent plane at $(0, 0)$ is given by

$$z = f(0, 0) + (x - 0) f_x(0, 0) + (y - 0) f_y(0, 0)$$

Since $f(0, 0) = 1$, $f_x(0, 0) = 0$, and $f_y(0, 0) = 0$, the equation of the tangent plane at $(0, 0)$ is $z = 1$, which shows that the tangent plane is horizontal.   ■

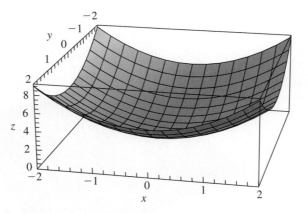

Figure 10.44 The graph of the function
$f(x, y) = x^2 + y^2 + 1$.

| EXAMPLE 2 | Find all critical points of |

$$f(x, y) = x^2 + y^2 + xy, \quad (x, y) \in \mathbf{R}^2$$

Solution   Since the function $f(x, y)$ is differentiable in $\mathbf{R}^2$, the only critical points are points that satisfy $\nabla f(x, y) = \begin{bmatrix} 0 \\ 0 \end{bmatrix}$. Now,

$$\nabla f(x, y) = \begin{bmatrix} 2x + y \\ 2y + x \end{bmatrix} = \begin{bmatrix} 0 \\ 0 \end{bmatrix}$$

We need to solve the system of linear equations

$$2x + y = 0$$
$$x + 2y = 0$$

It follows from the first equation that $y = -2x$. Substituting this into the second equation yields

$$x + 2(-2x) = 0, \quad \text{or} \quad -3x = 0, \quad \text{or} \quad x = 0$$

and, therefore, $y = 0$. The function thus has one critical point: $(0, 0)$.   ■

We now give a sufficient condition that will allow us to determine whether a candidate for a local extremum is indeed a local extremum and, if so, whether it is a local maximum or a local minimum. The proof of this condition is beyond the scope of this book.

Recall that in the case of a function of one variable we obtained the following sufficient condition for twice-differentiable functions: If $f'(x_0) = 0$ and $f''(x_0) > 0$ [$f''(x_0) < 0$], then $f(x)$ has a local minimum (a local maximum) at $x = x_0$. In the multivariable case, there is an analogous condition involving second partial derivatives.

Theorem Suppose the second partial derivatives of f are continuous in a disk centered at (x_0, y_0). Suppose also that $\nabla f(x_0, y_0) = \begin{bmatrix} 0 \\ 0 \end{bmatrix}$. Define

$$D = f_{xx}(x_0, y_0) f_{yy}(x_0, y_0) - \left(f_{xy}(x_0, y_0) \right)^2$$

1. If $D > 0$ and $f_{xx}(x_0, y_0) > 0$, then f has a local minimum at (x_0, y_0).
2. If $D > 0$ and $f_{xx}(x_0, y_0) < 0$, then f has a local maximum at (x_0, y_0).
3. If $D < 0$, then f does not have a local extremum at (x_0, y_0). The point (x_0, y_0) is then called a **saddle point**.

In all other cases, the test is inconclusive. We now return to Example 2 and determine whether $(0, 0)$ is a local extremum.

EXAMPLE 2

(continued) Determine whether the critical point $(0, 0)$ of $f(x, y) = x^2 + y^2 + xy$ in Example 2 is a local maximum or a local minimum (Figure 10.45).

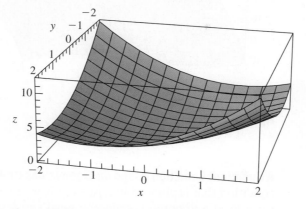

Figure 10.45 The graph of the function $f(x, y) = x^2 + y^2 + xy$.

Solution

We need to find all second partial derivatives. Since

$$\frac{\partial f(x, y)}{\partial x} = 2x + y \quad \text{and} \quad \frac{\partial f(x, y)}{\partial y} = x + 2y$$

we have

$$\frac{\partial^2 f}{\partial x^2} = 2, \quad \frac{\partial^2 f}{\partial y^2} = 2, \quad \frac{\partial^2 f}{\partial x \partial y} = 1$$

Hence,

$$D = f_{xx} f_{yy} - f_{xy}^2 = (2)(2) - (1)^2 = 3 > 0$$

Since $D > 0$ and $f_{xx} > 0$, we conclude that $(0, 0)$ is a local minimum (Figure 10.45). ■

EXAMPLE 3

Find all local extrema of

$$f(x, y) = 3xy - x^3 - y^3, \quad (x, y) \in \mathbf{R}^2$$

and classify them according to whether each is a local maximum, a local minimum, or neither.

Solution

The function $f(x, y)$ is differentiable on its domain. The critical points thus satisfy

$$\nabla f(x, y) = \begin{bmatrix} 3y - 3x^2 \\ 3x - 3y^2 \end{bmatrix} = \begin{bmatrix} 0 \\ 0 \end{bmatrix}$$

which yields $y = x^2$ and $x = y^2$. This set of equations has the solutions $(0, 0)$ and $(1, 1)$. Now,

$$\frac{\partial^2 f(x, y)}{\partial x^2} = -6x, \quad \frac{\partial^2 f(x, y)}{\partial y^2} = -6y, \quad \frac{\partial^2 f(x, y)}{\partial x \partial y} = 3$$

Therefore,

$$D = f_{xx} f_{yy} - (f_{xy})^2 = 36xy - 9$$

At $(1, 1)$, $D = 36 - 9 > 0$. Since $f_{xx}(1, 1) = -6 < 0$, $f(x, y)$ has a local maximum at $(1, 1)$. At $(0, 0)$, $D = -9 < 0$. The critical point $(0, 0)$ is neither a local maximum nor a local minimum, since $D < 0$ (Figure 10.46). ■

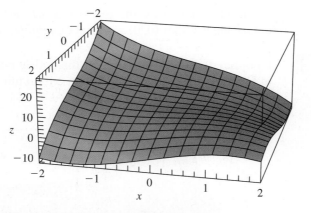

Figure 10.46 The graph of the function
$f(x, y) = 3xy - x^3 - y^3$.

A Sufficient Condition Based on Eigenvalues (Optional)   We will now give a sufficient condition that is phrased in terms of eigenvalues to determine whether a candidate for a local extremum is indeed a local extremum and, if so, what type (i.e., local maximum or local minimum). To motivate this condition, we will look at the following three functions defined for $(x, y) \in \mathbf{R}^2$:

$$f_1(x, y) = x^2 + y^2, \qquad f_2(x, y) = x^2 - y^2, \qquad f_3(x, y) = -x^2 - y^2$$

These functions are illustrated in Figures 10.47 through 10.49.

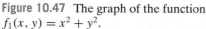

Figure 10.47 The graph of the function
$f_1(x, y) = x^2 + y^2$.

Figure 10.48 The graph of the function
$f_2(x, y) = x^2 - y^2$.

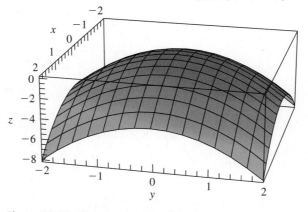

Figure 10.49 The graph of the function
$f_3(x, y) = -x^2 - y^2$.

Computing $\nabla f_i(x, y)$, $i = 1, 2$, and 3, we find that

$$\nabla f_1(x, y) = \begin{bmatrix} 2x \\ 2y \end{bmatrix}, \qquad \nabla f_2(x, y) = \begin{bmatrix} 2x \\ -2y \end{bmatrix}, \qquad \nabla f_3(x, y) = \begin{bmatrix} -2x \\ -2y \end{bmatrix}$$

It follows immediately that $(0, 0)$ is a candidate for a local extremum for all three functions.

The analogue of a second derivative of a function of one variable for a function of two variables with continuous second partial derivatives is the following second-derivative matrix, called the **Hessian matrix**:

$$\text{Hess } f(x, y) = \begin{bmatrix} \frac{\partial^2 f(x,y)}{\partial x^2} & \frac{\partial^2 f(x,y)}{\partial y \partial x} \\ \frac{\partial^2 f(x,y)}{\partial x \partial y} & \frac{\partial^2 f(x,y)}{\partial y^2} \end{bmatrix}$$

Computing this matrix for the functions $f_i(x, y)$, $i = 1, 2$, and 3, we obtain

$$\text{Hess } f_1(x, y) = \begin{bmatrix} 2 & 0 \\ 0 & 2 \end{bmatrix}, \qquad \text{Hess } f_2(x, y) = \begin{bmatrix} 2 & 0 \\ 0 & -2 \end{bmatrix},$$

$$\text{Hess } f_3(x, y) = \begin{bmatrix} -2 & 0 \\ 0 & -2 \end{bmatrix}$$

It now turns out that the eigenvalues of the second-derivative matrix provide a sufficient condition for determining whether a critical point is a local maximum, a local minimum, or neither. The following holds:

Suppose the second partial derivatives of f are continuous in a disk centered at (x_0, y_0). Suppose also that $\nabla f(x_0, y_0) = \begin{bmatrix} 0 \\ 0 \end{bmatrix}$. Then

1. If the two eigenvalues of the second-derivative matrix Hess $f(x_0, y_0)$ at (x_0, y_0) are positive, then f has a local minimum at (x_0, y_0).
2. If the two eigenvalues of Hess $f(x_0, y_0)$ are negative, then f has a local maximum at (x_0, y_0).
3. If the two eigenvalues of Hess $f(x_0, y_0)$ are of opposite signs, then f does not have a local extremum at (x_0, y_0). The point (x_0, y_0) is then called a **saddle point**.

In all other cases, the test is inconclusive. Returning to our example, we need to evaluate Hess $f_i(x, y)$, $i = 1, 2$, and 3, at $(0, 0)$. We see that Hess $f_i(x, y)$, $i = 1, 2$, and 3, do not depend on x or y; thus, Hess $f_i(0, 0) = $ Hess $f_i(x, y)$. We can read off the eigenvalues of each of the second-derivative matrices evaluated at $(0, 0)$, since they are in diagonal form. We find that the eigenvalues of Hess $f_1(0, 0)$ are both 2; hence, $f_1(0, 0)$ is a local minimum, which agrees with the graph in Figure 10.47. The eigenvalues of Hess $f_2(0, 0)$ are 2 and -2, and we conclude that $f_2(0, 0)$ is not a local extremum. (See Figure 10.48.) The graph of $f(x, y)$ resembles a saddle near $(0, 0)$ — hence the name "saddle point." The eigenvalues of Hess $f_3(0, 0)$ are both -2, and we conclude that $f_3(0, 0)$ is a local maximum. (See Figure 10.49.)

We assumed in the second-derivative criterion that all second partial derivatives are continuous in a disk centered at (x_0, y_0). This assumption implies that

$$\frac{\partial^2 f(x, y)}{\partial x\, \partial y} = \frac{\partial^2 f(x, y)}{\partial y\, \partial x}$$

That is, the off-diagonal elements of Hess $f(x, y)$ are identical and the Hessian matrix is of the form $\begin{bmatrix} a & c \\ c & b \end{bmatrix}$. Such a matrix is called **symmetric**. We can show that the eigenvalues of a symmetric matrix are always real. (See Problem 34.) This fact has an important consequence: If the second partial derivatives of f are continuous in a disk centered at (x_0, y_0), then the eigenvalues of Hess $f(x_0, y_0)$ are both real. Provided that neither eigenvalue is equal to zero, one of the three cases in our second-derivative criterion occurs, thereby allowing us to settle the question whether the candidate (x_0, y_0) for which $\nabla f(x_0, y_0) = \begin{bmatrix} 0 \\ 0 \end{bmatrix}$ is a local extremum and, if so, of what type it is. If one or both eigenvalues are equal to zero, we cannot say anything about the nature of the critical point on the basis of the Hessian matrix. This case is discussed in Problem 11.

EXAMPLE 4

Find the local extrema of

$$f(x, y) = 2x^2 - xy + y^4, \quad (x, y) \in \mathbf{R}^2$$

Solution

We compute

$$\nabla f(x, y) = \begin{bmatrix} 4x - y \\ -x + 4y^3 \end{bmatrix}$$

Setting both partial derivatives equal to 0, we find that

$$4x - y = 0 \quad \text{and} \quad -x + 4y^3 = 0$$

It follows from the first equation that $x = y/4$. Substituting this into the second equation, we obtain

$$-\frac{y}{4} + 4y^3 = 0$$

$$-\frac{y}{4}(1 - 16y^2) = 0$$

yielding

$$y_1 = 0, \quad y_2 = \frac{1}{4}, \quad \text{and} \quad y_3 = -\frac{1}{4}$$

Hence, the corresponding x-values are

$$x_1 = 0, \quad x_2 = \frac{1}{16}, \quad \text{and} \quad x_3 = -\frac{1}{16}$$

Since ∇f is defined for all $(x, y) \in \mathbf{R}^2$, there are no other critical points. The three candidates for local extrema are thus

$$(0, 0), \quad \left(\frac{1}{16}, \frac{1}{4}\right), \quad \text{and} \quad \left(-\frac{1}{16}, -\frac{1}{4}\right)$$

The Hessian matrix is of the form

$$\text{Hess } f(x, y) = \begin{bmatrix} 4 & -1 \\ -1 & 12y^2 \end{bmatrix}$$

We evaluate the Hessian matrix at each candidate and compute its eigenvalues:

(i) Hess $f(0, 0) = \begin{bmatrix} 4 & -1 \\ -1 & 0 \end{bmatrix}$

The eigenvalues satisfy

$$\det \begin{bmatrix} 4 - \lambda & -1 \\ -1 & -\lambda \end{bmatrix} = \lambda(\lambda - 4) - 1 = \lambda^2 - 4\lambda - 1 = 0$$

Thus,

$$\lambda_{1,2} = \frac{4 \pm \sqrt{16 + 4}}{2} = 2 \pm \sqrt{5} \approx \begin{cases} 4.2361 \\ -0.2361 \end{cases}$$

implying that f has a saddle point at $(0, 0)$.

(ii) Hess $f\left(\frac{1}{16},\frac{1}{4}\right) = \begin{bmatrix} 4 & -1 \\ -1 & \frac{3}{4} \end{bmatrix}$

The eigenvalues satisfy

$$\det\begin{bmatrix} 4-\lambda & -1 \\ -1 & \frac{3}{4}-\lambda \end{bmatrix} = (4-\lambda)(\frac{3}{4}-\lambda)-1 = 3-4\lambda-\frac{3}{4}\lambda+\lambda^2-1$$

$$= \lambda^2 - \frac{19}{4}\lambda + 2 = 0$$

Thus,

$$\lambda_{1,2} = \frac{\frac{19}{4} \pm \sqrt{\frac{361}{16}-8}}{2} = \frac{19}{8} \pm \frac{1}{8}\sqrt{233} \approx \begin{cases} 4.2830 \\ 0.4670 \end{cases}$$

implying that f has a local minimum at $(\frac{1}{16},\frac{1}{4})$.

(iii) Hess $f\left(-\frac{1}{16},-\frac{1}{4}\right) = \begin{bmatrix} 4 & -1 \\ -1 & \frac{3}{4} \end{bmatrix}$

This is the same matrix as that for $(\frac{1}{16},\frac{1}{4})$ [i.e., case (ii)]. We thus conclude that f has a local minimum at $(-\frac{1}{16},-\frac{1}{4})$ as well.

The graph of $f(x,y)$ is illustrated in Figure 10.50. ■

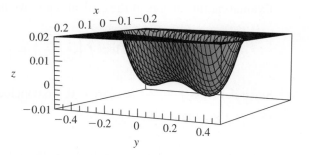

Figure 10.50 The graph of the function $f(x,y) = 2x^2 - xy + y^4$.

In the last example, we saw that finding the eigenvalues of the Hessian matrix can be time consuming. There is another criterion which follows from the relationship that expresses the determinant and the trace of a 2×2 matrix in terms of the eigenvalues of the matrix. Recall that if A is a 2×2 matrix with eigenvalues λ_1 and λ_2, then $\det A = \lambda_1\lambda_2$ and $\operatorname{tr} A = \lambda_1 + \lambda_2$. If the eigenvalues of A are both real (as is the case for a symmetric matrix), and if $\det A > 0$, then either both λ_1 and λ_2 are positive or both are negative. If, in addition, $\operatorname{tr} A > 0$, then both λ_1 and λ_2 are positive. We thus arrive at the following criterion:

Suppose the second partial derivatives of f are continuous in a disk centered at (x_0, y_0). Suppose also that $\nabla f(x_0, y_0) = \begin{bmatrix} 0 \\ 0 \end{bmatrix}$. Then

1. If $\det$ Hess $f(x_0, y_0) > 0$ and tr Hess $f(x_0, y_0) > 0$, then f has a local minimum at (x_0, y_0).
2. If $\det$ Hess $f(x_0, y_0) > 0$ and tr Hess $f(x_0, y_0) < 0$, then f has a local maximum at (x_0, y_0).
3. If $\det$ Hess $f(x_0, y_0) < 0$, then (x_0, y_0) is not a local extremum; instead, (x_0, y_0) is a saddle point.

Recall that if one of the eigenvalues of Hess $f(x_0, y_0)$ is equal to 0 [or, equivalently, if det Hess $f(x_0, y_0) = 0$], then we cannot say anything about the nature of the critical point on the basis of the Hessian matrix. (Such a case is explored in Problem 11.)

EXAMPLE 5   Find and classify the critical points of

$$f(x, y) = x^3 - 4xy + y, \quad (x, y) \in \mathbf{R}^2$$

Solution   We find that

$$\nabla f(x, y) = \begin{bmatrix} 3x^2 - 4y \\ -4x + 1 \end{bmatrix} = \begin{bmatrix} 0 \\ 0 \end{bmatrix}$$

when

$$3x^2 - 4y = 0 \quad \text{and} \quad -4x + 1 = 0$$

The second equation yields $x = 1/4$. Substituting this value into the first equation yields

$$\frac{3}{16} - 4y = 0$$

implying that $y = 3/64$. Since f is differentiable for all $(x, y) \in \mathbf{R}^2$, there is only one critical point: $(\frac{1}{4}, \frac{3}{64})$. To classify the critical point, we compute

$$\text{Hess } f(x, y) = \begin{bmatrix} 6x & -4 \\ -4 & 0 \end{bmatrix}$$

Evaluating this matrix at the critical point, we find that

$$\text{Hess } f\left(\frac{1}{4}, \frac{3}{64}\right) = \begin{bmatrix} \frac{3}{2} & -4 \\ -4 & 0 \end{bmatrix}$$

Since det Hess $f(\frac{1}{4}, \frac{3}{64}) = -16 < 0$, we conclude that $f(x, y)$ has a saddle point at $(\frac{1}{4}, \frac{3}{64})$. (See Figure 10.51.)   ■

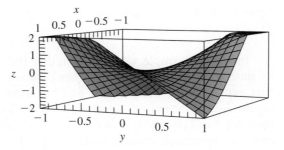

Figure 10.51 The graph of the function $f(x, y) = x^3 - 4xy + y$.

EXAMPLE 6   Find and classify the critical points of

$$f(x, y) = \sqrt{x^2 + y^2}, \quad (x, y) \in \mathbf{R}^2$$

Solution   We find that

$$\nabla f(x, y) = \begin{bmatrix} \frac{2x}{2\sqrt{x^2+y^2}} \\ \frac{2y}{2\sqrt{x^2+y^2}} \end{bmatrix} = \frac{1}{\sqrt{x^2 + y^2}} \begin{bmatrix} x \\ y \end{bmatrix}, \quad (x, y) \neq (0, 0)$$

Since the gradient of f is undefined at $(0, 0)$, the point $(0, 0)$ is a critical point. There are no other critical points, because $\nabla f(x, y) \neq \begin{bmatrix} 0 \\ 0 \end{bmatrix}$ for all $(x, y) \neq (0, 0)$. Now, $f(x, y) > 0$ for $(x, y) \neq (0, 0)$ and $f(x, y) = 0$ for $(x, y) = (0, 0)$. Therefore, $f(x, y)$ has a local minimum at $(0, 0)$. (See Figure 10.52.) Note that we cannot use the Hessian here to decide whether $(0, 0)$ is a local extremum and, if so, of what type it is, since the theorem requires that the gradient be zero at the point, but here the gradient is undefined at $(0, 0)$. [The Hessian is also not defined at $(0, 0)$.] ■

Figure 10.52 The graph of the function $f(x, y) = \sqrt{x^2 + y^2}$.

Global Extrema We now turn our discussion to global extrema. Recall that, for functions of one variable, the extreme-value theorem guarantees the existence of global extrema for functions defined on a *closed* interval. The analogue of closed intervals in the two-dimensional plane is a **closed** set; similarly, the analogue of an open interval is an **open** set. An example of a closed set is a closed disk; an example of an open set is an open disk. (See Section 10.2.)

To define these concepts more generally, we start with a set $D \subset \mathbf{R}^2$. A point (x, y) is called an **interior point** of D if there exists a $\delta > 0$ such that the disk centered at (x, y) with radius δ is contained in D—that is, if $B_\delta(x, y) \subset D$. (See Figure 10.53a.) A point (x, y) is a **boundary point** of D if every disk centered at (x, y) contains both points in D and points not in D; the boundary point (x, y) need not be contained in D. (See Figure 10.53b.) The **interior** of D consists of all interior points of D; the **boundary** of D consists of all boundary points of D. A set $D \subset \mathbf{R}^2$ is **open** if it consists only of interior points; a set $D \subset \mathbf{R}^2$ is **closed** if it contains all its boundary points as well as its interior points. (See Figure 10.54.)

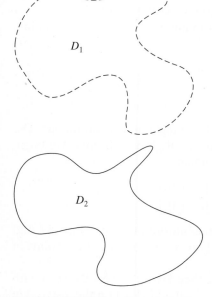

Figure 10.54 The set D_1 on the top is open. The set D_2 on the bottom is closed; the solid line is the boundary.

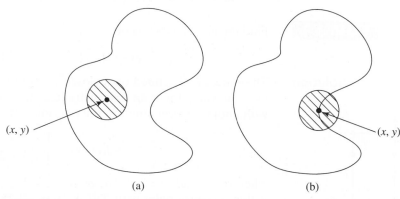

(a)　　　　(b)

Figure 10.53 The point (x, y) on the left is an interior point; the point (x, y) on the right is a boundary point.

Most of the time, the domains of our functions will be rectangles or disks. Figure 10.55 illustrates the concepts we just learned on the unit disk. We start with the open unit disk $\{(x, y) : x^2 + y^2 < 1\}$ (Figure 10.55a). Every point in this set is an interior point. The unit circle $\{(x, y) : x^2 + y^2 = 1\}$ is the boundary of the open unit disk (Figure 10.55b). If we combine the open disk and its boundary, we obtain the closed unit disk $\{(x, y) : x^2 + y^2 \leq 1\}$ (Figure 10.55c).

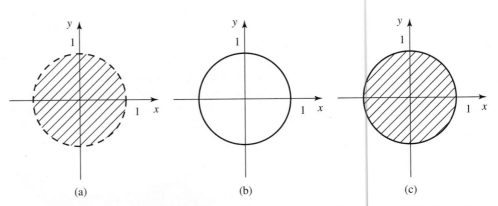

Figure 10.55 The set on the left is the open unit disk. The solid line in the middle figure is the boundary of the unit disk. The set on the right combines the open disk and its boundary; it is the closed unit disk.

To formulate the extreme-value theorem in $\mathbf{R}^2$, we also need the notion of a **bounded** set. A set is bounded if it is contained within some disk.

> **Extreme-Value Theorem in $\mathbf{R}^2$** If f is continuous on a closed and bounded set $D \subset \mathbf{R}^2$, then f has both a global maximum and a global minimum on D.

Global extrema can occur in the interior of D or on the boundary of D. We already discussed how to find candidates for local extrema in the interior. To find local extrema on the boundary, it is useful to think of $f(x, y)$ restricted to the boundary of D. This restriction often allows us to write the function that is so restricted as a function of just one variable; we can then use the tools of single-variable calculus to find all candidates for local extrema on the boundary of D. (See Example 7.) To find global extrema for continuous functions defined on a closed and bounded set, we thus proceed as follows:

1. Determine all candidates for local extrema in the interior of D.
2. Determine all candidates for local extrema on the boundary of D.
3. Select the global maximum and the global minimum from the set of points determined in steps 1 and 2.

EXAMPLE 7 Find the global extrema of

$$f(x, y) = x^2 - 3y + y^2, \quad -1 \leq x \leq 1, 0 \leq y \leq 2$$

Solution The function is defined on a closed and bounded rectangle and is continuous. The extreme-value theorem thus guarantees the existence of global extrema. We begin with finding critical points in the interior of the domain,

$$\nabla f(x, y) = \begin{bmatrix} 2x \\ -3 + 2y \end{bmatrix} = \begin{bmatrix} 0 \\ 0 \end{bmatrix}$$

when $x = 0$ and $y = 3/2$. The point $(0, 3/2)$ is in the interior of the domain of f and is thus a critical point with $f(0, 3/2) = -2.25$. There are no other critical points in the interior of the domain of f.

Next, we need to check the boundary values. (See Figure 10.56.) We start with the line segment C_1, which connects the points $(-1, 0)$ and $(1, 0)$ on the x-axis. On C_1, $y = 0$. Hence, on C_1, f is of the form

$$f(x, 0) = x^2, \quad -1 \leq x \leq 1$$

By restricting $f(x, y)$ to the curve $y = 0$, we obtained a function of just one variable. Using single-variable calculus, we find that $f'(x, 0) = 2x = 0$ for $x = 0$. The critical

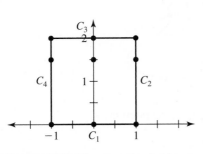

Figure 10.56 The domain of $f(x, y)$ in Example 7. The boundary consists of the line segments C_1, C_2, C_3, and C_4.

point on C_1 is thus $(0, 0)$, with $f(0, 0) = 0$; in addition, there are the two endpoints $(-1, 0)$, with $f(-1, 0) = 1$, and $(1, 0)$, with $f(1, 0) = 1$.

On C_2, we have $x = 1$, which yields $f(1, y) = 1 - 3y + y^2, 0 \leq y \leq 2$, which is again a function of just one variable. Now, $f'(1, y) = -3 + 2y = 0$ for $y = 3/2$. Hence, we find a candidate at $(1, 3/2)$, with $f(1, 3/2) = -1.25$; other candidates are the endpoints $(1, 0)$, with $f(1, 0) = 1$, and $(1, 2)$, with $f(1, 2) = -1$.

On C_3, we have $y = 2$, yielding $f(x, 2) = x^2 - 2$. Thus, $f'(x, 2) = 2x = 0$ for $x = 0$, giving the critical point $(0, 2)$, with $f(0, 2) = -2$. Other candidates are the endpoints $(-1, 2)$, with $f(-1, 2) = -1$, and $(1, 2)$, with $f(1, 2) = -1$.

On C_4, $x = -1$ and $f(-1, y) = 1 - 3y + y^2$, which is the same as on C_2. We thus have the additional points $(-1, 3/2)$, with $f(-1, 3/2) = -1.25$; $(-1, 0)$, with $f(-1, 0) = 1$; and $(-1, 2)$, with $f(-1, 2) = -1$.

Comparing all the values of $f(x, y)$ at the candidate points (see Figure 10.56 and the table that follows), we find that the global minimum is $f(0, 3/2) = -2.25$ and the global maxima are $f(-1, 0) = 1$ and $f(1, 0) = 1$.

(x, y)	$(0, 3/2)$	$(0, 0)$	$(-1, 0)$	$(1, 0)$	$(1, 3/2)$	$(1, 2)$	$(0, 2)$	$(-1, 2)$	$(1, 2)$	$(-1, 3/2)$
$f(x, y)$	-2.25	0	1	1	-1.25	-1	-2	-1	-1	-1.25

■

EXAMPLE 8

Find the absolute maxima and minima of $f(x, y) = x^2 + y^2 - 2x + 4$ on the disk $D = \{(x, y) : x^2 + y^2 \leq 4\}$.

Solution

The function is defined on a closed and bounded disk and is continuous. The extreme-value theorem thus guarantees global extrema. We begin with finding critical points in the interior of the domain,

$$\nabla f(x, y) = \begin{bmatrix} 2x - 2 \\ 2y \end{bmatrix} = \begin{bmatrix} 0 \\ 0 \end{bmatrix}$$

when $x = 1$ and $y = 0$. Since $x^2 + y^2 = 1 \leq 4$, the point $(1, 0)$ is in the interior of the domain of f and is thus a critical point, with $f(1, 0) = 3$. There are no other critical points in the interior of the domain of f.

Next, we seek extrema on the boundary of the domain: the circle $x^2 + y^2 = 4$. The circle is centered at the origin $(0, 0)$ and has radius 2. We need a mathematical description of the circle in terms of a function of just one variable so that we can use single-variable calculus to identify extrema on the boundary. Toward that end, every point (x, y) on the circle can be written as

$$x = 2 \cos \theta$$
$$y = 2 \sin \theta$$

for $0 \leq \theta < 2\pi$. This is called a **parameterization** of the circle.

On this circle,

$$f(x, y) = x^2 + y^2 - 2x + 4$$
$$= 4 \cos^2 \theta + 4 \sin^2 \theta - 4 \cos \theta + 4$$
$$= 4 - 4 \cos \theta + 4 = 8 - 4 \cos \theta$$

where we used $\sin^2 \theta + \cos^2 \theta = 1$. To find maxima and minima of the single-valued function $g(\theta) = 8 - 4 \cos \theta$, we need to differentiate $g(\theta)$ thus:

$$g'(\theta) = 4 \sin \theta$$

Then we solve $g'(\theta) = 0$ in $[0, 2\pi)$. We find the two angles $\theta = 0$ and $\theta = \pi$. Now,

$$g(0) = 8 - 4 = 4 \quad \text{and} \quad g(\pi) = 8 + 4 = 12$$

The maximum on the boundary is at $\theta = \pi$, which corresponds to the point $(-2, 0)$. The minimum on the boundary is at $\theta = 0$, which corresponds to the point $(2, 0)$.

Comparing the extrema on the boundary with the extremum in the interior of the set D, we find that the global minimum is in the interior at $(1, 0)$ and the global maximum is on the boundary at $(-2, 0)$ (Figure 10.57). ■

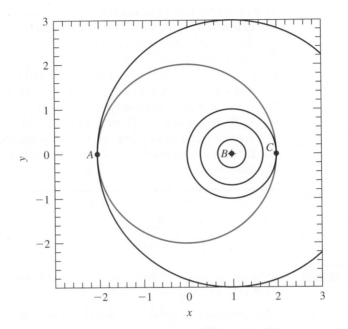

Figure 10.57 Shown are the contour lines of the function $f(x, y) = x^2 + y^2 - 2x + 4$ [in black for the values $f(x, y) = c$ with $c = 3.1, 3.5, 4$, and 12] and the boundary of the disk $x^2 + y^2 \leq 4$ (in blue). The point $B = (1, 0)$ is the local minimum in the interior of the disk $x^2 + y^2 \leq 4$, the point $A = (-2, 0)$ is the local maximum on the boundary of the disk, and the point $C = (2, 0)$ is the local minimum on the boundary of the disk.

We conclude this subsection with an application.

EXAMPLE 9 Determine the values of three nonnegative numbers whose sum is 90 and whose product is maximal.

Solution We denote the three numbers by x, y, and z, respectively. Then $x + y + z = 90$. Now, their product is xyz, and since $z = 90 - x - y$, we can write the product as $xyz = xy(90 - x - y)$. Our goal is to maximize this product. We define the function

$$f(x, y) = xy(90 - x - y), \quad x + y \leq 90, \ x \geq 0, \ y \geq 0$$

Since x, y, and z are nonnegative numbers and their sum is equal to 90, the domain is the set $\{(x, y) : x + y \leq 90, x \geq 0, y \geq 0\}$, which is the triangular region bounded by the lines $x = 0$, $y = 0$, and $y = 90 - x$.

We need to find (x, y) so that $f(x, y)$ is maximal. Now,

$$\nabla f(x, y) = \begin{bmatrix} 90y - 2xy - y^2 \\ 90x - x^2 - 2xy \end{bmatrix} = \begin{bmatrix} 0 \\ 0 \end{bmatrix}$$

when

$$y(90 - 2x - y) = 0 \quad \text{and} \quad x(90 - 2y - x) = 0$$

Solutions with $x = 0$ or $y = 0$ are points on the boundary. To find solutions in the interior of the domain, we need to solve

$$2x + y = 90$$
$$x + 2y = 90$$

Multiplying the second equation by 2 and subtracting the first equation from the resulting equation yields

$$3y = 90$$

or $y = 30$. Using the first equation, we obtain $2x = 90 - y = 60$, so $x = 30$, yielding the candidate $(30, 30)$, which is in the interior of the domain, with $f(30, 30) = 30^3 = 27,000$. There are no other candidates for local extrema in the interior of D.

The function $f(x, y)$ is continuous on a closed and bounded set, namely the triangle with corners $(0, 0)$, $(0, 90)$, and $(90, 0)$. The extreme-value theorem guarantees that f has a global maximum on the domain. We see that $f(x, y)$ takes on the value 0 on the boundary of the domain. Comparing the values of $f(x, y)$ on the boundary of D with the value at the candidate point $(30, 30)$, we conclude that the function $f(x, y)$ has the global maximum at the interior candidate $(30, 30)$.

The product xyz is therefore maximal when $x = y = z = 30$.   ■

■ 10.6.2 Extrema with Constraints

A number of studies have shown that, in butterflies which lay their eggs singly, egg size decreases with maternal age. Begon and Parker (1986) proposed a mathematical model to explain this decline in egg size in terms of a maternal strategy that would optimize reproductive fitness. The basic assumptions of their model are that all resources necessary for egg production are gathered before eggs are laid and that clutch size is fixed (e.g., a single egg per clutch). Under these assumptions, Begon and Parker were able to show that if egg fitness is an increasing and concave-down function of egg size, then a decline in egg size with maternal age is an optimal maternal strategy (i.e., a strategy that would maximize maternal reproductive fitness).

Mathematically, the problem can be phrased in terms of finding the maximum of a function that describes maternal reproductive fitness in terms of egg size per clutch during the lifetime of the individual under the constraint that the total amount of reproductive resources are fixed before reproduction starts. This type of problem falls into the category of finding extrema with constraints.

To have a concrete example at hand when we go through the discussion that follows on how to find such extrema, let's take the case of a female that has at most two clutches, each of size n, during her lifetime. We denote egg size of the first clutch by x_1, and egg size of the second clutch by x_2; we also assume that eggs in the same clutch have the same size. If the total amount of resources available for reproduction is R, then the constraint can be written as

$$nx_1 + nx_2 = R \tag{10.23}$$

[In order for the units to agree in (10.23), assume that both egg size and amount of resources are measured in the same units—say, calories.] We define a function $\rho(x)$ that describes egg fitness as a function of egg size x. If p_i is the probability that the female survives to lay her ith clutch ($i = 1, 2$), then the following function can be used to express maternal reproductive fitness:

$$f(x_1, x_2) = p_1 n \rho(x_1) + p_2 n \rho(x_2)$$

Our goal is thus to find extrema of the function $f(x_1, x_2)$ under the constraint $nx_1 + nx_2 = R$.

Finding extrema with constraints involves two functions: one describing the constraint, the other the function we wish to maximize. All of the constraints in this section will be of the form

$$g(x, y) = 0$$

For instance, the constraint (10.23) can be written as $g(x_1, x_2) = nx_1 + nx_2 - R = 0$.

We can illustrate the constraint $g(x, y) = 0$ as a set of points in the x–y plane: $\{(x, y) : g(x, y) = 0\}$. These will typically be curves of the sort shown in Figure 10.58. In the discussion that follows, it will be useful to think of traveling on this

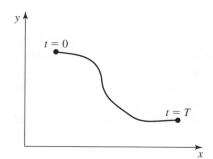

Figure 10.58 A constraint curve.

curve starting, say, at the point labeled $t = 0$ and ending at the point $t = T$. We can then refer to any point on the curve by giving the time t at which we pass through it. We say that we *parameterize* the curve by using the parameter t. The curve can thus be written as a vector-valued function

$$\mathbf{r}(t) = \begin{bmatrix} x(t) \\ y(t) \end{bmatrix}$$

where, at time t, we pass through the point $(x(t), y(t))$, just as in Subsection 9.4.3, where we parameterized lines. The constraint then satisfies the equation $g(\mathbf{r}(t)) = 0$. Let's look at two examples.

EXAMPLE 10 Let

$$g(x, y) = x + y - 1$$

Then the set $\{(x, y) : g(x, y) = 0\}$ describes a curve in the x–y plane, namely, the straight line $y = 1 - x$. We can parameterize this curve by using

$$x(t) = t \qquad \text{and, consequently,} \qquad y(t) = 1 - t$$

This is not the only parameterization, but it is the simplest. At time $t = 0$, for instance, we are at the point $(0, 1)$. At time $t = 1$, we are at $(1, 0)$; at time $t = 2$, we are at $(2, -1)$; and so on, as illustrated in Figure 10.59. This parameterization describes motion along the line to the right and downward, at the rate of one unit of x per unit of time. ■

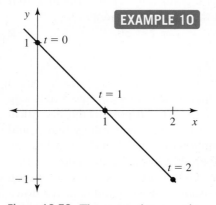

Figure 10.59 The constraint curve in Example 10.

Note that the parameterization $x = t$ and $y = f(t)$ always works when y is given explicitly as a function of x [i.e., $y = f(x)$], as in Example 10. The next example shows a parameterization of the unit circle.

EXAMPLE 11 Let

$$g(x, y) = x^2 + y^2 - 1$$

Then $g(x, y) = 0$ is the circle with radius 1 centered at the origin. A natural parameterization of the unit circle is

$$x(t) = \cos t \quad \text{and} \quad y(t) = \sin t, \quad 0 \le t < 2\pi$$

The parameter t describes the angle, as illustrated in Figure 10.60. That is, we move counterclockwise around the circle at 1 radian per unit time. ■

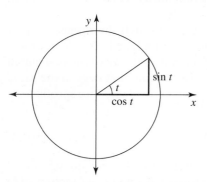

Figure 10.60 The constraint curve in Example 11.

What is the advantage of such a parameterization? The answer is that it will allow us to relate the gradient of g at (x, y), namely, $\nabla g(x, y)$, to the tangent at (x, y) on the graph of the constraint curve. Let's see how. Let $\mathbf{r}(t)$ be the vector from the origin to the point $(x(t), y(t))$ on the parameterized constraint curve $g(x, y) = 0$. Using the formal definition of the derivative, we define the derivative of $\mathbf{r}(t)$ with respect to t as

$$\mathbf{r}'(t) = \frac{d}{dt}\mathbf{r}(t) = \lim_{\Delta t \to 0} \frac{\mathbf{r}(t + \Delta t) - \mathbf{r}(t)}{\Delta t}$$

provided that the limit exists. The vector $\mathbf{r}(t + \Delta t) - \mathbf{r}(t)$ is illustrated in Figure 10.61. Dividing this vector by Δt changes its length, but not its direction. It thus appears that, in the limit $\Delta t \to 0$, the limiting vector $\mathbf{r}'(t)$, if it exists, is tangent to the curve at $(x(t), y(t))$. (See Figure 10.62.) We'll try to understand this in the case where the curve $g(x, y) = 0$ can be written as a function that gives y explicitly in terms of x, namely, $y = h(x)$, and where $h'(x)$ exists. Recall that the derivative of $\mathbf{r}(t)$ with respect to t is defined componentwise:

$$\frac{d}{dt}\mathbf{r}(t) = \frac{d}{dt}\begin{bmatrix} x(t) \\ y(t) \end{bmatrix} = \begin{bmatrix} \frac{dx(t)}{dt} \\ \frac{dy(t)}{dt} \end{bmatrix}$$

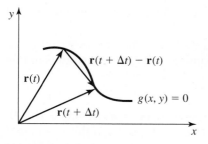

Figure 10.61 The vector $\mathbf{r}(t + \Delta t) - \mathbf{r}(t)$.

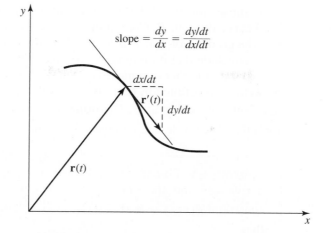

Figure 10.62 The vector $\mathbf{r}'(t)$.

The slope of the tangent line at a point (x, y) on the curve is given by $h'(x)$. Since

$$h'(x) = \frac{dy}{dx} = \frac{dy/dt}{dx/dt}$$

the slope of the tangent line can be expressed as the ratio of the components dx/dt and dy/dt of the vector $\mathbf{r}'(t)$, as illustrated in Figure 10.62, implying that $\mathbf{r}'(t)$ is tangential to the curve at $(x(t), y(t))$.

The preceding result has an important implication: Since $g(x, y) = 0$ is a level curve, the gradient of g at (x, y), namely, $\nabla g(x, y)$, is perpendicular to the level curve at (x, y) and thus to $\mathbf{r}'(t)$. (See Subsection 10.5.3.) How can we see this? Because the curve is given by the equation $g(x(t), y(t)) = 0$, it follows that $\frac{d}{dt} g(x(t), y(t)) = 0$. Now, using the chain rule, we find that

$$\frac{d}{dt} g(x(t), y(t)) = \frac{\partial g}{\partial x} \frac{dx}{dt} + \frac{\partial g}{\partial y} \frac{dy}{dt} = \nabla g(x, y) \cdot \mathbf{r}'(t)$$

where we used the definition of the dot product in the last step. We thus have

$$\nabla g(x, y) \cdot \mathbf{r}'(t) = 0$$

which implies that the gradient vector $\nabla g(x, y)$ is perpendicular to the tangent vector $\mathbf{r}'(t)$ at $(x(t), y(t))$, as claimed previously. This is an important fact that we will need next when we try to find extrema under constraints.

Now let's go back to the problem of finding extrema under constraints. We denote the function we wish to optimize by $f(x, y)$ and the constraint by $g(x, y) = 0$. Finding extrema with constraints amounts to restricting the function $f(x, y)$ to the constraint curve and seeking its extrema there. The constraint $g(x, y) = 0$ defines a set of points (x, y) in the x–y plane. The graph of $z = f(x, y)$ is a surface in three-dimensional x–y–z space. Using level curves for $f(x, y)$, we can represent $f(x, y)$ in the x–y plane. We can then graph both the level curves of $f(x, y)$ and the constraint $g(x, y) = 0$ in the same two-dimensional coordinate system. (See Figure 10.63.) We claim that $f(x, y)$ has a local extremum at the point P in the figure under the constraint $g(x, y) = 0$. To see this, imagine traveling on the curve $g(x, y) = 0$ starting at the point Q and traveling in the direction of the arrow. You first intersect the level curve $f(x, y) = c_1$ and, later, $f(x, y) = c_2$. To make the discussion that follows more concrete, assume that $c_1 < c_2 < c_3 < c_4$. (Other cases will be discussed in Problems 59 and 60.) The values of f along your travel route then increase until you reach the point P. Once you pass P, the values of f decrease again. Thus, along the curve $g(x, y) = 0$, the function f has a local extremum (in this case, a local maximum) at P.

What characterizes the point P? The level curve through P and the constraint curve touch each other at P; that is, they both have the same tangent line. Recall

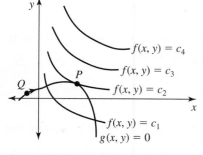

Figure 10.63 Level curves and constraints.

from Subsection 10.5.3 that the gradient of f at the point P is perpendicular to the level curve through P. Combining this observation with the previously derived fact that the gradient of g at P is perpendicular to the tangent line at P on the graph of g, we conclude that if there is an extremum at P, then ∇g and ∇f are parallel.

To state the theorem more generally [and not just in the case where $g(x, y)$ can be written as a function that gives y explicitly in terms of x], we need to require that $\nabla g(x, y) \neq \begin{bmatrix} 0 \\ 0 \end{bmatrix}$ at P. Denoting the coordinates of P by (x_0, y_0), we can then formulate the result as Lagrange's theorem:

Lagrange's Theorem Assume that f and g have continuous first partial derivatives and that $f(x, y)$ has an extremum at (x_0, y_0) subject to the constraint $g(x, y) = 0$. If $\nabla g(x_0, y_0) \neq \begin{bmatrix} 0 \\ 0 \end{bmatrix}$, then there exists a number λ such that

$$\nabla f(x_0, y_0) = \lambda \nabla g(x_0, y_0) \qquad (10.24)$$

The number λ is called a **Lagrange multiplier**. Using Lagrange multipliers to find candidates for extrema subject to a constraint is called the **method of Lagrange multipliers**. The condition (10.24) is a necessary condition. In the next example, we illustrate how to use Lagrange multipliers to find extrema subject to constraints.

EXAMPLE 12   Find all extrema of

$$f(x, y) = e^{-xy}$$

subject to the constraint $x^2 + 4y^2 = 1$.

Solution   We define $g(x, y) = x^2 + 4y^2 - 1$. Then the constraint is of the form $g(x, y) = 0$. Using the method of Lagrange multipliers, we are looking for (x, y) and λ such that

$$\nabla f(x, y) = \lambda \nabla g(x, y) \qquad \text{and} \qquad g(x, y) = 0$$

This translates into the set of equations

$$-ye^{-xy} = 2\lambda x, \quad -xe^{-xy} = 8\lambda y, \quad \text{and} \quad x^2 + 4y^2 = 1$$

since

$$\nabla f(x, y) = \begin{bmatrix} -ye^{-xy} \\ -xe^{-xy} \end{bmatrix} \qquad \text{and} \qquad \nabla g(x, y) = \begin{bmatrix} 2x \\ 8y \end{bmatrix}$$

We can eliminate λ from the first two equations. (Multiply the first equation by $4y$ and the second by x, and take the difference of the two equations.) We then find that

$$-4y^2 e^{-xy} + x^2 e^{-xy} = 0$$

Simplifying yields $e^{-xy}(x^2 - 4y^2) = 0$. Since $e^{-xy} \neq 0$, we obtain $x^2 - 4y^2 = 0$. Combining this with the constraint equation, we get the system

$$x^2 - 4y^2 = 0$$
$$x^2 + 4y^2 = 1$$

We leave the first equation and eliminate y from the second equation by adding the two equations. We then obtain

$$x^2 - 4y^2 = 0$$
$$2x^2 = 1$$

Thus, $x^2 = 1/2$ and $4y^2 = x^2 = 1/2$, implying that $y^2 = 1/8$. Simultaneously solving $x^2 = 1/2$ and $y^2 = 1/8$ gives the candidates

$$\left(\sqrt{\frac{1}{2}}, \frac{1}{2}\sqrt{\frac{1}{2}} \right), \quad \left(\sqrt{\frac{1}{2}}, -\frac{1}{2}\sqrt{\frac{1}{2}} \right), \quad \left(-\sqrt{\frac{1}{2}}, \frac{1}{2}\sqrt{\frac{1}{2}} \right), \quad \left(-\sqrt{\frac{1}{2}}, -\frac{1}{2}\sqrt{\frac{1}{2}} \right)$$

with

$$f\left(\sqrt{\frac{1}{2}},\frac{1}{2}\sqrt{\frac{1}{2}}\right)=e^{-1/4}, \quad f\left(\sqrt{\frac{1}{2}},-\frac{1}{2}\sqrt{\frac{1}{2}}\right)=e^{1/4},$$

$$f\left(-\sqrt{\frac{1}{2}},\frac{1}{2}\sqrt{\frac{1}{2}}\right)=e^{1/4}, \quad f\left(-\sqrt{\frac{1}{2}},-\frac{1}{2}\sqrt{\frac{1}{2}}\right)=e^{-1/4}$$

The extreme-value theorem applies to the constraint curve because that curve is closed and bounded; we can then conclude that maxima and minima exist on this curve, and we can select them from among our candidates. The maxima are

$$\left(-\sqrt{\frac{1}{2}},\frac{1}{2}\sqrt{\frac{1}{2}}\right) \quad \text{and} \quad \left(\sqrt{\frac{1}{2}},-\frac{1}{2}\sqrt{\frac{1}{2}}\right)$$

the minima are

$$\left(\sqrt{\frac{1}{2}},\frac{1}{2}\sqrt{\frac{1}{2}}\right) \quad \text{and} \quad \left(-\sqrt{\frac{1}{2}},-\frac{1}{2}\sqrt{\frac{1}{2}}\right)$$

Figure 10.64 The level curve of $f(x, y)$ that touches the constraint curve $g(x, y) = 0$ in Example 12.

(See Figure 10.64.) ■

Looking back at Example 12, we notice that we did not have to compute the actual value of λ to find extrema. This will typically be the case.

In the statement of the result, we mentioned that the condition $\nabla f = \lambda \nabla g$ is a necessary condition. This means that finding (x_0, y_0) so that $\nabla f(x_0, y_0) = \lambda \nabla g(x_0, y_0)$ only identifies candidates for local extrema. The next example illustrates this point.

EXAMPLE 13

Use Lagrange multipliers to identify candidates for local extrema of

$$f(x, y) = y$$

subject to the constraint $y - x^3 = 0$, and show that there is one such candidate that turns out not to be a local extremum. Furthermore, show that the function $f(x, y)$ subject to the constraint $y - x^3 = 0$ has no global extrema.

Solution

We define $g(x, y) = y - x^3$. Then

$$\nabla f(x, y) = \begin{bmatrix} 0 \\ 1 \end{bmatrix} \quad \text{and} \quad \nabla g(x, y) = \begin{bmatrix} -3x^2 \\ 1 \end{bmatrix}$$

With $y - x^3 = 0$, we obtain

$$0 = -3\lambda x^2 \quad \text{and} \quad 1 = \lambda \quad \text{and} \quad y = x^3$$

Eliminating λ, we find $x = 0$ and thus $y = 0$. We claim that $(0, 0)$ is *not* a local extremum. The easiest way to see this is to find out what $f(x, y)$ looks like along the constraint curve $y = x^3$. If we use $y = x^3$ to substitute y in the function $f(x, y)$, we obtain a single-variable function $h(x) = x^3$, $x \in \mathbf{R}$. We know from single-variable calculus that $h(x) = x^3$ has no local extrema on $\mathbf{R}$ even though, since $h'(x) = 0$ for $x = 0$, there is a candidate for a local extremum at $x = 0$. Because $\lim_{x \to \infty} h(x) = \infty$ and $\lim_{x \to -\infty} h(x) = -\infty$, the function $f(x, y)$ subject to the constraint $y - x^3 = 0$ has no global extrema. ■

The method of Lagrange multipliers only identifies candidates for local extrema, and as we saw in the previous example, these candidates may not turn out to be local extrema. Just finding candidates, however, is often good enough if we are interested in global extrema. This scenario is illustrated in the next example.

EXAMPLE 14

Suppose you wish to enclose a rectangular plot. You have 1600 ft of fencing. Using that material, what are the dimensions of the plot that will have the largest area? (See Figure 10.65.)

Solution

We wish to maximize

$$A = xy$$

subject to the constraint $2x + 2y = 1600$. We define

$$f(x, y) = xy \quad \text{and} \quad g(x, y) = 2x + 2y - 1600 = 0$$

Then

$$\nabla f(x, y) = \begin{bmatrix} y \\ x \end{bmatrix} \quad \text{and} \quad \nabla g(x, y) = \begin{bmatrix} 2 \\ 2 \end{bmatrix}$$

Using Lagrange multipliers, we need to find (x, y) and λ such that

$$\nabla f(x, y) = \lambda \nabla g(x, y) \quad \text{and} \quad 2x + 2y - 1600 = 0$$

This yields the system of equations

$$y = 2\lambda \qquad x = 2\lambda \qquad x + y = 800$$

Eliminating λ from the first two equations, we conclude that $x = y$ and thus $2x = 800$ or $x = 400$. For physical reasons, x and y can take only nonnegative values. The constraint thus restricts x and y to the line segment $y = 800 - x$, $0 \leq x \leq 800$. To see that $f(400, 400)$ gives us a maximum, we compare $f(400, 400)$ with the values at the endpoints of the line segment describing the constraint: $f(0, 800)$ and $f(800, 0)$. Since $f(400, 400) = 160,000$ and $f(0, 800) = f(800, 0) = 0$, $f(400, 400)$ is indeed the global maximum. ■

Figure 10.65 The rectangular plot in Example 14.

You probably remember the type of problem presented in Example 14 from Section 5.4. The method of Lagrange multipliers provides another method for solving the problems we discussed in that section. The method of Lagrange multipliers is more general than the method we learned there; it can be used even if we cannot solve the constraint for either x or y to eliminate one of the two variables, as we did in Section 5.4.

As the last example in this subsection, we return to the motivating example at the beginning of the subsection. There, we wished to maximize $f(x_1, x_2) = p_1 n \rho(x_1) + p_2 n \rho(x_2)$ subject to the constraint $nx_1 + nx_2 = R$. We now make the additional assumption that $\rho(x)$ increases at a decelerating rate and satisfies $\rho(0) = 0$ (see Figure 10.66), implying that there is a diminishing return to increasing egg size.

Figure 10.66 The function $\rho(x)$.

EXAMPLE 15

Assume that x_1 and x_2 are nonnegative. Maximize

$$f(x_1, x_2) = p_1 n \rho(x_1) + p_2 n \rho(x_2)$$

subject to the constraint $nx_1 + nx_2 = R$, and show that egg size should decline with maternal age.

Solution

We find that

$$\nabla f(x_1, x_2) = \begin{bmatrix} p_1 n \rho'(x_1) \\ p_2 n \rho'(x_2) \end{bmatrix} \quad \text{and} \quad \nabla g(x_1, x_2) = \begin{bmatrix} n \\ n \end{bmatrix}$$

Thus, we need to find (x_1, x_2) and λ such that

$$p_1 n \rho'(x_1) = n\lambda \qquad p_2 n \rho'(x_2) = n\lambda \qquad nx_1 + nx_2 = R$$

Eliminating λ, we obtain

$$p_1 \rho'(x_1) = p_2 \rho'(x_2) \tag{10.25}$$

Now, the constraint curve $nx_1 + nx_2 = R$ is a straight line. For biological reasons, we require that both x_1 and x_2 be nonnegative. Therefore, the constraint curve is the line segment with endpoints $(R/n, 0)$ and $(0, R/n)$.

On this line segment, there might not be a point (x_1, x_2) that satisfies (10.25). It is not too difficult to show, however, that there is at most one such point. To do so, we solve the constraint curve $x_2 = R/n - x_1$ for x_2 and substitute the result into (10.25), yielding

$$\frac{p_1}{p_2} = \frac{\rho'(R/n - x_1)}{\rho'(x_1)} \tag{10.26}$$

Since $\rho'(x) > 0$ and $\rho''(x) < 0$, we have

$$\frac{d}{dx_1} \frac{\rho'(R/n - x_1)}{\rho'(x_1)} = \frac{-\rho''(R/n - x_1)\rho'(x_1) - \rho'(R/n - x_1)\rho''(x_1)}{[\rho'(x_1)]^2} > 0$$

and it follows that there is at most one value of x_1 such that (10.26) [and hence (10.25)] holds.

We thus have the following situation: If there is a point (x_1, x_2) that satisfies (10.25) and lies on that line segment, then there are three candidates for global extrema, namely, (x_1, x_2), $(R/n, 0)$, and $(0, R/n)$; otherwise, there are only the two endpoints $(R/n, 0)$ and $(0, R/n)$. We need to select the global maximum from this set of candidates.

Now, $p_1 > p_2$ implies that $f(R/n, 0) > f(0, R/n)$. Thus, the global maximum cannot occur at the endpoint $(0, R/n)$. We claim that if there is a point (x_1, x_2) that satisfies (10.25) and lies on the segment with endpoints $(R/n, 0)$ and $(0, R/n)$, then the global maximum occurs at the point (x_1, x_2); otherwise, it occurs at the endpoint $(R/n, 0)$. Which of the two points yields the global maximum depends on the function ρ and the ratio of the survival probabilities p_1 and p_2. We claim that if

$$\frac{\rho'(R/n)}{\rho'(0)} < \frac{p_2}{p_1} \tag{10.27}$$

then there is a point (x_1, x_2) that satisfies (10.25) and lies on the segment with endpoints $(R/n, 0)$ and $(0, R/n)$. The global maximum then occurs at the point (x_1, x_2); otherwise, the global maximum occurs at the endpoint $(R/n, 0)$. How can we see this? Since $x_2 = R/n - x_1$, we can write the fitness function f as a function of x_1 alone and determine where the function is increasing and where it is decreasing. We find that

$$y = f(x_1, R/n - x_1) = np_1\rho(x_1) + np_2\rho(R/n - x_1)$$

Differentiating the right-hand side with respect to x_1 yields

$$y' = np_1\rho'(x_1) - np_2\rho'(R/n - x_1)$$

which is positive at $x_1 = 0$ since $p_1 > p_2$ and $\rho'(0) > \rho'(R/n)$. [Recall that $\rho'(x)$ is increasing at a decelerating rate.] The derivative y' is negative at $x_1 = R/n$ if (10.27) holds. [Recall that $\rho'(x) > 0$.] Therefore, if (10.27) holds, the function f has a maximum at some point (x_1, x_2) with $x_1 > 0$, $x_2 > 0$, and $nx_1 + nx_2 = R$. If (10.27) does not hold, the maximum is at the endpoint $(R/n, 0)$. That is, the strategy $(R/n, 0)$ can be improved by laying eggs in the second clutch if (10.27) holds. If (10.27) does not hold, the optimal strategy is laying all eggs in the first clutch and choosing egg size R/n for each of the n eggs.

To show that egg size should decline, we again use the assumption that $\rho(x)$ is a function with a diminishing return; that is, $\rho(x)$ is increasing at a decelerating rate. Since $p_1 > p_2$ (the probability of being alive at a later age is smaller than at an earlier age), it follows that $\rho'(x_1) < \rho'(x_2)$ and therefore $x_1 > x_2$ (see Figure 10.66), which implies that egg size should decline. If the optimal strategy is at the endpoint $(R/n, 0)$, then the egg size in the first clutch is R/n and in the second clutch it is 0, implying that egg size should decline in this case as well. ■

■ 10.6.3 Diffusion

Suppose that we place a sugar cube into a glass of water without stirring the water. The sugar dissolves and the sugar molecules move about randomly in the water. If we wait long enough, the sugar concentration will eventually be uniform throughout the water. This random movement of molecules is called **diffusion** and plays an important role in many processes of life. For instance, gas exchange in unicellular organisms and in many small multicellular organisms takes place by diffusion. Diffusion is a slow process, which means that cells have to be close to the surface if they want to exchange gas by diffusion. This limits the size and shape of organisms, unless they evolve different gas exchange mechanisms. (There are large organisms, such as kelp, that rely on diffusion for gas exchange, but their blades are extremely thin so that all of their cells are close to the surface.)

Derivation of the One-Dimensional Diffusion Equation We want to understand what type of microscopic description yields a diffusion equation. We assume that molecules move along the x-axis, and we denote the concentration of these molecules at x at time t by $c(x, t)$. That is, the number of molecules at time t in the interval $[x_1, x_2)$ is given by

$$N_{[x_1, x_2)}(t) = \int_{x_1}^{x_2} c(x, t)\, dx \tag{10.28}$$

Because the molecules move around, the number of molecules in a given interval changes over time. We will express this change as the difference between the net movement of molecules on the left and that on the right end of the interval. The quantity that describes this net movement is called the **flux** and is denoted by $J(x, t)$, where

$$J(x, t)\, \Delta t = \quad \text{the net number of molecules crossing } x \text{ from the} \\ \text{left to the right during an interval } \Delta t$$

That is, if we consider the change in the number of molecules in the interval $[x_0, x_0 + \Delta x)$ during the interval $[t, t + \Delta)$, then we have

$$N_{[x_0, x_0 + \Delta x)}(t + \Delta t) - N_{[x_0, x_0 + \Delta x)}(t) \\ = J(x_0, t)\, \Delta t - J(x_0 + \Delta x, t)\, \Delta t \tag{10.29}$$

Dividing both sides of (10.29) by Δt and letting $\Delta t \to 0$, we find that

$$\lim_{\Delta t \to 0} \frac{N_{[x_0, x_0 + \Delta x)}(t + \Delta t) - N_{[x_0, x_0 + \Delta x)}(t)}{\Delta t} \\ = J(x_0, t) - J(x_0 + \Delta x, t) \tag{10.30}$$

The left-hand side of (10.30) is equal to

$$\frac{d}{dt} N_{[x_0, x_0 + \Delta x)}(t) \tag{10.31}$$

Using (10.28), we can write (10.31) as

$$\frac{d}{dt} \int_{x_0}^{x_0 + \Delta x} c(x, t)\, dx$$

When $c(x, t)$ is sufficiently smooth, we can interchange differentiation and integration. (We cannot present a justification of this step here, but courses in real analysis do so.) We find that

$$\frac{d}{dt} \int_{x_0}^{x_0 + \Delta x} c(x, t)\, dx = \int_{x_0}^{x_0 + \Delta x} \frac{\partial c(x, t)}{\partial t}\, dx$$

Note that the d changed into a ∂ when we moved the derivative inside the integral. Before we moved it inside, we differentiated a function that depended only on t, but once we moved it inside, we differentiated a function that depends on two variables, namely, x and t. Summarizing, we arrive at the equation

$$\int_{x_0}^{x_0+\Delta x} \frac{\partial}{\partial t} c(x,t)\,dx = J(x_0, t) - J(x_0 + \Delta x, t) \tag{10.32}$$

To obtain the diffusion equation, we divide both sides of (10.32) by Δx and take the limit as $\Delta x \to 0$. On the left-hand side of (10.32), we have

$$\lim_{\Delta x \to 0} \frac{1}{\Delta x} \int_{x_0}^{x_0+\Delta x} \frac{\partial}{\partial t} c(x,t)\,dx = \frac{\partial c(x_0, t)}{\partial t}$$

On the right-hand side, we find that

$$\lim_{\Delta x \to 0} \frac{J(x_0, t) - J(x_0 + \Delta x, t)}{\Delta x} = -\frac{\partial J(x_0, t)}{\partial x}$$

Equating the two results, we arrive at

$$\frac{\partial c(x_0, t)}{\partial t} = -\frac{\partial J(x_0, t)}{\partial x} \tag{10.33}$$

A phenomenological law called *Fick's law* relates the flux to the change in concentration when molecules move around randomly in a solvent. Fick's law says that

$$J = -D\frac{\partial c}{\partial x} \tag{10.34}$$

where D is a positive constant called the **diffusion constant**. Equation (10.34) means that the flux is proportional to the change in concentration; the minus sign means that the net movement of molecules is from regions of high concentration to regions of low concentration. This movement agrees with our intuition: Going back to our example of sugar dissolving in water, we expect the net movement of sugar molecules to be from regions of high concentration to regions of low concentration so that ultimately the sugar concentration is uniform.

Combining (10.33) and (10.34), we arrive at the **diffusion equation**:

$$\frac{\partial c}{\partial t} = D\frac{\partial^2 c}{\partial x^2} \tag{10.35}$$

The diffusion approach is ubiquitous in biology. It is used not only in the description of the random movement of molecules, but in a wide array of applications, such as the change in allele frequencies due to random genetic drift, the invasion of alien species into virgin habitat, the directed movement of organisms along gradients of chemicals (chemotaxis), pattern formation, and many more phenomena. Equation (10.35) is the simplest form of an equation that incorporates diffusion. In physics, (10.35) is called the *heat equation*. It describes the diffusion of heat through a solid bar; in this case, $c(x, t)$ represents the temperature at point x at time t.

Equation (10.35) is an example of a **partial differential equation**, which is an equation that contains partial derivatives. The theory of partial differential equations is complex and well beyond the scope of this course. We will be able to discuss only some aspects of the diffusion equation.

Solving the Diffusion Equation For most partial differential equations, it is not possible to find an analytical solution; such equations often can be solved only numerically, and even this is typically not an easy task. Fortunately, (10.35) is simple enough that we can find a solution. We claim that

$$c(x,t) = \frac{1}{\sqrt{4\pi Dt}} \exp\left[-\frac{x^2}{4Dt}\right] \tag{10.36}$$

is a solution of (10.35). As in the case of ordinary differential equations, we can check this by computing the appropriate derivatives. For the left-hand side of (10.35), we need the first partial derivative of $c(x, t)$ with respect to t. We find that

$$
\begin{aligned}
\frac{\partial c(x, t)}{\partial t} &= \frac{\partial}{\partial t} \frac{1}{\sqrt{4\pi Dt}} \exp\left[-\frac{x^2}{4Dt}\right] \\
&= -\frac{1}{2} \frac{4\pi D}{(4\pi Dt)^{3/2}} \exp\left[-\frac{x^2}{4Dt}\right] + \frac{1}{\sqrt{4\pi Dt}} \exp\left[-\frac{x^2}{4Dt}\right] \frac{x^2}{4Dt^2} \\
&= \exp\left[-\frac{x^2}{4Dt}\right] \left\{ \frac{x^2}{4Dt^2\sqrt{4\pi Dt}} - \frac{2\pi D}{4\pi Dt\sqrt{4\pi Dt}} \right\} \\
&= \frac{1}{2t\sqrt{4\pi Dt}} \exp\left[-\frac{x^2}{4Dt}\right] \left\{ \frac{x^2}{2Dt} - 1 \right\}
\end{aligned}
\tag{10.37}
$$

On the right-hand side of (10.35), we need the second partial derivative of $c(x, t)$ with respect to x. We find that

$$
\begin{aligned}
\frac{\partial c(x, t)}{\partial x} &= \frac{\partial}{\partial x} \frac{1}{\sqrt{4\pi Dt}} \exp\left[-\frac{x^2}{4Dt}\right] \\
&= \frac{1}{\sqrt{4\pi Dt}} \exp\left[-\frac{x^2}{4Dt}\right] \left(-\frac{2x}{4Dt}\right)
\end{aligned}
$$

and

$$
\begin{aligned}
\frac{\partial^2 c(x, t)}{\partial x^2} &= \frac{-1}{\sqrt{4\pi Dt}} \frac{1}{2Dt} \exp\left[-\frac{x^2}{4Dt}\right] \left\{ 1 - x\frac{2x}{4Dt} \right\} \\
&= \frac{1}{2Dt\sqrt{4\pi Dt}} \exp\left[-\frac{x^2}{4Dt}\right] \left\{ \frac{x^2}{2Dt} - 1 \right\}
\end{aligned}
\tag{10.38}
$$

Putting things together, we see that (10.36) satisfies (10.35).

The function in (10.36) is called the **Gaussian density**. Figure 10.67 shows $c(x, t)$ for $t = 1, 2$, and 4; it clearly indicates that the concentration $c(x, t)$ becomes more uniform as time goes on.

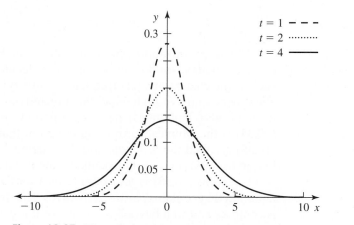

Figure 10.67 The solution of the diffusion equation at different times.

Diffusion is a very slow process. The diffusion constant D measures how quickly it proceeds. The larger D, the faster the concentration spreads out, and we can show that within t units of time, the bulk of the molecules spread over a region of length of order $\sqrt{t}$.

To give an idea of how slow diffusion is, here are a few examples taken from Yeargers, Shonkwiler, and Herod (1996). Oxygen in blood at 20°C has a diffusion constant of 10^{-5} cm^2/s, which means that it takes an oxygen molecule roughly 500 seconds to cross a distance of 1 mm by diffusion alone. Ribonuclease (an enzyme that hydrolyzes ribonucleic acid) in water at 20°C has a diffusion constant of 1.1×10^{-6} cm^2/s, which means that ribonuclease takes roughly 4672 seconds (or 1 hr, 18 min) to cross a distance of 1 mm by diffusion alone. These examples illustrate why organisms frequently rely on other active mechanisms to transport molecules.

The diffusion equation (10.35) can be generalized to higher dimensions. In that case, (10.33) becomes

$$\frac{\partial c}{\partial t} = -\nabla J \tag{10.39}$$

and (10.34) becomes

$$J = -D\nabla c \tag{10.40}$$

Combining (10.39) and (10.40), we find that

$$\frac{\partial c}{\partial t} = D\nabla \cdot (\nabla c)$$

where $\nabla \cdot (\nabla c)$ is to be interpreted as a dot product. That is, if $\mathbf{x} = (x_1, x_2, x_3) \in \mathbf{R}^3$, $t \in \mathbf{R}$, then

$$\frac{\partial c}{\partial t} = D\left(\frac{\partial^2 c}{\partial x_1^2} + \frac{\partial^2 c}{\partial x_2^2} + \frac{\partial^2 c}{\partial x_3^2}\right)$$

As a shorthand notation, we define

$$\Delta = \frac{\partial^2}{\partial x_1^2} + \frac{\partial^2}{\partial x_2^2} + \frac{\partial^2}{\partial x_3^2}$$

where Δ is called the **Laplace operator**. We then write

$$\frac{\partial c}{\partial t} = D\,\Delta c$$

More generally, if $\mathbf{x} = (x_1, x_2, \ldots, x_n) \in \mathbf{R}^n$, then

$$\Delta = \frac{\partial^2}{\partial x_1^2} + \frac{\partial^2}{\partial x_2^2} + \cdots + \frac{\partial^2}{\partial x_n^2}$$

(Δc is read "the Laplacian of c.")

Section 10.6 Problems

■ 10.6.1

In Problems 1–10, the functions are defined for all $(x, y) \in \mathbf{R}^2$. Find all candidates for local extrema, and use the Hessian matrix to determine the type (maximum, minimum, or saddle point).

1. $f(x, y) = x^2 + y^2 - 2x$ **2.** $f(x, y) = -2x^2 - y^2 + 3y$

3. $f(x, y) = x^2 y - 4x^2 - 4y$ **4.** $f(x, y) = xy - 2y^2$

5. $f(x, y) = -2x^2 + y^2 - 6y$ **6.** $f(x, y) = x(1 - x + y)$

7. $f(x, y) = e^{-x^2 - y^2}$ **8.** $f(x, y) = yxe^{-y}$

9. $f(x, y) = x \cos y$ **10.** $f(x, y) = y \sin x$

11. In this problem, we will illustrate that if one of the eigenvalues of the Hessian matrix at a point where the gradient vanishes is equal to 0, then we cannot make any statements about whether the point is a local extremum just on the basis of the Hessian matrix.

Consider the following functions:

$$f_1(x, y) = x^2$$
$$f_2(x, y) = x^2 + y^3$$
$$f_3(x, y) = x^2 + y^4$$

Figures 10.68 through 10.70 show graphs of the three functions.
(a) Show that, for $i = 1, 2,$ and 3,

$$\nabla f_i(0, 0) = \begin{bmatrix} 0 \\ 0 \end{bmatrix}$$

(b) Show that, for $i = 1, 2,$ and 3,

$$\text{Hess } f_i(0, 0) = \begin{bmatrix} 2 & 0 \\ 0 & 0 \end{bmatrix}$$

and determine the eigenvalues of Hess $f_i(0, 0)$.

Figure 10.68 $f_1(x, y)$ in Problem 11.

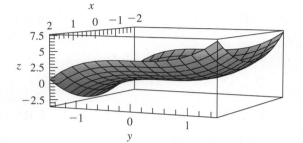

Figure 10.69 $f_2(x, y)$ in Problem 11.

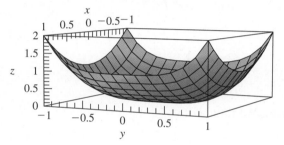

Figure 10.70 $f_3(x, y)$ in Problem 11.

(c) Since one of the eigenvalues of Hess $f_i(0, 0)$ is equal to 0, we cannot use the criterion stated in the text to determine the behavior of the three functions at $(0, 0)$. Use Figures 10.68 through 10.70 to describe what happens at $(0, 0)$ for each function.

12. Consider the function

$$f(x, y) = ax^2 + by^2$$

(a) Show that

$$\nabla f(0, 0) = \begin{bmatrix} 0 \\ 0 \end{bmatrix}$$

(b) Find values for a and b such that **(i)** $(0, 0)$ is a local minimum, **(ii)** $(0, 0)$ is a local maximum, and **(iii)** $(0, 0)$ is a saddle point.

In Problems 13–16, the functions are defined on the rectangular domain

$$D = \{(x, y) : -1 \le x \le 1, -1 \le y \le 1\}$$

Find the absolute maxima and minima of f on D.

13. $f(x, y) = 2x - y$ **14.** $f(x, y) = 3 - x + 2y$

15. $f(x, y) = x^2 - y^2$ **16.** $f(x, y) = x^2 + y^2$

17. Find the absolute maxima and minima of

$$f(x, y) = x^2 + y^2 - x + 2y$$

on the set

$$D = \{(x, y) = 0 \le x \le 1, -2 \le y \le 0\}$$

18. Find the absolute maxima and minima of

$$f(x, y) = x^2 - y^2 + 4x + y$$

on the set

$$D = \{(x, y) = -4 \le x \le 0, 0 \le y \le 1\}$$

19. Maximize the function

$$f(x, y) = 2xy - x^2y - xy^2$$

on the triangle bounded by the line $x + y = 2$, the x-axis, and the y-axis.

20. Maximize the function

$$f(x, y) = xy(15 - 5y - 3x)$$

on the triangle bounded by the line $5y + 3x = 15$, the x-axis, and the y-axis.

21. Find the absolute maxima and minima of

$$f(x, y) = x^2 + y^2 + 4x - 1$$

on the disk

$$D = \{(x, y) : x^2 + y^2 \le 9\}$$

22. Find the absolute maxima and minima of

$$f(x, y) = x^2 + y^2 - 6y + 3$$

on the disk

$$D = \{(x, y) : x^2 + y^2 \le 16\}$$

23. Find the absolute maxima and minima of

$$f(x, y) = x^2 + y^2 + x - y$$

on the disk

$$D = \{(x, y) : x^2 + y^2 \le 1\}$$

24. Find the absolute maxima and minima of

$$f(x, y) = x^2 + y^2 + x + 2y$$

on the disk

$$D = \{(x, y) : x^2 + y^2 \le 4\}$$

25. Can a continuous function of two variables have two maxima and no minima? Describe in words what the properties of such a function would be, and contrast this behavior with a function of one variable.

26. Suppose $f(x, y)$ has a horizontal tangent plane at $(0, 0)$. Can you conclude that f has a local extremum at $(0, 0)$?

27. Suppose crop yield Y depends on nitrogen (N) and phosphorus (P) concentrations as

$$Y(N, P) = NPe^{-(N+P)}$$

Find the value of (N, P) that maximizes crop yield.

28. Choose three numbers x, y, and z so that their sum is equal to 60 and their product is maximal.

29. Find the maximum volume of a rectangular closed (top, bottom, and four sides) box with surface area 48 m².

30. Find the maximum volume of a rectangular open (bottom and four sides, no top) box with surface area 75 m².

31. Find the minimum surface area of a rectangular closed (top, bottom, and four sides) box with volume 216 m³.

32. Find the minimum surface area of a rectangular open (bottom and four sides, no top) box with volume 256 m³.

33. The distance between the origin $(0, 0, 0)$ and the point (x, y, z) is

$$\sqrt{x^2 + y^2 + z^2}$$

Find the minimum distance between the origin and the plane $x + y + z = 1$. (*Hint*: Minimize the squared distance between the origin and the plane.)

34. Given the symmetric matrix

$$A = \begin{bmatrix} a & c \\ c & b \end{bmatrix}$$

where a, b, and c are real numbers, show that the eigenvalues of A are real. (*Hint*: Compute the eigenvalues.)

35. Understanding species richness and diversity is a major concern of ecological studies. A frequently used measure of diversity is the Shannon and Weaver index

$$H = -\sum_{i=1}^{n} p_i \ln p_i$$

where p_i is equal to the proportion of species i, $i = 1, 2, \ldots, n$, and n is the total number of species in the study area. Assume that a community consists of three species with relative proportions p_1, p_2, and p_3.
(a) Use the fact that $p_1 + p_2 + p_3 = 1$ to show that H is of the form

$$H(p_1, p_2) = -p_1 \ln p_1 - p_2 \ln p_2$$
$$- (1 - p_1 - p_2) \ln(1 - p_1 - p_2)$$

and that the domain of $H(p_1, p_2)$ is the triangular set in the p_1-p_2 plane bounded by the lines $p_1 = 0$, $p_2 = 0$, and $p_1 + p_2 = 1$.
(b) Show that H attains its absolute maximum when $p_1 = p_2 = p_3 = 1/3$.

■ **10.6.2**

In Problems 36–45, use Lagrange multipliers to find the maxima and minima of the functions under the given constraints.

36. $f(x, y) = 2x - y; x^2 + y^2 = 5$
37. $f(x, y) = 3x^2 + y; x^2 + y^2 = 1$
38. $f(x, y) = xy; x^2 + y^2 = 4$
39. $f(x, y) = xy; 2x - 4y = 1$
40. $f(x, y) = x^2 - y^2; 2x + y = 1$
41. $f(x, y) = x^2 + y^2; 3x - 2y = 4$
42. $f(x, y) = xy^2; x^2 - y = 0$
43. $f(x, y) = x^2 y; x^2 + 3y = 1$
44. $f(x, y) = x^2 y^2; 2x - 3y = 4$
45. $f(x, y) = x^2 y^2; x^2 - y^2 = 1$

In Problems 46–55, use Lagrange multipliers to find the answers to the indicated problems in Section 5.4.

46. Problem 1   **47.** Problem 2
48. Problem 3   **49.** Problem 4
50. Problem 5   **51.** Problem 6
52. Problem 7   **53.** Problem 9
54. Problem 12   **55.** Problem 18

56. Let

$$f(x, y) = x + y \quad (x, y) \in \mathbf{R}^2$$

with constraint function $xy = 1$.
(a) Use Lagrange multipliers to find all local extrema.
(b) Are there global extrema?

57. Let

$$f(x, y) = x + y$$

with constraint function

$$\frac{1}{x} + \frac{1}{y} = 1, \ x \neq 0, y \neq 0$$

(a) Use Lagrange multipliers to find all local extrema.
(b) Are there global extrema?
58. Let

$$f(x, y) = xy, \quad (x, y) \in \mathbf{R}^2$$

with constraint function $y - x^2 = 0$.
(a) Use Lagrange multipliers to find candidates for local extrema.
(b) Use the constraint $y - x^2 = 0$ to reduce $f(x, y)$ to a single-variable function, and then use this function to show that $f(x, y)$ has no local extrema on the constraint curve.
59. Explain why $f(x, y)$ has a local extremum at the point P in Figure 10.63 under the constraint $g(x, y) = 0$ if $c_1 > c_2 > c_3 > c_4$.
60. Explain why $f(x, y)$ has a local extremum at the point P in Figure 10.63 under the constraint $g(x, y) = 0$ if $c_1 < c_2$ and $c_2 > c_3 > c_4$.
61. In the introductory example, we discussed how egg size depends on maternal age. Assume now that the total amount of resources available is 10 (in appropriate units), the number of eggs per clutch is 3, the number of clutches is 2, and the egg size in clutch number i is denoted by x_i.
(a) Find the constraint function.
(b) Suppose the fitness function is given by

$$f(x_1, x_2) = \frac{3}{2}\rho(x_1) + \frac{3}{4}\rho(x_2)$$

where $\rho(x) = \frac{2x}{5+x}$. Find the optimal egg sizes for clutch 1 and clutch 2 under the constraint in (a).
62. In the introductory example in this subsection, we discussed how egg size depends on maternal age. Assume now that the fitness function is given by

$$f(x_1, x_2) = \frac{5}{3}\rho(x_1) + \frac{5}{6}\rho(x_2)$$

with

$$\rho(x) = \frac{3x}{4+x}$$

The constraint function is given by

$$5x_1 + 5x_2 = 7$$

(a) Compare the given functions with the corresponding ones in the text, and identify the parameters n, p_1, p_2, and R from the text.
(b) Solve the constraint function for x_2 and substitute your expression for x_2 into the function f. This then yields a function of one variable. Find the domain of this single-variable function and use single-variable calculus to determine optimal egg sizes for clutch 1 and clutch 2.

■ **10.6.3**

63. Show that

$$c(x, t) = \frac{1}{\sqrt{8\pi t}} \exp\left[-\frac{x^2}{8t}\right]$$

solves

$$\frac{\partial c(x, t)}{\partial t} = 2\frac{\partial^2 c(x, t)}{\partial x^2}$$

64. Show that

$$c(x, t) = \frac{1}{\sqrt{2\pi t}} \exp\left[-\frac{x^2}{2t}\right]$$

solves

$$\frac{\partial c(x, t)}{\partial t} = \frac{1}{2} \frac{\partial^2 c(x, t)}{\partial x^2}$$

65. A solution of

$$\frac{\partial c(x, t)}{\partial t} = D \frac{\partial^2 c(x, t)}{\partial x^2}$$

is the function

$$c(x, t) = \frac{1}{\sqrt{4\pi Dt}} \exp\left[-\frac{x^2}{4Dt}\right]$$

for $x \in \mathbf{R}$ and $t > 0$.
(a) Show that, as a function of x for fixed values of $t > 0$, $c(x, t)$ is **(i)** positive for all $x \in \mathbf{R}$, **(ii)** is increasing for $x < 0$ and decreasing for $x > 0$, **(iii)** has a local maximum at $x = 0$, and **(iv)** has inflection points at $x = \pm\sqrt{2Dt}$.
(b) Graph $c(x, t)$ as a function of x when $D = 1$ for $t = 0.01$, $t = 0.1$, and $t = 1$.

66. A solution of

$$\frac{\partial c(x, t)}{\partial t} = D \frac{\partial^2 c(x, t)}{\partial x^2}$$

is the function

$$c(x, t) = \frac{1}{\sqrt{4\pi Dt}} \exp\left[-\frac{x^2}{4Dt}\right]$$

for $x \in \mathbf{R}$ and $t > 0$.
(a) Show that a local maximum of $c(x, t)$ occurs at $x = 0$ for fixed t.
(b) Show that $c(0, t), t > 0$, is a decreasing function of t.
(c) Find

$$\lim_{t \to 0^+} c(x, t)$$

when $x = 0$ and when $x \neq 0$.

(d) Use the fact that

$$\int_{-\infty}^{\infty} e^{-u^2/2} \, du = \sqrt{2\pi}$$

to show that, for $t > 0$,

$$\int_{-\infty}^{\infty} c(x, t) \, dx = 1$$

(e) The function $c(x, t)$ can be interpreted as the concentration of a substance diffusing in space. Explain the meaning of

$$\int_{-\infty}^{\infty} c(x, t) \, dx = 1$$

and use your results in (c) and (d) to explain why this means that initially (i.e., at $t = 0$) the entire amount of the substance was released at the origin.

Mathematically, we can specify such an initial condition (in which the substance is concentrated at the origin at time 0) by the δ-function $\delta(x)$, with the property that

$$\delta(x) = 0, \quad \text{for } x \neq 0$$

and

$$\int_{-\infty}^{\infty} \delta(x) \, dx = 1$$

67. The two-dimensional diffusion equation

$$\frac{\partial n(\mathbf{r}, t)}{\partial t} = D\left(\frac{\partial^2 n(\mathbf{r}, t)}{\partial x^2} + \frac{\partial^2 n(\mathbf{r}, t)}{\partial y^2}\right) \qquad (10.41)$$

where $n(\mathbf{r}, t)$, $\mathbf{r} = (x, y)$, denotes the population density at the point $\mathbf{r} = (x, y)$ in the plane at time t, can be used to describe the spread of organisms. Assume that a large number of insects are released at time 0 at the point $(0, 0)$. Furthermore, assume that, at later times, the density of these insects can be described by the diffusion equation (10.41). Show that

$$n(x, y, t) = \frac{n_0}{4\pi Dt} \exp\left[-\frac{x^2 + y^2}{4Dt}\right]$$

satisfies (10.41).

■ 10.7 Systems of Difference Equations (Optional)

■ 10.7.1 A Biological Example

About 14% of all insect species (and thus about 10% of all species of multicellular animals) are estimated to belong to a group of insects called *parasitoids*. These are insects (mostly in the order Hymenoptera) that lay their eggs on, in, or near the (in most cases, immature) body of another arthropod, which serves as a host for the developing parasitoids. The eggs develop into free-living adults while consuming the host.

Parasitoids play an important role in biological control. A successful example is *Trichogramma* wasps, which parasitize insect eggs. These wasps are reared in factories for subsequent release to the field. Every year, millions of hectares of agricultural land are treated with released *Trichogramma* wasps, for instance, to protect sugar cane from the sugarcane borer, *Chilo* spp., in China, or to protect cornfields from the European corn borer, *Ostrinia nubilalis* (Hübner), in western Europe. Another successful example of biological control of an insect pest is the parasitoid wasp *Aphytis melinus*, which regulates red scale (*Aonidiella aurantii*), which damages citrus trees in California.

The importance of parasitoids in pest control stimulated both empirical and theoretical work. Theoretical studies of host–parasitoid interactions go back to Thompson (1924) and Nicholson and Bailey (1935). The work of Nicholson and Bailey was particularly influential. These researchers introduced discrete-generation, host–parasitoid models of the form

$$N_{t+1} = bN_t e^{-aP_t}$$
$$P_{t+1} = cN_t[1 - e^{-aP_t}]$$

for $t = 0, 1, 2, \ldots$. Here, N and P denote the population sizes of, respectively, susceptible hosts and searching adult female parasitoids at times t and $t + 1$. The parameter b is interpreted as the *net growth parameter*. We see from the first equation that hosts grow exponentially in the absence of parasitoids ($P = 0$). The term e^{-aP_t} is the fraction of hosts that are *not* parasitized (and thus $1 - e^{-aP_t}$ is the fraction of hosts that are parasitized) at generation t. Parasitized hosts produce parasitoids. The parameter c is equal to the number of parasitoids produced per parasitized host. Note that only nonparasitized hosts reproduce.

A numerical simulation (Figure 10.71) of the Nicholson–Bailey equation shows that population sizes oscillate with increasing amplitude until either the parasitoid becomes extinct, followed by an exponential increase of the host, or until the host becomes extinct, followed by the extinction of the parasitoid.

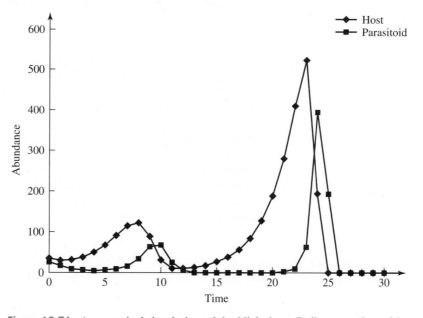

Figure 10.71 A numerical simulation of the Nicholson–Bailey equation with $a = 0.023$, $b = 1.5$, and $c = 1$. The population sizes oscillate until extinction occurs.

The behavior of the model disagrees with most empirical studies (although some laboratory experiments have produced such unstable behavior). The model has since been modified in a number of ways to stabilize the dynamics. One such attempt is called the *negative binomial model* (Griffiths, 1969; May, 1978), in which

$$N_{t+1} = bN_t \left(1 + \frac{aP}{k}\right)^{-k}$$

$$P_{t+1} = cN_t \left[1 - \left(1 + \frac{aP}{k}\right)^{-k}\right]$$

The form of this set of equations is quite similar to that of the Nicholson–Bailey equation, and the parameters b and c have the same interpretation as before. The main (and crucial) difference is the term $\left(1 + \frac{aP}{k}\right)^{-k}$, which replaces the term e^{-aP} in the Nicholson–Bailey model. It has the same interpretation, though, in both equations, denoting the fraction of hosts that escape parasitism.

The choices of parameter in the numerical simulation of the negative binomial model (Figure 10.72) show that host and parasitoids equilibrate so that they both have positive abundances. (We call this *coexistence*.)

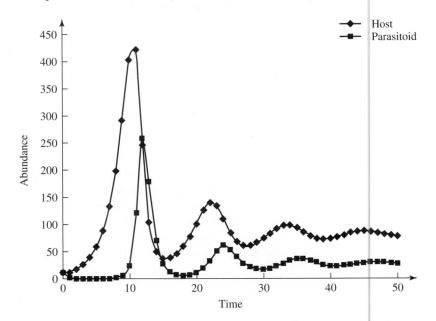

Figure 10.72 A numerical simulation of the negative binomial model with $a = 0.023$, $b = 1.5$, $c = 1$, and $k = 0.5$. The choices for the parameters a, b, and c are the same as in Figure 10.71. The choice for k stabilizes the host–parasitoid interactions, and coexistence occurs.

The two host–parasitoid examples just give a glimpse of the possible behavior of multispecies interactions that are modeled by discrete-generation difference equations. In what follows, we will concentrate on coexistence in two-species, discrete-time models. The analysis will parallel our discussion of difference equations in Section 5.6, where we examined the equilibria and stability of single-species, discrete-time models of the form

$$x_{t+1} = f(x_t)$$

There, we found that point equilibria satisfy the equation

$$x^* = f(x^*)$$

and that such point equilibria are locally stable if $|f'(x^*)| < 1$. We obtained this condition by linearizing $f(x)$ about the equilibrium x^*.

We will see in this section that point equilibria satisfy a similar condition in two-species models and that the same strategy of linearizing about the equilibrium will yield an analogous condition of local stability in two-species models. Since investigating nonlinear difference equations will lead us to the study of linear difference equations, we begin our discussion with the latter.

■ 10.7.2 Equilibria and Stability in Systems of Linear Difference Equations

Linear difference equations are of the form

$$x_1(t + 1) = a_{11}x_1(t) + a_{12}x_2(t) \tag{10.42}$$

$$x_2(t + 1) = a_{21}x_1(t) + a_{22}x_2(t) \tag{10.43}$$

where $t = 0, 1, 2, \ldots$. This set of equations can be written in matrix form

$$\underbrace{\begin{bmatrix} x_1(t+1) \\ x_2(t+1) \end{bmatrix}}_{\mathbf{x}(t+1)} = \underbrace{\begin{bmatrix} a_{11} & a_{12} \\ a_{21} & a_{22} \end{bmatrix}}_{A} \underbrace{\begin{bmatrix} x_1(t) \\ x_2(t) \end{bmatrix}}_{\mathbf{x}(t)}$$

which shows that linear difference equations are *linear maps*, which we discussed in Section 9.3. First note that if $\mathbf{x}(0) = \begin{bmatrix} 0 \\ 0 \end{bmatrix}$, then $\mathbf{x}(t) = \begin{bmatrix} 0 \\ 0 \end{bmatrix}$ for all $t = 1, 2, 3, \ldots$. We call $\begin{bmatrix} 0 \\ 0 \end{bmatrix}$ a **(point) equilibrium**. More generally, a point equilibrium satisfies the equation

$$\mathbf{x}^* = A\mathbf{x}^*$$

We see that $\begin{bmatrix} 0 \\ 0 \end{bmatrix}$ is a point equilibrium of the linear system $\mathbf{x}(t+1) = A\mathbf{x}(t)$. We will now investigate what happens when $\mathbf{x}(0) \neq \begin{bmatrix} 0 \\ 0 \end{bmatrix}$.

In Section 9.3, we learned how to compute $\mathbf{x}(t)$ for a given initial condition $\mathbf{x}(0)$ without computing the values of $\mathbf{x}(s)$ for all values of s between 0 and t. We found that if A has two real and distinct eigenvalues λ_1 and λ_2, then we can write any vector $\mathbf{x}(0)$ as a linear combination of its eigenvectors $\mathbf{u}_1$ and $\mathbf{u}_2$ (corresponding to λ_1 and λ_2, respectively); that is,

$$\mathbf{x}(0) = c_1\mathbf{u}_1 + c_2\mathbf{u}_2$$

where c_1 and c_2 are real numbers. Using this representation of $\mathbf{x}(0)$, we found that

$$\mathbf{x}(t) = c_1\lambda_1^t\mathbf{u}_1 + c_2\lambda_2^t\mathbf{u}_2 \tag{10.44}$$

which we can use to say something about the *long-term behavior* of $\mathbf{x}(t)$, or $\lim_{t\to\infty} \mathbf{x}(t)$. Let's return to the question of what happens to $\mathbf{x}(t)$ when $\mathbf{x}(t) \neq \begin{bmatrix} 0 \\ 0 \end{bmatrix}$. We see from the representation of $\mathbf{x}(t)$ in (10.44) that if $|\lambda_1| < 1$ and $|\lambda_2| < 1$, then $\lim_{t\to\infty} \mathbf{x}(t) = \begin{bmatrix} 0 \\ 0 \end{bmatrix}$, regardless of $\mathbf{x}(0)$. In this case, we say that $\begin{bmatrix} 0 \\ 0 \end{bmatrix}$ is a **stable equilibrium**. If either $|\lambda_1| > 1$ or $|\lambda_2| > 1$, then the equilibrium $\begin{bmatrix} 0 \\ 0 \end{bmatrix}$ is called **unstable**.

The stability condition for the equilibrium $\begin{bmatrix} 0 \\ 0 \end{bmatrix}$ of $\mathbf{x}(t+1) = A\mathbf{x}(t)$, where A is a 2×2 matrix, holds more generally (not just for λ_1 and λ_2 real):

The point $\begin{bmatrix} 0 \\ 0 \end{bmatrix}$ is a stable equilibrium of

$$\mathbf{x}(t+1) = A\mathbf{x}(t)$$

if both eigenvalues λ_1 and λ_2 of A satisfy

$$|\lambda_1| < 1 \quad \text{and} \quad |\lambda_2| < 1$$

If either $|\lambda_1| > 1$ or $|\lambda_2| > 1$, $\begin{bmatrix} 0 \\ 0 \end{bmatrix}$ is an unstable equilibrium.

In the preceding criterion, we no longer require the eigenvalues of A to be real and distinct. However, it is beyond the scope of this book to show the criterion for general λ_1 and λ_2.

EXAMPLE 1

Show that $\begin{bmatrix} 0 \\ 0 \end{bmatrix}$ is an equilibrium of

$$\begin{bmatrix} x_1(t+1) \\ x_2(t+1) \end{bmatrix} = \begin{bmatrix} -0.4 & 0.2 \\ -0.3 & 0.1 \end{bmatrix} \begin{bmatrix} x_1(t) \\ x_2(t) \end{bmatrix}$$

and determine its stability.

Solution   To check that $\begin{bmatrix} 0 \\ 0 \end{bmatrix}$ is an equilibrium, we need to show that $\begin{bmatrix} 0 \\ 0 \end{bmatrix}$ satisfies

$$\begin{bmatrix} 0 \\ 0 \end{bmatrix} = \begin{bmatrix} -0.4 & 0.2 \\ -0.3 & 0.1 \end{bmatrix} \begin{bmatrix} 0 \\ 0 \end{bmatrix}$$

But this is true, since $A\begin{bmatrix} 0 \\ 0 \end{bmatrix} = \begin{bmatrix} 0 \\ 0 \end{bmatrix}$ for any matrix A with constant entries. To determine the stability of A, we need to find the eigenvalues of A. That is, we need to solve

$$\det(A - \lambda I) = \det \begin{bmatrix} -0.4 - \lambda & 0.2 \\ -0.3 & 0.1 - \lambda \end{bmatrix}$$
$$= (-0.4 - \lambda)(0.1 - \lambda) + (0.2)(0.3)$$
$$= \lambda^2 + 0.3\lambda + 0.02 = 0$$

The solutions are

$$\lambda_{1,2} = \frac{-0.3 \pm \sqrt{0.09 - 0.08}}{2} = \frac{-0.3 \pm 0.1}{2}$$

Hence, $\lambda_1 = -0.1$ and $\lambda_2 = -0.2$. Since $|\lambda_1| = |-0.1| < 1$ and $|\lambda_2| = |-0.2| < 1$, it follows that $\begin{bmatrix} 0 \\ 0 \end{bmatrix}$ is stable.   ■

When the eigenvalues λ_1 and λ_2 are complex conjugate, the criterion for stability can be simplified. Specifically, if λ_1 and λ_2 are complex conjugates, then

$$|\lambda_1|^2 = |\lambda_2|^2 = \lambda_1 \lambda_2 \tag{10.45}$$

This identity is not obvious at first sight, but can be demonstrated when we graph λ_1 and λ_2 and determine their absolute values. Let

$$\lambda_1 = a + ib \quad \text{and} \quad \lambda_2 = a - ib$$

be the two complex conjugate eigenvalues of A. In Figure 10.73, we draw λ_1 and λ_2 when both a and b are positive. An application of the Pythagorean theorem shows that

$$|\lambda_1|^2 = a^2 + b^2 \quad \text{and} \quad |\lambda_2|^2 = a^2 + b^2 \tag{10.46}$$

Algebraically, we find that

$$\lambda_1 \lambda_2 = (a + ib)(a - ib) = a^2 - (ib)^2 = a^2 - i^2 b^2 = a^2 + b^2$$

since $i^2 = -1$. Combining the preceding equation for $\lambda_1 \lambda_2$ with (10.46) proves (10.45), which we use in the next example.

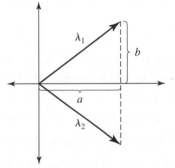

Figure 10.73 A graphical illustration of the identity $|\lambda_1|^2 = |\lambda_2|^2 = \lambda_1 \lambda_2$.

EXAMPLE 2   Show that the equilibrium $\begin{bmatrix} 0 \\ 0 \end{bmatrix}$ of

$$\begin{bmatrix} x_1(t+1) \\ x_2(t+1) \end{bmatrix} = \begin{bmatrix} -0.3 & -0.5 \\ 0.7 & 0.15 \end{bmatrix} \begin{bmatrix} x_1(t) \\ x_2(t) \end{bmatrix}$$

is stable.

Solution   To test the stability of $\begin{bmatrix} 0 \\ 0 \end{bmatrix}$, we need to solve

$$\det \begin{bmatrix} -0.3 - \lambda & -0.5 \\ 0.7 & 0.15 - \lambda \end{bmatrix} = 0$$

This amounts to solving the quadratic equation

$$(-0.3 - \lambda)(0.15 - \lambda) + (0.5)(0.7) = 0$$

or

$$\lambda^2 + 0.15\lambda + 0.305 = 0 \tag{10.47}$$

Since the discriminant $(0.15)^2 - (4)(1)(0.305) = -1.1975 < 0$, it follows that the two solutions λ_1 and λ_2 of (10.47) are complex conjugates. Without computing λ_1 and λ_2, we can check whether $|\lambda_1|$ and $|\lambda_2|$ are both less than 1, since $|\lambda_1|^2 = |\lambda_2|^2 = \lambda_1 \lambda_2 = \det A$. Now,

$$\det A = \begin{bmatrix} -0.3 & -0.5 \\ 0.7 & 0.15 \end{bmatrix}$$

$$= (-0.3)(0.15) - (0.7)(-0.5) = 0.305 < 1$$

Therefore, $|\lambda_1| < 1$ and $|\lambda_2| < 1$, and $\begin{bmatrix} 0 \\ 0 \end{bmatrix}$ is a stable equilibrium. ■

■ 10.7.3 Equilibria and Stability of Nonlinear Systems of Difference Equations

We saw examples of nonlinear systems of difference equations at the beginning of this section: the Nicholson–Bailey equation and the negative binomial model. The general form of a system of two nonlinear difference equations is

$$x_1(t+1) = F(x_1(t), x_2(t)) \tag{10.48}$$
$$x_2(t+1) = G(x_1(t), x_2(t)) \tag{10.49}$$

where F and G are (nonlinear) functions of the two variables x_1 and x_2 and $t = 0, 1, 2, \ldots$ is the independent variable that denotes time.

As in the previous subsection, we will be interested here in equilibria and their stability. We say that the point (x_1^*, x_2^*) is a (**point**) **equilibrium** of the system (10.48) and (10.49) if x_1^* and x_2^* simultaneously satisfy the two equations

$$x_1^* = F(x_1^*, x_2^*)$$
$$x_2^* = G(x_1^*, x_2^*)$$

EXAMPLE 3

Find all equilibria of

$$x_1(t+1) = 2x_1(t)[1 - x_1(t)]$$
$$x_2(t+1) = x_1(t)[1 - x_2(t)]$$

Solution

To find the equilibria, we need to solve

$$x_1 = 2x_1(1 - x_1)$$
$$x_2 = x_1(1 - x_2)$$

Multiplying out and rearranging terms yields

$$2x_1^2 - x_1 = 0$$
$$x_2 + x_1 x_2 - x_1 = 0$$

The first equation has the solution $x_1 = 0$ or $\frac{1}{2}$. Solving the second equation for x_2, we find that

$$x_2(1 + x_1) = x_1, \qquad \text{or} \qquad x_2 = \frac{x_1}{1 + x_1}$$

If $x_1 = 0$, then $x_2 = 0$; if $x_1 = 1/2$, then $x_2 = (1/2)/(3/2) = 1/3$. Summarizing our results, we found two point equilibria:

$$\begin{bmatrix} 0 \\ 0 \end{bmatrix} \qquad \text{and} \qquad \begin{bmatrix} 1/2 \\ 1/3 \end{bmatrix}$$

■

EXAMPLE 4   Find all biologically relevant equilibria of the Nicholson–Bailey model

$$N_{t+1} = bN_t e^{-aP_t}$$
$$P_{t+1} = cN_t[1 - e^{-aP_t}]$$

Solution   To find the equilibria in question, we need to solve

$$N = bNe^{-aP} \qquad (10.50)$$
$$P = cN[1 - e^{-aP}] \qquad (10.51)$$

The first equation is satisfied for $N = 0$. If we substitute $N = 0$ into the second equation, we obtain $P = 0$. The system thus has the trivial equilibrium $(N^*, P^*) = (0, 0)$, which corresponds to the state when both host and parasitoid are absent.

If $N \neq 0$, then we can cancel N in equation (10.50), resulting in

$$e^{aP^*} = b, \qquad \text{or} \qquad P^* = \frac{1}{a} \ln b$$

We see that, in order for P^* to be positive (this is required to be a biologically reasonable nontrivial equilibrium), we require b to be greater than 1. Now, using $e^{aP^*} = b$, we find that

$$P^* = cN^* \left[1 - \frac{1}{b}\right]$$

This equation reduces to

$$N^* = \frac{P^*}{c[1 - 1/b]} = \frac{\ln b}{ac[1 - 1/b]} = \frac{b}{b - 1} \frac{1}{ac} \ln b$$

when $P^* = \frac{1}{a} \ln b$. We see that, for $b > 1$, $N^* > 0$. We conclude that, in addition to the trivial equilibrium $(N^*, P^*) = (0, 0)$, if $b > 1$, then there exists a biologically reasonable, nontrivial equilibrium—that is, an equilibrium in which both the host and the parasitoid densities are positive. This equilibrium is given by

$$N^* = \frac{b}{b - 1} \frac{1}{ac} \ln b \qquad \text{and} \qquad P^* = \frac{1}{a} \ln b \qquad \blacksquare$$

To determine the stability of point equilibria, we proceed in the same way as in the single-species case: We linearize about the equilibria and use the linearized system and what we learned about linear maps in Chapter 9 to derive an analytical condition for local stability. Here is how this works: We start with the general system of difference equations

$$x_1(t + 1) = F(x_1(t), x_2(t)) \qquad (10.52)$$
$$x_2(t + 1) = G(x_1(t), x_2(t)) \qquad (10.53)$$

and assume that it has a point equilibrium (x_1^*, x_2^*) which simultaneously satisfies

$$x_1^* = F(x_1^*, x_2^*) \qquad \text{and} \qquad x_2^* = G(x_1^*, x_2^*)$$

To linearize about (x_1^*, x_2^*), we write

$$x_1(t) = x_1^* + z_1(t) \qquad \text{and} \qquad x_2(t) = x_2^* + z_2(t)$$

where we interpret $z_1(t)$ and $z_2(t)$ as small perturbations, just as in Chapter 5, where we discussed stability of equilibria in difference equations, or in Chapter 8, where we discussed the stability of equilibria in differential equations. Now linearizing $F(x_1(t), x_2(t))$ and $G(x_1(t), x_2(t))$ about the equilibrium (x_1^*, x_2^*), we find that

$$\text{linearization of } F(x_1(t), x_2(t)) \text{ is } F(x_1^*, x_2^*) + \left(\frac{\partial F}{\partial x_1}\right)^* z_1(t) + \left(\frac{\partial F}{\partial x_2}\right)^* z_2(t)$$

linearization of $G(x_1(t), x_2(t))$ is $G(x_1^*, x_2^*) + \left(\dfrac{\partial G}{\partial x_1}\right)^* z_1(t) + \left(\dfrac{\partial G}{\partial x_2}\right)^* z_2(t)$

where $(\cdot)^*$ means that we evaluate the expression in the parentheses at the equilibrium (x_1^*, x_2^*).

With $x_1(t) = x_1^* + z_1(t)$ and $x_2(t) = x_2^* + z_2(t)$, we find that

$$x_1^* + z_1(t+1) \approx \underbrace{F(x_1^*, x_2^*)}_{x_1^*} + \left(\frac{\partial F}{\partial x_1}\right)^* z_1(t) + \left(\frac{\partial F}{\partial x_2}\right)^* z_2(t)$$

$$x_2^* + z_2(t+1) \approx \underbrace{G(x_1^*, x_2^*)}_{x_2^*} + \left(\frac{\partial G}{\partial x_1}\right)^* z_1(t) + \left(\frac{\partial G}{\partial x_2}\right)^* z_2(t)$$

Canceling x_1^* from the first equation and x_2^* from the second equation and writing the resulting approximation in matrix form, we obtain

$$\begin{bmatrix} z_1(t+1) \\ z_2(t+1) \end{bmatrix} \approx \begin{bmatrix} \left(\frac{\partial F}{\partial x_1}\right)^* & \left(\frac{\partial F}{\partial x_2}\right)^* \\ \left(\frac{\partial G}{\partial x_1}\right)^* & \left(\frac{\partial G}{\partial x_2}\right)^* \end{bmatrix} \begin{bmatrix} z_1(t) \\ z_2(t) \end{bmatrix} \qquad (10.54)$$

We recognize the 2×2 matrix as the Jacobi matrix of the vector-valued function $\begin{bmatrix} F(x_1, x_2) \\ G(x_1, x_2) \end{bmatrix}$. The right-hand side of (10.54) is a linear map of the form $A\mathbf{x}$, where A is a 2×2 matrix and $\mathbf{x} = \begin{bmatrix} x_1 \\ x_2 \end{bmatrix}$ is a 2×1 vector. In the previous subsection, we found that a linear map

$$\mathbf{x}_{t+1} = A\mathbf{x}_t$$

has the equilibrium $(0, 0)$, which is stable if the absolute values of the two eigenvalues of A are each less than 1. This is the criterion we need to determine the stability of the equilibrium (x_1^*, x_2^*) of the system (10.52) and (10.53).

The point equilibrium $\begin{bmatrix} x_1^* \\ x_2^* \end{bmatrix}$ of the system (10.52) and (10.53) is locally stable if the two eigenvalues λ_1 and λ_2 of the Jacobi matrix

$$\begin{bmatrix} \left(\frac{\partial F}{\partial x_1}\right)^* & \left(\frac{\partial F}{\partial x_2}\right)^* \\ \left(\frac{\partial G}{\partial x_1}\right)^* & \left(\frac{\partial G}{\partial x_2}\right)^* \end{bmatrix}$$

evaluated at (x_1^*, x_2^*) satisfy

$$|\lambda_1| < 1 \qquad \text{and} \qquad |\lambda_2| < 1$$

If $|\lambda_1| > 1$ or $|\lambda_2| > 1$, the point equilibrium $\begin{bmatrix} x_1^* \\ x_2^* \end{bmatrix}$ is an unstable equilibrium.

Note that, as in Chapters 5 and 8, the stability analysis is only a local analysis and we must say that an equilibrium is *locally* stable; the local analysis does not reveal anything about global stability.

EXAMPLE 5

Discuss the stability of the equilibria of the system in Example 3.

Solution

In Example 3,

$$F(x_1, x_2) = 2x_1(1 - x_1)$$
$$G(x_1, x_2) = x_1(1 - x_2)$$

The Jacobi matrix is

$$J(x_1, x_2) = \begin{bmatrix} 2 - 4x_1 & 0 \\ 1 - x_2 & -x_1 \end{bmatrix}$$

Evaluating $J(x_1, x_2)$ at the equilibrium $(0, 0)$, we find that

$$J(0, 0) = \begin{bmatrix} 2 & 0 \\ 1 & 0 \end{bmatrix}$$

which has eigenvalues $\lambda_1 = 2$ and $\lambda_2 = 0$. Since $|\lambda_1| > 1$, it follows that $(0, 0)$ is an unstable equilibrium.

Evaluating $J(x_1, x_2)$ at the equilibrium $(1/2, 1/3)$, we obtain

$$J(1/2, 1/3) = \begin{bmatrix} 0 & 0 \\ 2/3 & -1/2 \end{bmatrix}$$

which has eigenvalues $\lambda_1 = 0$ and $\lambda_2 = -1/2$. Since $|\lambda_1| < 1$ and $|\lambda_2| < 1$, it follows that $(1/2, 1/3)$ is a locally stable equilibrium. ■

EXAMPLE 6 Show that the nontrivial equilibrium of the Nicholson–Bailey equation is unstable.

Solution The Nicholson–Bailey equation is of the form

$$F(N, P) = bNe^{-aP}$$
$$G(N, P) = cN[1 - e^{-aP}]$$

To find the Jacobi matrix evaluated at the nontrivial equilibrium, we differentiate F and G and then evaluate the derivatives at the equilibrium

$$(N^*, P^*) = \left(\frac{b}{b-1} \frac{1}{ac} \ln b, \frac{1}{a} \ln b \right)$$

which we computed in Example 3:

$$\left(\frac{\partial F}{\partial N} \right)^* = be^{-aP}\big|_{(N^*, P^*)} = 1$$

$$\left(\frac{\partial F}{\partial P} \right)^* = -abNe^{-aP}\big|_{(N^*, P^*)} = -aN^* \quad \text{(since } be^{-aP^*} = 1\text{)}$$

$$\left(\frac{\partial G}{\partial N} \right)^* = c[1 - e^{-aP}]\big|_{(N^*, P^*)} = c\left[1 - \frac{1}{b} \right] \quad \left(\text{since } e^{-aP^*} = \frac{1}{b} \right)$$

$$\left(\frac{\partial G}{\partial P} \right)^* = caNe^{-aP}\big|_{(N^*, P^*)} = acN^*\frac{1}{b}$$

The Jacobi matrix evaluated at (N^*, P^*) is then

$$J(N^*, P^*) = \begin{bmatrix} 1 & -aN^* \\ c\left[1 - \frac{1}{b} \right] & acN^*\frac{1}{b} \end{bmatrix}$$

Instead of computing the eigenvalues explicitly, we will first show that the two eigenvalues of the matrix J are complex conjugate if $b > 1$. (This was the condition we found in Example 3 that guaranteed a biologically reasonable, nontrivial equilibrium.) The eigenvalues of J satisfy the equation $\det(J - \lambda I) = 0$; that is,

$$(1 - \lambda)\left(acN^*\frac{1}{b} - \lambda \right) + acN^*\left(1 - \frac{1}{b} \right) = 0$$

which simplifies to

$$\lambda^2 - \left(1 + \frac{ac}{b}N^* \right)\lambda + acN^* = 0$$

The solutions of this equation are complex conjugate if the discriminant

$$\left(1 + \frac{ac}{b}N^*\right)^2 - 4acN^* < 0$$

With $N^* = \frac{b}{b-1}\frac{1}{ac}\ln b$, the discriminant is

$$f(b) = \left(1 + \frac{\ln b}{b-1}\right)^2 - \frac{4b}{b-1}\ln b$$

This function depends only on b. Graphing $f(b)$ (see Figure 10.74) shows that $f(b) < 0$ for $b > 1$, thus confirming that the two eigenvalues of J are complex conjugate if $b > 1$.

When we discussed linear systems of difference equations, we derived the identity

$$|\lambda_1|^2 = |\lambda_2|^2 = \lambda_1\lambda_2 = \det J$$

The determinant of J is given by

$$\det J = acN^*\frac{1}{b} + acN^*\left(1 - \frac{1}{b}\right) = acN^* = \frac{b\ln b}{b-1}$$

Graphing $g(b) = \frac{b\ln b}{b-1}$ as a function of b (see Figure 10.75), we see that $g(b) > 1$ for $b > 1$, from which we conclude that, for $b > 1$,

$$|\lambda_1|^2 = |\lambda_2|^2 = \lambda_1\lambda_2 > 1$$

implying that the nontrivial equilibrium is unstable. ■

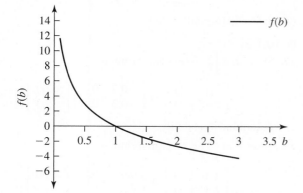

Figure 10.74 The graph of $f(b)$ confirms that $f(b) < 0$ for $b > 1$.

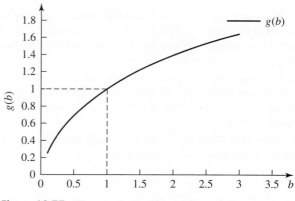

Figure 10.75 The graph of $g(b)$ confirms that $g(b) > 1$ for $b > 1$.

Section 10.7 Problems

■ 10.7.1

Problems 1–6 refer to the Nicholson–Bailey host–parasitoid model. Problems 1, 2, 5, and 6 are best done with the help of a spreadsheet, but can also be done with a calculator. Nicholson and Bailey introduced the discrete-generation host–parasitoid model of the form

$$N_{t+1} = bN_t e^{-aP_t}$$

$$P_{t+1} = cN_t[1 - e^{-aP_t}]$$

for $t = 0, 1, 2, \ldots$.

1. Evaluate the Nicholson–Bailey model for the first 10 generations when $a = 0.02$, $c = 3$, and $b = 1.5$. For the initial host density, choose $N_0 = 5$, and for the initial parasitoid density, choose $P_0 = 0$.

2. Evaluate the Nicholson–Bailey model for the first 10 generations when $a = 0.02$, $c = 3$, and $b = 0.5$. For the initial host density, choose $N_0 = 15$, and for the initial parasitoid density, choose $P_0 = 0$.

3. Show that when the initial parasitoid density is $P_0 = 0$, the Nicholson–Bailey model reduces to

$$N_{t+1} = bN_t$$

With N_0 denoting the initial host density, find an expression for N_t in terms of N_0 and the parameter b.

4. When the initial parasitoid density is $P_0 = 0$, the Nicholson–Bailey model reduces to

$$N_{t+1} = bN_t$$

as shown in the previous problem. For which values of b is the host density increasing if $N_0 > 0$? For which values of b is it decreasing? (Assume that $b > 0$.)

5. Evaluate the Nicholson–Bailey model for the first 15 generations when $a = 0.02$, $c = 3$, and $b = 1.5$. For the initial host density, choose $N_0 = 5$, and for the initial parasitoid density, choose $P_0 = 5$.

6. Evaluate the Nicholson–Bailey model for the first 25 generations when $a = 0.02$, $c = 3$, and $b = 1.5$. For the initial host density, choose $N_0 = 15$, and for the initial parasitoid density, choose $P_0 = 8$.

Problems 7–12 refer to the negative binomial host–parasitoid model. Problems 7, 8, 11, and 12 are best done with the help of a spreadsheet, but can also be done with a calculator. The negative binomial model is a discrete-generation host–parasitoid model of the form

$$N_{t+1} = bN_t \left(1 + \frac{aP_t}{k} \right)^{-k}$$

$$P_{t+1} = cN_t \left[1 - \left(1 + \frac{aP_t}{k} \right)^{-k} \right]$$

for $t = 0, 1, 2, \ldots$.

7. Evaluate the negative binomial model for the first 10 generations when $a = 0.02$, $c = 3$, $k = 0.75$, and $b = 1.5$. For the initial host density, choose $N_0 = 5$, and for the initial parasitoid density, choose $P_0 = 0$.

8. Evaluate the negative binomial model for the first 10 generations when $a = 0.02$, $c = 3$, $k = 0.75$, and $b = 0.5$. For the initial host density, choose $N_0 = 15$, and for the initial parasitoid density, choose $P_0 = 0$.

9. Show that when the initial parasitoid density is $P_0 = 0$, the negative binomial model reduces to

$$N_{t+1} = bN_t$$

With N_0 denoting the initial host density, find an expression for N_t in terms of N_0 and the parameter b.

10. When the initial parasitoid density is $P_0 = 0$, the negative binomial model reduces to

$$N_{t+1} = bN_t$$

as shown in the previous problem. For which values of b is the host density increasing if $N_0 > 0$? For which values of b is it decreasing? (Assume that $b > 0$.)

11. Evaluate the negative binomial model for the first 25 generations when $a = 0.02$, $c = 3$, $k = 0.75$, and $b = 1.5$. For the initial host density, choose $N_0 = 100$, and for the initial parasitoid density, choose $P_0 = 50$.

12. Evaluate the negative binomial model for the first 25 generations when $a = 0.02$, $c = 3$, $k = 0.75$, and $b = 0.5$. For the initial host density, choose $N_0 = 100$, and for the initial parasitoid density, choose $P_0 = 50$.

13. In the Nicholson–Bailey model, the fraction of hosts escaping parasitism is given by

$$f(P) = e^{-aP}$$

(a) Graph $f(P)$ as a function of P for $a = 0.1$ and $a = 0.01$.

(b) For a given value of P, how are the chances of escaping parasitism affected by increasing a?

14. In the negative binomial model, the fraction of hosts escaping parasitism is given by

$$f(P) = \left(1 + \frac{aP}{k} \right)^{-k}$$

(a) Graph $f(P)$ as a function of P for $a = 0.1$ and $a = 0.01$ when $k = 0.75$.

(b) For $k = 0.75$ and a given value of P, how are the chances of escaping parasitism affected by increasing a?

15. In the negative binomial model, the fraction of hosts escaping parasitism is given by

$$f(P) = \left(1 + \frac{aP}{k} \right)^{-k}$$

(a) Graph $f(P)$ as a function of P for $k = 0.75$ and $k = 0.5$ when $a = 0.02$.

(b) For $a = 0.02$ and a given value of P, how are the chances of escaping parasitism affected by increasing k?

16. The negative binomial model can be reduced to the Nicholson–Bailey model by letting the parameter k in the negative binomial model go to infinity. Show that

$$\lim_{k \to \infty} \left(1 + \frac{aP}{k} \right)^{-k} = e^{-aP}$$

(*Hint*: Use l'Hospital's rule.)

■ **10.7.2**

17. Show that $\begin{bmatrix} 0 \\ 0 \end{bmatrix}$ is an equilibrium of

$$\begin{bmatrix} x_1(t+1) \\ x_2(t+1) \end{bmatrix} = \begin{bmatrix} -0.7 & 0 \\ -0.3 & 0.2 \end{bmatrix} \begin{bmatrix} x_1(t) \\ x_2(t) \end{bmatrix}$$

and determine its stability.

18. Show that $\begin{bmatrix} 0 \\ 0 \end{bmatrix}$ is an equilibrium of

$$\begin{bmatrix} x_1(t+1) \\ x_2(t+1) \end{bmatrix} = \begin{bmatrix} 0.4 & 0.2 \\ 0 & -0.9 \end{bmatrix} \begin{bmatrix} x_1(t) \\ x_2(t) \end{bmatrix}$$

and determine its stability.

19. Show that $\begin{bmatrix} 0 \\ 0 \end{bmatrix}$ is an equilibrium of

$$\begin{bmatrix} x_1(t+1) \\ x_2(t+1) \end{bmatrix} = \begin{bmatrix} -1.4 & 0 \\ -0.5 & 0.1 \end{bmatrix} \begin{bmatrix} x_1(t) \\ x_2(t) \end{bmatrix}$$

and determine its stability.

20. Show that $\begin{bmatrix} 0 \\ 0 \end{bmatrix}$ is an equilibrium of

$$\begin{bmatrix} x_1(t+1) \\ x_2(t+1) \end{bmatrix} = \begin{bmatrix} 0.1 & 0.4 \\ 0.1 & -0.2 \end{bmatrix} \begin{bmatrix} x_1(t) \\ x_2(t) \end{bmatrix}$$

and determine its stability.

21. Show that $\begin{bmatrix} 0 \\ 0 \end{bmatrix}$ is an equilibrium of

$$\begin{bmatrix} x_1(t+1) \\ x_2(t+1) \end{bmatrix} = \begin{bmatrix} 1 & 2 \\ 3 & 2 \end{bmatrix} \begin{bmatrix} x_1(t) \\ x_2(t) \end{bmatrix}$$

and determine its stability.

22. Show that $\begin{bmatrix} 0 \\ 0 \end{bmatrix}$ is an equilibrium of

$$\begin{bmatrix} x_1(t+1) \\ x_2(t+1) \end{bmatrix} = \begin{bmatrix} 1.5 & 0.2 \\ 0.08 & 0 \end{bmatrix} \begin{bmatrix} x_1(t) \\ x_2(t) \end{bmatrix}$$

and determine its stability.

23. Show that the equilibrium $\begin{bmatrix} 0 \\ 0 \end{bmatrix}$ of

$$\begin{bmatrix} x_1(t+1) \\ x_2(t+1) \end{bmatrix} = \begin{bmatrix} -0.2 & -0.4 \\ 0.6 & 0.1 \end{bmatrix} \begin{bmatrix} x_1(t) \\ x_2(t) \end{bmatrix}$$

is stable.

24. Show that the equilibrium $\begin{bmatrix} 0 \\ 0 \end{bmatrix}$ of

$$\begin{bmatrix} x_1(t+1) \\ x_2(t+1) \end{bmatrix} = \begin{bmatrix} 0.2 & 0.3 \\ -0.5 & -0.4 \end{bmatrix} \begin{bmatrix} x_1(t) \\ x_2(t) \end{bmatrix}$$

is stable.

25. Show that the equilibrium $\begin{bmatrix} 0 \\ 0 \end{bmatrix}$ of

$$\begin{bmatrix} x_1(t+1) \\ x_2(t+1) \end{bmatrix} = \begin{bmatrix} 4.2 & -3.4 \\ 2.4 & -1.1 \end{bmatrix} \begin{bmatrix} x_1(t) \\ x_2(t) \end{bmatrix}$$

is unstable.

26. Show that the equilibrium $\begin{bmatrix} 0 \\ 0 \end{bmatrix}$ of

$$\begin{bmatrix} x_1(t+1) \\ x_2(t+1) \end{bmatrix} = \begin{bmatrix} 2 & -4 \\ 5 & -6 \end{bmatrix} \begin{bmatrix} x_1(t) \\ x_2(t) \end{bmatrix}$$

is stable.

■ **10.7.3**

27. Show that the equilibrium $\begin{bmatrix} 0 \\ 0 \end{bmatrix}$ of

$$x_1(t+1) = \frac{x_2(t)}{4(1+x_1^2(t))}$$

$$x_2(t+1) = \frac{2x_1(t)}{1+x_2^2(t)}$$

is locally stable.

28. Show that the equilibrium $\begin{bmatrix} 0 \\ 0 \end{bmatrix}$ of

$$x_1(t+1) = \frac{3x_2(t)}{1+x_1^2(t)}$$

$$x_2(t+1) = \frac{2x_1(t)}{1+x_2^2(t)}$$

is unstable.

29. Show that the equilibrium $\begin{bmatrix} 0 \\ 0 \end{bmatrix}$ of

$$x_1(t+1) = x_2(t)$$

$$x_2(t+1) = \frac{2x_2(t)-x_1(t)}{2+x_1(t)}$$

is locally stable.

30. Show that, for any $a > 1$, the equilibrium $\begin{bmatrix} 0 \\ 0 \end{bmatrix}$ of

$$x_1(t+1) = x_2(t)$$

$$x_2(t+1) = \frac{ax_2(t)-(a-1)x_1(t)}{a+x_1(t)}$$

is locally stable.

31. Show that $\begin{bmatrix} 0 \\ 0 \end{bmatrix}$ is an equilibrium point of

$$x_1(t+1) = ax_2(t)$$

$$x_2(t+1) = 2x_1(t) - \cos(x_2(t)) + 1$$

Assume that $a > 0$. For which values of a is $\begin{bmatrix} 0 \\ 0 \end{bmatrix}$ locally stable?

32. Show that $\begin{bmatrix} 0 \\ 0 \end{bmatrix}$ and $\begin{bmatrix} -\pi \\ \pi \end{bmatrix}$ are equilibria of

$$x_1(t+1) = -x_2(t)$$

$$x_2(t+1) = \sin(x_2(t)) - x_1(t)$$

and analyze their stability.

33. Find all nonnegative equilibria of

$$x_1(t+1) = x_2(t)$$

$$x_2(t+1) = \frac{1}{2}x_1(t) + \frac{2}{3}x_2(t) - x_2^2(t)$$

and analyze their stability.

34. Find all nonnegative equilibria of

$$x_1(t+1) = x_2(t)$$

$$x_2(t+1) = \frac{1}{2}x_1(t) + \frac{1}{3}x_2(t) - x_2^2(t)$$

and analyze their stability.

35. For which values of a is the equilibrium $\begin{bmatrix} 0 \\ 0 \end{bmatrix}$ of

$$x_1(t+1) = \frac{ax_2(t)}{1+x_1^2(t)}$$

$$x_2(t+1) = \frac{x_1(t)}{1+x_2^2(t)}$$

locally stable?

36. For which values of a is the equilibrium $\begin{bmatrix} 0 \\ 0 \end{bmatrix}$ of

$$x_1(t+1) = x_2(t)$$

$$x_2(t+1) = \frac{1}{2}x_1(t) + ax_2(t) - x_2^2(t)$$

locally stable?

37. Denote by $x_1(t)$ the number of juveniles, and by $x_2(t)$ the number of adults, at time t. Assume that $x_1(t)$ and $x_2(t)$ evolve according to

$$x_1(t+1) = x_2(t)$$

$$x_2(t+1) = \frac{1}{2}x_1(t) + rx_2(t) - x_2^2(t)$$

(a) Show that if $r > 1/2$, there exists an equilibrium $\begin{bmatrix} x_1^* \\ x_2^* \end{bmatrix}$ with $x_1^* > 0$ and $x_2^* > 0$. Find x_1^* and x_2^*.

(b) Determine the stability of the equilibrium found in (a) when $r > 1/2$.

38. Find all biologically relevant equilibria of the Nicholson–Bailey model

$$N_{t+1} = 2N_t e^{-0.2P_t}$$
$$P_{t+1} = N_t \left[1 - e^{-0.2P_t}\right]$$

and analyze their stability.

39. Find all biologically relevant equilibria of the Nicholson–Bailey model

$$N_{t+1} = 4N_t e^{-0.1P_t}$$
$$P_{t+1} = N_t \left[1 - e^{-0.1P_t}\right]$$

and analyze their stability.

40. Find all biologically relevant equilibria of the negative binomial host–parasitoid model

$$N_{t+1} = 4N_t \left(1 + \frac{0.01P_t}{2}\right)^{-2}$$
$$P_{t+1} = N_t \left[1 - \left(1 + \frac{0.01P_t}{2}\right)^{-2}\right]$$

and analyze their stability.

41. Find all biologically relevant equilibria of the negative binomial host–parasitoid model

$$N_{t+1} = 4N_t \left(1 + \frac{0.01P_t}{0.5}\right)^{-0.5}$$
$$P_{t+1} = N_t \left[1 - \left(1 + \frac{0.01P_t}{0.5}\right)^{-0.5}\right]$$

and analyze their stability.

Chapter 10 Key Terms

Discuss the following definitions and concepts:

1. Real-valued function
2. Function of two variables
3. Surface
4. Level curve
5. Limit
6. Limit laws
7. Continuity
8. Partial derivative
9. Geometric interpretation of a partial derivative
10. Mixed-derivative theorem
11. Tangent plane
12. Differentiability
13. Differentiability and continuity
14. Sufficient condition for differentiability
15. Standard linear approximation, tangent plane approximation
16. Vector-valued function
17. Jacobi matrix, derivative matrix
18. Chain rule
19. Implicit differentiation
20. Directional derivative
21. Gradient
22. Local extrema
23. Sufficient condition for finding local extrema
24. Hessian matrix
25. Global extrema
26. The extreme-value theorem
27. Diffusion
28. Systems of difference equations
29. Point equilibria and their stability
30. Nicholson–Bailey equation

Chapter 10 Review Problems

1. Germination Suppose that you conduct an experiment to measure the germination success of seeds of a certain plant as a function of temperature and humidity. You find that seeds don't germinate at all when the humidity is too low, regardless of temperature; germination success is highest for intermediate values of temperature; and seeds tend to germinate better when you increase humidity levels. Use the preceding information to sketch a graph of germination success as a function of temperature for different levels of humidity. Also, sketch the graph of germination success as a function of humidity for different temperature values.

2. Plant Physiology Gaastra (1959) measured the effects of atmospheric CO_2 enrichment on CO_2 fixation in sugar beet leaves at various light levels. He found that increasing CO_2 at fixed light levels increases the fixation rate and that increasing light levels at fixed atmospheric CO_2 concentration also increased fixation. If $F(A, I)$ denotes the fixation rate as a function of atmospheric CO_2 concentration (A) and light intensity (I), determine the signs of $\partial F/\partial A$ and $\partial F/\partial I$.

3. Plant Ecology In Burke and Grime (1996), a long-term field experiment in a limestone grassland was described.
(a) One of the experiments related total area covered by *indigenous* species to fertility and disturbance gradients. The experiment was designed so that the two variables (fertility and disturbance) could be altered independently. Burke and Grime found that the area covered by indigenous species generally increased with the amount of fertilizer added and decreased with the intensity of a disturbance. If $A_i(F, D)$ denotes the area covered by indigenous species as a function of the amount of fertilizer added (F) and the intensity of disturbance (D), determine the signs of $\partial A_i/\partial F$ and $\partial A_i/\partial D$ for Burke and Grime's experiment.

(b) In another experiment, Burke and Grime related the total area covered by *introduced* species to fertility and disturbance gradients. Let $A_e(F, D)$ denote the area covered by introduced species as a function of the amount of fertilizer added (F) and the intensity of disturbance (D). Burke and Grime found that

$$\frac{\partial A_e}{\partial F} > 0$$

and

$$\frac{\partial A_e}{\partial D} > 0$$

Explain in words what this means.

(c) Compare the responses to fertilization and disturbance with the area covered in the two experiments.

4. Plant Physiology Vitousek and Farrington (1997) investigated nutrient limitations in soils of different ages. In the abstract of their paper, they say,

> Walker and Syers (1976) proposed a conceptual model that describes the pattern and regulation of soil nutrient pools and availability during long-term soil and ecosystem development. Their model implies that plant production generally should be limited by N [nitrogen] on young soils and by P [phosphorus] on old soils; N and P supply should be more or less equilibrate on intermediate aged soils.

Vitousek and Farrington tested this hypothesis by conducting fertilizer experiments along a gradient of soil age, measuring the average increment in diameter (in mm/yr) of *Metrosideros polymorpha* trees.

Denote by $D(N, P, t)$ the diameter increment (in mm/yr) as a function of the amount of nitrogen (N) added, the amount of phosphorus (P) added, and the age (t) of the soil. Vitousek and Farrington's experiments showed that

$$\frac{\partial D}{\partial t}(N, 0, t) < 0$$

and

$$\frac{\partial D}{\partial t}(0, P, t) > 0$$

for their choices of N > 0 and P > 0. Explain why their results support the Walker and Syers hypothesis.

5. Find the Jacobi matrix

$$\mathbf{f}(x, y) = \begin{bmatrix} x^2 - y \\ x^3 - y^2 \end{bmatrix}$$

6. Find a linear approximation to

$$\mathbf{f}(x, y) = \begin{bmatrix} 2xy^2 \\ \frac{x}{y} \end{bmatrix}$$

at $(1, 1)$.

7. Mark–Recapture Experiment We can compute the average radius of spreading individuals at time t, denoted by r_{avg}. We find that

$$r_{avg} = \sqrt{\pi D t} \tag{10.55}$$

(a) Graph r_{avg} as a function of D for $t = 0.1$, $t = 1$, and $t = 5$. Describe in words how an increase in D affects the average radius of spread.

(b) Show that

$$D = \frac{(r_{avg})^2}{\pi t} \tag{10.56}$$

(c) Equation (10.56) can be used to measure D, the diffusion constant, from field data of mark–recapture experiments, taken from Kareiva (1983), as follows: Marked organisms are released from the release site and then recaptured after a certain amount of time t from the time of release. The distance of the recaptured organisms from the release site is measured.

If N denotes the total number of recaptured organisms, d_i denotes the distance of the ith recaptured organism from the release site, and t is the time between release and recapture, use (10.56) to explain why

$$D = \frac{1}{\pi t} \left(\frac{1}{N} \sum_{i=1}^{N} d_i \right)^2$$

can be used to measure D from field data. (Note that the time between release and recapture is the same for each individual in this study.)

11

Systems of Differential Equations

We first encountered systems of differential equations in Section 8.3. There, the emphasis was on understanding where such systems arise and on finding equilibria. (Section 8.3 is not a prerequisite for this chapter.) We will now give a systematic treatment of such systems. Suppose that we are given a set of variables $x_1, x_2, \ldots, x_n$, each depending on an independent variable, say, t, so that $x_1 = x_1(t), x_2 = x_2(t), \ldots,$ $x_n = x_n(t)$. Suppose also that the dynamics of the variables are linked by differential equations of the form

$$\frac{dx_1}{dt} = g_1(t, x_1, x_2, \ldots, x_n)$$

$$\frac{dx_2}{dt} = g_2(t, x_1, x_2, \ldots, x_n) \tag{11.1}$$

$$\vdots$$

$$\frac{dx_n}{dt} = g_n(t, x_1, x_2, \ldots, x_n)$$

This set of equations is called a *system of differential equations*. On the left-hand side are the derivatives of $x_i(t)$ with respect to t. On the right-hand side of each equation is a function g_i that depends on the variables $x_1, x_2, \ldots, x_n$ and on t. We will first look at the case when the functions g_i are linear in the variables $x_1, x_2, \ldots, x_n$—that is, when, for $i = 1, 2, \ldots, n,$

$$g_i(t, x_1, x_2, \ldots, x_n) = a_{i1}(t)x_1 + a_{i2}(t)x_2 + \ldots + a_{in}(t)x_n + f_i(t)$$

We can write the linear system in matrix form as

$$\frac{d\mathbf{x}}{dt} = A(t)\mathbf{x}(t) + \mathbf{f}(t) \tag{11.2}$$

where

$$\mathbf{x}(t) = \begin{bmatrix} x_1(t) \\ x_2(t) \\ \vdots \\ x_n(t) \end{bmatrix} \qquad A = \begin{bmatrix} a_{11}(t) & a_{12}(t) & \ldots & a_{1n}(t) \\ a_{21}(t) & a_{22}(t) & \ldots & a_{2n}(t) \\ \vdots & \vdots & \vdots & \vdots \\ a_{n1}(t) & a_{n2}(t) & \ldots & a_{nn}(t) \end{bmatrix}$$

and

$$\mathbf{f}(t) = \begin{bmatrix} f_1(t) \\ f_2(t) \\ \vdots \\ f_n(t) \end{bmatrix}$$

Equation (11.2) is called a system of **linear first-order equations** (first order because only first derivatives occur). We will investigate only the case when $\mathbf{f}(t) = \mathbf{0}$ and $A(t)$ does not depend on t. Equation (11.2) then reduces to

$$\frac{d\mathbf{x}}{dt} = A\mathbf{x}(t) \tag{11.3}$$

where

$$A = \begin{bmatrix} a_{11} & a_{12} & \cdots & a_{1n} \\ a_{21} & a_{22} & \cdots & a_{2n} \\ \vdots & \vdots & \vdots & \vdots \\ a_{n1} & a_{n2} & \cdots & a_{nn} \end{bmatrix}$$

Equation (11.3) is called a **homogeneous** linear first-order system with constant co-efficients. [*Homogeneous* refers to the fact that $\mathbf{f}(t) = \mathbf{0}$; when $\mathbf{f}(t) \neq \mathbf{0}$, the system is called **inhomogeneous**.] Since the matrix A does not depend on t, all the coefficients are constant; such a system is **autonomous**. (We encountered autonomous systems in Section 8.1.) Note that A is a square matrix.

In Section 11.1, we will present some of the theory for systems of the form (11.3). In Section 11.2, we will discuss some applications of linear systems. Section 11.3 is devoted to the theory of nonlinear systems, Section 11.4 to applications of nonlinear systems.

■ 11.1 Linear Systems: Theory

In this section, we will analyze homogeneous, linear first-order systems with constant coefficients—that is, systems of the form

$$\begin{aligned} \frac{dx_1}{dt} &= a_{11}x_1 + a_{12}x_2 + \ldots + a_{1n}x_n \\ \frac{dx_2}{dt} &= a_{21}x_1 + a_{22}x_2 + \ldots + a_{2n}x_n \\ &\vdots \\ \frac{dx_n}{dt} &= a_{n1}x_1 + a_{n2}x_2 + \ldots + a_{nn}x_n \end{aligned} \tag{11.4}$$

where the variables $x_1, x_2, \ldots, x_n$ are functions of t and the parameters $a_{ij}, 1 \leq i, j \leq n$, are constants. We can write this system in matrix form as

$$\frac{d\mathbf{x}}{dt} = A\mathbf{x}(t)$$

where

$$\mathbf{x}(t) = \begin{bmatrix} x_1(t) \\ x_2(t) \\ \vdots \\ x_n(t) \end{bmatrix} \qquad A = \begin{bmatrix} a_{11} & a_{12} & \cdots & a_{1n} \\ a_{21} & a_{22} & \cdots & a_{2n} \\ \vdots & \vdots & \vdots & \vdots \\ a_{n1} & a_{n2} & \cdots & a_{nn} \end{bmatrix}$$

Most of the time we will restrict ourselves to the case $n = 2$.

EXAMPLE 1   Write

$$\frac{dx_1}{dt} = 4x_1 - 2x_2$$

$$\frac{dx_2}{dt} = -3x_1 + x_2$$

in matrix notation.

Solution   We write $x_1 = x_1(t)$ and $x_2 = x_2(t)$. Using the rules for matrix multiplication, we find that

$$\begin{bmatrix} \frac{dx_1}{dt} \\ \frac{dx_2}{dt} \end{bmatrix} = \begin{bmatrix} 4x_1 - 2x_2 \\ -3x_1 + x_2 \end{bmatrix} = \begin{bmatrix} 4 & -2 \\ -3 & 1 \end{bmatrix} \begin{bmatrix} x_1(t) \\ x_2(t) \end{bmatrix}$$

That is,

$$\frac{d\mathbf{x}(t)}{dt} = \begin{bmatrix} 4 & -2 \\ -3 & 1 \end{bmatrix} \mathbf{x}(t)$$ ■

First, we will be concerned with solutions of (11.4): A *solution* is an ordered n-tuple of functions $(x_1(t), x_2(t), \ldots, x_n(t))$ that satisfies all n equations in (11.4). Second, we will discuss equilibria: An *equilibrium* is a point $\hat{\mathbf{x}} = (\hat{x}_1, \hat{x}_2, \ldots, \hat{x}_n)$ such that $A\hat{\mathbf{x}} = \mathbf{0}$. We begin with a graphical approach to visualizing solutions.

■ 11.1.1 The Direction Field

Before we turn to finding solutions for homogeneous, linear first-order systems with constant coefficients, we discuss an important property of solution curves that will allow us to sketch solutions graphically in the x_1–x_2 plane with the help of **direction fields**. Consider

$$\frac{dx_1}{dt} = x_1 - 2x_2$$

$$\frac{dx_2}{dt} = x_2$$ (11.5)

Imagine now that you are standing at a point (x_1, x_2) in the x_1–x_2 plane and the system (11.5) determines your future location. Where should you go next? The two differential equations tell you how your coordinates will change. To give a concrete example, look at the point $(2, -1)$. We claim that you move along a curve whose tangent line at the point $(2, -1)$ has slope

$$\frac{dx_2}{dx_1} = \frac{dx_2/dt}{dx_1/dt} = \frac{x_2}{x_1 - 2x_2} = \frac{-1}{2 - (2)(-1)} = -\frac{1}{4}$$

Why is this true? A solution of (11.5) that starts at time 0 at the point $(x_1(0), x_2(0))$ is given by points of the form $(x_1(t), x_2(t))$, $t \geq 0$, that satisfy (11.5). An example of a solution curve is shown in Figure 11.1. We see that the solution is a curve in the x_1–x_2 plane, and at each point, we can draw a tangent line whose slope is dx_2/dx_1. The tangent line at $(2, -1)$, for which we computed the slope (namely, $-1/4$), is also drawn in Figure 11.1.

We can draw tangent lines at each point (x_1, x_2) in the x_1–x_2 plane. Knowing all the tangent lines then allows us to sketch the corresponding solution curve. This is done by assigning each point (x_1, x_2) in the x_1–x_2 plane a vector $\begin{bmatrix} dx_1/dt \\ dx_2/dt \end{bmatrix}$ which has the property that it is tangential to the solution curve that passes through the point (x_1, x_2) and it points in the direction of the solution.

In our example, the vector is of the form $\begin{bmatrix} x_1 - 2x_2 \\ x_2 \end{bmatrix}$ and the slope of the solution curve that goes through (x_1, x_2) is

$$\frac{dx_2}{dx_1} = \frac{dx_2/dt}{dx_1/dt} = \frac{x_2}{x_1 - 2x_2}$$

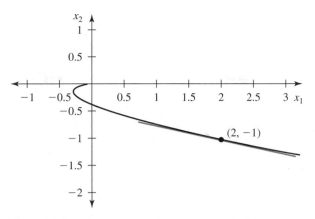

Figure 11.1 Solution curve through $(2, -1)$ with tangent line.

The collection of these vectors is called a **direction field** or **slope field**, and each vector of the direction field is called a **direction vector**. Since (11.5) is an autonomous system, the direction vector depends only on the location of the point (x_1, x_2) and not on t. This property implies that the direction field is the same for all times t.

The direction field for (11.5) is shown in Figure 11.2. [The figure also contains four solution curves that were computer generated; each curve starts at a different point very close to the origin $(0, 0)$.] The length of the direction vector at a given point tells us how quickly the solution curve passes through the point; the length is proportional to $\sqrt{(dx_1/dt)^2 + (dx_2/dt)^2}$. If we are interested only in the direction of the solution curves, we can indicate the direction by small line segments (as shown in Figure 11.3), which often results in a less cluttered picture. Like Figure 11.2, Figure 11.3 contains four solution curves that were computer generated; note that the direction vectors are always tangent to the curves. We can therefore use the direction field to sketch solution curves by drawing curves in such a way that the direction vectors are always tangent to the curve.

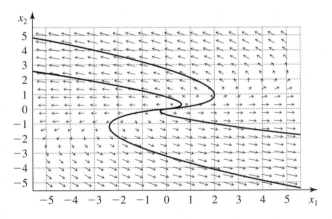

Figure 11.2 The direction field of the system (11.5) together with some solution curves.

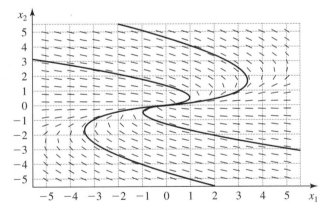

Figure 11.3 The direction field of the system (11.5) with some solution curves, where the direction vectors are not drawn to scale.

The point $(0, 0)$ is special: When we compute the direction vector at $(0, 0)$, we find that

$$\begin{bmatrix} \frac{dx_1}{dt} \\ \frac{dx_2}{dt} \end{bmatrix} = \begin{bmatrix} 0 \\ 0 \end{bmatrix}$$

That is, if we start at this point, neither $x_1(t)$ nor $x_2(t)$ will change. We call such points *equilibria*. We will discuss their significance in Subsection 11.1.3; they play a central role in how solutions behave as $t \to \infty$.

■ 11.1.2 Solving Linear Systems

Specific Solutions   Consider the following differential equation:

$$\frac{dx}{dt} = ax$$

This is a linear, first-order differential equation with a constant coefficient. [That is, the equation is of the form (11.4) with $n = 1$]. We can find a solution by integrating after separating variables (as we learned in Chapter 8). All solutions are of the form

$$x(t) = ce^{at}$$

where c is a constant that depends on the initial condition. An initial condition picks out a specific solution among the set of solutions. For instance, if $x(0) = x_0$, then $c = x_0$.

We will now show that exponential functions are also solutions of *systems* of linear differential equations. We restrict our discussion to systems of two differential equations. Consider the system

$$\frac{dx_1}{dt} = a_{11}x_1(t) + a_{12}x_2(t) \tag{11.6}$$

$$\frac{dx_2}{dt} = a_{21}x_1(t) + a_{22}x_2(t) \tag{11.7}$$

which, in matrix form, is written as

$$\frac{d\mathbf{x}}{dt} = A\mathbf{x}(t) \tag{11.8}$$

A solution of (11.8) is a vector-valued function. As in the example presented at the beginning of this subsection (i.e., $dx/dt = ax$), we will find that (11.8) admits a collection of solutions, and if we choose an initial condition, a particular solution will be picked out. We will get to initial conditions later; let's first see what the solutions look like.

We claim that the vector-valued function

$$\mathbf{x}(t) = \begin{bmatrix} u_1 e^{\lambda t} \\ u_2 e^{\lambda t} \end{bmatrix} = e^{\lambda t} \begin{bmatrix} u_1 \\ u_2 \end{bmatrix} \tag{11.9}$$

where λ, u_1, and u_2 are constants, is a solution of (11.8) for an appropriate choice of values for λ, u_1, and u_2. To see how we must choose these values, we differentiate $\mathbf{x}(t)$ in (11.9):

$$\frac{d\mathbf{x}}{dt} = \begin{bmatrix} u_1 \lambda e^{\lambda t} \\ u_2 \lambda e^{\lambda t} \end{bmatrix} = \lambda e^{\lambda t} \begin{bmatrix} u_1 \\ u_2 \end{bmatrix} \tag{11.10}$$

If $\mathbf{x}(t)$ solves (11.8), then $\mathbf{x}(t)$ must satisfy (11.8). Using (11.10) for the left-hand side of (11.8) and (11.9) for the right-hand side, we find that

$$\underbrace{\lambda e^{\lambda t} \begin{bmatrix} u_1 \\ u_2 \end{bmatrix}}_{\frac{d\mathbf{x}(t)}{dt}} = \underbrace{A e^{\lambda t} \begin{bmatrix} u_1 \\ u_2 \end{bmatrix}}_{A\mathbf{x}(t)}$$

or, after dividing both sides by $e^{\lambda t}$,

$$\lambda \begin{bmatrix} u_1 \\ u_2 \end{bmatrix} = A \begin{bmatrix} u_1 \\ u_2 \end{bmatrix}$$

This last expression should remind you of eigenvalues and eigenvectors that we encountered in Section 9.4: $\begin{bmatrix} u_1 \\ u_2 \end{bmatrix}$ is an eigenvector corresponding to an eigenvalue λ of A. With this choice, (11.9) is a solution of (11.8).

In this subsection, we will look only at differential equations of the form (11.8) for which the eigenvalues of A are both real and distinct. We will discuss complex eigenvalues in the next subsection. (We will not discuss the case when both eigenvalues are identical.) We use the following system to illustrate how to find specific solutions of a system of the form (11.8) in the case when both eigenvalues of A are real and distinct:

$$\frac{dx_1}{dt} = 2x_1 - 2x_2$$

$$\frac{dx_2}{dt} = 2x_1 - 3x_2$$

(11.11)

1. Finding eigenvalues The coefficient matrix of (11.11) is given by

$$A = \begin{bmatrix} 2 & -2 \\ 2 & -3 \end{bmatrix}$$

We first determine the eigenvalues and corresponding eigenvectors of A. To find the eigenvalues of A, we must solve

$$\det(A - \lambda I) = 0$$

That is,

$$\det \begin{bmatrix} 2 - \lambda & -2 \\ 2 & -3 - \lambda \end{bmatrix} = (2 - \lambda)(-3 - \lambda) + 4$$

$$= \lambda^2 + \lambda - 2 = (\lambda - 1)(\lambda + 2) = 0$$

which has solutions

$$\lambda_1 = 1 \quad \text{and} \quad \lambda_2 = -2$$

2. A solution corresponding to the eigenvalue λ_1 To find an eigenvector $\mathbf{u} = \begin{bmatrix} u_1 \\ u_2 \end{bmatrix}$ corresponding to $\lambda_1 = 1$, we solve

$$A \begin{bmatrix} u_1 \\ u_2 \end{bmatrix} = \lambda_1 \begin{bmatrix} u_1 \\ u_2 \end{bmatrix} \quad \text{with } \lambda_1 = 1$$

We find

$$\begin{bmatrix} 2u_1 - 2u_2 \\ 2u_1 - 3u_2 \end{bmatrix} = \begin{bmatrix} u_1 \\ u_2 \end{bmatrix}$$

which can be written as

$$u_1 - 2u_2 = 0$$
$$2u_1 - 4u_2 = 0$$

Both equations reduce to the same equation, namely,

$$u_1 = 2u_2$$

Setting $u_2 = 1$, for instance, we obtain $u_1 = 2$. An eigenvector corresponding to $\lambda_1 = 1$ is then

$$\mathbf{u} = \begin{bmatrix} 2 \\ 1 \end{bmatrix}$$

(Recall that any nonzero multiple of $\mathbf{u}$ is also an eigenvector corresponding to $\lambda_1 = 1$.)

We claim that

$$\mathbf{x}(t) = e^t \begin{bmatrix} 2 \\ 1 \end{bmatrix}$$

solves (11.11). Let's check. We find that

$$\frac{d\mathbf{x}}{dt} = e^t \begin{bmatrix} 2 \\ 1 \end{bmatrix} = \mathbf{x}(t)$$

Since $\begin{bmatrix} 2 \\ 1 \end{bmatrix}$ is an eigenvector corresponding to $\lambda_1 = 1$, $\mathbf{x}(t)$ also satisfies

$$A\mathbf{x}(t) = Ae^t \begin{bmatrix} 2 \\ 1 \end{bmatrix} = e^t A \begin{bmatrix} 2 \\ 1 \end{bmatrix} = e^t \begin{bmatrix} 2 \\ 1 \end{bmatrix} = \mathbf{x}(t)$$

where, for the same reason, we used the fact that

$$A \begin{bmatrix} 2 \\ 1 \end{bmatrix} = \lambda_1 \begin{bmatrix} 2 \\ 1 \end{bmatrix} = \begin{bmatrix} 2 \\ 1 \end{bmatrix}$$

Therefore,

$$\frac{d\mathbf{x}}{dt} = A\mathbf{x}(t)$$

and we conclude that $\mathbf{x}(t) = e^t \begin{bmatrix} 2 \\ 1 \end{bmatrix}$ is indeed a solution of (11.11).

3. A solution corresponding to the eigenvalue λ_2   We can now repeat the same steps for the eigenvalue $\lambda_2 = -2$. To find an eigenvector $\mathbf{v} = \begin{bmatrix} v_1 \\ v_2 \end{bmatrix}$ corresponding to the eigenvalue $\lambda_2 = -2$, we solve

$$\begin{bmatrix} 2v_1 - 2v_2 \\ 2v_1 - 3v_2 \end{bmatrix} = -2 \begin{bmatrix} v_1 \\ v_2 \end{bmatrix}$$

which can be written as

$$4v_1 - 2v_2 = 0$$
$$2v_1 - v_2 = 0$$

The two equations reduce to the same equation, namely,

$$2v_1 = v_2$$

Setting $v_1 = 1$, we find that $v_2 = 2$. An eigenvector corresponding to $\lambda_2 = -2$ is then

$$\mathbf{v} = \begin{bmatrix} 1 \\ 2 \end{bmatrix}$$

We set

$$\mathbf{x}(t) = e^{-2t} \begin{bmatrix} 1 \\ 2 \end{bmatrix}$$

and check that

$$\frac{d\mathbf{x}}{dt} = -2e^{-2t} \begin{bmatrix} 1 \\ 2 \end{bmatrix} = -2\mathbf{x}(t)$$

and

$$A\mathbf{x}(t) = Ae^{-2t} \begin{bmatrix} 1 \\ 2 \end{bmatrix} = e^{-2t} A \begin{bmatrix} 1 \\ 2 \end{bmatrix} = -2e^{-2t} \begin{bmatrix} 1 \\ 2 \end{bmatrix} = -2\mathbf{x}(t)$$

where we used the fact that

$$A \begin{bmatrix} 1 \\ 2 \end{bmatrix} = \lambda_2 \begin{bmatrix} 1 \\ 2 \end{bmatrix} = -2 \begin{bmatrix} 1 \\ 2 \end{bmatrix}$$

since $\begin{bmatrix} 1 \\ 2 \end{bmatrix}$ is an eigenvector corresponding to $\lambda_2 = -2$. This shows that $\mathbf{x}(t) = e^{-2t} \begin{bmatrix} 1 \\ 2 \end{bmatrix}$ is also a solution of (11.11).

4. *The direction field* Consider the two particular solutions we just found. We wish to illustrate them in the corresponding direction field. Recall that an eigenvector can be used to construct a straight line through the origin in the direction of the eigenvector. In Figure 11.4, we show the direction field together with the two lines in the direction of the eigenvectors.

Figure 11.4 The direction field of the system (11.11) with the lines in the direction of the eigenvectors.

If $\mathbf{x}(0)$ is a point on one of the lines defined by the eigenvectors, then the solution $\mathbf{x}(t)$ will remain on that line at *all* later times. Of course, the location of $\mathbf{x}(t)$ on the line will change with time. If the corresponding eigenvalue is positive, the solution will move away from the origin; if the eigenvalue is negative, it will move toward the origin, as can be seen from the direction of the direction vectors. The solid line in Figure 11.4 ($2x_1 - x_2 = 0$) corresponds to the eigenvalue $\lambda_2 = -2$, and we see that the direction vectors on this line point toward the origin. The dashed line in Figure 11.4 ($x_1 - 2x_2 = 0$) corresponds to the eigenvalue $\lambda_2 = 1$, and we see that the direction vectors on this line point away from the origin.

The General Solution Suppose that $\mathbf{y}(t)$ and $\mathbf{z}(t)$ are two solutions of

$$\frac{d\mathbf{x}}{dt} = A\mathbf{x}(t) \tag{11.12}$$

That is,

$$\frac{d\mathbf{y}}{dt} = A\mathbf{y}(t) \qquad \text{and} \qquad \frac{d\mathbf{z}}{dt} = A\mathbf{z}(t)$$

Then the linear combination

$$c_1\mathbf{y}(t) + c_2\mathbf{z}(t) \tag{11.13}$$

also solves (11.12). This can be seen as follows: First, note that

$$\frac{d}{dt}\left[c_1\mathbf{y}(t) + c_2\mathbf{z}(t)\right] = c_1\frac{d\mathbf{y}}{dt} + c_2\frac{d\mathbf{z}}{dt}$$

But since $\mathbf{y}(t)$ and $\mathbf{z}(t)$ are both solutions of (11.12), it follows that

$$c_1\frac{d\mathbf{y}}{dt} + c_2\frac{d\mathbf{z}}{dt} = c_1 A\mathbf{y}(t) + c_2 A\mathbf{z}(t) \tag{11.14}$$

Using the linearity properties of A, we can write the right-hand side of (11.14) as

$$A\left[c_1\mathbf{y}(t) + c_2\mathbf{z}(t)\right]$$

Summarizing, we find that

$$\frac{d}{dt}\underbrace{[c_1\mathbf{y}(t) + c_2\mathbf{z}(t)]}_{\mathbf{x}(t)} = A\underbrace{[c_1\mathbf{y}(t) + c_2\mathbf{z}(t)]}_{\mathbf{x}(t)}$$

which shows that the linear combination $c_1\mathbf{y}(t) + c_2\mathbf{z}(t)$ is also a solution.

Combining solutions, as in (11.13), illustrates the important **superposition principle**:

Superposition Principle Suppose that

$$\begin{bmatrix} \frac{dx_1}{dt} \\ \frac{dx_2}{dt} \end{bmatrix} = \begin{bmatrix} a_{11} & a_{12} \\ a_{21} & a_{22} \end{bmatrix} \begin{bmatrix} x_1(t) \\ x_2(t) \end{bmatrix} \tag{11.15}$$

If

$$\mathbf{y}(t) = \begin{bmatrix} y_1(t) \\ y_2(t) \end{bmatrix} \quad \text{and} \quad \mathbf{z}(t) = \begin{bmatrix} z_1(t) \\ z_2(t) \end{bmatrix}$$

are solutions of (11.15), then

$$\mathbf{x}(t) = c_1 \mathbf{y}(t) + c_2 \mathbf{z}(t)$$

is also a solution of (11.15).

We just saw that $e^{\lambda t} \begin{bmatrix} w_1 \\ w_2 \end{bmatrix}$, where $\begin{bmatrix} w_1 \\ w_2 \end{bmatrix}$ is an eigenvector corresponding to the real eigenvalue λ of A, solves (11.12). If we have two real and distinct eigenvalues λ_1 and λ_2 with corresponding eigenvectors $\mathbf{u}$ and $\mathbf{v}$, then, setting $\mathbf{y}(t) = e^{\lambda_1 t} \mathbf{u}$ and $\mathbf{z}(t) = e^{\lambda_2 t} \mathbf{v}$ and using the superposition principle, we find that the linear combination

$$\mathbf{x}(t) = c_1 e^{\lambda_1 t} \mathbf{u} + c_2 e^{\lambda_2 t} \mathbf{v} \tag{11.16}$$

where c_1 and c_2 are constants, is also a solution of (11.12). The constants c_1 and c_2 depend on the initial conditions. We can show that every solution of (11.2) can be written in the form (11.16); we therefore call a solution of the form (11.16) the *general solution*. (The situation is more complicated when A has repeated eigenvalues—that is, when $\lambda_1 = \lambda_2$. We do not give the general solution for that case here, but will discuss two such examples in Problems 27 and 28.) We summarize our findings as follows:

The General Solution Let

$$\frac{d\mathbf{x}}{dt} = A\mathbf{x}(t) \tag{11.17}$$

where A is a 2×2 matrix with two real and distinct eigenvalues λ_1 and λ_2 with corresponding eigenvectors $\mathbf{u}$ and $\mathbf{v}$. Then

$$\mathbf{x}(t) = c_1 e^{\lambda_1 t} \mathbf{u} + c_2 e^{\lambda_2 t} \mathbf{v} \tag{11.18}$$

is the general solution of (11.17). The constants c_1 and c_2 depend on the initial condition.

We will now check that (11.18) is indeed a solution of (11.17). To do so, we differentiate (11.18) with respect to t:

$$\frac{d\mathbf{x}}{dt} = \lambda_1 c_1 e^{\lambda_1 t} \mathbf{u} + \lambda_2 c_2 e^{\lambda_2 t} \mathbf{v}$$

Since $\mathbf{u}$ and $\mathbf{v}$ are eigenvectors corresponding to λ_1 and λ_2, it follows that

$$A\mathbf{x}(t) = A(c_1 e^{\lambda_1 t} \mathbf{u} + c_2 e^{\lambda_2 t} \mathbf{v})$$
$$= c_1 e^{\lambda_1 t} A\mathbf{u} + c_2 e^{\lambda_2 t} A\mathbf{v}$$
$$= c_1 e^{\lambda_1 t} \lambda_1 \mathbf{u} + c_2 e^{\lambda_2 t} \lambda_2 \mathbf{v}$$

Hence,

$$\frac{d\mathbf{x}}{dt} = A\mathbf{x}(t)$$

Going back to the example at the beginning of this subsection, we find that the general solution of (11.11), or

$$\frac{d\mathbf{x}}{dt} = \begin{bmatrix} 2 & -2 \\ 2 & -3 \end{bmatrix} \mathbf{x}(t)$$

is

$$\mathbf{x}(t) = c_1 e^t \begin{bmatrix} 2 \\ 1 \end{bmatrix} + c_2 e^{-2t} \begin{bmatrix} 1 \\ 2 \end{bmatrix} \tag{11.19}$$

5. An initial condition for (11.11)   Suppose we know that, at time 0,

$$\mathbf{x}(0) = \begin{bmatrix} -1 \\ 4 \end{bmatrix} \tag{11.20}$$

holds for (11.11). Then we can determine the constants c_1 and c_2 in (11.19) so that $\mathbf{x}(t)$ satisfies the initial condition (11.20), or

$$\mathbf{x}(0) = c_1 \begin{bmatrix} 2 \\ 1 \end{bmatrix} + c_2 \begin{bmatrix} 1 \\ 2 \end{bmatrix} = \begin{bmatrix} -1 \\ 4 \end{bmatrix}$$

To find c_1 and c_2, we must solve the system of linear equations

$$\begin{aligned} 2c_1 + c_2 &= -1 \\ c_1 + 2c_2 &= 4 \end{aligned}$$

Eliminating c_1 in the second equation yields

$$\begin{aligned} 2c_1 + c_2 &= -1 \\ 3c_2 &= 9 \end{aligned}$$

Hence, $c_2 = 3$, and therefore,

$$2c_1 = -1 - c_2 = -1 - 3 = -4 \qquad \text{or} \qquad c_1 = -2$$

Thus,

$$\mathbf{x}(t) = -2e^t \begin{bmatrix} 2 \\ 1 \end{bmatrix} + 3e^{-2t} \begin{bmatrix} 1 \\ 2 \end{bmatrix}$$

or, writing the vector equation out as separate equations,

$$\begin{aligned} x_1(t) &= -4e^t + 3e^{-2t} \\ x_2(t) &= -2e^t + 6e^{-2t} \end{aligned}$$

To summarize, this solution solves the system

$$\frac{dx_1}{dt} = 2x_1 - 2x_2$$

$$\frac{dx_2}{dt} = 2x_2 - 3x_1$$

which was given in (11.11) with initial condition $x_1(0) = -1$ and $x_2(0) = 4$.

We give one more example of an initial-value problem that illustrates all at once the different steps that must be carried out in order to obtain a solution in the case when A has two distinct and real eigenvalues.

EXAMPLE 2

Solve

$$\frac{d\mathbf{x}}{dt} = \begin{bmatrix} 2 & -3 \\ 1 & -2 \end{bmatrix} \mathbf{x}(t) \tag{11.21}$$

with the initial condition

$$\mathbf{x}(0) = \begin{bmatrix} 3 \\ -1 \end{bmatrix} \tag{11.22}$$

Solution

The first step is to find the eigenvalues and the corresponding eigenvectors. To find the eigenvalues, we must solve

$$\det(A - \lambda I) = \det \begin{bmatrix} 2 - \lambda & -3 \\ 1 & -2 - \lambda \end{bmatrix} = (2 - \lambda)(-2 - \lambda) + 3$$

$$= \lambda^2 - 1 = 0$$

which has solutions

$$\lambda_1 = 1 \quad \text{and} \quad \lambda_2 = -1$$

The eigenvector $\mathbf{u}$ corresponding to the eigenvalue $\lambda_1 = 1$ satisfies

$$\begin{bmatrix} 2 & -3 \\ 1 & -2 \end{bmatrix} \begin{bmatrix} u_1 \\ u_2 \end{bmatrix} = \begin{bmatrix} u_1 \\ u_2 \end{bmatrix}$$

Rewriting this matrix equation as a system of equations, we obtain

$$2u_1 - 3u_2 = u_1$$
$$u_1 - 2u_2 = u_2$$

These two equations reduce to the same equation, namely,

$$u_1 - 3u_2 = 0$$

If we set $u_2 = 1$, then $u_1 = 3$, and therefore,

$$\mathbf{u} = \begin{bmatrix} 3 \\ 1 \end{bmatrix}$$

is an eigenvector corresponding to $\lambda_1 = 1$.

The eigenvector $\mathbf{v}$ corresponding to the eigenvalue $\lambda_2 = -1$ satisfies

$$\begin{bmatrix} 2 & -3 \\ 1 & -2 \end{bmatrix} \begin{bmatrix} v_1 \\ v_2 \end{bmatrix} = - \begin{bmatrix} v_1 \\ v_2 \end{bmatrix}$$

Rewriting this as a system of equations, we find that

$$2v_1 - 3v_2 = -v_1$$
$$v_1 - 2v_2 = -v_2$$

These two equations reduce to the same equation, namely,

$$v_1 - v_2 = 0$$

If we set $v_1 = 1$, then $v_2 = 1$, and therefore,

$$\mathbf{v} = \begin{bmatrix} 1 \\ 1 \end{bmatrix}$$

is an eigenvector corresponding to $\lambda_2 = -1$. The general solution of (11.21) is thus

$$\mathbf{x}(t) = c_1 e^t \begin{bmatrix} 3 \\ 1 \end{bmatrix} + c_2 e^{-t} \begin{bmatrix} 1 \\ 1 \end{bmatrix}$$

where c_1 and c_2 are constants. The initial condition (11.22) will allow us to determine the constants c_1 and c_2:

$$\mathbf{x}(0) = c_1 \begin{bmatrix} 3 \\ 1 \end{bmatrix} + c_2 \begin{bmatrix} 1 \\ 1 \end{bmatrix} = \begin{bmatrix} 3 \\ -1 \end{bmatrix}$$

That is, c_1 and c_2 satisfy

$$3c_1 + c_2 = 3$$
$$c_1 + c_2 = -1$$

We solve this system by the standard elimination method:

$$3c_1 + c_2 = 3$$
$$2c_1 = 4$$

Hence, $c_1 = 2$ and $c_2 = 3 - 3c_1 = 3 - 6 = -3$. The solution of (11.21) that satisfies the initial condition (11.22) is therefore

$$\mathbf{x}(t) = 2e^t \begin{bmatrix} 3 \\ 1 \end{bmatrix} - 3e^{-t} \begin{bmatrix} 1 \\ 1 \end{bmatrix}$$

which can also be written in the form

$$x_1(t) = 6e^t - 3e^{-t}$$
$$x_2(t) = 2e^t - 3e^{-t}$$

The direction field, with two lines in the direction of the two eigenvectors and the solution, is shown in Figure 11.5. ■

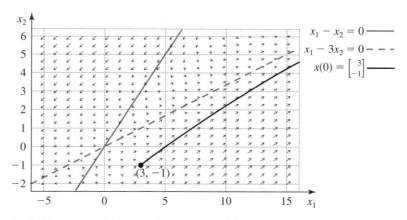

Figure 11.5 The direction field of the system (11.21) with the lines in the direction of the eigenvectors and the solution with initial condition $\begin{bmatrix} 3 \\ -1 \end{bmatrix}$.

You might have noticed that all initial conditions we have discussed thus far have been formulated at time $t = 0$. This is a natural choice for an initial condition; however, we could have chosen any other time—for instance, $t = 1$. Suppose that in Example 2 the initial condition had been

$$\mathbf{x}(1) = \begin{bmatrix} 2 \\ 1 \end{bmatrix} \tag{11.23}$$

Then the general solution would still be

$$\mathbf{x}(t) = c_1 e^t \begin{bmatrix} 3 \\ 1 \end{bmatrix} + c_2 e^{-t} \begin{bmatrix} 1 \\ 1 \end{bmatrix}$$

but the constants c_1 and c_2 would now satisfy

$$3ec_1 + e^{-1}c_2 = 2$$
$$ec_1 + e^{-1}c_2 = 1$$

We solve this system by the standard elimination method:

$$3ec_1 + e^{-1}c_2 = 2$$
$$2ec_1 = 1$$

Hence, $c_1 = \frac{1}{2e}$ and $c_2 = \frac{e}{2}$. The solution satisfying (11.23) is then

$$\mathbf{x}(t) = \frac{1}{2}e^{t-1}\begin{bmatrix} 3 \\ 1 \end{bmatrix} + \frac{1}{2}e^{1-t}\begin{bmatrix} 1 \\ 1 \end{bmatrix}$$

■ 11.1.3 Equilibria and Stability

In this subsection, we will be concerned with equilibria and stability, two concepts that we encountered in Section 8.2, where we discussed ordinary differential equations. Both concepts can be extended to systems of differential equations. We will restrict ourselves to the case

$$\frac{d\mathbf{x}}{dt} = A\mathbf{x}(t) \tag{11.24}$$

with

$$A = \begin{bmatrix} a_{11} & a_{12} \\ a_{21} & a_{22} \end{bmatrix} \quad \text{and} \quad \mathbf{x}(t) = \begin{bmatrix} x_1(t) \\ x_2(t) \end{bmatrix} \tag{11.25}$$

We say that a point

$$\hat{\mathbf{x}} = \begin{bmatrix} \hat{\mathbf{x}}_1 \\ \hat{\mathbf{x}}_2 \end{bmatrix}$$

is an equilibrium of (11.24) if

$$A\hat{\mathbf{x}} = \mathbf{0}$$

that is, if $\hat{\mathbf{x}}$ is a point at which the direction vector in the corresponding direction field has length 0. If we start a system of differential equations at an equilibrium point, it will remain there at all later times.

To find equilibria of (11.24), we must solve $A\mathbf{x} = \mathbf{0}$. We see immediately that $\hat{\mathbf{x}} = \mathbf{0}$ solves $A\mathbf{x} = \mathbf{0}$. It follows from results in Subsection 9.2.3 that if $\det A \neq 0$, then $(0, 0)$ is the only equilibrium of (11.24). If $\det A = 0$, then there will be other equilibria.

As in Chapter 8, the characteristic property of an equilibrium is that if we start a system in equilibrium, it will stay there at all future times. This does *not* mean that if the system is in equilibrium and is **perturbed** by a small amount (i.e., the solution is moved to a nearby point), it will return to the equilibrium (as we learned in Chapter 8). Whether or not a solution will return to an equilibrium after a small perturbation is addressed by the **stability** of the equilibrium. We saw in the previous subsection that, in the case when A has two real and distinct eigenvalues, the solution of (11.24) is given by

$$\mathbf{x}(t) = c_1 e^{\lambda_1 t}\mathbf{u} + c_2 e^{\lambda_2 t}\mathbf{v} \tag{11.26}$$

where $\mathbf{u}$ and $\mathbf{v}$ are the eigenvectors corresponding to the eigenvalues λ_1 and λ_2 of the matrix A and the constants c_1 and c_2 depend on the initial condition. Knowing the solution will allow us to study the behavior of the solution (11.26) as $t \to \infty$ and thus address the question of stability, at least when the eigenvalues are real and distinct.

Since A is a 2×2 matrix and all entries of A are real, the eigenvalues of A are either both real or both complex conjugate. We will treat these two cases separately. To simplify our discussion, we again assume that A has two distinct eigenvalues.

The equilibria of (11.24) can be found by solving

$$A\mathbf{x} = \mathbf{0} \tag{11.27}$$

If $\det A \neq 0$, then (11.27) has only one solution, namely, the trivial solution $(0, 0)$. (See Subsection 9.2.3.) Since $\det A = \lambda_1\lambda_2$, where λ_1 and λ_2 are the eigenvalues of A, we see that if λ_1 and λ_2 are both nonzero, then $\det A \neq 0$.

Case 1: A has two distinct real nonzero eigenvalues In this case, the equation $A\mathbf{x} = \mathbf{0}$ has only one solution, namely $(0, 0)$, and thus $(0, 0)$ is the only equilibrium. The general solution of (11.24) is given by (11.26), and we can therefore study the behavior of the solution of (11.24) directly by investigating (11.26). We are interested in determining

$$\lim_{t \to \infty} \mathbf{x}(t) = \lim_{t \to \infty} \left[c_1 e^{\lambda_1 t}\mathbf{u} + c_2 e^{\lambda_2 t}\mathbf{v} \right] \tag{11.28}$$

We see immediately that the behavior of $\mathbf{x}(t)$ is determined by $e^{\lambda_1 t}$ and $e^{\lambda_2 t}$. Recall that

$$\lim_{t \to \infty} e^{\lambda_i t} = \begin{cases} 0 & \text{if } \lambda_i < 0 \\ \infty & \text{if } \lambda_i > 0 \end{cases}$$

We distinguish the following three categories:

1. Both eigenvalues are negative: $\lambda_1 < 0$ and $\lambda_2 < 0$.
2. The eigenvalues are of opposite signs: $\lambda_1 < 0$ and $\lambda_2 > 0$ or $\lambda_1 > 0$ and $\lambda_2 < 0$.
3. Both eigenvalues are positive: $\lambda_1 > 0$ and $\lambda_2 > 0$.

Representative direction fields for each of the three categories are shown in Figures 11.6 through 11.8. We will discuss each category separately.

Figure 11.6 The direction field of a linear system where both eigenvalues are negative, together with the lines in the direction of the eigenvectors.

Category 1 When both eigenvalues are negative, we conclude from (11.28) that

$$\lim_{t \to \infty} \mathbf{x}(t) = \mathbf{0}$$

regardless of $(x_1(0), x_2(0))$. We say that the equilibrium $(0, 0)$ is **globally stable**, since the solution will approach the equilibrium $(0, 0)$ regardless of the starting point (and not just from nearby points). We call $(0, 0)$ a **sink** or a **stable node**. A direction field for this case is shown in Figure 11.6; the shape of the field explains why we call the equilibrium $(0, 0)$ a sink: All solutions "flow" into the origin.

The system of differential equations that gave rise to the direction field in Figure 11.6 is

$$\frac{d\mathbf{x}}{dt} = A\mathbf{x}(t) \quad \text{with } A = \begin{bmatrix} -2 & 0 \\ 0 & -1 \end{bmatrix}$$

The eigenvalues of A are $\lambda_1 = -2$ and $\lambda_2 = -1$. Both are negative. The two straight lines in the figure are the lines in the directions of the two eigenvectors. We see that, starting at a point on either straight line, the solution will approach the equilibrium $(0, 0)$ along the straight line. Furthermore, we see from the direction field that, starting from any other point, the solution will approach the equilibrium $(0, 0)$, as we concluded from the general solution and the fact that both eigenvalues are negative.

Category 2 When the eigenvalues are of opposite signs, we see from (11.28) that the component of the solution associated with the negative eigenvalue goes to 0 as $t \to \infty$ and the component associated with the positive eigenvalue goes to infinity. That is, unless we start in the direction of the eigenvector associated with the negative eigenvalue, the solution will not converge to the equilibrium $(0, 0)$. We say that the equilibrium $(0, 0)$ is **unstable** and call $(0, 0)$ a **saddle point**. A direction field for this case is shown in Figure 11.7; the shape of the field explains why we call this equilibrium a saddle point.

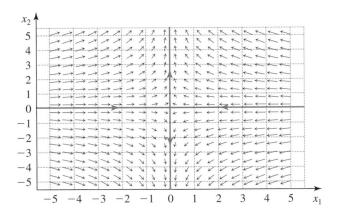

Figure 11.7 The direction field of a linear system where both eigenvalues are of opposite sign, together with the lines in the direction of the eigenvectors.

The system of differential equations that gave rise to this direction field is

$$\frac{d\mathbf{x}}{dt} = A\mathbf{x}(t) \quad \text{with } A = \begin{bmatrix} -2 & 0 \\ 0 & 1 \end{bmatrix}$$

The eigenvalues of A are $\lambda_1 = -2$ and $\lambda_2 = 1$. The two straight lines in Figure 11.7 are the lines in the directions of the two eigenvectors. We see that, starting at a point on the straight line corresponding to the negative eigenvalue (the horizontal line), the solution will approach the equilibrium $(0, 0)$ along the straight line. Starting at a point on the straight line corresponding to the positive eigenvalue (the vertical line), the solution will move away from the equilibrium $(0, 0)$ along the straight line. Furthermore, we see from the direction field that, starting from any other point, the solution will eventually move away from the equilibrium $(0, 0)$.

In sum, we see from the direction field that the solution can approach $(0, 0)$ from only one direction (the direction of the eigenvector associated with the negative eigenvalue); it gets pushed away from $(0, 0)$ elsewhere.

Category 3 Finally, if both eigenvalues are positive, we see from (11.28) that the solution will not converge to $(0, 0)$ unless we start at $(0, 0)$. We say that the equilibrium $(0, 0)$ is unstable, and we call $(0, 0)$ a **source** or an **unstable node**. A direction field for this case is shown in Figure 11.8; the shape of the field explains why we call this equilibrium a source.

The system of differential equations that gave rise to this direction field is

$$\frac{d\mathbf{x}}{dt} = A\mathbf{x}(t) \quad \text{with } A = \begin{bmatrix} 2 & 0 \\ 0 & 1 \end{bmatrix}$$

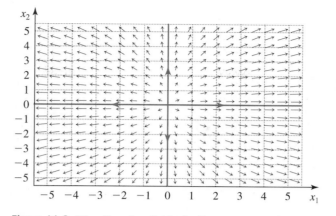

Figure 11.8 The direction field of a linear system where both eigenvalues are positive, together with the lines in the direction of the eigenvectors.

The eigenvalues of A are $\lambda_1 = 2$ and $\lambda_2 = 1$, which are both positive. The two straight lines in Figure 11.8 are the lines in the directions of the two eigenvectors. We see that, starting at a point on either straight line, the solution will move away from the equilibrium $(0, 0)$ along the straight line. Furthermore, we see from the direction field that, starting from any other point, the solution will move away from the equilibrium $(0, 0)$.

Case 2: A has complex conjugate eigenvalues We will not solve the system when A has complex conjugate eigenvalues. Instead, we will look at some examples to see what typical direction fields look like.

Category 1 Let

$$A = \begin{bmatrix} -1 & -1 \\ 1 & 0 \end{bmatrix} \tag{11.29}$$

Then

$$\det(A - \lambda I) = \det \begin{bmatrix} -1 - \lambda & -1 \\ 1 & -\lambda \end{bmatrix} = (-1 - \lambda)(-\lambda) + 1$$
$$= \lambda^2 + \lambda + 1 = 0$$

which gives

$$\lambda_{1,2} = \frac{-1 \pm \sqrt{1 - 4}}{2} = -\frac{1}{2} \pm \frac{i}{2}\sqrt{3}$$

Both eigenvalues are complex and form a conjugate pair. (Note that the real parts of both eigenvalues are negative.) To see what the solutions of

$$\frac{d\mathbf{x}}{dt} = A\mathbf{x}(t)$$

look like, we graph the direction field and solutions $x_1(t)$ and $x_2(t)$ in Figures 11.9 and 11.10, respectively.

We see from the direction field that, starting from any point other than $(0, 0)$, solutions spiral into the equilibrium $(0, 0)$. For this reason, the equilibrium $(0, 0)$ is called a **stable spiral**. When we plot solutions as functions of time, they show oscillations, as illustrated in Figure 11.10. The amplitude of the oscillations decreases over time. We therefore call the oscillations **damped**.

The oscillations are caused by the imaginary part of the eigenvalues; the damping of the oscillations is caused by the negative real part of the eigenvalues. Before we explain any further, we give two more examples, one in which the complex conjugate eigenvalues have positive real parts, the other in which the eigenvalues are purely imaginary.

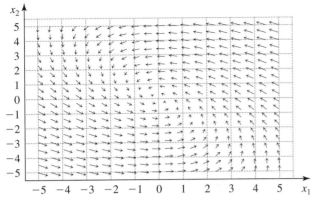

Figure 11.9 The direction field for the system with matrix A in (11.29).

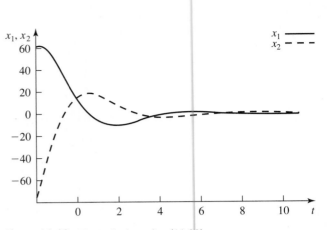

Figure 11.10 The solutions for (11.29).

Category 2 Let

$$B = \begin{bmatrix} 1 & -1 \\ 1 & 0 \end{bmatrix}$$

Then

$$\det(B - \lambda I) = \det \begin{bmatrix} 1 - \lambda & -1 \\ 1 & -\lambda \end{bmatrix} = (1 - \lambda)(-\lambda) + 1$$

$$= \lambda^2 - \lambda + 1 = 0$$

which has solutions

$$\lambda_{1,2} = \frac{1 \pm \sqrt{1 - 4}}{2} = \frac{1}{2} \pm \frac{i}{2}\sqrt{3}$$

Both eigenvalues of B are complex and form a complex conjugate pair, but their real parts are positive. To see what the solutions of

$$\frac{d\mathbf{x}}{dt} = B\mathbf{x}(t) \tag{11.30}$$

look like, we graph the direction field and solutions $x_1(t)$ and $x_2(t)$ in Figures 11.11 and 11.12, respectively.

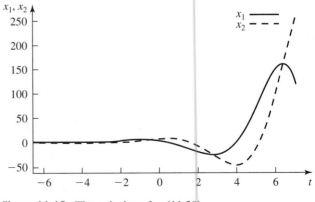

Figure 11.11 The direction field for the system with matrix B in (11.30).

Figure 11.12 The solutions for (11.30).

We see from the direction field that, starting from any point other than $(0, 0)$, the solutions spiral out from the equilibrium $(0, 0)$. For this reason, we call the equilibrium $(0, 0)$ an **unstable spiral**. When we plot solutions as functions of time, as in Figure 11.12, we see that the solutions show oscillations as before, but this time their amplitudes are increasing. The oscillations are again caused by the imaginary parts of the eigenvalues; the increase in amplitude is caused by the positive real parts.

Category 3 Here is an example where both eigenvalues are purely imaginary. Let

$$C = \begin{bmatrix} 0 & -1 \\ 1 & 0 \end{bmatrix}$$

Then

$$\det(C - \lambda I) = \det \begin{bmatrix} -\lambda & -1 \\ 1 & -\lambda \end{bmatrix} = \lambda^2 + 1 = 0$$

which gives

$$\lambda_{1,2} = \pm i$$

Both eigenvalues of C are complex and form a complex conjugate pair, but they are purely imaginary. (Their real parts are equal to 0.) To see what the solutions of

$$\frac{d\mathbf{x}}{dt} = C\mathbf{x}(t) \tag{11.31}$$

look like, we graph both the direction field and solutions $x_1(t)$ and $x_2(t)$ in Figures 11.13 and 11.14.

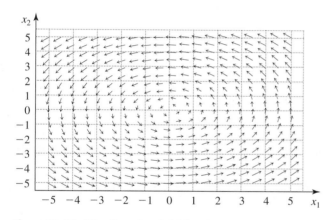

Figure 11.13 The direction field for the system with matrix C in (11.31).

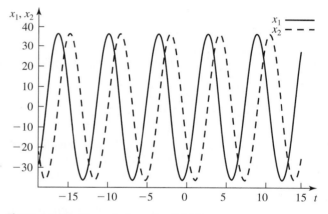

Figure 11.14 The solutions for (11.31).

Looking at solution curves in Figure 11.14, we see that the solutions oscillate as in the previous two examples, but this time the amplitude does not change with time. Looking at the direction field in Figure 11.13, we see that solutions spiral around the equilibrium $(0, 0)$, but since the amplitude of the solutions does not change, the solutions neither approach nor move away from the equilibrium. The equilibrium $(0, 0)$ is called a **neutral spiral** or a **center**. We can show that the solutions form closed curves. (We will analyze this case more closely in Example 3, where we will show the direction field with a solution curve.)

Where Do the Oscillations Come From? We can show that the solutions of (11.24) are given by (11.26), regardless of whether the eigenvalues are real or complex, as long as they are distinct. If the eigenvalues are complex, then the solution contains terms of the form e^z, where z is a complex number.

We can write the complex number z as

$$z = a + ib$$

where a and b are both real. Recall that the number a is called the *real part* of z, the number b the *imaginary part* of z.

To understand what e^z means when z is complex, we write

$$e^z = e^{a+ib} = e^a e^{ib}$$

The term e^a is a real number; we know what it means. The term e^{ib} is an exponential with a purely imaginary exponent. We have not encountered this kind of number before. The following formula, which we cannot prove here, explains its meaning:

Euler's Formula
$$e^{ib} = \cos b + i \sin b$$

Both the sine and the cosine functions show oscillations; this is the reason systems of differential equations with complex eigenvalues have solutions that oscillate, as in the next example, in which we can guess the answer.

EXAMPLE 3   Show that $x_1(t) = \cos t$ and $x_2(t) = \sin t$ solve

$$\frac{dx_1}{dt} = -x_2$$

$$\frac{dx_2}{dt} = x_1 \tag{11.32}$$

with $x_1(0) = 1$ and $x_2(0) = 0$.

Solution   The associated matrix of coefficients is

$$A = \begin{bmatrix} 0 & -1 \\ 1 & 0 \end{bmatrix}$$

which is the same as the matrix C that we saw in (11.31). We therefore know that

$$\lambda_1 = i \quad \text{and} \quad \lambda_2 = -i$$

We need to show that

$$x_1(t) = \cos t$$
$$x_2(t) = \sin t$$

solves (11.32). Let's check. We find that

$$\frac{dx_1}{dt} = -\sin t = -x_2(t)$$

and

$$\frac{dx_2}{dt} = \cos t = x_1(t)$$

In addition, we must check that our solution satisfies the initial condition. Indeed, $x_1(0) = \cos 0 = 1$ and $x_2(0) = \sin 0 = 0$. Thus, the solution of this system is a pair of real-valued functions, as claimed. The solution is shown in Figure 11.15, and the trajectory of $(x_1(t), x_2(t))$ in the x_1–x_2 plane is shown in Figure 11.16.

We see from Figure 11.15 that the solutions are periodic in this case. They show sustained oscillations; that is, the amplitudes of $x_1(t)$ and $x_2(t)$ do not change with time. In Figure 11.16, we see that the solution in the x_1–x_2 plane forms closed curves, indicating that the solutions are periodic.   ■

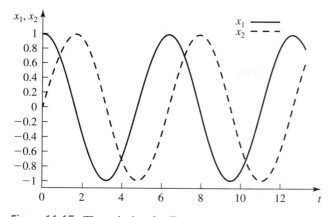

Figure 11.15 The solution for Example 3 as functions of t with initial condition $x_1(0) = 1$ and $x_2(0) = 0$.

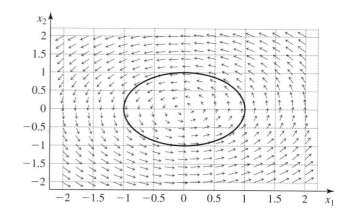

Figure 11.16 The trajectory for Example 3 in the direction field with initial condition $x_1(0) = 1$ and $x_2(0) = 0$. The trajectory is a closed curve.

Summary We close this section with a brief summary of the classification of the equilibrium $(0, 0)$ of

$$\frac{d\mathbf{x}}{dt} = A\mathbf{x}(t) \quad \text{with } A = \begin{bmatrix} a_{11} & a_{12} \\ a_{21} & a_{22} \end{bmatrix}$$

Recall that

The trace of A is $a_{11} + a_{22}$.

The determinant of A is $a_{11}a_{22} - a_{12}a_{21}$.

We denote the trace of A by τ and the determinant of A by Δ. The eigenvalues of A are found by solving

$$\det(A - \lambda I) = \det \begin{bmatrix} a_{11} - \lambda & a_{12} \\ a_{21} & a_{22} - \lambda \end{bmatrix}$$

$$= (a_{11} - \lambda)(a_{22} - \lambda) - a_{12}a_{21}$$

$$= \lambda^2 - (a_{11} + a_{22})\lambda + a_{11}a_{22} - a_{12}a_{21}$$

$$= \lambda^2 - \tau\lambda + \Delta = 0$$

where we used $\tau = a_{11} + a_{22}$ and $\Delta = a_{11}a_{22} - a_{12}a_{21}$. The last equation gives

$$\lambda_1 = \frac{\tau + \sqrt{\tau^2 - 4\Delta}}{2} \quad \text{and} \quad \lambda_2 = \frac{\tau - \sqrt{\tau^2 - 4\Delta}}{2} \tag{11.33}$$

We see immediately that

If $\tau^2 > 4\Delta$, both eigenvalues are real and distinct.

If $\tau^2 < 4\Delta$, the eigenvalues are complex conjugates.

When both eigenvalues are real and distinct—that is, when $\tau^2 > 4\Delta$—we can distinguish the following three cases:

1. $\Delta < 0$: $\lambda_1 > 0$, $\lambda_2 < 0$ (saddle point)
2. $\Delta > 0$, $\tau < 0$: $\lambda_1 < 0$, $\lambda_2 < 0$ (sink, or stable node)
3. $\Delta > 0$, $\tau > 0$: $\lambda_1 > 0$, $\lambda_2 > 0$ (source, or unstable node)

When both eigenvalues are complex conjugates—that is, when $\tau^2 < 4\Delta$—we can distinguish the following three cases:

1. $\tau < 0$: Both eigenvalues have negative real parts (stable spiral).
2. $\tau > 0$: Both eigenvalues have positive real parts (unstable spiral).
3. $\tau = 0$: Both eigenvalues are purely imaginary (center).

We can summarize all this graphically in the τ–Δ plane as shown in Figure 11.17. The parabola $4\Delta = \tau^2$ is the boundary line between oscillatory and nonoscillatory behavior. The line $\tau = 0$ divides the stable and the unstable regions. The line $\Delta = 0$ separates the saddle point from the node regions. The case in which the eigenvalues are identical resides on the boundary line $4\Delta = \tau^2$.

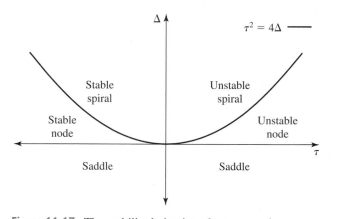

Figure 11.17 The stability behavior of a system of two linear, homogeneous differential equations with constant coefficients.

The line $\Delta = 0$ corresponds to the case in which one eigenvalue is equal to 0. As long as the other eigenvalue is not equal to 0, both eigenvalues are again distinct and the solution is of the form (11.26). However, in this case there are equilibria other than $(0, 0)$. We will discuss two such examples in Problems 67 and 68 and one in Section 11.2.

Section 11.1 Problems

■ 11.1.1

In Problems 1–4, write each system of differential equations in matrix form.

1.
$$\frac{dx_1}{dt} = 2x_1 + 3x_2$$
$$\frac{dx_2}{dt} = -4x_1 + x_2$$

2.
$$\frac{dx_1}{dt} = x_1 + x_2$$
$$\frac{dx_2}{dt} = -2x_2$$

3.
$$\frac{dx_1}{dt} = x_3 - 2x_1$$
$$\frac{dx_2}{dt} = -x_1$$
$$\frac{dx_3}{dt} = x_1 + x_2 + x_3$$

4.
$$\frac{dx_1}{dt} = 2x_2 - 3x_1 - x_3$$
$$\frac{dx_2}{dt} = -x_1 + x_2$$
$$\frac{dx_3}{dt} = 5x_1 + x_3$$

5. Consider
$$\frac{dx_1}{dt} = -x_1 + 2x_2$$
$$\frac{dx_2}{dt} = x_1$$

Determine the direction vectors associated with the following points in the x_1–x_2 plane, and graph the direction vectors in the x_1–x_2 plane: $(1, 0), (0, 1), (-1, 0), (0, -1), (1, 1), (0, 0)$, and $(-2, 1)$.

6. Consider
$$\frac{dx_1}{dt} = 2x_1 - x_2$$
$$\frac{dx_2}{dt} = -x_2$$

Determine the direction vectors associated with the following points in the x_1–x_2 plane, and graph the direction vectors in the x_1–x_2 plane: $(2, 0), (1.5, 1), (1, 0), (0, -1), (1, 1), (0, 0)$, and $(-2, -2)$.

7. Consider
$$\frac{dx_1}{dt} = x_1 + 3x_2$$
$$\frac{dx_2}{dt} = -x_1 + 2x_2$$

Determine the direction vectors associated with the following points in the x_1–x_2 plane, and graph the direction vectors in the x_1–x_2 plane: $(1, 0), (0, 1), (-1, 1), (0, -1), (-3, 1), (0, 0)$, and $(-2, 1)$.

8. Consider
$$\frac{dx_1}{dt} = -x_2$$
$$\frac{dx_2}{dt} = x_1 + x_2$$

Determine the direction vectors associated with the following points in the x_1–x_2 plane, and graph the direction vectors in the x_1–x_2 plane: $(1, 0), (0, 1), (-1, 0), (0, -1), (1, 1), (0, 0)$, and $(-2, -2)$.

9. In Figures 11.18 through 11.21, direction fields are given. Each of the following systems of differential equations corresponds to exactly one of the direction fields. Match the systems to the appropriate figures.

(a)
$$\frac{dx_1}{dt} = 2x_1$$
$$\frac{dx_2}{dt} = x_1 + x_2$$

(b)
$$\frac{dx_1}{dt} = x_1 + 2x_2$$
$$\frac{dx_2}{dt} = -2x_1$$

Figure 11.18

Figure 11.19

Figure 11.20

Figure 11.21

(c) $\dfrac{dx_1}{dt} = x_1$

$\dfrac{dx_2}{dt} = x_2$

(d) $\dfrac{dx_1}{dt} = -x_2$

$\dfrac{dx_2}{dt} = x_1$

10. The direction field of

$$\frac{dx_1}{dt} = x_1 + 3x_2$$

$$\frac{dx_2}{dt} = 2x_1 + 3x_2$$

is given in Figure 11.22. Sketch the solution curve that goes through the point $(1, 0)$.

11. The direction field of

$$\frac{dx_1}{dt} = 2x_1 + 3x_2$$

$$\frac{dx_2}{dt} = -x_1 + x_2$$

is given in Figure 11.23. Sketch the solution curve that goes through the point $(2, -1)$.

Figure 11.22

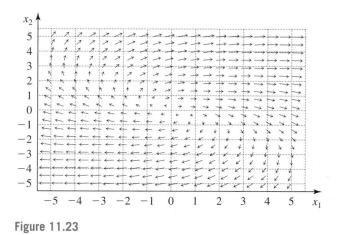

Figure 11.23

12. The direction field of

$$\frac{dx_1}{dt} = -x_1 - x_2$$

$$\frac{dx_2}{dt} = -2x_2$$

is given in Figure 11.24. Sketch the solution curve that goes through the point $(-3, -3)$.

Figure 11.24

■ **11.1.2**

In Problems 13–18, find the general solution of each given system of differential equations and sketch the lines in the direction of the eigenvectors. Indicate on each line the direction in which the solution would move if it starts on that line.

13. (Figure 11.25)

$$\begin{bmatrix} \frac{dx_1}{dt} \\ \frac{dx_2}{dt} \end{bmatrix} = \begin{bmatrix} 1 & 3 \\ 5 & 3 \end{bmatrix} \begin{bmatrix} x_1(t) \\ x_2(t) \end{bmatrix}$$

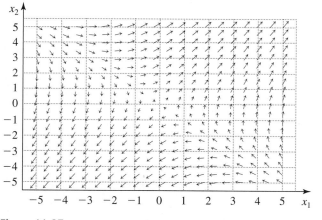

Figure 11.25

14. (Figure 11.26)

$$\begin{bmatrix} \frac{dx_1}{dt} \\ \frac{dx_2}{dt} \end{bmatrix} = \begin{bmatrix} 2 & 1 \\ 4 & -1 \end{bmatrix} \begin{bmatrix} x_1(t) \\ x_2(t) \end{bmatrix}$$

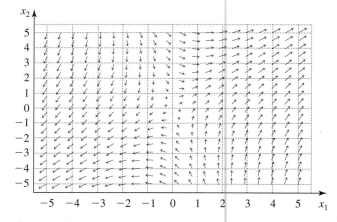

Figure 11.26

15. (Figure 11.27)

$$\begin{bmatrix} \frac{dx_1}{dt} \\ \frac{dx_2}{dt} \end{bmatrix} = \begin{bmatrix} -3 & 3 \\ 6 & 4 \end{bmatrix} \begin{bmatrix} x_1(t) \\ x_2(t) \end{bmatrix}$$

Figure 11.27

16. (Figure 11.28)

$$\begin{bmatrix} \frac{dx_1}{dt} \\ \frac{dx_2}{dt} \end{bmatrix} = \begin{bmatrix} -5 & 3 \\ -2 & 0 \end{bmatrix} \begin{bmatrix} x_1(t) \\ x_2(t) \end{bmatrix}$$

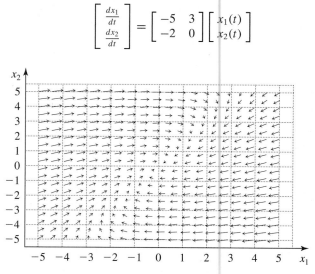

Figure 11.28

17. (Figure 11.29)

$$\begin{bmatrix} \frac{dx_1}{dt} \\ \frac{dx_2}{dt} \end{bmatrix} = \begin{bmatrix} -2 & 0 \\ -3 & 1 \end{bmatrix} \begin{bmatrix} x_1(t) \\ x_2(t) \end{bmatrix}$$

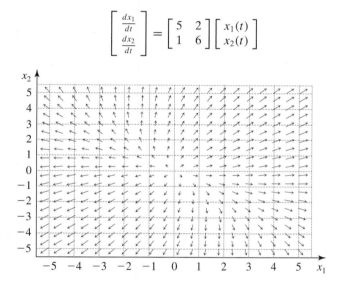

Figure 11.29

18. (Figure 11.30)

$$\begin{bmatrix} \frac{dx_1}{dt} \\ \frac{dx_2}{dt} \end{bmatrix} = \begin{bmatrix} 5 & 2 \\ 1 & 6 \end{bmatrix} \begin{bmatrix} x_1(t) \\ x_2(t) \end{bmatrix}$$

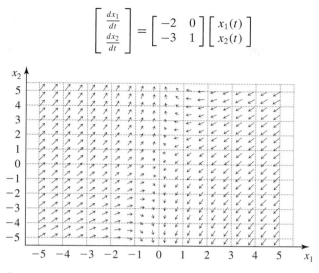

Figure 11.30

In Problems 19–26, solve the given initial-value problem.

19. $\begin{bmatrix} \frac{dx_1}{dt} \\ \frac{dx_2}{dt} \end{bmatrix} = \begin{bmatrix} -3 & 0 \\ 4 & 2 \end{bmatrix} \begin{bmatrix} x_1(t) \\ x_2(t) \end{bmatrix}$

with $x_1(0) = -5$ and $x_2(0) = 5$.

20. $\begin{bmatrix} \frac{dx_1}{dt} \\ \frac{dx_2}{dt} \end{bmatrix} = \begin{bmatrix} 1 & 3 \\ 0 & 2 \end{bmatrix} \begin{bmatrix} x_1(t) \\ x_2(t) \end{bmatrix}$

with $x_1(0) = 2$ and $x_2(0) = -1$.

21. $\begin{bmatrix} \frac{dx_1}{dt} \\ \frac{dx_2}{dt} \end{bmatrix} = \begin{bmatrix} 3 & -2 \\ 0 & 1 \end{bmatrix} \begin{bmatrix} x_1(t) \\ x_2(t) \end{bmatrix}$

with $x_1(0) = 1$ and $x_2(0) = 1$.

22. $\begin{bmatrix} \frac{dx_1}{dt} \\ \frac{dx_2}{dt} \end{bmatrix} = \begin{bmatrix} -1 & 0 \\ 1 & -2 \end{bmatrix} \begin{bmatrix} x_1(t) \\ x_2(t) \end{bmatrix}$

with $x_1(0) = -1$ and $x_2(0) = -2$.

23. $\begin{bmatrix} \frac{dx_1}{dt} \\ \frac{dx_2}{dt} \end{bmatrix} = \begin{bmatrix} 4 & -7 \\ 2 & -5 \end{bmatrix} \begin{bmatrix} x_1(t) \\ x_2(t) \end{bmatrix}$

with $x_1(0) = 13$ and $x_2(0) = 3$.

24. $\begin{bmatrix} \frac{dx_1}{dt} \\ \frac{dx_2}{dt} \end{bmatrix} = \begin{bmatrix} -3 & 4 \\ -1 & 2 \end{bmatrix} \begin{bmatrix} x_1(t) \\ x_2(t) \end{bmatrix}$

with $x_1(0) = 1$ and $x_2(0) = 2$.

25. $\begin{bmatrix} \frac{dx_1}{dt} \\ \frac{dx_2}{dt} \end{bmatrix} = \begin{bmatrix} 4 & 7 \\ 1 & -2 \end{bmatrix} \begin{bmatrix} x_1(t) \\ x_2(t) \end{bmatrix}$

with $x_1(0) = -1$ and $x_2(0) = -2$.

26. $\begin{bmatrix} \frac{dx_1}{dt} \\ \frac{dx_2}{dt} \end{bmatrix} = \begin{bmatrix} 2 & 6 \\ 1 & 3 \end{bmatrix} \begin{bmatrix} x_1(t) \\ x_2(t) \end{bmatrix}$

with $x_1(0) = -3$ and $x_2(0) = 1$.

In Problems 27 and 28, we discuss the case of repeated eigenvalues.

27. Let

$$\begin{bmatrix} \frac{dx_1}{dt} \\ \frac{dx_2}{dt} \end{bmatrix} = \begin{bmatrix} 1 & 0 \\ 0 & 1 \end{bmatrix} \begin{bmatrix} x_1(t) \\ x_2(t) \end{bmatrix} \qquad (11.34)$$

(a) Show that

$$A = \begin{bmatrix} 1 & 0 \\ 0 & 1 \end{bmatrix}$$

has the repeated eigenvalues $\lambda_1 = \lambda_2 = 1$.

(b) Show that $\begin{bmatrix} 1 \\ 0 \end{bmatrix}$ and $\begin{bmatrix} 0 \\ 1 \end{bmatrix}$ are eigenvectors of A and that any vector $\begin{bmatrix} c_1 \\ c_2 \end{bmatrix}$ can be written as

$$\begin{bmatrix} c_1 \\ c_2 \end{bmatrix} = c_1 \begin{bmatrix} 1 \\ 0 \end{bmatrix} + c_2 \begin{bmatrix} 0 \\ 1 \end{bmatrix}$$

(c) Show that

$$\mathbf{x}(t) = c_1 e^t \begin{bmatrix} 1 \\ 0 \end{bmatrix} + c_2 e^t \begin{bmatrix} 0 \\ 1 \end{bmatrix}$$

is a solution of (11.34) that satisfies the initial condition $x_1(0) = c_1$ and $x_2(0) = c_2$.

28. Let

$$\begin{bmatrix} \frac{dx_1}{dt} \\ \frac{dx_2}{dt} \end{bmatrix} = \begin{bmatrix} 1 & 2 \\ 0 & 1 \end{bmatrix} \begin{bmatrix} x_1(t) \\ x_2(t) \end{bmatrix} \qquad (11.35)$$

(a) Show that

$$A = \begin{bmatrix} 1 & 2 \\ 0 & 1 \end{bmatrix}$$

has the repeated eigenvalues $\lambda_1 = \lambda_2 = 1$.

(b) Show that every eigenvector of A is of the form

$$c_1 \begin{bmatrix} 1 \\ 0 \end{bmatrix}$$

where c_1 is a real number different from 0.

(c) Show that

$$\mathbf{x}_1(t) = e^t \begin{bmatrix} 1 \\ 0 \end{bmatrix}$$

is a solution of (11.35).

(d) Show that

$$\mathbf{x}_2(t) = te^t \begin{bmatrix} 1 \\ 0 \end{bmatrix} + e^t \begin{bmatrix} 0 \\ 0.5 \end{bmatrix}$$

is a solution of (11.35).

(e) Show that

$$c_1\mathbf{x}_1(t) + c_2\mathbf{x}_2(t)$$

is a solution of (11.35). (It turns out that this is the general solution.)

■ **11.1.3**

In Problems 29–42, we consider differential equations of the form

$$\frac{d\mathbf{x}}{dt} = A\mathbf{x}(t)$$

where

$$A = \begin{bmatrix} a_{11} & a_{12} \\ a_{21} & a_{22} \end{bmatrix}$$

The eigenvalues of A will be real, distinct, and nonzero. Analyze the stability of the equilibrium $(0,0)$, and classify the equilibrium according to whether it is a sink, a source, or a saddle point.

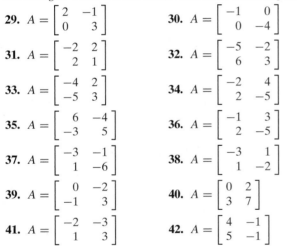

29. $A = \begin{bmatrix} 2 & -1 \\ 0 & 3 \end{bmatrix}$ **30.** $A = \begin{bmatrix} -1 & 0 \\ 0 & -4 \end{bmatrix}$

31. $A = \begin{bmatrix} -2 & 2 \\ 2 & 1 \end{bmatrix}$ **32.** $A = \begin{bmatrix} -5 & -2 \\ 6 & 3 \end{bmatrix}$

33. $A = \begin{bmatrix} -4 & 2 \\ -5 & 3 \end{bmatrix}$ **34.** $A = \begin{bmatrix} -2 & 4 \\ 2 & -5 \end{bmatrix}$

35. $A = \begin{bmatrix} 6 & -4 \\ -3 & 5 \end{bmatrix}$ **36.** $A = \begin{bmatrix} -1 & 3 \\ 2 & -5 \end{bmatrix}$

37. $A = \begin{bmatrix} -3 & -1 \\ 1 & -6 \end{bmatrix}$ **38.** $A = \begin{bmatrix} -3 & 1 \\ 1 & -2 \end{bmatrix}$

39. $A = \begin{bmatrix} 0 & -2 \\ -1 & 3 \end{bmatrix}$ **40.** $A = \begin{bmatrix} 0 & 2 \\ 3 & 7 \end{bmatrix}$

41. $A = \begin{bmatrix} -2 & -3 \\ 1 & 3 \end{bmatrix}$ **42.** $A = \begin{bmatrix} 4 & -1 \\ 5 & -1 \end{bmatrix}$

In Problems 43–56, we consider differential equations of the form

$$\frac{d\mathbf{x}}{dt} = A\mathbf{x}(t)$$

where

$$A = \begin{bmatrix} a_{11} & a_{12} \\ a_{21} & a_{22} \end{bmatrix}$$

The eigenvalues of A will be complex conjugates. Analyze the stability of the equilibrium $(0,0)$, and classify the equilibrium according to whether it is a stable spiral, an unstable spiral, or a center.

43. $A = \begin{bmatrix} 2 & -1 \\ 3 & 0 \end{bmatrix}$ **44.** $A = \begin{bmatrix} -1 & -5 \\ 4 & -3 \end{bmatrix}$

45. $A = \begin{bmatrix} -2 & 4 \\ -3 & -2 \end{bmatrix}$ **46.** $A = \begin{bmatrix} 1 & 3 \\ -2 & 1 \end{bmatrix}$

47. $A = \begin{bmatrix} 1 & 3 \\ -2 & -2 \end{bmatrix}$ **48.** $A = \begin{bmatrix} 2 & -3 \\ 2 & -1 \end{bmatrix}$

49. $A = \begin{bmatrix} 4 & 5 \\ -3 & -3 \end{bmatrix}$ **50.** $A = \begin{bmatrix} 2 & 2 \\ -6 & -4 \end{bmatrix}$

51. $A = \begin{bmatrix} -1 & 1 \\ -3 & 1 \end{bmatrix}$ **52.** $A = \begin{bmatrix} 3 & -2 \\ 1 & 3 \end{bmatrix}$

53. $A = \begin{bmatrix} 0 & -1 \\ 1 & 0 \end{bmatrix}$ **54.** $A = \begin{bmatrix} 0 & -3 \\ 2 & 2 \end{bmatrix}$

55. $A = \begin{bmatrix} 1 & 2 \\ -5 & -3 \end{bmatrix}$ **56.** $A = \begin{bmatrix} 2 & -3 \\ 3 & -2 \end{bmatrix}$

In Problems 57–66, we consider differential equations of the form

$$\frac{d\mathbf{x}}{dt} = A\mathbf{x}(t)$$

where

$$A = \begin{bmatrix} a_{11} & a_{12} \\ a_{21} & a_{22} \end{bmatrix}$$

Analyze the stability of the equilibrium $(0,0)$, and classify the equilibrium.

57. $A = \begin{bmatrix} -1 & -2 \\ 1 & 3 \end{bmatrix}$ **58.** $A = \begin{bmatrix} -2 & 2 \\ -4 & 3 \end{bmatrix}$

59. $A = \begin{bmatrix} -1 & -1 \\ 5 & -3 \end{bmatrix}$ **60.** $A = \begin{bmatrix} 2 & 2 \\ 2 & 3 \end{bmatrix}$

61. $A = \begin{bmatrix} 1 & 3 \\ 2 & 3 \end{bmatrix}$ **62.** $A = \begin{bmatrix} -1 & 5 \\ -3 & 0 \end{bmatrix}$

63. $A = \begin{bmatrix} -2 & 3 \\ 1 & -4 \end{bmatrix}$ **64.** $A = \begin{bmatrix} -2 & -7 \\ 1 & 2 \end{bmatrix}$

65. $A = \begin{bmatrix} 3 & -5 \\ 2 & -1 \end{bmatrix}$ **66.** $A = \begin{bmatrix} 3 & 6 \\ -1 & -4 \end{bmatrix}$

67. The following system has two distinct real eigenvalues, but one eigenvalue is equal to 0:

$$\frac{d\mathbf{x}}{dt} = \begin{bmatrix} 4 & 8 \\ 1 & 2 \end{bmatrix} \mathbf{x}(t) \tag{11.36}$$

(a) Find both eigenvalues and the associated eigenvectors.

(b) Use the general solution (11.26) to find $x_1(t)$ and $x_2(t)$.

(c) The direction field is shown in Figure 11.31. Sketch the lines corresponding to the eigenvectors. Compute dx_2/dx_1, and conclude that all direction vectors are parallel to the line in the direction of the eigenvector corresponding to the nonzero eigenvalue. Describe in words how solutions starting at different points behave.

Figure 11.31

68. The following system has two distinct real eigenvalues, but one eigenvalue is equal to 0:

$$\frac{d\mathbf{x}}{dt} = \begin{bmatrix} 2 & 4 \\ 3 & 6 \end{bmatrix} \mathbf{x}(t) \qquad (11.37)$$

(a) Find both eigenvalues and the associated eigenvectors.

(b) Use the general solution (11.26) to find $x_1(t)$ and $x_2(t)$.

(c) The direction field is shown in Figure 11.32. Sketch the lines corresponding to the eigenvectors. Compute dx_2/dx_1, and conclude that all direction vectors are parallel to the line in the direction of the eigenvector corresponding to the nonzero eigenvalue. Describe in words how solutions starting at different points behave.

Figure 11.32

■ 11.2 Linear Systems: Applications

■ 11.2.1 Compartment Models

Compartment models (which we encountered in Chapter 8) describe flow between compartments, such as nutrient flow between lakes or the flow of a radioactive tracer between different parts of an organism. In the simplest situations, the resulting model is a system of linear differential equations.

We will consider a general two-compartment model that can be described by a system of two linear differential equations. A schematic description of the model is given in Figure 11.33.

We denote by $x_1(t)$ the amount of matter in compartment 1 at time t and by $x_2(t)$ the amount of matter in compartment 2 at time t. To have a concrete example in mind, think of $x_1(t)$ and $x_2(t)$ as the amount of water in each of the two compartments, respectively. The direction of the flow of matter and the rates at which matter flows are shown in Figure 11.33. We see that matter enters compartment 1 at the constant rate I and moves from compartment 1 to compartment 2 at rate ax_1 if x_1 is the amount of matter in compartment 1. Matter in compartment 1 is lost at rate cx_1. In addition, matter flows from compartment 2 to compartment 1 at rate bx_2 if x_2 is the amount of matter in compartment 2. Matter in compartment 2 is lost at rate dx_2; there is no external input into compartment 2. The constants $I, a, b, c,$ and d are all nonnegative.

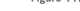

Figure 11.33 A schematic description of a general two-compartment model.

We describe the dynamics of $x_1(t)$ and $x_2(t)$ by the following system of differential equations:

$$\frac{dx_1}{dt} = I - (a + c)x_1 + bx_2$$
$$\frac{dx_2}{dt} = ax_1 - (b + d)x_2 \qquad (11.38)$$

If $I > 0$, then (11.38) is a system of inhomogeneous linear differential equations with constant coefficients. Constant input is often important in real situations, such as the flow of nutrients between soil and plants, in which nutrients might be added at a constant rate. In the discussion that follows, however, we will set $I = 0$, since this corresponds to the situation discussed in the previous section (i.e., no matter is added over time). It is not difficult to guess how the system behaves when $I = 0$: Either some matter is continually lost, so one or both compartments empty out, or no matter is lost, so at least one compartment will contain matter. We will discuss both cases.

When $I = 0$, (11.38) reduces to the linear system

$$\frac{d\mathbf{x}}{dt} = A\mathbf{x}(t) \quad \text{with} \quad A = \begin{bmatrix} -(a + c) & b \\ a & -(b + d) \end{bmatrix} \qquad (11.39)$$

To avoid trivial situations, we assume that at least one of the parameters $a, b, c,$ and d is positive. (Otherwise, no material would ever move in the system.)

To find the eigenvalues of A, we compute

$$\det \begin{bmatrix} -(a+c) - \lambda & b \\ a & -(b+d) - \lambda \end{bmatrix}$$
$$= (a + c + \lambda)(b + d + \lambda) - ab$$
$$= \lambda^2 + (a + b + c + d)\lambda + (a + c)(b + d) - ab$$
$$= \lambda^2 - \tau\lambda + \Delta = 0$$

where τ is the trace of A and Δ is the determinant of A:

$$\tau = -(a + b + c + d) < 0$$
$$\Delta = (a + c)(b + d) - ab = ad + bc + cd \geq 0$$

The eigenvalues λ_1 and λ_2 are the solutions of $\lambda^2 - \tau\lambda + \Delta = 0$:

$$\lambda_{1,2} = \frac{\tau \pm \sqrt{\tau^2 - 4\Delta}}{2}$$

To check whether the eigenvalues are real, we simplify the expression under the square root:

$$\tau^2 - 4\Delta = (a + b + c + d)^2 - 4(a + c)(b + d) + 4ab$$
$$= [(a + b) - (c + d)]^2 + 4ab \geq 0$$

From the last expression, we see that both eigenvalues are real.

Since $\Delta \geq 0$ and $\tau^2 - 4\Delta \geq 0$, it follows that $\sqrt{\tau^2 - 4\Delta} \leq |\tau|$. Since, in addition, $\tau < 0$, we conclude that both eigenvalues are less than or equal to 0. As long as $\Delta > 0$, both eigenvalues will be strictly negative; then $(0, 0)$ will be the only equilibrium and will be a stable node.

$\Delta > 0$: When $\Delta = ad + bc + cd > 0$, at least one of the terms in $ad + bc + cd$ must be positive; that is, either

$$a, d > 0 \quad \text{or} \quad b, c > 0 \quad \text{or} \quad c, d > 0$$

To see what this means, we return to Figure 11.33. When a and d are both positive, the matter in compartment 1 can move to compartment 2, and matter in compartment 2 can leave the system. Similarly, when both b and c are positive, matter from both compartments can leave the system through compartment 1. If c and d are both positive, then matter can leave both compartments. It follows that, in any of these three cases, all matter will eventually leave the system, implying that $(0, 0)$ is a stable equilibrium.

$\Delta = 0$: Then

$$\lambda_1 = \tau = -(a + b + c + d) \quad \text{and} \quad \lambda_2 = 0$$

Since we assume that at least one of the four parameters a, b, c, and d is positive, it follows that $\lambda_1 < 0$; thus, the eigenvalues are distinct. To have $\Delta = 0$, one of the following must hold:

$$c = a = 0 \quad \text{or} \quad d = b = 0 \quad \text{or} \quad c = d = 0$$

If $c = a = 0$, then matter gets stuck in compartment 1; if $d = b = 0$, then matter gets stuck in compartment 2; if $c = d = 0$, no matter will ever leave the system, and the amount of matter present at time t is equal to the amount of matter present at time 0. In other words, the total amount of matter, $x_1(t) + x_2(t)$, is constant (i.e., it does not depend on t) and is therefore called a **conserved quantity**. This conclusion follows from (11.38) as well, since, in this case,

$$\frac{dx_1}{dt} = -ax_1 + bx_2$$

$$\frac{dx_2}{dt} = ax_1 - bx_2$$

Adding the two equations, we find that

$$\frac{dx_1}{dt} + \frac{dx_2}{dt} = 0$$

Since

$$\frac{dx_1}{dt} + \frac{dx_2}{dt} = \frac{d}{dt}(x_1 + x_2)$$

it follows that $x_1(t) + x_2(t)$ is a constant.

Solving the system when $c = d = 0$ We can write the solution explicitly in order to find out the equilibrium state. To determine the solution, we must compute the eigenvectors corresponding to the eigenvalues λ_1 and λ_2.

The eigenvector **u** corresponding to $\lambda_1 = \tau$ satisfies

$$\begin{bmatrix} -a & b \\ a & -b \end{bmatrix} \begin{bmatrix} u_1 \\ u_2 \end{bmatrix} = -(a+b) \begin{bmatrix} u_1 \\ u_2 \end{bmatrix}$$

Writing this system out yields

$$-au_1 + bu_2 = -au_1 - bu_1$$
$$au_1 - bu_2 = -au_2 - bu_2$$

These equations simplify to one equation, namely,

$$u_1 = -u_2$$

If we set $u_1 = 1$, then $u_2 = -1$, and an eigenvector corresponding to $\lambda_1 = -(a+b)$ is

$$\mathbf{u} = \begin{bmatrix} 1 \\ -1 \end{bmatrix}$$

The eigenvector **v** corresponding to $\lambda_2 = 0$ satisfies

$$\begin{bmatrix} -a & b \\ a & -b \end{bmatrix} \begin{bmatrix} v_1 \\ v_2 \end{bmatrix} = \begin{bmatrix} 0 \\ 0 \end{bmatrix}$$

This matrix equation simplifies to one algebraic equation, namely,

$$-av_1 + bv_2 = 0$$

If we set $v_1 = b$, then $v_2 = a$, and an eigenvector corresponding to $\lambda_2 = 0$ is

$$\mathbf{v} = \begin{bmatrix} b \\ a \end{bmatrix}$$

The general solution when $\Delta = 0$ is, therefore,

$$\mathbf{x}(t) = c_1 e^{-(a+b)t} \begin{bmatrix} 1 \\ -1 \end{bmatrix} + c_2 \begin{bmatrix} b \\ a \end{bmatrix}$$

and

$$\lim_{t \to \infty} \mathbf{x}(t) = c_2 \begin{bmatrix} b \\ a \end{bmatrix}$$

At time 0,

$$\begin{bmatrix} x_1(0) \\ x_2(0) \end{bmatrix} = c_1 \begin{bmatrix} 1 \\ -1 \end{bmatrix} + c_2 \begin{bmatrix} b \\ a \end{bmatrix}$$

That is,

$$c_1 + bc_2 = x_1(0)$$
$$-c_1 + ac_2 = x_2(0)$$

Adding these two equations, we eliminate c_1 and find that

$$(a+b)c_2 = x_1(0) + x_2(0)$$

or

$$c_2 = \frac{x_1(0) + x_2(0)}{a+b}$$

Recall that the sum $x_1(t) + x_2(t)$ is a constant for all $t \geq 0$; we set $x_1(0) + x_2(0) = K$. Then

$$\lim_{t \to \infty} x_1(t) = K \frac{b}{a+b} \qquad \text{and} \qquad \lim_{t \to \infty} x_2(t) = K \frac{a}{a+b}$$

Together, these two limit equations mean that compartment 1 will eventually contain a fraction $b/(a+b)$ of the total amount of matter, and compartment 2 a fraction $a/(a+b)$ of the total amount. These are the relative rates at which matter enters the respective compartments.

EXAMPLE 1

Find the system of differential equations corresponding to the compartment diagram shown in Figure 11.34, and analyze the stability of the equilibrium $(0,0)$.

Solution

The compartment model is described by the following system of differential equations:

$$\frac{dx_1}{dt} = -(0.1 + 0.2)x_1 + 0.5x_2$$

$$\frac{dx_2}{dt} = 0.2x_1 - 0.5x_2$$

In matrix notation, this system is equal to

$$\frac{d\mathbf{x}}{dt} = \begin{bmatrix} -0.3 & 0.5 \\ 0.2 & -0.5 \end{bmatrix} \mathbf{x}(t)$$

Figure 11.34 The compartment diagram for Example 1.

To investigate the stability of $(0,0)$, we find the eigenvalues of the matrix describing the system. That is, we solve

$$\det \begin{bmatrix} -0.3 - \lambda & 0.5 \\ 0.2 & -0.5 - \lambda \end{bmatrix} = (-0.3 - \lambda)(-0.5 - \lambda) - (0.2)(0.5)$$

$$= (0.3)(0.5) + (0.3 + 0.5)\lambda + \lambda^2 - (0.2)(0.5)$$

$$= \lambda^2 + 0.8\lambda + 0.05 = 0$$

We obtain

$$\lambda_{1,2} = \frac{-0.8 \pm \sqrt{0.64 - 0.2}}{2}$$

$$= \begin{cases} -0.4 + \frac{1}{2}\sqrt{0.44} \approx -0.068 \\ -0.4 - \frac{1}{2}\sqrt{0.44} \approx -0.732 \end{cases}$$

We find that both eigenvalues are negative. Therefore, the equilibrium $(0,0)$ is a stable node. ∎

EXAMPLE 2 Given the system of differential equations

$$\frac{dx_1}{dt} = -0.7x_1 + 0.2x_2$$

$$\frac{dx_2}{dt} = 0.3x_1 - 0.2x_2$$

determine the parameter for the compartment diagram in Figure 11.33.

Solution   The general compartment diagram describing a linear system with two states is shown in Figure 11.33. Comparing the diagram and the system of equations, we find that $I = 0$ and

$$a + c = 0.7$$
$$b = 0.2$$
$$a = 0.3$$
$$b + d = 0.2$$

Solving this system of equations, we conclude that

$$a = 0.3, \qquad b = 0.2, \qquad c = 0.4, \qquad \text{and} \qquad d = 0 \qquad \blacksquare$$

Compartment models are used in pharmacology to study how drug concentrations change within a human being or other animal's body. In the simplest of such models, a drug is administered to a person in a single dose; we investigate this situation in the next example.

EXAMPLE 3 A drug is administered to a person in a single dose. We assume that the drug does not accumulate in body tissue, but is secreted through urine. We denote the amount of drug in the body at time t by $x_1(t)$ and in the urine at time t by $x_2(t)$. Initially,

$$x_1(0) = K \qquad \text{and} \qquad x_2(0) = 0$$

We describe the movement of the drug between the body and the urine by

$$\frac{dx_1}{dt} = -ax_1(t) \tag{11.40}$$

$$\frac{dx_2}{dt} = ax_1(t) \tag{11.41}$$

That is, the body's amount of the drug decreases at the same rate as the drug accumulates in the urine. We can solve (11.40) directly, to find that

$$x_1(t) = c_1 e^{-at}$$

Since $x_1(0) = K = c_1$, we have

$$x_1(t) = K e^{-at}$$

Plugging this into (11.41) yields

$$\frac{dx_2}{dt} = aK e^{-at}$$

or

$$\int dx_2 = \int aK e^{-at}\, dt$$

Hence,

$$x_2(t) = c_2 - K e^{-at}$$

With $x_2(0) = 0$, we find that $0 = c_2 - K$. Thus,

$$x_2(t) = K(1 - e^{-at})$$

The constant a is called the *excretion rate* in this application. Finding the excretion rate is important for determining how long a drug will remain in the body. We can find the excretion rate a by plotting $\ln(K - x_2(t))$ versus t:

$$\ln(K - x_2(t)) = \ln(Ke^{-at}) = \ln K - at$$

That is, plotting $\ln(K - x_2(t))$ versus t results in a straight line with slope $-a$, which allows us to determine a.   ■

■ 11.2.2 The Harmonic Oscillator (Optional)

Consider a particle moving along the x-axis. We assume that the acceleration is proportional to the distance to the origin and that the direction of the acceleration always points toward the origin. If $x(t)$ is the location of the particle at time t, then the second derivative of $x(t)$ denotes the acceleration of the particle, and we find that

$$\frac{d^2x}{dt^2} = -kx(t) \tag{11.42}$$

for some $k > 0$. This is a **second-order differential equation**, because the derivative of the highest-order derivative in the equation is of order 2. We will use this example to show how a second-order differential equation can be transformed into a system of first-order differential equations. To do so, we set

$$\frac{dx}{dt} = v(t)$$

Then

$$\frac{dv}{dt} = \frac{d^2x}{dt^2}$$

and, hence,

$$\frac{dv}{dt} = -kx(t)$$

We thus obtain the following system of first-order differential equations:

$$\frac{dx}{dt} = v$$

$$\frac{dv}{dt} = -kx$$

or

$$\frac{d}{dt}\begin{bmatrix} x \\ v \end{bmatrix} = \begin{bmatrix} 0 & 1 \\ -k & 0 \end{bmatrix}\begin{bmatrix} x \\ v \end{bmatrix}$$

If we denote the matrix by A, then

$$\text{tr } A = 0 \quad \text{and} \quad \det A = k > 0$$

which implies that the eigenvalues of A are complex conjugates with real parts equal to 0. We find that

$$\det(A - \lambda I) = \det\begin{bmatrix} -\lambda & 1 \\ -k & -\lambda \end{bmatrix} = \lambda^2 + k = 0$$

Hence,

$$\lambda_1 = i\sqrt{k} \quad \text{and} \quad \lambda_2 = -i\sqrt{k}$$

Thus, we expect the equilibrium $(0, 0)$ to be a neutral spiral (or center) and the solutions to exhibit oscillations whose amplitudes do not change with time.

We can solve (11.42) directly: Since

$$\frac{d}{dt}\sin(at) = a\cos(at)$$

$$\frac{d}{dt}\cos(at) = -a\sin(at)$$

it follows that

$$\frac{d^2}{dt^2}\sin(at) = -a^2\sin(at)$$

and

$$\frac{d^2}{dt^2}\cos(at) = -a^2\cos(at)$$

If we set $a = \sqrt{k}$, we see that $\cos(\sqrt{k}t)$ and $\sin(\sqrt{k}t)$ solve (11.42). Using the superposition principle, we therefore obtain the solution of (11.42) as

$$x(t) = c_1\sin(\sqrt{k}t) + c_2\cos(\sqrt{k}t)$$

To determine the constants c_1 and c_2, we must fix an initial condition. If we assume, for instance, that

$$x(0) = 0 \quad \text{and} \quad v(0) = v_0 \tag{11.43}$$

then

$$0 = c_2$$

Since $v(t) = dx/dt$, we have

$$v(t) = c_1\sqrt{k}\cos(\sqrt{k}t) - c_2\sqrt{k}\sin(\sqrt{k}t)$$

and, therefore,

$$v(0) = c_1\sqrt{k} = v_0$$

which implies that

$$c_1 = \frac{v_0}{\sqrt{k}}$$

Hence, the solution of (11.42) that satisfies the initial condition (11.43) is given by

$$x(t) = \frac{v_0}{\sqrt{k}}\sin(\sqrt{k}t)$$

The harmonic oscillator is quite important in physics. It describes, for instance, a frictionless pendulum when the displacement from the resting state is not too large.

Section 11.2 Problems

■ 11.2.1

In Problems 1–8, determine the system of differential equations corresponding to each compartment model and analyze the stability of the equilibrium $(0, 0)$. The parameters have the same meaning as in Figure 11.33.

1. $a = 0.5, b = 0.1, c = 0.05, d = 0.02$
2. $a = 0.4, b = 1.2, c = 0.3, d = 0$
3. $a = 2.5, b = 0.7, c = 0, d = 0.1$
4. $a = 1.7, b = 0.6, c = 0.1, d = 0.3$
5. $a = 0, b = 0.1, c = 0, d = 0.3$
6. $a = 0.2, b = 0.1, c = 0, d = 0$
7. $a = 0.1, b = 1.2, c = 0.5, d = 0.05$
8. $a = 0.2, b = 0, c = 0, d = 0.3$

In Problems 9–18, find the corresponding compartment diagram for each system of differential equations.

9. $\dfrac{dx_1}{dt} = -0.4x_1 + 0.3x_2$

$\dfrac{dx_2}{dt} = 0.1x_1 - 0.5x_2$

10. $\dfrac{dx_1}{dt} = -0.4x_1 + 3x_2$

$\dfrac{dx_2}{dt} = 0.2x_1 - 3x_2$

11. $\dfrac{dx_1}{dt} = -0.2x_1 + 0.1x_2$

$\dfrac{dx_2}{dt} = -0.1x_2$

12. $\dfrac{dx_1}{dt} = -0.2x_1 + 1.1x_2$

$\dfrac{dx_2}{dt} = 0.2x_1 - 1.1x_2$

13. $\dfrac{dx_1}{dt} = -2.3x_1 + 1.1x_2$

$\dfrac{dx_2}{dt} = 0.2x_1 - 2.3x_2$

14. $\dfrac{dx_1}{dt} = -1.6x_1 + 0.3x_2$

$\dfrac{dx_2}{dt} = 0.1x_1 - 0.5x_2$

15. $\dfrac{dx_1}{dt} = -1.2x_1$

$\dfrac{dx_2}{dt} = 0.3x_1 - 0.2x_2$

16. $\dfrac{dx_1}{dt} = -0.2x_1 + 0.4x_2$

$\dfrac{dx_2}{dt} = 0.2x_1 - 0.4x_2$

17. $\dfrac{dx_1}{dt} = -0.2x_1$

$\dfrac{dx_2}{dt} = -0.3x_2$

18. $\dfrac{dx_1}{dt} = -x_1$

$\dfrac{dx_2}{dt} = x_1 - 0.5x_2$

19. Suppose that a drug is administered to a person in a single dose, and assume that the drug does not accumulate in body tissue, but is excreted through urine. Denote the amount of drug in the body at time t by $x_1(t)$ and in the urine at time t by $x_2(t)$. If $x_1(0) = 4$ mg and $x_2(0) = 0$, find $x_1(t)$ and $x_2(t)$ if

$$\frac{dx_1}{dt} = -0.3x_1(t)$$

20. Suppose that a drug is administered to a person in a single dose, and assume that the drug does not accumulate in body tissue, but is excreted through urine. Denote the amount of drug in the body at time t by $x_1(t)$ and in the urine at time t by $x_2(t)$. If $x_1(0) = 6$ mg and $x_2(0) = 0$, find a system of differential equations for $x_1(t)$ and $x_2(t)$ if it takes 20 minutes for the drug to be at one-half of its initial amount in the body.

21. A very simple two-compartment model for gap dynamics in a forest assumes that gaps are created by disturbances (wind, fire, etc.) and that gaps revert to forest as trees grow in the gaps. We denote by $x_1(t)$ the area occupied by gaps and by $x_2(t)$ the area occupied by adult trees. We assume that the dynamics are given by

$$\frac{dx_1}{dt} = -0.2x_1 + 0.1x_2 \qquad (11.44)$$

$$\frac{dx_2}{dt} = 0.2x_1 - 0.1x_2 \qquad (11.45)$$

(a) Find the corresponding compartment diagram.

(b) Show that $x_1(t) + x_2(t)$ is a constant. Denote the constant by A and give its meaning. [*Hint*: Show that $\frac{d}{dt}(x_1 + x_2) = 0$.]

(c) Let $x_1(0) + x_2(0) = 20$. Use your answer in (b) to explain why this equation implies that $x_1(t) + x_2(t) = 20$ for all $t > 0$.

(d) Use your result in (c) to replace x_2 in (11.44) by $20 - x_1$, and show that doing so reduces the system (11.44) and (11.45) to

$$\frac{dx_1}{dt} = 2 - 0.3x_1 \qquad (11.46)$$

with $x_1(t) + x_2(t) = 20$ for all $t \geq 0$.

(e) Solve the system (11.44) and (11.45), and determine what fraction of the forest is occupied by adult trees at time t when $x_1(0) = 2$ and $x_2(0) = 18$. What happens as $t \to \infty$?

22. One simple model for forest succession is a three-compartment model. We assume that gaps in a forest are created by disturbances and are colonized by early successional species, which are then replaced by late successional species. We denote by $x_1(t)$ the total area occupied by gaps at time t, by $x_2(t)$ the total area occupied by early successional species at time t, and by $x_3(t)$ the total area occupied by late successional species at time t. The dynamics are given by

$$\frac{dx_1}{dt} = 0.2x_2 + x_3 - 2x_1$$

$$\frac{dx_2}{dt} = 2x_1 - 0.7x_2$$

$$\frac{dx_3}{dt} = 0.5x_2 - x_3$$

(a) Draw the corresponding compartment diagram.

(b) Show that

$$x_1(t) + x_2(t) + x_3(t) = A$$

where A is a constant, and give the meaning of A.

■ **11.2.2**

23. Solve

$$\frac{d^2x}{dt^2} = -4x$$

with $x(0) = 0$ and $\frac{dx(0)}{dt} = 6$.

24. Solve

$$\frac{d^2x}{dt^2} = -9x$$

with $x(0) = 0$ and $\frac{dx(0)}{dt} = 12$.

25. Transform the second-order differential equation

$$\frac{d^2x}{dt^2} = 3x$$

into a system of first-order differential equations.

26. Transform the second-order differential equation

$$\frac{d^2x}{dt^2} = -\frac{1}{2}x$$

into a system of first-order differential equations.

27. Transform the second-order differential equation

$$\frac{d^2x}{dt^2} + \frac{dx}{dt} = x$$

into a system of first-order differential equations.

28. Transform the second-order differential equation

$$\frac{d^2x}{dt^2} - 2\frac{dx}{dt} = 3x$$

into a system of first-order differential equations.

■ 11.3 Nonlinear Autonomous Systems: Theory

In this section, we will develop some of the theory needed to analyze systems of differential equations of the form

$$\frac{dx_1}{dt} = f_1(x_1, x_2, \ldots, x_n)$$

$$\frac{dx_2}{dt} = f_2(x_1, x_2, \ldots, x_n)$$

$$\vdots$$

$$\frac{dx_n}{dt} = f_n(x_1, x_2, \ldots, x_n)$$

(11.47)

where $f_i : \mathbf{R}^n \to \mathbf{R}$, for $i = 1, 2, \ldots, n$. We assume that the functions $f_i, i = 1, 2, \ldots, n$, do not explicitly depend on t; the system (11.47) is therefore called *autonomous*. We no longer assume that the functions f_i are linear, as in Section 11.1. Using vector notation, we can write the system (11.47) in the form

$$\frac{d\mathbf{x}}{dt} = \mathbf{f}(\mathbf{x})$$

where $\mathbf{x} = [x_1, x_2, \ldots, x_n]'$, and $\mathbf{f}(\mathbf{x})$ is a vector-valued function $\mathbf{f} : \mathbf{R}^n \to \mathbf{R}^n$ with components $f_i(x_1, x_2, \ldots, x_n) : \mathbf{R}^n \to \mathbf{R}, i = 1, 2, \ldots, n$. The function $\mathbf{f}(\mathbf{x})$ defines a direction field, just as in the linear case.

Unless the functions f_i are linear, it is typically not possible to find explicit solutions of systems of differential equations. If we want to solve such systems, we frequently must use numerical methods. Instead of trying to find solutions, we will focus on point equilibria and their stability, just as in Section 8.2.

The definition of a point equilibrium (as given in Section 8.2) must be extended to systems of the form (11.47). We say that a point

$$\hat{\mathbf{x}} = (\hat{x}_1, \hat{x}_2, \ldots, \hat{x}_n)$$

is a point equilibrium (or simply equilibrium) of (11.47) if

$$\mathbf{f}(\hat{\mathbf{x}}) = \mathbf{0}$$

An equilibrium is also called a **critical point**. As in the linear case, this is a point in the direction field at which the direction vector has length 0, implying that if we start a system of differential equations at an equilibrium point, it will stay there for all later times.

As in the linear case, a solution might not return to an equilibrium after a small perturbation; this possibility is addressed by the stability of the equilibrium. The theory of stability for systems of nonlinear autonomous differential equations is parallel to that in Section 8.2; there is both an analytical and graphical approach that reduces to the theory set forth there when there is a single differential equation. We will restrict our discussion to systems of two equations in two variables. (The concepts are the same when we have more than two equations, but the calculations become more involved.)

■ 11.3.1 Analytical Approach

A Single Autonomous Differential Equation

EXAMPLE 1   Find all equilibria of

$$\frac{dx}{dt} = x(1 - x)$$

(11.48)

and analyze their stability.

Solution
We developed the theory for single autonomous differential equations in Section 8.2. To find equilibria, we set

$$x(1 - x) = 0$$

which yields

$$\hat{x}_1 = 0 \quad \text{and} \quad \hat{x}_2 = 1$$

To analyze the stability of these equilibria, we linearize the differential equation (11.48) about each equilibrium and compute the corresponding eigenvalue. We set

$$f(x) = x(1 - x)$$

Then

$$f'(x) = 1 - 2x$$

The eigenvalue associated with the equilibrium $\hat{x}_1 = 0$ is

$$\lambda_1 = f'(0) = 1 > 0$$

which implies that $\hat{x}_1 = 0$ is unstable.

The eigenvalue associated with the equilibrium $\hat{x}_2 = 1$ is

$$\lambda_2 = f'(1) = -1 < 0$$

which implies that $\hat{x}_2 = 1$ is locally stable. ■

The eigenvalue corresponding to an equilibrium of the differential equation

$$\frac{dx}{dt} = f(x) \tag{11.49}$$

is the slope of the function $f(x)$ at the equilibrium value. The reason for this is discussed in detail in Section 8.2; we repeat the basic argument here. Suppose that $\hat{x}$ is an equilibrium of (11.49); that is, $f(\hat{x}) = 0$. If we perturb $\hat{x}$ slightly (i.e., if we look at $\hat{x} + z$ for small $|z|$), we can find out what happens to $\hat{x} + z$ by examining

$$\frac{dx}{dt} = \frac{d}{dt}(\hat{x} + z) = \frac{dz}{dt}$$

Since the perturbation is small, we can linearize

$$f(\hat{x} + z) \approx f(\hat{x}) + f'(\hat{x})(\hat{x} + z - \hat{x}) = f'(\hat{x})z$$

[In the last step, we used the fact that $f(\hat{x}) = 0$.] We find that

$$\frac{dz}{dt} \approx f'(\hat{x})z$$

which has the approximate solution

$$z(t) \approx z(0)e^{\lambda t} \quad \text{with } \lambda = f'(\hat{x})$$

Therefore, if $f'(\hat{x}) < 0$, then $z(t) \to 0$ as $t \to \infty$ and, hence, $x(t) = \hat{x} + z(t) \to \hat{x}$ as $t \to \infty$; that is, the solution will return to the equilibrium $\hat{x}$ after a small perturbation. In this case, $\hat{x}$ is locally stable. If $f'(\hat{x}) > 0$, then $z(t)$ will not go to 0, which implies that $\hat{x}$ is unstable. The linearization of $f(x)$ thus tells us whether an equilibrium is locally stable or unstable. We will use linearization as well to determine the stability of equilibria of systems of differential equations.

Systems of Two Differential Equations We consider differential equations of the form

$$\frac{dx_1}{dt} = f_1(x_1, x_2) \tag{11.50}$$

$$\frac{dx_2}{dt} = f_2(x_1, x_2) \tag{11.51}$$

or, in vector notation,

$$\frac{d\mathbf{x}}{dt} = \mathbf{f}(\mathbf{x}) \tag{11.52}$$

where $\mathbf{x} = \mathbf{x}(t) = (x_1(t), x_2(t))$ and $\mathbf{f}(\mathbf{x})$ is a vector-valued function $\mathbf{f}(\mathbf{x}) = (f_1(\mathbf{x}), f_2(\mathbf{x}))$ with $f_i(\mathbf{x}) : \mathbf{R}^2 \to \mathbf{R}$. An equilibrium or critical point, $\hat{\mathbf{x}}$, of (11.52) satisfies

$$\mathbf{f}(\hat{\mathbf{x}}) = \mathbf{0}$$

Suppose that $\hat{\mathbf{x}}$ is a point equilibrium. Then, as in the case of one nonlinear equation, we look at what happens to a small perturbation of $\hat{\mathbf{x}}$. We perturb $\hat{\mathbf{x}}$; that is, we look at how $\hat{\mathbf{x}} + \mathbf{z}$ changes under the dynamics described by (11.52):

$$\frac{d}{dt}(\hat{\mathbf{x}} + \mathbf{z}) = \frac{d}{dt}\mathbf{z} = \mathbf{f}(\hat{\mathbf{x}} + \mathbf{z})$$

The linearization of $\mathbf{f}(\mathbf{x})$ about $\mathbf{x} = \hat{\mathbf{x}}$ is

$$\mathbf{f}(\hat{\mathbf{x}}) + D\mathbf{f}(\hat{\mathbf{x}})\mathbf{z} = D\mathbf{f}(\hat{\mathbf{x}})\mathbf{z}$$

where we used the fact that $\mathbf{f}(\hat{\mathbf{x}}) = \mathbf{0}$. The matrix $D\mathbf{f}(\hat{\mathbf{x}})$ is the Jacobi matrix evaluated at $\hat{\mathbf{x}}$. If we approximate $\mathbf{f}(\hat{\mathbf{x}} + \mathbf{z})$ by its linearization $D\mathbf{f}(\hat{\mathbf{x}})$, then

$$\frac{d\mathbf{z}}{dt} = D\mathbf{f}(\hat{\mathbf{x}})\mathbf{z} \tag{11.53}$$

is the linear approximation of the dynamics of the perturbation $\mathbf{z}$.

We now have a system of linear differential equations that is a good approximation, provided that $\mathbf{z}$ is sufficiently close to $\mathbf{0}$. In Section 11.1, we learned how to analyze linear systems. We saw that eigenvalues of the matrix $D\mathbf{f}(\hat{\mathbf{x}})$ allowed us to determine the nature of the equilibrium. We will use the same approach here, but we emphasize that this is now a *local* analysis, just as in the case of a single differential equation, since we know that the linearization (11.53) is a good approximation only as long as we are sufficiently close to the point about which we linearized.

We return to our classification scheme for the linear case,

$$\frac{d\mathbf{x}}{dt} = A\mathbf{x}$$

where $\mathbf{x}(t) = (x_1(t), x_2(t))$ and A is a 2×2 matrix. We let

$$\Delta = \det A \quad \text{and} \quad \tau = \operatorname{tr} A$$

When we linearize a nonlinear system about an equilibrium, the matrix A is the Jacobi matrix evaluated at the equilibrium:

$$A = D\mathbf{f}(\hat{\mathbf{x}})$$

We exclude the following cases: (i) $\Delta = 0$ (when $\Delta = 0$, at least one eigenvalue is equal to 0), (ii) $\tau = 0$ and $\Delta > 0$ (when $\tau = 0$ and $\Delta > 0$, the two eigenvalues are purely imaginary), and (iii) $\tau^2 = 4\Delta$ (when $\tau^2 = 4\Delta$, the two eigenvalues are identical). Except in these three cases, we can use the same classification scheme as in the linear case (Figure 11.35).

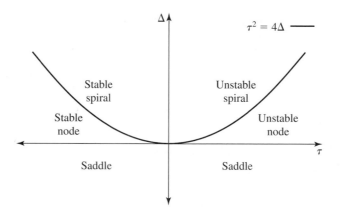

Figure 11.35 The stability behavior of a system of two autonomous equations.

The extension from the linear case is possible because the linearized vector field and the original vector field are geometrically similar close to an equilibrium point. (After all, that is the idea behind linearization.) This result is known as the *Hartman–Grobman theorem*, which says that as long as $D\mathbf{f}(\hat{\mathbf{x}})$ has no zero or purely imaginary eigenvalues, then the linearized and the original vector fields are similar close to the equilibrium. That is,

$$\frac{d\mathbf{x}}{dt} = \mathbf{f}(\mathbf{x}) \qquad \text{and} \qquad \frac{d\mathbf{z}}{dt} = D\mathbf{f}(\hat{\mathbf{x}})\mathbf{z}$$

behave similarly for $\mathbf{x} = \hat{\mathbf{x}} + \mathbf{z}$ with $\mathbf{z}$ close to $\mathbf{0}$.

We find the same classification scheme as in the linear case:

- The equilibrium $\hat{\mathbf{x}}$ is a *node* if both eigenvalues of $D\mathbf{f}(\hat{\mathbf{x}})$ are real, distinct, nonzero, and of the same sign. The node is locally stable if the eigenvalues are negative and unstable if the eigenvalues are positive.
- The equilibrium $\hat{\mathbf{x}}$ is a saddle point if both eigenvalues are real and nonzero but have opposite signs. A saddle point is unstable.
- The equilibrium $\hat{\mathbf{x}}$ is a spiral if both eigenvalues are complex conjugates with nonzero real parts. The spiral is locally stable if the real parts of the eigenvalues are negative and unstable if the real parts of the eigenvalues are positive.

When the two eigenvalues are purely imaginary, we cannot determine the stability by linearization.

EXAMPLE 2   Consider

$$\frac{dx_1}{dt} = x_1 - 2x_1^2 - 2x_1x_2$$

$$\frac{dx_2}{dt} = 4x_2 - 5x_2^2 - 7x_1x_2 \tag{11.54}$$

(a) Find all equilibria of (11.54) and (b) analyze their stability.

Solution   (a) To find equilibria, we set the right-hand side of (11.54) equal to 0:

$$x_1 - 2x_1^2 - 2x_1x_2 = 0 \tag{11.55}$$

$$4x_2 - 5x_2^2 - 7x_1x_2 = 0 \tag{11.56}$$

Factoring out x_1 in the first equation and x_2 in the second yields

$$x_1(1 - 2x_1 - 2x_2) = 0 \qquad \text{and} \qquad x_2(4 - 5x_2 - 7x_1) = 0$$

That is,

$$x_1 = 0 \qquad \text{or} \qquad 2x_1 + 2x_2 = 1$$

and

$$x_2 = 0 \quad \text{or} \quad 7x_1 + 5x_2 = 4$$

Combining the different solutions, we get the following four cases:

(i) $x_1 = 0$ and $x_2 = 0$
(ii) $x_1 = 0$ and $7x_1 + 5x_2 = 4$
(iii) $x_2 = 0$ and $2x_1 + 2x_2 = 1$
(iv) $2x_1 + 2x_2 = 1$ and $7x_1 + 5x_2 = 4$

First, we will compute the equilibria in these four cases:

Case (i) There is nothing to compute; the equilibrium is $(\hat{x}_1, \hat{x}_2) = (0, 0)$.

Case (ii) To find the equilibrium, we must solve the system

$$x_1 = 0$$
$$7x_1 + 5x_2 = 4$$

which has the solutions

$$x_1 = 0 \quad \text{and} \quad x_2 = \frac{4}{5}$$

Hence, the equilibrium is $(\hat{x}_1, \hat{x}_2) = (0, \frac{4}{5})$.

Case (iii) To find the equilibrium, we must solve the system

$$x_2 = 0$$
$$2x_1 + 2x_2 = 1$$

which has the solutions

$$x_2 = 0 \quad \text{and} \quad x_1 = \frac{1}{2}$$

Hence, the equilibrium is $(\hat{x}_1, \hat{x}_2) = (\frac{1}{2}, 0)$.

Case (iv) To find the equilibrium, we must solve the system

$$2x_1 + 2x_2 = 1$$
$$7x_1 + 5x_2 = 4$$

We use the standard elimination method: Leaving the first equation alone, changing the second by multiplying the first by 7 and the second by 2, and subtracting the second equation from the first, we find that this system is equivalent to

$$2x_1 + 2x_2 = 1$$
$$4x_2 = -1$$

which has the solutions

$$x_2 = -\frac{1}{4} \quad \text{and} \quad x_1 = \frac{1}{2} - x_2 = \frac{3}{4}$$

Hence, the equilibrium is $(\hat{x}_1, \hat{x}_2) = (\frac{3}{4}, -\frac{1}{4})$.

We can illustrate all equilibria in the direction field of (11.54), which is displayed in Figure 11.36. The equilibria are shown as dots.

(b) To analyze the stability of the equilibria, we compute the Jacobi matrix

$$\mathbf{Df} = \begin{bmatrix} \dfrac{\partial f_1}{\partial x_1} & \dfrac{\partial f_1}{\partial x_2} \\ \dfrac{\partial f_2}{\partial x_1} & \dfrac{\partial f_2}{\partial x_2} \end{bmatrix}$$

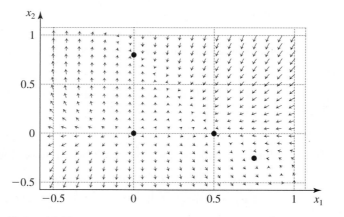

Figure 11.36 The direction field of (11.54) together with the equilibria.

With

$$f_1(x_1, x_2) = x_1 - 2x_1^2 - 2x_1x_2 \quad \text{and} \quad f_2(x_1, x_2) = 4x_2 - 5x_2^2 - 7x_1x_2$$

we find that

$$Df(x_1, x_2) = \begin{bmatrix} 1 - 4x_1 - 2x_2 & -2x_1 \\ -7x_2 & 4 - 10x_2 - 7x_1 \end{bmatrix}$$

We will now go through the four cases and analyze each equilibrium:

Case (i) The equilibrium is the point $(0, 0)$. The Jacobi matrix at $(0, 0)$ is

$$Df(0, 0) = \begin{bmatrix} 1 & 0 \\ 0 & 4 \end{bmatrix}$$

Since this matrix is in diagonal form, the eigenvalues are the diagonal elements, and we find that $\lambda_1 = 1$ and $\lambda_2 = 4$. Because both eigenvalues are positive, the equilibrium is unstable. Using the same classification as in the linear case, we say that $(0, 0)$ is an unstable node.

The linearization of the direction field about $(0, 0)$ is displayed in Figure 11.37, where we show the direction field of

$$\frac{d\mathbf{x}}{dt} = \begin{bmatrix} 1 & 0 \\ 0 & 4 \end{bmatrix} \mathbf{x}(t)$$

Figure 11.37 confirms that $(0, 0)$ is an unstable node (or source). If you now compare Figure 11.37 with the direction field of (11.54) shown in Figure 11.36,

Figure 11.37 The linearization of the direction field about $(0, 0)$.

you will find that the linearized direction field and the direction field of (11.54) close to the equilibrium $(0, 0)$ are similar.

Case (ii) The equilibrium is the point $(0, \frac{4}{5})$. The Jacobi matrix at $(0, \frac{4}{5})$ is

$$Df\left(0, \frac{4}{5}\right) = \begin{bmatrix} 1 - \frac{8}{5} & 0 \\ -\frac{28}{5} & 4 - 8 \end{bmatrix} = \begin{bmatrix} -\frac{3}{5} & 0 \\ -\frac{28}{5} & -4 \end{bmatrix}$$

Since this matrix is in lower triangular form, the eigenvalues are the diagonal elements, and we find that the eigenvalues of $Df(0, \frac{4}{5})$ are $\lambda_1 = -\frac{3}{5}$ and $\lambda_2 = -4$. Because both eigenvalues are negative, $(0, \frac{4}{5})$ is locally stable. Using the same classification as in the linear case, we say that $(0, \frac{4}{5})$ is a stable node.

The linearization of the direction field about $(0, \frac{4}{5})$ is displayed in Figure 11.38, which confirms that $(0, \frac{4}{5})$ is a stable node (or sink). If you compare Figure 11.38 with the direction field of (11.54) shown in Figure 11.36, you will find that the linearized direction field and the direction field of (11.54) close to the equilibrium $(0, \frac{4}{5})$ are similar.

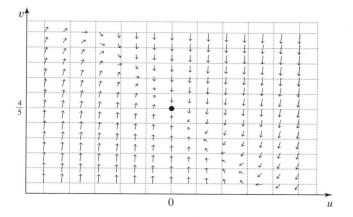

Figure 11.38 The linearization of the direction field about $(0, \frac{4}{5})$.

Case (iii) The equilibrium is the point $(\frac{1}{2}, 0)$. The Jacobi matrix at $(\frac{1}{2}, 0)$ is

$$Df\left(\frac{1}{2}, 0\right) = \begin{bmatrix} 1 - 2 & -1 \\ 0 & 4 - \frac{7}{2} \end{bmatrix} = \begin{bmatrix} -1 & -1 \\ 0 & \frac{1}{2} \end{bmatrix}$$

Since this matrix is in upper triangular form, the eigenvalues are simply the diagonal elements, and we find that the eigenvalues of $Df(\frac{1}{2}, 0)$ are $\lambda_1 = -1$ and $\lambda_2 = \frac{1}{2}$. Because one eigenvalue is positive and the other is negative, $(\frac{1}{2}, 0)$ is unstable. Using the same classification as in the linear case, we say that $(\frac{1}{2}, 0)$ is a saddle point.

The linearization of the direction field about $(\frac{1}{2}, 0)$ is displayed in Figure 11.39, which confirms that $(\frac{1}{2}, 0)$ is a saddle point. If you compare Figure 11.39 with the direction field of (11.54) shown in Figure 11.36, you will find that the linearized direction field and the direction field of (11.54) close to the equilibrium $(\frac{1}{2}, 0)$ are similar.

Case (iv) The Jacobi matrix at the equilibrium $(\frac{3}{4}, -\frac{1}{4})$ is

$$Df\left(\frac{3}{4}, -\frac{1}{4}\right) = \begin{bmatrix} 1 - 3 + \frac{1}{2} & -\frac{3}{2} \\ \frac{7}{4} & 4 + \frac{10}{4} - \frac{21}{4} \end{bmatrix} = \begin{bmatrix} -\frac{3}{2} & -\frac{3}{2} \\ \frac{7}{4} & \frac{5}{4} \end{bmatrix}$$

Figure 11.39 The linearization of the direction field about $(\frac{1}{2}, 0)$.

To find the eigenvalues, we must solve

$$\det \begin{bmatrix} -\frac{3}{2} - \lambda & -\frac{3}{2} \\ \frac{7}{4} & \frac{5}{4} - \lambda \end{bmatrix} = 0$$

Evaluating the determinant on the left-hand side and simplifying yields

$$\left(-\frac{3}{2} - \lambda\right)\left(\frac{5}{4} - \lambda\right) + \left(\frac{3}{2}\right)\left(\frac{7}{4}\right) = 0$$

$$\lambda^2 + \frac{3}{2}\lambda - \frac{5}{4}\lambda - \frac{15}{8} + \frac{21}{8} = 0$$

$$\lambda^2 + \frac{1}{4}\lambda + \frac{3}{4} = 0$$

Solving this quadratic equation, we find that

$$\lambda_{1,2} = \frac{-\frac{1}{4} \pm \sqrt{\frac{1}{16} - 3}}{2}$$

$$= -\frac{1}{8} \pm \frac{1}{8}\sqrt{-47} = -\frac{1}{8} \pm \frac{1}{8}i\sqrt{47}$$

That is,

$$\lambda_1 = -\frac{1}{8} + \frac{1}{8}i\sqrt{47} \qquad \text{and} \qquad \lambda_2 = -\frac{1}{8} - \frac{1}{8}i\sqrt{47}$$

The eigenvalues are complex conjugates with negative real parts. Thus, $(\frac{3}{4}, -\frac{1}{4})$ is locally stable, and we expect the solutions to spiral into the equilibrium when we start close to the equilibrium.

The linearization of the direction field about $(\frac{3}{4}, -\frac{1}{4})$ is displayed in Figure 11.40, which confirms that $(\frac{3}{4}, -\frac{1}{4})$ is a stable spiral. If you compare Figure 11.40 with the direction field of (11.54) shown in Figure 11.36, you will find that the linearized direction field and the direction field of (11.54) close to the equilibrium $(\frac{3}{4}, -\frac{1}{4})$ are similar. ∎

Figure 11.40 The linearization of the direction field about $(\frac{3}{4}, -\frac{1}{4})$.

■ 11.3.2 Graphical Approach for 2×2 Systems

In this subsection, we will discuss a graphical approach to systems of two autonomous differential equations. Suppose that

$$\frac{dx_1}{dt} = f_1(x_1, x_2)$$

$$\frac{dx_2}{dt} = f_2(x_1, x_2)$$

which in vector form is

$$\frac{d\mathbf{x}}{dt} = \mathbf{f}(\mathbf{x})$$

The curves

$$f_1(x_1, x_2) = 0$$
$$f_2(x_1, x_2) = 0$$

are called **zero isoclines** or **null clines**, and they represent the points in the x_1–x_2 plane where the growth rates of the respective quantities are equal to zero. This situation is illustrated in Figure 11.41 for a particular choice of f_1 and f_2. Let's assume that x_1 and x_2 are nonnegative; this restricts the discussion to the first quadrant of the x_1–x_2 plane. The two curves in Figure 11.41 divide the first quadrant into four regions, and we label each region according to whether dx_i/dt is positive or negative. Without specifying the signs of f_1 and f_2 any further, we make assumptions about the signs of dx_1/dt and dx_2/dt as indicated in Figure 11.41.

The point where both null clines in Figure 11.41 intersect is a point equilibrium or critical point, which we call $\hat{\mathbf{x}}$. We can use the graph to determine the signs of the entries in the Jacobi matrix

$$D\mathbf{f}(\hat{\mathbf{x}}) = \begin{bmatrix} a_{11} & a_{12} \\ a_{21} & a_{22} \end{bmatrix}$$

where $a_{ij} = \frac{\partial f_i}{\partial x_j}(\hat{\mathbf{x}})$. Clearly, the entry $a_{11} = \frac{\partial f_1}{\partial x_1}$ is the effect of a change in f_1 in the x_1-direction when we keep x_2 fixed. To determine the sign of a_{11}, follow the horizontal arrow in the figure: The arrow goes from a region where f_1 is positive to a region where f_1 is negative, implying that f_1 is decreasing and hence $\frac{\partial f_1}{\partial x_1}(\hat{\mathbf{x}}) = a_{11} < 0$. We conclude that the sign of a_{11} in $D\mathbf{f}(\hat{\mathbf{x}})$ is negative, which we indicate in the Jacobi matrix by a minus sign in place of a_{11}:

$$D\mathbf{f}(\hat{\mathbf{x}}) = \begin{bmatrix} - & a_{12} \\ a_{21} & a_{22} \end{bmatrix}$$

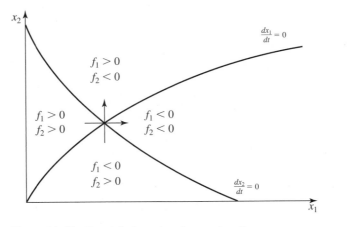

Figure 11.41 Graphical approach: zero isoclines.

Next, we determine the sign of $a_{12} = \frac{\partial f_1}{\partial x_2}(\hat{\mathbf{x}})$. This time, we want to know how f_1 changes at the equilibrium $\hat{\mathbf{x}}$ when we move in the x_2-direction and keep x_1 fixed. This is the direction of the vertical arrow through the equilibrium point. Since the arrow goes from a region where f_1 is negative to a region where f_1 is positive, f_1 increases in the direction of x_2 and, therefore, $a_{12} = \frac{\partial f_1}{\partial x_2}(\hat{\mathbf{x}}) > 0$.

The signs of the other two entries are found similarly, and we obtain

$$D\mathbf{f}(\hat{\mathbf{x}}) = \begin{bmatrix} - & + \\ - & - \end{bmatrix}$$

Thus, the trace of $D\mathbf{f}(\hat{\mathbf{x}})$ is negative and the determinant of $D\mathbf{f}(\hat{\mathbf{x}})$ is positive. Using the criterion stated in Subsection 9.4.2, we conclude that both eigenvalues have negative real parts and, therefore, that the equilibrium is locally stable.

EXAMPLE 3

Use the graphical approach to analyze the equilibrium $(3, 2)$ of

$$\frac{dx_1}{dt} = 5 - x_1 - x_1 x_2 + 2x_2$$

$$\frac{dx_2}{dt} = x_1 x_2 - 3x_2$$

Solution First, note that $(3, 2)$ is indeed an equilibrium of this system. Now, the zero isoclines satisfy

$$\frac{dx_1}{dt} = 0, \quad \text{which holds for } x_2 = \frac{5 - x_1}{x_1 - 2}$$

and

$$\frac{dx_2}{dt} = 0, \quad \text{which holds for } x_2 = 0 \text{ or } x_1 = 3$$

The zero isoclines in the x_1–x_2 plane are drawn in Figure 11.42. The equilibrium $(3, 2)$ is the point of intersection of the zero isoclines $x_1 = 3$ and $x_2 = \frac{5-x_1}{x_1-2}$. The signs of $\frac{dx_1}{dt}$ and $\frac{dx_2}{dt}$ are indicated in the figure as well. We claim that

$$D\mathbf{f}(\hat{\mathbf{x}}) = \begin{bmatrix} - & - \\ + & 0 \end{bmatrix}$$

Here is why: To find the sign of $a_{11} = \frac{\partial f}{\partial x_1}$, we need to determine how dx_1/dt changes as we cross the zero isocline of x_1 in the x_1-direction. We see from the graph that dx_1/dt changes from positive to negative when we follow the horizontal arrow while crossing the zero isocline of x_1. Therefore, a_{11} is negative. To see why $a_{22} = 0$, follow

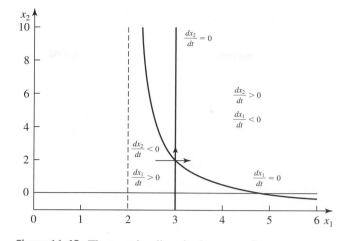

Figure 11.42 The zero isoclines in the x_1–x_2 plane.

the vertical arrow in the x_2-direction. Since the vertical arrow is on the zero isocline of x_2, the sign of dx_2/dt does not change as we cross the equilibrium in the x_2-direction. Therefore, $a_{22} = 0$. The signs of a_{12} and a_{21} follow from observing that if we cross the zero isocline of x_1 in the x_2-direction (the vertical arrow), then dx_1/dt changes from positive to negative, making $a_{12} < 0$. If we cross the zero isocline of x_2 in the direction of x_1 (the horizontal arrow), we see that dx_2/dt changes from negative to positive, making $a_{21} > 0$.

To determine the stability of $\hat{\mathbf{x}}$, we look at the trace and the determinant. Since the trace is negative and the determinant is positive, we conclude that the equilibrium is locally stable. ■

This simple graphical approach does not always give us the signs of the real parts of the eigenvalues, as illustrated in the following example: Suppose that we arrive at the Jacobi matrix in which the signs of the entries are

$$\begin{bmatrix} + & - \\ - & - \end{bmatrix}$$

The trace may now be positive or negative. Therefore, we cannot conclude anything about the eigenvalues. In this case, we would have to compute the eigenvalues or the trace and the determinant explicitly and cannot rely on the signs alone.

Section 11.3 Problems

■ 11.3.1

In Problems 1–6, the point $(0, 0)$ is always an equilibrium. Use the analytical approach to investigate its stability.

1. $\dfrac{dx_1}{dt} = x_1 - 2x_2 + x_1 x_2$

$\dfrac{dx_2}{dt} = -x_1 + x_2$

2. $\dfrac{dx_1}{dt} = -x_1 - x_2 + x_1^2$

$\dfrac{dx_2}{dt} = x_2 - x_1^2$

3. $\dfrac{dx_1}{dt} = x_1 + x_1^2 - 2x_1 x_2 + x_2$

$\dfrac{dx_2}{dt} = x_1$

4. $\dfrac{dx_1}{dt} = 3x_1 x_2 - x_1 + x_2$

$\dfrac{dx_2}{dt} = x_2^2 - x_1$

5. $\dfrac{dx_1}{dt} = x_1 e^{-x_2}$

$\dfrac{dx_2}{dt} = 2x_2 e^{x_1}$

6. $\dfrac{dx_1}{dt} = -2 \sin x_1$

$\dfrac{dx_2}{dt} = -x_2 e^{x_1}$

In Problems 7–12, find all equilibria of each system of differential equations and use the analytical approach to determine the stability of each equilibrium.

7. $\dfrac{dx_1}{dt} = -x_1 + 2x_1(1 - x_1)$

$\dfrac{dx_2}{dt} = -x_2 + 5x_2(1 - x_1 - x_2)$

8. $\dfrac{dx_1}{dt} = -x_1 + 3x_1(1 - x_1 - x_2)$

$\dfrac{dx_2}{dt} = -x_2 + 5x_2(1 - x_1 - x_2)$

9. $\dfrac{dx_1}{dt} = 4x_1(1 - x_1) - 2x_1 x_2$

$\dfrac{dx_2}{dt} = x_2(2 - x_2) - x_2$

10. $\dfrac{dx_1}{dt} = 2x_1(5 - x_1 - x_2)$

$\dfrac{dx_2}{dt} = 3x_2(7 - 3x_1 - x_2)$

11. $\dfrac{dx_1}{dt} = x_1 - x_2$

$\dfrac{dx_2}{dt} = x_1 x_2 - x_2$

12. $\dfrac{dx_1}{dt} = x_1 x_2 - x_2$

$\dfrac{dx_2}{dt} = x_1 + x_2$

13. For which value of a has

$$\frac{dx_1}{dt} = x_2(x_1 + a)$$

$$\frac{dx_2}{dt} = x_2^2 + x_2 - x_1$$

a unique equilibrium? Characterize its stability.

14. Assume that $a > 0$. Find all point equilibria of

$$\frac{dx_1}{dt} = 1 - ax_1 x_2$$

$$\frac{dx_2}{dt} = ax_1 x_2 - x_2$$

and characterize their stability.

■ **11.3.2**

15. Assume that

$$\frac{dx_1}{dt} = x_1(10 - 2x_1 - x_2)$$

$$\frac{dx_2}{dt} = x_2(10 - x_1 - 2x_2)$$

(a) Graph the zero isoclines.

(b) Show that $(\frac{10}{3}, \frac{10}{3})$ is an equilibrium, and use the analytical approach to determine its stability.

16. Assume that

$$\frac{dx_1}{dt} = x_1(10 - x_1 - 2x_2)$$

$$\frac{dx_2}{dt} = x_2(10 - 2x_1 - x_2)$$

(a) Graph the zero isoclines.

(b) Show that $(\frac{10}{3}, \frac{10}{3})$ is an equilibrium, and use the analytical approach to determine its stability.

In Problems 17–22, use the graphical approach for 2×2 systems to find the sign structure of the Jacobi matrix at the indicated equilibrium. If possible, determine the stability of the equilibrium. Assume that the system of differential equations is given by

$$\frac{dx_1}{dt} = f_1(x_1, x_2)$$

$$\frac{dx_2}{dt} = f_2(x_1, x_2)$$

Furthermore, assume that x_1 and x_2 are both nonnegative. In each problem, the zero isoclines are drawn and the equilibrium we want to investigate is indicated by a dot. Assume that both x_1 and x_2 increase close to the origin and that f_1 and f_2 change sign when crossing their zero isoclines.

17. See Figure 11.43.

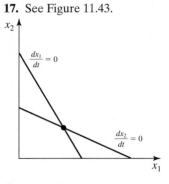

Figure 11.43

18. See Figure 11.44.

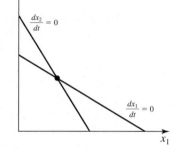

Figure 11.44

19. See Figure 11.45.

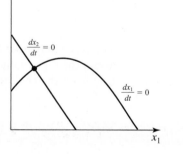

Figure 11.45

20. See Figure 11.46.

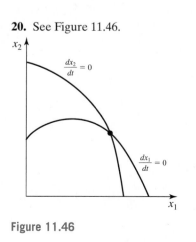

Figure 11.46

21. See Figure 11.47.

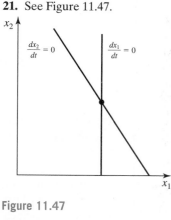

Figure 11.47

22. See Figure 11.48.

Figure 11.48

23. Let

$$\frac{dx_1}{dt} = x_1(2 - x_1) - x_1 x_2$$

$$\frac{dx_2}{dt} = x_1 x_2 - x_2$$

(a) Graph the zero isoclines.

(b) Show that $(1, 1)$ is an equilibrium. Use the graphical approach to determine its stability.

24. Let

$$\frac{dx_1}{dt} = x_1(2 - x_1^2) - x_1 x_2$$

$$\frac{dx_2}{dt} = x_1 x_2 - x_2$$

(a) Graph the zero isoclines.

(b) Show that $(1, 1)$ is an equilibrium. Use the graphical approach to determine its stability.

■ 11.4 Nonlinear Systems: Applications

■ 11.4.1 The Lotka–Volterra Model of Interspecific Competition

Imagine two species of plants growing together in the same plot. They both use similar resources: light, water, and nutrients. The use of these resources by one individual reduces their availability to other individuals. We call this type of interaction between individuals **competition**. **Intraspecific competition** occurs between individuals of the same species, **interspecific competition** between individuals of different species. Competition may result in reduced fecundity or reduced survivorship (or both). The effects of competition are often more pronounced when the number of competitors is higher.

In this subsection, we will discuss the Lotka–Volterra model of interspecific competition, which incorporates density-dependent effects of competition in the manner described previously. The model is an extension of the logistic equation to the case of two species. To describe it, we denote the population size of species 1 at time t by $N_1(t)$ and that of species 2 at time t by $N_2(t)$. Each species grows according to the logistic equation when the other species is absent. We denote their respective carrying capacities by K_1 and K_2, and their respective intrinsic rates of growth by r_1 and r_2. We assume that $K_1, K_2, r_1,$ and r_2 are positive. In addition, the two species may have inhibitory effects on each other. We measure the effect of species 1 on species 2 by the **competition coefficient** α_{21}; the effect of species 2 on species 1 is measured by the competition coefficient α_{12}. The Lotka–Volterra model of interspecific competition is then given by the following system of differential equations:

$$\frac{dN_1}{dt} = r_1 N_1 \left(1 - \frac{N_1}{K_1} - \alpha_{12} \frac{N_2}{K_1} \right)$$

$$\frac{dN_2}{dt} = r_2 N_2 \left(1 - \frac{N_2}{K_2} - \alpha_{21} \frac{N_1}{K_2} \right)$$

(11.57)

Let's look at the first equation. If $N_2 = 0$, the first equation reduces to the logistic equation $dN_1/dt = r_1 N_1 (1 - N_1/K_1)$, as mentioned. To understand the precise meaning of the competition coefficient α_{12}, observe that N_2 individuals of species 2 have the same effect on species 1 as $\alpha_{12} N_2$ individuals of species 1. The term $\alpha_{12} N_2$ thus converts the number of N_2 individuals into "N_1-equivalents." For instance, set $\alpha_{12} = 0.2$ and assume that $N_2 = 20$; then the effect of 20 individuals of species 2 on species 1 is the same as the effect of $(0.2)(20) = 4$ individuals of species 1 on species 1, because both reduce the growth rate by the same amount. A similar interpretation can be attached to the competition coefficient α_{21} in the second equation.

This model takes a simplistic view of competition; actual competitive interactions are more complicated. Nevertheless, it serves an important purpose: Because of its simple form, it allows us to study the consequences of competition, potentially giving us valuable insight into more complex situations.

We will use both zero isoclines and eigenvalues to analyze the model.

Zero Isoclines   The first step is to find the equations of the zero isoclines. To find the zero isocline for species 1, we set

$$r_1 N_1 \left(1 - \frac{N_1}{K_1} - \alpha_{12} \frac{N_2}{K_1} \right) = 0$$

The solutions are $N_1 = 0$ or

$$N_2 = \frac{K_1}{\alpha_{12}} - \frac{1}{\alpha_{12}} N_1$$

(11.58)

To find the zero isocline for species 2, we set

$$r_2 N_2 \left(1 - \frac{N_2}{K_2} - \alpha_{21} \frac{N_1}{K_2} \right) = 0$$

The solutions are $N_2 = 0$ or

$$N_2 = K_2 - \alpha_{21} N_1$$

(11.59)

The isocline $N_i = 0, i = 1, 2$, corresponds to the case in which species i is absent. This is biologically reasonable, because individuals of either species are not created spontaneously—once a species is absent, it remains absent.

The other two isoclines, given by (11.58) and (11.59), are of particular interest because they tell us whether the two species can coexist stably. Both isoclines are straight lines in the N_1–N_2 plane, and there are four ways these isoclines can be arranged. These are illustrated in Figures 11.49 through 11.52, together with the corresponding direction fields. The solid dots in each figure are the equilibria. We see that there are always the equilibria $(K_1, 0)$ and $(0, K_2)$. These are the equilibria representing the fact that only one species is present; they are referred to as **monoculture** equilibria. In addition, there are two cases where an equilibrium exists when both species are present; such an equilibrium is called a *nontrivial equilibrium*. We will now discuss each case separately.

Case 1: $K_1 > \alpha_{12} K_2$ and $K_2 < \alpha_{21} K_1$ When we look at the direction field in Figure 11.49, we see that species 1 drives species 2 to extinction. If both species are present initially, then the abundance of species 2 declines over time, and species 2 will eventually become extinct, whereas species 1 will reach its carrying capacity K_1. We say that species 1 outcompetes species 2 and refer to this as the case of **competitive exclusion**.

Figure 11.49 Case 1: Species 1 outcompetes species 2.

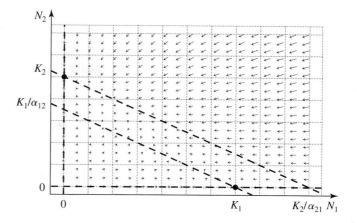

Figure 11.50 Case 2: Species 2 outcompetes species 1.

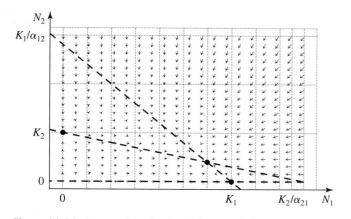

Figure 11.51 Case 3: Species 1 and 2 can coexist.

Figure 11.52 Case 4: Either species 1 or species 2 wins, depending on the initial condition.

We can understand why species 1 is the better competitor by looking at the inequalities $K_1 > \alpha_{12}K_2$ and $K_2 < \alpha_{21}K_1$. The carrying capacity K_1 is the abundance of species 1 that results in a zero per capita growth rate of species 1 in the absence of species 2. That is, if $N_1 = K_1$ and $N_2 = 0$, then $\frac{1}{N_1}\frac{dN_1}{dt} = 0$. The carrying capacity K_2 has a similar interpretation. To see the effect of species 2 on species 1, we look at the case when species 2 is at its carrying capacity K_2 and the abundance of species 1 is negligible. In this case, the per capita growth rate of species 1 is approximately equal to $r_1(1 - \alpha_{12}K_2/K_1)$, which is positive because $\alpha_{12}K_2/K_1$ is less than 1. On the one hand, this implies that species 2 cannot prevent species 1 from increasing its abundance when species 1 is rare; we say that species 1 can *invade* species 2 when species 1 is rare. On the other hand, since $K_2 < \alpha_{21}K_1$, species 2 cannot invade species 1. That is, if we set $N_1 = K_1$ and assume that the abundance of species 2 is negligible, then the per capita growth rate of species 2 is approximately $r_2(1 - \alpha_{21}K_1/K_2)$, which is negative. We say that species 1 is the *strong*, and species 2 the *weak*, interspecific competitor.

Case 2: $K_1 < \alpha_{12}K_2$ and $K_2 > \alpha_{21}K_1$ This is the same as case 1, but with the roles of species 1 and 2 interchanged. That is, species 2 is now the strong, and species 1 is the weak, interspecific competitor. We see from the direction field in Figure 11.50 that species 2 outcompetes species 1 and drives species 1 to extinction. The equilibrium $(K_1, 0)$ is therefore unstable, and the equilibrium $(0, K_2)$ is locally stable.

Case 3: $K_1 > \alpha_{12}K_2$ and $K_2 > \alpha_{21}K_1$ The two inequalities imply that each species can invade the other species. When we look at the direction field of Figure 11.51, we see that the interior equilibrium representing the fact that both species are present is locally stable. We say that **coexistence** is possible. The two monoculture equilibria $(K_1, 0)$ and $(0, K_2)$ are unstable.

Case 4: $K_1 < \alpha_{12} K_2$ and $K_2 < \alpha_{21} K_1$ In this case, neither species can invade the other. When we look at the direction field of Figure 11.52, we see that the interior equilibrium is a saddle point and, hence, unstable. The outcome of competition depends on the initial densities. For instance, if the densities of N_1 and N_2 are given initially by the point A in the figure, then species 1 will win and species 2 will become extinct (following the direction of the direction vectors in the region that contains the point A). If, however, the densities of N_1 and N_2 are given initially by the point B in the figure, then species 2 will win and species 1 will become extinct. Since the outcome of competition depends on initial abundances, we refer to this scenario as **founder control**. In this scenario, both monoculture equilibria $(K_1, 0)$ and $(0, K_2)$ are locally stable.

We see from the preceding analysis that the Lotka–Volterra model allows for three possible outcomes in a two-species interaction. Cases 1 and 2 show the possibility of competitive exclusion. Case 3 shows that coexistence is possible. Case 4, founder control, shows that, depending on the initial abundances, one or the other species eventually wins.

Although we stated that the system (11.57) describes a highly idealized situation of competition, there is a famous example that fits the equations very well. The Russian ecologist G. F. Gause (1934) studied competition between species of the protozoan *Paramecium*. When *P. aurelia* and *P. caudatum* were grown together, *P. aurelia* competitively excluded *P. caudatum*. When *P. caudatum* and *P. bursaria* were grown together, they coexisted stably . Solution curves from (11.57) were fitted to both sets of data by estimating the relevant parameters in (11.57), and an excellent fit was obtained.

We will now turn to using eigenvalues to analyze (11.57).

Eigenvalues We will first determine all possible equilibria. Setting $dN_1/dt = 0$, we find that

$$N_1 = 0 \quad \text{or} \quad N_1 + \alpha_{12} N_2 = K_1$$

Setting $dN_2/dt = 0$, we find that

$$N_2 = 0 \quad \text{or} \quad \alpha_{21} N_1 + N_2 = K_2$$

There are four possible combinations:

1. The equilibrium $(\hat{N}_1, \hat{N}_2) = (0, 0)$ represents the case when both species are absent. This is the trivial equilibrium.

2. The equilibrium $(\hat{N}_1, \hat{N}_2) = (K_1, 0)$ represents the case when species 2 is absent and species 1 is at its carrying capacity K_1.

3. The equilibrium $(\hat{N}_1, \hat{N}_2) = (0, K_2)$ represents the case when species 1 is absent and species 2 is at its carrying capacity K_2.

4. The fourth equilibrium is obtained by simultaneously solving

$$N_1 + \alpha_{12} N_2 = K_1$$
$$\alpha_{21} N_1 + N_2 = K_2$$

and requiring that both solutions be positive.

To analyze the stability of these equilibria, we must compute the Jacobi matrix associated with (11.57). We find that

$$D\mathbf{f}(N_1, N_2) = \begin{bmatrix} r_1 - 2\frac{r_1}{K_1} N_1 - \frac{r_1 \alpha_{12}}{K_1} N_2 & -\frac{r_1 \alpha_{12}}{K_1} N_1 \\ -\frac{r_2 \alpha_{21}}{K_2} N_2 & r_2 - 2\frac{r_2}{K_2} N_2 - \frac{r_2 \alpha_{21}}{K_2} N_1 \end{bmatrix}$$

We look at each equilibrium separately:

1. Evaluating the Jacobi matrix at the trivial equilibrium $(\hat{N}_1, \hat{N}_2) = (0, 0)$, we obtain

$$Df(0, 0) = \begin{bmatrix} r_1 & 0 \\ 0 & r_2 \end{bmatrix}$$

which is a diagonal matrix; the eigenvalues are therefore

$$\lambda_1 = r_1 \quad \text{and} \quad \lambda_2 = r_2$$

Since $r_1 > 0$ and $r_2 > 0$, the equilibrium $(0, 0)$ is unstable.

2. Evaluating the Jacobi matrix at the equilibrium $(\hat{N}_1, \hat{N}_2) = (K_1, 0)$, we obtain

$$Df(K_1, 0) = \begin{bmatrix} -r_1 & -r_1\alpha_{12} \\ 0 & r_2\left(1 - \alpha_{21}\frac{K_1}{K_2}\right) \end{bmatrix}$$

Because $Df(K_1, 0)$ is an upper triangular matrix, we can immediately read off the eigenvalues:

$$\lambda_1 = -r_1 \quad \text{and} \quad \lambda_2 = r_2\left(1 - \alpha_{21}\frac{K_1}{K_2}\right)$$

Since $r_1 > 0$, it follows that $\lambda_1 < 0$. The eigenvalue $\lambda_2 < 0$ when $K_2 < \alpha_{21}K_1$. Therefore, the equilibrium

$$(K_1, 0) \quad \text{is} \quad \begin{cases} \text{locally stable} & \text{for } K_2 < \alpha_{21}K_1 \\ \text{unstable} & \text{for } K_2 > \alpha_{21}K_1 \end{cases}$$

3. Evaluating the Jacobi matrix at the equilibrium $(\hat{N}_1, \hat{N}_2) = (0, K_2)$, we find that

$$Df(0, K_2) = \begin{bmatrix} r_1\left(1 - \alpha_{12}\frac{K_2}{K_1}\right) & 0 \\ -r_2\alpha_{21} & -r_2 \end{bmatrix}$$

Because $Df(0, K_2)$ is a lower triangular matrix, we can immediately read off the eigenvalues:

$$\lambda_1 = r_1\left(1 - \alpha_{12}\frac{K_2}{K_1}\right) \quad \text{and} \quad \lambda_2 = -r_2$$

Since $r_2 > 0$, it follows that $\lambda_2 < 0$. The eigenvalue $\lambda_1 < 0$ when $K_1 < \alpha_{12}K_2$. Therefore, the equilibrium

$$(0, K_2) \quad \text{is} \quad \begin{cases} \text{locally stable} & \text{for } K_1 < \alpha_{12}K_2 \\ \text{unstable} & \text{for } K_1 > \alpha_{12}K_2 \end{cases}$$

4. We stated that the fourth equilibrium can be obtained by simultaneously solving

$$N_1 + \alpha_{12}N_2 = K_1$$
$$\alpha_{21}N_1 + N_2 = K_2$$

Using the standard method of elimination, we get

$$N_1 + \alpha_{12}N_2 = K_1$$
$$(\alpha_{21}\alpha_{12} - 1)N_2 = \alpha_{21}K_1 - K_2$$

Therefore,

$$\hat{N}_2 = \frac{\alpha_{21}K_1 - K_2}{\alpha_{21}\alpha_{12} - 1}$$

and

$$\hat{N}_1 = K_1 - \alpha_{12}\frac{\alpha_{21}K_1 - K_2}{\alpha_{21}\alpha_{12} - 1} = \frac{\alpha_{12}K_2 - K_1}{\alpha_{21}\alpha_{12} - 1}$$

We require that both $\hat{N}_1$ and $\hat{N}_2$ be positive. (After all, these are population densities.) That is, we require that

$$\frac{\alpha_{21}K_1 - K_2}{\alpha_{21}\alpha_{12} - 1} > 0 \quad \text{and} \quad \frac{\alpha_{12}K_2 - K_1}{\alpha_{21}\alpha_{12} - 1} > 0 \qquad (11.60)$$

If $\alpha_{21}\alpha_{12} > 1$, then (11.60) reduces to

$$K_2 < \alpha_{21}K_1 \quad \text{and} \quad K_1 < \alpha_{12}K_2 \qquad (11.61)$$

If $\alpha_{21}\alpha_{12} < 1$, then (11.60) reduces to

$$K_2 > \alpha_{21}K_1 \quad \text{and} \quad K_1 > \alpha_{12}K_2 \qquad (11.62)$$

It turns out that finding the eigenvalues associated with this nontrivial equilibrium is algebraically rather involved if we use the Jacobi matrix associated with the original system of differential equations (11.57). However, if we investigate the effect of a perturbation directly, then the analysis becomes manageable. To demonstrate what we mean by this, we set

$$z_1 = N_1 - \hat{N}_1 \quad \text{and} \quad z_2 = N_2 - \hat{N}_2$$

Then (z_1, z_2) represents the deviation from the equilibrium. First, observe that

$$\frac{dz_1}{dt} = \frac{dN_1}{dt} \quad \text{and} \quad \frac{dz_2}{dt} = \frac{dN_2}{dt}$$

Substituting $z_1 + \hat{N}_1$ for N_1 and $z_2 + \hat{N}_2$ for N_2 in (11.57) yields

$$\frac{dz_1}{dt} = r_1(z_1 + \hat{N}_1)\left(1 - \frac{z_1 + \hat{N}_1}{K_1} - \alpha_{12}\frac{z_2 + \hat{N}_2}{K_1}\right)$$

$$= r_1(z_1 + \hat{N}_1)\left(\underbrace{1 - \frac{\hat{N}_1}{K_1} - \alpha_{12}\frac{\hat{N}_2}{K_1}}_{=0} - \frac{z_1}{K_1} - \alpha_{12}\frac{z_2}{K_1}\right)$$

and

$$\frac{dz_2}{dt} = r_2(z_2 + \hat{N}_2)\left(1 - \frac{z_2 + \hat{N}_2}{K_2} - \alpha_{21}\frac{z_1 + \hat{N}_1}{K_2}\right)$$

$$= r_2(z_2 + \hat{N}_2)\left(\underbrace{1 - \frac{\hat{N}_2}{K_2} - \alpha_{21}\frac{\hat{N}_1}{K_2}}_{=0} - \frac{z_2}{K_2} - \alpha_{21}\frac{z_1}{K_2}\right)$$

Instead of analyzing the Jacobi matrix associated with (11.57) evaluated at the nontrivial equilibrium $(\hat{N}_1, \hat{N}_1)$, we investigate the equilibrium $(z_1, z_2) = (0, 0)$ of the new system

$$\frac{dz_1}{dt} = -\frac{r_1}{K_1}(z_1 + \hat{N}_1)(z_1 + \alpha_{12}z_2)$$

$$\frac{dz_2}{dt} = -\frac{r_2}{K_2}(z_2 + \hat{N}_2)(z_2 + \alpha_{21}z_1)$$

The Jacobi matrix is

$$J(z_1, z_2) =$$

$$\begin{bmatrix} -\frac{r_1}{K_1}(z_1 + \alpha_{12}z_2) - \frac{r_1}{K_1}(z_1 + \hat{N}_1) & -\frac{r_1\alpha_{12}}{K_1}(z_1 + \hat{N}_1) \\ -\frac{r_2\alpha_{21}}{K_2}(z_2 + \hat{N}_2) & -\frac{r_2}{K_2}(z_2 + \alpha_{21}z_2) - \frac{r_2}{K_2}(z_2 + \hat{N}_2) \end{bmatrix}$$

Evaluating the matrix $J(z_1, z_2)$ at $(z_1, z_2) = (0, 0)$, we obtain

$$J(0, 0) = \begin{bmatrix} -\frac{r_1}{K_1}\hat{N}_1 & -\frac{r_1\alpha_{12}}{K_1}\hat{N}_1 \\ -\frac{r_2\alpha_{21}}{K_2}\hat{N}_2 & -\frac{r_2}{K_2}\hat{N}_2 \end{bmatrix}$$

Now,

$$\mathrm{tr}(J(0, 0)) = -\frac{r_1\hat{N}_1}{K_1} - \frac{r_2\hat{N}_2}{K_2}$$

$$\det(J(0, 0)) = \frac{r_1 r_2}{K_1 K_2}\hat{N}_1\hat{N}_2(1 - \alpha_{12}\alpha_{21})$$

The nontrivial equilibrium $(\hat{N}_1, \hat{N}_2)$ satisfies $\hat{N}_1 > 0$ and $\hat{N}_2 > 0$, implying that $\mathrm{tr}(J(0, 0)) < 0$. Since

$$\det(J(0, 0)) > 0 \quad \text{when } 1 - \alpha_{12}\alpha_{21} > 0$$

it follows that $(\hat{N}_1, \hat{N}_2)$ is unstable if (11.61) holds and is locally stable if (11.62) holds.

When we compare the conditions in Case 4 with those in Cases 2 and 3, we see that the interior equilibrium is locally stable when the two boundary (monoculture) equilibria $(K_1, 0)$ and $(0, K_2)$ are unstable and that the interior equilibrium is unstable when the two boundary equilibria are locally stable. We can also show that both eigenvalues are always real; that is, we do not expect oscillations.

Comparing the two approaches, we find the same results. The graphical approach gives qualitative answers only, albeit rather useful ones. The graphical approach also turns out to be easier. The eigenvalue approach allows us to obtain quantitative answers in terms of eigenvalues, which tell us something about how quickly the system returns to a stable equilibrium after a small perturbation. Such quantitative answers are important if we want to know how the system responds to small perturbations.

■ 11.4.2 A Predator–Prey Model

In this subsection, we investigate a simple model for predation, defined as the consumption of prey by a predator. We assume that the prey is alive when the predator attacks it and that the predator kills its prey and thus removes it from the population.

There are many patterns of abundance that result from predator–prey interactions—most notably, those situations in nature in which predator and prey abundances appear to be closely linked and show periodic fluctuations. A frequently cited example is the Canada lynx (*Lynx canadensis*) and snowshoe hare (*Lepus americanus*) system. The population abundances show periodic oscillations. A laboratory example of coupled predator–prey oscillations is the azuki bean weevil (*Callosbruchus chinensis*) and its parasitoid wasp (*Heterospilus prosopidis*).

The simplest predator–prey model that exhibits coupled oscillations is the Lotka–Volterra model (Lotka, 1932; Volterra, 1926). We describe this system in Volterra's own words:

The first case I have considered is that of two associated species, of which one, finding sufficient food in its environment, would multiply indefinitely when left to itself, while the other would perish for lack of nourishment if left alone; but the second feeds upon the first, and so the two species can coexist together.

The proportional rate of increase of the eaten species diminishes as the number of individuals of the eating species increases, while augmentation of the eating species increases with the increase of the number of individuals of the eaten species.

We will now translate this verbal formulation of the model into a system of differential equations. If we denote the abundance of the prey by $N(t)$ and the abundance of the predator by $P(t)$, then the model is given by the system of differential equations

$$\frac{dN}{dt} = rN(t) - aP(t)N(t)$$

$$\frac{dP}{dt} = baP(t)N(t) - dP(t)$$

(11.63)

where $r, a, b,$ and d are positive constants. Note that the prey increases exponentially in the absence of the predator ($P = 0$). The intrinsic rate of increase of the prey in the absence of the predator is r. The constant a denotes the attack rate, and the term aPN is the consumption rate of prey. Predators decline exponentially in the absence of prey; the rate of decline is given by the constant d. The consumption of prey results in an increase in predator abundance. The constant b describes how efficiently predator turn prey into predator offspring.

To find equilibria of (11.63), we set

$$N(r - aP) = 0$$
$$P(abN - d) = 0$$

This system yields the zero isoclines $N = 0$ and $P = r/a$ for $dN/dt = 0$ and $P = 0$ and $N = d/ab$ for $dP/dt = 0$. The zero isoclines are shown in Figure 11.53. The points of intersection of the zero isoclines for N and P are equilibria. We find that

$$(\hat{N}, \hat{P}) = (0, 0) \qquad \text{and} \qquad (\hat{N}, \hat{P}) = \left(\frac{d}{ab}, \frac{r}{a}\right)$$

To analyze the stability of these two equilibria, we linearize (11.63). The Jacobi matrix is given by

$$\mathbf{Df}(N, P) = \begin{bmatrix} r - aP & -aN \\ baP & baN - d \end{bmatrix}$$

When $(\hat{N}, \hat{P}) = (0, 0)$,

$$\mathbf{Df}(0, 0) = \begin{bmatrix} r & 0 \\ 0 & -d \end{bmatrix}$$

This is a diagonal matrix; hence, the eigenvalues are given by the diagonal elements

$$\lambda_1 = r > 0 \qquad \text{and} \qquad \lambda_2 = -d < 0$$

We conclude that $(0, 0)$ is unstable.

When $(\hat{N}, \hat{P}) = (\frac{d}{ab}, \frac{r}{a})$,

$$\mathbf{Df}\left(\frac{d}{ab}, \frac{r}{a}\right) = \begin{bmatrix} 0 & -\frac{d}{b} \\ rb & 0 \end{bmatrix}$$

To determine the eigenvalues of $\mathbf{Df}(\frac{d}{ab}, \frac{r}{a})$, we must solve

$$\det \begin{bmatrix} -\lambda & -\frac{d}{b} \\ rb & -\lambda \end{bmatrix} = \lambda^2 + rd = 0$$

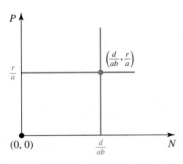

Figure 11.53   Zero isoclines.

Solving this equation, we find that

$$\lambda_1 = i\sqrt{rd} \quad \text{and} \quad \lambda_2 = -i\sqrt{rd}$$

That is, both eigenvalues are purely imaginary. Thus, we cannot determine the stability of this equilibrium by linearizing about the equilibrium, as was pointed out in Section 11.3. Fortunately, however, we can solve (11.63) exactly.

To solve (11.63) exactly, we divide dP/dt by dN/dt:

$$\frac{dP/dt}{dN/dt} = \frac{dP}{dN} = \frac{P(abN - d)}{N(r - aP)}$$

Separating variables and integrating, we obtain

$$\int \frac{r - aP}{P}\, dP = \int \frac{abN - d}{N}\, dN$$

Carrying out the integration gives

$$r \ln P - aP = abN - d \ln N + C$$

where C is the constant of integration. Rearranging terms and exponentiating yields

$$\left(N^d e^{-abN}\right)\left(P^r e^{-aP}\right) = K$$

where $K = e^C$ depends on the initial condition. We define the function

$$f(N, P) = \left(N^d e^{-abN}\right)\left(P^r e^{-aP}\right)$$

and set

$$g(N) = N^d e^{-abN} \quad \text{and} \quad h(P) = P^r e^{-aP}$$

Then we can show that $g(N)$ has its absolute maximum when $N = d/ab$ and $h(P)$ has its absolute maximum when $P = r/a$. The function $f(N, P)$ thus takes on its absolute maximum at the equilibrium point $(d/ab, r/a)$. We can therefore define level curves

$$f(N, P) = K$$

for $K \leq K_{max}$, where K_{max} is the value of f at the equilibrium $(d/ab, r/a)$. Although we will not be able to demonstrate it here, these level curves are closed curves. (We show such level curves in Figure 11.54, to convince you that they are indeed closed.) The level curves are solution curves.

Solutions of $N(t)$ and $P(t)$ as functions of time corresponding to two of the closed curves in Figure 11.54 are shown in Figures 11.55 and 11.56, respectively. When we plot $N(t)$ and $P(t)$ versus t, we see that the closed trajectories in the N–P plane correspond to periodic solutions for the predator and the prey. The amplitudes of the oscillations depend on the initial condition. Note that the amplitudes in the two figures are different.

The closed trajectories shown in Figures 11.55 and 11.56 are not stable under perturbations. That is, if a small perturbation changes the value of N or P, the solution will follow a *different* closed trajectory. This property is a major drawback of the model; it implies that if a natural population actually followed this simple model, its abundance would not exhibit regular cycles, because external factors would constantly shift the population to different trajectories. If a natural population exhibits regular cycles, we would expect these cycles to be stable; that is, the population would return to the same cycle after a small perturbation. Such cycles are called **stable limit cycles**. A locally stable equilibrium that is approached by oscillations can be obtained by modifying the original Lotka–Volterra model to include a nonlinear predator response to prey abundance. Such a modification is discussed in Problems 19 through 21.

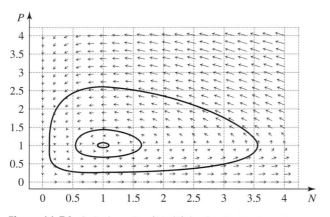

Figure 11.54   Solutions for (11.63) in the N–P plane.

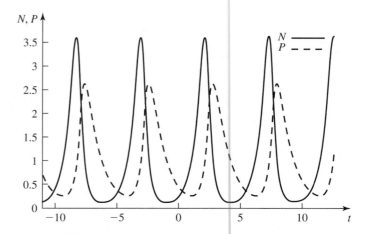

Figure 11.55   Solutions for (11.63) as functions of time.

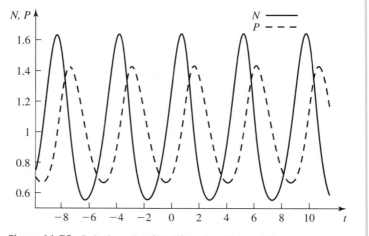

Figure 11.56   Solutions for (11.63) as functions of time.

■ 11.4.3 The Community Matrix

In this subsection, we consider a fairly general multispecies population model, initiated by Levins (1970) and further developed by May (1975). The goal is to determine how interactions between pairs of species influence the stability of the equilibria of an assemblage of species.

   We assume an assemblage of two species in which the abundance of species i at time t is given by $N_i(t)$. (At the end of this subsection, we will generalize this binary situation to an assemblage of m species.) Suppose that the following set of differential equations describes the dynamics of our two-species assemblage:

$$\frac{dN_1}{dt} = f_1(N_1(t), N_2(t))$$

$$\frac{dN_2}{dt} = f_2(N_1(t), N_2(t))$$

(11.64)

In vector notation,

$$\frac{d\mathbf{N}}{dt} = \mathbf{f}(\mathbf{N}(t))$$

(11.65)

To determine equilibria, we must solve the system of equations

$$f_1(N_1, N_2) = 0$$
$$f_2(N_1, N_2) = 0$$

(11.66)

We assume that (11.66) has a nontrivial solution $\hat{\mathbf{N}} = (\hat{N}_1, \hat{N}_2)$ with $\hat{N}_1 > 0$ and $\hat{N}_2 > 0$. We can now proceed to determine its stability. To do so, we must evaluate the Jacobi matrix associated with the system (11.64) at the equilibrium $(\hat{N}_1, \hat{N}_2)$. The Jacobi matrix at the equilibrium $\hat{\mathbf{N}}$ is given by

$$
D\mathbf{f}(\hat{\mathbf{N}}) = \begin{bmatrix} \dfrac{\partial f_1(\hat{N}_1, \hat{N}_2)}{\partial N_1} & \dfrac{\partial f_1(\hat{N}_1, \hat{N}_2)}{\partial N_2} \\[2mm] \dfrac{\partial f_2(\hat{N}_1, \hat{N}_2)}{\partial N_1} & \dfrac{\partial f_2(\hat{N}_1, \hat{N}_2)}{\partial N_2} \end{bmatrix}
$$

This 2×2 matrix is called the **community matrix**. Its elements

$$
a_{ij} = \frac{\partial f_i(\hat{N}_1, \hat{N}_2)}{\partial N_j}
$$

describe the effect of species j on species i at equilibrium, because the partial derivative $\frac{\partial f_i}{\partial N_j}$ tells us how the function f_i, which describes the growth of species i, changes when the abundance of species j changes.

The diagonal elements $a_{ii} = \frac{\partial f_i}{\partial N_i}$ measure the effect species i has on itself, whereas the off-diagonal elements $a_{ij} = \frac{\partial f_i}{\partial N_j}, i \neq j$, measure the effect species j has on species i. The signs of the elements a_{ij} thus tell us something about the pairwise effects the species in this assemblage have on each other at equilibrium.

The quantity a_{ij} can be negative, 0, or positive. If $a_{ij} < 0$, then the growth rate of species i is decreased if species j increases its abundance; we therefore say that species j has a negative, or *inhibitory*, effect on species i. If $a_{ij} = 0$, then changes in the abundance of species j have no effect on the growth rate of species i. If $a_{ij} > 0$, then the growth rate of species i is increased if species j increases its abundance; we then say that species j has a positive, or *facilitory*, effect on species i.

We look at the possible combinations of the pair (a_{21}, a_{12}). This pair describes the interactions *between* the two species in the assemblage. The following table lists all possible combinations:

		a_{12}		
		+	**0**	**−**
	+	$++$	$+0$	$--$
a_{21} **0**		$0+$	00	$0-$
	−	$-+$	-0	$--$

To interpret the table, take the pair (00), for instance. The pair (00) represents the case in which neither species has an effect on the other species at equilibrium. With another pair—for instance, $(0+)-a_{21} = 0$ and species 1 has no effect on species 2, but $a_{12} > 0$ and species 2 has a positive effect on species 1.

The case (00) is the simplest, and we will discuss it first. The community matrix in this case is

$$
\begin{bmatrix} a_{11} & 0 \\ 0 & a_{22} \end{bmatrix}
$$

Since the community matrix is in diagonal form, its eigenvalues are the diagonal elements a_{11} and a_{22}. Hence, the equilibrium $(\hat{N}_1, \hat{N}_2)$ is stable, provided that both a_{11} and a_{22} are negative. That is, if neither species has an effect on the other species, a locally stable nontrivial equilibrium in which both species coexist exists only if they each have a negative effect on themselves; this means that each species needs to regulate its own population size.

Following May (1975), the remaining eight combinations in the table can be categorized into five biologically distinct types of interactions:

Mutualism, or symbiosis (++): Each species has a positive effect on the other.

Competition (−−): Each species has a negative effect on the other.

Commensalism (+0): One species benefits from the interaction, whereas the other is unaffected.

Amensalism (−0): One species is harmed by the interaction, whereas the other is unaffected.

Predation (+−): One species benefits, whereas the other is harmed.

We will now discuss the stability of the nontrivial equilibrium $(\hat{N}_1, \hat{N}_2)$ in all five cases. Recall that we assumed that this equilibrium exists and that both $\hat{N}_1 > 0$ and $\hat{N}_2 > 0$. The community matrix [the Jacobi matrix of (11.64) at equilibrium] is of the form

$$A = \begin{bmatrix} a_{11} & a_{12} \\ a_{21} & a_{22} \end{bmatrix}$$

Both eigenvalues of A have negative real parts if and only if

$$\operatorname{tr} A = a_{11} + a_{22} < 0 \quad \text{and} \quad \det A = a_{11}a_{22} - a_{12}a_{21} > 0$$

In what follows, we assume that

$$a_{11} < 0 \quad \text{and} \quad a_{22} < 0 \tag{11.67}$$

so that the first condition $\operatorname{tr} A < 0$ is automatically satisfied. This has the same interpretation as discussed in the case (00): that both species have a negative effect on themselves or regulate their own population densities.

We will now go through all five cases and determine under which conditions the nontrivial equilibrium is stable:

Mutualism We assume (11.67). The sign structure of the community matrix at equilibrium in the case of mutualism is then of the form

$$A = \begin{bmatrix} - & + \\ + & - \end{bmatrix}$$

Since

$$\det A = \underbrace{a_{11}a_{22}}_{>0} - \underbrace{a_{12}a_{21}}_{>0}$$

the determinant of A may be either positive or negative. If $a_{12}a_{21}$ is sufficiently small compared with $a_{11}a_{22}$, then $\det A > 0$ and the equilibrium $(\hat{N}_1, \hat{N}_2)$ is locally stable. In other words, if the positive effects of the species on each other are sufficiently counteracted by their own population control (represented by a_{11} and a_{22}), then the equilibrium is locally stable.

Competition Again, we assume (11.67). The sign structure of the community matrix at equilibrium in the case of competition is then of the form

$$A = \begin{bmatrix} - & - \\ - & - \end{bmatrix}$$

Since

$$\det A = \underbrace{a_{11}a_{22}}_{>0} - \underbrace{a_{12}a_{21}}_{>0}$$

the determinant of A may be either positive or negative. Now, the equilibrium $(\hat{N}_1, \hat{N}_2)$ is locally stable if the negative effects each species has on the other are smaller than the effects each species has on itself. (In this case, $\det A > 0$ and, hence, the equilibrium is locally stable.)

Commensalism and Amensalism Once more, assume (11.67). The sign structure of the community matrices at equilibrium are then of the form

$$A = \begin{bmatrix} - & 0 \\ + & - \end{bmatrix} \quad \text{or} \quad A = \begin{bmatrix} - & 0 \\ - & - \end{bmatrix}$$

In either case, the determinant is positive; therefore, the equilibrium is locally stable.

Predation Yet again, we assume (11.67). The sign structure of the community matrix at equilibrium in the case of predation is then of the form

$$A = \begin{bmatrix} - & - \\ + & - \end{bmatrix}$$

and det A is always positive. That is, provided that a nontrivial equilibrium exists, it is locally stable.

The Multispecies Case We will now briefly look at the case in which more than two species form an assemblage. Suppose there are m species. We denote the density of species i at time t by $N_i(t)$. The dynamics are described by the following system of differential equations:

$$\frac{dN_1}{dt} = f_1(N_1(t), N_2(t), \ldots, N_m(t))$$

$$\frac{dN_2}{dt} = f_2(N_1(t), N_2(t), \ldots, N_m(t))$$

$$\vdots$$

$$\frac{dN_m}{dt} = f_m(N_1(t), N_2(t), \ldots, N_m(t))$$

The equilibria are found by solving the following system of equations:

$$f_1(\hat{N}_1, \hat{N}_2, \ldots, \hat{N}_m) = 0$$

$$f_2(\hat{N}_1, \hat{N}_2, \ldots, \hat{N}_m) = 0$$

$$\vdots$$

$$f_m(\hat{N}_1, \hat{N}_2, \ldots, \hat{N}_m) = 0$$

If we assume that $\hat{\mathbf{N}} = (\hat{N}_1, \hat{N}_2, \ldots, \hat{N}_m)$ is an equilibrium, then the community matrix at equilibrium is the Jacobi matrix at equilibrium, given by

$$D\mathbf{f}(\hat{\mathbf{N}}) = \begin{bmatrix} \frac{\partial f_1}{\partial N_1} & \cdots & \frac{\partial f_1}{\partial N_m} \\ \vdots & & \vdots \\ \frac{\partial f_m}{\partial N_1} & \cdots & \frac{\partial f_m}{\partial N_m} \end{bmatrix}$$

As in the case of two species, the element

$$a_{ij} = \frac{\partial f_i(\hat{\mathbf{N}})}{\partial N_j}$$

describes the effect of species j on species i at equilibrium.

■ 11.4.4 A Mathematical Model for Neuron Activity

The nervous system of an organism is a communication network that allows the rapid transmission of information between cells. The nervous system consists of nerve cells called **neurons**. A typical neuron has a cell body that contains the cell nucleus and nerve fibers. Nerve fibers that receive information are called **dendrites**, whereas those that transport information are called **axons**; the latter provide links to other neurons via **synapses**. A typical vertebrate neuron is shown in Figure 11.57.

Neurons respond to electrical stimuli, a property that is exploited by scientists studying them. When the cell body of an isolated neuron is stimulated with a very mild electrical shock, the neuron shows no response; increasing the intensity of

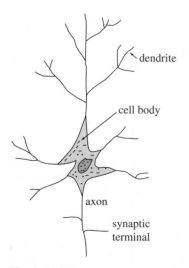

Figure 11.57 A typical vertebrate neuron.

the shock beyond a certain threshold, however, will trigger a response, namely, an impulse that travels along the axon. Increasing the intensity of the electrical shock further does not change the response. The impulse is thus an all-or-nothing response.

A. L. Hodgkin and A. F. Huxley studied the giant axon of a squid experimentally and developed a mathematical model for neuron activity. Their work, which appeared in a series of papers in 1952, is an excellent example of how experimental and theoretical research can be combined to gain a thorough understanding of a natural system. In 1963, Hodgkin and Huxley were awarded the Nobel Prize in Physiology or Medicine for their work on neurons.

Let us briefly examine how a neuron works. The main players in the functioning of a neuron are sodium (Na^+) and potassium (K^+) ions. The cell membrane of a neuron is impermeable to these ions when the cell is in a resting state. In a typical neuron in its resting state, the concentration of Na^+ in the interior of the cell is about one-tenth of the extracellular concentration of Na^+ and the concentration of K^+ in the interior of the cell is about 30 times the extracellular concentration of K^+. When the neuron is in its resting state, the interior of the cell is negatively charged (at -70 mV) relative to the exterior of the cell.

When a nerve cell is stimulated, its surface becomes permeable to Na^+ ions, which rush into the cell through sodium channels in the surface. This influx of Na^+ ions results in a reversal of polarization at the points where those ions entered the cell. The surface inside the cell is now positively charged relative to the outside of the cell and becomes permeable to K^+ ions, which rush outside through potassium channels. Because the potassium ions (K^+) are positively charged and move from the inside to the outside of the cell, the polarization at the surface of the cell is again reversed and is now below the polarization of the resting cell. To restore the original concentration of Na^+ and K^+ (and thus the original polarization), energy must be expended to run the so-called sodium and potassium pumps on the surface of the cell to pump the excess Na^+ from the interior to the exterior of the cell and to pump K^+ from the exterior to the interior of the cell.

To trigger such a reaction, the intensity of the stimulus must be above a certain threshold. The reaction described occurs locally on the surface of the cell. This large local change in polarization triggers the same reaction in the neighborhood of the cells, allowing the reaction to propagate along the nerve cell and thus creating the observed impulse that travels along the cell. The local change in polarization, called an **action potential**, is then reversed to the original polarization. An action potential is illustrated in Figure 11.58.

Figure 11.58 The action potential.

Hodgkin and Huxley measured sodium and potassium conductance and fitted curves to their data. The sodium curve is fitted by a cubic function, the potassium curve by a quartic function. Hodgkin and Huxley then developed a model for the action potential. The model, which consists of a system of four autonomous differential equations, is a phenomenological model; that is, the equations are based on fitting curves to experimental data for the various components of the model. One equation describes the change of voltage on the cell surface, two equations describe the sodium channel, and one equation describes the potassium channel. We will not give the details of this system, as it is far too complicated to analyze. (It is typically solved numerically.) Instead, we will present a simplified version of the model developed by Fitzhugh (1961) and Nagumo et al. (1962).

The Fitzhugh–Nagumo model is based on the fact that the time scales of the two channels are quite different. The sodium channel works on a much faster time scale than the potassium channel. This fact led Fitzhugh and Nagumo to assume that the sodium channel is essentially always in a steady state, an assumption that allowed them to reduce the four equations of the Hodgkin and Huxley model to two. The Fitzhugh–Nagumo model is thus an approximation to the Hodgkin and Huxley model, retaining the essential features of the action potential, but much easier to analyze.

The Fitzhugh–Nagumo model is described by two variables. One variable, denoted by V, describes the potential of the cell surface. The other variable, denoted by w, models the sodium and potassium channels. The equations are

$$\frac{dV}{dt} = -V(V-a)(V-1) - w$$

$$\frac{dw}{dt} = b(V - cw)$$

(11.68)

where a, b, and c are constants that satisfy $0 < a < 1, b > 0$, and $c > 0$.

We will analyze the system graphically. The zero isoclines of (11.68) are given by

$$w = -V(V-a)(V-1) \qquad \text{and} \qquad w = \frac{1}{c}V$$

The important feature of this model is that the zero isocline $dV/dt = 0$ is the graph of a cubic function in the $V-w$ plane. The zero isoclines are illustrated in Figure 11.59.

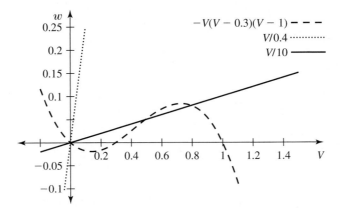

Figure 11.59 The zero isoclines of (11.68).

We see that if c is small, there is just one equilibrium, namely, $(0, 0)$, whereas when c is sufficiently large, the line $w = V/c$ intersects the graph of $dV/dt = 0$ three times.

The threshold phenomenon and the action potential are observed when c is small. In this case, there is just one equilibrium, again, $(0, 0)$, that corresponds to the resting state.

We can analyze the stability of $(0, 0)$ by linearizing the system about this equilibrium. We find that

$$\mathbf{Df}(V, w) = \begin{bmatrix} -3V^2 + 2V + 2aV - a & -1 \\ b & -bc \end{bmatrix}$$

Hence,

$$\mathbf{Df}(0, 0) = \begin{bmatrix} -a & -1 \\ b & -bc \end{bmatrix}$$

To find the eigenvalues, we compute

$$\det \begin{bmatrix} -a - \lambda & -1 \\ b & -bc - \lambda \end{bmatrix} = (-a - \lambda)(-bc - \lambda) + b = 0$$

That is, we must solve

$$\lambda^2 + (a + bc)\lambda + b(ac + 1) = 0$$

which has solutions

$$\lambda_{1,2} = \frac{-(a + bc) \pm \sqrt{(a + bc)^2 - 4b(ac + 1)}}{2}$$

$$= \frac{-(a + bc) \pm \sqrt{(a - bc)^2 - 4b}}{2}$$

Since a, b, and c are positive constants, it follows that the expression under the square root—that is, $(a + bc)^2 - 4b(ac + 1)$—is smaller than $(a + bc)^2$. Therefore, both λ_1 and λ_2 have negative real parts, which implies that $(0, 0)$ is locally stable. As long as $(a - bc)^2 > 4b$, the equilibrium is a stable sink. When $(a - bc)^2 < 4b$, the eigenvalues are complex conjugates and $(0, 0)$ becomes a stable spiral.

The system mimics the action potential when both eigenvalues are real and negative. If we apply a weak stimulus (i.e., if we increase V to a value less than a), then V will quickly return to 0. However, if we apply a strong enough stimulus [i.e., we let $V \in (a, 1)$], the trajectory will move away from the equilibrium point, as shown in Figure 11.60. A plot of voltage versus time reveals that if the stimulus is too weak, $V(t)$ will quickly return to the equilibrium, whereas if the stimulus is large enough, the solution curve of $V(t)$ resembles the action potential. In Figures 11.61 and 11.62, we present two solution curves $V(t)$. In either case, $w(0) = 0$. In Figure 11.61, $V(0) = 0.5 > a$, and we observe an action potential; in Figure 11.62, $V(0) = 0.2 < a$, and the initial stimulus is too small and dies away quickly.

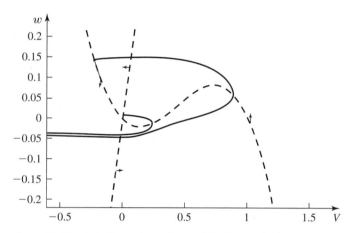

Figure 11.60 Solution curves for the Fitzhugh–Nagumo model.

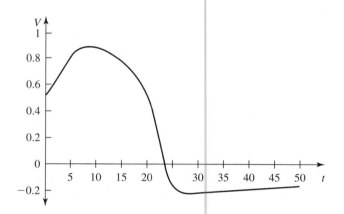

Figure 11.61 The action potential when $V(0) > a$.

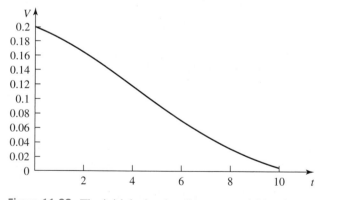

Figure 11.62 The initial stimulus dies away quickly when $V(0) < a$.

■ 11.4.5 A Mathematical Model for Enzymatic Reactions

The goal of this subsection is threefold. First, we will learn how to model biochemical reactions; second, we will see how mathematical models can be used to understand empirical observations; third, we will introduce the important idea that, by making appropriate assumptions, the number of variables in a model can sometimes be reduced, which typically facilitates the analysis of the model.

The class of biochemical reactions we will study are *enzymatic reactions*, which are ubiquitous in the living world. Enzymes are proteins that act as catalysts in chemical reactions by reducing the activation energy required to initiate the reaction.

Enzymes are not altered by the reaction; they aid in the initial steps of the reaction and control the rate of the reaction by binding the reactants (called substrates) to the active site of the enzyme, thus forming an enzyme–substrate complex that then allows the substrates to react and to form the product. This chain of steps is illustrated in Figure 11.63.

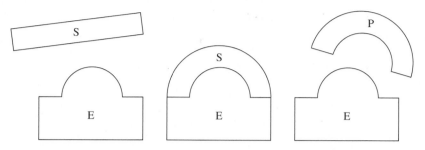

Figure 11.63 A schematic description of an enzymatic reaction.

If we denote the substrate by S, the enzyme by E, and the product of the reaction by P, then an enzymatic reaction can be described by

$$E + S \underset{k_-}{\overset{k_+}{\rightleftarrows}} ES \overset{k_2}{\longrightarrow} E + P \tag{11.69}$$

where k_+, k_-, and k_2 are the reaction rates in the corresponding reaction steps.

We can translate the schematic description of the reaction (11.69) into a system of differential equations. We use the following notation:

$$e(t) = [E] = \text{enzyme concentration at time } t$$
$$s(t) = [S] = \text{substrate concentration at time } t$$
$$c(t) = [ES] = \text{concentration of enzyme–substrate complex at time } t$$
$$p(t) = [P] = \text{product concentration at time } t$$

Using the mass action law gives

$$\frac{ds}{dt} = k_- c - k_+ se$$

$$\frac{de}{dt} = (k_- + k_2)c - k_+ se$$

$$\frac{dc}{dt} = k_+ se - (k_- + k_2)c \tag{11.70}$$

$$\frac{dp}{dt} = k_2 c$$

Michaelis and Menten (1913) were instrumental in the description of enzyme kinetics, through both experimental and theoretical work. On the experimental side, they developed techniques that allowed them to measure reaction rates under controlled conditions. Their experiments showed a hyperbolic relationship between the rate of the enzymatic reaction dp/dt and the substrate concentration s, which they described as

$$\frac{dp}{dt} = \frac{v_m s}{K_m + s} \tag{11.71}$$

where v_m is the saturation constant and K_m is the half-saturation constant (i.e., if $s = K_m$, then $dp/dt = v_m/2$).

On the theoretical side, they developed a mathematical model for enzyme kinetics that predicted the observed hyperbolic relationship between the substrate concentration and the initial rate at which the product is formed.

In what follows, we will analyze (11.70). This system of four equations is not easy to analyze, but we will simplify it, and in the end, we will obtain (11.71). We will also use (11.70) to illustrate that it is sometimes possible to reduce the number of equations in a system.

We claim that the system has a conserved quantity—that is, a quantity that does not depend on time and is therefore constant throughout the reaction. To find this quantity, note that

$$\frac{de}{dt} + \frac{dc}{dt} = 0$$

That is,

$$\frac{d}{dt}(e + c) = 0$$

which implies that

$$e(t) + c(t) = e_0 \tag{11.72}$$

where e_0 is a constant. Since $e(t) + c(t)$ is constant, we say that the sum $e(t) + c(t)$ is a conserved quantity. The advantage of having conserved quantities is that if we know the initial concentrations $e(0)$ and $c(0)$ and one of the two quantities $e(t)$ and $c(t)$, we immediately know the other, since $e(0) + c(0) = e(t) + c(t)$. This reduces the number of equations from four to three.

To reduce the number of equations even further, we make another assumption. Whereas the existence of a conserved quantity followed from the system of equations, and we could have obtained it without knowing the meaning of those equations, the next assumption requires a thorough understanding of the enzymatic reaction itself and cannot be deduced from the set of equations (11.70). Briggs and Haldane (1925), who had this understanding, proposed that the rate of formation balances the rate of breakdown of the complex; that is, they assumed that

$$\frac{dc}{dt} = 0$$

This assumption yields the equation

$$0 = k_+ se - (k_- + k_2)c$$

which can be rewritten as

$$\frac{se}{c} = \frac{k_- + k_2}{k_+}$$

We denote this ratio by K_m; that is,

$$K_m = \frac{k_- + k_2}{k_+}$$

and, therefore,

$$\frac{se}{c} = K_m \tag{11.73}$$

Solving (11.72) for e—that is, $e = e_0 - c$—and substituting the result into (11.73), we find that

$$\frac{s(e_0 - c)}{c} = K_m$$

which, when we solve for c, yields

$$c = \frac{e_0 s}{K_m + s} \tag{11.74}$$

allowing us to rewrite the equation for the rate at which the product is formed. Since $dp/dt = k_2 c$, it follows from (11.74) that

$$\frac{dp}{dt} = \frac{k_2 e_0 s}{K_m + s} \tag{11.75}$$

We can interpret the factor $k_2 e_0$ as follows: If all of the enzyme is complexed with the substrate, then $e = 0$ and therefore $c = e_0$ (since $e + c = e_0$ is a conserved quantity). This implies that the rate at which the product is formed, $dp/dt = k_2 c$, is fastest when $c = e_0$, in which case $dp/dt = k_2 e_0$. We can therefore interpret $k_2 e_0$ as the maximum rate at which this reaction can proceed. We introduce the notation

$$v_m = k_2 e_0$$

and rewrite (11.75) as

$$\frac{dp}{dt} = \frac{v_m s}{K_m + s} \tag{11.76}$$

This equation is known as the Michaelis–Menten law and describes the velocity of an enzymatic reaction. We see from Equation (11.76) that the reaction rate dp/dt is limited by the availability of the substrate S.

Equations (11.76) and (11.71) are the same. Whereas (11.71) was derived from fitting a curve to data points that related the measured substrate concentration s to the velocity dp/dt of the reaction, (11.76) is derived from a mathematical model. The mathematical model allows us to interpret the constants v_m and K_m in terms of the enzymatic reaction, and the experiments allow us to measure v_m and K_m.

Microbial Growth in a Chemostat: An Application of Substrate-Limited Growth

A rather simplistic view of microbial growth is that microbes convert a substrate through enzymatic reactions into products that are then converted into microbial biomass. In what follows, we will investigate a mathematical model for microbial growth in a chemostat. The growth of the microbes will be limited by the availability of the substrate.

A chemostat is a growth chamber in which sterile medium with concentration s_0 of the substrate enters the chamber at a constant rate D. Air is pumped into the chamber to mix and aerate the culture. To keep the volume in the chamber constant, the content of the chamber is removed at the same rate D as new medium enters. A sketch of a chemostat is shown in Figure 11.64.

nutrient → inlet

air escape

air →

overflow

Figure 11.64 A chemostat.

We denote the microbial biomass at time t by $x(t)$ and the substrate concentration at time t by $s(t)$. Jacques Lucien Monod was highly influential in the development of quantitative microbiology; in 1950, he derived the following system of differential equations to describe the growth of microbes in a chemostat:

$$\begin{aligned} \frac{ds}{dt} &= D(s_0 - s) - q(s)x \\ \frac{dx}{dt} &= Yq(s)x - Dx \end{aligned} \tag{11.77}$$

In these equations, $s_0 > 0$ is the substrate concentration of the entering medium, $D > 0$ is the rate at which medium enters or leaves the chemostat, and $Y > 0$ is the yield constant. The function $q(s)$ is the rate at which microbes consume the substrate; the argument s indicates that q depends on the substrate concentration. The yield factor Y can thus be interpreted as a conversion factor of substrate into biomass.

Monod (1942) showed empirically that the uptake rate $q(s)$ fits the hyperbolic relationship

$$q(s) = \frac{v_m s}{K_m + s} \tag{11.78}$$

where v_m is the saturation level and K_m is the half-saturation constant [i.e., $q(K_m) = v_m/2$]. A graph of $q(s)$ is shown in Figure 11.65. It later occurred to Monod that (11.78) is identical to the Michaelis–Menten law (11.76)—an identity which might suggest that microbial growth is governed by enzymatic reactions.

In what follows, we will determine possible equilibria of (11.77) and analyze their stability. There is always, of course, the trivial equilibrium, which is obtained when substrate enters a growth chamber that is devoid of microbes [i.e., when $x(0) = 0$].

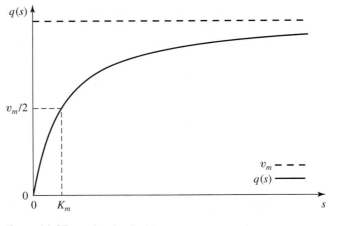

Figure 11.65 A graph of $q(s)$.

In that case, there will be no microbes at later times and, hence, $\frac{dx}{dt} = 0$ for all times $t \geq 0$. The substrate equilibrium is then found by setting $\frac{ds}{dt} = 0$ with $x = 0$:

$$0 = D(s_0 - s)$$

This equation has the solution $s = s_0$. Hence, one equilibrium is

$$(\hat{s}_1, \hat{x}_1) = (s_0, 0) \tag{11.79}$$

To obtain a nontrivial equilibrium $(\hat{s}_2, \hat{x}_2)$, we will look for an equilibrium with $\hat{x}_2 > 0$. To find this equilibrium, we solve the simultaneous equations

$$\frac{ds}{dt} = 0 \quad \text{and} \quad \frac{dx}{dt} = 0$$

It follows from $dx/dt = 0$ that

$$q(\hat{s}_2) = \frac{D}{Y} \tag{11.80}$$

We see immediately from Figure 11.65 that (11.80) has the solution $\hat{s}_2 > 0$ if $0 < D/Y < v_m$. Substituting (11.78) into (11.80), we can compute $\hat{s}_2$:

$$\frac{v_m \hat{s}_2}{K_m + \hat{s}_2} = \frac{D}{Y}$$

$$\hat{s}_2 = \frac{D K_m}{Y v_m - D} \tag{11.81}$$

Equation (11.81) is indeed positive, provided that $0 < D/Y < v_m$. Using (11.80) in $ds/dt = 0$, we find that

$$D(s_0 - \hat{s}_2) = \frac{D}{Y} \hat{x}_2$$

or

$$\hat{x}_2 = Y(s_0 - \hat{s}_2) \tag{11.82}$$

from which we can see that $\hat{x}_2 > 0$, provided that $\hat{s}_2 < s_0$. Under the assumption that $0 < D/Y < v_m$ and $\hat{s}_2 < s_0$, we therefore have the nontrivial equilibrium

$$(\hat{s}_2, \hat{x}_2) = \left(\frac{D K_m}{Y v_m - D}, Y(s_0 - \hat{s}_2) \right) \tag{11.83}$$

There are no other equilibria.

To analyze the stability of the two equilibria (11.79) and (11.83), we find the Jacobi matrix $D\mathbf{f}$ associated with the system (11.77):

$$D\mathbf{f}(s, x) = \begin{bmatrix} -D - q'(s)x & -q(s) \\ Yq'(s)x & Yq(s) - D \end{bmatrix}$$

We analyze the stability of the trivial equilibrium (11.79) first:

$$Df(s_0, 0) = \begin{bmatrix} -D & -q(s_0) \\ 0 & Yq(s_0) - D \end{bmatrix}$$

Since the Jacobi matrix is in upper triangular form, the eigenvalues are the diagonal elements, and we find that

$$\lambda_1 = -D < 0$$

$$\lambda_2 = Yq(s_0) - D < 0, \quad \text{provided that } \frac{D}{Y} > q(s_0)$$

Therefore, the equilibrium

$$(s_0, 0) \quad \text{is} \quad \begin{cases} \text{locally stable} & \text{if } \frac{D}{Y} > q(s_0) \\ \text{unstable} & \text{if } \frac{D}{Y} < q(s_0) \end{cases}$$

For the nontrivial equilibrium (11.83), we obtain

$$A = Df(\hat{s}_2, \hat{x}_2) = \begin{bmatrix} -D - q'(\hat{s}_2)\hat{x}_2 & -q(\hat{s}_2) \\ Yq'(\hat{s}_2)\hat{x}_2 & Yq(\hat{s}_2) - D \end{bmatrix}$$

Using (11.80), we see that this equation simplifies to

$$A = \begin{bmatrix} -D - q'(\hat{s}_2)\hat{x}_2 & -\frac{D}{Y} \\ Yq'(\hat{s}_2)\hat{x}_2 & 0 \end{bmatrix}$$

Now,

$$\text{tr } A = -D - q'(\hat{s}_2)\hat{x}_2 < 0$$

and

$$\det A = Dq'(\hat{s}_2)\hat{x}_2 > 0$$

for $\hat{x}_2 > 0$, since $q(s)$ is an increasing function. Therefore, if the nontrivial equilibrium exists (i.e., if both $\hat{s}_2 > 0$ and $\hat{x}_2 > 0$), then it is locally stable.

We saw that the nontrivial equilibrium does exist, provided that $0 < D/Y < v_m$ and $s_0 > \hat{s}_2$. These two conditions can be summarized as

$$0 < \frac{D}{Y} < v_m \quad \text{and} \quad \frac{DK_m}{Yv_m - D} < s_0$$

If the first inequality holds, then the denominator in the second inequality is positive. Solving the second inequality for D, we then find that

$$D < \frac{Ys_0v_m}{s_0 + K_m} = Yq(s_0) \tag{11.84}$$

We can now summarize our results. The chemostat has two equilibria: a trivial one in which microbes are absent and a nontrivial one that allows stable microbial growth. If $D > Yq(s_0)$, then the trivial equilibrium is the only biologically reasonable equilibrium and it is locally stable. If $D < Yq(s_0)$, both equilibria are biologically reasonable; the trivial one is now unstable and the nontrivial one is the locally stable one. Stable microbial growth is therefore possible, provided that the rate at which medium enters and leaves the growth chamber is between 0 and $Yq(s_0)$.

Section 11.4 Problems

■ 11.4.1

1. Suppose that the densities of two species evolve in accordance with the Lotka–Volterra model of interspecific competition. Assume that species 1 has intrinsic rate of growth $r_1 = 2$ and carrying capacity $K_1 = 20$ and that species 2 has intrinsic rate of growth $r_2 = 3$ and carrying capacity $K_2 = 15$. Furthermore, assume that 20 individuals of species 2 have the same effect on species 1 as 4 individuals of species 1 have on themselves and that 30 individuals of species 1 have the same effect on species 2 as 6 individuals of species 2 have on themselves. Find a system of differential equations that describes this situation.

2. Suppose the densities of two species evolve in accordance with the Lotka–Volterra model of interspecific competition. Assume that species 1 has intrinsic rate of growth $r_1 = 4$ and carrying capacity $K_1 = 17$ and that species 2 has intrinsic rate of growth $r_2 = 1.5$ and carrying capacity $K_2 = 32$. Furthermore, assume that 15 individuals of species 2 have the same effect on species 1 as 7 individuals of species 1 have on themselves and that 5 individuals of species 1 have the same effect on species 2 as 7 individuals of species 2 have on themselves. Find a system of differential equations that describes this situation.

In Problems 3–6, use the graphical approach to classify the following Lotka–Volterra models of interspecific competition according to "coexistence," "founder control," "species 1 excludes species 2," or "species 2 excludes species 1."

3. $\dfrac{dN_1}{dt} = 2N_1 \left(1 - \dfrac{N_1}{10} - 0.7\dfrac{N_2}{10} \right)$

$\dfrac{dN_2}{dt} = 5N_2 \left(1 - \dfrac{N_2}{15} - 0.3\dfrac{N_1}{15} \right)$

4. $\dfrac{dN_1}{dt} = 3N_1 \left(1 - \dfrac{N_1}{50} - 0.3\dfrac{N_2}{50} \right)$

$\dfrac{dN_2}{dt} = 4N_2 \left(1 - \dfrac{N_2}{30} - 0.8\dfrac{N_1}{30} \right)$

5. $\dfrac{dN_1}{dt} = N_1 \left(1 - \dfrac{N_1}{20} - \dfrac{N_2}{5} \right)$

$\dfrac{dN_2}{dt} = 2N_2 \left(1 - \dfrac{N_2}{15} - \dfrac{N_1}{3} \right)$

6. $\dfrac{dN_1}{dt} = 3N_1 \left(1 - \dfrac{N_1}{25} - 1.2\dfrac{N_2}{25} \right)$

$\dfrac{dN_2}{dt} = N_2 \left(1 - \dfrac{N_2}{30} - 0.8\dfrac{N_1}{30} \right)$

In Problems 7–10, use the eigenvalue approach to analyze all equilibria of the given Lotka–Volterra models of interspecific competition.

7. $\dfrac{dN_1}{dt} = 3N_1 \left(1 - \dfrac{N_1}{18} - 1.3\dfrac{N_2}{18} \right)$

$\dfrac{dN_2}{dt} = 2N_2 \left(1 - \dfrac{N_2}{20} - 0.6\dfrac{N_1}{20} \right)$

8. $\dfrac{dN_1}{dt} = 4N_1 \left(1 - \dfrac{N_1}{12} - 0.3\dfrac{N_2}{12} \right)$

$\dfrac{dN_2}{dt} = 5N_2 \left(1 - \dfrac{N_2}{15} - 0.2\dfrac{N_1}{15} \right)$

9. $\dfrac{dN_1}{dt} = N_1 \left(1 - \dfrac{N_1}{35} - 3\dfrac{N_2}{35} \right)$

$\dfrac{dN_2}{dt} = 3N_2 \left(1 - \dfrac{N_2}{40} - 4\dfrac{N_1}{40} \right)$

10. $\dfrac{dN_1}{dt} = N_1 \left(1 - \dfrac{N_1}{25} - 0.1\dfrac{N_2}{25} \right)$

$\dfrac{dN_2}{dt} = N_2 \left(1 - \dfrac{N_2}{28} - 1.2\dfrac{N_1}{28} \right)$

11. Suppose that two species of beetles are reared together in one experiment and separately in another. When species 1 is reared alone, it reaches an equilibrium of about 200. When species 2 is reared alone, it reaches an equilibrium of about 150. When both of them are reared together, they seem to be able to coexist: Species 1 reaches an equilibrium of about 180 and species 2 reaches an equilibrium of about 80. If their densities follow the Lotka–Volterra equation of interspecific competition, find α_{12} and α_{21}.

12. Suppose that two species of beetles are reared together. Species 1 wins if there are initially 100 individuals of species 1 and 20 individuals of species 2. But species 2 wins if there are initially 20 individuals of species 1 and 100 individuals of species 2. When the beetles are reared separately, both species seem to reach an equilibrium of about 120. On the basis of this information and assuming that the densities follow the Lotka–Volterra model of interspecific competition, can you give lower bounds on α_{12} and α_{21}?

■ 11.4.2

In Problems 13 and 14, use a graphing calculator to sketch solution curves of the given Lotka–Volterra predator–prey model in the N–P plane. Also graph $N(t)$ and $P(t)$ as functions of t.

13. $\dfrac{dN}{dt} = 2N - PN$

$\dfrac{dP}{dt} = \frac{1}{2}PN - P$

with initial conditions

(a) $(N(0), P(0)) = (2, 2)$ **(b)** $(N(0), P(0)) = (3, 3)$

(c) $(N(0), P(0)) = (4, 4)$

14. $\dfrac{dN}{dt} = 3N - 2PN$

$\dfrac{dP}{dt} = PN - P$

with initial conditions

(a) $(N(0), P(0)) = (1, 3/2)$ **(b)** $(N(0), P(0)) = (2, 2)$

(c) $(N(0), P(0)) = (3, 1)$

In Problems 15 and 16, we investigate the Lotka–Volterra predator–prey model.

15. Assume that

$$\dfrac{dN}{dt} = N - 4PN$$

$$\dfrac{dP}{dt} = 2PN - 3P$$

(a) Show that this system has two equilibria: the trivial equilibrium $(0, 0)$, and a nontrivial one in which both species have positive densities.

(b) Use the eigenvalue approach to show that the trivial equilibrium is unstable.

(c) Determine the eigenvalues corresponding to the nontrivial equilibrium. Does your analysis allow you to infer anything about the stability of this equilibrium?

(d) Use a graphing calculator to sketch curves in the N–P plane. Also, sketch solution curves of the prey and the predator densities as functions of time.

16. Assume that

$$\frac{dN}{dt} = 5N - PN$$

$$\frac{dP}{dt} = PN - P$$

(a) Show that this system has two equilibria: the trivial equilibrium $(0, 0)$, and a nontrivial one in which both species have positive densities.

(b) Use the eigenvalue approach to show that the trivial equilibrium is unstable.

(c) Determine the eigenvalues corresponding to the nontrivial equilibrium. Does your analysis allow you to infer anything about the stability of this equilibrium?

(d) Use a graphing calculator to sketch curves in the N–P plane. Also, sketch solution curves of the prey and the predator densities as functions of time.

17. Assume that $N(t)$ denotes the density of an insect species at time t and $P(t)$ denotes the density of its predator at time t. The insect species is an agricultural pest, and its predator is used as a biological control agent. Their dynamics are given by the system of differential equations

$$\frac{dN}{dt} = 5N - 3PN$$

$$\frac{dP}{dt} = 2PN - P$$

(a) Explain why

$$\frac{dN}{dt} = 5N \qquad (11.85)$$

describes the dynamics of the insect in the absence of the predator. Solve (11.85). Describe what happens to the insect population in the absence of the predator.

(b) Explain why introducing the insect predator into the system can help to control the density of the insect.

(c) Assume that at the beginning of the growing season the insect density is 0.5 and the predator density is 2. You decide to control the insects by using an insecticide in addition to the predator. You are careful and choose an insecticide that does not harm the predator. After you spray, the insect density drops to 0.01 and the predator density remains at 2. Use a graphing calculator to investigate the long-term implications of your decision to spray the field. In particular, investigate what would have happened to the insect densities if you had decided not to spray the field, and compare your results with the insect density over time that results from your application of the insecticide.

18. Assume that $N(t)$ denotes prey density at time t and $P(t)$ denotes predator density at time t. Their dynamics are given by the system of equations

$$\frac{dN}{dt} = 4N - 2PN$$

$$\frac{dP}{dt} = PN - 3P$$

Assume that initially $N(0) = 3$ and $P(0) = 2$.

(a) If you followed this predator–prey community over time, what would you observe?

(b) Suppose that bad weather kills 90% of the prey population and 67% of the predator population. If you continued to observe this predator–prey community, what would you expect to see?

19. An unrealistic feature of the Lotka–Volterra model is that the prey exhibits unlimited growth in the absence of the predator. The model described by the following system remedies this shortcoming (in the model, we assume that the prey evolves according to logistic growth in the absence of the predator; the other features of the model are retained):

$$\frac{dN}{dt} = 3N\left(1 - \frac{N}{10}\right) - 2PN$$

$$\frac{dP}{dt} = PN - 4P \qquad (11.86)$$

(a) Explain why the prey evolves according to

$$\frac{dN}{dt} = 3N\left(1 - \frac{N}{10}\right) \qquad (11.87)$$

in the absence of the predator. Investigate the long-term behavior of solutions to (11.87).

(b) Find all equilibria of (11.86), and use the eigenvalue approach to determine their stability.

(c) Use a graphing calculator to sketch the solution curve of (11.86) in the N–P plane when $N(0) = 2$ and $P(0) = 2$. Also, sketch $N(t)$ and $P(t)$ as functions of time, starting with $N(0) = 2$ and $P(0) = 2$.

20. An unrealistic feature of the Lotka–Volterra model is that the prey exhibits unlimited growth in the absence of the predator. The model described by the following system remedies this shortcoming (in the model, we assume that the prey evolves according to logistic growth in the absence of the predator; the other features of the model are retained):

$$\frac{dN}{dt} = N\left(1 - \frac{N}{K}\right) - 4PN$$

$$\frac{dP}{dt} = PN - 5P \qquad (11.88)$$

Here, $K > 0$ denotes the carrying capacity of the prey in the absence of the predator. In what follows, we will investigate how the carrying capacity affects the outcome of this predator–prey interaction.

(a) Draw the zero isoclines of (11.88) for (i) $K = 10$ and (ii) $K = 3$.

(b) When $K = 10$, the zero isoclines intersect, indicating the existence of a nontrivial equilibrium. Analyze the stability of this nontrivial equilibrium.

(c) Is there a minimum carrying capacity required in order to have a nontrivial equilibrium? If yes, find it and explain what happens when the carrying capacity is below this minimum and what happens when the carrying capacity is above this minimum.

21. An unrealistic feature of the Lotka–Volterra model is that the prey exhibits unlimited growth in the absence of the predator. The model described by the following system remedies this shortcoming (in the model, we assume that the prey evolves according to logistic growth in the absence of the predator; the other features of the model are retained):

$$\frac{dN}{dt} = N\left(1 - \frac{N}{20}\right) - 5PN$$

$$\frac{dP}{dt} = 2PN - 8P \qquad (11.89)$$

(a) Draw the zero isoclines of (11.89).

(b) Use the graphical approach of Subsection 11.3.2 to determine whether the nontrivial equilibrium is locally stable.

In Problems 22–26, we will analyze how a change in parameters in the modified Lotka–Volterra predator–prey model

$$\frac{dN}{dt} = aN\left(1 - \frac{N}{K}\right) - bPN$$

$$\frac{dP}{dt} = cPN - dP \qquad (11.90)$$

affects predator–prey interactions.

22. (a) Find the zero isoclines of (11.90), and determine conditions under which a nontrivial equilibrium (i.e., an equilibrium in which both prey and predator have positive densities) exists.

(b) Use the graphical approach of Subsection 11.3.2 to show that if a nontrivial equilibrium exists, it is locally stable.

In Problems 23–26, we use the results of Problem 22. Assume that the parameters are chosen so that a nontrivial equilibrium exists.

23. Use the results of Problem 22 to show that an increase in a (the intrinsic rate of growth of the prey) results in an increase in the predator density, but leaves the prey density unchanged.

24. Use the results of Problem 22 to show that an increase in b (the searching efficiency) reduces the predator density, but has no effect on the equilibrium abundance of the prey.

25. Use the results of Problem 22 to show that an increase in c (the predator growth efficiency) reduces the prey equilibrium abundance and increases the predator equilibrium abundance.

26. Use the results of Problem 22 to show that an increase in K (the prey carrying capacity in the absence of the predator) increases the predator equilibrium abundance, but has no effect on the prey equilibrium abundance.

■ **11.4.3**

In Problems 27–34, classify each community matrix at equilibrium according to the five cases considered in Subsection 11.4.3 and determine whether the equilibrium is stable. (Assume in each case that the equilibrium exists.)

27. $\begin{bmatrix} -1 & -1.3 \\ 0.3 & -2 \end{bmatrix}$ **28.** $\begin{bmatrix} -3 & -1.2 \\ -1 & -2 \end{bmatrix}$

29. $\begin{bmatrix} -1.5 & 1.6 \\ 2.3 & -5.1 \end{bmatrix}$ **30.** $\begin{bmatrix} -0.3 & 0 \\ 0.4 & -0.7 \end{bmatrix}$

31. $\begin{bmatrix} -1 & 1.3 \\ 2 & -1.5 \end{bmatrix}$ **32.** $\begin{bmatrix} -2.7 & 0 \\ -1.3 & -0.6 \end{bmatrix}$

33. $\begin{bmatrix} -5 & -1.7 \\ -2.3 & -0.2 \end{bmatrix}$ **34.** $\begin{bmatrix} -2.3 & -4.7 \\ 1.2 & -3.2 \end{bmatrix}$

In Problems 35–40, we consider communities composed of two species. The abundance of species 1 at time t is given by $N_1(t)$, the abundance of species 2 at time t by $N_2(t)$. Their dynamics are described by

$$\frac{dN_1}{dt} = f_1(N_1, N_2)$$

$$\frac{dN_2}{dt} = f_2(N_1, N_2)$$

Assume that when both species are at low abundances their abundances increase and that f_1 and f_2 change sign when crossing their zero isoclines. In each problem, determine the sign structure of the community matrix at the nontrivial equilibrium (indicated by a dot) on the basis of the graph of the zero isoclines. Determine the stability of the equilibria if possible.

35. See Figure 11.66.

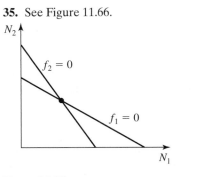

Figure 11.66

36. See Figure 11.67.

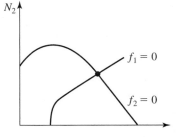

Figure 11.67

37. See Figure 11.68.

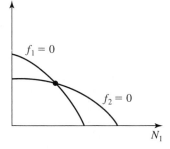

Figure 11.68

38. See Figure 11.69.

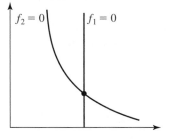

Figure 11.69

39. See Figure 11.70.

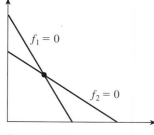

Figure 11.70

40. See Figure 11.71.

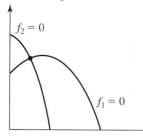

Figure 11.71

41. Assume that the diagonal elements a_{ii} of the community matrix of a species assemblage in equilibrium are negative. Explain why this assumption implies that species i exhibits self-regulation.

42. Consider a community composed of two species. Assume that both species inhibit themselves. Explain why mutualistic and competitive interactions lead to qualitatively similar predictions about the stability of the corresponding equilibria. That is, show that if $A = [a_{ij}]$ is the community matrix at equilibrium for the case of mutualism, and if $B = [b_{ij}]$ is the community matrix at equilibrium for the case of competition, then the following holds: If $|a_{ij}| = |b_{ij}|$ for $1 \leq i, j \leq 2$, then either both equilibria are locally stable or both are unstable.

43. The classical Lotka–Volterra model of predation is given by

$$\frac{dN}{dt} = aN - bNP$$

$$\frac{dP}{dt} = cNP - dP$$

where $N = N(t)$ is the prey density at time t and $P = P(t)$ is the predator density at time t. The constants $a, b, c,$ and d are all positive.

(a) Find the nontrivial equilibrium $(\hat{N}, \hat{P})$ with $\hat{N} > 0$ and $\hat{P} > 0$.

(b) Find the community matrix corresponding to the nontrivial equilibrium.

(c) Explain each entry of the community matrix found in (b) in terms of how individuals in this community affect each other.

44. The modified Lotka–Volterra model of predation is given by

$$\frac{dN}{dt} = aN\left(1 - \frac{N}{K}\right) - bNP$$

$$\frac{dP}{dt} = cNP - dP$$

where $N = N(t)$ is the prey density at time t and $P = P(t)$ is the predator density at time t. The constants $a, b, c, d,$ and K are positive. Assume that $d/c < K$.

(a) Find the nontrivial equilibrium $(\hat{N}, \hat{P})$ with $\hat{N} > 0$ and $\hat{P} > 0$.

(b) Find the community matrix corresponding to the nontrivial equilibrium.

(c) Explain each entry of the community matrix found in (b) in terms of how individuals in this community affect each other.

■ 11.4.4

45. Use a graphing calculator to study the following example of the Fitzhugh–Nagumo model:

$$\frac{dV}{dt} = -V(V - 0.3)(V - 1) - w$$

$$\frac{dw}{dt} = 0.01(V - 0.4w)$$

Sketch the graph of the solution curve in the V–w plane when **(i)** $(V(0), w(0)) = (0.4, 0)$ and **(ii)** $(V(0), w(0)) = (0.2, 0)$.

46. Use a graphing calculator to study the following example of the Fitzhugh–Nagumo model:

$$\frac{dV}{dt} = -V(V - 0.6)(V - 1) - w$$

$$\frac{dw}{dt} = 0.03(V - 0.6w)$$

Sketch the graph of the solution curve in the V–w plane when **(i)** $(V(0), w(0)) = (0.8, 0)$ and **(ii)** $(V(0), w(0)) = (0.4, 0)$.

47. Assume the following example of the Fitzhugh–Nagumo model:

$$\frac{dV}{dt} = -V(V - 0.3)(V - 1) - w$$

$$\frac{dw}{dt} = 0.01(V - 0.4w)$$

Assume that $w(0) = 0$. For which initial values of $V(0)$ can you observe an action potential?

48. Assume the following example of the Fitzhugh–Nagumo model:

$$\frac{dV}{dt} = -V(V - 0.6)(V - 1) - w$$

$$\frac{dw}{dt} = 0.03(V - 0.6w)$$

Assume that $w(0) = 0$. For which initial values of $V(0)$ can you observe an action potential?

■ 11.4.5

In Problems 49–52, use the mass action law to translate each chemical reaction into a system of differential equations.

49. $A + B \xrightarrow{k} C$ **50.** $A + B \underset{k_-}{\overset{k_+}{\rightleftharpoons}} C$

51. $E + S \xrightarrow{k_1} ES \xrightarrow{k_2} E + P$ **52.** $A + B \xrightarrow{k} A + C$

53. Show that the following system of differential equations has a conserved quantity, and find it:

$$\frac{dx}{dt} = 2x - 3y$$

$$\frac{dy}{dt} = 3y - 2x$$

54. Show that the following system of differential equations has a conserved quantity, and find it:

$$\frac{dx}{dt} = -4x + 2y$$

$$\frac{dy}{dt} = -y + 2x$$

55. Show that the following system of differential equations has a conserved quantity, and find it:

$$\frac{dx}{dt} = -x + 2xy + z$$

$$\frac{dy}{dt} = -2xy$$

$$\frac{dz}{dt} = x - z$$

56. Suppose that $x(t) + y(t)$ is a conserved quantity. If

$$\frac{dx}{dt} = -3x + 2xy$$

find the differential equation for $y(t)$.

57. The Michaelis–Menten law [Equation (11.76)] states that

$$\frac{dp}{dt} = \frac{v_m s}{K_m + s}$$

where $p = p(t)$ is the concentration of the product of the enzymatic reaction at time t, $s = s(t)$ is the concentration of the substrate at time t, and v_m and K_m are positive constants. Set

$$f(s) = \frac{v_m s}{K_m + s}$$

where v_m and K_m are positive constants.
(a) Show that

$$\lim_{s \to \infty} f(s) = v_m$$

(b) Show that

$$f(K_m) = \frac{v_m}{2}$$

(c) Show that, for $s \geq 0$, $f(s)$ is **(i)** nonnegative, **(ii)** increasing, and **(iii)** concave down. Sketch a graph of $f(s)$. Label v_m and K_m on your graph.

(d) Explain why we said that the reaction rate dp/dt is limited by the availability of the substrate.

58. The growth of microbes in a chemostat was described by (11.77). Using the notation of that equation, together with the relationship

$$q(s) = \frac{v_m s}{K_m + s}$$

where v_m and K_m are positive constants, we will investigate how the substrate concentration $\hat{s}$ in equilibrium depends on the uptake rate Y.
(a) Assume that the microbes have a positive equilibrium density. Find the equilibrium concentration $\hat{s}$ algebraically, and investigate how the uptake rate Y affects $\hat{s}$.

(b) Assume that the microbes have a positive equilibrium density. Sketch a graph of $q(s)$, and explain how you would determine $\hat{s}$ graphically. Use your graph to explain how the uptake rate Y affects $\hat{s}$.

59. The growth of microbes in a chemostat was described by (11.77). Using the notation of that equation, together with the relationship

$$q(s) = \frac{v_m s}{K_m + s}$$

we will investigate how the substrate concentration $\hat{s}$ in equilibrium depends on D, the rate at which the medium enters the chemostat.
(a) Assume that the microbes have a positive equilibrium density. Find the equilibrium concentration $\hat{s}$ algebraically. Investigate how the rate D affects $\hat{s}$.

(b) Assume that the microbes have a positive equilibrium density. Sketch a graph of $q(s)$, and explain how you would determine $\hat{s}$ graphically. Use your graph to explain how the rate D affects $\hat{s}$.

In Problems 60 and 61, we investigate specific examples of microbial growth described by (11.77). We use the notation of Subsection 11.4.5. In each case, determine all equilibria and their stability.

60. $\dfrac{ds}{dt} = 2(4 - s) - \dfrac{3s}{2 + s}x$ **61.** $\dfrac{ds}{dt} = 2(4 - s) - \dfrac{3s}{1 + s}x$

$\dfrac{dx}{dt} = \dfrac{sx}{2 + s} - 2x$ $\qquad$ $\dfrac{dx}{dt} = \dfrac{3sx}{1 + s} - 2x$

Chapter 11 Key Terms

Discuss the following definitions and concepts:

1. Linear first-order equation

2. Homogeneous

3. Direction field, slope field, direction vector

4. Solution of a system of linear differential equations

5. Eigenvalue, eigenvector

6. Superposition principle

7. General solution

8. Stability

9. Sink, or stable node

10. Saddle point

11. Source, or unstable node

12. Spiral

13. Euler's formula

14. Compartment model

15. Conserved quantity

16. Harmonic oscillator

17. Nonlinear autonomous system of differential equations

18. Critical point

19. Zero isoclines, or null clines

20. Graphical approach to stability

21. Lotka–Volterra model of interspecific competition

22. Intraspecific competition, interspecific competition

23. Competitive exclusion, founder control, coexistence

24. Lotka–Volterra predator–prey model

25. Community matrix

26. Fitzhugh–Nagumo model

27. Action potential

28. Michaelis–Menten law

Chapter 11 Review Problems

1. Population Growth Let $N_1(t)$ and $N_2(t)$ denote the respective sizes of two populations at time t, and assume that their dynamics are respectively given by

$$\frac{dN_1}{dt} = r_1 N_1$$

$$\frac{dN_2}{dt} = r_2 N_2$$

where r_1 and r_2 are positive constants denoting the intrinsic rate of growth of the two populations. Set $Z(t) = N_1(t)/N_2(t)$, and show that $Z(t)$ satisfies

$$\frac{d}{dt} \ln Z(t) = r_1 - r_2 \qquad (11.91)$$

Solve (11.91), and show that $\lim_{t \to \infty} Z(t) = \infty$ if $r_1 > r_2$. Conclude from this that population 1 becomes numerically dominant when $r_1 > r_2$.

2. Population Growth Let $N_1(t)$ and $N_2(t)$ denote the respective sizes of two populations at time t, and assume that their dynamics are respectively given by

$$\frac{dN_1}{dt} = r_1 N_1$$

$$\frac{dN_2}{dt} = r_2 N_2$$

where r_1 and r_2 are positive constants denoting the intrinsic rate of growth of the two populations. Denote the combined population size at time t by $N(t)$; that is, $N(t) = N_1(t) + N_2(t)$. Define the relative proportions

$$p_1 = \frac{N_1}{N} \quad \text{and} \quad p_2 = \frac{N_2}{N}$$

Use the fact that $p_1/p_2 = N_1/N_2$ to show that

$$\frac{dp_1}{dt} = p_1(1 - p_1)(r_1 - r_2)$$

Show that if $r_1 > r_2$ and $0 < p_1(0) < 1$, $p_1(t)$ will increase for $t > 0$ and population 1 will become numerically dominant.

3. Predator–Prey Interactions An unrealistic feature of the Lotka–Volterra model is that the prey exhibits unlimited growth in the absence of the predator. The model described by the following system remedies this shortcoming (in the model, we assume that the prey evolves according to logistic growth in the absence of the predator; the other features of the model are retained):

$$\frac{dN}{dt} = 2N\left(1 - \frac{N}{10}\right) - 3PN$$

$$\frac{dP}{dt} = PN - 3P \qquad (11.92)$$

(a) Draw the zero isoclines of (11.92).

(b) Use the graphical approach of Subsection 11.3.2 to determine whether the nontrivial equilibrium is locally stable.

4. Resource Competition Tilman (1982) developed a theoretical framework for studying resource competition in plants. In its simplest form, the theory posits that one species competes for a single resource—for instance, nitrogen. If $B(t)$ denotes the total biomass at time t and $R(t)$ is the amount of the resource available

at time t, then the dynamics are described by the following system of differential equations:

$$\frac{dB}{dt} = B[f(R) - m]$$

$$\frac{dR}{dt} = a(S - R) - cBf(R)$$

The first equation describes the rate of change of biomass, where the function $f(R)$ describes how the species growth rate depends on the resource, and m is the specific loss rate. The second equation describes the resource dynamics; the constant S is the maximal amount of the resource in a given habitat. The rate of resource supply (dR/dt) is assumed to be proportional to the difference between the current resource level and the maximal amount of the resource; the constant a is the constant of proportionality. The term $cBf(R)$ describes the resource uptake by the plants; the constant c can be considered a conversion factor.

In what follows, we assume that $f(R)$ follows the Monod growth function

$$f(R) = \frac{dR}{k + R}$$

where d and k are positive constants.

(a) Find all equilibria. Show that if $d > m$ and $S > mk/(d - m)$, then there exists a nontrivial equilibrium.

(b) Sketch the zero isoclines for the case in which the system admits a nontrivial equilibrium. Use the graphical approach to analyze the stability of the nontrivial equilibrium.

5. Plant Competition In this problem, we describe a simple competition model in which two species of plants compete for vacant space. Assume that the entire habitat is divided into a large number of patches. Each patch can be occupied by at most one species. We denote by $p_i(t)$ the fraction of patches occupied by species i. Note that $0 \leq p_1(t) + p_2(t) \leq 1$. The dynamics are described by

$$\frac{dp_1}{dt} = c_1 p_1(1 - p_1 - p_2) - m_1 p_1$$

$$\frac{dp_1}{dt} = c_2 p_2(1 - p_1 - p_2) - m_2 p_2$$

where c_1, c_2, m_1, and m_2 are positive constants. The first term on the right-hand side of each equation describes the colonization of vacant patches; the second term on the right-hand side of each equation describes how occupied patches become vacant.

(a) Show that the dynamics of species 1 in the absence of species 2 are given by

$$\frac{dp_1}{dt} = c_1 p_1(1 - p_1) - m_1 p_1 \qquad (11.93)$$

and find conditions on c_1 and m_1 so that (11.93) admits a nontrivial equilibrium (an equilibrium in which $0 < p_1 \leq 1$).

(b) Assume now that $c_1 > m_1$ and $c_2 > m_2$. Show that if

$$\frac{c_1}{m_1} > \frac{c_2}{m_2}$$

then species 1 will exclude species 2 if species 1 initially occupies a positive fraction of the patches.

6. Paradox of Enrichment Rosenzweig (1971) analyzed a number of predator–prey models and concluded that enriching the system by increasing the nutrient supply destabilizes the nontrivial equilibrium. We will think of the predator–prey model as a plant–herbivore system in which plants represent prey and herbivores represent predators. The models analyzed were of the form

$$\frac{dN}{dt} = f(N, P) \tag{11.94}$$

$$\frac{dP}{dt} = g(N, P) \tag{11.95}$$

where $N = N(t)$ is the plant abundance at time t and $P = P(t)$ is the herbivore abundance at time t. The models all shared the property that the zero isocline for the herbivore was a vertical line and the zero isocline for the plants was a hump-shaped curve. We will look at one of the models, namely,

$$\frac{dN}{dt} = aN\left(1 - \frac{N}{K}\right) - bP\left(1 - e^{-rN}\right)$$
$$\frac{dP}{dt} = cP\left(1 - e^{-rN}\right) - dP \tag{11.96}$$

(a) Find the zero isoclines for (11.96), and show that **(i)** the zero isocline of the herbivore ($dP/dt = 0$) is a vertical line in the N–P plane and **(ii)** the zero isocline for the plants ($dN/dt = 0$) intersects the N-axis at $N = K$.

(b) Plot the zero isoclines in the N–P plane for $a = b = c = r = 1$ and $d = 0.9$ and for three levels of the carrying capacity: **(i)** $K = 1$, **(ii)** $K = 4$, and **(iii)** $K = 10$.

(c) For each of the three carrying capacities, determine whether a nontrivial equilibrium exists.

(d) Use the graphical approach of Subsection 11.3.2 to determine the stability of the existing nontrivial equilibria in (c).

(e) Enriching the community could mean increasing the carrying capacity of the plants. For instance, adding nitrogen or phosphorus to plant communities frequently results in an increase in biomass, which can be interpreted as an increase in the carrying capacity of the plants (the K-value). On the basis of your answers in (d), explain why enriching the community (increasing the carrying capacity of the plants) can result in a destabilization of the nontrivial equilibrium. What are the consequences?

7. Microbial Growth The growth of microbes in a chemostat was described by Equation (11.77). We will investigate how the microbial abundance in equilibrium depends on the characteristics of the system.

(a) Assume that $q(s)$ is a nonnegative function. Show that the equilibrium abundance of the microbes is given by

$$\hat{x} = Y(s_0 - \hat{s})$$

where $\hat{s}$ is the substrate equilibrium abundance. When is $\hat{x} > 0$?

(b) Assume now that

$$q(s) = \frac{v_m s}{K_m + s}$$

Investigate how the uptake rate Y and the rate D at which new medium enters the chemostat affect the equilibrium abundance of the microbes.

8. Successional Niche Pacala and Rees (1998) discuss a simple mathematical model of competition to explain successional diversity by means of a successional niche mechanism. In this model, two species—an early successional and a late successional—occupy discrete patches. Each patch experiences disturbances (such as fire) at rate D. After a patch is disturbed, both species are present. Over time, however, the late successional species outcompetes the early successional species, causing the early successional species to become extinct. This change, from a patch that is occupied by both species to a patch that is occupied by the late successional species only, happens at rate a. We keep track of the number of patches occupied by both species at time t, denoted by $x(t)$, and the number of patches occupied by just the late successional species at time t, denoted by $y(t)$. The dynamics are given by the system of linear differential equations

$$\frac{dx}{dt} = -ax + Dy$$
$$\frac{dy}{dt} = ax - Dy \tag{11.97}$$

where a and D are positive constants.
(a) Show that all equilibria are of the form $(x, ax/D)$.
(b) Find the eigenvalues and eigenvectors corresponding to each equilibrium.
(c) Show that

$$\begin{bmatrix} x(t) \\ y(t) \end{bmatrix} = C_1 \begin{bmatrix} u_1 \\ u_2 \end{bmatrix} + C_2 e^{\lambda_2 t} \begin{bmatrix} v_1 \\ v_2 \end{bmatrix}$$

where $\begin{bmatrix} u_1 \\ u_2 \end{bmatrix}$ is the eigenvector corresponding to the zero eigenvalue and $\begin{bmatrix} v_1 \\ v_2 \end{bmatrix}$ is the eigenvector corresponding to the nonzero eigenvalue λ_2, is a solution of (11.97).
(d) Show that $x(t) + y(t)$ does not depend on t. [*Hint*: Show that $\frac{d}{dt}(x(t)+y(t)) = 0$.] Show also that the line $x+y = A$ (where A is a constant) is parallel to the line in the direction of the eigenvector corresponding to the nonzero eigenvalue.
(e) Show that the zero isoclines of (11.97) are given by

$$y = \frac{a}{D}x$$

and that this line is the line in the direction of the eigenvector corresponding to the zero eigenvalue.

(f) Suppose now that $x(t) + y(t) = c$, where c is a positive constant. Show that (11.97) can be reduced to just one equation, namely,

$$\frac{dx}{dt} = -(a + D)x + Dc$$

Show that $\hat{x} = c\frac{D}{D+a}$ is the only equilibrium, and determine its stability.

Probability and Statistics

12

LEARNING OBJECTIVES

This chapter develops the principles of probability theory and statistics. Specifically, we will learn how to

- apply the principles of counting;
- define and calculate probabilities for discrete and continuous random variables;
- analyze the behavior of averages; and
- describe data, estimate parameters, and find statistical relationships.

We conclude this book with a chapter on probability and statistics. Although neither field is part of calculus, they both rely on calculus in providing indispensable tools for life scientists.

Many phenomena in nature are not deterministic. To give just a few examples, consider the number of eggs laid by a bird, the life span of an organism, the inheritance of genes, and the number of people infected during an outbreak of a disease. To deal with the inherent randomness (or stochasticity) of natural phenomena, we need to develop special tools; these tools are supplied by the disciplines of probability and statistics.

A short description of the role of probability and statistics in the life sciences might be as follows: Probability theory provides tools for modeling randomness and forms the foundation of statistics. Statistics provides tools for analyzing data from scientific experiments.

12.1 Counting

It is often necessary to count the ways in which a certain task can be performed. The mathematical field of **combinatorics** deals with such enumeration problems. Frequently, the total number of possible ways is very large, making it impractical to write down all possible choices. There are three basic counting principles that will help us to count in a more systematic way. The first is the multiplication principle; the other two, which follow from it, are rules concerning permutations and combinations.

■ 12.1.1 The Multiplication Principle

To illustrate the multiplication principle, consider the next example.

EXAMPLE 1   Imagine that we wish to experimentally manipulate growth conditions for plants—say, the grass species big bluestem, *Andropogon gerardi*. We want to grow plants in pots in a greenhouse at two different levels of fertilizer (low and high) and four different temperatures (10°C, 15°C, 20°C, and 25°C). If we want three replicates of each possible combination of fertilizer and temperature treatment, how many pots will we need?

Solution   We can answer this question with the help of a tree diagram, as shown in Figure 12.1. We see from the tree that we will need

$$2 \cdot 4 \cdot 3 = 24$$

pots for our experiment.   ■

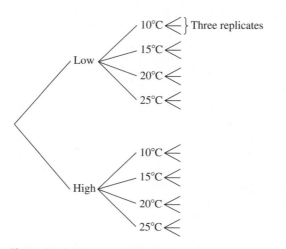

Figure 12.1   The tree diagram illustrates how many pots are needed in the experiment described in Example 1.

The counting principle that we just used is called the *multiplication principle*, which we can summarize as follows:

Multiplication Principle Suppose that an experiment consists of m ordered tasks. Task 1 has n_1 possible outcomes, task 2 has n_2 possible outcomes, ..., and task m has n_m possible outcomes. Then the total number of possible outcomes of the experiment is

$$n_1 \cdot n_2 \cdot n_3 \cdots n_m$$

Looking back at Example 1, we see that the experiment consisted of three tasks: first, to select the fertilizer level; second, to select the temperature; and third, to replicate each combination of fertilizer and temperature three times. The successive tasks are illustrated in the tree diagram, and the total number of pots required for the experiment can be obtained by counting the number of tips of the tree.

We present one more example that illustrates this counting principle.

EXAMPLE 2   Suppose that after a long day in the greenhouse you decide to order pizza. You call a local pizza parlor and learn that there are three choices of crust and five choices of toppings and that you can order the pizza with or without cheese. If you want only one topping, how many different choices do you have for selecting a pizza?

Solution   Your "experiment," which consists of ordering a pizza, involves three tasks. The first task is to choose a crust, the second is to choose the topping, and the third is to decide whether or not you want cheese. Using the multiplication principle, we find that there are

$$3 \cdot 5 \cdot 2 = 30$$

different pizzas that you could order. ■

■ 12.1.2 Permutations

EXAMPLE 3   Suppose that you grow plants in a greenhouse. To control for spatially varying environmental conditions, you rearrange the pots every other day. If you have six pots arranged in a row on a bench, in how many ways can you arrange the pots?

Solution   To answer this question, imagine that you arrange the pots on the bench from left to right: You have six choices for the leftmost position on the bench, for the next position you can choose any of the remaining five pots, for the third position you can choose any of the remaining four pots, and so on, until there is one pot left that must go into the rightmost position. Using the multiplication principle, we find that there are

$$6 \cdot 5 \cdot 4 \cdot 3 \cdot 2 \cdot 1 = 720$$

ways to arrange the six pots on the bench. ■

As shorthand notation for the type of descending products of positive integers we encountered in Example 3, we define

$$n! = n(n-1)(n-2)\cdots 3 \cdot 2 \cdot 1$$

and read $n!$ as "n factorial." We can now write $6!$ instead of $6 \cdot 5 \cdot 4 \cdot 3 \cdot 2 \cdot 1$.
   We define

$$0! = 1$$

Then, for $n = 0, 1, 2, \ldots,$

$$(n+1)! = (n+1)n!$$

The quantity $n!$ grows very quickly. Suppose that instead of 6 pots you have 7; then there are $7! = 5040$ ways to arrange the seven pots. With 12 pots, there are $12! = 479,001,600$ possible ways to arrange them.
   We will look at another example and then state a general principle.

EXAMPLE 4   Suppose that a track team has 10 sprinters, any 4 of whom can form a relay team. Assume that each person can run in any position on the team. How many teams can be formed if teams that consist of the same 4 people in different running orders are considered different teams?

Solution   We select the members of the team in the order in which they run. There are 10 available sprinters for the first position. After having chosen a person for the first position, there are 9 left, and we can choose any of the 9 for the second position. For the third position, we can choose among the 8 remaining people and, finally, for the fourth position, we can select a person from the remaining 7. Using the multiplication principle, we find that there are

$$10 \cdot 9 \cdot 8 \cdot 7 = 5040$$

different relay teams. ■

In Example 3, we selected $k = 6$ objects from a set of $n = 6$; the order of selection was important. In Example 4, we selected $k = 4$ objects from a set of $n = 10$, where again the order of selection was important. Such selections are called **permutations**. Using the multiplication principle, we can find the number of possible permutations of a given number of objects.

> **Permutations** A **permutation** of n different objects taken k at a time is an *ordered* subset of k out of the n objects. The number of ways that this can be done is denoted by $P(n, k)$ and is given by
>
> $$P(n, k) = n(n-1)(n-2)\cdots(n-k+1)$$

Note that the last term in the product defining $P(n, k)$ is of the form $(n-k+1)$, because there are k descending factors and the first factor is n.

Returning to Example 3, where we wanted to select six out of six objects in an ordered arrangement, we can now use our permutation rule to compute the number of ways that we can make the selection. Setting $n = 6$ and $k = 6$, we find that

$$P(6, 6) = 6 \cdot 5 \cdot 4 \cdot 3 \cdot 2 \cdot 1$$

for the number of different ways to arrange the pots. Since the product consists of six terms in descending order, starting with 6, the last term is $n - k + 1 = 6 - 6 + 1 = 1$.

Returning to Example 4, in which we wanted to select 4 out of 10 objects in an ordered arrangement, we can now use our permutation rule to compute the number of ways that we can make the selection. Setting $n = 10$ and $k = 4$, we obtain

$$P(10, 4) = 10 \cdot 9 \cdot 8 \cdot 7$$

for the number of different relay teams. Since the product consists of four terms in descending order, starting with 10, the last term is $n - k + 1 = 10 - 4 + 1 = 7$.

Another way to compute $P(n, k)$ follows from the calculation

$$P(n, k) = n(n-1)(n-2)\cdot(n-k+1)\frac{(n-k)(n-k-1)\cdots 3 \cdot 2 \cdot 1}{(n-k)(n-k-1)\cdots 3 \cdot 2 \cdot 1}$$

which, after simplification, yields

$$P(n, k) = \frac{n!}{(n-k)!} \tag{12.1}$$

Let us consider one more example before we introduce the third counting principle.

EXAMPLE 5

How many 5-letter words with no repeated letters can you form out of the 26 letters of the alphabet? (Note that a "word" here need not be in the dictionary.)

Solution

This task amounts to choosing 5 letters from 26, where the order is important. Hence, there are

$$P(26, 5) = \frac{26!}{21!} = 26 \cdot 25 \cdot 24 \cdot 23 \cdot 22 = 7{,}893{,}600$$

different words.   ■

■ 12.1.3 Combinations

In choosing a permutation, the order of the objects is important. But what if the order is not important? How can we then compute the number of arrangements?

We return to Example 4, in which we chose a relay team. The order on the team is important when the members on the team actually run. But if we wanted to know only who was on the team, the order would no longer be important. We saw that there are $10 \cdot 9 \cdot 8 \cdot 7$ different relay teams. But since 4 people can be arranged in $4! = 4 \cdot 3 \cdot 2 \cdot 1 = 24$ different ways, each choice of 4 people appears in 24 different teams. If we divide $10 \cdot 9 \cdot 8 \cdot 7$ by $4 \cdot 3 \cdot 2 \cdot 1$, we obtain the number of ways that we can choose 4 people from a group of 10 if the order does not matter. We find that this number is

$$\frac{10 \cdot 9 \cdot 8 \cdot 7}{4 \cdot 3 \cdot 2 \cdot 1} = 210$$

Such unordered selections are called **combinations**. The approach we just used gives us a general formula for the number of combinations, summarized as follows:

> **Combinations** A **combination** of n different objects taken k at a time is an *unordered* subset of k out of n objects. The number of ways that this can be done is denoted by $C(n, k)$ and is given by
>
> $$C(n, k) = \frac{n(n-1)(n-2)\cdots(n-k+1)}{k!}$$

Note that it follows from the probabilistic meaning of $C(n, k)$ that $C(n, k)$ is always an integer. Instead of $C(n, k)$, we often write $\binom{n}{k}$, which we read "n choose k." The symbol $\binom{n}{k}$ is called a **binomial coefficient**. Using (12.1), we find that

$$C(n, k) = \binom{n}{k} = \frac{P(n, k)}{k!} = \frac{n!}{k!\,(n-k)!} \tag{12.2}$$

Looking at the rightmost expression, we see that it is equal to $\binom{n}{n-k}$. We therefore find the identity

$$\binom{n}{k} = \binom{n}{n-k} \tag{12.3}$$

The following counting argument also explains this identity: The expression $\binom{n}{k}$ denotes the number of ways that we can select an unordered subset of size k from a set of size n. Instead of choosing the elements that go into the set, we could choose the elements that do not go into the set. There are $n-k$ such elements, and we can select them in $\binom{n}{n-k}$ different ways.

Another identity follows from setting $k = 0$ in (12.3) and using (12.2) and $0! = 1$:

$$\binom{n}{0} = \binom{n}{n} = \frac{n!}{0!\,n!} = 1$$

This identity can also be understood from the following counting argument: The expression $\binom{n}{0}$ means that we select a subset of size 0 from a set of size n, where the order is not important. But there is only one such set: the empty set. Similarly, $\binom{n}{n}$ means that we select a subset of n objects from a set of size n where the order is not important. There is only one way to do this: We must take the entire set of n objects.

We can use similar reasoning to argue that $\binom{n}{1} = n$; this represents the number of ways that we can choose subsets of size 1 where the order is not important. There are n such subsets: all the singletons. [Actually, the order does not play a role when we consider sets with one element, as is reflected in the fact that $P(n, 1) = n$ as well.]

In the next example, we also use the rule for counting combinations.

EXAMPLE 6

Suppose that you wish to plant 5 grass species in a plot. You can choose among 12 different species. How many choices do you have?

Solution

Since the order is not important for this selection, there are

$$C(12, 5) = \frac{12!}{5!\,7!} = \frac{12 \cdot 11 \cdot 10 \cdot 9 \cdot 8}{5 \cdot 4 \cdot 3 \cdot 2 \cdot 1} = 792$$

different ways to make the selection. Alternatively, we could have written

$$C(12, 5) = \binom{12}{5} = \frac{P(12, 5)}{5!}$$

These expressions are equivalent and are evaluated as before. ■

As a last example in this subsection, we prove the binomial theorem, which we encountered in Chapter 4 when we used the formal definition of derivatives to prove the rule for differentiating the function $f(x) = x^n$ for n a positive integer.

EXAMPLE 7

Show that if n is a positive integer, then

$$(x + y)^n = \sum_{k=0}^{n} \binom{n}{k} x^k y^{n-k} \tag{12.4}$$

Solution

The term $(x + y)^n$ consists of n factors, each of the form $x + y$. When we multiply out, each factor contributes either an x or a y. Thus, the product consists of sums that contain terms of the form $x^k y^{n-k}$ for $k = 0, 1, 2, \ldots, n$. A term of the form $x^k y^{n-k}$ occurs $\binom{n}{k}$ times, since there are $\binom{n}{k}$ ways of selecting the factor x k times from among the n factors $(x + y)$. Equation (12.4) then follows. ■

■ 12.1.4 Combining the Counting Principles

The difficult part of counting is to decide which rule to use. To gain experience with this decision, we discuss several examples in which we combine the three counting principles.

EXAMPLE 8

How many different 11-letter words can be formed from the letters in the word MISSISSIPPI?

Solution

There are four S's, four I's, two P's, and one M. There are 11! ways to arrange the letters, but some of the resulting words will be indistinguishable from one another, since letters that repeat in a word can be swapped without creating a new word. Therefore, we need to divide by the order of the repeated letters. We then find that there are

$$\frac{11!}{4! \, 4! \, 2! \, 1!} = 34,650$$

different words. ■

EXAMPLE 9

Returning to Example 2, suppose that you now want *two* different toppings on your pizza. How does this change affect your answer?

Solution

Since there are five toppings and the order in which we choose them is not important, we have $\binom{5}{2}$ choices for the toppings. Everything else remains the same, and we find that there are

$$3 \cdot \binom{5}{2} \cdot 2 = 60$$

different pizzas to choose from. ■

EXAMPLE 10

Suppose that a license plate consists of three letters followed by three digits. How many license plates can there be if repetition of letters, but not of digits, is allowed?

Solution

The order is important in this case. For each letter, there are 26 choices, since repetition is allowed. There are 10 choices for the first digit, 9 choices for the second digit, and 8 choices for the third digit, since repetition of digits is not allowed. Hence, there are

$$26 \cdot 26 \cdot 26 \cdot 10 \cdot 9 \cdot 8 = 12,654,720$$

different license plates with the aforementioned restriction. ■

EXAMPLE 11

An urn contains six green and four blue balls. You take out three balls at random without replacement. How many different selections contain exactly two green balls and one blue ball? (Assume that the balls are distinguishable.)

Solution   The order in this selection is not important. To obtain two green balls and one blue ball, we select two of the six green and one of the four blue balls and then combine our choices. That is, there are

$$\binom{6}{2}\binom{4}{1} = 60$$

different selections.   ■

In the preceding example, we explicitly stated that all of the balls were distinguishable, and from now on we will always assume that the objects we select are distinguishable without explicitly stating this assumption. Our counting principles apply only to distinguishable objects, as assumed in the definitions ("*n* different objects"). There are cases in physics where objects are indistinguishable—for instance, electrons—but we will not deal with this possibility here.

EXAMPLE 12   A collection contains seeds for five different annual plants; two produce yellow flowers, and the other three produce blue flowers. You plan a garden bed with three different annual plants from this selection, but you do not want both of the plants with yellow flowers in the bed. How many different selections can you make?

Solution   Possible selections contain either plants with blue flowers only or one plant with yellow and two with blue flowers. We must choose three different plants out of the available five. There are $\binom{3}{3}\binom{2}{0}$ choices for a flower bed with blue flowers only and $\binom{3}{2}\binom{2}{1}$ choices for a flower bed with exactly one plant with yellow flowers. Hence, adding up the two cases, there are

$$\binom{3}{3}\binom{2}{0} + \binom{3}{2}\binom{2}{1} = (1)(1) + (3)(2) = 7$$

different selections.

Alternatively, we could have approached this problem in the following way: There are $\binom{5}{3}$ ways to select three plants from the available five. Choices with two plants with yellow flowers and one plant with blue flowers are not acceptable; there are $\binom{3}{1}\binom{2}{2}$ such choices. There are no choices of plants with three yellow and no blue flowers, since there are only two plants with yellow flowers. All other choices are acceptable. Hence, there are

$$\binom{5}{3} - \binom{3}{1}\binom{2}{2} = 10 - (3)(1) = 7$$

different selections.   ■

EXAMPLE 13   A standard deck of cards consists of 52 cards, arranged in four suits, each with 13 different values. In the game of poker, a hand consists of 5 cards drawn at random from the deck without replacement.

(a) How many hands are possible?

(b) How many hands consist of exactly one pair (i.e., two cards of equal value, with the three other cards having different values)?

Solution   (a) There are

$$\binom{52}{5} = 2,598,960$$

different ways of choosing 5 cards from a deck of 52 cards.

(b) To pick exactly one pair, we first assign the value to the pair (13 ways) and then choose their suits [$\binom{4}{2}$ ways]. The 3 remaining cards all have different values that are different from the pair and from each other [$\binom{12}{3}$ ways]. There are four ways to assign a suit to each card and thus a total of 4^3 ways. Combining the different steps, we find that there are

$$13 \cdot \binom{4}{2}\binom{12}{3} \cdot 4^3 = 1,098,240$$

ways to pick exactly one pair. ■

Section 12.1 Problems

■ 12.1.1

1. Suppose that you want to investigate the influence of light and fertilizer levels on plant performance. You plan to use five fertilizer and two light levels. For each combination of fertilizer and light level, you want four replicates. What is the total number of replicates?

2. Suppose that you want to investigate the effects of leaf damage on the performance of drought-stressed plants. You plan to use three levels of leaf damage and four different watering protocols. For each combination of leaf damage and watering protocol, you plan to have three replicates. What is the total number of replicates?

3. *Coleomegilla maculata*, a lady beetle, is an important predator of egg masses of *Ostrinia nubialis*, the European corn borer. *C. maculata* also feeds on aphids and maize pollen. To study its food preferences, you choose two satiation levels for *C. maculata* and combinations of two of the three food sources (i.e., either egg masses and aphids, egg masses and pollen, or aphids and pollen). For each experimental protocol, you want 20 replicates. What is the total number of replicates?

4. To test the effects of a new drug, you plan the following clinical trial: Each patient receives the new drug, an established drug, or a placebo. You enroll 50 patients. In how many ways can you assign them to the three treatments?

5. The Muesli-Mix is a popular breakfast hangout near a campus. A typical breakfast there consists of one beverage, one bowl of cereal, and a piece of fruit. If you can choose among three different beverages, seven different cereals, and four different types of fruit, how many choices for breakfast do you have?

6. To study sex differences in food preferences in rats, you offer one of three choices of food to each rat. You plan to have 12 rats for each food-and-sex combination. How many rats will you need?

7. The genome of the HIV virus consists of 9749 nucleotides. There are four different types of nucleotides. Determine the total number of different genomes of size 9749 nucleotides.

8. Automated chemical synthesis of DNA has made it possible to custom-order moderate-length DNA sequences from commercial suppliers. Assume that a single nucleotides weighs about 5.6×10^{-22} gram and that there are four kinds of nucleotides. If you wish to order all possible DNA sequences of a fixed length, at what length will your order exceed **(a)** 100 kg and **(b)** the mass of the Earth (5.9736×10^{24} kg)?

■ 12.1.2

9. You plan a trip to Europe during which you wish to visit London, Paris, Amsterdam, Rome, and Heidelberg. Because you want to buy a railway ticket before you leave, you must decide on the order in which you will visit these five cities. How many different routes are there?

10. Five people line up for a photograph. How many different lineups are possible?

11. You have just bought seven different books. In how many ways can they be arranged on your bookshelf?

12. Four cars arrive simultaneously at an intersection. Only one car can go through at a time. In how many different ways can they leave the intersection?

13. How many four-letter words with no repeated letters can you form from the 26 letters of the alphabet?

14. A committee of 3 people must be chosen from a group of 10. The committee consists of a president, a vice president, and a treasurer. How many committees can be selected?

15. Three different awards are to be given to a class of 15 students. Each student can receive at most one award. Count the number of ways these awards can be given out.

16. You have just enough time to play 4 songs out of 10 from your favorite CD. In how many ways can you program your CD player to play the 4 songs?

17. Six customers arrive at a bank at the same time. Only one customer at a time can be served. In how many ways can the six customers be served?

18. An amino acid is encoded by triplet nucleotides. How many different amino acids are possible if there are four different nucleotides that can be chosen for a triplet?

■ 12.1.3

19. A bag contains 10 different candy bars. You are allowed to choose 3. How many choices do you have?

20. During International Movie Week, 60 movies are shown. You have time to see 5 movies. How many different plans can you make?

21. A committee of 3 people must be formed from a group of 10. How many committees can there be if no specific tasks are assigned to the members?

22. A standard deck contains 52 different cards. In how many ways can you select 5 cards from the deck?

23. An urn contains 15 different balls. In how many ways can you select 4 balls without replacement?

24. Twelve people wait in front of an elevator that has room for only 5. Count the number of ways that the first group of people to take the elevator can be chosen.

25. Four A's and five B's are to be arranged into a nine-letter word. How many different words can you form?

26. Suppose that you want to plant a flower bed with four different plants. You can choose from among eight plants. How may different choices do you have?

27. Amin owns a 4-GB music storage device that holds 1000 songs. How many different playlists of 20 songs are there if the order of the songs is important?

28. A bookstore has 300 science fiction books. Molly wants to buy 5 of the 300 science fiction books. How many selections are there?

■ 12.1.4

29. A box contains five red and four blue balls. You choose two balls.

(a) How many possible selections contain exactly two red balls, how many exactly two blue balls, and how many exactly one of each color?

(b) Show that the sum of the number of choices for the three cases in (a) is equal to the number of ways that you can select two balls out of the nine balls in the box.

30. Twelve children are divided up into three groups, of five, four, and three children, respectively. In how many ways can this be done if the order within each group is not important?

31. Five A's, three B's, and six C's are to be arranged into a 14-letter word. How many different words can you form?

32. A bag contains 45 beans of three different varieties. Each variety is represented 15 times in the bag. You grab 9 beans out of the bag.

(a) Count the number of ways that each variety can be represented exactly three times in your sample.

(b) Count the number of ways that only one variety appears in your sample.

33. Let $S = \{a, b, c\}$. List all possible subsets, and argue that the total number of subsets is $2^3 = 8$.

34. Suppose that a set contains n elements. Argue that the total number of subsets of this set is 2^n.

35. In how many ways can Brian, Hilary, Peter, and Melissa sit on a bench if Peter and Melissa want to be next to each other?

36. Paula, Cindy, Gloria, and Jenny have dinner at a round table. In how many ways can they sit around the table if Cindy wants to sit to the left of Paula?

37. In how many ways can you form a committee of three people from a group of seven if two of the people do not want to serve together?

38. In how many ways can you form two committees of three people each from a group of nine if

(a) no person is allowed to serve on more than one committee?

(b) people can serve on both committees simultaneously?

39. A collection contains seeds for four different annual and three different perennial plants. You plan a garden bed with three different plants, and you want to include at least one perennial. How many different selections can you make?

40. In diploid organisms, chromosomes appear in pairs in the nuclei of all cells except gametes (sperm or ovum). Gametes are formed during meiosis, a process in which the number of chromosomes in the nucleus is halved; that is, only one member of each pair of chromosomes ends up in a gamete. Humans have 23 pairs of chromosomes. How may kinds of gametes can a human produce?

41. Sixty patients are enrolled in a small clinical trial to test the efficacy of a new drug against a placebo and the currently used drug. The patients are divided into 3 groups of 20 each. Each group is assigned one of the three treatments. In how many ways can the patients be assigned?

42. One hundred patients wish to enroll in a small study in which patients are divided into four groups of 25 patients each. In how many ways can this be done if no patient is to be assigned to more than one group?

43. Expand $(x + y)^4$.   **44.** Expand $(2x - 3y)^5$.

45. In how many ways can four red and five black cards be selected from a standard deck of cards if cards are drawn without replacement?

46. In how many ways can two aces and three kings be selected from a standard deck of cards if cards are drawn without replacement?

47. In the game of poker, determine the number of ways exactly two pairs can be picked.

48. In the game of poker, determine the number of ways a *flush* (five cards of the same suit) can be picked.

49. In the game of poker, determine the number of ways *four of a kind* (four cards of the same value, plus one other cards) can be picked.

50. In the game of poker, determine the number of ways a *straight* (five cards with consecutive values, such as A 2 3 4 5 or 7 8 9 10 J or 10 J Q K A, but not all of the same suit) can be picked.

51. Counterpoint   *Counterpoint* is a musical term that means the combination of simultaneous voices; it is synonymous with *polyphony*. In *triple counterpoint*, three voices are arranged such that any voice can take any place of the three possible positions: highest, intermediate, and lowest voice. In how many ways can the three voices be arranged?

52. Counterpoint   *Counterpoint* is a musical term that means the combination of simultaneous voices; it is synonymous with *polyphony*. In *quintuple counterpoint*, five voices are arranged such that any voice can take any place of the five possible positions: from highest to lowest voice. In how many ways can the five voices be arranged?

■ 12.2 What Is Probability?

■ 12.2.1 Basic Definitions

A **random experiment** is a repeatable experiment in which the outcome is uncertain. Tossing a coin and rolling a die are examples of random experiments. The set of all possible outcomes of a random experiment is called the **sample space** and is often denoted by Ω (uppercase Greek omega). We look at some examples in which we describe random experiments and give the associated sample space.

EXAMPLE 1   Suppose that we toss a coin labeled heads (H) on one side and tails (T) on the other. If we toss the coin once, the possible outcomes are H and T, and the sample space is therefore

$$\Omega = \{H, T\}$$

If we toss the coin twice in a row, then each outcome is an ordered pair describing the outcome of the first toss followed by the outcome of the second toss, such as HT, which means heads followed by tails. The sample space is

$$\Omega = \{HH, HT, TH, TT\}$$ ■

EXAMPLE 2

Consider a population for which we keep track of the genotype at one locus. We assume that the genes at this locus occur in three different forms, called alleles and denoted by A_1, A_2, and A_3. Furthermore, we assume that the individuals in the population are diploid; that is, the chromosomes occur in pairs. This means that a genotype is described by a pair of genes, such as A_1A_1. Since the order of the chromosomes is not important, the genotype A_1A_2 is the same as A_2A_1. If our random experiment consists of picking one individual out of the population and noting the genotype, then the sample space is given by

$$\Omega = \{A_1A_1, A_1A_2, A_1A_3, A_2A_2, A_2A_3, A_3A_3\}$$ ■

EXAMPLE 3

An urn contains five balls, numbered 1–5, respectively. We draw two balls from the urn without replacement and note the numbers drawn.

There is some ambiguity in the formulation of this experiment. We can draw the two balls one after the other (without replacing the first in the urn after having noted its number), or we can draw the two balls simultaneously. In the first case, the sample space consists of ordered pairs (i, j), where the first entry is the number on the first ball and the second entry is the number on the second ball. Because the sampling is done without replacement, the two numbers are different. The sample space can then be written as

$$\begin{aligned} \Omega = \{&(1, 2), \ (1, 3), \ (1, 4), \ (1, 5), \\ &(2, 1), \ (2, 3), \ (2, 4), \ (2, 5), \\ &(3, 1), \ (3, 2), \ (3, 4), \ (3, 5), \\ &(4, 1), \ (4, 2), \ (4, 3), \ (4, 5), \\ &(5, 1), \ (5, 2), \ (5, 3), \ (5, 4)\} \end{aligned}$$

or, in short,

$$\Omega = \{(i, j) : 1 \leq i \leq 5, 1 \leq j \leq 5, i \neq j\}$$

In the second case, there is no first or second ball, because we draw the balls simultaneously. We can write this sample space as

$$\begin{aligned} \Omega = \{&(1, 2), \ (1, 3), \ (1, 4), \ (1, 5), \\ &(2, 3), \ (2, 4), \ (2, 5), \\ &(3, 4), \ (3, 5), \\ &(4, 5)\} \end{aligned}$$

or, in short,

$$\Omega = \{(i, j) : 1 \leq i < j \leq 5\}$$

where the first entry of (i, j) represents the smaller of the two numbers on the balls in our sample.

The specifics of the random experiment determine which description of the sample space we prefer. ■

When we perform random experiments, we often consider a particular outcome, or, more formally, a particular subset of the sample space. We call subsets of the sample space **events**. Since an outcome is an element of the sample space, an outcome is an event as well, namely, a subset that consists of just one element. We will use the basic set operations to deal with events.

Basic Set Operations   Suppose that A and B are events of the sample space Ω. The **union** of A and B, denoted by $A \cup B$ (read "A union B"), is the set of all outcomes that belong to either A or B (or both). The **intersection** of A and B, denoted by $A \cap B$ (read "A intersected with B"), is the set of all outcomes that belong to both A and B. Figures 12.2 and 12.3 show these first two set operations. These figures, in which sets are visualized as "bubbles," are called **Venn diagrams**.

We can generalize the union and intersection of two events to a finite number of events. Let $A_1, A_2, \ldots, A_n$ be a finite number of events. Then

$$\bigcup_{i=1}^{n} A_i = A_1 \cup A_2 \cup \cdots \cup A_n = (A_1 \cup A_2 \cup \cdots \cup A_{n-1}) \cup A_n$$

$$= \left[\begin{array}{l} \text{the set of all outcomes that} \\ \text{belong to at least one set } A_i \end{array} \right]$$

$$\bigcap_{i=1}^{n} A_i = A_1 \cap A_2 \cap \cdots \cap A_n = (A_1 \cap A_2 \cap \cdots \cap A_{n-1}) \cap A_n$$

$$= \left[\begin{array}{l} \text{the set of all outcomes that} \\ \text{belong to all sets } A_i, i = 1, 2, \ldots, n \end{array} \right]$$

The **complement** of A, denoted by A^c, is the set of all outcomes contained in Ω that are not in A. (See Figure 12.4.) It follows that

$$\Omega^c = \emptyset \qquad \text{and} \qquad \emptyset^c = \Omega$$

where $\emptyset$ denotes the empty set. Furthermore,

$$(A^c)^c = A$$

When we take complements of unions or intersections, the following two identities are useful (Figure 12.5):

De Morgan's Laws

$$(A \cup B)^c = A^c \cap B^c$$
$$(A \cap B)^c = A^c \cup B^c$$

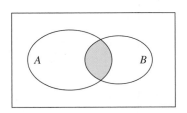

Figure 12.2   The union of A and B, $A \cup B$, illustrated in a Venn diagram. The rectangle represents the set Ω.

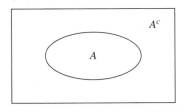

Figure 12.3   The intersection of A and B, $A \cap B$, illustrated in a Venn diagram.

Figure 12.4   The complement of A is A^c.

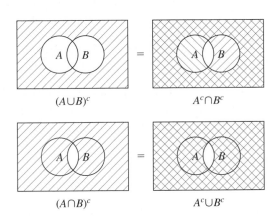

Figure 12.5   De Morgan's laws.

EXAMPLE 4

Let us consider the experiment described in Example 1, in which we tossed a coin twice. The sample space is given by

$$\Omega = \{HH, HT, TH, TT\}$$

We denote by A the event that at least one head occurred:

$$A = \{HH, HT, TH\}$$

Let B denote the event that the first toss resulted in tails:

$$B = \{TH, TT\}$$

We see that

$$A \cup B = \{HH, HT, TH, TT\} \quad \text{and} \quad A \cap B = \{TH\}$$

Furthermore,

$$A^c = \{TT\} \quad \text{and} \quad B^c = \{HH, HT\}$$

To see how De Morgan's laws work, we compute both $(A \cup B)^c$ and $A^c \cap B^c$. We find that

$$(A \cup B)^c = \emptyset \quad \text{and} \quad A^c \cap B^c = \emptyset$$

which is consistent with De Morgan's first law. The second De Morgan's law claims that $(A \cap B)^c$ and $A^c \cup B^c$ are the same. We find that, indeed,

$$(A \cap B)^c = \{HH, HT, TT\} = A^c \cup B^c \qquad \blacksquare$$

We say that $A_1, A_2, \ldots, A_n$ are **pairwise disjoint** (or, simply, disjoint) if

$$A_i \cap A_j = \emptyset \quad \text{whenever} \quad i \neq j$$

This situation is illustrated for four sets in the Venn diagram in Figure 12.6.

Figure 12.6 The four sets A_1, A_2, A_3, and A_4 are disjoint.

EXAMPLE 5

Is it true that if $A_1 \cap A_2 \cap A_3 = \emptyset$, then A_1, A_2, and A_3 are pairwise disjoint?

Solution

No. Figure 12.7 shows a counterexample: $A_1 \cap A_2 \cap A_3 = \emptyset$, but $A_2 \cap A_3 \neq \emptyset$, implying that A_1, A_2, and A_3 are not pairwise disjoint. $\blacksquare$

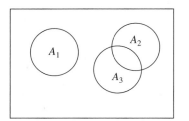

Figure 12.7 A counterexample to Example 5.

The Definition of Probability In the following definition, we assume that the sample space Ω has finitely many elements:

> **Definition** Let Ω be a finite sample space and A and B be events in Ω. A **probability** is a function that assigns values between 0 and 1 to events. The probability of an event A, denoted by $P(A)$, satisfies the following properties:
>
> 1. For any event $A, 0 \leq P(A) \leq 1$.
> 2. $P(\emptyset) = 0$ and $P(\Omega) = 1$.
> 3. For two disjoint events A and B,
>
> $$P(A \cup B) = P(A) + P(B)$$

Note that a probability is a number that is *always* between 0 and 1. If you compute a probability and get either a negative number or a number greater than 1, you know immediately that your answer must be wrong.

EXAMPLE 6

Assume that $\Omega = \{1, 2, 3, 4, 5\}$ and that

$$P(1) = P(2) = 0.2, \quad P(3) = P(4) = 0.1, \quad \text{and} \quad P(5) = 0.4$$

where we wrote $P(i)$ for $P(\{i\})$. Set $A = \{1, 2\}$ and $B = \{4, 5\}$. Find $P(A \cup B)$, and show that $P(\Omega) = 1$.

Solution

Since A and B are disjoint ($A \cap B = \emptyset$), it follows that

$$P(A \cup B) = P(A) + P(B) = P(\{1, 2\}) + P(\{4, 5\})$$

Also, since $\{1, 2\} = \{1\} \cup \{2\}$ and $\{4, 5\} = \{4\} \cup \{5\}$, and both are unions of disjoint sets,

$$P(\{1, 2\}) = P(1) + P(2)$$
$$P(\{3, 4\}) = P(3) + P(4)$$

Hence,

$$P(A \cup B) = P(1) + P(2) + P(4) + P(5)$$
$$= 0.2 + 0.2 + 0.1 + 0.4 = 0.9$$

To show that $P(\Omega) = 1$, we observe that

$$\Omega = \{1, 2, 3, 4, 5\} = \{1\} \cup \{2\} \cup \{3\} \cup \{4\} \cup \{5\}$$

Because this is a union of disjoint sets,

$$P(\Omega) = P(1) + P(2) + P(3) + P(4) + P(5)$$
$$= 0.2 + 0.2 + 0.1 + 0.1 + 0.4 = 1$$ ■

We next derive two additional basic properties of probabilities. The first is

$$P(A^c) = 1 - P(A)$$

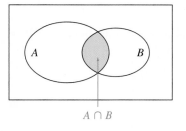

Figure 12.8 The set Ω, represented by the rectangle, can be written as a disjoint union of the sets A and A^c.

To see why this is true, we observe in Figure 12.8 that $\Omega = A \cup A^c$ and that A and A^c are disjoint. Therefore,

$$1 = P(\Omega) = P(A \cup A^c) = P(A) + P(A^c)$$

from which the claim follows after a rearrangement of terms.

Note that in Property 2 of the definition we wrote $P(\emptyset) = 0$ *and* $P(\Omega) = 1$. It would have been sufficient to require just one of these two identities. For instance, since $\Omega^c = \emptyset$, we can write $\Omega = \Omega \cup \emptyset$. Now, Ω and $\emptyset$ are disjoint and $P(\Omega) = 1$. It then follows that

$$1 = P(\Omega) = P(\Omega \cup \emptyset) = P(\Omega) + P(\emptyset)$$

and therefore,

$$P(\emptyset) = 1 - P(\Omega) = 0$$

The second property allows us to compute probabilities of unions of two sets (which are not necessarily disjoint, counter to Property 3 of the definition):

For any sets A and B,

$$P(A \cup B) = P(A) + P(B) - P(A \cap B)$$

Figure 12.9 To compute $P(A \cup B)$, we add $P(A)$ and $P(B)$, but since we count $A \cap B$ twice, we need to subtract $P(A \cap B)$.

The preceding equation is illustrated in Figure 12.9 and follows from the fact that we count $A \cap B$ twice when we compute $P(A) + P(B)$. The proof of this property is left to the reader in Problem 14. Here, we give an example in which we use both of the two additional properties we just described.

EXAMPLE 7

Assume that $\Omega = \{1, 2, 3, 4, 5\}$ and that

$$P(1) = P(2) = 0.2, \quad P(3) = P(4) = 0.1, \quad \text{and} \quad P(5) = 0.4$$

Set $A = \{1, 3, 4\}$ and $B = \{4, 5\}$. Find $P(A \cup B)$.

Solution

Observe that A and B are not disjoint. We find that

$$A \cap B = \{4\}$$

Using the second of the additional properties, we obtain

$$P(A \cup B) = P(A) + P(B) - P(A \cap B)$$
$$= P(\{1, 3, 4\}) + P(\{4, 5\}) - P(\{4\})$$
$$= (0.2 + 0.1 + 0.1) + (0.1 + 0.4) - (0.1) = 0.8$$

We could have gotten the same result by observing that

$$A \cup B = \{1, 3, 4, 5\}, \quad \text{and therefore,} \quad A \cup B = \{2\}^c$$

which yields

$$P(A \cup B) = P(\{2\}^c) = 1 - P(\{2\}) = 1 - 0.2 = 0.8$$ ■

■ 12.2.2 Equally Likely Outcomes

An important class of random experiments with finite sample spaces is that in which all outcomes are equally likely. That is, if $\Omega = \{1, 2, \ldots, n\}$, then $P(1) = P(2) = \cdots = P(n)$, where we wrote $P(i)$ for $P(\{i\})$. Then

$$1 = P(\Omega) = \sum_{i=1}^{n} P(i) = nP(1)$$

which implies that

$$P(1) = P(2) = \cdots = P(n) = \frac{1}{n}$$

If we denote the number of elements in A by $|A|$, and if $A \subset \Omega$ with $|A| = k$, then

$$P(A) = \frac{|A|}{|\Omega|} = \frac{k}{n}$$

In the next two examples, we will discuss random experiments in which all outcomes are equally likely. You should pay particular attention to the sample space, to make sure that you understand that all of its elements are indeed equally likely.

EXAMPLE 8

Toss a fair coin three times and find the probability of the event $A = \{$at least two heads$\}$.

Solution

The sample space in this case is

$$\Omega = \{HHH, HHT, HTH, THH, HTT, THT, TTH, TTT\}$$

All outcomes are equally likely, since we assumed that the coin is fair. Because $|\Omega| = 8$, it follows that each possible outcome has probability $1/8$. Thus,

$$P(A) = P(HHH, HHT, HTH, THH)$$
$$= \frac{|A|}{|\Omega|} = \frac{4}{8} = \frac{1}{2}$$ ■

EXAMPLE 9

An urn contains five blue and six green balls. We draw two balls from the urn without replacement.

(a) Determine the sample space Ω and find $|\Omega|$.

(b) What is the probability that the two balls are of a different color?

(c) What is the probability that at least one of the two balls is green?

Solution

(a) As physical objects, the balls are distinguishable and we can imagine them being numbered from 1 to 11, where we assign the first 5 numbers to the 5 blue balls and the remaining 6 numbers to the 6 green balls. The sample space for this random experiment then consists of all subsets of size 2 that can be drawn from the set of 11 balls. Each subset of size 2 is then equally likely. Using the counting techniques from Section 12.1, we find that the size of the sample space is

$$|\Omega| = \binom{11}{2}$$

since the order in which the balls are removed from the urn is not important and sampling is done without replacement.

(b) Let A denote the event that the two balls are of a different color. To obtain an outcome in which one ball is blue and the other is green, we must select one blue ball from the five blue balls, which can be done in $\binom{5}{1}$ different ways, and one green ball from the six green balls, which can be done in $\binom{6}{1}$ ways. Using the multiplication rule from Section 12.1, we find that

$$|A| = \binom{5}{1}\binom{6}{1}$$

Hence,

$$P(A) = \frac{|A|}{|\Omega|} = \frac{\binom{5}{1}\binom{6}{1}}{\binom{11}{2}} = \frac{5 \cdot 6}{\frac{11 \cdot 10}{2}} = \frac{6}{11}$$

(c) Let B denote the event that at least one ball is green. This event can be written as a union of the following two disjoint sets:

$$B_1 = \{\text{exactly one ball is green}\}$$
$$B_2 = \{\text{both balls are green}\}$$

Since $B = B_1 \cup B_2$ with $B_1 \cap B_2 = \emptyset$, it follows that

$$P(B) = P(B_1) + P(B_2)$$

Using a similar argument as in (b), we find that

$$P(B) = \frac{\binom{5}{1}\binom{6}{1}}{\binom{11}{2}} + \frac{\binom{5}{0}\binom{6}{2}}{\binom{11}{2}} = \frac{5 \cdot 6}{55} + \frac{15}{55} = \frac{9}{11} \qquad ■$$

EXAMPLE 10

Four cards are drawn at random and without replacement from a standard deck of 52 cards. What is the probability of at least two kings?

Solution

We find
$$P(\text{at least two kings}) = 1 - [P(\text{no kings}) + P(\text{one king})]$$

There are $\binom{52}{4}$ ways of selecting four cards. There are four kings in a standard deck of cards. There are $\binom{48}{4}$ ways of selecting a hand of four cards that does not contain any kings and there are $\binom{4}{1}\binom{48}{3}$ ways of selecting a hand of four cards that contains exactly one king. Hence,

$$P(\text{at least two kings}) = 1 - \frac{\binom{48}{4}}{\binom{52}{4}} - \frac{\binom{4}{1}\binom{48}{3}}{\binom{52}{4}}$$

$$= 1 - \frac{48 \cdot 47 \cdot 46 \cdot 45}{52 \cdot 51 \cdot 50 \cdot 49} - \frac{16 \cdot 48 \cdot 47 \cdot 46}{52 \cdot 51 \cdot 50 \cdot 49} \approx 0.0257 \qquad ■$$

An Application from Genetics Gregor Mendel, an Austrian monk, experimented with peas to study the laws of inheritance. He started his experiments in 1856. His work was fundamental in understanding the laws of inheritance. It took over 35 more years until Mendel's original work was publicized and his conclusions were confirmed in additional experiments.

We will use the current knowledge about inheritance to determine the likelihood of outcomes of certain crossings. We describe one of Mendel's experiments that studies the inheritance of flower color in peas. Mendel had seeds that produced plants with either red or white flowers. Flower color in Mendel's peas is determined by a single locus on the chromosome. The genes at this locus occur in two forms, called alleles, which we denote by C and c, respectively. Since pea plants are diploid organisms, each plant has two genes that determine flower color, one from each parent plant. The following genotypes are thus possible: CC, Cc, and cc. The genotypes CC and Cc have red flowers, whereas the genotype cc has white flowers.

EXAMPLE 11

Suppose that you cross two pea plants, both of type Cc. Determine the probability of each genotype occurring in the next generation. What is the probability that a randomly chosen seed from this crossing results in a plant with red flowers?

Solution

We denote the offspring of the crossing as a pair Cc whose first entry is the maternal contribution and second entry is the paternal contribution. For instance, an offspring of type (c, C) inherited a c from the mother and a C from the father. (See Figure 12.10.) We list all possible outcomes of this crossing in the sample space

$$\Omega = \{(C, C), (C, c), (c, C), (c, c)\}$$

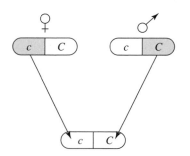

Figure 12.10 The crossing $Cc \times Cc$ resulted in an offspring of type Cc.

The laws of inheritance imply that gametes form at random and that, therefore, all outcomes in Ω are equally likely. (This is Mendel's first law.) Since $|\Omega| = 4$, it follows that each outcome has probability 1/4.

Although the sample space has four different outcomes, there are only three different genotypes, since (C, c) and (c, C) denote the same genotype, Cc. In what follows, we will often denote the event $\{(C, c), (c, C)\}$ simply by Cc, as is customary in genetics. We therefore find that

$$P(CC) = \frac{1}{4}, \quad P(Cc) = \frac{1}{2}, \quad \text{and} \quad P(cc) = \frac{1}{4}$$

Since the two genotypes CC and Cc result in red flowers, it follows that

$$P(\text{red}) = P(\{(C, C), (C, c), (c, C)\}) = \frac{3}{4}$$ ■

The Mark–Recapture Method   The mark–recapture method is commonly used to estimate population sizes. We illustrate the method with a fish population. Suppose that N fish are in a lake, where N is unknown. To get an idea of how big N is, we capture K fish, mark them, and subsequently release them. We wait until the marked fish in the lake have had sufficient time to mix with the other fish. We then capture n fish. Suppose that k of the n fish are marked. (Assume that $k > 0$.) Then if the fish are mixed well again, the ratio of the marked to unmarked fish in the sample of size n should approximately be equal to the ratio of marked to unmarked fish in the lake; that is,

$$\frac{k}{n} \approx \frac{K}{N}$$

We might therefore conclude that there are about

$$N \approx K\frac{n}{k}$$

fish in the lake. We will explain in the next two examples why this approach makes sense.

EXAMPLE 12

Given the mark–recapture experiment, compute the probability of finding k marked fish in a sample of size n.

Solution

There are N fish in the lake, K of which are marked. We choose a sample of size n. Each outcome is therefore a subset of size n, and all outcomes are equally likely. Using the counting techniques from Section 12.1, we find that

$$|\Omega| = \binom{N}{n}$$

since the order in the sample is not important.

We denote by A the event that the sample of size n contains exactly k marked fish. To determine how many outcomes contain exactly k marked fish, we argue as follows: We need to select k fish from the K marked ones and $n-k$ fish from the $N-K$

unmarked ones. Selecting the k marked fish can be done in $\binom{K}{k}$ ways; selecting the $n - k$ unmarked fish can be done in $\binom{N-K}{n-k}$ ways. Since each choice of k marked fish can be combined with any choice of the $n-k$ unmarked fish, we use the multiplication principle to find the total number of ways of obtaining a sample of size n with exactly k marked fish. We obtain

$$|A| = \binom{K}{k}\binom{N-K}{n-k}$$

Therefore,

$$P(A) = \frac{|A|}{|\Omega|} = \frac{\binom{K}{k}\binom{N-K}{n-k}}{\binom{N}{n}}$$

(This example is of the same basic type as the urn problem in Example 9.) ■

We will now give an argument that explains why the total number of fish in the lake can be estimated from the formula $N \approx Kn/k$.

EXAMPLE 13

Assume that there are K marked fish in the lake. We take a sample of size n and observe k marked fish. Show that the value of N which maximizes the probability of finding k marked fish in a sample of size n is the largest integer less than or equal to Kn/k. We use this value as our estimate for the population size N. Since this estimate of N maximizes the probability of what we observe, it is called a **maximum likelihood estimate**.

Solution

We denote by A the event that the sample of size n contains exactly k marked fish. We found in Example 12 that

$$P(A) = \frac{|A|}{|\Omega|} = \frac{\binom{K}{k}\binom{N-K}{n-k}}{\binom{N}{n}}$$

We now consider $P(A)$ as a function of N and denote it by p_N. To find the value of N that maximizes p_N, we look at the ratio p_N/p_{N-1}. (The function p_N is not continuous, since it is defined only for integer values of N; therefore, we cannot differentiate p_N to find its maximum.) The ratio is given by

$$\frac{p_N}{p_{N-1}} = \frac{\dfrac{\binom{K}{k}\binom{N-K}{n-k}}{\binom{N}{n}}}{\dfrac{\binom{K}{k}\binom{N-1-K}{n-k}}{\binom{N-1}{n}}} = \frac{\binom{N-K}{n-k}}{\binom{N-1-K}{n-k}} \, \frac{\binom{N-1}{n}}{\binom{N}{n}}$$

$$= \frac{(N-K)!\,(n-k)!\,(N-1-K-n+k)!}{(n-k)!\,(N-K-n+k)!\,(N-1-K)!} \, \frac{(N-1)!\,n!\,(N-n)!}{n!\,(N-1-n)!\,N!}$$

When we cancel terms, we find that

$$\frac{p_N}{p_{N-1}} = \frac{N-K}{N-K-n+k} \, \frac{N-n}{N}$$

We will now investigate when this ratio is greater than or equal to 1, since, when it is, we can find the values of N for which p_N exceeds p_{N-1}. Values of N for which p_N exceeds both p_{N-1} and p_{N+1} are local maxima. The ratio p_N/p_{N-1} is greater than or equal to 1 if

$$(N-K)(N-n) \geq N(N-K-n+k)$$

Multiplying out both sides of this inequality gives

$$N^2 - Nn - KN + Kn \geq N^2 - NK - Nn + Nk$$

Simplifying yields

$$Kn \geq kN$$

or

$$N \leq K \frac{n}{k}$$

Thus, $p_N \geq p_{N-1}$ as long as $N \leq Kn/k$. If Kn/k is an integer, then $p_N = p_{N-1}$ for $N = Kn/k$ and both Kn/k and $Kn/k-1$ maximize the probability of observing k fish in the sample of size n. Either of the two values can then be chosen as estimates for the number of fish in the lake. If Kn/k is not an integer, then the largest integer less than Kn/k maximizes the probability p_N. To arrive at just one value, we will always use the largest integer less than or equal to Kn/k to estimate the total number of fish in the lake. ■

EXAMPLE 14

Assume that there are 15 marked fish in a lake. We take a sample of size 10 and observe 4 marked fish. Find an estimate of the number of fish in the lake on the basis of Example 13.

Solution

It follows from Example 13 that an estimate for the number of fish in the lake, denoted by N, is the largest integer less than or equal to Kn/k, where, in this example, $K = 15, n = 10$, and $k = 4$. Since

$$K \frac{n}{k} = 15 \cdot \frac{10}{4} = 37.5$$

we estimate that there are 37 fish in the lake.

To see that this value indeed maximizes

$$p_N = \frac{\binom{K}{k} \binom{N-K}{n-k}}{\binom{N}{n}}$$

(defined in Example 13), we graph p_N as a function of N for $K = 15, n = 10$, and $k = 4$ (Figure 12.11). Since there are 15 marked fish in the lake and we sample 6 unmarked fish, the number of fish in the lake must be at least 21. Therefore, $p_N = 0$ for $N < 21$. ■

Figure 12.11 The function p_N for different values of N when $K = 15, n = 10$, and $k = 4$. The graph shows that p_N is maximal for $N = 37$.

Section 12.2 Problems

■ 12.2.1

In Problems 1–4, determine the sample space for each random experiment.

1. The random experiment consisting of tossing a coin three times.

2. The random experiment consisting of rolling a six-sided die twice.

3. An urn contains five balls numbered 1–5, respectively. The random experiment consists of selecting two balls simultaneously without replacement.

4. An urn contains six balls numbered 1–6, respectively. The random experiment consists of selecting five balls simultaneously without replacement.

In Problems 5–8, assume that

$$\Omega = \{1, 2, 3, 4, 5, 6\}$$

$A = \{1, 3, 5\}$, *and* $B = \{1, 2, 3\}$.

5. Find $A \cup B$ and $A \cap B$.

6. Find A^c and show that $(A^c)^c = A$.

7. Find $(A \cup B)^c$.

8. Are A and B disjoint?

In Problems 9–12, assume that

$$\Omega = \{1, 2, 3, 4, 5\}$$

$P(1) = 0.1$, $P(2) = 0.2$, *and* $P(3) = P(4) = 0.05$. *Furthermore, assume that* $A = \{1, 3, 5\}$ *and* $B = \{2, 3, 4\}$.

9. Find $P(5)$. **10.** Find $P(A)$ and $P(B)$.

11. Find $P(A^c)$. **12.** Find $P(A \cup B)$.

In Problems 13–15, assume that

$$\Omega = \{1, 2, 3, 4\}$$

and $P(1) = 0.1$. *Furthermore, assume that* $A = \{2, 3\}$ *and* $B = \{3, 4\}$, $P(A) = 0.7$, *and* $P(B) = 0.5$.

13. Find $P(3)$. **14.** Set $C = \{1, 2\}$. Find $P(C)$.

15. Find $P((A \cap B)^c)$.

16. Assume that $P(A \cap B^c) = 0.1$, $P(B \cap A^c) = 0.5$, and $P((A \cup B)^c) = 0.2$. Find $P(A \cap B)$.

17. Assume that $P(A \cap B) = 0.1$, $P(A) = 0.4$, and $P(A^c \cap B^c) = 0.2$. Find $P(B)$.

18. Assume that $P(A) = 0.4$, $P(B) = 0.4$, and $P(A \cup B) = 0.7$. Find $P(A \cap B)$ and $P(A^c \cap B^c)$.

19. Show the second of the additional properties, namely,

$$P(A \cup B) = P(A) + P(B) - P(A \cap B) \qquad (12.5)$$

(a) Use a diagram to show that B can be written as a disjoint union of the sets $A \cap B$ and $B \cap A^c$.

(b) Use a diagram to show that $A \cup B$ can be written as a disjoint union of the sets A and $B \cap A^c$.

(c) Use your results in (a) and (b) to show that

$$P(A \cup B) = P(A) + P(B \cap A^c)$$

and

$$P(B \cap A^c) = P(B) - P(A \cap B)$$

Conclude from these two equations that (12.5) holds.

20. If $A \subset B$, we can define the difference between the two sets A and B, denoted by $B - A$ (read "B minus A"),

$$B - A = B \cap A^c$$

as illustrated in Figure 12.12.

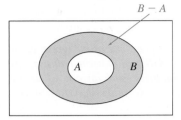

Figure 12.12 The set A is contained in B. The shaded area is the difference of A and B, $B - A$.

Go through the following steps to show that the difference rule

$$P(B - A) = P(B) - P(A) \qquad (12.6)$$

holds:

(a) Use the diagram in Figure 12.12 to show that B can be written as a disjoint union of A and $B - A$.

(b) Use your result in (a) to conclude that

$$P(B) = P(A) + P(B - A)$$

and show that (12.6) follows from this equation.

(c) An immediate consequence of (12.6) is the result that if $A \subset B$, then

$$P(A) \leq P(B)$$

Use (12.6) to show this inequality.

■ 12.2.2

21. Toss two fair coins and find the probability of at least one head.

22. Toss three fair coins and find the probability of no heads.

23. Toss four fair coins and find the probability of exactly two heads.

24. Toss four fair coins and find the probability of three or more heads.

25. Roll a fair die twice and find the probability of at least one 4.

26. Roll two fair dice and find the probability that the sum of the two numbers is even.

27. Roll two fair dice, one after the other, and find the probability that the first number is larger than the second number.

28. Roll two fair dice and find the probability that the minimum of the two numbers will be greater than 4.

29. In Example 11, we considered a cross between two pea plants, each of genotype Cc. Find the probability that a randomly chosen seed from this cross has white flowers.

30. In Example 11, we considered a cross between two pea plants, each of genotype Cc. Now we cross a pea plant of genotype cc with a pea plant of genotype Cc.

(a) What are the possible outcomes of this crossing?

(b) Find the probability that a randomly chosen seed from this crossing results in red flowers.

31. Suppose that two parents are of genotype Aa. What is the probability that their offspring is of genotype Aa? (Assume Mendel's first law.)

32. Suppose that one parent is of genotype AA and the other is of genotype Aa. What is the probability that their offspring is of genotype AA? (Assume Mendel's first law.)

33. A family has three children. Assuming a 1:1 sex ratio, what is the probability that all of the children are girls?

34. A family has three children. Assuming a 1:1 sex ratio, what is the probability that at least one child is a boy?

35. A family has four children. Assuming a 1:1 sex ratio, what is the probability that no more than two children are girls?

In Problems 36–37, we discuss the inheritance of red–green color blindness. Color blindness is an X-linked inherited disease. A woman who carries the color blindness gene on one of her X chromosomes, but not on the other, has normal vision. A man who carries the gene on his only X chromosome is color blind.

36. If a woman with normal vision who carries the color blindness gene on one of her X chromosomes has a child with a man who has normal vision, what is the probability that their child will be color blind?

37. If a woman with normal vision who carries the color blindness gene on one of her X chromosomes has a child with a man who is red–green color blind, what is the probability that their child has normal vision?

38. Cystic fibrosis is an autosomal recessive disease, which means that two copies of the gene must be mutated for a person to be affected. Assume that two unaffected parents who each carry a single copy of the mutated gene have a child. What is the probability that the child is affected?

39. An urn contains three red and two blue balls. You remove two balls without replacement. What is the probability that the two balls are of a different color?

40. An urn contains five blue and three green balls. You remove three balls from the urn without replacement. What is the probability that at least two out of the three balls are green?

41. You select 2 cards without replacement from a standard deck of 52 cards. What is the probability that both cards are spades?

42. You select 5 cards without replacement from a standard deck of 52 cards. What is the probability that you get four aces?

43. An urn contains four green, six blue, and two red balls. You take three balls out of the urn without replacement. What is the probability that all three balls are of different colors?

44. An urn contains three green, five blue, and four red balls. You take three balls out of the urn without replacement. What is the probability that all three balls are of the same color?

45. Four cards are drawn at random without replacement from a standard deck of 52 cards. What is the probability of at least one ace?

46. Four cards are drawn at random without replacement from a standard deck of 52 cards. What is the probability of exactly one pair?

47. Thirteen cards are drawn at random without replacement from a standard deck of 52 cards. What is the probability that all are red?

48. Four cards are drawn at random without replacement from a standard deck of 52 cards. What is the probability that all are of different suits?

49. Five cards are drawn at random without replacement from a standard deck of 52 cards. What is the probability of exactly two pairs?

50. Five cards are drawn at random without replacement from a standard deck of 52 cards. What is the probability of three of a kind and a pair (for instance, Q Q Q 3 3)? (This is called a *full house* in poker.)

51. A lake contains an unknown number of fish, denoted by N. You capture 100 fish, mark them, and subsequently release them. Later, you return and catch 10 fish, 3 of which are marked.

(a) Find the probability that exactly 3 out of 10 fish you just caught will be marked. This probability will be a function of N, the unknown number of fish in the lake.

(b) Find the value of N that maximizes the probability you computed in (a), and show that this value agrees with the value we computed in Example 13.

■ 12.3 Conditional Probability and Independence

Before we define *conditional probability* and *independence*, we will illustrate these concepts by using the Mendelian crossing of peas that we considered in the previous section to study flower color inheritance.

Assume that two parent pea plants are of genotype Cc. Suppose you know that the offspring of the crossing $Cc \times Cc$ has red flowers. What is the probability that it is of genotype CC? We can find this probability by noting that one of the three equally likely possibilities that produce red flowers [namely, (C, C), (C, c), and (c, C) if we list the types according to maternal and paternal contributions as in Example 11 of the previous section] is of type CC. Hence, the probability that the offspring is of genotype CC is 1/3. Such a probability, conditioned on some prior knowledge (such as flower color of offspring), is called a **conditional probability**.

Suppose now that the paternally transmitted gene in the offspring of the crossing $Cc \times Cc$ is of type C. What is the probability that the maternally transmitted gene in the offspring is of type c? To answer this question, we note that the paternal gene has no impact on the choice of the maternal gene in this case. The probability that the maternal gene is of type c is therefore 1/2. We say that the maternal gene is **independent** of the paternal gene: Knowing which of the paternal genes was chosen does not change the probability of the maternal gene.

■ 12.3.1 Conditional Probability

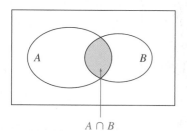

Figure 12.13 The conditional probability of A given B is the proportion of A in the set B, $A \cap B$, relative to the set B.

$A \cap B$

As illustrated in the introduction of this section, conditional probabilities have something to do with prior knowledge. Suppose we know that the event B has occurred and that $P(B) > 0$. Then the conditional probability of the event A given B, denoted by $P(A \mid B)$, is the probability that A will occur given the fact that B has occurred. This probability is defined as

$$P(A \mid B) = \frac{P(A \cap B)}{P(B)} \tag{12.7}$$

To explain this definition, we look at Figure 12.13. The probability of A given B is the proportion of A in the set B relative to B.

In the next example, we will use the definition (12.7) to repeat the introductory example and to find the probability that an offspring is of genotype CC given that its flower color is red.

EXAMPLE 1   Find the probability that the offspring of a $Cc \times Cc$ crossing of pea plants is of type CC given that its flowers are red.

Solution   Let A denote the event that the offspring is of genotype CC and B represent the event that the flower color of the offspring is red. We want to find $P(A \mid B)$. Using (12.7), we have

$$P(A \mid B) = \frac{P(A \cap B)}{P(B)}$$

The *unconditional* probabilities $P(B)$ and $P(A \cap B)$ are computed with the use of the sample space $\Omega = \{(C, C), (C, c), (c, C), (c, c)\}$, whose outcomes all have the same probability. The probability $P(B)$ is the probability that the genotype of the offspring is in the set $\{(C, C), (C, c), (c, C)\}$. Since the sample space has equally likely outcomes, $P(B) = 3/4$. To compute $P(A \cap B)$, we note that $A \cap B$ is the event that the offspring is of genotype CC. Using the sample space Ω, we find that $P(A \cap B) = 1/4$. Hence,

$$P(A \mid B) = \frac{P(A \cap B)}{P(B)} = \frac{\frac{1}{4}}{\frac{3}{4}} = \frac{1}{3}$$

which is the same answer that we obtained before.   ■

As we have seen in Example 1, (12.7) can be used to compute conditional probabilities. By rearranging terms, we can also use (12.7) to compute probabilities of the intersection of events. All we need do is multiply both sides of that equation by $P(B)$, to obtain

$$P(A \cap B) = P(A \mid B)P(B) \tag{12.8}$$

In Equation (12.7), we conditioned on the event B. If we condition on A instead, we have the following identity:

$$P(B \mid A) = \frac{P(A \cap B)}{P(A)}$$

Rearranging terms as in (12.8), we find that

$$P(A \cap B) = P(B \mid A)P(A) \tag{12.9}$$

Formulas (12.8) and (12.9) are particularly useful for computing probabilities in two-stage experiments. We illustrate the use of these identities in the next example. We will also see that there is a natural choice for which of the two events to condition on.

EXAMPLE 2   Suppose that we draw 2 cards at random without replacement from a standard deck of 52 cards. Compute the probability that both cards are diamonds.

Solution

This example can be thought of as a two-stage experiment: We first draw one card and then, without replacing the first one, we draw a second card. We define the two events

$$A = \{\text{the first card is diamond}\}$$
$$B = \{\text{the second card is diamond}\}$$

Then

$$A \cap B = \{\text{both cards are diamonds}\}$$

Now, should we use (12.8) or (12.9)? Since the first card is drawn first, it will be easier to condition on the outcome of the first draw than on the second draw; that is, we will compute $P(B \mid A)$ rather than $P(A \mid B)$ and then use (12.9). We have

$$P(A) = \frac{13}{52}$$

since 13 out of the 52 cards are diamonds and each card has the same probability of being drawn. To compute $P(B \mid A)$, we note that if the first card is a diamond, then there are 12 diamonds left in the deck of the remaining 51 cards. Therefore,

$$P(B \mid A) = \frac{12}{51}$$

Using (12.9), we find that

$$P(A \cap B) = P(B \mid A) P(A) = \frac{12}{51} \cdot \frac{13}{52} = \frac{1}{17}$$

$\blacksquare$

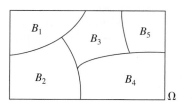

Figure 12.14 The sets $B_1, B_2, \ldots, B_5$ form a partition of Ω.

■ 12.3.2 The Law of Total Probability

We begin this subsection by defining a partition of a sample space. Suppose the sample space Ω is written as a union of n disjoint sets $B_1, B_2, \ldots, B_n$. That is,

(i) $B_i \cap B_j = \emptyset$ whenever $i \neq j$
(ii) $\Omega = \bigcup_{i=1}^{n} B_i$

We then say that the sets $B_1, B_2, \ldots, B_n$ form a **partition** of the sample space Ω. (See Figure 12.14.)

Now, let A be an event. We can use our newly defined partition of Ω to write A as a union of disjoint sets:

$$A = (A \cap B_1) \cup (A \cap B_2) \cup \cdots \cup (A \cap B_n)$$

We illustrate this union in Figure 12.15.

Since the sets $A \cap B_i$, $i = 1, 2, \ldots, n$, are disjoint, it follows that

$$P(A) = \sum_{i=1}^{n} P(A \cap B_i)$$

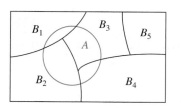

Figure 12.15 The set A is a disjoint union of sets of the form $A \cap B_i$.

To evaluate $P(A \cap B_i)$, we might find it useful to condition on B_i; that is, $P(A \cap B_i) = P(A \mid B_i) P(B_i)$. Then

$$P(A) = \sum_{i=1}^{n} P(A \mid B_i) P(B_i) \tag{12.10}$$

Equation (12.10) is known as the **law of total probability**.

 EXAMPLE 3

A test for the HIV virus shows a positive result in 99% of all cases when the virus is actually present and in 5% of all cases when the virus is not present (a *false positive* result). If such a test is administered to a randomly chosen individual, what is the probability that the test result is positive? Assume that the prevalence of the virus in the population is 1/200.

Solution   We set

$$A = \{\text{test result is positive}\}$$

Individuals in this population fall into two sets: those who are infected with the HIV virus and those who are not. These two sets form a partition of the population. If we pick an individual at random from the population, then the person belongs to one of the two sets. We define

$$B_1 = \{\text{person is infected}\}$$
$$B_2 = \{\text{person is not infected}\}$$

Using (12.10), we can write

$$P(A) = P(A \mid B_1)P(B_1) + P(A \mid B_2)P(B_2)$$

Now, $P(B_1) = 1/200$ and $P(B_2) = 199/200$. Furthermore, $P(A \mid B_1) = 0.99$ and $P(A \mid B_2) = 0.05$. Hence,

$$P(A) = (0.99)\frac{1}{200} + (0.05)\frac{199}{200} = 0.0547 \qquad ■$$

We can illustrated the last example by a **tree diagram**, as shown in Figure 12.16.

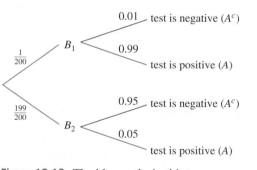

Figure 12.16 The blue paths in this tree diagram lead to the event A. The numbers on the branches represent the respective probabilities.

The numbers on the branches represent the respective probabilities. To find the probability of a positive test result, we multiply the probabilities along the paths that lead to tips labeled "test positive." (The two paths are marked in the tree.) Adding the results along the different paths then yields the desired probability—that is,

$$P(\text{test positive}) = \frac{1}{200}(0.99) + \frac{199}{200}(0.05) = 0.0547$$

as in Example 3. Tree diagrams are quite useful when used together with the law of total probability.

In the next example, we will return to our pea plants. Recall that red-flowering pea plants are of genotype CC or Cc and that white-flowering pea plants are of genotype cc.

EXAMPLE 4   Suppose that you have a batch of red-flowering pea plants, 20% of which are of genotype Cc and 80% of which are of genotype CC. You pick one of the red-flowering plants at random and cross it with a white-flowering plant. Find the probability that the offspring will produce red flowers.

Solution   A white-flowering plant is of genotype cc. If the red-flowering parent plant is of genotype CC (probability 0.8) and is crossed with a white-flowering plant, then all offspring are of genotype Cc and therefore produce red flowers. If the red-flowering parent plant is of genotype Cc (probability 0.2) and is crossed with a white-flowering plant, then, with probability 0.5, an offspring is of genotype Cc (and therefore red flowering) and, with probability 0.5, an offspring is of genotype cc (and therefore white flowering). We use a tree diagram to illustrate the computation of the probability of a red-flowering offspring (Figure 12.17).

The paths that lead to a red-flowering offspring are marked. Using the tree diagram (or the law of total probability), we find that

$$P(\text{red-flowering offspring}) = (0.8)(1) + (0.2)(0.5) = 0.9 \qquad \blacksquare$$

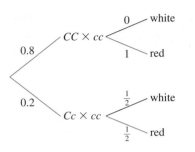

Figure 12.17 The tree diagram for Example 4: The blue paths lead to red flowering offspring.

■ 12.3.3 Independence

Suppose that you toss a fair coin twice. Let A be the event that the first toss results in heads and B the event that the second toss results in heads. Suppose that A occurs. Does this change the probability that B will occur? The answer is obviously no. The outcome of the first toss does not influence the outcome of the second toss. We can express this fact mathematically with conditional probabilities:

$$P(B \mid A) = P(B) \qquad (12.11)$$

Although we say that A and B are *independent*, we will not use (12.11) as the definition of independence; rather, we will use the definition of conditional probabilities to rewrite (12.11). Since $P(B \mid A) = P(A \cap B)/P(A)$, (12.11) can be written as

$$\frac{P(A \cap B)}{P(A)} = P(B)$$

Multiplying both sides by $P(A)$, we obtain $P(A \cap B) = P(A)P(B)$. We use this formula as our definition:

> Two events A and B are **independent** if
> $$P(A \cap B) = P(A)P(B)$$

EXAMPLE 5

Suppose you draw 1 card from a standard deck of 52 cards. Let

$$A = \{\text{card is spade}\}$$
$$B = \{\text{card is king}\}$$

Show that A and B are independent.

Solution

To show that A and B are independent, we compute

$$P(A) = \frac{13}{52} \qquad P(B) = \frac{4}{52}$$

and

$$P(A \cap B) = P(\text{card is king of spade}) = \frac{1}{52}$$

Since

$$P(A)P(B) = \frac{13}{52} \cdot \frac{4}{52} = \frac{1}{52} = P(A \cap B)$$

it follows that A and B are independent. $\qquad \blacksquare$

We now return to our pea plant example to illustrate how we can compute probabilities of intersections of events when we know that the events are independent.

EXAMPLE 6

What is the probability that the offspring of a $Cc \times Cc$ crossing is of genotype cc?

Solution

Previously, we used a sample space with equally likely outcomes to compute the answer. But we can also use independence to compute this probability.

In order for the offspring to be of genotype cc, both parents must contribute a c gene. Let

$$A = \{\text{paternal gene is } c\}$$
$$B = \{\text{maternal gene is } c\}$$

Now, it follows from the laws of inheritance that A and B are independent and that $P(A) = P(B) = 1/2$. Hence,

$$P(cc) = P(A \cap B) = P(A)P(B) = \frac{1}{2} \cdot \frac{1}{2} = \frac{1}{4}$$

This is the same result that we obtained when we looked at the sample space $\Omega = \{(C, C), (C, c), (c, C), (c, c)\}$ and observed that all four possible types were equally likely. ■

We can extend independence to more than two events. We say that events $A_1, A_2, \ldots, A_n$ are independent if, for any $1 \le i_1 < i_2 < \cdots < i_k \le n$,

$$P(A_{i_1} \cap A_{i_2} \cap \cdots \cap A_{i_k}) = P(A_{i_1})P(A_{i_2}) \cdots P(A_{i_k}) \qquad (12.12)$$

To see what (12.12) means when we have three events A, B, and C, we write the conditions explicitly: Three events A, B, and C are independent if

$$\left.\begin{array}{l} P(A \cap B) = P(A)P(B) \\ P(A \cap C) = P(A)P(C) \\ P(B \cap C) = P(B)P(C) \end{array}\right\} \qquad (12.13)$$

and

$$P(A \cap B \cap C) = P(A)P(B)P(C) \qquad (12.14)$$

That is, both (12.13) *and* (12.14) must hold.

The number of conditions we must verify increases quickly with the number of events. When there are just 2 sets, only one condition must be checked, namely, $P(A \cap B) = P(A)P(B)$. When there are 3 sets, as we just saw, there are four conditions: $\binom{3}{2}$ conditions involving pairs of sets and $\binom{3}{3}$ conditions involving all 3 sets. With 4 sets, there is a total of $\binom{4}{2} + \binom{4}{3} + \binom{4}{4} = 11$ conditions; with 5 sets, 26 conditions; and with 10 sets, 1013 conditions.

We emphasize that it is *not* enough to check independence between pairs of events to determine whether a collection of sets is independent. However, independence between pairs of sets is an important property itself, and we wish to define it. We say that events $A_1, A_2, \ldots, A_n$ are **pairwise independent** if

$$P(A_i \cap A_j) = P(A_i)P(A_j) \qquad \text{whenever } i \neq j$$

The next example presents a situation in which events are *pairwise* independent but not *independent*.

EXAMPLE 7

Roll two dice. Let

$$A = \{\text{the first die shows an even number}\}$$
$$B = \{\text{the second die shows an even number}\}$$
$$C = \{\text{the sum of the two dice is odd}\}$$

Show that A, B, and C are pairwise independent but not independent.

Solution   To show pairwise independence, we compute

$$P(A) = \frac{18}{36} = \frac{1}{2} \qquad P(B) = \frac{18}{36} = \frac{1}{2} \qquad P(C) = \frac{18}{36} = \frac{1}{2}$$

$$P(A \cap B) = \frac{9}{36} = \frac{1}{4} = P(A)P(B)$$

$$P(A \cap C) = \frac{9}{36} = \frac{1}{4} = P(A)P(C)$$

$$P(B \cap C) = \frac{9}{36} = \frac{1}{4} = P(B)P(C)$$

However, the event $A \cap B \cap C = \emptyset$, since if both dice show even numbers, then the sum of two even numbers cannot be odd. Therefore,

$$P(A \cap B \cap C) = P(\emptyset) = 0 \neq P(A)P(B)P(C)$$

which shows that A, B, and C are not independent.   ■

When events are independent, we can use (12.12) to compute the probability of their intersections, as illustrated in the next example.

EXAMPLE 8   Assume a 1:1 sex ratio. A family has five children. Find the probability that at least one of the children is a girl.

Solution   Instead of computing the probability that at least one of the children is a girl, we will look at the complement of this event. This is a particularly useful trick when we look at events that ask for the probability of "at least one." We denote by

$$A_i = \{\text{the } i\text{th child is a boy}\}$$

Then the events $A_1, A_2, \ldots, A_5$ are independent. Let

$$B = \{\text{at least one of the children is a girl}\}$$

Instead of computing the probability of B directly, we compute the complement of B. Now, B^c is the event that all children are boys, which is expressed as

$$B^c = A_1 \cap A_2 \cap \cdots \cap A_5$$

It follows that

$$\begin{aligned} P(B) &= 1 - P(B^c) = 1 - P(A_1 \cap A_2 \cap \cdots \cap A_5) \\ &= 1 - P(A_1)P(A_2)\cdots P(A_5) \\ &= 1 - \left(\frac{1}{2}\right)^5 = \frac{31}{32} \end{aligned}$$

If we had tried to compute $P(B)$ directly, we would have needed to compute the probability of exactly one girl, exactly two girls, and so on, and then add all the probabilities. It is quicker and easier to compute the probability of the complement.   ■

■ 12.3.4 The Bayes Formula

In Example 3 of this section, we computed the probability that the result of an HIV test of a randomly chosen individual is positive. For the individual, however, it is much more important to know whether a positive test result actually means that he or she is infected. Recall that we defined

$$A = \{\text{test result is positive}\}$$
$$B_1 = \{\text{person is infected}\}$$
$$B_2 = \{\text{person is not infected}\}$$

We are interested in $P(B_1 | A)$—that is, the probability that a person is infected given that the result is positive. We saw in Example 3 that $P(A | B_1)$ and $P(A | B_2)$ followed immediately from the characteristics of the test. Now we wish to compute a conditional probability such that the roles of A and B_1 are reversed.

Before we compute the probability for this specific example, we look at the general case. We assume that the sets $B_1, B_2, \ldots, B_n$ form a partition of the sample space Ω, A is an event, and the probabilities $P(A | B_i)$, $i = 1, 2, \ldots, n$ are known. We are interested in computing $P(B_i | A)$. We can accomplish this as follows: Using the definition of conditional probabilities, we find that

$$P(B_i | A) = \frac{P(A \cap B_i)}{P(A)} \tag{12.15}$$

To compute $P(A \cap B_i)$, we now condition on B_i; that is,

$$P(A \cap B_i) = P(A | B_i) P(B_i) \tag{12.16}$$

To evaluate the denominator $P(A)$, we use the law of total probability:

$$P(A) = \sum_{j=1}^{n} P(A | B_j) P(B_j) \tag{12.17}$$

Combining (12.15), (12.16), and (12.17), we arrive at the following definition, known as the Bayes formula:

The Bayes Formula Let $B_1, B_2, \ldots, B_n$ form a partition of Ω, and let A be an event. Then

$$P(B_i | A) = \frac{P(A | B_i) P(B_i)}{\sum_{j=1}^{n} P(A | B_j) P(B_j)}$$

We will now return to our example. We wish to find $P(B_1 | A)$—that is, the probability that a person is infected given a positive result. We partition the population into the two sets B_1 and B_2. Then, using the Bayes formula, we find that

$$P(B_1 | A) = \frac{P(A | B_1) P(B_1)}{P(A | B_1) P(B_1) + P(A | B_2) P(B_2)}$$

$$= \frac{(0.99) \frac{1}{200}}{(0.99) \frac{1}{200} + (0.05) \frac{199}{200}} \approx 0.090$$

The denominator is equal to $P(A)$, which we already computed in Example 3.

This example is worth discussing in more detail. The probability is quite small that a person is infected given a positive result. But you should compare this probability with the unconditional probability that a person is infected, namely, $P(B_1)$. The ratio of the conditional to the unconditional probability is

$$\frac{P(B_1 | A)}{P(B_1)} \approx 18.1$$

That is, if a test result is positive, the chance of actually being infected increases by a factor of 18 compared with the chance that a randomly chosen individual in the population who has not been tested is infected. (In practice, if a test result is positive, more than one test is performed to see whether a person is indeed infected or whether the first result was a false positive.)

If a test result is negative, we can also use the Bayes formula to compute the probability that the individual is not infected. We find that

$$P(B_2 \mid A^c) = \frac{P(B_2 \cap A^c)}{P(A^c)} = \frac{P(A^c \mid B_2)P(B_2)}{P(A^c)}$$

$$= \frac{(0.95)\frac{199}{200}}{1 - 0.0547} \approx 0.999947$$

where we used $P(A) = 0.0547$, which we computed in Example 3. This result is rather reassuring.

The reason that the probability of being infected given a positive result is so small comes from the fact that the prevalence of the disease is relatively low (1 in 200). To illustrate, we treat the prevalence of the disease as a variable and compute $P(B_1 \mid A)$ as a function of the prevalence of the disease. That is, we set

$$p = P(\text{a randomly chosen individual is infected})$$

Using the same test characteristics as before, we obtain

$$f(p) = P(B_1 \mid A) = \frac{p(0.99)}{p(0.99) + (1-p)(0.05)} = \frac{0.99p}{0.05 + 0.94p}$$

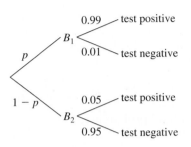

Figure 12.18 The tree diagram for computing the probability that a person is infected given that the test came back positive when the prevalence of the disease is p.

(See Figure 12.18.) A graph of $f(p)$ is shown in Figure 12.19. We see that, for small p, $f(p)$ (the probability of being infected given a positive result) is quite small. The ratio $f(p)/p$ is shown in Figure 12.20, from which we conclude that, although $f(p)$ is small when p is small, the ratio $f(p)/p$ is quite large for small p.

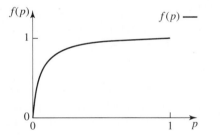

Figure 12.19 The probability of being infected given that the test came back positive as a function of the prevalence of the disease.

Figure 12.20 The ratio $f(p)/p$ as a function of p illustrates by what factor the probability of being infected increases when the test result is positive compared with the prevalence of the disease in the population.

The Bayes formula is also important in genetic counseling. Hemophilia is a blood disorder that is characterized by a deficiency of a blood-clotting factor. Individuals afflicted with this disease suffer from excessive bleeding. The disease is caused by an abnormal gene that resides on the X chromosome. A female who carries the abnormal gene on one of her X chromosomes, but not on the other, is a carrier of the disease but will not develop symptoms. A male who carries the abnormal gene on his (only) X chromosome will develop symptoms of the disease. Almost all symptomatic individuals are males.

In what follows, we assume that only one parent carries the abnormal gene. If the father carries the abnormal gene (and thus suffers from hemophilia), all his daughters will be carriers, since they inherit their father's X chromosome; but all his sons will be disease free, since they inherit their father's Y chromosome. If the mother carries the abnormal gene, then her daughters have a 50% chance of being carriers and her sons have a 50% chance of suffering from the disease.

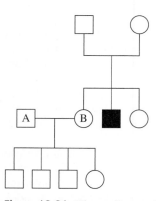

Pedigrees of families show family relationships among individuals and are indispensable tools for tracing diseases of genetic origin. In a pedigree, males are denoted by squares, females by circles; blackened symbols denote individuals who suffer from the disease that is tracked by the pedigree. Figure 12.21 shows the pedigree of a family in which one male (the black square) suffers from hemophilia. We will use this pedigree to determine the probability that individual B is a carrier of the disease given that all three sons of A and B are disease free.

We see from the pedigree that B has a hemophilic brother. Therefore, B's mother must be a carrier. There is a 50% chance that a sister of the affected individual is a carrier. We denote the event that B is a carrier by E. Then $P(E) = 1/2$. Now, assume that we are told that B has three sons with an unaffected male (A). If F denotes the event that all three sons are healthy, then

Figure 12.21 The pedigree of a family in which one member suffers from hemophilia. Squares indicate males, circles females. The black square shows an afflicted individual.

$$P(F \mid E) = \frac{1}{2} \cdot \frac{1}{2} \cdot \frac{1}{2} = \frac{1}{8}$$

since if B is a carrier, each son has probability 1/2 of not inheriting the disease gene and thus being healthy.

We can use the Bayes formula to compute the probability that B is a carrier given that none of her three sons has the disease:

$$P(E \mid F) = \frac{P(E \cap F)}{P(F)} = \frac{P(F \mid E)P(E)}{P(F)}$$

To compute the denominator, we must use the law of total probability, as illustrated in the tree diagram in Figure 12.22.

We find that

$$P(F) = \frac{1}{2} \cdot \frac{1}{8} + \frac{1}{2} \cdot 1 = \frac{9}{16}$$

Therefore, using the Bayes formula, we obtain

$$P(E \mid F) = \frac{\frac{1}{8} \cdot \frac{1}{2}}{\frac{9}{16}} = \frac{1}{9}$$

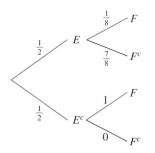

Figure 12.22 The sample space is partitioned into two sets—E and E^c—where E is the event that the individual B is a carrier of the hemophilia gene. Based on whether or not B is a carrier, the probability of the event that all three sons are healthy (F) can be computed as shown.

or, in words, on the basis of the pedigree, the probability that B is a carrier of the gene causing hemophilia given that none of her three sons is symptomatic for the disease is 1/9.

Section 12.3 Problems

■ 12.3.1

1. Suppose you draw 2 cards from a standard deck of 52 cards. Find the probability that the second card is a spade given that the first card is a club.

2. Suppose you draw 2 cards from a standard deck of 52 cards. Find the probability that the second card is a spade given that the first card is a spade.

3. Suppose you draw 3 cards from a standard deck of 52 cards. Find the probability that the third card is a club given that the first two cards are spades.

4. Suppose you draw 3 cards from a standard deck of 52 cards. Find the probability that the third card is a club given that the first two cards are clubs.

5. An urn contains five blue and six green balls. You take two balls out of the urn, one after the other, without replacement. Find the probability that the second ball is green given that the first ball is blue.

6. An urn contains five green, six blue, and four red balls. You take three balls out of the urn, one after the other, without replacement. Find the probability that the third ball is green given that the first two balls were red.

7. A family has two children. The younger one is a girl. Find the probability that the other child is a girl as well.

8. A family has two children. One of their children is a girl. Find the probability that both children are girls.

9. You roll two fair dice. Find the probability that the first die is a 4 given that the sum is 7.

10. You roll two fair dice. Find the probability that the first die is a 5 given that the minimum of the two numbers is a 3.

11. You toss a fair coin three times. Find the probability that the first coin is heads given that at least one head occurred.

12. You toss a fair coin three times. Find the probability that at least two heads occurred given that the second toss resulted in heads.

13. You toss a fair coin four times. Find the probability that four heads occurred given that the first toss and the third toss resulted in heads.

14. You toss a fair coin four times. Find the proability of no more than three heads given that at least one toss resulted in heads.

■ 12.3.2

15. A screening test for a disease shows a positive test result in 90% of all cases when the disease is actually present and in 15% of all cases when it is not. Assume that the prevalence of the disease is 1 in 100. If the test is administered to a randomly chosen individual, what is the probability that the result is negative?

16. A screening test for a disease shows a positive result in 92% of all cases when the disease is actually present and in 7% of all cases when it is not. Assume that the prevalence of the disease is 1 in 600. If the test is administered to a randomly chosen individual, what is the probability that the result is positive?

17. A patient underwent a diagnostic test for hypothyroidism. The diagnostic test correctly identifies patients who in fact have the disease in 93% of the cases and correctly identifies healthy patients in 81% of the cases. If 4 in 100 individuals have the disease, what is the probability that a test comes back negative?

18. A screening test for a disease shows a positive test result in 95% of all cases when the disease is actually present and in 20% of all cases when it is not. When the test was administered to a large number of people, 21.5% of the results were positive. What is the prevalence of the disease?

19. A drawer contains three bags numbered 1–3, respectively. Bag 1 contains three blue balls, bag 2 contains four green balls, and bag 3 contains two blue balls and one green ball. You choose one bag at random and take out one ball. Find the probability that the ball is blue.

20. A drawer contains six bags numbered 1–6, respectively. Bag i contains i blue balls and 2 green balls. You roll a fair die and then pick a ball out of the bag with the number shown on the die. What is the probability that the ball is blue?

21. You pick 2 cards from a standard deck of 52 cards. Find the probability that the second card is an ace. Compare this with the probability that the first card is an ace.

22. You pick 3 cards from a standard deck of 52 cards. Find the probability that the third card is an ace. Compare this with the probability that the first card is an ace.

23. Suppose that you have a batch of red-flowering pea plants of which 40% are of genotype CC and 60% of genotype Cc. You pick one plant at random and cross it with a white-flowering pea plant. Find the probability that the offspring of this crossing will have white flowers.

24. Suppose that you have a batch of red- and white-flowering pea plants, and suppose also that all three genotypes CC, Cc, and cc are equally represented in the batch. You pick one plant at random and cross it with a white-flowering pea plant. What is the probability that the offspring will have red flowers?

25. A bag contains two coins, one fair and the other with two heads. You pick one coin at random and flip it. Find the probability that the outcome is heads.

26. A drug company claims that a new headache drug will bring instant relief in 90% of all cases. If a person is treated with a placebo, there is a 20% chance that the person will feel instant relief. In a clinical trial, half the subjects are treated with the new drug and the other half receive the placebo. If an individual from this trial is chosen at random, what is the probability that the person will have experienced instant relief?

■ 12.3.3

27. You are dealt 1 card from a standard deck of 52 cards. If A denotes the event that the card is a spade and if B denotes the event that the card is an ace, determine whether A and B are independent.

28. You are dealt 2 cards from a standard deck of 52 cards. If A denotes the event that the first card is an ace and B denotes the event that the second card is an ace, determine whether A and B are independent.

29. An urn contains five green and six blue balls. You take two balls out of the urn, one after the other, without replacement. If A denotes the event that the first ball is green and B denotes the event that the second ball is green, determine whether A and B are independent.

30. An urn contains four green and three blue balls. You take one ball out of the urn, note its color, and replace it. You then take a second ball out of the urn, note its color, and replace it. If A denotes the event that the first ball is green and B denotes the event that the second ball is green, determine whether A and B are independent.

31. Assume a 1:1 sex ratio. A family has three children. Find the probability of the event

(a) $A = \{$all children are girls$\}$ **(b)** $B = \{$at least one boy$\}$
(c) $C = \{$at least two girls$\}$ **(d)** $D = \{$at most two boys$\}$

32. Assume that 20% of a very common insect species in your study area is parasitized. Assume that insects are parasitized independently of each other. If you collect 10 specimens of this species, what is the probability that no more than 2 specimens in your sample are parasitized?

33. A multiple-choice question has four choices, and a test has a total of 10 multiple-choice questions. A student passes the test only if he or she answers all questions correctly. If the student guesses the answers to all questions randomly, find the probability that he or she will pass.

34. Assume that A and B are disjoint and that both events have positive probability. Are they independent?

35. Assume that the probability that an insect species lives more than five days is 0.1. Find the probability that, in a sample of size 10 of this species, at least one insect will still be alive after five days.

36. **(a)** Use a Venn diagram to show that

$$(A \cup B)^c = A^c \cap B^c$$

(b) Use your result in (a) to show that if A and B are independent, then A^c and B^c are independent.

(c) Use your result in (b) to show that if A and B are independent, then

$$P(A \cup B) = 1 - P(A^c)P(B^c)$$

■ 12.3.4

37. A screening test for a disease shows a positive result in 95% of all cases when the disease is actually present and in 10% of all cases when it is not. If the prevalence of the disease is 1 in 50 and an individual tests positive, what is the probability that the individual actually has the disease?

38. A screening test for a disease shows a positive result in 95% of all cases when the disease is actually present and in 10% of all cases when it is not. If a result is positive, the test is repeated. Assume that the second test is independent of the first test. If the prevalence of the disease is 1 in 50 and an individual tests positive twice, what is the probability that the individual actually has the disease?

39. A bag contains two coins, one fair and the other with two heads. You pick one coin at random and flip it. What is the probability that you picked the fair coin given that the outcome of the toss was heads?

40. You pick 2 cards from a standard deck of 52 cards. Find the probability that the first card was a spade given that the second card was a spade.

41. Suppose a woman has a hemophilic brother and one healthy son. Suppose furthermore that neither her mother nor her father were hemophilic but that her mother was a carrier for hemophilia. Find the probability that she is a carrier of the hemophilia gene.

The pedigree in Figure 12.23 shows a family in which one member (III-4) is hemophilic. In Problems 42 and 43, refer to this pedigree.

42. (a) Given the pedigree, find the probability that the individual I-2 is a carrier of the hemophilia gene.

(b) Given the pedigree, find the probability that II-3 is a carrier of the hemophilia gene.

43. (a) Given the pedigree, find the probability that II-3 is a carrier of the hemophilia gene.

(b) Given the pedigree, find the probability that III-2 is a carrier of the hemophilia gene.

(c) Given the pedigree, find the probability that II-2 is a carrier of the hemophilia gene.

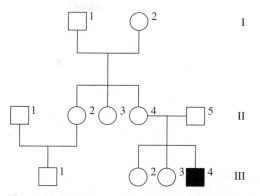

Figure 12.23 The pedigree for Problems 42 and 43. The solid black square (individual III-4) represents an afflicted male.

■ 12.4 Discrete Random Variables and Discrete Distributions

Outcomes of random experiments frequently are real numbers, such as the number of heads in a coin-tossing experiment, the number of seeds produced in a crossing between two plants, or the life span of an insect. Such numerical outcomes can be described by **random variables**. A random variable is a function from the sample space Ω into the set of real numbers. Random variables are typically denoted by X, Y, or Z, or other capital letters chosen from the end of the alphabet. For instance,

$$X : \Omega \to \mathbf{R}$$

describes the random variable X as a map from the sample space Ω into the set of real numbers.

Random variables are classified according to their range. If X takes on a discrete set of values (finite or infinite), X is called a **discrete random variable**. If X takes on a continuous range of values—for instance, values that range over an interval—X is called a **continuous random variable**. Discrete random variables are the topic of this section; continuous random variables are the topic of the next section.

■ 12.4.1 Discrete Distributions

In the first two examples in this section, we look at random variables that take on a discrete set of values. In the first example, this set is finite; in the second example, the set is infinite.

EXAMPLE 1

Toss a fair coin three times. Let X be a random variable that counts the number of heads in each outcome. The sample space is

$$\Omega = \{HHH, HHT, HTH, THH, HTT, THT, TTH, TTT\}$$

and the random variable

$$X : \Omega \to \mathbf{R}$$

takes on values 0, 1, 2, or 3. For instance,

$$X(HHH) = 3 \quad \text{or} \quad X(TTH) = 1 \quad \text{or} \quad X(TTT) = 0 \qquad ■$$

EXAMPLE 2

Toss a fair coin repeatedly until the first time heads appears. Let Y be a random variable that counts the number of trials until the first time heads shows up. The sample space is

$$\Omega = \{H, TH, TTH, TTTH, \ldots\}$$

and the random variable

$$Y : \Omega \to \mathbf{R}$$

takes on values $1, 2, 3, \ldots$. For instance,

$$Y(H) = 1 \qquad Y(TH) = 2 \qquad Y(TTH) = 3 \qquad \cdots \qquad \blacksquare$$

We will now turn to the problem of how to assign probabilities to the different values of a random variable X. For the moment, we will restrict the discussion to the case when the range of X is finite.

Let's go back to Example 1. The coin in Example 1 is fair. This means that each outcome in Ω has the same probability, namely, 1/8. We can translate this set of probabilities into probabilities for X. For instance,

$$
\begin{aligned}
P(X = 1) &= P(\{HTT, THT, TTH\}) \\
&= P(HTT) + P(THT) + P(TTH) \\
&= \frac{1}{8} + \frac{1}{8} + \frac{1}{8} = \frac{3}{8}
\end{aligned}
$$

x	$P(X = x)$
0	1/8
1	3/8
2	3/8
3	1/8

We can perform similar computations for all other values of X. The table on the left summarizes the results.

The function $p(x) = P(X = x)$ is called a **probability mass function**. Note that $p(x) \geq 0$ and $\sum_x p(x) = 1$; these are defining properties of a probability mass function.

> **Definition** A random variable is called a **discrete random variable** if it takes on a finite or infinite set of discrete values. The probability distribution of X can be described by the probability mass function $p(x) = P(X = x)$, which has the following properties:
>
> 1. $p(x) \geq 0$
> 2. $\sum_x p(x) = 1$, where the sum is over all values of X with $P(X = x) > 0$

The probability mass function is one way to describe the probability distribution of a discrete random variable. Another important function that describes the probability distribution of a random variable X is the (cumulative) distribution function $F(x) = P(X \leq x)$. This function is defined for *any* random variable, not just discrete ones.

> **Definition** The **(cumulative) distribution function** $F(x)$ of a random variable X is defined as
>
> $$F(x) = P(X \leq x)$$

Instead of "cumulative distribution function," we will simply say "distribution function."

The probability mass function and the distribution function are equivalent ways of describing the probability distribution of a discrete random variable, and we can obtain one from the other, as illustrated in the next two examples.

EXAMPLE 3

Suppose that the probability mass function of a discrete random variable X is given by the table on the left on the following page. Find and graph the corresponding distribution function $F(x)$.

Solution

The function $F(x)$ is defined for all values of $x \in \mathbf{R}$. For instance, $F(-2.3) = P(X \leq -2.3) = P(\emptyset) = 0$ or $F(1) = P(X \leq 1) = P(X = -1 \text{ or } 0) = 0.3$. Since $F(x) =$

x	$P(X = x)$
-1	0.1
0	0.2
1.5	0.05
3	0.15
5	0.5

$P(X \le x)$, we must be particularly careful when x is in the range of X. To illustrate, we compute $F(1.4)$ and $F(1.5)$. We find that

$$F(1.4) = P(X \le 1.4) = P(X = -1 \text{ or } 0) = 0.1 + 0.2 = 0.3$$
$$F(1.5) = P(X \le 1.5) = P(X = -1, 0, \text{ or } 1.5) = 0.1 + 0.2 + 0.05 = 0.35$$

The distribution function $F(x)$ is a piecewise-defined function. We have

$$F(x) = \begin{cases} 0 & \text{for } x < -1 \\ 0.1 & \text{for } -1 \le x < 0 \\ 0.3 & \text{for } 0 \le x < 1.5 \\ 0.35 & \text{for } 1.5 \le x < 3 \\ 0.5 & \text{for } 3 \le x < 5 \\ 1 & \text{for } x \ge 5 \end{cases}$$

The graph of $F(x)$ is shown in Figure 12.24. ■

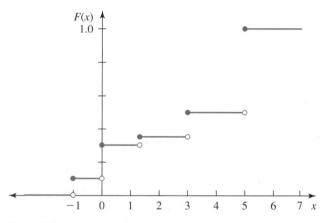

Figure 12.24 The distribution function $F(x)$ of Example 3. The solid circles on the left ends of the line segments indicate that the distribution function takes on this value at the points where the function jumps.

Looking at Figure 12.24, we see that the graph of $F(x)$ is a nondecreasing and piecewise-constant function that takes jumps at those values x where $P(X = x) > 0$. The function $F(x)$ is right continuous; that is, for any $c \in \mathbf{R}$,

$$\lim_{x \to c^+} F(x) = F(c)$$

It is not left continuous everywhere, since, at values $c \in \mathbf{R}$ where $P(X = c) > 0$,

$$\lim_{x \to c^-} F(x) \ne F(c)$$

For instance, when $c = 3$,

$$\lim_{x \to 3^-} F(x) = 0.35 \ne F(3) = 0.5$$

Furthermore, a distribution function has the following additional characteristics:

$$\lim_{x \to -\infty} F(x) = 0 \quad \text{and} \quad \lim_{x \to \infty} F(x) = 1$$

It is possible to obtain the probability mass function from the distribution function. Let's look at the distribution function of Example 3. The function jumps at $x = 3$

and the jump height is 0.15. Since $F(x) = P(X \leq x)$, it follows that

$$p(3) = P(X = 3) = P(X \leq 3) - P(X < 3)$$
$$= F(3) - \lim_{x \to 3^-} F(x) = 0.5 - 0.35 = 0.15$$

We see that the distribution function jumps at the values of X for which $P(X = x) > 0$. The jump height is then equal to the probability that X takes on this value.

EXAMPLE 4

Suppose the distribution function of a discrete random variable X is given by

$$F(x) = \begin{cases} 0 & \text{for } x < -5 \\ 0.2 & \text{for } -5 \leq x < 2 \\ 0.6 & \text{for } 2 \leq x < 3 \\ 0.7 & \text{for } 3 \leq x < 6.5 \\ 1.0 & \text{for } x \geq 6.5 \end{cases}$$

Find the corresponding probability mass function.

Solution

We need to look at the points $x \in \mathbf{R}$ where $F(x)$ jumps. Those are the points where $p(x) = P(X = x) > 0$. The jump height is equal to the probability that X takes on this value. We find that

$$p(-5) = P(X = -5) = P(X \leq -5) - P(X < -5)$$
$$= F(-5) - \lim_{x \to 5^-} F(x) = 0.2 - 0.0 = 0.2$$

Likewise,

$$p(2) = P(X = 2) = 0.6 - 0.2 = 0.4$$
$$p(3) = P(X = 3) = 0.7 - 0.6 = 0.1$$
$$p(6.5) = P(X = 6.5) = 1.0 - 0.7 = 0.3$$

There are no other values of x where $P(X = x) > 0$. We can check our result by adding up the probabilities we just found:

$$p(-5) + p(2) + p(3) + P(6.5) = 0.2 + 0.4 + 0.1 + 0.3 = 1.0$$

The sum adds to 1, which indicates that there cannot be any other values x where $P(X = x) > 0$. ■

■ 12.4.2 Mean and Variance

Knowing the distribution of a random variable tells us everything about the random variable. In practice, however, it is often impossible or unnecessary to know the full probability distribution of a random variable that describes a particular random experiment. Instead, it might suffice to determine a few characteristic quantities, such as the average value and a measure that describes the spread around the average value.

The Average Value, or the Mean, of a Discrete Random Variable

EXAMPLE 5

x	$P(X = x)$
1	0.05
2	0.1
3	0.2
4	0.3
5	0.25
6	0.1

Clutch size can be thought of as a random variable. Let X denote the number of eggs per clutch laid by a certain species of bird, and assume that the distribution of X is described by the probability mass function on the left. The average number of eggs per clutch is computed as the weighted sum

$$\text{average value} = \sum_x x P(X = x)$$
$$= (1)(0.05) + (2)(0.1) + (3)(0.2)$$
$$+ (4)(0.3) + (5)(0.25) + (6)(0.1) = 3.9$$

and we find that the average clutch size is 3.9. ■

The average value of X is called the **expected value**, or **mean**, of X and is denoted by $E(X)$. The expected value is a very important quantity. Here is its definition:

If X is a discrete random variable, then the expected value, or mean, of X is

$$E(X) = \sum_x x P(X = x)$$

where the sum is over all values of x with $P(X = x) > 0$.

When the range of X is finite, the sum in the definition is always defined. When the range of X is countably infinite, we must sum an infinite number of terms. Such sums can be finite or infinite, depending on the distribution of X. The expected value of X is defined only if both $\sum_{x<0} x P(X = x)$ and $\sum_{x \geq 0} x P(X = x)$ are finite. Determining whether such infinite sums are finite is beyond the scope of this book, and we will therefore restrict the discussion to cases in which these sums are finite.

The next example shows that the definition of the mean of a discrete random variable coincides with our everyday notion of average values.

EXAMPLE 6

On a winter day somewhere in southern Minnesota, the following temperature readings T_k (in Fahrenheit) at hour k were obtained:

k	0	1	2	3	4	5	6	7	8	9	10	11
T_k	6	6	6	5	5	5	5	5	8	10	12	12

k	12	13	14	15	16	17	18	19	20	21	22	23
T_k	12	12	12	12	10	8	8	8	5	5	3	3

On the basis of these hourly observations, the average temperature on that day, denoted by $\overline{T}$, is

$$\overline{T} = \frac{1}{24}(6 + 6 + 6 + 5 + 5 + 5 + 5 + 5 + 8 + 10 + 12$$

$$+ 12 + 12 + 12 + 12 + 12 + 10 + 8 + 8 + 8 + 5 + 5 + 3 + 3)$$

$$= \frac{183}{24} = 7.625$$

Rearranging these values according to size, we get

$$\overline{T} = \frac{1}{24}[(3 + 3) + (5 + 5 + 5 + 5 + 5 + 5 + 5) + (6 + 6 + 6)$$

$$+ (8 + 8 + 8 + 8) + (10 + 10) + (12 + 12 + 12 + 12 + 12 + 12)]$$

$$= \frac{1}{24}[(3)(2) + (5)(7) + (6)(3) + (8)(4) + (10)(2) + (12)(6)]$$

$$= 3 \cdot \frac{2}{24} + 5 \cdot \frac{7}{24} + 6 \cdot \frac{3}{24} + 8 \cdot \frac{4}{24} + 10 \cdot \frac{2}{24} + 12 \cdot \frac{6}{24}$$

$$= \sum [\text{temperature}] \times [\text{relative frequency of that temperature}]$$

$$= \frac{183}{24} = 7.625$$

■

In Example 6, we introduced the notion of a **relative frequency**, which tells us how often a value appears in a sample relative to the total sample size. For instance, 3°F appears twice in the sample of 24 measurements, so the relative frequency of 3°F is 2/24.

If we interpret the relative frequencies as probabilities, we see that $\overline{T}$ in Example 6 is indeed the expected value of the temperature T on that day.

EXAMPLE 7 The following table contains the number of leaves per basil plant in a sample of 25 basil plants:

16	15	13	16	16
14	16	15	18	17
16	18	16	13	16
16	16	15	15	16
15	18	16	16	15

To find the relative frequency distribution, we must count how often each value occurs and then divide by the sample size, which is 25 in this case. The result is summarized in the following table:

No. of leaves	13	14	15	16	17	18
Relative frequency	$\frac{2}{25}$	$\frac{1}{25}$	$\frac{6}{25}$	$\frac{12}{25}$	$\frac{1}{25}$	$\frac{3}{25}$

We interpret relative frequencies as probabilities. If the random variable X denotes the number of leaves per plant with probability distribution given by the relative frequency distribution, then the expected value of the number of leaves per plant is

$$E(X) = 13 \cdot \frac{2}{25} + 14 \cdot \frac{1}{25} + 15 \cdot \frac{6}{25} + 16 \cdot \frac{12}{25} + 17 \cdot \frac{1}{25} + 18 \cdot \frac{3}{25}$$

$$= 393 \cdot \frac{1}{25} = 15.72$$

Note that although the number of leaves per plant is an integer, the average number of leaves per plant is not. You would actually lose valuable information if you rounded the average number to the closest integer. ■

It is important to understand that the expected value of an integer-valued random variable need not be an integer. To emphasize this point, consider the average number of lifetime births expected by women 18 to 34 years old in 1992. (The data that follow are data from the U.S. Census Bureau, published in 1994.) The number of lifetime births expected by a woman who is not a high school graduate is 2.393, whereas the corresponding number for a woman with a graduate or professional degree is 1.990. If we rounded these numbers to the closest integer, they would be the same, namely, 2; we would no longer see the difference between the two groups of women.

We can extend the definition of the expected value of X to the expected value of a function of X. Let $g(x)$ be a function of x. Then

$$E[g(X)] = \sum_x g(x)P(X = x) \tag{12.18}$$

EXAMPLE 8 Compute $E(X^2)$ for the random variable X in Example 5.

Solution Using the probability mass function given in Example 5, we find that

$$E(X^2) = \sum_x x^2 P(X = x)$$

$$= (1)^2(0.05) + (2)^2(0.1) + (3)^2(0.2) + (4)^2(0.3) + (5)^2(0.25) + (6)^2(0.1)$$

$$= 16.9$$

■

k	$P(X = k)$	$P(Y = k)$
−10	0	0.2
−1	0.2	0
0	0.6	0.6
1	0.2	0
10	0	0.2

The Variance of a Discrete Random Variable   Another important quantity that characterizes the distribution of a random variable, the **variance** describes how spread out the range of the random variable is. To motivate the definition, let's look at the two random variables X and Y, with probability mass functions as shown on the left. We illustrate these two distributions in Figure 12.25. Both random variables have mean 0, but the range of Y is much more spread out than the range of X.

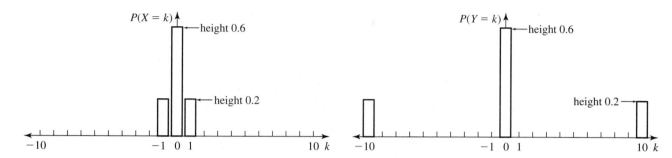

Figure 12.25 The probability mass functions of X and Y. The distribution of Y is more spread out than the distribution of X.

To capture this idea in a single quantity, we will compute the variance, which is defined as a weighted average of the squared distances to the mean:

For any random variable X with mean μ, the **variance** of X is defined as

$$\text{var}(X) = E[(X - \mu)^2]$$

If X is a discrete random variable, then

$$\text{var}(X) = \sum_x (x - \mu)^2 P(X = x)$$

Since the variance is an average value of a squared quantity, it is always nonnegative.

Let's return to the random variables X and Y. Their means are both equal to 0, so their variances are

$$\text{var}(X) = (-1 - 0)^2(0.2) + (0 - 0)^2(0.6) + (1 - 0)^2(0.2) = 0.4$$
$$\text{var}(Y) = (-10 - 0)^2(0.2) + (0 - 0)^2(0.6) + (10 - 0)^2(0.2) = 40$$

We see that the variance of Y is larger than the variance of X, reflecting the fact that the range of Y is more spread out than the range of X.

The variance of X is often denoted by σ^2 (read "sigma squared"). A quantity that is closely related to the variance is the **standard deviation**, denoted by s.d. or σ. The standard deviation is defined as the square root of the variance:

$$\text{s.d.} = \sigma = \sqrt{\text{var}(X)}$$

The standard deviation has the advantage that it has the same units as the mean and so can be interpreted more easily than the variance.

EXAMPLE 9

Compute the variance and the standard deviation of the number of leaves per plant in Example 7.

Solution

Denote by X the random variable that counts the number of leaves per plant with probability distribution given in the table of Example 7. In that example, we found that $E(X) = 15.72$. Therefore, the variance of X is

$$\text{var}(X) = (13 - 15.72)^2 \frac{2}{25} + (14 - 15.72)^2 \frac{1}{25} + (15 - 15.72)^2 \frac{6}{25}$$

$$+ (16 - 15.72)^2 \frac{12}{25} + (17 - 15.72)^2 \frac{1}{25} + (18 - 15.72)^2 \frac{3}{25}$$

$$= 1.5616$$

and the standard deviation of X is

$$\text{s.d.}(X) = \sqrt{\text{var}(X)} = \sqrt{1.5616} \approx 1.2496 \qquad \blacksquare$$

We will now collect some important rules regarding expected values and variances. The first rule tells us how to compute the expected value and the variance of a linear transformation of X. This rule holds for any random variable, not just discrete ones.

Let a and b be constants. Then

$$E(aX + b) = a[E(X)] + b$$
$$\text{var}(aX + b) = a^2 \, \text{var}(X)$$

The first property says that the expected value of a linear function of X is the linear function evaluated at the expected value of X. The second property tells us what happens to the variance when we multiply a random variable by a constant factor; it is important to note that the constant factor is squared when we pull it out of the variance. Furthermore, we see that the variance is unchanged when we add a constant term to a random variable. The latter fact can be understood intuitively: Adding a constant term merely shifts the distribution without changing its shape. We will prove these two properties in Problems 25 and 26 for the case when X is a discrete random variable.

EXAMPLE 10 Suppose the average minimum temperature, measured in degrees Fahrenheit, in Minneapolis, Minnesota, in January is 2°F. Find the average minimum temperature in degrees Celsius.

Solution The linear transformation

$$C = \frac{5(F - 32)}{9}$$

converts temperature measured in degrees Fahrenheit (F) into temperature measured in degrees Celsius (C). Hence,

$$E(C) = E\left[\frac{5(F - 32)}{9}\right] = 5\frac{EF - 32}{9}$$

$$= 5\frac{2 - 32}{9} = -\frac{150}{9} \approx -16.67$$

and we find that the average minimum temperature in January in Minneapolis is about $-16.67°$C. $\qquad \blacksquare$

EXAMPLE 11 Find a formula that converts the variance of a temperature measured in degrees Celsius into the variance of the temperature measured in degrees Fahrenheit.

Solution We use the linear transformation of Example 10. This transformation relates a temperature measured in degrees Fahrenheit (F) to the temperature measured in degrees Celsius (C):

$$C = \frac{5(F - 32)}{9}$$

Solving this equation for F, we obtain

$$F = \frac{9}{5}C + 32$$

Therefore,

$$\text{var}(F) = \text{var}\left(\frac{9}{5}C + 32\right) = \left(\frac{9}{5}\right)^2 \text{var}(C)$$

$$= \frac{81}{25}\text{var}(C) = (3.24)\,\text{var}(C) \qquad ■$$

It is often necessary to look at sums of random variables. We collect some rules without proof. Let X and Y be two random variables. Then $X + Y$ is also a random variable, and we have

$$\boxed{E(X + Y) = E(X) + E(Y)}$$

This formula holds for any random variables, not just discrete ones.

EXAMPLE 12

Suppose the average number of women who enter a coffee shop during lunch hour is 52.2 and the average number of men is 47.3. Find the average total number of people entering the coffee shop during lunch hour.

Solution

If we denote the number of women by X and the number of men by Y, then we are interested in finding $E(X + Y)$. With $E(X) = 52.2$ and $E(Y) = 47.3$, we have

$$E(X + Y) = E(X) + E(Y) = 52.2 + 47.3 = 99.5 \qquad ■$$

We can use our rules to find an alternative formula for the variance. We start with

$$(X - \mu)^2 = X^2 - 2X\mu + \mu^2$$

Taking expectations on both sides, we find that

$$E(X - \mu)^2 = E(X^2 - 2\mu X + \mu^2)$$

Since the expectation of a sum is the sum of the expectations, the right-hand side simplifies to

$$E(X^2) - E(2\mu X) + E(\mu^2) = E(X^2) - 2\mu E(X) + \mu^2 = E(X^2) - [E(X)]^2$$

because $\mu = E(X)$, $E(\mu^2) = \mu^2 = [E(X)]^2$, and $E(2\mu X) = 2\mu E(X) = 2[E(X)]^2$. With $E(X - \mu)^2 = \text{var}(X)$, we have

$$\boxed{\text{var}(X) = E(X^2) - [E(X)]^2}$$

This formula for the variance is often more convenient to use, since it leads to algebraically simpler expressions. Note that $E(X^2) \neq [E(X)]^2$, unless $\text{var}(X) = 0$, and that $E(X^2) \geq [E(X)]^2$, since $\text{var}(X) \geq 0$. In the next example, we apply this formula to the random variable X in Example 5.

EXAMPLE 13

Use the random variable X in Example 5, the result of Example 8, and the preceding formula to compute the variance of X.

Solution

In Example 5, we found that $E(X) = 3.9$. In Example 8, we computed $E(X^2)$ and obtained

$$E(X^2) = 16.9$$

Hence,

$$\text{var}(X) = 16.9 - (3.9)^2 = 1.69 \qquad ■$$

Joint Distributions   It is often important to investigate the relationship between random variables.

EXAMPLE 14   Gout is a type of arthritis in which uric acid is deposited in crystalline form within joints. A medical study might focus on whether the prevalence of the disease is dependent on gender. A survey in 1986 revealed that about 13.6 per 1000 men and 6.4 per 1000 women are affected. We can treat gender as one random variable and the presence of gout as another by defining

$$X = \begin{cases} 1 & \text{if male} \\ 0 & \text{if female} \end{cases}$$

and

$$Y = \begin{cases} 1 & \text{if gout is present} \\ 0 & \text{if gout is not present} \end{cases}$$

The following table lists the number of individuals in each of the four combinations in a study of 10,000 men and 10,000 women:

	$X = 0$	$X = 1$	Total
$Y = 0$	9936	9864	19,800
$Y = 1$	64	136	200
Total	10,000	10,000	20,000

We see that the fraction of individuals in this study that are both male and affected by gout is $136/20,000 = 0.0068$. If we interpret this ratio as a probability, we could write

$$P(X = 1, Y = 1) = 0.0068$$

Converting all numbers in the table into relative frequencies and interpreting them as probabilities produces the **joint probability distribution** of X and Y:

	$X = 0$	$X = 1$	Total
$Y = 0$	0.4968	0.4932	0.99
$Y = 1$	0.0032	0.0068	0.01
Total	0.5	0.5	1.0

■

In general, when X and Y are discrete random variables, we define the joint probability distribution of X and Y by

$$p(x, y) = P(X = x, Y = y)$$

for all values of x in the range of X and all values of y in the range of Y. We can then obtain the distribution of X or of Y called the **marginal distribution** as follows:

$$p_X(x) = P(X = x) = \sum_y P(X = x, Y = y)$$

$$p_Y(y) = P(Y = y) = \sum_x P(X = x, Y = y)$$

EXAMPLE 15   Use the data from Example 14 to determine the probability that a randomly chosen person in the study described there has gout.

Solution   We want to find the probability that $Y = 1$:

$$P(Y = 1) = P(X = 0, Y = 1) + P(X = 1, Y = 1)$$
$$= 0.0032 + 0.0068 = 0.01$$

■

We can define conditional probabilities as in Section 12.3, or

$$P(X = x \mid Y = y) = \frac{P(X = x, Y = y)}{P(Y = y)} \tag{12.19}$$

provided that $P(Y = y) > 0$.

EXAMPLE 16   Use the data of Example 14 to determine the probability that a randomly chosen man in the study described there has gout.

Solution   We want to find $P(Y = 1 \mid X = 1)$. Using (12.19), we find that

$$P(Y = 1 \mid X = 1) = \frac{P(X = 1, Y = 1)}{P(X = 1)} = \frac{0.0068}{0.5} = 0.0136$$

or 13.6 per 1000 men.   ■

   Whether or not gender influences the prevalence of gout leads us to the concept of **independence** of random variables. The definition of independence of random variables follows from that of the independence of events. Recall that events X and Y are **independent** if, for any two sets A and B,

$$P(X \in A, Y \in B) = P(X \in A)P(Y \in B)$$

When X and Y are discrete random variables, this equation simplifies to

$$P(X = x, Y = y) = P(X = x)P(Y = y) \tag{12.20}$$

for all values of x in the range of X and for all values of y in the range of Y.

EXAMPLE 17   Use the data of Example 14 to determine whether X and Y are independent.

Solution   We check whether $P(X = 1, Y = 1)$ is equal to $P(X = 1)P(Y = 1)$. We find that $P(X = 1, Y = 1) = 0.0068$, $P(X = 1) = 0.5$, and $P(Y = 1) = 0.01$. Consequently, $P(X = 1)P(Y = 1) = (0.5)(0.01) = 0.005$, which is different from $P(X = 1, Y = 1)$. We conclude that X and Y are not independent; they are then said to be *dependent*. Note that to show that X and Y are *not* independent, it is enough to find one pair (x, y) for which (12.20) does not hold.   ■

EXAMPLE 18   Suppose that X and Y are two independent discrete random variables with probability mass functions as listed in the following table:

k	$P(X = k)$	$P(Y = k)$
−1	0.1	0.3
0	0.0	0.2
1	0.7	0.1
2	0.2	0.4

(a) Find the probability that X takes on the value −1 and Y takes on the value 2.
(b) Find the probability that X is negative and Y is positive.

Solution   **(a)** We want to find $P(X = -1, Y = 2)$. Since X and Y are independent, we have

$$P(X = -1, Y = 2) = P(X = -1)P(Y = 2) = (0.1)(0.4) = 0.04$$

(b) We want to find the probability of the event that X is negative and Y is positive. This is the event $\{X = -1 \text{ and } Y = 1 \text{ or } 2\}$. Since X and Y are independent, we have

$$P(X = -1 \text{ and } Y = 1 \text{ or } 2) = P(X = -1)P(Y = 1 \text{ or } 2)$$
$$= (0.1)(0.1 + 0.4) = (0.1)(0.5) = 0.05 \quad ■$$

700 Chapter 12 ■ Probability and Statistics

The definition of independence in (12.20) allows us to find the expected value of a product of independent discrete random variables. The following calculation shows the result when X and Y have finite ranges:

$$E(XY) = \sum_{x,y} xy P(X = x, Y = y) = \sum_{x,y} xy P(X = x) P(Y = y)$$
$$= \sum_{x} x P(X = x) \sum_{y} y P(Y = y) = E(X)E(Y)$$

This relationship holds more generally:

> If X and Y are two independent random variables, then
> $$E(XY) = E(X)E(Y)$$

EXAMPLE 19

For the random variables X and Y in Example 18, find $E(XY)$.

Solution

Since X and Y are independent, we have $E(XY) = E(X)E(Y)$. Now,

$$E(X) = (-1)(0.1) + (0)(0.0) + (1)(0.7) + (2)(0.2) = 1.0$$

and

$$E(Y) = (-1)(0.3) + (0)(0.2) + (1)(0.1) + (2)(0.4) = 0.6$$

Hence,

$$E(XY) = E(X)E(Y) = (1.0)(0.6) = 0.6 \qquad ■$$

We can use the rule about the expected value of a product of independent random variables to compute the variance of the sum of two independent random variables. Suppose that X and Y are independent. Then

$$\text{var}(X + Y) = E(X + Y)^2 - [E(X + Y)]^2$$
$$= E(X^2 + 2XY + Y^2) - [E(X) + E(Y)]^2$$
$$= E(X^2) + 2E(XY) + E(Y^2) - [E(X)]^2 - 2E(X)E(Y) - [E(Y)]^2$$

Since X and Y are independent, $E(XY) = E(X)E(Y)$, and the sum simplifies to

$$E(X^2) - [E(X)]^2 + E(Y^2) - [E(Y)]^2$$

But we recognize this quantity as $\text{var}(X) + \text{var}(Y)$. Hence,

> If X and Y are independent random variables, then
> $$\text{var}(X + Y) = \text{var}(X) + \text{var}(Y)$$

Both formulas, $E(XY) = E(X)E(Y)$ and $\text{var}(X + Y) = \text{var}(X) + \text{var}(Y)$, hold for any independent random variables X and Y, not just for discrete ones. However, we will use these identities only in the context of discrete random variables. It is important to keep in mind that these two formulas only hold when X and Y are independent.

In the remaining subsections, we introduce a number of important discrete distributions.

■ 12.4.3 The Binomial Distribution

In this subsection, we will discuss a discrete random variable that models the number of successes among a fixed number of trials. Suppose that you perform a random experiment of repeated trials in which each trial has two possible outcomes: success or failure. Each trial is called a **Bernoulli trial**. The trials are independent and the probability of success in each trial is p. We define the random variables X_k, $k = 1, 2, \ldots, n$, as

$$X_k = \begin{cases} 1 & \text{if the } k\text{th trial is successful} \\ 0 & \text{otherwise} \end{cases}$$

Then $P(X_k = 1) = p = 1 - P(X_k = 0)$ for $k = 1, 2, \ldots, n$.

If we repeat these trials n times, we might want to know the total number of successes. We set

$$S_n = \text{number of successes in } n \text{ trials}$$

We can define S_n in terms of the random variables X_k as

$$S_n = \sum_{k=1}^{n} X_k \tag{12.21}$$

Since the trials are independent, this representation shows that S_n can be written as a sum of independent random variables, all having the same distribution. We will use (12.21) subsequently.

The random variable S_n is discrete and takes on values $0, 1, 2, \ldots, n$. To find its probability mass function $p(k) = P(S_n = k)$, we argue as follows: The event $\{S_n = k\}$ can be represented as a string of zeros and ones of length n, where 0 represents failure and 1 represents success. For instance, if $n = 5$ and $k = 3$, then 01101 could be interpreted as the outcome of five trials, the first resulting in failure, followed by two successes, then a failure, and finally a success. The probability of this particular outcome is easy to compute, since the trials are independent. We obtain

$$P(01101) = (1 - p)pp(1 - p)p = p^3(1 - p)^2$$

The outcome 01101 is not the only one with three successes in five trials: Any other string of length 5 with exactly three ones has the same probability. To determine the number of different strings with this property, note that there are $\binom{5}{3}$ different ways of placing the three ones in the five possible positions and there is exactly one way to place the zeros in the remaining two positions. Hence, there are $\binom{5}{3} \cdot 1 = \binom{5}{3}$ different strings of length 5 with exactly three ones. There is another way to find this; namely, there are 5! ways of arranging the three ones and the two zeros if the zeros and the ones are distinguishable. Since the zeros and the ones can be rearranged among themselves without changing the outcome, we must divide by the order. We then find that there are

$$\frac{5!}{3!2!} = \binom{5}{3}$$

different outcomes. As all outcomes are equally likely, we have

$$P(S_5 = 3) = \binom{5}{3} p^3(1 - p)^2$$

We can use similar reasoning to derive the general formula, which we summarize as follows:

Binomial Distribution Let S_n be a random variable that counts the number of successes in n independent trials, each having probability p of success. Then S_n is said to be **binomially distributed** with parameters n and p, and

$$P(S_n = k) = \binom{n}{k} p^k (1 - p)^{n-k}, \quad k = 0, 1, 2, \ldots, n$$

The random variable S_n is called a **binomial** random variable, and its distribution is called the **binomial distribution**.

EXAMPLE 20

Toss a fair coin four times. Find the probability that there are exactly three heads.

Solution

Let S_4 denote the number of heads. If heads denote success, then the probability of success is $p = 1/2$. S_4 is thus binomially distributed with parameters $n = 4$ and $p = 1/2$. Therefore,

$$P(S_4 = 3) = \binom{4}{3}\left(\frac{1}{2}\right)^3 \left(1 - \frac{1}{2}\right) = 4 \cdot \frac{1}{16} = \frac{1}{4}$$ ∎

EXAMPLE 21

In a shipment of 10 boxes, each box has probability 0.2 of being damaged. Find the probability of having two or more damaged boxes in the shipment.

Solution

Let S_{10} denote the number of damaged boxes in the shipment. S_{10} is binomially distributed with parameters $n = 10$ and $p = 0.2$. The event of two or more damaged boxes can then be written as $S_{10} \geq 2$. To compute $P(S_{10} \geq 2)$, we use the formula

$$P(S_{10} \geq 2) = 1 - P(S_{10} < 2) = 1 - \left[P(S_{10} = 0) + P(S_{10} = 1)\right]$$
$$= 1 - \left[\binom{10}{0}(0.2)^0(0.8)^{10} + \binom{10}{1}(0.2)(0.8)^9\right]$$
$$\approx 0.6242$$ ∎

EXAMPLE 22

Down syndrome, or trisomy 21, is a genetic disorder in which three copies of chromosome 21 instead of two copies are present. In the United States, the prevalence is about 1 in 700 pregnancies. What is the probability that at least 1 in 100 pregnancies is affected?

Solution

If S_{100} is the number of pregnancies and $p = 1/700$ is the probability of a pregnancy being affected, then S_{100} is binomially distributed with parameters $n = 100$ and $p = 1/700$. Thus,

$$P(S_{100} \geq 1) = 1 - P(S_{100} = 0)$$
$$= 1 - \left(1 - \frac{1}{700}\right)^{100} \approx 0.1332$$ ∎

If we use the representation $S_n = \sum_{k=1}^n X_k$ from (12.21) for the binomial random variable S_n, it is straightforward to compute its mean and its variance. We find that

$$E(X_1) = (1)p + (0)(1 - p) = p$$

and, with $E(X_1^2) = (1)^2 p + (0)^2(1 - p) = p$, we have

$$\text{var}(X_1) = E(X_1^2) - [E(X_1)]^2 = p - p^2 = p(1 - p)$$

Since all $X_k, k = 1, 2, \ldots, n$, have the same distribution, it follows that

$$E(S_n) = E\left(\sum_{k=1}^n X_k\right) = \sum_{k=1}^n E(X_k) = np \qquad (12.22)$$

In addition, because the X_k are independent,

$$\text{var}(S_n) = \text{var}\left(\sum_{k=1}^n X_k\right) = \sum_{k=1}^n \text{var}(X_k) = np(1 - p) \qquad (12.23)$$

We summarize these results as follows:

If S_n is binomially distributed with parameters n and p, then

$$E(S_n) = np \qquad \text{and} \qquad \text{var}(S_n) = np(1 - p)$$

We present two more applications of the binomial distribution.

EXAMPLE 23

We consider the flowering pea plants again. Suppose that 20 independent offspring result from $Cc \times Cc$ crossings. Find the probability that at most two offspring have white flowers, and compute the expected value and the variance of the number of offspring that have white flowers.

Solution

In a $Cc \times Cc$ crossing, the probability of a white-flowering offspring (genotype cc) is 1/4 and the probability of a red-flowering offspring (genotype CC or Cc) is 3/4. The flower colors of different offspring are independent. We can therefore think of this experiment as one with 20 trials, each having a probability of success of 1/4. We want to know the probability of at most two successes. With $n = 20$, $k \le 2$, and $p = 1/4$, we find that

$$P(S_{20} \le 2) = P(S_{20} = 0) + P(S_{20} = 1) + P(S_{20} = 2)$$

$$= \binom{20}{0}\left(\frac{1}{4}\right)^0\left(\frac{3}{4}\right)^{20} + \binom{20}{1}\left(\frac{1}{4}\right)^1\left(\frac{3}{4}\right)^{19} + \binom{20}{2}\left(\frac{1}{4}\right)^2\left(\frac{3}{4}\right)^{18}$$

$$\approx 0.0913$$

The expected value of the number of white-flowering offspring is

$$E(S_{20}) = (20)\left(\frac{1}{4}\right) = 5$$

and the variance is

$$\text{var}(S_{20}) = (20)\left(\frac{1}{4}\right)\left(\frac{3}{4}\right) = \frac{15}{4} = 3.75 \qquad ■$$

EXAMPLE 24

Suppose that a woman who is a carrier for hemophilia has four daughters with a man who is not hemophilic. Find the probability that at least one daughter carries the hemophilia gene.

Solution

Each daughter has probability 1/2 of carrying the disease gene, independently of all others. We can think of this experiment as one with four independent trials and a probability of success of 1/2. ("Success" in this case is being a carrier.) Therefore,

$$P(S_4 \ge 1) = 1 - P(S_4 = 0)$$

$$= 1 - \binom{4}{0}\left(\frac{1}{2}\right)^0\left(\frac{1}{2}\right)^4$$

$$= 1 - \frac{1}{16} = \frac{15}{16} = 0.9375 \qquad ■$$

Sampling with and without Replacement Consider an urn with 10 green and 15 blue balls. Sampling balls from this urn can be done with or without replacement. If we sample with replacement, we take out a ball, note its color, and then place the ball back into the urn. The number of balls of a specific color is then binomially distributed.

EXAMPLE 25

If we sample five balls with replacement from the aforementioned urn, what is the probability of three blue balls in the sample?

Solution

Denote the number of blue balls in the sample by S_5. Then S_5 is binomially distributed, with n, the number of trials, equal to 5 and probability of success $p = 15/(15 + 10) = 3/5$. We find that

$$P(S_5 = 3) = \binom{5}{3}\left(\frac{3}{5}\right)^3\left(\frac{2}{5}\right)^2 = 0.3456 \qquad ■$$

If we sample without replacement, we take the balls out one after the other without putting them back into the urn and note their colors. If we take the balls out one after the other without replacing them, then the composition of the urn changes every time we remove a ball and the number of balls of a certain color is no longer binomially distributed.

EXAMPLE 26

If we sample five balls without replacement from the urn, what is the probability of three blue balls in the sample?

Solution

(We encountered a similar problem in Example 9 of Section 12.2.) There are $\binom{25}{5}$ ways of sampling 5 balls from this urn, so $\binom{25}{5}$ is the size of the sample space. Each outcome in this sample space has the same probability. In order to have 3 blue balls, we need to select 3 out of the 15 blue balls, which can be done in $\binom{15}{3}$ ways. Since we want a total of 5 balls, we also need to select 2 out of the 10 green balls, which can be done in $\binom{10}{2}$ ways. Combining the blue and the green balls, we find that there are $\binom{15}{3}\binom{10}{2}$ ways of selecting 3 blue and 2 green balls from the urn. Therefore, the probability of obtaining 3 blue balls in a sample of size 5 when sampling is done without replacement is

$$\frac{\binom{15}{3}\binom{10}{2}}{\binom{25}{5}} = \frac{455 \cdot 45}{53130} \approx 0.3854$$

Note that the answer is different from that in Example 25. ■

The probability distribution in Example 26 is called the **hypergeometric distribution**. The hypergeometric distribution describes sampling without replacement if two types of objects are in the urn. Suppose the urn has M green and N blue balls, and a sample of size n is taken from the urn without replacement. If X denotes the number of blue balls in the sample, then

$$P(X = k) = \frac{\binom{N}{k}\binom{M}{n-k}}{\binom{M+N}{n}}, \quad k = 0, 1, 2, \ldots, n$$

■ 12.4.4 The Multinomial Distribution

In the previous subsection, we considered experiments in which each trial resulted in exactly one of two possible outcomes. We will now extend this situation to more than two possible outcomes. The distribution is then called the **multinomial distribution**.

EXAMPLE 27

To study food preferences in the lady beetle *Coleomegilla maculata*, we present each beetle with three different food choices: maize pollen, egg masses of the European corn borer, and aphids. We suspect that 20% of the time the beetle prefers the aphids, 35% of the time egg masses, and 45% of the time pollen. We carry out this experiment with 30 beetles and find that 8 beetles prefer aphids, 10 egg masses, and 12 pollen. Compute the probability of this event, assuming that the trials are independent.

Solution

We define the random variables

$$N_1 = \text{number of beetles that prefer aphids}$$
$$N_2 = \text{number of beetles that prefer egg masses}$$
$$N_3 = \text{number of beetles that prefer pollen}$$

and the probabilities

$$p_1 = P(\text{beetle prefers aphids}) = 0.2$$
$$p_2 = P(\text{beetle prefers egg masses}) = 0.35$$
$$p_3 = P(\text{beetle prefers pollen}) = 0.45$$

We claim that

$$P(N_1 = 8, N_2 = 10, N_3 = 12) = \frac{30!}{8!\,10!\,12!}(0.2)^8(0.35)^{10}(0.45)^{12}$$

The term $\frac{30!}{8!\,10!\,12!}$ counts the number of ways we can arrange 30 objects—8 of one type, 10 of another, and 12 of a third—that represent the beetles preferring aphids, egg masses, and pollen, respectively. The term $(0.2)^8(0.35)^{10}(0.45)^{12}$ comes from the probability of a particular arrangement, just as in the binomial case. ■

A more involved example for the multinomial distribution is another of Mendel's experiments, one in which he crossed pea plants that had round, yellow seeds with plants that had green, wrinkled seeds. Roundness and yellow color are dominant traits, and greenness and wrinkled texture are recessive traits. We denote the allele for round seeds by R, the allele for wrinkled seeds by r, the allele for yellow seeds by Y, and the allele for green seeds by y. Then a crossing between plants that are homozygous for round, yellow seeds (genotype RR/YY) and plants that are homozygous for wrinkled, green seeds (genotype rr/yy) is written as

$$RR/YY \times rr/yy$$

This crossing results in offspring of type Rr/Yy. That is, all offspring are heterozygous with round, yellow seeds. Crossing plants from this offspring generation then results in all possible combinations, as illustrated in the following table:

	RY	Ry	rY	ry
RY	RR/YY round, yellow	RR/Yy round, yellow	Rr/YY round, yellow	Rr/Yy round, yellow
Ry	RR/Yy round, yellow	RR/yy round, green	Rr/yY round, yellow	Rr/yy round, green
rY	rR/YY round, yellow	rR/Yy round, yellow	rr/YY wrinkled, yellow	rr/Yy wrinkled, yellow
ry	rR/yY round, yellow	rR/yy round, green	rr/yY wrinkled, yellow	rr/yy wrinkled, green

The laws of inheritance tell us that each outcome of this crossing is equally likely. There are 16 different genotypes, some of which give rise to the same morphological type of seed (e.g., round, yellow). The morphological type of the seed is called the **phenotype**. We summarize the phenotypes and their probability distribution in the table on the left.

	Yellow	Green
Round	9/16	3/16
Wrinkled	3/16	1/16

EXAMPLE 28

Suppose that you obtain 50 independent offspring from the crossing

$$Rr/Yy \times Rr/Yy$$

where 25 seeds are round and yellow, 9 are round and green, 12 are wrinkled and yellow, and 4 are wrinkled and green. Find the probability of this outcome.

Solution

This is another application of the multinomial distribution. Arguing as in the previous example, we find that the probability of this outcome is

$$\frac{50!}{25!\,9!\,12!\,4!}\left(\frac{9}{16}\right)^{25}\left(\frac{3}{16}\right)^{9}\left(\frac{3}{16}\right)^{12}\left(\frac{1}{16}\right)^{4}$$

■

■ 12.4.5 Geometric Distribution

We again consider a sequence of independent Bernoulli trials, namely, a random experiment of repeated trials where each trial has two possible outcomes—success or failure—and the trials are independent. As in Subsection 12.4.3, we denote the probability of success by p. This time, however, we define a random variable X that counts the number of trials until the first success. The random variable X takes on values $1, 2, 3, \ldots$ and is therefore a discrete random variable. Its probability distribution is called the **geometric distribution** and is given by

$$P(X = k) = (1 - p)^{k-1}p, \quad k = 1, 2, 3, \ldots \tag{12.24}$$

since the event $\{X = k\}$ means that the first $k - 1$ trials resulted in failure (each one has probability $1 - p$ and the trials are independent) and were followed by a trial that resulted in a success (with probability p and independently of all other trials).

The range of X is the set of all positive integers; thus, X takes on infinitely (though still countably) many values. This is the first time we encounter a random variable of this kind. There are some issues we need to discuss that pertain to a countably infinite range. For instance, to show that $P(X = k)$ in (12.24) is indeed a probability mass function, we will need to sum up the probabilities from $k = 1$ to $k = \infty$. This means summing up an infinite number of terms, and we will need to explain what that means.

To show that (12.24) is a probability mass function, we need to check that $P(X = k) \geq 0$ and that it sums to 1. The first part is straightforward: Since p is a probability, it follows that $0 \leq p \leq 1$, making $(1 - p)^{k-1}p \geq 0$ for all $k = 1, 2, 3, \ldots$. For the second part, we need to show that

$$\sum_k P(X = k) = 1$$

where the sum ranges over all values of k in the range of X, namely, $k = 1, 2, 3, \ldots$. Mathematically, we write this as

$$\sum_{k=1}^{\infty} P(X = k)$$

and define this **infinite sum** as

$$\sum_{k=1}^{\infty} P(X = k) = \lim_{n \to \infty} \sum_{k=1}^{n} P(X = k)$$

To compute the infinite sum (and later the mean and the variance), we introduce the **geometric series**: the infinite sum

$$\sum_{k=0}^{\infty} q^k = 1 + q + q^2 + q^3 + \ldots$$

The finite sum

$$S_n = \sum_{k=0}^{n} q^k = 1 + q + q^2 + \cdots + q^n$$

can be computed with the use of the following computational "trick": Write

$$S_n = 1 + q + q^2 + \cdots + q^n$$
$$q S_n = q + q^2 + q^3 + \cdots + q^n + q^{n+1}$$

and then subtract $q S_n$ from S_n. Most terms cancel, and we find that

$$S_n - q S_n = 1 - q^{n+1}$$

Factoring out S_n on the left-hand side and solving for S_n yields

$$(1 - q)S_n = 1 - q^{n+1}$$

$$S_n = \frac{1 - q^{n+1}}{1 - q}$$

provided that $q \neq 1$.

If $|q| < 1$, then $\lim_{n \to \infty} q^{n+1} = 0$, and therefore,

$$\sum_{k=0}^{\infty} q^k = \lim_{n \to \infty} \sum_{k=0}^{\infty} q^k = \lim_{n \to \infty} \frac{1 - q^{n+1}}{1 - q} = \frac{1}{1 - q} \quad \text{for } |q| < 1$$

These are important results, which we summarize as follows:

For $q \neq 1$,

$$\sum_{k=0}^{n} q^k = \frac{1 - q^{n+1}}{1 - q} \tag{12.25}$$

For $|q| < 1$,

$$\sum_{k=0}^{\infty} q^k = \frac{1}{1 - q} \tag{12.26}$$

We can use the preceding results to check that the probability mass function of the geometric distribution adds up to 1:

$$\sum_{k=1}^{\infty} P(X = k) = \sum_{k=1}^{\infty} (1 - p)^{k-1} p$$

$$= p \sum_{l=0}^{\infty} (1 - p)^l = p \frac{1}{1 - (1 - p)} = p \cdot \frac{1}{p} = 1$$

In the summation, we made the substitution $l = k - 1$. Then the summation range $k = 1, 2, 3, \ldots$ changes to $l = 0, 1, 2, \ldots$, allowing us to apply the results for the geometric series we derived in (12.26).

EXAMPLE 29

A random experiment consists of rolling a fair die until the first time a six appears. Find the probability that the first six appears at the fifth trial.

Solution

Denote by X the first time a six appears and by p the probability that the die shows a six in a single trial (the success in this experiment). Since the die is fair, all six numbers on the die are equally likely and we find that $p = 1/6$. Then

$$P(X = 5) = \left(1 - \frac{1}{6}\right)^4 \frac{1}{6} \approx 0.0804 \qquad \blacksquare$$

EXAMPLE 30

Consider a sequence of independent Bernoulli trials with probability of success p. Find the probability of no success in the first k trials.

Solution

Denote by X the number of trials until the first success. We want to find the event $\{X > k\}$. Now, this event can be phrased in terms of a binomial random variable S_k that counts the number of successes in the first k trials. The event $\{X > k\}$ is equivalent to the event $\{S_k = 0\}$. Therefore,

$$P(X > k) = P(S_k = 0) = (1 - p)^k \qquad \blacksquare$$

EXAMPLE 31

Compare the probability of no success in the first k trials of independent Bernoulli trials with the probability of no success in k trials following n unsuccessful trials.

Solution

If X denotes the number of trials with probability of success p, then we want to compare $P(X > k)$ with $P(X > n + k \mid X > n)$. From Example 30, we conclude that

$$P(X > k) = (1 - p)^k$$

To compute the conditional probability $P(X > n + k \,|\, X > n)$, we use

$$P(X > n + k \,|\, X > n) = \frac{P(X > n + k, X > n)}{P(X > n)}$$

Since the event $\{X > n + k\}$ is contained in the event $\{X > n\}$, it follows that

$$P(X > n + k, X > n) = P(\{X > n + k\} \cap \{X > n\}) = P(X > n + k)$$

and

$$\frac{P(X > n + k, X > n)}{P(X > n)} = \frac{P(X > n + k)}{P(X > n)} = \frac{(1 - p)^{n+k}}{(1 - p)^n} = (1 - p)^k$$

We then find that

$$P(X > k) = P(X > n + k \,|\, X > n)$$

That is, not having had a success in the first n trials does not change the probability of not having successes in the following k trials compared with not having k successes in the first k trials. This result is a consequence of the independence of trials.   ■

EXAMPLE 32   If both parents are carriers of a recessive autosomal disease, but are not symptomatic for the disease, then there is a 25% chance that a child of theirs will be symptomatic for the disease. Suppose the parents have three asymptomatic children and plan on having a fourth child. What is the probability that the fourth child will not be symptomatic for the disease?

Solution   Denote by X the waiting time for the first symptomatic child in this family. With probability of "success" $p = 1/4$, we find, from Example 31, that

$$P(X > 4 \,|\, X > 3) = P(X > 1) = 1 - P(X = 1) = 1 - \frac{1}{4} = \frac{3}{4}$$

We can also argue as follows: The fact that the first three children are asymptomatic for the disease does not change the probability that their next child will be asymptomatic for the disease, since these events are independent.   ■

We will now compute the mean and the variance of the geometric distribution. If X is a geometrically distributed random variable with $P(X = k) = (1 - p)^{k-1} p$, then

$$E(X) = \sum_{k=1}^{\infty} k(1 - p)^{k-1} p = p \sum_{k=1}^{\infty} k(1 - p)^{k-1}$$

To compute this infinite sum, we need (12.26) and the following result, which we cannot prove here: If $|q| < 1$, then

$$\frac{d}{dq} \sum_{k=0}^{\infty} q^k = \sum_{k=0}^{\infty} \frac{d}{dq} q^k$$

In words, the derivative of this infinite sum can be obtained by differentiating each term separately and then taking the sum of all the derivatives. Interchanging differentiation and summation when the sum is an infinite sum cannot always be done, but can be justified in this case. Using this result, we find that

$$\sum_{k=0}^{\infty} \frac{d}{dq} q^k = \sum_{k=0}^{\infty} k q^{k-1} = \sum_{k=1}^{\infty} k q^{k-1}$$

where, in the last step, we used the fact that the term with $k = 0$ is equal to 0. Since $\sum_{k=0}^{\infty} q^k = \frac{1}{1-q}$ when $|q| < 1$, it follows that

$$\frac{d}{dq} \sum_{k=0}^{\infty} q^k = \frac{d}{dq} \frac{1}{1 - q} = \frac{1}{(1 - q)^2}$$

and hence,

$$\sum_{k=1}^{\infty} kq^{k-1} = \frac{1}{(1-q)^2}$$

With $q = 1 - p$,

$$E(X) = p \sum_{k=1}^{\infty} k(1-p)^{k-1} = p\frac{1}{(1-(1-p))^2} = \frac{p}{p^2} = \frac{1}{p}$$

To compute the variance of X, we employ a similar argument. We first derive the following result, which we will need to carry out the calculation of the variance: If $|q| < 1$, then

$$\frac{d^2}{dq^2} \sum_{k=0}^{\infty} q^k = \sum_{k=0}^{\infty} \frac{d^2}{dq^2} q^k = \sum_{k=2}^{\infty} k(k-1)q^{k-2}$$

Now, since

$$\frac{d^2}{dq^2} \frac{1}{1-q} = \frac{2}{(1-q)^3}$$

we have

$$\sum_{k=2}^{\infty} k(k-1)q^{k-2} = \frac{2}{(1-q)^3}$$

To compute the variance, it is useful to first compute

$$E[X(X-1)] = \sum_{k=2}^{\infty} k(k-1)P(X=k) = \sum_{k=2}^{\infty} k(k-1)(1-p)^{k-1}p$$

$$= (1-p)p \sum_{k=2}^{\infty} k(k-1)(1-p)^{k-2} = (1-p)p\frac{2}{(1-(1-p))^3}$$

$$= \frac{2p(1-p)}{p^3} = \frac{2(1-p)}{p^2}$$

Since $E[X(X-1)] = E(X^2) - E(X)$, it follows that

$$\text{var}(X) = E(X^2) - [E(X)]^2 = E[X(X-1)] + E(X) - [E(X)]^2$$
$$= \frac{2(1-p)}{p^2} + \frac{1}{p} - \frac{1}{p^2} = \frac{1-p}{p^2}$$

We summarize our results as follows:

If X is geometrically distributed with $P(X=k) = (1-p)^{k-1}p$, then

$$E(X) = \frac{1}{p} \quad \text{and} \quad \text{var}(X) = \frac{1-p}{p^2}$$

EXAMPLE 33 You roll a fair die until the first time a 6 appears. How long do you have to wait, on average?

Solution To answer this question, we need to find $E(X)$, where X is geometrically distributed with probability of success $p = 1/6$. Therefore,

$$E(X) = \frac{1}{p} = 6$$

In words, you have to roll a die six times, on average, until the first time a 6 appears. ∎

■ **12.4.6 The Poisson Distribution**

The Poisson distribution is one of the most important probability distributions. It is used to model, for instance, amino acid substitutions in proteins, the escape probability of hosts from parasitism, and spatial distributions of plants. It often models "rare events," as we will see.

We say that X is **Poisson** distributed with parameter $\lambda > 0$ if

$$P(X = k) = e^{-\lambda}\frac{\lambda^k}{k!}, \quad k = 0, 1, 2, \ldots$$

The random variable X is a discrete random variable whose range is the set of all nonnegative integers. Thus, the range is infinite but still countable.

To show that the probability distribution we defined sums to 1, or to find the mean and the variance of X, we need some additional results. Recall that the Taylor polynomial of order n of $f(x) = e^x$ is

$$P_n(x) = \sum_{k=0}^{n} \frac{x^k}{k!}$$

It turns out (but we cannot prove this here) that, in the limit as $n \to \infty$,

$$e^x = \lim_{n\to\infty} P_n(x) = \lim_{n\to\infty} \sum_{k=0}^{n} \frac{x^k}{k!}$$

for all $x \in \mathbf{R}$. On the right-hand side, we have an infinite sum, and we will use the notation (as we did in the previous section)

$$\sum_{k=0}^{\infty} \frac{x^k}{k!} = \lim_{n\to\infty} \sum_{k=0}^{n} \frac{x^k}{k!}$$

to denote this infinite sum. Thus, for any $x \in \mathbf{R}$,

$$e^x = \sum_{k=0}^{\infty} \frac{x^k}{k!} \tag{12.27}$$

We can use (12.27) to show that the probability mass function for the Poisson distribution indeed sums to 1:

$$\sum_{k=0}^{\infty} P(X = k) = \sum_{k=0}^{\infty} e^{-\lambda}\frac{\lambda^k}{k!} = e^{-\lambda} \sum_{k=0}^{\infty} \frac{\lambda^k}{k!} = e^{-\lambda}e^{\lambda} = 1$$

Furthermore, since $\lambda > 0$, $P(X = k) \geq 0$. Thus, the probability mass function for the Poisson distribution (of course) satisfies the two conditions of a probability mass function.

EXAMPLE 34

Suppose the number of plants per hectare of a certain species is Poisson distributed with parameter $\lambda = 3$ plants per hectare. Find the probability that there are **(a)** no plants in a given hectare and **(b)** at least two plants in a given hectare.

Solution

Denote by X the number of plants in a given hectare. Then X is Poisson distributed with parameter $\lambda = 3$.

(a) The probability that there are no plants in a given hectare is

$$P(X = 0) = e^{-\lambda} = e^{-3} \approx 0.0498$$

(b) The probability that there are at least two plants in a given hectare is

$$P(X \geq 2) = 1 - P(X \leq 1) = 1 - [P(X = 0) + P(X = 1)]$$
$$= 1 - e^{-\lambda}(1 + \lambda) = 1 - e^{-3}(1 + 3) \approx 0.8009 \qquad ■$$

To find the mean and the variance of a Poisson-distributed random variable, we need to use (12.27) repeatedly. Let X be Poisson distributed with parameter λ. Then, formally,

$$E(X) = \sum_{k=0}^{\infty} k P(X = k) = \sum_{k=0}^{\infty} k e^{-\lambda} \frac{\lambda^k}{k!}$$

$$= e^{-\lambda} \lambda \sum_{k=1}^{\infty} \frac{\lambda^{k-1}}{(k-1)!} = e^{-\lambda} \lambda \sum_{k=0}^{\infty} \frac{\lambda^k}{k!} = \lambda e^{-\lambda} e^{\lambda} = \lambda$$

To compute the variance of X, we begin by computing

$$E[X(X-1)] = \sum_{k=0}^{\infty} k(k-1) P(X=k) = \sum_{k=2}^{\infty} k(k-1) P(X=k)$$

$$= \sum_{k=2}^{\infty} k(k-1) e^{-\lambda} \frac{\lambda^k}{k!} = e^{-\lambda} \lambda^2 \sum_{k=2}^{\infty} \frac{\lambda^{k-2}}{(k-2)!}$$

$$= e^{-\lambda} \lambda^2 \sum_{k=0}^{\infty} \frac{\lambda^k}{k!} = e^{-\lambda} \lambda^2 e^{\lambda} = \lambda^2$$

Therefore,

$$\operatorname{var}(X) = E(X^2) - [E(X)]^2 = E[X(X-1)] + E(X) - [E(X)]^2$$
$$= \lambda^2 + \lambda - \lambda^2 = \lambda$$

We summarize our findings as follows:

If X is Poisson distributed with parameter $\lambda > 0$, then

$$E(X) = \lambda \qquad \text{and} \qquad \operatorname{var}(X) = \lambda$$

EXAMPLE 35

The number of substitutions on a given amino acid sequence during a fixed period is modeled by a Poisson distribution. Suppose the number of substitutions on a sequence of 100 amino acids over a period of 1 million years is Poisson distributed with average number of substitutions equal to 1. What is the probability that at least one substitution occurred?

Solution   If X denotes the number of substitutions, then X is Poisson distributed with mean 1. Since the mean of a Poisson distribution is equal to its parameter, we find that $\lambda = 1$. Hence,

$$P(X \geq 1) = 1 - P(X = 0) = 1 - e^{-\lambda} = 1 - e^{-1} \approx 0.6321 \qquad ■$$

We mentioned previously that the Poisson distribution frequently models rare events. The next result makes this precise. Consider a sequence of independent Bernoulli trials with probability of success p. The number of successes among n trials is binomially distributed. We denote the number of successes in n trials by S_n. We consider the case when the number of trials n is very large but the probability of success p is very small, so successes are rare. To make this concept mathematically precise, we will need to take the limit as n tends to infinity such that the product np, which denotes the expected number of successes among n trials, approaches a constant. To do so, we need to let p tend to 0 as n tends to infinity. To indicate that the probability of success depends on n, we will denote it by p_n. The following result says that the number of successes among a large number of trials is approximately Poisson distributed if the probability of success is small:

> **Poisson Approximation to the Binomial Distribution**   Suppose S_n is binomially distributed with parameters n and p_n. If $p_n \to 0$ as $n \to \infty$ such that $\lim_{n\to\infty} np_n = \lambda > 0$, then
>
> $$\lim_{n\to\infty} P(S_n = k) = e^{-\lambda}\frac{\lambda^k}{k!}$$

In all the examples of the binomial distribution we have seen thus far, the probability of success p was fixed and did not depend on n. How, then, should we interpret the preceding result? It is the precise mathematical formulation; it allows us to use a Poisson distribution as an approximation to the binomial distribution when the number of trials n is large and the probability of success p is small. Let's use the result to compare the Poisson approximation to the exact results of a binomial distribution. This will illustrate how to use the result. We will prove it afterward.

EXAMPLE 36

Suppose we toss a biased coin 100 times and denote the number of heads by S_{100}. If the probability of heads is $1/50$, compute $P(S_{100} = k)$ for $k = 0, 1,$ and 2 exactly and compare your answer with the Poisson approximation.

Solution

Since $n = 100$ and $p = 1/50$, we compare the distribution of S_{100} with a Poisson distribution with parameter $\lambda = np = 100/50 = 2$. We find that

$$P(S_{100} = k) = \binom{100}{k}\left(\frac{1}{50}\right)^k\left(\frac{49}{50}\right)^{100-k} \approx e^{-2}\frac{2^k}{k!}$$

For $k = 0$,

$$P(S_{100} = 0) = \left(\frac{49}{50}\right)^{100} \approx 0.1326$$

$$e^{-2} \approx 0.1353$$

For $k = 1$,

$$P(S_{100} = 1) = 100 \cdot \frac{1}{50}\left(\frac{49}{50}\right)^{99} \approx 0.2707$$

$$2e^{-2} \approx 0.2707$$

For $k = 2$,

$$P(S_{100} = 2) = \frac{100 \cdot 99}{2}\left(\frac{1}{50}\right)^2\left(\frac{49}{50}\right)^{98} \approx 0.2734$$

$$e^{-2}\frac{2^2}{2} \approx 0.2707$$

In each case, we see that the approximate value is quite close to the exact value. The advantage is that the Poisson distribution is much easier to calculate than the binomial distribution, since the binomial coefficients $\binom{n}{k}$ are computationally intensive.

■

We will now show that

$$\lim_{n\to\infty} P(S_n = k) = e^{-\lambda}\frac{\lambda^k}{k!}$$

We will need the following result: If $\lim_{n\to\infty} x_n = x$, then

$$\lim_{n\to\infty} \left(1 + \frac{x_n}{n}\right)^n = e^x$$

To prove this result, we show that

$$\lim_{n\to\infty} \ln\left(1 + \frac{x_n}{n}\right)^n = x$$

Now,

$$\lim_{n\to\infty} \ln\left(1 + \frac{x_n}{n}\right)^n = \lim_{n\to\infty} n \ln\left(1 + \frac{x_n}{n}\right)$$

This limit is of the form $\infty \cdot 0$, which suggests that we should rewrite the limit in the form $\frac{0}{0}$ or $\frac{\infty}{\infty}$ and use l'Hospital's rule. We rewrite it in the form

$$\lim_{n\to\infty} n \ln\left(1 + \frac{x_n}{n}\right) = \lim_{n\to\infty} \frac{\ln\left(1 + \frac{x_n}{n}\right)}{\frac{1}{n}} = \lim_{n\to\infty} \frac{\ln\left(1 + \frac{x_n}{n}\right)}{\frac{x_n}{n}} x_n$$

To evaluate this limit, we use l'Hospital's rule to compute

$$\lim_{y\to 0} \frac{\ln(1 + y)}{y}$$

We find that

$$\lim_{y\to 0} \frac{\ln(1 + y)}{y} = \lim_{y\to 0} \frac{\frac{1}{1+y}}{1} = 1$$

This result, together with $\lim_{n\to\infty} x_n = x$ and $\lim_{n\to\infty} \frac{x_n}{n} = 0$, yields

$$\lim_{n\to\infty} \frac{\ln\left(1 + \frac{x_n}{n}\right)}{\frac{x_n}{n}} x_n = (1)(x) = x$$

We can now prove the Poisson approximation. Observe that

$$P(S_n = k) = \binom{n}{k} p_n^k (1 - p_n)^{n-k}$$

We define $\lambda_n = np_n$, so $p_n = \frac{\lambda_n}{n}$. Then

$$P(S_n = k) = \binom{n}{k} \left(\frac{\lambda_n}{n}\right)^k \left(1 - \frac{\lambda_n}{n}\right)^{n-k}$$

$$= \frac{n(n-1)(n-2)\cdots(n-k+1)}{n^k} \frac{\lambda_n^k}{k!} \left(1 - \frac{\lambda_n}{n}\right)^n \left(1 - \frac{\lambda_n}{n}\right)^{-k}$$

Since $\lim_{n\to\infty} np_n = \lim_{n\to\infty} \lambda_n = \lambda$,

$$\lim_{n\to\infty} \frac{n(n-1)(n-2)\cdots(n-k+1)}{n^k} = \lim_{n\to\infty} \frac{n}{n}\frac{n-1}{n}\frac{n-2}{n}\cdots\frac{n-k+1}{n} = 1$$

$$\lim_{n\to\infty} \frac{\lambda_n^k}{k!} = \frac{\lambda^k}{k!}, \qquad \lim_{n\to\infty} \left(1 - \frac{\lambda_n}{n}\right)^{-k} = 1$$

and

$$\lim_{n\to\infty} \left(1 - \frac{\lambda_n}{n}\right)^n = \lim_{n\to\infty} \left(1 + \frac{-\lambda_n}{n}\right)^n = e^{-\lambda}$$

it follows that

$$\lim_{n\to\infty} P(S_n = k) = (1)\frac{\lambda^k}{k!} e^{-\lambda} (1) = e^{-\lambda} \frac{\lambda^k}{k!}$$

which is the Poisson approximation to the binomial distribution.

In Section 10.7, we discussed host–parasitoid models. Parasitoids are insects that lay their eggs on, in, or near the (in most cases, immature) body of another arthropod, which serves as the host for the developing parasitoid. The eggs develop into free-living adults while consuming the host. Parasitoids make up about 14% of all insect species. A key component in modeling host–parasitoid interactions is the probability of a host escaping parasitism. The first models were developed by Nicholson and Bailey (1935), who assumed that the probability that a host escapes parasitism is given by e^{-aP}, where P is the parasitoid density and a is a positive parameter called the *search efficiency*. The next example explains where this functional form of the escape probability comes from.

EXAMPLE 37

Parasitoid encounters with their hosts are sometimes modeled with a Poisson distribution. Suppose a host is surrounded by P parasitoids. Each parasitoid, independently of all others, has a probability a of encountering the host. We can consider this probability as a sequence of P Bernoulli trials with probability of success a. If no parasitoid encounters the host, the host will escape parasitism. If P is large and a is small, we can use the Poisson approximation. The number of encounters is then approximately Poisson distributed with parameter aP, and we find that

$$P(\text{host escapes parasitism}) = e^{-aP}$$

This is the escape probability used in the Nicholson–Bailey host–parasitoid model we discussed in Section 10.7. It is the zeroth term of a Poisson distribution and comes about because parasitoids are assumed to search randomly. ■

The Poisson approximation plays a crucial role in using amino acid sequence data to estimate the time of divergence of species. Sequences of the same protein across different species are compared, and the number of pairwise amino acid differences gives an indication of the evolutionary distance between each pair of species. The simplest mathematical model for estimating times of divergence based on amino acid sequences assumes that the probability of a substitution at a given site in the amino acid sequence is the same for all sites and depends only on the time since divergence. Furthermore, all sites are assumed to be independent. The number of amino acid substitutions along a sequence of length n is then binomially distributed with probability of success equal to the probability of a substitution at the given site, provided that multiple substitutions at the site can be ignored and the time since divergence is not too long. If the sequence is sufficiently long and the time since divergence is not too long, so that the probability of substitution is small, then the number of substitutions is equal to the number of differences between the two sequences and can be approximated by a Poisson distribution.

EXAMPLE 38

Suppose you compare the hemoglobin-α chain (length 140 amino acids) of two vertebrate species that diverged about 10 million years ago. A previous study found that, for any amino acid along this chain, the probability of an amino acid difference is about 0.014.

(a) How many amino acid differences would you expect when comparing the two sequences?

(b) What is the probability of finding at least three sites with amino acid differences?

Solution

The number of amino acid differences is approximately Poisson distributed with parameters equal to the product of the length of the sequence and the probability of finding a difference at a given site. We find that $\lambda = 140 \cdot 0.014 = 1.96$.

(a) The expected number of amino acid differences is equal to the parameter of the distribution, 1.96 in this case.

(b) The probability of finding at least three differences is equal to

$$1 - e^{-\lambda}(1 + \lambda + \lambda^2/2) = 1 - e^{-1.96}(1 + 1.96 + 1.96^2/2) \approx 0.3125 \qquad ■$$

Sums of Poisson Random Variables Suppose X is a Poisson random variable with parameter λ and Y is a Poisson random variable with parameter μ. Then, if X and Y are independent, it follows that $X + Y$ is Poisson distributed with parameter $\lambda + \mu$. To prove this result, we calculate $P(X + Y = n)$ for $n = 0, 1, 2, \ldots$.

First, note that the event $\{X+Y = n\}$ can be decomposed into mutually exclusive events of the form $\{X = k, Y = n - k\}$ for $k = 0, 1, 2, \ldots, n$, so that

$$P(X + Y = n) = \sum_{k=0}^{n} P(X = k, Y = n - k)$$

Next, because of independence,

$$P(X = k, Y = n - k) = P(X = k)P(Y = n - k)$$

Therefore,

$$P(X + Y = n) = \sum_{k=0}^{n} P(X = k)P(Y = n - k)$$

$$= \sum_{k=0}^{n} e^{-\lambda}\frac{\lambda^k}{k!}e^{-\mu}\frac{\mu^{n-k}}{(n-k)!}$$

$$= e^{-(\lambda+\mu)}\frac{1}{n!}\sum_{k=0}^{n}\frac{n!}{k!\,(n-k)!}\lambda^k\mu^{n-k}$$

Recall from Example 7 in Section 12.1 that

$$(x + y)^n = \sum_{k=0}^{n}\binom{n}{k}x^k y^{n-k}$$

Hence,

$$P(X + Y = n) = e^{-(\lambda+\mu)}\frac{(\lambda + \mu)^n}{n!}$$

EXAMPLE 39 The number of annual cancer deaths in a population is sometimes modeled with a Poisson distribution. In Ireland in the mid-1990s, the number of deaths from lung cancer was about 71 per 100,000 men and 34 per 100,000 women. Use these data to find the probability that no lung cancer deaths occurred during a given year in a group of 1000 men and 1000 women.

Solution We model the number of annual deaths with a Poisson distribution. We denote the number of deaths in men by X and use $\lambda = 0.71$ for the parameter of the Poisson distribution. We denote the number of deaths in women by Y and use $\mu = 0.34$ for the parameter of the Poisson distribution. The number of annual deaths in this group of 1000 men and 1000 women is then Poisson distributed with parameter

$$\lambda + \mu = 0.71 + 0.34 = 1.05$$

and it follows that

$$P(X + Y = 0) = e^{-(\lambda+\mu)} = e^{-1.05} \approx 0.3499$$ ■

■ **12.4.1**

1. Toss a fair coin twice. Let X be the random variable that counts the number of tails in each outcome. Find the probability mass function describing the distribution of X.

2. Toss a fair coin four times. Let X be the random variable that counts the number of heads. Find the probability mass function describing the distribution of X.

3. Roll a fair die twice. Let X be the random variable that gives the absolute value of the differences betwen the two numbers. Find the probability mass function describing the distribution of X.

4. Roll a fair die twice. Let X be the random variable that gives the maximum of the two numbers. Find the probability mass function describing the distribution of X.

5. An urn contains three green and two blue balls. You remove two balls at random without replacement. Let X denote the number of green balls in your sample. Find the probability mass function describing the distribution of X.

6. An urn contains five green balls, two blue balls, and three red balls. You remove three balls at random without replacement. Let X denote the number of red balls. Find the probability mass function describing the distribution of X.

7. You draw 3 cards from a standard deck of 52 cards without replacement. Let X denote the number of spades in your hand. Find the probability mass function describing the distribution of X.

8. You draw 5 cards from a standard deck of 52 cards without replacement. Let X denote the number of aces in your hand. Find the probability mass function describing the distribution of X.

9. Suppose that the probability mass function of a discrete random variable X is given by the following table:

x	$P(X = x)$
-3	0.2
-1	0.3
1.5	0.4
2	0.1

Find and graph the corresponding distribution function $F(x)$.

10. Suppose the probability mass function of a discrete random variable X is given by the following table:

x	$P(X = x)$
-1	0.2
-0.5	0.25
0.1	0.1
0.5	0.1
1	0.35

Find and graph the corresponding distribution function $F(x)$.

11. Let X be a random variable with distribution function

$$F(x) = \begin{cases} 0 & x < -2 \\ 0.2 & -2 \le x < 0 \\ 0.3 & 0 \le x < 1 \\ 0.7 & 1 \le x < 2 \\ 1 & x \ge 2 \end{cases}$$

Determine the probability mass function of X.

12. Let X be a random variable with distribution function

$$F(x) = \begin{cases} 0 & x < 0 \\ 0.05 & 0 \le x < 1.3 \\ 0.30 & 1.3 \le x < 1.7 \\ 0.85 & 1.7 \le x < 1.9 \\ 0.90 & 1.9 \le x < 2 \\ 1.0 & x \ge 2 \end{cases}$$

Determine the probability mass function of X.

13. Let $S = \{1, 2, 3, \ldots, 10\}$, and assume that

$$p(k) = \frac{k}{N}, \; k \in S$$

where N is a constant.

(a) Determine N so that $p(k)$, $k \in S$, is a probability mass function.

(b) Let X be a discrete random variable with $P(X = k) = p(k)$. Find the probability that X is less than 8.

14. Geometric Distribution In Example 2, we tossed a coin repeatedly until the first heads showed up. Assume that the probability of heads is p, where $p \in (0, 1)$. Let Y be a random variable that counts the number of trials until the first heads shows up.

(a) Show that $P(Y = 1) = p$, $P(Y = 2) = (1 - p)p$, and $P(Y = 3) = (1 - p)^2 p$.

(b) Explain why

$$P(Y = j) = (1 - p)^{j-1} p$$

for $j = 1, 2, \ldots$. This equation is called the **geometric distribution**.

(c) Prove that

$$\sum_{j \ge 1} P(Y = j) = 1$$

as follows:

(i) For $0 \le q < 1$, define

$$S_n = 1 + q + q^2 + \cdots + q^n$$

Show that

$$S_n - q S_n = 1 - q^{n+1}$$

and conclude from this equation that

$$S_n = \frac{1 - q^{n+1}}{1 - q}$$

(ii) Show that

$$P(Y \le k) = \sum_{j=1}^{k} P(Y = j) = p \sum_{j=1}^{k} (1 - p)^{j-1}$$

Use your results in (i) to show that this formula simplifies to

$$1 - (1 - p)^k$$

and conclude from this equation that

$$\lim_{k \to \infty} P(Y \le k) = 1$$

which is equivalent to

$$\sum_{j \ge 1} P(Y = j) = 1$$

■ 12.4.2

15. The following table contains the number of leaves per basil plant in a sample of size 25:

19	21	20	13	18
14	17	14	17	17
13	15	12	15	17
15	16	18	17	14
14	14	13	20	13

(a) Find the relative frequency distribution.
(b) Compute the average value by **(i)** averaging the values in the table directly and **(ii)** using the relative frequency distribution obtained in (a).

16. The following table contains the number of aphids per plant in a sample of size 30:

15	27	13	2	0	16
26	0	2	1	17	15
21	13	5	0	19	25
12	11	0	16	22	1
28	9	0	0	1	17

(a) Find the relative frequency distribution.
(b) Compute the average value by **(i)** averaging the values in the table directly and **(ii)** using the relative frequency distribution obtained in (a).

17. The following table contains the scores of 25 students on a certain exam:

7	8	8	3	2
5	6	9	10	6
8	8	7	6	9
10	4	4	8	6
9	10	5	5	8

(a) Find the relative frequency distribution.
(b) Compute the average value by **(i)** averaging the values in the table directly and **(ii)** using the relative frequency distribution obtained in (a).

18. The following table contains the number of flower heads per plant in a sample of size 20:

15	17	19	18	15
17	18	15	14	19
17	15	15	18	19
20	17	14	17	18

(a) Find the relative frequency distribution.
(b) Compute the average value by **(i)** averaging the values in the table directly and **(ii)** using the relative frequency distribution obtained in (a).

19. Suppose that the probability mass function of a discrete random variable X is given by the following table:

x	$P(X = x)$
-2	0.1
-1	0.4
0	0.3
1	0.2

(a) Find $E(X)$. **(b)** Find $E(X^2)$. **(c)** Find $E[X(X-1)]$.

20. Suppose that the probability mass function of a discrete random variable X is given by the following table:

x	$P(X = x)$
0	0.3
1	0.3
2	0.1
3	0.1
4	0.2

(a) Find $E(X)$. **(b)** Find $E(X^2)$. **(c)** Find $E(2X-1)$.

21. Suppose that the probability mass function of a discrete random variable X is given by the following table:

x	$P(X = x)$
-3	0.2
-1	0.3
1.5	0.4
2	0.1

Find the mean, the variance, and the standard deviation of X.

22. Suppose that the probability mass function of a discrete random variable X is given by the following table:

x	$P(X = x)$
-1	0.1
-0.5	0.2
0.1	0.1
0.5	0.25
1	0.35

Find the mean, the variance, and the standard deviation of X.

23. Let X be **uniformly distributed** on the set

$$S = \{1, 2, 3, \ldots, 10\}$$

That is,

$$P(X = k) = \frac{1}{10}, \quad k \in S$$

(a) Find $E(X)$. **(b)** Find var(X).

24. Let X be uniformly distributed on the set

$$S = \{1, 2, 3, \ldots, n\}$$

where n is a positive integer; that is,

$$P(X = k) = \frac{1}{n}, \quad k \in S$$

(a) Find $E(X)$. **(b)** Find var(X).
Hint: Recall that

$$\sum_{k=1}^{n} k = \frac{n(n+1)}{2}$$

and

$$\sum_{k=1}^{n} k^2 = \frac{n(n+1)(2n+1)}{6}$$

25. Assume that X is a discrete random variable with finite range, and set

$$p(x) = P(X = x)$$

(a) Show that

$$E(aX + b) = \sum_x (ax + b)p(x)$$

(b) Use your result in (a) and the rules for finite sums to conclude that

$$E(aX + b) = aE(X) + b$$

26. Assume that X is a discrete random variable with finite range, and set

$$p(x) = P(X = x)$$

(a) Show that

$$\text{var}(aX + b) = a^2 \sum_x [x - E(X)]^2 p(x)$$

(b) Use your result in (a) and the rules for finite sums to conclude that

$$\text{var}(aX + b) = a^2 \, \text{var}(X)$$

27. Let X and Y be two random variables with the following joint distribution:

	X = 0	X = 1
Y = 0	0.3	0.1
Y = 1	0.2	0.4

(a) Find $P(X = 1, Y = 0)$.   **(b)** Find $P(X = 1)$.
(c) Find $P(Y = 0)$.   **(d)** Find $P(Y = 0 | X = 1)$.

28. Let X and Y be two random variables with the following joint distribution:

	X = 0	X = 1
Y = 0	0.2	0.0
Y = 1	0.3	0.5

(a) Find $P(X = 0, Y = 1)$.   **(b)** Find $P(X = 0)$.
(c) Find $P(Y = 1)$.   **(d)** Find $P(X = 0 | Y = 0)$.

29. Let X and Y be two independent random variables with probability mass function described by the following table:

k	P(X = k)	P(Y = k)
−2	0.1	0.2
−1	0	0.2
0	0.3	0.1
1	0.4	0.3
2	0.05	0
3	0.15	0.2

(a) Find $E(X)$ and $E(Y)$.   **(b)** Find $E(X + Y)$.
(c) Find $\text{var}(X)$ and $\text{var}(Y)$.   **(d)** Find $\text{var}(X + Y)$.

30. Let X and Y be two independent random variables with probability mass function described by the following table:

k	P(X = k)	P(Y = k)
−3	0.1	0.1
−1	0.1	0.2
0	0.2	0.1
0.5	0.3	0.3
2	0.15	0.1
2.5	0.15	0.2

(a) Find $E(X)$ and $E(Y)$.   **(b)** Find $E(X + Y)$.
(c) Find $\text{var}(X)$ and $\text{var}(Y)$.   **(d)** Find $\text{var}(X + Y)$.

31. We have two formulas for computing the variance of X, namely,

$$\text{var}(X) = E[X - E(X)]^2$$

and

$$\text{var}(X) = E(X^2) - [E(X)]^2$$

(a) Explain why $\text{var}(X) \geq 0$.
(b) Use your results in (a) to explain why

$$E(X^2) \geq [E(X)]^2$$

32. Assume that X is a discrete random variable with finite range. Show that if $\text{var}(X) = 0$, then $P(X = E(X)) = 1$.

■ 12.4.3

33. Toss a fair coin 10 times. Let X be the number of heads. Find
(a) $P(X = 5)$.   **(b)** $P(X \geq 8)$.   **(c)** $P(X \leq 9)$.

34. Toss a coin with probability of heads 0.3 five times. Let X be the number of tails. Find
(a) $P(X = 2)$.   **(b)** $P(X \geq 1)$.

35. Roll a fair die six times. Let X be the number of times you roll a 6. Find the probability mass function.

36. A loaded die has probability 0.5 of rolling a 6 and probability 0.1 of rolling each of the other five numbers. Find the probability of rolling a 6 three times in a row.

37. A loaded die is weighted so that rolling a 4 is three times as likely as rolling any of the other numbers. You roll the die twice and record the sum of the two numbers. What is the probability that the sum is equal to 7.

38. An urn contains four green and six blue balls. You draw a ball at random, note its color, and replace it. You repeat these steps four times. Let X denote the total number of green balls you obtain. Find the probability mass function of X.

39. An urn contains three blue and two white balls. You draw a ball at random, note its color, and replace it. You repeat these steps three times. Let X denote the total number of white balls. Find $P(X \leq 1)$.

40. An urn contains four red, seven green, and two white balls. You draw a ball at random, note its color, and replace it. You repeat these steps four times. Let X denote the number of red balls and Y the number of green balls. Find $P(X + Y) = 2$.

41. Assume that 20% of all plants in a field are infested with aphids. Suppose that you pick 20 plants at random. What is the probability that none of them carried aphids?

42. To test for a disease that has a prevalence of 1 in 100 in a population, blood samples of 10 individuals are pooled and the pooled blood is then tested. What is the probability that the test result is negative (the disease is not present in the pooled blood sample)?

43. Suppose that a box contains 10 apples. The probability that any one apple is spoiled is 0.1. (Assume that spoilage of the apples is an independent phenomenon.)
(a) Find the expected number of spoiled apples per box.
(b) A shipment contains 10 boxes of apples. Find the expected number of boxes that contain no spoiled apples.

44. Toss a fair coin 10 times. Let X denote the number of heads. What is the probability that X is within one standard deviation of its mean?

45. A multiple-choice exam contains 50 questions. Each question has four choices. Find the expected number of correct answers if a student guesses the answers at random.

46. A true–false exam has 20 questions. Find the expected number of correct answers if a student guesses the answers at random.

47. Sampling with and without Replacement An urn contains 12 green and 24 blue balls.

(a) You take 10 balls out of the urn. Find the probability that 6 of the 10 balls are blue.

(b) You take a ball out of the urn, note its color, and replace it. You repeat these steps 10 times. Find the probability that 6 of the 10 balls are blue.

48. Sampling with and without Replacement An urn contains K green and $N - K$ blue balls.

(a) You take n balls out of the urn. Find the probability that k of the n balls are green.

(b) You take a ball out of the urn, note its color, and replace it. You repeat these steps n times. Find the probability that k of the n balls are green.

■ 12.4.4

50. Repeat Example 27 when $N_1 = 10$, $N_2 = 14$, and $N_3 = 6$.

51. Repeat Example 27 when $N_1 = 5$, $N_2 = 15$, and $N_3 = 10$.

52. Repeat Example 28 when 20 seeds are round and yellow, 10 are round and green, 8 are wrinkled and yellow, and 2 are wrinkled and green.

53. Repeat Example 28 when 17 seeds are round and yellow, 22 are round and green, 13 are wrinkled and yellow, and 8 are wrinkled and green.

54. An urn contains six green, eight blue, and 10 red balls. You take one ball out of the urn, note its color, and replace it. You repeat these steps six times. What is the probability that you sampled two of each color?

55. An urn contains eight green, four blue, and six red balls. You take one ball out of the urn, note its color, and replace it. You repeat these steps four times. What is the probability that you sampled two green, one blue, and one red ball?

56. In a $Cc \times Cc$ crossing of peas, 5 offspring are of genotype CC, 12 are of genotype Cc, and 6 are of genotype cc. What is the probability of this event?

57. In a $Cc \times Cc$ crossing of peas, two offspring are of genotype CC, three are of genotype Cc, and one is of genotype cc. What is the probability of this event?

A number of human traits are caused by a single pair of recessive genes and thus manifest themselves only in individuals who are homozygous for the mutant gene. An individual with one normal and one mutant gene is a carrier, but does not exhibit the trait. In Problems 57–59, calculate each of the probabilites.

57. The inability to roll one's tongue is caused by a single pair of recessive genes (rr). For a couple consisting of a heterozygote individual (Rr) and an affected person (rr), what is the probability that, among their four children, at most one child is unable to roll his or her tongue?

58. An attached earlobe is caused by a single pair of recessive genes (aa). For a couple consisting of a heterozygous individual (Aa) and an affected person (aa), what is the probability that a child has an unattached earlobe?

59. Tay–Sachs disease is caused by a single pair of recessive genes. If both parents are carriers of the mutant gene, what is the likelihood that none of their four children will be affected?

60. Assume a 1:1 sex ratio. A woman who is a carrier of hemophilia has two daughters and two sons with a man who is

not hemophilic. What is the probability that one daughter is not a carrier, one daughter is a carrier, one son is hemophilic, and one son is not hemophilic?

■ 12.4.5

61. A random experiment consists of flipping a fair coin until the first time heads appears. Find the probability that the first heads appears on the kth trial for $k = 1, 2$, and 3.

62. A random experiment consists of flipping a biased coin with probability 0.3 of heads until the first time heads appears. Find the probability that heads appears for the first time on the fifth trial.

63. A random experiment consists of rolling a fair die until the first time an even number appears. Find the probability that the first even number appears on the third trial.

64. A random experiment consists of rolling a fair die until the first time a five or a six appears. Find the probability that the first five or six appears on the kth trial for $k = 1, 2, \ldots, 5$.

65. A random experiment consists of flipping a fair coin until the first time heads appears. Find the probability that the first heads appears after the third trial.

66. A random experiment consists of rolling a fair die until the first six appears. Find the probability that the first six appears after the seventh trial.

67. A random experiment consists of flipping a fair coin until the first time heads appears. Find the probability that the first heads appears within the first four trials.

68. A random experiment consists of rolling a fair die until the first time a 1 or a 2 appears. Find the probability that the first 1 or 2 appears within the first five trials.

69. An urn contains one black and 14 white balls. Balls are drawn at random, one at a time, until the black ball is selected. Each ball is replaced before the next ball is drawn. Find the probability that at least 20 draws are needed.

70. An urn contains one black and $n - 1$ white balls. Balls are drawn at random, one at a time, until the black ball is selected. Each ball is replaced before the next ball is drawn. Find the probability that at least n draws are needed. What happens as $n \to \infty$?

71. An urn contains five green and 25 blue balls. Balls are drawn at random, one at a time, until a green ball is selected. Each ball is replaced before the next ball is drawn. Let T denote the first time until a green ball is drawn. Find $E(T)$ and $\mathrm{var}(T)$.

72. An urn contains 10 green and 20 blue balls. Balls are drawn at random, one at a time, until a green ball is selected. Each ball is replaced before the next ball is drawn. Let T denote the first time until a green ball is drawn. Find $E(T)$ and $\mathrm{var}(T)$.

73. An urn contains one black and nine white balls. Balls are drawn at random until the black ball is selected. Find the probability that exactly six white balls will be drawn before the black one is if **(a)** each ball is replaced before the next ball is drawn and **(b)** balls are not replaced.

74. An urn contains one black and $n - 1$ white balls. Balls are drawn at random until the black ball is selected. Find the probability that exactly k white balls will be drawn before the black one is if **(a)** each ball is replaced before the next ball is drawn and **(b)** balls are not replaced.

75. Suppose the waiting time for the first success in an experiment is geometrically distributed with mean $1/p$.

(a) Find the probability that the first success occurs on the kth trial.

(b) The experiment is repeated after the first success. Assume that the waiting time for the second success has the same distribution as the waiting time for the first success. Find the probability mass function for the distribution of the second success.

76. A Bernoulli experiment with probability of success p is repeated until the nth success. Assume that each trial is independent of all others. Find the probability mass function of the distribution of the nth success. (This distribution is called the *negative binomial distribution*.)

■ **12.4.6**

77. Suppose X is Poisson distributed with parameter $\lambda = 2$. Find $P(X = k)$ for $k = 0, 1, 2,$ and 3.

78. Suppose X is Poisson distributed with parameter $\lambda = 0.5$. Find $P(X = k)$ for $k = 0, 1, 2,$ and 3.

79. Suppose X is Poisson distributed with parameter $\lambda = 1$.
(a) Find $P(X \geq 2)$.   **(b)** Find $P(1 \leq X \leq 3)$.

80. Suppose X is Poisson distributed with parameter $\lambda = 0.2$.
(a) Find $P(X < 3)$.   **(b)** Find $P(2 \leq X \leq 4)$.

81. Suppose X is Poisson distributed with parameter $\lambda = 1.5$. Find the probability that X exceeds 3.

82. Suppose X is Poisson distributed with parameter $\lambda = 1.2$. Find the probability that X is at most 3.

83. Suppose X is Poisson distributed with parameter $\lambda = 2$. Find the probability that X is at least 2.

84. Suppose X is Poisson distributed with parameter $\lambda = 0.6$. Find the probability that X is less than 3.

85. Suppose the number of phone calls arriving at a switchboard per hour is Poisson distributed with mean 7 calls per hour. Find the probability that no phone calls arrive during a certain hour.

86. Suppose the number of phone calls arriving at a switchboard per hour is Poisson distributed with mean 3 calls per hour.
(a) Find the probability that at least one phone call arrives between noon and 1 P.M.
(b) Assuming that phone calls in different hours are independent of each other, find the probability that no phone calls arrive between noon and 2 P.M.

87. Suppose the number of typos on a book page is Poisson distributed with mean 0.5. Find the probability that there is at least one typo on a given page.

88. Suppose the number of typos on a book page is Poisson distributed with mean 0.1.
(a) Find the probability that there are no typos on a page.

(b) How many pages with typos do you expect in a 200-page book?

89. The number of amino acid substitutions on a given amino acid sequence is Poisson distributed with mean 3. What is the probability of at least two substitutions?

90. The number of amino acid substitutions on a given amino acid sequence is Poisson distributed with mean 2. Given that there are substitutions on the sequence, what is the probability that there are at least two substitutions?

91. X and Y are independent and Poisson with mean 3.
(a) Find $P(X + Y = 2)$.
(b) Given that $X + Y = 2$, find the probability that $X = k$ for $k = 0, 1,$ and 2.

92. X is Poisson distributed with mean 2, and Y is Poisson distributed with mean 3.
(a) Find $P(X + Y = 4)$
(b) Given that $X + Y = 1$, find the probability that $X = 1$.

93. Let X be Poisson distributed with mean 4 and Y be Poisson distributed with mean 2. Calculate $P(X = 2 | X + Y = 3)$.

94. Suppose X and Y are independent and Poisson with mean λ. Given that $X + Y = n$, find the probability that $X = k$ for $k = 0, 1, 2, \ldots, n$.

In Problems 95–99, use the Poisson approximation.

95. For a certain vaccine, 1 in 1000 individuals experiences some side effects. Find the probability that, in a group of 500 people, nobody experiences side effects.

96. For a certain vaccine, 1 in 500 individuals experiences some side effects. Find the probability that, in a group of 200 people, at least 1 person experiences side effects.

97. About 1 in 700 births in the United States is affected by Down syndrome, a chromosomal disorder. Find the probability that there is at most one case of Down syndrome among 1000 births by **(a)** computing the exact probability and **(b)** using a Poisson approximation.

98. About 1 in 1000 boys is affected by fragile X syndrome, a genetic disorder that causes learning difficulties. Find the probability that, in a group of 500 boys, nobody is affected by this disorder by **(a)** computing the exact probability and **(b)** using a Poisson approximation.

99. (Refer to Example 37.) Suppose a parasitoid has a probability of 0.03 of detecting a given host. If 50 parasitoids are trying to find a particular host, what is the probability that the host will avoid detection?

■ 12.5 Continuous Distributions

■ 12.5.1 Density Functions

In the previous section, we discussed random variables that took on a discrete number of values. In this section, we will discuss random variables that take on a continuum of values. Called **continuous random variables**, they arise, for instance, when we consider the length distribution of an organism: Within an appropriate (species-specific) interval, an individual's length can take on any value.

To illustrate, consider the following example adapted from de Roos (1996): The water flea *Daphnia pulex* feeds on the alga *Chlamydomonas rheinhardii*. An important component of the feeding behavior of *Daphnia* is the strong dependence of the amount of food consumed on the size of the individual. To model the feeding behavior of *Daphnia*, we would need to describe the size structure of the population.

We view size as a random variable, denoted by X, and determine, for all possible values of x, the fraction of *Daphnia* whose size is less than or equal to x. If the population size is large, this fraction can be well approximated by a continuous function, which we denote by $F(x)$. This function serves the role of a distribution function, so $F(x) = P(X \leq x)$.

A distribution function completely characterizes the probability distribution of a random variable, as we saw in the previous section. This property is no different for continuous random variables. To describe the probability distribution of a continuous random variable X, we will therefore use its distribution function $F(x)$. The distribution function for a continuous random variable has the same definition as the one for a discrete random variable:

$$F(x) = P(X \leq x)$$

$F(x)$ has the following properties (see Figure 12.26), some of which are absent in the case of a discrete random variable:

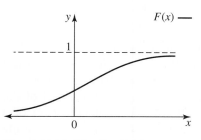

Figure 12.26 The distribution function of a continuous random variable.

1. $0 \leq F(x) \leq 1$.
2. $\lim_{x \to -\infty} F(x) = 0$ and $\lim_{x \to \infty} F(x) = 1$.
3. $F(x)$ is nondecreasing and continuous.

Note on the one hand, that the distribution function of a continuous random variable is continuous, unlike the distribution function of a discrete random variable, which is piecewise constant and takes jumps at those values of x where $P(X = x) > 0$. On the other hand, both distribution functions are nondecreasing and take on values between 0 and 1.

EXAMPLE 1   Show that

$$F(x) = \begin{cases} 1 - e^{-2x} & \text{for } x > 0 \\ 0 & \text{for } x \leq 0 \end{cases}$$

is a distribution function of a continuous random variable.

Solution   A graph of $F(x)$ is shown in Figure 12.27. We must check the three properties of distribution functions of continuous random variables:

1. Since $0 \leq 1 - e^{-2x} \leq 1$ for $x > 0$ and $F(x) = 0$ for $x \leq 0$, it follows that $0 \leq F(x) \leq 1$ for all $x \in \mathbf{R}$.
2. Since $F(x) = 0$ for $x \leq 0$ and $\lim_{x \to \infty} e^{-2x} = 0$, it also follows that

$$\lim_{x \to -\infty} F(x) = 0 \quad \text{and} \quad \lim_{x \to \infty} F(x) = 1$$

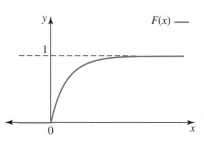

Figure 12.27 The distribution function in Example 1.

3. To show that $F(x)$ is continuous for all $x \in \mathbf{R}$, note that $F(x)$ is continuous for both $x > 0$ and $x < 0$. To check continuity at $x = 0$, we compute

$$\lim_{x \to 0^+} F(x) = \lim_{x \to 0^+} (1 - e^{-2x}) = 0 = \lim_{x \to 0^-} F(x)$$

which is equal to $F(0) = 0$. Hence, $F(x)$ is continuous at $x = 0$.

To show that $F(x)$ is nondecreasing, we compute $F'(x)$ for $x > 0$:

$$F'(x) = 2e^{-2x} > 0 \text{ for } x > 0$$

This equation implies that $F(x)$ is increasing for $x > 0$. Since $F(x)$ is continuous for all $x \in \mathbf{R}$ and equal to 0 for $x \leq 0$, it follows that $F(x)$ is nondecreasing for all $x \in \mathbf{R}$.   ■

If there is a nonnegative function $f(x)$ such that the distribution function $F(x)$ of a random variable X has the representation

$$F(x) = \int_{-\infty}^{x} f(u)\, du$$

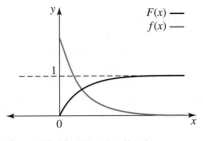

Figure 12.28 The distribution function $F(x)$ and corresponding density function $f(x)$ of a continuous random variable.

we say that X is a continuous random variable with **(probability) density function** $f(x)$. (See Figure 12.28.)

Since $F(x)$ is a distribution function, it follows that

$$\int_{-\infty}^{\infty} f(x)\,dx = 1 \tag{12.28}$$

Any nonnegative function that satisfies (12.28) defines a density function. The function $f(x)$ need not be continuous, although in all of our applications $f(x)$ will be continuous, except perhaps for a finite number of points. The function $f(x)$ will frequently be defined as a piecewise continuous function. Using part I of the fundamental theorem of calculus, we can obtain the density function $f(x)$ of a continuous random variable from the distribution function $F(x)$ by differentiating the distribution function. Thus, $f(x) = F'(x)$ at all points x where $F(x)$ is differentiable. At points where $F(x)$ is not differentiable, we set $f(x) = 0$, a strategy which ensures that the density function is defined everywhere.

Furthermore, since

$$P(a < X \le b) = P(X \le b) - P(X \le a) = F(b) - F(a)$$

it follows that

$$P(a < X \le b) = \int_{a}^{b} f(x)\,dx \tag{12.29}$$

That is, the area under the curve $y = f(x)$ between a and b represents the probability that the random variable takes on values between a and b, as illustrated in Figure 12.29. It is important to realize that, in (12.29), probabilities are represented by areas; in particular, the *integral* of $f(x)$—not $f(x)$ itself—has this physical interpretation. The density function $f(x)$ does not have an immediate physical interpretation. Later in the section, we will explain how to find $f(x)$ empirically.

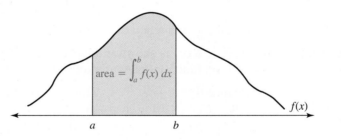

Figure 12.29 The area between the density function $f(x)$ and the x-axis between a and b represents the probability that the random variable X lies between a and b.

In contrast to discrete random variables, for which $P(X \le b)$ and $P(X < b)$ can differ, there is no difference for continuous random variables, since

$$P(X = b) = \int_{b}^{b} f(x)\,dx = 0$$

Therefore,

$$\int_{a}^{b} f(x)\,dx = P(a < X \le b) = P(a \le X \le b)$$
$$= P(a < X < b) = P(a \le X < b)$$

EXAMPLE 2   The distribution function of a continuous random variable X is given by

$$F(x) = \begin{cases} 0 & \text{for } x \le 0 \\ x^2 & \text{for } 0 < x < 1 \\ 1 & \text{for } x \ge 1 \end{cases}$$

(a) Find the corresponding density function.

(b) Compute $P(-1 \le X \le 1/2)$.

Solution

(a) To find the density function, we need to invoke part I of the fundamental theorem of calculus. Thus, we have

$$F(x) = \int_a^x f(u) \, du \qquad \text{implies} \qquad F'(x) = f(x)$$

The density function $f(x)$ can be found by differentiating the distribution function $F(x)$ in the intervals $(-\infty, 0)$, $(0, 1)$, and $(1, \infty)$:

$$f(x) = F'(x) = \begin{cases} 2x & \text{for } 0 < x < 1 \\ 0 & \text{for } x < 0 \text{ or } x > 1 \end{cases}$$

To define the density function everywhere, we set $f(0) = f(1) = 0$. The distribution function, together with its density function, is shown in Figure 12.30.

(b) Using the distribution function, we immediately find that

$$P\left(-1 \le X \le \frac{1}{2}\right) = F\left(\frac{1}{2}\right) - F(-1) = \frac{1}{4} - 0 = \frac{1}{4}$$

If, instead, we use the density function, we must evaluate

$$P\left(-1 \le X \le \frac{1}{2}\right) = \int_{-1}^{1/2} f(x) \, dx = \int_0^{1/2} 2x \, dx$$

$$= x^2 \Big]_0^{1/2} = \frac{1}{4}$$ ■

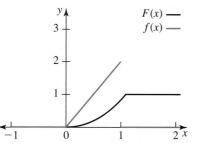

Figure 12.30 The distribution function $F(x)$ and corresponding density function $f(x)$ in Example 2.

The formulas for the mean and the variance of a continuous random variable are analogous to those for the discrete case. The expected value, or mean, $E(X)$ of a continuous random variable X with density function $f(x)$ is defined as

$$E(X) = \int_{-\infty}^{\infty} x f(x) \, dx$$

The expected value of a function of a random variable $g(X)$ is

$$E[g(X)] = \int_{-\infty}^{\infty} g(x) f(x) \, dx$$

where $f(x)$ is the density function of X.

The variance of a continuous random variable X with mean μ is defined as

$$\text{var}(X) = E(X - \mu)^2 = \int_{-\infty}^{\infty} (x - \mu)^2 f(x) \, dx$$

The alternative formula that we gave in the previous section holds as well:

$$\text{var}(X) = E(X^2) - [E(X)]^2 = \int_{-\infty}^{\infty} x^2 f(x) \, dx - \left(\int_{-\infty}^{\infty} x f(x) \, dx\right)^2$$

Recall that these integrals are defined on unbounded intervals and $f(x)$ might be discontinuous. To evaluate such integrals, we must use the methods developed in Section 7.4.

EXAMPLE 3   The density function of a random variable X is given by

$$f(x) = \begin{cases} 3x^2 & \text{for } 0 < x < 1 \\ 0 & \text{otherwise} \end{cases}$$

(See Figure 12.31.) Compute the mean and the variance of X.

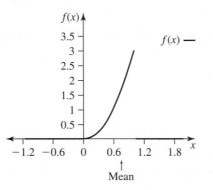

Figure 12.31   The density function $f(x)$ in Example 3, together with the location of the mean of X.

Solution   To compute the mean, we must evaluate

$$E(X) = \int_{-\infty}^{\infty} x f(x)\, dx = \int_0^1 3x^3\, dx = \left. \frac{3}{4}x^4 \right]_0^1 = \frac{3}{4}$$

The mean is indicated in Figure 12.31. To compute the variance, we first evaluate

$$E(X^2) = \int_{-\infty}^{\infty} x^2 f(x)\, dx = \int_0^1 3x^4\, dx = \left. \frac{3}{5}x^5 \right]_0^1 = \frac{3}{5}$$

The variance of X is then given by

$$\text{var}(X) = E(X^2) - [E(X)]^2 = \frac{3}{5} - \left(\frac{3}{4}\right)^2 = \frac{3}{80} \qquad ■$$

EXAMPLE 4   The exponential function

$$f(r) = \begin{cases} ae^{-ar} & \text{for } r > 0 \\ 0 & \text{for } r \le 0 \end{cases}$$

where $a > 0$ is a constant, is frequently used to model seed dispersal. The function $f(r)$ is a density function, and $\int_a^b f(r)\, dr$ describes the fraction of seeds dispersed between distances a and b from the source at 0. Find the average dispersal distance.

Solution   We use the formula for the average value, namely,

$$\text{average dispersal distance} = \int_{-\infty}^{\infty} r f(r)\, dr = \int_0^{\infty} rae^{-ar}\, dr$$

since $f(r) = 0$ for $r \le 0$. To evaluate this integral, we must integrate by parts:

$$\text{average dispersal distance} = \int_0^{\infty} rae^{-ar}\, dr = \lim_{z \to \infty} \int_0^z rae^{-ar}\, dr$$

$$= \lim_{z \to \infty} \left[r(-e^{-ar}) \right]_0^z + \lim_{z \to \infty} \int_0^z e^{-ar}\, dr$$

The first expression on the right is equal to

$$\lim_{z \to \infty} [-ze^{-az} + 0] = -\lim_{z \to \infty} \frac{z}{e^{az}}$$

This limit is of the form $\frac{\infty}{\infty}$. Using l'Hospital's rule, we find that

$$\lim_{z \to \infty} \frac{z}{e^{az}} = \lim_{z \to \infty} \frac{1}{ae^{az}} = 0$$

since $a > 0$. The second expression is

$$\lim_{z \to \infty} \int_0^z e^{-ar} dr = \lim_{z \to \infty} \left[-\frac{1}{a} e^{-ar} \right]_0^z = \frac{1}{a}$$

Therefore,

$$\text{average dispersal distance} = \frac{1}{a} \qquad\qquad ■$$

We will now discuss how to determine $f(x)$ empirically. We set

$$F(x) = \int_{-\infty}^x f(t) \, dt$$

Then

$$F(x + \Delta x) - F(x) = \int_{-\infty}^{x+\Delta x} f(t) \, dt - \int_{-\infty}^x f(t) \, dt = \int_x^{x+\Delta x} f(t) \, dt$$

If Δx is sufficiently small, then $f(t)$ will not vary much over the interval $[x, x + \Delta x)$, and we approximate $f(t)$ by $f(x)$ over the interval $[x, x + \Delta x)$. Hence,

$$\int_x^{x+\Delta x} f(t) \, dt \approx f(x) \Delta x \qquad (12.30)$$

for Δx sufficiently small, and we can think of $f(x)\Delta x$ as approximately representing the fraction that falls into the interval $[x, x + \Delta x)$. This is an important interpretation that will help us to determine density functions empirically. It should remind you of the Riemann sum approximation that we discussed in Section 6.1. For Δx small, just one rectangle gives a good approximation, as illustrated in Figure 12.32.

To determine the density $f(x)$ empirically, we will use the approximation (12.30). We take a sufficiently large sample from the population and measure the quantity of interest of each individual sampled. We partition the interval over which the quantity of interest varies into subintervals of length Δx_i. For each subinterval, we count the number of sample points that fall into the respective subintervals. To display the data graphically, we use a **histogram**, which consists of rectangles whose widths are equal to the lengths of the corresponding subintervals and whose *areas* are equal to the number of sample points that fall into the corresponding subintervals. This approach is analogous to approximating areas under curves by rectangles; we illustrate it in the example that follows.

Brachiopods form a marine invertebrate phylum whose soft body parts are enclosed in shells. These organisms were the dominant seabed shelled animals in the Paleozoic era, but suffered greatly during the Permian–Triassic mass extinction.[1] They are still present today (with approximately 120 genera) and occupy a diverse range of habitats, but they are no longer the dominant seabed shelled animal, their place having been taken by bivalve mollusks. (The brachiopod story is described in Ward, 1992.)

Figure 12.32 The area of the rectangle is $f(x)\Delta x$, a good approximation of $\int_x^{x+\Delta x} f(t)\, dt$, which is the area under the curve between x and $x + \Delta x$.

---

(1) The Permian geological period lasted from 286 million to 248 million years ago; the Triassic followed the Permian and lasted from 248 million to 213 million years ago. The Permian–Triassic mass extinction is believed to have been the most severe mass extinction of life that has ever occurred.

Figure 12.33 The saggital length of a shell.

Brachiopod *Dielasma* fossils are common in Permian reef deposits in the north of England. The following table (adapted from Benton and Harper, 1997) represents measurements of the sagittal length of the animal's shell, measured in mm (see Figure 12.33):

Length	5	7	9	11	13	15	17	19	21	23	25
Frequency	3	28	12	2	4	4	6	6	5	3	1

The length measurements are divided into classes $[0, 2)$, $[2, 4)$, $[4, 6)$, ..., $[26, 28)$. The midpoint of each subinterval represents the size class. For instance, length 11 in the table corresponds to the size class of lengths between 10 mm and 12 mm. The number below each size class—the frequency—represents the number of brachiopods in the sample whose lengths fell into the corresponding size class. For instance, there were two brachiopods in the sample whose lengths fell into the size class $[10, 12)$.

To display this data set graphically, we use a histogram, as shown in Figure 12.34. The horizontal axis shows the midpoint of each size class. The graph consists of rectangles whose widths are equal to the length of the corresponding size class and whose *area* is proportional to the number of specimens in the corresponding class. It is very important to note that a histogram represents numbers by area, not by height. For instance, the number of specimens in size class $[8, 10)$ is equal to 12. This size class is represented by 9, the midpoint of the interval $[8, 10)$. Because the width of the size class is 2, the height of the rectangle must be equal to 6 units so that the area of the corresponding rectangle is equal to 12 units. In our example, all size classes are of the same length; of course, this need not be the case in general.

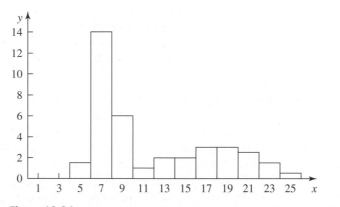

Figure 12.34 The histogram for the frequency distribution of the saggital length.

Displaying data graphically has certain advantages. For instance, we see immediately that the size distribution is biased toward smaller sizes, because the rectangles in the histogram that correspond to smaller shell lengths have larger areas. This bias toward smaller sizes might indicate, for instance, that brachiopods suffered a high juvenile mortality.

It is often convenient to scale the vertical axis of the histogram so that the total area of the histogram is equal to 1. One of the advantages of this approach is that the histogram then does not directly reflect the sample size, since only proportions are represented. Consequently, it is easier to compare histograms from different samples. For instance, if someone else had obtained a different sample of this type of brachiopod in the same location, and if both samples are representative of its length distribution, then both histograms should look similar.

If we scale the total area of the histogram to 1, then the area of each rectangle in the histogram represents the fraction of the sample in the corresponding class. To obtain the fraction of sample points in a certain class, we divide the number of sample points in that class by the total sample size. In our example, the sample size was 74. The fraction of the sample in size class [8, 10), for instance, would then be $12/74 = 0.16$, or 16%.

The choice of the widths of the classes in the histogram is somewhat arbitrary. The goal is to obtain an informative graph. Typically, the larger the sample size, the smaller the widths of the classes can be, and the better the approximation of $f(x)$. The outline of the normalized histogram (i.e., the total area of the histogram is equal to 1) can then be used as an approximation to the density function $f(x)$. (See Figure 12.35.)

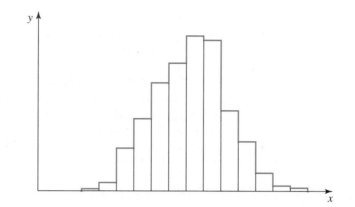

Figure 12.35 The histogram as an approximation of the density function.

In the subsections that follow, we will introduce some continuous distributions and their applications.

■ 12.5.2 The Normal Distribution

The normal distribution was first introduced by Abraham De Moivre (1667–1754) in the context of computing probabilities in binomial experiments when the number of trials is large. Later, Gauss showed that this distribution was important in the error analysis of measurements. It is the most important continuous distribution, and we discuss an application first.

Quantitative genetics is concerned with metric characters, such as plant height, litter size, body weight, and so on. Such characters are called **quantitative characters**. There are many quantitative characters whose frequency distributions follow a bell-shaped curve. For instance, counting the bristles on some particular part of the abdomen (fifth sternite) of a strain of *Drosophila melanogaster*, Mackay (1984) found that the number of bristles varied according to a bell-shaped curve. (This curve is shown in Figure 12.36, which is adapted from Hartl and Clark, 1989.)

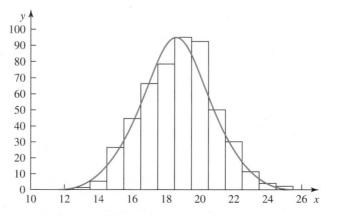

Figure 12.36 The number of abdominal bristles.

The smooth curve in Figure 12.36 that is fitted to the histogram is proportional to the density function of a **normal distribution**. (The curve is not scaled, so the area under the curve is not equal to 1.) The density function of the normal distribution is described by just two parameters, called μ and σ, which can be estimated from data. The parameter μ can be any real number; the parameter σ is a positive real number. The density function of a normal distribution is described as follows:

> A continuous random variable X is normally distributed with parameters μ and σ if it has density function
>
> $$f(x) = \frac{1}{\sigma\sqrt{2\pi}}e^{-(x-\mu)^2/2\sigma^2}, \quad -\infty < x < \infty$$

The parameter μ is the mean and the parameter σ is the standard deviation of the normal distribution. In Problem 11, we will investigate the shape of the density function of the normal distribution. (A graph is shown in Figure 12.37.) Following are the properties of $f(x)$:

1. $f(x)$ is symmetric about $x = \mu$.
2. The maximum of $f(x)$ is at $x = \mu$.
3. The inflection points of $f(x)$ are at $x = \mu - \sigma$ and $x = \mu + \sigma$.

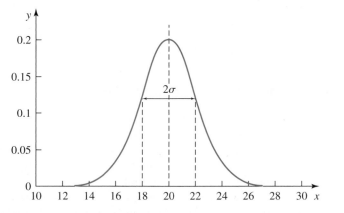

Figure 12.37 The graph of the normal density with $\mu = 20$ and $\sigma = 2$. The maximum is at $\mu = 20$; the two inflection points are at 18 and 22, respectively.

Since $f(x)$ is a density function,

$$f(x) \geq 0 \quad \text{and} \quad \int_{-\infty}^{\infty} f(x)\, dx = 1$$

With the tools we have so far, we cannot show that the density function is normalized to 1. In Problem 12, we will show that the mean μ is indeed the expected value of X; that is,

$$\mu = \int_{-\infty}^{\infty} x f(x)\, dx$$

Furthermore, if a quantity X is normally distributed with parameters μ and σ, then

$$P(a \leq X \leq b) = \int_{a}^{b} \frac{1}{\sigma\sqrt{2\pi}}e^{-(x-\mu)^2/2\sigma^2}\, dx$$

It is not possible to evaluate this integral by using elementary functions; it can be evaluated only numerically. There are tables for the normal distribution with parameters $\mu = 0$ and $\sigma = 1$ that list values for

$$F(x) = P(X \le x) = \int_{-\infty}^{x} \frac{1}{\sqrt{2\pi}} e^{-z^2/2}\, dz$$

A table for $F(x)$ is reproduced in the Appendix B and can be used to obtain probabilities for general μ and σ. Later, we will see how to do this.

At the moment, we will only need a few values. The area $A(k)$ under the density function of the normal distribution with mean μ and standard deviation σ between $\mu - k\sigma$ and $\mu + k\sigma$, for $k = 1, 2$, and 3, is shown in the table on the left. (See Figure 12.38.) That is, if a certain quantity X of a population is normally distributed with mean μ and standard deviation σ, then 68% of the population falls within one standard deviation of the mean, 95% falls within two standard deviations of the mean, and 99% falls within three standard deviations of the mean.

k	A(k)
1	68%
2	95%
3	99%

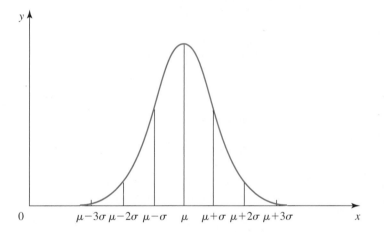

Figure 12.38 The density of the normal distribution with mean μ and standard deviation σ.

These percentages can also be interpreted in the following way: Suppose that a quantity X in a population is normally distributed with mean μ and standard deviation σ. If we sampled from this population—that is, if we picked one individual at random from the population—then there is a 68% chance that the observation would fall within one standard deviation of the mean. We can therefore say that the probability that X is in the interval $[\mu - \sigma, \mu + \sigma]$ is equal to 0.68, which we write as

$$P(X \in [\mu - \sigma, \mu + \sigma]) = 0.68$$

Likewise,

$$P(X \in [\mu - 2\sigma, \mu + 2\sigma]) = 0.95 \quad \text{and} \quad P(X \in [\mu - 3\sigma, \mu + 3\sigma]) = 0.99$$

Finding Probabilities by Using the Mean and the Standard Deviation

EXAMPLE 5 Assume that a certain quantitative character X is normally distributed with mean $\mu = 4$ and standard deviation $\sigma = 1.5$. Find an interval centered at the mean such that there is a 95% chance that an observation will fall into this interval. Then do the same for a 99% chance.

Solution Since 95% corresponds to a range within two standard deviations of the mean (see Figure 12.39), the resulting interval is

$$[4 - (2)(1.5), 4 + (2)(1.5)] = [1, 7]$$

We can therefore write

$$P(X \in [1, 7]) = 0.95$$

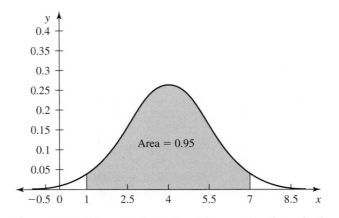

Figure 12.39 The normal density with mean 4 and standard deviation 1.5 in Example 5; 95% of the observations fall into the interval [1, 7].

Similarly, 99% corresponds to a range within three standard deviations of the mean, resulting in an interval of the form

$$[4 - (3)(1.5), 4 + (3)(1.5)] = [-0.5, 8.5]$$

We can therefore write

$$P(X \in [-0.5, 8.5]) = 0.99$$ ■

EXAMPLE 6

Assume that a certain quantitative character X is normally distributed with mean $\mu = 3$ and standard deviation $\sigma = 2$. We take a sample of size 1. What is the chance that we observe a value greater than 9?

Solution

Since we know that $9 = 3 + (3)(2)$, we want to find the chance that the observation is three standard deviations above the mean. (See Figure 12.40.) Now, 99% of the population is within three standard deviations of the mean; therefore, 1% is outside of the interval $[\mu - 3\sigma, \mu + 3\sigma]$. Because the density function of the normal distribution is symmetric about the mean, the area to the left of $\mu - 3\sigma$ and the area to the right of $\mu + 3\sigma$ are the same. Hence, there is a $(1\%)/2 = 0.5\%$ chance that the observation is above 9. We can therefore write

$$P(X > 9) = 0.005$$ ■

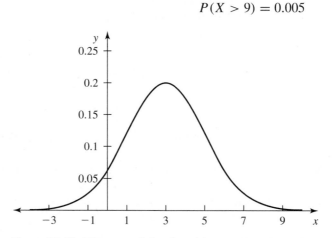

Figure 12.40 The normal density with mean 3 and standard deviation 2 in Example 6; 0.5% of the observations are greater than 9.

EXAMPLE 7

Assume that a certain quantitative character X is normally distributed with parameters μ and σ. What is the probability that an observation lies below $\mu + \sigma$?

Solution   Since 68% of the population falls within one standard deviation of the mean, and since the density curve is symmetric about the mean, it follows that 34% of the population falls into the interval $[\mu, \mu+\sigma]$. Furthermore, because of the symmetry of the density function, 50% of the population is below the mean. Hence, 50% + 34% = 84% of the population lie below $\mu+\sigma$, as illustrated in Figure 12.41. We can therefore write

$$P(X < \mu + \sigma) = 0.84$$ ■

Figure 12.41   The area for Example 7.

Using the Table to Find Probabilities   The table for a normal distribution with mean 0 and standard deviation 1 (see Appendix B) can be used to compute probabilities when the distribution is normal with mean μ and standard deviation σ.

We begin by explaining how to use the table for the normal distribution with mean 0 and standard deviation 1, called the **standard normal distribution**, whose density is given by

$$f(u) = \frac{1}{\sqrt{2\pi}} e^{-u^2/2} \qquad \text{for } -\infty < u < \infty$$

The table lists values for

$$F(z) = \int_{-\infty}^{z} \frac{1}{\sqrt{2\pi}} e^{-u^2/2} \, du$$

Geometrically, $F(z)$ is the area to the left of the line $x = z$ under the graph of the density function, as illustrated in Figure 12.42.

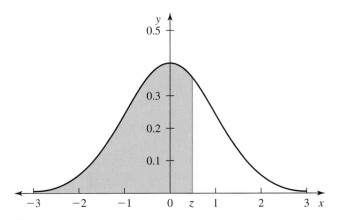

Figure 12.42   The area to the left of $x = z$ under the graph of the normal density $F(z)$, which is listed in the table for the normal distribution.

We interpret $F(z)$ as the probability that an observation is to the left of z. For instance, when $z = 1$, $F(1) = 0.8413$, and we say that the probability that an observation has a value less than or equal to 1 is 0.8413. In other words, 84.13% of the population has a value less than or equal to 1.

As you can see, the table does not provide entries for negative values of z. To compute such values, we take advantage of the symmetries of the density function. For instance, we see from the graph of the function that the area to the left of -1 is the same as the area to the right of 1. (See Figure 12.43.) Thus, if we wish to compute $F(-1)$, we write

$$F(-1) = \int_{-\infty}^{-1} \frac{1}{\sqrt{2\pi}} e^{-u^2/2}\, du = \int_{1}^{\infty} \frac{1}{\sqrt{2\pi}} e^{-u^2/2}\, du$$

$$= \int_{-\infty}^{\infty} \frac{1}{\sqrt{2\pi}} e^{-u^2/2}\, du - \int_{-\infty}^{1} \frac{1}{\sqrt{2\pi}} e^{-u^2/2}\, du$$

$$= 1 - F(1) = 1 - 0.8413 = 0.1587$$

Here, we used the fact that the total area is equal to 1.

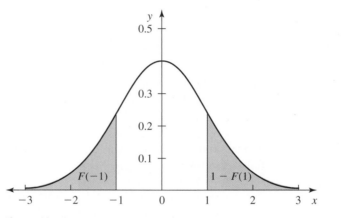

Figure 12.43 The area of the shaded region $F(-1)$ is the same as the area of the shaded region $1 - F(1)$.

We can use the table to compute areas under the graph of a normal density with arbitrary mean μ and standard deviation σ. Thus, to find the value of

$$\int_{a}^{b} \frac{1}{\sigma\sqrt{2\pi}} e^{-(x-\mu)^2/2\sigma^2}\, dx$$

where $-\infty < a < b < \infty$, we use the substitution

$$u = \frac{x - \mu}{\sigma} \quad \text{with} \quad \frac{du}{dx} = \frac{1}{\sigma}$$

which yields

$$\int_{a}^{b} \frac{1}{\sigma\sqrt{2\pi}} e^{-(x-\mu)^2/2\sigma^2}\, dx = \int_{(a-\mu)/\sigma}^{(b-\mu)/\sigma} \frac{1}{\sqrt{2\pi}} e^{-u^2/2}\, du$$

We recognize the right-hand side as the area under the *standard normal density* between $(a - \mu)/\sigma$ and $(b - \mu)/\sigma$. Therefore, the area under the normal density with mean μ and standard deviation σ between a and b is the same as the area under the standard normal density between $(a - \mu)/\sigma$ and $(b - \mu)/\sigma$. (See Figure 12.44.) We illustrate this equality in the next example.

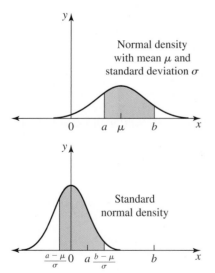

Figure 12.44 The area under the normal density with mean μ and standard deviation σ between a and b is the same as the area under the standard normal density between $(a - \mu)/\sigma$ and $(b - \mu)/\sigma$.

EXAMPLE 8

Suppose that a quantity X is normally distributed with mean 3 and standard deviation 2. Find the fraction of the population that falls into the interval $[2, 5]$; that is, find $P(X \in [2, 5])$.

Solution   To solve this problem, we must compute

$$\int_2^5 \frac{1}{2\sqrt{2\pi}} e^{(x-3)^2/8}\,dx \tag{12.31}$$

Using the transformation $u = (x - 3)/2$, we find that when

$$x = 2, \quad u = \frac{2 - 3}{2} = -\frac{1}{2}$$

and when

$$x = 5, \quad u = \frac{5 - 3}{2} = 1$$

Therefore, the area under the normal density with mean 3 and standard deviation 2 between 2 and 5 is the same as the area under the standard normal density between $-1/2$ and 1. Hence, the integral in (12.31) is equal to

$$\int_{-1/2}^1 \frac{1}{\sqrt{2\pi}} e^{-u^2/2}\,du = \int_{-\infty}^1 \frac{1}{\sqrt{2\pi}} e^{-u^2/2}\,du - \int_{-\infty}^{-1/2} \frac{1}{\sqrt{2\pi}} e^{-u^2/2}\,du$$

$$= F(1) - F\left(-\frac{1}{2}\right) = F(1) - \left(1 - F\left(\frac{1}{2}\right)\right)$$

$$= F(1) + F\left(\frac{1}{2}\right) - 1 = 0.8413 + 0.6915 - 1 = 0.5328$$

and it follows that $P(X \in [2, 5]) = 0.5328$.

Instead of writing out these integrals, it is easier to determine what we need to compute when we sketch the relevant area under the standard normal curve. From Figure 12.45, we see that we need to compute $F(1) - F(-\frac{1}{2})$. Since $F(-\frac{1}{2}) = 1 - F(\frac{1}{2})$, we need to find $F(1) - 1 + F(\frac{1}{2})$, which we computed previously. ∎

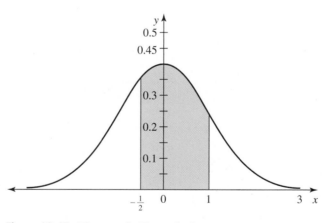

Figure 12.45 The area in Example 8.

EXAMPLE 9   Let X be normally distributed with mean 3 and variance 4. Find $P(1 \le X \le 6)$.

Solution   We apply the transformation $z = (x - \mu)/\sigma$ with $\mu = 3$ and $\sigma = 2$ and denote a standard normally distributed random variable by Z:

$$P(1 \le X \le 6) = P\left(\frac{1 - 3}{2} \le \frac{X - \mu}{\sigma} \le \frac{6 - 3}{2}\right)$$

$$= P\left(-1 \le Z \le \frac{3}{2}\right) = F\left(\frac{3}{2}\right) - F(-1)$$

$$= F(1.5) - (1 - F(1)) = 0.9332 - 1 + 0.8413 = 0.7745 \quad ∎$$

EXAMPLE 10

Suppose that a quantitative character X is normally distributed with mean 2 and standard deviation 1/2. Find x such that 30% of the population is above x.

Solution

We need to find x such that $P(X > x) = 0.3$. Using the transformation $z = (x-\mu)/\sigma$ and letting Z denote a quantity that is normally distributed with mean 0 and standard deviation 1, we obtain

$$P(X > x) = P\left(\frac{X - \mu}{\sigma} > \frac{x - \mu}{\sigma}\right) = P\left(Z > \frac{x - 2}{1/2}\right)$$
$$= P(Z > 2(x - 2)) = 0.3$$

Now, $P(Z > z) = 1 - F(z) = 0.3$; hence, $F(z) = 0.7$. We then find that $F(0.52) = 0.6985 \approx 0.7$. Thus,

$$2(x - 2) = 0.52, \qquad \text{or} \qquad x = 2.26$$

That is, $P(X > 2.26) = 0.3$. Therefore, the value of x that we are seeking is 2.26. ■

A Note on Samples To obtain information about a quantity, such as size or bristle number, we cannot survey an entire population. Instead, we take a sample from the population and find the distribution of the quantity of interest in the sample. We must choose the sample so that it is representative of the population; this is a difficult problem, which we cannot discuss here. Even if we assume that the sample is representative of the population, it will still be the case that samples differ.

EXAMPLE 11

The numbers in the following table, representing two samples, each from the same population, show values of a quantity that is normally distributed with mean $\mu = 0$ and standard deviation $\sigma = 1$:

Sample 1				
−1.633	0.542	0.250	−0.166	0.032
1.114	0.882	1.265	−0.202	0.151
1.151	−1.210	−0.927	0.425	0.290
−1.939	0.891	−0.227	0.602	0.873
0.385	−0.649	−0.577	0.237	−0.289

Sample 2				
−0.157	0.693	1.710	0.800	−0.265
1.492	−0.713	0.821	−0.031	−0.780
−0.042	1.615	−1.440	−0.989	−0.580
0.289	−0.904	0.259	−0.600	−1.635
0.721	−1.117	0.635	0.592	−1.362

Both samples are obtained from a table of random numbers that are normally distributed with mean 0 and standard deviation 1 (Beyer, 1991).

(a) Count the number of observations in each sample that fall below the mean $\mu = 0$, and compare the number with what you would expect on the basis of properties of the normal distribution.

(b) Count the number of observations in each sample that fall within one standard deviation of the mean, and compare the number with what you would expect on the basis of properties of the normal distribution.

Solution

(a) Since the mean is equal to 0, to find the number of observations that are below the mean, we simply count the number of observations that are negative. In the first sample, 10 observations are below the mean; in the second sample, 14 are. We expect that half of the sample points are below the mean. Because each sample is of size 25, we expect about 12 or 13 sample points to be below the mean.

(b) Since the standard deviation is 1, we count the number of observations that fall into the interval $[-1, 1]$. In the first sample, there are 19 such observations; in the second sample, there are 18. To compare these with the theoretical value, note that 68% of the population falls within one standard deviation of the mean. Since the sample size is equal to 25, and since $(0.68)(25) = 17$, we expect about 17 observations to fall into the interval $[-1, 1]$.   ∎

The preceding example illustrates an important point: Even if random samples are taken from the same population, they are not identical. For instance, in the preceding example we expect that half of the observations will be below the mean. In the first sample less than half of the observations will be below the mean, whereas in the second sample more than half of the observations will be below the mean.

As the sample size increases, however, the sample will reflect the population increasingly more faithfully. That is, in order to determine the distribution of a quantitative character, such as the number of bristles in *D. melanogaster*, you would take a sample and find, for instance, the histogram associated with the quantity of interest. Then, if the sample is large enough, the histogram will reflect the population distribution quite well. But if you repeat the experiment, you should not expect the two histograms to be exactly the same. If the sample size is large enough, however, they will be close.

The importance of the normal distribution cannot be overstated. A large part of statistics is based on the assumption that observed quantities are normally distributed. You will probably ask why we can assume the normal distribution in the first place. The reason for this is quite deep, and we will discuss some of it in the next section. At this point, we wish to just give you the gist of it.

Many quantities can be thought of as a sum of a large number of small contributions. We can show that the distribution of any sum of independent random variables that all have the same distribution with a finite mean and a finite variance converges to a normally distributed random variable when the number of terms in the sum increases. This result is known as the **central limit theorem**. (See Section 12.6.)

The central limit theorem is evoked, for instance, in quantitative genetics. Many quantitative traits (such as the height or birth weight of an organism) are thought of as resulting from numerous genetic and environmental factors that all act in an additive or multiplicative way. If these factors do in fact act in an additive way and are independent, the central limit theorem can be applied directly and the distribution of values of the trait will resemble a normal distribution. If the factors act in a multiplicative way, then a logarithmic transformation reduces this case to the additive one.

The same reasoning is used when we consider measurement errors. A measurement error is frequently thought of as a sum of a large number of independent contributions from different sources that act additively. This model for measurement errors is based on empirical evidence, so such errors are often assumed to be normally distributed.

■ 12.5.3 The Uniform Distribution

The uniform distribution is in some ways the simplest continuous distribution. We say that a random variable U is **uniformly distributed** over the interval (a, b) if its density function is given by

$$f(x) = \begin{cases} \frac{1}{b-a} & \text{for } x \in (a, b) \\ 0 & \text{otherwise} \end{cases}$$

as illustrated in Figure 12.46.

The reason for the term *uniform* can be seen when we compute the probability that the random variable U falls into the interval $(x_1, x_2) \subset (a, b)$. To compute probabilities of events for a uniformly distributed random variable, we compute the area of a rectangle. We therefore use the simple geometric formula asserting that the

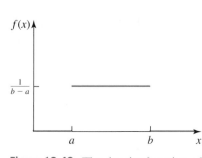

Figure 12.46 The density function of a uniformly distributed random variable over the interval (a, b).

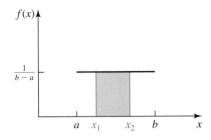

Figure 12.47 The probability $P(U \in (x_1, x_2))$ is equal to the area of the shaded region.

area is equal to the product of height and length, instead of formally integrating the density function. (See Figure 12.47.) When we do, we find that

$$P(U \in (x_1, x_2)) = \left[\begin{array}{c} \text{area under } f(x) = \frac{1}{b-a} \\ \text{between } x_1 \text{ and } x_2 \end{array} \right] = \frac{x_2 - x_1}{b - a}$$

We see that this probability depends only on the length of the interval (x_1, x_2) relative to the length of the interval (a, b), and *not* on the location of (x_1, x_2), provided that (x_1, x_2) is a subset of (a, b). Therefore, intervals of equal lengths that are contained in (a, b) have equal chances of containing U.

To find the mean of a uniformly distributed random variable, we evaluate

$$E(U) = \int_{-\infty}^{\infty} x f(x) \, dx = \int_a^b x \frac{1}{b-a} \, dx = \frac{1}{b-a} \left[\frac{1}{2} x^2 \right]_a^b$$

$$= \frac{1}{b-a} \frac{b^2 - a^2}{2} = \frac{a+b}{2}$$

The last identity comes about because $b^2 - a^2 = (b - a)(b + a)$. The mean of a uniformly distributed random variable over the interval (a, b) is therefore the midpoint of the interval (a, b). To find the variance, we first compute

$$E(U^2) = \int_{-\infty}^{\infty} x^2 f(x) \, dx = \int_a^b x^2 \frac{1}{b-a} \, dx = \frac{1}{b-a} \left[\frac{1}{3} x^3 \right]_a^b$$

$$= \frac{1}{b-a} \frac{b^3 - a^3}{3} = \frac{b^2 + ab + a^2}{3}$$

The last identity comes about because $b^3 - a^3$ factors into $(b - a)(b^2 + ab + a^2)$. Then,

$$\text{var}(U) = E(U^2) - [E(U)]^2 = \frac{b^2 + ab + a^2}{3} - \frac{(a + b)^2}{4}$$

$$= \frac{4b^2 + 4ab + 4a^2 - 3a^2 - 6ab - 3b^2}{12} = \frac{b^2 - 2ab + a^2}{12}$$

$$= \frac{(b - a)^2}{12}$$

Uniform distributions are frequently used in computer simulations of random experiments, as in the next example.

EXAMPLE 12

Suppose that you wish to simulate on the computer a random experiment that consists of tossing a coin five times with probability 0.3 of heads. The software that you want to use can generate uniformly distributed random variables in the interval $(0, 1)$. How do you proceed?

Solution

Each trial consists of flipping the coin and then recording the outcome. To simulate a coin flip with a probability of heads of 0.3, we draw a uniformly distributed random variable U from the interval $(0, 1)$. The computer then returns a number u in the interval $(0, 1)$. Since $P(U \leq 0.3) = 0.3$, if

$$u = \begin{cases} \leq 0.3, & \text{we record heads} \\ > 0.3, & \text{we record tails} \end{cases}$$

We repeat this experiment five times.

To be concrete, assume that successive values of the uniform random variable are $0.2859, 0.9233, 0.5187, 0.8124$, and 0.0913. These numbers are then translated into $HTTTH$, where H stands for heads and T for tails. ∎

In the next example, we compute the distribution function of a uniformly distributed random variable.

EXAMPLE 13

(a) Find the density and distribution functions of a uniformly distributed random variable on the interval $(1, 5)$, and graph both in the same coordinate system.

(b) Suppose that we draw a uniformly distributed random variable from the interval $(1, 5)$. Compute the probability that the first digit after the decimal point is a 2.

Solution

Figure 12.48 The density function $f(x)$ and the distribution function $F(x)$ of a uniformly distributed random variable over the interval $(1, 5)$. (See Example 13.)

(a) Since the length of the interval $(1, 5)$ is 4, the density function is given by

$$f(x) = \begin{cases} \frac{1}{4} & \text{for } 1 < x < 5 \\ 0 & \text{otherwise} \end{cases}$$

The distribution function $F(x) = P(X \le x)$ is given by

$$F(x) = \int_{-\infty}^{x} f(u)\, du = \begin{cases} 0 & \text{for } x \le 1 \\ \int_{1}^{x} \frac{1}{4}\, du = \frac{1}{4}x - \frac{1}{4} & \text{for } 1 < x < 5 \\ 1 & \text{for } x \ge 5 \end{cases}$$

Graphs of $f(x)$ and $F(x)$ are shown in Figure 12.48.

(b) The event that the first digit after the decimal point is a 2 is the event that the random variable U falls into the set

$$A = [1.2, 1.3) \cup [2.2, 2.3) \cup [3.2, 3.3) \cup [4.2, 4.3)$$

Therefore,

$$P(U \in A) = (4)(0.1)\left(\frac{1}{4}\right) = 0.1$$

as illustrated in Figure 12.49. ■

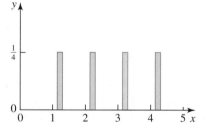

Figure 12.49 The probability $P(U \in A)$ in Example 13 is equal to the sum of the areas of the shaded regions.

■ 12.5.4 The Exponential Distribution

We give the density function of the exponential distribution first and then explain where this distribution plays a role. We say that a random variable X is **exponentially distributed** with parameter $\lambda > 0$ if its density function is given by

$$f(x) = \begin{cases} \lambda e^{-\lambda x} & \text{for } x > 0 \\ 0 & \text{for } x \le 0 \end{cases}$$

Since

$$\int_{0}^{x} \lambda e^{-\lambda u}\, du = -e^{-\lambda u}\Big|_{0}^{x} = 1 - e^{-\lambda x}$$

its distribution function $F(x) = P(X \le x)$ is given by

$$F(x) = \begin{cases} 1 - e^{-\lambda x} & \text{for } x > 0 \\ 0 & \text{for } x \le 0 \end{cases}$$

The expected value of X is

$$E(X) = \int_{0}^{\infty} x\lambda e^{-\lambda x}\, dx = \frac{1}{\lambda}$$

and the variance of X is

$$\text{var}(X) = \int_{0}^{\infty} \left(x - \frac{1}{\lambda}\right)^2 \lambda e^{-\lambda x}\, dx = \frac{1}{\lambda^2}$$

(The mean and the variance will be calculated in Problems 47 and 48.)

The exponential distribution is frequently used to model waiting times or lifetimes.

EXAMPLE 14

Suppose that the time between arrivals of insect pollinators to a flowering plant is exponentially distributed with parameter $\lambda = 0.3$/hr.

(a) Find the mean and the standard deviation of the waiting time between successive pollinator arrivals.

(b) If a pollinator just left the plant, what is the probability that you will have to wait for more than three hours before the next pollinator arrives?

Solution

(a) Since $\lambda = 0.3$/hr, the mean is equal to $1/\lambda = 10/3$ hours and the standard deviation, which is the square root of the variance, is equal to $1/\lambda = 10/3$ hours.

(b) If we denote the waiting time by T, then

$$P(T > 3) = 1 - F(3) = e^{-(0.3)(3)} \approx 0.4066$$

■

EXAMPLE 15

Suppose that the lifetime of an organism is exponentially distributed with parameter $\lambda = (1/200)$ yr^{-1}.

(a) Find the probability that the organism will live for more than 50 years.

(b) Given that the organism is 100 years old, find the probability that it will live for at least another 50 years.

Solution

We denote the lifetime of the organism by T, measured in units of years. Then T is exponentially distributed with parameter $\lambda = (1/200)$ yr^{-1}.

(a) We want to find the probability that T exceeds 50 years. We compute

$$P(T > 50) = 1 - P(T \leq 50) = 1 - (1 - e^{-50/200})$$
$$= e^{-50/200} = e^{-1/4} \approx 0.7788$$

Note that the units in the exponent canceled out.

(b) We want to find $P(T > 150 \,|\, T > 100)$. This is a conditional probability. We evaluate it in the following way:

$$P(T > 150 \,|\, T > 100) = \frac{P(T > 150 \text{ and } T > 100)}{P(T > 100)}$$

Since $\{T > 150\} \subset \{T > 100\}$, it follows that

$$\{T > 150\} \cap \{T > 100\} = \{T > 150\}$$

Therefore,
$$P(T > 150 \text{ and } T > 100) = P(T > 150)$$

We can now continue to evaluate

$$\frac{P(T > 150 \text{ and } T > 100)}{P(T > 100)} = \frac{P(T > 150)}{P(T > 100)} = \frac{e^{-150/200}}{e^{-100/200}}$$
$$= e^{-3/4 + 1/2} = e^{-1/4} \approx 0.7788$$

This is the same answer as that in (a). The fact that the organism has lived for 100 years does not change its probability of living for another 50 years. We say that the organism does not age. This nonaging property is a characteristic feature of the exponential distribution. Of course, most organisms do age. Nevertheless, the exponential distribution is still frequently used to model lifetimes—even if the organism ages, in which case the distribution should be considered as an approximation of the real situation. ■

Let us examine the nonaging property in more detail. If T is an exponentially distributed lifetime, then T satisfies the equation

$$P(T > t + h \,|\, T > t) = P(T > h)$$

In words, if the organism is still alive after t units of time, then the probability that it will live for at least another h units of time is the same as the probability that the organism survived the first h units of time. This implies that death does not become more (or less) likely with age.

The nonaging property follows immediately from the calculation

$$P(T > t + h \mid T > t) = \frac{P(T > t + h \text{ and } T > t)}{P(T > t)}$$

$$= \frac{P(T > t + h)}{P(T > t)} = \frac{e^{-\lambda(t+h)}}{e^{-\lambda t}}$$

$$= e^{-\lambda h} = P(T > h)$$

which is the same as the one that we carried out in Example 15. Note that, for most organisms or objects, the exponential distribution is a poor lifetime model. However, as the next example illustrates, it is the correct distribution to model radioactive decay.

EXAMPLE 16

Assume that the lifetime of a radioactive atom is exponentially distributed with parameter $\lambda = 3/\text{days}$.

(a) Find the average lifetime of this atom.

(b) Find the time T_h such that the probability that the atom will not have decayed at time t is equal to $1/2$. (The time T_h is called the *half-life*.)

Solution

(a) The average lifetime is $1/\lambda = (1/3)$ days.

(b) If the random variable T denotes the lifetime of the atom, then T_h satisfies

$$P(T > T_h) = \frac{1}{2}$$

That is,

$$e^{-\lambda T_h} = \frac{1}{2} \quad \text{or} \quad T_h = \frac{\ln 2}{\lambda}$$

With $\lambda = 3/\text{days}$, we find that

$$T_h = \frac{\ln 2}{3} \text{ days} \approx 0.2310 \text{ days} \qquad ■$$

EXAMPLE 17

In Example 4, we used the exponential function

$$f(r) = \begin{cases} \lambda e^{-\lambda r} & \text{for } r > 0 \\ 0 & \text{for } r \leq 0 \end{cases}$$

where $\lambda > 0$ is a constant, to model seed dispersal. The function $f(r)$ is a density function, and, for $0 < a < b$, $\int_a^b f(r)\, dr$ describes the fraction of seeds dispersed between distances a and b from the source at 0. We recognize this function as the density function of an exponentially distributed random variable with parameter λ.

(a) Show that $f(r)$ is a density function.

(b) Show that the fraction of seeds that are dispersed a distance R or more declines exponentially with R.

(c) Find R such that 60% of the seeds are dispersed within distance R of the source. How does R depend on λ?

Solution

(a) To show that $f(r)$ is a density function, we need to show that $f(r) \geq 0$ for all $r \in \mathbf{R}$ and that $\int_{-\infty}^{\infty} f(r)\, dr = 1$. Since $\lambda > 0$ and $e^{-\lambda r} > 0$, it follows immediately that $f(r) \geq 0$ for $r > 0$. Combining this result with $f(r) = 0$ for $r \leq 0$, we find that $f(r) \geq 0$ for all $r \in \mathbf{R}$. To check the second criterion, we need to carry out the integration. Since the function $f(r)$ is a piecewise-defined function, we need to split the integral into two parts:

$$\int_{-\infty}^{\infty} f(r)\,dr = \int_{-\infty}^{0} f(r)\,dr + \int_{0}^{\infty} f(r)\,dr$$

$$= \int_{-\infty}^{0} 0\,dr + \int_{0}^{\infty} \lambda e^{-\lambda r}\,dr$$

The term $\int_{-\infty}^{0} 0\,dr$ is equal to 0. An antiderivative of $\lambda e^{-\lambda r}$ is $-e^{-\lambda r}$. Hence,

$$\int_{0}^{\infty} \lambda e^{-\lambda r}\,dr = \lim_{z\to\infty}\int_{0}^{z} \lambda e^{-\lambda r}\,dr = \lim_{z\to\infty}\left[-e^{-\lambda r}\right]_{0}^{z}$$

$$= \lim_{z\to\infty}\left[-e^{-\lambda z} - (-1)\right] = 1$$

since $\lim_{z\to\infty} e^{-\lambda z} = 0$.

(b) For $R > 0$, let $G(R)$ denote the fraction of seeds that are dispersed a distance R or more. Then

$$G(R) = \int_{R}^{\infty} f(r)\,dr = \int_{R}^{\infty} \lambda e^{-\lambda r}\,dr = \lim_{z\to\infty}\int_{R}^{z}\lambda e^{-\lambda r}\,dr$$

$$= \lim_{z\to\infty}\left[-e^{-\lambda r}\right]_{R}^{z} = \lim_{z\to\infty}(-e^{-\lambda z} + e^{-\lambda R}) = e^{-\lambda R}$$

This result shows that $G(R)$ declines exponentially with R.

(c) The number R satisfies

$$0.6 = \int_{0}^{R} \lambda e^{-\lambda r}\,dr$$

Carrying out the integration, we obtain

$$0.6 = \left[-e^{-\lambda r}\right]_{0}^{R} = 1 - e^{-\lambda R}$$

To find R, we need to solve

$$e^{-\lambda R} = 0.4$$
$$-\lambda R = \ln 0.4$$
$$R = -\frac{\ln 0.4}{\lambda} = \frac{1}{\lambda}\ln\frac{5}{2}$$

where, in the last step, we used the fact that $-\ln 0.4 = \ln\frac{1}{0.4} = \ln\frac{5}{2}$. We see that $R \propto 1/\lambda$ (i.e., the bigger λ, the smaller R is), which means that seeds tend to be dispersed more closely to the source for larger values of λ. ■

EXAMPLE 18

Suppose that you wish to use a computer to generate exponentially distributed random variables, but the computer's software can generate only uniformly distributed random variables in the interval (0, 1). How do you proceed?

Solution

The key ingredient to solving this problem is the following result: If X is a continuous random variable with a strictly increasing distribution function $F(x)$, then $F(X)$ is uniformly distributed in the interval (0, 1). To prove this result, we need to show that

$$P(F(X) \le u) = u \quad \text{for } 0 < u < 1$$

Now, the event $\{F(X) \le u\}$ is equivalent to the event $\{X \le F^{-1}(u)\}$, where $F^{-1}(\cdot)$ is the inverse function of $F(\cdot)$. That the inverse function exists follows from the assumption that the distribution function is strictly increasing, which implies that $F(x)$ is one to one. Therefore,

$$P(F(X) \leq u) = P(X \leq F^{-1}(u))$$

Since $F(x) = P(X \leq x)$, we have

$$P(X \leq F^{-1}(u)) = F(F^{-1}(u)) = u$$

where the last identity follows from the properties of the inverse function.

How can we use the preceding result to generate exponentially distributed random variables from uniformly distributed random variables? The computer generates a uniformly distributed random variable U, which we interpret as $F(X)$, where X is distributed according to the distribution function $F(x)$. Since

$$U = F(X) \quad \text{is equivalent to} \quad X = F^{-1}(U)$$

we need to find the inverse function of $F(x)$ and compute $F^{-1}(U)$. The result of our computation is then the random variable X.

In the case of an exponential distribution with parameter λ, the distribution function $F(x)$ is

$$F(x) = 1 - e^{-\lambda x} \quad \text{for } x \geq 0$$

which is strictly increasing on $[0, \infty)$. Set $u = 1 - e^{-\lambda x}$ and solve for x:

$$1 - u = e^{-\lambda x}$$

$$-\frac{1}{\lambda} \ln(1 - u) = x$$

Therefore,

$$F^{-1}(u) = -\frac{1}{\lambda} \ln(1 - u)$$

To be concrete, assume that $\lambda = 2$ and that a computer generated the following three random variables, distributed uniformly in $(0, 1)$:

$$u_1 = 0.8890, \quad u_2 = 0.9394, \quad u_3 = 0.3586$$

Then

$$x_1 = F^{-1}(u_1) = -\frac{1}{2} \ln(1 - 0.8890) \approx 1.099$$

$$x_2 = F^{-1}(u_2) = -\frac{1}{2} \ln(1 - 0.9394) \approx 1.402$$

$$x_3 = F^{-1}(u_3) = -\frac{1}{2} \ln(1 - 0.3586) \approx 0.2221$$

are the corresponding realizations of the exponentially distributed random variable.

■

■ 12.5.5 The Poisson Process

In Example 14, we modeled the time between arrivals of insect pollinators to a flowering plant. Our model was an exponential distribution with parameter λ. Suppose we start observing at time 0 and insects arrive at times $T_1, T_2, T_3, \ldots$. The interarrival times $T_1 - 0, T_2 - T_1, T_3 - T_2, \ldots$ are assumed to be independent and exponentially distributed with parameter λ. One can show that the time T_n of the nth arrival is a continous random variable whose distribution is given by the density function

$$f_{n,\lambda}(x) = \frac{\lambda^n x^{n-1}}{(n-1)!} e^{-\lambda x} \quad \text{for } x \geq 0$$

This density function is called the *gamma* (n, λ) *density function.* We will not derive it. Note, however, that if $n = 1$, then

$$f_{1,\lambda} = \lambda e^{-\lambda x} \quad \text{for } x > 0$$

which is indeed the density of an exponential distribution.

Instead of asking when the nth insect pollinator arrived, we can count the number of arrivals up to time t, which we denote by $N(t)$. Now, $N(t)$ is a discrete random variable that takes on values $0, 1, 2, \ldots$.

To calculate the probability mass function for $N(t)$, we begin with the event $\{N(t) = 0\}$. Note that $\{N(t) = 0\}$ is equivalent to $\{T_1 > t\}$. Since T_1 is exponentially distributed with parameter λ, it follows that

$$P(N(t) = 0) = P(T_1 > t) = e^{-\lambda t} \tag{12.32}$$

Generalizing this argument to arbitrary n, we find that the event $\{N(t) < n\}$ is equivalent to $\{T_n > t\}$. The event $\{T_n > t\}$ can be calculated with the use of the gamma (n, λ) density function

$$P(T_n > t) = \int_t^\infty f_{n,\lambda}(x)\,dx = \int_t^\infty \frac{\lambda^n x^{n-1}}{(n-1)!} e^{-\lambda x}\,dx \tag{12.33}$$

Using integration by parts, with $u = \frac{x^{n-1}}{(n-1)!}$ and $v' = \lambda^n e^{-\lambda x}$, on the right-hand side of (12.33), we obtain

$$P(T_n > t) = -\frac{x^{n-1}}{(n-1)!} \lambda^{n-1} e^{-\lambda x} \Big|_t^\infty + \int_t^\infty \frac{x^{n-2}}{(n-2)!} \lambda^{n-1} e^{-\lambda x}\,dx$$

$$= \frac{\lambda^{n-1} t^{n-1}}{(n-1)!} e^{-\lambda t} + P(T_{n-1} > t)$$

Using (12.32), we can calculate

$$P(N(t) < 2) = P(T_2 > t) = \lambda t e^{-\lambda t} + e^{-\lambda t}$$

$$P(N(t) < 3) = P(T_3 > t) = \frac{(\lambda t)^2}{2!} e^{-\lambda t} + \lambda t e^{-\lambda t} + e^{-\lambda t}$$

$$\vdots$$

$$P(N(t) < n) = P(T_n > t) = \sum_{k=0}^{n-1} e^{-\lambda t} \frac{(\lambda t)^k}{k!}$$

The last equation shows that $N(t)$ is Poisson distributed with parameter λt. We say that $N(t)$ is a **Poisson process** with rate λ and summarize as follows:

If the interarrival times are independent and exponentially distributed with parameter λ, then the number of arrivals up to time t is a Poisson process with rate λ and

$$P(N(t) = n) = e^{-\lambda t} \frac{(\lambda t)^n}{n!} \quad \text{for } n = 0, 1, 2, \ldots$$

EXAMPLE 19

Continuation of Example 14 Suppose that the time between arrivals of insect pollinators to a flowering plant is exponentially distributed with parameter $\lambda = 0.3$/hour. Find the probability of fewer than two arrivals within four hours of observation.

Solution

If $N(t)$ denotes the number of arrivals within t hours of observation, then $N(t)$ is a Poisson process with rate $\lambda = 0.3$/hour. Hence,

$$P(N(4) < 2) = e^{-(0.3)(4)}[1 + (0.3)(4)]$$

$$= 2.2 e^{-1.2} \approx 0.6626$$

■

It follows from the properties of the Poisson distribution that

$$E[N(t)] = \lambda t \quad \text{and} \quad \text{var}[N(t)] = \lambda t$$

The Poisson process has three important properties:

1. Nonoverlapping intervals are independent. That is, if $s < t$, then $N(s)$ and $N(t) - N(s)$ are independent.

2. For a small time interval Δt, the probability of an arrival occurring in the interval $[t + \Delta t)$ is approximately proportional to the length of the interval Δt:

$$\lim_{\Delta t \to 0} \frac{P(N([t, t + \Delta t)) = 1)}{\Delta t} = \lim_{\Delta t \to 0} \frac{1}{\Delta t} e^{-\lambda \Delta t} \lambda \Delta t = \lambda$$

3. For a small time interval of length Δt, the probability of more than one arrival is negligible:

$$\lim_{\Delta t \to 0} \frac{P(N([t, t + \Delta t)) > 1)}{\Delta t} = \lim_{\Delta t \to 0} \frac{1 - e^{-\lambda \Delta t} - \lambda \Delta t e^{-\lambda \Delta t}}{\Delta t}$$

$$= \lim_{\Delta t \to 0} \frac{\lambda e^{-\lambda \Delta t} - \lambda e^{-\lambda \Delta t} + \lambda^2 \Delta t e^{-\lambda \Delta t}}{1} = 0$$

Note that we used l'Hospital's rule in the penultimate step.

These three properties characterize the Poisson process and are used to derive its distribution.

■ 12.5.6 Aging

Aging is a universal feature of both living organisms and mechanical devices; it is described by a progressive loss of vitality or reliability. A number of mathematical models are used to describe the phenomenon. The starting point for these models is the **survival function** $S(x)$, which is defined as the probability that the individual or device is still alive or functioning at age x. If X is the lifetime, then

$$S(x) = P(X > x) = 1 - P(X \le x) = 1 - F(x)$$

where $F(x)$ is the distribution function of X. We assume now that X is a nonegative, continuous random variable with density function $f(x) > 0$ for $x > 0$. This assumption implies that $F(0) = 0$ and that $F(x)$ is strictly increasing for $x > 0$. The **failure-rate**, or **hazard-rate**, function $\lambda(x)$ is defined as the relative rate of decline of the survival function:

$$\lambda(x) = -\frac{1}{S(x)} \frac{dS}{dx}$$

Note that $\lambda(x)$ is positive for $x > 0$, since $S(x)$ is strictly decreasing for $x > 0$. Also, because

$$S(x) = P(X > x) = \int_x^\infty f(u) \, du$$

for $x > 0$, it follows that

$$\lambda(x) = -\frac{1}{S(x)} \frac{dS}{dx} = \frac{1}{P(X > x)} f(x)$$

Consequently, $\lambda(x) \, dx$ can be interpreted as the conditional probability of dying within the age interval $[x, x + dx)$, given that the individual is still alive at age $x > 0$. Mathematically,

$$\lambda(x) \, dx = P(X \in [x, x + dx) \mid X > x) \quad \text{for } x > 0$$

Non-aging   Following the preceding interpretation of $\lambda(x)\,dx$, we say that a system does not age if the failure rate $\lambda(x)$ is constant. In this case, for $x > 0$,

$$-\frac{1}{S(x)}\frac{dS}{dx} = \lambda = \text{constant}$$

Separating variables and integrating yields

$$\int \frac{dS}{S} = -\int \lambda\,dx$$

or

$$\ln|S(x)| = -\lambda x + C_1$$

Thus,

$$S(x) = C_2 e^{-\lambda x}$$

with $C_2 = \pm e^{C_1}$. Since X is a nonnegative continuous random variable with $S(0) = P(X > 0) = 1$, it follows that $C_2 = 1$. Therefore,

$$S(x) = e^{-\lambda x} = 1 - F(x) \qquad \text{or} \qquad F(x) = 1 - e^{-\lambda x} \tag{12.34}$$

We conclude that X is exponentially distributed with parameter λ. We showed earlier that the exponential distribution has the non-aging property. Equation (12.34) shows that the reverse holds as well: If a device has the non-aging property (i.e., a constant failure rate), then its lifetime distribution is exponential.

EXAMPLE 20   Suppose the hazard-rate function $\lambda(x) = 3/\text{year}$ for $x \geq 0$. Find the probability that an individual will die before age one year.

Solution   If $\lambda(x) = 3/\text{year}$, then $S(x) = e^{-3x}$ for $x \geq 0$, where x is measured in years. If X denotes the lifetime of the individual, then

$$P(X \leq 1) = 1 - S(1) = 1 - e^{-3} \approx 0.9502$$

Thus, there is about a 95% chance that the individual will die before age one year.   ■

Aging   When the hazard-rate function increases with age, then an older device has a higher probability of dying than a younger device, and we say that the system is an **aging system**. Figure 12.50 shows an empirical hazard-rate function based on 8926 males from an inbred line of *Drosophila melanogaster* obtained in the lab of Professor Jim Curtsinger at the University of Minnesota. The smoothed line is a curve fitted to the data, which are based on daily measurements of survival. The horizontal axis lists

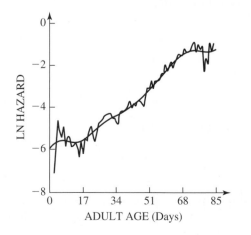

Figure 12.50   An empirical hazard-rate function (courtesy of Dr. Jim Curtsinger).

the age of individuals. The vertical axis lists $-\ln(N_{x+1}/N_x)$, where N_x is the number of adults alive at age x. This quantity can be interpreted as the hazard-rate function, averaged over the interval $[x, x + 1]$. (See Problem 71.) We see from the graph that the hazard-rate function increases with age, but seems to level off at very old ages. This is the typical pattern one observes in these kinds of studies.

We again assume that the the lifetime X is a nonnegative, continuous random variable with hazard-rate function $\lambda(x) > 0$ for $x > 0$ and survival function $S(x)$ for $x \geq 0$, with $S(0) = 1$.

Given the hazard-rate function $\lambda(x)$, we can find the survival function $S(x)$ by integration. For $x > 0$,

$$-\frac{1}{S(x)}\frac{dS}{dx} = \lambda(x)$$

$$\frac{dS}{S} = -\lambda(x)\,dx$$

$$\ln|S(x)| = -\int_0^x \lambda(u)\,du + C_1$$

and therefore,

$$S(x) = C\exp\left[-\int_0^x \lambda(u)\,du\right]$$

with $C = \pm e^{C_1}$. Since $S(0) = 1$, it follows that $C = 1$. Therefore,

$$S(x) = \exp\left[-\int_0^x \lambda(u)\,du\right]$$

The two most prominent hazard-rate functions that model aging are the **Gompertz law**, in which the function increases exponentially with age, and the **Weibull law**, in which the function increases according to a power law.

Gompertz Law:
$$\lambda(x) = A + Be^{\alpha x}, \ x \geq 0$$
where A, B, and α are positive constants.

Weibull Law:
$$\lambda(x) = Cx^{\beta}, \ x \geq 0$$
where C and β are positive parameters.

The Weibull law is often used to calculate the reliability of technical devices, whereas the Gompertz law is used with biological systems. (See the review article by Gavrilov and Gavrilova, 2001.)

EXAMPLE 21

Suppose the lifetime of an organism follows the Gompertz law with hazard-rate function
$$\lambda(x) = 1.5 + 0.3e^{0.1x}, \ x \geq 0$$

where x is measured in years. Find the probability that the organism will live for more than one year.

Solution

The survival function is given by

$$S(x) = \exp\left[-\int_0^x (1.5 + 0.3e^{0.1u})\,du\right], \quad x \geq 0$$

We evaluate the integral first. For $x \geq 0$,

$$\int_0^x (1.5 + 0.3e^{0.1u}) \, du = 1.5u + \frac{0.3}{0.1}e^{0.1u} \Big|_0^x$$

$$= (1.5x + 3e^{0.1x}) - (0 + 3) = 1.5x + 3e^{0.1x} - 3$$

Therefore,

$$S(x) = \exp[-(1.5x + 3e^{0.1x} - 3)], \quad x \geq 0$$

If X denotes the lifetime of the organism, then the probability that the organism will live for more than one year is

$$P(X > 1) = S(1) = \exp\left[-(1.5 + 3e^{0.1} - 3)\right] \approx 0.1628$$

The organism has about a 16% chance of living for more than one year. ■

EXAMPLE 22 Mortality data from *Drosophila melanogaster* were fitted to a Weibull law. It was found that the hazard-rate function

$$\lambda(x) = (3 \times 10^{-6})x^{2.5}, \quad x \geq 0$$

where x is measured in days, provided the best fit.

(a) Find the probability that an individual will die within the first 20 days.

(b) Find the age at which the probability of still being alive is 0.5.

Solution **(a)** The survival function is

$$S(x) = \exp\left[-\int_0^x (3 \times 10^{-6})u^{2.5} \, du\right], \quad x \geq 0$$

We evaluate the integral first. For $x \geq 0$,

$$\int_0^x (3 \times 10^{-6})u^{2.5} \, du = (3 \times 10^{-6})\frac{1}{3.5}u^{3.5} \Big|_0^x$$

$$= (3 \times 10^{-6})\frac{1}{3.5}x^{3.5}$$

Thus, the probability that an individual will die within the first 20 days is

$$1 - S(20) = 1 - \exp\left[-(3 \times 10^{-6})\frac{1}{3.5}(20)^{3.5}\right] \approx 0.0302$$

(b) We need to find x such that

$$S(x) = 0.5$$

We solve for x:

$$\exp\left[-(3 \times 10^{-6})\frac{1}{3.5}x^{3.5}\right] = \frac{1}{2}$$

$$-(3 \times 10^{-6})\frac{1}{3.5}x^{3.5} = \ln\frac{1}{2}$$

$$x^{3.5} = \frac{(3.5)(\ln 2)}{3 \times 10^{-6}}$$

$$x = \left(\frac{(3.5)(\ln 2)}{3 \times 10^{-6}}\right)^{1/3.5}$$

$$x \approx 48.746$$

The age at which the probability of survival is 0.5 is about 48.7 days. ■

Section 12.5 Problems

■ 12.5.1

1. Show that
$$f(x) = \begin{cases} 3e^{-3x} & \text{for } x > 0 \\ 0 & \text{for } x \le 0 \end{cases}$$
is a density function. Find the corresponding distribution function.

2. Show that
$$f(x) = \begin{cases} \frac{1}{2} & \text{for } 0 < x < 2 \\ 0 & \text{otherwise} \end{cases}$$
is a density function. Find the corresponding distribution function.

3. Determine c such that
$$f(x) = \frac{c}{1+x^2}, \quad x \in \mathbf{R}$$
is a density function.

4. Determine c such that
$$f(x) = \begin{cases} \frac{c}{x^2} & \text{for } x > 1 \\ 0 & \text{for } x \le 1 \end{cases}$$
is a density function.

5. Let X be a continuous random variable with density function
$$f(x) = \begin{cases} 2e^{-2x} & \text{for } x > 0 \\ 0 & \text{for } x \le 0 \end{cases}$$
Find $E(X)$ and var(X).

6. Let X be a continuous random variable with density function
$$f(x) = \frac{1}{2}e^{-|x|}$$
for $x \in \mathbf{R}$. Find $E(X)$ and var(X).

7. Let X be a continuous random variable with distribution function
$$F(x) = \begin{cases} 1 - \frac{1}{x^3} & \text{for } x > 1 \\ 0 & \text{for } x \le 1 \end{cases}$$
Find $E(X)$ and var(X).

8. Let X be a continuous random variable with
$$P(X > x) = e^{-ax}, \quad x \ge 0$$
where a is a positive constant. Find $E(X)$ and var(X).

9. Let X be a continuous random variable with density function
$$f(x) = \begin{cases} (a-1)x^{-a} & \text{for } x > 1 \\ 0 & \text{for } x \le 1 \end{cases}$$
(a) Show that $E(X) = \infty$ when $a \le 2$.
(b) Compute $E(X)$ when $a > 2$.

10. Suppose that X is a continuous random variable that takes on only nonnegative values. Set
$$G(x) = P(X > x)$$
(a) Show that
$$G'(x) = -f(x)$$
where $f(x)$ is the corresponding density function.

(b) Assume that
$$\lim_{x \to \infty} xG(x) = 0$$
and use integration by parts and (a) to show that
$$E(X) = \int_0^\infty G(x)\, dx \qquad (12.35)$$

(c) Let X be a continuous random variable with
$$P(X > x) = e^{-ax}, \quad x > 0$$
where a is a positive constant. Use (12.35) to find $E(X)$. (If you did Problem 8, compare your answers.)

■ 12.5.2

11. Denote by the density of a normal distribution with mean μ and standard deviation σ
$$f(x) = \frac{1}{\sigma\sqrt{2\pi}}e^{-(x-\mu)^2/2\sigma^2}$$
for $-\infty < x < \infty$.
(a) Show that $f(x)$ is symmetric about $x = \mu$.
(b) Show that the maximum of $f(x)$ is at $x = \mu$.
(c) Show that the inflection points of $f(x)$ are at $x = \mu - \sigma$ and $x = \mu + \sigma$.
(d) Graph $f(x)$ for $\mu = 2$ and $\sigma = 1$.

12. Suppose that $f(x)$ is the density function of a normal distribution with mean μ and standard deviation σ. Show that
$$\mu = \int_{-\infty}^\infty xf(x)\, dx$$
is the mean of this distribution. (*Hint*: Use substitution.)

13. Suppose that a quantitative character is normally distributed with mean $\mu = 12.8$ and standard deviation $\sigma = 2.7$. Find an interval centered at the mean such that 95% of the population falls into the interval. Do the same for 99% of the population.

14. Suppose a quantitative character is normally distributed with mean $\mu = 15.4$ and standard deviation $\sigma = 3.1$. Find an interval centered at the mean such that 95% of the population falls into this interval. Do the same for 99% of the population.

In Problems 15–20, assume that a quantitative character is normally distributed with mean μ and standard deviation σ. Determine what fraction of the population falls into the given interval.

15. $[\mu, \infty)$ **16.** $[\mu - 2\sigma, \mu + \sigma]$
17. $(-\infty, \mu + 3\sigma]$ **18.** $[\mu + \sigma, \mu + 2\sigma]$
19. $(-\infty, \mu - 2\sigma]$ **20.** $[\mu - 3\sigma, \mu]$

21. Suppose that X is normally distributed with mean $\mu = 3$ and standard deviation $\sigma = 2$. Use the table in Appendix B to find the following:
(a) $P(X \le 4)$ **(b)** $P(2 \le X \le 4)$
(c) $P(X > 5)$ **(d)** $P(X \le 0)$

22. Suppose that X is normally distributed with mean $\mu = -1$ and standard deviation $\sigma = 1$. Use the table in Appendix B to find the following:
(a) $P(X > 0)$ **(b)** $P(0 < X < 1)$
(c) $P(-1.5 < X < 2.5)$ **(d)** $P(X > 1.5)$

23. Suppose that X is normally distributed with mean $\mu = 1$ and standard deviation $\sigma = 2$. Use the table in Appendix B to find x such that the following hold:
(a) $P(X \le x) = 0.9$ **(b)** $P(X > x) = 0.4$

(c) $P(X \leq x) = 0.4$ (d) $P(|X - 1| < x) = 0.5$

24. Suppose that X is normally distributed with mean $\mu = -2$ and standard deviation $\sigma = 1$. Use the table in Appendix B to find x such that the following hold.

(a) $P(X \geq x) = 0.8$ (b) $P(X < 2x + 1) = 0.5$

(c) $P(X \leq x) = 0.1$ (d) $P(|X - 2| > x) = 0.4$

25. Assume that the mathematics score X on the Scholastic Aptitude Test (SAT) is normally distributed with mean 500 and standard deviation 100.

(a) Find the probability that an individual's score exceeds 700.

(b) Find the math SAT score so that 10% of the students who took the test have that score or greater.

26. In a study of *Drosophila melanogaster* by Mackey (1984), the number of bristles on the fifth abdominal sternite in males was shown to follow a normal distribution with mean 18.7 and standard deviation 2.1.

(a) What percentage of the male population has fewer than 17 abdominal bristles?

(b) Find an interval centered at the mean so that 90% of the male population have bristle numbers that fall into this interval.

27. Suppose the weight of an animal is normally distributed with mean 3720 g and standard deviation 527 g. What percentage of the population has a weight that exceeds 5000 g?

28. Suppose the height of an adult animal is normally distributed with mean 17.2 in. Find the standard deviation if 10% of the animals have a height that exceeds 19 in.

29. Suppose that X is normally distributed with mean 2 and standard deviation 1. Find $P(0 \leq X \leq 3)$.

30. Suppose that X is normally distributed with mean -1 and standard deviation 2. Find $P(-3.5 \leq X \leq 0.5)$.

31. Suppose that X is normally distributed with mean μ and standard deviation σ. Show that $E(X) = \mu$. [You may use the fact that if Z is standard normally distributed, then $E(Z) = 0$ and $\text{var}(X) = 1$.]

32. Suppose that X is normally distributed with mean μ and standard deviation σ. Show that $\text{var}(X) = \sigma^2$. [You may use the fact that if Z is standard normally distributed, then $E(Z) = 0$ and $\text{var}(X) = 1$.]

33. Suppose that X is standard normally distributed. Find $E(|X|)$.

34. Suppose that the number of seeds a plant produces is normally distributed, with mean 142 and standard deviation 31. Find the probability that in a sample of five plants, at least one produces more than 200 seeds. Assume that the plants are independent.

35. The total maximum score on a calculus exam was 100 points. The mean score was 74 and the standard deviation was 11. Assume that the scores are normally distributed.

(a) Determine the percentage of students scoring 90 or above.

(b) Determine the percentage of students scoring between 60 and 80 (inclusive).

(c) Determine the minimum score of the highest 10% of the class.

(d) Determine the maximum score of the lowest 5% of the class.

36. The mean weight of female students at a small college is 123 lb, and the standard deviation is 9 lb. If the weights are normally distributed, determine what percentage of female students weigh
(a) between 110 and 130 lb, (b) less than 100 lb, and (c) more than 150 lb.

■ **12.5.3**

37. Suppose that you pick a number at random from the interval $(0, 4)$. What is the probability that the first digit after the decimal point is a 3?

38. Suppose that you pick a number X at random from the interval $(0, a)$. If $P(X \geq 1) = 0.2$, find a.

39. Suppose that you pick a number X at random from the interval (a, b). If $E(X) = 4$ and $\text{var}(X) = 3$, find a and b.

40. Suppose that you pick five numbers at random from the interval $(0, 1)$. Assume that the numbers are independent. What is the probability that all numbers are greater than 0.7?

41. Suppose that X_1, X_2, and X_3 are independent and uniformly distributed over $(0, 1)$. Define

$$Y = \max(X_1, X_2, X_3)$$

Find $E(Y)$. [*Hint*: Compute $P(Y \leq y)$, and use it to deduce the density of Y.]

42. Suppose that X_1, X_2, and X_3 are independent and uniformly distributed over $(0, 1)$. Define

$$Y = \min(X_1, X_2, X_3)$$

Find $E(Y)$. [*Hint*: Compute $P(Y > y)$, and use it to deduce the density of Y.]

43. Suppose that you wish to simulate a random experiment that consists of tossing a coin with probability 0.6 of heads 10 times. The computer generates the following 10 random variables: 0.1905, 0.4285, 0.9963, 0.1666, 0.2223, 0.6885, 0.0489, 0.3567, 0.0719, 0.8661. Find the corresponding sequence of heads and tails.

44. Suppose that you wish to simulate a random experiment that consists of rolling a fair die. The computer generates the following 10 random variables: 0.7198, 0.2759, 0.4108, 0.7780, 0.2149, 0.0348, 0.5673, 0.0014, 0.3249, 0.6630. Describe how you would find the corresponding sequence of numbers on the die, and find them.

45. Suppose $X_1, X_2, \ldots, X_n$ are independent random variables with uniform distribution on $(0, 1)$. Define $X = \min(X_1, X_2, \ldots, X_n)$.

(a) Compute $P(X > x)$.

(b) Show that $P(X > x/n) \to e^{-x}$ as $n \to \infty$.

46. Suppose $X_1, X_2, \ldots, X_n$ are independent random variables with uniform distribution on $(0, 1)$. Define $X = \max(X_1, X_2, \ldots, X_n)$.

(a) Find the distribution function of X.

(b) Use Problem 10 to compute $E(X)$.

■ **12.5.4**

47. Let X be exponentially distributed with parameter λ. Find $E(X)$.

48. Let X be exponentially distributed with parameter λ. Find $\text{var}(X)$.

49. Suppose that the lifetime of a battery is exponentially distributed with an average life span of three months. What is the probability that the battery will last for more than four months?

50. Suppose that the lifetime of a battery is exponentially distributed with an average life span of two months. You buy six batteries. What is the probability that none of them will last more than two months? (Assume that the batteries are independent.)

51. Suppose that the lifetime of a radioactive atom is exponentially distributed with an average life span of 27 days.

(a) Find the probability that the atom will not decay during the first 20 days after you start to observe it.

(b) Suppose that the atom does not decay during the first 20 days that you observe it. What is the probability that it will not decay during the next 20 days?

52. If X has distribution function $F(x)$, we can show that $F(X)$ is uniformly distributed over the interval $(0, 1)$. Use this fact to generate exponentially distributed random variables with mean 1. [Assume that a computer generated the following four uniformly distributed random variables on the interval $(0, 1)$: $0.0371, 0.5123, 0.1370, 0.9865.$]

■ **12.5.5**

53. Suppose the number of customers per hour arriving at the post office is a Poisson process with an average of four customers per hour.
(a) Find the probability that no customer arrives between 2 and 3 P.M.
(b) Find the probability that exactly two customers arrive between 3 and 4 P.M.
(c) Assuming that the number of customers arriving between 2 and 3 P.M. is independent of the number of customers arriving between 3 and 4 P.M., find the probability that exactly two customers arrive between 2 and 4 P.M.
(d) Assume that the number of customers arriving between 2 and 3 P.M. is independent of the number of customers arriving between 3 and 4 P.M. Given that exactly two customers arrive between 2 and 4 P.M., what is the probability that both arrive between 3 and 4 P.M.?

54. Suppose the number of customers per hour arriving at the post office is a Poisson process with an average of five customers per hour.
(a) Find the probability that exactly one customer arrives between 2 and 3 P.M.
(b) Find the probability that exactly two customers arrive between 3 and 4 P.M.
(c) Assuming that the number of customers arriving between 2 and 3 P.M. is independent of the number of customers arriving between 3 and 4 P.M., find the probability that exactly three customers arrive between 2 and 4 P.M.
(d) Assume that the number of customers arriving between 2 and 3 P.M. is independent of the number of customers arriving between 3 and 4 P.M. Given that exactly three customers arrive between 2 and 4 P.M., what is the probability that one arrives between 2 and 3 P.M. and two between 3 and 4 P.M.?

55. You arrive at a bus stop at a random time. Assuming that busses arrive according to a Poisson process with rate 4/hr, what is the expected time to the next arrival?

56. Assume that $N(t)$ is a Poisson process with rate λ and T_1 is the time of the first arrival. Show that, for $s < t$,

$$P(T < s \mid N(t) = 1) = \frac{s}{t}$$

That is, show that, given that an arrival occurred in the interval $[0, t)$, the time of occurrence is uniform over the interval.

57. Suppose the lifetime of a technical device is exponentially distributed with mean 3 years. The device is instantly replaced upon failure.
(a) Find the probability that the device will have failed after two years.
(b) What is the probability that, over a period of five years, the device was replaced only once?

58. Suppose the lifetime of a light bulb is exponentially distributed with mean 1 year. The light bulb is instantly replaced upon failure. What is the probability that, over a period of five years, at most five light bulbs are needed?

■ **12.5.6**

59. Suppose the lifetime of a technical device is exponentially distributed with mean five years.
(a) Find the probability that the device will have failed after three years.
(b) Given that the device has worked for six years, find the probability that it will work for another year.

60. Suppose the lifetime of an organism is exponentially distributed with hazard rate function $\lambda(x) = 2/\text{day}$.
(a) Find the probability that an individual of this species lives for more than three days.
(b) What is the expected lifetime?

61. Suppose the lifetime of a technical device is exponentially distributed with parameter $\lambda = 0.2/\text{year}$.
(a) What is the expected lifetime?
(b) The **median lifetime** is defined as the age x_m at which the probability of not having failed by age x_m is 0.5. Find x_m.

62. The median lifetime is defined as the age x_m at which the probability of not having failed by age x_m is 0.5. If the life span of an organism is exponentially distributed, and if $x_m = 4$ years, what is the hazard-rate function?

63. The hazard-rate function of an organism is given by

$$\lambda(x) = 0.3 + 0.1e^{0.01x}, \quad x \geq 0$$

where x is measured in days.
(a) What is the probability that the organism will live for more than five days?
(b) What is the probability that the organism will live between 7 and 10 days?

64. The hazard-rate function of an organism is given by

$$\lambda(x) = 0.1 + 0.5e^{0.02x}, \quad x \geq 0$$

where x is measured in days.
(a) What is the probability that the organism will live less than 10 days?
(b) What is the probability that the organism will live for another five days, given that it survived the first five days?

65. The median lifetime is defined as the age x_m at which the probability of not having failed by age x_m is 0.5. Use a graphing calculator to numerically approximate the median lifetime if the hazard-rate function is

$$\lambda(x) = 1.2 + 0.3e^{0.5x}, \quad x \geq 0$$

66. The median lifetime is defined as the age x_m at which the probability of not having failed by age x_m is 0.5. Use a graphing calculator to numerically approximate the median lifetime if the hazard-rate function is

$$\lambda(x) = 0.5 + 0.1e^{0.2x}, \quad x \geq 0$$

67. The hazard-rate function of an organism is given by

$$\lambda(x) = (2 \times 10^{-5})x^{1.5}, \quad x \geq 0$$

where x is measured in days.
(a) What is the probability that the organism will live for more than 50 days?
(b) What is the probability that the organism will live between 50 and 70 days?

68. The hazard-rate function of an organism is given by

$$\lambda(x) = 0.04x^{3.1}, \quad x \geq 0$$

where x is measured in years.

(a) What is the probability that the organism will live for more than three years?

(b) What is the probability that the organism will live for another three years, given that it survived the first three years?

69. The median lifetime is defined as the age x_m at which the probability of not having failed by age x_m is 0.5. Find the median lifetime if the hazard-rate function is

$$\lambda(x) = (4 \times 10^{-5})x^{2.2}, \quad x \geq 0$$

70. The median lifetime is defined as the age x_m at which the probability of not having failed by age x_m is 0.5. Find the median lifetime if the hazard-rate function is

$$\lambda(x) = (3.7 \times 10^{-6})x^{2.7}, \quad x \geq 0$$

71. Let N_x be the number of individuals that are still alive at age x. Show that

$$-\ln \frac{N_{x+1}}{N_x}$$

can be estimated by

$$\int_x^{x+1} \lambda(u)\, du$$

where $\lambda(x)$ is the hazard-rate function at age x.

■ 12.6 Limit Theorems

■ 12.6.1 The Law of Large Numbers

Prostate-specific antigen (PSA) levels are a diagnostic tool for detecting prostate cancer. They are also used to screen for "biochemical failure" after surgical removal of the prostate. Biochemical failure is defined as a PSA level that exceeds 0.5 ng/ml; it can be an indication that prostate cancer cells are still in the body of the patient. In a study by Iselin et al. (1999), the PSA levels of 817 men with prostate cancer were followed after surgical removal of the prostate. In 429 of the men, the disease was found to be confined to the prostate. After five years, 8% of the men in the group whose cancer was confined to the prostate had biochemical failure.

Suppose now that you heard of a small study of 30 men whose prostate was surgically removed and in whom the disease was confined to the prostate. After five years, 3 out of the 30 men, or 10%, experienced biochemical failure, as defined previously. Which of the two figures, 8% or 10% would you deem more reliable? Your answer will likely be 8%. We tend to trust larger studies more than smaller ones. The reason is found in a mathematical result known as the **law of large numbers**, which implies that estimates of proportions become more reliable as the sample size increases.

To state the law of large numbers, we consider a sequence of independent random variables $X_1, X_2, \ldots, X_n$, all with the same distribution. We say that $X_1, X_2, \ldots, X_n$ are **independent and identically distributed** (i.i.d., for short). We assume that $E(X_i) = \mu$ and $\text{var}(X_i) = \sigma^2$. We can think of the sequence as coming from some random experiment that we repeat n times and X_i is the outcome on the ith trial. For instance, the X_i's could denote the successive outcomes of tossing a coin n times, with $X_i = 1$ if the ith toss results in heads and $X_i = 0$ otherwise.

We are interested in the (arithmetic) **average** of $X_1, X_2, \ldots, X_n$, denoted by

$$\overline{X}_n = \frac{1}{n}\sum_{i=1}^{n} X_i$$

In the case of the coin-tossing example, $\overline{X}_n$ would be the fraction of heads in n trials.

To state the law of large numbers, we need a type of convergence called **convergence in probability**. We say that a random variable Z_n converges to a constant γ in probability as $n \to \infty$ if, for any $\epsilon > 0$,

$$\lim_{n \to \infty} P(|Z_n - \gamma| \geq \epsilon) = 0$$

Now we state the law:

> **Weak Law of Large Numbers**[2] If $X_1, X_2, \ldots, X_n$ are i.i.d. with $E(|X_i|) < \infty$, then, as $n \to \infty$, $\overline{X}_n$ converges to $E(X_1)$ in probability.

The weak law of large numbers explains why taking a larger sample improves the reliability of estimates of proportions. In the prostate cancer example set forth at the beginning of this subsection, we were interested in the likelihood of biochemical failure after five years in men who underwent prostate surgery when the cancer was confined to the prostate. We can think of this likelihood as a (to us unknown) probability that we estimate by taking a large sample in which all individuals are considered independent. (Such a sample is called a *random sample*.) If we set

$$X_i = \begin{cases} 1 & \text{if biochemical failure appears after five years in individual } i \\ 0 & \text{otherwise} \end{cases}$$

then $\overline{X}_n = \frac{1}{n}\sum_{i=1}^{n} X_i$ is the fraction of men in the sample who experience biochemical failure after five years. Now, if we set

$$\mu = E(X_i) = P(X_i = 1)$$

then the law of large numbers tells us that $\overline{X}_n \to \mu$ in probability as $n \to \infty$. Thus, if the sample size is sufficiently large, $\overline{X}_n$ provides a good estimate of the probability of biochemical failure, in the sense that, with high probability, $\overline{X}_n$ will be close to μ.

Two inequalities help us to prove the weak law of large numbers.

> **Markov's Inequality** If X is a nonnegative random variable with $E(X) < \infty$, then, for any $a > 0$,
> $$P(X \geq a) \leq \frac{E(X)}{a}$$

Proof We prove Markov's inequality when X is a nonnegative, continuous random variable with density function $f(x)$. Let $a > 0$. Then

$$E(X) = \int_0^\infty xf(x)\,dx = \int_0^a xf(x)\,dx + \int_a^\infty xf(x)\,dx \qquad (12.36)$$

Since $\int_0^a xf(x)\,dx \geq 0$, it follows that

$$E(X) \geq \int_a^\infty xf(x)\,dx \qquad (12.37)$$

Using $x \geq a$, we find that the right-hand side of (12.37) is bounded as follows:

$$\int_a^\infty xf(x)\,dx \geq \int_a^\infty af(x)\,dx = a\int_a^\infty f(x)\,dx = aP(X \geq a)$$

Therefore,

$$E(X) \geq aP(X \geq a) \qquad \text{or} \qquad P(X \geq a) \leq \frac{E(X)}{a} \qquad ■$$

The next inequality is a consequence of Markov's inequality.

---

(2) In addition to the weak law of large numbers, there is also a strong law of large numbers. The weak law does not exclude the possibility that the averages $\overline{X}_n$ may occasionally be quite different from $E(X_1)$, even for large n. The strong law of large numbers excludes this possibility. Because it requires additional theoretical background to state, the strong law of large numbers will not be discussed in this book.

> **Chebyshev's Inequality** If X is a random variable with finite mean μ and finite variance σ^2, then, for $c > 0$,
>
> $$P(|X - \mu| \geq c) \leq \frac{\sigma^2}{c^2}$$

Proof   The events $\{|X - \mu| \geq c\}$ and $\{(X - \mu)^2 \geq c^2\}$ are the same. Therefore,

$$P(|X - \mu| \geq c) = P((X - \mu)^2 \geq c^2)$$

The random variable $(X - \mu)^2$ is nonnegative, and $E(X - \mu)^2 = \sigma^2 < \infty$ by assumption. We can thus apply Markov's inequality and obtain

$$P((X - \mu)^2 \geq c^2) \leq \frac{E(X - \mu)^2}{c^2} = \frac{\sigma^2}{c^2} \qquad ■$$

To prove the weak law of large numbers, we need to find the mean and the variance of $\overline{X}_n$. First, the mean is

$$E(\overline{X}_n) = E\left(\frac{1}{n}\sum_{i=1}^{n} X_i\right) = \frac{1}{n}\sum_{i=1}^{n} E(X_i) = \frac{1}{n}\sum_{i=1}^{n} \mu = \frac{1}{n} \cdot n\mu = \mu$$

Then, using independence yields the variance:

$$\text{var}(\overline{X}_n) = \text{var}\left(\frac{1}{n}\sum_{i=1}^{n} X_i\right) = \frac{1}{n^2}\sum_{i=1}^{n} \text{var}(X_i) = \frac{1}{n^2}\sum_{i=1}^{n} \sigma^2 = \frac{1}{n^2} \cdot n\sigma^2 = \frac{\sigma^2}{n}$$

Applying Chebyshev's inequality with $c > 0$ to $\overline{X}_n$ gives

$$P(|\overline{X}_n - \mu| \geq c) \leq \frac{\text{var}(\overline{X}_n)}{c^2} = \frac{\sigma^2}{nc^2}$$

If we let $n \to \infty$, the right-hand side tends to 0. Since probabilities are (always) nonnegative, we can conclude that

$$\lim_{n \to \infty} P(|\overline{X}_n - \mu| \geq c) = 0 \qquad (12.38)$$

The result about the limit in (12.38) can thus be expressed as "$\overline{X}_n$ converges to μ in probability." This is the weak law of large numbers.

We proved the weak law under the additional assumption that $\text{var}(X_i) < \infty$. The weak law of large numbers still holds if $\text{var}(X_i) = \infty$, provided that $E(|X_i|) < \infty$, but the proof of this result would be quite a bit more complicated (and we won't do it).

EXAMPLE 1   Suppose $X_1, X_2, \ldots, X_n$ are i.i.d. with

$$X_i = \begin{cases} 1 & \text{with probability } p \\ 0 & \text{with probability } 1 - p \end{cases}$$

Set $\overline{X}_n = \frac{1}{n}\sum_{i=1}^{n} X_i$, and show that $\overline{X}_n$ converges to p in probability as $n \to \infty$.

Solution   Since

$$E(|X_i|) = |1|(p) + |0|(1 - p) = p < \infty \quad \text{and} \quad E(X_i) = (1)(p) + (0)(1 - p) = p$$

we can invoke the law of large numbers and conclude that

$$\overline{X}_n = \frac{1}{n}\sum_{i=1}^{n} X_i \to E(X_i) = p$$

in probability as $n \to \infty$. ■

EXAMPLE 2

Monte Carlo Integration Suppose that $f(x)$ is an integrable function on $[0, 1]$ with $f(x) \geq 0$ for $x \in [0, 1]$. Let $U_1, U_2, \ldots, U_n$ be i.i.d. with U_i uniformly distributed on $(0, 1)$. Show that

$$\lim_{n \to \infty} \frac{1}{n} \sum_{i=1}^{n} f(U_i) = \int_0^1 f(x)\, dx \qquad \text{in probability}$$

Solution

Define $X_i = f(U_i)$. Then, using the fact that the density function of a uniform distribution on $(0, 1)$ is equal to 1 on that same interval, we obtain

$$E(|X_i|) = E(|f(U_i)|) = \int_0^1 |f(x)|\, dx = \int_0^1 f(x)\, dx < \infty$$

and

$$E(X_i) = E[f(U_i)] = \int_0^1 f(x)\, dx$$

The X_i's are i.i.d. We can therefore apply the law of large numbers and conclude that

$$\lim_{n \to \infty} \frac{1}{n} \sum_{i=1}^{n} f(U_i) = \lim_{n \to \infty} \frac{1}{n} \sum_{i=1}^{n} X_i = E(X_i) = \int_0^1 f(x)\, dx$$

in probability. ■

We can use Chebyshev's inequality to get an estimate of sample size.

EXAMPLE 3

Suppose $X_1, X_2, \ldots, X_n$ are i.i.d. with

$$X_i = \begin{cases} 1 & \text{with probability } p \\ 0 & \text{with probability } 1 - p \end{cases}$$

Use Chebyshev's inequality to find n such that $\overline{X}_n$ will differ from p by less than 0.01 with probability at least 0.95.

Solution

We know from Example 1 that $\overline{X}_n$ converges to p in probability as $n \to \infty$. We want to investigate how fast the convergence is. More precisely, we want to find n such that

$$P(|\overline{X}_n - p| < 0.01) \geq 0.95$$

or, taking complements,

$$P(|\overline{X}_n - p| \geq 0.01) \leq 0.05$$

Using Chebyshev's inequality, we find that

$$P(|\overline{X}_n - p| \geq 0.01) \leq \frac{\operatorname{var}(\overline{X}_n)}{(0.01)^2}$$

Since the X_i's are independent,

$$\operatorname{var}(\overline{X}_n) = \operatorname{var}\left(\frac{1}{n} \sum_{i=1}^{n} X_i\right) = \frac{1}{n^2} \sum_{i=1}^{n} \operatorname{var}(X_i)$$

$$= \frac{1}{n^2} \sum_{i=1}^{n} p(1 - p) = \frac{p(1 - p)}{n}$$

Therefore,

$$\frac{\operatorname{var}(\overline{X}_n)}{(0.01)^2} = 10{,}000 \frac{p(1 - p)}{n}$$

We want this expression to be less than or equal to 0.05. However, we don't know the value of p. Fortunately, the following reasoning allows us to find a bound on $p(1-p)$: The function $f(p) = p(1-p)$ is an upside-down parabola with roots at $p = 0$ and $p = 1$. Its maximum is at $p = 1/2$ and is $1/4$; that is, $f(1/2) = 1/4$. Therefore,

$$p(1-p) \leq \frac{1}{4} \quad \text{for all } 0 \leq p \leq 1$$

We thus obtain

$$10{,}000\frac{p(1-p)}{n} \leq 10{,}000\frac{1}{4n} \leq 0.05, \quad \text{or} \quad n \geq 50{,}000$$

We conclude that a sample size of 50,000 would suffice to estimate p within an error of 0.01 with 95% probability. It turns out that Chebyshev's inequality does not give very good estimates and this lower bound on the sample size is much larger than what we would really need. In the next subsection, we will learn a better way to estimate sample sizes. ■

■ 12.6.2 The Central Limit Theorem

We now come to a result that is indeed central to probability theory. It says that if we add up a large number of independent and identically distributed random variables with finite mean and variance, then, after suitable scaling, the distribution of the resulting quantity is approximately normally distributed. We will not be able to prove the theorem here, but we will be able to explore some of its implications.

> **Central Limit Theorem** Suppose $X_1, X_2, \ldots, X_n$ are i.i.d. with mean $E(X_i) = \mu$ and variance $\mathrm{var}(X_i) = \sigma^2 < \infty$. Define $S_n = \sum_{i=1}^{n} X_i$. Then, as $n \to \infty$,
>
> $$P\left(\frac{S_n - n\mu}{\sqrt{n\sigma^2}} \leq x\right) \to F(x)$$
>
> where $F(x)$ is the distribution function of the standard normal distribution.

EXAMPLE 4

Toss a fair coin 500 times. Use the central limit theorem to find an approximation for the probability of at least 265 heads.

Solution

We define

$$X_i = \begin{cases} 1 & \text{if } i\text{th toss results in heads} \\ 0 & \text{otherwise} \end{cases}$$

Then

$$\mu = E(X_i) = \frac{1}{2} \quad \text{and} \quad \sigma^2 = \mathrm{var}(X_i) = \frac{1}{2}\left(1 - \frac{1}{2}\right) = \frac{1}{4}$$

Denote the number of heads among the first 500 tosses by $S_{500} = \sum_{i=1}^{500} X_i$. Then

$$P(S_{500} \geq 265) = P\left(\frac{S_{500} - 500\mu}{\sqrt{500\sigma^2}} \geq \frac{265 - 250}{\sqrt{125}}\right)$$

$$\approx 1 - F\left(\frac{15}{\sqrt{125}}\right) = 1 - F(1.34)$$

$$= 1 - 0.9099 = 0.0901$$

where $F(x)$ is the distribution function of a standard normally distributed random variable and the values of $F(x)$ are obtained from the table in Appendix B. ■

When the central limit theorem is applied to integer-valued random variables, a correction is used to get a better approximation. The correction, called the **histogram correction**, is explained in the next example.

EXAMPLE 5

For S_{500} defined in Example 4, use the central limit theorem to find an approximation for $P(S_{500} = 250)$.

Solution

If we applied the central limit theorem without any corrections, we would find that

$$P(S_{500} = 250) = P\left(\frac{S_{500} - 500\mu}{\sqrt{500\sigma^2}} = \frac{250 - 250}{\sqrt{125}}\right) \approx P(Z = 0) = 0$$

where Z is a standard normally distributed random variable. We can compare this result with the exact value. In that case, S_{500} is binomially distributed with parameters $n = 500$ and $p = 1/2$. We find that

$$P(S_{500} = 250) = \binom{500}{250}\left(\frac{1}{2}\right)^{500} \approx 0.036$$

The central limit theorem does not give a good approximation. We can do better by writing the event $\{S_{500} = 250\}$ as $\{249.5 \le S_{500} \le 250.5\}$. (See Figure 12.51.) Then

$$P(S_{500} = 250) = P(249.5 \le S_{500} \le 250.5)$$

$$= P\left(\frac{249.5 - 250}{\sqrt{125}} \le \frac{S_{500} - 500\mu}{\sqrt{500\sigma^2}} \le \frac{250.5 - 250}{\sqrt{125}}\right)$$

$$= P(-0.04 \le Z \le 0.04)$$

where Z is standard normally distributed. It then follows that

$$P(-0.04 \le Z \le 0.04) = 2F(0.04) - 1 = (2)(0.5160) - 1 = 0.032$$

where $F(x)$ is the distribution function of a standard normal distribution. ■

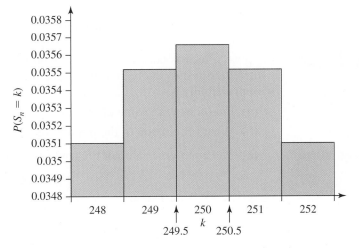

Figure 12.51 The histogram correction for Example 5.

EXAMPLE 6

Redo Example 4 with the histogram correction.

Solution

We write the event $\{S_{500} \ge 265\}$ as $\{S_{500} \ge 264.5\}$. (See Figure 12.52.) Then

$$P(S_{500} \ge 265) = P(S_{500} \ge 264.5) = P\left(\frac{S_{500} - 500\mu}{\sqrt{500\sigma^2}} \ge \frac{264.5 - 250}{\sqrt{125}}\right)$$

$$\approx P(Z \ge 1.30) = 1 - F(1.30) = 1 - 0.9032 = 0.0968$$

where Z is a standard normally distributed random variable with distribution function $F(x)$. ■

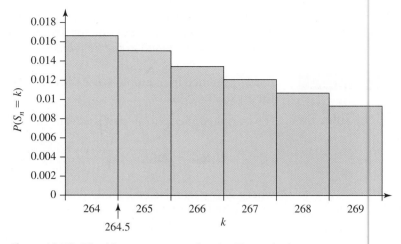

Figure 12.52 The histogram correction for Example 6.

Quantitative genetics is a field in biology that attempts to explain quantitative differences between individuals that are either of genotypic or environmental origin, such as differences in height, litter size, number of abdominal bristles in *Drosophila*, and so on. Estimates on the number of loci involved in a quantitative trait range from very few, such as five loci for skull length in rabbits (Wright, 1968), to very many, such as 98 loci for abdominal bristles in *Drosophila* (Falconer, 1989).

When many loci are involved in a quantitative trait, the **infinitesimal model** is used to model the genotypic value of that trait. The genotypic value G of a trait is considered to be a sum of the contributions of each of the loci involved:

$$G = X_1 + X_2 + \cdots + X_n$$

In the simplest case, the X_i's are assumed to be independent and identically distributed and represent the contribution of locus i to the genotypic value. If the X_i's have finite mean and variance, and if n is large, the distribution of G can be approximated by a normal distribution.

EXAMPLE 7

Suppose a trait is controlled by 100 loci. Each locus, independently of all others, contributes to the genotypic value of the trait either $+1$ with probability 0.6 or -0.7 with probability 0.4.

(a) Find the mean value of the trait.

(b) What proportion of the population has a trait value greater than 40?

Solution

(a) The genotypic value of the trait can be written as

$$S_{100} = \sum_{i=1}^{100} X_i$$

with

$$X_i = \begin{cases} 1 & \text{with probability } 0.6 \\ -0.7 & \text{with probability } 0.4 \end{cases}$$

Hence,

$$E(S_{100}) = \sum_{i=1}^{100} E(X_i) = \sum_{i=1}^{100} [(1)(0.6) + (-0.7)(0.4)] = \sum_{i=1}^{100} 0.32 = 32$$

Hence, the mean value of the trait is 32.

(b) To find the proportion of the population that has a trait value greater than 40, we employ the central limit theorem. We compute the variance of X_i first:

$$E(X_i^2) = (1)^2(0.6) + (-0.7)^2(0.4) = 0.796$$

Thus,

$$\text{var}(X_i) = E(X_i^2) - [E(X_i)]^2 = 0.796 - (0.32)^2 = 0.6936$$

Now,

$$P(S_{100} > 40) = P\left(\frac{S_{100} - 32}{\sqrt{(100)(0.6936)}} > \frac{40.5 - 32}{\sqrt{69.36}}\right)$$
$$\approx 1 - F(1.02) = 1 - 0.8461 = 0.1539$$

where $F(x)$ is the distribution function of a standard normal distribution. Consequently, about 15% of the population has trait value greater than 40. ■

EXAMPLE 8

Estimating Sample Sizes Suppose you wish to conduct a medical study to determine the fraction of people in the general population whose total cholesterol level is above 220 g/dl. How large a sample size would you need to estimate the proportion within 0.01 of the true value with probability at least 0.95?

Solution

Define

$$X_i = \begin{cases} 1 & \text{if } i\text{th individual has cholesterol} \geq 220 \text{ mg/dl} \\ 0 & \text{otherwise} \end{cases}$$

and assume that the individuals are selected so that the X_i's are i.i.d. Then

$$\overline{X}_n = \frac{1}{n}\sum_{i=1}^{n} X_i$$

is an estimate of the proportion of individuals whose cholesterol level exceeds 220 mg/dl.

If we set $S_n = \sum_{i=1}^{n} X_i$, then, with $p = E(X_i)$ and $\sigma^2 = \text{var}(X_i)$,

$$\frac{S_n - np}{\sqrt{n\sigma^2}} \quad \text{is approximately standard normally distributed}$$

Dividing both numerator and denominator by n, we find that, with $\sigma = \sqrt{\text{var}(X_i)} = \sqrt{p(1-p)}$,

$$\sqrt{n}\frac{\overline{X}_n - p}{\sqrt{p(1-p)}} \quad \text{is approximately standard normally distributed}$$

We are interested in finding n in order to estimate the proportion within 0.01 of the true value p with probability at least 0.95. That is,

$$P(|\overline{X}_n - p| \leq 0.01) \geq 0.95$$

We rewrite this inequality as

$$P(-0.01 \leq \overline{X}_n - p \leq 0.01) \geq 0.95$$

or

$$P\left(\sqrt{n}\frac{-0.01}{\sqrt{p(1-p)}} \leq \sqrt{n}\frac{\overline{X}_n - p}{\sqrt{p(1-p)}} \leq \sqrt{n}\frac{0.01}{\sqrt{p(1-p)}}\right) \geq 0.95$$

Since $\sqrt{n}\frac{\overline{X}_n - p}{\sqrt{p(1-p)}}$ is approximately standard normally distributed, we find that the left-hand side is approximately

$$2F\left(\sqrt{n}\frac{0.01}{\sqrt{p(1-p)}}\right) - 1$$

This is ≥ 0.95 if

$$F\left(\sqrt{n}\,\frac{0.01}{\sqrt{p(1-p)}}\right) \geq 0.975$$

or

$$\sqrt{n}\,\frac{0.01}{\sqrt{p(1-p)}} \geq 1.96$$

Solving for n, we obtain

$$n \geq (196)^2\, p(1-p)$$

Since we do not know p, we take the worst possible case that maximizes $p(1-p)$. As we saw in Example 3, this occurs for $p = 1/2$. Therefore,

$$n \geq (196)^2\frac{1}{2}\left(1-\frac{1}{2}\right) = 9604$$

Thus, about 9604 individuals would suffice for this study. You should compare this result with that in Example 3, where we solved the same problem (in a different application) by using Chebyshev's inequality instead of the central limit theorem. ■

Remark. Both the normal and the Poisson distribution serve as approximations to the binomial distribution. As a rule of thumb, the approximations are reasonably good when $n \geq 40$. When $np \leq 5$, the Poisson approximation should be used; when $np \geq 5$, the normal approximation should be used.

Section 12.6 Problems

■ 12.6.1

1. Let X be exponentially distributed with parameter $\lambda = 1/2$. Use Markov's inequality to estimate $P(X \geq 3)$, and compare your estimate with the exact answer.

2. Let X be uniformly distributed over $(1, 4)$.
(a) Use Markov's inequality to estimate $P(X \geq a)$, $1 \leq a \leq 4$, and compare your estimate with the exact answer.
(b) Find the value of $a \in (1, 4)$ that minimizes the difference between the bound and the exact probability computed in (a).

3. Prove Markov's inequality when X is a nonnegative discrete random variable with $E(X) < \infty$.

4. Let X be a continuous random variable with density $f(x)$, and assume that $X \geq 2$. Why is $E(X) \geq 2$?

5. Let X be uniformly distributed over $(-2, 2)$. Use Chebyshev's inequality to estimate $P(|X| \geq 1)$, and compare your estimate with the exact answer.

6. Let X be standard normally distributed. Use Chebyshev's inequality to estimate **(a)** $P(|X| \geq 1)$, **(b)** $P(|X| \geq 2)$, and **(c)** $P(|X| \geq 3)$. Compare each estimate with the exact answer.

7. Suppose X is a random variable with mean 10 and variance 9. What can you say about $P(|X - 10| \geq 5)$?

8. Suppose X is a random variable with mean -5 and variance 2. What can you say about the probability that X deviates from its mean by at least 4?

9. Suppose $X_1, X_2, \ldots, X_n$ are i.i.d. with

$$X_i = \begin{cases} -1 & \text{with probability } 0.2 \\ 1 & \text{with probability } 0.5 \\ 2 & \text{with probability } 0.3 \end{cases}$$

What can you say about $\frac{1}{n}\sum_{i=1}^{n} X_i$ as $n \to \infty$?

10. Suppose $X_1, X_2, \ldots, X_n$ are independent random variables with $P(X_i > x) = e^{-2x}$. What can you say about $\frac{1}{n}\sum_{i=1}^{n} X_i$ as $n \to \infty$?

11. Suppose $X_1, X_2, \ldots, X_n$ are independent random variables with density function

$$f(x) = \frac{1}{\pi(1 + x^2)}, \quad x \in \mathbf{R}$$

Can you apply the law of large numbers to $\frac{1}{n}\sum_{i=1}^{n} X_i$? If so, what can you say about $\frac{1}{n}\sum_{i=1}^{n} X_i$ as $n \to \infty$?

12. How often do you have to toss a coin to determine $P(\text{heads})$ within 0.1 of its true value with probability at least 0.9?

13. A certain study showed that less than 5% of the population suffers from a certain disorder. To get a more accurate estimate of this proportion, you plan to conduct another study. What sample size should you choose if you want to be at least 95% sure that your estimate is within 0.05 of the true value?

14. Assume that $E(e^{cX}) < \infty$ for $c > 0$. Use Markov's inequality to prove **Bernstein's inequality**,

$$P(X \geq x) \leq e^{-cx}\, E(e^{cX})$$

for $c > 0$.

■ 12.6.2

15. Toss a fair coin 400 times. Use the central limit theorem and the histogram correction to find an approximation for the probability of getting at most 190 heads.

16. Toss a fair coin 150 times. Use the central limit theorem and the histogram correction to find an approximation for the probability that the number of heads is at least 70.

17. Toss a fair coin 200 times.

(a) Use the central limit theorem and the histogram correction to find an approximation for the probability that the number of heads is at least 120.

(b) Use Markov's inequality to find an estimate for the event in (a), and compare your estimate with that in (a).

18. Toss a fair coin 300 times.

(a) Use the central limit theorem and the histogram correction to find an approximation for the probability that the number of heads is between 140 and 160.

(b) Use Chebyshev's inequality to find an estimate for the event in (a), and compare your estimate with that in (a).

19. Suppose S_n is binomially distributed with parameters $n = 200$ and $p = 0.3$. Use the central limit theorem to find an approximation for $P(99 \leq S_n \leq 101)$ **(a)** without the histogram correction and **(b)** with the histogram correction. **(c)** Use a graphing calculator to compute the exact probabilities, and compare your answers with those in (a) and (b).

20. Suppose S_n is binomially distributed with parameters $n = 150$ and $p = 0.4$. Use the central limit theorem to find an approximation for $P(S_n = 60)$ **(a)** without the histogram correction and **(b)** with the histogram correction. **(c)** Use a graphing calculator to compute the exact probabilities and compare your answers with those in (a) and (b).

21. Suppose a genotypic trait is controlled by 80 loci. Each locus, independently of all others, contributes to the genotypic value of the trait either $+0.3$ with probability 0.2, -0.1 with probability 0.5, or -0.5 with probability 0.3.

(a) Find the mean value of the trait.

(b) What proportion of the population has a trait value between -12 and -7?

22. Suppose a genotypic trait is controlled by 90 loci. Each locus, independently of all others, contributes to the genotypic value of the trait either 1.1 with probability 0.7, 0.9 with probability 0.1, or 0.1 with probability 0.2.

(a) Find the mean value of the trait.

(b) What proportion of the population has a trait value less than 72?

23. How often should you toss a coin to be at least 90% certain that your estimate of $P(\text{heads})$ is within 0.1 of its true value?

24. How often should you toss a coin to be at least 90% certain that your estimate of $P(\text{heads})$ is within 0.01 of its true value?

25. To forecast the outcome of a presidential election in which two candidates run for office, a telephone poll is conducted. How many people should be surveyed to be at least 95% sure that the estimate is within 0.05 of the true value? (Assume that there are no undecided people in the survey.)

26. A medical study is conducted to estimate the proportion of people suffering from seasonal affective disorder. How many people should be surveyed to be at least 99% sure that the estimate is within 0.02 of the true value?

In Problems 27–30, S_n is binomially distributed with parameters n and p.

27. For $n = 100$ and $p = 0.01$, compute $P(S_n = 0)$ **(a)** exactly, **(b)** by using a Poisson approximation, and **(c)** by using a normal approximation.

28. For $n = 100$ and $p = 0.1$, compute $P(S_n = 10)$ **(a)** exactly, **(b)** by using a Poisson approximation, and **(c)** by using a normal approximation.

29. For $n = 50$ and $p = 0.1$, compute $P(S_n = 5)$ **(a)** exactly, **(b)** by using a Poisson approximation, and **(c)** by using a normal approximation.

30. For $n = 50$ and $p = 0.5$, compute $P(S_n = 25)$ **(a)** exactly, **(b)** by using a Poisson approximation, and **(c)** by using a normal approximation.

31. Suppose you want to estimate the proportion of people in the United States who do not believe in evolution. You happen to take a class on evolutionary theory at a U.S. college that is attended by 200 students, all of whom are biology majors. Do you think you would get an accurate estimate if you asked all 200 students in your class? Discuss.

32. A soft-drink company introduces a new beverage. One month later, the company wants to know whether its marketing strategies have reached young adults of ages 18–20. You happen to work part time for the marketing company that is conducting the survey. At the same time, you are taking a calculus class that is attended by 250 students. It would be easy for you to hand out a survey in class. Would you suggest this to your supervisor in the marketing company? Discuss.

33. Clementine oranges are sold in boxes. Each box contains 50 clementines. The probability that a clementine in a box is spoiled is 0.01.

(a) Use an appropriate approximation to determine the probability that a box contains 0, 1, or at least 2 spoiled clementines.

(b) A shipment of clementines (said to be hybrid crossings between oranges and tangerines) with 100 boxes is considered unacceptable if 35% or more of the boxes contain spoiled clementines. What is the probability that a shipment is unacceptable?

34. Turner's syndrome is a rare chromosomal disorder in which girls have only one X chromosome. The condition affects about 1 in 2000 girls in the United States. About 1 in 10 girls with Turner's syndrome suffers from an abnormal narrowing of the aorta.

(a) In a group of 4000 girls, what is the probability that no girls are affected with Turner's syndrome? That one girl is affected? Two? At least three?

(b) In a group of 170 girls affected with Turner's syndrome, what is the probability that at least 20 of them suffer from an abnormal narrowing of the aorta?

In Problems 35–37, use the following facts: Cystic fibrosis is an inherited disorder that causes abnormally thick body secretions. About 1 in 2500 white babies in the United States has this disorder. About 3 in 100 children with cystic fibrosis develop diabetes mellitus, and about 1 in 5 females with cystic fibrosis is infertile.

35. Find the probability that, in a group of 5000 newborn white babies in the United States, at least 4 babies suffer from cystic fibrosis.

36. Find the probability that, in a group of 1000 children with cystic fibrosis, at least 25 will develop diabetes mellitus.

37. Find the probability that, in a group of 250 women with cystic fibrosis, no more than 60 are infertile.

■ 12.7 Statistical Tools

In the preceding sections, we learned how to model various random experiments. Using the underlying probability distribution of the model, we were able to compute the probabilities of events, such as the probability of obtaining white-flowering pea plants in Mendel's pea experiment.

To understand observations or outcomes of experiments in the biological or medical sciences, however, we often take the reverse approach: We infer the underlying probability distribution from events we have observed. On the basis of a collection of observations, called **data**, we estimate characteristics of the underlying probability distribution. For instance, in the case of the normal distribution, our goal might be to estimate the mean and the variance, which are parameters that describe a normal distribution.

This section provides an introduction to data, parameters estimation, confidence intervals, and linear regression. Throughout, we assume that all observations are expressed numerically.

■ 12.7.1 Describing Univariate Data

Observations that are expressed numerically can be described by **variables**. We can measure one or more variables in an experiment or an observational study. **Univariate** data refers to data obtained from measuring a single variable. **Bivariate** data consists of pairs of observations for each sample point. If more than two variables are measured, we call the data **multivariate**. Data in medical studies are typically multivariate. For instance, Dyck et al. (1999) measured 26 variables in a cohort of patients with diabetes in Rochester, Minnesota, to understand the effects of chronic hyperglycemia (high blood sugar levels) on diabetic neuropathy (a neurological disorder that often accompanies long-term diabetes). Examples of variables included in the study are age, height, weight, duration of diabetes, cholesterol levels, and percent glycosylated hemoglobin. In this section, we will encounter only univariate data.

To learn something about the distribution of a character in a population (severity of diabetic neuropathy, clutch size, plant height, life span, effect of a new drug, and so forth), we cannot measure the occurrence of that character on every individual in the population. Instead, we take a subset of the population, called a **sample**, obtain individual observations on the character of interest, which are assumed to be independent of each other, and then infer the distribution of the character in the population from its distribution in the sample. A sample that is representative of the population and whose individual observations are independent of each other and have the same distribution is called a **random sample**.

To illustrate the importance of choosing a random sample, let's look at the recent debate regarding the benefit of hormone replacement therapy for postmenopausal women. For several decades, physicians recommended hormone replacement therapy for these women. The recommendation was based on observational studies that indicated health benefits, such as a decrease in coronary heart disease (CHD). Three well-designed and large drug trials—the Heart and Estrogen/Progestin Replacement Study (HERS), the Women's Health Initiative (WHI) estrogen-plus-progestin trial, and the WHI estrogen-alone trial—found an increase in CHD in addition to other harmful effects. In 2002, after the conclusion of the first two studies, the U.S. Food and Drug Administration issued a warning about the potential harm of this treatment. An editorial by Hulley and Grady (2004) that accompanied the research article of the third study (the WHI estrogen-alone trial) stated, "Given the absence of evidence in all 3 trials that these hormone regimens prevent CHD in these populations..., it is now clear that previously available evidence was misleading. Observational studies were probably confounded by the tendency of healthier women to seek and comply with hormone treatment."

In what follows, we will always assume that our sample is a random sample and thus representative of the population. By definition, all observations in a random sample are independent and come from the same distribution (the distribution of the quantity in the entire population). A typical scheme to obtain a random sample of size n is to pick an individual at random from the population, record the quantity of interest, replace the individual in the population, and then select the next individual. This procedure is repeated until a sample of size n is obtained. Replacing the sampled individual after recording the quantity of interest ensures that the population always has the same composition and, hence, that all observations come from the same distribution. It also means that an individual may be chosen more than once, unless the population size is much larger than the sample size. In practice, individuals are often chosen without replacement; for instance, in medical studies, when a representative group of participants is chosen, each individual is represented only once. When the sample size is much smaller than the population size, the difference between sampling with replacement and without replacement is negligible. We denote the sample by the vector $(X_1, X_2, \ldots, X_n)$, where X_k is the kth observation. The X_k are independent random variables that all have the same distribution. We say that the X_k, $k = 1, 2, \ldots, n$, are **independent and identically distributed**.

We use the data set on the left, which describes the number of lions at 17 locations in the Serengeti over a three-day period in late October 1990, to introduce some important definitions. These data were collected by Dr. Craig Packer of the University of Minnesota and a group of collaborators working with him.

One way to summarize data is to give the number of times a certain category (in this case, number of lions) occurs. These numbers are called *frequencies*. If we divide the frequencies by the total number of observations, we obtain *relative frequencies*. The list of (relative) frequencies is called a (relative) frequency distribution.

To obtain the frequency distribution of the number of lions, we count how often each of the values in the category "number of lions" appears. For instance, the value 4 appears five times, so its frequency is 5; the value 7 appears once, so its frequency is 1, and so on.

The frequencies appear in the following table, which shows that the range of values is between 0 and 7:

Location	Number of Lions
Kenyangaga	0
Lemai	1
Kogatende	6
Bolagonja	4
Klein's Camp	7
Tabora B	0
Lobo	6
Tagoro	6
Kirawira A	3
Kirawira B	0
Ikimoa	6
Ndabaka	0
Handajega	4
Mereo	4
Seronera	4
Maswa HQ	0
Mamarehe	4

Number of Lions	0	1	2	3	4	5	6	7
Number of Locations	5	1	0	1	5	0	4	1

To obtain relative frequencies, we divide each frequency by 17, because there are 17 locations. The relative frequencies appear in the following table.

Number of Lions	0	1	2	3	4	5	6	7
Relative Frequency of Locations	5/17	1/17	0	1/17	5/17	0	4/17	1/17

A quantity that is computed from observations in a sample is called a **statistic**.

The first statistic that we define is the **sample median**: the middle of the observations when we order the data according to size. If the number of observations is odd, there is one data point in the middle of the ordered data. If the number of observations is even, we take the average of the two observations in the middle. The number of data points in our example is odd. The list of the ordered data is as follows:

Ordered Lion Data: 0, 0, 0, 0, 0, 1, 3, 4, **4**, 4, 4, 4, 6, 6, 6, 6, 7

After ordering the data points, we find that the middle of the observations is the ninth data point, which is the median. Thus, the median is 4.

The two most frequently used statistics are the **sample mean** and the **sample variance**. In addition, the sample standard deviation is used. To define these quantities, we recall that we denoted a sample of size n by the vector $(X_1, X_2, \ldots, X_n)$, where X_k is the kth observation. The sample mean, the sample variance, and the sample standard deviation are defined as follows:

$$\text{Sample mean:} \quad \overline{X}_n = \frac{1}{n} \sum_{k=1}^{n} X_k$$

$$\text{Sample variance:} \quad S_n^2 = \frac{1}{n-1} \sum_{k=1}^{n} (X_k - \overline{X}_n)^2$$

$$\text{Sample standard deviation:} \quad S_n = \sqrt{S_n^2}$$

The sample mean is thus the arithmetic average of the observations. The sample variance is the sum of the squared deviations from the sample mean, divided by $n-1$. (We will explain in the next subsection why we divide by $n-1$ rather than n.) The sample standard deviation is the square root of the sample variance.

The preceding definition of the sample variance is not very convenient for computation. An alternative form that is typically easier to use follows from an algebraic manipulation of the definition of the sample variance:

$$S_n^2 = \frac{1}{n-1} \sum_{k=1}^{n} (X_k - \overline{X}_n)^2$$

$$= \frac{1}{n-1} \sum_{k=1}^{n} (X_k^2 - 2X_k\overline{X}_n + \overline{X}_n^2)$$

$$= \frac{1}{n-1} \left[\sum_{k=1}^{n} X_k^2 - 2\overline{X}_n \sum_{k=1}^{n} X_k + n\overline{X}_n^2 \right]$$

Using the fact that $\sum_{k=1}^{n} X_k = n\overline{X}_n$, we simplify the equation for S_n^2 to

$$S_n^2 = \frac{1}{n-1} \left[\sum_{k=1}^{n} X_k^2 - n\overline{X}_n^2 \right]$$

$$= \frac{1}{n-1} \left[\sum_{k=1}^{n} X_k^2 - \frac{1}{n} \left(\sum_{k=1}^{n} X_k \right)^2 \right]$$

EXAMPLE 1

Compute the sample mean, the sample variance, and the sample standard deviation for the data on lions.

Solution

We find that

$$\sum_{k=1}^{17} X_k = 55 \quad \text{and} \quad \sum_{k=1}^{17} X_k^2 = 283$$

Hence,

$$\overline{X}_{17} = \frac{55}{17} \approx 3.2,$$

$$S_{17}^2 = \frac{1}{16} \left[\sum_{k=1}^{17} X_k^2 - \frac{1}{17} \left(\sum_{k=1}^{17} X_k \right)^2 \right]$$

$$= \frac{1}{16} \left(283 - \frac{1}{17}(55)^2 \right) \approx 6.57$$

and

$$S_{17} = \sqrt{S_{17}^2} \approx 2.56$$

The sample standard deviation (S.D.) is an estimate of the variance of the population. When we report the sample mean and wish to give an indication of the variance of the population, we report the sample mean and the sample standard deviation as

$$\text{Mean} \pm \text{S.D.}$$

In Example 1, we would thus report 3.2 ± 2.6.

Reporting the sample mean and the variance of the population is common practice in scientific publications. These statistics frequently are reported under a heading of the form "Mean ± S.D." The expression "Mean" stands for the sample mean $\overline{X}_n$. The abbreviation "S.D." stands for the sample standard deviation, an estimate of the variance of the population. For instance, in the study by Dyck et al. (1999), mentioned earlier, the baseline characteristics of 149 Type 2 diabetes patients are listed in a table as "Mean ± S.D." The age of this group is listed as 69.7 ± 9.7, the height (in cm) as 166.2 ± 4.4, and the weight (in kg) as 84.8 ± 16.9.

If the sample distribution is summarized in a frequency distribution, the sample mean and the sample variance take on the following form: Assume that a sample of size n has l distinct values $x_1, x_2, \ldots, x_l$, where x_k occurs f_k times in the sample. Then the sample mean is given by the formula

$$\overline{X}_n = \frac{1}{n} \sum_{k=1}^{l} x_k f_k$$

and the sample variance has the form

$$S_n^2 = \frac{1}{n-1} \left[\sum_{k=1}^{l} x_k^2 f_k - \frac{1}{n} \left(\sum_{k=1}^{l} x_k f_k \right)^2 \right]$$

EXAMPLE 2

Use the frequency distribution of the data on lions to calculate the sample mean and the sample variance.

Solution

We expect to find the same answers as in Example 1. We have

$$\overline{X}_n = \frac{1}{n} \sum_{k=1}^{l} x_k f_k$$

$$= \frac{1}{17} [(0)(5) + (1)(1) + (2)(0) + (3)(1) + (4)(5) + (5)(0) + (6)(4) + (7)(1)]$$

$$= \frac{55}{17}$$

and

$$S_n^2 = \frac{1}{n-1} \left[\sum_{k=1}^{l} x_k^2 f_k - \frac{1}{n} \left(\sum_{k=1}^{l} x_k f_k \right)^2 \right]$$

$$= \frac{1}{16} [(0)(5) + (1)(1) + (4)(0) + (9)(1) + (16)(5) + (25)(0) + (36)(4) + (49)(1)$$

$$- \frac{1}{17} (55)^2]$$

$$= \frac{1}{16} \left[283 - \frac{1}{17} (55^2) \right] \approx 6.57 \qquad \blacksquare$$

It is important to realize that any statistic computed from a sample will vary from sample to sample, since the samples are random subsets of the population. Statistics are therefore random variables with their own probability distributions.

EXAMPLE 3

Assume that a population consists of the three numbers 2, 4, and 7. List all samples of size 2 that can be drawn from this population with replacement, and find the sample mean of each sample.

Solution

There are nine equally likely samples. We list them in the table on the left, together with the sample mean of each sample. We see from the table that the sample mean 3.0 occurs twice. This means that if we drew a sample of size 2 from this population, we would obtain a sample mean 3.0 with probability 2/9. ■

Sample	Sample Mean
(2, 2)	2.0
(2, 4)	3.0
(2, 7)	4.5
(4, 2)	3.0
(4, 4)	4.0
(4, 7)	5.5
(7, 2)	4.5
(7, 4)	5.5
(7, 7)	7.0

To illustrate the point further, we look at simulated data from a normal distribution with mean 5 and variance 2. We take 5000 samples from this population, each of size 20, and record the sample mean and the sample median for each sample. Let's denote the vector of sample means by $(y_1, y_2, \ldots, y_{5000})$ and the vector sample medians by $(z_1, z_2, \ldots, z_{5000})$.

This simulation generated a table filled with random numbers from a normal distribution with mean 5 and variance 2. Although we cannot reproduce the full matrix, the following table lists the first four and the last two samples, together with the sample means and sample medians of each of the samples:

	Sample 1	Sample 2	Sample 3	Sample 4	...	Sample 4999	Sample 5000
1	5.4920	3.6955	4.8933	3.4264	...	2.4151	6.1035
2	5.4436	6.0239	6.7905	4.8145	...	3.2596	3.7615
3	6.3139	4.2353	5.8876	4.9372	...	3.6499	8.5641
4	6.5158	4.7598	3.5983	3.9115	...	2.0195	2.3705
5	3.5321	5.1131	4.9673	4.5547	...	5.0145	6.6010
6	4.2789	6.0736	6.3267	8.3319	...	3.5154	4.8617
7	4.0582	4.4552	8.2660	5.1155	...	5.2439	6.9706
8	5.0427	7.4530	4.7665	4.3044	...	5.6628	4.5673
9	4.6626	9.5493	6.6253	3.6156	...	4.0330	4.0917
10	4.1105	5.4313	4.1620	5.8632	...	6.0880	4.5681
11	6.0461	5.7836	5.0604	3.6048	...	5.8100	5.0355
12	7.4127	5.9725	3.8046	5.4154	...	6.0803	7.7348
13	3.2608	5.2567	5.9747	5.3955	...	6.3623	5.0953
14	4.3100	5.6717	4.2149	7.2308	...	6.3486	7.3613
15	4.4968	5.3302	7.0006	6.7701	...	2.1907	4.7502
16	2.9105	4.3761	6.8677	6.9040	...	4.6863	3.3855
17	5.2193	5.9124	5.2160	5.4927	...	5.1773	4.1588
18	6.5259	0.8479	6.3996	5.4561	...	4.3043	3.6232
19	3.0414	4.8115	6.5505	4.7408	...	7.5806	4.6489
20	3.5006	3.2167	5.0989	5.8786	...	4.9495	5.9608
Mean	4.8087	5.1985	5.6236	5.2882	...	4.7196	5.2107
Median	4.5797	5.2935	5.5518	5.2555	...	4.9820	4.8060

Histograms of the values of the sample means and sample medians are shown in Figure 12.53.

We can treat the sample means $(y_1, y_2, \ldots, y_{5000})$ and the sample medians $(z_1, z_2, \ldots, z_{5000})$ as data and calculate the sample means and sample variances of these two statistics.

For the simulated data that generated the histograms in Figure 12.53, we find, for the sample mean and the sample variance of the sample mean,

$$\bar{y} = \frac{1}{5000} \sum_{k=1}^{5000} y_k = 4.9966 \qquad S_{\bar{y}}^2 = \frac{1}{4999} \sum_{k=1}^{5000} (y_k - \bar{y})^2 = 0.0979$$

Figure 12.53 (a) The histogram of the sample means. (b) The histogram of the sample medians.

and, for the sample mean and the sample variance of the sample median,

$$\bar{z} = \frac{1}{5000} \sum_{k=1}^{5000} z_k = 5.0012 \qquad S_{\bar{z}}^2 = \frac{1}{4999} \sum_{k=1}^{5000} (z_k - \bar{z})^2 = 0.1442$$

As we said earlier, statistics are random variables with their own probability distributions. We know that the distribution of the population in our example is a normal distribution with mean 5 and variance 2. It can be shown that the distribution of the sample mean of normally distributed random variables is also normally distributed. The mean of the distribution of the sample mean is equal to the mean of the population distribution, and the variance of the sample mean is equal to the variance of the population divided by the sample size. In our example with samples of size 20, the mean is therefore 5 and the variance is $2/20 = 0.1$. The simulated data agree with the theoretical predictions.

The histogram for the sample median (Figure 12.53b) suggests that the sampling distribution for the median of a sample from a normal distribution is approximately normal, and this is correct even for relatively small sample sizes. The mean of the distribution of the sample median is the same as the population distribution, namely, 5. However, it appears that the sample variance of the sample mean is smaller than that of the sample median (cf. Figures 12.53a and b). If the population is normally distributed with mean μ and variance σ^2, then the variance of the sample median of a sample of size n, denoted by σ_m^2, has the following relationship with σ^2 for large values of n:

$$\sigma_m^2 \approx \sigma^2 \frac{\pi}{2n}$$

When the population distribution is not normal, the sample mean is still approximately normally distributed for a large enough sample size, provided that the population distribution has a finite mean and a finite variance. This statement follows from the central limit theorem. The distribution of the sample *median* is often more complicated and may not even be approximately normally distributed.

It is important to note that a new simulation would generate different samples, and thus different values, for the various quantities we have computed. Because of the large number of samples, however, the sample means and sample variances of the two statistics would not differ much from simulation run to simulation run.

■ 12.7.2 Estimating Parameters

We take random samples to learn something about the distribution of a variable in a population. For instance, we might be interested in knowing what the average cholesterol level in 50-year-old white males is. To estimate this level, we would take a random sample of 50-year-old white males, measure their cholesterol levels, and then compute the average of the measurements. This is an example of a *point estimate*. More generally, when estimating a parameter of a distribution, we can either give a single number, called a **point estimate**, or give a range, called an **interval estimate**.

In 1994, the National Institute of Standards and Technology (NIST)[3] published *Guidelines for Evaluating and Expressing the Uncertainty of NIST Measurement Results*. These guidelines are valuable for measurements in general. They state, "The result of a measurement is only an approximation or estimate of the value of the specific quantity subject to measurement, that is, the *measurand*, and thus the result is complete only when accompanied by a quantitative statement of uncertainty." The mean of a population is a parameter of particular interest. We will see in what follows how to provide point and interval estimates for this parameter and how to assess uncertainty in its measurement.

Estimates for parameters rely on outcomes of measurements. We expect measurements to be **accurate** and **precise**. *Accuracy* refers to how close measurements are to the true value, and *precision* refers to how close repeated measurements are to each other.

Point Estimates of Means   We assume that the population distribution has a finite mean μ and that the value of this parameter is unknown to us. For reasons that will become clear shortly, we will also assume that the distribution has a finite variance σ^2. We wish to estimate the mean by taking a random sample $(X_1, X_2, \ldots, X_n)$ of size n from the population. The X_k are independent and identically distributed according to the distribution of the population, with

$$E(X_k) = \mu \quad \text{and} \quad \text{var}(X_k) = \sigma^2 \quad \text{for } k = 1, 2, \ldots, n \quad (12.39)$$

To estimate the mean of the population distribution, we will use the sample mean defined in the previous section. The sample mean is the arithmetic average

$$\overline{X}_n = \frac{1}{n} \sum_{k=1}^{n} X_k$$

We stated that statistics are random variables. Hence, we can compute their mean and variance. Using (12.39), we find that the mean of the sample mean is

$$E(\overline{X}_n) = E\left(\frac{1}{n} \sum_{k=1}^{n} X_k\right) = \frac{1}{n} \sum_{k=1}^{n} E(X_k) = \frac{1}{n}(n\mu) = \mu$$

From the independence of the observations, in addition to (12.39), the variance of the sample mean is

$$\text{var}(\overline{X}_n) = \text{var}\left(\frac{1}{n} \sum_{k=1}^{n} X_k\right) = \frac{1}{n^2} \sum_{k=1}^{n} \text{var}(X_k) = \frac{1}{n^2}(n\sigma^2) = \frac{\sigma^2}{n}$$

We see from these equations that the expected value of the sample mean is equal to the population mean. The spread of the distribution of $\overline{X}_n$ is described by the variance of $\overline{X}_n$. Since that variance becomes smaller as the sample size increases ($\sigma^2/n \to 0$ as $n \to \infty$), we conclude that the sample mean of large samples shows less variation about its mean than does the sample mean of small samples. This conclusion implies that the larger the sample size, the more accurately the mean of the population can be estimated. In fact, invoking the weak law of large numbers from the previous section, we find that

$$\overline{X}_n \to \mu \quad \text{in probability as } n \to \infty$$

---

(3) NIST, founded in 1901, is a nonregulatory federal agency within the U.S. Commerce Department's Technology Administration. NIST's mission is "to promote U.S. innovation and industrial competitiveness by advancing measurement science, standards, and technology in ways that enhance economic security and improve quality of life."

This relationship justifies the use of $\overline{X}_n$ as an estimate for the mean of the distribution. Since $E(\overline{X}_n) = \mu$, we say that $\overline{X}_n$ is an **unbiased estimator** for μ.

We illustrate the behavior of the sample mean as a function of sample size in the following simulation, where we draw random samples from a standard normal distribution (i.e., $\mu = 0$ and $\sigma^2 = 1$): Figure 12.54(a) shows a histogram for the sample means of 1000 random samples, each of size 10; Figure 12.54(b) shows a histogram for the sample means of 1000 random samples, each of size 50. The simulations confirm that as the sample size increases, the variance of the sample mean decreases, resulting in a narrower histogram. If the sample is drawn from a normal distribution with mean 0 and variance 1, then the sample mean is normally distributed with mean 0 and variance $1/n$, where n is the sample size. When the distribution of the population is not normal, but has a finite mean and variance, we can invoke the central limit theorem to conclude that the sample mean is approximately normal. We can view each histogram as an approximation to the theoretical distribution of the sample mean.

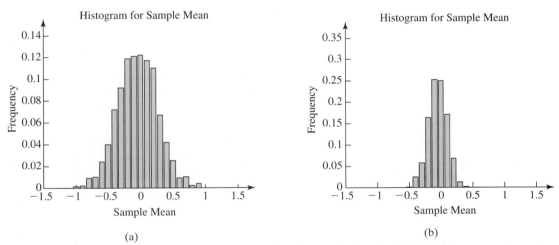

Figure 12.54 (a) The histogram of sample means when the sample size is 10. (b) The histogram of sample means when the sample size is 50.

EXAMPLE 4

Assume that we draw a random sample from a population that consists of individuals whose lifetimes are exponentially distributed with a mean of five years.

(a) Denote by X the random variable that is exponentially distributed with a mean of five years. What is the probability density function of the random variable that describes the lifetime of an individual in this population? What are the mean and the variance of this random variable?

(b) If you took a large number of random samples, each of size 50, from this population and graphed the histogram for the sample mean of these random samples, what do you expect the histogram to look like? Compute the mean and the variance of the sample mean.

Solution

(a) The probability density of an exponential distribution with mean 5 has parameter $\lambda = 0.2$ and is given by

$$f(x) = 0.2 \exp(-0.2x)$$

The mean of an exponentially distributed random variable is $1/\lambda$ and the variance is $1/\lambda^2$. Hence, $E(X) = 5$ and $\mathrm{var}(X) = 25$. The graph of the probability density is shown in Figure 12.55(a).

(b) Since the mean of the population distribution are finite, the sample mean of a large sample is approximately normally distributed with mean equal to the population mean and variance equal to the population variance divided by the sample size. The sample size is 50, so the mean of the sample mean is 5 and the variance of the sample mean is $25/50 = 0.5$. We simulated 10,000 samples, each of size 50, and displayed the sample means in the histogram in Figure 12.55(b). Note that the histogram resembles a normal distribution and that it is centered around 5, the mean of the sample mean. ■

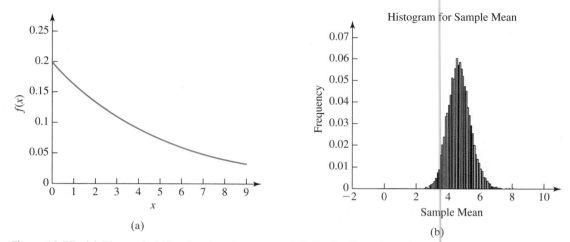

(a)

(b)

Figure 12.55 (a) The probability density of an exponentially distributed random variable with mean 5 years. (b) The histogram of sample means when the sample size is 50.

A Remark on Using the Sample Mean to Estimate the Mean You might wonder why we used the sample mean, and not some other quantity, to estimate the mean of the population distribution. Statisticians have established criteria to assess the quality of an estimator. We mentioned one such criterion already: It is desirable to use unbiased estimators. The sample mean is unbiased, since its expected value is equal to the parameter it is supposed to estimate, the population mean. A biased estimator introduces a systematic error. Later we will see an example in which we will define an estimator that will turn out to be biased, but then we will find a way to remove the bias.

Another criterion is as follows: Choose an estimator with as small a variance as possible. We illustrate the application of this criterion in a population distribution that is normal with mean μ and variance σ^2. Suppose we wish to estimate the mean of the distribution. We already know that the sample mean is an unbiased estimator of the mean. In Subsection 12.7.1, we mentioned that the sample median is also an unbiased estimator of the mean (although not necessarily if the population distribution is not normal). If we know that the population distribution is normal, why don't we choose the sample median instead of the sample mean? After all, the sample median seems easier to compute. We saw in Subsection 12.7.1 that the variance of the sample mean is smaller than the variance of the sample median. If the sample size is denoted by n, then the former is equal to σ^2/n, and the latter is equal to $\pi\sigma^2/(2n) \approx 1.508\sigma^2/n$, if n is sufficiently large. A smaller variance increases the precision of our estimate, so an estimator with a smaller variance is preferable.

Point Estimates of Proportions Estimating proportions is a special case of estimating means. Examples of estimating proportions are, for instance, estimating the proportion of white-flowering pea plants in a $Cc \times Cc$ crossing or estimating the proportion of patients in a clinical study who had a recurrence of a disease after a fixed length of time. If we take a random sample of size n from a population of which a proportion p has a certain characteristic, then the number of observations in the sample with this characteristic is binomially distributed with parameters n and p. If we denote this quantity by B_n, then we have

$$P(B_n = k) = \binom{n}{k} p^k (1-p)^{n-k} \quad \text{for } k = 0, 1, 2, \ldots, n$$

To derive an estimator for the parameter p, we consider each single observation in the sample as a success or a failure, depending on whether the observation has the characteristic under investigation. We set

$$X_k = \begin{cases} 1 & \text{if the } k\text{th observation is a success} \\ 0 & \text{if the } k\text{th observation is a failure} \end{cases}$$

Then $B_n = \sum_{k=1}^{n} X_k$ is the total number of successes in the sample and p is the probability of success. The sample mean $\overline{X}_n = \frac{1}{n} \sum_{k=1}^{n} X_k$ is then the fraction of successes in the sample. We find that

$$E(\overline{X}_n) = \frac{1}{n} \sum_{k=1}^{n} E(X_k) = \frac{1}{n}(np) = p$$

and

$$\text{var}(\overline{X}_n) = \text{var}\left(\frac{1}{n} \sum_{k=1}^{n} X_k\right) = \frac{1}{n^2} n(\text{var}(X_1))$$

$$= \frac{np(1-p)}{n^2} = \frac{p(1-p)}{n}$$

The sample mean $\overline{X}_n$ will serve as an estimator for the probability of success p. It is customary to use $\hat{p}$ instead of $\overline{X}_n$; that is, we denote the estimate for p by $\hat{p}$ (read "p hat"). If we observe k successes in a sample of size n, then

$$\hat{p} = \frac{k}{n}$$

with

$$E(\hat{p}) = p \qquad \text{and} \qquad \text{var}(\hat{p}) = \frac{p(1-p)}{n}$$

The next example considers Mendel's experiment of crossing pea plants.

EXAMPLE 5

To estimate the probability of white-flowering pea plants in a $Cc \times Cc$ crossing, Mendel randomly crossed red-flowering pea plants of genotype Cc. He obtained 705 plants with red flowers and 224 plants with white flowers. Estimate the probability of a white-flowering pea plant in a $Cc \times Cc$ crossing.

Solution

The sample mean is

$$\overline{X}_n = \frac{224}{224 + 705} \approx 0.24$$

The estimate for the probability of a white-flowering pea plant in a $Cc \times Cc$ crossing is therefore $\hat{p} = 0.24$. We know from the laws of inheritance that the expected value of $\overline{X}_n$ is $p = 0.25$. ■

The next example illustrates the use of estimating proportions in a clinical trial.

EXAMPLE 6

The Women's Health Initiative Steering Committee (2004) conducted a clinical study on the effects of estrogen-alone therapy in postmenopausal women who had had a hysterectomy. The study was halted in July 2002 because the risks exceeded the benefits. Of 5310 patients who had been randomly assigned to receive the hormone therapy, 158 suffered strokes during the 81.6 months of follow-up, whereas 118 of 5429 patients in the placebo group suffered strokes during the 81.9 months of follow-up. Estimate the number of strokes per 10,000 person-years in each group. Which group had the higher incidence of stroke?

Solution We first need to convert the number of patients in each group into person-years. The hormone therapy group has $(81.6)(5310)/12 = 36,108$ person-years; the placebo group has $(81.9)(5429)/12 = 37,053$ person-years. The incidence in the hormone therapy group is therefore $158/36,108 \approx 0.0044$, or 44 per 10,000 person-years. The incidence in the placebo group is $118/37,053 \approx 0.0032$, or 32 per 10,000 person-years. The group that received hormone therapy had a higher risk of stroke. ■

Point Estimates of Variances Recall the sample variance

$$S_n^2 = \frac{1}{n-1} \sum_{k=1}^{n} (X_k - \overline{X}_n)^2$$

We will show that dividing by $n-1$ (instead of by n) yields an unbiased estimator for the variance. To compute the mean of the sample variance, we need the following identity: For any $c \in \mathbf{R}$,

$$\sum_{k=1}^{n} (X_k - c)^2 = \sum_{k=1}^{n} (X_k - \overline{X}_n)^2 + n(\overline{X}_n - c)^2 \qquad (12.40)$$

To see why (12.40), we expand its left-hand side after adding $0 = \overline{X}_n - \overline{X}_n$ inside the parentheses:

$$\sum_{k=1}^{n} (X_k - c)^2 = \sum_{k=1}^{n} (X_k - \overline{X}_n + \overline{X}_n - c)^2$$

$$= \sum_{k=1}^{n} [(X_k - \overline{X}_n)^2 + 2(X_k - \overline{X}_n)(\overline{X}_n - c) + (\overline{X}_n - c)^2]$$

$$= \sum_{k=1}^{n} (X_k - \overline{X}_n)^2 + 2(\overline{X}_n - c) \sum_{k=1}^{n} (X_k - \overline{X}_n) + n(\overline{X}_n - c)^2$$

Since $\sum_{k=1}^{n} (X_k - \overline{X}_n) = 0$, the middle term is equal to 0, and (12.40) follows.

If we set $c = \mu$ in (12.40) and rearrange the equation, we obtain

$$\sum_{k=1}^{n} (X_k - \overline{X}_n)^2 = \sum_{k=1}^{n} (X_k - \mu)^2 - n(\overline{X}_n - \mu)^2$$

Taking expectations on both sides and using, on the right-hand side, the fact that the expectation of a sum is the sum of the expectations, we find that

$$E\left(\sum_{k=1}^{n} (X_k - \overline{X}_n)^2\right) = \sum_{k=1}^{n} E(X_k - \mu)^2 - n E(\overline{X}_n - \mu)^2$$

Now, $\sum_{k=1}^{n} (X_k - \overline{X}_n)^2 = (n-1)S_n^2$, $E(X_k - \mu)^2 = \sigma^2$, and $E(\overline{X}_n - \mu)^2 = \text{var}(\overline{X}_n) = \frac{1}{n}\sigma^2$. Hence,

$$(n-1)E(S_n^2) = n\sigma^2 - \sigma^2 = (n-1)\sigma^2$$

and, therefore,

$$E(S_n^2) = \sigma^2$$

which shows that S_n^2 is an unbiased estimate of the variance of the population. This is the reason that we divided by $n-1$ instead of n when we computed the sample variance. We will not compute the variance of the sample variance; it is given by a complicated formula. One can show, however, that the variance of the sample variance goes to 0 as the sample size becomes infinite. Hence, invoking the weak law of large numbers, we find that

$$S_n^2 \to \sigma^2 \quad \text{in probability as } n \to \infty$$

The next example illustrates how we would estimate the mean and the variance of a characteristic of a population.

EXAMPLE 7

Suppose that a computer generates the following sample of independent observations from a population:

$$0.0201, 0.8918, 0.9619, 0.1713, 0.0357,$$
$$0.6325, 0.4276, 0.2517, 0.2330, 0.6754$$

Estimate the mean and the variance of these observations.

Solution

To estimate the mean, we compute the sample mean $\overline{X}_n$. We sum the 10 numbers in the sample and divide the result by 10, which yields

$$\overline{X}_n = 0.4301$$

Thus, our estimate of the mean is 0.4301.

To estimate the variance, we compute the sample variance S_n^2. We square the difference between each sample point and the sample mean and add the results. We then divide the resulting number by 9 to obtain

$$S_n^2 = 0.1176$$

Thus, our estimate of the variance is 0.1176. ■

Confidence Intervals   Earlier, we learned that the sample mean varies from sample to sample. The variation of the distribution of the sample mean is described by the **standard error** (S.E.), also denoted by $S_{\overline{x}}$ to indicate that it is the standard deviation of the sample mean.

The definition of the standard error is motivated by the following considerations (again, we assume that the mean and the variance of the population distribution are finite): To estimate the population mean μ, we use the sample mean $\overline{X}_n$. The variance of $\overline{X}_n$ gives us an idea as to how much the distribution of $\overline{X}_n$ varies. Now,

$$\text{var}(\overline{X}_n) = \text{var}\left(\frac{1}{n}\sum_{k=1}^{n} X_k\right) = \frac{1}{n^2}\text{var}\left(\sum_{k=1}^{n} X_k\right)$$

Since the X_k are independent, the variance of the sum is the sum of the variances; moreover, all X_k have the same distribution. Hence,

$$\text{var}(\overline{X}_n) = \frac{1}{n^2}n\sigma^2 = \frac{\sigma^2}{n} \tag{12.41}$$

The variance of $\overline{X}_n$ thus depends on another population parameter: the variance σ^2. In problems where we wish to estimate the mean, we typically don't know the variance either. If the sample size is large, however, the sample variance will be close to the population variance. We can therefore approximate the variance of $\overline{X}_n$ by replacing σ^2 in (12.41) by S_n^2, giving S_n^2/n. The standard error is then the square root of that expression:

$$\text{S.E.} = S_{\overline{x}} = \frac{S_n}{\sqrt{n}}$$

The next example illustrates how to determine a sample mean and its standard error.

EXAMPLE 8

In a sample of six leaves from a morning glory plant that is infested with aphids, the following numbers of aphids per leaf are found: 12, 27, 17, 35, 14, and 18. Find the sample mean, the sample variance, and the standard error.

Solution

The sample size is $n = 6$. We use the formulas

$$\overline{X}_n = \frac{1}{n}\sum_{k=1}^{n} X_k \quad \text{and} \quad S_n^2 = \frac{1}{n-1}\left[\sum_{k=1}^{n} X_k^2 - \frac{1}{n}\left(\sum_{k=1}^{n} X_k\right)^2\right]$$

with $n = 6$ to construct the following table:

$\sum_{k=1}^{n} X_k$	$\sum_{k=1}^{n} X_k^2$	$\overline{X}_n$	S_n^2	$\frac{S_n}{\sqrt{n}}$
123	2907	20.5	77.1	3.58

We thus find that S.E. = 3.58.   ■

To find the standard error in the case of k successes in a sample of size n, note that

$$S_n^2 = \frac{1}{n-1}\left[\sum_{j=1}^{n} X_j^2 - \frac{1}{n}\left(\sum_{j=1}^{n} X_j\right)^2\right]$$

$$= \frac{1}{n-1}\left(k - \frac{k^2}{n}\right)$$

$$= \frac{n}{n-1}\frac{k}{n}\left(1 - \frac{k}{n}\right) = \frac{n\hat{p}(1-\hat{p})}{n-1}$$

The standard error of the sample mean p is therefore

$$S_{\hat{p}} = \frac{S_n}{\sqrt{n}} = \sqrt{\frac{\hat{p}(1-\hat{p})}{n-1}} \qquad (12.42)$$

In the literature, you will often find the standard error for proportions stated as

$$S_{\hat{p}} = \sqrt{\frac{\hat{p}(1-\hat{p})}{n}} \qquad (12.43)$$

For n large, the two numbers are very close, so that it does not matter which one you use.

EXAMPLE 5

(continued) In Example 5, we estimated the probability of white-flowering pea plants in Mendel's $Cc \times Cc$ crossing that produced 705 plants with red flowers and 224 plants with white flowers. We obtained $\hat{p} = 0.24$ as an estimate for the probability of a white-flowering pea plant in this crossing. Find the standard error of the sample mean p.

Solution

The standard error of the sample mean is

$$\text{S.E.} = \sqrt{\frac{\hat{p}(1-\hat{p})}{n}} = \sqrt{\frac{(0.24)(0.76)}{705 + 224}} \approx 0.014$$

We can thus report the result as 0.24 ± 0.014.   ■

In the scientific literature, variability is often reported as either "Mean ± S.D." or "Mean ± S.E." Since S.E. = S.D./$\sqrt{n}$, the two ways are not the same. It must be made clear which is used.

EXAMPLE 9

During 2000–2003, the temperature in Medicine Lake, Minnesota, was measured repeatedly at a 3-m depth. The following table lists the mean temperature (in degree Celsius) and standard error for every other month from June through October, together with number of sample points (n) for each month:

Month	Mean ± S.E.	n
June	21.5 ± 1.52	6
August	25.1 ± 1.45	9
October	13.5 ± 1.42	7

Convert "Mean ± S.E." to "Mean ± S.D."

Solution   Since S.E. = S.D./$\sqrt{n}$, we multiply the S.E. given in the table by $\sqrt{n}$. For instance, to obtain the S.D. for June, we calculate S.D. = $1.52\sqrt{6} = 3.72$. We get the following results:

Month	Mean ± S.D.	n
June	21.5 ± 3.72	6
August	25.1 ± 4.35	9
October	13.5 ± 3.76	7

■

Interpreting Mean ± S.E.   What does "Mean ± S.E." mean? When we write "Mean ± S.E.," we specify an interval, namely, [Mean − S.E., Mean + S.E.]. Since we use "Mean" (= $\overline{X}_n$) as an estimate for the population mean μ, we would like this interval to contain μ. Surely, since $\overline{X}_n$ is a random variable, if we took repeated samples and computed such intervals for each sample, not *all* the intervals would contain μ. But maybe we can at least find out what fraction of these intervals are likely to contain the population mean μ. In other words, before taking the sample, we might wish to know what the probability is that the interval [Mean − S.E., Mean + S.E.], or, more generally, [Mean − aS.E., Mean + aS.E.], where a is a positive constant, will contain the actual value of μ. To be concrete, we will try to determine a so that this probability is equal to 0.95.

If the sample size n is large, it follows from the central limit theorem that

$$\frac{\overline{X}_n - \mu}{\sigma/\sqrt{n}}$$

is approximately standard normally distributed. If Z is standard normally distributed, then

$$P(-1.96 \le Z \le 1.96) = 0.95$$

Hence, the event

$$-1.96 \le \frac{\overline{X}_n - \mu}{\sigma/\sqrt{n}} \le 1.96 \tag{12.44}$$

has a probability of approximately 0.95 for n sufficiently large.

Rearranging terms in (12.44), we find that

$$-1.96\frac{\sigma}{\sqrt{n}} \le \overline{X}_n - \mu \le 1.96\frac{\sigma}{\sqrt{n}}$$

or

$$\overline{X}_n - 1.96\frac{\sigma}{\sqrt{n}} \le \mu \le \overline{X}_n + 1.96\frac{\sigma}{\sqrt{n}}$$

We can thus write

$$P\left(\overline{X}_n - 1.96\frac{\sigma}{\sqrt{n}} \le \mu \le \overline{X}_n + 1.96\frac{\sigma}{\sqrt{n}}\right) \approx 0.95 \quad \text{for large } n \tag{12.45}$$

This equation tells us that if we repeatedly draw random samples of size n from a population with mean μ and standard deviation σ, then, in about 95% of the samples, the interval $\left[\overline{X}_n - 1.96\frac{\sigma}{\sqrt{n}}, \overline{X}_n + 1.96\frac{\sigma}{\sqrt{n}}\right]$ will contain the true mean μ. Such an interval is referred to as a **95% confidence interval**.

Notice that this interval contains the parameter σ. If we do not know μ, we probably do not know σ either. We might then wish to replace σ by the square root of the sample variance, denoted by S_n. Fortunately, when n is large, S_n will be very close to σ and (12.45) holds approximately when σ is replaced by S_n.

We thus find that, for large n,

$$P\left(\overline{X}_n - 1.96\frac{S_n}{\sqrt{n}} \le \mu \le \overline{X}_n + 1.96\frac{S_n}{\sqrt{n}}\right) \approx 0.95$$

or, with the event rewritten in interval notation,

$$P\left(\mu \in \left[\overline{X}_n - 1.96\frac{S_n}{\sqrt{n}}, \overline{X}_n + 1.96\frac{S_n}{\sqrt{n}}\right]\right) \approx 0.95$$

The interval in this expression is of the form

$$[\text{Mean} - (1.96)\text{S.E.}, \text{Mean} + (1.96)\text{S.E.}] \tag{12.46}$$

We have succeeded in determining what fraction of intervals of the form (12.46) contain the mean. That is, if n is large and we take repeated samples each of size n, then approximately 95% of the intervals [Mean − (1.96)S.E., Mean + (1.96)S.E.] would contain the true mean. [Or, equivalently, before we take the sample, the probability that the interval in (12.46) will contain the actual value of μ is 0.95.] If we wanted 99% of such intervals to contain the true mean, we would need to replace the factor 1.96 by 2.58, since if Z is standard normally distributed, then $P(-2.58 \le Z \le 2.58) = 0.99$ (95% and 99% are the most frequently used percentages for confidence intervals). Because $P(-1 \le Z \le 1) = 0.68$, we conclude that if the sample size is large, approximately 68% of intervals of the form [Mean − S.E., Mean + S.E.] contain the population mean μ.

EXAMPLE 10

In the study by Dyck et al. (1999) mentioned at the beginning of Section 12.7, the weight (in kg) of the group of 149 Type 2 diabetes patients was reported as Mean ± S.D. = 84.4 ± 16.9. Find a 99% confidence interval for the mean.

Solution

If Z is a standard normally distributed random variable, we need to find z such that

$$P(-z \le Z \le z) = 0.99$$

Now,

$$P(-z \le Z \le z) = 2\Phi(z) - 1 = 0.99$$

where $\Phi(z)$ denotes the distribution function of a standard normal distribution. Hence,

$$\Phi(z) = \frac{1.99}{2} = 0.995$$

and it follows that $z = 2.58$. With S.D. = 16.9 and $n = 149$, we find that

$$\text{S.E.} = \frac{16.9}{\sqrt{149}} \approx 1.38$$

Therefore, the 99% confidence interval is of the form

$$[84.8 - (2.58)(1.38), 84.8 + (2.58)(1.38)] = [81.2, 88.4] \qquad \blacksquare$$

■ 12.7.3 Linear Regression

In textbooks or in scientific literature, you will frequently see plots that fit a straight line to data (as shown in Figure 12.56). The quantities on the horizontal and the vertical axes are linearly related, and a linear model is used to describe the relationship. We denote the quantity on the horizontal axis by x and the quantity on the vertical axis by Y. We think of x as a particular treatment that is under the control of the experimenter and of Y as the response to that treatment. In measurements of Y, errors are typically present, so that the data points will not lie exactly on the straight line (even if the linear model is correct) but will be scattered around it; that is, Y is not completely determined by x. The degree of scatter is an indication of how much random variation there is. In what follows, we will see how to separate the random variation from the actual relationship between the two quantities in the case when they are linearly related. We will discuss one particular model.

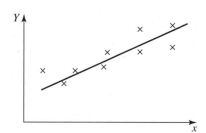

Figure 12.56 A straight line fitted to data.

We assume that x is under the control of the experimenter to the extent that it can be measured without error. The response Y, however, shows random variation. We assume the linear model

$$Y = a + bx + \epsilon$$

where ϵ is a normal random variable representing the **error**, which has mean 0 and standard deviation σ. The standard deviation of the error does not depend on x and is thus the same for all values of x.

Our goal is to estimate a and b from data that consist of the points (x_i, y_i), $i = 1, 2, \ldots, n$. The approach will be to choose a and b such that the sum of the squared deviations

$$h(a, b) = \sum_{k=1}^{n} [y_k - (a + bx_k)]^2$$

Figure 12.57 The line $Y = a + bx$ is chosen so that the sum of the squared residuals is minimized.

is minimized. The deviations $y_k - (a + bx_k)$ are called **residuals**. The procedure of finding a and b is called the **method of least squares** and is illustrated in Figure 12.57. The resulting straight line is called the **least square line** (or **linear regression line**).

EXAMPLE 11

Given the three points $(0, 2)$, $(1, 0)$, and $(2, 1)$, use the method of least squares to find the least square line.

Solution

We wish to find a straight line of the form $y = a + bx$. For given values of a and b, the residuals are

$$2 - (a + 0b) \qquad 0 - (a + b) \qquad 1 - (a + 2b)$$

and the sum of their squares is

$$\begin{aligned}
(2 - a)^2 &+ (a + b)^2 + (1 - a - 2b)^2 \\
&= (4 - 4a + a^2) + (a^2 + 2ab + b^2) + (1 + a^2 + 4b^2 - 2a - 4b + 4ab) \\
&= 5 - 6a + 3a^2 + 6ab + 5b^2 - 4b \\
&= (2b^2 + 2b) + (3 + 3b^2 + 3a^2 + 6ab - 6a - 6b) + 2 \\
&= 2(b^2 + b) + 3(1 + b^2 + a^2 + 2ab - 2a - 2b) + 2
\end{aligned}$$

Grouping the terms in this way allows us to complete the squares; we find that the sum of the squares is then equal to

$$2\left(b + \frac{1}{2}\right)^2 + 3(1 - a - b)^2 + \frac{3}{2} \tag{12.47}$$

(If this looks like magic, don't worry; we will derive a general formula for a and b shortly.) Since (12.47) consists of two squares (plus a constant term), the expression is minimized when the two squares are both equal to 0. Accordingly, we solve

$$b + \frac{1}{2} = 0$$

$$1 - a - b = 0$$

which yields $b = -1/2$ and $a = 3/2$. Therefore, the least square line is of the form

$$y = \frac{3}{2} - \frac{1}{2}x$$

This line, together with the given three points, is shown in Figure 12.58. ■

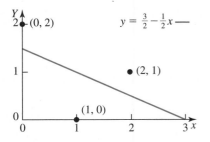

Figure 12.58 The least square line together with the three data points of Example 11.

We will now derive the general formula for finding a and b. The basic steps will be similar to those in Example 11. We will first rewrite the residuals. We set

$$\bar{x} = \frac{1}{n}\sum_{k=1}^{n} x_k \qquad \text{and} \qquad \bar{y} = \frac{1}{n}\sum_{k=1}^{n} y_k$$

and

$$y_k - (a + bx_k) = (y_k - \overline{y}) + (\overline{y} - a - b\overline{x}) - b(x_k - \overline{x})$$

In what follows, we will simply write $\sum$ instead of $\sum_{k=1}^{n}$. If we square the expression and sum over k, we find that

$$\begin{aligned}
\sum [y_k - (a + bx_k)]^2 = {} & \sum (y_k - \overline{y})^2 + n(\overline{y} - a - b\overline{x})^2 \\
& + b^2 \sum (x_k - \overline{x})^2 - 2b \sum (x_k - \overline{x})(y_k - \overline{y}) \\
& + 2(\overline{y} - a - b\overline{x}) \sum (y_k - \overline{y}) \\
& - 2b(\overline{y} - a - b\overline{x}) \sum (x_k - \overline{x})
\end{aligned} \qquad (12.48)$$

The last two terms are equal to 0.

Next, we introduce notation to simplify our derivation. Let

$$SS_{xx} = \sum (x_k - \overline{x})^2 = \sum x_k^2 - \frac{(\sum x_k)^2}{n}$$

$$SS_{yy} = \sum (y_k - \overline{y})^2 = \sum y_k^2 - \frac{(\sum y_k)^2}{n}$$

$$SS_{xy} = \sum (x_k - \overline{x})(y_k - \overline{y}) = \sum x_k y_k - \frac{(\sum x_k)(\sum y_k)}{n}$$

Using this notation, we can write the right-hand side of (12.48) as

$$SS_{yy} + n(\overline{y} - a - b\overline{x})^2 + b^2 SS_{xx} - 2b SS_{xy}$$

The last two terms suggest that we should complete the square:

$$SS_{yy} + n(\overline{y} - a - b\overline{x})^2 + SS_{xx}\left(b^2 - 2b\frac{SS_{xy}}{SS_{xx}} + \left(\frac{SS_{xy}}{SS_{xx}}\right)^2\right) - \frac{(SS_{xy})^2}{SS_{xx}}$$

$$= n(\overline{y} - a - b\overline{x})^2 + SS_{xx}\left(b - \frac{SS_{xy}}{SS_{xx}}\right)^2 + SS_{yy} - \frac{(SS_{xy})^2}{SS_{xx}}$$

As in Example 11, we succeeded in writing the sum of the squared deviations as a sum of two squares plus an additional term. We can minimize this expression by setting each squared expression equal to 0:

$$\overline{y} - a - b\overline{x} = 0$$

$$b - \frac{SS_{xy}}{SS_{xx}} = 0$$

Solving for a and b yields

$$b = \frac{SS_{xy}}{SS_{xx}}$$

$$a = \overline{y} - b\overline{x}$$

The right-hand sides serve as estimates of a and b, respectively denoted by $\hat{a}$ and $\hat{b}$. Summarizing, we have the following result:

The least square line (or linear regression line) is given by

$$y = \hat{a} + \hat{b}x$$

with

$$\hat{b} = \frac{\sum_{k=1}^{n}(x_k - \overline{x})(y_k - \overline{y})}{\sum_{k=1}^{n}(x_k - \overline{x})^2} \qquad (12.49)$$

$$\hat{a} = \overline{y} - \hat{b}\overline{x} \qquad (12.50)$$

We illustrate finding $\hat{a}$ and $\hat{b}$ in the next example.

EXAMPLE 12   Fit a linear regression line through the points

$$(1, 1.62), (2, 3.31), (3, 4.57), (4, 5.42), (5, 6.71)$$

Solution   To facilitate the computation, we construct the following table:

x_k	y_k	$x_k - \overline{x}$	$y_k - \overline{y}$	$(x_k - \overline{x})(y_k - \overline{y})$
1	1.62	−2	−2.706	5.412
2	3.31	−1	−1.016	1.016
3	4.57	0	0.244	0
4	5.42	1	1.094	1.094
5	6.71	2	2.384	4.768

Now,

$$\hat{b} = \frac{12.29}{10} = 1.229$$

$$\hat{a} = 4.326 - (1.229)(3) = 0.639$$

Hence, the linear regression line is given by

$$y = 1.23x + 0.64$$

This line and the given data points are shown in Figure 12.59.   ■

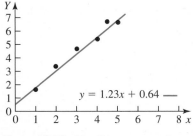

Figure 12.59 The linear regression line and the data points of Example 12.

$y = 1.23x + 0.64$

Now that we know how to fit a straight line to a set of points, we might want to know how good the fit is. To this end, we will define a quantity known as the **coefficient of determination**. We motivate its definition as follows: We start with a set of observations $(x_k, y_k), k = 1, 2, \ldots, n$, and assume the linear model $Y = a + bx + \epsilon$. We set

$$\hat{y}_k = \hat{a} + \hat{b}x_k \qquad \text{and} \qquad \overline{y} = \frac{1}{n}\sum_{k=1}^{n} y_k$$

We think of $\hat{y}_k$ as the expected response under the linear model if $x = x_k$. Now, $y_k - \overline{y}$ is the deviation of the observation from the sample mean, $y_k - \hat{y}_k$ is the deviation of the observation from the expected response under the linear model, and $\hat{y}_k - \overline{y}$ is the deviation of the expected response under the linear model from the sample mean. The deviation $\hat{y}_k - \overline{y}$ can be thought of as being explained by the model, and the deviation $y_k - \hat{y}_k$ can be thought of as the unexplained part due to random variation (the stochastic error). We can write

$$y_k - \overline{y} = (\hat{y}_k - \overline{y}) + (y_k - \hat{y}_k) \tag{12.51}$$

If we look at $\sum(y_k - \overline{y})^2$, which is the total sum of the squared deviations, and use (12.51), we find

$$\sum(y_k - \overline{y})^2 = \sum[(\hat{y}_k - \overline{y}) + (y_k - \hat{y}_k)]^2$$

$$= \sum(\hat{y}_k - \overline{y})^2 + 2\sum(\hat{y}_k - \overline{y})(y_k - \hat{y}) \tag{12.52}$$

$$+ \sum(y_k - \hat{y}_k)^2$$

We want to show that $\sum(\hat{y}_k - \overline{y})(y_k - \hat{y}) = 0$. To do so, we observe that

$$\hat{y}_k - \overline{y} = (\hat{a} + \hat{b}x_k) - (\hat{a} + \hat{b}\overline{x}) = \hat{b}(x_k - \overline{x})$$

$$y_k - \hat{y}_k = (y_k - \overline{y}) - (\hat{y}_k - \overline{y}) = (y_k - \overline{y}) - \hat{b}(x_k - \overline{x})$$

Therefore,

$$\sum(\hat{y}_k - \overline{y})(y_k - \hat{y}_k) = \sum \hat{b}(x_k - \overline{x})[(y_k - \overline{y}) - \hat{b}(x_k - \overline{x})]$$

$$= \hat{b}\sum(x_k - \overline{x})(y_k - \overline{y}) - \hat{b}^2 \sum(x_k - \overline{x})^2$$

Using (12.49) to substitute for one of the $\hat{b}$'s in the term $\hat{b}^2$, we obtain

$$\sum(\hat{y}_k - \overline{y})(y_k - \hat{y}_k) = \hat{b}\sum(x_k - \overline{x})(y_k - \overline{y})$$

$$- \hat{b}\frac{\sum(x_k - \overline{x})(y_k - \overline{y})}{\sum(x_k - \overline{x})^2} \sum(x_k - \overline{x})^2 \qquad (12.53)$$

$$= 0$$

This equation allows us to partition the total sum of squares into the explained and the unexplained sums of squares. Accordingly, continuing with (12.52) and using (12.53), we find that

$$\underbrace{\sum(y_k - \overline{y})^2}_{\text{total}} = \underbrace{\sum(\hat{y}_k - \overline{y})^2}_{\text{explained}} + \underbrace{\sum(y_k - \hat{y}_k)^2}_{\text{unexplained}}$$

The ratio

$$\frac{\text{explained}}{\text{total}} = \frac{\sum(\hat{y}_k - \overline{y})^2}{\sum(y_k - \overline{y})^2}$$

is therefore the proportion of variation that is explained by the model. It is denoted by r^2 and is called the *coefficient of determination*. With $\hat{y}_k - \overline{y} = \hat{b}(x_k - \overline{x})$ and $\hat{b}$ given in (12.49), the coefficient of determination can be written as

$$r^2 = (\hat{b})^2\frac{\sum(x_k - \overline{x})^2}{\sum(y_k - \overline{y})^2} = \left(\frac{\sum(x_k - \overline{x})(y_k - \overline{y})}{\sum(x_k - \overline{x})^2}\right)^2 \frac{\sum(x_k - \overline{x})^2}{\sum(y_k - \overline{y})^2}$$

$$= \frac{\left[\sum(x_k - \overline{x})(y_k - \overline{y})\right]^2}{\sum(x_k - \overline{x})^2 \sum(y_k - \overline{y})^2}$$

We summarize this result as follows:

The coefficient of determination is given by

$$r^2 = \frac{\left[\sum(x_k - \overline{x})(y_k - \overline{y})\right]^2}{\sum(x_k - \overline{x})^2 \sum(y_k - \overline{y})^2}$$

and represents the proportion of variation that is explained by the model.

Returning to Example 12, we find that

$$r^2 = \frac{(12.29)^2}{(10)(15.29)} = 0.988$$

That is, 98.8% of the variation is explained by the model.

Since r^2 is the ratio of explained to total variation, it follows that $r^2 \leq 1$. Furthermore, since r^2 is the square of an expression, it is always nonnegative. That is, we have

$$0 \leq r^2 \leq 1$$

The closer r^2 is to 1, the more closely the data points follow the straight line resulting from the linear model. In the extreme case, when $r^2 = 1$, all points lie on the line; there is no random variation.

Section 12.7 Problems

■ 12.7.1

1. The following data represent the number of aphids per plant found in a sample of 10 plants:

$$17, 13, 21, 47, 3, 6, 12, 25, 0, 18$$

Find the median, the sample mean, and the sample variance.

2. The following data represent the number of seeds per flower head in a sample of nine flowering plants:

$$27, 39, 42, 18, 21, 33, 45, 37, 21$$

Find the median, the sample mean, and the sample variance.

3. The following data represent the age of patients in a clinical trial:

$$28, 45, 34, 36, 30, 42, 35, 45, 38, 27$$

Find the median, the sample mean, and the sample variance.

4. The following data represent blood cholesterol levels, in mg/dL, of patients in a clinical trial:

$$174, 138, 212, 203, 194, 245, 146, 149, 164, 209, 158$$

Find the median, the sample mean, and the sample variance.

5. The following data represent the frequency distribution of seed numbers per flower head in a flowering plant:

Seed Number	Frequency
9	37
10	48
11	53
12	49
13	61
14	42
15	31

Calculate the sample mean and the sample variance.

6. The following data represent the frequency distribution of the numbers of days that it took a certain ointment to clear up a skin rash:

Number of Days	Frequency
1	2
2	7
3	9
4	27
5	11
6	5

Calculate the sample mean and the sample variance.

7. The following data represent the relative frequency distribution of clutch size in a sample of 300 laboratory guinea pigs:

Clutch Size	Relative Frequency
2	0.05
3	0.09
4	0.12
5	0.19
6	0.23
7	0.12
8	0.13
9	0.07

Calculate the sample mean and the sample variance.

8. The following data represent the relative frequency distribution of clutch size in a sample of 42 mallards:

Clutch Size	Relative Frequency
6	0.10
7	0.24
8	0.29
9	0.21
10	0.16

Calculate the sample mean and the sample variance.

9. Assume that a population consists of the three numbers 1, 6, and 8. List all samples of size 2 that can be drawn from this population with replacement, and find the sample mean of each sample.

10. Use a graphing calculator to generate five samples, each of size 6, from a uniform distribution over the interval $(0,1)$. Compute the sample means of each sample.

11. Let $(X_1, X_2, \ldots, X_n)$ denote a sample of size n. Show that

$$\sum_{k=1}^{n}(X_k - \overline{X}) = 0$$

where $\overline{X}$ is the sample mean.

12. Let $(X_1, X_2, \ldots, X_n)$ denote a sample of size n. Show that

$$n\overline{X}^2 = \frac{1}{n}\left(\sum_{k=1}^{n} X_k\right)^2$$

where $\overline{X}$ is the sample mean.

13. Assume that a sample of size n has l distinct values $x_1, x_2, \ldots, x_l$, where x_k occurs f_k times in the sample. Explain why the sample mean is given by the formula

$$\overline{X} = \frac{1}{n}\sum_{k=1}^{l} x_k f_k$$

14. Assume that a sample of size n has l distinct values $x_1, x_2, \ldots, x_l$, where x_k occurs f_k times in the sample. Explain why the sample variance is given by the formula

$$S^2 = \frac{1}{n-1}\left[\sum_{k=1}^{l} x_k^2 f_k - \frac{1}{n}\left(\sum_{k=1}^{l} x_k f_k\right)^2\right]$$

15. Assume that X is exponentially distributed with parameter $\lambda = 3.0$.

(a) Assume that a sample of size 50 is taken from this population. What is the approximate distribution of the sample mean?

(b) Assume now that 1000 samples, each of size 50, are taken from this population and a histogram of the sample means of each of the samples is produced. What shape will the histogram be approximately?

16. Assume that X is exponentially distributed with parameter $\lambda = 3.0$. Assume that a sample of size 50 is taken from this population and that the sample mean of this sample is calculated. How likely is it that the sample mean will exceed 0.43?

■ **12.7.2**

17. Use the random-number generator on a graphing calculator to generate three samples, each of size 10, from a uniform distribution over the interval $(0, 1)$.

(a) Compute the sample mean and the sample variance of each sample.

(b) Combine all three samples, and compute the mean and the sample variance of the combined sample.

(c) Compare your answers in (a) and (b) with the true values of the mean and the variance.

18. Suppose that X is exponentially distributed with mean 1. A computer generates the following sample of independent observations from the population X:

$$0.3169, 0.5531, 2.376, 1.150, 0.6174,$$
$$0.1563, 2.936, 1.778, 0.7357, 0.1024$$

Find the sample mean and the sample variance, and compare them with the corresponding population parameters.

19. Compute the sample mean and the standard error for the sample in Problem 1.

20. Compute the sample mean and the standard error for the sample in Problem 2.

21. The following data represent a sample from a normal distribution with mean 0 and variance 1:

$$-0.68, 1.22, 1.33, -0.84, -0.06,$$
$$0.50, 0.03, -0.13, -0.29, -0.47$$

Construct a 95% confidence interval.

22. The following data represent a sample from a normal distribution with mean 0 and variance 1:

$$-1.18, 0.52, 0.36, -0.16, 0.92,$$
$$0.68, -0.61, -0.54, 0.15, 1.04$$

Construct a 95% confidence interval.

23. Use a graphing calculator to construct a 95% confidence interval for a sample of size 30 from a uniform distribution over the interval $(0, 1)$. Take a class poll to determine the percentage of confidence intervals that contain the true mean. Discuss the result in class.

24. (a) If X has distribution function $F(x)$, we can show that $F(X)$ is uniformly distributed over the interval $(0, 1)$. Use this fact, a graphing calculator, and the table for the standard normal distribution to generate 15 standard normally distributed random variables.

(b) Use your data from (a) to construct a 95% percent confidence interval. Take a class poll to determine the percentage of confidence intervals that contain the true mean. Discuss the result in class.

25. To determine the germination success of seeds of a certain plant, you plant 162 seeds. You find that 117 of the seeds germinate. Estimate the probability of germination and give a 95% confidence interval.

26. To test a new drug for lowering cholesterol, 72 people with elevated cholesterol receive the drug; 51 of them show reduced cholesterol levels. Estimate the probability that the drug lowers cholesterol, and construct a 95% confidence interval.

■ **12.7.3**

In Problems 27 and 28, fit a linear regression line through the given points and compute the coefficient of determination.

27. $(-3, -6.3), (-2, -5.6), (-1, -3.3), (0, 0.1), (1, 1.7), (2, 2.1)$

28. $(0, 0.1), (1, -1.3), (2, -3.5), (3, -5.7), (4, -5.8)$

29. Show that the sum of the residuals about any linear regression line is equal to 0.

30. Show that the last two terms in (12.48), namely

$$2(\overline{y} - a - b\overline{x}) \sum (y_k - \overline{y})$$

and

$$2(\overline{y} - a - b\overline{x}) \sum (x_k - \overline{x})$$

are equal to 0.

31. To determine whether the frequency of chirping crickets depends on temperature, the following data were obtained (Pierce, 1949):

Temperature (°F)	Chirps/s
69	15
70	15
72	16
75	16
81	17
82	17
83	16
84	18
89	20
93	20

Fit a linear regression line to the data, and compute the coefficient of determination.

32. The initial velocity v of an enzymatic reaction that follows Michaelis–Menten kinetics is given by

$$v = \frac{v_{max}s}{K_m + s} \qquad (12.54)$$

where s is the substrate concentration and v_{max} and K_m are two parameters that characterize the reaction. The following computer-generated table contains values of the initial velocity v when the substrate concentration s was varied:

s	v
1	4.1
2.5	6.1
5	9.3
10	12.9
20	17.1

(a) Invert (12.54) and show that

$$\frac{1}{v} = \frac{K_m}{v_{max}}\frac{1}{s} + \frac{1}{v_{max}} \qquad (12.55)$$

This is the Lineweaver–Burk equation. If we plot $1/v$ as a function of $1/s$, a straight line with slope K_m/v_{max} and intercept $1/v_{max}$ results. Use (12.55) to transform the data, and fit a linear regression line to the transformed data. Find the slope and the intercept of the linear regression line, and determine K_m and v_{max}.

(b) Dowd and Riggs (1965) proposed to use the transformation

$$v = v_{max} - K_m\frac{v}{s} \qquad (12.56)$$

and then plot v against v/s. The resulting straight line has slope $-K_m$ and intercept v_{max}. Use (12.56) to transform the data, and fit a linear regression line to the transformed data. Find the slope and the intercept of the linear regression line, and determine K_m and v_{max}.

Chapter 12 Key Terms

Discuss the following definitions and concepts:

1. Multiplication principle
2. Permutation
3. Combination
4. Random experiment
5. Sample space
6. Basic set operations, Venn diagram, De Morgan's laws
7. Definition of probability
8. Equally likely outcomes
9. Mendel's pea experiments
10. Mark–recapture method
11. Maximum likelihood estimate
12. Conditional probability
13. Partition of sample space
14. Law of total probability
15. Independence
16. Bayes formula
17. Random variable
18. Discrete distribution
19. Probability mass function
20. Distribution function of a disctete random variable
21. Mean and variance
22. Joint distributions
23. Binomial distribution
24. Multinomial distribution
25. Geometric distribution
26. Poisson distribution
27. Poisson approximation to the binomial distribution
28. Continuous random variable
29. Density function
30. Distribution function of a continuous random variable
31. Mean and variance of a continuous random variable
32. Histogram
33. Normal distribution
34. Uniform distribution
35. Exponential distribution
36. Aging
37. Gompertz law
38. Weibull law
39. Law of large numbers
40. Markov's inequality
41. Chebyshev's inequality
42. Central limit theorem
43. Histogram correction
44. Sample
45. Statistic
46. Sample median, sample mean, sample variance, standard error
47. Confidence interval
48. Estimating proportions
49. Linear regression line
50. Coefficient of determination

Chapter 12 Review Problems

1. (a) There are 25 students in a calculus class. What is the probability that no two students have the same birthday?

(b) Let p_n denote the probability that, in a group of n people, no two people have the same birthday. Show that

$$p_1 = 1 \quad \text{and} \quad p_{n+1} = p_n\frac{365-n}{365}$$

Use this formula to generate a table of p_n for $1 \le n \le 25$.

2. Thirty patients are to be randomly assigned to two different treatment groups. How many ways can this be done?

3. Fifteen different plants are to be equally divided among five plots. How many ways can this be done?

4. Assume that a certain disease either is caused by a genetic mutation or appears spontaneously. The disease will appear in 67% of all people with the mutation and in 23% of all people without the mutation. Assume that 3% of the population carries the disease gene.

(a) What is the probability that a randomly chosen individual will develop the disease?

(b) Given an individual who suffers from the disease, what is the probability that he or she has the genetic mutation?

5. Suppose that 42% of the seeds of a certain plant germinate.

(a) What is the expected number of germinating seeds in a sample of 10 seeds?

(b) You plant 10 seeds in one pot. What is the probability that none of the seeds will germinate?

(c) You plant five pots with 10 seeds each. What is the expected number of pots with no germinating seeds?

(d) You plant five pots with 10 seeds each. What is the probability that at least one pot has no germinating seeds?

6. Suppose that the amount of yearly rainfall in a certain area is normally distributed with mean 27 and standard deviation 5.7 (measured in inches).

(a) What is the probability that, in a given year, the rainfall will exceed 35 inches?

(b) What is the probability that, in five consecutive years, the rainfall will exceed 35 inches in each year?

(c) What is the probability that, in at least 1 out of 10 years, the rainfall will exceed 35 inches per year?

7. Suppose that, each time a student takes a particular test, he or she has a 20% chance of passing. (Assume that consecutive trials are independent.)

(a) What are the chances of passing the test on the second trial?

(b) Given that a student failed the test the first time, what are the chances that he or she will pass the test on the second trial?

8. Explain why

$$2^n = \sum_{k=1}^{n} \binom{n}{k}$$

9. A bag contains 170 chocolate-covered raisins, on average. Production standards require that, in 95% of all bags, the number of raisins does not deviate from 170 by more than 10. Assume that the number of raisins is normally distributed with mean μ and variance σ^2.

(a) Determine μ and σ.

(b) A shipment contains 100 bags. What is the probability that no bag contains fewer than 160 raisins?

10. Suppose that two parents are carriers of a recessive gene causing a metabolic disorder. Neither parent has the disease. If they have three children, what is the probability that none of the children will be afflicted with the disease? (Note that a recessive gene causes a disorder only if an individual has two copies of the gene.)

11. Suppose that you choose a plant from a large batch of red-flowering pea plants and cross it with a white-flowering pea plant. What percentage of the red-flowering parent plants are of genotype Cc if 90% of the offspring have red flowers?

12. Suppose that a random variable is normally distributed with mean μ and variance σ^2. How would you estimate μ and σ?

13. Let $(X_1, X_2, \ldots, X_n)$ be a sample of size n from a population with mean μ and variance σ^2. Define

$$V = \frac{1}{n} \sum_{k=1}^{n} (X_k - \overline{X})^2$$

where $\overline{X}$ is the sample mean.

(a) If S^2 is the sample variance, show that

$$V = \frac{n-1}{n} S^2$$

(b) Compute $E(V)$.

14. Assume that the weight of a certain species is normally distributed with mean μ and variance σ^2. The following data represent the weight (measured in grams) of 10 individuals:

$$171, 168, 151, 192, 175, 163, 182, 157, 177, 169$$

(a) Find the median, the sample mean, and the sample variance.

(b) Construct a 95% confidence interval for the population mean.

15. (a) Generate five observations (x, y) from a random experiment, where

$$y = 2x + 1 + \epsilon$$

$x = 1, 2, 3, 4, 5$, and ϵ is normally distributed with mean 0 and variance 1.

(b) Use your data from (a) to find the least square line, and compare your results with the linear model that describes this experiment.

(c) What proportion of your data is explained by the model?

16. Suppose X is a continuous random variable with density function

$$f(x) = \begin{cases} 0 & \text{for } x < 1 \\ (r-1)x^{-r} & \text{for } x \geq 1 \end{cases}$$

where r is a constant greater than 1.

(a) For which values of r is $E(X) = \infty$?

(b) Compute $E(X)$ for those values of r for which $E(X) < \infty$.

Appendices

Appendix A Frequently Used Symbols

A.1 Greek Letters

Lowercase Letters

α	alpha	η	eta	ν	nu	τ	tau
β	beta	θ	theta	ξ	xi	υ	upsilon
γ	gamma	ι	iota	o	omicron	ϕ	phi
δ	delta	κ	kappa	π	pi	χ	chi
ϵ	epsilon	λ	lambda	ρ	rho	ψ	psi
ζ	zeta	μ	mu	σ	sigma	ω	omega

Uppercase Letters

Γ	Gamma	Λ	Lambda	Σ	Sigma
Δ	Delta	Π	Pi	Ω	Omega

A.2 Mathematical Symbols

$<$	less than	$\subset$	subset	$=$	equal to
$\leq$	less than or equal to	$\in$	element of	$\neq$	not equal to
$>$	greater than	$\perp$	perpendicular	$\approx$	approximately
$\geq$	greater than or equal to	$\parallel$	parallel	$\propto$	proportional to
$\cup$	union	$\cap$	intersection	Σ	sum

■ Appendix B Table of the Standard Normal Distribution

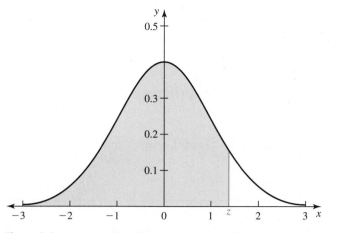

Figure B.1 Areas under the standard normal curve from $-\infty$ to z.

z	0	1	2	3	4	5	6	7	8	9
0.0	.5000	.5040	.5080	.5120	.5160	.5199	.5239	.5279	.5319	.5359
0.1	.5398	.5438	.5478	.5517	.5557	.5596	.5636	.5675	.5714	.5754
0.2	.5793	.5832	.5871	.5910	.5948	.5987	.6026	.6064	.6103	.6141
0.3	.6179	.6217	.6255	.6293	.6331	.6368	.6406	.6443	.6480	.6517
0.4	.6554	.6591	.6628	.6664	.6700	.6736	.6772	.6808	.6844	.6879
0.5	.6915	.6950	.6985	.7019	.7054	.7088	.7123	.7157	.7190	.7224
0.6	.7258	.7291	.7324	.7357	.7389	.7422	.7454	.7486	.7518	.7549
0.7	.7580	.7612	.7642	.7673	.7704	.7734	.7764	.7794	.7823	.7852
0.8	.7881	.7910	.7939	.7967	.7996	.8023	.8051	.8078	.8106	.8133
0.9	.8159	.8186	.8212	.8238	.8264	.8289	.8315	.8340	.8365	.8389
1.0	.8413	.8438	.8461	.8485	.8508	.8531	.8554	.8577	.8599	.8621
1.1	.8643	.8665	.8686	.8708	.8729	.8749	.8770	.8790	.8810	.8830
1.2	.8849	.8869	.8888	.8907	.8925	.8944	.8962	.8980	.8997	.9015
1.3	.9032	.9049	.9066	.9082	.9099	.9115	.9131	.9147	.9162	.9177
1.4	.9192	.9207	.9222	.9236	.9251	.9265	.9279	.9292	.9306	.9319
1.5	.9332	.9345	.9357	.9370	.9382	.9394	.9406	.9418	.9429	.9441
1.6	.9452	.9463	.9474	.9484	.9495	.9505	.9515	.9525	.9535	.9545
1.7	.9554	.9564	.9573	.9582	.9591	.9599	.9608	.9616	.9625	.9633
1.8	.9641	.9649	.9656	.9664	.9671	.9678	.9686	.9693	.9699	.9706
1.9	.9713	.9719	.9726	.9732	.9738	.9744	.9750	.9756	.9761	.9767
2.0	.9772	.9778	.9783	.9788	.9793	.9798	.9803	.9808	.9812	.9817
2.1	.9821	.9826	.9830	.9834	.9838	.9842	.9846	.9850	.9854	.9857
2.2	.9861	.9864	.9868	.9871	.9875	.9878	.9881	.9884	.9887	.9890
2.3	.9893	.9896	.9898	.9901	.9904	.9906	.9909	.9911	.9913	.9916
2.4	.9918	.9920	.9922	.9925	.9927	.9929	.9931	.9932	.9934	.9936
2.5	.9938	.9940	.9941	.9943	.9945	.9946	.9948	.9949	.9951	.9952
2.6	.9953	.9955	.9956	.9957	.9959	.9960	.9961	.9962	.9963	.9964
2.7	.9965	.9966	.9967	.9968	.9969	.9970	.9971	.9972	.9973	.9974
2.8	.9974	.9975	.9976	.9977	.9977	.9978	.9979	.9979	.9980	.9981
2.9	.9981	.9982	.9982	.9983	.9984	.9984	.9985	.9985	.9986	.9986

Answers to Odd-Numbered Problems

■ **Section 1.1**

1. (a) $\{-4, 2\}$ (b) $\{-4, 2\}$ **3.** (a) $\{-1, 5\}$ (b) $\{1, 5\}$ (c) $\{-4, 1\}$
(d) no solution **5.** (a) $[-\frac{2}{5}, \frac{6}{5}]$, (b) $(-\infty, -\frac{7}{3}) \cup (3, \infty)$,
(c) $(-\infty, -1] \cup [-\frac{1}{7}, \infty)$, (d) $(-\frac{1}{5}, \frac{13}{5})$ **7.** $x + 3y - 14 = 0$
9. $3x + y + 2 = 0$ **11.** $7x - 3y + 5 = 0$ **13.** $4x + 3y - 12 = 0$
15. $2y - 3 = 0$ **17.** $x + 1 = 0$ **19.** $3x - y + 2 = 0$
21. $2y - x - 4 = 0$ **23.** $y + 2x - 2 = 0$ **25.** $x + 4y - 3 = 0$
27. $x + 2y + 4 = 0$ **29.** $x + 3y + 4 = 0$ **31.** $2x + 5y - 22 = 0$
33. $y - x + 6 = 0$ **35.** $y = 2$ **37.** $x = -1$ **39.** $x = 1$
41. $y = 3$ **43.** (a) $y = kx, [x] = $ ft, $[y] = $ cm, $k = \frac{30.5\,\text{cm}}{1\,\text{ft}}$ implies
$y = (30.5 \frac{\text{cm}}{\text{ft}})x$ (b) (i) 183 cm, (ii) $\frac{1159}{12}$ cm, (iii) $\frac{1159}{24}$ cm
(c) (i) $\frac{346}{61}$ ft, (ii) $\frac{150}{61}$ ft, (iii) $\frac{96}{61}$ ft **45.** $s(t) = (40\,\text{mi/hr})t$,
$k = 40\,\text{mi/hr}$ **47.** $\frac{1}{(0.305)^2}$ ft^2 **49.** (a) $[y] = $ liter, $[x] = $ ounces,
$y = \left(\frac{1}{33.81} \frac{\text{liter}}{\text{ounces}}\right)x$ (b) $\frac{12}{33.81}$ liters **51.** (a) 88 km/hr, (b) 81 mi/hr
53. (a) $C = K - 273.15$, (b) $77.4\,\text{K} = -195.75°\text{C} = -320.35°\text{F}$,
$90.2\,\text{K} = -182.95°\text{C} = -297.31°\text{F}$; nitrogen gets distilled first,
since it has the lower boiling point. **55.** $(x + 1)^2 + (y - 4)^2 = 9$
57. (a) $(x - 2)^2 + (y - 5)^2 = 9$ (b) $y = 5 + \sqrt{5}$ or $y = 5 - \sqrt{5}$
(c) No **59.** center: $(2, 0)$; radius: 4 **61.** center: $(2, -1)$; radius: 4
63. (a) $\frac{5}{12}\pi$, (b) 255° **65.** (a) $\frac{1}{2}\sqrt{2}$, (b) $-\frac{1}{2}\sqrt{3}$, (c) $\sqrt{3}$
67. (a) $\alpha = \frac{5\pi}{3}$ or $\alpha = \frac{4\pi}{3}$ (b) $\alpha = \frac{\pi}{3}$ or $\alpha = \frac{4\pi}{3}$ **69.** Divide both
left and right side by $\cos^2\theta$. **71.** $\{0, \frac{\pi}{3}, \pi, \frac{5\pi}{3}\}$ **73.** (a) $4^{7/3}$ (b) 27
(c) 5^{4k-2} **75.** (a) $x = \frac{1}{16}$ (b) $x = 27$ (c) $x = \frac{1}{100}$ **77.** (a) $x = -5$
(b) $x = -4$ (c) $x = -3$ **79.** (a) $\ln 3$ (b) $\log_4(x - 2) + \log_4(x + 2)$
(c) $6x - 2$ **81.** (a) $x = \frac{1}{3}(\ln 2 + 1)$ (b) $x = -\frac{1}{2}\ln 10$
(c) $x = \pm\sqrt{1 + \ln 10}$ **83.** (a) $x = 3 + e^5$ (b) $x = \sqrt{4 + e}$
(c) $x = 18$ **85.** $5 - 7i$ **87.** $13 + 2i$ **89.** $15 + 9i$ **91.** 37
93. $3 + 2i$ **95.** $6 - 3i$ **97.** $8 - 2i$ **99.** $z + \bar{z} = 2a, z - \bar{z} = 2bi$
101. $x_1 = \frac{3}{4} + i\frac{\sqrt{7}}{4}, x_2 = \frac{3}{4} - i\frac{\sqrt{7}}{4}$ **103.** $x_1 = -1, x_2 = 2$
105. $x_1 = \frac{3}{8} + i\frac{\sqrt{7}}{8}, x_2 = \frac{3}{8} - i\frac{\sqrt{7}}{8}$ **107.** $x_1 = \frac{7}{3}, x_2 = -1$
109. $x_1 = x_2 = 1$ **111.** $x_1 = \frac{5 + i\sqrt{47}}{6}, x_1 = \frac{5 - i\sqrt{47}}{6}$

■ **Section 1.2**

1. range: $y \geq 0$ **3.** range: $y \in [0, 1)$

5. (b) No, their domains are different.

7. $f(x)$ is odd. $f(-x) = -2x - f(x)$

9. $f(x)$ is even. $f(-x) = |3(-x)| = |3x| = f(x)$

11. $f(x)$ is even. $f(-x) = -|-x| = -|x| = f(x)$

15. (a) $(f \circ g)(x) = 1 - 4x^2, x \geq 0$ (b) $(g \circ f)(x) = 2(1 - x^2)$,
$x \in \mathbf{R}$

17. $(f \circ g)(x) = 3x, x \geq 9$

19. $(f \circ g)(x) = x, x \geq 0; (g \circ f)(x) = x, x \geq 0$

21. $x^2 > x^4$ for $0 < x < 1; x^2 < x^4$ for $x > 1$

23. They intersect at $x = 0$ or 1. **25.** (a)

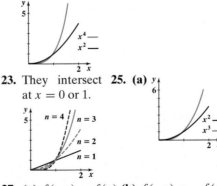

27. (a) $f(-x) = f(x)$ (b) $f(-x) = -f(x)$

29. (a) $k = \frac{3}{2}$ (b) domain: $0 \leq x \leq 3$

31. $s(t) = t$, polynomial of degree 1   **33.** domain: $x \neq 1$;
range: $y \neq 0$ **35.** domain: $x \neq -3, 3$; range: $\mathbf{R}$

37. $\frac{1}{x} < \frac{1}{x^2}$ for $0 < x < 1$; $\frac{1}{x} > \frac{1}{x^2}$ for $x > 1$;
they intersect at $x = 1$.

39. (a)   **(b)** range of $f(x)$ is $(0, \infty)$ **(c)** $x = -\frac{1}{2}$

(d) exactly one.

41. (a)   **(b)** range of $f(x)$ is $[0, 3)$, **(c)** $x = 2$,

(d) exactly one, $x = \frac{a}{3-a}$.   **43.** 83.3 4.76

45. (a)   **(b)** range of $f(x)$ is $[0, 1)$,

(c) $f(x)$ approaches 1.

47.   **49.**   **51. (a)**

53. increases **55.** increases

57. (a) 1, 2, 4, 8, 16 **(b)**

59. $20 \exp\left[-\frac{\ln 2}{5730} 2000\right]$ **61.** $\lambda = \frac{\ln 2}{7\,\text{days}}$

63. (a) $W(t) = (300\,\text{gr}) \exp\left[-\frac{\ln 2}{140\,\text{days}} t\right]$
(b) $t = \frac{\ln 5}{\ln 2} 140\,\text{days} \approx 325\,\text{days}$ **(c)**

65. $\frac{W(t)}{W(0)} = \exp\left[-\frac{\ln 2}{5730} 15,000\right] \approx 16.3\%$ **67. (a)** $r = 3$
(b) $r = \ln 1.25$ **69. (a)** yes **(b)** no **(c)** yes **(d)** yes **(e)** no **(f)** yes

71. (a) $f^{-1}(x) = \sqrt{x - 1}, x \geq 1$ **(b)**

73. $f^{-1}(x) = \sqrt[3]{\frac{1}{x}}, x > 0$

75. $f^{-1}(x) = \log_3 x, x > 0$ **77.** $f^{-1}(x) = \log_{1/4} x, x > 0$

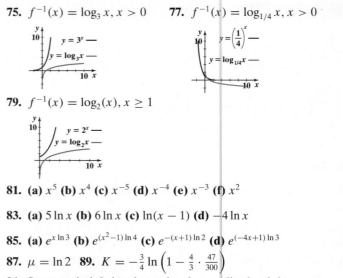

79. $f^{-1}(x) = \log_2(x), x \geq 1$

81. (a) x^5 **(b)** x^4 **(c)** x^{-5} **(d)** x^{-4} **(e)** x^{-3} **(f)** x^2

83. (a) $5 \ln x$ **(b)** $6 \ln x$ **(c)** $\ln(x - 1)$ **(d)** $-4 \ln x$

85. (a) $e^{x \ln 3}$ **(b)** $e^{(x^2-1) \ln 4}$ **(c)** $e^{-(x+1) \ln 2}$ **(d)** $e^{(-4x+1) \ln 3}$

87. $\mu = \ln 2$ **89.** $K = -\frac{3}{4} \ln\left(1 - \frac{4}{3} \cdot \frac{47}{300}\right)$

91. Same period; $2 \sin x$ has twice the amplitude of $\sin x$.

93. Same period; $2 \cos x$ has twice the amplitude of $\cos x$.

95. Same period; $y = 2 \tan x$ is stretched by a factor of 2.

97. amplitude: 3; period: $\frac{\pi}{2}$ **99.** amplitude: 4; period: 1

101. amplitude: 4; period: 8π **103.** amplitude: 3; period: 10

■ **Section 1.3**

1. **3.** **5.** $y = -2x^2 - 3$

7. $y = 3 + 1/x$ **9.** $y = 1/(x - 1)$ **11.** $y = \exp(x - 2)$

13. $y = e^{-(x + 3)}$ **15.** **17.** $y = -\ln(x - 1) + 1$

19. $2\sin\left(x+\frac{\pi}{4}\right)$ —   **21.** $y=-\sin(\pi x/2)$ —

23. **(a)** Shift two units down. **(b)** Shift $y=x^2$ one unit to the right and then one unit up. **(c)** Shift $y=x^2$ two units to the left, stretch by a factor of 2, and reflect about the x-axis.   **25. (a)** 1. Reflect $\frac{1}{x}$ about the x-axis. 2. Shift up one unit. **(b)** 1. Shift $\frac{1}{x}$ one unit to the right. 2. Reflect about the x-axis. **(c)** 1. Shift $y=\frac{1}{x}$ one unit to the left. 2. Reflect about x-axis. 3. Shift up one unit.

27. (a) Stretch $y=e^x$ by a factor of 2, and then shift one unit down. **(b)** Reflect $y=e^x$ about the y-axis, and then reflect about the x-axis. **(c)** Shift $y=e^x$ two units to the right, and then shift one unit up.   **29. (a)** Shift $y=\ln x$ one unit to the right.
(b) Reflect $y=\ln x$ about the x-axis, and then shift up one unit.
(c) Shift $y=\ln x$ three units to the left, then down one unit.
31. (a) Reflect $y=\sin x$ about the x-axis, then one unit up.
(b) Shift $y=\sin x$ by $\pi/4$ units to the right. **(c)** Shift $y=\sin x$ by $\pi/3$ units to the left, and then reflect about the x-axis.
33. Calculate log of each number, for instance, $\log 100=2$.
35. (b) No **(c)** No   **37.** four   **39.** one, three   **41.** six to seven
43. $y=5\times(0.58)^x$ **45.** $y=3^{1/3}\times(3^{-1/3})^x$
47. $\log y=\log 3-2x$ **49.** $\log y=\log 2-(1.2)(\log e)x$
51. $\log y=\log 5+(4\log 2)x$ **53.** $\log y=\log 4+(2\log 3)x$
55. $y=(2)x^{-(\log 2)/\log 5}$ **57.** $y=\frac{1}{8}x^2$ **59.** $\log y=\log 2+5\log x$
61. $\log y=6\log x$ **63.** $\log y=-2\log x$
65. $\log y=\log 4-3\log x$ **67.** $\log y=\log 3+1.7\log x$, log-log transformation   **69.** $\log N(t)=\log 130+(1.2t)\log 2$, log-linear transformation   **71.** $\log R(t)=\log 3.6+1.2\log t$, log-log transformation   **73.** $y=1.8x^{0.2}$ **75.** $y=4\times 10^x$
77. $y=(5.7)x^{2.1}$ **79.** $\log_2 y=x$ **81.** $\log_2 y=-x$
83. (a) $\log N=\log 2+3t\log e$ **(b)** slope: $3\log e\approx 1.303$
85. $\log S=\log C+z\log A$, $z=$ slope of straight line
87. $v_{max}=$ horizontal-line intercept, $\frac{v_{max}}{K_m}=$ vertical-line intercept
89. (a) $\log S=\log 1.162+0.93\log B$ **91. (a)** $\alpha=-\ln 0.9/\text{m}$
(b) 10% **(c)** 1 m: 90%, 2 m: 81%, 3 m: 72.9% **(e)** slope
$=\log 0.9=-\alpha/\ln 10$ **(f)** $z=-\frac{1}{\alpha}\ln(0.01)=\frac{\ln(0.01)}{\ln(0.9)}$ **(g)** Clear lake: small α; milky lake: large α **93.** $y=(100)(10^{1/3})^x$
95. $y=(2^{1/3})(2^{2/3})^x$ **97.** $y=\log x$

99.
$$\Delta G=2.303RT\log\frac{C_2}{C_1}\text{ —}$$

101.

103.

105.

107.

109.

111.

113.

115.

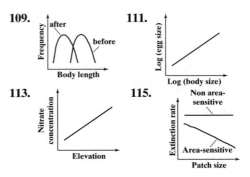

■ Chapter 1 Review Problems

1. (a) 10^4, 1.1×10^4, 1.22×10^4, 1.35×10^4, 1.49×10^4
(b) $t=10\ln 10\approx 23.0$
3. (b) $R(x)=-4kx^3+4k(a+b)x^2-kb(4a+b)x+kab^2$, polynomial of degree 3
(c) $R(x)=(0.3)(5-x)(6-2x)^2$, $0\le x\le 3$

5. (a) $L(t)=0.69, 2.40, 4.62, 6.91$, $E(t)=1.72, 2.20\times 10^4$, 2.69×10^{43}, 1.97×10^{434} **(b)** 20.93 years, 3.09 ft **(c)** $10^{536,000,000}$ years, 21.98 ft **(d)** $L=19.92$ ft, $E=10^{195\times 10^6}$ ft   **7.** $T=\frac{\ln 2}{\ln(1+\frac{q}{100})}$, T goes to infinity as q gets closer to 0.   **9. (a)**

(b) $Y=C^{1.17}10^{-1.92}$ **(c)** $Y_p=2.25Y_c$ **(d)** 8.5%   **11. (a)** $21.\overline{8}$ hours per day, 400 days per year **(b)** line through $(0,24)$ and $(380, 21.\overline{8})$: $y=4320-180x$ **(c)** 376×10^6 to 563×10^6 years ago
13. (a) males: $S(t)=\exp[-(0.019t)^{3.41}]$; females: $S(t)=\exp[-(0.022t)^{3.24}]$ **(b)** males: 47.27 days; females: 40.59 days **(c)** males should live longer   **15. (a)** $x=k$, $v=\frac{a}{2}$
(b) $x_{0.9}=81x_{0.1}$ **17.**

19. $g(s)=\frac{v_{max}}{S_k}S$

21. (a) $\alpha=-\frac{\ln(0.01)}{18\text{m}}\approx 0.25\frac{1}{\text{m}}$ **(b)** 4.87 m

■ Section 2.1

1. $3, 9, 27, 81, 243, 729$ **3.** $25, 25/4, 25/16, 25/64, 25/256, 25/1024$
5. $N_t=2\cdot 2^t$, $t=0,1,2,\ldots$ **7.** $N_t=4^t$, $t=0,1,2,\ldots$
9. $N_t=2\cdot 2^t$, $t=0,1,2,\ldots$ **11.** 1.5 hrs   **13.** $1,2,4,8,16,32$
15. 161 minutes   **17.** 50 minutes   **19.** $N(t)=(40)(2^t)$,
$t=0,1,2,\ldots$ **21.** $N(t)=(20)(3^t)$, $t=0,1,2,\ldots$
23. $N(t)=(5)(4^t)$, $t=0,1,2,\ldots$ **25.** $N(t+1)=2N(t)$,
$N(0)=20$ **27.** $N(t+1)=3N(t)$, $N(0)=10$
29. $N(t+1)=4N(t)$, $N(0)=30$

31. **33.** **35.** 3, 6, 12, 24, 48, 96

37. 2, 6, 18, 54, 162, 486 **39.** 1, 5, 25, 125, 625, 3125 **41.** 1024,
512, 256, 128, 64, 32 **43.** 729, 243, 81, 27, 9, 3 **45.** 31250, 6250,
1250, 250, 50, 10 **47.** $N(t) = (15)(2^t), t = 0, 1, 2, \ldots$
49. $N(t) = (12)(3^t), t = 0, 1, 2, \ldots$ **51.** $N(t) = (24)(4^t)$,
$t = 0, 1, 2, \ldots$ **53.** $N(t) = (5000)\left(\frac{1}{2}\right)^t, t = 0, 1, 2, \ldots$

55. $N(t) = (8000)\left(\frac{1}{3}\right)^t, t = 0, 1, 2, \ldots$

57. $N(t) = (1200)\left(\frac{1}{5}\right)^t, t = 0, 1, 2, \ldots$

59. **61.**

63. **65.**

67. Parent–offspring ratio: 1/2 **69.** Parent–offspring ratio: 1/3

71. Parent–offspring ratio: 2 **73.** Parent–offspring ratio: 3

75. (a) No **(b)** Yes **(c)** Can argue either way. **77.** Limited food
resources; limited habitat; limited nesting sites.

■ **Section 2.2**

1. 0, 1, 2, 3, 4, 5 **3.** $\frac{1}{2}, \frac{1}{3}, \frac{1}{4}, \frac{1}{5}, \frac{1}{6}, \frac{1}{7}$ **5.** 1, 0.25, 0.11, 0.063, 0.04,
0.028 **7.** −1, 0, 3, 8, 15, 24 **9.** 0, −1, 2, −3, 4, −5 **11.** 0, $\frac{1}{2}, \frac{4}{3}, \frac{9}{4}$,
$\frac{16}{5}, \frac{25}{6}$ **13.** 1, 2.72, 4.11, 5.65, 7.39, 9.36 **15.** 1, 0.33, 0.11, 0.037,
0.012, 0.0041 **17.** 6, 7, 8, 9 **19.** $\frac{1}{36}, \frac{1}{49}, \frac{1}{64}, \frac{1}{81}$ **21.** $\frac{6}{7}, \frac{7}{8}, \frac{8}{9}, \frac{9}{10}$
23. $\sqrt{6 + e^6}, \sqrt{7 + e^7}, \sqrt{8 + e^8}, \sqrt{9 + e^9}$ **25.** $a_n = n$,
$n = 0, 1, 2, \ldots$ **27.** $a_n = 2^n, n = 0, 1, 2, \ldots$ **29.** $a_n = \frac{1}{3^n}$,
$n = 0, 1, 2, \ldots$ **31.** $a_n = (-1)^{n+1}(n + 1), n = 0, 1, 2, \ldots$
33. $a_n = (-1)^{n+1}\frac{1}{(n+2)}, n = 0, 1, 2, \ldots$ **35.** $a_n = \sin[(n + 1)\pi]$,
$n = 0, 1, 2, \ldots$ **37.** $\frac{1}{2}, \frac{1}{3}, \frac{1}{4}, \frac{1}{5}, \frac{1}{6}; 0$ **39.** 0, $\frac{1}{2}, \frac{2}{3}, \frac{3}{4}, \frac{4}{5}; 1$ **41.** 1, $\frac{1}{2}$,
$\frac{1}{5}, \frac{1}{10}, \frac{1}{17}; 0$ **43.** 1, $-\frac{1}{2}, \frac{1}{3}, -\frac{1}{4}, \frac{1}{5}; 0$ **45.** 0, $\frac{1}{2}, \frac{4}{3}, \frac{9}{4}, \frac{16}{5}$; limit does
not exist. **47.** 0, 1, $\sqrt{2}, \sqrt{3}, \sqrt{4}$; limit does not exist. **49.** 1, 2, 4,
8, 16; limit does not exist. **51.** 1, 3, 9, 27, 64; limit does not exist.
53. $a = 0, N = 100$ **55.** $a = 0, N = 10$ **57.** $a = 0, N = 100$
59. $a = 0, N = 100$ **61.** $a = 1, N = 99$ **63.** $a = 1, N = 9$
65. N is the largest integer less than or equal to $1/\epsilon$. **67.** N is
the largest integer less than or equal to $\sqrt{1/\epsilon}$. **69.** N is the
largest integer less than or equal to $1/\epsilon$. **71.** 0 **73.** 1 **75.** 1
77. 0 **79.** 0 **81.** 1 **83.** 2, 4, 8, 16, 32 **85.** 1, 1, 1, 1, 1

87. −6, 16, −28, 60, −116 **89.** $\frac{1}{2}, \frac{1}{3}, \frac{1}{4}, \frac{1}{5}, \frac{1}{6}$ **91.** 2, $\frac{5}{2}, \frac{29}{10}, \frac{941}{290}$,
$\frac{969,581}{272,890}$ **93.** 4 **95.** −3 **97.** 2, −2 **99.** $-1 + \sqrt{3}, -1 - \sqrt{3}$
101. 0, 5 **103.** 5; 5 **105.** 0, 2; 2 **107.** 0, $\frac{1}{2}; \frac{1}{2}$ **109.** 2, −2; 2

■ **Section 2.3**

1. $N_t = \frac{2N_t}{1 + \frac{1}{15}N_t}$ **3.** $N_t = \frac{1.5N_t}{1 + \frac{0.5}{40}N_t}$

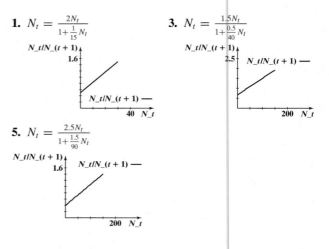

5. $N_t = \frac{2.5N_t}{1 + \frac{1.5}{90}N_t}$

7. $R = 2, K = 20$ **9.** $R = 1.5, K = 30$ **11.** $R = 4, K = 450$

13. 0, 90 **15.** 0, 30 **17.** 0, 60 **19.** Limiting population size: 10

21. Limiting population size: 15 **23.** Limiting population size:

40 **25.** $r = 2, x_t = \frac{1}{20}N_t$ **27.** $r = 3, x_t = \frac{2}{45}N_t$ **29.** $r = 3.5$,
$x_t = \frac{2.5}{105}N_t$ **31. (c)** $N_t = 1000M_t, K = 1000L$, **(d)** $M_t = 20$,

$L = 5$ **33.** $z = 6$ **35.**

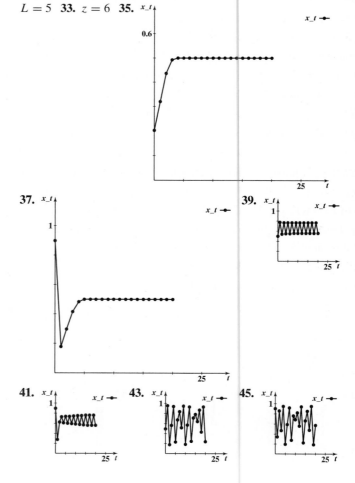

37. 39.

41. 43. 45.

47. Points of intersection: $N = 0$ and $N = 10$

49. Points of intersection: $N = 0$ and $N = 12$

51.

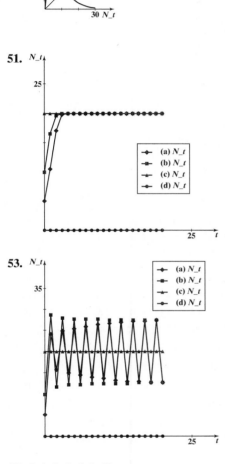

53.

55. $1, 1, 2, 3, 5, 8, 13, \ldots$

57. One-month-old rabbit pairs produce one pair of rabbits; two-month-old rabbit pairs produce two pairs of rabbits.

■ Chapter 2 Review Problems

1. 0 **3.** 40 **5.** ∞ **7.** 1 **9.** 0 **11.** $a_n = \frac{2n+1}{2n+2}, n = 0, 1, 2, \ldots$

13. $a_n = \frac{n+1}{(n+1)^2+1}, n = 0, 1, 2, \ldots$ **15.**

17. $\hat{R}_t \approx 0.8$ for t large; extinction will occur.

19.

■ Section 3.1

1. -3 **3.** -1 **5.** $\frac{3}{2}\sqrt{2}$ **7.** $\frac{4}{3}\sqrt{3}$ **9.** e^{-2} **11.** 0 **13.** 7 **15.** 0
17. 1 **19.** 0 **21.** $-\infty$ **23.** ∞ **25.** ∞ **27.** ∞ **29.** $\frac{1}{6}$ **31.** $\frac{1}{2}$
33. $\lim_{x\to\infty} f(x) = \lim_{x\to-\infty} f(x) = 0$; $\lim_{x\to 0} f(x) = \infty$
35. divergence by oscillations **37.** -9 **39.** 54 **41.** $\frac{53}{3}$ **43.** $\frac{47}{2}$
45. $\frac{28}{3}$ **47.** 2 **49.** 4 **51.** $-\frac{1}{4}$ **53.** -5

■ Section 3.2

1. 1 **3.** 5 **5.** $f(2) = 3$ **7.** $a = 6$ **9.** $x = 3$ **11.** $x = 1, x = 2$
13. $f(5/2) = 5/2$. See Example 3 for $k = 3$. **15.** $x \in \mathbf{R}$
17. $x \neq 1$ **19.** $x \in \mathbf{R}$ **21.** $\{x : x < -1 \text{ or } x > 0\}$
23. $\{x \in \mathbf{R} : x \neq \frac{1}{4} + \frac{k}{2}, k \in \mathbf{Z}\}$
25. (a) $f(x)$ is not continuous at $x = 0$

(b) $c = 2$ **27. (b)**   **(c)** No **29.** $\frac{1}{2}$ **31.** 1 **33.** 3

35. 1 **37.** 1 **39.** 1 **41.** 2 **43.** $\frac{1}{4}$ **45.** $\frac{1}{6}$ **47.** 0

■ Section 3.3

1. 0 **3.** ∞ **5.** 2 **7.** ∞ **9.** ∞ **11.** -1 **13.** 4 **15.** 2 **17.** 0
19. $\frac{3}{2}$ **21.** $\frac{3}{2}$ **23.** 0 **25.** a **27. (a)**   **(b)** 500

(c) 250 **29. (a)**   **(b)** 50

■ Section 3.4

1. (a)

3. (a) [graph of $\frac{\ln x}{x}$, 0.5, 14 x] **(b)** $x \geq e$ **5.** 1 **7.** 5 **9.** π **11.** 0 **13.** 1

15. 0 **17.** 0 **19.** 0

■ **Section 3.5**

1. (a) [graph $f(x)$, 3.5, 2.5 x] **(b)** $f(0) = -1$, $f(2) = 3$

3. (a) [graph $f(x)$, 3, 2.5 x] **(b)** $f(1) = \sqrt{3} < 2 < f(2) = \sqrt{6}$

5. $f(0) = 1 > 0$, $f(1) = e^{-1} - 1 < 0$ **7.** $x \approx 0.57$
9. (a) $x \approx -0.67$ **(b)** [graph $f(x)$, 6, 2 x] **(c)** No

11. (a) $N(10) = 23$ **(b)** $N(10) = 23$ (in millions)

■ **Section 3.6**

1. $0.495 < x < 0.505$ **3.** $2.99 < x < 3.01$ or $-3.01 < x < -2.99$
5. (b) $1.95 < x < 2.05$ **7. (a)** [graph $\frac{1}{x}$, 5, 5 x] **(b)** $0 < x < \frac{1}{4}$

9. $\delta = \frac{\epsilon}{2}$ **11.** $\delta = \epsilon^{1/5}$ **13.** $\delta = \frac{2}{\sqrt{M}}$ **15.** $\delta = \frac{1}{\sqrt[4]{M}}$ **17.** $x_0 = \sqrt{\frac{2}{\epsilon}}$
19. $x_0 = \frac{1}{\epsilon}$ **21.** $\delta = \frac{\epsilon}{|m|}$

■ **Chapter 3 Review Problems**

1. $x \in \mathbf{R}$ **3.** $x \in \mathbf{R}$ **5.** [graph $f(x)$, 1 x] **7.** [graph $f(x)$, 1, x]

9. $f(-2) = -2$, $\lim_{x \to -2^+} f(x) = -2$, $\lim_{x \to -2^-} f(x) = -3$
11. $a = 1.24 \times 10^6$, $k = 5$
13. (a) $g(t) = \begin{cases} 1 & \text{for } \frac{1}{6} + 2k \leq x \leq \frac{5}{6} + 2k, k = 0, 1, 2, \dots \\ 0 & \text{otherwise} \end{cases}$
(b) $s(t)$ is continuous, $g(t)$ is not continuous
15. (c) $T_h = 0$: $g(N) = aTN = f(N)$ **(d)** $\lim_{N \to \infty}(aTN) = \infty$;
$\lim_{N \to \infty} \frac{aTN}{1 + aT_h N} = \frac{T}{T_h}$ **17. (a)** [graph, 5, 4 x, sinh x, cosh x, tanh x]
(b) $\lim_{x \to \infty} \sinh x = \infty$, $\lim_{x \to -\infty} \sinh x = -\infty$,
$\lim_{x \to \infty} \cosh x = \infty$, $\lim_{x \to -\infty} \cosh x = \infty$, $\lim_{x \to \infty} \tanh x = 1$,
$\lim_{x \to -\infty} \tanh x = -1$

■ **Section 4.1**

1. 0 **3.** 4 **5.** 0 **7.** 0 **9.** 0 **11.** 2 **13.** 3 **15.** $c = 2k + 1, k \in \mathbf{Z}$
17. $-2h$ **19.** $\sqrt{4+h} - 2$ **21. (a)** $f'(-1) = -10$
(b) $y = -10x - 5$ **23. (a)** $f'(2) = -12$ **(b)** $y = \frac{1}{12}x - \frac{43}{6}$

25. $f'(x) = \frac{1}{2\sqrt{x}}$ **27.** $y = 6x - 3$ **29.** $y = \frac{1}{4}x + 1$
31. $y = -\frac{1}{6}x - \frac{19}{6}$ **33.** $y = -\frac{1}{4}x + \frac{5}{4}$ **35.** $f(x) = 2x^2$ and
$x = a$ **37.** $f(x) = \frac{1}{x^2+1}, a = 2$ **39. (a)** [graph $s(t)$, 100, 2 t]
(b) 40 km/hr **(c)** 40 km/hr **41. (a)** $s\left(\frac{3}{4}\right) = 30$, $s(1) = \frac{160}{3}$ **(b)** $\frac{280}{3}$
(c) $v\left(\frac{3}{4}\right) = 80$, $\left|v\left(\frac{3}{4}\right)\right| = 80$
43. $R^* = 1.25$ [graph $f(R)$, 200, $\frac{200R}{5+R}$, 40, 1.25, 15 R]
45. $f(N) = 0$ for $N = 0$ and $N = 20$ [graph $\frac{dN}{dt}$, $3N\left(1 - \frac{N}{20}\right)$, 15, 20 N]

47. $x = 7$ or $x = 4$; reaction ceases when $x = 4$.
49. $\frac{dN}{dt} = 0$ for $N = 0$ or $N = K$ **51.** B **53.** [graph, a c b x]
55. No, $f(x)$ could have a discontinuity at $x = c$. **57.** $x = -5$
59. $x = -2$ **61.** $x = 3$ **63.** $x = -1$ **65.** $x = \sqrt{1/2}$ and
$x = -\sqrt{1/2}$ **67.** $x = 1$ **69.** $x = 0$

■ **Section 4.2**

1. $12x^2 - 7$ **3.** $-10x^4 + 7$ **5.** $-4 - 10x$ **7.** $35s^6 + 6s^2 - 5$
9. $-\frac{4}{3}t^3 + 4$ **11.** $2x \sin \frac{\pi}{3}$ **13.** $-12x^3 \tan \frac{\pi}{6}$ **15.** $3t^2 e^{-2} + 1$
17. $3s^2 e^3$ **19.** $60x^2 - 24x^5 + 72x^7$ **21.** $3\pi x^2 + \frac{1}{\pi}$ **23.** $3ax^2$
25. $2ax$ **27.** $2rs$ **29.** $3rs^2x^2 - r$ **31.** $4(b-1)N^3 - \frac{2N}{b}$
33. $a^3 - 3at^2$ **35.** $V_0\gamma$ **37.** $1 - \frac{2N}{K}$ **39.** $2rN - 3\frac{r}{K}N^2$
41. $4\frac{2\pi^5}{15}\frac{k^4}{c^2h^3}T^3$ **43.** $y - 191x - 377 = 0$ **45.** $y - 3x + 6 = 0$
47. $\sqrt{2}y - 8x + 18 = 0$ **49.** $2y - x - 7 = 0$
51. $24y - x - 73\sqrt{3} = 0$ **53.** $3y + x + 5 = 0$
55. $2ax - y - a = 0$ **57.** $(a^2 + 2)y - 4ax + 4a = 0$
59. $\frac{1}{3a}x + y + a + \frac{1}{3a} = 0$
61. $2a(a+1)y + \frac{1}{2}(a+1)^2x - 8a^2 - (a+1)^2 = 0$ **63.** $(0,0)$
65. $\left(\frac{3}{2}, \frac{9}{4}\right)$ **67.** $(0,0)$ and $\left(\frac{2}{9}, -\frac{4}{243}\right)$ **69.** $(0,0)$, $\left(-\frac{1}{2}, -\frac{17}{96}\right)$, and
$\left(4, -\frac{160}{3}\right)$ **71.** $(0, 4)$; only point **73.** $\left(-\frac{1}{4}, -\frac{3}{8}\right)$; only point
75. $\left(\frac{1}{3}\sqrt{3}, \frac{7}{9}\sqrt{3} + 2\right)$, $\left(-\frac{1}{3}\sqrt{3}, -\frac{7}{9}\sqrt{3} + 2\right)$
77. Tangent line: $y = 2x - 1$ **79.** $y = 2ax - a^2$ and
$y = -2ax - a^2$ **81.** $P'(x)$ is a polynomial of degree 3.

■ **Section 4.3**

1. $f'(x) = 3x^2 + 10x - 3$ **3.** $f'(x) = -105x^6 + 30x^4 + 75x^2 - 10$
5. $f'(x) = x(2x + 3x^2) + (\frac{1}{2}x^2 - 1)(2 + 6x)$ **7.** $f'(x) = \frac{4}{5}x^3$
9. $f'(x) = 6(3x - 1)$ **11.** $f'(x) = -12(1 - 2x)$
13. $g'(s) = 2(4s - 5)(2s^2 - 5s)$
15. $g'(t) = 6(4t - 20t^3)(2t^2 - 5t^4)$ **17.** $y = x - 1$
19. $y = -56x - 64$ **21.** $y = -\frac{1}{6}x + \frac{7}{3}$ **23.** $y = \frac{1}{5}x + 2$
25. $f'(x) = (12x + 5)(1 - x) - (2x - 1)(3x + 4)$

27. $f'(x) = (6x^2 - 12x + 1)(1 - x^2) - 2x(x - 3)(2x^2 + 1)$
29. $f'(x) = a(4x - 3)$ **31.** $f'(x) = 8ax(x^2 - a)$
33. $g'(t) = 2a(at + 1)$ **35.** 11 **37.** 2 **39.** -27
41. $y' = 3f'(x)g(x) + 3f(x)g'(x)$
43. $y' = f'(x)g(x) + f(x)g'(x) + 4g(x)g'(x)$
45. $f(B) = Bg(B)$ with $f'(0) = g(0)$
47. $f'(N) = r\left[(a - 2N)\left(1 - \frac{N}{K}\right) - \frac{1}{K}(aN - N^2)\right]$
49. $f'(x) = \frac{4}{(x+1)^2}$ **51.** $2\frac{3x^2+3x-2}{(2x+1)^2}$ **53.** $f'(x) = \frac{2x^3-3x^2+3}{(1-x)^2}$
55. $h'(t) = \frac{t^2+2t-4}{(t+1)^2}$ **57.** $f'(s) = \frac{2s^2-4s+4}{(1-s)^2}$
59. $f'(x) = \frac{1}{2\sqrt{x}}(x - 1) + \sqrt{x}$
61. $f'(x) = \frac{\sqrt{3}}{2\sqrt{x}}(x^2 - 1) + 2x\sqrt{3x}$ **63.** $f'(x) = 3x^2 + \frac{3}{x^4}$
65. $f'(x) = 4x - \frac{3-6x}{x^4}$ **67.** $g'(s) = \frac{2s^{-1/3}-s^{-2/3}-1}{3(s^{2/3}-1)^2}$
69. $f'(x) = (-2)\left(\sqrt{2x} + \frac{2}{\sqrt{x}}\right) + (1 - 2x)\left(\frac{1}{\sqrt{2x}} - \frac{1}{x^{3/2}}\right)$
71. $y = -8x - \frac{55}{3}$ **73.** $y = \frac{7}{16}x - 1$ **75.** $f'(x) = \frac{3a}{(3+x)^2}$
77. $f'(x) = \frac{8ax}{(4+x^2)^2}$ **79.** $f'(R) = \frac{nR^{n-1}k^n}{(k^n+R^n)^2}$
81. $h'(t) = \frac{\sqrt{a}}{2\sqrt{t}}(t - a) + \sqrt{at} + a$ **83.** $-\frac{5}{18}$ **85.** -8 **87.** 3
89. $y' = \frac{f'(x)g(x)-2f(x)g'(x)}{[g(x)]^3}$
91. $y' = \frac{1}{2\sqrt{x}}f(x)g(x) + \sqrt{x}f'(x)g(x) + \sqrt{x}f(x)g'(x)$
93. $y = -\frac{c}{x_1^2}x + 2\frac{c}{x_1}; x = 2x_1$

■ **Section 4.4**

1. $2(x - 3)$ **3.** $-24x(1 - 3x^2)^3$ **5.** $\frac{x}{\sqrt{x^2+3}}$ **7.** $\frac{-3x^2}{2\sqrt{3-x^3}}$
9. $-\frac{12x^2}{(x^3-2)^5}$ **11.** $\frac{2x-3}{(2x^2-1)^{3/2}}$ **13.** $\frac{1-3x}{\sqrt{2x-1}(x-1)^3}$ **15.** $\frac{2\sqrt{s}+1}{4\sqrt{s}\sqrt{s+\sqrt{s}}}$
17. $\frac{-9t^2}{(t-3)^4}$
19. $(r^2 - r)^2(r + 3r^3)^{-5}[3(2r - 1)(r + 3r^3) - 4(1 + 9r^2)(r^2 - r)]$
21. $-\frac{4}{5}x^3(3 - x^4)^{-4/5}$ **23.** $\frac{2x-2}{7(x^2-2x+1)^{6/7}}$
25. $g'(s) = \frac{3}{2}(3s^7 - 7s)^{1/2}(21s^6 - 7)$ **27.** $\frac{2}{5}(3t + \frac{3}{t})^{-3/5}(3 - \frac{3}{t^2})$
29. $3a(ax + 1)^2$ **31.** $g'(N) = \frac{bk-bN}{(k+N)^3}$ **33.** $g'(T) = -3a(T_0 - T)^2$
35. (a) $\frac{2x}{x^2+3}$ (b) $\frac{1}{2(x-1)}$ **37.** $2\left(\frac{f(x)}{g(x)} + 1\right)\frac{f'(x)g(x)-f(x)g'(x)}{[g(x)]^2}$
39. $\frac{2f(x)f'(x)[g(2x)+2x]-[f(x)]^2(g'(2x)\cdot2+2)}{[g(2x)+2x]^2}$
41. $y' = 4(\sqrt{x^3 - 3x} + 3x)^3\left(\frac{3x^2-3}{2\sqrt{x^3-3x}} + 3\right)$
43. $y' = 36x(3x^2 - 1)(1 + (3x^2 - 1)^3)$
45. $y' = 3\left(\frac{2x+1}{3(x^3-1)^3-1}\right)^2\frac{6(x^3-1)^3-2-27x^2(2x+1)(x^3-1)^2}{(3(x^3-1)^3-1)^2}$ **47.** $\frac{dy}{dx} = -\frac{x}{y}$
49. $\frac{dy}{dx} = -\left(\frac{y}{x}\right)^{1/4}$ **51.** $\frac{dy}{dx} = 4\sqrt{xy} - \frac{y}{x}$ **53.** $\frac{dy}{dx} = \frac{x}{y}$
55. (a) $y = \frac{4}{3}x - \frac{25}{3}$ (b) $y = -\frac{3}{4}x$ **57.** (a) $y = \frac{3}{4}x - \frac{9}{4}$
(b) $y = -\frac{4}{3}x + \frac{136}{9}$ **59.** (a) $(27)^{1/6} = \sqrt{3}$ (b)

61. $-\frac{2}{3}\sqrt{3}$ **63.** $-\frac{3}{4}$ **65.** $\frac{dV}{dt} = 3x^2\frac{dx}{dt}$ **67.** $\frac{dS}{dt} = 8\pi r\frac{dr}{dt}$
69. $\frac{dh}{dt} = \frac{1}{100\pi}\frac{m}{min}$ **71.** $\frac{183}{\sqrt{61}}\frac{mi}{hr}$ for both $t = 20$ min and $t = 40$ min
73. $f'(x) = 3x^2 - 6x$, $f''(x) = 6x - 6$ **75.** $g'(x) = 2(x + 1)^{-2}$,
$g''(x) = -\frac{4}{(x+1)^3}$ **77.** $g'(t) = \frac{9t^2+2}{2\sqrt{3t^3+2t}}$, $g''(t) = \frac{27t^4+36t^2-4}{4(3t^3+2t)^{3/2}}$
79. $f'(s) = \frac{3}{4}\left(s^{1/2} - \frac{1}{s}\right)^{-1/2}$, $f''(s) = -\frac{3}{8}\frac{\frac{1}{2}s^{3/2}+1}{\sqrt{s}(s^{3/2}-1)^{3/2}}$
81. $g'(t) = -\frac{5}{2}t^{-7/2} - \frac{1}{2}t^{-1/2}$, $g''(t) = \frac{35}{4}t^{-9/2} + \frac{1}{4}t^{-3/2}$

83. $f(x) = x^5$, $f'(x) = 5x^4$, $f''(x) = 20x^3$, $f'''(x) = 60x^2$,
$f^{(4)}(x) = 120x$, $f^{(5)}(x) = 120$, $f^{(6)} = \cdots = f^{(10)}(x) = 0$
85. $p(x) = 3x^2 + 2x + 3$
87. (a) velocity: $v_0 - gt$, acceleration: $-g$ (b) $t = \frac{v_0}{g}$

■ **Section 4.5**

1. $f'(x) = 2\cos x + \sin x$
3. $f'(x) = 3\cos x - 5\sin x - 2\sec x \tan x$
5. $f'(x) = \sec^2 x + \csc^2 x$ **7.** $f'(x) = 3\cos(3x)$
9. $f'(x) = 6\cos(3x + 1)$ **11.** $f'(x) = 4\sec^2(4x)$
13. $f'(x) = 4\sec(1 + 2x)\tan(1 + 2x)$ **15.** $f'(x) = 6x\cos(x^2)$
17. $f'(x) = 6x\sin^2(x^2 - 3)\cos(x^2 - 3)$
19. $f'(x) = 12x\sin x^2\cos x^2$
21. $f'(x) = -8x\sin x^2 + 4\sin x \cos x$
23. $f'(x) = -8\sin x \cos x - 8x^3\sin x^4$
25. $f'(x) = -4x\sec^2(1 - x^2)$
27. $f'(x) = -18\tan^2(3x - 1)\sec^2(3x - 1)$
29. $f'(x) = \frac{2x\cos(2x^2-1)}{\sqrt{\sin(2x^2-1)}}$ **31.** $g'(s) = \frac{-\sin s}{2\sqrt{\cos s}} + \frac{1}{2\sqrt{s}}\sin\sqrt{s}$
33. $g'(t) = \frac{2\cos(2t)[\cos(6t)-1]+6\sin(6t)[\sin(2t)+1]}{[\cos(6t)-1]^2}$
35. $f'(x) = \frac{2x\sec(x^2-1)[\tan(x^2-1)+\cot(x^2+1)]}{\csc(x^2+1)}$
37. $f'(x) = 2\cos(2x - 1)\cos(3x + 1) - 3\sin(2x - 1)\sin(3x + 1)$
39. $f'(x) = 6x\sec^2(3x^2 - 1)\cot(3x^2 + 1)$
  $- 6x\csc^2(3x^2 + 1)\tan(3x^2 - 1)$ **41.** $f'(x) = \sec^2 x$
43. $f'(x) = 0$ **45.** $g'(x) = \frac{-6x\cos(3x^2-1)}{\sin^2(3x^2-1)}$
47. $g'(x) = -30x\sin^2(1 - 5x^2)\cos(1 - 5x^2)$
49. $h'(x) = -\frac{6\sec^2(2x)-3}{(\tan(2x)-x)^2}$ **51.** $h'(s) = 3\sin s \cos s(\sin s - \cos s)$
53. $f'(x) = \frac{2(1+x^2)\cos(2x)-2x\sin(2x)}{(1+x^2)^2}$ **55.** $f'(x) = -\frac{1}{x^2}\sec^2(\frac{1}{x})$
57. $f'(x) = \frac{2(\sec x^2)[x(\tan x^2)-\tan x]}{\sec^2 x}$ **59.** $x = \frac{3}{2} + 3k, k \in \mathbf{Z}$
61. Write $\frac{d}{dx}\cos x = \lim_{h\to0}\frac{\cos(x+h)-\cos x}{h}$ and use the identity for
$\cos(x + h)$. **63.** Use quotient rule. **65.** $f'(x) = \frac{x\cos\sqrt{x^2+1}}{\sqrt{x^2+1}}$
67. $f'(x) = (\cos\sqrt{3x^3 + 3x})\frac{9x^2+3}{2\sqrt{3x^3+3x}}$
69. $f'(x) = 4x\sin(x^2 - 1)\cos(x^2 - 1)$
71. $f'(x) = 27x^2\tan^2(3x^3 - 3)\sec^2(3x^3 - 3)$
73. (a) $\frac{dc}{dt} = \frac{\pi}{2}\cos(\frac{\pi}{2}t)$ (b)

(c) (i) $\frac{dc}{dt} = 0$ (ii) increasing (iii) $c(t)$ has a horizontal tangent and
either a maximum or a minimum.

■ **Section 4.6**

1. $f'(x) = 3e^{3x}$ **3.** $f'(x) = -12e^{1-3x}$
5. $f'(x) = (-4x + 3)e^{-2x^2+3x-1}$ **7.** $f'(x) = 28x(x^2 + 1)e^{7(x^2+1)^2}$
9. $f'(x) = e^x(1 + x)$ **11.** $f'(x) = xe^{-x}(2 - x)$
13. $f'(x) = \frac{e^x(1+x^2-2x)-2x}{(1+x^2)^2}$ **15.** $f'(x) = \frac{2e^x-2e^{-x}-2}{(2+e^x)^2}$
17. $f'(x) = 3\cos(3x)e^{\sin(3x)}$ **19.** $f'(x) = e^{\sin(x^2-1)}\cos(x^2 - 1)2x$
21. $f'(x) = e^x\cos(e^x)$ **23.** $f'(x) = (2e^{2x} + 1)\cos(e^{2x} + x)$
25. $f'(x) = (1 - \cos x)\exp(x - \sin x)$
27. $g'(s) = \exp[\sec s^2](\sec s^2)(\tan s^2)(2s)$
29. $f'(x) = (\sin x + x\cos x)e^{x\sin x}$
31. $f'(x) = (-3)(2x + \sec^2 x)e^{x^2+\tan x}$ **33.** $f'(x) = (\ln 2)2^x$
35. $f'(x) = (\ln 2)2^{x+1}$ **37.** $f'(x) = \frac{\ln 5}{\sqrt{2x-1}}5^{\sqrt{2x-1}}$

39. $f'(x) = 2x(\ln 2)2^{x^2+1}$ **41.** $h'(t) = (2t)(\ln 2)2^{t^2-1}$

43. $f'(x) = (\ln 2)\frac{1}{2\sqrt{x}}2^{\sqrt{x}}$ **45.** $f'(x) = (\ln 2)\frac{x}{\sqrt{x^2-1}}2^{\sqrt{x^2-1}}$

47. $h'(t) = \frac{\ln 5}{2\sqrt{t}}5^{\sqrt{t}}$ **49.** $g'(x) = -2(\ln 2)(\sin x)2^{2\cos x}$

51. $g'(r) = \frac{\ln 3}{5r^{4/5}}3^{r^{1/5}}$ **53.** 2 **55.** 0 **57.** $\frac{1}{\ln 2}$ **59. (a)** $N(0) = 1$

61. $\frac{dN}{dt} = (\ln 2)N(t)$ which implies that $\frac{dN}{dt}$ is proportional to

$N(t)$ **63. (a)** $\frac{dN}{dt} = \frac{rK(\frac{K}{N(0)}-1)e^{-rt}}{[1+(\frac{K}{N(0)}-1)e^{-rt}]^2}$ **(c)** $dN/(Ndt)$

65. (a) $L(x)$

(b) L_∞ is the limiting size and L_0 is the initial size. **(c)** The fish with $k = 1$ reaches $L = 5$ more quickly. **(d)** With age, the rate of growth decreases. **(e)** The larger the value of k, the more quickly the fish grows and reaches its limiting size.

67. $\frac{dW}{dt} = -4W(t)\frac{1}{\text{days}}$ **69.** $\frac{dW}{dt} = -\frac{\ln 2}{5}W(t)\frac{1}{\text{days}}$

71. (a) $W(4) = 6e^{-12}$ **(b)** half-life: $\frac{\ln 2}{3}$

73. (a) $\frac{dW}{dt} = -(\ln \frac{5}{2})W(t)$ **(b)** $W(3) = 5(\frac{5}{2})^{-3}$ **(c)** half-life: $\frac{\ln 2}{\ln 5 - \ln 2}$

■ **Section 4.7**

1. $\frac{d}{dx}f^{-1}(x) = x$ **3.** $\frac{d}{dx}f^{-1}(x) = \frac{1}{\sqrt{8(x+1)}}$

5. $f^{-1}(x) = \left(\frac{3-x}{2}\right)^{1/3}$, $x \le 3$, $\frac{d}{dx}f^{-1}(x) = -\frac{1}{6}\left(\frac{2}{3-x}\right)^{2/3}$

7. $\frac{d}{dx}f^{-1}(0) = \frac{1}{4}$ **9.** $\frac{d}{dx}f^{-1}(2) = 4$ **11.** $\frac{d}{dx}f^{-1}(1) = \frac{1}{2}$

13. $\frac{d}{dx}f^{-1}(\pi) = \frac{1}{2}$ **15.** $\frac{d}{dx}f^{-1}(0) = 1$ **17.** $\frac{d}{dx}f^{-1}(-\ln 2) = \frac{1}{\sqrt{3}}$

19. $\frac{d}{dx}f^{-1}(1) = 1$ **21.** $\frac{d}{dx}f^{-1}(1) = \frac{1}{2}$ **23.** $f'(x) = \frac{1}{x+1}$

25. $f'(x) = \frac{-2}{1-2x}$ **27.** $f'(x) = \frac{2}{x}$ **29.** $f'(x) = \frac{6x^2-1}{2x^3-x}$

31. $f'(x) = 2(\ln x)\frac{1}{x}$ **33.** $f'(x) = \frac{8\ln x}{x}$ **35.** $f'(x) = \frac{x}{x^2+1}$

37. $f'(x) = \frac{1}{x(x+1)}$ **39.** $f'(x) = \frac{-1}{1-x} - \frac{2}{1+2x}$

41. $f'(x) = (1-\frac{1}{x})\exp[x - \ln x]$ **43.** $f'(x) = \cot x$

45. $f'(x) = \frac{2x\sec^2(x^2)}{\tan(x^2)}$ **47.** $f'(x) = \ln x + 1$ **49.** $f'(x) = \frac{1-\ln x}{x^2}$

51. $h'(t)) = \cos(\ln(3t))\frac{1}{t}$ **53.** $f'(x) = \frac{2x}{x^2-3}$ for $|x| \ne \sqrt{3}$

55. $f'(x) = \frac{-2x}{(\ln 10)(1-x^2)}$ **57.** $f'(x) = \frac{3x^2-3}{(\ln 10)(x^3-3x)}$

59. $f'(u) = \frac{4u^3}{(\ln 3)(3+u^4)}$ **63.** $\frac{dy}{dx} = 2x^x(\ln x + 1)$

65. $\frac{dy}{dx} = (\ln x)^x[\ln(\ln x) + \frac{1}{\ln x}]$ **67.** $\frac{dy}{dx} = x^{\ln x}2(\ln x)\frac{1}{x}$

69. $\frac{dy}{dx} = x^{1/x-2}(1 - \ln x)$

71. $\frac{dy}{dx} = [x^x(\ln x + 1)\ln x + x^{x-1}]x^{x^x}$

73. $\frac{dy}{dx} = x^{\cos x}\left[\frac{\cos x}{x} - (\sin x)(\ln x)\right]$

75. $\frac{1}{y}\frac{dy}{dx} = 2 + \frac{27}{9x-2} - \frac{x}{2(x^2+1)} - \frac{9x^2}{4(3x^3-7)}$

■ **Section 4.8**

1. $\sqrt{65} \approx 8.0625$, error $= 2.42 \times 10^{-4}$ **3.** $\sqrt[3]{124} \approx 5 - \frac{1}{75}$,

error $\approx 3.57 \times 10^{-5}$ **5.** $(0.99)^{25} \approx 0.75$, error ≈ 0.0278

7. $\sin\left(\frac{\pi}{2} + 0.02\right) \approx 1$, error $\approx 2.00 \times 10^{-4}$ **9.** $\ln(1.01) \approx 0.01$,

error $\approx 4.97 \times 10^{-5}$ **11.** $L(x) = 1 - x$ **13.** $L(x) = \frac{3}{2} - \frac{1}{2}x$

15. $L(x) = 1 - 2x$ **17.** $L(x) = x$ **19.** $L(x) = \frac{1}{\ln 10}(x - 1)$

21. $L(x) = 1 + x$ **23.** $L(x) = 1 - x$ **25.** $L(x) = x$

27. $L(x) = 1 - nx$ **29.** $L(x) = 1$ **31.** 100.3 **33.** $B(1.1) \approx 5.005$

35. [1.8, 2.2] **37.** [10.8, 13.2] **39.** [5.91, 8.87] **41.** ±6%

43. ±0.668% **45.** ±9% **47.** ±2.4% **49.** $\pm\frac{(a+b-2x)x}{(a-x)(b-x)}\left(100\frac{\Delta x}{x}\right)$

■ **Chapter 4 Review Problems**

1. $f'(x) = -12x^3 - \frac{1}{x^{3/2}}$ **3.** $h'(t) = \frac{1}{3}\left(\frac{1+t}{1-t}\right)^{2/3}\frac{-2}{(1+t)^2}$

5. $f'(x) = 2e^{2x}\sin\left(\frac{\pi}{2}x\right) + e^{2x}\frac{\pi}{2}\cos\left(\frac{\pi}{2}x\right)$

7. $f'(x) = \frac{\frac{\ln x}{x+1} - \frac{\ln(x+1)}{x}}{(\ln x)^2}$ **9.** $f'(x) = -xe^{-x^2/2}$,

$f''(x) = e^{-x^2/2}(x^2 - 1)$ **11.** $h'(x) = \frac{1}{(x+1)^2}$, $h''(x) = -\frac{2}{(x+1)^3}$

13. $\frac{dy}{dx} = \frac{\cos x + y^2 - 2xy}{x^2 - 2xy}$ **15.** $\frac{dy}{dx} = 1 - 2(x - y)$ **17.** $\frac{dy}{dx} = -\frac{x}{y}$,

$\frac{d^2y}{dx^2} = -\frac{16}{y^3}$ **19.** $\frac{dy}{dx} = \frac{1}{x\ln x}$, $\frac{d^2y}{dx^2} = -\frac{\ln x + 1}{(x\ln x)^2}$ **21.** $5.70\frac{\text{ft}}{\text{sec}}$

23. (a) $\frac{dy}{dx} = f'(x)e^{f(x)}$ **(b)** $\frac{dy}{dx} = \frac{f'(x)}{f(x)}$ **(c)** $\frac{dy}{dx} = 2f(x)f'(x)$

25. (a) **(b)** $c = 1$, $y = \frac{1}{2}x$

27. $y - \frac{1}{2}e^{-(\pi/3)^2} = -e^{-(\pi/3)^2}\left(\frac{1}{2}\sqrt{3} + \frac{\pi}{3}\right)\left(x - \frac{\pi}{3}\right)$ **29.** $y = x$

31. $p(x) = 2x^2 + 4x + 8$ **33. (a)** $s(5.5) \approx 54.84$ miles

(b) $v(t) = \frac{ds}{dt} = 3\pi + 3\pi\sin(\pi t)$, $a(t) = \frac{dv}{dt} = 3\pi^2\cos(\pi t)$

(c) $v(t) \ge 0$ **(d)** Three valleys and three hills

35. (b) $N(t) = 100 + 50\cos\left(\frac{\pi}{2}t\right)$

(c) The population size shows oscillations. **37.** 9.33%

■ **Section 5.1**

1. absolute minimum: $(0, -1)$; absolute maximum: $(1, 1)$

3. absolute minimum: $(\frac{3\pi}{4}, -1)$; absolute maximum: $(\frac{\pi}{4}, 1)$

5. absolute minimum: $(0, 0)$; absolute maximum: $(-1, 1)$ and $(1, 1)$ **7.** absolute minima: $(1, e^{-1})$ and $(-1, e^{-1})$; absolute maximum: $(0, 1)$ **9.**   **11.**

13. local maximum = global maximum = $(-1, 4)$, no local and global minima

15. local minimum = global minimum = $(0, -2)$, local maximum = global maximum = $(-1, -1)$ and $(1, -1)$

17. local minimum = $(-2, -3)$ and $(1, 0)$, global minimum = $(-2, -3)$, local maximum = global maximum = $(0, 1)$ **19.** $f'(0) = 0$; $f(x)$ has a local minimum at $x = 0$ **21.** $f'(0) = 0$; $f(x) = -x^2$ has a local maximum at $x = 0$ **23.** $f'(0) = 0$, but $x = 0$ is not a local extremum: $f(x) < 0$ for $x < 0$ and $f(x) > 0$ for $x > 0$

25. $f'(-1) = 0$, but $x = -1$ is not a local extremum: $f(x) < 0$ for $x < -1$ and $f(x) > 0$ for $x > -1$ **27.** $f(0) = 0$ and $f(x) > 0$ for $x \ne 0$; $\lim_{x\to 0^-} f'(x) = -1$, $\lim_{x\to 0^+} f'(x) = 1$.

29. $f(1) = f(-1) = 0$ and $f(x) > 0$ for $x \ne 1, -1$; $\lim_{x\to 1^-} f'(x) = -2 \ne \lim_{x\to 1^+} f'(x) = 2$ and $\lim_{x\to -1^-} f'(x) = -2 \ne \lim_{x\to -1^+} f'(x) = 2$.

31. loc min = glob min = $(-1, 0)$ and $(1, 0)$; loc max = glob max = $(0, 1)$ and $(2, 1)$

33. (a)

$\frac{dN}{dt}$ is maximal for $N = 50$ **(b)** $f'(N) = r - \frac{2r}{K}N$

35. (a) Slope: 2 **(b)** $c = 1$; guaranteed by the mean-value theorem.   **37.** Slope: 0; $c = 0$ **39.** $[0, 1]$

41. $f(-1) = 1$, $f(2) = -2$, $\frac{f(2)-f(-1)}{2-(-1)} = -1$

43. The mean-value theorem guarantees such a point.

45. (a) Slope: $a + b$

(b) Apply the mean-value theorem. Midpoint $= a + \frac{b-a}{2} = \frac{a+b}{2}$.

47. Use the mean-value theorem and argue that the slope of the secant line connecting $(a, f(a))$ and $(c, f(c))$ and the slope of the secant line connecting $(b, f(b))$ and $(c, f(c))$ have opposite signs, where $c \in (a, b)$ with $f(c) \neq 0$.   **49. (a)** 0.25 m/s **(b)** $\frac{3}{100}t^2$, $0 < t < 5$ **(c)** $t = \frac{5}{3}\sqrt{3}$ s   **51.** $0 \leq B(3) \leq 6$ **53.** $f(x) = 3, x \in \mathbf{R}$

■ **Section 5.2**

1. $y' = 3 - 2x$, $y'' = -2$; $y' > 0$ and y is increasing on $(-\infty, 3/2)$; $y' < 0$ and y is decreasing on $(3/2, \infty)$; $y'' < 0$ and y is concave down.

3. $y' = 2x + 1$, $y'' = 2$; $y' > 0$ and y is increasing on $(-1/2, \infty)$; $y' < 0$ and y is decreasing on $(-\infty, -1/2)$; $y'' > 0$ and y is concave up.

5. $y' = -2x^2 + 7x - 3$, $y'' = -4x + 7$; $y' > 0$ and y is increasing on $(1/2, 3)$; $y' < 0$ and y is decreasing on $(-\infty, 1/2) \cup (3, \infty)$; $y'' > 0$ and y is concave up on $(-\infty, 7/4)$ and $y'' < 0$ and y is concave down on $(7/4, \infty)$.

7. $y' = \frac{1}{2\sqrt{x+1}}$, $x > -1$; $y'' = -\frac{1}{4(x-1)^{3/2}}$, $x > -1$; $y' > 0$ and y is increasing; $y'' < 0$ and y is concave down.

9. $y' = -\frac{1}{x^2}$, $x \neq 0$; $y'' = \frac{2}{x^3}$, $x \neq 0$; $y' < 0$ and y is decreasing for $x \neq 0$; $y'' < 0$ and y is concave down for $x < 0$; $y'' > 0$ and y is concave up for $x > 0$.

11. $y' = \frac{1}{3}(x^2 + 1)^{-2/3}2x$, $y'' = \frac{6-2x^2}{9(x^2+1)^{5/3}}$; $y' > 0$ and y is increasing on $(0, \infty)$; $y' < 0$ and y is decreasing on $(-\infty, 0)$; $y'' > 0$ and y is concave up on $(-\sqrt{3}, \sqrt{3})$; $y'' < 0$ and y is concave down on $(-\infty, -\sqrt{3}) \cup (\sqrt{3}, \infty)$.

13. $y' = -\frac{2}{(1+x)^3}$, $y'' = \frac{6}{(1+x)^4}$; $y' > 0$ and y is increasing on $(-\infty, -1)$; $y' < 0$ and y is decreasing on $(-1, \infty)$; $y'' > 0$ and y is concave up for $x \neq -1$.

15. $y' = \cos x$, $y'' = -\sin x$; $y' > 0$ and y is increasing on $(0, \pi/2) \cup (3\pi/2, 2\pi)$; $y' < 0$ and y is decreasing on $(\pi/2, 3\pi/2)$; $y'' > 0$ and y is concave up on $(\pi, 2\pi)$; $y'' < 0$ and y is concave down on $(0, \pi)$.

17. $y' = e^x$, $y'' = e^x$; $y' > 0$ and y is increasing for $x \in \mathbf{R}$; $y'' > 0$ and y is concave up for $x \in \mathbf{R}$.

19. $y' = -xe^{-x^2/2}$, $y'' = e^{-x^2/2}(x^2 - 1)$; $y' > 0$ and y is increasing for $x < 0$; $y' < 0$ and y is decreasing for $x > 0$; $y'' > 0$ and y is concave up for $x < -1$ and $x > 1$; $y'' < 0$ and y is concave down for $-1 < x < 1$.

21. (a) **(b)**

(c) In (a), $y' > 0$ and $y'' > 0$; in (b), $y' > 0$ and $y'' < 0$.
23. (b) $f(-1) > 0$ and $f(1) < 0$ **25.** $f'(x)$ is decreasing; use the definition of concave down. **27. (a)**

(b) $g'(N) = -\frac{r}{K} < 0$ **29.** $f'(N) = 1 - \left(\frac{N}{K}\right)^\theta (1 + \theta)$; $f(N)$ is increasing for $0 < N < N^*$ and decreasing for $N > N^*$, where $N^* = K\left(\frac{1}{1+\theta}\right)^{1/\theta}$. **31.** The probability of escaping decreases with parasitoid density. **33. (a)** $y' = 117e^{-10/x}\frac{10}{x^2} > 0$; maximum attainable height is 117. **(b)** $y(x)$ is concave up on $(0, 5)$; $y(x)$ is concave down on $(5, \infty)$

(c) **(d)**

35. For $0 < \gamma < 1$, the average number of pollinator visits increases with the number of flowers on the plant, but at a decelerating rate.
37. (a) $N^* = e^{-aN^*}$ **(b)** $\frac{dN^*}{da} = -N^*e^{-aN^*}/(1 + ae^{-aN^*}) < 0$
39. (a) $\frac{d}{dN}\left(\frac{A}{N}\right) = -S[1 + (aN)^b]^{-2}b(aN)^{b-1}a < 0$ **(b) (v)** The number of surviving plants in the next year is the same as the number of plants this year. **41. (a)** For $0 < a < 1$, Y is an increasing function of X but $\frac{Y}{X}$ is a decreasing function of X. $Y(X)$ is concave down. **(b)** Juveniles have relatively larger heads than adults. **43.** $y = f(x)$ is concave up for $k > 1$ and concave down for $0 < k < 1$.

■ **Section 5.3**

1. local max: $(-2, 16)$ and $(3, 1)$; absolute max: $(-2, 16)$; local min: $(2, 0)$; absolute min: $(2, 0)$; y is increasing on $(2, 3)$ and decreasing on $(-2, 2)$. **3.** local max: $(2, \ln 3)$; absolute max: $(2, \ln 3)$; local min: $(1, 0)$; absolute min: $(1, 0)$; y is increasing on $(1, 2)$. **5.** local min: $(0, 0)$; absolute min $(0, 0)$; local max:

$(1, e^{-1})$; absolute max: $(1, e^{-1})$; y is increasing on $[0, 1]$ **7.** no extrema; y is increasing for all $x \in \mathbf{R}$. **9.** local max: $(0, 1)$; absolute max: $(0, 1)$; local min: $(-1, -1)$ and $(1, -1)$; absolute min: $(-1, -1)$ and $(1, -1)$; y is increasing on $(-1, 0)$ and decreasing on $(0, 1)$. **11.** local max: $(0, 1)$; absolute max: $(0, 1)$; y is increasing on $(-\infty, 0)$, and decreasing on $(0, \infty)$. **13.** local max: $(-3, 15.5)$; local min: $(2, -16/3)$; y is increasing on $(-\infty, -3) \cup (2, \infty)$; y is decreasing on $(-3, 2)$. **15.** y is increasing on $\mathbf{R}$. No extrema. **17.** $f'(x) = 3x^2 > 0$ for $x \neq 0$
19. $(0, -2)$ **21.** $(\sqrt{2}/2, e^{-1/2})$ **23.** $(0, 0)$ **25.** $f'(x) = 4x^3$, $f''(x) = 12x^2 > 0$ for all $x \neq 0$ **27.** local min: $(-2, 2/3)$, $(3, -16)$; local max: $(-1, 16/3)$, $(5, 16/3)$; absolute min: $(3, -16)$; absolute max: $(-1, 16/3)$, $(5, 16/3)$; inflection point: $(1, -16/3)$
y is increasing on $(-2, -1) \cup (3, 5)$; y is decreasing on $(-1, 3)$; y is concave up on $(1, 5)$; y is concave down on $(-2, 1)$;

29. local min: $(-3, 0)$, $(3, 0)$; local max: $(-4, 5)$, $(0, 9)$, $(5, 16)$; absolute min: $(-3, 0)$, $(3, 0)$; absolute max: $(5, 16)$; inflection points: $(-3, 0)$, $(3, 0)$; y is increasing on $(-3, 0) \cup (3, 5)$; y is decreasing on $(-4, -3) \cup (0, 3)$; y is concave up on $(-4, -3) \cup (3, 5)$; y is concave down on $(-3, 3)$;

31. no extrema; inflection points: $x = \pi/2 + k\pi$, $k \in \mathbf{Z}$; y is increasing on $\mathbf{R}$; y is concave up on $(\pi/2 + 2k\pi, 3\pi/2 + 2k\pi)$, $k \in \mathbf{Z}$; y is concave down on $(-\pi/2 + 2k\pi, \pi/2 + 2k\pi)$, $k \in \mathbf{Z}$;

33. local min: $(0, -1)$; absolute min: $(0, -1)$; inflection points at $x = -\frac{\sqrt{3}}{3}$ and $x = \frac{\sqrt{3}}{3}$; y is increasing on $(0, \infty)$; y is decreasing on $(-\infty, 0)$; y is concave up on $(-\frac{\sqrt{3}}{3}, \frac{\sqrt{3}}{3})$; y is concave down on $(-\infty, -\frac{\sqrt{3}}{3}) \cup (\frac{\sqrt{3}}{3}, \infty)$;

35. (c) decreasing for all $x \neq 1$; no extrema **(d)** $f(x)$ is concave down on $(-\infty, 1)$ and concave up on $(1, \infty)$; no inflection points **(e)**

37. (a) $\lim_{x \to -2^+} f(x) = \infty$, $\lim_{x \to -2^-} f(x) = -\infty$ **(b)** $f(x)$ is increasing on $(-\infty, -2 - \sqrt{6}/2) \cup (-2 + \sqrt{6}/2, \infty)$; $f(x)$ is decreasing on $(-2 - \sqrt{6}/2, -2) \cup (-2, -2 + \sqrt{6}/2)$; local max at $x = -2 - \sqrt{6}/2$; local min at $-2 + \sqrt{6}/2$ **(c)** $f(x)$ is concave up on $(-2, \infty)$; $f(x)$ is concave down on $(-\infty, -2)$ **(d)** $y = 2x - 4$ **(e)**

39. (a) $f'(x) = \frac{2x}{(1+x^2)^2}$ and $f(x)$ is increasing for $x > 0$ and decreasing for $x < 0$ **(b)** $f(x)$ is concave up for $-1/\sqrt{3} < x < 1/\sqrt{3}$ and concave down for $x < -1/\sqrt{3}$ or $x > 1/\sqrt{3}$. $f(x)$ has two inflection points at $x = \pm 1/\sqrt{3}$ **(c)** $\lim_{x \to \infty} = 1$, $\lim_{x \to -\infty} = 1$ **(d)**

41. (a) $f'(x) = \frac{a}{(a+x)^2}$; $f(x)$ is increasing for $x > 0$ **(b)** $f''(x) = -\frac{2a}{(a+x)^3} < 0$; f is concave down for all x; there are no inflection points **(c)** $\lim_{x \to \infty} \frac{x}{a+x} = 1$; there is a horizontal asymptote at $y = 1$ **(d)**

43. The growth rate is maximal for $N = K \left(\frac{1}{1+\theta}\right)^{1/\theta}$.

■ **Section 5.4**

1. 20 in. **3.** 4 **5.** The field is 80 ft by 160 ft, where the length along the river is 160 ft. **7.** $5(1 + \sqrt{2})$ **9.** 4 **11. (b)** $(6/5, \sqrt{1.6})$ **(c)** local minimum at $x = 6/5$ as in (b). **13.** $\sqrt{2}$

15. $g'(x) = 2f(x)f'(x)$ has the same sign change as $f'(x)$.

17. The height is equal to the diameter, namely $2\left(\frac{500}{\pi}\right)^{1/3}$.

19. (a) $r = \sqrt{2}, \theta = 2$ **(b)** $r = \sqrt{10}, \theta = 2$ **21.** $r = \left(\frac{355}{4\pi}\right)^{1/3}$
23. $a = 2, b = -2$ **25. (a)** Local maximum at t^* satisfies $f'(t^*) = \frac{f(t^*)}{C+t^*}$, which is the slope of the straight line through $(-C, 0)$ and the point with $t = t^*$.
(b) $y = \frac{1}{(1+\sqrt{2})^2}t + \frac{2}{(1+\sqrt{2})^2}$

27. (a) $-x\frac{dr}{dx} - r(x) - L + \frac{3ke^{-kx}}{1-e^{-kx}} - \frac{r'(x)e^{-(r(x)+L)}}{1-e^{-(r(x)+L)}} = 0$

■ **Section 5.5**

1. 10 **3.** -7 **5.** $\frac{1}{2}$ **7.** 1 **9.** $\frac{1}{2}$ **11.** ∞ **13.** 0 **15.** $\frac{\ln 2}{\ln 3}$ **17.** $-\frac{\ln 3}{\ln 2}$
19. $\frac{1}{2}$ **21.** 0 **23.** 0 **25.** 0 **27.** 0 **29.** 0 **31.** 0 **33.** 1 **35.** 0
37. 0 **39.** 0 **41.** 1 **43.** 1 **45.** e^3 **47.** e^{-2} **49.** e^{-1} **51.** 0

53. ∞ **55.** 0 **57.** 0 **59.** 1 **61.** $\frac{\ln a}{\ln b}$

63. $\lim_{x \to \infty} \left(1 + \frac{c}{x^p}\right) = \begin{cases} \infty & \text{if } 0 < p < 1 \\ e^c & \text{if } p = 1 \\ 1 & \text{if } p > 1 \end{cases}$

65. Apply l'Hospital's rule. **67.** Let $y = 121e^{-17/x}$.
(a) $\lim_{x \to 0^+} \frac{dy}{dx} = 0$, $\lim_{x \to \infty} \frac{dy}{dx} = 0$ **(b)** $\frac{17}{2}$ **(c)** Height is increasing at an accelerating rate for $0 < x < \frac{17}{2}$ and at a decelerating rate for $x > \frac{17}{2}$. **(d)**

■ **Section 5.6**

1. (a) $N_5 = (10)(1.03)^5$ **(b)** $t = \frac{\ln 2}{\ln 1.03}$ **3. (a)** $b = 1.02$
(b) $N_{10} = (20)(1.02)^{10}$ **(c)** $t = \frac{\ln 2}{\ln 1.02}$ **5. (a)** $b = 1 + \frac{x}{100}$
(b) $t = \frac{\ln 2}{\ln(1+x/100)}$; 693.5, 139.0, 70.0, 35.0, 14.2, 7.3 **7. (a)** 0
(b) stable **9. (a)** $x = 1/2$ and $x = -2$ **(b)** $x = 1/2$ is locally stable, $x = -2$ is unstable. **11.** $x = 0$ is unstable, $x = 0.5$ is locally stable. **13. (a)** $x = 0$ is locally stable, $x = 1$ is unstable, $x = 4$ is locally stable **(b) (i)** 0 **(ii)** 4 **15. (b)** 0 **(c)** $1/\beta$
(d) $P = 2/\beta$ is an inflection point **(e)**

17. (b) To find equilibria, solve $N = 10Ne^{-0.01N}$; $N = 0$ is another equilibrium. **(c)** Oscillations seem to appear, and the system does not seem to converge to the nontrivial equilibrium.
19. Equilibria: $N = 0$ and $N = 100$. **(b)** Starting from $N = 10$, it appears that the limiting population size is 100. $N = 100$ is locally stable. **21.** Fixed points are 0 and $1 - \frac{1}{r}$. The nonzero fixed point is locally stable for $1 < r < 3$. **23.** $N^* = r^{1/(\gamma-1)}$; locally stable for $1 < r < 3$ **25.** $N^* = K$; locally stable for $0 < r < 2$

■ **Section 5.7**

1. $\sqrt{7} = 2.645751$ **3.** 0.6529186 **5.** 1.895494
7. (a) $|x_n| = 2^n x_0$ **(b)** ∞
9. (a) $x_0 = 3, x_1 = 4.166667, x_2 = 4.003333, x_3 = 4.000001$
(b) $x_0 = x_1 = x_2 = \cdots = 4$

■ **Section 5.8**

1. $F(x) = \frac{4}{3}x^3 - \frac{1}{2}x + C$ **3.** $F(x) = \frac{1}{3}x^3 + \frac{3}{2}x^2 - 4x + C$
5. $F(x) = \frac{1}{5}x^5 - x^3 + x + C$ **7.** $F(x) = x^4 - x^2 + 3x + C$
9. $F(x) = \frac{x^2}{2} + \ln|x| - \frac{1}{x} + C$ **11.** $F(x) = x + \frac{1}{x} + C$
13. $F(x) = \ln|1+x| + C$ **15.** $F(x) = x^5 - \frac{5}{3x^3} + C$
17. $F(x) = \frac{1}{2}\ln|1+2x| + C$ **19.** $F(x) = -\frac{1}{3}e^{-3x} + C$
21. $F(x) = e^{2x} + C$ **23.** $F(x) = -\frac{1}{2}e^{-2x} + C$
25. $F(x) = -\frac{1}{2}\cos(2x) + C$
27. $F(x) = -3\cos(x/3) + 3\sin(x/3) + C$
29. $F(x) = -\frac{4}{\pi}\cos(\pi x/2) - \frac{6}{\pi}\sin(\pi x/2) + C$
31. $F(x) = \frac{1}{2}\tan(2x) + C$ **33.** $F(x) = 3\tan(x/3) + C$
35. $F(x) = \tan x + x + C$ **37.** $-\frac{1}{6}x^{-6} + \frac{1}{2}x^6 - \frac{1}{2}\cos(2x) + C$
39. $\frac{1}{3}\tan(3x - 1) + \frac{1}{2}x^2 - 3\ln x + C$ **41.** $\frac{1}{a(a+1)}e^{(a+1)x}$
43. $\frac{1}{a}\ln|ax + 3|$ **45.** $\frac{1}{a+3}x^{a+3} - \frac{1}{\ln a}a^{x+2}$
47. $y = 2\ln|x| - \frac{1}{2}x^2 + C$ **49.** $y = \frac{1}{2}x^2 + \frac{1}{3}x^3 + C$

51. $y = \frac{1}{2}t^2 - \frac{1}{3}t^3 + C$ **53.** $y = -2e^{-t/2} + C$

55. $y = -\frac{1}{\pi}\cos(\pi s) + C$ **57.** $y = 2\tan(x/2) + C$

59. $y = x^3 + 1, x \geq 0$ **61.** $y = \frac{4}{3}x^{3/2} + \frac{2}{3}$

63. $N(t) = \ln t + 10, t \geq 1$ **65.** $W(t) = e^t, t \geq 0$

67. $W(t) = 1 - \frac{1}{3}e^{-3t}$ **69.** $T(t) = 3 + \frac{1}{\pi} - \frac{1}{\pi}\cos(\pi t)$

71. $y = \frac{e^x - e^{-x}}{2}$ **73.** $L(x) = 25 - 10e^{-0.1x}, L(0) = 15$

75. $t = 2.5$ s, $v(2.5) = 80\frac{\text{ft}}{\text{s}}$ **77. (a)** The first term on the right-hand side describes evaporation; the second one describes watering. **(b)** $a = \frac{1}{24}$

■ Chapter 5 Review Problems

1. (b) Absolute maximum at $(1, e^{-1})$ **(c)** inflection point at $(2, 2e^{-2})$ **(d)**

3. (a) $f'(x) = \frac{4}{(e^x + e^{-x})^2} > 0$, hence, $f(x)$ is strictly increasing.

5. (e)

7. (a) $\lim_{x \to \infty} f(x) = c$ **(b)** $f'(x) = \frac{ck}{(k+x)^2} > 0$, $f''(x) = -\frac{2ck}{(k+x)^3} < 0$ **(c)** $f(k) = \frac{c}{2}$ **(d)**

9. (a) $L(\theta) = (\theta^2)^8(2\theta(1-\theta))^6((1-\theta)^2)^3$ **(b)** $\frac{d}{d\theta}\ln L(\theta) = \frac{L'(\theta)}{L(\theta)}$ and $L(\theta) > 0$. **(c)** $\hat{\theta} = \frac{11}{17}$ **11. (a)** $c(t) = \frac{1}{3}e^{-0.3t}, t \geq 0$

(b) $t = \frac{\ln 2}{0.3}$ **13. (a)** $t = v_0/g$ **(b)** $\frac{v_0^2}{2g}$ **(c)** 0 **(d)** $t = \frac{2v_0}{g}$

■ Section 6.1

1. 0.21875 **3.** 0.46875 **5.** $\sqrt{1} + \sqrt{2} + \sqrt{3} + \sqrt{4}$

7. $3^2 + 3^3 + 3^4 + 3^5 + 3^6$ **9.** $1 + (x+1) + (x+1)^2 + (x+1)^3$

11. $-1 + 1 - 1 + 1$ **13.** $\left(\frac{1}{n}\right)^2\frac{1}{n} + \left(\frac{2}{n}\right)^2\frac{1}{n} + \left(\frac{3}{n}\right)^2\frac{1}{n} + \cdots + \left(\frac{n}{n}\right)^2\frac{1}{n}$

15. $\sum_{k=1}^{n} 2k$ **17.** $\sum_{k=2}^{5} \ln k$ **19.** $\sum_{k=2}^{6} \frac{k-3}{k+2}$ **21.** $\sum_{k=1}^{n} q^{k-1}$

23. 285 **25.** 112 **27.** $\frac{2n(n-1)(2n-1)}{3}$ **29.** 0 **33.** 1.36 **35.** 14

37. 0 **39.** $\int_a^b x \, dx = \frac{1}{2}b^2 - \frac{1}{2}a^2$

41. $2\int_1^2 x^3 \, dx$ **43.** $\int_{-3}^2 (2x-1)\, dx$ **45.** $\int_2^3 \frac{x-1}{x+2}\, dx$ **47.** $\int_{-5}^2 e^x \, dx$

49. $\lim_{\|P\| \to 0} \sum_{k=1}^{n} (c_k + 1)^{1/3}\Delta x_k$, where $\|P\|$ is a partition of $[2, 6]$, $c_k \in [x_{k-1}, x_k]$, and $\Delta x_k = x_k - x_{k-1}$

51. $\lim_{\|P\| \to 0} \sum_{k=1}^{n} \ln c_k \Delta x_k$, where $\|P\|$ is a partition of $[1, e]$, $c_k \in [x_{k-1}, x_k]$, and $\Delta x_k = x_k - x_{k-1}$

53. $\lim_{\|P\| \to 0} \sum_{k=1}^{n} g(c_k)\Delta x_k$, where $\|P\|$ is a partition of $[0, 5]$, $c_k \in [x_{k-1}, x_k]$, and $\Delta x_k = x_k - x_{k-1}$

55. $\int_{-1}^2 (x^2-1)\, dx = -A_- + A_+$

57. $\int_0^5 e^{-x}\, dx = A$

59. $\int_{1/2}^4 \ln x \, dx = A_2 - A_1$

61. $\int_{-2}^3 |x|\, dx = \frac{13}{2}$

63. $\int_2^5 \left(\frac{1}{2}x - 4\right) dx = -\frac{27}{4}$

65. $\int_{-2}^2 \left(\sqrt{4 - x^2} - 2\right) dx = 2\pi - 8$

67. $\int_{-3}^0 \left(4 - \sqrt{9 - x^2}\right) dx = 12 - \frac{9}{4}\pi$ **69.** 0

71. 0 **73.** 0 **75.** $x \geq x^2$ for $0 \leq x \leq 1$ **77.** $\sqrt{x} \geq 0$ for $x \geq 0$ and $\sqrt{x} \leq 2$ for $0 \leq x \leq 4$

79.

The rectangle with height $1/2$ from $\pi/6$ to $5\pi/6$ is contained in the area under $y = \sin x$ from $\pi/6$ to $5\pi/6$, which is contained in the rectangle with height 1 from $\pi/6$ to $5\pi/6$. **81.** $a = \pi/2$

83. $a = 3$

■ Section 6.2

1. $\frac{dy}{dx} = 2x^3$ **3.** $\frac{dy}{dx} = 4x^2 - 3$ **5.** $\frac{dy}{dx} = \sqrt{1 + 2x}$

7. $\frac{dy}{dx} = \sqrt{1 + \sin^2 x}$ **9.** $\frac{dy}{dx} = xe^{4x}$ **11.** $\frac{dy}{dx} = \frac{1}{x+3}$

13. $\frac{dy}{dx} = \sin(x^2 + 1)$ **15.** $\frac{dy}{dx} = 3(1 + 9x^2)$

17. $\frac{dy}{dx} = [2(1 - 4x)^2 + 1](-4)$ **19.** $\frac{dy}{dx} = 2x\sqrt{x^2 + 1}$

21. $\frac{dy}{dx} = 3(1 + e^{3x})$ **23.** $\frac{dy}{dx} = (6x + 1)[1 + (3x^2 + x)e^{3x^2 + x}]$

25. $\frac{dy}{dx} = -(1 + x)$ **27.** $\frac{dy}{dx} = -2[1 + \sin(2x)]$ **29.** $\frac{dy}{dx} = -\frac{1}{x^2}$

31. $\frac{dy}{dx} = -2x\sec(x^2)$ **33.** $\frac{dy}{dx} = 2[1 + (2x)^2] - (1 + x^2)$

35. $\frac{dy}{dx} = 3x^2 \ln(x^3 - 3) - 2x \ln(x^2 - 3)$

37. $\frac{dy}{dx} = (1 + 3x^2)\sin(x + x^3) + 2x\sin(2 - x^2)$ **39.** $x + x^3 + C$
41. $\frac{1}{9}x^3 - \frac{1}{4}x^2 + C$ **43.** $\frac{1}{6}x^3 + \frac{3}{2}x^2 - \frac{2}{3}x + C$
45. $x^{3/2}\left(\frac{4}{5}x - \frac{2}{3}\right) + C$ **47.** $\frac{2}{7}x^{7/2} + C$ **49.** $\frac{2}{9}x^{9/2} + \frac{7}{9}x^{9/7} + C$
51. $\frac{2}{3}x^{3/2} + 2\sqrt{x} + C$ **53.** $\frac{1}{3}x^3 - x + C$
55. $-\frac{1}{3}x^3 + \frac{5}{2}x^2 - 6x + C$ **57.** $\frac{1}{2}e^{2x} + C$ **59.** $-3e^{-x} + C$
61. $-e^{-x^2/2} + C$ **63.** $-\frac{1}{2}\cos(2x) + C$ **65.** $\frac{1}{3}\sin(3x) + C$
67. $\frac{1}{3}\tan(3x) + C$ **69.** $\sec x + C$ **71.** $\frac{1}{2}\ln|\sec(2x)| + C$
73. $\tan x + \ln|\sec x| + C$ **75.** $4\tan^{-1}x + C$ **77.** $\sin^{-1}x + C$
79. $\ln|x + 2| + C$ **81.** $\frac{2}{3}x - \frac{1}{3}\ln|x| + C$ **83.** $\ln|x - 3| + C$
85. $-\ln|x + 3| + C$ **87.** $5(x - \tan^{-1}x) + C$ **89.** $\frac{3^x}{\ln 3} + C$
91. $-\frac{9^{-x}}{\ln 9} + C$ **93.** $\frac{1}{3}x^3 + \frac{2^x}{\ln 2} + C$ **95.** $\frac{2}{3}x^{3/2} + 2e^{x/2} + C$
97. -6 **99.** $-\frac{1}{2}$ **101.** 3 **103.** $\frac{28}{3}$ **105.** $\frac{1}{2}$ **107.** $\frac{1}{2}$ **109.** $\frac{\pi}{4}$
111. $\frac{\pi}{6}$ **113.** $\frac{1}{2}\ln 2$ **115.** $\frac{1}{3}(1 - e^{-3})$ **117.** 1 **119.** 1 **121.** $\ln\frac{3}{2}$
123. $\frac{1}{2}$ **125.** $f(x) = 4x$

■ **Section 6.3**

1. $\frac{125}{6}$ **3.** $2e$ **5.** $\frac{4}{3}$ **7.** $\frac{14}{3} - \ln 4$ **9.** $\frac{\pi}{2} - 1$ **11.** $\frac{3}{2}$ **13.** $\frac{2}{3}$ **15.** $\frac{16}{3}$
17. (a) $N(t) = 101 - e^{-t}$ **(b)** $1 - e^{-5}$ **(c)** $N(5) - N(0) = \int_0^5 e^{-t}\,dt$;

shaded area.

19. (a)

(b) Particle moves to the left for $0 \le t \le 1$ and $3 \le t \le 5$, and to the right for $1 \le t \le 3$. **(c)** $s(t) = 2t^2 - \frac{1}{3}t^3 - 3t$; signed area between $v(u)$ and the horizontal axis from 0 to t

(d)

The rightmost position: $s(0) = s(3) = 0$; the leftmost position: $s(5) = -\frac{20}{3}$. **21.** Cumulative growth between $t = 2$ and $t = 7$.
23. Cumulative change in biomass between $t = 1$ and $t = 6$.
25. $-\frac{2}{3}$ **27. (a)** $68 + \sin\left(\frac{\pi}{R}t\right)$ **(b)** 68

29. $f(x)$ is symmetric about the origin. **31.** average value of $f(x) = 2$; $f(1) = 2$ **33.** $\frac{1}{3}\pi hr^2$ **35.** $\frac{256}{15}\pi$ **37.** 2π **39.** $2\pi\sqrt{3}$
41. $\frac{2}{15}\pi$ **43.** $\frac{\pi}{2}\left(e^4 + e^{-4} - 2\right)$ **45.** $\pi\left(\frac{\pi}{2} - 1\right)$ **47.** $\frac{32}{5}\pi$
49. $\pi\ln 3$ **51.** $\frac{3}{10}\pi$ **53. (a)** $2\sqrt{5}$ **(b)** $2\sqrt{5}$
55. $\frac{8}{27}\left[(10)^{3/2} - \left(\frac{13}{4}\right)^{3/2}\right]$ **57.** $\frac{14}{3}$ **59.** $\int_{-1}^{1}\sqrt{1 + 4x^2}\,dx$
61. $\int_0^1\sqrt{1 + e^{-2x}}\,dx$ **63. (a)** $\frac{\pi}{2}$ **(b)** $\frac{\pi}{2}$ **65.** $f'(a) = \frac{1}{2}(e^a - e^{-a})$

■ **Chapter 6 Review Problems**

1. $3.38\frac{m^3}{s}$ **3. (a)**

(figure: $v(d)$ vs d; curves labeled $(3-d)^{1/5}$ and $(3-d)^{1/7}$, axis marks 1.2 and 4)

(b) $v(D) = 0$, $v'(d) < 0$ for $d \in [0, D)$ **5. (a)** $\bar{v} = v(d_1)$

■ **Section 7.1**

1. $\frac{2}{3}(x^2 + 3)^{3/2} + C$ **3.** $-\frac{6}{5}(1 - x^2)^{5/4} + C$ **5.** $\frac{5}{3}\sin(3x) + C$
7. $-\frac{7}{12}\cos(4x^3) + C$ **9.** $\frac{1}{2}e^{2x+3} + C$ **11.** $-e^{-x^2/2} + C$
13. $\frac{1}{2}\ln|x^2 + 4x| + C$ **15.** $3(x + 4) - 12\ln|x + 4| + C$
17. $\frac{2}{3}(x + 3)^{3/2} + C$ **19.** $\frac{2}{3}(2x^2 - 3x + 2)^{3/2} + C$
21. $-\frac{1}{4}\ln|1 + 4x - 2x^2| + C$ **23.** $\frac{1}{2}\ln|1 + 2x^2| + C$
25. $\frac{3}{2}e^{x^2} + C$ **27.** $-\cot(\ln x) + C$ **29.** $-\frac{2}{3\pi}\cos\left(\frac{3\pi}{2}x + \frac{\pi}{4}\right) + C$
31. $\frac{1}{2}\tan^2 x + C$ **33.** $\frac{1}{3}(\ln x)^3 + C$
35. $\frac{1}{15}(5 + x^2)^{3/2}(3x^2 - 10) + C$ **37.** $\ln|ax^2 + bx + c| + C$
39. $\frac{1}{n+1}[g(x)]^{n+1} + C$ **41.** $-e^{-g(x)} + C$ **43.** $\frac{1}{3}(10^{3/2} - 1)$
45. $\frac{7}{2025}$ **47.** $-e^{-9/2} + 1$ **49.** $\frac{3}{8}$ **51.** $\frac{1}{2}$ **53.** $4 + 3\ln 3$ **55.** $\frac{1}{2}$
57. $2(e^{-1} - e^{-3})$ **59.** $\ln|\sin x| + C$

■ **Section 7.2**

1. $x\sin x + \cos x + C$ **3.** $\frac{2}{3}x\sin(3x - 1) + \frac{2}{9}\cos(3x - 1) + C$
5. $-2x\cos(x - 1) + 2\sin(x - 1) + C$ **7.** $xe^x - e^x + C$
9. $x^2 e^x - 2xe^x + 2e^x + C$ **11.** $\frac{1}{2}x^2\ln|x| - \frac{1}{4}x^2 + C$
13. $\frac{1}{2}x^2\ln(3x) - \frac{1}{4}x^2 + C$ **15.** $x\tan x + \ln|\cos x| + C$
17. $\frac{1}{2}\left(\sqrt{3} - \frac{\pi}{3}\right)$ **19.** $2\ln 2 - 1$ **21.** $2\ln 4 - \frac{3}{2}$ **23.** $1 - 2e^{-1}$
25. $\frac{1}{2} + \frac{1}{4}(\sqrt{3} - 1)e^{\pi/3}$ **27.** $\frac{2e^{-3x}}{36+\pi^2}\left[\pi\sin\left(\frac{\pi}{2}x\right) - 6\cos\left(\frac{\pi}{2}x\right)\right] + C$
29. $\frac{1}{2}x[\sin(\ln x) - \cos(\ln x)] + C$
31. $\int\cos^2 x\,dx = \frac{1}{2}\sin x\cos x + \frac{1}{2}x + C$
33. (b) $\int\arcsin x\,dx = x\arcsin x + \sqrt{1 - x^2} + C$
35. (b) $\frac{1}{2}(\ln x)^2 + C$ **37. (b)** $-\frac{1}{3}x^2 e^{-3x} - \frac{2}{9}xe^{-3x} - \frac{2}{27}e^{-3x} + C$
39. $2\sqrt{x}\sin(\sqrt{x}) + 2\cos(\sqrt{x}) + C$ **41.** $-e^{-x^2/2}(2 + x^2) + C$
43. $\sin xe^{\sin x} - e^{\sin x} + C$ **45.** 2 **47.** $-\frac{1}{2} + 3\ln 3$
49. $-\frac{1}{2}xe^{-2x} - \frac{1}{4}e^{-2x} + C$ **51.** $\ln|\sin x| + C$ **53.** $-\cos(x^2) + C$
55. $\frac{1}{4}\tan^{-1}\left(\frac{x}{4}\right) + C$ **57.** $x - 3\ln|x + 3| + C$
59. $\frac{1}{2}\ln|x^2 + 3| + C$ **61.** $\int\ln x\,dx = x\ln x - x + C$
63. $\frac{4}{9}(x - 2)^{9/4} + \frac{8}{5}(x - 2)^{5/4} + C$ **65.** $2e^2$ **67.** $\frac{\pi}{2}$ **69.** $\frac{1}{2}$

■ **Section 7.3**

1. $2x + 1 - \frac{3}{x+1}$ **3.** $3x - 2 + \frac{2x}{x^2+1}$ **5.** $\frac{5}{x+1} - \frac{3}{x}$ **7.** $\frac{2}{x} - \frac{1}{x-3} + \frac{3}{x+1}$
9. $\frac{2}{x-1} + \frac{3}{x+1}$ **11.** $\frac{3}{x-5} + \frac{1}{x+2}$ **13.** $\frac{1}{2}\ln|x - 2| - \frac{1}{2}\ln|x| + C$
15. $\frac{1}{4}\ln|x - 3| - \frac{1}{4}\ln|x + 1| + C$
17. $\frac{3}{2}\ln|x + 2| + \frac{1}{x} - \frac{1}{2}\ln|x| + C$
19. $-\tan^{-1}x + \frac{1}{2}\ln|x^2 + 4| + C$ **21.** $2\tan^{-1}x + \frac{3}{2(x^2+1)} + C$
23. $\tan^{-1}(x - 1) + C$ **25.** $\frac{1}{3}\tan^{-1}\left(\frac{x-2}{3}\right) + C$ **27.** $\frac{1}{5}\ln\left|\frac{x-3}{x+2}\right| + C$
29. $\frac{1}{6}\ln\left|\frac{x-3}{x+3}\right| + C$ **31.** $\frac{1}{3}\ln\left|\frac{x-2}{x+1}\right| + C$
33. $x - 5\ln|x + 2| + 2\ln|x + 1| + C$ **35.** $x + 2\ln\left|\frac{x-2}{x+2}\right| + C$
37. $2 - \ln 5 + \ln 3$ **39.** $\frac{1}{2}\ln 2$ **41.** $-\ln 2$ **43.** $\frac{\pi}{4} - \frac{1}{2}\ln 2$

45. $\frac{1}{1+x} + \ln\left|\frac{x}{x+1}\right|$ **47.** $-\frac{2}{x+1} + \ln\left|\frac{x+1}{x-1}\right|$

49. $\frac{1}{108} \ln\left|\frac{x+3}{x-3}\right| - \frac{1}{36}\left(\frac{1}{x+3} + \frac{1}{x-3}\right) + C$ **51.** $-\frac{1}{x} - \tan^{-1} x$

■ Section 7.4

1. infinite interval; $\frac{1}{2}$ **3.** infinite interval; π **5.** infinite interval; 2 **7.** infinite interval; 2 **9.** infinite interval; 0 **11.** integrand discontinuous; 6 **13.** integrand discontinuous; 2 **15.** integrand discontinuous at $x = 0$; -2 **17.** infinite interval; $\frac{1}{2}$
19. integrand discontinuous at $x = 0$; integral divergent
21. integrand discontinuous at $x = 1$; 0 **23.** infinite interval; integral divergent **25.** infinite interval; integral divergent
27. integrand discontiuous at $x = \pm 1$; 0 **29.** integrand discontiuous at $x = \pm 1$; integral divergent **31.** $c = 3$
35. **(b)** $0 \le \int_1^\infty e^{-x^2}\, dx \le \lim_{z\to\infty} \int_1^z e^{-x}\, dx = e < \infty$
37. **(b)** $\int_1^\infty \frac{1}{\sqrt{1+x^2}}\, dx \ge \lim_{z\to\infty} \frac{1}{2}\int_1^z \frac{1}{x}\, dx = \infty$; divergent
39. For $x \ge 1$: $0 \le e^{-x^2/2} \le e^{-x}$; convergent **41.** For $x \ge 1$: $\frac{1}{\sqrt{x+1}} \ge \frac{1}{2\sqrt{x}}$; divergent **43.** **(a)** Use l'Hospital's rule **(b)** Show that $\lim_{x\to\infty} \frac{\ln x}{\sqrt{x}} = 0$ and use a graphing calculator to show that, for $x > 74.2$, $2\ln x \le \sqrt{x}$ **(c)** For $x > 74.2$, use the fact that $e^{-\sqrt{x}} \le e^{-2\ln x} = \frac{1}{x^2}$ and conclude that the integral is convergent

■ Section 7.5

1. 2.328 **3.** 0.6292 **5.** $M_4 \approx 0.6912$; error ≈ 0.0019
7. $M_4 \approx 5.3838$; error ≈ 0.0505 **9.** $T_4 \approx 2.3438$
11. $T_3 \approx 0.6380$ **13.** $T_5 = 20.32$; error ≈ 0.32 **15.** $T_4 \approx 1.8195$; error ≈ 0.0661 **17.** $n = 82$ **19.** $n = 58$ **21.** $n = 92$ **23.** $n = 50$
25. **(a)** $M_5 = 0.245$; $T_5 = 0.26$; $\left|\int_0^1 x^3\, dx - M_5\right| = 0.005$;
$\left|\int_0^1 x^3\, dx - T_5\right| = 0.01$ **(c)** $0.6433 \le \int_0^1 \sqrt{x}\, dx \le 0.6730$

■ Section 7.6

1. $L(x) = 1 + 2x$ **3.** $L(x) = 1 + x$ **5.** $L(x) = \ln 2$
7. $P_5(x) = 1 - \frac{x^2}{2!} + \frac{x^4}{4!}$ **9.** $P_6(x) = \frac{120}{5!}x^5 = x^5$
11. $P_3(x) = \sqrt{2} + \frac{1}{2\sqrt{2}}x - \frac{1}{16\sqrt{2}}x^2 + \frac{1}{64\sqrt{2}}x^3$; $P_3(0.1) \approx 1.4491$;
$f(0.1) = \sqrt{2.1}$; $|f(0.1) - P_3(0.1)| \approx 3.34 \times 10^{-7}$
13. $P_5(x) = x - \frac{x^3}{3!} + \frac{x^5}{5!}$; $P_5(1) \approx 0.8417$; $f(1) \approx 0.8415$;
$|P_5(1) - f(1)| \approx 1.96 \times 10^{-4}$ **15.** $P_2(x) = x$; $P_2(0.1) = 0.1$;
$f(0.1) \approx 0.10033$; $|P_2(0.1) - f(0.1)| \approx 3.35 \times 10^{-4}$
17. **(a)** $P_3(x) = x - \frac{x^3}{3!}$ **(b)** $\lim_{x\to 0} \frac{P_3(x)}{x} = 0$ and $P_3(x)$ approximates $f(x) = \sin x$ at $x = 0$
19. $P_3(x) = 1 + \frac{1}{2}(x-1) - \frac{1}{8}(x-1)^2 + \frac{1}{16}(x-1)^3$;
$P_3(2) \approx 1.4375$; $f(2) \approx 1.4142$; $|P_3(2) - f(2)| \approx 0.023$
21. $P_3(x) = \frac{1}{2}\sqrt{3} - \frac{1}{2}(x - \frac{\pi}{6}) - \frac{1}{4}\sqrt{3}(x - \frac{\pi}{6})^2 + \frac{1}{12}(x - \frac{\pi}{6})^3$;
$P_3(\frac{\pi}{7}) \approx 0.9010$; $f(\frac{\pi}{7}) \approx 0.9010$; $|P_3(\frac{\pi}{7}) - f(\frac{\pi}{7})| \approx 6.861 \times 10^{-5}$
23. $P_3(x) = e^2 + e^2(x-2) + \frac{e^2}{2}(x-2)^2 + \frac{e^2}{6}(x-2)^3$;
$P_3(2.1) \approx 8.1661$; $f(2.1) \approx 8.1662$;
$|P_3(2.1) - f(2.1)| \approx 3.14 \times 10^{-5}$ **25.** $\int_0^1 x^3\, dx = 0.25$;
$M_5 \approx 0.245$; $T_5 \approx 0.26$ **27.** $n = 10$ **29.** $n = 2$ **31.** $P_2(x) = 0$;
error term: $f(x) - P_2(x) = f(x)$ **33.** **(b)** Use $x = 1$: $\tan^{-1} = \frac{\pi}{4}$

■ Section 7.7

1. $\frac{x}{2} + \frac{3}{4}\ln|2x - 3| + C$

3. $\frac{1}{2}\left(x\sqrt{x^2 - 16} - 16\ln|x + \sqrt{x^2 - 16}|\right) + C$ **5.** $6 - 16e^{-1}$
7. $\frac{2}{9}e^3 + \frac{1}{9}$ **9.** $\frac{1}{2}e^{\pi/6} - \frac{1}{4}(\sqrt{3} - 1)$ **11.** $-2e^{-x/2}(x^2 + 4x + 7) + C$
13. $\frac{1}{20}[10x - 6 + \sin(10x - 6)] + C$

15. $x\sqrt{\frac{9}{4} + x^2} + \frac{9}{4}\left(\ln|x + \sqrt{\frac{9}{4} + x^2}|\right) + C$
17. $\frac{4e^{2x+1}}{16+\pi^2}\left[2\sin\left(\frac{\pi}{2}x\right) - \frac{\pi}{2}\cos\left(\frac{\pi}{2}x\right)\right] + C$ **19.** $2\ln 2$
21. $\frac{x}{2}(\sin(\ln(3x)) + \cos(\ln(3x))) + C$

■ Chapter 7 Review Problems

1. $-\frac{1}{9}(1 - x^3)^3 + C$ **3.** $-2e^{-x^2} + C$
5. $\frac{6}{7}(1 + \sqrt{x})^{7/3} - \frac{3}{2}(1 + \sqrt{x})^{4/3} + C$ **7.** $\frac{1}{6}\tan(3x^2) + C$
9. $\frac{1}{2}x^2 \ln x - \frac{1}{4}x^2 + C$ **11.** $\tan x \ln(\tan x) - \tan x + C$
13. $\frac{1}{2}\tan^{-1}\left(\frac{x}{2}\right) + C$ **15.** $-\ln|\cos x| + C$
17. $\frac{e^{2x}}{5}(2\sin x - \cos x) + C$ **19.** $2e^{x/2} + C$
21. $-\frac{1}{2}\sin x \cos x + \frac{x}{2} + C$ **23.** $\ln|\frac{x-1}{x}| + C$
25. $x - 5\ln|x + 5| + C$ **27.** $\ln|x + 5| + C$
29. $\frac{1}{2}x^2 + 3x + 4\ln|x - 1| + C$ **31.** $4 + \ln 3$ **33.** $1 - e^{-1/2}$
35. $\frac{\pi}{8}$ **37.** 4 **39.** $\frac{\pi}{6}$ **41.** divergent **43.** divergent **45.** 2
47. $-\frac{1}{4}$ **49.** $e - e^{\sqrt{2}/2}$ **51.** **(a)** $M_4 = 0.625$ **(b)** $T_4 = 0.75$
53. **(a)** $M_5 \approx 0.6311$ **(b)** $T_5 \approx 0.6342$ **55.** $P_3(x) = 2x - \frac{4}{3}x^3$
57. $P_3(x) = (x - 1) - \frac{1}{2}(x - 1)^2 + \frac{1}{3}(x - 1)^3$
59.

■ Section 8.1

1. $y = \frac{1}{2}x^2 - \cos x + 1$ **3.** $y = \ln x$ for $x > 0$
5. $x(t) = 2 - \ln(1 - t)$ for $t < 1$ **7.** $s(t) = \frac{2}{9}(3t + 1)^{3/2} + \frac{7}{9}$ for $t \ge -\frac{1}{3}$ **9.** $V(t) = \sin t + 5$ **11.** $y = 2e^{3x}$ **13.** $x(t) = 5e^{2-2t}$
15. $h(s) = \frac{1}{2}(9e^{2s} - 1)$ **17.** $N(t) = 20e^{0.3t}$, $N(5) = 20e^{1.5} \approx 90$
19. **(a)** $N(t) = Ce^{rt}$ **(b)** $\log N(t) = \log C + (r \log e)t$. To determine r, graph $N(t)$ on a semilog graph; the slope is then $r \log e$. **(c)** 1. Obtain data at various points in time. 2. Plot on semilog paper. 3. Determine the slope of the resulting straight line. 4. Divide the slope by $\log e$; this yields r.
21. **(a)** $N(t) = \frac{100}{10-t}$ **(b)** $\lim_{t\to 10^-} N(t) = \infty$

23. **(a)** $L_\infty = 123$; $k = \frac{1}{27}\ln\frac{244}{123} \approx 0.0254$
(b) $L(10) = 123 - 122e^{-0.254} \approx 28.37$ in.
(c) $t = \frac{1}{0.0254}\ln\frac{122}{12.3} \approx 90.33$ months **25.** $y = \frac{2}{3e^{-x}-2}$
27. $y = \frac{5}{1+4e^{5x}}$ **29.** $y = \frac{3}{1-\frac{2}{3}e^{-6(x-1)}}$ **31.** $y = \frac{1}{\frac{1}{C}e^{-x}-1}$
33. $y = -1 \pm (-2(x + C))^{-1/2}$ **35.** **(b) (i)** $y = \frac{2-2e^{4x}}{1+e^{4x}}$ **(ii)** $y = 2$
(iii) $y = \frac{\frac{2}{3}e^{4x}+2}{1-\frac{1}{3}e^{4x}}$ **37.** $N(t) = \frac{200}{1+3e^{-0.34t}}$, $\lim_{t\to\infty} N(t) = 200$
39. **(a)** $N(t) = \frac{50}{1+4e^{-1.5t}}$ **(b)** $N(t) = \frac{50}{1-\frac{4}{9}e^{-1.5t}}$ **(c)**

(d) $\lim_{t\to\infty} N(t) = 50$ in both (a) and (b)
41. **(a)** $\frac{dN}{dt} = 5N\left(1 - \frac{N}{30}\right)$ **(b)**

43. **(a)** $p(t) = \frac{1}{\frac{1-p_0}{p_0}e^{-st/2}+1}$, $t \ge 0$ **(b)** $t = \frac{2}{0.01}\ln 9 \approx 439.4$
(c) $\lim_{t\to\infty} p(t) = 1$, which means that eventually the population will consist only of A_1A_1 types. **45.** $y = \sqrt{x^2 + 2x + 4}$

47. $y = -1 + 3\exp[1 - e^{-x}]$ **49.** $y = 6x - 7$

51. $r(t) = \exp[1 - e^{-t}]$ **53.** $\frac{dc}{c} = k\frac{dm}{m}$ **55.** $\frac{dy}{dx} = \frac{1}{7.7}\frac{y}{x}$

57. $N(t) = 5\exp\left(\frac{1}{\pi} + 2t - \frac{1}{\pi}\cos(2\pi t)\right)$

■ Section 8.2

1. (a) $y = 0, 2$ **(b)**

$y(2 - y)$ — $y = 0$ is unstable;

$y = 2$ is locally stable. **(c)** Eigenvalue associated with $y = 0$ is
$2 > 0$; hence, $y = 0$ is unstable. Eigenvalue associated with $y = 2$
is $-2 < 0$; hence, $y = 2$ is locally stable.   **3. (a)** $y = 0, 1, 2$

(b) $y(y - 1)(y - 2)$ —

$g(y)$

$y = 0$ and $y = 2$ are unstable; $y = 1$ is locally stable.
(c) Eigenvalue associated with $y = 0$ is $2 > 0$; hence, $y = 0$ is
unstable. Eigenvalue associated with $y = 1$ is $-1 < 0$; hence,
$y = 1$ is locally stable. Eigenvalue associated with $y = 2$ is $2 > 0$;
hence, $y = 2$ is unstable.   **5. (a)** $\frac{dN}{dt} = 1.5N\left(1 - \frac{N}{100}\right)$

(b)

$1.5N\left(1 - \frac{N}{100}\right)$ —

$N = 0$ is unstable; $N = 100$ is locally stable **(c)** Eigenvalue
associated with $N = 0$ is $1.5 > 0$; hence, $N = 0$ is unstable.
Eigenvalue associated with $N = 100$ is $-1.5 < 0$; hence,
$N = 100$ is locally stable. Same results as in (b).

7. (a) $K = 2000$ **(b)** $t = \frac{1}{2}\ln 199 \approx 2.65$ **(c)** 2000

9. (a) $N \approx 52.79$ is unstable; $N \approx 947.21$ is locally stable
(b) The maximal harvesting rate is $rK/4$.

11.

$g(C)$ $\frac{q}{V}(C_I - C)$ —

Locally stable

C_I C

The equilibrium C_I is locally stable.   **13. (a)** $\frac{dC}{dt} = \frac{0.2}{400}(3 - C)$
(b) $C(t) = 3 - 3e^{-t/2000}, t \geq 0; \lim_{t\to\infty} C(t) = 3$ **(c)** $C = 3$ is
locally stable.   **15. (a)** Equilibrium concentration: $C_I = 254$
(b) $T_R = \frac{1}{0.37} \approx 2.703$ **(c)** $T_R = \frac{1}{0.37} \approx 2.703$ **(d)** They are the
same.   **17.** Use $T_R = \frac{V}{q}$.   **19.** $T_R = \frac{12.3 \times 10^9}{220}$ seconds ≈ 647.1 days;
$C(T_R) \approx 0.806 \frac{mg}{l}$ **21. (a)** $0.5p(1 - p) - 1.5p$ —

$g(p)$

$1\ p$

(b) $p = 0$ is locally stable **(c)** $g'(0) = -1 < 0$, which implies that
0 is locally stable.   **23. (a)** $\frac{dp}{dt}$ describes the rate of change of $p(t)$;
$cp(1 - p - D)$ describes the colonization of vacant undestroyed
patches; $-mp$ describes extinction.   **(b)**

p_2

$1 - D - \frac{m}{c}$ —

$1 - \frac{m}{c}$

$1 - \frac{m}{c}$ D

(c) $D < 1 - \frac{m}{c}$; $p_1 = 0$ is unstable; $p_2 = 1 - D - \frac{m}{c}$ is locally
stable.   **25. (a)** $N = 0$, $N = 17$, and $N = 200$ **(b)** $N = 0$ is locally

stable; $N = 17$ is unstable; $N = 200$ is locally stable.

(c)

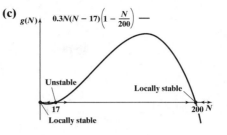

$g(N)$ $0.3N(N - 17)\left(1 - \frac{N}{200}\right)$ —

Unstable   Locally stable

17 $200\ N$

Locally stable

■ Section 8.3

1. (a) $R_0 = 1.5 > 1$; the disease will spread. **(b)** $R_0 = \frac{1}{2} < 1$; the
disease will not spread.   **3.** $R_0 = 0.9999 < 1$; the disease will not
spread.

5. (a) $\frac{dN}{dt} = N_I - 5N - 0.02NX + X$

$\frac{dX}{dt} = 0.02NX - 2X$

(b) equilibrium: $(\hat{N}, \hat{X}) = (100, N_I - 500)$; this is a nontrivial
equilibrium, provided that $N_I > 500$.

7. (a) $\frac{dN}{dt} = 200 - N - 0.01NX + 2X$ **(b)**

$\frac{dX}{dt} = 0.01NX - 3X$

X $\frac{dX}{dt} = 0$

100

300 N

$\frac{dN}{dt} = 0$

(c) No nontrivial equilibria   **9. (a)** Equilibria: $(0, 0), (0, 2/3),$
$(1/2, 0)$ **(b)** Since $\frac{dp_2}{dt} < 0$ when $p_1 = 1/2$ and p_2 is small, species
2 cannot invade.   **11. (a)**

p_2

$\frac{4}{5}$

$\frac{1}{2}$ $\frac{2}{3}$ p_1

(b) equilibria: $(0, 0), (2/3, 0), (0, 4/5)$

■ Chapter 8 Review Problems

1. (a) $\frac{dT}{dt}$ is proportional to the difference between the
temperature of the object and the temperature of the
surrounding medium. **(b)** $t = \frac{1}{0.013}\ln\frac{9}{4} \approx 62.38$ minutes

3. (a) $N(t) = N(0)e^{r_e t}$ **(b)** $N(t) = \frac{K}{1 + \left(\frac{K}{N_0} - 1\right)e^{-r_I t}}$ **(c)** $r_e \approx 0.691;$
$K = 1001: r_I \approx 1.382; K = 10,000: r_I \approx 0.701$

5. (a) $1000\exp[-(\ln 10)e^{-2t}]$ — **(c)** $2N(\ln 1000 - \ln N)$ —

$N(t)$ $g(N)$

1000

100 $1000\ N$

t

$N = 0$ is unstable; $N = 1000$ is locally stable; K is the carrying
capacity.   **7. (b)**

X $\frac{dX}{dt} = 0$

$\frac{qk}{b-q}$ N_0 N

$\frac{dN}{dt} = 0$

(c) $\hat{N} = \frac{qk}{b-q} > 0$ if $b > q$, $\hat{X} = \left(k + \frac{qk}{b-q}\right)\frac{q}{b}\left(\frac{N_0(b-q)}{qk} - 1\right)$

■ Section 9.1

1. $\{(4, 3)\}$

3. No solution; the lines are parallel.

5. (a) $c = 10$ (b) $c \neq 10$ (c) No

7. Eliminate x_1 from the second equation and solve the system.
9. $x = 2/5, y = -11/5$ **11.** $x = 9/17, y = -5/17$
13. No solution. **15.** Infinitely many solutions:
$\{(x, y) : x = t, y = \frac{3}{2} - \frac{1}{2}t; t \in \mathbf{R}\}$ **17.** Zach bought five fish
and six plants. **19.** Eliminate x_1 from the second equation and
solve the system. **21.** $x = 1, y = -1, z = 0$ **23.** $x = 2, y = 0$,
$z = -3$ **25.** $x = 1, y = 1, z = -2$
27. $\{(x, y, z) : x = 2 - t, y = 1 + t, z = t, t \in \mathbf{R}\}$
29. underdetermined;
$\{(x, y, z) : x = 7 + t, y = t + 2, z = t, t \in \mathbf{R}\}$
31. overdetermined; no solution **33.** underdetermined;
$\{(x, y, z) : x = \frac{10}{3} + \frac{13}{9}t, y = \frac{2}{3} + \frac{5}{9}t, z = t, t \in \mathbf{R}\}$ **35.** 750 gr of
SL 24–4–8; 1000 gr of SL 21–7–12; $\frac{11,000}{17}$ gr of SL 17–0–0.

■ Section 9.2

1. $\begin{bmatrix} 1 & -3 \\ 0 & -9 \end{bmatrix}$ **3.** $D = \begin{bmatrix} 1 & 0 \\ 4 & 11 \end{bmatrix}$ **5.** Use the rules of matrix
addition to calculate the left-hand side and the right-hand side.
Then compare.

7. $\begin{bmatrix} 19 & 3 & 6 \\ 9 & 11 & 4 \\ 3 & -13 & -7 \end{bmatrix}$ **9.** $D = \begin{bmatrix} -4 & -2 & -7 \\ -5 & -1 & -3 \\ -1 & 5 & -1 \end{bmatrix}$

11. Use the rules of matrix addition to calculate the left-hand
side and the right-hand side. Then compare. **13.** Write A, B,
and C as generic $m \times n$ matrices, and use the rules of matrix
addition and the fact that $a_{ij} + b_{ij} = c_{ij}$ implies $a_{ij} = c_{ij} - b_{ij}$.

15. $A' = \begin{bmatrix} -1 & 2 \\ 0 & 1 \\ 3 & -4 \end{bmatrix}$ **17.** Write A and B as generic $m \times n$

matrices, and use the rules of transposition to show that the
left-hand side is equal to the right-hand side. **19.** Write A as a
generic $m \times n$ matrix, and use the rules of transposition to show
that the left-hand side is equal to the right-hand side.

21. (a) $\begin{bmatrix} -2 & -3 \\ 0 & 5 \end{bmatrix}$ (b) $\begin{bmatrix} 1 & 6 \\ 2 & 2 \end{bmatrix}$

23. $AC = \begin{bmatrix} -1 & -2 \\ 1 & 0 \end{bmatrix} CA = \begin{bmatrix} 1 & 4 \\ -1 & -2 \end{bmatrix} AC \neq CA$

25. $(A + B)C = AC + BC = \begin{bmatrix} 1 & -1 \\ 0 & -3 \end{bmatrix}$ **27.** 3×2

29. (a) 1×4 (b) 3×3 (c) 4×3

31. (a) $\begin{bmatrix} 7 & 5 & 9 & -1 \\ -4 & -2 & -6 & 0 \end{bmatrix}$ (b) $\begin{bmatrix} 1 & -1 \\ 2 & 4 \\ 0 & -6 \\ -1 & -3 \end{bmatrix}$

33. $A^2 = \begin{bmatrix} 3 & -1 \\ 1 & 8 \end{bmatrix}, A^3 = \begin{bmatrix} 7 & 6 \\ -6 & -23 \end{bmatrix}, A^4 = \begin{bmatrix} 8 & -11 \\ 11 & 63 \end{bmatrix}$

35. The powers alternate between B and I_2. **37.** Calculate AI_2
and I_2A. Compare to A.

39. $\begin{bmatrix} 2 & 3 & -1 \\ 0 & 2 & 1 \\ 1 & 0 & -2 \end{bmatrix} \begin{bmatrix} x_1 \\ x_2 \\ x_3 \end{bmatrix} = \begin{bmatrix} 0 \\ 1 \\ 2 \end{bmatrix}$

41. $\begin{bmatrix} 2 & -3 \\ -1 & 1 \\ 3 & 0 \end{bmatrix} \begin{bmatrix} x_1 \\ x_2 \end{bmatrix} = \begin{bmatrix} 4 \\ 3 \\ 4 \end{bmatrix}$

43. $AB = \begin{bmatrix} 1 & 0 \\ 0 & 1 \end{bmatrix}, BA = \begin{bmatrix} 1 & 0 \\ 0 & 1 \end{bmatrix}$ **45.** $A^{-1} = \begin{bmatrix} -3/5 & 1/5 \\ 2/5 & 1/5 \end{bmatrix}$

47. $A^{-1} = \begin{bmatrix} -3/5 & 1/5 \\ 2/5 & 1/5 \end{bmatrix}$ and show that $(A^{-1})^{-1} = A$.

49. C does not have an inverse. **51.** (a) $x_1 = 2, x_2 = 3$

(b) $A^{-1} = \begin{bmatrix} -1 & 0 \\ -2/3 & -1/3 \end{bmatrix}, x_1 = 2, x_2 = 3$

53. $\det A = 7$, A is invertible **55.** $\det A = 0$, A is not invertible
57. (a) $\det A = 0$, A is not invertible (b) $\begin{aligned} 2x + 4y &= b_1 \\ 3x + 6y &= b_2 \end{aligned}$

(c) solution: $\{(x, y) : x = \frac{3}{2} - 2t, y = t, t \in \mathbf{R}\}$ (d) The system

has no solutions when $\frac{b_1}{2} \neq \frac{b_2}{3}$

59. $A^{-1} = \begin{bmatrix} 1/5 & 1/5 \\ 3/5 & -2/5 \end{bmatrix}$ **61.** $A^{-1} = \begin{bmatrix} -1/21 & 4/21 \\ 5/21 & 1/21 \end{bmatrix}$

63. $\det A = 2$, A is invertible; $A^{-1} = \begin{bmatrix} 1 & 1/2 \\ 0 & 1/2 \end{bmatrix}, X = \begin{bmatrix} 0 \\ 0 \end{bmatrix}$

65. C^{-1} does not exist. $\{(x, y) : x = -3t, y = t, t \in \mathbf{R}\}$

67. $\begin{bmatrix} 1/4 & 1/4 & 0 \\ -1/8 & 3/8 & 1/2 \\ -3/8 & 1/8 & -1/2 \end{bmatrix}$ **69.** $\begin{bmatrix} -2/3 & -1/6 & -1/3 \\ 0 & -1/2 & 0 \\ -1/3 & 1/6 & 1/3 \end{bmatrix}$

71. $L = \begin{bmatrix} 0 & 3.2 & 1.7 \\ 0.2 & 0 & 0 \\ 0 & 0.7 & 0 \end{bmatrix}, N(2) = \begin{bmatrix} 2232 \\ 580 \\ 280 \end{bmatrix}$

73. $L = \begin{bmatrix} 0 & 0 & 4.6 & 3.7 \\ 0.7 & 0 & 0 & 0 \\ 0 & 0.5 & 0 & 0 \\ 0 & 0 & 0.1 & 0 \end{bmatrix}, N(2) = \begin{bmatrix} 1242 \\ 934 \\ 525 \\ 25 \end{bmatrix}$

75. four age classes; 60% of one-year olds survive until the end
of the next breeding season; 2 is the average number of female
offspring of a two-year-old. **77.** four age classes; 20% of
two-year olds survive until the end of the next breeding season;
2.5 is the average number of female offspring of a one-year-old.

79. $q_0(t)$ and $q_1(t)$ seem to converge to 2.3; it appears that 74%
of females will be age 0 in the stable age distribution.

81. $q_0(t)$ and $q_1(t)$ oscillate between 0.4 and 3.

■ Section 9.3

1. (a) $\begin{bmatrix} 2(x_1 + y_1) + (x_2 + y_2) \\ 3(x_1 + y_1) + 4(x_2 + y_2) \end{bmatrix}$ (b) $\begin{bmatrix} 2\lambda x_1 + \lambda x_2 \\ 3\lambda x_1 + 4\lambda x_2 \end{bmatrix}$

3. length: $2\sqrt{2}$, angle: $\frac{\pi}{4}$

5. length: 3, angle: $\frac{\pi}{2}$

7. length: 2, angle: $\frac{5\pi}{6}$ **9.** $\begin{bmatrix} \sqrt{3} \\ 1 \end{bmatrix}$

11. $\begin{bmatrix} \cos 120° \\ \sin 120° \end{bmatrix}$ **13.** $x_1 = 3\cos 15°,\ x_2 = -3\sin 15°$

15. $x_1 = 5\cos 115°,\ x_2 = 5\sin 115°$ **17.** $\begin{bmatrix} 4 \\ 4 \end{bmatrix}$

19. $\begin{bmatrix} 1 \\ -3 \end{bmatrix}$ **21.** $\begin{bmatrix} 0 \\ 0 \end{bmatrix}$

23. $\begin{bmatrix} -4 \\ 2 \end{bmatrix}$ **25.** $\begin{bmatrix} 0 \\ -1 \end{bmatrix}$

27. $\begin{bmatrix} -1 \\ 1/4 \end{bmatrix}$ **29.** $\begin{bmatrix} 2 \\ 2 \end{bmatrix}$

31. $\begin{bmatrix} -2 \\ -6 \end{bmatrix}$ **33.** $\begin{bmatrix} 3 \\ 0 \end{bmatrix}$

35. leaves **x** unchanged **37.** counterclockwise rotation by $\theta = \frac{\pi}{2}$

39. counterclockwise rotation by $\theta = \frac{\pi}{6}$ **41.** $\begin{bmatrix} -\frac{1}{2}\sqrt{3} - 1 \\ -\frac{1}{2} + \sqrt{3} \end{bmatrix}$

43. $\begin{bmatrix} 5\cos(\pi/12) - 2\sin(\pi/12) \\ 5\sin(\pi/12) + 2\cos(\pi/12) \end{bmatrix}$ **45.** $\begin{bmatrix} \sqrt{2} + \sqrt{2}/2 \\ -\sqrt{2} + \sqrt{2}/2 \end{bmatrix}$

47. $\begin{bmatrix} 5\cos(-\pi/7) + 3\sin(-\pi/7) \\ 5\sin(-\pi/7) - 3\cos(-\pi/7) \end{bmatrix}$

49. $\lambda_1 = 2,\ \mathbf{v}_1 = \begin{bmatrix} 1 \\ 0 \end{bmatrix},\ \lambda_2 = -1,\ \mathbf{v}_2 = \begin{bmatrix} -1 \\ 1 \end{bmatrix}$

51. $\lambda_1 = 1,\ \mathbf{v}_1 = \begin{bmatrix} 1 \\ 0 \end{bmatrix},\ \lambda_2 = -1,\ \mathbf{v}_2 = \begin{bmatrix} 0 \\ 1 \end{bmatrix}$

53. $\lambda_1 = -2,\ \mathbf{v}_1 = \begin{bmatrix} 1 \\ 1 \end{bmatrix},\ \lambda_2 = -1,\ \mathbf{v}_2 = \begin{bmatrix} 2 \\ 3 \end{bmatrix}$

55. $\lambda_1 = 1,\ \mathbf{v}_1 = \begin{bmatrix} 1 \\ -1 \end{bmatrix},\ \lambda_2 = 4,\ \mathbf{v}_2 = \begin{bmatrix} 1 \\ 2 \end{bmatrix}$

57. $\lambda_1 = 4,\ \lambda_2 = 3$ **59.** $\lambda_1 = 1,\ \lambda_2 = 2$ **61.** $\lambda_1 = a,\ \lambda_2 = b$
63. The real parts of both eigenvalues are negative. **65.** The real parts of both eigenvalues are not negative. **67.** The real parts of both eigenvalues are negative. **69. (a)** $l_1: x_2 = 0$; l_2: $-3x_1 + x_2 = 0$; since l_1 and l_2 are not identical, $\mathbf{u}_1$ and $\mathbf{u}_2$ are linearly independent. **(b)** $\mathbf{x} = 2\mathbf{u}_1 - \mathbf{u}_2$ **(c)** $A^{20}\mathbf{x} = \begin{bmatrix} -1048574 \\ -3145728 \end{bmatrix}$

71. $\begin{bmatrix} -2 \\ 6 \end{bmatrix}$ **73.** $\begin{bmatrix} (-7)3^{20} + 4(-2)^{20} \\ (2)3^{20} - 4(-2)^{20} \end{bmatrix}$

75. (a) $\lambda_1 = 1 + \sqrt{2.2},\ \lambda_2 = 1 - \sqrt{2.2}$ **(b)** The larger eigenvalue corresponds to the growth rate. **(c)** 89.2% are in age class 0, and 10.8% are in age class 1 in the stable age distribution.
77. (a) $\lambda_1 = \frac{1}{2}(7 + \sqrt{50.2});\ \lambda_2 = \frac{1}{2}(7 - \sqrt{50.2})$ **(b)** The larger eigenvalue corresponds to the growth rate. **(c)** 98.6% are in age class 0, and 1.4% are in age class 1 in the stable age distribution.
79. (a) $\lambda_1 = \sqrt{0.45},\ \lambda_2 = -\sqrt{0.45}$ **(b)** The larger eigenvalue corresponds to the growth rate. **(c)** 88.17% are in age class 0, and 11.83% are in age class 1 in the stable age distribution.

■ **Section 9.4**

1. (a) $\begin{bmatrix} -1 \\ 5 \\ -1 \end{bmatrix}$ **(b)** $\begin{bmatrix} 2 \\ 8 \\ -2 \end{bmatrix}$ **(c)** $\begin{bmatrix} 6 \\ -3 \\ 0 \end{bmatrix}$ **3.** $\begin{bmatrix} 2 \\ -2 \end{bmatrix}$ **5.** $\begin{bmatrix} -1 \\ -2 \\ 5 \end{bmatrix}$

7. $\sqrt{10}$ **9.** $\sqrt{26}$ **11.** $\begin{bmatrix} 1/\sqrt{11} \\ 3/\sqrt{11} \\ -1/\sqrt{11} \end{bmatrix}$ **13.** $\begin{bmatrix} 1 \\ 0 \\ 0 \end{bmatrix}$ **15.** 1 **17.** 2

19. $\sqrt{5}$ **21.** $\sqrt{30}$ **23.** $\cos\theta = 1/\sqrt{50},\ \theta \approx 1.429$

25. $\cos\theta = 2/\sqrt{110},\ \theta \approx 1.379$ **27.** $\begin{bmatrix} 1 \\ 1 \\ 1 \end{bmatrix}$ **29.** $\begin{bmatrix} -2 \\ 1 \\ 1 \end{bmatrix}$

31. (a,b) $\overline{PQ} = 4,\ \overline{QR} = 3,\ \overline{PR} = 5$,
angle $QPR = \tan^{-1}(3/4) \approx 36.9°$,
angle $PRQ = 90° - \tan^{-1}(3/4) \approx 53.1°$,
angle $RQP = 90°$. **33. (a)** $\overline{PQ} = \sqrt{10},\ \overline{QR} = \sqrt{2},\ \overline{PR} = \sqrt{6}$
(b) angle $QPR = \cos^{-1}(7/\sqrt{60}) \approx 25.4° = 0.442$,
angle $PRQ = \cos^{-1}(-1/\sqrt{12}) \approx 106.8° = 1.864$,
angle $RQP = \cos^{-1}(3/\sqrt{20}) \approx 47.9° = 0.835$. **35.** $x + 2y = 4$
37. $4x + y = 2$ **39.** $-y + z = 1$ **41.** $x = 0$ **43.** $x = 1 + 2t$ and
$y = -1 + t$ for $t \in \mathbf{R}$ **45.** $x = -1 + t$ and $y = -2 - 3t$ for $t \in \mathbf{R}$
47. $2y - x - 5 = 0$ **49.** $y - x + 4 = 0$

51. $\begin{bmatrix} x \\ y \end{bmatrix} = \begin{bmatrix} 0 \\ 1/4 \end{bmatrix} + t\begin{bmatrix} 1 \\ -3/4 \end{bmatrix},\ t \in \mathbf{R}$

53. $\begin{bmatrix} x \\ y \end{bmatrix} = \begin{bmatrix} 0 \\ 3 \end{bmatrix} + t\begin{bmatrix} 1 \\ -2 \end{bmatrix},\ t \in \mathbf{R}$

55. $\begin{bmatrix} x \\ y \\ z \end{bmatrix} = \begin{bmatrix} 1 \\ -1 \\ 2 \end{bmatrix} + t\begin{bmatrix} 1 \\ -2 \\ 1 \end{bmatrix},\ t \in \mathbf{R}$

57. $\begin{bmatrix} x \\ y \\ z \end{bmatrix} = \begin{bmatrix} -1 \\ 3 \\ -2 \end{bmatrix} + t\begin{bmatrix} -1 \\ -2 \\ 4 \end{bmatrix},\ t \in \mathbf{R}$

59. $\begin{bmatrix} x \\ y \\ z \end{bmatrix} = \begin{bmatrix} 5 \\ 4 \\ -1 \end{bmatrix} + t\begin{bmatrix} 3 \\ 4 \\ -4 \end{bmatrix},\ t \in \mathbf{R}$

61. $\begin{bmatrix} x \\ y \\ z \end{bmatrix} = \begin{bmatrix} 2 \\ -3 \\ 1 \end{bmatrix} + t \begin{bmatrix} 7 \\ -5 \\ 0 \end{bmatrix}, t \in \mathbf{R}$ **63.** $(-5/2, -1/2, 9/2)$

65. $\begin{bmatrix} x \\ y \\ z \end{bmatrix} = \begin{bmatrix} 5 \\ -1 \\ 0 \end{bmatrix} + t \begin{bmatrix} 1 \\ 1 \\ 0 \end{bmatrix}, t \in \mathbf{R}$

■ Chapter 9 Review Problems

1. (a) $A\mathbf{x} = \begin{bmatrix} 0 \\ 2 \end{bmatrix} x_2$

(b) $\lambda_1 = -1, \mathbf{u}_1 = \begin{bmatrix} 1 \\ 0 \end{bmatrix}, \lambda_2 = 2, \mathbf{u}_1 = \begin{bmatrix} 1 \\ 3 \end{bmatrix}$

(c)   **(d)** $a_1 = 2/3, a_2 = 1/3$

3. Growth rate: $\lambda_1 = 1.75$; a stable age distribution: $\begin{bmatrix} 35 \\ 10 \end{bmatrix}$

5. $\begin{bmatrix} -2 & 5 \\ -2 & 9 \end{bmatrix}$ **7.** 1. Gaussian elimination; 2. Write in matrix form $AX = B$ and find the inverse of A. Then compute $X = A^{-1}B$ **9.** $a = -3$ **11.** For $\frac{5}{23} < a \le 1$, the population will grow.

■ Section 10.1

1. $CO = HR \times SV$; $[CO] = $ liter; domain: $\{(HR, SV) : HR \ge 0, SV \ge 0\}$; range: $\{CO : CO \ge 0\}$.

3.   **5.** $\frac{4}{13}$

7. (a) -14, **(b)** 1   **9.** $e^{-1/10}$ **11.** $-e^2$
13. domain: $\{(x, y) : x \in \mathbf{R}, y \in \mathbf{R}\}$; range: $\{z : z \ge 0\}$; level curves: $x^2 + y^2 = c$, circle with radius $\sqrt{c}$ centered at $(0, 0)$
15. domain: $\{(x, y) : y > x^2, x \in \mathbf{R}\}$; range: $\{z : z \in \mathbf{R}\}$; level curves: $y = e^c + x^2$, parabolas shifted in $y > 0$ direction by e^c
17. domain: $\{(x, y) : x \in \mathbf{R}, y \in \mathbf{R}, x + y \ne 0\}$; range: $\mathbf{R}$; level curves: $y = \frac{1-c}{1+c}x$, straight lines through the origin with slope $(1 - c)/(1 + c)$ for $c \ne -1$. When $c = -1$, level curve: $x = 0$.
19. Figure 10.23   **21.** Figure 10.24   **23. (a)** level curve: $x^2 + y^2 = c$, circle centered at origin with radius $\sqrt{c}$; intersection with x–z plane: $z = x^2$, intersection with y–z plane: $z = y^2$
(b) $c = 3$ intersection with x–z plane: $z = 4x^2$,
$c = 2$
$c = 1$ intersection with y–z plane: $z = y^2$

(c) $c = 3$ intersection with x–z plane: $z = \frac{1}{4}x^2$,
$c = 2$
$c = 1$ intersection with y–z plane: $z = y^2$

(d) The intersection with the y–z plane is always $z = y^2$. When $a = 1$, the resulting surface could be obtained by rotating the curve $z = y^2$ about the z-axis. When $0 < a < 1$, the resulting surface is still a paraboloid, but longer along the x-axis than along the y-axis. When $a > 1$, the paraboloid is longer along the y-axis than along the x-axis.   **25.** Day 180: 22 m; day 200: 18 m; day 220: 14 m

■ Section 10.2

1. 1   **3.** 2   **5.** 18   **7.** $-\frac{1}{2}$ **9.** 1   **11.** $-\frac{3}{2}$ **13.** $\frac{2}{3}$ **15.** Along positive x-axis: 1; along positive y-axis: -2 **17.** Along x-axis: 0; along y-axis: 0; along $y = x$: 2   **19.** Along $y = mx, m \ne 0$: 2; along $y = x^2$: 1; the limit does not exist.
21. 1. $f(x, y)$ is defined at $(0, 0)$. 2. $\lim_{(x,y)\to(0,0)}(x^2 + y^2)$ exists. 3. $f(0, 0) = 0 = \lim_{(x,y)\to(0,0)}(x^2 + y^2)$ **23.** In Problem 17, we showed that $\lim_{(x,y)\to(0,0)} \frac{4xy}{x^2+y^2}$ does not exist. Hence, $f(x, y)$ is discontinuous at $(0, 0)$ **25.** In Problem 19, we showed that $\lim_{(x,y)\to(0,0)} \frac{2xy}{x^3+yx}$ does not exist. Hence, $f(x, y)$ is discontinuous at $(0, 0)$ **27. (a)** $h(x, y) = g[f(x, y)]$ with $f(x, y) = x^2 + y^2$ and $g(z) = \sin z$; **(b)** the function is continuous for all $(x, y) \in \mathbf{R}^2$ **29. (a)** $h(x, y) = g[f(x, y)]$ with $f(x, y) = xy$ and $g(z) = e^z$; **(b)** the function is continuous for all $(x, y) \in \mathbf{R}^2$
31. $\{(x, y) : (x - 1)^2 + (y + 1)^2 < 4\}$

33. The boundary is a circle with radius 3, centered at $(0, 2)$. The boundary is not included.   **35.** Choose $2\delta^2 = \epsilon$.

■ Section 10.3

1. $\frac{\partial f}{\partial x} = 2xy + y^2, \frac{\partial f}{\partial y} = x^2 + 2xy$ **3.** $\frac{\partial f}{\partial x} = \frac{3}{2}y\sqrt{xy} - \frac{2y}{3(xy)^{1/3}}$, $\frac{\partial f}{\partial y} = \frac{3}{2}x\sqrt{xy} - \frac{2x}{3(xy)^{1/3}}$ **5.** $\frac{\partial f}{\partial x} = \cos(x + y), \frac{\partial f}{\partial y} = \cos(x + y)$
7. $\frac{\partial f}{\partial x} = -4x\cos(x^2 - 2y)\sin(x^2 - 2y)$, $\frac{\partial f}{\partial y} = 4\cos(x^2 - 2y)\sin(x^2 - 2y)$ **9.** $\frac{\partial f}{\partial x} = e^{\sqrt{x+y}}\frac{1}{2\sqrt{x+y}}$, $\frac{\partial f}{\partial y} = e^{\sqrt{x+y}}\frac{1}{2\sqrt{x+y}}$ **11.** $\frac{\partial f}{\partial x} = e^x\sin(xy) + ye^x\cos(xy)$, $\frac{\partial f}{\partial y} = xe^x\cos(xy)$ **13.** $\frac{\partial f}{\partial x} = \frac{2}{2x+y}, \frac{\partial f}{\partial y} = \frac{1}{2x+y}$
15. $\frac{\partial f}{\partial x} = \frac{-2x}{(\ln 3)(y^2-x^2)}, \frac{\partial f}{\partial y} = \frac{2y}{(\ln 3)(y^2-x^2)}$ **17.** 6   **19.** $3e^5$ **21.** 1
23. $\frac{2}{9}$ **25.** $f_x(1, 1) = -2, f_y(1, 1) = -2$ **27.** $f_x(-2, 1) = -4$, $f_y(-2, 1) = 4$ **29. (a)** $\frac{\partial P_e}{\partial a} > 0$: the number of prey items eaten increases with increasing attack rate. **(b)** $\frac{\partial P_e}{\partial T} > 0$: the number of prey items eaten increases with increasing T.
31. $\frac{\partial f}{\partial x} = 2xz - y, \frac{\partial f}{\partial y} = z^2 - x, \frac{\partial f}{\partial z} = 2yz + x^2$
33. $\frac{\partial f}{\partial x} = 3x^2y^2z + \frac{1}{yz}, \frac{\partial f}{\partial y} = 2x^3yz - \frac{x}{y^2z}, \frac{\partial f}{\partial z} = x^3y^2 - \frac{x}{yz^2}$
35. $\frac{\partial f}{\partial x} = e^{x+y+z}, \frac{\partial f}{\partial y} = e^{x+y+z}, \frac{\partial f}{\partial z} = e^{x+y+z}$
37. $\frac{\partial f}{\partial x} = \frac{1}{x+y+z}, \frac{\partial f}{\partial y} = \frac{1}{x+y+z}, \frac{\partial f}{\partial z} = \frac{1}{x+y+z}$
39. $2y$ **41.** e^y **43.** $2\sec^2(u + w)\tan(u + w)$ **45.** $-6x\sin y$
47. $\frac{2}{(x+y)^3}$ **49. (a)** $\frac{\partial f}{\partial N} > 0$: the number of prey encounters per

predator increases as the prey density increases. **(b)** $\frac{\partial f}{\partial T} > 0$: the function increases as the time for search increases. **(c)** $\frac{\partial f}{\partial T_h} < 0$: the function decreases as the handling time T_h increases.

(d)

$y = f(N, 2.4, 0.2) = \dfrac{1.536\,N^2}{1 + 0.5N + 0.16N^2}$

51.

■ Section 10.4

1. $8 = 6x + 4y - z$ **3.** $z = -2x - y - 2$ **5.** $z - y = 0$
7. $z - 2ex = -e$ **9.** $z = x + y - 1$ **11.** $f(x, y)$ is defined in an open disk centered at $(1, 1)$ and is continuous at $(1, 1)$.
13. $f(x, y)$ is defined in an open disk centered at $(0, 0)$ and is continuous at $(0, 0)$.   **15.** $f(x, y)$ is defined in an open disk centered at $(-1, 2)$ and is continuous at $(-1, 2)$.
17. $L(x, y) = x - 3y$ **19.** $L(x, y) = \frac{1}{2}x + 2y + \frac{1}{2}$
21. $L(x, y) = x + y$ **23.** $L(x, y) = x + \frac{1}{2}y - \frac{3}{2} + \ln 2$
25. $L(x, y) = 1 + x + y$, $L(0.1, 0.05) = 1.15$, $f(0.1, 0.05) \approx 1.1618$ **27.** $L(x, y) = 2x - 3y - 2$, $L(1.1, 0.1) = -0.1$, $f(1.1, 0.1) \approx -0.0943$
29. $Df(x, y) = \begin{bmatrix} 1 & 1 \\ 2x & -2y \end{bmatrix}$ **31.** $Df(x, y) = \begin{bmatrix} e^{x-y} & -e^{x-y} \\ e^{x+y} & e^{x+y} \end{bmatrix}$
33. $Df(x, y) = \begin{bmatrix} -\sin(x - y) & \sin(x - y) \\ -\sin(x + y) & -\sin(x + y) \end{bmatrix}$
35. $Df(x, y) = \begin{bmatrix} 4xy + 1 & 2x^2 - 3 \\ e^x \sin y & e^x \cos y \end{bmatrix}$
37. $L(x, y) = \begin{bmatrix} 4x + 2y - 4 \\ -x - y + 3 \end{bmatrix}$ **39.** $L(x, y) = \begin{bmatrix} e(2x - y) \\ 2x - y - 1 \end{bmatrix}$
41. $L(x, y) = \begin{bmatrix} x - y + 1 \\ y - x + 1 \end{bmatrix}$
43. $L(1.1, 1.9) = \begin{bmatrix} -0.9 \\ 9.8 \end{bmatrix}$, $f(1.1, 1.9) \approx \begin{bmatrix} -0.88 \\ 9.83 \end{bmatrix}$
45. $L(1.9, -3.1) = \begin{bmatrix} 25 \\ -22.4 \end{bmatrix}$, $f(1.9, -3.1) \approx \begin{bmatrix} 25 \\ -22.382 \end{bmatrix}$

■ Section 10.5

1. $18 \ln 2 + 8$ **3.** $\dfrac{\frac{\pi}{3} + \frac{\sqrt{3}}{4}}{\sqrt{\frac{\pi^2}{9} + \frac{3}{4}}}$ **5.** 0 **7.** $\frac{dz}{dt} = \frac{\partial f}{\partial x} u'(t) + \frac{\partial f}{\partial y} v'(t)$
9. $-\dfrac{2x}{2y + x^2 + y^2}$ **11.** $-\dfrac{2x - 3y(x^2 + y^2)}{2y - 3x(x^2 + y^2)}$
13. $\frac{dy}{dx} = -\dfrac{1}{\sqrt{1 - x^2}}$ for $-1 \le x \le 1$
15. The growth rate decreases over time.
17. $\operatorname{grad} f = \begin{bmatrix} 3x^2 y^2 \\ 2x^3 y \end{bmatrix}$ **19.** $\operatorname{grad} f = \dfrac{1}{2\sqrt{x^3 - 3xy}} \begin{bmatrix} 3x^2 - 3y \\ -3x \end{bmatrix}$
21. $\operatorname{grad} f = \dfrac{\exp[\sqrt{x^2 + y^2}]}{\sqrt{x^2 + y^2}} \begin{bmatrix} x \\ y \end{bmatrix}$ **23.** $\operatorname{grad} f = \dfrac{x^2 - y^2}{x^2 + y^2} \begin{bmatrix} \frac{1}{x} \\ -\frac{1}{y} \end{bmatrix}$
25. $\frac{2}{3}\sqrt{3}$ **27.** $-\sqrt{2}$ **29.** $\frac{3}{2}\sqrt{10}$ **31.** $D_{\mathbf{u}} f(2, 1) = \dfrac{13}{\sqrt{2}}$
33. $D_{\mathbf{u}} f(1, 6) = -\dfrac{1}{4\sqrt{29}}$ **35.** $f(x, y)$ increases most rapidly in the direction $\begin{bmatrix} 5 \\ -3 \end{bmatrix}$ at the point $(-1, 1)$.   **37.** $f(x, y)$ increases most rapidly in the direction $\begin{bmatrix} 5/4 \\ -3/4 \end{bmatrix}$ at the point $(5, 3)$.
39. $\begin{bmatrix} 3/5 \\ 4/5 \end{bmatrix}$ **41.** $\dfrac{1}{\sqrt{733}} \begin{bmatrix} 2 \\ -27 \end{bmatrix}$
43. The amoeba will move in the direction $\begin{bmatrix} -4/25 \\ -4/25 \end{bmatrix}$.

■ Section 10.6

1. $f(x, y)$ has a local minimum at $(1, 0)$ **3.** $f(x, y)$ has saddle points at $(2, 4)$ and $(-2, 4)$ **5.** $f(x, y)$ has a saddle point at $(0, 3)$ **7.** $f(x, y)$ has a local maximum at $(0, 0)$.   **9.** $f(x, y)$ has saddle points at $(0, \pi/2 + k\pi)$ for $k \in \mathbf{Z}$ **11. (c)** Figure 10.65: $f(x, y)$ stays constant for fixed x; there is neither a maximum nor a minimum at $(0, 0)$. Figure 10.66: saddle point at $(0, 0)$, Figure 10.67: local minimum at $(0, 0)$.   **13.** Absolute maximum: $(1, -1)$; absolute minimum: $(-1, 1)$ **15.** Absolute maxima: $(1, 0)$ and $(-1, 0)$; absolute minima: $(0, 1)$ and $(0, -1)$
17. Absolute maxima: $(0, 0)$, $(1, 0)$, $(1, -2)$, and $(0, -2)$; absolute minimum: $(1/2, -1)$ **19.** Absolute maximum at $(2/3, 2/3)$; absolute minima occur at all points along the boundary of the domain.   **21.** Absolute minimum: $(-2, 0)$; absolute maximum: $(3, 0)$ **23.** Absolute minimum: $(-1/2, 1/2)$; absolute maximum: $(1/\sqrt{2}, -1/\sqrt{2})$ **25.** Yes.   **27.** Absolute maximum at $(N, P) = (1, 1)$.   **29.** Maximum volume is $(2\sqrt{2})^3$ m^3.   **31.** The minimum surface area is 216 m^2.   **33.** The minimum distance is $1/\sqrt{3}$.   **35. (a)** Use $p_3 = 1 - p_1 - p_2$ and $0 \le p_3 \le 1$.   **37.** Absolute maxima: $(-\sqrt{35}/6, 1/6)$, $(\sqrt{35}/6, 1/6)$; absolute minimum: $(0, -1)$ **39.** Absolute minimum: $(1/4, -1/8)$; no maxima   **41.** Absolute minimum: $(12/13, -8/13)$; no maxima.   **43.** Local minimum: $(0, 1/3)$; no absolute minima; absolute maxima: $(1/\sqrt{2}, 1/6)$, $(-1/\sqrt{2}, 1/6)$.
45. Absolute minima: $(1, 0)$ and $(-1, 0)$; no maxima.   **47.** Set $f(x, y) = xy$ and $g(x, y) = x + y - c$. Then $y = \lambda$ and $x = \lambda$ implies $x = y$.   **49.** The total length of the fence is 96 ft.
51. Largest possible area is 4.   **53.** Smallest perimeter is 4.
55. $r = \sqrt{A}$, $\theta = 2$, perimeter is $4\sqrt{A}$ **57.** Local minimum at $(2, 2)$; no absolute extrema.   **59.** *Hint*: Look at the relative positions of level curves and constraint.   **61. (a)** $3x_1 + 3x_2 = 10$,
(b) absolute maximum at $\left(\dfrac{65 - 40\sqrt{2}}{3}, \dfrac{40\sqrt{2} - 55}{3} \right)$.

■ Section 10.7

1. $N_t = 5, 7.5, 11.25, 16.875, 25.31, 37.97, 56.95, 85.43, 128.14, 192.22, 288.33$; $P_t = 0$ for $t = 0, 1, 2, \ldots, 10$ **3.** $N_t = b^t N_0$
5. $N_t = 5, 6.79, 9.89, 14.67, 21.86, 32.59, 48.51, 71.67, 102.92, 128.47, 68.40, 0.71, 0.02, 0.03, 0.04, 0.06$; $P_t = 5, 1.43, 0.57, 0.34, 0.30, 0.39, 0.76, 2.18, 9.18, 51.81, 248.67, 203.69, 2.09, 0.0022, 0, 0$
7. $N_t = 5, 7.5, 11.25, 16.88, 25.31, 37.97, 56.95, 85.43, 128.14, 192.22, 288.33$; $P_t = 0$ for $t = 0, 1, 2, \ldots 10$ **9.** $N_t = b^t N_0$
11. (rounded to the closest integer) $N_t = 100, 79, 37, 16, 10, 10, 13, 17, 24, 34, 47, 61, 67, 54, 32, 18, 13, 13, 16, 20, 27, 36, 45, 51, 48, 36$; $P_t = 50, 141, 164, 80, 27, 10, 5, 3, 3, 4, 8, 19, 49, 94, 99, 59, 27, 13, 8, 6, 7, 10, 17, 33, 58, 72$ **13. (a)**

(b) Increasing a increases the chances of escaping parasitism.
15. (a)

(b) Increasing k increases the chances of escaping parasitism.
17. Stable   **19.** Unstable   **21.** Unstable   **23.** Stable

25. Unstable **27.** Eigenvalues: $1/\sqrt{2}, -1/\sqrt{2}$ **29.** Eigenvalues: $0.5 + i0.5, 0.5 - i0.5$ **31.** For $0 < a < 1/2$, $(0, 0)$ is locally stable. **33.** $(0, 0)$ is unstable; $(1/6, 1/6)$ is locally stable. **35.** $(0, 0)$ is locally stable if $-1 < a < 1$. **37. (a)** If $r > 1/2$, then $(r - 1/2, r - 1/2)$ is an equilibrium. **(b)** For $1/2 < r < 3/2$, the equilibrium $(r - 1/2, r - 1/2)$ is locally stable. **39.** $(0, 0)$ is unstable; $((40 \ln 4)/3, 10 \ln 4)$ is unstable. **41.** $(0, 0)$ is unstable; $(1000, 750)$ is locally stable.

■ **Chapter 10 Review Problems**

1.

3. (a) $\frac{\partial A_i}{\partial F} > 0, \frac{\partial A_i}{\partial D} < 0$ **(b)** $\frac{\partial A_e}{\partial F} > 0$: area covered by introduced species increased with the amount of fertilizer added.
$\frac{\partial A_e}{\partial D} > 0$: area covered by introduced species increased with intensity of disturbance [note that this is the opposite of (a)].
(c) Fertilization had a positive effect in both cases. That is, fertilization increased the total area covered of both introduced and indigenous species. Intensity of disturbance had a negative effect on indigenous species and a positive effect on introduced species. **5.** $D\mathbf{f}(x, y) = \begin{bmatrix} 2x & -1 \\ 3x^2 & -2y \end{bmatrix}$ **7. (a)**

(b) Use (a): $r_{avg}^2 = \pi Dt$ and solve for D.
(c) r_{avg} = arithmetic average = $\frac{1}{N} \sum_{i=1}^{N} d_i$, and use formula for D in (b).

■ **Section 11.1**

1. $\frac{d\mathbf{x}}{dt} = \begin{bmatrix} 2 & 3 \\ -4 & 1 \end{bmatrix} \mathbf{x}(t)$ **3.** $\frac{d\mathbf{x}}{dt} = \begin{bmatrix} -2 & 0 & 1 \\ -1 & 0 & 0 \\ 1 & 1 & 1 \end{bmatrix} \mathbf{x}(t)$

5. $(1, 0): \begin{bmatrix} -1 \\ 1 \end{bmatrix}$; $(0, 1): \begin{bmatrix} 2 \\ 0 \end{bmatrix}$; $(-1, 0): \begin{bmatrix} 1 \\ -1 \end{bmatrix}$; $(0, -1): \begin{bmatrix} -2 \\ 0 \end{bmatrix}$;
$(1, 1): \begin{bmatrix} 1 \\ 1 \end{bmatrix}$; $(0, 0): \begin{bmatrix} 0 \\ 0 \end{bmatrix}$; $(-2, 1): \begin{bmatrix} 4 \\ -2 \end{bmatrix}$;

7. $(1, 0): \begin{bmatrix} 1 \\ -1 \end{bmatrix}$; $(0, 1): \begin{bmatrix} 3 \\ 2 \end{bmatrix}$; $(-1, 1): \begin{bmatrix} 2 \\ 3 \end{bmatrix}$; $(0, -1): \begin{bmatrix} -3 \\ -2 \end{bmatrix}$;
$(-3, 1): \begin{bmatrix} 0 \\ 5 \end{bmatrix}$; $(0, 0): \begin{bmatrix} 0 \\ 0 \end{bmatrix}$; $(-2, 1): \begin{bmatrix} 1 \\ 4 \end{bmatrix}$;

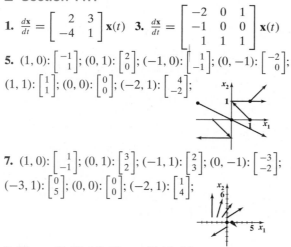

9. Figure 11.18: (d); Figure 11.19: (c); Figure 11.20: (b); Figure 11.21: (a)
11.

13. $\mathbf{x}(t) = c_1 e^{6t} \begin{bmatrix} 3 \\ 5 \end{bmatrix} + c_2 e^{-2t} \begin{bmatrix} 1 \\ -1 \end{bmatrix}$

15. $\mathbf{x}(t) = c_1 e^{-5t} \begin{bmatrix} -3 \\ 2 \end{bmatrix} + c_2 e^{6t} \begin{bmatrix} 1 \\ 3 \end{bmatrix}$

17. $\mathbf{x}(t) = c_1 e^{-2t} \begin{bmatrix} 1 \\ 1 \end{bmatrix} + c_2 e^t \begin{bmatrix} 0 \\ 1 \end{bmatrix}$

19. $\mathbf{x}(t) = e^{-3t} \begin{bmatrix} -5 \\ 4 \end{bmatrix} + e^{2t} \begin{bmatrix} 0 \\ 1 \end{bmatrix}$ **21.** $\mathbf{x}(t) = e^t \begin{bmatrix} 1 \\ 1 \end{bmatrix}$

23. $\mathbf{x}(t) = 2e^{2t} \begin{bmatrix} 7 \\ 2 \end{bmatrix} - e^{-3t} \begin{bmatrix} 1 \\ 1 \end{bmatrix}$ **25.** $\mathbf{x}(t) = \frac{13}{8} e^{-3t} \begin{bmatrix} 1 \\ -1 \end{bmatrix} - \frac{3}{8} e^{5t} \begin{bmatrix} 7 \\ 1 \end{bmatrix}$
27. (c) Use differentiation to find dx_1/dt and dx_2/dt.

29. unstable node **31.** saddle **33.** saddle **35.** unstable node
37. stable node **39.** saddle **41.** saddle **43.** unstable spiral
45. stable spiral **47.** stable spiral **49.** unstable spiral
51. neutral spiral **53.** neutral spiral **55.** stable spiral
57. saddle **59.** stable spiral **61.** saddle **63.** stable node
65. unstable spiral **67. (a)** $\lambda_1 = 0, \mathbf{v}_1 = \begin{bmatrix} 2 \\ -1 \end{bmatrix}$; $\lambda_2 = 6, \mathbf{v}_1 = \begin{bmatrix} 4 \\ 1 \end{bmatrix}$
(b) $\mathbf{x}(t) = c_1 \begin{bmatrix} 2 \\ -1 \end{bmatrix} + c_2 e^{6t} \begin{bmatrix} 4 \\ 1 \end{bmatrix}$
(c)

The direction vectors approach the eigenvector associated with the eigenvalue $\lambda_2 = 6$.

Section 11.2

1. $\frac{dx_1}{dt} = -0.55x_1 + 0.1x_2$ $(0, 0)$ is a stable node.

$\frac{dx_2}{dt} = 0.5x_1 - 0.12x_2$

3. $\frac{dx_1}{dt} = -2.5x_1 + 0.7x_2$ $(0, 0)$ is a stable node.

$\frac{dx_2}{dt} = 2.5x_1 - 0.8x_2$

5. $\frac{dx_1}{dt} = 0.1x_2$ All material gets stuck in compartment 1.

$\frac{dx_2}{dt} = -0.4x_2$

7. $\frac{dx_1}{dt} = -0.6x_1 + 1.2x_2$ $(0, 0)$ is a stable node.

$\frac{dx_2}{dt} = 0.1x_1 - 1.25x_2$

9. $a = 0.1, b = 0.3, c = 0.3, d = 0.2$

11. $a = 0, b = 0.1, c = 0.2, d = 0$

13. $a = 0.2, b = 1.1, c = 2.1, d = 1.2$

15. $a = 0.3, b = 0, c = 0.9, d = 0.2$

17. $a = 0, b = 0, c = 0.2, d = 0.3$ **19.** $x_1(t) = 4e^{-0.3t}$,

$x_2(t) = 4(1 - e^{-0.3t})$ **21. (a)** $a = 0.2, b = 0.1, c = 0, d = 0$

(b) The constant is the total area.

(d) $x_2(t) = 20 - x_1(t) = \frac{40}{3} + \frac{14}{3}e^{-0.3t}$, $\lim_{t\to\infty} x_1(t) = \frac{20}{3}$,

$\lim_{t\to\infty} x_2(t) = \frac{40}{3}$ **23.** $x(t) = 3\sin(2t)$ **25.** $\frac{dx}{dt} = v, \frac{dv}{dt} = 3x$

27. $\frac{dx}{dt} = v, \frac{dv}{dt} = x - v$

Section 11.3

1. saddle **3.** saddle **5.** unstable node **7.** $(0, 0)$: unstable node;

$(0, 4/5)$: saddle; $(1/2, 0)$: saddle; $(0.5, 0.3)$: stable node **9.** $(0, 0)$:

unstable node; $(0, 1)$: saddle; $(1, 0)$: saddle; $(1/2, 1)$: stable node

11. $(0, 0)$: saddle; $(1, 1)$: unstable spiral **13.** $a \geq 1/4$. Unstable

spiral for $a > 1/4$. **15. (a)**

(b) $(10/3, 10/3)$ is a stable node. **17.** $\mathrm{tr} < 0, \det = ?$ **19.** $\mathrm{tr} = ?$,

$\det < 0$ **21.** $\mathrm{tr} < 0, \det > 0$; equilibrium is locally stable

23. (a)

(b) $\mathrm{tr} < 0, \det > 0$: $(1, 1)$ is locally stable

Section 11.4

1. $\frac{dN_1}{dt} = 2N_1 \left(1 - \frac{N_1}{20} - \frac{N_2}{100}\right)$

$\frac{dN_2}{dt} = 3N_2 \left(1 - \frac{N_2}{15} - \frac{N_1}{75}\right)$

3. Species 2 excludes species 1. **5.** Founder control **7.** $(0, 0)$:

unstable (source); $(18, 0)$: unstable (saddle); $(0, 20)$: stable (sink)

9. $(0, 0)$: unstable (source); $(35, 0)$: stable (sink);

$(0, 40)$: stable (sink); $(85/11, 100/11)$: unstable (saddle)

11. $(\alpha_{12}, \alpha_{21}) = (1/4, 7/18)$

13.

15. (a) trivial equilibrium: $(0, 0)$; nontrivialequilibrium:

$(3/2, 1/4)$ **(b)** $\lambda_1 = 1, \lambda_2 = -3$: $(0, 0)$ is an unstable saddle

(c) $\lambda_1 = i\sqrt{3}, \lambda_2 = -i\sqrt{3}$: purely imaginary eigenvalues, linear

stability analysis cannot be used to infer stability of equilibrium.

(d)

17. (a) $\frac{dN}{dt} = 5N, N(t) = N(0)e^{5t}$; in the absence of the

predator, the insect species grows exponentially fast. **(b)** If

$P(t) > 0$, then $N(t)$ stays bounded.

(c)

By spraying the field, the solution moves to a different cycle; this

results in a much larger insect outbreak later in the year

compared to before the spraying. **19. (a)** When $P = 0$, then

$\frac{dN}{dt} = 3N\left(1 - \frac{N}{10}\right)$; equilibria: $\hat{N} = 0$ (unstable) and $\hat{N} = 10$

(locally stable). If $N(0) > 0$, then $\lim_{t\to\infty} N(t) = 10$ **(b)** $(0, 0)$:

unstable (saddle); $(10, 0)$: unstable (saddle); $(4, 0.9)$: stable spiral

(c)

21. (a)

(b) tr $Df(\hat{N}, \hat{P}) < 0$ and det $Df(\hat{N}, \hat{P}) > 0$: the nontrivial equilibrium is locally stable. **23.** $\hat{N} = \frac{d}{c}$ does not depend on a; hence, it remains unchanged if a changes. $\hat{P} = \frac{a}{b}(1 - \frac{d/c}{K})$ is an increasing function of a; hence, the predator equilibrium increases when a increases. **25.** $\hat{N} = \frac{d}{c}$ is a decreasing function of c; hence, the prey abundance decreases as c increases. $\hat{P} = \frac{a}{b}(1 - \frac{d/c}{K})$ is an increasing function of c; hence, the predator abundance increases as c increases. **27.** predation; locally stable **29.** mutualism; locally stable **31.** mutualism; unstable **33.** competition; unstable

35. $\begin{bmatrix} - & - \\ - & - \end{bmatrix}$; trace negative; determinant undetermined

37. $\begin{bmatrix} - & - \\ - & - \end{bmatrix}$; trace negative; determinant undetermined

39. $\begin{bmatrix} - & - \\ - & - \end{bmatrix}$; trace negative; determinant undetermined

41. If $a_{ii} < 0$, then the growth rate of species i is negatively affected by an increase in the density of species i. This is referred to as self regulation. **43. (a)** $(\hat{N}, \hat{P}) = (\frac{d}{c}, \frac{a}{b})$ **(b)** $\begin{bmatrix} 0 & -b\frac{d}{c} \\ c\frac{a}{b} & 0 \end{bmatrix}$
(c) $a_{11} = a_{22} = 0$: neither species has an effect on itself; $a_{12} = -b\frac{d}{c} < 0$: prey is affected negatively by predators; $a_{21} = c\frac{a}{b} > 0$: predators are affected positively by prey.

45.

47. $V(0) > 0.3$ **49.** $\frac{dc}{dt} = kab$
51. $e = [E], s = [S], c = [ES], p = [P], \frac{de}{dt} = -k_1es + k_2c,$
$\frac{ds}{dt} = -k_1es, \frac{dc}{dt} = k_1es - k_2c, \frac{dp}{dt} = k_2c$ **53.** $\frac{dx}{dt} + \frac{dy}{dt} = 0,$
$x(t) + y(t)$ is constant. **55.** $\frac{dx}{dt} + \frac{dy}{dt} + \frac{dz}{dt} = 0, x(t) + y(t) + z(t)$
is constant. **57. (c)**

(d) Since $\frac{dp}{dt} = f(s)$, the reaction rate is a function of s and hence, the availability of s determines the reaction rate.

59. (a) $\hat{s} = \frac{DK_m}{Yv_m - D}$; $\hat{s}$ is an increasing function of D.

(b) $q(s)$

The s-coordinate of the point of intersection of the graph of $q(s)$ and the horizontal line $f(s) = D/Y$ is the equilibrium.

61. $(4, 0)$: unstable; $(2, 2)$: stable

■ Chapter 11 Review Problems

1. $Z(t) = Z(0)e^{(r_1-r_2)t}$ **3. (a)**

(b) The nontrivial equilibrium is locally stable. **5. (a)** If $c_1 > m_1$, then there exists a nontrivial equilibrium in which species 1 has positive density and species 2 is absent.
7. (a) $\hat{x} = Y(s_0 - \hat{s}) > 0$ when $\hat{s} < s_0$ **(b)** $\frac{\partial \hat{s}}{\partial D} > 0; \frac{\partial \hat{s}}{\partial Y} < 0$

■ Section 12.1

1. 40 **3.** 120 **5.** 84 **7.** $4^{9749} \approx 3.04 \times 10^{5869}$ **9.** 120 **11.** 5040 **13.** 358,800 **15.** 2730 **17.** 6! **19.** 120 **21.** 120 **23.** 1365 **25.** 126 **27.** $\binom{1000}{20} \approx 3.4 \times 10^{41}$ **29. (a)** exactly two red balls: 10; exactly two blue balls: 6; one of each: 20 **(b)** total: 36 **31.** 168, 168 **33.** Ø, $\{a\}, \{b\}, \{c\}, \{a, b\}, \{a, c\}, \{b, c\}, \{a, b, c\}$; to form a subset, for each element, we need to decide whether it should be in the subset. There are $2^3 = 8$ choices. **35.** 12 **37.** 30 **39.** 31 **41.** $\binom{60}{20}\binom{40}{20}\binom{20}{20}$ **43.** $x^4 + 4x^3y + 6x^2y^2 + 4xy^3 + y^4$ **45.** $\binom{26}{4}\binom{26}{5}$ **47.** $4\binom{13}{2}\binom{4}{2}\binom{4}{2}\binom{11}{1}$ **49.** $\binom{13}{1}\binom{4}{4}\binom{12}{1}\binom{4}{1}$ **51.** 3!

■ Section 12.2

1. $\Omega = \{HHH, HHT, HTH, THH, HTT, THT, TTH, TTT\}$
3. $\Omega = \{(i, j) : 1 \le i < j \le 5\}$
5. $A \cup B = \{1, 2, 3, 5\}, A \cap B = \{1, 3\}$ **7.** $\{4, 6\}$ **9.** 0.6
11. 0.25 **13.** 0.3 **15.** 0.7 **17.** 0.5
19. (a) **(b)**

21. $\frac{3}{4}$ **23.** 3/8 **25.** $\frac{11}{36}$ **27.** $\frac{5}{12}$ **29.** $\frac{1}{4}$ **31.** $\frac{1}{2}$ **33.** $\frac{1}{8}$ **35.** 11/16
37. 1/4 **39.** 3/5 **41.** $\frac{1}{17}$ **43.** $\frac{12}{55}$ **45.** $1 - \frac{\binom{48}{4}}{\binom{52}{4}}$ **47.** $\frac{\binom{26}{13}}{\binom{52}{13}}$
49. $\frac{\binom{13}{2}\binom{4}{2}\binom{4}{2}\binom{11}{1}\binom{4}{1}}{\binom{52}{5}}$ **51. (a)** $\frac{\binom{N-100}{7}\binom{100}{3}}{\binom{N}{10}}$ **(b)** 333

■ Section 12.3

1. $\frac{13}{51}$ **3.** 13/50 **5.** $\frac{3}{5}$ **7.** $\frac{1}{2}$ **9.** $\frac{1}{6}$ **11.** $\frac{4}{7}$ **13.** 1/4 **15.** 0.8425
17. 0.7804 **19.** $\frac{5}{9}$
21. P(first card is an ace) $= P$(second card is an ace) $= \frac{1}{13}$
23. 0.3 **25.** $\frac{3}{4}$ **27.** A and B are independent. **29.** A and B are not independent. **31. (a)** $\frac{1}{8}$ **(b)** $\frac{7}{8}$ **(c)** $\frac{1}{2}$ **(d)** $\frac{7}{8}$ **33.** $\left(\frac{1}{4}\right)^{10}$
35. $1 - (0.9)^{10}$ **37.** 0.1624 **39.** $\frac{1}{3}$ **41.** $\frac{1}{3}$ **43. (a)** $\frac{1}{2}$ **(b)** $\frac{1}{2}$ **(c)** $\frac{1}{3}$

■ Section 12.4

1. $P(X = 0) = \frac{1}{4}, P(X = 1) = \frac{1}{2}, P(X = 2) = \frac{1}{4}$
3. $P(X = 0) = \frac{6}{36}, P(X = 1) = \frac{10}{36}, P(X = 2) = \frac{8}{36},$
$P(X = 3) = \frac{6}{36}, P(X = 4) = \frac{4}{36}, P(X = 5) = \frac{2}{36}$
5. $P(X = 0) = \frac{\binom{3}{0}\binom{2}{2}}{\binom{5}{2}}, P(X = 1) = \frac{\binom{3}{1}\binom{2}{1}}{\binom{5}{2}}, P(X = 2) = \frac{\binom{3}{2}\binom{2}{0}}{\binom{5}{2}}$
7. $P(X = 0) = \frac{\binom{13}{0}\binom{39}{3}}{\binom{52}{3}}, P(X = 1) = \frac{\binom{13}{1}\binom{39}{2}}{\binom{52}{3}},$
$P(X = 2) = \frac{\binom{13}{2}\binom{39}{1}}{\binom{52}{3}}, P(X = 3) = \frac{\binom{13}{3}\binom{39}{0}}{\binom{52}{3}}$

9. $F(x) = \begin{cases} 0, & x < -3 \\ 0.2, & -3 \le x < -1 \\ 0.5, & -1 \le x < 1.5 \\ 0.9, & 1.5 \le x < 2 \\ 1, & x \ge 2 \end{cases}$

11. $P(X = -1) = 0.2$, $P(X = 0) = 0.1$, $P(X = 1) = 0.4$, $P(X = 2) = 0.3$ **13. (a)** $N = 55$ **(b)** $\frac{28}{55}$

15. (a)

k	12	13	14	15	16	17	18	19	20	21
p_k	1/25	4/25	5/25	3/25	1/25	5/25	2/25	1/25	2/25	1/25

(b) 15.84

17. (a)

k	2	3	4	5	6	7	8	9	10
p_k	1/25	1/25	2/25	3/25	4/25	2/25	6/25	3/25	3/25

(b) 6.84

19. (a) -0.4 **(b)** 1.0 **(c)** 1.4 **21.** $E(X) = -0.1$, $\mathrm{var}(X) = 3.39$, s.d. $= \sqrt{3.39}$ **23. (a)** $E(X) = \frac{55}{10}$ **(b)** $\mathrm{var}(X) = 8.25$ **27. (a)** 0.1 **(b)** 0.5 **(c)** 0.4 **(d)** 0.2 **29. (a)** $E(X) = 0.75$, $EY = 0.3$ **(b)** $E(X + Y) = 1.05$ **(c)** $\mathrm{var}(X) = 1.7875$, $\mathrm{var}(Y) = 3.01$ **(d)** $\mathrm{var}(X + Y) = \mathrm{var}(X) + \mathrm{var}(Y) = 4.7975$ **31. (a)** Since $[X - E(X)]^2 \ge 0$, it follows that $E[X - E(X)]^2 \ge 0$; therefore, $\mathrm{var}(X) \ge 0$. **(b)** Since $\mathrm{var}(X) = E(X^2) - [E(X)]^2 \ge 0$, it follows that $E(X^2) \ge [E(X)]^2$. **33. (a)** $\binom{10}{5}(0.5)^{10}$ **(b)** $(0.5)^{10}\left[\binom{10}{8} + \binom{10}{9} + \binom{10}{10}\right]$ **(c)** $1 - (0.5)^{10}$

35. $P(X = k) = \binom{6}{k}\left(\frac{1}{6}\right)^k\left(\frac{5}{6}\right)^{6-k}$, $k = 0, 1, 2, \ldots, 6$ **37.** $14/64$

39. $\left(\frac{3}{5}\right)^3 + 3\left(\frac{3}{5}\right)\left(\frac{3}{5}\right)^2$ **41.** $(0.8)^{20}$ **43. (a)** 1 **(b)** $(10)(0.9)^{10}$

45. 12.5 **47. (a)** $\dfrac{\binom{24}{6}\binom{12}{4}}{\binom{36}{10}}$ **(b)** $\binom{10}{6}\left(\frac{2}{3}\right)^6\left(\frac{1}{3}\right)^4$

49. $\frac{30!}{10!14!6!}(0.2)^{10}(0.35)^{14}(0.45)^6$

51. $\frac{40!}{20!10!8!2!}\left(\frac{9}{16}\right)^{20}\left(\frac{3}{16}\right)^{10}\left(\frac{3}{16}\right)^8\left(\frac{1}{16}\right)^2$

53. $\frac{6!}{2!2!2!}\left(\frac{6}{24}\right)^2\left(\frac{8}{24}\right)^2\left(\frac{10}{24}\right)^2$ **55.** $\frac{23!}{5!12!6!}\left(\frac{1}{4}\right)^5\left(\frac{1}{2}\right)^{12}\left(\frac{1}{4}\right)^6$ **57.** $5/16$

59. $(3/4)^4$ **61.** $1/2, 1/4, 1/8$ **63.** $1/8$ **65.** $1/8$ **67.** $15/16$

69. $\left(\frac{14}{15}\right)^{19}$ **71.** $E(T) = 6$, $\mathrm{var}(T) = 30$ **73. (a)** $\left(\frac{9}{10}\right)^5\frac{1}{10}$ **(b)** $\frac{9}{10}\frac{8}{9}\frac{7}{8}\frac{6}{7}\frac{5}{6}\frac{1}{5}$ **75. (a)** $(1 - p)^{k-1}p$ **(b)** $\binom{k-1}{1}p^2(1 - p)^{k-2}$

77.

k	0	1	2	3
$P(X = k)$	e^{-2}	$2e^{-2}$	$2e^{-2}$	$\frac{4}{3}e^{-2}$

79. (a) $1 - 2e^{-1}$ **(b)** $e^{-1}\left(1 + \frac{1}{2} + \frac{1}{6}\right)$

81. $1 - e^{-1.5}\left[1 + 1.5 + \frac{(1.5)^2}{2} + \frac{(1.5)^3}{6}\right]$ **83.** $1 - 3e^{-2}$ **85.** e^{-7} **87.** $1 - e^{-0.5}$ **89.** $1 - 4e^{-3}$ **91. (a)** $18e^{-6}$ **(b)** $1/4, 1/2, 1/4$ **93.** $(2/3)^2$ **95.** $P(X = 0) \approx e^{-0.5}$ **97. (a)** 0.5819 **(b)** 0.5820 **99.** $e^{-1.5}$

■ Section 12.5

1. distribution function: $F(x) = 1 - e^{-3x}$ for $x \ge 0$, $F(x) = 0$ for $x \le 0$ **3.** $c = \frac{1}{\pi}$ **5.** $E(X) = \frac{1}{2}$, $\mathrm{var}(X) = \frac{1}{4}$ **7.** $E(X) = \frac{3}{2}$, $\mathrm{var}(X) = \frac{3}{4}$ **9. (b)** $E(X) = \frac{a-1}{a-2}$ **11. (d)**

$\frac{1}{\sqrt{2\pi}}e^{-(x-2)^2/2}$

13. 95%: $(7.4, 18.2)$, 99%: $(4.7, 20.9)$ **15.** 50% **17.** 99.5% **19.** 2.5% **21. (a)** 0.6915 **(b)** 0.383 **(c)** 0.1587 **(d)** 0.0668 **23. (a)** $x = 3.56$ **(b)** $x = 1.5$ **(c)** $x = 0.5$ **(d)** $x = 1.34$ **25. (a)** 0.0228 **(b)** $x = 628$ **27.** 0.76% **29.** 0.8185 **33.** $E(|X|) = 2/\sqrt{2\pi}$ **35. (a)** 0.0735 **(b)** 0.6068 **(c)** 88 **(d)** 56 **37.** 0.1 **39.** $a = 1, b = 7$ **41.** $E(Y) = \frac{3}{4}$ **43.** $HHTHHTHHHT$ **45. (a)** $(1 - x)^n$ **(b)** Use l'Hospital's rule on $\lim_{n \to \infty} \ln(1 - \frac{x}{n})^n$. **47.** $E(X) = \frac{1}{\lambda}$ **49.** $e^{-4/3}$ **51. (a)** $e^{-20/27}$ **(b)** $e^{-20/27}$ **53. (a)** e^{-4} **(b)** $8e^{-4}$ **(c)** $32e^{-8}$ **(d)** $\frac{1}{4}$ **55.** 0.25 hour **57. (a)** $1 - e^{-2/3}$ **(b)** $P(N(5) = 1) = \frac{5}{3}e^{-5/3}$ **59. (a)** $1 - e^{-3/5}$ **(b)** $e^{-1/5}$ **61. (a)** 5 years **(b)** $\ln 2/0.2$ years **63. (a)** $\exp\left[-(1.5 + 10e^{0.05} - 10)\right]$ **(b)** $\exp\left[-(2.1 + 10e^{0.07} - 10)\right] - \exp\left[-(3 + 10e^{0.1} - 10)\right]$ **65.** Solution of $1.2x + (0.6)e^{0.5x} - 0.6 - \ln 2 = 0$ is approximately 0.451. **67. (a)** $\exp\left[-(2 \times 10^{-5})\frac{(50)^{2.5}}{2.5}\right]$ **(b)** 0.1477 **69.** $x_m \approx 30.4$

■ Section 12.6

1. Exact probability: $e^{-3/2}$; Markov's inequality: $P(X \ge 3) \le \frac{2}{3}$ **5.** Exact: $P(|X| \ge 1) = 1/2$; Chebyshev's inequality: $P(|X| \ge 1) \le \frac{4}{3}$ **7.** $\frac{9}{25}$ **9.** $\frac{1}{n}\sum_{i=1}^n X_i$ converges to 0.9 as $n \to \infty$ **11.** Since $E(|X_i|) = \infty$, we cannot apply the law of large numbers as stated in Section 12.6. **13.** The sample size should be at least 380. **15.** 0.1587 **17. (a)** 0.0023 **(b)** 0.83 **19. (a)** 0.1114 **(b)** 0.1664 **(c)** 0.1679 **21. (a)** -11.2 **(b)** 0.579 **23.** 69 **25.** 385 **27. (a)** 0.3660 **(b)** 0.3679 **(c)** 0.243 **29. (a)** 0.1849 **(b)** 0.1755 **(c)** 0.1896 **31.** Likely not. **33. (a)** $0.6065, 0.3033, 0.0758$ **(b)** 0.8391 **35.** 0.1429 **37.** 0.9515

■ Section 12.7

1. median: 15; sample mean: 16.2; sample variance: 180.2 **3.** median: 35.5; sample mean: 36; sample variance: 43.1 **5.** $\overline{X} = 11.93$; $S^2 = 3.389$ **7.** $\overline{X} = 5.69$; $S^2 = 3.465$ **9.**

Sample	Sample Mean
(1, 1)	1.0
(1, 6)	3.5
(1, 8)	4.5
(6, 1)	3.5
(6, 6)	6.0
(6, 8)	7.0
(8, 1)	4.5
(8, 6)	7.0
(8, 8)	8.0

15. (a) approximately normal with mean 1/3 and variance 1/450 **(b)** approximately normal **17. (c)** true values: $\mu = 0.5$, $\sigma^2 = \frac{1}{12}$ **19.** $\overline{X} = 16.2$; S.E. $= 4.245$ **21.** $[-0.3993, 0.5213]$ **23.** $\hat{p} = 0.72$; $[0.651, 0.789]$ **25.** $y = 1.92x - 0.92$; $r^2 = 0.9521$ **31.** $y = 0.201x + 0.481$; $r^2 = 0.841$

■ Chapter 12 Review Problems

1. (a) 0.431 **3.** $168,168,000$ **5. (a)** $E(X) = 4.2$ **(b)** 0.0043 **(c)** 0.0215 **(d)** $1 - (0.0043)^5$ **7. (a)** 0.16 **(b)** 0.2 **9. (a)** $\mu = 170$, $\sigma = 6.098$ **(b)** 0.4013 **11.** 20% **13. (b)** $E(V) = \frac{n-1}{n}\sigma^2$

References

W. C. Allee (1931). *Animal Aggregations: A Study in General Sociology*. University of Chicago Press.

H. G. Andrewartha and L. C. Birch (1954). *The Distribution and Abundance of Animals*. University of Chicago Press.

M. Begon and G. A. Parker (1986). Should egg size and clutch size decrease with age? *Oikos* **47**: 293–302.

T. S. Bellows, Jr. (1981). The descriptive properties of some models for density dependence. *Journal of Animal Ecology* **50**: 139–156.

M. J. Benton (1997). Models for the diversification of life. *TREE* **12**: 490–495.

M. Benton and D. Harper (1997). *Basic Paleontology*. Addison-Wesley and Longman.

William H. Beyer, ed. (1991). *CRC Standard Probability and Statistics Tables and Formulae*. CRC Press.

P. J. Bohlen, P. M. Groffman, C. T. Driscoll, T. J. Fahey, and T. G. Siccama (2001). Plant-soil-microbial interactions in a northern hardwood forest. *Ecology* **82**: 965–978.

R. Borchert (1994). Soil and stem water storage determine phenology and distribution of tropical dry forest trees. *Ecology* **75**: 1437–1449.

T. Boulinier, J. D. Nichols, and J. E. Hines (2001). Forest fragmentation and bird community dynamics: Inference at regional scales. *Ecology* **82**: 1159–1169.

G. E. Briggs and J. B. S. Haldane (1925). A note on the kinematics of enzyme actions. *Biochemical Journal* **19**: 338–339.

J. L. Brooks and S. I. Dodson (1965). Predation, body-size and composition of plankton. *Science* **150**: 28–35.

M. J. W. Burke and J. P. Grime (1996). An experimental study of plant community invasibility. *Ecology* **77**: 776–790.

M. L. Cain, S. W. Pacala, J. A. Silander, Jr., and M.-J. Fortin (1995). Neighborhood models of clonal growth in the white clover *Trifolium repens*. *American Naturalist* **145**: 888–917.

G. S. Campbell (1986). *An Introduction to Environmental Biophysics*. Springer.

M. J. Crawley (1997). Plant-herbivore dynamics. In *Plant Ecology*, ed. M. J. Crawley, pp. 401–474. Blackwell Science.

J. W. Dalling, K. Winter, and J. D. Nason (2001). The unusual life history of *Alseis blackiana*: A shade-persistent pioneer tree? *Ecology* **82**: 933–945.

D. P. Dalzell (1944). On 22/7. *Journal of the London Mathematical Society* **19**: 133–134.

D. P. Dalzell (1971). On 22/7 and 355/113. *Eureka: The Archimedians' Journal* **34**: 1013.

C. Darwin (1859). *On the Origin of Species by Means of Natural Selection*. John Murray.

C. Darwin (1959). *The Origin of Species*. Reprinted by Penguin Classics (1985). Penguin Books.

D. L. DeAngelis (1992). *Dynamics of Nutrient Cycling and Food Webs*. Chapman & Hall.

W. R. DeMott, R. D. Gulati, and K. Siewertsen (1998). Effects of phosphorus-deficient diets on the carbon and phosphorus balance of *Daphnia magna*. *Limnology and Oceanography* **43**: 1147–1161.

A. M. de Roos (1996). A gentle introduction to physiologically structured population models. In *Structured-Population Models in Marine, Terrestrial, and Freshwater Systems*, eds. S. Tuljapurkar and H. Caswell. Chapman & Hall.

R. H. Dott, Jr., and R. L. Batten (1976). *Evolution of the Earth*, 2nd edition. McGraw-Hill.

J. E. Dowd and D. S. Riggs (1965). A comparison of estimates of Michaelis-Menten kinetic constants from various linear transformations. *Journal of Biological Chemistry* **240**: 863–869.

C. M. Duarte and S. Agustí (1998). The CO_2 balance of unproductive aquatic ecosystems. *Science* **281**: 234–236.

P. J. Dyck, J. L. Davies, D. M. Wilson, F. J. Service, L. J. Melton III, and P. C. O'Brien (1999). Risk factors for severity of diabetic polyneuropathy. *Diabetes Care* **22**: 1479–1486.

D. S. Falconer (1989). *Introduction to Quantitative Genetics*. Longman.

D. S. Falconer and T. F. C. Mackay (1996). *Introduction to Quantitative Genetics*, 4th edition. Longman.

R. Fitzhugh (1961). Impulses and physiological states in theoretical models of nerve membrane. *Biophysics Journal* **1**: 445–466.

D. J. Futuyama (1995). *Science on Trial*, 2nd edition. Sinauer.

P. Gaastra (1959). Mededelinger van de Landbouwhogeschool Te Wageningen, The Netherlands. In *Plant Physiology* (1978), eds. F. B. Salisbury and C. W. Ross. Wadsworth.

G. F. Gause (1934). *The Struggle for Existence*. Reprinted by Hafner (1964). Williams & Wilkins.

L. A. Gavrilov and N. S. Gavrilova (2001). The reliability theory of aging and longevity. *Journal of Theoretical Biology* **213**: 527–545.

Z. M. Gliwicz (1990). Food threshold and body size in caldocerans. *Nature* **343**: 683–740.

P. R. Grant (1982). Variation in the size and shape of Darwin's finch eggs. *Auk* **99**: 15–23.

P. R. Grant, I. Abbott, D. Schluter, R. L. Curry, and L. K. Abbott (1985). Variation in the size and shape of Darwin's finches. *Biological Journal of the Linnean Society* **25**: 1–39.

K. J. Griffiths (1969). The importance of coincidence in the functional and numerical responses of two parasites of the European pine sawfly, *Neodiprion sertifer*. *Canadian Entomologist* **101**: 673–713.

J. B. S. Haldane (1957). The cost of natural selection. *Journal of Genetics* **55**: 511–524.

D. J. Hall (1964). An experimental approach to the dynamics of a natural population of *Daphnia galeata mendota*. *Ecology* **45**: 94–111.

D. L. Hartl and A. G. Clark (1989). *Principles of Population Genetics*, 2nd edition. Sinauer.

R. W. Herschy (1995). *Streamflow Measurement*, 2nd edition. Taylor & Francis.

T. Hojo and K. Pearson (1931). Distribution of the median, quartiles and interquartile distance in samples from a normal population. *Biometrika* **23**: 315–363.

L. Holdridge, W. Grenke, W. Hatheway, T. Liang, and J. Tosi, Jr. (1971). *Forest Environments in Tropical Life Zones: A Pilot Study*. Pergamon Press.

C. S. Holling (1959). Some characteristics of simple types of predation and parasitism. *Canadian Entomologist* **91**: 385–398.

H. S. Horn (1971). The adaptive geometry of trees. *Monographs in Population Biology 3*. Princeton University Press.

S. B. Hulley and D. Grady (2004). The WHI estrogen-alone trial—do things look any better? *JAMA* **14**: 1769–1771.

C. E. Iselin, J. E. Robertson, and D. F. Paulson (1999). Radical perineal prostatectomy: Oncological outcome during a 20-year period. *Journal of Urology* **161**: 163–168.

Y. Iwasa, T. J. de Jong and P. G. L. Klinkhamer (1995). Why pollinators visit only a fraction of the open flowers on a plant and its consequence for fitness curves of plants. *Journal of Evolutionary Biology* **8**: 439–453.

T. H. Jukes and C. R. Cantor (1969). Evolution of protein molecules. In *Mammalian Protein Metabolism III*, ed. H. N. Munro, pp. 21–132. Academic Press.

P. M. Kareiva (1983). Local movement in herbivorous insects: Applying a passive diffusion model to mark-recapture field experiments. *Oecologia* **57**: 322–327.

W. O. Kermack and A. G. McKendrick (1927). Contributions to the mathematical theory of epidemics. *Proceedings of the Royal Society A* **115**: 700–721.

W. O. Kermack and A. G. McKendrick (1932). Contributions to the mathematical theory of epidemics. *Proceedings of the Royal Society A* **138**: 55–83.

W. O. Kermack and A. G. McKendrick (1933). Contributions to the mathematical theory of epidemics. *Proceedings of the Royal Society A* **141**: 94–122.

K. B. Krauskopf and D. K. Bird (1995). *Introduction to Geochemistry*, p. 268. McGraw-Hill.

H. E. Landsberg (1969). *Weather and Health, an Introduction to Biometerology*. Doubleday.

A. C. Leopold and P. E. Kriedemann (1975). *Plant Growth and Development*. McGraw-Hill.

P. Leslie (1945). On the use of matrices in certain population mathematics. *Biometrika* **33**: 183–212.

R. Levins (1969). Some demographic and genetic consequences of environmental heterogeneity for biological control. *Bulletin of the Entomological Society of America* **15**: 237–240.

R. Levins (1970). Community equilibria and stability, and an extension of the competitive exclusion principle. *American Naturalist* **104**: 413–423.

D. C. Lloyd (1987). Selection of offspring size at independence and other size-versus-number strategies. *American Naturalist* **129**: 800–817.

A. J. Lotka (1932). The growth of mixed populations: Two species competing for a common food supply. *Journal of the Washington Academy of Sciences* **22**: 461–469.

R. H. MacArthur and E. O. Wilson (1963). An equilibrium theory of insular zoogeography. *Evolution* **17**: 373–387.

T. F. C. Mackey (1984) Jumping genes meet abdominal bristles: Hybrid dysgenesis-induced quantitative variation in *Drosophila melanogaster*. *Genetics Research* **44**: 231–237.

R. M. May (1975). *Stability and Complexity in Model Ecosystems*. Princeton University Press.

R. M. May (1976). Simple mathematical models with very complicated dynamics. *Nature* **261**: 459–461.

R. M. May (1978). Host–parasitoid systems in patchy environments: A phenomenological model. *Journal of Animal Ecology*, **47**: 833–843.

L. Michaelis and M. I. Menten (1913). Die Kinetik der Invertinwirkung. *Biochemische Zeitschrift* **49**: 333–369.

J. L. Monod (1942). *Recherches sur la Croissance des Cultures Bacteriennes*. Hermann.

J. L. Monod (1950). La technique de culture continué: Théorie et applications. *Annales de l'Institut Pasteur* **79**: 390–410.

B. Moss (1980). *Ecology of Freshwaters*. Blackwell Scientific.

J. S. Nagumo, S. Arimoto, and S. Yoshizawa (1962). An active pulse transmission line simulating nerve axon. *Proceedings of the Institute of Radio Engineers* **50**: 2061–2071.

S. Nee and R. M. May (1992). Dynamics of metapopulation: Habitat destruction and competitive coexistence. *Journal of Animal Ecology* **61**: 37–40.

A. J. Nicholson (1933). The balance of animal populations. *Journal of Animal Ecology* **2**: 131–178.

A. J. Nicholson and V. A. Bailey (1935). The balance of animal populations. *Proceedings of the Zoological Society of London* **3**: 551–598.

K. J. Niklas (1994). *Plant Allometry: The Scaling of Form and Process*. University of Chicago Press.

R. T. Oglesby (1977). Relationships of fish yield to lake phytoplankton, standing crop, production, and morphoedaphic factors. *Journal of the Fisheries Reserve Board of Canada* **34**: 2271–2279.

J. G. Owen (1988). On productivity as a predictor of rodent and carnivore density. *Ecology* **69**: 1161–1165.

S. W. Pacala and M. Rees (1998). Models suggesting field experiments to test two hypotheses explaining successional diversity. *American Naturalist* **152**: 729–737.

R. Pearl and L. J. Reed (1920). On the rate of growth of the population of the United States since 1790 and its mathematical representation. *Proceedings of the National Academy of Sciences* **6**: 275–288.

G. W. Pierce (1949). *The Songs of Insects*. Harvard University Press.

A. Pisek, W. Larcher, W. Moser, and I. Pack (1969). Kardinale Temperaturbereiche der Photosynthese und Grenztemperaturen des Lebens der Blätter verschiedener Spermatophyten. III. Temperaturabhängigkeit und optimaler Temperaturbereich der Netto-Photosynthese. *Flora, Jena* **158**: 608–630.

S. D. Pletcher (1998). Mutation and the evolution of age-specific mortality rates: Experimental results and statistical developments. Ph.D. thesis, University of Minnesota.

F. W. Preston (1962). The canonical distribution of commonness and rarity. *Ecology* **43**: 185–215, 410–431.

M. J. Reiss (1989). *The Allometry of Growth and Reproduction*. Cambridge University Press.

D. A. Roff (1992). *The Evolution of Life Histories*. Chapman & Hall.

M. L. Rosenzweig (1971). Paradox of enrichment: Destabilization of exploitation ecosystems in ecological time. *Science* **171**: 385–387.

M. L. Rosenzweig and Z. Abramsky (1993). How are diversity and productivity related? In *Species Diversity in Ecological Communities*, eds. R. E. Ricklefs and D. Schluter, pp. 52–65. University of Chicago Press.

J. Roughgarden (1996). *Theory of Population Genetics and Evolutionary Ecology*, 2nd edition. Prentice Hall.

S. M. Stanley (1979). *Macroevolution: Pattern and Process*, p. 332. W. H. Freeman.

R. W. Sterner (1997). Modeling interactions between food quality and quantity in homeostatic consumers. *Freshwater Biology* **38**: 473–482.

R. W. Sterner and J. J. Elser (2002). *Ecological Stoichiometry: The Biology of Elements from Molecules to the Biosphere.* Princeton University Press.

C. R. Taylor, G. M. O. Maloiy, E. R. Weibel, U. A. Langman, J. M. Z. Kaman, H. J. Seeherman, and B. N. C. Heglund (1980). Design of the mammalian respiratory system. III. Scaling maximum aerobic capacity to body mass: Wild and domestic mammals. *Respiration Physiology* **44**: 25–37.

W. R. Thompson (1924). La théorie mathématique de l'action des parasites entomophages et le facteur du hasard. *Annales de la Faculté des Sciences de Marseille* **2**: 69–89.

D. Tilman (1982). *Resource Competition and Community Structure.* Princeton University Press.

D. Tilman (1994). Competition and biodiversity in spatially structured habitats. *Ecology* **75**: 2–16.

M. M. Tilzer, W. Geller, U. Sommer, and H. H. Stable (1982). Kohlenstoffkreislauf und Nahrungsketten in der Freiwasserzone des Bodensees. *Konstanzer Blätter für Hochschulfragen* **73**: 51.

U.S. Bureau of the Census (1994). *Statistical Abstract of the United States*, 14th edition. Government Printing Office.

J. W. Valentine (1985). Biotic diversity and clade diversity. In *Phanerozoic Diversity Patterns: Profiles in Macroevolution*, ed. J. W. Valentine, pp. 419–424. Princeton University Press.

P. M. Vitousek and H. Farrington (1997). Nutrient limitation and soil development: Experimental test of a biogeochemical theory. *Biogeochemistry* **37**: 63–75.

V. Volterra (1926). Fluctuations in the abundance of species considered mathematically. *Nature* **118**: 558–560.

T. D. Walker (1985). Diversification functions and the rate of taxonomic evolution. In *Phanerozoic Diversity Patterns: Profiles in Macroevolution*, ed. J. W. Valentine, pp. 311–334. Princeton University Press.

T. W. Walker and J. K. Syers (1976). The fate of phosphorus during pedogenesis. *Geoderma* **15**: 1–19.

T. D. Walker and J. W. Valentine (1984). Equilibrium models of evolutionary species diversity and the number of empty niches. *American Naturalist* **124**: 887–899.

P. D. Ward (1992). *On Methuselah's Trail—Living Fossils and the Great Extinctions.* Freeman.

Women's Health Initiative Steering Committee (2004). Effects of conjugated equine estrogen in postmenopausal women with hysterectomy: The Women's Health Initiative Randomized Controlled Trial. *JAMA* **291** (14), 1701–1712.

S. Wright (1968). *Evolution of Populations, Vol. 1. Genetics and Biometrik Foundation.* University of Chicago Press.

E. K. Yeargers, R. W. Shonkwiler, and J. V. Herod (1996). *An Introduction to the Mathematics of Biology: With Computer Algebra Models.* Birkhäuser.

Photo Credits

Front Endpapers

Descartes	Library of Congress
Newton	Library of Congress
Leibniz	Wikemedia Commons
Euler	Library of Congress
Kepler	Library of Congress
Pascal	Courtesy of IBM Archives
L'Hospital	The Granger Collection, NY
Bernoulli	Jupiter Unlimited
Lagrange	Library of Congress

Gauss	Wikemedia Commons
Cauchy	C. H. Reutlinger/Smithsonian Institution Archives
Riemann	Library of Congress
Lebesgue	The Granger Collection, NY
Agnesi	Library of Congress
Weierstrass	Smithsonian Institution Archives
Kovalevsky	Library of Congress
Gibbs	Wikemedia Commons

About the Author

Neuhauser	Dawn Villella/AP, © HHMI

Chapter Openers

Photodisc

Index